HARTMANN & KESTER'S
PLANT PROPAGATION
PRINCIPLES AND PRACTICES

EIGHTH EDITION

Hudson T. Hartmann, PhD
University of California, Davis

Dale E. Kester, PhD
University of California, Davis

Fred T. Davies, Jr., PhD
Texas A&M University College Station

Robert L. Geneve, PhD
University of Kentucky, Lexington

Prentice Hall

Boston   Columbus   Indianapolis   New York   San Francisco   Upper Saddle River
Amsterdam   Cape Town   Dubai   London   Madrid   Milan   Munich   Paris   Montreal   Toronto
Delhi   Mexico City   São Paulo   Sydney   Hong Kong   Seoul   Singapore   Taipei   Tokyo

Editorial Director: Vernon Anthony
Acquisitions Editor: William Lawrensen
Editorial Assistant: Lara Dimmick
Director of Marketing: David Gesell
Senior Marketing Coordinator: Alicia Wozniak
Marketing Assistant: Les Roberts
Associate Managing Editor: Alexandrina Benedicto Wolf
Inhouse Production Liaison: Alicia Ritchey
Operations Specialist: Laura Weaver
Art Director: Diane Ernsberger

Cover Designer: Anne Demarinis
Cover Art: Robert L. Geneve
Lead Media Project Manager: Karen Bretz
Full-Service Project Management: Douglas Bell/PreMediaGlobal
Composition: PreMediaGlobal
Printer/Binder: Courier/Kendallville
Cover Printer: Lehigh-Phoenix Color
Text Font: AGaramond 11/13

Credits and acknowledgments borrowed from other sources and reproduced, with permission, in this textbook appear on appropriate page within text.

Copyright © 2011, 2002, 1997, 1990, 1983, 1975, 1968, 1959 Pearson Education, Inc., Publishing as Prentice Hall, One Lake Street, Upper Saddle River, NJ 07458. All rights reserved. Manufactured in the United States of America. This publication is protected by copyright, and permission should be obtained from the publisher prior to any prohibited reproduction, storage in a retrieval system, or transmission in any form or by any means, electronic, mechanical, photocopying, recording, or likewise. To obtain permission(s) to use material from this work, please submit a written request to Pearson Education, Inc., Permissions Department, Prentice Hall, One Lake Street, Upper Saddle River, NJ 07458.

Many of the designations by manufacturers and seller to distinguish their products are claimed as trademarks. Where those designations appear in this book, and the publisher was aware of a trademark claim, the designations have been printed in initial caps or all caps.

Library of Congress Cataloging-in-Publication Data

Davies, Fred T., 1949-
   Hartmann and Kester's plant propagation : principles and practice / Fred T. Davies, Robert L. Geneve, Dale E. Kester.—8th ed.
   p. cm.
   Revision of: 7th ed. 2002.
   Includes bibliographical references and index.
   ISBN 978-0-13-501449-3 (alk. paper)
   1. Plant propagation.   I. Geneve, R. L.   II. Kester, Dale E.   III. Hartmann, Hudson Thomas.
   IV. Title.   V. Title: Plant propagation.
   SB119.P55 2011
   631.5'3—dc22

2010012518

10 9 8 7 6 5 4 3 2 1

www.pearsonhighered.com

Paper bound	ISBN 10:	0-13-501449-2
	ISBN 13:	978-0-13-501449-3
Loose leaf	ISBN 10:	0-13-505441-9
	ISBN 13:	978-0-13-505441-3

Dedications

7 July 2009

The eighth edition of *Plant Propagation* is dedicated to Dr. Dale Emmert Kester, Professor Emeritus for the University of California, Davis. Dale passed away on November 21, 2003.

His lifelong interest in horticulture led Dale to enroll as a horticulture student at Iowa State University in Ames, Iowa in 1941. His college career was interrupted in 1943 when Dale joined the war effort as a U.S. Air Force P-51 Mustang pilot. As a World War II pilot, he escorted bombers on 28 missions over Italy and Central Europe. Dale met his future wife, Daphne Dougherty, while he was stationed in Baton Rouge, Louisiana. Daphne was a USO dancer at the time. Following the war, he returned to Iowa State University and completed his horticulture degree in 1947.

Dale was the first PhD graduate from the University of California, Davis Pomology Department following the war. His dissertation concerned embryo culture of peaches. In 1951, he was offered an Assistant Professor position in the Department of Pomology at UC Davis where his work was to focus on almond production and breeding. This was the position he would hold until his retirement 40 years later in 1991. He taught undergraduate plant propagation and pomology courses. Early in his career, he partnered with Dr. Hudson Hartmann to publish the first edition of "Plant Propagation—Principles and Practices" in 1959.

Along with Hudson Hartmann and others, Dale was a founding member of the Western Region of the International Plant Propagators' Society. He served that organization as Vice-President, program chair in 1996 and President in 1997. Dale received the Curtis J. Alley Award in 1999 for his lifetime service to the International Plant Propagators' Society. In 2002, shortly before his death, he received the society's highest award, the International Award of Honor. With this award, he was recognized for "his long-standing reputation as a dedicated teacher of students interested in plant propagation, his service to the International Plant Propagators' Society and especially, for his seminal textbook on plant propagation used the world over."

Dale was a longtime member of the American Society for Horticultural Science and was recognized as a Fellow in 1977. He served as the first chair of the Propagation Working Group and received the Stark Award in 1980. In 1998, he was the Spenser Ambrose Beach Lecturer at Iowa State University. He published over 120 research papers in journals and conference proceedings. His research efforts in almond led to numerous root stock introductions, as well as the cause for noninfectious bud failure in almond.

Dale Kester was one of the most internationally recognized horticulturists of his generation, but remained a very unpretentious man. He was easy-going, good humored and appeared more impressed with his colleagues' achievements than his own. Dale was a mentor, role model, and a friend. He will be greatly missed by the horticultural community.

The seventh edition of *Plant Propagation* was dedicated to Dr. Hudson T. Hartmann. Dr. Hartmann died March 2, 1994 just as plans for the sixth edition were getting underway. He is remembered as a dedicated, hard-working, conscientious scientist, teacher, and human being. He conceived of the writing of this text about 1955 and asked the second author, Dr. Dale E. Kester, to join him. Dr. Hartmann taught Plant Propagation at the University of California at Davis from 1945 to his retirement in 1980. His research in propagation involved early studies on hormones, mist propagation, and other aspects of cutting propagation particularly as they applied to fruit trees. He was also a specialist in olive research and development, attaining a worldwide reputation for this crop.

One of his primary accomplishments was his activity with the International Plant Propagation Society. He became a member in 1953 and then was instrumental in initiating the Western Region of the Society in 1960. He served as Western Region Editor for the Society from 1960 to 1993, serving also as International Editor from 1970 until 1991. During his career he published many scientific papers and popular articles. As well as the present text, he was senior author of *Plant Science: Growth, Development and Utilization of Cultivated Plants,* first edition (1981), second edition (1988) published by Prentice Hall.

Dr. Hartmann was a member of the American Society for Horticultural Science, becoming a Fellow in 1974. As an undergraduate he was a member of Gamma Sigma Delta and Alpha Zeta. He received many awards, including the Charles G. Woodbury Award (1960), Joseph H. Gourley Award (1962), and Stark Brothers Award (1964) from ASHS. The American Association of Nurserymen awarded him its Norman J. Coleman Award (1970), The California Association of Nurserymen presented him with its Research award (1977), and Pi Alpha Xi made him an honorary member (1981). The Western Region IPPS awarded Dr. Hartmann its Merit award (1979), Honorary Membership (1983), and established the Hudson T. Hartmann Western Region Research Grant in his honor. The International IPPS Board of Directors awarded him the International Award of Honor in 1990.

Dr. Hartmann was a close personal friend, a collaborator who made working together a pleasure, and a respected peer whose guidance and insight are missed.

brief contents

Preface x
About the Authors xii

part one
General Aspects of Propagation

1. How Plant Propagation Evolved in Human Society 2
2. Biology of Plant Propagation 14
3. The Propagation Environment 49

part two
Seed Propagation

4. Seed Development 110
5. Principles and Practices of Seed Selection 140
6. Techniques of Seed Production and Handling 162
7. Principles of Propagation from Seeds 200
8. Techniques of Propagation by Seed 250

part three
Vegetative Propagation

9. Principles of Propagation by Cuttings 280
10. Techniques of Propagation by Cuttings 344
11. Principles of Grafting and Budding 415
12. Techniques of Grafting 464
13. Techniques of Budding 512
14. Layering and Its Natural Modifications 537
15. Propagation by Specialized Stems and Roots 561
16. Principles and Practices of Clonal Selection 594

part four
Cell and Tissue Culture Propagation

17. Principles of Tissue Culture and Micropropagation 644
18. Techniques for Micropropagation 699

part five
Propagation of Selected Plant Species

19. Propagation Methods and Rootstocks for Fruit and Nut Species 728
20. Propagation of Ornamental Trees, Shrubs, and Woody Vines 774
21. Propagation of Selected Annuals and Herbaceous Perennials Used as Ornamentals 840

Subject Index 881
Plant Index: Scientific Names 898
Plant Index: Common Names 908

contents

Preface x
About the Authors xii

part one
General Aspects of Propagation 1

1
How Plant Propagation Evolved in Human Society 2

Introduction 2
Learning Objectives 2
Stages of Agricultural Development 3
Organization of Human Societies 4
Exploration, Science, and Learning 5
The Development of Nurseries 8
The Modern Plant Propagation Industry 12
Discussion Items 12
References 12

2
Biology of Plant Propagation 14

Introduction 14
Learning Objectives 14
Biological Life Cycles in Plants 14
Taxonomy 18
Legal Protection of Cultivars 21
Genetic Basis for Plant Propagation 21
Genetic Inheritance 27
Gene Structure and Activity 30
Plant Hormones and Plant Development 38
Discussion Items 45
References 45

3
The Propagation Environment 49

Introduction 49
Learning Objectives 49
Environmental Factors Affecting Propagation 50
Physical Structures for Managing the Propagation Environment 54
Containers for Propagating and Growing Young Liner Plants 70
Management of Media and Nutrition in Propagation and Liner Production 77
Management of Microclimatic Conditions in Propagation and Liner Production 85
Biotic Factors—Pathogen and Pest Management in Plant Propagation 90
Post-Propagation Care of Liners 100
Discussion Items 102
References 103

part two
Seed Propagation 109

4
Seed Development 110

Introduction 110
Learning Objectives 110
Reproductive Life Cycles of Vascular Plants 110
Characteristics of a Seed 112
Reproductive Parts of the Flower 117
Relationship Between Flower and Seed Parts 118
Stages of Seed Development 122
Unusual Types of Seed Development 130
Plant Hormones and Seed Development 133
Ripening and Dissemination 136
Discussion Items 137
References 137

5 Principles and Practices of Seed Selection 140

Introduction 140

Learning Objectives 140

Breeding Systems 140

Categories of Seed-Propagated Cultivars and Species 147

Control of Genetic Variability During Seed Production 150

Seed Production Systems 153

Discussion Items 159

References 159

6 Techniques of Seed Production and Handling 162

Introduction 162

Learning Objectives 162

Sources For Seeds 162

Harvesting and Processing Seeds 166

Seed Testing 175

Seed Treatments to Improve Germination 184

Seed Storage 189

Discussion Items 195

References 195

7 Principles of Propagation from Seeds 200

Introduction 200

Learning Objectives 200

The Germination Process 200

Dormancy: Regulation of Germination 218

Kinds of Primary Seed Dormancy 220

Secondary Dormancy 235

Dormancy Control by Plant Hormones 236

Discussion Items 240

References 240

8 Techniques of Propagation by Seed 250

Introduction 250

Learning Objectives 250

Seedling Production Systems 250

Discussion Items 276

References 276

part three
Vegetative Propagation 279

9 Principles of Propagation by Cuttings 280

Introduction 280

Learning Objectives 280

Descriptive Observations of Adventitious Root and Bud (and Shoot) Formation 281

Correlative Effects: How Hormonal Control Affects Adventitious Root and Bud (and Shoot) Formation 293

The Biochemical Basis for Adventitious Root Formation 299

Molecular/Biotechnological Advances in Asexual Propagation 304

Management and Manipulation of Adventitious Root and Shoot Formation 305

Management of Stock Plants to Maximize Cutting Propagation 307

Treatment of Cuttings 318

Environmental Manipulation of Cuttings 323

Discussion Items 331

References 332

10 Techniques of Propagation by Cuttings 344

Introduction 344

Learning Objectives 344

Types of Cuttings 344
Sources of Cutting Material 363
Rooting Media 367
Wounding 373
Treating Cuttings with Auxins 373
Preventative Disease Control 381
Environmental Conditions for Rooting Leafy Cuttings 383
Preparing the Propagation Bed, Bench, Rooting Flats, and Containers, and Inserting the Cuttings 393
Preventing Operation Problems with Mist and Fog Propagation Systems 395
Management Practices 396
Care of Cuttings During Rooting 401
Hardening-Off and Post-Propagation Care 403
Handling Field-Propagated Plants 406
Container-Grown Plants and Alternative Field Production Systems 409
Discussion Items 409
References 409

Principles of Grafting and Budding 415

Introduction 415
Learning Objectives 415
The History of Grafting 415
Terminology 417
Seedling and Clonal Rootstock Systems 419
Reasons for Grafting and Budding 419
Natural Grafting 424
Formation of the Graft Union 425
Graft Union Formation in T- and Chip Budding 432
Factors Influencing Graft Union Success 433
Genetic Limits of Grafting 439
Graft Incompatibility 441
Scion-Rootstock (Shoot-Root) Relationships 450
Discussion Items 457
References 457

12
Techniques of Grafting 464

Introduction 464
Learning Objectives 464
Requirements for Successful Grafting 464

Types of Grafts 465
Production Processes of Graftage 491
Aftercare of Grafted Plants 502
Field, Bench, and Miscellaneous Grafting Systems 504
Discussion Items 509
References 509

Techniques of Budding 512

Introduction 512
Learning Objectives 512
Importance and Utilization of Budding 512
Rootstocks for Budding 513
Time of Budding—Summer, Spring, or June 513
Types of Budding 519
Top-Budding (Topworking) 532
Double-Working by Budding 533
Microbudding 534
Discussion Items 535
References 536

Layering and Its Natural Modifications 537

Introduction 537
Learning Objectives 537
Reasons for Layering Success 537
Management of Plants During Layering 539
Procedures in Layering 539
Plant Modifications Resulting in Natural Layering 551
Discussion Items 558
References 558

Propagation by Specialized Stems and Roots 561

Introduction 561
Learning Objectives 561
Bulbs 563
Corms 577
Tubers 579

Tuberous Roots and Stems 581

Rhizomes 584

Pseudobulbs 587

Discussion Items 590

References 590

16
Principles and Practices of Clonal Selection 594

Introduction 594

Learning Objectives 594

History 594

Using Clones as Cultivars 595

Origin of Clones as Cultivars 597

Phenotypic Variations Within Clones 601

Patterns of Genetic Chimeras Within Clones 603

Management of Phase Variation During Vegetative Propagation 613

Pathogens and Plant Propagation 619

Selection and Management of Propagation Sources 623

Propagation Sources and Their Management 630

Discussion Items 635

References 636

part four
Cell and Tissue Culture Propagation 643

17
Principles of Tissue Culture and Micropropagation 644

Introduction 644

Learning Objectives 644

A Brief History of Tissue Culture and Micropropagation 644

Types of Tissue Culture Systems 649

Control of the Tissue Culture Environment 679

Special Problems Encountered by *In Vitro* Culture 681

Variation in Micropropagated Plants 684

Discussion Items 687

References 687

18
Techniques for Micropropagation 699

Introduction 699

Learning Objectives 699

Uses for Micropropagation 699

Disadvantages of Micropropagation 701

General Laboratory Facilities and Procedures 702

Micropropagation Procedures 712

Stage I—Establishment 713

Stage II—Shoot Multiplication 716

Stage III—Root Formation 717

Stage IV—Acclimatization to Greenhouse Conditions 718

Discussion Items 724

References 724

part five
Propagation of Selected Plant Species 727

19
Propagation Methods and Rootstocks for Fruit and Nut Species 728

Introduction 728

References 766

20
Propagation of Ornamental Trees, Shrubs, and Woody Vines 774

Introduction 774

References 825

21
Propagation of Selected Annuals and Herbaceous Perennials Used as Ornamentals 840

Introduction 840

References 869

Subject Index 881
Plant Index: Scientific Names 898
Plant Index: Common Names 908

Preface

The eighth edition of *Plant Propagation: Principles and Practices* continues the legacy of updating the ever-changing principles and practices associated with plant propagation, but it is also the first edition with expanded color figures throughout the text. This is an exciting prospect that the co-authors hope will enhance student learning. Some 90 percent or more of the images and illustrations are either new or enhanced.

The eighth edition is published a half-century after the initial printing of *Plant Propagation: Principles and Practices* in 1959, but still continues the tradition of presenting paired chapters where the principles underlying the science of propagation alternate with the technical practices and skills utilized for commercial plant propagation. As with previous editions, the amount of material between editions has increased at an incredible rate and many aspects of growth and development have expanded beyond the wildest forecasts in 1959. We have tried to integrate the most current commercial techniques and understanding of the biology of propagation into current chapters. We have substantially updated the references and sections on "Getting More in Depth on the Subject" to help the reader delve deeper into these subjects than the general scope of this textbook.

As in previous editions, the book is organized into four basic parts. The initial three chapters are general chapters meant to support general aspects of propagation including a historical perspective, basic plant biology concepts and the impact and control of the environment as it affects propagation and nursery practices. Chapter 2 has been significantly revised to reflect the significant progress in plant hormone biology and the molecular advances in plant growth and development. We hope that it serves as background support for understanding the concepts described in the Principles chapters, and provides a foundation for students to pursue these fascinating subjects in the literature. Chapter 3 continues the integration of concepts and application to control the propagation environment, which is of major importance in commercial propagation. The latest engineering, computerization, and mechanization systems for propagation are included. The next two sections describe seed and vegetative propagation, respectively. Each revised section provides a chapter on the concepts behind genetic selection for either sexual or clonal plants, and then specific chapters for the principles and practices. The final section is an updated compilation of propagation techniques for specific crops.

New with this edition is the inclusion of study questions at the end of each chapter to compliment the keywords provided in the page margins, and web-based student resources available through www.pearsonhighered.com/hartmann. There is also an instructors' resource website at www.pearsonhighered.com/hartmann. Propagation instructors are encouraged to contact their local Prentice Hall representative for a complimentary copy of the textbook.

A substantial increase in the number of figures was used to support the text for the eighth edition. The majority of these images have been taken by the co-authors while visiting commercial producers and research labs throughout the world. This opportunity was only possible because of the generosity of companies and individuals associated with those organizations. These groups are too numerous to acknowledge here, but the authors would like to express our sincere appreciation for the access granted to us that has made it possible to illustrate commercial plant propagation techniques to students. Additional images were taken while using the library resources of the Lloyd Library in

Cincinnati, and the rare book collections at the Missouri Botanical Garden and the University of Kentucky. We would also like to express our appreciation to those colleagues who have generously supplied images to enhance this and previous editions.

Mention or photographs of any products or techniques are for information purposes only, and are not intended as endorsements; neither is criticism implied for products not mentioned. Always follow instructions on product labels, and be aware that regulations may vary by country, state, and region. In any commercial propagation system it is important to conduct small trials before propagating on a large scale. Any propagation techniques and references listed are to serve as a guide. Propagators must develop their own procedures and chemical treatments that work best for their particular propagation system.

In preparing the eighth edition of this book, we have depended upon the assistance of authorities in the various fields of propagation and related subjects. We thank them for their critical evaluation and suggestions. We also thank our wives, Maritza Davies and Pat Geneve, and families for their support, encouragement, and patience during the writing and production of this edition. We thank Mike Geneve for preparing selected illustrations used in the text.

Finally we acknowledge the skill and professionalism of the Prentice Hall and associated editors who made this production possible including: Stephanie Kelly, William Lawrensen, Alicia Ritchey, Laura Weaver, Lara Dimmick, and Alex Wolf.

Download Instructor Resources from the Instructor Resource Center

To access supplementary materials online, instructors need to request an instructor access mode. Go to www.pearsonhighered.com/irc to register for an instructor access code. Within 48 hours of registering, you will receive a confirming e-mail including an instructor access code. Once you have received your code, locate your text in the online catalog and click the Instructor Resources button on the left side of the catalog product page. Select a supplement, and a login page will appear. Once you have logged in, you can access instructor material for all Prentice Hall textbooks. If you have any difficulties accessing the site or downloading a supplement, please contact Customer Service at http://247.prenhall.com.

Acknowledgments

The authors and publisher would like to thank the following reviewers for their time and content expertise:

R. Lee Ivy, *Associate Professor*
Landscape Gardening
Sandhills Community College
Pinehurst, NC

G. N. Mohan Kumar, *Associate Professor*
Horticulture and Landscape Architecture
Washington State University
Pullman, WA

Mark J. Schusler, *Assistant Professor*
Horticulture
Tarrant County College
Fort Worth, TX

Todd P. West, PhD, *Assistant Professor*
Horticulture
West Virginia University
Morgantown, WV

Susan Wiegrefe, PhD, *Contributing Faculty*
Plant and Earth Science
University of Wisconsin—River Falls
River Falls, WI

Sandra B. Wilson, *Associate Professor*
Environmental Horticulture
University of Florida
Fort Pierce, FL

About the Authors

Fred T. Davies, Jr., Regents Professor of Horticultural Sciences, and Molecular & Environmental Plant Sciences, and TAES Research Faculty Fellow, Texas A&M University, has taught courses in plant propagation and nursery production and management since 1979. He has co-authored over 150 research and technical publications. He was a J. S. Guggenheim Fellow (1999), and a Fulbright Senior Fellow to Mexico (1993) and Peru (1999). He is a Fellow of the American Society for Horticultural Sciences (ASHS) (2003) and the International Plant Propagators' Society (IPPS). He received the Distinguished Achievement Award for Nursery Crops from the ASHS (1989), L.M. Ware Distinguished Research Award—ASHS—SR (1995), and S. B. Meadows Award of Merit—IPPS (1994). He is a recipient of the Association of Former Students Distinguished Achievement Award for Teaching—TAMU (1997), Chancellor of Agriculture's Award in Excellence in Undergraduate Teaching—TAMU (1998), L.M. Ware Distinguished Teaching Award, ASHS—SR (1998), and L.C. Chadwick Educator's Award, American Nursery and Landscape Association (1999). He was the International Division Vice-President—ASHS. He was President, and is currently Editor, of the IPPS—SR. He is President of the ASHS.

Robert L. Geneve is a Professor in the Department of Horticulture at the University of Kentucky. He teaches courses in plant propagation and seed biology. He has co-authored over 100 scientific and technical articles in seed biology, cutting propagation, and tissue culture. He is also the co-editor of the book *Biotechnology of Ornamental Plants* and author of *A Book of Blue Flowers*. He has served as a Vice-President, program chair and President for the International Plant Propagators' Society–Eastern Region. He has served as the Editor for the international horticulture journal, *Scientia Horticulturae* from 2001 to 2008 and is currently on the editorial boards of the Propagation of Ornamental Plants and the Journal of Seed Technology. He is a recipient of the University of Kentucky, George E. Mitchell Jr. Award for Outstanding Faculty Service to Graduate Students (2006), and is a Fellow the American Society for Horticultural Science (2005), and the International Plant Propagators' Society–Eastern Region (2003).

Fred T. Davies, Jr.
Robert L. Geneve

part one
General Aspects of Propagation

CHAPTER 1 **How Plant Propagation Evolved in Human Society**
CHAPTER 2 **Biology of Plant Propagation**
CHAPTER 3 **The Propagation Environment**

This book about **plant propagation** not only describes procedures originating thousands of years ago, but also the application of recent scientific advances. Plant propagation can be described as *the purposeful act of reproducing plants.* It has been practiced for perhaps the past 10,000 years, and its beginning probably marks the start of civilization. The traditional concept of a propagator is a skilled technician who loves plants and who acquired the art from traditional skills learned by experience, or whose knowledge was handed down from one generation to another. Today, propagation may be carried out by an array of general and specialized industries that produce plants to feed the world; to provide fiber, building materials, and pharmaceuticals; and to enhance the world's beauty.

*1 How Plant Propagation Evolved in Human Society

learning objectives

- Describe the evolution of plant propagation during human history.
- Describe aspects of modern plant propagation activities.

"And the earth brought forth grass, and herb yielding seed after his kind, and the tree yielding fruit, whose seed was in itself, after his kind: and God saw that it was good."

Genesis 1:12.

"Man has become so utterly dependent on the plants he grows for food that, in a sense, the plants have 'domesticated him.' A fully domesticated plant cannot survive without the aid of man, but only a minute fraction of the human population could survive without cultivated plants."

from: J. R. Harlan, *Crops and Man*, 2nd edition. Madison, WI: Amer. Soc. of Agron. 1992.

INTRODUCTION

The propagation of plants is a fundamental occupation of humankind. Its discovery began what we now refer to as civilization and initiated human dominion over the earth. Agriculture began some 10,000 years ago when ancient peoples, who lived by hunting and gathering, began to cultivate plants and domesticate animals. These activities led to stable communities where people began to select and propagate the kinds of plants that provided a greater and more convenient food supply, as well as other products for themselves and their animals (21, 35). Once this process began, humans could remain at the same site for long periods of time, thus creating centers of activity that eventually would become cities and countries.

agriculture The deliberate practice of propagating and growing plants for human use.

Agriculture is the deliberate cultivation of crops and animals for use by humans and involves five fundamental activities:

1. Plant selection—selecting and (or) developing specific kinds of plants.
2. Plant propagation—multiplying plants and preserving their unique qualities.
3. Crop production—growing plants under more controlled conditions for maximum yield.
4. Crop handling and storage—preserving crop products for long-term usage and transport to other areas.
5. Food technology—transforming and preserving crop products for food or other uses (e.g., making bread, pressing oil, preparing wine, dehydration, etc.).

STAGES OF AGRICULTURAL DEVELOPMENT

The pivotal role of plant propagation in the evolution of human society can be seen in terms of particular stages of agricultural development.

Hunting and Gathering

Most of the millions of years of human existence as hunters and gatherers were related to the presence of specific food resources including seeds, fruits, roots, and tubers, as well as animals that fed on the plants. The distribution and the characteristics of plant species were determined by the environment; that is, both the physical world (climate, soil, topography) and the biological interactions of plant, animal, and human populations (21, 32, 35). Humans have existed for millions of years, spreading from their presumed place of origin in western Africa into Asia, Europe, and, eventually, into North and South America. Food supplies were abundant in the native vegetation, although quite variable in different parts of the world. Apparently, early humans were quite effective in searching out those that were useful, as well as in developing processes to utilize and preserve them.

What motivated humans to begin to propagate and grow specific kinds of plants near their homes has been the subject of much scientific debate (21, 35). It is clear that the development of agriculture forever changed the relationship between humans and their surrounding environment. This event occurred in separate areas of the world, more or less simultaneously within a relatively short period of a few thousand years nearly 10,000 years ago. These areas included the Near East fertile crescent of Southwest Asia and Northeast Africa, extending from the valley of the Euphrates and Tigris Rivers along the coasts of Syria, Turkey, and Israel to the Nile Valley of Egypt; China, including a northern and a tropical southern area; and Central and South America, including areas in Mexico, and the coastal lowlands and highlands of Peru (21, 23).

The key activity bringing about this change must have been the deliberate selection and propagation and cultivation of specific kinds of plants that were particularly useful to humans. As a result, a larger and more stable population could be supported, which evolved into cities and countries. Human organization changed from subsistence existence, where everyone participated in the production of food and other items, to a division of labor between agricultural and non-agricultural segments of the population, and even to specialization within the agricultural segment. In this context, the plant propagator, who possessed specific knowledge and skills, had to assume a key role.

Domestication

Early civilization developed with relatively few domesticated plant species, determined both by their usefulness in the primitive economy and the ease with which they could be propagated. The lists differed in the separate areas of the world where human societies evolved (21, 32, 34, 35). In the Near East, the earliest domesticated food crops included wheat, barley, peas, and lentil. In the Far East, millet appears to be the first domesticated crop, followed by rice. In Central and South America, the first food crops domesticated were apparently squash and avocado, followed by such important modern-day food crops as corn, bean, pepper, tomato, and potato. Many of the early food crops were seed plants (cereals, such as wheat, barley, rice), which provided carbohydrates, and legumes (beans, peas), which provided protein. These seed-propagated plants could be subjected to genetic selection in consecutive propagation cycles for such agricultural characteristics as high yield, "non-shattering," large seed size, and reduced seed dormancy. These species were maintained more or less "fixed" because of their genetic tolerance to inbreeding (see Chapter 5). Highly desirable single plants of certain species, such as grape, fig, olive, pomegranate, potato, yam, banana, and pineapple (39) could be selected directly from wild populations and "fixed" through vegetative propagation (see Chapters 2 and 9). Domestication of fruit plants, such as apple, pear, peach, apricot, citrus, and others occurred with the discovery of grafting methods (see Chapters 12, 13, and 14). By the time of recorded history (or that which can be reconstructed), most of the basic methods of propagation had been discovered. During domestication, crop plants had evolved beyond anything that existed in nature.

> **domestication** The process of selecting specific kinds of wild plants and adapting them to human use.

The establishment of specific crops and cropping systems resulted in some side effects that have continued to create problems (21). As the fields used to grow plants near human sites were disturbed and became depleted, certain aggressive plant species also were spontaneously established in these sites. These so-called weedy species have become a part of the agricultural system and more or less evolved along with cultivated plants.

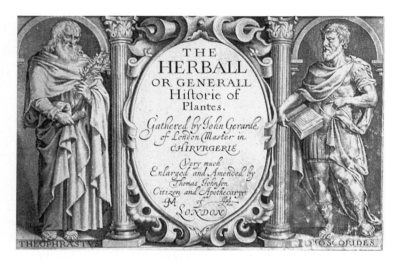

Figure 1–1
Theophrastus (300 BC) was an important influence on Renaissance agriculture, as indicated by his being depicted and commemorated on the front page of John Gerard's influential herbal, published in 1597. His image is in the left panel opposite his Greek counterpart Dioscorides (1 AD), renowned as an authority on the medicinal use of plants.

ORGANIZATION OF HUMAN SOCIETIES

Ancient (7000 BC)

The initial phases of domestication probably involved plant selection, plant propagation, and plant production. With an increase in food supply, a larger population could be supported and division of labor began to occur. Classes of individuals may have included laborers, manufacturers, artisans, government bureaucrats associated with irrigation systems, religious groups, and soldiers, as well as farmers and herdsmen. Historical records of early civilizations in Egypt and the Middle East (as well as archaeological investigations) have shown that the agricultural sector was well organized to produce food (cereals, vegetables, fruits, dates), fiber (flax, cotton), and other items for the non-agricultural components of society (25). Early Chinese writings indicate the knowledge of grafting, layering, and other techniques, although rice and millet were the principal food sources. In the Americas, seed-propagated crops (maize, beans, cucurbits, squash), as well as vegetatively propagated crops (potato, cassava, sweet potato, pineapple), were developed and grown.

Greek and Roman (500 BC to AD 1000)

Early writings described the agricultural world in detail with accounts of propagation techniques much as we know them today. Control of land and agricultural surplus was the key to power and wealth (35). Small and large farms existed. Olive oil and wine were exported, and grains were imported. Vegetables were grown near the home as were many fruits (fig, apple, pear, cherry, plum). Not only were food plants essential, but Romans developed ornamental gardening to a high level (21).

Some of the earliest references to plant propagation come from Theophrastus, a Greek philosopher (circa 300 BC) and disciple of Aristotle (Fig. 1–1). He described many aspects of plant propagation including seeds, cuttings, layering, and grafting in his two books *Historia de Plantis* and *De Causis Plantarum* (36, 37). An example from the translation of *De Causis Plantarum* (37) illustrates his understanding of propagation: "while all the trees which are propagated by some kind of slip seem to be alike in their fruits to the original tree, those raised from the fruit . . . are nearly all inferior, while some quite lose the character of their kin, as vine, apple, fig, pomegranate, pear."

Additional information on propagation can be seen in surviving works from Romans Pliny the Elder and Columella (circa 1 AD). For example, Pliny recommends that cabbage seeds be soaked in the juice of houseleek before being sown so that they will be "immune to all kinds of insects" (30), and Columella describes taking leafless, mallet stem cuttings in grape (12).

Medieval Period (AD 750 to 1500)

Society was organized around large estates, manor houses, and castles with landlords providing protection. Large areas of forest were kept as game preserves. Equally important were the monasteries that acted as independent agricultural and industrial organizations and preserved a great deal of the written and unwritten knowledge (Fig. 1–2). In both kinds of institutions, a separation developed among those involved in the production of cereals, fibers, and forages grown extensively in large fields (agronomy); vegetables, fruits, herbs, and flowers grown in "kitchen gardens" and orchards near the home (horticulture); and woody plants grown for lumber, fuel, and game preserves (forestry) (25).

Figure 1–2
The monastic garden was an enclosed area of medicinal and edible plants. The Cloisters in New York has several representative enclosed period gardens.

The end of the medieval period and the beginnings of modern Europe brought a shift from a subsistence existence to a market economy and the emergence of land ownership (35). In Western Europe, both large landowners and owners of smaller individual plots emerged. In Eastern Europe, the shift was toward large wealthy estates with the populace being largely serfs.

Through these periods, the specific skills and knowledge of the plant propagator were possessed by specific individuals. These skills, considered "trade secrets," were passed from father to son or to specific individuals. Often this knowledge was accompanied by superstition and, sometimes, attained religious significance.

EXPLORATION, SCIENCE, AND LEARNING

Plant Exchanges

The plant material exchange from the area of origin to other countries of the world has been one of the major aspects of human development. Not only did the range of plants available for food, medicine, industrial uses, and gardening expand, but plant propagation methods to reproduce them were required. Early movement of useful plants often followed military expansion into different countries when the invading soldier brought plants from his home country into a new land. Conversely, returning soldiers introduced to their homelands new plants they found while on a military campaign. There are numerous examples of this type of exchange taking place during the Roman conquests of northern Europe. Similarly, Islamic expansion in the 9th Century introduced citrus and rice to southern Europe, along with new concepts of cultivation and the use of irrigation. The voyages of Columbus opened the world to exploration and the interchange of plant materials from continent to continent. Such food staples as potatoes, tomatoes, beans, corn, squash and peppers all became available to Europe in the 16th and 17th centuries after voyages to the new world.

In addition to edible food crops, new and exotic plants were being sought out for introduction. Centers of learning in which scientific investigations began on all aspects of the biological and physical world were established in many countries. Linnaeus established the binomial system of nomenclature, and botanists began to catalog the plants of the world. Exploration trips were initiated where the primary mission was plant introduction, such as the voyages of Captain Cook in 1768, which included the plant explorers Sir Joseph Banks and Francis Masson who brought large numbers of exotic plants to England for the Royal Botanic Garden, established at Kew, outside of London (23, 31). Nathanial Ward, a London physician and amateur horticulturist, invented the Wardian case early in the 1800s to help preserve plant material on these long expeditions (Fig. 1–3) (38).

plant exchange The movement of plants from their place of origin to their place of use.

Wardian case A glazed wooden cabinet designed to keep high humidity inside and salt water spray outside the case on long sea voyages.

Plant-collecting trips continued throughout the world: from Europe (David Douglas, Joseph D. Hooker,

Figure 1–3
The Wardian case was invented by N. B. Ward in the early 19th Century to use when transporting plants over long ocean voyages.

Figure 1–4
Herbals were produced soon after the invention of the printing press to describe the utility of local and introduced plants. Plants such as this pea in Matthioli's herbal (*Commentarii*, 1564) were depicted from woodcuts on blocks.

Robert Fortune, George Forrest, Frank Kingdon Ward) and from the United States (David Fairchild, Frank Meyer, Joseph Rock, Charles Sargent, Ernest Wilson) (13, 18, 23, 31). Significant ornamental species that are mainstays of modern gardens were collected: from the Orient (rhododendron, primula, lily, rose, chrysanthemum), Middle East (tulips, many bulb crops), and North America (evergreen and deciduous trees and shrubs). "Orangeries" and glasshouses (greenhouses) were expanded to grow the exotic species being collected from India, Africa, and South and Central America.

Scientific and Horticultural Literature

The first important written works on agriculture, plant medicinal uses, and propagation that shaped western society came from the early Greek, Roman, and Arab writers between 300 BC and AD 2. Although many works were undoubtedly lost, many survive today because they were preserved in Arab libraries and passed on though medieval monasteries. Following the invention of the printing press in 1436, there was resurgence in the production of books called herbals (Fig. 1–4) describing and illustrating plants with medicinal properties. Much of the information came from older first century Greek literature, especially Dioscorides (Fig. 1–2). These early works were written in Latin, but eventually works began to appear in local languages (2), making plant information available to a wider audience.

The Renaissance heralded the appearance of scientific enquiry that relied heavily on meticulous observation of plant morphology and behavior. This is wonderfully shown in the illustrations from Marcello Malpighi (29) on plant anatomy in 1675 (Fig. 1–5).

In the late 1800s, the concepts of natural selection and genetics made a big impact on scientific advancement. Charles Darwin and his *Origin of Species* (14) as well as its important contemporary *The Variation of Animals and Plants Under Domestication* (15) introduced the concept of evolution and set the stage for the genetic discoveries following the rediscovery of Mendel's papers in 1900. The subsequent explosion in knowledge and application provided the framework on which present-day plant propagation is based, as did the increase in knowledge of plant growth, anatomy, physiology, and other basics of biological science (31).

Books and articles on gardening and propagation began to appear (16). The first book on nurseries, *Seminarium*, was written by Charles Estienne in 1530. Later, Charles Baltet, a practical nurseryman, published a famous book, *The Art of Grafting and Budding*, in 1821, describing 180 methods of grafting (see Figs. 1–6 and 1–7) (11). A book by Andrew J. Fuller—*Propagation of Plants*—was published in 1885 (19).

The Morrill Act

The passage of the Morrill Act by the United States Congress in 1862 was a landmark event that established land-grant colleges and fostered the scientific investigation of agriculture and mechanical arts.

> **Morrill Act** An act of Congress in 1862 that established land-grant universities for scientific study and teaching of agriculture and mechanical arts.

Departments of agronomy, horticulture, pomology, and related fields were established, which became centers of scientific investigation, teaching, and extension. Liberty Hyde Bailey (33), a product of this system, published his

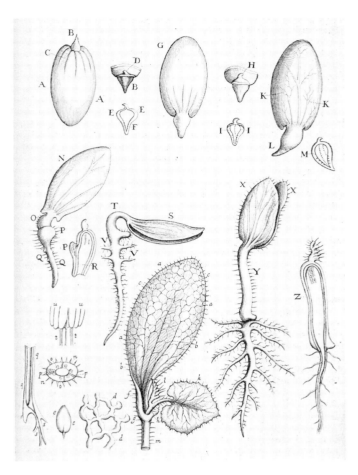

Figure 1–5
With the Renaissance, there was a resurgence in scientific inquiry. Malpighi was a keen observer of plants, as seen in his depiction of this germinating cucumber in his wonderfully illustrated *Anatome Plantarum*, 1675.

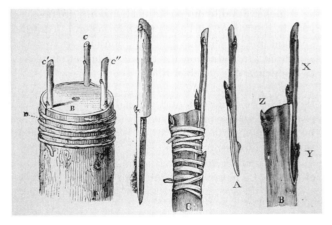

Figure 1–6
Bark grafting as illustrated in *The Art of Grafting and Budding* (1910) by Baltet.

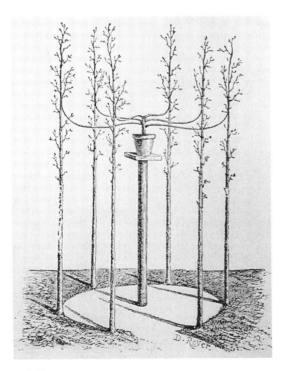

Figure 1–7
Approach grafting was a more important propagation technique before the introduction of mist propagation (11).

first edition of *The Nursery Book* (3) later revised as the *Nursery Manual* in 1920 (6), which cataloged what was known about plant propagation and the production of plants in the nursery (Fig. 1–8). His *Cyclopedia of American Horticulture* (4) in 1900–1902, *Standard Cyclopedia of Horticulture* (5) in 1914–1917, *Hortus* (7) in 1930, *Hortus Second* (8) in 1941, and *Manual of*

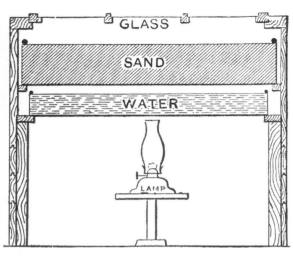

Figure 1–8
Liberty Hyde Bailey is considered the Father of American Horticulture (Seeley, 1990). He provides an interesting version of bottom heat for germination and cutting propagation in the *Nursery Book* (3), one of his 63 published books on horticulture.

Cultivated Plants (9) in 1940 and 1949 described the known plants in cultivation. An update, *Hortus Third* (10), is a classic in the field.

M. G. Kains of Pennsylvania State College and, later, Columbia University in New York, published *Plant Propagation* (26), later revised by Kains and McQuesten (27), which remained a standard text for many years (Fig. 1–9). Several other books were written during this period including titles by Adriance and Brison (1), Duruz (17), Hottes (24), and Mahlstede and Haber (28). The first edition of *Plant Propagation: Principles and Practices* (22) was published in 1959 and has continued through eight editions.

THE DEVELOPMENT OF NURSERIES

The concept of the nursery, where plants are propagated to be transplanted to their permanent site either as part of the agricultural unit or to be sold to others, has likely been a part of agriculture since its beginning. Nevertheless, the development of commercial nurseries is probably something that has developed largely within the recent era (16). Most agronomic crops (wheat, corn, etc.) and many vegetables were grown by seed. A portion of the seed was retained each year to supply the seed for the next cycle. In regions with cold winters, starting vegetables and flowers in protected structures

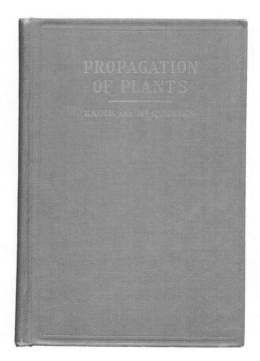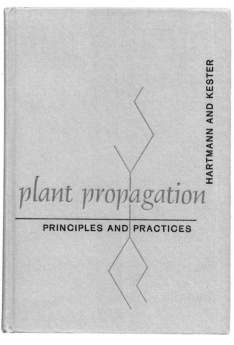

Figure 1–9
Early books for students and nursery professionals include *Propagation of Plants* by Kains and McQuesten (1938) and the first edition of *Plant Propagation: Principles and Practices* by Hartmann and Kester (1959).

(cold frames, hotbeds) and later transplanting them to the open was an important part of production, because doing so extended the length of the growing season.

A number of important nurseries existed in France during the 16th and 17th centuries and, eventually, throughout Europe (17). Ghent, Belgium, had a gardener's guild as early as 1366. The first glass house (greenhouse) was built in 1598. The Vilmorin family established a seed house and nursery business in 1815, which was maintained through seven generations.

Early plant breeding was often combined with a nursery, as exemplified by Victor Lemoine (1850) who specialized in tuberous begonias, lilies, gladiolus, and other garden flowers. Nickolas Hardenpont and Jean Baptiste van Mons specialized in fruits, particularly pears. The Veitch family started a major nursery in England in 1832. Thomas Andrew Knight, a famous hybridizer of fruits, established the Royal Horticultural Society in 1804.

Early colonists brought seeds, scion, and plants to the United States from Europe, and Spanish priests brought material to the West Coast. John Bartram is credited with providing a major impetus with his Botanical Garden in Philadelphia in 1728. The first nursery, however, was credited to William Prince and Son in 1730 on Long Island (Fig. 1–10). These were followed by the expansion of nurseries throughout the eastern United States during the 19th Century. To a large extent, the early nurseries specialized in selecting and grafting fruit trees, although ornamentals and forest trees also began to be produced.

David Landreth established a seed company, and the seed industry in the Philadelphia area, in 1784. He offered seeds internationally and later distributed seeds collected during the Lewis and Clark expedition. In 1906, Bernard McMahon produced the American Gardener's Calendar, which was reprinted through

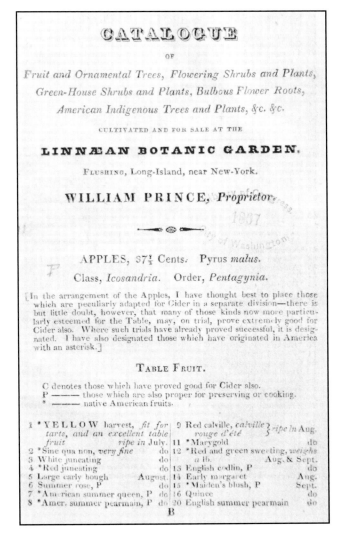

Figure 1–10
The first established nursery in the United States was begun in New York in 1730 by William Prince.

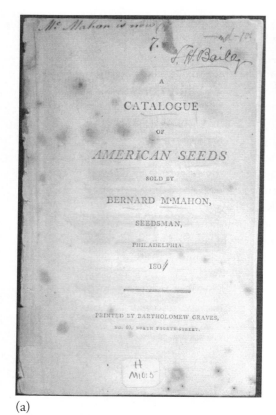

(a)

(b)

Figure 1–11
Seeds were offered through the mail by placing orders through seed catalogs. (a) Liberty Hyde Bailey's copy of Bernard McMahon's *Catalogue of American Seeds*. (b) The Shakers from Mount Lebanon, New York, pioneered the use of retain seed packets.

eleven editions. His Philadelphia seed house sold over 1,000 species of plants (Fig. 1–11a). The Shakers in Mount Lebanon, New York, began packaging seeds in individual envelopes for local retail sales in the early 1800s (Fig. 1–11b). The first seed catalog in color was produced in 1853 by B. K. Bliss. At the turn of the 20th Century, these mail order catalogs became wonderful lithographic works of art (Fig. 1–12).

Figure 1–12
The seed business was competitive, so companies produced colorful mail order seed catalogs to attract potential customers.

> BOX 1.1 **GETTING MORE IN DEPTH ON THE SUBJECT**
> PLANT PROPAGATION ORGANIZATIONS
>
> **American Seed Trade Association (ASTA)** This organization of seed companies has been serving the industry since 1883. ASTA holds a general meeting each year and sponsors conferences on specific crops. It publishes a newsletter, an annual yearbook, and proceedings of individual conferences. It participates in regulatory activities that affect the seed industry. (http://amseed.com)
>
> **American Society for Horticultural Science (ASHS)** This organization has a membership of public and private scientists, educators, extension personnel, and industry members with an interest in horticulture. The organization holds annual national and regional meetings and publishes scientific reports in the *Journal of American Society for Horticultural Science, HortScience,* and *HortTechnology*. It includes working groups in all propagation areas. (http://www.ashs.org)
>
> **Association of Official Seed Analysts, Inc. (AOSA)** Membership is seed laboratories, both private and governmental, mostly in the continental United States. The association holds an annual meeting and publishes the *Journal of Seed Technology*. They provide numerous handbooks on the rules for seed testing, seed sampling, purity analysis, etc. They also provide a seed technologist's training manual. (http://www.aosaseed.com/)
>
> **Association of Official Seed Certifying Agencies (AOSCA)** Originally organized in 1919 as the International Crop Improvement Association, membership includes United States and Canadian agencies responsible for seed certification in their respective areas. These agencies maintain a close working relationship with the seed industry, seed regulatory agencies, governmental agencies involved in international seed market development and movement, and agricultural research and extension services. (http://aosca.org/)
>
> **International Fruit Tree Association** This organization is for members interested in fruit tree rootstocks and propagation but also includes cultural aspects. An annual meeting is held, and the proceedings are published in *Compact Fruit Tree*. (http://www.ifruittree.org/)
>
> **International Plant Propagators Society (IPPS)** The society was organized in 1951 to recognize the special skills of the plant propagator and to foster the exchange of information among propagators. The organization has expanded to include Eastern, Western, and Southern Regions of the United States; Great Britain and Ireland; Australia; New Zealand; Japan; and a Southern African Region. Each region holds an annual meeting, and their papers are published in a *Combined Proceedings*. (http://www.ipps.org)
>
> **International Seed Testing Association (ISTA)** This is an intergovernmental association with worldwide membership accredited by the governments of 59 countries and involving 137 official seed-testing associations. The primary purpose is to develop, adopt, and publish standard procedures for sampling and testing seeds and to promote uniform application of these procedures for evaluation of seeds moving in international trade. Secondary purposes are to promote research in all areas of seed science and technology, to encourage cultivar certification, and to participate in conferences and training courses promoting these activities. They hold an annual conference and publish the *Seed Science and Technology* journal, as well as a newsletter, bulletins, and technical handbooks on seed testing. (http://www.seedtest.org/en/home.html)
>
> **International Society for Horticultural Science (ISHS)** This organization is an international society for horticultural scientists, educators, extension, and industry personnel. It sponsors an International Horticultural Congress every four years as well as numerous workshops and symposia. Proceedings are published in *Acta Horticulture*. A newsletter, *Chronica Horticulturae*, is published four times per year. (http://www.ishs.org)
>
> **American Nursery and Landscape Association (ANLA)** Organized in 1875 as the American Association of Nurserymen, this association is a national trade organization of the United States nursery and landscape industry. It serves member firms involved in the nursery business—wholesale growers, garden center retailers, landscape firms, mail-order nurseries, and allied suppliers to the horticultural community. (http://www.anla.org/)
>
> **Society for In Vitro Biology (SIVB)** This organization is composed of biologists, both plant and animal, who do research on plant cellular and developmental biology, including the use of plant tissue culture techniques. The organization publishes the journal *In Vitro Cellular and Developmental Biology—Plant* and holds an annual meeting. (http://www.sivb.org/)
>
> **Southern Nursery Association (SNA)** An organization of nurseries in the southeastern United States, this trade organization has annual conferences and publishes newsletters and conference proceedings. (http://www.sna.org)

The establishment of the nursery industry in the Pacific Northwest was a unique accomplishment (17). In the summer of 1847, Henderson Lewelling of Salem, Iowa, established a traveling nursery of grafted nursery stock growing in a mixture of soil and charcoal in boxes on heavy wagons pulled by oxen, which crossed the Great Plains, covering 2,000 miles to Portland, Oregon. The 350 surviving trees were used to establish a nursery at Milwaukee, Oregon.

THE MODERN PLANT PROPAGATION INDUSTRY

The present-day plant propagation industry is large and complex, and involves not only the group that multiplies plants for sale and distribution, but also a large group of industries that provides services, sells the product, is involved in regulation, provides consultation, carries on research, or is involved in teaching. The key person within this complex is the plant propagator who possesses the knowledge and skills either to perform or to supervise the essential propagation task for specific plants. In 1951, the Plant Propagator's Society was established to provide the nursery profession with knowledge and research support.

DISCUSSION ITEMS

Modern day plant propagation is a complex, many faceted industry that represents a synthesis of different skills. Underlying these skills is a love and appreciation for the rich history and importance plant propagation has played in agriculture development.

1. Discuss how the relationship between the domestication of plants has been symbiotic with human development.
2. Discuss the relationship between plant selection and domestication with methods of plant propagation.
3. The number of plant species used for food is relatively small. Speculate on some of the reasons why.
4. The terms "agriculture," "forestry," and "horticulture" became distinct disciplines during the medieval period of human history. What do you see as the differences in these disciplines that led to their separation in medieval times, and does this relate to our modern views of these disciplines?
5. Why do you think the "modern" nursery developed and how did the period of plant exploration relate to nurseries?
6. Visit the web site of a professional organization and discuss why you think membership would be important to a person working in plant propagation or horticulture.

REFERENCES

1. Adriance, G. W., and F. R. Brison. 1955. *Propagation of horticultural plants.* New York: McGraw Hill.
2. Agricola, G. A. 1716. A philosophical treatise of husbandry and gardening.
3. Bailey, L. H. 1891 (revised, 1896). *The nursery book.* Harrisburg, PA: Mount Pleasant Press, J. Horace McFarland.
4. Bailey, L. H. 1900–1902. 4th ed. 1906. *Cyclopedia of American horticulture.* New York: Macmillan.
5. Bailey, L. H. 1914–1917. *Standard cyclopedia of horticulture.* 3 vols. New York: Macmillan.
6. Bailey, L. H. 1920 (revised). *The nursery manual.* New York: Macmillan.
7. Bailey, L. H. 1930. *Hortus.* New York: Macmillan.
8. Bailey, L. H., and E. Z. Bailey. 1941. *Hortus second.* New York: Macmillan.
9. Bailey, L. H., E. Z. Bailey, and staff of Bailey Hortorium. 1940, 1949. *Manual of cultivated plants.* New York: Macmillan.
10. Bailey, L. H., E. Z. Bailey, and staff of Bailey Hortorium. 1976. *Hortus third.* New York: Macmillan.
11. Baltet, C. 1910. *The art of grafting and budding.* 6th ed. London: Crosby Lockwood (quoted by Hottes, 1922).
12. Columella, L. J. M. 1948. *De re rustica.* Loeb classical library. William Heinemann Ltd. London and Harvard University Press. Boston.
13. Cunningham, I. S. 1984. *Frank N. Meyer: Plant hunter in Asia.* Ames, IA: Iowa State University Press.
14. Darwin, C. 1859. *The origin of species by means of natural selection, or the preservation of favoured races in the struggle for life.* London: J. Murray.
15. Darwin, C. 1868. *The variation of animals and plants under domestication.* London: J. Murray.
16. Davidson, H., R. Mecklenburg, and C. Peterson. 2000. *Nursery management.* 4th ed. Upper Saddle River, NJ: Prentice Hall.
17. Duruz, W. P. 1st ed. 1949, 2nd ed. 1953. *The principles of nursery management.* New York: A. T. de la Mare Co.

18. Fairchild, D. 1938. *The world was my garden.* New York: Scribner's.

19. Fuller, A. S. 1887. *Propagation of plants* (quoted by Hottes, 1922).

20. Gerard, J. 1597. *The Herball or General Historie of Plants.*

21. Harlan, J. R. 1992. *Crops and man.* 2nd ed. Madison, WI: Amer. Soc. of Agron., Inc. Crop Science of America.

22. Hartmann, H. T., and D. E. Kester. 1959. *Plant propagation: principles and practices.* Englewood Cliffs, NJ: Prentice-Hall.

23. Hartmann, H. T., A. M. Kofranek, V. E. Rubatsky, and W. J. Flocker. 1988. *Plant science: Growth, development and utilization of cultivated plants.* 2nd ed. Englewood Cliffs, NJ: Prentice Hall.

24. Hottes, A. C. 1917, 1922 (revised). *Practical plant propagation.* New York: A. T. de la Mare Co.

25. Janick, J., R. W. Shery, F. W. Woods, and V. W. Ruttan. 1969. *Plant science.* San Francisco: W. H. Freeman.

26. Kains, M. G. 1916, 1920. *Plant propagation: Greenhouse and nursery practice.* New York: Orange Judd Publishing Co.

27. Kains, M. G., and L. M. McQuesten. 1938, 1942, 1947. *Propagation of plants.* New York: Orange Judd Publishing Co.

28. Mahlstede, J. P., and E. S. Haber. 1957. *Plant propagation.* New York: Wiley.

29. Malpighi, M. 1675. *Anatome plantarum.* London.

30. Plinius Secondus. 1962. *The history of the world.* Carbondale, IL: Southern Illinois University Press.

31. Reed, H. S. 1942. *A short history of the plant sciences.* New York: The Ronald Press Co.

32. Sauer, C. O. 1969. *Agricultural origins and dispersal.* 2nd ed. Cambridge, MA: Massachusetts Institute of Technology Press.

33. Seeley, J. G. 1990. Liberty Hyde Bailey—Father of Modern Horticulture. *HortScience* 25:1204–9.

34. Simmonds, N. W., ed. 1976. *Evolution of crop plants.* London: Longman Group Limited.

35. Solbrig, O. T., and D. J. Solbrig. 1994. *So shall you reap: Farming and crops in human affairs.* Washington, DC: Island Press.

36. Theophrastus. 1961. *De causis plantarum.* Loeb classical library. William Heinemann Ltd. London and Harvard University Press. Boston.

37. Theophrastus. 1961. *De historia plantarum.* Loeb classical library. William Heinemann Ltd. London and Harvard University Press. Boston.

38. Ward, N. B. 1842. *On the growing of plants in closely glazed cases.* 2nd ed. London: J. van Voorst.

39. Zohary, D., and P. Spiegel-Roy. 1975. Beginnings of fruit growing in the old world. *Science* 187(4174):319–27.

*2 Biology of Plant Propagation

learning objectives

- Describe the basic life cycles of plants as related to sexual (seed) and asexual (vegetative) propagation.
- Explain the rules for naming plants.
- Describe how ownership of cultivars can be controlled.
- Explain the difference between mitosis and meiosis.
- Describe how genes and gene expression impact plant growth and development.
- Identify plant hormones and their role in plant development.

INTRODUCTION

The natural world is covered by populations of many different kinds of plants that have evolved over eons of time. We identify these as **species**, although there are other divisions that will be described in this text. These populations can more or less maintain themselves from generation to generation because of their natural genetic characteristics. If not, they evolve into other variants or become extinct.

In agriculture and horticulture, on the other hand, propagators primarily deal with special kinds of plants, which are defined as **cultivars (varieties)** (9). We buy 'Thompson Seedless' grapes and 'Elberta' peaches for our table, grow 'Queen Elizabeth' roses and 'Bradford' pear trees in our landscape, and plant 'Hybrid Yellow Granex' onion seed and 'Marquis' wheat in our fields. All of these represent populations of plants that are unique and only exist in cultivation. These plants would likely change drastically, or disappear altogether, if not maintained by genetic selection during propagation. Chapters 19, 20, and 21 describe the range of cultivated plants and their propagation.

Plant propagation and plant breeding both involve genetic selection. The role of the plant breeder is *to recreate patterns of genetic variation in its many forms from which to select new kinds of plants useful to humans.* The role of the plant propagator, on the other hand, is *to multiply these selected cultivars and to do it in such a manner as to maintain the genetic characteristics of the original population.* To do both requires an understanding of genetic principles and procedures.

BIOLOGICAL LIFE CYCLES IN PLANTS

Plant Life Cycles

In natural systems, plant life cycles can be described based on their life span and reproductive pattern. Therefore, they are referred to as annuals, biennials, or perennials:

1. **Annuals** are plants that complete the entire sequence from germination to seed dissemination and death in one growing season. Technically, annuals are monocarpic, meaning that they die after reproducing. However, "annuals" also refers to plants that may be perennial in mild climates but are not winter hardy, and so die after the first growing season due to cold temperatures.

2. **Biennials** are plants that require two growing seasons to complete their life cycle. During the first year, the plants are vegetative and grow as low clumps or a rosette of leaves. These plants usually need a period of cold weather for **vernalization** of the shoot meristem before they can become reproductive. During the second season, biennial plants bolt, producing a fast-growing flowering spike, flower, produce seeds, and then die. Although the terminology is confusing, winter annuals fit into this category. Seeds germinate in late summer, forming a seedling with numerous rosette leaves that hug the ground. After winter vernalization, the meristem bolts, flowers, sets seeds, and dies before summer (less than 12 months).

3. **Perennials** are plants that live for more than 2 years and repeat the vegetative-reproductive cycle annually. Perennial cycles tend to be related to seasonal cycles of warm-cold (temperature climates) or wet-dry periods (tropical climates). Both herbaceous and woody plants can be perennial:

   a. **Herbaceous perennials** produce shoots that grow during one season and die back during the winter or periods of drought. It may take herbaceous perennials several growing cycles before they become reproductive, and they may not flower every year, depending on the plant's accumulation of resources during the growing cycle. Plants survive during adverse conditions as specialized underground structures with roots and crown that remain perennial. Geophytes (bulbs, corms, rhizomes, tubers; see Chapters 15) are included in this group.

   b. **Woody perennials** develop permanent aboveground woody stems that continue to increase annually from apical and lateral buds with characteristic growth and dormancy periods. Woody perennials are trees and shrubs.

clonal propagation
A group of plants originating from a single source plant by vegetative propagation.

In horticultural systems, plant life cycles can also be described based on their propagation methods. Here they can be described based on the seedling, **clonal**, and **apomictic** life cycles.

Life Cycles of Seedling Cultivars

In propagation, an individual plant that develops from a seed is referred to as a **seedling** whether it is an **annual, biennial, herbaceous perennial,** or **woody perennial.** During the **life cycle of a seedling,** the sequence of growth and development is separated into four broad **phases** (Fig. 2–1a) (10, 25, 29, 46).

seedling life cycle
Growth and development of a plant when propagated from a seed.

Phase I Embryonic This phase begins with the formation of a zygote. This cell grows into an **embryo,** which receives nourishment from the mother plant through physiological stages of development. At first, growth involves cell division of the entire embryo as it increases in size. Later, growth potential develops with a polar orientation as the embryo develops its characteristic structure. These embryonic changes are described in detail in Chapter 4.

Phase II Juvenile Seed germination initiates a dramatic change from the embryonic pattern to the developmental pattern of the young seedling. Vegetative growth is now polar, extending in two directions via the shoot and root axis. Cell division is concentrated in the root tips, shoot tips, and axillary growing points. Subsequently, the extension of the root and shoot is accompanied by an increase in volume. New nodes are continually laid down as leaves and axillary growing points are produced. Lateral growing points produce only shoots that are not competent to flower. The juvenile period is the growth stage where plants cannot flower even though the inductive flowering signals are present in the environment (33, 61).

Phase III Transition The vegetative period at the end of the juvenile phase and prior to the reproductive stage is marked by subtle changes in growth and morphology. Growth tends to decrease as the plants enter the reproductive period when flowering occurs. The important point is that the developmental potential of the growing points is sensitive to particular signals, partly internal, although often dictated by cues from the environment such as changes in day length and chilling.

Phase IV Adult (or Mature) During this phase, shoot meristems have the potential to develop flower buds, and the plant produces flowers, fruits, and seeds.

The duration and expression of these phases represent fundamental variation in plant development, which is analogous to comparable phases in animal development. The most conspicuous expression of phases occurs in long-lived perennial plants, such as trees and shrubs, where conspicuous differences in juvenile and mature traits may be observed in the same plant. Nevertheless, phase changes have been

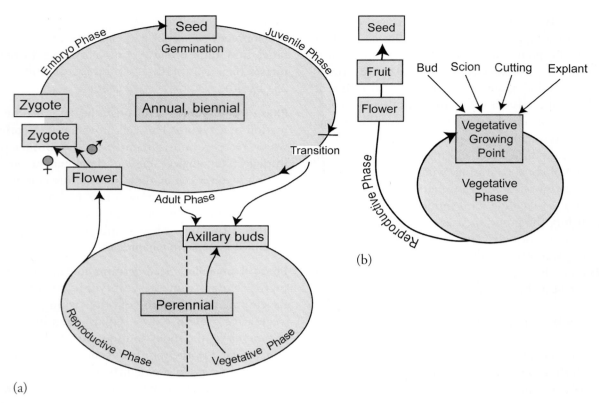

Figure 2–1
Seedling and clonal life cycles. (a) Seedling cycle in plants. Model illustrates epigenetic changes involving embryo development, juvenile, transition, and adult phases. In the annual or biennial, the apical meristem progresses more or less continuously through one (*annual*) or two (*biennial*) growing seasons (*top circle*). In herbaceous and woody perennials (*bottom circle*), the adult vegetative meristem is renewed continuously by seasonal cycles of growth and development. (b) A clonal life cycle results when a plant originates by vegetative propagation. The type of growth, time to flower, and other characteristics may vary among different propagules depending on the location on the seedling plant from which the propagule was taken. With continued vegetative propagation, the clone is stabilized at its mature form by characteristic consecutive vegetative and reproductive phases.

identified in annual plants, such as maize (61), and must be recognized as a fundamental aspect of all plant development.

The following characteristics of plant development are associated with phase change:

- *Time of flowering* (52, 79, 85). The age when flowering begins is the most characteristic aspect of phase change. Time of first flowering varies from days to a few months in some annuals to as much as 50 years in some perennials (Table 2–1). Usually, flowering begins in the upper and peripheral parts of the tree where shoots and branches have attained the prerequisite phase.
- *Morphological expression of leaves and other structures.* Leaf form in the juvenile phase sometimes differs radically from that of the adult phase (Fig. 2–2). English ivy is a classic example of phase change, as illustrated in Figures 2–3 and 2–4. Juvenile parts of apple, pear, and citrus seedlings may be very thorny, although the trait disappears in the adult phase (33, 80).
- *Potential for regeneration* (34, 80). Each phase tends to have a differing potential for regeneration. For instance, cuttings taken from the juvenile phase usually have a higher potential for rooting than do cuttings from the adult phase. An in-depth discussion of the impact of phase change on propagation is found in Chapter 16.

Life Cycles of Apomictic Cultivars

Apomixis is a natural reproductive process possessed by some species of plants in which the embryo develops directly from specific vegetative cells of some part of the reproductive structure that has not undergone meiosis (50). The result is that an asexual process has replaced the normal sexual process.

| **apomixis** Reproduction in which vegetative cells in the flower develop into zygotes to create seeds by a clonal reproduction process.

Table 2–1
AGE OF FLOWER DEVELOPMENT IN SOME WOODY PLANTS

Species	Length of juvenile period
Rose (*Rosa* spp.)	20–30 days
Grape (*Vitis*)	1 year
Stone fruits (*Prunus* spp.)	2–8 years
Apple (*Malus* spp.)	4–8 years
Citrus (*Citrus* spp.)	5–8 years
Scotch pine (*Pinus sylvestris*)	5–10 years
Ivy (*Hedera helix*)	5–10 years
Birch (*Betula pubescens*)	5–10 years
Sequoia (*Sequoia sempervirens*)	5–10 years
Pear (*Pyrus* spp.)	6–10 years
Pine (*Pinus monticola*)	7–20 years
Larch (*Larix decidua*)	10–15 years
Ash (*Franxinus excelsior*)	15–20 years
Maple (*Acer pseudoplatanus*)	15–20 years
Douglas-Fir (*Pseudotsuga menziesii*)	20 years
Bristlecone pine (*Pinus aristata*)	20 years
Redwood (*Sequoiadendron giganteum*)	20 years
Norway spruce (*Picea abies*)	20–25 years
Hemlock (*Tsuga heterophylla*)	20–30 years
Sitka spruce (*Picea sitchensis*)	20–35 years
Oak (*Quercus robur*)	25–30 years
Fir (*Abies amabilis*)	30 years
Beech (*Fagus sylvatica*)	30–40 years

(a) (b)

Figure 2–2
In some woody plants, there is a dramatic change in leaf shape (foliar dimorphism) that accompanies the change from juvenile (red arrows) to mature phase (white arrows).
(a) *Eucalyptus*; (b) *Pseudopanax*.

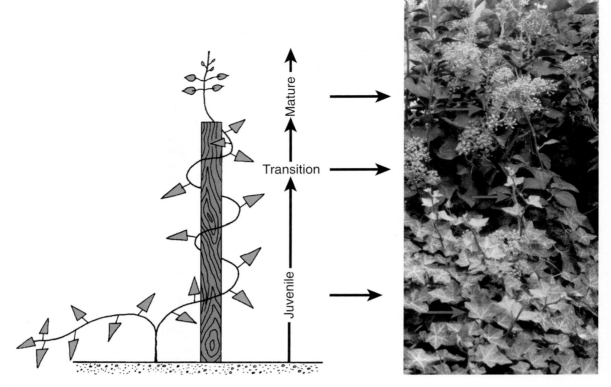

Figure 2–3
Phase change in Ivy (*Hedera helix*) in which the *juvenile* (non-flowering) phase is a vine which, as it grows into a vertical form, undergoes a *transition* into the *mature* (adult) flowering and fruiting phase.

apomictic life cycle
Growth and development of a plant when propagated from an apomictic embryo.

Different types of these phenomena are described in Chapter 4. The **apomictic life cycle** (not shown) is the same as the seedling cycle, except that the embryo is essentially a clone since it is produced as a result of mitosis and is asexual. Plants of the apomictic cycle go through the same phase changes as the sexual life cycle. The significance of this phenomenon is described in Chapter 5.

Life Cycles of Clonal Cultivars

clonal life cycle Growth and development of a plant when propagated vegetatively from a particular propagule of an individual plant.

Two essential aspects characterize **clonal life cycles** (Fig. 2–1b) (46):

- A clone originates by vegetative propagation from an individual plant using various types of **vegetative propagules.** The basic kinds are bud, scion, cutting, layer, bulb, corm, tuber, and explant. Depending on their history and origin, each of these propagules may represent a different phase of the seedling cycle.

- The phase-potential of the propagule is maintained during vegetative propagation such that the progeny plants may vary significantly in their morphological characteristics. For instance, Figure 2–4 compares the appearance of a plant propagated from the juvenile and mature phase of English ivy and *Chamaecyparis*.

TAXONOMY

Organisms are named in a hierarchical system described as their taxonomy. A sample hierarchy is provided for apple (Table 2–2).

The basic system for naming plants was introduced by Linnaeus (Fig. 2–4) as the Latin system of binomial nomenclature using a **genus** and **species** name for each plant (each of which are italicized). The genus describes a group of plants that are similar in morphological, biochemical, and genetic properties. The species is used to designate a

species The natural grouping of plants that have common characteristics in appearance, adaptation, and breeding behavior (i.e., can freely interbreed with each other).

Figure 2–4
The juvenile or mature phase may be retained by vegetative propagation. (a) Juvenile and (b) mature forms of English ivy (*Hedera helix*). The juvenile form is a vine, while the mature form is a three-foot shrub with terminal inflorescences. (c) Mature and (d) juvenile foliage forms of false cypress (*Chamaecyparis*).

Table 2–2
THE TAXONOMIC HIERARCHY FOR APPLE

Classification

Kingdom: Plantae
Division: Spermatophyta
  Subdivision: Angiospermae
    Class: Dicotyledonae
      Order: Rosales
        Family: Rosaceae
          Genus: *Malus*
            species: *domestica*

population of plants within a genus that can be recognized and reproduced as a unit (51). The rules for naming plants are maintained by the International Association of Plant Taxonomists under the longstanding International Code of Botanical Nomenclature (http://ibot.sav.sk/icbn/main). In nature, individuals within one species normally interbreed freely but do not interbreed well with members of another species. Geographical isolation or some physiological, morphological, or genetic barrier prevents gene exchange between them. A true species can usually be propagated and maintained by seed but may require some control during propagation.

Cultivated plants may also be designated by binomial name even though they may be a complex hybrid rather than a distinct "natural" species (51, 72). For example, peach cultivars are variations within a recognized species *Prunus persica* L., but the European prune (*Prunus domestica* L.) is a complex hybrid that apparently developed in cultivation. **Cultivars** may also be derived from repeated vegetative propagation of an initial desirable mutation. The rules for naming cultivated plants are spelled out in the *International Code of Nomenclature for Cultivated Plants* (9).

cultivar A group of plants that have originated in cultivation, are unique and similar in appearance, and whose essential characteristics are maintained during propagation.

Eastern redbud illustrates the various subgroups occurring in selected or natural populations within a species (Fig. 2–6, page 21):

Genus and species: *Cercis canadensis* L.
Subspecies: *Cercis canadensis* subsp. *texensis*.

 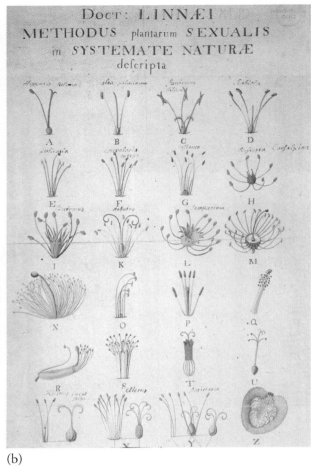

Figure 2–5
Linnaeus was important in championing the binomial system for naming plants. (a) Portrait as a young man in Sweden. (b) The Linnean system grouped plants based on the number of male and female parts of the flower as illustrated in this old plate, "The Sexual System of Linneaus." Ehret, 1736.

Botanical variety (*varietas* in Latin): *Cercis canadensis* var. *alba*

Cultivar: *Cercis canadensis* cv. Forest Pansy or 'Forest Pansy'

In some cases, breeders have been able to make genetic crosses between different species or even between genera. Interspecific hybrids within a genus are designated with an "x" between the genus and species (i.e., *Viburnum xburkwoodii*, which is a hybrid between *V. carlesii* and *V. utile*). Intergeneric hybrids are formed between genera within a family and are designated with an "x" before the new genus name, which is a contraction of the two genera names (i.e., *xFatshedera lizei* is an intergeneric hybrid between *Fatsia japonica* and *Hedera helix*).

There are a number of web sites that provide information on current taxonomy for plant names:

International Plant Names Index	http://www.ipni.org/index.html
USDA PLANTS database	http://plants.usda.gov/
USDA Germplasm Resources Information Network (GRIN) Taxonomy for Plants	http://www.ars-grin.gov/cgi-bin/npgs/html/index.pl
eFloras.org	http://www.efloras.org/index.aspx
World checklist of plant families	http://www.kew.org/ (choose Scientific Research & Data, and in the search box enter World Checklist of Selected Plant Families)

(a) **Genus and species:** *Cercis canadensis* L. The authority indicates who is responsible for giving this plant its name. In this case "L." is for Linnaeus.

(b) **Botanical variety:** *Cercis canadensis* var. *alba* (white-flowered eastern redbud). A botanical variety is considered a variant that occurs in the wild, but its differences from the species are less distinct compared to a subspecies.

(c) **Subspecies:** *Cercis canadensis* subsp. *texensis*. A subspecies is a group of variants that occur consistently in nature. They can be viewed as the beginning of a new species. They are often geographically isolated from the main species.

(d) **Cultivar:** *Cercis canadensis* cv. Forest Pansy or 'Forest Pansy'. A cultivar (cultivated variety) can be set off from the species by the "cv." abbreviation or by single quotes.

Figure 2–6
Major categories for naming plants include Genus, species, botanical variety, subspecies, and cultivar.

Names for new plants should be registered with the proper registration authority. The International Society for Horticultural Sciences provides a home for the Commission for Nomenclature and Cultivar Registration (http://www.ishs.org/sci/icra.htm). They provide a link to individuals or organizations that maintain the registry for a single genus or group of plants. For example, the registry for English ivy (*Hedera*) is maintained by the American Ivy Society, while woody plants without specific registries are handled by the American Public Gardens Association.

LEGAL PROTECTION OF CULTIVARS

In modern agricultural and horticultural industries, individual cultivars and breeding materials have commercial value and, according to law, are entitled to legal protection as is any invention made by humans (17, 40, 42, 59). The right to propagate specific cultivars that are developed through controlled selection and/or breeding programs can be protected by a number of legal devices. These allow the originator to control their distribution and receive monetary awards for their efforts.

Legal protection has been available in the United States with the passage of the Townsend-Purnell Act in 1930, which added vegetatively propagated plants to the general patenting law for inventions. Protection was provided to seed-propagated cultivars by the 1970 Plant Variety Protection Act, revised in 1994 (4). Many countries of the world have legal systems that grant protection to patents and breeders' rights, and a large network of such programs have developed. Guidelines have been produced by the International Union for the Protection of New Varieties of Plants (http://www.upov.int/index_en.html) in 1961, 1972, 1978, and 1991 (77) and the Food and Agriculture Organization of the United Nations (38). Propagators need to be aware of the rights and obligations under these particular conditions (see Box 2.1, page 22).

GENETIC BASIS FOR PLANT PROPAGATION

The life cycle of plants begins with a single cell known as a **zygote.** This cell is the result of the union of male and female gametes. From this initial cell, additional cells

> **BOX 2.1 GETTING MORE IN DEPTH ON THE SUBJECT**
> **LEGAL PROTECTION OF CULTIVARS**
>
> **Patent** A **plant patent** is a grant from the United States Patent and Trademark Office, which extends patent protection to plants. Exclusive rights are given to the inventor of a "distinct and new" kind of plant (cultivar) for a 20-year period. Only vegetatively propagated cultivars are covered—not tuber-propagated plants. A plant growing wild is not considered patentable. There is no necessity to prove that the cultivar is superior, only that it is "new and different." To obtain information, contact the United States Patent and Trademark Office, Washington, DC 20231 (http://www.uspto.gov).
>
> *plant patent* Legal protection of a vegetatively propagated cultivar (except tuber) granted by the United States Patent and Trademark Office to allow the inventor of the plant to control its propagation.
>
> **Plant Variety Protection** The United States **Plant Variety Protection Act (PVPA)** extends plant patent protection to seed-propagated cultivars that can be maintained as "lines," including F_1 hybrids. Tuber-propagated plants are also protected. The new cultivar must be novel, distinctive, and stable. A plant-breeding certificate allows breeders propagation protection for many agricultural and horticultural crops propagated by seed, including such crops as cotton, alfalfa, soybean, and marigolds. The length of time is 20 years for most plants, but 25 for trees, shrubs, and vines. These rights may be sold or licensed. To obtain information, contact the Plant Variety Protection Office, USDA National Agricultural Library Building, Room 500, 10301 Baltimore Blvd., Beltsville, MD 20705, USA. It is also available at the USDA's web site in PDF form (http://www.ams.usda.gov/AMSv1.0/; Type "Plant Variety Protection Act" in the Search box, choose the link for "Plant Variety Protection Act [PDF]")
>
> *plant variety protection* Legal protection granted by the United States Plant Variety Protection Act for a seed-propagated cultivar; a plant-breeding certificate allows the inventor of the plant to control its propagation.
>
> **Trademarks** A registered trademark offers protection for a name that indicates the specific origin of a plant (or product). The trademark is any word, symbol, device, logo, or distinguishing mark. It is granted for 10 years but can be renewed indefinitely as long as it remains in use. The trademark is distinct from the cultivar name and both identities should be provided. Unfortunately, the ways nurseries are using trademark names can confuse and even mislead consumers. For example, *Acer rubrum* 'Franks Red' is the cultivar name for the popular Red Sunset® maple, although most consumers assume Red Sunset is the cultivar name. The owner of the Red Sunset trademark has every right to use that name for a different red maple cultivar if he chose to make that change because the trademark is a company mark that is not permanently linked to *Acer rubrum* 'Frank's Red'. There are also examples where the same cultivar is being sold under numerous trademark names by different companies. This is the case for *Loropetalum chinensis* 'Hines Purple Leaf' that is being sold under the trademark names Plum Delight and Pizzaz even though they are the same plant.
>
> **Utility Patents** This protection is under the general patent law, which uses the criteria of novelty and utility. An application requires the same full description as a plant patent. It may include more than one claim that involves specific uses of the plant. Utility patents are used by commercial biotechnology and engineering firms to control the use of specific genes and technologies.
>
> **Other Methods** *Contracts* can be used to control the propagation of specific plants as well as the selling of their fruit or other products. Enforcement comes under contract law. Trade secrets are protected by law and can provide some protection for disclosure of certain technology. This may include information that is not disclosed to the public, or temporary protection prior to disclosure for patent application. *Copyrights* have the purpose of preventing unauthorized reproduction or copies of printed material. Although this device could apply to plant materials, copyrights are usually used to control reproduction of pictures or printed material about the plant that is used in brochures or catalogs.

multiply and develop the body of the plant. Living plant cells contain a **nucleus** embedded within the **cytoplasm,** all enclosed within a **cell wall** (Fig. 2–7). The nucleus contains the genetic material that directs growth and development by determining when particular **RNAs (ribonucleic acid)** and proteins are made by a cell. **Chromosomes** within the nucleus contain **DNA (deoxyribonucleic acid)** that forms the genetic blueprint for heredity. DNA is present in two other structures of the cell—**chloroplasts** and **mitochondria.** Individual characteristics and traits are associated with sequences of DNA nucleotides coded on the chromosome as **genes.** Genetic information is passed along from cell to cell during cell division.

chromosome Structures within the nucleus of a cell that contain the genes.

DNA (deoxyribonucleic acid) The basic biochemical compound that makes up the gene.

Gene Hereditary unit of inheritance now known to be composed of specific arrangements of nucleotides to make up a genetic code.

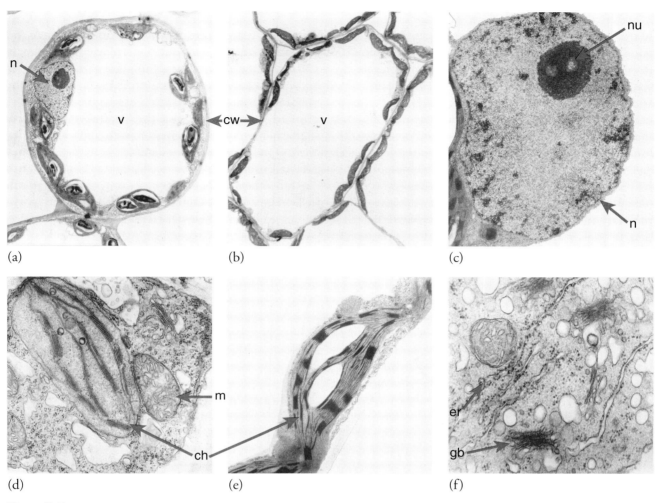

Figure 2–7
Electron micrographs of cells and cell components. (a) A mesophyll cell; (b) parenchyma cell with a large central vacuole and cytoplasm and organells pushed against the cell wall; (c) nucleus and nucleoli; (d) chloroplast and mitochondria; (e) mature chloroplast with starch; (f) Golgi body and endoplasmic reticulum. Abbreviations: n—nucleus; nu—nucleolus; cw—cell wall; ch—chloroplast; m—mitochondria; gb—Golgi body; er—endoplasmic reticulum; v—vacuole.

Cell Division

There are two types of cell division in plants—mitosis and meiosis. **Mitosis** is cell division in vegetative tissue used for growth, while **meiosis** is a reductive division used during the sexual reproductive cycle to produce gametes.

mitosis The special kind of cell division that results in vegetative propagation.

meiosis The special kind of cell division that results in sex cells, which are utilized in sexual reproduction.

Mitosis The **cell cycle** (24) is the period from the beginning of one cell division to the next (Fig. 2–8). The cell cycle is divided into a two parts: interphase and mitosis. Interphase is composed of three phases: G_1, S, and G_2. During the G_1 (G stands for gap) phase, there are active biochemical processes that increase the internal contents of the cell as well as its size. Cells that are not

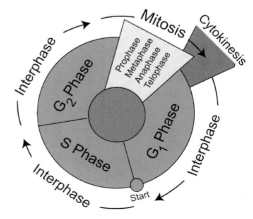

Figure 2–8
Cell cycle – see text for details.

preparing for cell division are arrested in the G_1 phase. In order for the cell cycle to proceed, there is a critical point referred to as the "start" where the cell commits to cell division. Progression through the cell cycle is controlled by proteins called cyclin-dependent protein kinases. The S (synthesis) phase involves DNA replication and synthesis. During the second gap phase (G_2), the cell, which now has replicated sets of chromosomes, prepares to partition these into two identical daughter cells during the cell division phase of mitosis.

Mitosis is separated into four phases (prophase, metaphase, anaphase, and telophase) related to the way the chromosomes appear within the dividing cell (Fig. 2–9). During ***prophase,*** chromosomes condense and appear as short, thickened structures with distinctive morphology, size, and number. The chromosomes exist as homologous pairs of chromatids attached together at their centers by centromeres. After the nuclear envelope disappears, **metaphase** spindle fibers form and the chromosomes migrate to the center of the cell. In **anaphase,** the mitotic spindle fiber microtubules attached to each chromosome pair at the centromere contract, pulling the chromosomes apart. The daughter chromosomes move to opposite ends of the cell in preparation for division. Nuclear envelopes reform around the separated daughter chromosomes during **telophase.** The phragmoplast forms at the cell's center. The phragmoplast is the initial formation of the cell plate, which will eventually form the new cell wall. The chromosomes again become less distinct within the nuclear matrix as the cell cycle proceeds from mitosis to interphase. Cell division ends with

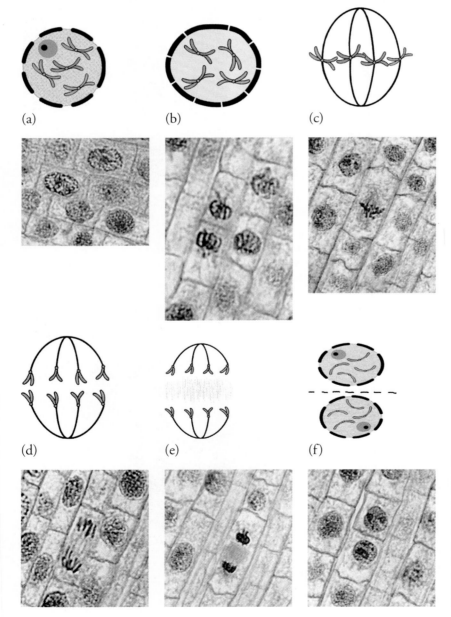

Figure 2–9
Stages in mitosis. (a) Early prophase, chromosomes begin to condense as nuclear envelope and nucleolus begin to deteriorate. (b) Prophase, chromosomes thicken and become conspicuous. (c) Metaphase, chromosomes line up across the center. (d) Anaphase, chromosomes separate. (e and f) Early and late Telophase, cell plate is laid down to produce two new cells.

cytokinesis, which is the division of the cytoplasm by the completed new cell wall. The result is the production of two new cells identical in genotype to the original cell.

Growth by mitosis increases the vegetative size of the plant. Cells may undergo enlargement, differentiation, and development into different kinds of cells (e.g., **parenchyma, collenchyma, fibers, and sclereids**) (Fig. 2–10). Parenchyma cells represent the basic living cell type. It is a living cell with a primary cell wall that is metabolically active and capable of differentiating into specific cell types. These may be for reserve storage as in endosperm cells or specialized for photosynthesis as the palisade and spongy mesophyll layers of the leaf. They may also develop into cells that provide structural support for stems and leaves or protective layers for seeds. These include collenchyma cells that are living cells with thickened primary cell walls. Collenchyma is usually found just below the epidermis in herbaceous and woody stems. Fibers and sclereids are examples of sclerenchyma cells that are nonliving at maturity. These have thick secondary walls that provide strength and structural support.

Eventually, cells differentiate into **tissues** (e.g., **xylem, phloem**) and **organs** such as stems, roots, leaves, and fruit (Fig. 2–11). Cells capable of dividing are referred to as **meristematic** and are located in primary or apical meristems (shoot and root tips) and secondary

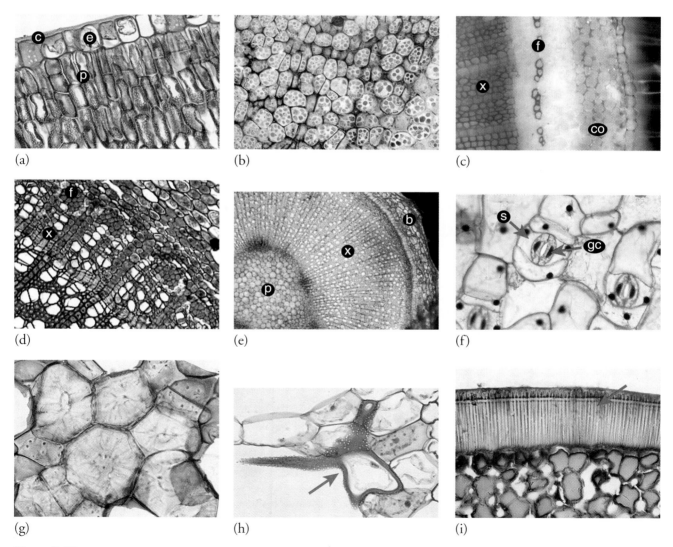

Figure 2–10
Different cell types in plants. (a) Cross-section of the adaxial portion of a leaf showing cuticle—c, epidermis—e, and palisade—p cells. (b) Parenchyma cells in an endosperm with storage bodies. (c) Cross-section of tomato stem showing xylem—x, phloem fibers—f, and collenchyma—co cells. (d) Cross-section of a woody plant stem showing xylem—x, and fibers—f. (e) Cross-section of azalea stem showing pith—p, xylem—x, and bark—b. (f) Lower (abaxial) surface of a leaf showing stomates with guard cells. (g, h, and i) Three types of sclereid cells: (g) brachysclereids, or stone cells, in pear fruit, (h) trichosclereids in water lily, (i) macrosclereids in a legume seed.

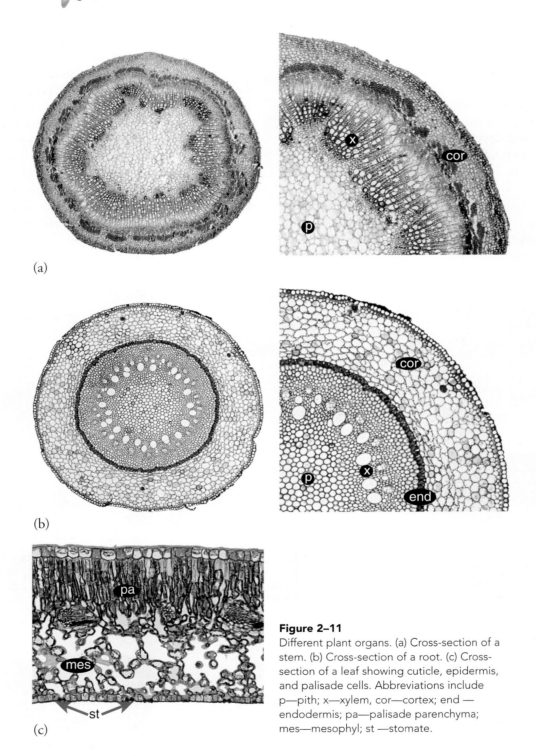

Figure 2–11
Different plant organs. (a) Cross-section of a stem. (b) Cross-section of a root. (c) Cross-section of a leaf showing cuticle, epidermis, and palisade cells. Abbreviations include p—pith; x—xylem, cor—cortex; end—endodermis; pa—palisade parenchyma; mes—mesophyl; st—stomate.

growing points (vascular cambium, cork cambium, leaf marginal meristems) (Fig. 2–12).

Meiosis The key feature of sexual reproduction is cell division through *meiosis* (64). Meiosis takes place within mother cells **(microspore mother cells** and **megaspore mother cells)** of the flower to produce **pollen** (male) and the **embryo sac** (female). Meiosis is the division of the nucleus that results in a reduction in the chromosome number by one-half, producing the haploid ($1n$) condition. Eventually, successful fertilization between haploid male and female gametes restores the diploid ($2n$) zygote leading to seed formation (see Chapter 4). Meiosis (Fig. 2–13, page 28) is separated into two parts: meiosis I and II. Each part of meiosis I and II includes prophase, metaphase, anaphase, and telophase stages.

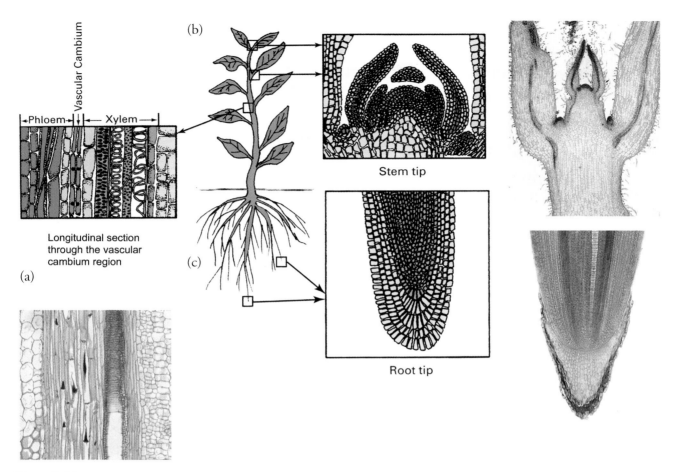

Figure 2–12
Location of growing points where meristematic cells occur and mitosis takes place. The points are located in the (a) cambium, (b) shoot tip, and (c) root tip.

Meiosis differs from mitosis in several important aspects:

1. Mitosis results in two genetically identical diploid cells, while meiosis results in four genetically different haploid cells.
2. There is only one division cycle for mitosis, while meiosis requires two division cycles.

Just as in mitosis, cells preparing for meiosis duplicate and double their chromosome number during interphase in preparation for division. During prophase I, the chromosomes become visible as centromeric chromatids and are arranged into homologous pairs. Then a remarkable process begins as the homologous chromosomes pairs exchange parts (**crossing-over**) of individual chromatids. Attached pairs of chromosomes then separate during metaphase, anaphase, and telophase to generate two new cells to complete meiosis I.

In meiosis II, each pair of chromosomes separates at the centromere and produces two daughter cells (four **gametes**), each with a haploid (n) number and genetically different from the parent cell and from each other. During sexual reproduction, a haploid gamete from a pollen unites with the haploid gamete from the embryo sac to produce a diploid zygote.

The consequence of meiosis is the creation of new patterns of genetic variation. Three opportunities for variation exist: (a) **crossing-over** (i.e., the interchange of genetic information during the early stages of meiosis I), (b) the **independent assortment** of the chromosomes during the later stages of meiosis II, and (c) the **recombination** of (haploid) male and female gametes in the creation of new zygotes during fertilization.

GENETIC INHERITANCE

Because of the exchange of genetic material during crossing-over, the independent assortment of chromosomes during meiosis, and the chance recombination during fertilization, patterns of genetic variation may appear in seedling populations that can be expressed in mathematical ratios of individual traits (see Figs. 2–14 and 2–15).

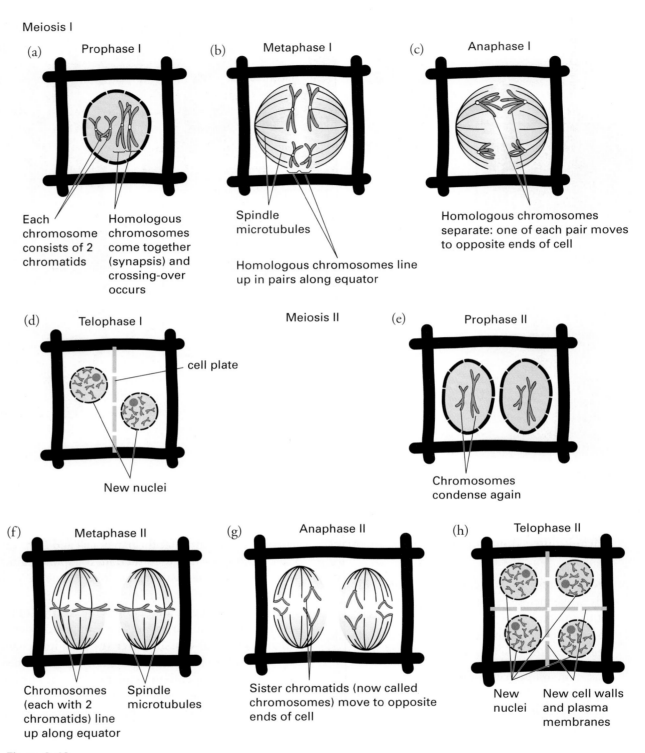

Figure 2–13

Stages of meiosis. **Meiosis I** - Interphase, in between divisions (not shown). However, chromosomes divide in preparation for division but remain attached at the centromere. (a) Prophase I, each shortened and conspicous chromosome has two chromatids attached at centromere. Chromosomes pair and exchange segments (crossing-over or synapsis). (b) Metaphase I, pairs line up along the center of the cell. (c) Anaphase I, pairs separate and move to opposite ends. (d) Telophase I, chromosomes disperse to form two nuclei. **Meiosis II** - (e) Prophase II, chromosomes again condense to form conspicuous pairs. (f) Metaphase II, chromosomes line up across the center of each cell. (g) Anaphase II, chromosomes separate into chromatids and move to opposite ends. (h) Telophase II, cell walls laid down to produce four haploid (n) gametes. Adapted from Linda R. Berg. 1997. *Introductory Botany.* Saunders College Publishing.

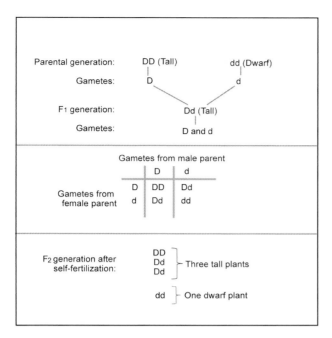

Figure 2–14
Inheritance involving a single pair of alleles in the gene controlling height in the garden pea. *Tallness* (D) is dominant over *dwarf* (d). A tall pea plant is either homozygous (DD) or heterozygous (Dd). Segregation occurs in the F₂ generation to produce three genotypes (DD, Dd, or dd) and the two phenotypes *tall* and *dwarf*.

These phenotypic distributions will be affected by whether the two genes are **dominant** or **recessive** and whether they are present as **homozygous** or **heterozygous** pairs. Many traits, however, are determined **quantitatively** by the interactions of a large number of genes that may be expressed uniquely in different environments (Fig. 2–16). In nature, seedling variability provides the opportunities for selection so that new genotypes can evolve that are adapted to specific environmental niches. Over time, genotypes tend to become more or less stabilized, or **"fixed,"** when grown over a long period in the same environment. This genotype–environment interaction is the basis for the origin of species (21, 70). In cultivation, seedling variation provides the opportunity for plant breeders to develop new kinds of plants that have special traits useful for humans but whose genotype must be maintained by special techniques of seed production. In general, plant breeding includes transferring genes from desirable parents to their offspring by crossing and then stabilizing (**fixing**) the genotype of the offspring population for propagation (1, 12, 37, 71).

> **"fixing"** The process of genetically stabilizing the genotype so that the cultivar will breed true from seed.

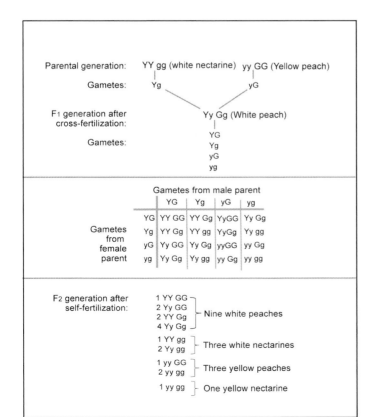

Figure 2–15
Simultaneous inheritance of two genes in a cross involving peach and nectarine (*Prunus persica*). *Fuzzy skin* (G) of a peach is dominant over the *smooth skin* (g) of a nectarine. *White flesh color* (Y) is dominant over *yellow flesh color* (y). In the example shown, the phenotype of the F₁ generation is different from either parent. Segregation in the F₂ generation produces nine genotypes and four phenotypes.

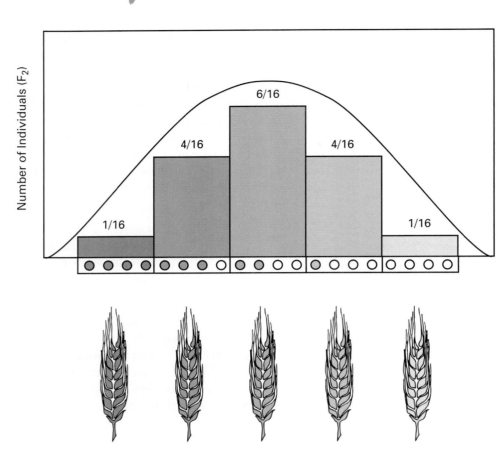

Figure 2–16
Quantitative genetic distribution is illustrated by the continuous varying pattern of wheat grain color. This makes a normal distribution curve, which indicates that many genes contribute to this phenotype. Adapted from Linda R. Berg. 1997. *Introductory Botany*. Saunders College Publishing.

GENE STRUCTURE AND ACTIVITY

Genes play a dual role in all organisms (6, 49, 64). First, they provide the physical mechanism by which individual traits and characteristics are reproduced from generation to generation both by seed (meiosis) or vegetative propagation (mitosis). Second, genes contain the specific directions for regulating the chain of morphological and physiological events that determine the expression of specific traits and characteristics of the phenotype. The central dogma of this process is that genetic information flows (with some exceptions) from deoxyribonucleic acid (DNA) to ribonucleic acid (RNA) to proteins through processes of transcription and translation.

Genes as Structural Units of Inheritance

Pre-Mendel The concepts and practices of plant (and animal) selection has a long and progressive history (see Chapter 1). Prior to 1900, plant breeders and plant propagators (often the same individual) carried on selection by visual inspection of specific traits and characteristics. That is, in seed propagation, the phenotypes of parents were compared with their seedling offspring; in vegetative propagation, the clonal source plant was compared with its vegetative progeny. Improvement was through mass selection in which the "best" phenotypes of one generation were chosen as parents for the next.

Mendelian Genetics The rediscovery in 1900 of Gregor Mendel's (56) paper published in 1866 marked the start of a new era in which selection became based on experimentally determined hereditary principles under the term *Mendelian genetics*. The concept of *gene* emerged as well as the term *genotype*. Chromosomes, which had been discovered about 50 years earlier, were found to be related to patterns of gene inheritance (20). Chromosomes were found to be composed of DNA and proteins. A basic question was whether proteins or DNA were responsible for inheritance. The answer obtained from studies with specific bacterial viruses called **bacteriophages** (39) showed that DNA was responsible for inheritance.

DNA-Based Genetics The studies of Watson and Crick published in 1953 on the structure of the DNA molecule (81) not only provided the biochemical model for DNA duplication during mitosis and meiosis but ushered in a new era of genetic research (Fig. 2–17). Subsequent

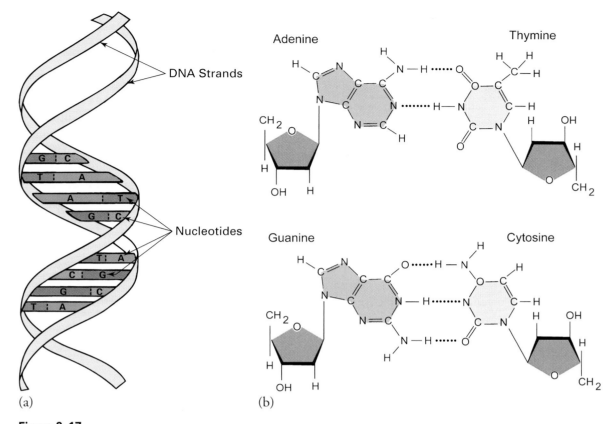

Figure 2–17
DNA structure. (a) Double helix of DNA strands made up of alternating sequences of ribose sugar joined by phosphate (PO₄) radicles. Nucleotides are made up of four possible bases—identified as A, T, G, C—that are joined at one end to a sugar molecule in the strand and loosely joined on the other to a complementary base (i.e., A with T, G with C). Combinations of base-pairs make up the genetic code. Adapted from Linda R. Berg. 1997. *Introductory Botany*. Saunders College Publishing. (b) Molecular structure of the binding pairs of nucleotide bases.

genetic code
Combinations of base pairs that create a code for different amino acids, whose combination in turn creates different proteins.

nucleotide A component of the DNA molecule whose important component is one of the four bases identified as T, G, A, or C; particular combinations of the complementary bases on homologous chromosomes pair G with C, and A with T.

studies identified the **genetic code** used to translate genetic information into functional proteins. This universal code was first identified in bacteria and then confirmed as a universal code for all organisms. This identification of genetic code was accompanied by the elucidation of **gene regulation** through the processes of transcription and translation, which regulate the expression of individual genes.

The structure of a chromosome consists of two strands of DNA in combination with various structural proteins called histones. The essential components of the DNA structure are **nucleotides,** which are combinations of one of four possible chemical nitrogenous bases **(thymine, adenine, guanine, cytosine),** a five-carbon sugar molecule **(deoxyribose),** and **phosphate (PO₄)** (Fig. 2–17). Nucleotides are attached to long chemical strands made up of phosphate (PO_4^-) radicals that connect the 5' (five-prime) position of one sugar molecule to the 3' (three-prime) position on the next sugar molecule. A **base** is attached at one end to a sugar molecule and loosely attracted by a **hydrogen bond** to a different, but complementary, base on the other DNA strand. Guanine (G) pairs with cytosine (C), and adenine (A) with thymine (T). The result is a **double-helix** structure of long, double chains of repeating nucleotides. This structure gives DNA a unique capacity to replicate itself during mitosis and meiosis when catalyzed by the enzyme **DNA polymerase.** The specific sequence of nucleotide bases, i.e., **base pairs,** provides the genetic information that determines inheritance, establishes

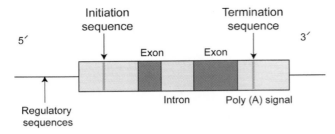

Figure 2–18
Schematic drawing of the structure of a gene. See text for details.

the specific genotype of the organism, and directs the pattern of gene expression. A coding unit consists of a specific sequence of three of the four bases and is known as a **codon.** Codons translate into one of twenty **amino acids** used to make proteins. From a molecular standpoint, a gene can be described as a linear piece of DNA (65), which includes the following: (a) coding regions, known as **exons,** that contain the genetic instructions, (b) noncoding regions known as **introns,** (c) an **initiation codon** (also known as a **promoter**), (d) a **termination codon,** and (e) a **regulator sequence** adjoining the gene on the 5′ end that determines when a gene is turned on and off (Fig. 2–18) (6, 64).

Gene Expression

Transcription Genetic information is copied from one of the strands of DNA onto similar macromolecules called **RNA (ribonucleic acid).** The structure of RNA differs from nuclear plant DNA in that it has only a single strand, a different sugar **(ribose),** and includes **uracil** instead of thymine. RNA exists in several forms. At the transcription stage it is called **messenger RNA** or **mRNA.** The process begins with a signal **recognition** process that involves various environmental, physiological, or hormonal cues to turn on the gene. This is followed by the initiation of **transcription** of specific DNA sequences to make single-stranded mRNA molecules. A specific enzyme **(RNA polymerase)** mediates transcription that results

RNA (ribonucleic acid) Biochemical compound that functions to transcribe genetic code information from the chromosome to mRNA where it is translated into protein synthesis.

transcription The process by which the genetic code of genes present in the DNA is enzymatically transcribed to a strand of RNA.

RNA polymerase An enzyme within the nucleus that mediates the transcription of DNA codes to tRNA.

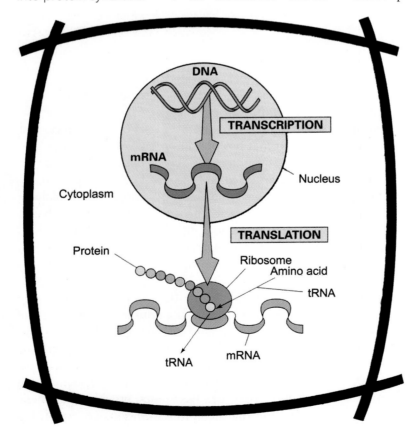

Figure 2–19
Transcription and translation. The diagram illustrates the fact that transcription is carried out in the nucleus in which mRNA transcribes the nucleotide sequences for a specific gene on one of the strands and then migrates to the cytoplasm. Here the message is used to manufacture specific proteins within the ribosomes with the help of tRNA and rRNA.
Adapted from Linda R. Berg. 1997. *Introductory Botany.* Saunders College Publishing.

translation The process by which the genetic code from genes is translated from mRNA by ribosomal RNA to combine amino acids to create peptides, polypeptides, and, eventually, proteins.

ribosome A structure within the cytoplasm composed of protein and ribosomal RNA (rRNA) within which translation takes place.

in the synthesis of mRNA molecules—which may vary from 200 to 10,000 nucleotides in size. These move from the nucleus across the nuclear membrane into the cytoplasm (Fig. 2–19).

Translation Translation is the process of building a protein based on the genetic code sequenced on the mRNA. Translation is a coordinated effort among mRNA, **ribosomes, transfer RNA (tRNA),** and amino acids. Proteins are made at the ribosome where the mRNA passes between the two ribosome subunits. Transfer RNA brings the appropriate amino acid to the ribosomal complex for translation into the protein called for by the codon on the mRNA. Amino acids become linked together in chains first as **peptides,** then **polypeptides,** and, eventually, specific **proteins.** Proteins expressed after translation can be visualized using gel electrophoresis (Fig. 2–20).

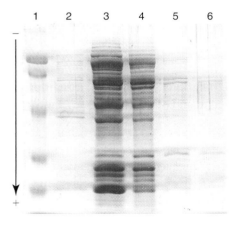

Figure 2–20
Gel electrophoresis showing the migration of proteins down the gel. Gel electrophoresis uses an electric current to move molecules from the top of the gel toward the base. Different sized molecules move at different rates. Gel electrophoresis can also be used to visually separate DNA and RNA. In this gel, proteins are stained with Coomassie blue to visualize the proteins. Lane 1 is a molecular weight ladder used as a reference. The five other lanes represent treatments with different protein level expression. Lanes 3 and 4 qualitatively have very similar protein profiles, but the treatment represented in lane 3 has more protein being expressed.

Regulating Gene Expression

Proteins are large, complex macromolecules, many of which function as enzymes that regulate the biochemical reactions controlling metabolic and developmental plant processes. The types and functionality of proteins produced by the cell determines plant growth and development. Therefore, regulation of gene transcription is an important component of determining a cell's developmental fate.

Regulation of gene transcription involves effector and repressor molecule interactions at the regulatory sequences found at the three-prime portion of the gene called the promoter region. This type of gene regulation can be illustrated by the repressor/de-repressor model for auxin action (Fig. 2–21) (83). The auxin responsive gene has a sequence in the promoter region called the auxin response element (AuxRE). The promoter protein called auxin response factor (ARF) physically interacts with this regulatory element to promote gene expression. However, when auxin is not present, the repressor molecule (AUX/IAA) interacts with ARF in such a way that it is unable to promote transcription. When auxin is present in the cell, auxin binds to its receptor moleculre (TIR1) to initiate degradation of the Aux/IAA repressor. This releases ARF from its repression by AUX/IAA to promote gene transcription. This type of repressor/de-repressor interaction seems to be a common mechanism controlling gene expression.

Gene expression can also be regulated after transcription is complete and mRNA is made. One example of this control is by small, nontranslating RNA molecules such as microRNAs. **MicroRNAs** function in translational repression and are important for controlling development in plants and

microRNA (miRNA) A small RNA molecule involved in post-translational control of gene expression.

animals. They are small ~22 nucleotide RNAs that are components of a RNA-induced silencing complex. MicroRNAs seek out complementary mRNA, bind to them, and target them for enzymatic degradation. Using auxin-induced gene expression again as an example, several microRNAs that are developmental regulated target ARF mRNA for silencing. These microRNAs prevent ARF mRNA translation, which, in turn, eliminates ARF as a promoter of auxin-responsive genes.

Post-translational control is also an important regulatory mechanism for growth and development. Proteins made through the gene expression pathway may not have regulatory function until they are modified. A common protein modification is through phosphorylation by kinase enzymes. This sets up "kinase

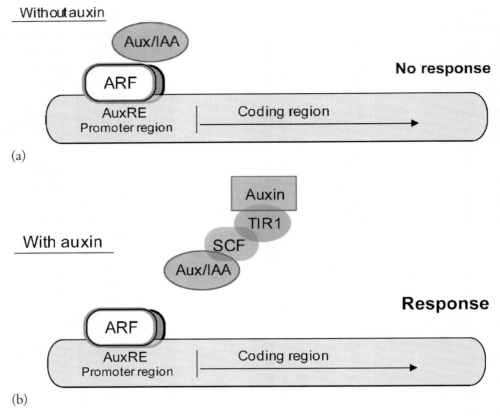

Figure 2–21
Presented is a model for auxin hormone action related to gene expression. (a) The gene being controlled by auxin as an auxin response element (AuxRE) in its promoter region. A transcription regulator called auxin response factor (ARF) is required for gene expression. It is available to bind the promoter region even in the absence of auxin; however, a repressor molecule (Aux/IAA) binds to ARF to inhibit gene expression. (b) When auxin is present, it binds to its receptor (TIR1) and initiates a ubiquitan-ligase complex (SCF) that targets AUX/IAA for destruction. With the repressor removed, ARF can initiate gene transcription.

cascades" that are important consequenses of hormone-receptor binding and downstream hormone activity.

Biotechnology

A long sequence of basic laboratory studies has led to a revolution in genetic research which is described under the umbrella term of **biotechnology.** These have begun to have far-reaching applications not only in propagation but across the whole range of applied biology.

Cell and Tissue Culture Technology This term refers to an array of concepts and procedures involving the propagation and culture of cells, tissues, and individual plant organs in aseptic closed systems. Among the culture systems developed are those for embryos, ovules, shoot apices, callus, protoplasts, and cell suspensions. These concepts and procedures, covered in this text in Chapters 17 and 18, are powerful tools that have revolutionized many aspects of plant physiology, genetics, and propagation. Some procedures are used commercially in nursery operations, others are primarily for genetic improvement, and others are for scientific investigations.

DNA-Based Marker Technology This category refers to the group of laboratory procedures that utilize the nucleotide sequences present on small DNA fragments produced artificially from chromosomes by specific enzyme treatments to identify and label specific locations in the **genome.** With appropriate procedures, the sequences on these segments can be used as **DNA markers,** which are visually observed as bands on an electrophoresis plate (Fig. 2–22). This technique

genome All of the genetic material (i.e., genes) present in the chromosomes of an organism; some DNA may be present in chloroplasts and mitochondria as well.

DNA markers Specific combinations of base pairs (bp) that are used to identify genes and genotypes in the laboratory.

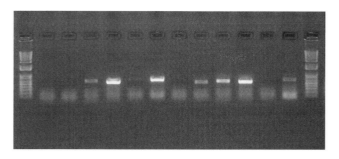

Figure 2–22
DNA visualization on an agarose electrophoretic gel. The first and last lanes are the DNA size markers. The DNA is visualized on an X-ray film taking advantage of the radiolabeled phosphorus that was added to mark the DNA during the PCR reaction.

makes it possible to identify specific genes and, eventually, to characterize whole genomes. Nucleotide sequencing techniques are used to monitor and predict variation during breeding operations (marker-assisted gene linkage maps) and to identify specific cultivars (DNA fingerprinting) (7, 78, 84). They also provide data to investigate botanical and evolutionary relationships by creating cladograms that show genetic similarities among members of a genus or plant family (23, 27, 63).

Recombinant DNA Technology This term includes a group of procedures in which the nucleotide sequences of the DNA molecule representing a gene can be isolated, cloned, and hybridized with other DNA fragments to produce what is known as **recombinant DNA.** These hybrid DNA clones can be used as **genetic probes** to identify and characterize gene expression. Also, by using appropriate methods, DNA from a donor organism can be introduced into cells of another organism to become part of its genome (19, 30, 44). Plants transformed using recombinant DNA techniques are called **transgenic** and are popularly

> recombinant DNA The combination of DNA representing a particular gene cloned with other DNA fragments in the laboratory in order to be inserted into the genome of another organism.

BOX 2.2 GETTING MORE IN DEPTH ON THE SUBJECT
TECHNIQUES USED TO STUDY GENE EXPRESSION

DNA Marker Technology
Fragmentation Restriction Enzymes have been discovered in bacteria that cause chromosomal DNA to split into small fragments at specific nucleotide sequences and with different numbers of nucleotides. Under various laboratory procedures, large numbers of these **restriction fragments** can be generated, which, taken together, represent pieces of the entire genome of individual organisms. These fragments then become markers of specific segments of the genome representing specific genes and can be stored as **genomic libraries** in the laboratory. These fragments become the working tools of various procedures described in the subsequent text.

Amplification Treating DNA fragments with the bacterial enzyme **Taq polymerase** under appropriate temperature sequences causes single DNA strands to replicate (up to 1 million times in a few hours). The process known as **polymerase chain reaction (PCR)** is a form of cloning. As a result, large quantities of specific DNA clones can be produced.

polymerase chain reaction (PCR) DNA fragments can be caused to replicate in order to produce large amounts of a specific DNA clone.

Visualization and Separation DNA fragments are identified by the pattern of consecutive bands in the gel on an electrophoresis plate (Fig. 2–22). Mixtures of fragments are placed at one end of the gel and individual segments migrate to the other end in response to an electric current. The location of the segments differs primarily because of fragment size (i.e., numbers of base pairs). To visualize the pattern, the gel is treated by appropriate indicators (stains, ultraviolet light, radioactivity). The gel can be sliced into sections to isolate specific DNA fragments.

DNA Sequencing A DNA sample is divided into parts, each to be treated separately by different restriction enzymes, which recognize different nucleotide pairings. The samples are amplified (cloned) by the PCR reaction, electrophoresed, and analyzed for nucleotide sequences. The latter is done automatically by a DNA sequencing machine that utilizes different colored fluorescent dyes for visualization. Because the base-pair patterns of different fragments overlap, a complete "map" of an entire genome or individual gene location can be produced.

Molecular Genetics
Molecular genetics is the study of the function of genes at the molecular level. One important tool in the study of molecular genetics is to generate mutants impaired in an area of growth and development.

Mutant Generation and Analysis Mutant plants are usually generated by chemical (EMS) or radiation exposure. Mutagenesis directly impacts the DNA sequence, altering a gene's ability to be transcribed and translated into a viable protein. Mutant screens must be developed to visualize the few desired mutants in the thousands of treated seeds or plant parts. For example, seedlings germinated

(Continued)

in the presence of ethylene will develop a classical triple response (short, thickened stems growing horizontally). Mutant seedlings impaired for ethylene action were discovered because they grew as tall, upright seedlings seemingly immune to ethylene.

Transformation Technology (41). Foreign DNA can be introduced into a plant's genome using *Agrobacterium*-mediated transformation or particle bombardment. *Agrobacterium tumefaciens* is a bacteria that uses a circular piece of DNA called a plasmid to integrate a portion of its DNA into the plant's genome to create a plant tumor and facilitate bacterial replication (8). Researchers can modify this plasmid to replace bacterial genes with novel genes for plant improvement or basic plant science studies. Plant tissue cultures or intact flowers are exposed to the engineered bacteria for gene insertion (Fig. 2–23). A second transformation method is particle bombardment, sometimes called biolistics. This method uses microprojectiles (gold or tungsten particles) coated with DNA that are shot into plant tissue where they enter dividing plant cells and become integrated into the plant's genome (Fig. 2–23c). Following gene transfer, seedlings or plant tissue are placed on a selection medium where transformed individuals can be identified and raised into reproductive whole plants (Fig. 2–23d).

Plants are transformed to up-regulate a gene's activity, introduce a novel gene product (like herbicide resistance), or suppress or silence a gene (see later in this chapter).

Plants may also be randomly transformed with short pieces of DNA that can insert into a gene to disrupt its activity; this is called tDNA insertional mutagenesis. Plants can be screened for activity in a similar way to mutants generated by chemicals or radiation. However, because the tDNA insertion has a known DNA nucleotide sequence, the disrupted gene can usually be more easily identified compared to other mutants.

Gene Silencing Technology

Once a gene is suspected of having regulatory properties, gene silencing technology can evaluate the importance of that gene in growth and development. Gene silencing significantly knocks down or eliminates the gene product (protein) from being produced and should impair the growth and development process being evaluated (like seed germination or flowering). Commercial plant cultivars can be developed so that a particular gene has been silenced in order to influence production of a biochemical product or slow a process like fruit ripening or flower senescence.

Antisense Technology (26) DNA consists of two complementary strands of nucleotides. Only one of these two strands of DNA serves as a template for mRNA formation and is called the **sense strand**. It is possible to reverse the order of a particular segment controlling a particular gene copy of the sense strand within the chromosome, which is

(a)

(b)

(c)

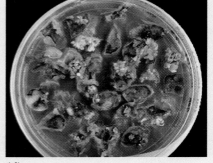
(d)

Figure 2–23
The most common forms of genetic transformation use *Agrobacterium*-mediated transformation or biolistics. (a) A solution of bio-engineered *Agrobacterium* designed to integrate new DNA into the plant's genome (b) *Arabidopsis* at the proper flowering stage to be dipped in the diluted *Agrobacterium* solution. (c) Technician placing a sample into the biolistics machine, which will shoot DNA-coated particles into the plant sample. (d) Leaf pieces on a selection medium after being transformed. Green, new shoots represent plants that were transformed, while non-transformed leaf pieces do not survive on the selection medium (i.e., antibiotic medium).

now called **antisense**. The nucleotides within the specific segment are not copied now, and in effect become nonfunctional, effectively turning off the gene associated with the segment. This antisense feature is then inherited like any gene.

RNA Interference (RNAi) Small RNA molecules (fewer than 20 nucleotides) have recently been discovered as important for plant defense (disease resistance) and for control of growth and developmental processes (43, 47). These include small interfering RNA (siRNA) and microRNA (miRNA). These small RNA molecules attach to complementary sequences on mRNA to prevent translation. These are natural processes of control in plants and animals. Researchers take advantage of this technology by inducing RNAi silencing of a gene of interest to investigate the gene's function.

Genome-Wide Gene Expression

Techniques have become available to do global gene expression profiling that measures the activity of thousands of genes at once. These techniques provide a huge amount of data that has lead to the development of a new field of study called bioinformatics, which aids in gene discovery experiments.

Transcriptome Analysis The transcriptome represents the mRNA being produced by a cell or plant tissue at a given time during growth and development. This is a measure of the gene expression at that particular developmental event in time (i.e. radicle protrusion during seed germination). The identification of these mRNA has been greatly enhanced by the availability of the gene sequences for entire genomes in plants such as *Arabidopsis*, poplar (*Populus*), rice, and *Medicago truncatula*. For plants like corn or tomato where genome sequencing is still under development, expressed sequence tag (EST) libraries have been developed that contain information about mRNA expression. Microarray (also called a gene chip) technology has been developed to measure global gene expression (Fig. 2–24). A microarray contains thousands of partial DNA sequences arranged on a slide or platform (62). These sequences will hybridize to cDNA (complementary DNA) synthesized from the mRNA extracted from the plant tissue. A positive interaction leads to a fluorescent label being activated that indicates the relative abundance of the mRNA signal.

Although microarray analysis reveals the different mRNAs being transcribed in the cell, that information does not necessarily give a full profile of the functional proteins being translated from those mRNA. Therefore, a second complementary technique called proteomics has been developed to measure all of the proteins made during that same developmental time.

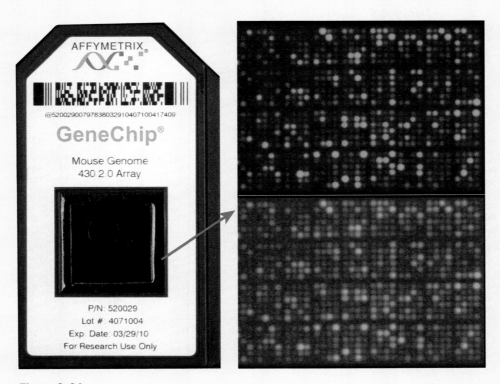

Figure 2–24
Microarray chips contain thousands of gene sequences as individual microscopic spots. They act as probes to visualize gene expression. Positive interactions can be seen by color and intensity on the chip.

known as **genetically modified organisms (GMOs)**. Examples of economic traits being engineered include various seed components (13), flower longevity (31), and disease (53) and insect resistance (74).

PLANT HORMONES AND PLANT DEVELOPMENT

phytohormones (plant hormones) Organic chemicals that regulate growth and development.

Plant hormones (phytohormones) are naturally occurring organic chemicals of relatively low molecular weight, active in small concentrations. The classic definition of a hormone is that they are synthesized at a given site and translocated to their site of action; however, there are some exceptions for plant hormones. They are specific molecules involved in the induction and regulation of growth and development. The five major plant hormones are **auxin, cytokinin, gibberellin, abscisic acid,** and **ethylene**. Additional compounds considered hormones include **brassinosteroids, jasmonates, salicylic acid, polyamines,** and **peptide hormones**. Plant hormones have great importance in propagation because they not only are part of the internal mechanism that regulates plant function, but they also can induce specific responses such as root initiation in cuttings and dormancy release in seeds.

In addition to these substances, certain chemicals—some natural, others synthetic—show hormonal effects to plants. Both natural and synthetic types are classed together as **plant growth regulators (PGRs)**. Table 2–3 lists the characteristics of important PGRs used in propagation. Their usage will be further described in subsequent chapters.

plant growth regulators (PGRs) Any natural and synthetic chemical that shows hormonal effects.

Here is a usual set of events that occurs during hormone-induced growth and development:

1. Biosynthesis of the hormone
2. Transport or distribution to its site of action

Table 2–3
CHARACTERISTICS OF IMPORTANT PLANT GROWTH REGULATORS AND HORMONES. THOSE MARKED WITH ASTERISK (*) OCCUR NATURALLY

Name	Chemical name	Mol. Wt.	Solvent	Sterilization[1]	Storage Powder	Storage Liquid
	A. Auxins					
IAA*	indole-3-acetic acid	175.2	EtOH or 1N NaOH	CA/F	–0°C	–0°C
IBA*	indole-3-butyric acid	203.2	EtOH or 1N NaOH	CA/F	0–5°C	–0°C
K-IBA	indole-3-butyric acid-potassium salt	241.3	Water	CA/F	0–5°C	–0°C
NAA	α-naphthaleneacetic acid	186.2	EtOH or 1N NaOH	CA	RT	0–5°C
2,4-D	2,4-dichloro-phenoxy-acetic acid	221.0	EtOH or 1N NaOH	CA	RT	0–5°C
	B. Cytokinins					
BA	6-benzyl-amino-purine	225.3	1N NaOH	CA/F	RT	0–5°C
2iP*	6(di-methyl-allyl-amino) purine	203.2	1N NaOH	CA/F	–0°C	–0°C
Kinetin		215.2	1N NaOH	CA/F	–0°C	–0°C
TDZ	Thidiazuron	220.2	DMSO or EtOH	CA	RT	0–5°C
Zeatin*		219.2	1N NaOH	CA/F	–0°C	–0°C
	C. Gibberellins					
GA$_3$*	gibberellic acid	346.4	EtOH	F	RT	0–5°C
K-GA$_3$	gibberellic acid potassium salt	384.5	water	F	0–5°C	–0°C
	D. Inhibitors					
ABA*	Abscisic acid	264.3	1N NaOH	CA/F	–0°C	–0°C

[1]CA = coautoclavable with other media; F = filter sterilize; CA/F = autoclavable with other components but some loss in activity may occur.
Source: Adapted from *Plant Cell Culture* 1993 catalog. Sigma Chemical Co., St. Louis, Mo.

3. Perception of the hormone signal by its cellular receptor
4. Signal transduction leading to downstream events often at the molecular (gene expression) level

It has become evident that many types of growth and development are not controlled by a single hormone; rather there is considerable interaction and "cross-talk" often between several hormones. Often there is one principle hormone controlling development with other hormones modifying its action (45). For example, as discussed in Chapter 7, abscisic acid's control over seed dormancy is modulated by gibberellin, cytokinin, ethylene, and brassinosteroid. Some of the plant hormones are present in active and conjugated forms. Conjugation is the addition of a sugar or amino acid to the chemical structure of the hormone. Conjugation may inactivate the hormone permanently, or enzymes can interconvert the hormone between conjugated and free forms through a process called homeostatic control.

Auxins

Auxin was the first plant hormone discovered by plant scientists. Phototropism, where uni-directional light altered the growth of plant coleoptiles, in grass seedlings was one of the first biological systems studied by botanists including Charles Darwin (22). Fritz Went, Kenneth Thimann (82), and a number of other researchers showed that these effects could be induced by plant extracts, which were subsequently shown to contain the chemical **indole-3-acetic acid (IAA).**

There are two biosynthetic pathways for IAA in plants (5). Primary auxin biosynthesis is via the amino acid L-tryptophan, but IAA can also be synthesized by a tryptophan-independent pathway. Most of the IAA in plant tissue is in the conjugated form using both amino acids and sugars for conjugation. Free, active IAA comprises approximately 1 percent of the total auxin content, with the remaining portion in the conjugated form. Primary sites of auxin biosynthesis include root and shoot meristems, young leaf primordia, vascular tissue, and reproductive organs including developing seeds (Fig. 2–25).

Auxin movement from cell to cell requires efflux carriers located on the plant membrane (Fig. 2–26) (83). They control polar auxin movement from plant tips (distal ends) to their base (proximal end). Cellular auxin movement and the subsequent polar gradient established between cells is important for normal development of the plant embryo as well as the shoot apical meristem (57).

Auxin has a major role for controlling phototrophism, inhibition of lateral buds by terminal buds (apical dominance), formation of abscission layer on leaves and fruit, activation of cambial growth, and adventitious root initiation. Auxin is the most widely used hormone in plant propagation because of its impact on adventitious rooting in cuttings (see Chapter 10) and its control of morphogenesis during micropropagation (see Chapters 17 and 18).

IAA degrades in the light, and exogenously applied IAA is quickly degraded by the enzyme IAA-oxidase.

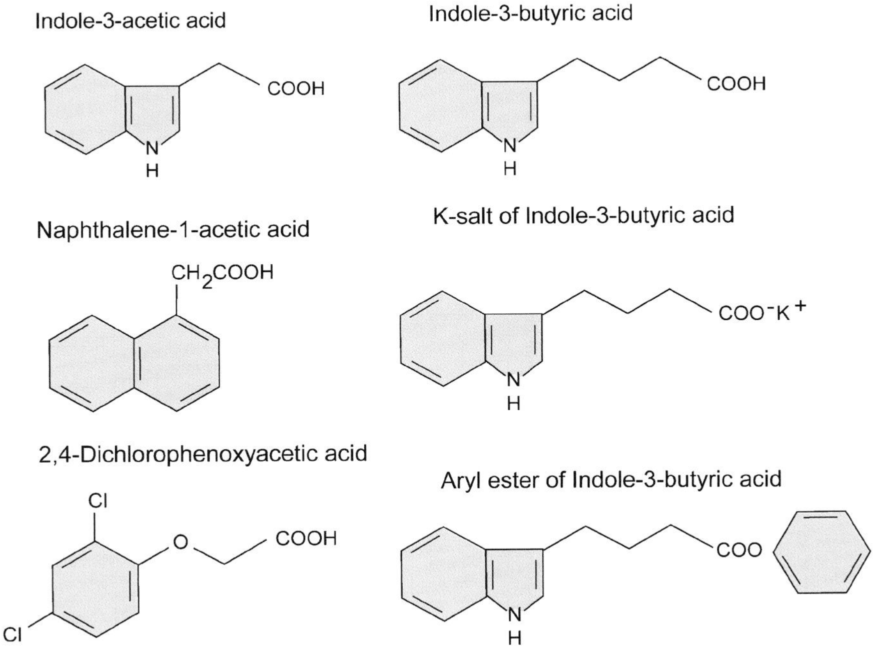

Figure 2–25
Chemical structures of various auxins.

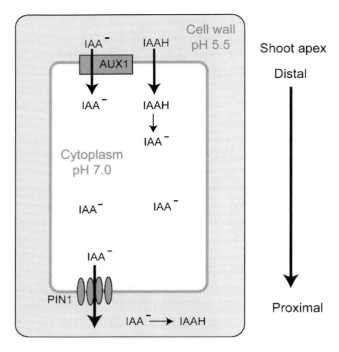

Figure 2–26
The chemiosmotic model for polar auxin transport. Auxin in the protonated form at the low cell wall pH can pass through the cell membrane or it may be transported by an influx carrier (AUX1). At the higher cytoplasmic pH, IAA dissociates. In this state, auxin can only move back into the cell wall by active transport using efflux carriers (PIN1). Since efflux carriers are only located at the base (proximal) end of the cell, auxin moves in a polar fashion from shoot to the root–shoot junction.

Synthetic auxins are less susceptible to IAA-oxidase degradation and are, therefore, used more often for commercial applications. The most useful synthetic auxins, discovered about 1935, are **indole-3-butyric acid (IBA)** and 1-**naphthalene acetic acid (NAA)**. IBA has been subsequently found to occur naturally, but in less abundance compared to IAA. IBA must be converted by plant tissue into IAA to function. The herbicide, 2,4-D (2,4-dichlorophenoxyacetic acid) has auxin activity and is an important inducer of somatic embryogenesis in tissue culture (see Chapter 17). Various synthetic IBA conjugates (such as its aryl ester PITB—Fig. 2–25) have been developed with good auxin activity but are not widely available or used (35). Auxins are not readily dissolved in water and must be dissolved in a solvent (ethanol, DMSO) or a base (1N NaOH) before being quickly added to water. Potassium salts of IBA and NAA (K-IBA, K-NAA) are auxin formulations that easily dissolve in water and are available commercially.

Cytokinins

Cytokinins were discovered by Miller and Skoog at the University of Wisconsin in efforts to develop methods for growing plant cells in tissue culture (68). Through the 1940s and 1950s, researchers were frustrated because isolated plant cells and tissues grew poorly or not at all in tissue culture. At that time, tissue culture media supplemented with coconut milk (liquid endosperm) had the most stimulating effect on cell division compared to other compounds evaluated. Then Miller and Skoog inadvertently discovered that an extract from autoclaved fish sperm DNA yielded a compound that greatly stimulated cell division. This synthetic compound was called **kinetin** and the hormone class was called cytokinins because of their ability to stimulate cell division. Subsequently the naturally occurring cytokinins **zeatin** (isolated from corn endosperm) and **isopentenyladenine (2iP)** were found in seeds and other plant parts. These previously mentioned cytokinins along with the naturally occurring **dihydrozeatin** and the synthetic **benzyladenine (BA or BAP)** represent the aminopurine type cytokinins (Fig. 2–27). Another class of compounds—the dipheylureas—displays potent cytokinin activity but are structurally dissimilar to natural occurring cytokinins, including thiourea, diphenylurea, **thidizuron (TDZ),** and N-(2–chloro-4-pyridyl) n'-phenylurea (CPPU).

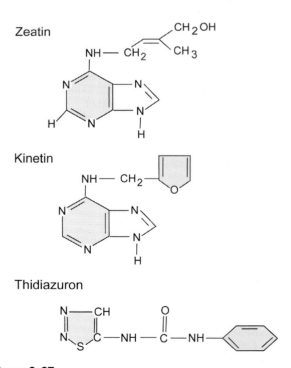

Figure 2–27
Chemical structure of cytokinin.

The major route for cytokinin biosynthesis is via the isoprenoid pathway with isopenteyltransferase (*ipt*) being the key regulatory enzyme (48). The root tip is a primary source of cytokinin, but biosynthesis also occurs in seeds (embryos) and developing leaves. In addition to free forms of cytokinin, conjugated derivatives include ribosides, ribotides, aminoacids and sugars—many of which freely interconvert. The major enzyme for cytokinin destruction is cytokinin-oxidase.

Cytokinins are thought to play a regulatory role in cell division, shoot initiation and development, senescence, photomorphogenesis, and apical dominance. Cytokinins play a key role in regulating various aspects of the cell cycle and mitosis. Transgenic plants over-expressing the *ipt* gene show elevated cytokinin levels, reduced height, increased lateral branching, and reduced chlorophyll destruction leading to a deep green color. Tissue infected with *Agrobacterium tumifaciens* grow and proliferate in tissue culture independent of growth regulator application. This is because it induces elevated cytokinin levels by inserting an *ipt* gene from its plasmid into the plant's genome.

The interaction of auxin and cytokinin is one of the primary hormonal relationships in plant growth and development as well as plant propagation. A high auxin:cytokinin ratio favors rooting, a high cytokinin:auxin ratio favors shoot formation, and a high level of both favors callus development (see Chapter 9 on rooting and Chapter 17 on micropropagation).

Gibberellins

Gibberellins (69) were discovered before World War II by Japanese scientists trying to explain the abnormally tall growth and reduced yield of rice infected by the fungi *Gibberella fukikuori* (perfect stage) or *Fusarium moniliformne* (imperfect stage). An active ingredient was extracted from the fungus and its chemical structure was determined as gibberellins (named after the fungus). Subsequently, gibberellins were found to be naturally occurring hormones in plants. All gibberellins are cyclic diterpenoids and named for their structure not their activity. More than 100 forms of gibberellins have been found in plants but only a few are physiologically active. The most important naturally occurring active gibberellins include GA_1, GA_4, GA_7 (Fig. 2–28). Depending on the plant, they will tend to make either GA_1 or GA_4 as their primary gibberellin. Gibberellic acid (GA_3) is the gibberellin found in fungi and is the most important commercial product.

Biosynthesis of gibberellins (73, 76) starts with mevalonate (an important precursor for many secondary compounds in plants) and proceeds via the iosprenoid pathway. Its biosynthesis is a coordinated process involving the plastids, endoplasmic reticulum, and cytosol. Numerous enzymes are regulated during gibberellin biosynthesis, but GA_{20} oxidase appears to especially important. Active gibberellins are inactivated by GA_2 oxidase. Gibberellins can also be sugar conjugated as previous discussed with other hormones.

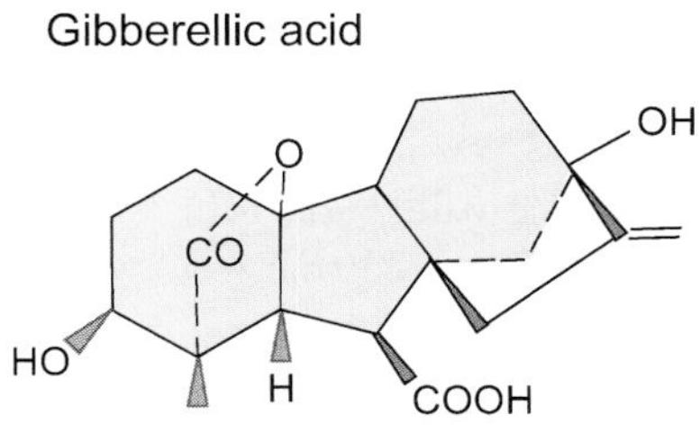

Figure 2–28
Chemical structure of gibberellic acid.

Gibberellins are made in developing seeds and fruits, elongating shoots, and roots. Gibberellins are the primary hormone controlling plant height. Gibberellin mutants impaired for gibberellin biosynthesis are dwarfed compared to wild type plants, demonstrating the importance of gibberellins for shoot elongation. Several commercially available gibberellin biosynthesis inhibitors, including ancymidol, cycocel, paclobutrazol (Bonzi), and uniconizole (Sumagic), are important plant growth regulators used to control plant height during greenhouse pot and bedding plant production. Gibberellins also play a role in plant maturation and in triggering flowering. Gibberellins are particularly important during seed germination, where the antagonistic interactions between gibberellin and abscisic acid are involved in dormancy release and germination.

Abscisic Acid (ABA)

Abscisic acid was originally discovered during the 1960s in studies searching for hormonal control of leaf abscission and bud dormancy (Fig. 2–29). These studies

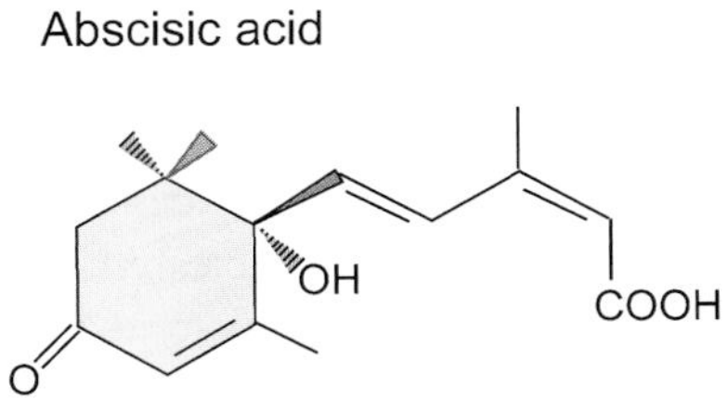

Figure 2–29
Chemical structure of abscisic acid (ABA).

suggested that ABA was involved in abscission, and the isolated compound was called "Abscisin II." Studies also suggested that ABA was involved in bud dormancy, and that compound was called "dormin." However, subsequent analyses determined that ABA was not a major factor in leaf abscission (2) but may be involved in bud dormancy. ABA's major role in plant growth and development is to modulate environmental stresses, especially water stress. ABA regulates stomatal opening and closure as an indicator of plant water status and promotes root growth under water stress. ABA's other major roles are as a major determinant of zygotic embryo growth during seed development (see Chapter 4) and in maintaining seed dormancy (see Chapter 7). ABA mutants typically show reduced seed dormancy, increased precocious germination, and wilted leaves at the whole plant level.

ABA is a sesquiterpene synthesized directly from carotenoids (ß-carotene and zeaxanthin) rather than the usual mevalonate pathway observed for gibberellins. Biosynthesis occurs in coordination between enzymes in the plastids and cytosol (67). The ABA molecule has two isomeric forms, *cis* and *trans*. The trans form is the more active and common form in plants. The chemical structure also has a (+) and (−) form that cannot be interconverted. The (+) form is active and occurs in nature. Commercial products are mixtures of both (+) and (−) forms. Fluridone is a carotenoid biosynthesis inhibitor that chemically reduces ABA levels in plants. Cellular ABA concentrations are important for controlling ABA action. Important regulated enzymes in the biosynthetic pathway appear to be 9-*cis*-epoxy-carotenoid dioxygenase (NCED) and xeaxanthin epoxidase (ZEP) for increased ABA levels, while cytochrome P450 707A (CYP707A) is the major enzyme reducing ABA levels (28).

Ethylene

Dimitry Neljubow, a Russian scientist, is credited with the first report of the effects of ethylene on plants in 1901 (66). He demonstrated that ethylene was the agent from illuminating gas used in street lamps that caused plant damage. He also used etiolated pea seedlings to study the effects of ethylene on plant growth and identified the triple response in ethylene-treated seedlings. Seedlings displaying the triple respone show inhibition of stem elongation, increased radial swelling in the hypocotyl, and horizontal stem orientation to gravity.

Ethylene is a gas with a very simple hydrocarbon structure (Fig. 2–30). However, it can have profound effects on plant growth, including epinasty at high

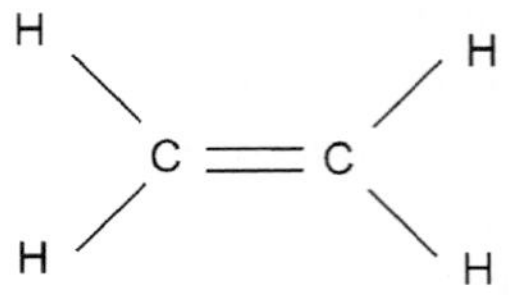

Figure 2–30
Chemical structure of ethylene.

concentrations, senescence and abscission in leaves and fruit, flowering, apical dominance, latex production, and flower induction. In propagation, ethylene can induce adventitious roots, stimulate germination, and overcome dormancy. Wounding, stress, and auxin usually stimulate increased ethylene production. Naturally occurring ethylene is involved in the maturity of certain fruits and is widely used to induce ripening in commercial storage. Ethephon (2–chloroethylphosphoric acid) is absorbed by plant tissue where it breaks down to ethylene. It is used on some crops to promote ripening, to act as a thinning agent, to promote or reduce flowering, and to reduce apical dominance. Ethylene gas is a natural by-product of combustible fuels, and escaping fumes can cause damage in commercial storage and greenhouse production. Likewise, ethylene from ripening fruit causes damage to other plant material in common storage.

Ethylene is synthesized from the amino acid methionine via a pathway that includes S-adenosylmethionine and l-aminocyclopropane-l-carboxylic acid (ACC) as precursors. Key regulated enzymes in the pathway include ACC-synthase and ACC-oxidase. Ethylene inhibitors are used commercially to inhibit flower senescence and delay fruit ripening. Aminoethoxyvinylglycine (AVG; Retain) inhibits ethylene biosynthesis, while silver thiosulfate, silver nitrate, and 1-methylcyclopropene (MCP) inhibit ethylene action by altering ethylene's ability to bind to its receptor.

Additional Plant Hormones

Certain other naturally occurring substances are considered by some to show hormonal action. These include brassinosteroids, jasmonates, salicylic acid, polyamines, and peptide hormones.

Brassinosteroids Brassinosteroids were originally extracted from *Brassica napus* pollen and called "Brassins" (15). They were shown to have growth-regulator activity in seedling bioassays, and the active components were identified as brassinolide. Brassinosteroid's importance as a new plant hormone on plant growth was demonstrated when brassinosteroid-deficient mutants were discovered in *Arabidopsis* that showed extreme dwarf plant growth

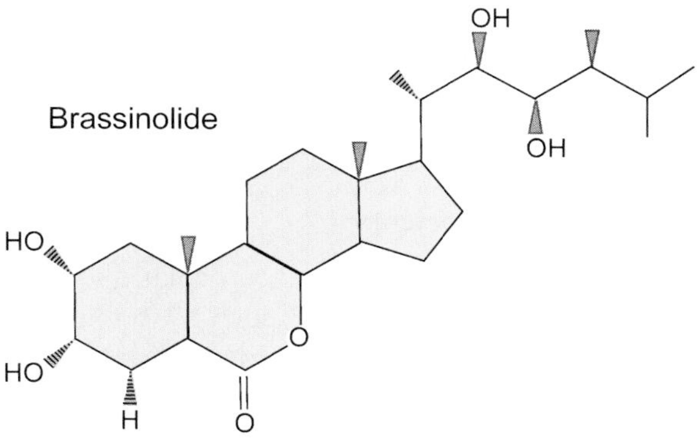

Figure 2–31
Chemical structure of brassinolide.

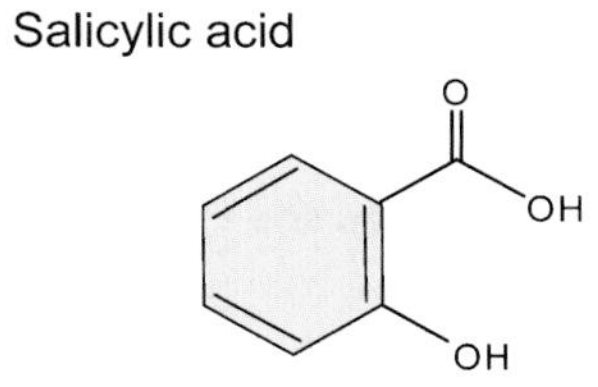

Figure 2–32
Chemical structure of salicylic acid.

that was recovered to wild type growth with exogenous brassinosteroid application.

Brassinosteroids are a class of plant steroid hormones that include over forty members including brassinolide (Fig. 2–31). Biosynthesis of brassinosteroids is from plant sterols (cycloartenol, campestrol) derived from the mevalonate pathway. Brassinosteroid-deficient mutants show reduced shoot growth, reduced fertility, and vascular development. Brassinosteroids complement auxin and cytokinin for cell division, gibberellin for seed germination, and are involved in phytochrome-mediated photomorphogenesis.

Jasmonates Jasmonic acid and methyl jasmonate are collectively called jasomnates and are members of the oxylipins derived from the oxidation of fatty acids starting with membrane linolenic acid (Fig. 2–33) (11, 18). The name jasmonate is in reference to its first discovery as a component in *Jasminum grandiflora* oil. In the 1980s, jasmonate was found to naturally occur as a germination inhibitor in bean seeds, and was thought to have similar properties to ABA. However, their primary roles are in plant defense, abiotic stress, and plant developmental process like wounding and senescence. Jasmonate levels increase with wounding and are important for inducing systemic wound responses, and exogenous jasmonate accelerates senescence. They interact with salicylic acid and ethylene as part of regulatory systems involved in plant defense. Methyl-jasmonate is a volatile compound thought to move within and between plants as a form of communication that can induce defense genes and compounds in plants prior to being exposed to the invading organism.

Salicylic Acid Salicylic acid is a plant phenolic compound derived from the shikimic acid pathway. Its name comes from its discovery in willow (*Salix*) bark (Fig. 2–32). It is a precursor to aspirin (acetylsalicylic acid). Salicylic acid has a major role in plant defense and is a critical component in systemic acquired resistance against pathogen attack (36). Salicylic acid may also be involved in plant growth via photosynthesis, flowering, and mineral nutrition. One interesting role for salicylic acid is as part of the heat-generating system found in themogenic aroid and cycad plants. Application of salicylic acid to voodoo lily led to temperature increases of as much as 12°C. As previously indicated, salicylic acid interacts with jasmonates and ethylene for plant defense.

Polyamines Polyamines, (putrescine, cadaverine, spermidine, and spermine) are synthesized from the amino acids arginine and ornithine and are widespread in animals and plants (3). Polyamines are required for cell growth and can function to stabilize DNA. In 1678, Antoni van Leeuwenhoek using the recently invented microscope found stellate crystals in human semen, which were later identified and named spermine. Key enzymes in the pathway include ornithine decarboxylase, arginine decarboxylase, spermine synthase, and spermidine synthase. Inhibitors are available for each of these enzymes. Spermine and spermidine synthase share S-adenosylmethionine as a precursor with the ethylene biosynthetic pathway. Competition for this precursor has been shown to be important for a number of plant processes, including seed germination, senescence, fruit ripening, and adventitious root formation. Additional processes where polyamines appear important or essential include seed development, somatic embryogenesis, flower initiation, and plant stress. Inclusion of polyamines in tissue culture systems has enhanced both of these processes.

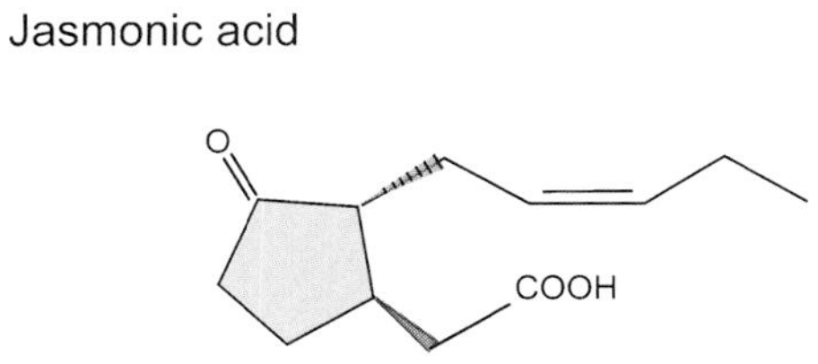

Figure 2–33
Chemical structure of jasmonic acid.

Flower initiation in thin layer culture systems of tobacco have been directly related to polyamines.

Plant Peptide and Polypeptide Hormones Peptide hormones have an established role in animal physiology, but it has only been recently that small peptide molecules have been discovered that influence plant growth in development (54). Systemin was the first peptide hormone discovered as an 18-amino acid peptide involved in long-distance communication in response to insect attack. Other peptide hormones include SCR/SP11, involved in pollen/stigma self-incompatibility; ENOD40, involved in *Rhizobium*-induced nodule formation in legumes; IDA, involved in flower petal abscission; and phytosulfokines, involved in cell proliferation during carrot tissue culture. **Florigen,** the long-sought-after factor promoting plant flowering, may be a polypeptide transcription factor called the FT (FLOWERING LOCUS T) protein (16, 75).

Plant Development, Competency, and Determinism

One of the principles in biology is that each living plant cell has the potential to reproduce an entire organism since it possesses all of the necessary genetic information in its genes to reproduce all the characteristics of the plant. This concept is known as **totipotency** (32).

totipotency The concept that a single cell has the necessary genetic factors to reproduce all of the characteristics of the plant.

The basic concepts of competency and determination for plant organ formation were developed from a series of experiments inducing shoot and root regeneration in field bindweed (*Convolvulus arvensis*) leaf explants (14). **Competency** was described as the potential of a cell(s) or tissue to develop in a particular direction; for example, the initiation of adventitious roots on a stem cutting or the change from a vegetative to a flowering meristem (55, 58, 60). At some point of development, the process becomes irreversible and the cells are said to be determined. Therefore, **determination** describes the degree to which cells are committed toward a specific organ formation. A general scheme for adventitious organ formation is shown in Figure 2–34.

competence The potential of a cell(s) to develop in a particular direction, such as forming adventitious roots.

determination The degree that a cell(s) is committed toward a given developmental direction at a given stage of development.

Development of competency into a particular kind of cell, tissue, or organ also may require a special signal. These may originate internally within the plant or externally as an environment signal. For example, the internal change in type of growing point, such as vegetative to flowering, may be associated with a shift in the activity of specific hormones. External parameters, including the application of specific growth regulators or the subjection of the plant to various environments, may bring about

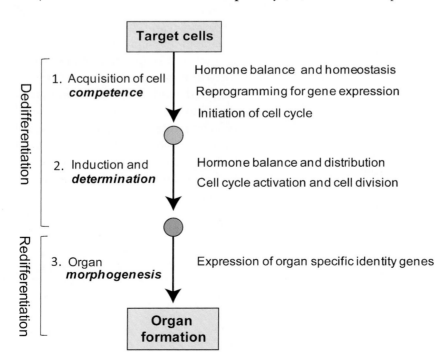

Figure 2–34
A generalized scheme for organ formation from target cells, such as parenchyma cells, during adventitious rooting or shoot formation. Target cells must acquire cell competency and become determined during dedifferentiation in order to redifferentiate into an organ (root or shoot).

changes. The signal may be applied exogenously, such as hormone application for the induction of adventitious roots, buds, shoots, or somatic embryos. These events take place if specific cells retain the potential for regeneration during development. In addition, specific cells can be induced to "dedifferentiate" and develop the capacity to regenerate. This potential to regenerate is the basis of propagation by cuttings (see Chapters 10 and 11), layering (see Chapter 15), specialized stems and roots (see Chapter 16), and tissue culture (see Chapters 17 and 18).

DISCUSSION ITEMS

This chapter covers areas of biology that are fundemantal to understanding plant propagation. These include plant nomenclature, plant life cycles, cell division, genes and gene expression, and plant hormones.

1. Compare and contrast mitosis and meiosis, and discuss how they function during sexual and asexual (vegetative) propagation.
2. How does phase change impact the seedling and clonal plant life cycles?
3. How do trademarks seem to be in contradiction to the rules for naming plants as set forward in the Botanical Code of Nomenclature for Cultivated plants?
4. Compare gene silencing used by the plant to regulate gene expression with gene silencing used as a biotechnology tool by the scientist.
5. How do plant scientists use mutants to understand growth and development? What kind of mutants might be important to better understand plant propagation?
6. What is hormone cross-talk, and why is it important for understanding plant propagation?

REFERENCES

1. Acquaah, G. 2006. *Principles of plant genetics and breeding.* Hoboken, NJ: Wiley-Blackwell Press.
2. Addicott, F. 1992. *Abscission.* Berkeley: Univ. of Calif. Press.
3. Baron, K., and C. Stasolla. 2008. The role of polyamines during *in vivo* and *in vitro* development. *In Vitro Cell. Dev. Bio.—Plant* 44:384–95.
4. Barton, J. 1997. Intellectual property and regulatory requirements affecting the commercialization of transgenic plants. In E. Galun and A. Breiman, eds. *Transgenic plants.* London: Imperial Press. pp. 254–76.
5. Benjamins, R., and B. Scheres. 2008. Auxin: The looping star in plant development. *Annu. Rev. Plant Biol.* 59:443–65.
6. Berg, L. R. 1997. *Introductory botany.* Orlando, FL: Saunders College Publications.
7. Bhattacharya, A., and J. A. Teixeira da Silva. 2004. Molecular systematic in *Chrysanthemum x grandiflorum* (Ramat.) Kitamura. *Scientia Hort.* 109:379–84.
8. Braun, A. C. 1959. A demonstration of the recovery of crown-gall tumor cell with the use of complex tumors and single-cell origin. *Proc. Natl. Acad. Sci. USA* 45:932–38.
9. Brickell, C. D. (Commission Chairman), B. R. Baum, W. L. A. Hetterscheid, A. C. Leslie, J. McNeill, P. Trehane, F. Vrugtman, and J. H. Wiersema (eds.). 2004. *New edition of the international code of nomenclature for cultivated plants.* (Obtainable from International Society for Horticultural Sciences.)
10. Brink, R. A. 1962. Phase change in higher plants and somatic cell heredity. *Quart. Rev. Bio.* 37(1):1–22.
11. Browse, J. 2009. Jasmonate passes muster: A receptor and targets for the defense hormone. *Annu. Rev. Plant Biol.* 60:183–205.
12. Callaway, D., and M. B. Callaway. 2000. *Breeding ornamental plants.* Portland, OR: Timber Press.
13. Chrispeels, M. J., and D. E. Sadava. 2002. *Plants, genes, and crop biotechnology.* 2nd ed. London, UK: Jones and Bartlett Pub.
14. Christianson, M. L., and D. A. Warnick. 1983. Competence and determination in the process of *in vitro* shoot organogenesis. *Develop. Bio.* 95:288–93.
15. Clouse, S. D. 2002. Brassinosteroids. *The Arabidopsis book.* American Society Plant Biologists (http://www.aspb.org/publications/arabidopsis/).
16. Corbesier, L., C. Vincent, S. Jang, F. Fornara, and Q. Fan. 2007. FT protein movement contributes to long distance signaling in floral induction of *Arabidopsis. Science* 316:1030–3.
17. Craig, R. 1994. Intellectual property protection of *Pelargoniums. HortTechnology* 3:284–90.

18. Creelmana, R. A., and R. Mulpuri. 2002. The oxylipin pathway in *Arabidopsis. The Arabidopsis book.* American Society Plant Biologists (http://www.aspb.org/publications/arabidopsis/).

19. Dandekar, A. M., G. H. McGranahan, and D. J. James. 1993. Transgenic woody plants. In S. D. Kung and R. Wu, eds. *Transgenic plants, Vol. 2: Present status and social and economic impacts.* New York: Academic Press. pp. 129–51.

20. Darlington, C. D. 1956. *Chromosome botany.* London: George Allen & Unwin.

21. Darwin, C. 1975. *On the origin of species by means of natural selection, or the preservation of the favored races in the struggle for life.* New York: Cambridge University Press.

22. Darwin, C., and F. Darwin. 1881. *The power of movement in plants.* New York: Appleton Century Crofts.

23. Davis, C. C., P. W. Fritsch, J. Li, and M. J. Donoghue. 2002. Phylogeny and biogeography of *Cercis* (Fabaceae): Evidence from nuclear ribosomal ITS and chloroplast *ndhF* sequence data. *Systematic Botany* 27:289–302.

24. Dewitte, W., and J. A. Murray. 2003. The plant cell cycle. *Annu. Rev. Plant Biol.* 54:235–64.

25. Doorenbos, J. 1965. Juvenile and adult phases in woody plants. In W. Ruhland, ed. *Handbuch der Pflanzenphysiologie,* Vol. 15 (Part 1). Berlin: Springer-Verlag. pp. 1222–35.

26. Ecker, J. R., and R. W. Davis. 1986. Inhibition of gene expression in plant cells by expression of antisense RNA. *Proc. Natl. Acad. Sci. USA* 83:5372–6.

27. Fang, D., R. R. Krueger, and M. L. Roose. 1998. Phylogenetic relationships among selected Citrus germplasm accessions revealed by intersimple sequence repeat (ISSR) markers. *J. Amer. Soc. Hort. Sci.* 124(4):612–17.

28. Finkelstein R. R., and C. D. Rock. 2002. Abscisic acid biosynthesis and response. *The Arabidopsis book.* American Society Plant Biologists (http://www.aspb.org/publications/arabidopsis/).

29. Fortanier, E. J., and H. Jonkers. 1976. Juvenility and maturity as influenced by their ontogenetical and physiological aging. *Acta Hort.* 56:37–44.

30. Galun, E., and A. Breiman. 1997. *Transgenic plants.* London: Imperial College Press.

31. Gubrium, E. K., D. J. Clevenger, D. G. Clark, J. E. Barrett, and T. A. Nell. 2000. Reproduction and horticultural performance of transgenic ethylene-insensitive petunias. *J. Amer. Soc. Hort. Sci.* 125(3):277–81.

32. Haberlandt, G. 1902. Kulturversuche mit isolierten Pflanzellen. *Sitzungsber. Akad. der Wiss. Wien, Math.—Naturwiss.* Ki. 111–69–92.

33. Hackett, W. P. 1985. Juvenility, maturation and rejuvenation in woody plants. *Hort. Rev.* 7:109–55.

34. Hackett, W. P., and J. R. Murray. 1997. Approaches to understanding maturation or phase change. In R. L. Geneve, J. E. Preece, and S. A. Merkle, ed. *Biotechnology of ornamental plants.* Wallingford, UK: CAB International.

35. Haissig, B. E. 1979. Influence of aryl esters of indole-3-acetic and indole-3-butyric acids on adventitious root primordium development. *Physiol. Plant.* 57:435–40.

36. Hayat, S., and A. Ahmad. 2007. *Salicylic acid—A plant hormone.* Dordrecht: Springer.

37. Hayward, M. D., N. O. Bosemark, and I. Romagoza, eds. 1993. *Plant breeding: Principles and prospects.* London: Chapman Hall.

38. Helfer, L. R. 2004. *Intellectual property rights in plant varieties: International legal regimes and policy options for national governments.* FAO Legislative study 85.

39. Hershey, A. D., and M. Chase. 1952. Independent functions of viral protein and nucleic acid in growth of bacteriophage. *J. Gen. Phys.* 36:39–56.

40. Hutton, R. J. 1994. Plant patents—a boon to horticulture and the American public. *HortTech.* 2:280–83.

41. Jackson, J. F., and H. F. Linskens. 2003. *Genetic transformation of plants.* Dordrecht: Springer.

42. Jondle, R. J. 1993. Legal protection for plant intellectual property. *HortTech.* 3:301–7.

43. Jones-Rhoades, M., D. Bartel, and B. Bartel. 2006. MicroRNAs and their regulatory roles in plants. *Annu. Rev. Plant Biol.* 57:19–53.

44. Kays, S. J. 1991. Biotechnology—a horticulturist's perspective: Introduction to the colloquium. *HortScience* 26:1020–21.

45. Kepinski, S. 2006. Integrating hormone signaling and patterning mechanisms in plant development. *Curr. Opin. Plant Bio.* 9:28–34.

46. Kester, D. E. 1983. The clone in horticulture. *HortScience* 18(6):831–37.

47. Kidner, C. A., and R. A. Martienssen. 2005. The developmental role of microRNA in plants. *Curr. Opin. Plant Bio.* 8:38–44.

48. Kieber, J. J. 2002. Cytokinins. *The Arabidopsis book.* American Society Plant Biologists (http://www.aspb.org/publications/arabidopsis/).

49. Klug, W. S., and M. R. Cummings. 2000. *Concepts of Genetics.* 6th ed. Upper Saddle River, NJ: Prentice Hall.

50. Koltunow, A. M., and U. Grossniklaus. 2003. APOMIXIS: A developmental perspective. *Annu. Rev. Plant Biol.* 54:547–74.

51. Lanjouw, J., ed. 1966. International code of botanical nomenclature. *Regnum Vegetabile* 46:402.

52. Lavi, U., E. Lahav, C. Degani, and S. Gazit. 1992. The genetics of the juvenile phase in avocado and its applications for breeding. *J. Amer. Soc. Hort. Sci.* 117(6):981–84.

53. Li, W., K. A. Zarka, D. S. Douches, J. J. Coombs, W. L. Pett, and E. J. Graflus. 1999. Coexpression of potato PVY° coat protein and *cry*V-Bt genes in potato. *J. Amer. Soc. Hort. Sci.* 124(3):218–23.

54. Matsubayashi, Y., and Y. Sakagami. 2006. Peptide hormones in plants. *Annu. Rev. Plant Biol.* 57:649–74.

55. Meins, F. 1986. Determination and morphogenetic competence in plant tissue culture. In M. M. Yeoman, ed. *Plant cell culture technology.* Boston: Blackwell. pp. 43–75.

56. Mendel, G. 1866. Versuche uber pflanzenhybriden. *Verhandlungen des Naturforschenden den Vereines in Brunn* 4:3–47.

57. Michniewicz, M., P. B. Brewer, and J. Friml. 2007. Polar auxin transport and asymmetric auxin distribution. *The Arabidopsis book.* American Society Plant Biologists (http://www.aspb.org/publications/arabidopsis/).

58. Mohnen, D. 1994. Novel experimental systems for determining cellular competence and determination. In T. D. Davis and B. E. Haissig, eds. *Biology of adventitious root formation.* New York and London: Plenum Press. pp. 87–98.

59. Moore, J. N. 1993. Plant patenting: A public fruit breeder's assessment. *HortTech.* 3:262–66.

60. Murray, J. R., M. C. Sanchez, A. G. Smith, and W. P. Hackett. 1994. Differential competence for adventitious root formation in histologically similar cell types. In T. D. Davis and B. E. Haissig, eds. *Biology of adventitious root formation.* New York: Plenum Press. pp. 99–110.

61. Poethig, R. S. 1990. Phase change and the regulation of shoot morphogenesis in plants. *Science* 250:923–29.

62. Pollack, J. R., C. M. Perou, A. A. Alizadeh, M. B. Eisen, A. Pergamenschikov, C. F. Williams, S. S. Jeffrey, D. Botstein, and P. O. Brown. 1999. Genome-wide analysis of DNA copy-number changes using cDNA microarrays. *Nature Genetics* 23:41–6.

63. Potter, D., J. J. Luby, and R. E. Harrison. 2000. Phylogenetic relationships among species of Fragaria (*Rosaceae*) inferred from non-coding nuclear and chloroplast DNA sequences. *Systematic Botany* 25(2):337–48.

64. Raven, P. H., R. F. Evert, and S. E. Eichhorn. 2005. *Biology of plants,* 7th ed. New York: Freeman\Worth.

65. Schaff, D. A. 1992. Biotechnology—gene transfer: Terminology, techniques and problems involved. *HortScience* 26:1021–24.

66. Schallera, G. E., and J. J. Kieber. 2002. Ethylene. *The Arabidopsis book.* American Society Plant Biologists (http://www.aspb.org/publications/arabidopsis/).

67. Seo, M., and T. Koshiba. 2002. Complex regulation of ABA biosynthesis in plants. *Trends Plant Sci.* 7:41–8.

68. Skoog, F., and C. O. Miller. 1957. Chemical regulation of growth and organ formation in plant tissues cultured in vitro. *Symp. Soc. for Exp. Biol.* 11:118–31.

69. Sponsel, V. M. 1995. Gibberellin biosynthesis and metabolism. In P. J. Davies, ed. *Plant hormones: Physiology, biochemistry and molecular biology.* Dordrecht: Kluwer Acad. Pub. pp. 43–75.

70. Stebbins, G. L. 1950. *Variation and evolution in plants.* New York: Columbia University Press.

71. Stoskopf, N. C., D. T. Tomes, and B. D. Christie. 1993. *Plant breeding theory and practice.* Boulder, CO: Westview Press.

72. Stout, A. B. 1940. The nomenclature of cultivated plants. *Amer. J. Bot.* 27:339–47.

73. Sun, T. 2008. Gibberellin metabolism, perception and signaling pathways. *The Arabidopsis book.* American Society Plant Biologists (http://www.aspb.org/publications/arabidopsis/).

74. Tao, R., A. M. Dandekar, S. L. Uratsu, P. V. Vail, and J. S. Tebbets. 1997. Engineering genetic resistance against insects in Japanese persimmon using the *cry*IA gene of *Bacillus thuringiensis. J. Amer. Soc. Hort. Sci.* 122(6):764–71.

75. Turck, F., F. Fornara, and G. Coupland. 2008. Regulation and identity of Florigen: FLOWERING LOCUS T moves centre stage. *Annu. Rev. Plant Biol.* 59:573–94.

76. Ueguchi-Tanaka, M., M. Nakajima, A. Motoyuki, and M. Matsuoka. 2007. Gibberellin receptor and its role in gibberellin signaling in plants. *Annu. Rev. Plant Biol.* 58:183–98.

77. UPOV. 1991. *International convention for the protection of new varieties of plants.* UPOV Publ. No. 221(E).

78. Vainstein A. 2002. *Breeding for ornamentals: Classical and molecular approaches.* Dordrecht: Kluwer Acad. Pub.

79. Visser, T., J. J. Verhaegh, and D. P. de Vries. 1976. A comparison of apple and pear seedlings with reference to the juvenile period. I. Seedling growth and yield. *Acta Hort.* 56:205–14.

80. Wareing, P. F., and V. M. Frydman. 1976. General aspects of phase change with special reference to *Hedera helix* L. *Acta Hort.* 56:57–68.

81. Watson, J. D., and F. C. Crick. 1953. Molecular structure for nucleic acids: A structure for deoxyribose nucleic acid. *Nature* 171:737–38.

82. Went, F. W. 1937. *Phytohormones.* New York: Macmillan.

83. Woodward, A. W., and B. Bartel. 2005. Auxin: regulation, action, and interaction. *Ann. Bot.* 95:707–35.

84. Xiang, N., and Y. Hong. 2000. The AFLP technique and its applications for plant study. *Comm. of Plant Physiol.* 36:236–40.

85. Zimmerman, R. 1972. Juvenility and flowering in woody plants: A review. *HortScience* 7:447–55.

3
The Propagation Environment

INTRODUCTION

Propagation can be done in the field, orchard, forest, outdoor raised beds, and in protected culture environments such as greenhouses, poly-covered houses, and tissue culture laboratories. The plant propagation period is generally a very narrow segment of a plant's life, ranging from several weeks for fast-growing herbaceous plants to one to two years for woody perennials. Following propagation, the rooted cuttings, seedlings **(plugs), layers,** or tissue culture produced plants are transplanted as **liner plants.** The liner plants are grown in small pots and then transplanted into larger containers or directly transplanted into field production. In other production systems plants may be propagated and produced in the same container or field location without going through a liner stage.

To enhance the propagation of plants, commercial producers manipulate the environment of **propagules** (cuttings, seeds) by managing:

a. **microclimatic conditions** (light, water-relative humidity, temperature, and gases)
b. **edaphic factors** (propagation medium or soil, mineral nutrition and water), and
c. **biotic factors**—interaction of propagules with other organisms (such as beneficial bacteria, mycorrhizal fungi, pathogens, insect pests, etc.) (Fig. 3–1).

Unique ecological conditions exist during propagation. Commercial propagators may have to compromise to obtain an "average environment" in

plugs   Small seedling plants.

layers   Plants produced asexually from layering, such as air layering or stooling.

propagule   A plant structure used for regenerating plants, which can include cuttings, seeds, grafts, layers, tissue culture explants, and single cells.

microclimatic conditions   Any environmental factors (relative humidity, temperature, light, gases, etc.) in the immediate vicinity of the propagule during propagation.

edaphic factors   Any factors influenced by the soil or propagation medium (substrate).

learning objectives

- Identify the environmental factors affecting propagation.
- Describe the physical structures for managing the propagation environment.
- Describe the containers for propagating and growing young liner pots.
- Discuss the management of media and nutrients in propagation and liner production.
- Discuss the management of microclimatic conditions in propagation and liner production.
- Discuss the management of biotic factors—pathogens and pests—in plant propagation.
- Explain the post-propagation care of liners.

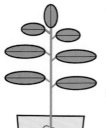

Atmospheric
- Light
- Temperature
- Gas exchange, Oxygen, Carbon Dioxide, Ethylene
- Water and Humidity

Biotic Pathogens and Pests

Edaphic
- Temperature
- Gas exchange, Oxygen, Carbon Dioxide, Ethylene
- Water
- Propagation Substrate or Medium
- Propagation Flat and Liner Container Dynamics
- Mineral Nutrients

Biotic
- Rhizophere Microbes -- Mycorrhizal Fungi Beneficial Bacteria, etc.
- Pathogens, Pests, and Weeds

Figure 3–1
The propagation environment: Manipulation of microclimatic, edaphic, and biotic factors. Modified from Landis (69).

Shading Partial reduction of light to 100 percent light exclusion that can occur during stock plant manipulation and/or propagation

hardening-off The stress adaptation process or **acclimation** that occurs as a propagule, such as a cutting, is gradually weaned from a high to a low relative humidity environment during rooting; in **micropropagation** (tissue culture) acclimation is referred to as **acclimatization**.

which a whole range of species are propagated by cuttings, seed, and/or tissue culture explants (69). The environmental conditions that are optimum for plant propagation are frequently conducive for pests (pathogenic fungi, viruses, bacteria, insect, and mite development). Astute propagators not only manage the environment during propagation, but also manipulate the environment of **stock plants** prior to selecting propagules, such as **shading** and stooling to maximize rooting potential of a propagule; and post propagation—**hardening-off** (weaning rooted cuttings from the mist system and changing fertility regimes) to assure growth and survival of tender-rooted liner plants after propagation.

ENVIRONMENTAL FACTORS AFFECTING PROPAGATION

In propagating and growing young nursery plants, facilities and procedures are designed to optimize the response of plants to environmental factors influencing their growth and development, such as **light, water, temperature, gases,** and **mineral nutrition.** In addition, young nursery plants require protection from pathogens and other pests, as well as control of salinity levels in the growing media. The propagation structures, equipment, and procedures described in this chapter, if handled properly, maximize the plants' growth and development by controlling their environment.

BOX 3.1 GETTING MORE IN DEPTH ON THE SUBJECT
LINER PRODUCTION

A **liner** traditionally refers to lining out nursery stock in a field row. The term has evolved to mean a small plant produced from a rooted cutting, seedling, plug, or tissue culture plantlet. **Direct sticking** or **direct rooting** into smaller **liner pots** is commonly done in United States propagation nurseries. Seedlings and rooted cuttings can also be transplanted into small liner pots and allowed to become established during liner production, before being transplanted to larger containers (**upcanned**) or outplanted into the field.

BOX 3.2 GETTING MORE IN DEPTH ON THE SUBJECT
MEASUREMENT OF LIGHT

Irradiance is the relative amount of light as measured by radiant energy per unit area. Irradiance, intensity, and photon flux all measure the amount of light very differently; they are not interchangeable terms. **Photosynthetic photon flux (PPF)** is the best light measurement for plant propagation, since the process of photosynthesis relies on the number of photons intercepted, not light given off by a point source (intensity) or energy content (irradiance). **Photosynthetic active radiation (PAR)** is measured in the 400 to 700 nanometer (nm) waveband as PPF in micromoles of photons per unit area per time ($\mu mol\ m^{-2}s^{-1}$) with a **quantum sensor** or as watts per square meter (W/m^2) with a **pyranometric sensor**. Some propagators still measure light intensity with a **photometric sensor**, which determines foot-candles or lux (1 foot-candle = 10.8 lux). A photometric sensor is relatively insensitive to wavelengths that are important for plant growth; that is, it may record high light intensity from an artificial electric light source, but it does *not* take into account if the light source is rich in green and yellow, or poor in red and blue light—which would lead to poor plant growth. Quantum and radiometric (pyranometer) sensors can be purchased from instrument companies (i.e., LI-COR Biosciences, www.licor.com; or Apogee Instruments, Inc., www.apogee-inst.com). For determining **light quality** or **wavelength**, the spectral distribution is measured with a portable spectroradiometer, which is a very expensive piece of equipment.

Light

Light is important for photosynthesis as a source of radiant energy. Light also generates a heat load that needs to be controlled (i.e., too high a temperature can quickly desiccate and kill cuttings). The management of light can be critical for rooting cuttings, germinating seeds, growing seedlings, or shoot multiplication of **explants** during tissue culture propagation. Light can be manipulated by controlling *irradiance* (see Box 3.2), *light duration* (daylength, photoperiod), and *light quality* (wavelength). For a relative comparison of light units for propagation, see Box 3.3 on page 52.

Irradiance While many propagators still measure light intensity, determining the photon flux of light is more accurate because the process of photosynthesis depends on the number of photons intercepted (*photosynthetic photon flux*), not just the light given off by a point source (*intensity*).

Daylength (Photoperiod) Higher plants are classified as long-day, short-day, or day-neutral, based on the effect of photoperiod on initiation of reproductive growth. **Long-day** plants, which flower chiefly in the summer, will flower when the **critical photoperiod** of light is equaled or exceeded; **short-day** plants, such as chrysanthemums, flower when the critical photoperiod is not exceeded. Reproductive growth in **day-neutral** plants, such as roses, is not triggered by photoperiod. The discovery of *photoperiodism* by Garner and Allard demonstrated that the dark period, not the light period, is most critical to initiation of reproductive growth, even though light cycles are traditionally used to denote a plant's photoperiod. In propagation, fresh seed collected in the fall from selected woody plant species, such as *Larix,* need long-day conditions to germinate. Dahlia cuttings need short-day conditions to trigger tuberous root formation.

Photoperiod can be extended under short-day conditions of late fall and early winter by lighting with incandescent lights, or high intensity discharge lights (HID) (Fig. 3–14, page 65). Conversely, photoperiod can be shortened under the long-day conditions of late spring and summer by covering stock plants and cuttings with black cloth or plastic that eliminates all light. See the in-depth discussion of phytochrome and photoperiodism in Chapter 7.

Light Quality Light quality is perceived by the human eye as color, and corresponds to a specific range of wavelengths. Red light is known to enhance seed germination of selected lettuce cultivars, while far-red light inhibits germination. Far-red light can promote bulb formation on long-day plants, such as onion (*Allium cepa*). Blue light enhances in vitro bud regeneration of tomato (77). Using greenhouse covering materials with different spectral light-transmitting characteristics, researchers at Clemson University (97) have been able to control the height and development of greenhouse-grown plants, rather than relying on the chemical application of growth regulators for height control. This has application for plant propagation, liner production, and plant tissue culture systems. Red shade cloth shifts light quality towards the blue/green and is being used to enhance root development of cuttings (Fig. 3–11, page 62). Red shade cloth can also be used to increase leaf surface and branching, which is important in liner development (111).

BOX 3.3 GETTING MORE IN DEPTH ON THE SUBJECT
RELATIVE COMPARISON OF LIGHT UNITS FOR SOLAR RADIATION AND ARTIFICIAL LIGHTING (67, 72, 117)*

Light Source	Energy [Photosynthetic photon plux] (μmol m^{-2}s^{-1})	Radiation [Irradiance] (watts m^{-2})	Illumination [Light intensity] (lux)	(ft-candles)
Solar Radiation				
Full sunlight	2,000	450	108,000	10,037
Heavy overcast	60	15	3,200	297
Artificial Light Source				
Metal halide (400 W) lamp @ 2 m height	19	4	1,330	124

* Photosynthetically active radiation (PAR): 400 to 700 nm. Conversions between energy, radiation, and illumination units are complicated and will be different for each light source. The spectral distribution curve of the radiant output must be known in order to make conversions.

Water-Humidity Control

Water management and humidity control are critical in propagation. Water management is one of the most effective tools for regulating plant growth. Evaporative cooling of an **intermittent mist** system can help control the propagation house microenvironment and reduce the heat load on cuttings, thereby permitting utilization of high light conditions to increase photosynthesis and encourage subsequent root development. A solid support medium, such as peat-perlite, is not always necessary to propagate plants;

intermittent mist
A thin film of water produced through a pressurized irrigation system that cools the atmosphere and leaf surface of cuttings.

peach cuttings can be rooted under aeroponic systems, while woody and herbaceous ornamentals can be rooted in modified, aero-hydroponic systems without relying on overhead mist (108). Tissue culture explants are often grown in a liquid phase rather than on a solid agar media.

While leaf **water potential** (Ψ_{leaf}) is an important parameter for measuring water status of seedlings and cuttings, and influences rooting of cuttings, **turgor** (Ψ_p) is physiologically more important for growth processes. The water status of seedlings and cuttings is a balance between transpirational losses and uptake of water. Later in this chapter the methods to control water loss of leaves of cuttings, seedlings, and containerized grafted plants are discussed.

BOX 3.4 GETTING MORE IN DEPTH ON THE SUBJECT
PLANT WATER MEASUREMENTS IN PROPAGATION

Water potential (Ψ_{water}) refers to the difference between the activity of water molecules in pure distilled water and the activity of water molecules in any other system in the plant. Pure water has a water potential of zero. Since the activity of water in a cell is usually less than that of pure water, the water potential in a cell is usually a negative number. The magnitude of water potential is expressed in megapascals [1 megapascal (MPa) = 10 bars = 9.87 atmospheres]. Propagators can determine water potential by using a pressure chamber (pressure bomb) manufactured by PMS Instrument Company (www.pmsinstrument.com) or Soil Moisture Corporation (www.soilmoisture.com). A psychrometer with a microvolt meter (LiCor, www.licor.com) can also be used. Estimation of **turgor** (Ψ_p) (or pressure potential) requires measurement of **water potential** (Ψ_{water}) minus the **osmotic potential** (Ψ_π), which is based on the formula $\Psi_{water} = \Psi_p + \Psi_\pi$. Osmotic potential can also be determined by either a pressure chamber or a psychrometer. The matrix potential (Ψ_m) is generally insignificant in determining Ψ_{water} but is important in seed germination. See the discussion on water potential and seed germination in Chapter 7.

Temperature

Temperature affects plant propagation in many ways. Seed dormancy is broken in some woody species by cool-moist stratification conditions that allow the germination process to proceed. Temperature of the propagation medium can be suboptimal for seed germination or rooting due to seasonally related ambient air temperature or the cooling effect of mist. In grafting, heating devices are sometimes placed in the graft union area to speed up graft union formation, while the rest of the rootstock is kept dormant under cooler conditions (see Fig. 12–48).

It is often more satisfactory and cost-effective to manipulate temperature by bottom heating at the propagation bench level, rather than heating the entire propagation house (Fig. 3–2). The use of heating and cooling systems in propagation structures is discussed further in this chapter (see Chapter 10 for heating equipment and sensors).

Gases and Gas Exchange

High respiration rates occur with seed germination and plug development, and during adventitious root formation at the base of a cutting. These aerobic processes require that O_2 be consumed and CO_2 be given off by the propagule. Seed germination is impeded when a hard seed coat restricts gas exchange. Likewise, gas exchange at the site of root initiation and subsequent rooting are reduced when cuttings are stuck in highly water-saturated propagation media with small air pore spaces. In leaves of droughted propagules, stomata are closed, gas exchange is limited, and suboptimal rates of photosynthesis occur. During propagation in enclosed greenhouses, ambient CO_2 levels can drop to suboptimal levels, limiting photosynthesis and propagule development. The buildup of ethylene gas (C_2H_4) can be deleterious to propagules during storage, shipping, and propagation conditions. Ethylene also plays a role in plant respiration, rooting of cuttings, and seed propagation.

Mineral Nutrition

To avoid stress and poor development during propagation, it is important that the stock plants be maintained under optimal nutrition—prior to harvesting propagules. During propagation, nutrients are generally applied to seedlings and plugs by **fertigation** (soluble fertilizers added to irrigation water) or with controlled-release fertilizers that are either

> fertigation The application of soluble fertilizer during the irrigation of a seedling or rooted cutting.

(a) (b) (c) (d)

Figure 3–2
Propagation house heating systems. (a) Gas-fired infrared or vacuum-operated radiant heaters (arrow). (b) Forced hot air heating system. (c) Greenhouse, hot water boilers. (d) Heating below the bench for better control of root zone temperature.

preincorporated into the propagation medium or broadcast (**top-dressed**) across the medium surface. Cuttings are normally fertilized with a controlled-release fertilizer preincorporated into the propagation medium (which is discussed later in this chapter and in Chapter 10) or with soluble fertilizer applied *after* roots are initiated. The development of intermittent mist revolutionized propagation, but the mist can severely leach cuttings of nutrients. This is a particular problem with cuttings of difficult-to-root species that have long propagation periods.

PHYSICAL STRUCTURES FOR MANAGING THE PROPAGATION ENVIRONMENT

Propagation Structures

Facilities required for propagating plants by seed, cuttings, and grafting, and other methods include two basic units. One is a structure with temperature control and ample light, such as a greenhouse, modified quonset house, or hotbed—where seeds can be germinated, or cuttings rooted, or tissue culture microplants rooted and acclimatized. The second unit is a structure into which the young, tender plants (liners) can be moved for hardening, which is preparatory to transplanting outdoors. Cold frames, low polyethylene tunnels or sun tunnels covered by Saran, and lathhouses are useful for this purpose. Any of these structures may, at certain times of the year and for certain species, serve as a propagation and acclimation structure. A synopsis of how structures are utilized in propagation is presented in Table 3–1.

Aseptic Micropropagation Facilities

Aseptic micropropagation facilities are described in Chapter 18.

Greenhouses

Greenhouses have a long history of use by horticulturists as a means of forcing more rapid growth of plants (11, 41, 55, 75, 122). Most of the greenhouse area in

Table 3–1
UTILIZATION OF PROPAGATION STRUCTURES

Propagation structure	Micropropagation	Cuttings	Seedlings/ Plugs	Grafting	Layering	Liner production and hardening-off
Micropropagation facilities (indoor)	Yes	No; except microcuttings	No	No; except micrografting	No	No
Greenhouses	Yes; during acclimatization	Yes	Yes	Yes	Yes; air layering	Yes
Closed-case propagation Hot frames (hotbeds) Heated sun tunnels	No	Yes	Yes	Yes	No	Yes
Closed-case propagation Cold frames Unheated sun tunnels	No; except acclimatization	Yes; hardwood and semi-hardwood cuttings	Yes	Yes	Yes	Yes
Lathhouses (shade houses)	No; except acclimatization	Yes; hardwood and semi-hardwood cuttings	Yes	Yes	Yes	Yes; used extensively for this
Miscellaneous closed-case propagation systems in greenhouses: (a) Propagating frames (b) Contact polyethylene systems	No; except acclimatization	Yes; hardwood and semi-hardwood cuttings	Yes	Yes; sometimes with bench grafting and acclimation	No	Yes

the United States is used for the wholesale propagation and production of floricultural crops, such as pot plants, foliage plants, bedding plants, and cut flowers; fewer are used for nursery stock and vegetable crops (104).

Greenhouse structures vary from elementary, home-constructed to elaborate commercial installations. Commercial greenhouses are usually independent structures of even-span, **gable-roof construction,** proportioned so that the space is well utilized for convenient walkways and propagating benches (55). In larger propagation operations, several single greenhouse units are often attached side by side, eliminating the cost of covering the adjoining walls with glass or polyethylene (Fig. 3–3). These gutter-connected houses, while more expensive to construct than independent ground-to-ground structures, allow easy access between houses and decrease the square footage (meters) of land needed for propagation houses. Heating and cooling equipment is more economical to install and operate, since a large growing area can share the same equipment (62). Greenhouses with double-tiered, moveable benches that can be rolled outside, and **retractable roof** greenhouses reduce energy costs (Figs. 3–4 and 3–5); they are being used in cutting and

gable-roof constructed greenhouse   A unit that has more expensive, reinforced upper support for hanging mist systems, supplementary lights, or additional tiers of potted plants.

retractable roof greenhouse   A unit with a roof that can be opened during the day and closed at night.

Figure 3–3
Gutter-connected propagation greenhouses. (a) A series of gutter-connected propagation houses. (b) The basic types of gutter-connected propagation greenhouses: bow or truss. Bows are less expensive, but offer less structural strength. Trusses make for a stronger house, while giving propagators the ability to hang plants and equipment, such as monorails, curtain systems, and irrigation booms. (c) Non—load-carrying bow propagation house. (d) Load-bearing, gutter-connected truss house (arrow).

Figure 3–4
(a and b) Instead of a movable bench, propagation trays are placed on rollers; notice how all trays on rollers slant toward the middle of the propagation house for easier movement of materials. (c) Movable benches for seedling plug production. (d and e). Propagation house with retractable benches, which can be rolled from the greenhouse structure to the outdoors, have reduced energy costs. (d) Inside of house with double-tiered benches that can be brought in at night and during inclement weather. Benches slide through opening of greenhouse and can be left outside under full sun conditions.

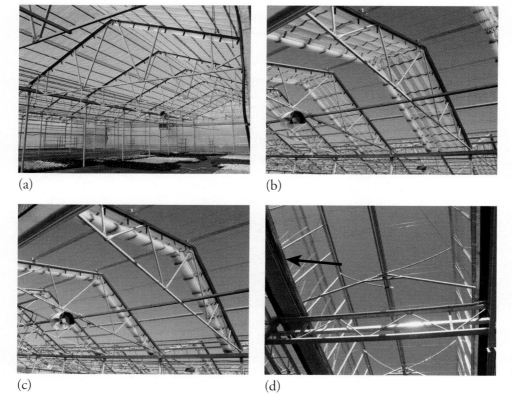

Figure 3–5
(a, b, and c) Retractable roof greenhouse for reducing heat load during propagation and liner production, and (d) a top-vented Dutch-style glasshouse with thermal curtains (arrow) for shade and trapping heat during winter nights.

seed propagation, and seedling plug production. Since the liner seedlings are partly produced under full sun conditions, they are better acclimatized for the consumer (8).

Quonset-type greenhouse An inexpensive propagation house made of bent tubing or PVC frame that is covered with polyethylene plastic.

Quonset-type construction is very popular. Such houses are inexpensive to build, usually consisting of a framework of piping, and are easily covered with one or two layers of polyethylene (Fig. 3–6).

Arrangement of benches in greenhouses varies considerably. Some propagation installations do not have permanently attached benches, their placement varying according to the type of equipment, such as lift trucks or electric carts, used to move flats and plants. The correct bench system can increase production efficiency and reduce labor costs (124). Rolling benches can reduce aisle space and increase the usable space by 30 percent in a propagation greenhouse. The benches are pushed together until one needs to get between them, and then rolled apart (Fig. 3–4). With rolling benches, propagation work can be done in an ergonomically correct fashion, making workers more comfortable, efficient, and productive (118). Besides increased propagation production numbers, rolling benches allow other automation features to be added (Fig. 3–7). Conversely, to reduce costs, many propagation houses are designed not to use benches, but rather cutting flats or small liner containers are placed on the gravel or Saran-covered floor (Figs. 3–6 and 3–7). It all depends on the propagation system and units to be produced.

In an **floor ebb and flood system (flood floor)**, greenhouse benches are eliminated and plants are produced with an automated floor watering and fertility system. There are below-ground floor-heating pipes and irrigation lines, a system of runoff-capturing tanks

Figure 3–6
Versatility of a polyethylene, saran-shaded quonset house. (a) Propagators sticking cuttings into rooting media floor beds previously prepared and sterilized with methyl bromide. (b) Cuttings in small liner rooting pots under mist.
(c) Rooted liner crop protected under saran shade with poly sidewalls, and (d) shade removed and rooted liner crop ready for transplanting and finishing off in larger container pots.

Figure 3–7
For more efficient use of costly greenhouse propagation space, movable benches on rollers have been installed to reduce aisle space. (a and b) Hydraulic lift system (arrow) to pick up and move benches. (c) Movable benches for maintaining coleus stock plants. (d) To eliminate bench space, cuttings in liner pots are placed on the cement propagation house floor and intermittent mist is applied from mist nozzles suspended from the ceiling.

with filters, and computer-controlled return of appropriate levels of irrigation water mixed with soluble fertilizer to the floor growing area (9, 89). While this has received limited use in the propagation of plants, it does have application for liner stock plant production of seedling plugs, rooted cuttings, and tissue culture produced plantlets (Fig. 3–8). Flood floor systems are more efficient than conventional bench greenhouses. They are highly automated, require less labor, and are environmentally friendly—since irrigation runoff, including nutrients and pesticides, is recaptured and recycled. The drawback of these benchless systems is the potential for rapid disease spread.

Greenhouse construction begins with a metal framework covered with polycarbonate, acrylic, glass, or poly (plastic) material. Gutter-connected greenhouses can be constructed as bow-style houses, which are less expensive and offer less structural strength, or as load-bearing truss-style houses, which give propagators the ability to hang mist and irrigation booms, install ceiling curtains for temperature and light control, and so on (Fig. 3–3). All-metal prefabricated greenhouses with prewelded or prebolted trusses are also widely used and are available from several manufacturers.

In any type of greenhouse or bench construction using wood, the wood should be pressure-treated with a preservative such as chromatid copper arsenate (CCA), which will add many years to its life (5). The two most common structural materials for greenhouses are steel and aluminum. Most greenhouses are made from galvanized steel, which is cheaper, stronger, lighter, and smaller than an aluminum member of equal strength. Aluminum has rust and corrosion resistance, and can be painted or anodized in various colors (62). With the high cost of

---

BOX 3.5 GETTING MORE IN DEPTH ON THE SUBJECT
SOURCES OF COMMERCIAL GREENHOUSES

For sources of commercial greenhouses, contact the National Greenhouse Manufacturers Association (www.ngma.com). A number of trade journals such as *GrowerTalks* (www.ballpublishing.com, choose the link for GrowerTalks) and *Greenhouse Beam Pro* (www.greenbeampro.com) list commercial greenhouse manufacturers and suppliers that include greenhouse structures, shade and heat retention systems, cooling and ventilation, environmental control computers, bench systems, and internal transport systems in greenhouses.

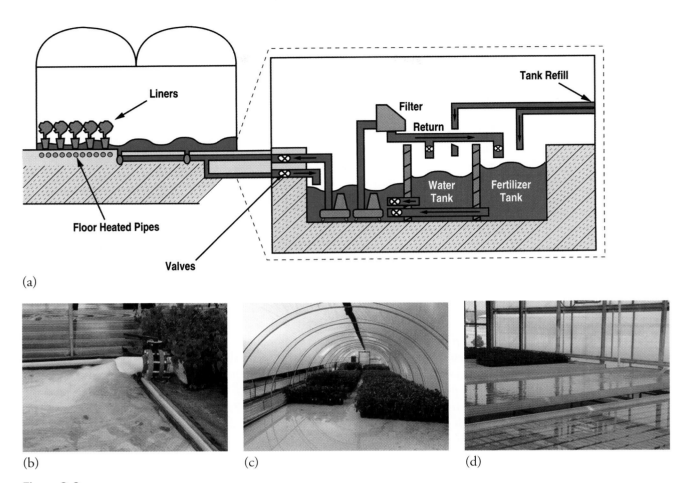

Figure 3–8
(a, b, and c) An ebb and flood or flood floor system. No benches are used and stock plants are produced with an automated floor watering and fertility system. There are below-ground floor heating pipes and irrigation lines, a system of runoff-capturing tanks with filters, and computer-controlled return of appropriate levels of irrigation water mixed with soluble fertilizer to the floor growing area. (a) Schematic of ebb and flood system with liner plants. (b and c) Flood floor system for maintaining stock plants. (d) Ebb and flood bench system.

lumber, fewer greenhouses are constructed with wood, and traditional wooden benches are being replaced by rigid plastics, metal benches, and other synthetic materials.

Greenhouse Heating and Cooling Systems

Ventilation, to provide air movement and air exchange with the outside, is necessary in all greenhouses to aid in controlling temperature and humidity. A mechanism for manual opening of panels at the ridge and sides or with passive ventilation can be used in smaller greenhouses, but most larger installations use a forced-air fan and pad-cooling ventilation system either regulated by thermostats or controlled by computer (42, 89).

Traditionally, greenhouses have been heated by steam or hot water from a central boiler through banks of pipes (some finned to increase radiation surface) suitably located in the greenhouse (Fig. 3–2). Unit heaters for each house, with fans for improved air circulation, are also used. If oil or gas heaters are used, they must be vented to the outside because the combustion products are toxic to plants (and people!), and ethylene gas generated can adversely affect plant growth. In large greenhouses, heated air is often blown into large—30 to 60 cm (12 to 24 in)—4-mil convection polyethylene tubes hung overhead. These extend the length of the greenhouse. Small—5 to 7.5 cm (2 to 3 in)—holes spaced throughout the length of these tubes allow the hot air to escape, thus giving uniform heating throughout the house. These same convection tubes can be used for forced-air ventilation and cooling in summer, eliminating the need for manual side and top vents.

Gas-Fired Infrared Heaters Gas-fired infrared heaters are vacuum-operated radiant heaters that are sometimes installed in the ridges of greenhouses with the concept of heating the plants but not the air mass. Infrared heaters consist of several lines of radiant tubing running the length of the house, with reflective shielding above the tubes installed at a height of 1.8 to 3.7 m (6 to 12 ft) above the plants (Fig. 3–2). The principal advantage of infrared heating systems in greenhouses is lower energy use. Cultural practices may need to be changed because infrared heating heats the plant but not the soil underneath.

> **gas-fired infrared heaters** Vacuum-operated radiant heaters installed in the ridges of greenhouses with the concept of heating the plants but not the air mass.

Root Zone Heating In contrast to infrared heating, root zone heating is done by placing pipes on or below the soil surface in the floor of the greenhouse, or on the benches, with recirculating hot water—controlled by a thermostat—circulating through the pipes. This places the heat below the plants, which hastens the germination of seeds, rooting of cuttings, or growth of liner plants. This popular system has been very satisfactory in many installations, heating the plants' roots and tops, but not the entire air mass in the greenhouse, yielding substantial fuel savings. It is also excellent for controlling foliage diseases. The majority of propagation (seed germination, rooted cuttings, and plug growing) is done with some form of root zone heat (Figs. 3–2 and 3–9) (55).

Solar Heating Conservation of energy in the greenhouse is important (83). In greenhouses, solar heating occurs naturally. The cost of fossil fuels has evoked considerable interest in methods of conserving daytime solar heat for night heating (50, 64). Conservation methods need to be developed and utilized; otherwise, high heating costs may eventually make winter use of greenhouses in colder regions economically unfeasible—relegating greenhouse operations to areas with relatively mild winters (89, 122).

Most heat loss in greenhouses occurs through the roof. One method of reducing heat loss in winter is to install sealed polyethylene sheeting outside over the glass or fiberglass covered structure, or to use two layers of polyethylene sheeting, as in a quonset house. This double-poly method of insulation is very effective. The two layers are kept separate by an air cushion from a low-pressure blower. Energy savings from the use of this system are substantial—more than 50 percent reduction in fuel compared to conventional glass greenhouses—but the greatly lowered light intensity with the double-layer plastic cover can lower yields of many greenhouse crops.

(a)

(b)

(c)

(d)

Figure 3–9
Hot water, root zone heating of propagation flats. (a) Biotherm tubing heating root zone of the plug tray. (b) Notice the probe (arrow) for regulating temperature. (c) The flexible hot water tubing is hooked into larger PVC pipes at set distances to assure more uniform heating. (d) Cuttings in propagation flats placed over white PVC hot water tubing; in milder climates, the ground hot water tubing may be all that is used to control root zone temperature and the air temperature of the propagation house.

Figure 3–10
(a) Prop house with thermal and shade curtains (arrow) to reduce winter heating costs and reduce light irradiance and greenhouse cooling expenses during summer months. (b) Thermal screen for energy conservation, made of woven aluminized polyester fabric, covering for propagation house with 46 percent light transmission; (c and d) the fabric is placed on top of polyethylene propagation house the covered house.

movable thermal curtains A device that reduces heat loss at night by creating a barrier between the crop and greenhouse roof and walls.

black clothing A curtain that is drawn over plants to exclude light for manipulating photoperiod.

Another device that reduces heat loss dramatically is a **movable thermal curtain** (Fig. 3–10), which, at night, is placed between the crop and the propagation house roof and walls (119). Winter heating bills are reduced as much as 30 percent, since the peak of the propagation house is not heated (67). During summer, automated curtains also reduce heat stress on propagules and workers, and less energy is needed to run fans for cooling. Modified curtains can be used for light reduction during the day and **"black clothing"** for light exclusion during photoperiod manipulation of plants. Curtains range from 20 percent shade reduction to complete blackout curtains—ULS Obscura A + B (67). Curtain fibers are available in white, black, with aluminum coated fibers, and/or with strips of aluminum sewn in. Black shade cloth reduces light to the plants, but absorbs heat and emits heat back into the propagation house. Aluminum-coated curtain fabrics are good reflectors of light, but poor absorbers of heat (Fig. 3–10). Some curtain materials come with a top side for reflecting heat and reducing condensation and a bottom side for heat retention. Insulating the north wall reduces heat loss without appreciably lowering the available light. Heat reduction also occurs with red and blue shade cloth used for control of plant growth (Fig. 3–11).

Greenhouses can be cooled mechanically in the summer by the use of large evaporative cooling units, as

(a)

(b)

(c)

Figure 3–11
(a and b) Propagation houses covered with red shade cloth for enhanced root initiation and development. The red netting increases the red, while reducing the blue and green spectra. (c) Shading seed propagation flats to reduce light irradiance and heat load.

pad and fan system
A system commonly used in greenhouse cooling to reduce the air temperature by raising the relative humidity and circulating air.

shown in Figure 3–12. The **"pad and fan" system,** in which a wet pad of material, such as special honeycombed cellulose, aluminum mesh, or plastic fiber, is installed at one side (or end) of a greenhouse with large exhaust fans at the other, has proved to be the best method of cooling greenhouses, especially in low-humidity climates (6). Fog can be used to cool greenhouses, but is more expensive than conventional pad and fan systems, and is inefficient in climates with high relative humidity (e.g., the Texas Gulf Coast).

(a)

(b)

Figure 3–12
Fully automated polycarbonate-covered greenhouse. (a) Air is pulled by exhaust fans (black arrows) to vent and cool. Components of both heating and cooling systems are electronically controlled via a weather monitoring station (white arrow) that feeds environmental inputs to computerized controls. (b) Cool cells (wettable pads) through which cooler, moist air is pulled across the propagation house by exhaust fans.

> **BOX 3.6 GETTING MORE IN DEPTH ON THE SUBJECT**
> **ENVIRONMENTAL CONTROL EQUIPMENT**
>
> Environmental control equipment is available from such companies as Priva (www.priva.nl), Wadsworth Control Systems, Inc., (www.wadsworthcontrols.com), and HortiMaX USA Inc. (www.qcom-controls.com).

Greenhouses are often sprayed on the outside at the onset of warm spring weather with a thin layer of *whitewash* or a white cold-water paint. This coating reflects much of the heat from the sun, thus preventing excessively high temperatures in the greenhouse during summer. The whitewash is removed in the fall. Too heavy a coating of whitewash, however, can reduce the light irradiance to undesirably low levels. Aluminized polyester fabric coverings are used for reducing heat load and can be placed on top of polyethylene-covered propagation houses (Fig. 3–10).

Environmental Controls

Controls are needed for greenhouse heating and evaporative cooling systems. Although varying with the plant species, a minimum night temperature of 13 to 15.5°C (55 to 60°F) is common. Thermostats for evaporative cooling are generally set to start the fans at about 24°C (75°F). In the early days of greenhouse operation, light, temperature, and humidity were about the only environmental controls attempted. Spraying the greenhouse with whitewash in summer and opening and closing side and ridge vents with a crank to control temperatures, along with turning on steam valves at night to prevent freezing, constituted environmental control. Humidity was increased by spraying the walks and benches by hand at least once a day. Later, it was found that thermostats, operating solenoid valves, could activate electric motors to raise and lower vents, and to open and close steam and water valves, thus giving some degree of automatic control. Most environmental controllers of greenhouse environments are now analog or computerized systems.

Analog Environmental Controls Analog controls (i.e., Wadsworth Step 500) have evolved for controlling the greenhouse environment. They use proportioning thermostats or electronic sensors to gather temperature information. This information drives amplifiers and electronic logic (i.e., decision making) circuitry (55). Essentially, they combine functions of several thermostats into one unit (10). Analog controls cost more than thermostats, but are more versatile and offer better performance.

Computerized Environmental Controls The advent of computer technology (i.e., Wadsworth EnviroSTEP) has replaced the amplifiers and logic circuits of an analog control with a microprocessor "computer on a chip" (Figs. 3–13 and 3–14). Computer controls are quicker and more precise in combining information from a variety of sensors (temperature, relative humidity, light intensity, wind direction) to make complex judgments about how to control the propagation environment. Computers can be utilized as zone controllers or in more expensive integrated computer systems (10, 55).

Although more costly than thermostats or analogs, computer controls offer significant energy and labor savings and improved production efficiency in propagation. Not only can temperature, ventilation, and humidity be controlled, but many other factors, such as propagating bed temperatures, application of liquid fertilizers through the irrigation system, daylength lighting, light-intensity regulation with mechanically operated shade cloth (and thermal sheets or curtains), operation of a mist or fog system, and CO_2 enrichment—all can be varied for different times of the day and night and for different banks of propagation units (7, 47, 56, 124). Computers can be programmed so that alarms are triggered or propagators paged by phone if deviations from preset levels occur—such as a heating failure on a cold winter night or a mist system failure on cuttings on a hot summer day. Some of these operations are shown in Figures 3–12, 3–13, 3–14, and 3–15. Most importantly, the computer can provide data on all factors being controlled for review to determine if changes are needed. This makes it easier for the propagator to make management decisions based on factual information (42).

Greenhouse Covering Materials

Common greenhouse covering materials include (54, 103):

- Glass
- Flexible covering materials
- Rigid covering materials

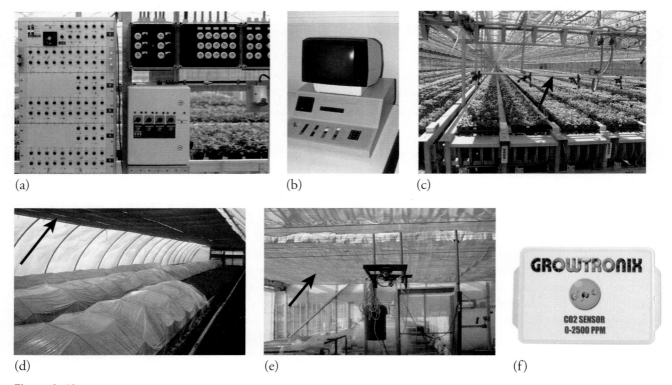

Figure 3–13
(a and b) Computer-controlled environmental manipulation of propagation facilities including (c) a mechanized traveling mist boom for irrigating flats on moveable benches. (d and e) Automated shade material programmed to close along the top of the propagation house when preset radiant energy levels are reached; this system works well with contact polyethylene propagation systems for rooting cuttings. (f) Automated metering system for monitoring CO_2 injection in propagation house.

Glass Glass-covered greenhouses are expensive, but for a permanent long-term installation under low-light winter conditions, glass may be more satisfactory than the popular, low-cost polyethylene (poly)-covered houses. Due to economics and the revolution in greenhouse covering materials from polyethylene to polycarbonates, glass greenhouses are no longer dominant. Glass is still used, due in part to its superior light transmitting properties and less excessive relative humidity problems. Glass "breathes" (the glass laps between panes allow air to enter), whereas polyethylene, acrylic, and polycarbonate-structured sheet houses are airtight, which can result in excessive humidity and undesirable water drip on the plants if not properly controlled. This problem can be overcome, however, by maintaining adequate ventilation and heating. Some of the newer greenhouse covering materials are designed to channel condensation to gutters, avoiding water dripping onto plant foliage. Control of high relative humidity is a key cultural technique to manage plant pathogens, since water can both disseminate pathogens and encourage plant infection. See the section on cultural controls in propagation under integrated pest management, later in the chapter.

Flexible Covering Materials are Categorized as Follows

Polyethylene (Polythene, Poly). Over half of the greenhouse area in the United States is covered with low-cost **polyethylene (poly)**, most with inflated double layers, giving good insulating properties. Poly is the most popular covering for propagation houses. Several types of plastic are available, but most propagators use either single- or double-layered polyethylene. Poly materials are lightweight and relatively inexpensive compared with glass. Their light weight also permits a less expensive supporting framework than is required for glass. Polyethylene has a relatively short life. It breaks down in sunlight and must be replaced after one or two years, generally in the fall in preparation for winter. The new polys, with ultraviolet (UV) inhibitors, can last three to four

| polyethylene (poly) A plastic covering used to cover propagation greenhouses. |

Figure 3–14
Manipulating the propagation environment. (a) Greenhouse sensors that are connected to an analog or computer-controlled environmental system. (b) Analog-type controller. (c) High vapor pressure sodium lighting for propagating plants during low-light conditions. (d and e) Lighting to extend photoperiod, which encourages (e) Japanese maple cuttings to avoid dormancy.

years, but in the southern United States where UV levels are higher, poly deteriorates more quickly and propagation houses need to be recovered more frequently.

A thickness of 4 to 6 mils (1 mil = 0.001 in) is recommended. For better insulation and lowered winter heating costs, a double layer of UV-inhibited copolymer material is used with a 2.5-cm (1-in) air gap between layers, kept separated by air pressure from a small blower.

Single-layer polyethylene-covered greenhouses lose more heat at night or in winter than a glass-covered house since polyethylene allows passage of heat energy from the soil and plants inside the greenhouse much more readily than glass. There are some newer infrared reflective polys, which save fuel but have lower light penetration than regular poly. Glass traps most infrared radiation, whereas polyethylene is transparent to it. However, double layer poly-covered greenhouses retain more heat than glass because the houses are more airtight and less infrared radiation escapes.

Only materials especially prepared for greenhouse covering should be used. Many installations, especially in windy areas, use a supporting material, usually welded wire mesh, for the polyethylene film. Occasionally, other supporting materials, such as Saran cloth, are used.

Polyethylene transmits about 85 percent of the sun's light, which is low compared with glass, but it passes all wavelengths of light required for plant growth. A tough, white, opaque film consisting of a mixture of polyethylene and vinyl plastic is available.

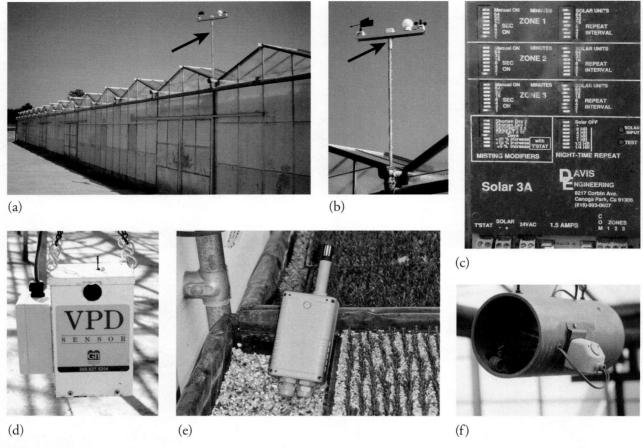

Figure 3–15
Environmental sensors for propagation. (a and b) A propagation house with a weather station for detecting light intensity, wind speed and direction, external temperature; this helps regulate temperature control and the fog propagation system. (c) Measurement of solar light allows for better mist control. (d, e, and f) Relative humidity sensors are needed to determine vapor pressure deficit (VPD) for critical fog propagation control.

This film stays more flexible under low winter temperatures than does clear polyethylene, but is more expensive. Because temperature fluctuates less under opaque film than under clear plastic, it is suitable for winter protection of field-bed or container-grown, liner plants (Fig. 3–16). Polyethylene permits the passage of oxygen and carbon dioxide, necessary for the growth processes of plants, while reducing the passage of water vapor.

For covering lath and shade structures, there are a number of satisfactory plastic materials prepared for the horticultural industry. Some commercially available materials include UV-treated cross-woven polyethylene and polypropylene fabric that resists ripping and tearing, and knitted high-density UV polyethylene shade cloth and Saran cloth that is strong and has greater longevity.

Rigid Covering (Structured Sheet) Materials Rigid Covering (Structured Sheet) Materials are Categorized as Follows

Acrylic (Plexiglass, Lucite, Exolite). Acrylic is highly weather resistant, does not yellow with age, has excellent light transmission properties, retains twice the heat of glass, and is very resistant to impact, but is brittle. It is somewhat more expensive and nearly as combustible as fiberglass. It is available in twin-wall construction which gives good insulation properties, and has a no-drip construction that channels condensation to run down to the gutters, rather than dripping on plants.

Polycarbonate (Polygal, Lexan, Cyroflex, Dynaglas). Polycarbonate is probably the most widely used structured sheet material today (55). Similar to acrylic in heat retention properties, it allows about 90 percent of

(a)

(b)

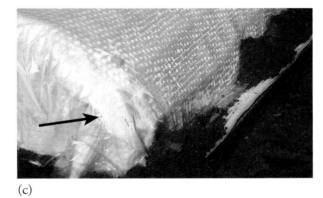
(c)

Figure 3–16
Low polyethylene tunnel or sun tunnel that is covered with polyethylene. (a) Sometimes a white poly material is used to avoid the higher temperature buildup and temperature fluctuation of clear poly. Propagation flats are placed on top of hot-water tubing or electric heating cables (b) Saran shade cloth can be used to cover the poly to reduce the heat load. (c) Winterization of sun tunnels can be done with white microfoam insulation covered with a clear poly or opaque poly (see arrow).

the light transmission of glass. Polycarbonate has high impact strength—about 200 times that of glass. It is lightweight, about one-sixth that of glass, making it easy to install. Polycarbonate's textured surface diffuses light and reduces condensation drip. It is available in twin-wall construction, which gives good insulation properties. Polycarbonate can be cut, sawn, drilled, or nailed, and is much more user-friendly than acrylic, which can shatter if nails or screws are driven into it. It is UV stabilized and will resist long outdoor exposure (some polycarbonates are guaranteed for ten years), but will eventually yellow with age (11, 90).

Fiberglass. Rigid panels, corrugated or flat, of polyester resin reinforced with fiberglass have been widely used for greenhouse construction. This material is strong, long-lasting, lightweight, and easily applied, and comes in a variety of dimensions (width, length, and thickness), but is not as permanent as glass. Only the clear material—especially made for greenhouses and in a thickness of 0.096 cm (0.038 in) or more and weighing 4 to 5 oz per square foot—should be used. New material transmits about 80 to 90 percent of the available light, but light transmission decreases over the years due to yellowing, which is a serious problem. Since fiberglass burns rapidly, an entire greenhouse may quickly be consumed by fire, so insurance costs can be higher. Fiberglass is more expensive than polyethylene, and is not as widely used as it once was.

The economics of using these greenhouse covering materials must be considered carefully before a decision is made. New materials are continually coming onto the market.

Closed-Case Propagation Systems

Hot Frames (Hotbeds) and Heated Sun Tunnels The **hot frame (hotbed)** is a small, low structure used for many of the same purposes as a propagation house. Traditionally, the hotbed is a large wooden box or frame with a sloping, tight-fitting lid made of window sash. Hotbeds can be used throughout the year, except in areas with severe winters where their use may be restricted to spring, summer, and fall. Another form of a hotbed is a **heated, low polyethylene tunnel or sun tunnel** that is made from hooped metal tubing or bent PVC pipe, which is covered with polyethylene (sometimes a white poly material is used to avoid the higher temperature buildup and temperature fluctuations of clear poly) (Fig. 3–16).

> **hot frames (hotbeds)** Propagation structures that are covered with poly and heated in the winter.

Traditionally, the size of the frame conforms to the size of the glass sash available—a standard size is 0.9 by 1.8 m (3 by 6 ft) (Fig. 3–17). If polyethylene is used as the covering, any convenient dimensions can be

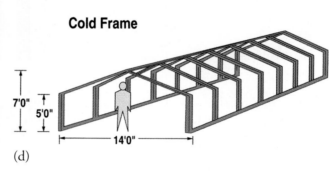

Figure 3–17
Traditional cold frames were used for propagating tender plants. Frames are opened after protection is no longer required. (a) Older commercial use of glass-covered cold frames in propagating ground cover plants by cuttings. (b) Wood sash used for liner production in a cold frame. Glass and lath coverings are rarely used due to the high labor costs in moving the heavy sash. Plastic coverings are more suitable. (c and d) Today a cold frame is most commonly a very low cost, budget, unheated poly-covered hoop or galvanized steel bow house.

used. The frame can be easily built with 3-cm (1-in) or 6-cm (2-in) lumber nailed to 4-by-4 corner posts set in the ground. Decay-resistant wood such as redwood, cypress, or cedar should be used, and preferably pressure-treated with wood preservatives, such as chromated copper arsenate (CCA). This compound retards decay for many years and does not give off fumes toxic to plants. Creosote must not be used on wood structures in which plants will be grown, since the fumes released, particularly on hot days, are toxic to plants.

Plastic or PVC tubing with recirculating hot water is quite satisfactory for providing bottom heat in hotbeds. The hotbed is filled with 10 to 15 cm (4 to 6 in) of a rooting or seed-germinating medium over the hot-water tubing. Alternatively, community propagation flats or flats with liner pots containing the medium can be used. These are placed directly on a thin layer of sand covering the hot-water tubing.

Seedlings can be started and leafy cuttings rooted in hotbeds early in the season. As in the greenhouse, close attention must be paid to shading and ventilation, as well as to temperature and humidity control. For small propagation operations, hotbed structures are suitable for producing many thousands of nursery plants without the higher construction expenditure for larger, walk-in propagation houses (60).

Cold Frames and Unheated Sun Tunnels
A primary use of **cold frames** is conditioning or hardening

| cold frames Propagation structures covered with poly, lath, or other covering material and which are not heated in the winter.

rooted cuttings or young seedlings (liners) preceding field, nursery-row, or container planting. Cold frames and unheated sun tunnels can be used for starting new plants in late spring, summer, or fall when no external supply of heat is necessary (129). Today, cold frames include not only low polyethylene-covered wood frames or unheated sun tunnels that people cannot walk within (Fig. 3–17), but also low-cost, poly-covered hoop houses (Fig. 3–17). The covered frames should fit tightly in order to retain heat and obtain high humidity. Cold frames should be placed in locations protected from winds, with the sash cover sloping down from north to south (south to north in the Southern Hemisphere).

Low-cost cold frame construction (Fig. 3–17) is the same as for hotbeds, except that no provision is made for supplying bottom heat. With older-style cold frames, sometimes a lath covering with open spaces between the lath boards is used to cover the cold frame. This does not prevent freezing temperatures from occurring, but does reduce high and low temperature fluctuations.

In these structures, only the heat of the sun, retained by the transparent or opaque white polyethylene coverings, is utilized. Close attention to ventilation, shading, watering, and winter protection is necessary for success with cold frames. When young, tender plants are first placed in a cold frame, the covers are generally kept tightly closed to maintain a high humidity, but as the plants become acclimated, the sash frames are gradually raised or the ends of the hoop house or sun tunnels opened to permit more ventilation and drier conditions.

The installation of a mist line or frequent irrigation of plants in a cold frame is essential to maintain humid conditions. During sunny days temperatures can build up to excessively high levels in closed frames unless ventilation and shading are provided. Spaced lath, Saran or poly shade cloth-covered frames, or reed mats are useful to lay over the sash to provide protection from the sun. In areas where extremely low temperatures occur, plants being overwintered in cold frames may require additional protective coverings.

Lathhouses Lathhouses or shade houses (Figs. 3–6 and 3–11) provide outdoor shade and protect container-grown plants from high summer temperatures and high light irradiance (50). They reduce moisture stress and decrease the water requirements of plants. Lathhouses have many uses in propagation, particularly in conjunction with the hardening-off and acclimation of liner plants prior to transplanting, and with maintenance of shade-requiring or tender plants. At times a lathhouse is used by nurseries simply to hold plants for sale. In mild climates, they are used for propagation, along with a mist facility, and can also be used as an overwintering structure for liner plants. Snow load can cause problems in higher latitude regions.

Lathhouse construction varies widely. Aluminum prefabricated lathhouses are available but may be more costly than wood structures. More commonly, pipe or wood supports are used, set in concrete with the necessary supporting cross-members. Today, most lathhouses are covered with high-density, woven, plastic materials, such as Saran, polypropylene fabric, and UV-treated polyethylene shade cloth, which come in varying shade percentages and colors. These materials are available in different densities, thus allowing lower irradiance of light, such as 50 percent sunlight, to the plants. They are lightweight and can be attached to heavy wire fastened to supporting posts. The shade cloth is resistant to ripping, and has an optimum life of 10 to 15 years, depending on climate and quality of material. For winterization in less temperate areas, producers will cover the shade cloth with polyethylene. Sometimes shade is provided by thin wood strips about 5 cm (2 in) wide, placed to give one-third to two-thirds cover, depending on the need. Both sides and the top are usually covered. Rolls of snow fencing attached to a supporting framework can be utilized for inexpensive construction.

Miscellaneous Closed-Case Systems There are a number of closed-case propagation systems that are used in the rooting of cuttings, acclimatization and rooting of tissue culture microcuttings, and propagation of seedlings. Besides the sun tunnels or cold frames previously described, closed-case propagation systems include nonmisted enclosures in glasshouses or polyhouses (shading, tent and contact polyethylene systems, wet tents, inverted glass jars).

Propagating Frames. Even in a greenhouse, humidity is not always high enough to permit satisfactory rooting of certain kinds of leafy cuttings. Enclosed frames covered with poly or glass may be necessary for successful rooting (see Figs. 3–18 and 10–36). There are many variations of such devices. Small ones were called Wardian cases in earlier days (see Fig. 1–3). Such enclosed frames are also useful for graft union formation of small potted nursery stock, since they retain high humidity.

Sometimes in cool summer climates (as far south as Virginia in the United States), when fall semi-hardwood cuttings are taken, a layer of very thin (1 or 2 mils) polyethylene laid directly on top of a bed of newly prepared leafy cuttings in a greenhouse or lathhouse will provide a sufficient increase in relative humidity to give good rooting. This is sometimes referred to as a **contact polyethylene system** (see Fig. 10–36). Good shade control to reduce light irradiance is essential for this system.

Figure 3–18
(a and b) Polyethylene-covered beds used in a greenhouse to maintain high humidity surrounding the cuttings during rooting. Propagation flats can be placed on beds or cuttings stuck directly into the mist beds and covered with poly. (c) Using shade (arrow) for light/temperature control. (d) Partially vented polycovered mist-bed under a quonset house for shade.

On a more limited scale, bell jars (large inverted glass jars) can be set over a container of unrooted cuttings or freshly grafted containerized plants to speed up graft union formation (see Fig. 12–49). Humidity is kept high in such devices, but some shading is necessary to control temperature.

In using all such structures, care is necessary to avoid the buildup of pathogenic organisms. The warm, humid conditions, combined with lack of air movement and relatively low light intensity, provide excellent conditions for the growth of various pathogenic fungi and bacteria. Cleanliness of all materials placed in such units is important; however, use of fungicides is sometimes necessary (see the section on **integrated pest management** later in the chapter).

Enclosed Poly Sweat Tent—Hydroponic System. An Australian producer of chrysanthemums uses a modified nutrient film technique (NFT) for growing greenhouse stock plants and propagating cuttings (58). Unrooted cuttings are stuck in Oasis root cubes and placed in mist propagation benches containing a reservoir of water, maintained with a float valve. The system is initially enclosed in a clear poly sweat tent. Once root initiation takes place, the mist is turned off and the poly tent lifted. Cuttings are then supplied with nutrient solution in the NFT system on the propagation bench and later transplanted with the roots intact and undisturbed in the root cube. Stock plants are also maintained in the NFT system and supported in root cubes, thus allowing more precise nutritional control and reduction in environmental stress to the stock plant.

CONTAINERS FOR PROPAGATING AND GROWING YOUNG LINER PLANTS

New types of containers for propagating and growing young liner plants are continually being developed, usually with a goal of reducing handling costs. Direct sticking of unrooted cuttings into small liner containers, as opposed to sticking into conventional propagation trays, saves a production step and later avoids root disturbance of cuttings, which can lead to transplant shock (Figs. 3–19, 3–20, and 3–21) (31).

Flats

Flats are shallow plastic, Styrofoam, wooden, or metal trays, with drainage holes in the bottom. They are useful for germinating seeds or rooting cuttings, since they permit young plants to be moved easily. In the past, durable kinds of wood, such as cypress, cedar, or redwood, were preferred for flats. The most popular flats are made of rigid plastic (polyethylene, polystyrene) and come in all shapes and sizes. The 28 × 53 cm (11 × 21 in) *1020* plastic flats are the industry standard. The number of cells or compartments per tray may range from 1 cell for a community rooting flat or seed germination tray, to 18 or more cells for a rooted liner tray, to 100 to 400 cells for a seedling plug tray. Trays also can be fitted with removable sheet inserts containing the cells. Plastic flats will nest, and thus require relatively little storage space. The costs of producing plastic for flats and containers and for disposing of used plastic have led to increased plastic recycling programs in horticulture and biodegradable paper tube liner pots (Fig. 3–19).

Plastic Pots

Plastic containers, round and square, have numerous advantages: they are nonporous, reusable, lightweight, and use little storage space because they will nest. Some types are fragile, however, and require careful handling,

(a)

(b)

(c)

(d)

Figure 3–19
(a and b) A paper pot system direct sticking (direct rooting) liner plants in paper tubes filled with peat-lite media. (b) Paper pot sleeve liner (arrow) inserted in plastic tray. (c) Rooted poinsettia in paper sleeve tube. (d) Plastic rooting tray with ribs (arrow) to reduce root circling of poinsettias during propagation and rooted liner development.

Figure 3–20
(a) Air-root pruning system for direct sticking (direct rooting) tree liners to minimize root circling, encourage more fibrous root development, and increase root surface area. (b) Direct rooting poinsettia cuttings in paper sleeves inserted in ribbed plastic liner pots.

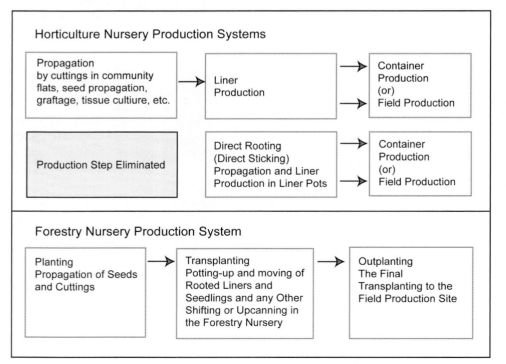

Figure 3–21
Flow diagram of a Horticulture Nursery Production System starting with propagation by rooted cuttings, seedlings, graftage, or tissue culture-produced plantlets—followed by transplanting into liner pots and final transplanting into larger containers or into nursery field production. Direct rooting (direct sticking) eliminates a production step, since both propagation and liner production occur in the same liner pot. A Forestry Nursery Production System of planting, transplanting, and outplanting is also described.

Figure 3–22
(a) Plastic (Roottainer) container made of preformed, hinged sheets for propagating seedling liners. (b) synthetic fiber media (Rockwool) blocks for inserting seedling plugs and growing in greenhouse. (c and d) ridged containers for minimizing root circling.

although other types, made from polyethylene, are flexible and quite sturdy. Small liner pots for direct rooting of cuttings, seedling propagation, and tissue culture plantlet acclimatization and production have gained considerable popularity.

Many of these small containers have rib-like structures to redirect root growth and prevent girdling (Figs. 3–19, 3–20, and 3–22). In forestry seedling production, ribbed book or sleeve containers are used, which consist of two matched sections of molded plastic that fit together to form a row of rectangular cells (Fig. 3–22). The inner walls of small propagation containers and liner pots can also be treated with chemical root pruning agents, such as copper hydroxide ($CuOH_2$), which chemically prune liner roots at the root-wall interface (71). The chemically pruned lateral roots become suberized but will begin to grow again after transplanting, which results in a well-distributed root system that helps minimize transplant shock (Fig. 3–23) (71).

Plastic pots (and flats) cannot be steam sterilized, but some of the more common plant pathogens can be controlled by a hot water dip, 70°C (158°F), for 3 minutes followed by a rinse in a dilute bleach solution (i.e., Clorox, Purex, etc.). Ultraviolet light inhibitors are sometimes incorporated in the plastic resin to prevent UV degradation of plastic pots under full sun conditions (Fig. 3–24, page 75).

Fiber Pots

Containers of various sizes, round or square, are pressed into shape from peat plus wood fiber, with fertilizer added. Dry, they will keep indefinitely. Since these pots are biodegradable, they are set in the soil along with the plants. Peat pots find their best use where plants are to be held for a relatively short time and then put in a larger container or in the field. During outplanting in the field, any portion of the fiber pot transplanted above the surface of the soil will act like a wick and quickly dry out the transplant.

During production, small peat pots with plants growing in them eventually deteriorate because of constant moisture, and may fall apart when moved. On the other hand, unless the pots are kept moist, roots will fail to penetrate the walls of the pot and will grow into

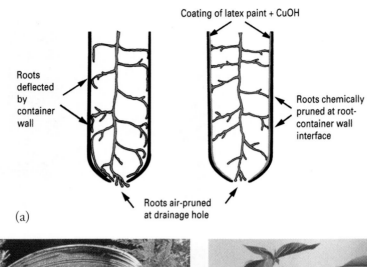

Figure 3–23
Chemical root pruning involves treating the interior container wall with a growth-inhibiting chemical such as copper hydroxide. This causes the lateral roots to be chemically pruned at the container wall. A well-branched root system occurs, which enhances transplant establishment. (a) Schematic of nonpruned versus chemically pruned seedling container roots. (b) Copper hydroxide-treated container. (c) Copper hydroxide-treated *Acalpha hispida* (see arrow) without visible surface roots. Photo courtesy of M. Arnold.

an undesirable spiral pattern. Units of 6 or 12 square peat pots fastened together are available. When large numbers of plants are involved, using peat pots results in time and labor savings.

Paper Pots

Paper pots or paper tube pots are more popular with seed plug and cutting propagation of ornamentals, vegetable and forestry species. They allow for greater mechanization with pot-filling machines, automatic seeders, and wire benches that allow air pruning of the root system. Typically, paper pots consist of a series of interconnected paper cells arranged in a honeycomb pattern that can be separated before outplanting (71). An advantage of the paper pot system is that pots are biodegradable, and the seedling plug can be planted intact into a larger container or into the ground without disturbing the root system. Some papier-mâché

(a) (b)

Figure 3–24
(a) Colorful, labeled, rigid-plastic containers are used for growing and merchandising landscape and garden plants. Frequently, inhibitors are incorporated with the plastic resin to prevent ultraviolet breakdown of the containers under full sun conditions. (b) Flexible poly container bags are used for nursery production in Europe, England, and Australia, where petroleum-based products are more costly than in the United States.

pots (paper, wax, asphalt) come treated with copper hydroxide, which enhances root development and retards deterioration of the pot.

In Europe and the United States, paper tube pots with predictable degradation rates are produced by machine (39). The propagation medium is formed into a continuous cylinder and wrapped with a length of paper or cellulose skin that is glued and heat sealed (Fig. 3–19).

Peat, Fiber, Expanded Foam, and Rockwool Blocks

Blocks of solid material, sometimes with a prepunched hole (Fig. 3–22), have become popular as a germinating medium for seeds and as a rooting medium for cuttings, especially for such plants as chrysanthemums and poinsettias. Sometimes fertilizers are incorporated into the material. One type is made of highly compressed peat which, when water is added, swells to its usable size and is soft enough for the cutting or seed to be inserted. Such blocks become a part of the plant unit and are set in the soil along with the plant. These blocks replace not only the pot but also the propagating mix.

Synthetic rooting blocks (oasis, rockwool) are becoming more widely used in the nursery industry (and forestry industry for seed propagation), and are well adapted to automation (Fig. 3–22). Other advantages are their light weight, consistent quality, reproducibility, and clean condition. Watering must be carefully controlled to provide constant moisture, while maintaining adequate aeration.

Plastic Growing Containers for Post-Liner Production

Many millions of nursery plants are grown and marketed each year in 3.8-liter (1-gal) and—to a lesser extent—11-liter (3-gal), 19-liter (5-gal), and larger containers. They are tapered for nesting and have drainage holes. Heavy-wall, injection-molded plastic containers are used extensively in the United States. Machine planters have been developed utilizing containers in which rooted cuttings or seedlings can be transplanted as rapidly as 10,000 or more a day. See the horticulture and forestry nursery production flow diagrams (Fig. 3–21). Plants are easily removed from tapered containers by inverting and tapping. Some plastic containers are made of preformed, hinged plastic sheets that can be separated for easy removal of the liner (Fig. 3–22).

In areas with high summer temperatures, use of light-colored (white or silver) containers may improve root growth by reducing heat damage to the roots, which is often encountered in dark-colored containers that absorb considerable heat when exposed to the sun. However, light-colored containers show dirt marks (as opposed to black or dark green containers) and must be cleaned prior to shipping. More and more colorful, labeled containers are being used for growing and merchandising landscape and garden plants (Fig. 3–24). A pot-in-pot system, in which a containerized plant is inserted into a hole in the ground lined with a plastic sleeve pot, helps moderate both high and low rootball temperatures (Fig. 3–25).

Polyethylene Bags and Plant Rolls

Polyethylene bags are widely used in Europe, Australia, New Zealand, and in less developed countries in the tropics—but rarely in North America—for growing rooted cuttings or seedling liners to a salable size. They are considerably less expensive than rigid plastic containers and seem to be satisfactory (Fig. 3–24), but some types deteriorate rapidly. They are usually black, but some are black on the inside and light-colored on the outside. The lighter color reflects heat and lowers the root temperature. Polybags do not prohibit root spiraling or allow air pruning, which is a drawback to their use in propagation and liner production; however,

Figure 3–25

Alternatives to traditional field production. (a) In-ground fabric containers or grow bags. (b) The pot-in-pot (P&P) system with individual pot, drip irrigation. (c) Copper-treated wall of outside sleeve containers (arrow) to prevent root penetration from the inner pots. (d and e) P&P containers. (f) The roots of the inside containers are very susceptible to heat stress when they are removed from the field. Here they are wrapped with an insulating packing fabric for shipping.

Figure 3–26
(a) Redwood containers used for large nursery specimen tree production. (b) Wood containerized tree and heavy equipment required to lift it. (c) A large, 8- to 9-year-old specimen tree produced in a 183 cm (72-in) box, weighing in excess of 3700 kg (8100 lbs). The enormous weight of the rootball will require a crane for lifting at the landscape site. The box is easier for landscapers to handle than heavy-duty plastic container that would need to be cut up.

poly tubes are open-ended, which reduces girdling problems. After planting, they cannot be stacked as easily as the rigid containers for truck transportation—the polybags often break, and the root system of the plant is more easily damaged.

A low-cost method of propagating some easy-to-root species is with a polyethylene plant roll (see Fig. 10–58). The basal ends of the cuttings are inserted in damp peat moss or sphagnum and rolled into the doubled-over plastic sheeting. The roll of cuttings is then set upright in a humid location for rooting. Polyethylene starter pouches with an absorbent paper inserted in the pouch are used for germinating selected seed lots.

Wood Containers

Large cedar-wood containers or boxes are used for growing large specimen trees and shrubs to provide "instant" landscaping for the customer. Some of the specimen trees are 8 to 9 years old and weigh up to 3700 kg (8100 lbs). Heavy moving equipment is required for handling such large nursery stock (Fig. 3–26).

MANAGEMENT OF MEDIA AND NUTRITION IN PROPAGATION AND LINER PRODUCTION

Media and Mixes for Propagating and Growing Young Liner Plants

Various substrates and mixtures of materials are used for germinating seeds and rooting cuttings. For good results, the following characteristics of the medium are required (51):

- The **medium must be sufficiently firm and dense** to hold the cuttings or seeds in place during rooting or germination. Its volume must be fairly constant when either wet or dry; excessive shrinkage after drying is undesirable.
- It should be **highly decomposed and stable** (preferably with a 20C:1N ratio) to prevent N immobilization and excessive shrinkage during production.
- It must be **easy to wet** (not too hydrophobic) and retain enough moisture to reduce frequent watering.

- It must be **sufficiently porous** so that excess water drains away, permitting adequate penetration of oxygen to the roots—all containers produce a perched water table that creates a zone of saturated growing medium at the bottom of the container.
- It must be **free from pests:** weed seeds, nematodes, and various pathogens.
- It must have a **low salinity** level.
- It should be **capable of being steam-pasteurized or chemically treated** without harmful effects.
- It should have a **high cation exchange capacity (CEC)** for retention of nutrients that may be applied preincorporated and/or in a supplementary soluble and/or controlled-release fertilizer program.
- It should be of **consistent quality** from batch to batch, and reproducible.
- It should be readily **available,** and **economical.**

Propagation media used in horticulture and forestry consist of a mixture of organic and inorganic components that have different but complementary properties. The **organic component** generally includes peat, softwood and hardwood barks, or sphagnum moss. Sawdust and rice hulls should be avoided since they oxidize readily and compact easily, which decreases pore space and aeration, and they have a high C:N ratio, which can result in nutritional problems for the propagule. A **coarse mineral component** is used to improve drainage and aeration by increasing the proportion of large, air-filled pores. A variety of mineral components include sand (avoid fine particle sands), grit, pumice, scoria, expanded shale, perlite, vermiculite, polystyrene, clay granules, and rockwool.

There is no single, ideal mix. An appropriate propagation medium depends on the species, propagule type, season, and propagation system (i.e., with fog, a waterlogged medium is less of a problem than with mist); cost and availability of the medium components are other considerations. The following media components can be used in propagation systems.

Soil A mineral soil is composed of materials in the solid, liquid, and gaseous states. For satisfactory plant growth, these materials must exist in the proper proportions. The solid portion of a soil is comprised of both inorganic and organic components. The inorganic part consists of the residue from parent rock after decomposition, resulting from the chemical and physical process of weathering. Such inorganic components vary in size from gravel down to extremely minute colloidal particles of clay, the texture of the soil being determined by the relative proportions of these particle sizes. The coarser particles serve mainly as a supporting framework for the remainder of the soil, whereas the colloidal clay fractions of the soil serve as storehouses for nutrients that are released and absorbed by plants. The organic portion of the soil consists of both living and dead organisms. Insects, worms, fungi, bacteria, and plant roots generally constitute the living organic matter, whereas the remains of such animal and plant life in various stages of decay make up the dead organic material. The residue from such decay (termed **humus**) is largely colloidal and assists in holding water and plant nutrients.

The liquid part of the soil, the soil solution, is made up of water that contains dissolved salts in various quantities, along with dissolved oxygen and carbon dioxide. Mineral elements, water, and some carbon dioxide enter the plant from the soil solution.

The gaseous portion of the soil is important to good plant growth. In poorly drained, waterlogged soils, water replaces the air, thus depriving plant roots as well as certain desirable aerobic microorganisms of the oxygen necessary for their existence.

The **texture** of a mineral soil depends upon the relative proportions of sand (0.05 to 2 mm particle diameter), silt (0.05 to 0.002 mm particle diameter), and clay (less than 0.002 mm particle diameter). In contrast to soil texture, which refers to the proportions of individual soil particles, **soil structure** refers to the arrangement of those particles in the entire soil mass. These individual soil grains are held together in aggregates of various sizes and shapes.

Propagation in commercial horticulture is generally done with flats, containers, and/or pot systems using *"soilless" media.* Some exceptions to this are field budding and grafting systems, stooling and layering systems, field propagation of hardwood cuttings without intermittent mist (see Fig. 10–2), direct seeding of crops, and utilizing outdoor seedbeds. With the greater reliance on containerized systems for propagation, mineral soils are either unsuitable or must be amended with other components to improve aeration and prevent the compaction that occurs with the structural changes of mineral soils in a container.

Sand Sand consists of small rock particles, 0.05 to 2.0 mm in diameter, formed as the result of the weathering of various rocks. The mineral composition of sand depends upon the type of rock. Quartz sand, consisting chiefly of a silica complex, is generally used for propagation purposes. Sand is the heaviest of all rooting media used, with a cubic foot of dry sand weighing about

Figure 3–27
Propagation medium. (a) Various types of propagation media components and mixes. (b) Sphagnum peat moss—excellent quality, but expensive. (c) A specialized azalea propagation mix composed of peat, bark, and perlite. (d) Media in bins used to fill propagation and liner flats inside the propagation house.

45 kg (100 lb). Preferably, it should be fumigated or steam-pasteurized before use, as it may contain weed seeds and various harmful pathogens. Sand contains virtually no mineral nutrients and has no buffering capacity or cation exchange capacity (CEC). It is used mostly in combination with organic materials. Sand collected near the ocean (beach sand) may be too high in salts. Calcareous sand will raise media pH and should be tested prior to mixing with vinegar or a dilute acid.

Peat Peat consists of the remains of aquatic, marsh, bog, or swamp vegetation that has been preserved under water in a partially decomposed state. The lack of oxygen in the bog slows bacterial and chemical decomposition of the plant material. Composition of different peat deposits varies widely, depending upon the vegetation from which it originated, state of decomposition, mineral content, and degree of acidity (82).

There are three types of peat as classified by the United States Bureau of Mines: moss peat, reed sedge, and peat humus. *Moss peat* (usually referred to in the market as **peat** or **peat moss**) is the least decomposed of the three types and is derived from sphagnum or other mosses. It varies in color from light tan to dark brown. It has a high moisture-holding capacity (15 times its dry weight), has a high acidity (pH of 3.2 to 4.5), and contains a small amount of nitrogen (about 1 percent) but little or no phosphorus or potassium. This type of peat generally comes from Canada, Ireland, or Germany, although some is produced in the northern United States. *Peat moss is the most commonly used peat in horticulture, the coarse grade being the best* (Fig. 3–27).

When peat moss is to be used in mixes, it should be broken apart and moistened before being added to the mix. Continued addition of coarse organic materials such as peat moss or sphagnum moss to greenhouse media can initially cause a decrease in wettability. Water will not penetrate easily, and many of the peat particles will remain dry even after watering. There is no good method for preventing this nonwettability, although the repeated use of commercial wetting agents, such as Aqua-Gro, can improve water penetration (12). Peat is not a uniform product and can be a source of weed seed, insects, and disease inoculum. Peat moss is relatively expensive so it is used less in nursery propagation and production mixes. It is gradually being replaced by other components, such as pulverized or shredded bark. However, peat is still the main organic ingredient in propagation and greenhouse mixes.

Sphagnum Moss Peat Commercial sphagnum moss peat or sphagnum peat is the dehydrated young residue or living portions of acid-bog plants in the genus *Sphagnum*, such as *S. papillosum*, *S. capillaceum*, and *S. palustre*. It is the most desirable peat for horticultural purposes, but its high cost limits its commercial use. It is relatively pathogen-free, light in weight, and has a very high water-holding capacity, able to absorb 10 to

20 times its weight in water. This material is generally shredded, either mechanically or by hand, before it is used in a propagating or growing media. It contains small amounts of minerals, but plants grown in it for any length of time require added nutrients. Sphagnum moss has a pH of about 3.5 to 4.0. It may contain specific fungistatic substances, including a strain of *Streptomyces* bacteria, which can inhibit damping-off of seedlings (2, 63).

Vermiculite Vermiculite is a micaceous mineral that expands markedly when heated. Extensive deposits are found in Montana, North Carolina, and South Africa. Chemically, it is a hydrated magnesium-aluminum-iron silicate. When expanded, vermiculite is very light in weight [90 to 150 kg per cubic meter (6 to 10 lbs per cubic foot)], neutral in reaction with good buffering properties, and insoluble in water. It is able to absorb large quantities of water—40 to 54 liters per cubic meter (3 to 4 gal per cubic foot). Vermiculite has a relatively high cation-exchange capacity and, thus, can hold nutrients in reserve for later release. It contains magnesium and potassium, but supplementary amounts are needed from other fertilizer sources.

In crude vermiculite ore, the particles consist of many thin, separate layers with microscopic quantities of water trapped between them. When run through furnaces at temperatures near 1090°C (1994°F), the water turns to steam, popping the layers apart and forming small, porous, spongelike kernels. Heating to this temperature provides complete sterilization. Horticultural vermiculite is graded to four sizes: No. 1 has particles from 5 to 8 mm in diameter; No. 2, the regular horticultural grade, from 2 to 3 mm; No. 3, from 1 to 2 mm; No. 4, which is most useful as a seed-germinating medium, from 0.75 to 1 mm. Expanded vermiculite should not be compacted when wet, as pressing destroys its desirable porous structure. Do not use nonhorticultural (construction grade) vermiculite, as it is treated with chemicals toxic to plant tissues.

Perlite Perlite, a gray-white silicaceous material, is of volcanic origin, mined from lava flows. The crude ore is crushed and screened, then heated in furnaces to about 760°C (1400°F), at which temperature the small amount of moisture in the particles changes to steam, expanding the particles to small, spongelike kernels that are very light, weighing only 80 to 100 kg per cubic meter (5 to 6.5 lbs per cubic foot). The high processing temperature provides a sterile product. Usually, a particle size of 1.6 to 3.0 mm (1/16 to 1/8 in) in diameter is used in horticultural applications (Fig. 3–27). Perlite holds three to four times its weight of water. It is essentially neutral with a pH of 6.0 to 8.0 but with no buffering capacity. Unlike vermiculite, it has no cation exchange capacity and contains no mineral nutrients. Perlite presents some problems with fluoride-sensitive plants, but fluoride can be leached out by watering heavily. It is most useful in increasing aeration in a mix. Perlite, in combination with peat moss, is a very popular rooting medium for cuttings (85). Perlite dust is a respiratory irritant. Perlite should be moistened to minimize dust, and workers should use respirators.

Calcined Clay and Other Aggregates Stable aggregates can be produced when minerals such as clay, shales, and pulverized fuel ash are heated (calcined) at high temperatures. They have no fertilizer value, are porous, are resistant to breakdown, and absorb water. The main purpose of these materials is to change the physical characteristics of a propagation or liner potting mix. Examples of commercial materials made from clay include Leca, Terragreen, and Turface. Haydite is a combination of clay and shale, while Hortag (used in the UK) is made from pulverized fuel ash (16). Clay-type kitty litter is also a calcined clay, but contains perfumes that are not desirable for propagation.

Pumice Chemically, pumice is mostly silicon dioxide and aluminum oxide, with small amounts of iron, calcium, magnesium, and sodium in the oxide form. It is of volcanic origin and is mined in several regions in the western United States. Pumice is screened to different-size grades, but is not heat-treated. It increases aeration and drainage in a propagation mix and can be used alone or mixed with peat moss.

Rockwool (Mineral Wool) This material is used as a rooting and growing medium in Europe, Australia, and the United States (Figs. 3–22 and 3–27). It is prepared from various rock sources, such as basalt rock, melted at a temperature of about 1600°C. As it cools, a binder is added, and it is spun into fibers and pressed into blocks. Horticultural rockwool is available in several forms—shredded, prills (pellets), slabs, blocks, cubes, or combined with peat moss as a mixture. Rockwool will hold a considerable amount of water, yet retains good oxygen levels. With the addition of fertilizers it can be used in place of the Peat-Lite mixes. Before switching from more traditional media mixes, it is best to initially conduct small-scale propagation trials with rockwool and other new media components as they become commercially available (51).

Shredded Bark Shredded or pulverized softwood bark from redwood, cedar, fir, pine, hemlock, or various hardwood bark species, such as oaks and maples, can be used as an organic component in propagation and growing mixes and are frequently substituted for peat moss at a lower cost (89, 91, 102, 112, 128). Before it is used as a growing medium, pine bark is hammer-milled into smaller component pieces, stockpiled in the open, and often composted by turning the piles and watering as needed. Fresh barks may contain materials toxic to plants, such as phenols, resins, terpenes, and tannins. Composting for 10 to 14 weeks before using reduces phenolic levels in bark and improves its wettability as media, and the higher bark pile temperatures help reduce insect and pathogen levels (16). Because of their moderate cost, light weight, and availability, barks are very popular and widely used in mixes for propagation and container-grown plants (Fig. 3–27). Wetting agents and gels increase available water content in pine bark and may play a greater role in helping propagators reduce irrigation frequency or the volume of water required during each irrigation (12).

Coconut Fiber/Coir   Coconut fiber (coir) is an economical peat substitute that can be mixed with a mineral component as propagation media. It is derived from coconut husks.

Compost   In some countries, compost is synonymous with container media for propagation and plant growth; however, we define *compost* (composting) as the product of biological decomposition of bulk organic wastes under controlled conditions, which takes place in piles or bins. The process occurs in three steps:

a. an initial stage lasting a few days in which decomposition of easily degradable soluble materials occurs;
b. a second stage lasting several months, during which high temperatures occur and cellulose compounds are broken down; and
c. a final stabilization stage when decomposition decreases, temperatures lower, and microorganisms recolonize the material.

Microorganisms include bacteria, fungi, and nematodes; larger organisms, such as millipedes, soil mites, beetles, springtails, earthworms, earwigs, slugs, and sowbugs, can often be found in compost piles in great numbers. Compost prepared largely from leaves may have a high soluble salt content, which will inhibit plant growth, but salinity can be lowered by leaching with water before use.

In the future, with dwindling landfill sites and environmental pressures to recycle organic scrapage materials, the use of composted yard wastes, chicken and cow manure, organic sludge from municipal sewage treatment plants, and so on will play a greater role as media components in the propagation and production of small liner plants. Many nurseries recycle culled, containerized plants and shred the plant and soil as compost or as a medium component to be mixed with fresh container medium. Composted sewage sludge not only provides organic matter, but nearly all the essential trace elements, and a large percentage of major elements needed by plants in a slowly available form (53). Mixes should always be analyzed for heavy metals and soluble salt levels. The usual recommended rate is that compost not comprise more than 30 percent of the volume of the mix (16).

Suggested Mixes—Media and Preplant Granular Fertilizers for Container Growing During Propagation and Liner Production

Following propagation, young seedlings, rooted cuttings, or acclimatized tissue culture plantlets (liners) are sometimes planted directly in the field but frequently are started in a blended, soilless mix in some type of container. Container growing of young seedlings and rooted cuttings has become an important alternative for field growers. In the southern and western United States, more than 80 percent of nursery plants are container produced (35). For this purpose, special growing mixes are needed (99, 128). It is sometimes more economical for a propagator to buy bags or bulk forms of premixed media. Typically, they are composed of a peat or peat-vermiculite, peat-perlite, hammer-milled and composted bark, rockwool, and other combinations. Preplant amendments in these mixes normally include dolomitic limestone, wetting agents (surfactants) to improve water retention and drainage of the peat or bark, starter fertilizers, trace elements, and sometimes gypsum and a pH buffer.

In preparing container mixes, the media should be screened for uniformity to eliminate excessively large particles. If the materials are very dry, they should be moistened slightly; this applies particularly to peat and bark, which, if mixed when dry, absorb moisture very slowly. In mixing, the various ingredients may be arranged in layers in a pile and turned with a shovel. A power-driven cement mixer, soil shredder, or front-end loader is used in large-scale operations. Most nurseries omit mineral soil from their mixes. The majority of container mixes for propagation and liner production use an organic component such as a bark or peat, which solely or in combination is mixed with mineral components such as sand, vermiculite, or pumice, depending on their availability and cost.

Preparation of the mixture should preferably take place at least a day prior to use. During the ensuing 24 hours, the moisture tends to become equalized throughout the mixture. The mixture should be just slightly moist at the time of use so that it does not crumble; on the other hand, it should not be sufficiently wet to form a ball when squeezed in the hand (44). With barks and other organic matter and supplementary components, particularly rice hulls and sugarcane begasse, it is necessary to compost the material for a period of months before using it as a container medium component.

Container mixes require fertilizer supplements and continued feeding of the plants until they become established in their permanent locations (132). For example, one successful mix for small seedlings, rooted cuttings, and bedding plants consists of one part each of shredded fir or hammer-milled pine bark, peat moss, perlite, and sand. To this mixture is added **preplant fertilizers**—gypsum, dolomitic limestone, microelements and sometimes controlled-release fertilizer. **Postplant fertilizers**—soluble forms of nitrogen, phosphorus, and potassium—are added later to the irrigation water (*fertigation*), or as a top dressing of controlled-release fertilizer, such as Osmocote or Nutricote.

preplant amendments/fertilizers Mineral nutrients that are applied to or incorporated in the propagation or container production media, prior to propagating propagules or transplanting liner plants into containers or into the field.

postplant amendments/fertilizers Mineral nutrients that are applied as a broadcast or liquid application during propagation or production of a containerized or field-grown plant.

In summary, nurseries have changed from loam-based growing media, as exemplified by the John Innes composts developed in England in the 1900s, to soilless mixes incorporating such materials as finely shredded bark, peat, sand, perlite, vermiculite, and pumice in varying proportions. The trend away from loam-based mixes is due to a lack of suitable uniform soils, the added costs of having to pasteurize soil mixes, and the costs of handling and shipping the heavier soils compared with lighter media materials. Much experimentation takes place in trying to develop other low-cost, readily available bulk material to be used as a component of growing mixes such as spent mushroom compost, papermill sludge (21, 26), composted sewage sludge (53), and other materials.

The Cornell Peat-Lite Mixes

The Cornell Peat-Lite mixes, like the earlier University of California (UC) potting mixes, are soilless media. First developed in the mid-1960s, they are used primarily for seed germination and for container growing of bedding plants, annuals, and flowering potted plants. The components are lightweight, uniform, readily available, and have chemical and physical characteristics suitable for the growth of plants. Excellent results have been obtained with these mixes. It may be desirable, however, to pasteurize the peat moss before use to eliminate any disease inoculum or other plant pests. Finely shredded bark is often substituted for the peat moss.

The term **peat-lite** refers to peat-based media containing perlite or vermiculite.

Peat-Lite Mix C (for germinating seeds): To Make 0.76 m^3 (1 cubic yard):

- 0.035 m^3 (1.2 ft^3) shredded German or Canadian sphagnum peat moss
- 0.035 m^3 (1.2 ft^3) horticultural grade vermiculite No. 4 (fine)
- 42 g (1.5 oz)—4 level tbsp ammonium nitrate
- 42 g (1.5 oz)—2 level tbsp superphosphate (20 percent), powdered
- 210 g (7.5 oz)—10 level tbsp finely ground dolomitic limestone

The materials should be mixed thoroughly, with special attention to wetting the peat moss during mixing. Adding a nonionic wetting agent, such as Aqua-Gro [(28 g (1 oz) per 23 liter (6 gal) of water)] usually aids in wetting the peat moss.

Many commercial ready-mixed preparations, based on the original Cornell peat-lite mixes, are available in bulk or bags and are widely used by propagators and producers. Some mixes are prefilled into cell packs, seed trays, or pots that are ready to be planted. Some soilless proprietary mixes are very sophisticated, containing peat moss, vermiculite, and perlite, plus a nutrient charge of nitrogen, potassium, phosphorus, dolomitic limestone, micronutrients, and a wetting agent with the pH adjusted to about 6.5.

Proprietary micronutrient materials, such as Esmigran, FTE 503, or Micromax, consisting of combinations of minor elements, are available for adding to growing media. Adding a controlled-release fertilizer such as Osmocote, MagAmp, Nutriform, Nutricote, or Polyon to the basic Peat-Lite mix is useful if the plants are to be grown in it for an extended period of time.

> **BOX 3.7 GETTING MORE IN DEPTH ON THE SUBJECT**
> **SOME SUPPLIERS OF COMMERCIAL MIXES IN NORTH AMERICA**
>
> Sun Gro Horticulture (www.sungro.com)
> Premier Horticulture (www.premierhort.com)
>
> Scotts Professional Horticulture Solutions (www.scottspro.com)
> Ball Horticultural Company (www.ballhort.com)

Managing Plant Nutrition with Postplant Fertilization During and After the Propagation Cycle

Developing an *efficient fertilizer program* for container plants for the 21st Century depends on (a) minimizing the loss of fertilizer from the production area and (b) increasing the amount of fertilizer utilized or taken up by the plant (133, 134). Suggested levels of *preincorporated* (*preplant*) granular fertilizers were discussed in the previous section on container media for propagation and small linear production. This section discusses some general fertilization practices for management of plant nutrition during **propagation and liner production** (Fig. 3–21). Both soluble and slow-release fertilizers are utilized.

Liquid Fertilizers For large-scale greenhouse and nursery operations, it is more practical to prepare a liquid concentrate and inject it into the regular watering or irrigating system by the use of a proportioner—**fertigation.** The most economical source of fertilizers to be applied through the irrigation water is from dealers who manufacture soluble liquid fertilizer for field crops. It is no longer recommended to use superphosphate in soilless mixes with outdoor container production because of the phosphorus leaching that occurs. Hence, more efficient, soluble forms of phosphorus are used, such as phosphoric acid or ammonium phosphate, in liquid feed programs. Potassium is typically applied as potassium chloride, or potassium nitrate, and nitrogen as Uran 30 (15 percent urea, 15 percent NH_4NO_3) or ammonium nitrate in the liquid concentrate.

An example of a liquid fertilizer system for production of containerized plants is the Virginia Tech System (VTS). With the VTS, all nutrients are supplied to the container by injecting liquid fertilizers into the irrigation water (131, 132). A $10N-4P_2O_5-6K_2O$ analysis liquid fertilizer is applied five times per week, 1.3 cm (0.5 in) each irrigation at an application rate of 100 to 80 ppm N, 15 to 10 ppm P, and 50 to 40 ppm K. Sometimes higher nitrogen levels are applied (200–300 ppm N), depending on the time of the year, plant growth conditions, or plant species. It is critical to regularly monitor soluble salt levels of the medium prior to fertigation. Supplemental micronutrients are also applied in a liquid form but from separate tanks and with separate injectors to prevent fertilizer precipitation. It is best to monitor soluble salt levels of the irrigation water by measuring electrical conductivity (EC) with a conductivity meter; that is, to apply 100 ppm N, the injector is set so that the conductivity of the irrigation water—minus the conductivity of the water before the fertilizer was injected—reads 0.55 mS/cm (millisiemens per cm or dS per m are the same units of measure) (132–134).

Controlled-Release Fertilizers (CRF) Controlled-release fertilizers (CRF) provide nutrients to the plants gradually over a long period and reduce the possibility of injury from excessive applications (127). There has been a long-term trend of nurseries in the southern United States incorporating CRF in propagation, liner and production media, and spot-fertilizing via liquid fertilizer (fertigation) or top-dressing with CRF. CRFs are some of the most cost-effective and ecologically friendly ways to fertilize plants because fertilizer is applied directly to the pots. In contrast, overhead fertigation with rainbird sprinkler-type systems is only about 30 percent efficient, and greater fertilizer runoff occurs from the container production area. Examples of CRF include Osmocote, Phycote, Nutricote, and Polyon, and some are available with micronutrients incorporated in the pellets. As previously described, for both cutting and seed propagation, a low concentration of macro and micro CRF can be included in the propagation mix, so the newly formed roots can have nutrients available for absorption (37). This is particularly important with mist propagation where nutrients can be leached out from both the plant and the medium.

Two types of CRF include coated water-soluble pellets or granules and inorganic materials that are slowly soluble, while slow-release, organic fertilizer includes organic materials of low solubility that gradually decompose by biological breakdown or by chemical hydrolysis.

Examples of the resin-coated-type pellets are (a) Osmocote, whose release rate depends on the thickness of the coating, and (b) Nutricote (105), whose release rate depends on a release agent in the coating. After a period of time the fertilizer will have completely diffused out of the pellets (130). Another kind of controlled-release fertilizer is the sulfur-coated urea granules, consisting of urea coated with a sulfur-wax mixture so that the final product is made up of about 82 percent urea, 13 percent sulfur, 2 percent wax, 2 percent diatomaceous earth, and 1 percent clay conditioner.

An example of the slowly soluble, inorganic type CRF is MagAmp (magnesium ammonium phosphate), an inorganic material of low water solubility. Added to the soilless mix, it supplies nutrients slowly for up to 2 years. MagAmp may be incorporated into media prior to steam pasteurization without toxic effects. On the other hand, steam pasteurization and sand abrasion in the preparation of mixes containing resin-coated, slow-release fertilizers, such as Osmocote, can lead to premature breakdown of the pellets and high soluble salt toxicity.

An example of the slow-release, organic, low-solubility type is urea-formaldehyde (UF), which will supply nitrogen slowly over a long period of time. Another organic slow-release fertilizer is isobutylidene diurea (IBDU), which is a condensation product of urea and isobutylaldehyde, having 31 percent nitrogen.

Fertilizer Systems for Propagation Commercial propagators often apply moderate levels of controlled-release macro and micro elements to the propagation media—preincorporated into the media—prior to sticking cuttings and starting seed germination and seedling plug production. During propagation, supplemental fertilizer is added by top dressing (broadcasting) with controlled-release fertilizer or by injecting gradually increasing concentrations of liquid fertilizer (fertigation). These supplementary nutrients do not promote root initiation (30, 66) in cuttings, but rather enhance root development after root primordia initiation has occurred. Hence, supplementary fertilization is generally delayed until cuttings have begun to root. Propagation turnover occurs more quickly and plant growth is maintained by producing rooted liners and plugs that are more nutritionally fit.

Some recommended levels of CRF for propagation are:

- 3.6 kg/m^3 (6 lb/yd^3) 18-6-12 Osmocote (or comparable product)
- 0.6 kg/m^3 (1 lb/yd^3) Micromax or other trace element mixtures—Perk, Esmigran, or FTE 503
- For unrooted cuttings, fast-germinating seeds, and tissue culture liners, CRF are preincorporated in the propagation media. For slower rooting or seed-germinating species, use Osmocote 153 g/m^2 (0.5 oz/ft^2).
- Nutricote and others are top-dressed on the media after rooting or seed germination starts to occur. Determining optimum levels of fertilization for propagation depends on the propagule system, and needs to be determined on a species basis (30).

Fertilizer Systems for Liner Production Soilless mixes must have fertilizers added (107, 132). Irrigation water and the container medium should be thoroughly analyzed for soluble salts, pH, and macro- and microelements before a fertility program can be established. It is always wise to conduct small trials before initiating large-scale fertility programs during propagation and liner production.

A satisfactory feeding program for growing liner plants is to combine a slowly available dry, granular fertilizer (preplant) in the original mix, with a (postplant) liquid fertilizer applied at frequent intervals during the growing season or with CRF added as top dressings, as needed (49).

Of the three major elements—nitrogen, phosphorus, and potassium—nitrogen has the most control on the amount of vegetative shoot growth. Phosphorus is very important, too, for root development, plant energy reactions, and photosynthesis. Potassium is important for plant water relations and enhanced drought resistance (40).

Nitrogen and potassium are usually supplied by CRF or fertigation—*100 to 80 ppm nitrogen* and *50 to 40 ppm potassium* are optional container medium levels when the Virginia Tech Extraction Method (VTEM) is used (134).

Negatively charged ions, such as phosphorus, leach from soilless media, so small amounts of phosphorus must be added to the media frequently. Past research indicates that *15 to 10 ppm phosphorus* should be maintained in container medium as determined by the saturated paste or VTEM (131, 132). Phosphorus from superphosphate leaches rapidly; so in order to maintain 10 ppm in the medium, CRF is used or small amounts of phosphorus in soluble form are applied.

Calcium and magnesium are supplied as a preplant amendment in dolomitic limestone and may naturally be supplied by irrigation water. Limestone is primarily added to adjust the pH of the media. It is important to have the irrigation water checked to determine the level of dolomitic limestone needed, if any. VTEM levels of

40 ppm calcium and *20 ppm magnesium* in the container medium are adequate.

MANAGEMENT OF MICROCLIMATIC CONDITIONS IN PROPAGATION AND LINER PRODUCTION

Water

Quality (Salinity) of Irrigation Water Good water quality is essential for propagating quality plants (78). The salt tolerance of unrooted cuttings, germinating seeds, and tissue culture explants is much lower than that of established plants, which can be grown under minor irrigation salinity by modifying cultural conditions.

water quality The amount of soluble salts (**salinity**) in irrigation water, which is measured with an electrical conductivity meter.

Water quality for propagation is considered good when the electrical conductivity (EC) reading is 0.75 mS (millisiemens) per cm or dS (decisiemens) per m (less than 525 ppm total soluble salts in ppm), and the sodium absorption ratio (SAR) is 5. Except for the most salt-tolerant plants, irrigation water with total soluble salts in excess of 1,400 ppm (approximately 2 mS/cm) (ocean water averages about 35,000 ppm) would be unsuitable for propagation. Salts are combinations of such cations as sodium, calcium, and magnesium with such anions as sulfate, chloride, and bicarbonate. Water containing a high proportion of sodium to calcium and magnesium can adversely affect the physical properties and water-absorption rates of propagation media and should not be used for irrigation purposes. It is prudent to have nursery irrigation water tested at least twice a year by a reputable laboratory that is prepared to evaluate all the elements in the water affecting plant growth. Most producers regularly monitor EC and pH of their irrigation water and container mix with inexpensive instruments. Some producers test and monitor their own container media nutrients, whereas plant leaf tissue is generally sent off to plant laboratories for nutrient analysis.

Although not itself detrimental to plant tissue, so-called hard water contains relatively high amounts of calcium and magnesium (as bicarbonates and sulfates) and can be a problem in mist-propagating units or in evaporative water cooling systems because deposits build up wherever evaporation occurs, which reduces the photosynthetic levels of cuttings, seedlings, and tissue culture plantlets. When *hard* water is run through a water softener, some types of exchange units replace the calcium and magnesium in the water with sodium ions. Misting and irrigating with such *soft*, high-sodium water can injure plant tissue.

A better, but more costly, method of improving water quality is using **deionization (DI).** Water passes over an absorptive cation resin to filter positively charged ions such as calcium and other ions in exchange for hydrogen. For further deionization, the water is passed through a second anion resin to filter out negatively charged ions such as carbonates, sulfates, and chlorides in exchange for hydroxyl (OH) ions.

Boron salts are not removed by deionization units, and, if present in water in excess of 1 ppm, they can cause plant injury. There is no satisfactory method for removing excess boron from water. The best solution is to acquire another water source and to use customized non-boron-blended fertilizers.

Another good, but expensive, method for improving water quality is **reverse osmosis (RO)** (Fig. 3–28), a process in which pressure applied to irrigation water forces it through a semipermeable membrane from a more concentrated solution to a less concentrated solution, eliminating unwanted salts from an otherwise good water source. There are combination RO/DI units, but they are cost-prohibitive for most propagation systems.

Municipal treatment of water supplies with chlorine (0.1 to 0.6 ppm) is not sufficiently high to cause plant injury. However, the addition of fluoride to water supplies at 1 ppm can cause leaf damage to a few tropical foliage plant species.

BOX 3.8 GETTING MORE IN DEPTH ON THE SUBJECT
MEASURING SALINITY

Salinity levels from irrigation water, and from water extracts of growing media (saturation-extract method) can be measured by electrical conductivity (EC) using a Solubridge. Various portable meters for testing salinity, as well as soil and water testing kits, are available through commercial greenhouse supply companies.

(a)

(b)

(c)

Figure 3–28
Good water quality is imperative for propagation. (a) A reverse osmosis system is shown for removing salts in commercial propagation. (b and c) Deionizing columns for removing salts.

When the water source is a pond, well, lake, or river, contamination by weed seeds, mosses, or algae can be a problem. Chemical contamination from drainage into the water source from herbicides applied to adjoining fields or from excess fertilizers on crop fields can also damage nursery plants. Recycled water, which is discussed in the section "Best Management Practices (BMP) (see page 98)," is used in nursery and greenhouse production, and is being evaluated for general propagation in some nurseries.

The pH of Irrigation Water and Substrate Media The pH is a measure of the concentration of hydrogen ions and can affect the rooting of cuttings, germination of

---

BOX 3.9 GETTING MORE IN DEPTH ON THE SUBJECT
TREATING RECYCLED IRRIGATION WATER

Nurseries using **recycled irrigation water** (Fig. 3–29) should treat the water before use. A good procedure is to:

- Initially utilize aquatic plants in runoff catchment ponds to reduce pollutants and sediments reentering the recycling system (113, 133).
- Add chlorine or bromine to suppress algae and plant pathogens as water is pumped from the catchment pond.
- Use strainers to remove large debris, then run the water through sand or mechanical filters with automatic back flushing to remove coarse particles and weed seed.
- Consider running the water through an activated charcoal tank to remove soluble herbicides and other residual chemicals.
- If the water has a high salt content, it can be improved by the use of deionization or reverse osmosis, but the processes are very expensive.

- Water can be treated with ultraviolet irradiation to reduce pathogens. Generally, all precipitate down to at least 20 µm is filtered out in order for UV light to be effective.
- Recycled water is acidified (to lower the pH, if necessary) and repumped into holding ponds with plastic liners and weed-free perimeters.
- Fresh well water is pumped into the holding pond and mixed with the recycled water. This allows for pumping from wells during the night to meet daily irrigation needs and dilutes soluble salts of recycled water.
- This water can then be used for field watering of container nursery plants and slow-release fertilizer incorporated into containers or soluble fertilizer injected into the irrigation system.

Figure 3–29
Systems for capturing, treating, and recycling irrigation water in commercial nurseries. (a and b) Irrigation water either drains into or is (c) pumped into a holding pond. (d) Irrigation water is treated with chlorine (sodium hypochlorite solution is one of the safest forms) as it is pumped from the holding pond into the irrigation lines system lines. (e) Filtration tanks for removing weed seed and particulate-suspended matter down to 20 mm (this is important if irrigation water is to be treated with ultraviolet light); some nurseries use tanks of activated charcoal to trap soluble herbicides and other undesirable chemicals. (f) Ultraviolet treatment of irrigation water with a UVS Ultra Pure model 5000. Scoresby, Victoria, Australia. (g) Bromination of water; some nurseries will inject acid at this point to lower water pH, if needed. (h) Monitoring water leaving the water treatment facility for pH and soluble salts or electrical conductivity (EC).

seeds, and micropropagation of explants. Liner production is also affected by pH influence on nutrient availability and activity of beneficial microorganisms in the container medium. A pH range of 5.5 to 7.0 is best for the growth of most plants (7.0 is neutral—below this level is acid and above is basic or alkaline). Nurseries may control carbonate problems by injecting sulfuric or phosphoric acid into the irrigation water supplies. Softwood bark and peat-based container mixes are acid and will lower irrigation water pH. Dolomitic limestone raises soil pH and is the primary source of Ca and Mg in many propagation and liner mixes. While pH is important, alkalinity has a greater impact on water quality (133).

Water-Humidity Control

Good water management is important to limiting plant stress. Care must be exercised to avoid overmisting and overirrigation, because too much water can be just as stressful as too little water. Root rots and damping-off organisms are favored by standing water and poor media drainage conditions.

Maintaining proper atmospheric humidity in the propagation house beds is important because low humidity can increase transpiration and subject unrooted cuttings and seedlings to water stress. Adequate humidity allows optimum growth, whereas extreme humidity promotes fungal pathogen, moss, and liverwort pests. Air always contains some water vapor, but at any given temperature it can hold only a finite amount. When the physical limit is reached, the air is **saturated,** and when it is exceeded, **condensation** occurs (72). The unique physical properties of water affect the propagation environment. When water is converted from a liquid to a gas (water vapor), a large amount of thermal energy (540 cal/g) is required. The cooling effect of mist irrigation results as heat is absorbed and the increased relative humidity minimizes plant transpiration. A heavy mist, which condenses and forms droplets of water, should be avoided because it leaches foliage of nutrients, saturates propagation media, and can promote disease problems.

Current systems used to control water loss of plant leaves (74) are:

1. **Enclosure Systems:** outdoor propagation under low tunnels or cold frames, or nonmisted enclosures in a glasshouse or polyhouse (shading, tent and contact polyethylene systems, wet tents).
2. **Intermittent Mist:** open and enclosed mist systems.
3. **Fogging Systems**

The effect of these systems on propagation environmental conditions and water relations of the propagule is discussed in greater detail in Chapter 9.

Temperature Control

As indicated in earlier sections, temperature is modified by environmental controls in the propagation structure and the type of propagation system that is used (see Chapter 9 for greater details). There is no environmental factor more critical than optimal temperature control for propagation. Optimal seed germination, rooting of cuttings, development of tissue culture plantlets, graft union formation, and specialized structure development are all temperature-driven plant responses. Hot air convection, infrared radiation, and hot water distribution systems are the three most viable ways to heat plants (Figs. 3–2 and 3–9). Of the three, hot water is the most flexible and commonly used system in propagation houses (98). It allows efficient root zone heating in the form of bottom heat. Some examples include Biotherm tubing and Delta tubes, which are used to maintain optimum propagation temperatures. A mist system accelerates root development of cuttings under high light irradiance, by evaporative cooling, which reduces the heat load on plant foliage. In fog systems, the fog particles remain suspended and reduce the light intensity, while a zero-transpiration environment is maintained, without the overwetting (condensation) that can occur with mist. Since only minimal condensation occurs, leaf and media temperatures are warmer with fog than mist. In liner production, DIF systems (cooler days and warmer night temperature) produce more compact plants. This works well for seedling plugs, bedding plants, and greenhouse crops under controlled environmental conditions (55).

Light Manipulation

The importance of light manipulation in propagation (irradiance, photoperiod, quality) was discussed earlier in the chapter and is covered in greater detail in later chapters on seed and cutting propagation, micropropagation, and specialized structure development and propagation. Light quality (which is commercially manipulated through greenhouse spectral filters, greenhouse coverings, and varying supplementary light sources) plays an important role in seed germination, and shoot development in macro- and micropropagation (Fig. 3–11). Photoperiod can be manipulated to delay bud dormancy and extend accelerated plant growth. Photoperiod can be utilized not only to enhance root initiation, but also to increase carbohydrate reserves of deciduous, rooted cuttings (liners) for better winter survival and subsequent vigorous spring growth (Fig. 3–14) (79).

Supplemental Photosynthetic Lighting in the Propagation House Plant growth in the winter in propagation houses can be slow due to the lack of sufficient light for photosynthesis, especially in the higher latitudes (19). This is due to several reasons:

- Low number of daily light hours
- Low angle of the sun, resulting in more of the earth's atmosphere that the sun's rays must penetrate
- Many cloudy and overcast days in the winter
- Shading by the greenhouse structure itself and dirt accumulating on the poly or glass or other covering materials

To overcome the problem of low natural winter light and reduced plant growth, supplemental light can be used over the plants (Fig. 3–14). The best light source for greenhouse lighting is high-pressure sodium vapor lamps. Most of the radiation from these lamps is in the red and yellow wavelengths and is very deficient in blue. However, when used in conjunction with the natural daylight radiation, these lamps are quite satisfactory.

The high-pressure sodium vapor lamps emit more photosynthetically active radiation (PAR) for each input watt of electricity than any other lamp that is commercially available. Sodium vapor lamps are long-lasting and degrade very slowly. They emit a considerable amount of heat that can be a benefit in the greenhouse in winter. They use a smaller fixture than fluorescent lamps, thus avoiding the substantial shading effect from the fluorescent lamp fixture itself. The installation should provide a minimum of about 65 µmol m^2 s^{-1} or 13 W/m^2 PAR at the plant level with a 16-hour photoperiod. For large greenhouses, the services of a lighting consultant should be used in designing the installation.

In the future, expect to see greater use of light-emitting diodes (LED) (86) with spectral qualities based on propagation needs under controlled environmental agriculture (CEA). The LED has no filament, just a microchip, and is extremely energy-efficient.

Photosynthetic lighting with high intensity discharge lights (HID) in more overcast climates has greatly expanded the production window for cuttings and seed propagation. Supplementary lighting is an important component in accelerated growth techniques (AGT) in propagating plants (Fig. 3–30).

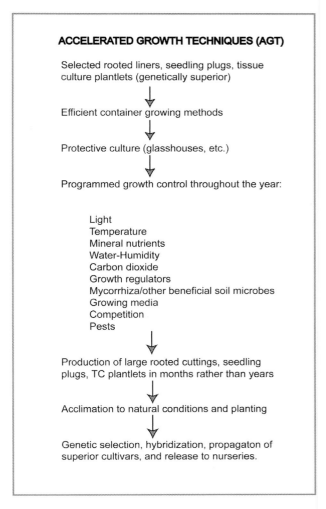

Figure 3–30
Components of accelerated growth techniques used in speeding up vegetative and seed propagation in the production of marketable liners.

Carbon Dioxide (CO$_2$) Enrichment in the Propagation House

Carbon dioxide is one of the required ingredients for the basic photosynthetic process that accounts for the dry-weight materials produced by the plant (59, 87, 94):

$$6H_2O + 6CO_2 + \text{Light Energy/Chlorophyll} \rightarrow C_6H_{12}O_6 + 6O_2$$

Ambient carbon dioxide (CO$_2$) in the atmosphere is around 380 ppm. Sometimes the concentration in winter in closed greenhouses may drop to 200 ppm, or lower, during the sunlight hours, owing to its use by plants (94). Under adequate light and temperature, but when low CO$_2$ concentration limits photosynthesis, a supplementary increase in CO$_2$ concentration 1,000 to 2,400 ppm can result in a 200 percent increase in photosynthesis. To take full advantage of this potential increase in dry-weight production, plant spacing must avoid shading of overcrowded leaves. When supplementary CO_2 is used during periods of sunny weather, the temperature in the greenhouse should be kept relatively high. Adding CO_2 at night is of no value. However, CO_2 generators can be turned on before dawn to increase photosynthesis early in the day. A tightly closed greenhouse is necessary to be able to increase the ambient CO_2.

Sources of CO_2 for greenhouses are either burners using kerosene, propane, or natural gas, or liquid CO_2. Liquid CO_2 is expensive but almost risk free. Kerosene burners must use high-quality, low-sulfur kerosene or SO_2 pollution can occur. With propane or natural gas, incomplete combustion is possible—by-products include carbon monoxide (dangerous to humans) and ethylene (harmful to plants). The flames should be a solid blue color. Monitoring of the CO_2 level in the greenhouse is very important. Accurate, inexpensive sensors are available and should be used (Fig. 3–13). With the newer computer technology, sensors in different parts of the greenhouse can give excellent control of the CO_2 levels. Excessively high levels of CO_2 in the greenhouse (over 5,000 ppm) can be dangerous to humans.

New tissue culture systems are utilizing high CO_2 enrichment and high light levels for *autotrophic micropropagation* (65). The plantlets are cultured without sugar in the culture medium as an energy and carbon source, and are stimulated by enriched atmospheric CO_2 and elevated light irradiance to photosynthesize and become autotrophic. The CO_2 is supplied either directly to the tissue culture vessel or indirectly via increased ambient CO_2 to permeable culture vessels. Autotrophic micropropagation improves plantlet growth and development, simplifies procedures, reduces contamination, and lowers production costs (see Chapter 18).

Accelerated Growth Techniques (AGT)

The forestry industry developed accelerated growth systems to speed up the production of liners from cutting and seed propagation. Woody perennial plants undergo cyclic (episodic) growth, and many tree species experience dormancy. Liners are grown in protective culture facilities where photoperiod is extended and water, temperature, carbon dioxide, nutrition, mycorrhizal fungi, and growing media are optimized for each woody species at different growth phases (Fig. 3–30).

This concept is also being used in propagation of horticultural crops where supplementary lighting with high-pressure sodium vapor lamps and injection of CO_2 gas into mist water are used to enhance seed germination, plug development, acclimation of tissue culture plantlets, and rooting of cuttings. The promotive effects of AGT on rooting of *Ilex aquifolium* (holly) cuttings has been attributed, in part, to enhanced photosynthesis.

Modeling in Plant Propagation Closely linked to AGT is the modeling of propagation environments to determine optimal light, temperature, water, CO_2, and nutritional regimes (125, 126). Computer technology allows the propagator to monitor and program the propagation environment and adjust environmental conditions as needed through automated environmental control systems (see Figs. 3–13, 3–14, 3–15, 10–42, and 10–43).

BIOTIC FACTORS—PATHOGEN AND PEST MANAGEMENT IN PLANT PROPAGATION

Pathogen and pest management begins *prior to propagation* with the proper manipulation of stock plants or the container plants from which the propagules are harvested, as well as with management of propagation beds and media preparation. If pathogens and pests are not checked during propagation, an inferior plant is produced and later production phases for finishing and selling the crop will be delayed, causing profit losses.

Pests are broadly defined as all biological organisms (bacteria, viruses, viroids, phytoplasma, fungi, insects, mites, nematodes, weeds, parasitic higher plants, birds, and mammals) that interfere with plant production (57). *Insect pests,* such as aphids, mealy bugs, thrips, white flies, and fire ants, actively seek out the plant host by migrating (flying, walking). When an infection can be spread from plant to plant, it is referred to as an infectious disease. *Infectious plant diseases* are caused by different pathogens *(infectious agents),* including pathogenic fungi, bacteria, viruses, viroids, and phytoplasma. Specific pathogens may infect only certain plant species or cultivars, or specific organs or tissue, which varies with the stage of development of the plant.

The pathogenic fungi most likely to cause disease development during propagation are species of *Pythium, Phytophthora, Fusarium, Cylindrocladium, Thielaviopsis, Sclerotinia, Rhizoctonia,* and *Botrytis* (27). These are all soil-borne or aerial organisms (*Botrytis*) that infect plant roots, stems, crowns, or foliage. The so-called damping-off commonly encountered in seedbeds is caused by soil fungi, such as species of *Pythium, Phytophthora, Rhizoctonia,* and *Fusarium.* Suppressing pathogens in propagation water is critical—*Phytophthora, Pythium,* and *Rhizoctonia* are readily disseminated in surface water.

Conversely, intermittent mist can wash off germinating fungal spores. Mist inhibits the spore germination of powdery mildew (*Sphaerotheca pannosa*) on leaves of cuttings, and it may be that other disease organisms are held in check in the same manner. However, mist propagation is highly conducive to diseases such as aerial *Rhizoctonia* blight, *Cylindrocladium,* bacterial soft rots, and so on.

A goal in propagation is to keep stock plants and propagules as clean and pest-free as possible and to suppress pathogenic fungi, viruses, nematodes, and weed seed from the propagation media. Optimum pest management depends on a thorough knowledge of the pest life cycle, as well as environmental conditions, cultural practices, and minimizing host plant stress—the rooting of a cutting and germination of a seed are vulnerable periods of plant growth. A stressed propagule is much more susceptible to pest problems. The management of pests through integrated pest management (IPM) is discussed in this section.

Preventive Measures

Cultivar Resistance Avoid producing crops that are susceptible to certain diseases and pests. A susceptible crop means more time, chemicals, and money spent to control the problem. In addition, the problem is passed on from the propagator to the consumer (3). By choosing a resistant cultivar, efforts are concentrated on propagating and producing the plant, rather than trying to control the pest (i.e., propagate disease-resistant crab apple cultivars, rather than disease susceptible *Malus* cultivars such as 'Hopa' and 'Mary Potter'). In the southern United States, Helleri hollies (*Ilex* 'Helleri') are plagued by southern red mites, root-knot nematodes, and black root rot—so why propagate them when other holly cultivars are more resistant (3)?

Scouting System All propagators should practice pest scouting. Early detection provides more effective pest and pathogen control with less reliance on pesticides. Propagation houses should be scouted on a regular basis and all propagation employees trained to recognize and report disease and insect pests. Workers are the ones in daily contact with plants and are an invaluable resource for early detection. Some large nurseries have detailed pest management programs with crews supervised by trained plant pathologists and entomologists (23). Such programs involve the proactive prevention of plant diseases and the avoidance of insects, mites, and weed problems. Serological test kits—ELISA (enzyme-linked immunosorbent assays)—are commercially available to propagators for the early detection of certain pathogens and viruses (88, 95). The user-friendly Alert Diagnostic Kits can rapidly identify the damping-off organisms—*Pythium, Rhizoctonia, Phytophthora,* and *Botrytis* (2). (Keeping viruses in check is considered in Chapter 16.)

Integrated Pest Management in Plant Propagation

Integrated Pest Management (IPM) is the most efficient, most economical, most environmentally safe system for managing pests in propagation and liner production systems. The components of IPM are divided into three management areas:

- Chemical
- Biological
- Cultural

Total elimination of a pest is not always feasible—nor is it biologically desirable if the process is environmentally damaging or leads to new, more resistant pests and eliminates beneficial fungi and insects. In the production of clean stock plants and propagules, pest-free plants may be a requirement, but this should be accomplished by using a variety of pest management methods without an overdependence on just one method (i.e., solely using chemical control). **Pest control** differs from **pest management** in that an individual pest control technique is used in isolation to eliminate a pest and *all* pest-related damage (57). Conversely, IPM uses as many *management* (control) methods as possible in a systematic program of suppressing pests (not necessarily annihilating) to a commercially acceptable level, which is a more ecologically sound system.

Chemical Control in IPM Chemical control methods in IPM include the use of:

- Fumigation
- Fungicides
- Insecticides

IPM does not imply that no chemicals are used in the control of pests and pathogens. Rather, better-targeted control with less chemical usage occurs because of the integration of additional biological and cultural management measures (109). IPM in propagation means that actions must be thoughtfully considered and carried out in ways that will ensure favorable economic, ecological, and sociological consequences (52, 93).

In the treatment of seeds, bulbs, corms, tubers, and roots, pesticides are sometime used in combination with cultural techniques, such as hot water soaks [43 to 57°C (110 to 135°F)]. The hot water temperature and duration is dependent on the species and propagule type being treated. For many ornamental plants, to control decay and damping-off, seeds are treated with fungicidal slurries or dusts of thiram, zineb, and so on. Seeds of California poppy, and *Strelitzia* (bird of paradise) are given hot water soaks to control pathogenic fungi, while *Delphinium* (larkspur) and *Digitalis* (foxglove) seeds are given a hot water soak and then dusted with thiram to control anthracnose. Bulbs and corms of many species are treated for nematodes and pathogenic fungi with hot water soaks and/or chemical treatment.

When using pesticides, it is important that propagators follow the **Worker Protection Standard (WPS)** rules and regulations to reduce pesticide-related illnesses and injuries (45). The WPS can complicate many jobs in propagation. Scheduling has become more critical so pesticide restricted entry intervals **(REI)** do not interfere with normal propagation assignments of workers. The United States Environmental Protection Agency (EPA) has a monthly updated bulletin that details WPS implementation information on reentry rules and times; see their web site (www.epa.gov/pesticides).

Fumigation with Chemicals Chemical fumigation kills organisms in the propagating mixes without disrupting the physical and chemical characteristics of the mixes, to the extent occurring with heat treatments. (In all cases, recommendations on pesticide labels must be followed to conform to permitted usages.) The mixes should be moist (between 40 and 80 percent of field capacity) and at temperatures of 18 to 24°C (65 to 75°F) for satisfactory results. Before using the mixture and after chemical fumigation, allow a waiting period of 2 days to 2 weeks, depending on the material, for dissipation of the fumes. A problem with chemically sterile media is that there are no competing microorganisms to limit the rapid recolonization of fungi and bacterium, which may create media aeration and pest problems.

Methyl Bromide (MB). MB is a highly effective fumigant for propagation. It is odorless, very volatile, and quite **toxic to animals and humans.** Because it contributes to the reduction of the earth's ozone layer, developing countries are limiting the use of MB with a complete phase out in 2015. The U.S. EPA is currently revising the reregistration of methyl bromide. The USDA has a special web site on MB alternatives, including methyl iodide and metam sodium (18), for agriculture (http://www.ars.usda.gov/is/np/mba/mebrhp.htm). It should be mixed with other materials and applied only by those trained in its use. Most nematodes, insects, weed seeds, and fungi are killed by methyl bromide. Methyl bromide is most often used by injecting the material from pressurized containers into an open vessel placed under a plastic sheet that covers the soil to be treated (Fig. 3–31). The cover is sealed around the edges with soil and should be kept in place for 48 hours. Penetration is very good and its effect extends to a depth of about 30 cm (12 in).

Methyl Bromide and Chloropicrin Mixtures. Proprietary materials are available that contain both methyl bromide and chloropicrin. Such combinations are more effective than either material alone in controlling weeds, insects, nematodes, and soil-borne pathogens. The

---

BOX 3.10 GETTING MORE IN DEPTH ON THE SUBJECT
IPM IN THE CULTURAL CONTROL OF APHIDS

An example of IPM is the **cultural control of aphids** in propagation by installing *microscreening* that covers vents and doorways of a propagation house, thereby reducing the movement of insects and the need for insecticides (48). Early detection of winged aphids with *yellow sticky cards* that are hung in the propagation house can alert personnel to monitor plants near cards for the presence of wingless females. The option to use **biological control** is possible with an efficient scouting system that detects controllable, low aphid levels. A beneficial midge, *Aphidoletes aphidimyza*, has been used to biologically control aphid colonies. If the aphid colony is small, other *biorational products* can be used such as insecticidal soap (M-Pede), horticultural oils (UltraFine SunSpray spray oil), *botanical insecticides* such as neem (Azatin and Margosan-O), and natural pyrethrums. *Insect growth regulators* such as kinoprene (Enstar II) and methoprene give safe, effective control of immature aphids. For large populations of aphids that were not detected early enough, **chemical control** with traditional pesticides are sometimes used, such as diazinon, bendiocarb, methiocarb, acephate; or the synthetic pyrethroids, such as fluvalinate (Mavrik), bifenthrin (Talstar), and fenpropathrin (Tame) (48).

Figure 3–31
Chemical and heat treatment of propagation mixes. (a) Methyl bromide (MB) being applied to propagation medium. (b) Methyl bromide is injected into media covered with poly. (c) Methyl bromide is extremely toxic; during soil treatment it is important to use warning signs and restrict the movement of personnel. (d, e, and f) Heat pasteurization with aerated steam.

addition of chloropicrin (tear gas) to methyl bromide was primarily so that humans could detect gas leaks and evacuate before being poisoned by methyl bromide. Aeration for 10 to 14 days is required following applications of methyl bromide-chloropicrin mixtures.

Fungicidal Soil Drenches. Fungicidal soil drenches can be applied to the container media in which young plants are growing or are to be grown to suppress growth of many soil-borne fungi. These materials may be applied either to media or to the plants. Preferably, a wetting agent should be added to the chemicals before application. It is very important when using such chemicals to read and follow the manufacturer's directions and prepare dilutions carefully, and to try the chemicals on a limited number of plants before going to large-scale applications. As with insect pests, pathogens can build up resistance to fungicides, so it is important to rotate fungicides and use mixtures with good residual action (63).

BOX 3.11 GETTING MORE IN DEPTH ON THE SUBJECT
MINOR-USE CHEMICALS

Chemicals used in propagation and horticulture are considered **minor use,** as opposed to pesticides used for large commodity crops such as cotton, soybean, corn, and others. The cost for chemical companies to develop new or to reregister specialty or minor-use chemicals is often prohibitive. Hence, more than 1,000 minor uses of agricultural chemicals are currently at risk, and another 2,600 newly sought minor uses may never come to fruition because of the 1988 Federal Insecticide, Fungicide and Rodenticide Act (FIFRA) (38) (see http://www.epa.gov/oecaagct/lfra.html).

Examples of fungicidal drench materials are Quintozene (PCNB, Terraclor), which controls *Rhizoctonia*, *Sclerotina*, and *Sclerotium*. Etridiazole (Terrazole, Truban) are incorporated into the propagating medium, which suppresses the water molds *Pythium* and *Phytophthora*. Banrot is a broad-spectrum fungicide that suppresses the damping-off organisms of *Pythium*, *Phytophthora*, and *Rhizoctonia*, as well as *Fusarium* and *Thielaviopsis*. Subdue and Heritage are some of the systemic fungicides used in propagation for control of root rots and foliar pathogens (27).

Propagators are adapting IPM systems—utilizing disease-free propagules, clean propagation media, disinfesting propagation facilities and incorporating beneficial rhizosphere organisms such as mycorrhiza; hence fungicidal sprays are applied only as needed and not as weekly preventive sprays (32). Some propagators dip the bottom 5 cm (2 in) of cuttings into Zerotol (hydrogen dioxide) to disinfect cuttings of potential pathogens; cuttings are then quick-dipped into auxin solutions for rooting (100).

Insecticidal Sprays and Drenches. An example of insecticidal spray and drench usage is in the control of fire ants, which are a major pest in the southern United States. The USDA implemented the Imported Fire Ant Quarantine and Imported Fire Ant Free Nursery program in 1958 to prevent the spread of fire ants, which infest twelve southern and western states and Puerto Rico. The ants are spread easily by accidentally shipping them with nursery stock and small liner plants. The ants do not directly harm plants and propagules (they will tend plants with aphids, and harvest the honeydew of the aphids from the plants' leaves)—but they do damage land and livestock, have killed people, and are a nuisance to propagation workers and the public. For short-term, small-container crops, such as liners, producers will drench plant containers with Dursban, Talstar 10WP, and Diazinon (in certain states). For propagation mixes and large-container crops, producers use soil-incorporated granular insecticide, such as Talstar and Dursban (22). Chemical baits are also effective for long-term fire ant control, but are slower acting than spray/drench applications; see the fire ant web site for the latest recommendations (http://fireant.tamu.edu).

Biological Control in IPM Biological control in IPM includes:

- Predator insects and mites
- Beneficial nematodes
- Beneficial fungi and bacteria

More and more insect pests and pathogens are being managed by biological methods. This is due in part to increased mite and insect resistance to pesticides, the fact that biological control can be cheaper and more effective than chemical control (i.e., two-spotted mite is effectively controlled by the Chilean predatory mite), increasing concern for environmental issues (contamination of groundwater, etc.), and worker safety (i.e., reentry times of workers after pesticide application, etc.). In the United States, there is the Association of Natural Biocontrol Producers (ANBP; www.anbp.org) for the production and utilization of beneficial insects and organisms.

In propagation, the bacterium *Bacillus thuringiensis* (BT) infects and controls most caterpillars and fungal gnat larvae but has little effect on other insects or the environment. Strains of this naturally occurring bacterium have been formulated into the biological control insecticides Dipel, Thuricide, Bactospeine, and so on.

Biofungicides are preventive, rather than curative, and must be applied or incorporated before disease onset to work properly. For example, the beneficial fungus *Trichoderma virens* (Soil-Gard) comes in an easy-to-apply granular form that is added to the propagation media. It has been cleared by the EPA for biological control of *Rhizoctonia colani* and *Pythium ultimum*, which are two of the principal pathogens causing damping-off diseases (31). Mycostop, a strain of *Streptomyces* bacteria isolated from Finnish peat, is used in propagation as a drench; dip for transplants, seeds, and cuttings; or as a foliar spray. It controls *Fusarium*, *Alternaria*, and *Phomopsis,* and suppresses *Botrytis*, *Pythium,* and *Phytophthora* (2, 63).

As higher plants have evolved, so have beneficial below-ground organisms interacting with the plant root system (the plant **rhizosphere**). Examples of this include symbiotic nitrogen-fixing bacteria, which are important for leguminous plants, and selected nematodes that control fungal gnats (i.e., X-Gnat from Biosys).

> rhizosphere The zone of soil immediately adjacent to plant roots in which the kinds, numbers, or activities of microorganisms differ from that of the bulk soil.

The nematodes come in water-dispersible granules, are applied with overhead irrigation equipment, and attack gnats in the larval stage in the container medium. It is well known that beneficial **mycorrhizal fungi** (which naturally colonize the root systems of *most* major horticulture, forestry, and agronomic plants) can increase plant disease resistance and help alleviate plant stress by enhancing the host plant water and nutrient uptake (32, 73). Mycorrhizae can also benefit propagation of cuttings, seedlings, and transplanting of liner plants (25, 33, 34, 110).

The use of **biocontrol agents** (beneficial bacteria, actinomycetes, or fungi living and functioning on or near roots in the rhizosphere soil) to control plant pathogens in propagation is gradually occuring (73). These beneficial microorganisms suppress fungal root pathogens by antibiosis (production of antibiotic chemicals), by parasitism (direct attack and killing of pathogen hyphae or spores), or by competing with the pathogen for space or nutrients, sometimes by producing chemicals such as siderophores, which bind nutrients (such as iron) needed by the pathogen for its disease-causing activities. The inhibitory capacity of these biocontrol antagonists increases in the presence of mycorrhizal fungi, and in the absence of plant pathogens there is a stimulation of plant growth by bacterial antagonists; somehow these bacteria stimulate plant growth, but the mechanism is not known. Perhaps in the future, plant protection during propagation will be done by inoculation of bacteria or combinations of bacteria with mycorrhizal fungi, which come closest to simulating natural conditions of the plant rhizosphere (73). For some commercial nurseries, incorporating mycorrhizal fungi during propagation is now standard procedure (32).

Cultural Control in IPM Cultural management continues to become more important in modern propagation systems with the loss of minor-use chemicals. In propagation, cultural control begins with the preplant treatment of soil mixes to suppress pathogens and pests. Other cultural control techniques include:

- sanitizing of propagation facilities
- suppressing pathogens and insect pests of stock blocks
- harvesting cuttings from stock blocks or containerized plants that are nutritionally fit and not drought stressed
- providing good water drainage to reduce the potential of *Phytophthora* root rot and other damping-off organisms
- reducing humidity to control *Botrytis*
- minimizing the spread of pathogens by quickly disposing of diseased plants from the propagation area, and
- hardening-off established propagules (96).

Cultural control in IPM includes:

- Stock plant management
- Media pasteurization
- Sanitation

Suppressing pathogens in propagation water is critical, since *Phytophthora, Pythium,* and *Rhizoctonia* are readily disseminated in surface water. Checking pathogens starts with the initial removal of suspended silt and solids, which can tie up chemicals being used to treat the water supply, a task most commonly accomplished by using a sand filterUltraviolet light irradiation is a nonchemical method of controlling pathogens, but water needs to be free of turbidity (suspended materials) that will shield some of the pathogens from the UV (Fig. 3–29). The most commonly used chemical treatments of irrigation water are with chlorination or bromination; one Australian nursery aims for a 4 ppm residual chlorine at the discharge of the irrigation water. They use a swimming pool chlorine test kit (15), such as easy-to-use, portable DPD color-indicator test kits (13). Current recommendations for chlorinated irrigation systems is to maintain a free chlorine level of 2 ppm (2 mg/liter) to kill *Phythiaceae* pathogens, and to increase the contact time to kill *Fusarium* and *Rhizoctonia* (20). A "free chlorine" level of 2.9 ppm is generally considered safe for most plants (106).

Selectrocide (chlorine dioxide) is also used for the control of algae and other microbial pests in greenhouse propagation irrrigation lines (68).

Preplanting Treatments of Mixes—Heat Treatment of Propagation and Liner Media Various Replanting Treatments of Mixes are Categorized as Follows

Pasteurization of Propagation Media. Propagation mixes such as bark, sand, and peat moss (14, 24) can contain pathogens and, ideally, should be pasteurized. The containers (bins, flats, pots) for such pasteurized mixes should, of course, have been treated to eliminate pathogens. Never put pasteurized mixes into dirty containers. New materials such as vermiculite, perlite, pumice, and rockwool, which have been heat-treated

---

BOX 3.12 GETTING MORE IN DEPTH ON THE SUBJECT
BENEFICIAL TRICHODERMA FUNGI

Trichoderma fungal species, which have plant growth-enhancing effects, independent of their biocontrol of root pathogens, have been reported to enhance the rooting of chrysanthemum cuttings, possibly by producing growth-regulating substances (76).

during their manufacture, need not be pasteurized unless they are reused.

Although the term soil *sterilization* has been commonly used, a more desirable process is **pasteurization,** since the recommended heating processes do not kill all organisms (Fig. 3–31). True sterilization would require heating the propagation media to a minimum temperature of 100°C (212°F) for a sufficient period to kill all pests and pathogenic organisms; all beneficial rhizosphere organisms are also killed by the process. Pasteurization of propagation media at lower temperatures with aerated steam is generally preferable to fumigation with chemicals.

After treatment with steam, the medium can be used much sooner. Steam is nonselective for pests, whereas chemicals may be selective. Aerated steam, when properly used, is much less dangerous to use than fumigant chemicals, for both plants and the operator. Chemicals do not vaporize well at low temperatures, but steam pasteurization can be used for cold, wet media.

Moist heat can be injected directly into the soil in covered bins or benches from perforated pipes placed 15 to 20 cm (6 to 8 in) below the surface. In heating the soil, which should be moist but not wet, a temperature of 82°C (180°F) for 30 minutes has been a standard recommendation because this procedure kills most harmful bacteria and fungi as well as nematodes, insects, and most weed seeds, as indicated in Figure 3–32. However, a lower temperature, such as 60°C (140°F) for 30 minutes, is more desirable since it kills pathogens but leaves many beneficial organisms that prevent explosive growth of harmful organisms if recontamination occurs. The lower temperature also tends to avoid toxicity problems, such as the release of excess ammonia and nitrite, as well as manganese injury, which can occur at high steam temperatures.

Electric Heat Pasteurizers. are in use for amounts of soil up to 0.4 m³ (0.5 yd³). *Microwave ovens* can be used effectively for small quantities of soil. They do not have the undesirable drying effect of conventional oven heating and will kill insects, disease organisms, weed seed, and nematodes.

Sanitation in Propagation In recent years, the importance of sanitation during propagation and growing has become widely recognized as an essential part of nursery operations. During propagation, losses of young seedlings, rooted cuttings, tissue-cultured rooted plants, and grafted nursery plants to various pathogens and insect pests can sometimes be devastating, especially under the warm, humid conditions found in propagation houses (80, 84). Ideally, sanitation strategies should be considered even in the construction phase of propagation structures (92).

Harmful pathogens and other pests are best managed by dealing with the three situations where they can enter and become a problem during propagation procedures:

- The propagation facilities: propagating room, containers, pots, flats, knives, shears, working surfaces, hoses, greenhouse benches, and the like
- The propagation media: rooting and growing mixes for cuttings, seedlings, and tissue culture plantlets
- The stock plant material: seeds, cutting material, scion, stock material for grafting, and tissue culture

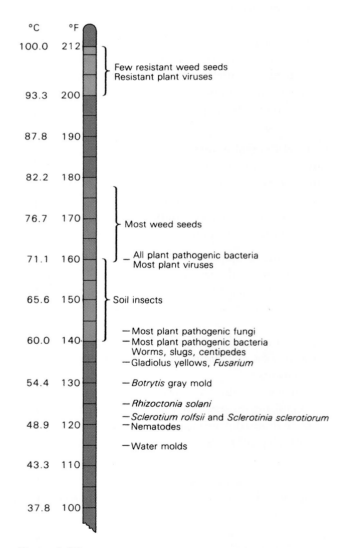

Figure 3–32
Soil temperatures required to kill weed seeds, insects, and various plant pathogens. Temperatures given are for 30 minutes under moist conditions.

If pathogens and other pests are suppressed in each of these areas, it is likely that the young plants can be propagated and grown to a salable size with minimal disease, insect, or mite infestations. Pathogenic fungi can best be controlled by using soilless mixes, pasteurizing propagation and growing mixes, considering general hygiene of the plants and facilities, avoiding overwatering, assuring good drainage of excess water, and using fungicides properly (81, 120).

Disinfection and Sanitation of Physical Propagation Facilities. Disinfection refers to the reduction of pathogens and algae, while sanitation refers to the level of cleanliness. The space where the actual propagation (making cuttings, planting seeds, grafting) takes place should be a light, very clean, cool room, completely separated from areas where the soil mixing, pot and flat storage, growing, and other operations take place. Traffic and visitors in this room should be kept to a minimum. At the end of each working day, all plant debris and soil should be cleaned out, the floors hosed down, and working surfaces washed with disinfectant solutions of sodium hypochlorite solution (Clorox), chlorine dioxide (Selectrocide), benzylkonium chloride (Physan 20, Green-shield), or pine disinfectant—diluted according to directions. Benzylkonium chloride is long-lasting and can be used for several days. Hydrogen dioxide (Zerotol, Oxidate) is a strong oxidizing agent used in sanitation of propagation facilities for the control of algae and pathogens (Fig. 3–33). Diluted household vinegar gives good control of algae and moss along walkways.

Flats and pots coming into this room should have been washed thoroughly and, if used previously, should be heat-treated or disinfected with chemicals (i.e., a 30-minute soak in sodium hypochlorite (Clorox) diluted one to nine). No dirty flats or pots should be allowed in the propagation area. Knives, shears, and

(a) (b) (c) (d)

(e)

Figure 3–33
Some common chemicals for disinfecting propagation facilities and propagules (a) Benzylkonium chloride, (b) hydrogen dioxide, (c) bromine and (d) diluted sodium hypochlorite solution (household bleach) can be used for (e) disinfesting both propagation facilities and propagules. Diluted household vinegar can control algae and moss along walkways. Always follow directions and try small trials first.

other equipment used in propagation should be sterilized periodically during the day by dipping in a disinfectant such as Physan or Zerotol.

Mist propagating and growing areas in greenhouses, cold frames, and lathhouses should be kept clean, and diseased and dead plant debris should be removed daily. Water to be used for misting should be free of pathogens. Water from ponds or reservoirs to be used for propagation purposes should be chlorinated to kill algae and pathogens. Proper chlorination will control *Phytophthora* and *Pythium* in irrigation water and can help reduce the cost of preventive fungicide programs (13, 20, 28).

Maintaining Clean Plant Material. In selecting propagating material, use only seed and those source plants that are disease- and insect-free. Some nurseries maintain stock plant blocks, which are kept meticulously "clean." However, stock plants of particularly disease-prone plants, such as *Euonymus,* might well be sprayed with a suitable fungicide several days before cuttings are taken. Drenches of fungicides and/or Agribrom (oxidizing biocide) are sometimes applied to stock plants in the greenhouse prior to selecting explants for tissue culture.

It is best to select cutting material from the upper portion of stock plants rather than from near the ground where the plant tissue could possibly be contaminated with soil pathogens. As cutting material is being collected, it should be placed in new plastic bags.

After the cuttings have been made and before sticking them in flats, they can be dipped in a dilute bleach solution, or treated with Zerotol, Agribrom, Physan 20, or various fungicides for broad-spectrum control of damping-off organisms—before any auxin treatment. One Oregon nursery disinfects Rhododendron cuttings with Consan, followed by washing in chlorinated water (46). Agri-strep (agricultural streptomycin) helps suppress bacterial problems, and one biological control, *Agrobacterium* spp., helps prevent crown gall of hardwood rose cuttings (31). However, once a cutting or seedling *becomes infected with a bacterium, there is no effective control* other than rouging-out and destroying the plant propagule.

Best Management Practices (BMP)

To a very limited degree, through some improper pesticide usage and inefficient irrigation and fertility systems, the nursery and greenhouse industries have been nonpoint source polluters of the environment. As a whole, the horticultural industries are good stewards of the environment. The environmentally friendly plants they produce are critical to the well-being, nutrition, and welfare of people, and are vital to enhancing the environment (reduced air and noise pollution, reduced heat loads around houses and urban areas, which lower utility cooling bills, adding O_2 to the air, and contributing to the abatement of current high global CO_2 conditions, etc.).

With the increased environmental regulations facing plant propagators and as an offshoot of integrated pest management programs, the development of Best Management Practices or BMP has occurred (61, 133). To help preserve the environment and head off additional state and federal regulation, BMP are being developed by the nursery industry, governmental agencies, and universities. Plans are for the nursery and greenhouse industries to self-regulate by adapting BMP, which many propagators have already been practicing for years. The above list of the ten best management practices applies to nursery propagation and liner production systems. To date, recycled water is generally not used to propagate plants (liners and container plants are irrigated with recycled water mixed with purer well or surface collected water), but, in the future with the scarcity of irrigation water and increased urban population pressure to use limited water supplies, more nurseries will have to develop propagation systems that utilize recycled water. Recycled water can present considerable challenges, since high salinity, trace levels of herbicides, pesticides, and pathogens such as *Phytophthora* can occur (Fig. 3–29).

BOX 3.13 GETTING MORE IN DEPTH ON THE SUBJECT
THE USE OF CHLORINE IN PROPAGATION

Chlorine can be used as a **sterilant,** which destroys all organisms, and as a **disinfectant,** which selectively destroys pests (70). When chlorine is used as a pesticide, it prevents pests from entering the propagation environment and minimizes the need for more toxic pesticides. Pest reduction or elimination is a cornerstone of IPM programs.

Chlorine, in the form of laundry bleach (Clorox, etc.), is one of the most affordable and readily available chemicals (36). Chlorine is used to sterilize greenhouse benches, floors, and other surfaces in the propagation area. Chlorination is being increasingly used in recycled irrigation water for controlling pathogenic fungi, algae, and other pests.

Chlorine is available as:

a. a gas (Cl_2), which is liquefied in pressurized metal containers and bubbled as a gas into water, but Cl_2 gas is very toxic and its corrosive nature makes it very hazardous to handle

b. calcium hypochlorite [$Ca(OCl)_2$] is used for domestic water treatment and is commercially available as granulated powder, large tablets, or liquid solutions; and

c. sodium hypochlorite (NaOCl), the active ingredient of household bleach, which is the most common form of chlorine used in propagation. When a continuous supply of chlorinated water is needed, concentrated solutions of sodium or calcium hypochlorite are injected. Chlorine injectors must be installed with an approved check-valve arrangement to prevent back flow into the fresh water system (13, 70). Bleach solutions are generally calculated as percent bleach or percent sodium hypochlorite; but these are not the same, since a **10 percent bleach solution** (which contains one part bleach to nine parts water) is 10 percent of 5.25 percent sodium hypochlorite or equivalent to **0.52 percent sodium hypochlorite**. Household bleach is commonly used as a disinfectant by diluting one part bleach to nine parts water.

Many chemicals, as well as organic residue from plants and propagation medium, react with chlorine and reduce its effectiveness. Enough chlorine must be added to produce an effective concentration of "free residual" chlorine (Fig. 3–34). Factors affecting chlorine activity include:

a. concentration—water treatment requires around 2 ppm free residual chlorine (20), and the bleaching of propagation benches and containers requires a 10 percent bleach solution or 5,250 ppm

b. exposure time

c. organic matter—contaminated water containing residual from soaking propagation containers or dipping propagules uses up available chlorine more rapidly than a clean solution

d. water temperature

e. pathogen growth stage—chlorine kills fungal mycelium on contact but is not systemic so fungal spores and pathogens embedded in roots and walls of Styrofoam containers are much more difficult to kill; soaking materials before treating with bleach allows spores to germinate and mycelia to grow, making pathogens easier to kill, and

f. a pH—around 6.5 is most effective, (70). At pH 6.0 to 7.5 total chlorine is predominately in the form of hypochlorous acid (strong sanitizer), whereas at pH 7.5 and above, hypoclorite is dominant, which is a weak, ineffective sanitizer.

For successful chlorination, clean the container, bed, and propagule materials prior to chlorinating, monitor the chlorine solution, and ventilate the work area. Dilute chlorine solutions irritate skin and chlorine vaporization irritates eyes, nose, and throat. It is important that propagation managers know the legal exposure limits (OSHA) that workers can be exposed to chlorine.

There are some environmental concerns about the use of bleach as a disinfectant to surface disinfect cuttings and for the sterilization of tools and propagation work surfaces. The hypochlorite ion from bleach attaches to organic compounds in the soil and forms very stable chlorinated organic compounds. These compounds can be taken up by plant roots, get into the food chain, and may bioaccumulate in the body fat of animals and humans (80). An alternative disinfectant for propagation is hydrogen peroxide (H_2O_2). It can be used as a sterilant for both fungi and bacteria, has no toxic by-products (it breaks down to water and oxygen), and it has no residual effect in water or soil. Hydrogen peroxide can be purchased in bulk form (35 percent concentration rate). A recommended rate for surface disinfestation of plant material is 1 part

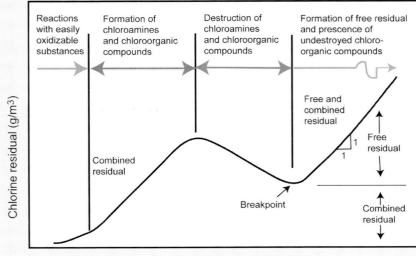

Figure 3–34
Many substances combine with chlorine to reduce its activity in solution, thus enough must be added to produce an effective concentration of "free residual" chlorine (70).

(Continued)

H₂O₂ (35 percent) to 100 parts water (80). Clorox (bleach) was found to be superior to hydrogen peroxide, Agrimycin 17 (agricultural streptomycin), or rubbing alcohol (isopropyl) in preventing the transmission of fire blight bacteria in pear trees (116).

Chlorine will continue to be used as an important disinfectant in propagation. Bleach is considerably cheaper than hydrogen peroxide, and with the dilute bleach solutions typically used in propagation, there should be little if any chlorine residual in tank solutions that are allowed to sit for several days (70). To be environmentally safe before discharging spent chlorinated water, test kits should be used to monitor residual chlorine levels, and local water quality officials can also be contacted.

POST-PROPAGATION CARE OF LINERS

Hardening-Off Liner Plants

Hardening-off or acclimating rooted propagules, seedlings, and tissue culture plantlets is critical for plant survival and growth. In commercial production, it assures a smooth transition and efficient turnover of plant product from propagation to liner production (Fig. 3–21) to finished plants in protected culture (greenhouses, etc.) or containerization and field production. This smooth transition and turnover of plant production units is essential in the marketing, sales, and profitability of plant manufacturing companies.

It is important to wean rooted cuttings from the mist system as quickly as possible (29). Reduction of irrigation and fertility in seedlings and plugs is done several weeks prior to shipping and/or transplanting to harden-off and ensure survival of the crop. Likewise, with acclimation of tissue culture-produced plantlets, light irradiance is increased and relative humidity is gradually reduced to stimulate the plantlet to increase photosynthetic rates and have better stomatal control. All of these ensure plant survival and a speedy transition when the acclimatized plant is shifted up and finished-off as a container or field crop. Acclimatization of liners is discussed in greater detail in Chapters 10 and 18.

Handling Container-Grown Plants

Irrigation Watering of container nursery stock is a major expense and environmental concern. In most operations, overhead sprinklers (i.e., Rainbird-type impact sprinklers) are used, although much runoff waste occurs. Watering of container plants by trickle, drip irrigation or low-volume emitters, results in less waste (121), and is becoming more widely used, particularly with plants produced in larger containers (Figs. 3–25 and 3–35). The development of solid-state soil tensiometers for the computer control of irrigation systems of containerized plants may help to increase water use efficiencies and decrease off-site pollution from runoff (17).

As part of Best Management Practices (BMP), many nurseries are switching to computer-controlled **cyclic** or **interval irrigation (pulse irrigation)** with impact sprinklers. Rather than manually turning on valves to run irrigation for 60 minutes, an environmental-control computer is programmed to precisely run the irrigation system three times daily at 5 to 10 minutes per cycle (123). Since most water is absorbed by the containers within the first 5 minutes, cyclic or pulse irrigation uses less water, greatly reduces water and fertility runoff, and lowers the amount of fertilizer needed in the fertigation system.

Flood floor systems for producing containerized plants and stock-plants for cuttings was discussed earlier in this chapter (Fig. 3–8).

BOX 3.14 GETTING MORE IN DEPTH ON THE SUBJECT
NURSERY BEST MANAGEMENT PRACTICES (BMP) (133)

- Collect runoff water when injecting fertilizer.
- Do not broadcast fertilizer on spaced containers.
- Do not top-dress fertilizer on containers prone to blow over.
- Water and fertilize according to plant needs.
- Group plants in a nursery according to water and fertilizer needs to minimize runoff.

- Monitor quantity of irrigation applied to prevent overwatering.
- Maintain minimal spacing between containers receiving overhead irrigation.
- Use low-volume irrigation for containers larger than 26 liter (7 gallon).
- Recycle runoff water.

Figure 3–35
Automatic watering system for container-grown plants. (a) Overhead sprinkler irrigation system for container crops. (b) Trickle irrigation can efficiently irrigate container plants with less water than overhead sprinkler irrigation systems. (c and d) Automated irrigation triggered by electronic eye (arrow) that turns on water as plants pass by on overhead conveyor system.

Fertilization Fertilizer solutions are usually injected into the irrigation system (fertigation) in commercial nurseries. Fertilizer may be supplied solely in controlled-release forms (Osmocote, Phycote, Nutricote, Polyon, etc.), or used in combination with fertigation (see page 53). After the container stock leaves the wholesale nursery, the retailer should maintain the stock with adequate irrigation until the plants have been purchased by the consumer. Controlled-release fertilizers added to containers leave a residual fertilizer supply (most retailers do not add supplementary fertilizer), and help maintain the plants until they are purchased by the consumer and planted in the landscape.

Root Development in Containerized Plants

When trees and shrubs from seedlings or rooted cuttings are grown in containers, roots often begin to circle on the outside of the rootball against the slick, smooth plastic container walls. If not mechanically controlled when the trees or shrubs are transplanted, circling roots may enlarge to the point of stressing or killing the plant by girdling (1). Internal walls of containers can be coated with copper compounds such as Spin Out, which is a latex-based paint containing copper hydroxide and a special formulated carrier (Figs. 3–23 and 3–25) that enhances root absorption of copper and temporarily inhibits root elongation (115), or containers can have special wall modifications as a means to reduce or prevent root circling during liner production and later container production. As shown in Figure 3–36, plants not properly air root-pruned or that are kept in containers too long will form an undesirable constricted root system from which they may never recover when planted in their permanent location. The plants should be shifted to larger containers before such "root spiraling" occurs.

The Ohio Production System (**OPS**), a system for rapidly producing container-grown shade trees (whips) in 1 year, compared with 3 years, also relies on copper-treated containers to control root growth. This eliminates or greatly reduces the need to root-prune when plants are upcanned to larger containers (114).

Using bottomless propagation and liner pots to **"air prune"** roots, judicious root pruning, early transplanting, and careful potting during the early transplanting stages can do much to encourage the development of a good root system by the time the young plant is ready

(a) (b)

Figure 3–36
One disadvantage of growing trees and shrubs in containers is the possibility of producing poorly shaped root systems. (a) Here a defective, twisted root system resulted from holding the young nursery tree too long in a container before transplanting. (b) Such spiraling roots retain this shape after planting and unacceptable tree growth occurs. This is avoided by proper root training, beginning with air-root pruning seed flats during propagation.

for transfer to its permanent location. Plastic containers with vertical grooves along the sides tend to prevent horizontal spiraling of the roots (Figs. 3–20 and 3–22).

Alternatives to Traditional Production Systems

Several in-ground alternatives to container production in the field and conventional field production of bare-root and B & B (balled-in-burlap) trees and shrubs have been developed, including (a) the **pot-in-pot system** (43), in which a container is inserted into an in-ground plastic sleeve container, and (b) **in-ground fabric containers (grow bags)** (see Fig. 3–25). Each of these methods can influence directional root development (1). The pot-in-pot, in-ground system involves sinking an outer or sleeve pot into the ground and inserting a second pot, which is the production pot that is harvested with the plant. The production container may have vertical ribs, or the interior walls are treated with copper to reduce root circling. The in-ground container system is a single container (unlike the pot-in-pot system) with rows of small holes along the container sides and bottom to enhance drainage.

In-ground fabric containers or grow bags are flexible, synthetic bags, which are filled with mineral soil and placed in predug holes in the field. The synthetic woven material of the bags limits most root penetration, and directs root growth to occur within the bag [more than 90 percent of the root system of conventional bare-root and balled and burlapped (B & B) plants are lost during digging]. Since the bag is placed in the ground, there is greater insulation of the root system against high and low temperatures (versus above-ground containerized crops), and the bag can be pulled out of the field, potentially reducing labor cost of traditional field techniques (101). This system does not work with all species, but has merits.

DISCUSSION ITEMS

1. What are some fundamental microclimatic and edaphic factors in the propagation environment?
2. How is light measured, and how is light manipulated in plant propagation?
3. Discuss the advantages and disadvantages of different types of plant propagation structures.
4. How does root zone heating save energy costs in propagation houses and enhance the rooting of cuttings?
5. Compare and contrast analog and computerized environmental controls of greenhouse propagation facilities.
6. What are some of the more popular covering materials for propagation houses?
7. What is closed-case propagation?
8. What kinds of containers are used for propagation and growing young liner plants?

9. Why is mineral soil rarely used in propagation and production of containerized plants?
10. Compare organic and inorganic media components used for propagation. What are peat-lite mixes?
11. How are pre-plant (preincorporated) and post-plant fertilization programs used in propagation and liner production systems?
12. How is salinity measured and controlled in irrigation water and container media used in propagation?
13. What are some potential problems in using recycled irrigation water for propagation?
14. How are accelerated growth techniques (AGT) used to enhance propagation?
15. Compare the broad definition of "pests" with insect pests.
16. What are "damping-off" pathogenic fungi? Give examples and indicate how they are disseminated.
17. How can integrated pest management (IPM) be utilized in propagation? Include the different areas of IMP and discuss the importance of the scouting system.
18. How are propagation equipment and facilities sanitized?
19. Why are best management practices (BMP) critical for environmental stewardship and the long-term profitability of the nursery industry?
20. What are some methods to "harden-off" liner plants during propagation and liner production?

REFERENCES

1. Appleton, B. L. 1994. Elimination of circling tree roots during nursery production. In G. A. Watson and D. Neely, eds. *The landscape below ground: Proceedings of an international workshop on tree root development in urban soils.* Savoy, IL: Intl. Soc. Arboricul.

2. Armstrong, M. 1992. Integrated disease management—research and development using new techniques and bioremediation at Vans pines. *Comb. Proc. Intl. Plant Prop. Soc.* 42:503.

3. Baker, J. R., and R. K. Jones. 1995. Pests of the south. *Amer. Nurser.* 181:78–89.

4. Ballester, A., M. C. San-Jose, N. Vidal, J. L. Fernández-Lorenzo, and A. M. Vieitez. 1999. Anatomical and biochemical events during in vitro rooting of microcuttings from juvenile and mature phases of chestnut. *Annals Bot.* 83:619–29.

5. Bartok, J. W. 1987. Chemical pressure treatment helps wood last longer. *Greenhouse Manag.* 6(2):151–52.

6. Bartok, J. W. 1988. Horizontal air flow. *Greenhouse Manag.* 6:197–212.

7. Bayles, E. 1988. Computers. *Greenhouse Manag.* 7:90–6.

8. Beytes, C. 1994. The sky's the limit with retractable roof greenhouses. *GrowerTalks* 58(4):22–6.

9. Beytes, C. 1995. Flood floor construction in 12 steps. *GrowerTalks* 58:40–6.

10. Beytes, C. 1998. Four levels of greenhouse climate control. *GrowerTalks* 61:38–48.

11. Beytes, C., ed. 2003. *The Ball red book: Crop production,* 17th ed. Vol. 2. Chicago, IL: Ball Publ.

12. Bilderback, T. 1993. Wetting agents and gels—where they have a purpose. *Comb. Proc. Intl. Plant Prop. Soc.* 43:421–23.

13. Black, R. 2008. Recycled irrigation water chlorination and pathogen prevention. *Comb. Proc. Intl. Plant Prop. Soc.* 58:508–11.

14. Bluhm, W. L. 1978. Peat, pests, and propagation. *Comb. Proc. Intl. Plant Prop. Soc.* 28:66–70.

15. Bunker, E. 1992. Water quality in propagation. *Comb. Proc. Intl. Plant Prop. Soc.* 42:81–4.

16. Bunt, A. C. 1988. *Media and mixes for container-grown plants,* 2nd ed. Modern potting composts. London: G. Allen & Unwin.

17. Burger, D. W. 1992. Water conserving irrigation systems. *Comb. Proc. Intl. Plant Prop. Soc.* 42:260–66.

18. Buzzo, R. J. 2007. Soil fumigation with metam sodium at lawyer nursery. *Comb. Proc. Intl. Plant Prop. Soc.* 57:553–56.

19. Cathey, H. M., and L. E. Campbell. 1980. Light and lighting systems for horticultural plants. *Hort. Rev.* 2:491–537.

20. Cayanan, D., P. Zhang, W. Liu, M. Dixon, and Y. Zheng. 2009. Efficacy of chlorine in controlling five common plant pathogens. *HortScience* 44:157–63.

21. Chong, C., R. A. Cline, and D. L. Rinker. 1988. Spent mushroom compost and papermill sludge as soil amendments for containerized nursery crops. *Comb. Proc. Intl. Plant Prop. Soc.* 37:347–53.

22. Collins, H. 1995. Fighting fire ants. *GrowerTalks* 59:48.

23. Connor, D. 1977. Propagation at Monrovia Nursery Company: Sanitation. *Comb. Proc. Intl. Plant Prop. Soc.* 27:102–6.

24. Coyier, D. L. 1978. Pathogens associated with peat moss used for propagation. *Comb. Proc. Intl. Plant Prop. Soc.* 28:70–2.

25. Cuny, H. 1995. Fungi lend a hand: Rooting out mycorrhizae's place in the nursery. *Nurs. Manag.* 10:45–9.

26. Dallon, J., Jr. 1988. Effects of spent mushroom compost on the production of greenhouse-grown crops. *Comb. Proc. Intl. Plant Prop. Soc.* 37:323–29.

27. Daughtrey, M. 2007. Propagation pathology: The basics. *Comb Proc. Intl. Plant Prop. Soc.* 57:293–95.

28. Daughtry, B. 1988. Control of *Phytophthora* and *Pythium* by chlorination of irrigation water. *Comb. Proc. Intl. Plant Prop. Soc.* 38:420–22.

29. Davies, F. T. 2005. Optimizing the water relations of cuttings during propagation. *Comb. Proc. Intl. Plant Prop. Soc.* 55:585–92.

30. Davies, F. T., Jr. 1988. The influence of nutrition and carbohydrates on rooting of cuttings. *Comb. Proc. Intl. Plant Prop. Soc.* 38:432–37.

31. Davies, F. T., Jr. 1991. Back to the basics in propagation. *Comb. Proc. Intl. Plant Prop. Soc.* 41:338–42.

32. Davies, F. T., Jr. 2008. Opportunities from down under—How mycorrhizal fungi can benefit nursery propagation and production systems. *Comb. Proc. Intl. Plant Prop. Soc.* 58:539–48.

33. Davies, F. T., Jr., J. A. Saraiva Grossi, L. Carpio, and A. A. Estrada-Luna. 2000. Colonization and growth effects of the mycorrhizal fungus *Glomus intraradicies* in a commercial nursery container production system. *J. Environ. Hort.* 18(4):247–51.

34. Davies, F. T., Jr., and C. A. Call. 1990. Mycorrhizae, survival and growth of selected woody plant species in lignite overburden in Texas. *Agric. Ecosystems Environ.* 31:243–52.

35. Davies, F. T., Jr., D. E. Kester, and T. D. Davis. 1994. Commercial importance of adventitious rooting: Horticulture. In T. D. Davis and B. E. Haissig, eds. *Biology of adventitious root formation.* New York: Plenum Press.

36. De Fossard, R. A. 1992. Treatment of plants with hypochlorite solutions. *Comb. Proc. Intl. Plant Prop. Soc.* 42:65–7.

37. Drahn, S. R. 2007. Propagating with controlled-release fertilizers. *Comb. Proc. Intl. Plant Prop. Soc.* 57:521–22.

38. Dysart, J. 1994. Minor use chemical loss likely—but not inevitable. *GrowerTalks* 58(4):94–100.

39. Edmonds, J. 1994. Europeans embrace new container and tunnel technologies. *Amer. Nurs.* 179(12):23–4.

40. Egilla, J. N., F. T. Davies, Jr., and M. C. Drew. 2000. Effect of potassium on drought resistance of *Hibiscus rosa-sinensis* cv. Leprechaun: Plant growth, leaf macro- and micronutrient content and root longevity. *Plant Soil* 16:1–12.

41. Falland, J. 1987. Greenhouses in plant propagation: A historical perspective. *Comb. Proc. Intl. Plant Prop. Soc.* 36:158–64.

42. Fessler, T. R. 1995. The propagation environment with a computer. *Comb. Proc. Intl. Plant Prop. Soc.* 45:320–21.

43. Fidler, A. R. 1999. Pot-in-pot system: A container-grown in the ground approach for diverse crops. *Comb. Proc. Intl. Plant Prop. Soc.* 49:297–99.

44. Fonteno, W. C. 1988. Know your media: The air, water, and container connection. *GrowerTalks* 51:110–11.

45. Geistlinger, L. 1994. The Worker Protection Standard. *Amer. Nurs.* 179:46–57.

46. George, C. 1993. Rhododendron propagation—methods and techniques carried out in the Pacific northwest of the USA. *Comb. Proc. Intl. Plant Prop. Soc.* 43:178–82.

47. Germing, G. H., ed. 1986. Symposium on greenhouse climate and its control. *Acta Hort.* 174:1–563.

48. Gill, S. 1995. Early detection key to aphid control. *GrowerTalks* 59(2):94.

49. Gilliam, C. H., and E. M. Smith. 1980. How and when to fertilize container nursery stock. *Amer. Nurs.* 151(2):7, 117–27.

50. Gordon, I. 1988. Structures used in Australia for plant propagation. *Comb. Proc. Intl. Plant Prop. Soc.* 37:482–89.

51. Gordon, I. 1992. A review of materials for propagation media. *Comb. Proc. Intl. Plant Prop. Soc.* 42:85–90.

52. Gough, N. 1992. Prospects for IPM in greenhouse ornamentals in Australia. *Comb. Proc. Intl. Plant Prop. Soc.* 42:103–7.

53. Gouin, F. R. 1989. Composted sewage sludge: An aid in plant propagation. *Comb. Proc. Intl. Plant Prop. Soc.* 39:489–93.

54. Hamrick, D. 1988. The covering choice. *GrowerTalks* 51(12):64–74.

55. Hamrick, D., ed. 2003. *Ball RedBook: Greenhouses and Equipment.* 17th ed. Chicago, IL: Ball Publ.

56. Hanan, J. 1987. A climate control system for greenhouse research. *HortScience* 22:704–8.

57. Harden, J. 1992. Plant protection—management of pest control techniques. *Comb. Proc. Intl. Plant Prop. Soc.* 42:99–102.

58. Herve, A. J. 1989. Production of spray chrysanthemums in a hydroponic system. *Comb. Proc. Intl. Plant Prop. Soc.* 36:66–70.

59. Hicklenton, P. R., and A. M. Armitage. 1988. *CO_2 enrichment in the greenhouse.* Portland, OR: Timber Press.

60. Hildebrandt, C. A. 1987. Economical propagation structures for the small grower. *Comb. Proc. Intl. Plant Prop. Soc.* 36:506–10.

61. Hottovy, S. A. 1994. Monrovia nursery's response to new environmental restrictions *Comb. Proc. Intl. Plant Prop. Soc.* 44:161–64.

62. Humphrey, C. 1995. Basics of greenhouse design and construction. *GrowerTalks* 58:60–4.

63. James, B. L. 1993. Update on fungicides. *Comb. Proc. Intl. Plant Prop. Soc.* 43:373–75.

64. Jensen, M. H. 1977. Energy alternative and conservation for greenhouses. *HortScience* 12:14–24.

65. Jeong, B. R., K. Fujiwara, and T. Kozai 1993. Carbon dioxide enrichment in autotrophic micropropagation: Methods and advantages. *HortTechnology.* 3(3):332–34.

66. Johnson, C. R., and D. F. Hamilton. 1977. Effects of media and controlled-release fertilizers on rooting and leaf nutrient composition of *Juniperus conferta* and *Ligustrum japonicum* cuttings. *J. Amer. Soc. Hort. Sci.* 102:320–22.

67. Kelly, M. H. 1995. Get a grip on the greenhouse climate with automated curtains. *GrowerTalks* 59(2):76–83.

68. Konjoian, P. 2006. Selectrocide chlorine dioxide as a new product for the control of algae and other microbial pests in greenhouse irrigation systems. *Comb. Proc. Intl. Plant Prop. Soc.* 56:342–53.

69. Landis, T. D. 1993. Using "limiting factors" to design and manage propagation environments. *Comb. Proc. Intl. Plant Prop. Soc.* 43:213–18.

70. Landis, T. D. 1994. Using chlorine to prevent nursery diseases. *Forest Nurs. Notes.* For. Serv., USDA, Pacific N. W. Region, RG-CP-TP-08-94.

71. Landis, T. D., R. W. Tinus, S. E. McDonald, and J. P. Barnett. 1990. Containers and growing media. *Container Tree Manual.* Vol. 2. Agri. Handbk. 674. Washington, DC: For. Serv., USDA.

72. Landis, T .D., R. W. Tinus, S. E. McDonald, and J. P. Barnett. 1992. Atmospheric environment. *Container Tree Manual.* Vol. 3. Agri. Handbk. 674. Washington, DC: For. Serv., USDA.

73. Linderman, R. G. 1993. Effects of biocontrol agents on plant growth. *Comb. Proc. Intl. Plant Prop. Soc.* 43:249–52.

74. Loach, K. 1989. Controlling environmental conditions to improve adventitious rooting. In B. E. Haissig, T. D. Davis, and N. Sankhla, eds. *Adventitious root formation in cuttings.* Portland, OR: Dioscorides Press.

75. MacDonald, A. B. 1986. Propagation facilities—past and present. *Comb. Proc. Intl. Plant Prop. Soc.* 35:170–75.

76. MacKenzie, A. J., T. W. Starmann, and M. T. Windham. 1995. Enhanced root and shoot growth of chrysanthemum cuttings propagated with the fungus *Trichoderma harzianum.* *HortScience* 30:496–98.

77. Marcenaro, S., C. Voyiatzi, and B. Lercari. 1994. Photocontrol of *in vitro* bud regeneration: A comparative study of the interaction between light and IAA in a wild type and an aurea mutant of *Lycopersicon esculentum. Physiol. Plant.* 91:329–33.

78. Mathers, H. 1999. Effect of water quality on plant propagation. *Comb. Proc. Intl. Plant Prop. Soc.* 49:535–40.

79. Maynard, B. K. 1993. Basics of propagation by cuttings: Light. *Comb. Proc. Intl. Plant Prop. Soc.* 43:445–49.

80. McClelland, M. T., and M. A. L. Smith. 1993. Alternative methods for sterilization and cutting disinfestation. *Comb. Proc. Intl. Plant Prop. Soc.* 43:526–30.

81. McCully, A. J., and M. B. Thomas. 1977. Soilborne diseases and their role in propagation. *Comb. Proc. Intl. Plant Prop. Soc.* 27:339–50.

82. Miller, N. 1981. Bogs, bales, and BTU'S: A primer on peat. *Horticulture* 49:38–45.

83. Monk, G. J., and J. M. Molnar. 1987. Energy efficient greenhouses. *Hort. Rev.* 9:1–52.

84. Moody, E. H., Sr. 1983. Sanitation: A deliberate, essential exercise in plant disease control. *Comb. Proc. Intl. Plant Prop. Soc.* 33:608–13.

85. Moore, G. 1988. Perlite: Start to finish. *Comb. Proc. Intl. Plant Prop. Soc.* 37:48–52.

86. Morrow, R. C. 2008. LED lighting in horticulture. *HortScience* 43:1947–50.

87. Mortensen, L. M. 1987. Review: CO_2 enrichment in greenhouses—crop responses. *Scientia. Hort.* 33:1–25.

88. Munro, D. 1989. Virus testing of perennial propagating stock. *Comb. Proc. Intl. Plant Prop. Soc.* 39:43–7.

89. Nelson, P. V. 2009. *Greenhouse Operation and Management*. Englewood Cliffs, NJ: Prentice Hall.

90. O'Donnell, K. 1988. Polycarbonates gain as growers seek high performance coverings. *GrowerTalks* 51:60–2.

91. Ogdon, R. J., F. A. Pokorny, and M. G. Dunavent. 1987. Elemental status of pine bark-based potting media. *Hort. Rev.* 9:103–31.

92. Orndorff, C. 1983. Constructing and maintaining disease-free propagation structures. *Comb. Proc. Intl. Plant Prop. Soc.* 32:599–605.

93. Parrella, M. P. 1992. An overview of integrated pest management for plant propagation. *Comb. Proc. Intl. Plant Prop. Soc.* 42:242–45.

94. Porter, M. A., and B. Grodzinski. 1985. CO_2 enrichment of protected crops. *Hort. Rev.* 7:345–98.

95. Pscheidt, J. W. 1991. Diagnosis of *Phytophthora* using ELISA test kits. *Comb. Proc. Intl. Plant Prop. Soc.* 41:2514.

96. Pyle, K. 1991. Fungus control in the post-benlate era. *GrowerTalks* 55:19–21.

97. Rajapakse, N. C., R. E. Young, M. J. McMahon, and R. Oi. 1999. Plant height control by photoselective filters: Current status and future prospects. *HortTechnology*. 9:618–24.

98. Reardon, J. 1994. Motivating plant growth with your heating system. *Comb. Proc. Intl. Plant Prop. Soc.* 44:364–66.

99. Sabalka, D. 1987. Propagation media for flats and for direct sticking: What works? *Comb. Proc. Intl. Plant Prop. Soc.* 36:409–13.

100. Scagel, C. F., K. Reddy, and J. M. Armstrong. 2003. Mycorrhizal fungi in rooting substrate influences the quantity and quality of roots on stem cuttings of Hick's yew. *HortTechnology* 13:62–6.

101. Self, R. 1977. Winter protection of nursery plants. *Comb. Proc. Intl. Plant Prop. Soc.* 77:303–7.

102. Self, R. 1978. Pine bark in potting mixes: Grades and age, disease and fertility problems. *Comb. Proc. Intl. Plant Prop. Soc.* 28:363–68.

103. Sherry, W. J. 1986. Greenhouse covering materials: Optical, thermal, and physical properties. *GrowerTalks* 49(12):53–8.

104. Sherry, W. J. 1988. Technology in the greenhouse: State of the industry, a survey report. *GrowerTalks* 51(9):48–57.

105. Shibata, A., T. Fujita, and S. Maeda. 1980. Nutricote coated fertilizers processed with polyolefin resins. *Acta Hort.* 99:179–86.

106. Skiminia, C. A. 1992. Recycling water, nutrients and waste in the nursery industry. *HortScience* 27:968–72.

107. Smith, E. M. 1980. How and when to fertilize container nursery stock. *Amer. Nurs.* 15:365–68.

108. Soffer, H., and D. W. Burger. 1989. Plant propagation using an aero-hydroponics system. *HortScience* 24:154.

109. Spooner-Hart, R. N. 1988. Integrated pest management with reference to plant propagation. *Comb. Proc. Intl. Plant Prop. Soc.* 38:119–25.

110. St. John, T. V., and J. M. Evans. 1990. Mycorrhizal inoculation of container plants. *Comb. Proc. Intl. Plant Prop. Soc.* 40:222–32.

111. Stephens, W. 2007. Colored shade cloth and plant growth. *Comb. Proc. Intl. Plant Prop. Soc.* 57:212–17.

112. Stewart, N. 1986. Production of bark for composts. *Comb. Proc. Intl. Plant Prop. Soc.* 35:454–58.

113. Street, C. 1994. Propagation of wetland species. *Comb. Proc. Intl. Plant Prop. Soc.* 44: 468–73.

114. Struve, D. K. 1990. Container production of hard-to-find or hard-to-transplant species. *Comb. Proc. Intl. Plant Prop. Soc.* 40:608–12.

115. Struve, D. K., M. A. Arnold, R. Beeson Jr., J. M. Ruter, S. Svenson, and W. T. Witte. 1994. The copper connection: The benefits of growing woody ornamentals in copper-treated containers. *Amer. Nurs.* 179:52–61.

116. Teviotdale, B. L., M. F. Wiley, and D. H. Harper. 1991. How disinfectants compare in pre-venting transmission of fire blight. *Calif. Agri.* 45:21–3.

117. Thimijan, R. W., and R. D. Heins. 1993. Photometric, radiometric, and quantum units of measure: A review of procedures for interconversion. *HortScience* 18:818–32.

118. Van Belle, B. 2004. Efficiencies from rolling bench propagation. *Comb Proc. Intl. Plant Prop. Soc.* 54:465–71.

119. Vollebregt, R. 1990. Analysis of greenhouse curtain systems for shading, cooling and heat retention. *Comb. Proc. Intl. Plant Prop. Soc.* 40:166–76.

120. Ward, J. 1980. An approach to the control of *Phytophthora cinnamomi*. *Comb. Proc. Intl. Plant Prop. Soc.* 30:230–37.

121. Weatherspoon, D. M., and C. C. Harrell. 1980. Evaluation of drip irrigation for container production of woody landscape plants. *HortScience* 15:488–89.

122. White, J. W. 1979. Energy efficient growing structures for controlled environment agriculture. *Hort. Rev.* 1:141–71.

123. Whitten, M. 1995. Water-tight irrigation. *Nurs. Manag.* 10(6):45–7.

124. Wilkerson, D. C. 1993. Improve your bottom line with innovative bench systems. *GrowerTalks* 57(1):23–31.

125. Wilkerson, E. G., R. S. Gates, S. Zolnier, S. T. Kester, and R. L. Geneve. 2005a. Transpiration capacity in poinsettia cuttings at different rooting stages and the development of a cutting coefficient for scheduling mist. *J. Amer. Soc. Hort. Sci.* 130:295–301.

126. Wilkerson, E. G., R. S. Gates, S. Zolnier, S. T. Kester, and R. L. Geneve. 2005b. Prediciting rooting stages in poinsettia cuttings using root zone temperature-based models. *J. Amer. Soc. Hort. Sci.* 130:302–7.

127. Williams, D. J. 1980. How slow-release fertilizers work. *Amer. Nurs.* 151(6):90–7.

128. Wilson, G. C. S. 1980. Symposium on substrates in horticulture other than soils in situ. *Acta Hort.* 99:1–248.

129. Wood, J. S. 1985. Sun frame propagation. *Comb. Proc. Intl. Plant Prop. Soc.* 34:306–11.

130. Worrall, R. 1982. High temperature release characteristics of resin-coated slow release fertilizers. *Comb. Proc. Intl. Plant Prop. Soc.* 31:176–81.

131. Wright, R. D. 1987. *The Virginia Tech liquid fertilizer system for container-grown plants.* Inform. Ser. #86-5. Blacksburg, VA: College of Agriculture and Life Sciences, Virginia Tech University.

132. Wright, R. D., and A. X. Niemiera. 1987. Nutrition of container-grown woody nursery crops. *Hort. Rev.* 9:75–101.

133. Yeager, T., T. Bilderback, D. Fare, C. Gilliam, J. Lea-Cox, A. Niemiera, J. Ruter, K. Tilt, S. Warren, T. Whitwell, and R. Wright. 2007. *Best management practices: Guide for producing nursery crops,* 2nd ed. Marietta, GA: South. Nurser. Assoc.

134. Yeager, T. H. 1989. Developing an efficient fertilizer program for container plants. *Woody Ornamentalist* 14:1–3.

part two
Seed Propagation

CHAPTER 4 **Seed Development**
CHAPTER 5 **Principles and Practices of Seed Selection**
CHAPTER 6 **Techniques of Seed Production and Handling**
CHAPTER 7 **Principles of Propagation from Seeds**
CHAPTER 8 **Techniques of Propagation by Seed**

During much of human existence, special kinds of crops, referred to as **landraces**, were maintained by farmers who kept a portion of each year's seed to produce the crops for the following year. These landraces received local names and represented some of our most important agricultural crops coming from Asia (rice, millet, soybean, many vegetables), southwest Asia (wheat, barley, oats, rye), Africa (rice, sorghum, watermelon), and the Americas (corn, squash, beans, pepper, potato, sunflower, cotton, tobacco).

Modern agriculture (agronomy, horticulture, and forestry) relies on seeds and seedlings to produce most of the world's food and fiber resources. Great advances have been made in the past century that permit seed companies to provide high quality seeds with superior genetics. Public and private plant breeders use the principles and practices of genetic research to breed new seedling cultivars that have superior growth characteristics, crop yields, pest resistance, and nutrition. Seed companies maintain germplasm for parental seed stocks and are responsible for production, storage, and distribution of seeds to producers. Millions of kilograms (pounds) of seeds are produced each year for use by propagators worldwide. This section deals with all aspects of the seed industry including genetic selection, seed production, and germination.

*4
Seed Development

learning objectives

- Trace the origin of seeds.
- Follow the relationship between flower parts and seed parts.
- Explain the general parts of a seed.
- Describe the stages of seed development.
- Explain unusual types of seed formation.
- Observe how plant hormones are important to seed development.
- Describe ripening and dissemination of fruits and seeds.

INTRODUCTION

Four hundred million years ago, plants moved out of the oceans to colonize land. Two major adaptations made this possible. The first was the evolution of the root. The root not only anchored the plant in soil but also allowed the plant to obtain water and minerals no longer brought to the plant by ocean water. A second adaptation that increased a plant's success on land was the development of a vascular system. This allowed materials obtained by the root system to be efficiently transported to the leafy photosynthetic parts of the plant. However, the price of these adaptations to land habitation was relative immobility. The first vascular plants (e.g., ferns) used spores to spread the result of sexual reproduction. However, plants that used spores to reproduce required a wet environment to allow male sperm to swim to fertilize the female egg. The development of the seed habit (dispersal of seeds rather than spores for reproduction) permitted plants to move away from perpetually wet environments and colonize areas with drier climates. This initiated the proliferation of the marvelous diversity found in seeds and their accompanying fruit structures. Seed-producing plants (especially angiosperms) became incredibly successful, and it is estimated that there are currently over 250,000 species of flowering plants, easily the most diverse group found in the plant kingdom.

Propagation by seeds is the major method by which plants reproduce in nature, and one of the most efficient and widely used propagation methods for cultivated crops. Plants produced from seeds are referred to as **seedlings.** Sowing seeds is the physical beginning of seedling propagation. The seed itself, however, is the end product of a process of growth and development within the parent plant, which is described in this chapter.

REPRODUCTIVE LIFE CYCLES OF VASCULAR PLANTS

Plant life cycles are characterized by alternate sporophytic and gametophytic generations. The sporophyte is usually plant-like in appearance with a diploid genetic composition. The sporophyte produces specialized reproductive structures that facilitate gamete production through meiosis. This initiates the gametophytic generation. Male and female gametes have a haploid genetic composition, and fusion of these gametes **(fertilization)** results in a reproductive zygote (embryo) that restarts the sporophytic generation.

fertilization The sexual union of a male and female gamete.

Vascular plants are separated into those that disseminate the next generation by spores or those who do so with seeds.

Seedless Vascular Plants

Seedless vascular plants reproduce from spores, and include horsetails (*Equisetum*), wiskferns (*Psilotum*), lycopods (*Lycopodium*), *Selaginella*, and ferns. The spore is a protective structure that is tolerant of environmental conditions, germinating when conditions are conducive for the gametophytic generation (usually wet conditions). The life cycle of a fern is depicted in Figure 4–1. Spores are produced in sporangia within a sorus produced on the underside or edge of the fern frond (sporophyte). The spore (1n) germinates and produces a small leafy structure called a prothallus. On the mature prothallus, male (antheridia) and female (archegonia) are formed. The male antheridium releases the motile sperm (1n) that swims into the archegonium uniting with a single egg cell (1n). Following fertilization, the zygote develops into a new fern.

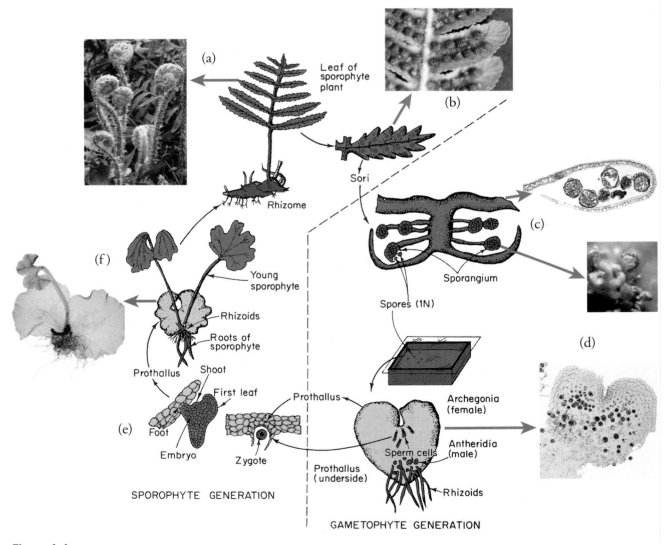

Figure 4–1
A representative fern life cycle includes alternate sporophytic and gametophytic generations. (a) A mature fern sporophyte produces fronds that typically produce (b) sori (spore producing structures) on the underside of the leaf-like frond. (c) Within the sori are sporangia that contain the spores that initiate the gametophytic gerneration. (d) When the spore germinates it produces a leaf-like gametophyte called the prothallus. On the prothallus, several female archegonia and many male antheridia are formed. (e) Fertilization occurs when the male sperm unites with the female egg within the archegonium. (f) The resultant young sporophyte becomes the long-lived fern. Adapted from Linda R. Berg. 1997. *Introductory Botany*. Saunders College Publishing.

Seed Plants

The seed habit developed during the Devonian period about 350 to 385 million years ago in an extinct group of plants called the progymnosperms (47). Progymnosperms are only known from the fossil record (Fig. 4–2) and produced seedlike structures enclosed in female tissue called cupules. They are considered the progenitors to our current-day gymnosperms and angiosperms. The seed habit is characterized by several anatomical features that differentiate them from spore-producing plants:

1. Rather than producing a single spore type (homospory), seed plants produce a separate female megaspore and male microspore (heterospory).
2. The female gametophyte is retained on the mother plant (sporophyte) and is enclosed within a protective maternal seed coat.
3. The ovule has an opening designed to receive pollen that does not depend on water for male gamete transfer.

Seed plants are separated into gymnosperms and angiosperms. Gymnosperms include the cycads, ginkgo, gnetophytes (*Ephedra*, *Gnetum*) and the conifers (like pine, fir, and hemlock). The term gymnosperm means "naked seeds" and refers to the absence of ovary tissue covering the seeds, which is a characteristic of angiosperms (flowering plants). Pine is representative of a gymnosperm life cycle (Fig. 4–3). Conifers produce separate male and female reproductive cones (strobili) on the same plant. Male cones produce winged pollen that is dispersed by wind. Egg cells are produced within the female megagametophyte located between the scales of the female cones. Haploid male and female gametes fuse to form a diploid zygote that develops into the embryo within the seed. Storage tissue (**endosperm**) in a gymnosperm seed is from the haploid female gametophyte.

> **endosperm** The major storage tissue in seeds. It is derived from the haploid female gametophyte in gymnosperms, while in angiosperms it is the result of gamete fusion that forms a triploid (3n) storage tissue.

Angiosperms are true flowering plants. The term angiosperm means "enclosed seeds" and refers to the female ovary tissue (carpels) that forms the fruit surrounding angiosperm seeds. Angiosperms are the dominant plant type on Earth with approximately 250,000 species, compared with only about 8,000 living species of gymnosperms. One reason for angiosperm success and diversity is the mutualistic co-evolution of animals (especially insects) as pollinators and seed dispersers. A representative angiosperm life cycle is presented in Figure 4–4 (page 114). A key development in the angiospermic life cycle is the presentation of the female megagametophyte as a multi-celled (usually 8) embryo sac within the ovule. Male gametes from the pollen enter the ovule. One gamete fuses with the egg cell to form a zygote, and the second fuses with two polar nuclei to form the endosperm. This double fertilization is a characteristic of angiosperms and leads to a triploid endosperm rather than the haploid endosperm seen in gymnosperms. Based on seedling morphology, angiosperms can be separated into **dicotyledonous** (seedlings with two cotyledons) and **monocotledonous** (seedlings with one cotyledon) plants.

> **dicots** Produce seedlings with two cotyledons.
>
> **monocots** Produce only a single modified cotyledon.

CHARACTERISTICS OF A SEED

A **seed** (20, 21) is a matured ovule containing an embryo, storage reserve tissue, and a protective outer covering (Figs. 4–5, 4–6, page 115). Seeds are the sexual reproductive unit in a plant.

Embryo

The **embryo** represents the new plant generation and develops after the sexual union of the male and female gametes during fertilization. Its basic structure is an embryo axis with growing points at each end—one for

Figure 4–2
Seed-producing plants evolved approximately 360 million years ago, but most were not successful and became extinct. Progymnosperms developed seeds enclosed within a cupule (arrow) and are thought to be the progenitors of the gymnosperms.

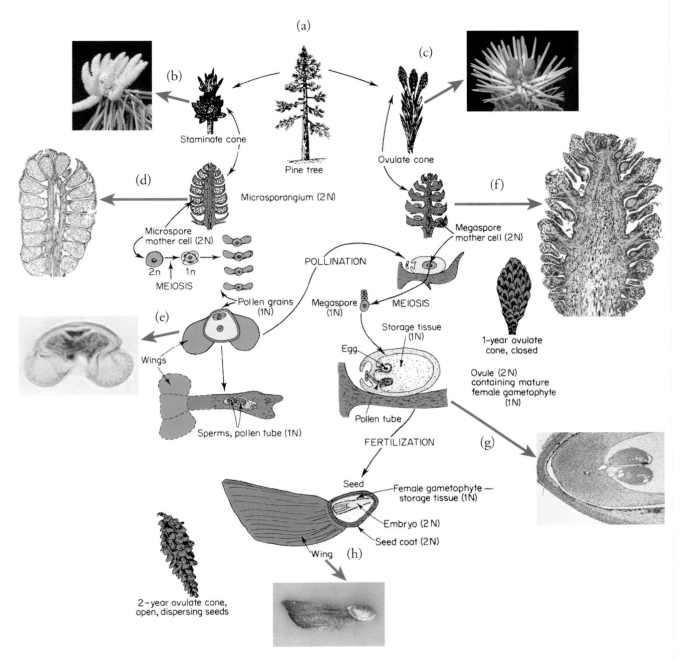

Figure 4–3
A representative gymnosperm life cycle. (a) A pine tree is a mature sporophyte. It produces separate male (b) and female (c) reproductive structures. The male gametophytes are produced in a (d) staminate cone as winged pollen grains (e) spread by the wind. The female gametophyte is produced within the female ovulate cone (f). The female egg cell (g) is fertilized by the male sperm to produce a seed (h)—the next sporophytic generation.

the shoot and one for the root—and one or more cotyledons attached to the embryo axis. The basic embryo types relative to the seed's storage tissue is represented in Figure 4–5.

The number of cotyledons in the embryo is used to classify plants. **Monocotyledonous** plants (such as coconut palm or grasses) have a single cotyledon, **dicotyledonous** plants (such as bean or peach) have two, and gymnosperms (such as pine or ginkgo) may have as many as fifteen. Embryo size in relation to the seed varies considerably (3, 48). In many seeds, the embryo occupies the entire inner seed (Figs. 4–5d, 4–6e), while others have small to miniature embryos (Figs. 4–5c, 4–6c).

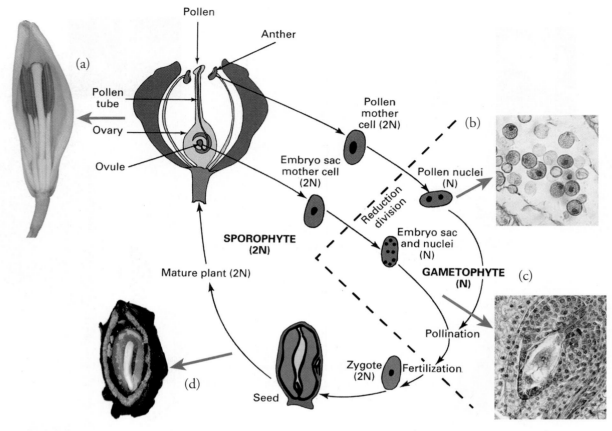

Figure 4–4
A representative angiosperm life cycle. (a) Flowers are formed during the sporophytic generation. In the gametophytic generation, (b) male gametophytes are produced within the anther as pollen grains and (c) the female gametophyte is produced in the ovule within the ovary. (d) The seed is formed following male and female gamete fusion (fertilization), which reinitiates the sporophytic generation.

Storage Reserves

storage and food reserves High-energy macromolecules like oils, carbohydrates, and protein that are produced during seed development and used for the early stages of seed germination and seedling emergence.

Storage tissue is designed to sustain the germinating embryo until the seedling can produce its own resources through photosynthesis. For dicots, storage materials are contained in the endosperm, cotyledons, and perisperm tissue. The endosperm is usually the result of the fusion of two female and one male nuclei during double fertilization and is triploid (3n). However, in some plants, the endosperm ploidy level may be higher (e.g., five-ploid in some members of the lily family and nine-ploid in peperomia). Storage tissue for monocots is the starchy endosperm (3n), and for gymnosperms, the storage tissue is an endosperm consisting of haploid (1n) female gametophytic tissue (7). **Perisperm** is nucellar tissue from the female plant and is diploid (2n).

Seeds can be separated into three basic storage reserve types that occur in **endospermic, non-endospermic,** or **unclassified** seeds (Table 4–1 and Figs. 4–5, 4–6).

In **endospermic seeds,** cotyledon growth is arrested in dicots at different stages of development such that the embryo may be only one-third to one-half the size of the seed at the time it is ripe. The remainder of the seed cavity contains large amounts of endosperm or perisperm depending on the species. Although the origin of the endosperm tissue is different, most monocot and gymnosperm seeds are endospermic.

The pattern for reserve metabolism in **non-endospermic** dicot seeds begins with an initial rapid

perisperm Nucellus tissue that remains in the mature seed and is used as storage tissue.

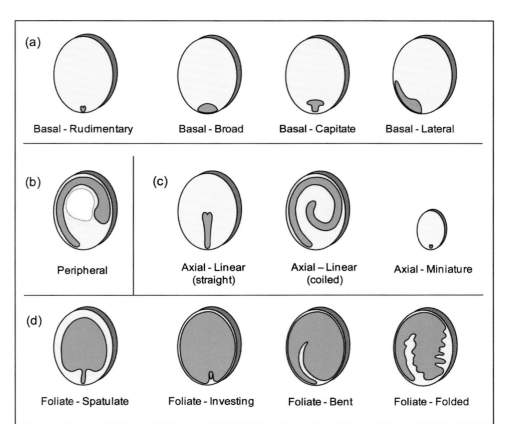

Figure 4–5
The basic embryo types found in seeds. Major forms include: (a) Basal embryos that have a high endosperm to embryo ratio. This is considered a more primitive evolutionary condition; (b) Peripheral embryos surround and inner mass of perisperm storage tissue; (c) Axial embryos occupy the center of the seed and contain a significant amount of endosperm; and (d) Foliate embryos where the cotyledons develop to occupy most of the seed and function as storage reserve tissue. Color codes for these images have the embryo in green, endosperm in yellow, perisperm in white, and seed coverings are brown. Adapted from Martin, A. C. 1946.

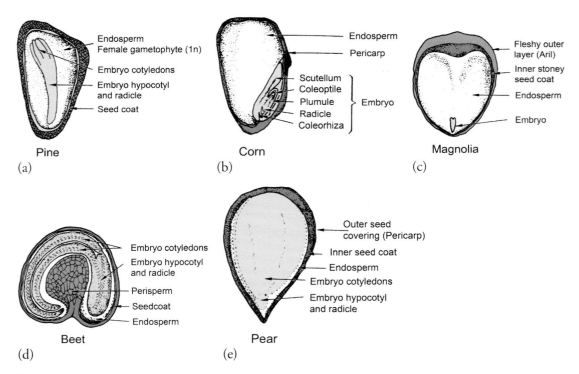

Figure 4–6
Representative seed morphologies. (a) Gymnosperm (conifer) seeds have embryos with multiple cotyledons and use the female gametophyte as reserve material. (b) Corn is an example of a monocot in the grass family. It has a peripheral embryo and a large endosperm reserve. The outer protective layer is fruit tissue—pericarp. (c, d, and e) Each of the representative dicots has embryos with two cotyledons. Magnolia has a small embryo and a large endosperm reserve. The fleshy outer covering is an aril derived from the funiculus. Beet seeds have a curved embryo and utilize perisperm derived from nucellar tissue. In pear, the cotyledons fill the seed and are used for storage reserve. The nutritive reserves in the endosperm have been transferred to the cotyledons, so there is only a small remnant endosperm between the embryo and seed coat. The outer layer is fruit (pericarp) tissue.

> **Table 4–1**
> **CLASSIFICATION OF SEEDS**
>
> The following classification is based upon morphology of embryo and seed coverings. It includes, as examples, families of herbaceous plants.
>
> I. Seeds with dominant endosperm (or perisperm) as seed storage organs (endospermic).
>   A. Rudimentary embryo. Embryo is very small and undeveloped but undergoes further increase at germination (see Fig. 4–5a, 4–6c Magnolia).
>   1. Ranunculaceae (*Aquilegia, Delphinium*), Papaveraceae (*Eschscholtzia, Papaver*), Fumariaceae (*Dicentra*), Araliaceae (*Fatsia*), Magnoliaceae (*Magnolia*), Aquifoliaceae (*Ilex*).
>   B. Linear embryo. Embryo is more developed than those in (A) and enlarges further at germination (Fig. 4–5c).
>   1. Apiaceae (*Daucus*), Ericaceae (*Calluna, Rhododendron*), Primulaceae (*Cyclamen, Primula*), Gentianaceae (*Gentiana*), Solanaceae (*Datura, Solanum*), Oleaceae (*Fraxinus*).
>   C. Miniature embryo. Embryo fills more than half the seed (Fig. 4–4c).
>   1. Crassulaceae (*Sedum, Heuchera, Hypericum*), Begoniaceae (*Begonia*), Solanaceae (*Nicotiana, Petunia, Salpiglossis*), Scrophulariaceae (*Antirrhinum, Linaria, Mimulus, Nemesia, Penstemon*), Lobeliaceae (*Lobelia*).
>   D. Peripheral embryo. Embryo encloses endosperm or perisperm tissue (Fig. 4–4b).
>   1. Polygonaceae (*Eriogonum*), Chenopodiaceae (*Kochia*), Amaranthaceae (*Amaranthus, Celosia, Gomphrena*), Nyctaginaceae (*Abronia, Mirabilis*).
> II. Seeds with embryo dominant (nonendospermic); classified according to the type of seed covering (Fig. 4–4d).
>   A. Hard seed coats restricting water entry.
>   1. Fabaceae (*Cercis, Gymnocladus, Gleditsia*), Geraniaceae (*Pelargonium*), Anacardiaceae (*Rhus*), Rhamnaceae (*Ceanothus*), Malvaceae (*Abutilon, Altea*), Convolvulaceae (*Convolvulus*).
>   B. Thin seed coats with mucilaginous layer.
>   1. Brassicaceae (*Arabis, Iberis, Lobularia, Mathiola*), Linaceae (*Linum*), Violaceae (*Viola*), Lamiaceae (*Lavandula*).
>   C. Woody outer seed coverings with inner semipermeable layer.
>   1. Rosaceae (*Geum, Potentilla*), Zygophyllaceae (*Larrea*), Balsaminaceae (*Impatiens*), Cistaceae (*Cistus, Helianthemum*), Onagraceae (*Clarkia, Oenothera*), Plumbaginaceae (*Armeria*), Apocynaceae, Polemoniaceae (*Phlox*), Hydrophyllaceae (*Nemophila, Phacelia*), Boraginaceae (*Anchusa*), Verbenaceae (*Lantana, Verbena*), Labiateae (*Coleus, Moluccela*), Dipsacaceae (*Dipsacus, Scabiosa*).
>   D. Fibrous outer seed covering with more or less semipermeable membranous layer, including endosperm remnant.
>   1. Asteraceae (many species).
> III. Unclassified
>   A. Rudimentary embryo with no food storage.
>   1. Orchidaceae (orchids, in general).
>   B. Modified miniature embryo located on periphery of seed (Fig. 4–6b).
>   1. Poaceae (grasses).
>   C. Axillary miniature embryo surrounded by gametophyte tissue (Fig. 4–6a).
>   1. Gymnosperms (in particular, conifers).
>
> Source: After Atwater (1).

nucellus Maternal tissue in which the megaspore mother cell (also called megasporocyte) undergoes meiosis and forms the embryo sac.

growth of the embryo that digests the enclosing **nucellus**. This is followed by expansion of the embryo through cell division at the periphery of the cotyledons that digests the developed endosperm. In these seeds, the endosperm and/or the nucellus is reduced to a remnant between the embryo and the seed coat (integuments), and the cotyledons function as the major storage tissue. Although the reduced endosperm may be only few cell layers thick, it can still play an important role in controlling seed germination and dormancy (see Chapter 7).

The third storage reserve type occurs in **unclassified seeds.** These are seeds that have negligible seed storage reserves like orchids. These tiny seeds rely on a fungal **(mycorrhiza)** symbiosis during germination to provide the nutrition required for development and growth (63).

Protective Seed Coverings

The **protective seed covering layer** surrounds the seed and provides physical protection; it may act to exclude water and gases. Seed coverings may consist of the seed coat, the remains of the nucellus and endosperm, and, sometimes, parts of the fruit. The seed coat, also termed the testa, is derived from the integuments of the ovule. During development, the seed coat becomes modified so that at maturity it presents an appearance often characteristic of the plant family (18). Usually, the outer layer of the seed coat becomes dry, somewhat hardened and thickened, and brownish in color. In particular families, it becomes hard and impervious to water. On the other hand, the inner seed coat layers are usually thin, transparent, and membranous. Remnants of the endosperm and nucellus are sometimes found within the inner seed coat, sometimes making a distinct, continuous layer around the embryo.

In some plants, parts of the fruit remain attached to the seed so that the fruit and seed are commonly handled together as the "seed." In fruits such as achenes, caryopsis, samaras, and schizocarps, the pericarp and seed coat layers are contiguous (Fig. 4–7a). In others, such as the acorn, the pericarp and seed coverings separate, but the fruit covering is indehiscent. In still others, such as the "pit" of stone fruits (Fig. 4–7b) or the shell of walnuts, the covering is a hardened portion of the pericarp, but it is dehiscent (splits along an existing suture line) and usually can be removed without much difficulty.

Figure 4–7
Fruit structures included as the "seed" unit. (a) Sunflower "seeds" actually include the entire fruit, called an achene. (b) Plum is an example of a pome (stone fruit) where the inner part of the fruit (endocarp) adheres to the seed and usually part of the seed unit.

Seed coverings provide mechanical protection for the embryo, making it possible to handle seeds without injury, and, thus, permitting transportation for long distances and storage for long periods of time. The seed coverings can also contribute to seed dormancy and control germination, as discussed in Chapter 7.

Seeds may also contain additional surface structures that usually aid in seed dispersal. These include arils, elaiosomes, caruncles, wings, and various plumes of hairs (Fig. 4–8, page 118). Elaiosomes are particularly interesting because they are nutrient-containing organs (especially oils) specifically designed to attract ants (4). The ants use the elaiosome as a food source, and the plant benefits by ant dispersal of the seeds.

REPRODUCTIVE PARTS OF THE FLOWER

Sexual reproduction (fusion of male and female gametes) occurs in the flower. The sexual cycle of plant reproduction starts with meiotic cell divisions that halve the number of chromosomes in male pollen cells and female cells in the embryo sac. Meiosis is described in detail in Chapter 2.

Pollen Development (Microsporogenesis)

Male gametes are formed in the pollen grains (microspores) that are produced within the stamen of the flower (Fig. 4–9, page 118). Pollen or microspore mother cells located within the stamen divide meiotically to form tetrads (four haploid microspores). These are surrounded by a nutritive cell layer called the tapetum. The exine is the outer pollen layer that provides protection for the pollen grain. The exine tends to be smooth in wind-pollinated plants and rough or spiked in insect-pollinated plants. A mature pollen grain typically is two or three celled—one or two generative cells and a tube cell (Fig. 4–9). The tube cell functions during pollen tube growth and the two generative cells are involved in fertilization.

Ovule Development (Megasporogenesis)

The ovule begins development within the nucellus of the female cones (gymnosperms) or flower (angiosperms) (Figs. 4–10 and 4–11, page 119). The nucellus is surrounded by one or two integuments that grow to eventually cover the nucellus. A megaspore mother cell is initiated in the nucellus that divides and begins meiosis. There are four linear nuclei formed at the end of meiosis. Only one nucleus survives to divide

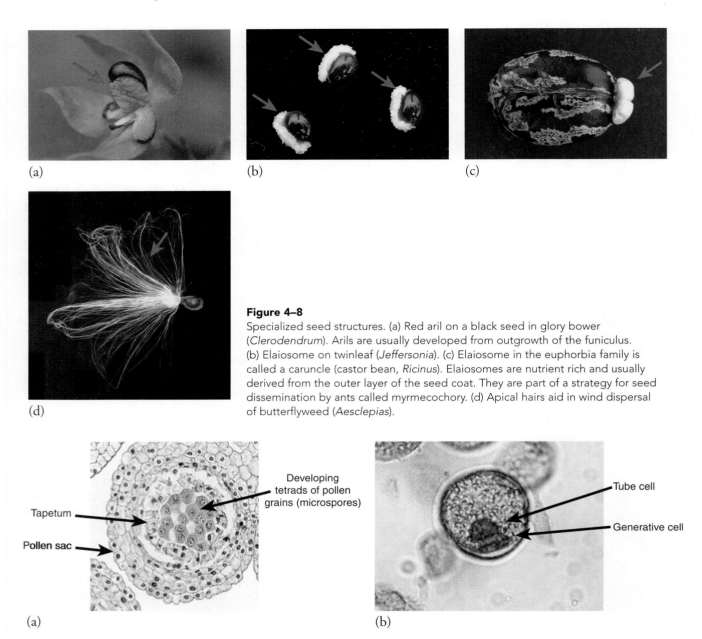

Figure 4–8
Specialized seed structures. (a) Red aril on a black seed in glory bower (*Clerodendrum*). Arils are usually developed from outgrowth of the funiculus. (b) Elaiosome on twinleaf (*Jeffersonia*). (c) Elaiosome in the euphorbia family is called a caruncle (castor bean, *Ricinus*). Elaiosomes are nutrient rich and usually derived from the outer layer of the seed coat. They are part of a strategy for seed dissemination by ants called myrmecochory. (d) Apical hairs aid in wind dispersal of butterflyweed (*Asclepias*).

Figure 4–9
Pollen development in a typical angiosperm. (a) Within the pollen sac, meiotic divisions give rise to the male gametes contained within a pollen grain. The tapetum is a nutritive layer of cells enclosing the pollen grains. (b) Mature pollen grain containing a tube and generative cell.

and form the archegonia in gymnosperms or the contents of the embryo sac in angiosperms. In angiosperms, the most common arrangement of cells in the embryo sac is called the *Polygonum* type and occurs in about two-thirds of flowering plants (Fig. 4–12, page 120). This type of embryo sac has seven cells (eight nuclei) that occupy specific locations that dictate their function (72). These cells include the egg apparatus consisting of a single egg and two synergid cells located at the micropylar end of the embryo sac, three antipodal cells at the opposite end of the embryo sac, and the central cell with two polar nuclei.

RELATIONSHIP BETWEEN FLOWER AND SEED PARTS

The initiation of seed formation generally requires two processes—**pollination** and **fertilization.** Pollination is the transfer of pollen within

pollination The transfer of male pollen to the female stigma.

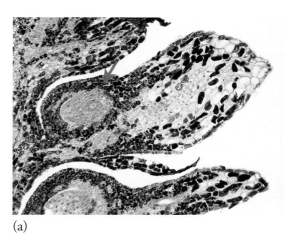
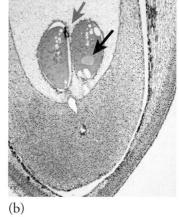

Figure 4–10
Development of the female gametophyte in a representative gymnosperm (pine). (a) The megaspore mother cell (arrow) develops in the female nucellar tissue. (b) Two archegonia (red arrow) form, each containing a female egg cell (black arrow).

a single flower (self-pollination) or from separate flowers (cross-pollination) to a receptive stigma. Pollen is transferred to the stigma by a variety of means including wind, insects, and, in some cases, mammals. The basic parts of an angiosperm flower are illustrated in Figure 4–13 (page 124). The pollen grain interacts with a receptive stigma and germinates. A pollen tube grows down specialized cells in the style called **transmitting cells** toward the embryo sac. The pollen tube contains three nuclei: one **tube nucleus** and two **generative nuclei** (Fig. 4–14, page 121). The tube nucleus acts to guide the pollen tube, while the generative nuclei will eventually fuse with female egg cells. The pollen tube enters the **micropyle** (a natural opening between the **integuments**) releasing the generative nuclei into the embryo sac.

Fertilization is the fusion of haploid (1n) male and female gametes inside the ovule. In gymnosperms, there is a single fertilization between the

transmitting cells Specialized cells in the style that conduct the pollen tube to the ovule.

micropyle An opening between the integuments through which the pollen tube enters the ovule.

integuments Two layers of cells that develop between the nucellus and embryo sac and become the seed coat.

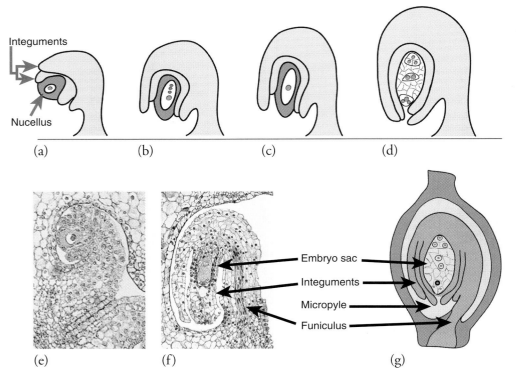

Figure 4–11
Development of the embryo sac in a representative angiosperm (lily). (a) The megaspore mother cell develops in the flower's nucellar tissue. (b) Meiosis results in one viable and three degenerative nuclei. (c and e) Progenitor nucleus for the embryo sac. (d, f, and g) Embryo sac within the ovule bounded by the integuments and attached to the ovary by the funiculus. It is common for the ovule to turn during development. The orientation illustrated is the most common form, called anatropous.

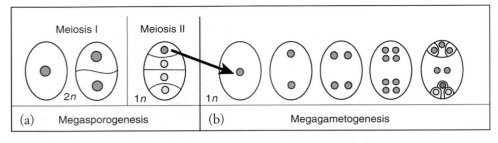

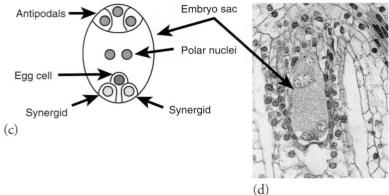

Figure 4–12
Development of the most common form of embryo sac (*Polygonum* type). (a) Initially a mother spore or mother cell develops in the nucellar tissue of the flower. Four haploid cells are formed during meiosis, but only one is retained. (b) It then divides to form the cell in the embryo sac. Each cell has a distinct role. (c, d) Three become antipodals, one is the central cell with two polar nuclei, two become synergids, and one becomes the egg cell.

sperm and egg cells. In angiosperms, double fertilization occurs.

Double fertilization (5) occurs when one generative nucleus fuses with the egg cell to form the **zygote** (2n embryo), while the second generative nucleus fuses with the central cell and its two polar nuclei to form the 3n **endosperm** (Fig. 4–15). The female synergid cells are closely associated with the egg cell and function to attract and guide male nuclei to the egg cell for fertilization (31). Synergids produce a chemical that attracts the pollen tube to the micropyle, arrests its growth, and ensures the proper release of the sperm cells into the ovule. Evidence suggests that the central cell signals the

zygote The result of sexual reproduction, which forms the embryo.

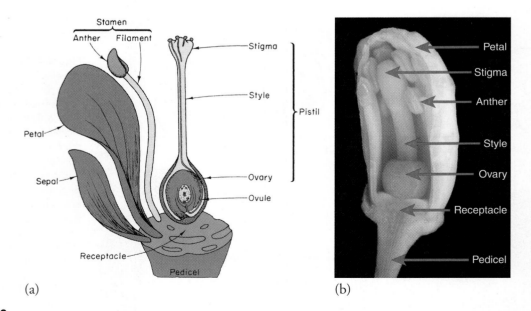

Figure 4–13
In a typical angiosperm flower, floral organs are produced in separate whorls. The outermost whorl are the sepals (caylx), the next are the petals (corolla), inside the petals are the male stamens, and innermost is the female pistil. Pollination occurs with the transfer of pollen from the stamens to the stigma of the pistil. The pollen grain germinates and the pollen tube grows down the style. Eventually, the pollen tube enters the ovule through the micropyle and deposits two male sperm cells. Fertilization involves the fusion of the male and female cells in the embryo sac.

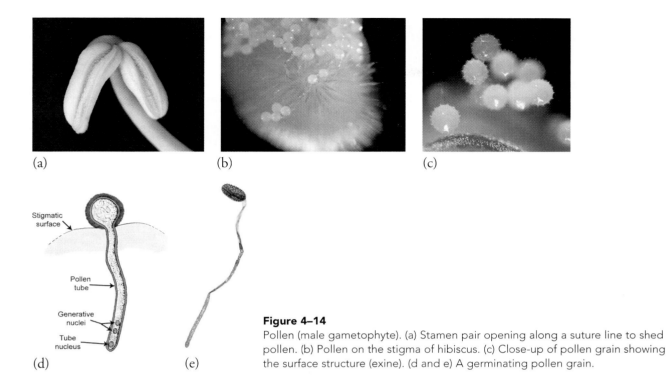

Figure 4–14
Pollen (male gametophyte). (a) Stamen pair opening along a suture line to shed pollen. (b) Pollen on the stigma of hibiscus. (c) Close-up of pollen grain showing the surface structure (exine). (d and e) A germinating pollen grain.

synergids to release the chemical attractant. The synergid cell degenerates soon after sperm cell release, permitting sperm cell access to the egg cell for fertilization and release of the second sperm cell to migrate to the central cell. The exact function of antipodal cells is not completely understood, but they disintegrate soon after fertilization of the egg cell.

The relationship between flower tissue and subsequent parts of the fruit and seed for a typical angiosperm species is outlined as follows:

1. Ovary grows into fruit tissue.
2. Ovule becomes the mature seed.
3. Embryo sac is the inner part of the seed.
4. Polar nuclei plus a generative nucleus become the endosperm.
5. Egg cell fuses with one generative nucleus to form embryo.
6. Integuments form the layers of the seed coat (also called testa).

Fertilization in gymnosperms differs from angiosperms because they do not produce elaborate flower parts. There is no true stigma in gymnosperms. Rather, there is either a stigmatic surface on the opening of the **ovule** or a sugary pollination

> **ovule** Develops in the nucellus and is enclosed by the integuments.

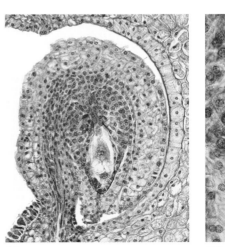
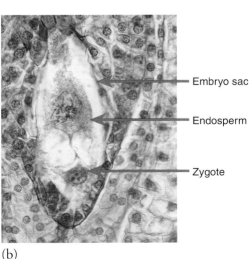

Figure 4–15
Double fertilization in lily. One sperm nucleus fuses with the egg cell to form the zygote and the other male nucleus fuses with the polar nuclei to form the triploid endosperm. (a) Shows the embryo sac within the developing ovule. (b) Is a close up of the embryo sac showing the onset of cell division following double fertilization.

> **BOX 4.1 GETTING MORE IN DEPTH ON THE SUBJECT**
> **PLOIDY LEVELS IN PLANTS**
>
> With many angiosperms, the zygote is diploid (2n) and divides to become the embryo; the endosperm is triploid (3n) and develops into nutritive tissue for the developing embryo. Terminology for ploidy levels in plants can be confusing. Ploidy indicates the number of sets of chromosomes in a plant. Not all plant species are diploid. Several important crop plants, like potato, are tetraploid or even octaploid like strawberry. However, the product of normal meiosis is still to produce gametes with half the original number of chromosomes. Therefore, current terminology for a tetraploid species is 2n = 4x, where "x" is the number of pairs of chromosomes. For example, in potato the ploidy level of the diploid would be expressed as 2n = 2x = 24, while the tetraploid would be 2n = 4x = 48. This means that there are 24 chromosomes in the diploid and 48 in the tetraploid.
>
> Crosses between species may fail to produce viable seed because the species have different ploidy levels. Failure of the endosperm to develop properly can also result in retardation or arrest of embryo development, and embryo abortion can result. This phenomenon is called somatoplastic sterility and commonly occurs when two genetically different individuals are hybridized, either from different species (15, 16, 17) or from two individuals of different ploidy constitution. It can be a barrier to hybridization in angiosperms but not in gymnosperms (62), since the "endosperm" in these plants is haploid female gametophytic tissue. Embryos that show some growth from these types of crosses can be "rescued" by isolating these embryos and placing them in tissue culture. Details on embryo rescue are found in Chapter 17.

drop exudes from the ovule to collect wind-borne pollen (62). In some species, like *Ginkgo*, the male gametes can be motile, but, in most cases, the pollination droplet pulls the pollen into the ovule and a pollen tube is formed. Double fertilization does not occur in gymnosperms. However, the gymnosperm, *Ephedra*, (in the Gnetophte group, which is possibly the progenitor line for the angiosperms) has a form of double fertilization, but no endosperm results from the second fertilization. Only angiosperms produce a true triploid (3n) endosperm. In gymnosperms, haploid female gametophyte tissue surrounds the developing embryo and performs the function of the endosperm.

STAGES OF SEED DEVELOPMENT

Three physiological stages of development are recognized in most seeds (Fig. 4–16). These include **histodifferentiation, cell expansion** (food reserve deposits), and **maturation drying.** Figure 4–17 shows the relative growth and development in lettuce seed (fruit), showing the physiological stages of seed development and days post-pollination.

Stage I Histodifferentiation (Embryo Differentiation)

Stage I is characterized by the differentiation of the embryo and endosperm mostly due to cell division. In Stage I, the embryo reaches the beginning of the cotyledon stage of development. There is rapid increase in both fresh and dry weight. There are characteristic stages of embryogenesis that occur during Stage I and these are distinct for dicots, monocots, and gymnosperms.

Embryo Differentiation in Dicots Although there are several variations on the types of angiosperm

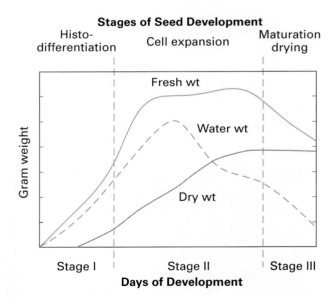

Figure 4–16
The stages of seed development. The stages include histodifferentiation (rapid increase in seed size due predominantly to cell division), cell expansion (largest increase in seed size for deposition of food reserves), and maturation drying (dramatic loss in seed fresh weight due to water loss). Redrawn from Bewley and Black, 1994.

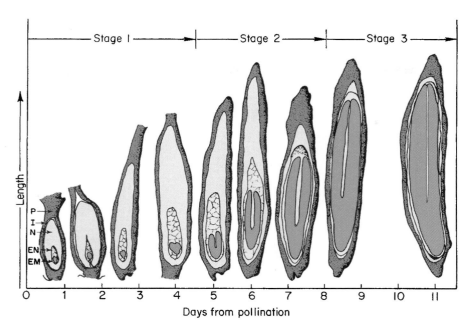

Figure 4–17
Growth and development of the fruit and seed in lettuce showing the relative changes in seed size during the three stages of seed development. P, pericarp; I, integuments; N, nucellus; EN, endosperm; EM, embryo. Redrawn from Jones 1927.

embryogenesis (8, 51, 59, 72), embryo formation in shepherd's purse (*Capsella bursa-pastoris*) has served as a good model for dicot embryogenesis and is very similar to *Arabidopsis*. Embryogenesis in dicots proceeds through the characteristic stages of development. These include the **proembryo, globular, heart, torpedo,** and **cotyledon** stages (Fig. 4–18, page 124).

proembryo The earliest stages of embryo development before the embryo and suspensor become easily recognized.

Following fertilization of the egg and sperm nuclei, a proembryo is initiated by a transverse cell division to form an apical and basal cell (Figure 4–18a–c). The basal cell forms the suspensor, while the apical cell forms the embryo. The suspensor in dicots is usually a column of single or multiple cells. The suspensor functions to push the proembryo into the embryo sac cavity and to absorb and transmit nutrients to the proembryo. The embryo is supplied with nutrients for growth via the suspensor until later stages of embryo development when the embryo is nourished by material from the endosperm. There is also hormone signaling between the suspensor and embryo. In shepherd's purse, basal cell derivatives in the globular embryo form the hypophysis that goes on to develop into the radicle (Fig. 4–18d). Tissue differentiation becomes evident in the sixteen-celled globular embryo (Fig. 4–18d–f). An outer layer of cells (protoderm) will develop into epidermal cells of the embryo. The inner cell layers will develop into the procambium and ground meristem.

As the embryo enters the cotyledon stage, the cotyledon primordia are evident in the heart-shaped stage of embryogenesis (Fig. 4–18g–i). These primordia elongate to give a typical torpedo stage embryo (Fig. 4–18g). In the torpedo stage, the embryo has organized to form an apical meristem, radicle, cotyledons, and hypocotyl. The endosperm has been developing along with the embryo and providing nutrition for its growth. When the embryo reaches the mature stage (Fig. 4–18j–l) in shepherd's purse, the major storage tissue is the cotyledons, which now occupy most of the seed cavity.

Embryo Differentiation in Monocots Monocots have a more complex embryo structure in the mature seed compared with dicots, but early embryo development is similar (60). Embryogenesis in monocots includes the **proembryo, globular, scutellar,** and **coleoptilar** stages (Fig. 4–19, page 125).

Following fertilization, an apical and basal cell is visible in corn (*Zea mays*) that initiates the proembryo stage (Fig. 4–19a). The proembryo and globular stages are similar to dicots, except that the suspensor is not a single or double row of cells and is less differentiated (Fig. 4–19b). In the late globular stage, the outer epidermal layer is evident and a group of cells on one side of the proembryo divides more rapidly; these will give rise to the embryo axis.

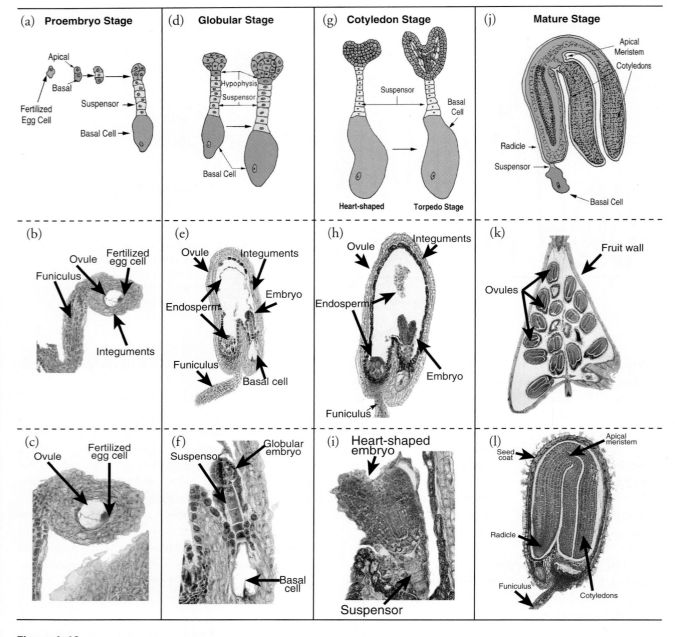

Figure 4–18
Embryo development in a typical dicot (shepherd's purse) showing the proembryo (a–c), globular (d–f), cotyledon (g–i), and mature (j–l) stages. See text for detailed description of each stage.

The remnant of the cotyledon can be seen in the scutellar stage of development. Monocots have reduced the pair of cotyledons represented in dicot embryos to a single modified cotyledon termed the **scutellum** (Fig. 4–9c). The scutellum acts as conductive tissue between the endosperm and embryo axis (Fig. 4–9d–e).

Finally, the embryo axis differentiates into the plumule (shoot) and radicle in the coleoptilar stage (Fig. 4–9d). In monocots, the embryo axis also has a specialized tissue surrounding the shoot and root tissue to aid in emergence during germination. These are the **coleoptile** and **coleorhiza,** respectively (Fig. 4–9d–e).

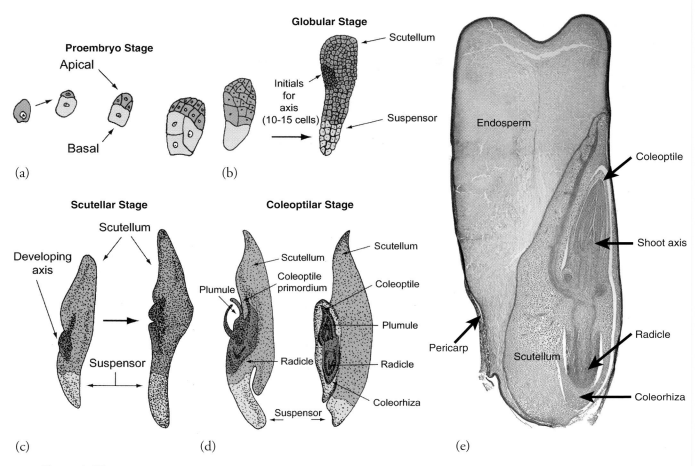

Figure 4–19
Embryo development in a typical monocot (corn). See text for description of figure. (e) Cross section of a mature seed of corn showing basic anatomical features.

Embryo Differentiation in Gymnosperms Compared with the more evolutionarily advanced angiosperms, embryo formation in gymnosperms (62) differs in several important ways (Fig. 4–20, page 126). Most conspicuous is that seeds of gymnosperms are not contained within a carpel or ovary (fruit). The term *gymnosperm* means "naked seeded." Only a single fertilization occurs in gymnosperms (Fig. 4–20a). Therefore, there is also no true triploid endosperm in gymnosperms. Rather, the developing embryo is nourished by haploid female gametophyte tissue also referred to as an endosperm (Fig. 4–20e). Pollination and fertilization may be separated by months (up to 12 months in pine), and seed formation can take two seasons in some species. The pollen tube germinates soon after pollination but must wait for the female gametophyte to complete development before fertilization can proceed. After fertilization, several embryos begin development within a single gymnosperm seed but rarely does more than one of these embryos mature.

In pine (*Pinus* sp.), the fertilized egg cell divides to form a free nuclear stage without cell walls between nuclei (Fig. 4–20b). Following cell wall formation, cells organize to form an embryo tier of cells and a suspensor tier (Fig. 4–20c). The suspensor differentiates into a set of primary suspensor cells (rosette cells) and embryonal suspensor tubes. The suspensor cells elongate and there are several cleavage events to give multiple embryos (polyembryos) inside a single seed (Fig. 4–20d). Usually, only one of these embryos continues to develop. The proembryo differentiates an epidermal layer (Fig. 4–20d) prior to the cotyledon primordia becoming evident. The mature pine embryo has multiple (usually eight) cotyledons compared to two or one in the dicots and monocots (Fig. 4–20d).

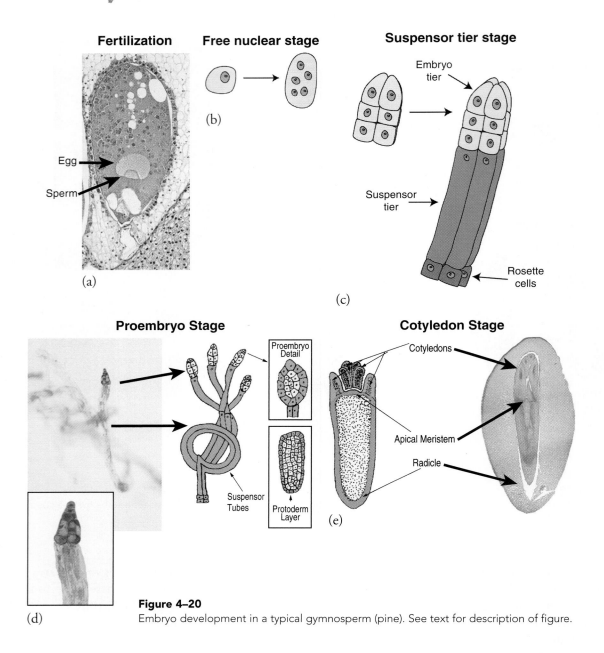

Figure 4–20
Embryo development in a typical gymnosperm (pine). See text for description of figure.

Stage II Cell Expansion

Stage II is a period of rapid cell enlargement—often called seed filling—due to the accumulation of food reserves (Fig. 4–16). This is an active period with large increases in DNA, RNA, and protein synthesis in the seed (7). The major food reserves include carbohydrates (starch), storage proteins, and lipids (oils or fats). Although different species may predominantly store a particular food reserve (i.e., cereal grains store starch, legumes store protein, and sunflower stores oil), most seeds contain all three types of food reserves (Table 4–2). Such substances not only provide essential energy substrates to ensure survival of the germinating seedling, but also provide essential food for humans and animals.

Food reserves are manufactured in the developing seed from photosynthate being "loaded" or moved into the seed from the mother plant. The process of seed reserve accumulation requires the translocation of small molecular weight compounds, such as sucrose, asparagine, glutamine, and minerals, into the seed. In dicot seeds, there is a direct vascular connection (phloem, xylem) between the mother plant and the seed through the funiculus (Fig. 4–21).

Table 4–2
FOOD RESERVES FOUND IN VARIOUS PLANT SPECIES

Species	Average percent composition			Major storage organ
	Protein	Oils	Starch	
Cereals	10–13%	2–8%	66–80%	Endosperm
Oil palm	9%	49%	28%	Endosperm
Legumes	23–37%	1–48%	12–56%	Cotyledons
Rape seed	21%	48%	19%	Cotyledons
Pine	35%	48%	6%	Female gametophyte

Source: From (7, 18, 62).

A vascular strand usually runs through the funiculus and down one side of the integuments (seed coat), allowing transfer of photosynthate and water into the developing seed (30). There is no direct vascular connection from the seed coat to the nucellus, endosperm, or embryo, and assimilates must reach the embryo by diffusion (75). Most viruses and large complex molecules are effectively screened from the embryo in this process, but may accumulate in the outer layers of the seed. There is no vascular connection between the mother plant and developing seed in monocots. Rather, there is a group of cells at the seed and mother plant interface called transfer cells that facilitate the passage of photosynthate into the endosperm (61).

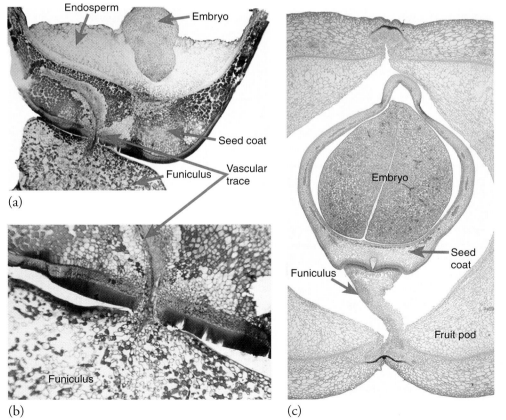

Figure 4–21
(a) Longitudinal section through a developing ovule of eastern redbud (*Cercis canadensis*) about 57 days post-anthesis (pollen shedding) showing the vascular connection between the funiculus and the ovule. (b) Close-up of the vascular trace. Note typical xylem cells in the vascular trace. (c) Bean seed with funiculus attached to the pod.
From Jones and Geneve (36).

BOX 4.2 GETTING MORE IN DEPTH ON THE SUBJECT
GENE EXPRESSION DURING SEED FILLING (69)

Specific mRNAs are required for the synthesis of storage compounds (7, 26, 70). The pattern of mRNA for storage protein accumulation is similar for a number of proteins and mRNAs including phaseolin, legumin, and vicilin in legumes; cruciferin in rape seed; and zein and hordein in cereals. A typical pattern for storage protein accumulation is illustrated in Figure 4–22a for broad bean (*Vicia faba*). This increase in storage protein is coincident with the increase in dry weight accumulation in Stage II embryos.

Figure 4–22b shows the increase in mRNA that precedes the accumulation of the storage protein, cruciferin, in rape seed (*Brassica napus*) (24).

Very specific genes are "turned on" during this stage of embryo growth (26, 67). These genes are only expressed during the embryogenesis stage of a plant's life cycle. The mRNAs for storage proteins are no longer translated after maturation drying and cannot be detected in germinating seeds.

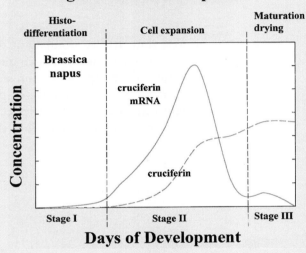
(a)

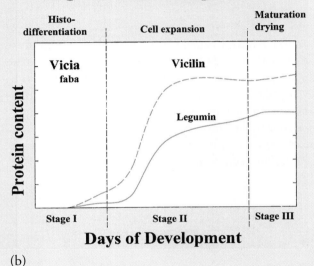
(b)

Figure 4–22
Accumulation of storage proteins related to the stages of seed development. (a) Pattern of protein accumulation in broad bean (*Vicia faba*) for vicilin and legumin, two major seed storage proteins in beans. (b) Accumulation of cruciferin protein and its mRNA in rape seed (*Brassica napus*). Note how the mRNA for the protein is only expressed at high levels during Stage II of seed development and is not detectable following maturation drying. Redrawn from Finkelstein and Crouch, 1987.

BOX 4.3 GETTING MORE IN DEPTH ON THE SUBJECT
BIOTECHNOLOGY OF SEED RESERVES (14)

The food reserves in seeds make up a major part of the world's diet both for human and livestock consumption. The nutritional quality of seeds can be improved by understanding the molecular genetics responsible for food reserve production. There are efforts through genetic engineering to improve the amino acid content of storage proteins in seeds (44, 64).

Cereals and legumes are important to worldwide diets, and their yield and nutrition have been improved significantly over years of conventional breeding. However, most cereal proteins are nutritionally low in the essential lysine-containing amino acids, and legume seeds produce storage proteins low in essential sulfur amino acids. The amino acid profile of these seeds can be improved using transformation technology (see Chapter 2) to insert new genes into crop plants to produce storage proteins high in lysine or sulfur. New germplasm is being developed that will increase the nutritional yield of some of our major crop plants. For example, genetic engineering of rice has resulted in grains containing beta-carotene that could serve as a major source of this essential nutrient for a large portion of the world's population (27).

Plants that store oils in seeds are also the target of increased efforts to produce novel oils that can be used for detergents, lubricants, and cooking oil that is healthier by producing lipids low in unsaturated fat (32). Canola (rape seed) and soybeans are major crops being bioengineered to produce novel oils.

Stage III Maturation Drying

Seeds at the end of Stage II of development have reached **physiological maturity** (also called mass maturity). Physiological maturity is the time prior to maturation drying when the seed has reached maximum dry weight through reserve accumulation. Seeds at physiological maturity can be removed from the fruit and show high germination potential as measured by seed viability and vigor (52). Seeds that do not tolerate desiccation drying are called **recalcitrant seeds** (see Box 4.4) and are usually shed from the plant at this stage without entering Stage III: maturation drying.

perisperm Nucellus tissue that remains in the mature seed and is used as storage tissue.

recalcitrant seeds Seeds that are unable to withstand maturation drying.

Orthodox seeds tolerate maturation drying and represent the condition of most crop seeds. Seeds in the maturation drying stage are characterized by rapid water loss (Fig. 4–23). There is no longer a vascular connection with the mother plant through the **funiculus.** Water loss occurs throughout the seed coat but may be more rapid where there are natural openings at the **hilum** (scar left on the seed coat after funiculus detachment) and micropyle. In species that develop impermeable seed coats as a form of dormancy (see Chapter 8), the final quantity of water leaves the seed at the hilum (34).

funiculus The attachment between the ovary and the ovule.

hilum The scar left on the seed coat after the funiculus abscises.

The low moisture level attained by dry seeds is a remarkable plant condition (9). Many plant tissues cannot tolerate moisture levels much below ~20 percent on a fresh weight basis for a prolonged time. Dry orthodox seeds can usually remain viable at 3 percent to 5 percent moisture. Orthodox seeds prepare for maturation drying towards the end of Stage II prior to physiological maturity. Abscisic acid (ABA) is the main signal for induction of desiccation tolerance. The physiological mechanisms for tolerating very dry conditions are not totally understood, but they are

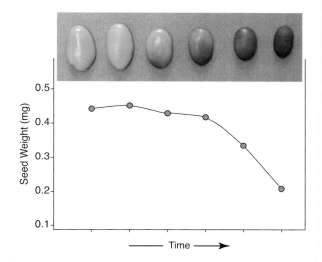

Figure 4–23
Water loss in honeylocust (*Gleditsia triacanthos*) seeds during development. Note the typical loss of chlorophyll during maturation drying and the overall reduction in seed size.

BOX 4.4 GETTING MORE IN DEPTH ON THE SUBJECT
RECALCITRANT SEEDS

After developing seeds reach physiological maturity, they proceed to desiccate (**orthodox seeds**), germinate on the plant (vivipary), or bypass complete desiccation (recalcitrant seeds). By definition, a recalcitrant seed loses viability after drying, while orthodox seeds tolerate drying (7). Germination in recalcitrant seeds must proceed soon after maturity or the seeds must be stored under conditions that prevent drying. Examples of storage life for some recalcitrant seeds stored at high humidity include coffee (*Coffee arabica*) for 10 months, coconut (*Cocus nucifera*) for 16 months, and oak (*Quercus*) for 20 months, compared to decades or years for many orthodox seeds. Recalcitrant seeds present challenges for propagators and limit germ plasm conservation because of their inability to store.

orthodox seeds Seeds that tolerate maturation drying and survive at less than 10 percent moisture.

The biological basis for this inability in recalcitrant seeds to tolerate drying is not well understood (6). Arabidopsis is an orthodox seeded species and its mutants have been a very useful tool for physiologists trying to study a variety of processes in plants. Arabidopsis mutants have been found with reduced levels of ABA, LEA proteins, and carbohydrates, and these mutants are impaired for tolerance to drying. These substances are thought to be critical for survival in orthodox seeds during desiccation drying. It would seem logical that recalcitrant seeds have reduced ABA levels or that they are impaired for the production of LEA proteins or some carbohydrates. However, most recalcitrant species produce these substances at almost normal levels. The true nature of recalcitrance drying remains to be found for this interesting group of seeds (23).

correlated to an increase in sugars (especially di- and oligosaccharides) and LEA (late embryogenesis abundant) proteins (see Box 4.5) (33). These are thought to preserve proteins and membranes by replacing the water function as cells become dry and enter a highly viscous state termed glassy (water replacement theory).

As indicated earlier, seeds also acquire the ability to germinate in Stage II prior to maturation drying. Usually, this potential to germinate is not expressed in orthodox seeds unless the fruit is removed from the plant and the seeds are gradually dried (38, 39). Germination of seeds prematurely on the plant without desiccation drying is termed precocious germination or **vivipary** (Fig. 4–24 and Box 4.6). It is usually the result of a mutation in the ability to produce or perceive ABA. During normal seed development, the seed does not germinate prior to maturation drying because of high ABA content in the seed and, for some seeds, the low water potential in the fruit coverings caused by high salt and sugar content.

Following maturation drying, the seed can be considered in a quiescent or dormant condition. Quiescent seeds fail to germinate because they are dry. Exposing quiescent seeds to a favorable environment will induce them to germinate. Dormant seeds fail to germinate even under favorable environmental conditions. There are several ecological advantages to seed dormancy and it is a common feature of many seeds. Over years of selection, dormancy has been bred out of most economically important crop species. Seed dormancy is discussed in detail in Chapter 7.

Figure 4–24

Precocious or viviparous germination occurs when the seed prematurely germinates in the fruit. This is the result of the developing seed not completing the third stage of development—maturation drying. The cause of precocious germination is usually the inability of the embryo to produce or perceive abscisic acid (ABA). ABA is a potent germination inhibitor and one of its roles during seed development is to prevent precocious germination. The tomato illustrated here is most likely an ABA production mutant.

UNUSUAL TYPES OF SEED DEVELOPMENT

Apomixis and polyembryony represent variations from the normal pattern of zygote formation and embryogenesis. Although related, they are not necessarily the same phenomenon. **Apomixis** is the asexual development of seeds that represent clonal duplicates of the mother plant.

apomixis
Asexual seed production.

BOX 4.5 GETTING MORE IN DEPTH ON THE SUBJECT
GENE EXPRESSION DURING MATURATION DRYING

Maturation drying can be considered a "switch," ending the seed's developmental program and preparing the seed for germination (22, 37, 40). Synthesis of developmental proteins stops prior to drying, and a new set of proteins is synthesized (38, 39); a major set of these proteins is called LEA (late embryogenesis abundant) proteins (33). LEA proteins are synthesized in response to water loss in the seed. LEA proteins are very stable and hydrophilic (attracts water), and possibly function as desiccation protectants by stabilizing membranes and proteins as the seed dries. There are many ecological advantages to the production of a dry seed for seed dissemination and seed survival. However, there are few living organisms that can survive drying below 15 percent moisture (55). LEA proteins appear to help the seed adjust to a dry condition. In addition, the seed is also protected during desiccation by an increase in certain sugars and oligosaccharides that also provide stabilization to proteins and membranes (2, 10).

Also during maturation drying, mRNAs for early germination are produced (19, 67). These are called conserved mRNA because they are stored in the dry seed and expressed early in germination. Although conserved mRNAs are lost in the first few hours of germination, they allow the seed to produce proteins essential for germination before the embryo regains the capacity to synthesize new mRNAs.

BOX 4.6 GETTING MORE IN DEPTH ON THE SUBJECT
PRECOCIOUS GERMINATION OR VIVIPARY

Precocious germination or vivipary is the phenomenon in which seeds precociously germinate without maturation drying. These seeds germinate in the fruit while still attached to the plant (Fig. 4–24). Precocious germination occurs naturally in some species like mangrove (*Rhizophora mangle*). In mangrove, precocious germination is an adaptation to growing in a wet (swampy) environment. Embryos germinate directly on the tree to produce seedlings with a long, javelin-shaped root (Fig. 4–25). The seedling eventually falls and becomes embedded in the mud below (65).

vivipary Germination of a seed while it is still attached to the mother plant.

For most plant species, however, precocious germination is undesirable. Premature seed sprouting occurs in many species including cereal grains (wheat and corn), fleshy fruits (citrus and tomato), and nuts (pecan). Precocious germination is considered a genetic mutation, but occurrence of precocious germination can be modified by the environment (71). Expect increased precocious germination in susceptible species during periods of wet weather (7).

The genetics of viviparous mutants in corn has been most extensively studied (50). Up to nine genes have been associated with precocious germination in corn. The common feature in viviparous mutants is reduced production, or insensitivity to abscisic acid (ABA). This supports the role for ABA in maintaining the embryo in the developmental mode through maturation drying.

(a)

(b)

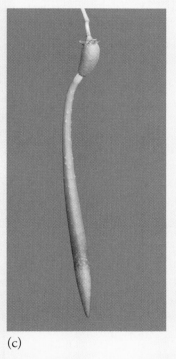
(c)

Figure 4–25
Precocious (viviparous) germination in mangrove (*Rhizophora mangle*). (a and b) Note the protrusion of the radicle from the fruit while it is still attached to the plant. (c) After sufficient radicle growth the fruit will fall from the plant and embed in the soft marshy soil around the mother plant.

---

polyembryony
The development of multiple embryos within the same seed.

Two types of apomixis are known: gametophytic and sporophytic apomixis (41, 68). **Polyembryony** means that more than one embryo develops within a single seed, sometimes many (Fig. 4–26, page 132).

Apomixis

Apomixis (53, 54, 59) is the production of an embryo that bypasses the usual process of **meiosis** and fertilization. The genotype of the embryo and resulting plant will be the same as the seed parent. **Seed production via apomixis is asexual.** Such clonal seedling plants are known as **apomicts**. Some species or individuals produce only apomictic embryos and are known as **obligate apomicts;** however, the majority of apomictic species produce both apomictic and sexual embryos on the same plant and are known as **facultative apomicts** (46).

Apomixis can be further divided into **gametophytic** versus **sporophytic apomixis** (see Box 4.7, page 133). From a horticultural production standpoint, sporophytic apomixis is the most significant because it is the type of

Figure 4–26
Polyembryony in trifoliate orange (*Poncirus trifoliata*) seeds as shown by the several seedlings arising from each seed. One seedling, usually the weakest, may be sexual; the others arise apomictically from cells in the nucellus and are diploid copies of the mother plant.

seed production that predominates in *Citrus*, mango (*Mangifera*), and mangosteen (*Garcinia*) and allows for clonal understock production from seeds for grafting or budding (11). Gametophytic apomixis results in multiple clonal embryos developing from nucellar (rarely, integument tissue) surrounding a normally developing sexual embryo sac. The seed usually contains one sexual embryo and multiple asexual embryos (Fig. 4–27). Often the seedling developing from the sexual embryo is easily identified as the weakest seedling in the group. This type of apomixis is a form of polyembryony and is termed **adventitious embryony** (also nucellar embryony and nucellar budding).

Nonrecurrent Apomixis In nonrecurrent apomixis, meiosis does occur and an embryo arises directly from the egg nucleus without fertilization. Since the egg is haploid, the resulting embryo will also be haploid. This case is rare and primarily of genetic interest. It does not consistently occur in any particular kind of plant, as do recurrent apomixis and adventitious embryony.

Polyembryony (46)

In 1719, Leeuwenhoek reported the first account of polyembryony in plants when he observed the production of twin embryos in *Citrus*. *Polyembryony* is the production of additional embryos within a seed other than the normal sexual embryo (43). The multiple embryos could be all sexual or a mixture of sexual and asexual (apomictic) embryos. Four types of polyembryony are recognized in angiosperms:

1. After the normal sexual embryo begins to form, additional embryos can "bud-off" from the proembryo (found in *Asparagus*, *Tulipa gesneriana*, and *Hamamelis*) or suspensor cells (found in *Acanthus*). The result is a sexual embryo and multiple copies of that sexual embryo.
2. Adventive embryony results in additional embryos formed from cells in nucellar (found in *Citrus*, *Mangifera*, and *Garcinia*) or integuments (found in *Spiranthes cernua*). The result is one sexual and multiple asexual embryos.
3. Multiple embryo sacs may be formed within a single ovule (seed). This has been observed in species of cotton (*Gossypium*). The result can be multiple sexual embryos from separate fertilizations or multiple asexual embryos from aposporic apomixis (Box 4.7).
4. Additional embryos may result from a synergid cell functioning as an egg cell. This can result from fertilization of the synergid by a male sperm cell or

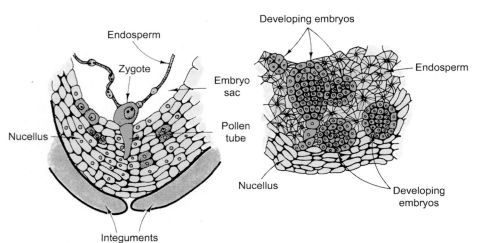

Figure 4–27
Development of nucellar embryos in *Citrus*. *Left:* Stage of development just after fertilization showing zygote and remains of pollen tube. Note individual active cells (shaded) of the nucellus, which are in the initial stages of nucellar embryony. *Right:* A later stage showing developing nucellar embryos. The large one may be the sexual embryo. Redrawn from Gustafsson, 1946.

from autonomous cell divisions in reduced synergids resulting in a haploid apomictic embryo. These types are found in *Pennisetum*, *Tamarix*, and *Solanum*.

Vegetative Apomixis

The term *apomixis* has been used in the past for any form of vegetative propagation. Today, its usage has been restricted to asexual production of an embryo within the ovule of flowering plants. However, some references still include the term *vegetative apomixis* to describe the production of other structures besides an embryo. In some cases, vegetative buds or **bulbils** are produced in the inflorescence in place of flowers. This occurs in *Poa bulbosa* and some *Allium*, *Agave*, and grass species.

PLANT HORMONES AND SEED DEVELOPMENT

In general, concentrations of plant hormones are high in seeds compared with other parts of the plant (7). Seeds were the first tissue where several of the plant hormones were discovered and studied in detail. All of the major hormones have been associated with seed development (57, 58). Plant hormones are involved in seed development in several ways:

1. growth and differentiation of the embryo
2. accumulation of food reserves
3. storage for use during germination and early seedling growth, and
4. growth and development of fruit tissue

BOX 4.7 GETTING MORE IN DEPTH ON THE SUBJECT
APOMIXIS (42)

Most seed plants produce embryos after the fusion of haploid female and male gametes. However, in a small proportion of plants, embryos form spontaneously from cells that bypass meiosis. These embryos have the same genotype as the mother plant, and this process is called **apomixis.**

The term *apomixis* is from the Greek *apo*, meaning from or away from, and *mixis*, to mingle or mix. Apomixis literally means "away from mixing" and refers to the production of new plants without mixing of gametes. Apomixis occurs in over 300 species from at least 35 plant families (12). Most display facultative apomixis. It is most common in the daisy (Asteraceae), grass (Graminaceae), rose (Rosaceae), and citrus (Rutaceae) families.

There are two general categories for apomixis. These are **gametophytic** and **sporophytic** apomixis.

Gametophytic Apomixis
During normal embryogenesis, female egg cells are derived from the megaspore mother cell found in the nucellus tissue of the flower. The megaspore mother cell undergoes reduction division (meiosis) to produce haploid nuclei that take up typical locations in the embryo sac (see page 119). In gametophytic apomixis, the megaspore mother cell degenerates or bypasses meiosis. Unreduced cell(s) divide to produce reproductive cells that take their normal location in the embryo sac. The unreduced nucleus corresponding to the egg cell undergoes spontaneous or parthenogenic divisions to form an embryo without fertilization from a male gamete. In some cases, polar nuclei can autonomously form endosperm, but, in many cases, an unreduced polar nuclei will fuse with a male sperm cell to form the endosperm. This process is known as pseudogamy.

There are two types of gametophytic apomixis called **diplospory** and **apospory.** These differ because of the location in the nucellus where apomictic cells arise. In diplospory, the megaspore mother cell does not undergo or complete meiosis. Rather, it divides to produce 2n rather than the normal 1n cells found in the embryo sac. Otherwise, the embryo sac has a normal appearance. This type of apomixis is common in the Asteraceae and in some grass species. In apospory, the megaspore mother cell undergoes normal meiosis, but the resulting cells usually degrade before they are fertilized. Additional cells in the nucellus become densely cytoplasmic and take on the role of apomictic initials bypassing meiosis to form cells in their own embryo sacs within the same ovule. Aposporous apomicts are found in the Rosaceae, Asteraceae, and in some grasses.

In some apomictic species, such as *Crepis*, dandelion (*Taraxacum*), bluegrass (*Poa*), and onion (*Allium*), there is no need for the stimulus of pollination; in others [e.g., species of guayule (*Parthenium*), raspberry (*Rubus*), apple (*Malus*), some grasses (*Poa* species), and coneflower (*Rudbeckia*)], pollination appears to be necessary, either to stimulate embryo development or to produce a viable endosperm.

Sporophytic Apomixis
Sporophytic apomixis is also known as **adventitious embryony.** It has been extensively studied because it is the type of apomixis that occurs in *Citrus* and mango (*Mangifera*) (Figs. 4–26 and 4–27). In adventitious embryony, the megaspore mother cell undergoes normal meiosis and forms a normal sexual embryo sac. These cells are fertilized by male sperm cells as in normal embryogenesis. However, at about the time the first divisions begin in the sexual embryo, cells in the nucellus begin abnormal cell divisions leading to multiple embryos forming in the micropylar region of the ovule. Since these are derived from mother plant cells, the resulting embryos are asexual. These asexual embryos do not produce their own embryo sac (thus the term sporophytic apomixis). Rather, they grow into the embryo sac of the sexual embryo and share its sexually derived endosperm. The result is a single seed that can contain a single sexual and multiple asexual embryos.

> **BOX 4.8 GETTING MORE IN DEPTH ON THE SUBJECT**
> **SIGNIFICANCE OF APOMIXIS**
>
> Apomixis is significant in agriculture and horticulture because the seedling plants have the same genotype as the mother plant (28). This asexual process eliminates variability and "fixes" the characteristics of any cultivar immediately. However, the apomictic life cycle has the same juvenile period found in sexually derived seeds (see Chapter 2).
>
> Only a few economically important food crops exhibit apomixis. These include *Citrus*, mango (*Mangifera*), and mangosteen (*Garcinia*). All three have adventitious embryony. They have mainly been exploited as clonal seedling understocks for grafting and budding because they are virus-free, show seedling vigor, and are uniform. However, sexual embryos can also be produced and can exhibit unwanted variability. Use of DNA fingerprinting is being used to separate sexual and asexual embryos for understock production.
>
> Several grass species and cultivars are facultative gametophytic apomicts. These include Kentucky bluegrass (*Poa pratensis*), 'King Ranch' bluestem (*Andropogun*), 'Argentine' Bahia grass (*Paspalum notatum*), and 'Tucson' side oats grama (*Bouteloua curtipendula*).
>
> There has been a recent resurgence in research concerning apomixis. It has been known for some time that apomixis is an inherited trait and that the gene maps to a single chromosome (68). This indicates that there is a single apomixis gene and that if that gene is isolated it could be used to genetically engineer apomixis into important crop plants. The major benefit would be that apomixis would fix hybrid vigor (heterosis) in crops that now require costly crossing between inbred parents (see Chapter 5). It would also be a simple way to eliminate virus in traditionally vegetatively propagated crops like potato. The seed produced would have the same genetics of the parents, but because the embryo is derived from a single cell, it should be virus-free.

Auxin

Free and conjugated forms of indoleacetic acid (IAA) are abundant in developing seeds. Free IAA is high during cell division stages of development (Stages I and II) and is essential for normal embryo and endosperm development. An auxin gradient is required to establish appropriate bipolar symmetry during embryo development. Mutations that cause seeds to have low auxin production or reduced auxin transport generally result in malformed embryos with fused cotyledons and poor endosperm development (74).

Conjugated forms of IAA (see Chapter 2) are abundant in mature seeds and during germination. Free IAA is released from the conjugated forms for utilization during early seedling growth. There is evidence that auxin from the developing seed signals the fruit to continue to develop (Fig. 4–28). Fruits usually abscise if seeds abort or are unfertilized. Auxin applied to tomato or strawberry can induce parthenocarpic fruit development (see Box 4.9, page 135).

Gibberellins

Various forms of gibberellins are abundant during seed development (Stages I and II). Most of the biochemistry known about gibberellins was first investigated in developing seeds. Active forms decline at seed maturity and are replaced by conjugated forms of gibberellins. Like auxin, these conjugated forms of gibberellins are utilized during germination.

Gibberellins were originally thought to play only a minor role in seed development. Gibberellin-deficient mutants in tomato and *Arabidopsis* generally show normal seed development only affecting final seed size. However in pea, gibberellins are required for embryo growth (66). In gibberellin-deficient mutants that show reduced gibberellin biosynthesis, gibberellin is required to sustain

Figure 4–28
Strawberry "fruit" (receptacle) enlargement requires auxin from the developing seed (actually the fruit-achene). Notice how the only swelling in the receptacle tissue is around the developing achenes (red arrow). The black arrow shows a non-fertilized seed where you can still see the style and stigma attached. There is no swelling in this area because there is no developing seed to provide the auxin.

> **BOX 4.9 GETTING MORE IN DEPTH ON THE SUBJECT**
> **PARTHENOCARPY**
>
> For many plant species, pollination is the stimulus for the beginning of fruit development. Continued fruit growth depends on seed formation. The number of seeds within a fruit strongly affects fruit size in species like apple and strawberry. Fruit that develop without seed formation (seedless) are called **parthenocarpic** fruit. Two types of parthenocarpy are recognized in plants (73). *Vegetative parthenocarpy* takes place in species like pear or fig, where the fruit develops even without pollination. *Stimulative parthenocarpy* takes place only after pollination but does not require fertilization or seed set for continued fruit growth. Grapes can form seedless fruit by stimulative parthenocarpy.
>
> parthenocarpy The formation of fruit without seeds.
>
> A number of species have been bred to naturally form parthenocarpic fruit. For example, parthenocarpy is essential for greenhouse cucumber fruit production because there are no reliable insect pollinators in the greenhouse. Other species (tomato, grape, some tree fruits) will form parthenocarpic fruit if sprayed with auxin or gibberellin. Interestingly, some species will only form parthenocarpic fruit if treated with auxin (tomato), while others require gibberellin (grape). The developing seed is the normal source for auxin and gibberellin for fruit growth. Both gibberellin and auxin are factors critical to fruit growth and interact during normal fruit development.

embryo growth in the first few days following pollination. It appears that the suspensor is the source for this gibberellin and that gibberellin from the suspensor is required for further development until the embryo grows sufficiently to receive hormones and nutrition from the endosperm.

Like auxin, gibberellins produced from the seed may also signal fruit development (56). Pea pods containing aborted seeds can continue development following application of gibberellic acid. Gibberellins can also induce **parthenocarpic** fruit development in crops like grapes (see Box 4.9).

Cytokinins

Several free and conjugated forms of cytokinins are high in developing seeds. The highest concentration of cytokinins is found during the cell division stages of embryogenesis (Stage I and early Stage II). Cytokinins appear to be supplied by the suspensor during histodifferentiation. The cytokinins-to-auxin ratio plays a key role in controlling shoot apical meristem formation, and this association appears to be important in the differentiation phase of Stage I embryos.

Abscisic Acid (ABA)

ABA levels are high in the maturation phase of developing seeds (Stage II). ABA has been shown to have a major role in all the major features of seed maturation. ABA mutants typically show reduced storage reserve synthesis, reduced tolerance to drying, and premature germination prior to maturation drying. ABA has a major influence on all four major genes (ABI3, FUS3, LEC1, and LEC2) that code transcription factors thought to be master regulators of seed maturation (29). However, there is cross talk among auxin, ABA, and gibberellin via these four regulator genes.

Ethylene

Significant amounts of ethylene are produced throughout seed development as seen in *Brassica* species (35, 49). Although the role of ethylene during seed development has not been extensively studied, it is interesting that ethylene production is high in developing *Brassica* embryos when embryos begin to "degreen" during maturation drying. In most seeds, embryos contain chlorophyll and are green during Stages I and II of development. There is a dramatic loss in chlorophyll during maturation drying while embryos "degreen" and appear yellow. Ethylene has a documented role in leaf senescence and could support embryo "de-greening." Ethylene probably plays only a minor role during seed development. Ethylene mutants of several species produce apparently normal seeds.

There is an interesting interaction between ethylene and ABA in controlling programmed cell death in corn endosperm (76). In corn, endosperm cells die prior to maturation drying. There must be a mechanism in place that programs these cells to die, while adjacent aleurone and embryo cells continue the maturation process. It appears that ethylene is differentially produced in the endosperm cells, and the response to that ethylene induces a senescence response in endosperm cells but not the other cell types. This differential response is partly due to the ability of the cells to respond to ABA.

RIPENING AND DISSEMINATION

Specific physical and chemical changes that take place during maturation and ripening of the fruit lead to fruit senescence and dissemination of the seed. One of the most obvious changes is the drying of the pericarp tissues. In certain species, this leads to dehiscence and the discharge of the seeds from the fruit. Changes may take place in the color of the fruit and the seed coats, and softening of the fruit may occur.

Seeds of most species dehydrate at ripening and prior to dissemination. Moisture content drops to 30 percent or less on the plant. The seed dries further during harvest, usually to about 4 percent to 6 percent for storage. Germination cannot take place at this level of dryness, so it is an important basis for maintaining viability and controlling germination.

In certain other species, seeds must not dry below about 30 percent to 50 percent or they will lose their ability to germinate (13). These plants include

a. species whose fruits ripen early in summer, drop to the ground, and contain seeds that germinate immediately (some maples, poplar, elm)
b. species whose seeds mature in autumn and remain in moist soil over winter (oak); and
c. species from warm, humid tropics (citrus). These are called **recalcitrant seeds** (see Box 4.6, page 131), which produce special problems in handling.

Seeds of species with fleshy fruits may become dry but are enclosed with soft flesh that can decay and cause injury. In most species, this fleshy tissue should be removed to prevent damage from spontaneous heating or an inhibiting substance. In some species, however (e.g., *Mahonia* and *Berberis*), the fruits and seeds may be dried together (45).

Many agents accomplish seed dispersal. Fish, birds, rodents, and bats consume and carry seeds in their digestive tracts (25). This is often a function of the type of fruit produced by that species (Table 4–3).

Table 4–3
DIFFERENT TYPES OF FRUITS

Type of fruit		Description	Example
Dry Fruits	Indehiscent Fruits		
	1. Caryopsis	Pericarp and seed coat are fused forming a single seed.	Most often in monocots like corn and wheat
	2. Samara	A one-seeded fruit with a specialized wing for wind dissemination.	Maple, ash, and elm
	3. Achene	A one-seeded fruit.	Strawberry, sunflower, and clematis
	4. Nut	Fruit develops from an ovary with multiple carpels, but only one survives.	Walnut and hazelnut
	5. Utricle	Single-seeded fruit with inflated pericarp.	Chenopodium
	Dehiscent Fruits		
	1. Follicle	Pod-like fruit from a single carpel that splits on one side.	Delphinium and columbine
	2. Legume	Pod that opens on both sides.	Bean, locust, and pea
	3. Capsule	There are numerous types of dry capsules that open along different suture lines near top of fruit.	Poppy, iris, and lily
	4. Silique	Develops from two carpels and opens along two suture lines.	Cabbage and arabidopsis
Fleshy Fruits	1. Berry	A fleshy fruit with many seeds with an endocarp, mesocarp, and exocarp that are soft.	Tomato and grape
	2. Drupe	Has a hard endocarp.	Peach, cherry, and fringe tree
	3. Pome	Has a papery endocarp.	Apple and pear
	4. Pepo	Outer endocarp forming hard rind.	Squash and pumpkin
	5. Hesperidium	Similar to a pepo but endocarp is not hard.	Orange and lemon
	6. Multiple fruits	Several fruits aggregated into a single structure.	Blackberry (multiple drupes), pineapple, and mulberry
Schizocarpic Fruits	Schizocarp	Fruits develop so that locules in an ovary separate into separate single-seeded units.	Sycamore, carrot, and parsley

Fruits with spines or hooks become attached to the fur of animals and are often moved considerable distances. Wind dispersal of seed is facilitated in many plant groups by "wings" on dry fruits; tumbleweeds can move long distances by rolling in the wind. Seeds carried by moving water, streams, or irrigation canals can be taken great distances and often become a source of weeds in cultivated fields. Some plants (e.g., *Impatiens* and *Oxalis*) have mechanisms for short-distance dispersal, such as explosive liberation of seeds. Human activities in purposeful shipment of seed lots all over the world are, of course, effective in seed dispersal.

DISCUSSION ITEMS

Knowledge of seed development is most important for understanding various aspects of seed quality discussed in Chapter 6. The environment during seed development and the conditions during seed harvest are critical to producing quality seeds. To evaluate problems related to seed quality, a fundamental understanding of seed development, especially seed filling (deposition of food reserves) and seed desiccation (maturation drying), are most important.

1. What are the three differences between pollination and fertilization?
2. How does the seed storage tissue differ among a monocot, dicot, and gymnosperm?
3. Compare zygotic and apomictic seed development.
4. How are the stages of embryogenesis similar and different in shepherd's purse vs. corn?
5. What might be the ecological advantages of vivipary as demonstrated by mangrove plants?
6. How is the scutellum of a monocot similar to and/or different from the cotyledons in a dicot?

REFERENCES

1. Atwater, B. R. 1980. Germination, dormancy and morphology of the seeds of herbaceous ornamental plants. *Seed Sci. and Tech.* 8:523–73.

2. Bailly C., C. Audigier, F. Ladonne, M. H. Wagner, F. Coste, F. Corineau, and D. Côme. 2001. Changes in oligosaccharide content and antioxidant enzyme activities in developing bean seeds as related to acquisition of drying tolerance and seed quality. *J. Exp. Bot.* 52:701–8.

3. Baskin, C. C., and J. M. Baskin. 2007. A revision of Martin's seed classification system, with particular reference to his dwarf-seed type. *Seed Sci. Res.* 17:11–20.

4. Beattie, A. J. 1985. *The evolutionary ecology of ant-plant mutualisms.* Cambridge, UK: Cambridge University Press.

5. Berger, F., Y. Hamamura, M. Ingouff, and T. Higashiyama. 2008. Double fertilization—caught in the act. *Trends Plant Sci.* 13:437–43.

6. Berjak, P. L., J. M. Farrant, and N. W. Pammenter. 1989. The basis of recalcitrant seed behavior. In R. B. Taylorson, ed. *Recent advances in the development and germination of seeds.* New York: Plenum Press. pp. 89–108.

7. Bewley, J. D., and M. Black. 1994. *Seeds: Physiology of development and germination.* New York: Plenum Press.

8. Bhatnager, S. P., and B. M. Johri. 1972. Development of angiosperm seeds. In T. T. Kozlowski, ed. *Seed biology,* Vol. 1. New York: Academic Press. pp. 77–149.

9. Black, M., and H. W. Pritchard. 2002. *Desiccation and survival in plants: Drying without dying.* Wallingford, UK: CABI Pub.

10. Blackman, S. A., R. L. Obendorf, and A. C. Leopold. 1992. Maturation proteins and sugars in desiccation tolerance of developing soybean seeds. *Plant Physiol.* 100:225–30.

11. Campbell, A. J., and D. Wilson. 1962. Apomictic seedling rootstocks for apples: Progress report, III. *Ann. Rpt. Long Ashton Hort. Res. Sta.* (1961):68–70.

12. Carman, J. G. 1997. Asynchronous expression of duplicate genes in angiosperms may cause apomixis, bispory, tetraspory, and polyembryony. *Biological J. Linnaen Soc.* 61:51–94.

13. Chin, H. F., and E. H. Roberts. 1980. *Recalcitrant crop seeds.* Kuala Lumpur: Tropical Press.

14. Chrispeels, M. J., and D. E. Sadava. 2002. *Plants, genes, and crop biotechnology*, 2nd ed. London: Jones and Bartlett Pub.

15. Collins, G. B., and J. W. Grosser. 1984. Culture of embryos. In I. K. Vail, ed. *Cell culture and somatic cell genetics of plants*, Vol. 1. New York: Academic Press. pp. 241–57.

16. Cooper, D. C., and R. A. Brink. 1940. Somatoplastic sterility as a cause of seed failure after interspecific hybridization. *Genetics* 25:593–617.

17. Cooper, D. C., and R. A. Brink. 1945. Seed collapse following matings between diploid and tetraploid races of *Lycopersicon pimpinellifollium*. *Genetics* 30:375–401.

18. Corner, E. J. H. 1976. *The seeds of dicotyledons*. Cambridge: Cambridge Univ. Press.

19. Dure, L. S., III. 1997. Lea proteins and the desiccation tolerance of seeds. *Cellular and molecular biology of plant seed development*. Boston, MA: Kluwer Acad. Pub. pp. 525–43.

20. Esau, K. 1977. *Anatomy of seed plants*. New York: John Wiley & Sons.

21. Fahn, A. 1982. *Plant anatomy*. New York: Pergamon Press.

22. Fait, A., R. Angelovici, H. Less, I. Ohad, E. Urbanczyk-Wochniak, A. R. Fernie, and G. Galili. 2006. Arabidopsis seed development and germination is associated with temporally distinct metabolic switches. *Plant Physiol.* 142:839–54.

23. Finch-Savage, W. E., S. K. Pramanik, and J. D. Bewley. 1994. The expression of dehydrin proteins in desiccation-sensitive (recalcitrant) seeds of temperate trees. *Planta* 193:478–85.

24. Finkelstein, R. R., and M. L. Crouch. 1987. Hormonal and osmotic effects on developmental potential of maturing rapeseed. *HortScience* 22:797–800.

25. Fordham, A. J. 1984. Seed dispersal as it concerns the propagator. *Comb. Proc. Intl. Plant Prop. Soc.* 34:531–34.

26. Goldberg, R. B., S. J. Barker, and L. Perez-Grau. 1989. Regulation of gene expression during plant embryogenesis. *Cell* 56:149–60.

27. Guerinot, M. L. 2000. The green revolution strikes gold. *Science* 287:241–43.

28. Gustafsson, A. 1946–1947. *Apomixis in higher plants*, Parts I–III. Lunds Univ. Arsskrift, N. F. Avid. 2 Bd 42, Nr. 3:42(2); 43(2); 43(12).

29. Gutierrez, L., O. Van Wuytswinkel, M. Castelain, and C. Bellini. 2007. Combined networks regulating seed maturation. *Trends Plant Sci.* 12:294–300.

30. Hardham, A. R. 1976. Structural aspects of the pathways of nutrient flow to the developing embryo and cotyledons of *Pisum sativum* L. *Aust. J. Bot.* 24:711–21.

31. Higashiyama, T., H. Kuroiwa, S. Kawano, and T. Kuroiwa. 1998. Guidance *in vitro* of the pollen tube to the naked embryo sac of *Torenia fournieri*. *Plant Cell* 10:2019–31.

32. Hills, M. J., and D. J. Murphy. 1991. Biotechnology of oil seeds. *Biotechnol. Genet. Eng. Rev.* 9:1–46.

33. Hoekstra F. A., E. A. Golovina, and J. Buitink. 2001. Mechanisms of plant desiccation tolerance. *Trends Plant Sci.* 6:431–48.

34. Hyde, E. O. 1954. The function of the hilum in some Papilionaceae in relation to the ripening of the seed and the permeability of the testa. *Ann. Bot.* 18:241–56.

35. Johnson-Flanagan, A. M., and M. S. Spencer. 1994. Ethylene production during development of mustard (*Brassica juncea*) and canola (*Brassica napus*) seed. *Plant Physiol.* 106:601–6.

36. Jones, R. E., and R. L. Geneve. 1995. Seed coat structure related to germination in eastern redbud (*Cercis canadensis* L.). *J. Amer. Soc. Hort. Sci.* 129:123–27.

37. Kermode, A. R. 1990. Regulatory mechanisms involved in the transition from seed development to germination. *Crit. Rev. Plant Sci.* 9:155–95.

38. Kermode, A. R., and J. D. Bewley. 1985a. The role of maturation drying in the transition from seed to germination. I. Acquisition of desiccation-tolerance and germinability during development of *Ricinus communis* L. seeds. *J. Exp. Bot.* 36:1906–15.

39. Kermode, A. R., and J. D. Bewley. 1985b. The role of maturation drying in the transition from seed to germination. II. Post-germinative enzyme production and soluble protein synthetic pattern changes within the endosperm of *Ricinus communis* L. seeds. *J. Exp. Bot.* 36:1916–27.

40. Kermode, A. R., J. D. Bewley, J. Dasgupta, and S. Misra. 1986. The transition from seed development to germination: A key role for desiccation? *HortScience* 21:1113–18.

41. Koltunow, A. M. 1993. Apomixis: Embryo sacs and embryos formed without meiosis or fertilization in ovules. *Plant Cell* 5:1425–37.

42. Koltunow, A. M., and U. Grossniklaus. 2003. APOMIXIS: A developmental perspective. *Annu. Rev. Plant Biol.* 54:547–74.

43. Lakshmanan, K. K., and K. B. Ambegaokar. 1984. Polyembryony. In B. M. Johri, ed. *Embryology of angiosperms*. Berlin: Springer-Verlag. pp. 445–74.

44. Larkins, B. A., C. R. Lending, and J. C. Wallace. 1993. Modification of maize-seed-protein quality. *Amer. J. Clin. Nutr.* 58:264S–9S.

45. MacDonald, B. 1986. *Practical woody plant propagation for nursery growers,* Vol. 1. Portland, OR: Timber Press.

46. Maheshwari, P., and R. C. Sachar. 1963. Polyembryony. In P. Maheshwari, ed. *Recent advances in the embryology of angiosperm.* Delhi, India: Univ. of Delhi, *Intl. Soc. of Plant Morph.* pp. 265–96.

47. Marshall, J. E. A., and A. R. Hemsley. 2003. A Mid-Devonian seed-megaspore from east Greenland and the origin of the seed plants. *Palaeontology* 46:647–70.

48. Martin, A. C. 1946. The comparative internal morphology of seeds. *Amer. Midland Nat.* 36:5126–60.

49. Matilla, A. J. 2000. Ethylene in seed formation and germination. *Seed Sci. Res.* 10:111–26.

50. McCarty, D. R., and C. B. Carlson. 1991. The molecular genetics of seed maturation in maize. *Physiol. Plant.* 81:267–72.

51. Meinke, D. W. 1991. Perspectives on genetic analysis of plant embryogenesis. *Plant Cell* 3:857–66.

52. Miles, D. F., D. M. TeKrony, and D. B. Egli. 1988. Changes in viability, germination, and respiration of freshly harvested soybean seed. *Crop Sci.* 28:700–4.

53. Naumova, T. N. 1993. Apomixis. In *Angiosperms: Nucellar and integumentary embryony.* Boca Raton, FL: CRC Press.

54. Nygren, A. 1954. Apomixis in the angiosperms II. *Bot. Rev.* 20:577–649.

55. Oliver, M. J., and J. D. Bewley. 1992. Desiccation tolerance in plants. In G. N. Somero, C. B. Osmond, and C. L. Bolis, eds. *Water and life.* Berlin: Springer-Verlag. pp. 141–60.

56. Ozaga, J. A., M. L. Brenner, and O. M. Reinecke. 1992. Seed effects on gibberellin metabolism in pea pericarp. *Plant Physiol.* 100:88–94.

57. Quatrano, R. S. 1987. The role of hormones during seed development. In P. J. Davies, ed. *Plant hormones and their role in plant growth and development.* Boston: Marinus Nijhoff Publishers. pp. 494–514.

58. Radley, M. 1979. The role of gibberellin, abscisic acid, and auxin in the regulation of developing wheat grains. *J. Exp. Bot.* 30:381–89.

59. Raghaven, V. 1986. *Embryogenesis in angiosperms.* Cambridge: Cambridge Univ. Press.

60. Randolf, L. F. 1936. Developmental morphology of caryopsis of maize. *J. Agric. Res.* 53:881–97.

61. Shannon, J. C. 1972. Movement of ^{14}C-labeled assimilates into kernels of *Zea mays* L.I. pattern and rate of sugar movement. *Plant Physiol.* 49:198–202.

62. Singh, H., and B. M. Johri. 1972. Development of gymnosperm seeds. In T. T. Kozlowski, ed. *Seed biology,* Vol. 1. New York: Academic Press. pp. 22–77.

63. Smith, S. E., and D. J. Read. 2008. *Mycorrhizal symbiosis.* 3rd ed. London, UK: Academic Press.

64. Sommerville, C. R. 1993. Future prospects for genetic modification of the composition of edible oils from higher plants. *Amer. J. Clin. Nutr.* 58:270S–5S.

65. Stephens, W. 1969. The mangrove. *Oceans* 2:51–5.

66. Swain, S. M., J. J. Ross, J. B. Reid, and Y. Kamiya. 1995. Gibberellins and pea seed development: Expression of the $lh\text{-}1$, ls and le^{5839} mutations. *Planta* 191:426–33.

67. Thomas, T. L. 1993. Gene expression during plant embryogenesis and germination: An overview. *Plant Cell* 5:1401–10.

68. Van Dijk, P., and J. van Damme. 2000. Apomixis technology and the paradox of sex. *Trends in Plant Sci.* 5:81–4.

69. Vicente-Carbajosa, J., and P. Carbonero. 2005. Seed maturation: Developing an intrusive phase to accomplish a quiescent state. *Intl. J. Dev. Biol.* 49:645–51.

70. Weber, H., L. Borisjuk, and U. Wobus. 2005. Molecular physiology of legume seed development. *Annu. Rev. Plant Biol.* 56:253–79.

71. Wellington, P. S., and V. W. Durham. 1958. Varietal differences in the tendency of wheat to sprout in the ear. *Empire J. Exp. Agr.* 26:47–54.

72. West, M. A. L., and J. J. Harada. 1993. Embryogenesis in higher plants: An overview. *Plant Cell* 5:1361–9.

73. Westwood, M. N. 1993. *Temperate-zone pomology: Physiology and culture.* Portland, OR: Timber Press.

74. Wijers, D., and G. Jurgens. 2005. Auxin and embryo axis formation the ends in sight? *Current Opinions Plant Biol.* 8:32–7.

75. Wolswinkel, P. 1992. Transport of nutrients into developing seeds: A review of physiological mechanisms. *Seed Sci. Res.* 2:59–73.

76. Young, T. E., and D. R. Gallie. 2000. Regulation of programmed cell death in maize endosperm by abscisic acid. *Plant Mol. Biol.* 42:397–414.

*5
Principles and Practices of Seed Selection

learning objectives

- Define breeding systems.
- Categorize seed-propagated cultivars and species.
- Define procedures to control genetic variability.
- Describe systems of seed selection and production.
- Define legal controls on genetic purity.

INTRODUCTION

Many annual and biennial crop, forage, vegetable, and ornamental selections are produced by plant breeding to be propagated by seed (7, 8, 29, 34). Breeding involves selection of parents, specific breeding procedures, and genotype stabilization (1, 3, 47). The last process is sometimes referred to as **"fixing the genotype."** Seed is used to reproduce most woody perennial plants in forestry as well as in the landscape. Propagation of many ornamental, fruit, and nut trees utilizes seedlings for rootstocks that are then grafted (49, 58). However, characteristics important in agriculture, horticulture, and forestry may not be consistently perpetuated into the next seedling generation unless appropriate principles and procedures are followed. This chapter deals with seed selection and the management of genetic variability in seedling populations in both herbaceous and perennial plant species for the purposes of propagation.

"fixing" The process of stabilizing the genotype of a seedling population to make it homozygous so that it will "breed true."

BREEDING SYSTEMS

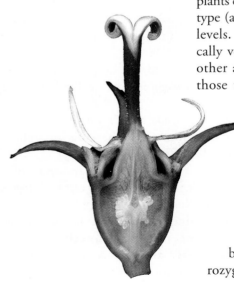

The main objective of a breeding program is to use the observed variability available within a particular genus or species to create new, stable populations with improved plant characteristics. Variability in seed-propagated plants can be described both at the phenotype (appearance) and genotype (genetic) levels. Seedlings that are phenotypically very similar in appearance to each other are termed **homogeneous**, while those that are dissimilar are described as **heterogeneous**. When more specific information is known about the seedling population's genetic makeup, they can be described as **homozygous** or **heterozygous**. Homozygous populations share many common paired alleles (genes) at each chromosome loci and breed true-to-type offspring. Heterozygous populations have dissimilar

homogenous
A population of seedlings that are phenotypically similar.

heterogeneous
A population of seedlings that are phenotypically dissimilar.

homozygous
A population of seedlings whose genotypes are very similar.

heterozygous
A population of seedlings whose genotypes are dissimilar.

self-pollination
A breeding system in which the plant flower is pollinated by itself because of flower structure or isolation.

cross-pollination
A breeding system in which the plant is pollinated by pollen from a separate genotype either because of flower structure or artificial control during pollination.

apomixis A breeding system in which the embryo is apomictic (i.e., produced from a vegetative cell and not as a result of reduction division and fertilization).

paired alleles at many chromosome loci and generally lead to diverse genetic offspring. These characteristics are determined by the breeding system, characteristics of the crop species, and management conditions under which seed populations are grown (1, 3, 22). Three important considerations for determining a plant breeding system are whether the plants reproduce primarily from **self-pollination, cross-pollination,** and **apomixis** (22).

Self-Pollination

Self-pollination occurs when pollen germinates on the stigma and the pollen tube grows down the style to fertilize the same flower or a flower of the same plant or clone. Self-pollination is a natural condition in some species because of flower structure. The extreme case is when pollination occurs before the flower opens (Fig. 5–1). This type of behavior is called cleistogamy and occurs in some crop plants like peanuts (*Arachis*). A wonderful example of this reproductive strategy is found in several types of violets (*Viola*). Violets can produce two types of flowers. Chasmogamous (open) flowers are produced in the spring or summer when pollinators are plentiful and active. Chasmogamous flowers open to permit cross-pollination between flowers and produce offspring (seeds) with generous genetic diversity. These same plants also produce underground cleistogamous flowers in the autumn that never open and self-pollinate. Although this restricts genetic diversity, it does not require the same level of plant resources for seed production and provides insurance against poor seed production from earlier out-crossing flowers.

The degree to which self-pollination occurs can vary among species. Some are highly self-pollinated (i.e., less than 4 percent cross-pollinated) such as cereal grains [barley (*Hordeum*), oats (*Avena*), wheat (*Triticum*), rice (*Oryza*)], legumes [field pea (*Pisum*), and garden bean (*Phaseolus*)], flax (*Linum*), and some grasses. There are also those that are self-fertile but can cross-pollinate at more than 4 percent, including cotton (*Gossypium*), pepper (*Capsicum*), and tomato (*Solanum*). Self-pollination is not typically found in most woody plant species, but some exceptions occur, such as peach (*Prunus*) (58).

Homozygosity in a self-pollinated herbaceous cultivar is **"fixed"** by consecutive generations of self-fertilizations (Table 5–1) (1, 22, 47). To produce a "true-breeding" homogeneous and homozygous cultivar, plant breeders will start with a single plant and then eliminate the off-type plants each generation for a period of six to ten generations. If one assumes a more or less homogeneous population with individuals possessing homozygous traits, self-pollination will result in a population of individuals that will remain homogeneous and homozygous. If a mutation occurs in one of the alleles and is recessive, the genotype for that trait becomes heterozygous. Then the next generation will produce homozygous plants that are similar in appearance but genetically heterozygous for the mutant allele. The proportion of homozygous individuals with the two traits will increase in consecutive generations, while the proportion with heterozygous genotypes will decrease by a factor of one-half each generation. The group of descendants of the original parent will segregate into a heterogeneous mixture of more or less true-breeding lines.

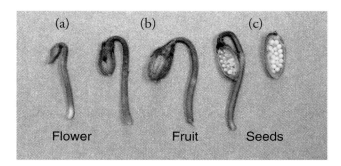

Figure 5–1
A cleistogomous flower in violet. The flower remains underground and never opens, forcing self-pollination. (a) Unopened cleistogomous flower. (b) Fruit with developing seeds. (c) Fruit with ovary wall removed to show the seeds.

Cross-Pollination

In nature, many, if not most, species are naturally cross-pollinated, a trait that seems to be desirable both for the individual and its population. Not only does the increased heterozygosity provide the opportunity for evolutionary adaptation within the population confronted with environmental change, but plant vigor

Table 5–1
EFFECT OF SELF-POLLINATION AND ROGUING FOLLOWING CROSSING OF A TALL (DD) PEA AND DWARF (dd) PEA (SEE FIG. 2-14).
"Fixing" of the two parental phenotypes can be observed in succeeding generations in the proportion of tall and dwarf plants. Continuous roguing for the recessive trait never quite eliminates its segregation from residual heterozygous individuals.

	A. Continuing self-pollination proportions				B. Roguing of all dwarfed plants		
	DD	Dd	dd	Percent homozygous	Tall	Dwarf	%dd
P_1	1		1	100			
F_1		1		0	all		
F_2	1	2	1	50	3	1	25
F_3	3	2	3	75	14	1	7.1
F_4	7	2	7	87.5	35	1	2.8
F_5	15	2	15	93.75	143	1	0.7
F_6	31	2	31	96.88	535	1	0.2
F_7	126	2	126	98.44	2143	1	0.05

also tends to be enhanced. Enforced self-pollination of naturally cross-pollinated plants through consecutive generations may result in homozygous plants and a homogeneous population (**inbred line**), but vigor, size, and productivity may be reduced, a condition described as **inbreeding depression**. If, however, two inbred lines are crossed, the vigor of the plants of the resulting population may not only be restored but may show more size and vigor than either parent, a phenomenon known as **heterosis** or **hybrid vigor**. In this case, the individual plants will be heterozygous, but the population is likely to be homogeneous and have uniform characteristics.

inbred line A population of seedlings that produced a consecutive series of self-pollinations.

hybrid vigor Vigor expressed by a seedling population that exceeds that of either of the parents.

Many species have also developed morphological or genetic mechanisms to prevent self-pollination and promote cross-pollination.

Here are four illustrations of morphological adaptations to facilitate cross-pollination (6):

- **Dioecy. Dioecious** plants have pistillate (female) and staminate (male) flowers present in separate plants, such as asparagus (*Asparagus*), pistachio (*Pistacia*) and holly (*Ilex*) (Fig. 5–2). Plants with only female flowers are called gynoecious, and those with only male flowers are androecious. This type of flower arrangement usually forces cross-pollination.

dioecious Plant trait in which male and female flowers are produced on different plants.

- **Monoecy. Monoecious** plants have pistillate (female) and staminate (male) flowers in separate flowers on the same plant. This system occurs in cucurbits (*Cucurbita*), corn (*Zea*), walnut (*Juglans*) (Fig. 5–3), oak (*Fagus*), and many conifers. Although this facilitates cross-pollination,

monoecious Plant trait in which the male and female parts are in different flowers but on the same plant.

(a)   (b)

Figure 5–2
Holly (*Ilex*) plants are dioecious, producing female (a) and male (b) flowers on separate plants, forcing cross-pollination. Many flowers in dioecious plants produce remnant female and male parts that are usually non-functional. Note the non-functional male stamens present in the female flowers.

Figure 5–3
Some nut-producing tree species have pollination systems that ensure cross-pollination. Walnuts (*Juglans*) are monoecious with female (a) and male (b) flowers produced separately on the same plant.

self-pollination is usually possible in monoecious plants unless another barrier to self-pollination is present.
- **Dichogamy.** **Dichogamy** is the separation of female and male flower function in time (50). There are two types of dichogamy depending on whether the female becomes receptive before the male sheds pollen (protogyny) or the male sheds pollen before the female is receptive (protandry). There are numerous examples of this type of flowering including carnation (*Dianthus*) (Fig. 5–4). Dichogamy does not ensure cross-pollination but reduces the ratio of self- to cross-pollinated flowers (40).
- **Polymorphism.** Floral polymorphisms refer to different arrangements of flower parts in flowers from the same or different plants within the same species. Many of these adaptations are designed to alter the ratio of self- to cross-pollination. A range of flower structures is illustrated in asparagus (*Asparagus*) (Fig. 5–5, page 144). These types of polymorphisms were of particular interest to Darwin (14) as he described the different flower forms in primrose (*Primula*) referred to as heterostyly. Plants exhibiting heterostyly have two or three different flower morphologies where the style of the female and the filaments of the male are produced at differernt lengths (Fig. 5–6, page 144). In addition to the different heterostylous morphologies, each style and filament length combination may be linked to a sexual incompatibility system to limit which flowers can cross with each other (23).

> **dichogamy** Genetic trait in which male and female flowers on the same plant bloom at different times.

Sexual incompatibility (10, 15) is a general term that describes the inability of plants that are not genetically related to cross and produce offspring. **Self-incompatibility** is a form of sexual incompatibility that has evolved to prevent self-pollination within closely related species and has been found in over 250 plant genera from at least 70 families. Some horticulturally important plants showing self-incompatibility include lily (*Lilium*), cabbage (*Brassica*), *Petunia*, almond (*Prunus dulcis*), apple (*Malus*), cherry, and plum (*Prunus*).

> **sexual incompatibility** Genetic trait in which the pollen either fails to grow down the style or does not germinate on the stigma of a plant with the same incompatibility alleles.

Self-incompatible crosses are characterized by a lack of pollen germination or arrested pollen-tube growth (53). Self-incompatibility is a genetic mechanism controlled by a single gene locus (in diploids) with several different S alleles. It is controlled by protein-to-protein recognition determined by the type of S allele in the male and female partners. The two most common forms of self-incompatibility are gametophytic and sporophytic (Fig. 5–7, page 145). **Gametophytic self-incompatibility** is the most common form of self-incompatibility, and the interaction between the male and female partners is determined by a single S-allele derived from the haploid genetics within the pollen grain. Recognition only occurs after pollen germination and tube growth. When the male and female share a common S-allele genotype, there is a protein-to-protein interaction that stops pollen-tube

Figure 5–4
Sweet William carnation (*Dianthus*) flowers show dichogamy. Note how the flower on the left has anthers (an) shedding pollen before the style (s) has fully developed and the flower on the right that has fully receptive female parts after the anthers have withered.

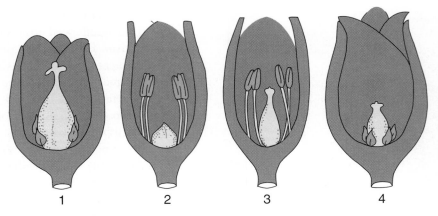

Figure 5–5
Range of flower structure types expressed in different asparagus flowers of individual plants. *Type 1.* Completely female. Dioecious. Flowers contain only the pistil; stamens (male) are reduced and nonfunctioning. *Type 2.* Completely male. Dioecious. Flowers only contain stamens. The pistil is reduced and nonfunctioning. *Type 3.* Both male and female structures are functioning. Perfect. *Type 4.* Both male and female structures are nonfunctioning. Sterile. Commercial seed production of asparagus results from growing Type 1 and Type 2 plants together to enforce cross-pollination and produce the desirable hybrid plants.
Courtesy Bryan Benson.

(a)

(b)

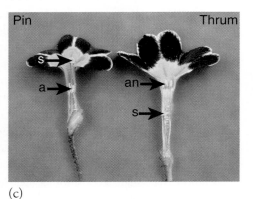
(c)

Figure 5–6
Examples of heterosyly in primrose (*Primula*). On the left are "pin" flowers where the stigma (s) is elevated above the corolla and the anthers (an) held on a short filament. On the right are "thrum" flowers with elongated filaments exposing the anthers above the corolla, and a shortened style, keeping the stigma within the corolla tube.

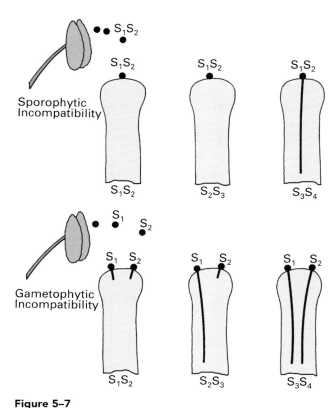

Figure 5–7
Incompatibility mechanisms prevent self-pollination in some species. *Top* (*cabbage*): Sporophytic incompatibility. Each pollen contains genes of both S_1 and S_2 alleles, and the pollen tube will only grow down a style with a different genotype. *Bottom* (*clover*): Gametophytic incompatibility. Each pollen grain has a single S allele. A pollen tube will not grow down a style where that allele is represented. Redrawn with permission from Stoskopf, et al. *Plant Breeding Theory and Practice*. Westview Press: Boulder, CO.

growth. In some families (i.e., Papaveraceae), the pollen tube stops growing soon after initial germination, while in others (i.e., Solanaceae, Rosaceae) the pollen tube will growth a considerable distance down the style before its growth is arrested (21). A unique breeding technique to bypass incompatibility in lily (*Lilium*) is to remove the upper two-thirds of the style (including the stigma) before applying pollen. This allows time for the pollen tube to reach the embryo sac before being arrested by the incompatibility reaction.

Sporophytic self-incompatibility differs from gametophytic self-incompatibility because it is the diploid S-allele pair from the male and female parents that determines compatibility (32). Protein types determined by different S alleles are deposited on the surface of the pollen grain where they interact with proteins on the stigmatic surface to determine whether the pollen grain will germinate and initiate tube growth. Examples of plants with this type of incompatibility are found in the Brassicaceae, Asteraceae, and Convolvulaceae families. Because multiple S alleles are involved in this recognition system, pollen/stigma interactions can be complex (Fig. 5–7).

Cross-pollination is mostly carried out by the movement of pollen by wind or insects. Also, pollination is sometimes by bats, birds, and water (48). Insect pollination is the rule for plants with white or brightly colored, fragrant, and otherwise conspicuous flowers that attract insects. The honeybee is one of the most important pollinating insects, although wild bees, butterflies, moths, and flies also obtain pollen and nectar from the flower (Fig. 5–8). Generally, pollen is heavy, sticky, and adheres to the body of the insect. Some important seed crops that require cross-pollination are alfalfa (*Trifolium*), birdsfoot trefoil (*Lotus*), red clover (*Trifolium pratense*), white clover (*Trifolium repens*), onion (*Allium*), watermelon (*Citrullis*) and sunflower (*Helianthus*) (Fig. 5–8c). In addition, many flower and vegetable crops are insect pollinated as are many fruit plants, ornamental plants, and deciduous and broad-leaved evergreens used in the landscape.

(a)   (b)   (c)

Figure 5–8
Important insect pollinators include (a) bees and (b) butterflies. (c) Bee hives are included in production fields to help pollination.

Wind pollination is the rule for many plants that have inconspicuous flowers, or those with monoecious, dioecious, or dichogamous flowers. Examples are grasses, corn, olive, and catkin-bearing trees such as the walnut (*Juglans*), oak (*Quercus*), alder (*Alnus*), cottonwood (*Populus*), and conifers (Fig. 5–9). The pollen produced from such plants is generally light and dry and, in some cases, carried long distances in wind currents.

Most trees and shrub species are both heterozygous and cross-pollinated such that considerable potential for genetic variability exists among the seedling progeny. Selection of seed source plants must take into account not only the characteristics of the plant itself but also the potential for cross-pollination with other species in the surrounding population. For example, the presence of off-type individuals in seedlings propagated from imported seed of *Eucalyptus* from Australia and pear (*Pyrus*) species from China and Japan (30) could be traced to hybridization with other species nearby.

Apomixis

Apomixis occurs when an embryo is asexually produced from a single cell of the sporophyte and does not develop from fertilization of two gametes (28). This new "vegetative" embryo may arise by mechanisms that were described in Chapter 4. In each case, the effect is that seed production becomes asexual and seed reproduction results in a clone. In some species, both apomictic and sexual seeds are produced, sometimes within the same ovule (**facultative**); bluegrass (*Poa pratensis*) falls into this category. Other species are essentially 100 percent apomictic (**obligate**); for example, Bahia grass (*Paspalum notatum*) and buffelgrass (*Pennisetum ciliare*).

> **facultative apomictic** A plant in which both sexual and asexual embryos are produced by the same plant.
>
> **obligate apomictic** A plant in which all the embryos are apomictic.

Breeding of apomictic cultivars requires that a genetic source for apomictic reproduction be found within that species. This trait is not identifiable by visual inspection of the parent plant but by its genetic performance (i.e., unexpected uniformity of its progeny from among normally variable populations). Apomixis has been most important in the breeding of grasses, forage crops, and sorghum. Introduced cultivars have included 'King Ranch' bluestem, 'Argentine' Bahia grass (*Paspalum*), and 'Tucson' side oats grama, 'Bonnyblue' and 'Adelphi' Kentucky bluegrass (*Poa*) (24), and buffelgrass (*Pennisetum*) (28). Relatively few genes apparently control apomixes, and breeding systems have been described to incorporate this trait into cultivars and particular species.

Apomictic reproduction in woody plant species and cultivars is found in many *Citrus* (9), mango (*Mangifera*), and some apple (*Malus*) species (52). Although apomixis produces genetically uniform seedlings, it is not necessarily useful for growing specific fruit cultivars because of undesirable juvenile tendencies, such as **thorniness, excess vigor,** and **delayed fruiting.** On the other hand, these characteristics make apomictic seedlings useful as rootstocks, characteristics exploited extensively in *Citrus*.

In apomixes, the seedling population is immediately stabilized as a "true-breeding" line without seedling variation. Such plants exhibit the apomictic cycle and express typical juvenile traits of the seedling population. Apomixis is particularly appropriate for

Figure 5–9
Conifers are usually wind pollinated. Male strobili (a) release pollen that is deposited on the female cone (b). True flowering plants (Angiosperms) developed along with insect pollinators. Wind pollination is a derived character that is usually associated with reduced flower parts (i.e., no petals) and unisexual flowers (c) as illustrated for chestnut (*Castanea*).

plants whose value lies in their vegetative characteristics—as occurs in forages and grasses—rather than in plants whose value depends on fruiting characteristics.

CATEGORIES OF SEED-PROPAGATED CULTIVARS AND SPECIES

Herbaceous Annual, Biennial, and Perennial Plants

Landraces Historically, farmers throughout the world have maintained seed-propagated plants by saving selected portions of the crop to be used to produce the next cycle. These populations, called **landraces,** evolved along with human societies and are still found in some parts of the world (56). These populations are variable but identifiable and have local names. This practice results in genetic populations adapted to a localized environment. Their inherent variability provides a buffer against environmental catastrophe and preserves a great deal of genetic diversity (Fig. 5–10).

> landrace Primitive varieties developed and maintained before the modern era of genetics.

Changes in cropping patterns have occurred during the 20th Century, particularly since about 1960. Many of the older populations around the world are being replaced by modern cultivars, which tend to be uniform and high yielding, particularly when grown in conjunction with high irrigation and fertility inputs. Sometimes, new cultivars lack adaptation to local environments. Although the trend has been to increase the world supply of essential food crops, concerns have been raised that a parallel loss of genetic diversity and germplasm has occurred. Exploration and conservation efforts have expanded to maintain these important raw materials for future use (19).

Cultivars A **cultivar** is a uniform and stable plant population that possesses recognizably distinct characteristics. Stated another way, a cultivar is a plant population that shows a minimum of variation, that can be propagated true-to-type for at least one characteristic, and is unique compared to the wild species or other cultivars. The term **variety** is often used interchangeably with cultivar especially when describing flower and vegetable populations. Care should be taken not to confuse variety with the concept of a true **botanical variety** (*varietas* or *var.*) that describes a type of naturally occurring population.

> botanical variety A population of plants originating in nature that are within one species but are phenotypically distinct.

Categories of seed-propagated cultivars include open-pollinated, lines, hybrids, synthetic, F_2, and clonal cultivars.

Open-pollinated cultivars can be maintained in cross-pollinated species that produce a relatively homogeneous population for specific traits important for production of that crop. Open-pollinated seed is often cheaper to produce compared to hybrid seed because

Figure 5–10

A landrace of soybeans (*Glycine*) in Africa showing the diversity inherent in seeds saved over many generations.

they do not require hand pollination to maintain the cultivar. However, because open-pollinated cultivars are a genetically heterogenic population, they can be more variable than hybrids (41).

Historically, many open-pollinated vegetable and flower varieties were maintained by families in their "kitchen gardens." Many of these varieties have since been maintained by generations of gardeners and local farmers and are being offered as **heirloom varieties.** The preservation and distribution of information concerning these varieties has been an objective of certain groups including Seed Savers Exchange, Inc., in Decorah, Iowa (4, 54, 59). There are also numerous commercial flower and vegetable crops produced as open-pollinated cultivars including *Begonia*, marigold (*Tagetes*), cucumber (*Cucumis*), and squash (*Cucurbita*).

line   A population of seedling plants whose genotype is maintained to a specific standard in consecutive generations.

Lines result in seedling populations whose genotype is maintained relatively intact during consecutive generations. These may be maintained as self- or cross-pollinated lines. An important type of seed population in this category is the **inbred** line, which are mainly used as parents for later production of F_1 hybrids (55).

Hybrid Cultivars include groups of individuals reconstituted each generation from specific parents. **F_1 hybrids** are the first generation of a planned cross. For seed production, they result from the cross between seedling populations of two or more inbred lines. When crossed with another inbred line, the result is a population of uniform, but heterozygous, plants. Often these populations exhibit greater vigor than the parents due to hybrid vigor (heterosis), depending on the combining ability of the parents. Hybridization is a means of "fixing" the genotype of the population similar to that described for self-pollinated lines. **Hybrid lines** were first produced in corn (*Zea mays*) (55) but have since been applied to many agronomic, vegetable, and flower crops (1, 26).

hybrid line   A seedling population that is produced by cross-pollinating two or more parental lines.

Hybrids may be produced between two inbred lines (**single-cross**), two single-crosses (**double-cross**), an inbred line and an open pollinated cultivar (**top-cross**), or between a single-cross and an inbred line (**three-way cross**) (55). Seeds saved from the hybrid population normally are not used for propagation because in the next generation, variability in size, vigor, and other characteristics may appear.

Synthetic cultivars are derived from the first generation of the open cross-fertilization of several lines or clones. For example, 'Ranger' alfalfa seed is made from inter-cropping five seed-propagated lines that results in genetically distinct but phenotypically similar seedlings in the seeded crop. Other crops in this category include pearl millet (*Pennisetum glaucus*), bromegrass (*Bromus*) and orchard grass (*Dactylis*).

synthetic line   A cultivar seedling population that is produced by combining a number of separately developing lines to produce a heterozygous but homozygous Cultivar.

F_2 cultivars are derived for open-pollination of an F_1 hybrid. Some flower crops, (*Petunia,* pansy (*Viola*), and *Cyclamen*) and vegetables (tomato and melon) can be maintained as F_2 populations.

Clonal seed cultivars are maintained through apomictic seed production (25, 51). Apomixis occurs when an embryo is asexually produced and does not develop from fertilization of two gametes (28). The result is a clonal copy of the parent plant. Apomixis is discussed in detail in Chapter 4. The degree of clonal seed production depends on whether the species has a **facultative** or **obligate** form of apomixis. In species with facultative apomixis, both apomictic and sexual seeds are produced, sometimes within the same seed. Bluegrass (*Poa pratensis*) falls into this category. Other species show essentially 100 percent obligate apomictic seed production. Examples include Bahia grass (*Paspalum notatum*) and buffelgrass (*Pennisetum ciliare*).

Woody Perennial Plants

Wild Populations   In nature, most species can be recognized as a more or less phenotypically (and genotypically) uniform seedling population that has evolved over time through consecutive generations to be adapted to the environment at a particular site. If the species covers a wide area, local variation in environment can result in different populations becoming adapted to different areas even though the plants may appear phenotypically similar. Plants within the species that show morphological differences compared to the species, but that are reproduced by seed, may be designated as **botanical varieties** or *varietas* or *var.* The term **form** indicates a particular

> **ecotype** A genetically distinct group of plants within a species that is adapted to a specific ecological location.
>
> **cline** Continuous genetic variation from one area to another in ecological adaptation.

phenotypic difference, as a blue or white color. Subgroups of a particular species that are morphologically similar but specifically adapted to a particular environmental niche are known as **ecotypes.** Variations that occur continuously between locations are known as **clines** (35, 43).

Provenance The climatic and geographical locality where seed is produced is referred to as its **seed origin** or **provenance** (2, 3, 38, 54). Variation can occur among plants associated with latitude, longitude, and elevation. Differences may be shown by morphology, physiology, adaptation to climate and to soil, and in resistance to diseases and insects. Natural plant populations growing within a given geographical area over a long period of time evolve so that they become adapted to the environmental conditions at that site.

> **provenance** A forestry term used to indicate the climatic and geographical locality where the seed originated.

Consequently, seeds of a given species collected in one locality may produce plants that are completely inappropriate to another locality. For example, seeds collected from trees in warm climates or at low altitudes are likely to produce seedlings that will not stop growing sufficiently early in the fall to escape freezing when grown in colder regions. The reverse situation—collecting seed from colder areas for growth in warmer regions—might be more satisfactory, but it also could result in a net reduction in growth resulting from the inability of the trees to fully utilize the growing season because of differences in the response to photoperiod (61).

Distinct ecotypes have been identified by means of **seedling progeny tests** in various native forest tree species, including Douglas-fir (*Pseudotsuga menziesii*), ponderosa pine (*Pinus ponderosa*), lodgepole pine (*Pinus murrayana*), eastern white pine (*Pinus strobus*), slash pine (*Pinus caribaea*), loblolly pine (*Pinus* sp.), shortleaf pine (*Pinus echinata*), and white spruce (*Picea alba*) (66). Other examples include the Baltic race of Scotch pine (*Pinus sylvestris*), the Hartz Mountain source of Norway spruce (*Picea abies*), the Sudeten (Germany) strain of European larch (*Larix*), the Burmese race of teak, Douglas-fir (*Pseudtxuga menziesii*) from the Palmer area in Oregon, ponderosa pine (*Pinus ponderosa*) from the Lolo Mountains in Montana, and white spruce (*Picea alba*) from the Pembroke, Ontario (Canada) area (16, 20, 31). Douglas-fir (*Pseudotsuga menziesii*) has at least three recognized races—*viridis, caesia,* and *glauca*—with various geographical strains within them that show different adaptations. For instance, progeny tests showed that a *viridis* strain from the United States West Coast was not winter-hardy in New York but was well suited to Western Europe. Those from Montana and Wyoming were very slow growing. Trees of the *glauca* (blue) strain from the Rocky Mountain region were winter-hardy but varied in growth rate and appearance. Strains collected farther inland were winter-hardy and vigorous; similar differences occurred in Scotch pine (*Pinus sylvestris*), mugho pine (*P. mugho*), Norway spruce (*Picea abies*), and others.

Improved Seed Sources Nursery propagation by woody plant seed can be upgraded significantly by the selection and development of improved seed sources. This practice applies to the production of rootstocks for fruit and nut trees (49, 58), shade trees (20, 43), and trees in the landscape. Likewise, foresters have been engaged in recent years in "domesticating" and upgrading forest tree production over that of "local" seed (38, 40, 60).

Elite Trees. Foresters refer to single seed-source trees with a superior genotype, as demonstrated by a progeny test, as **elite trees.** Nursery progeny tests can identify and characterize specific seed sources (e.g., for landscape or Christmas tree uses).

> **elite tree** An individual tree with outstanding phenotypic characteristics to be used as a seed source.

Clonal Seed Sources. Superior (elite) seed-source trees can be maintained as clones in seed orchards to preserve the genotype of the parent. Seeds from this clonal source are then used to produce seedling trees in the nursery. This procedure is used to produce rootstock seed for fruit and nut cultivars. For instance, 'Nemaguard' is a peach hybrid whose nematode-resistant seedling progeny are used for almond, peach, plum, and apricot trees in central California (49). Named cultivars of ornamental trees have been

> **clonal seed source** Cultivar maintained as a clone selected for producing outstanding seedlings.

identified as producing uniform, superior seedling progeny (16).

selected families Consecutive groups of progeny trees related by origin and showing superior characteristics.

Selected Families. Genetic improvement of forest-tree species has brought about **family selection** by growing progeny trees either from controlled crossing or selection from single open-pollinated superior (elite) trees. Seed orchards then may be established either from seedlings of these trees or established by grafting the parent trees. A minimum number of individual genotypes are selected—usually around twenty-five—to avoid the dangers of inbreeding and limits to the genetic range. Progeny trees are planted in test sites and evaluated for various forestry characteristics. Over time, superior sources are identified and preserved as parents to produce the next generation of new families of improved seed genotypes. Inferior seed sources are identified and eliminated.

Hybrid Seed Sources. F_1 hybrids of two species usually produce uniform populations of plants in the same manner as hybrid seeds of corn and other inbred lines (see page 157). For example, hybridization has been valuable in producing vigorous almond × peach hybrids for almond and peach rootstocks (33, 34), Paradox hybrids (*Juglans hindsii* × *J. regia*) for walnuts and fast-growing poplars from crosses with North American and European poplars (like *Populus trichocarpa.* × *P. deltoides*). Forest tree hybrids, such as *Pinus rigida* × *P. taeda* in Korea and *Larix decidua* × *L. leptolepsis* in Europe are not necessarily uniform, however, but have been a focus of improved forests. Because of expense and uncertainty of production, seeds of F_1 hybrids of the forest trees have been used to produce F_2 seedling populations. The more vigorous hybrid plants dominate and the weaker trees are crowded out.

CONTROL OF GENETIC VARIABILITY DURING SEED PRODUCTION

Herbaceous Annual, Biennial, and Perennial Plants

Isolation Isolation is used to prevent mechanical mixing of the seed during harvest and to prevent contamination by unwanted cross-pollination with a different but related cultivar. Isolation is achieved primarily through distance, but it can also be attained by enclosing plants or groups of plants in cages, enclosing individual flowers, or removing male flower parts (i.e., de-tasseling corn) and then manually applying pollen of a known source by hand or various other devices (Fig. 5–11). On a large scale, this goal can be achieved by using male-sterile parents (36). In a number of crop species [e.g., tobacco *(Nicotiana)* and onion *(Allium)*], specific genes have been identified that prevent normal formation of the male (pollen) reproductive structures (Fig. 5–12). This means that no viable pollen is produced. The most common form of **pollen sterility** is cytoplasmic male sterility, which is a complex interaction between nuclear and mitochondrial plant genes (12). Such traits can be bred into parental lines of specific cultivars for the production of **hybrid seed.** Using molecular biology to induce male sterility has also become a potential strategy to limit gene flow from transgenic plants into the environment (11).

pollen sterility Genetic phenomenon in which the pollen is nonviable.

hybrid seed Seed produced by the crossing of two dissimilar parents, usually produced when made between species.

(a)

(b)

Figure 5–11
Isolation is used to prevent unwanted cross-pollination during seed production. (a) An onion hybrid cross being isolated with an individual sac. (b) A small cage where flies will be introduced to pollinate onions for hybrid production. Female plants in the cage are made sterile to prevent self-pollination.

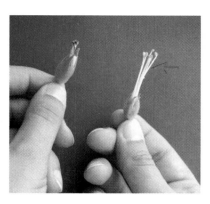

Figure 5–12
Male sterility in tobacco (*Nicotiana*). Note how the flower on the left lacks developed anthers compared to the perfect flower on the right with elongated and fertile anthers (arrow).

Self-pollinated cultivars of herbaceous plant species need only to be separated to prevent mechanical mixing of seed of different cultivars during harvest. The minimum distance usually specified between plots is 3 m (10 ft), but may be up to 50 to 65 m (150 to 200 ft) depending on the degree of cross-pollination capacity in the crop. For example, bell pepper is a self-pollinating crop but, given the opportunity, will cross-pollinate to a high degree from bee pollinators. Careful cleaning of the harvesting equipment is required when a change is made from one cultivar to another. Sacks and other containers used to hold the seed must be cleaned carefully to remove any seed that has remained from previous lots.

More isolation is needed to separate cultivars cross-pollinated by wind or insects. The minimum distance depends on a number of factors:

- the degree of natural cross-pollination
- the relative number of pollen-shedding plants
- the number of insects present
- the direction of prevailing winds

The minimum distance recommended for insect-pollinated herbaceous plant species is 0.4 km (1/4 mi) to 1.6 km (1 mi). The distance for wind-pollinated plants is 0.2 km (1/8 mi) to 3.2 km (2 mi), depending on species.

Effective cross-pollination usually can take place between cultivars of the same species; it may also occur between cultivars of a different species but in the same genus; rarely will it occur between cultivars belonging to another genus. Since the horticultural classification may not indicate taxonomic relationships, seed producers should be familiar with the botanical relationships among the cultivars they grow. It is also important to isolate GMO (genetically modified organisms) crops from non-GMO seed crops of the same cultivar or species.

In seed production areas, such as regions of Oregon, Washington, and California, seed companies cooperate to locate seed production fields at appropriate isolation distances from each other. The fields are located on "pinning" maps (each colored pin indicating a field and crop type), and the maps are located in County Extension offices within the production areas. Recently, these have also become available as virtual maps on the Internet, as is the case in California.

Roguing The removal of off-type plants, plants of other cultivars, and weeds in the seed production field is known as **roguing** (37). During the development of a seed-propagated cultivar, positive selection is practiced to retain a small portion of desirable plants and to maximize the frequency of desirable alleles in the population. During seed production, roguing following visual inspection exerts selection by eliminating the relatively small population that is not "true to type," thus keeping the cultivar "genetically pure."

> **roguing** The act of removing off-type plants, weeds, and plants of other cultivars in seed production fields.

Off-type characteristics (i.e., those that do not conform to the cultivar description) may arise because recessive genes may be present in a heterozygous condition even in homozygous cultivars. Recessive genes arising by mutation would not be immediately observed in the plant in which they occur. Instead, the plant becomes heterozygous for that gene, and, in a later generation, the gene segregates and the character appears in the offspring. Some cultivars have mutable genes that continuously produce specific off-type individuals (45). Off-type individual plants should be rogued out of the seed production fields before pollination occurs. Systematic inspection of the seed-producing fields by trained personnel is required.

Other sources of off-type plants include contamination by unwanted pollen due to inadequate isolation or volunteer plants arising from accidentally planted seed or from seed produced by earlier crops. Seed production fields of a particular cultivar should not have grown a potentially contaminating cultivar for a number of preceding years.

Weeds are plant species that have been associated with agriculture as a consequence of their ability to exploit disturbed land areas when cultivation occurs (29). Some weed species have evolved seed types that closely resemble crop seeds and are difficult to screen out during seed production.

Seedling Progeny Tests Planting representative seeds in a test plot or garden may be desirable to test for trueness-to-type. This procedure is used in the development of a cultivar to test its adaptability to various environments. The same method may be necessary to test whether changes have occurred in the frequency of particular genes or new gene combinations may have developed during seed increase generations. These changes can result from selection pressure exerted by management practices or environmental interaction. For example, intensive roguing may result in a **genetic drift** due to changes in the frequency of particular genes or gene combinations (24). Shifts may also occur due to environmental exposure in a growing area which is different from the initial selection area. Seedlings of particular genotypes may survive better than others and contribute more to the next generation. If sufficiently extensive, genetic drift could produce populations of progeny plants that differ somewhat from those of the same cultivar grown by other producers. Or the cultivar may have changed from the original breeder's seed.

> **genetic drift** Change in the frequency of specific genes as a result of environmental or other types of selection.

Problems can result if seed crops of particular perennial cultivars are grown in one environment (such as a mild winter area) to produce seed to be used in a different and more severe environment (such as an area requiring cold-hardiness). This situation has occurred, for example, with alfalfa (24) where rules for production of forage crop seed in a mild winter area can specify only one seedling generation of increase.

Woody Perennial Plants

Use of Local Seed "Local seed" means seed from a natural area subjected to a restricted range of climatic and soil influences. As applied to forest tree seed, this usually means that the collection site should be within 160 km (100 mi) of the planting site and within 305 m (1,000 ft) of its elevation. In the absence of these requirements, seed could be used from a region having as nearly as possible the same climatic characteristics. The reason for the historical emphasis on local seed is the phenomenon previously defined as seed origin or provenance (27). The use of local seed for herbaceous and woody plants is particularly important in the effort to restore any native ecosystem where the use of exotic species would be inappropriate (43).

Pure Stands. **Pure stand** refers to a group of phenotypically similar seedling plants of the same kind. This concept could apply both to plants growing in a natural environment or in a planting such as a wood lot. These populations are useful in seed collection because cross-pollination would likely occur from among this group of plants and one can judge both the female and the male parents. Although the individuals are likely to be heterozygous, they should produce good seeds and vigorous seedlings. The population should be homogeneous and reproduce the parental characteristics.

> **pure stand** An interbreeding group of phenotypically similar plants of the same kind growing in a given site.

Phenotypic Tree Selection Versus Genotypic Tree Selection. **Phenotypic selection** refers to evaluation of a seed source through visual inspection of the source plant(s). The basis of this procedure is that many important traits in forestry—such as stem form, branching habit, growth rate, resistance to diseases and insects, presence of surface defects, and other qualities—are inherited quantitatively. Geneticists refer to this relationship as high phenotypic correlation between parents and offspring. In practical terms, this means that the parental performance can be a good indicator of the performance of the offspring (57).

> **phenotypic selection** Selection of a seed source based on the phenotypic appearance of the source tree.

When individual trees in native stands show a superior phenotype, foresters call them "plus" or elite trees and sometimes leave them for natural reseeding or as seed sources. Such dominant seed trees may contribute the bulk of natural reseeding in a given area.

Genotypic selection refers to evaluation of a seed source based on the performance of their seedling progeny test (39). Seeds may be produced by open pollination (OP) where only one parent is known. Or the test may be made from a controlled cross, where both parents are known and the contributions of each can be evaluated. A progeny test establishes the breeding value of a particular seed source (5) because genetic potential is based on actual performance of the progeny.

> **genotypic selection** Selection of a seed source based on the phenotypic appearance of the seedling progeny.

A representative sample of seeds is collected, planted under test conditions, and the progeny observed over a period of years. A high correlation between the average phenotypic traits of the parent(s)

high additive heritability High correlation between phenotypic traits of offspring with the phenotypic traits of the parents.

low additive heritability Low correlation between phenotypic traits of offspring with the phenotypic traits of the parents.

and the average phenotypic response of the offspring is referred to as **high additive heritability** and justifies using the "best" trees for seed sources of the next generation (38). A low statistical correlation between parent and progeny characteristics is referred to as **low additive heritability** in that the desired traits of the progeny cannot be predicted from inspection of the parents.

Progeny testing is useful to verify the suitability of individual seed sources for future seed collecting. The procedure is an important component to the improvement of woody plants whether in forestry or horticulture.

SEED PRODUCTION SYSTEMS

Herbaceous Annual, Biennial, and Perennial Plants

Commercial Plantings Traditional seed selection of herbaceous plants utilized a portion of the seed from one year's crop to plant a crop for the next year. This system would be satisfactory for self-pollinated cultivars that are easy to maintain genetically. For cross-pollinated cultivars, knowledge of the production requirements of individual crops is needed and specific conditions are practiced depending upon the plant (3). Note, however, that inadequacies of this method led to its replacement by the pedigree system.

Pedigreed Stock System (4, 42) Commercial seed production of most self-pollinated and cross-pollinated lines is carried out in three steps (Fig. 5–13). The purpose of a **pedigreed stock system** is to maintain genetic purity

pedigreed stock system A controlled seed-production system of consecutive generations with standards to maintain genetic purity leading to commercial distribution.

through consecutive seed generations following appropriate standards of isolation, inspection, and roguing (with high costs) at the initial release with decreased standards (and lower costs) in the distribution of commercial seed. The overall program includes three phases. Phase 1 includes the *development* phase, which ends

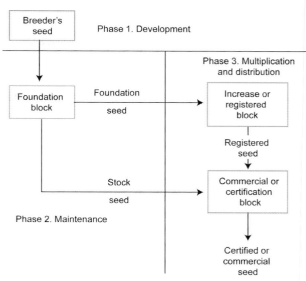

Figure 5–13
Pedigree system for seed production. See text for discussion.

with the production of a small quantity of seeds (**breeder's seed**) that is maintained by the originating institution as the primary reference for the cultivar. Phase 2 is a *maintenance* phase in which a quantity of seed called **foundation seed** (for certified seed classes; see Box 5.1, page 154) or **stock seed** (in commercial enterprises) is maintained under high standards of isolation, inspection, and roguing. Phase 3 is the *distribution* phase, which may include two steps: a *second-generation increase* block and a *third-generation* block to produce commercial seed for distribution to the public. A foundation planting originates only from breeder's seed or another foundation planting. An increase block originates only from a foundation seed or another increase planting. A seed production planting originates from foundation seed or increase block seed. This entire production process is carried out either by large commercial firms or groups of independent growers joined within a Crop Improvement Association to produce certified seed.

Seed Certification (4, 13, 17, 42) **Seed certification** is a legalized program that applies the previously mentioned principles to the production of specific seed-propagated plant cultivars to ensure the maintenance of seed purity. The system was established in the United States and Canada during the early 1920s to regulate the

seed certification A system of seed production utilizing pedigreed stock principles, which provides for legally enforceable standards of quality and genetic purity.

BOX 5.1 GETTING MORE IN DEPTH ON THE SUBJECT
CLASSES OF CERTIFIED SEEDS

Breeder's seed: that which originates with the sponsoring plant breeder or institution and provides the initial source of all the certified classes. **Foundation seed:** progeny of breeder's seed that is handled to maintain the highest standard of genetic identity and purity. It is the source of all other certified seed classes but can also be used to produce additional foundation plants. (**Select seed** is a comparable seed class used in Canada.) Foundation seed is labeled with a *white* tag or a certified seed tag with the word "foundation." **Registered seed:** progeny of foundation seed (or sometimes of breeder's seed or other registered seed) produced under specified standards approved and certified by the certifying agency and designed to maintain satisfactory genetic identity and purity. Bags of registered seed are labeled with a *purple* tag or with a blue tag marked with the word "registered." **Certified seed:** progeny of registered seed (or sometimes of breeder's, foundation, or other certified seed) that is produced in the largest volume and sold to crop producers. It is produced under specified standards designed to maintain a satisfactory level of genetic identity and purity and is approved and certified by the certifying agency. Bags of certified seed have a *blue* tag distributed by the seed-certifying agency.

---

commercial production of new cultivars of agricultural crops then being introduced in large numbers by state and federal plant breeders. The principles (as described for the pedigreed stock system) and accompanying regulations of seed maintenance were established through the cooperative efforts of public research, extension, regulatory agencies, and seed-certifying agencies known as Crop Improvement Associations, whose membership included commercial producers. These organizations were designated by law through the Federal Seed Act (1939) to conduct research, establish production standards, and certify seeds that are produced under these standards. Individual state organizations are coordinated through the Association of Official Seed Certifying Agencies (AOSCA) (4) in the United States and Canada. Similar programs exist at the international level where certification is regulated through the Organization for Economic Cooperation and Development (OECD).

The principal objective of seed certification is to provide standards to preserve the genetic qualities of a cultivar. Other requirements of seed quality also may be enforced as well as the eligibility of individual cultivars. The seed-certifying agency may determine production standards for isolation, maximum percentage of off-type plants, and quality of harvested seed; make regular inspections of the production fields to see that the standards are being maintained; and monitor seed processing.

The international OECD scheme includes similar classes but uses different terms. These include **basic** (equivalent to either foundation or registered seed), **certified first-generation** (blue tag) seed, and **second-generation** (red tag) seed.

Hybrid Seed Production (1) Hybrid cultivars (Fig. 5–14) are the F_1 progeny of two or more parental lines. Parent plants are maintained either as inbred lines (corn, onion) or as vegetatively propagated clones (asparagus). The same standards of isolation as for nonhybrid seed production may be required. To mass-produce hybrids, some system must be used to prevent self-pollination and to enforce cross-pollination. Hand pollination is sometimes practiced to produce seed in crops or situations in which the production of seed per flower is very high and/or the high value of the seed justifies the expense (Fig. 5–15, page 156). Hand pollination is used to produce some hybrid flower seeds and in breeding new cultivars (Fig. 5–16, page 157).

Perennial Sources

Commercial Sources Seeds for fruit and nut crops historically have been collected more or less successfully from commercial orchards particularly where the specific cultivar or origin is known. Fruit tree seeds such as apple (*Malus*), pear (*Pyrus*), and peach (*Prunus*) have been collected from canneries and dry yards where specific commercial cultivars, such as 'Lovell' peach, are used. Pure stands of local seedling landscape trees might be used. In several forest-tree species, seed has been collected from phenotypically above-average trees in commercial plantations. In New Zealand, seed from such trees is designated "CS" ("climbing selects") and rated higher in value than the seed from the surrounding trees, but below that of seed orchards or from well-tested families.

Seed-Collection Zones A **seed-collection zone** for forest trees is an area with defined boundaries and altitudinal limits in which soil and climate

seed-collection zone
Naturally occurring zone (forest plants) designated by elevation, latitude, and longitude that identifies a specific seed source.

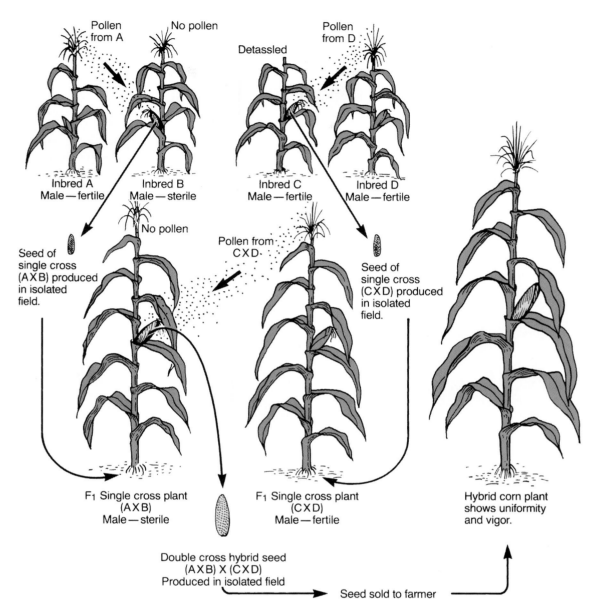

Figure 5–14
Hybrid seed corn production. Four inbred lines are produced to be used as parents for cross-pollination utilizing either detasseling (removal of male flower) or a pollen-sterile parent. The resulting F_1 plants are used as parents of the next (F_2) generation which are then sold for commercial crop production. The individual progeny plants are highly heterozygous, but the population is highly homogeneous, showing high vigor and production. Redrawn from USDA Yearbook of Agriculture 1937, Washington, D.C.: U.S. Government Printing Office.

are sufficiently uniform to indicate a high probability of reproducing a single ecotype. Seed-collection zones, designating particular climatic and geographical areas, have been established in most of the forest-tree–growing areas in the world (61). California, for example, is divided into six physiographic and climatic regions, 32 subregions, and 85 seed-collection zones (Fig. 5–17, page 158). Similar zones are established in Washington and Oregon and in the central region of the United States.

Seed-Production Areas (43)
A **seed-production area** contains a group of plants that can be identified and set aside specifically as a seed

seed-production area An area of source plants specifically utilized for seed collection.

Figure 5–15
(a and b) Hybrid pollination in petunia. Hand pollination involves removal of the male anthers (emasculation) before the flower opens; (c and d) followed by transfer of the pollen to the receptive stigma. Pollen is collected and stored dry in small vials at cold temperature. Pollen is transferred to the stigma using a small transfer stick or brush. (e) A successful pollination/fertilization is evident by continued development (red arrow). (f) Seed yield per capsule is high and the seed is very valuable justifying the use of hand labor.

source. Seed plants within the area are selected for their desirable characteristics. The value of the area might be improved by removing off-type plants, those that do not meet desired standards, and other trees or shrub species that would interfere with the operations. Competing trees can be eliminated to provide adequate space for tree development and seed production. In forestry, an isolation zone at least 120 m (400 ft) wide

(a)

(c)

(b)

Figure 5–16
F_1 hybrid cultivars of many vegetable and flowering annuals are created following large-scale hand pollination in a controlled environment. (a) Removing male parts in snapdragon prior to hand pollination. (b) Hand pollination. (c) Pepper fruit prior to harvest for seed extraction.

seed orchard
A planting used in forestry or in fruit tree nurseries to maintain seed sources as seedling populations of selected seed families or of a clone (fruit and nut trees) or collections of clones (forestry).

from which off-type plants are removed should be established around the area.

Seed Orchards Seed orchards are established to produce tree seeds of a particular origin or source. For example, fruit tree nurseries maintain seed orchards to produce seeds of specific rootstock cultivars under conditions that will prevent cross-pollination and the spread of pollen-borne viruses. A clonal cultivar such as 'Nemaguard' peach is budded to a rootstock, planted in isolation to avoid chance cross-pollination by virus-infected commercial cultivars, and grown specifically for rootstock seed production as part of the nursery operations.

BOX 5.2 GETTING MORE IN DEPTH ON THE SUBJECT
TREE CERTIFICATION CLASSES

Certification of forest-tree seeds is available in some states and European countries similar to that for crop seed (17, 18, 61). Recommended minimum standards are given by the Association of Official Seed Certifying Agencies (4). Forest-tree seed have different standards than agricultural seeds.

Tree certification classes are defined as **source-identified**: tree seed collected from natural stands where the geographic origin (source and elevation) is known and specified or from seed orchards or plantations of known provenance, specified by seed-certifying agencies. These seeds carry a *yellow* tag. **Selected**: tree seed collected from trees that have been selected for promising phenotypic characteristics but have not yet been adequately progeny-tested. The source and elevation must be stated (44). These seeds are given a *green* label. **Certified**: two types of seed are recognized. Seeds are from trees growing in a seed orchard whose genetic potential is based on phenotypic superiority. These are identified by a *pink* tag. When seedlings or seeds have been proven to be genetically superior in a progeny test, they are classified as **tested** and identified by a *blue* tag.

Fruit-tree rootstock clones that are self-pollinated are planted in solid blocks. An isolation zone 120 m (400 ft) wide should be established around the orchard to reduce pollen contamination from other cultivar sources. The size of this zone can be reduced if a buffer area of the same kind of tree is present around the orchard. Hybrid seed production involves planting both the parental clones in adjoining rows under the same condition.

Three general types of seed orchards are used for forest trees (42, 61): (a) seedling trees produced from selected parents through natural or controlled pollination; (b) clonal seed orchards in which selected clones are propagated by grafting, budding, or rooting cuttings; and (c) seedling-clonal seed orchards in which certain clones are grafted onto branches of some of the trees. The choice depends on the particular strategy used in the seed improvement program.

A site should be selected for good seed production. Forest trees and most other native species should include a range of genotypes in a suitable arrangement to ensure cross-pollination and to decrease the effects of inbreeding. Seven to thirty unrelated genotypes have been recommended to avoid this problem in a purely clonal orchard.

Nursery Row Selection Sometimes phenotypically unique individuals appear in nursery seed populations planted in the nursery row and can be identified visually. This is referred to as **nursery row selection.** Identification requires that the character be distinctive in vigor, appearance, or both. For example, Paradox hybrid walnut seedlings (*Juglans regia* × *J. hindsii*) used as a rootstock in California are produced by planting seeds of specific seed tree sources of black walnut (*Juglans hindsii*). Once germination takes place, leaf characteristics, bark color, and greater vigor of the

> **nursery row selection**
> A system of selection where specific progeny trees can be identified in the nursery row due to phenotype.

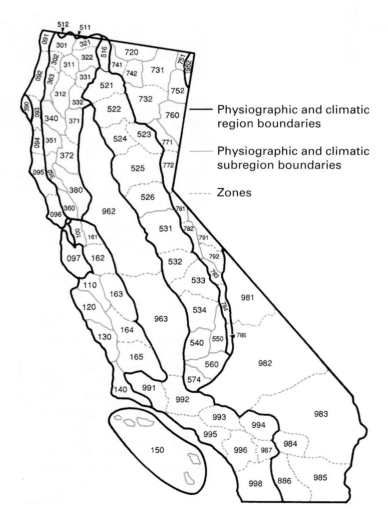

Figure 5–17
Seed collection zones in California. The 85 zones are identified by a three-digit number. The first gives 1 of the 6 major physiographic and climatic regions, the second gives 1 of the subregions, of which there are 32, and the third gives the individual zone. 0 = Coast; 1 = South coast; 3 = North coast; 5 = Mountains of Sierra Nevada and Cascade ranges; 7 = Northeast interior; 9 = Valley and desert areas, further divided into the central valley (6), southern California (9), and the desert areas (7). Redrawn from G. H. Schubert and R. S. Adams. 1971. Reforestation Practices for Conifers in California. Sacramento: Division of Forestry.

> **BOX 5.3   GETTING MORE IN DEPTH ON THE SUBJECT**
> **PLANT VARIETY PROTECTION ACT**
>
> Breeders of a new seed-reproduced plant variety (cultivar) in the United States may retain exclusive propagation rights through the Plant Variety Protection Act, established in 1970 (7), and revised in 1994 (46). The breeder applies to the U.S. Department of Agriculture for a Plant Variety Protection Certificate. For one to be granted, the cultivar must be "novel": it must differ from all known cultivars by one or more morphological, physiological, or other characteristics. It must be uniform; any variation must be describable, predictable, and acceptable. It must be stable (i.e., essential characteristics must remain unchanged during seed propagation). A certificate is good for 20 years. The applicant may designate that the cultivar be certified and that reproduction continue only for a given number of seed generations from the breeder or foundation stock. If designated that the cultivar be certified, it becomes unlawful under the Federal Seed Act to market seed by cultivar name unless it is certified. The passage of this law has greatly stimulated commercial cultivar development.

hybrid seedlings allow identification of the desired hybrid seedlings, whereas the black walnut seedlings are rogued out of the nursery row or separated at a later date. Genotypic progeny tests based on previous commercial experience indicate which walnut tree sources produce the highest percentages of hybrids, presumably from natural crossing with surrounding Persian walnut (*J. regia*) orchards.

Sometimes the phenotype of the propagated plant is sufficiently striking as to allow for selection in the nursery. For example, blue seedlings of the Colorado spruce (*Picea pungens*) tend to appear among seedling populations having the usual green form. Differences in fall coloring among seedlings of *Liquidambar* and *Pistacia chinensis* necessitate fall selection of individual trees for landscaping.

DISCUSSION ITEMS

Propagators of many plant species and cultivars may not be involved directly in the selection and handling of the seeds used but depend on the skill and knowledge of the specialized seed industry. Nothing is more important, however, than using seeds that are true-to-type and true-to-name. Consequently, knowledge of the basic principles and practices that are required to produce genetically pure seeds is important to propagators whether or not they are directly involved in seed selection.

1. From the propagator's standpoint, why do you think crop plants such as wheat, rye, and barley played such an important role in human history?
2. What are some major reasons why seed producers like to produce hybrid seed lines?
3. What are the differences and similarities among apomictic, inbred, and hybrid seed lines?
4. What is the function of seedling progeny tests in seed production?
5. Why is the seed origin (provenance) important to users of tree crops?
6. Would it be better to collect seeds from a single woody plant or from multiple plants?

REFERENCES

1. Acquaah, G. 2008. *Principles of plant genetics and breeding.* Wiley Press.
2. Ahuja, M. R., and W. L. Libby. 1993. *Clonal Forestry: I. Genetics and biotechnology. II. Conservation and application.* New York: Springer-Verlag.
3. Allard, R. W. 1999. *Principles of plant breeding.* 2nd ed. New York: John Wiley & Sons.
4. Assoc. Off. Seed Cert. Agencies. *Seed certification.* http://www.aosca.org
5. Barker, S. C. 1964. Progeny testing forest trees for seed certification programs. *Ann. Rpt. Inter. Crop Imp. Assn.* 46:83–7.
6. Barrett, S. C. H. 2002. The evolution of plant sexual diversity. *Nature Reviews Genetics* 3:274–84.
7. Barton, J. H. 1982. The international breeder's rights system and crop plant innovation. *Science* 216:1071–5.

8. Callaway, D. J., and M. B. Callaway. 2000. *Breeding ornamental plants.* Portland, OR: Timber Press.

9. Cameron, J. W., R. K. Soost, and H. B. Frost. 1957. The horticultural significance of nucellar embryony. In J. M. Wallace, ed. *Citrus virus diseases.* Berkeley: Univ. Calif. Div. Agr. Sci. pp. 191–96.

10. Charlesworth, D., X. Vekemans, V. Castric, and S. Glemin. 2005. Plant self-incompatibility systems: A molecular evolutionary perspective. *New Phytologist* 168:61–9.

11. Chase, C. D. 2006a. Cytoplasmic male sterility: A window to the world of plant mitochondrial–nuclear interactions. *Trends Genetics* 23:81–90.

12. Chase, C. D. 2006b. Genetically engineered male sterility. *Trends Plant Sci.* 11:7–9.

13. Cowan, J. R. 1972. Seed certification. In T. T. Kozlowski, ed. *Seed biology,* Vol. 3. New York: Academic Press. pp. 371–97.

14. Darwin, C. 1862. On the two forms, or dimorphic condition, in the species of *Primula,* and on their remarkable sexual relations. *J. Proc. Linnaean Soc. (Botany)* 6:77–96.

15. De Nettancourt, D. 1993. Self- and cross-incompatibility systems. In M. D. Hayward, N. O. Bosemark, and I. Romagosa, eds. *Plant breeding: Principles and prospects.* London: Chapman and Hall.

16. Dirr, M. A., and C. W. Heuser, Jr. 1987. *The reference manual of woody plant propagation.* Athens, Ga.: Varsity Press.

17. Edwards, D. G. W., and F. T. Portlock. 1986. Expansion of Canadian tree seed certification. *For. Chron.* 62:461–66.

18. Ehrenberg, C., A. Gustafsson, G. P. Forshell, and M. Simak. 1955. Seed quality and the principles of forest genetics. *Heredity* 41:291–366.

19. Esquinas-Akacazar, J. T. Plant genetic resources. In M. D. Hayward, N. O. Bosemark, and I. Romagosa, eds. *Plant breeding: Principles and prospects.* London: Chapman and Hall.

20. Flint, H. 1970. Importance of seed source to propagation. *Proc. Intl. Plant Prop. Soc.* 20:171–78.

21. Franklin-Tong, V. E., and F. C. H. Franklin. 2003. The different mechanisms of gametophytic self-incompatibility. *Philos. Trans. R. Soc. Lond. B. Biol. Sci.* 358:1025–32.

22. Frey, K. J. 1983. Plant population management and breeding. In D. R. Wood, K. M. Rawal, and M. N. Wood, eds. *Crop breeding.* Madison, Wis.: Amer. Soc. Agron. and Crop Sci. Soc. Amer. pp. 55–88.

23. Ganders, F. R. 1979. The biology of heterostyly. *New Zealand J. Bot.* 17:607–35.

24. Garrison, C. S., and R. J. Bula. 1961. Growing seeds of forages outside their regions of use. In *Seed yearbook of agriculture.* Washington, DC: U.S. Govt. Printing Office, pp. 401–6.

25. Geneve, R. L. 2006. Alternative strategies for clonal plant reproduction. *Comb. Proc. Intl. Plant Prop. Soc.* 56:269–73.

26. Goldsmith, G. A. 1976. The creative search for new F_1 hybrid flowers. *Proc. Intl. Plant Prop. Soc.* 26:100–3.

27. Haddock, P. G. 1968. The importance of provenance in forestry. *Proc. Intl. Plant Prop. Soc.* 17:91–8.

28. Hanna, W. W., and E. C. Bashaw. 1987. Apomixis: Its identification and use in plant breeding. *Crop Science* 27:1136–9.

29. Harlan, J. R. 1992. *Crops and man,* 2nd ed. Madison, Wis.: Amer. Soc. Agron. Crop Sci.

30. Hartmann, H. 1961. Historical facts pertaining to root and trunkstocks for pear trees. *Oreg. State Univ. Agr. Exp. Sta. Misc. Paper* 109:1–38.

31. Heit, C. E. 1964. The importance of quality, germinative characteristics and source for successful seed propagation and plant production. *Proc. Intl. Plant Prop. Soc.* 14:74–85.

32. Hiscock, S. J., and D. A. Tabah. 2003. The different mechanisms of sporophytic self-incompatibility. *Philos. Trans. R. Soc. Lond. B. Biol. Sci.* 358:1037–45.

33. Jones, R. W. 1969. Selection of intercompatible almond and root knot nematode resistant peach rootstocks as parents for production of hybrid rootstock seed. *J. Amer. Soc. Hort. Sci.* 94:89–91.

34. Kester, D. E., and C. Grasselly. 1987. Almond. In R. C. Rom, and R. Carlson, eds. *Rootstocks for fruit trees.* New York: John Wiley & Sons.

35. Langlet, O. 1962. Ecological variability and taxonomy of forest trees. In T. T. Kozlowski, ed. *Tree growth.* New York: Ronald Press pp. 357–69.

36. Lasa, L. M., and N. O. Bosemark. 1993. Male Sterility. In M. D. Hayward, N. O. Bosemark, and I. Romagosa, eds. *Plant breeding. Principles and prospects.* London: Chapman and Hall.

37. Laverack, G. K., and M. R. Turner. 1995. Roguing seed crops for genetic purity: A review. *Plant Varieties and Seeds* 8:29–45.

38. Libby, W. J., and R. M. Rauter. 1984. Advantages of clonal forestry. *For. Chron.* pp. 145–49.

39. Lindgren, D., J. Cui, S. G. Son, and J. Sonesson. 2004. Balancing seed yield and breeding value in clonal seed orchards. *New Forests* 28:11–22.

40. Macdonald, B. 1986. *Practical woody plant propagation for nursery growers,* Vol. 1. Portland, OR: Timber Press.

41. Maynard, D. N., and G. J. Hochmuth. 1997. *Knott's handbook for vegetable growers.* New York: John Wiley & Sons.

42. McDonald, M. B., Jr., and W. D. Pardee, eds. 1985. *The role of seed certification in the seed industry.* CSSA Spec. pub. 10. Madison, WI: Crop Sci. Soc. Amer., ASA.

43. Millar, C. I., and W. J. Libby. 1991. Strategies for conserving clinal, ecotypic, and disjunct population diversity in widespread species. In D. A. Falk and K. E. Holsinger, eds. *Genetics and conservation of rare plants.* Oxford: Oxford University Press.

44. National Tree Seed Laboratory. 2001. *Seed certification.* http://www.ntsl.net/Ntsl_dcert.htm.

45. Pearson, O. H. 1968. Unstable gene systems in vegetable crops and implications for selection. *HortScience* 3(4):271–74.

46. Plant Variety Protection Office. 2001. *Mission and general information.* Web site http://www.as.usda.gov./science/pvp.htm.

47. Poehlman, J. M. 1995. *Breeding field crops*, 4th ed. Westport, CT: *AVI*.

48. Raven, P. H., R. F. Evert, and S. E. Eichhorn. 2005. *Biology of plants,* 7th ed. New York: Freeman/Worth.

49. Rom, R. C., and R. F. Carlson, eds. 1987. *Rootstocks for fruit crops.* New York: John Wiley & Sons.

50. Routley, M. B., R. I. Bertin, and B. C. Husband. 2004. Correlated Evolution of Dichogamy and Self-Incompatibility: A phylogenetic perspective. *Intl. J. Plant Sci.* 165:983–93.

51. Savidan, Y. 2000. Apomixis: Genetics and breeding. *Plant Breeding Rev.* 18:13–86.

52. Sax, K. 1949. The use of *Malus* species for apple rootstocks. *Proc. Amer. Soc. Hort. Sci.* 53:219–20.

53. Schopfer, C. R., M. E. Nasrallah, and J. B. Nasrallah. 1999. The male determinant of self-incompatibility in *Brassica. Science* 266:1697–700.

54. Seed Savers Exchange. 2001. Web site http://www.seedsavers.com.

55. Sprague, G. F. 1950. Production of hybrid corn. *Iowa Agr. Exp. Sta. Bul. P48.* pp. 556–82.

56. Teshome, A., A. H. D. Brown, and T. Hodgkin. 2001. Diversity in landraces of cereal and legume crops. *Plant Breeding Rev.* 21:221–61.

57. Weng, Y. H., K. Tosh, G. Adam, M. S. Fullarton, C. Norfolk, and Y. S. Park. 2008. Realized genetic gains observed in a first generation seedling seed orchard for jack pine in New Brunswick, Canada. *New Forests* 36:285–98.

58. Westwood, M. N. 1994. *Temperate zone pomology,* 3rd ed. Portland, OR: Timber Press.

59. Whealy, K. 1992. *Garden seed inventory,* 3rd ed. Decorah, IA: Seed Saver Publications. RR 3, Box 239.

60. Young, J. A., and C. G. Young. 1992. *Seeds of woody plants in North America.* Portland, OR: Dioscorides Press.

61. Young, J. A., and C. G. Young. 1986. *Seeds of wild land plants.* Portland, OR: Timber Press.

*6
Techniques of Seed Production and Handling

learning objectives

- Determine different sources for seeds.
- Describe harvesting and processing of different seeds.
- Explain seed tests and their uses.
- Characterize different seed treatments to improve germination.
- Describe principles and procedures for seed storage.

INTRODUCTION

More plants are propagated for food, fiber, and ornamentals from seeds than any other method of propagation. Seed propagation is the cornerstone for producing agronomic, vegetable, forestry, and many ornamental plants. The production of high-quality seeds is of prime importance to propagators. In the production of any crop, the cost of the seed is usually minor compared with other production costs, yet, no single factor is as important in determining the success of the operation. Most crop plant seeds are produced by companies that specialize in both plant breeding and seed production. Growers expect these companies to introduce improved cultivars, as well as to produce high-quality seeds that have good germination characteristics and are true-to-type. To produce high-quality seeds, companies must not only pay close attention to the environment where seeds are produced, but must also have the means to test the quality of those seeds. This chapter discusses various aspects of seed production, testing, and storage. The steps taken to produce, clean, and store seeds for commercial crop production are summarized for a variety of crops in Table 6–1.

SOURCES FOR SEEDS

Commercial Seed Production

Commercial seed production is a specialized intensive industry with its own technology geared to the requirements of individual species (Fig. 6–1, page 164). This section on sources for seeds will be separated into herbaceous and woody plant seeds.

Agricultural, Vegetable, and Flower Seed (35, 50, 98)   Historically, seeds for next season's crop were collected as a by-product of production. Although some seeds may still be produced in this manner (e.g., some Third World production), modern seed production has become a very specialized industry (32, 134). A scheme for producing quality seed is included in Figure 6–2 (page 165).

Some agricultural seeds—such as corn, wheat, small grains, and grasses—are produced in the area where the crops are grown. The advantages for producing seeds in their production area include reduced transportation and handling costs as well as reduced potential for genetic shift (see Chapter 5). These are important considerations for agronomic crops where large amounts of seeds are required to produce a crop. However, crop production

Table 6–1
STEPS FOR PRODUCING, CLEANING, AND STORAGE OF SEEDS FOR COMMERCIAL CROP PRODUCTION

Crop	Production practices	Seed conditioning	Seed storage	Seed treatments
Sweet corn	Hybrid seed production from two inbred parents by wind pollination. Female parent requires detasseling before pollen is shed and is interplanted with rows of the male pollen parent.	Corn cobs are harvested when the seeds are between 35 and 45% moisture to avoid mechanical injury during harvest. Cobs are force-air–dried to about 12 to 13% moisture where the seeds are mechanically removed from the cob. Final moisture is removed in a drying oven (35 to 40°C).	Stored at 10% moisture at 10°C.	Usually treated with fungicide and/or insecticide. Often applied in a polymer film coating.
Tomato	Hybrid seed from inbred parents by hand pollination. Seed parent may be male-sterile, or hand emasculation of anthers is required.	Fruit pulp is separated from the seeds by juice extracting equipment. Extracts can be fermented for 2 to 3 days until the seeds separate from fruit gel and sink. Treatment with HCl acid (5%) is also used to extract seeds after several hours. Excessive fermentation or chemical treatment reduces seed quality. Seed drying should not exceed 43°C.	Stored at 6% moisture at 5 to 10°C.	Can be treated with a fungicide or, in some cases, primed.
Onion	Hybrid onion seed is produced by insect pollination between inbred parents. The female seed parent is male-sterile. Plants flower (bolt) after the second season. It is common to plant seed at close spacing the first year to produce small bulbs that are replanted at the appropriate spacing for seed production the second spring.	Seed maturity can vary because flowering umbels are not all initiated at the same time on the plant. Harvest the entire umbel when the first individual fruits begin to crack and show black seeds. Umbels are naturally air-dried for 2 to 3 weeks on open benches. These are threshed and seeds are separated by screens, air, and gravity separation.	Seeds of onion are short-lived in storage. Stored at 6% moisture at 5°C.	No special seed treatments.
Impatiens	Hybrid seeds are produced in the greenhouse by hand pollination between inbred parents. Seed parent is pollinated as soon as the stigma is receptive to prevent self-pollination.	Fruit of impatiens explodes when ripe, expelling seeds. Therefore, fruits are harvested prior to expulsion and placed on frames for several days until seeds are shed. Seeds are then air-dried or dried under gentle heat.	Stored between 3 and 5% moisture at 5°C.	Impatiens are a high-value seed crop. Seeds may be primed, pelleted, or pregerminated.
Pawpaw (*Asimina*)	Pawpaw understocks are produced from seeds. Hand pollination between trees with different genetic backgrounds will increase fruit and seed set.	In most cases, seeds are a by-product of fruit processing. Pulp can be removed by fermentation and washing.	Pawpaw seeds are recalcitrant and cannot withstand seed moisture below 35%. Seeds can be stored moist at 5°C for 2 years.	Stratification (moist, cold storage) for 8 to 10 weeks to relieve dormancy.

(*Continued*)

Table 6–1 Continued				
Crop	Production practices	Seed conditioning	Seed storage	Seed treatments
Pine	Seed orchards are established with elite trees with superior growth characteristics. Seed production takes 18 months and trees take between 2 and 10 years to bear a crop.	For some species, seed is collected on nets under trees after the cones naturally shed seeds. For most, cones are harvested and placed on wire benches where the cones air dry and shed seeds in 2 to 8 weeks. Some cones require oven drying at about 50°C to open cones. Seeds are collected and mechanically dewinged, followed by flotation or gravity separation to get viable seeds.	Stored at 6% moisture and 0 to 5°C.	Stratification (moist, cold storage) for 2 to 12 weeks to relieve dormancy.

Source: Adapted from Desai et al., 1997 (35); McDonald and Copeland, 1997 (98).

areas may not provide the best conditions for producing high-quality, disease-free seeds. Therefore, large amounts of high-value seeds such as forage, vegetable, and flowers are produced in specialized growing areas.

The major considerations for selecting areas to produce seeds are environmental conditions and cost of production (34, 94, 143). Large quantities of grass, vegetable, and flower seeds are produced in areas characterized by low summer rainfall, low humidity, and limited rainfall during the seed harvest season (11, 144). These conditions provide good seed yields and reduce disease problems, especially during harvest when seeds must dry before being handled. There are also crops that require special environmental conditions to flower and set seeds. These include the biennial vegetable and flower crops that require **vernalization** (a period of cold temperature) to flower (143). Examples are onion and carrot seed production. One-year-old biennial plants used for seed production have been called **stecklings** (65). Plants may be chilled by overwintering in the field, or in some cases, stecklings are brought into a cooler (5°C, 40°F) to satisfy vernalization requirements and shorten the seed-production cycle.

> vernalization A period of cold temperature required by plants to induce flowering. In natural systems, these crops grow in late summer, are chilled over winter, and then flower in early spring.

Major production areas for high-value seed production in the United States that meet these important environmental conditions include grass and forage seed production in the Pacific Northwest and vegetable and flower seed production in the Pacific Northwest down to the central, coastal valleys of California (Fig. 6–3). Increasingly, seed production has become an international industry. For example, the United States, Netherlands, and Japan provide over half of the world's flower seeds (61). Hybrid seed production that requires hand pollination has moved to areas of the world with reduced labor costs. These include Central and South America, Southeast Asia, India, and Africa. The advantages to producing seeds in the Southern Hemisphere include a reduced cost of production, and seed production in the season prior to planting in northern crop production areas, which reduces storage time and cost.

Regardless of the country where seeds are produced, there are several important considerations that

Figure 6–1
A majority of important agronomic, horticultural, and forestry crops are propagated by seeds that come in a large diversity of seed size and shape, resulting in diverse requirements for seed production, extraction, and conditioning.

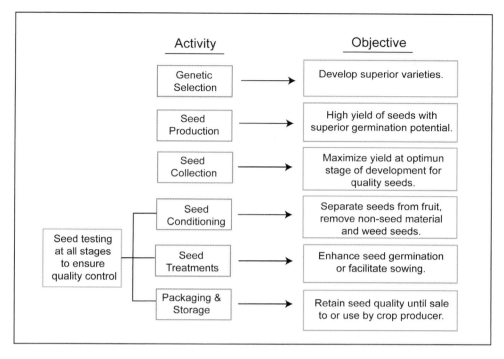

Figure 6–2
Procedures for producing and handling a commercial seed lot.

Figure 6–3
Seed production fields (a) Mallow produced as wildflower seed in Oregon. (b) Wildflowers (coneflower in forefront and grasses behind) production in Wisconsin. California production of (c) cucumber and (d) sunflower with bee hives for pollination.

must be satisfied when selecting specific sites for **seed production** (143, 144):

1. Appropriate **soil type** and **fertility** for good seed yields.
2. A detailed **cropping history** to avoid disease or herbicide carryover.
3. Adequate **soil moisture** or availability of supplemental irrigation.
4. A **dry environment** during seed harvest.
5. Ability to isolate **open or cross-pollinated crops.** For example, self-pollinated tomato plants require only 50 feet of separation between varieties, while some insect- or wind-pollinated crops require up to a mile of separation between varieties to avoid unwanted cross-pollination (65, 98).

Additional requirements for high-quality seed production are the selection of planting density, pest control, and availability of insect pollinators (144). In many cases, conditions for seed production and crop production are very similar.

Woody Plant Seed A number of commercial and professional seed-collecting firms exist that collect and sell seeds of certain timber, ornamental, or fruit species. Lists of such producers are available (36, 88, 90). Such seeds should be properly labeled as to their origin or provenance (see Chapter 5). Some tree seeds can be obtained as certified seeds.

Seed Exchanges. Many arboreta and plant societies have seed exchanges or will provide small amounts of specialty seed.

Seed Collecting. Propagators at individual nurseries may collect tree and shrub seeds (77, 119, 128, 148). These may be collected from specific seed-collection zones or from seed-production areas (see Chapter 5). Seeds may be collected from standing trees, trees felled for logging, or from squirrel caches. They might be collected from parks, roadways, streets, or wood lots. Seed collecting has the advantage of being under the control of the propagator, but requires intimate knowledge of each species and the proper method of handling. Most important, the collector should be aware of the importance of the selection principles described in Chapter 5.

Seed Orchards. Seed orchards or plantations are used to maintain seed source trees of particularly valuable species (23). They are extensively used by nurseries in the production of rootstock seeds of certain species or cultivars and for forest tree improvement. The major advantage to a seed orchard is that it is a consistent source of seeds from a known (often genetically superior) parentage (90). They also allow the seed producer to maximize seed harvest by reducing loss due to environmental conditions or animals. Such seed orchards are described in Chapter 5.

Fruit-Processing Industries. Historically, many of the fruit tree rootstock seeds were obtained as by-products of fruit-processing industries such as canneries, cider presses, and dry yards. Examples include peach and apricot in California, as well as pears in the Pacific Northwest. The procedure is satisfactory if the correct cultivar is used (see Chapter 5). In some cases, seed-borne viruses might be present in certain seed sources.

HARVESTING AND PROCESSING SEEDS

Maturity and Ripening

Each crop and plant species undergoes characteristic changes leading to seed ripening that must be known to establish the best time to harvest (35, 91, 147, 149). A seed is ready to harvest when it can be removed from the plant without impairing germination and subsequent seed vigor. This is called **harvest maturity.** In many cases, a balance must be made between late and early harvest to obtain the maximum number of high-quality seeds. If harvesting is delayed too long, the fruit may **dehisce**

> **harvest maturity**
> The time during seed development when the seeds can be harvested for germination without a significant reduction in seed quality.

("split open" or "shatter"), drop to the ground, or be eaten or carried off by birds or animals. If the fruit is harvested too soon when the embryo is insufficiently developed, seeds are apt to be thin, light in weight, shriveled, poor in quality, and short-lived (34). Some seeds that are mechanically harvested (i.e., sweet corn) can be damaged if the seed moisture at harvest is too dry. Therefore, developing seeds are sampled often to determine their stage of maturity. Seed moisture percentages are used as an indicator of seed maturity to determine the proper harvest time (see Box 6.1). Early seed harvest may also be desirable for seeds of some species of woody plants that produce a hard seed covering in addition to a dormant embryo. If seeds become dry and the seed coats harden, the seeds may not germinate until the second spring (146), whereas they would have germinated the first spring if harvested early.

> **BOX 6.1   GETTING MORE IN DEPTH ON THE SUBJECT**
> **TESTING SEED MOISTURE**
>
> Moisture content is found by the loss of weight when a sample is dried under standardized conditions (40). Oven drying at 130°C (266°F) for 1 to 4 hours is used for many kinds of seeds. For oily seeds, 103°C (217°F) for 17 hours is used, and for some seeds that lose oil at these temperatures (e.g., fir, cedar, beech, spruce, pine, hemlock) a toluene distillation method is used. Various kinds of electronic meters can be used for quick moisture tests (22, 35, 98).

Harvesting and Handling Procedures

Plants can be divided into three types for seed extraction, according to their fruit type:

1. Dry fruits that do not dehisce at maturity
2. Dry fruits that dehisce at maturity
3. Plants with fleshy fruits

Type 1: Dry Fruits That Do Not Dehisce at Maturity
Plants in this group have seed and fruit covers that adhere to each other at maturity. These are dry fruits that do not dehisce (open), and the seeds are not disseminated immediately upon maturity. This group includes most of the agricultural crops, such as corn, wheat, and other grains. Many of these have undergone considerable selection during domestication for ease of harvest and handling. This group also contains the nut crops like oak (*Quercus*), hazel (*Corylus*), and chestnut (*Castanea*).

Field-grown crops (cereals, grasses, corn) can be mechanically harvested using a combine, a machine that cuts and threshes the standing plant in a single operation (Fig. 6–4). Plants that tend to fall over or "lodge" are cut, piled, or windrowed for drying and curing. Low humidity is important during harvest. Rain damage results in seeds that show low vigor. The force required to dislodge seeds may result in mechanical damage, can reduce viability, and result in abnormal seedlings. Some of these injuries are internal and not noticeable, but they result in low viability after storage (3, 66, 109). Damage is most likely to occur if seed moisture is too high or low, or if the machinery is not properly adjusted. Usually less injury occurs if seeds are somewhat moist at harvest (i.e., up to 45 percent for corn).

Nut crops usually have an involucre covering (i.e., the cup of an acorn) that should be separated from the nut at harvest. Floatation is a common method for separating viable from non-viable seeds (Fig. 6–5, page 168). Floating seeds are more buoyant usually because of insect infestation and are discarded.

(a)

(b)

(c)

(d)

Figure 6–4
Corn seed is actually a fruit (caryopsis) and is an example of a crop with dry non-dehiscent fruits. (a) Corn seed is harvested with a picker, leaving the kernels attached to the cob. Although corn used for grain is combined (harvested and shelled in one operation), corn for seed is usually not shelled until it is allowed to dry further to prevent mechanical injury. (b) Corn dehusker. (c) Dehusked corn cobs. (d) Shelled kernels (seeds) ready for storage.

Figure 6–5
Non-viable oak nuts (acorns) float in water, while viable seeds are more dense and sink.

Type 2: Dry Fruits That Dehisce at Maturity These plants produce seeds from fruits that dehisce readily at maturity. This type includes seeds in follicles, pods, capsules, siliques, and cones. Crops of this group include many annual or biennial flowers (delphinium, pansy, petunia) (94) and various vegetables (onion, cabbage, other cruciferous crops, and okra). In most cases, these fruits must be harvested before they are fully mature, and then dried or cured before extraction. Consequently, some seeds will be underdeveloped and immature at the time of harvest. For example, in onion seed production there is a difference of up to 20 days between the opening of the first and last flowers on a plant. From a practical standpoint, onion seeds are harvested when the first fruits begin to open and seeds have turned black, which corresponds to about 50 days after flowers first open and begin shedding pollen (52).

In addition, many tree and shrub plants also have fruits that fall into this group and are handled with similar procedures. The steps for handling these types of seeds:

1. **Drying.** Plants are cut (sometimes by hand) or dry fruits may be windrowed in the field (Fig. 6–6), or placed on a canvas, tray, or screen (Fig. 6–7) to dry for 1 to 3 weeks. If there are only a few plants, they can be cut and hung upside down in a paper bag to dry. Some crops may need the benefit of forced air drying units for quick dryings, especially in harvest areas with high humidity at the time of harvest (Fig. 6–7).

2. **Extraction.** Commercial seeds may be harvested and extracted in a single operation (Fig. 6–8) with a combine or dried fruits may be passed through threshing machines that extract seeds by beating, flailing, or rolling dry fruit followed by separation of seeds from fruit parts, dirt, and other debris (Fig. 6–9, page 170). Seeds from small seed lots are extracted by hand.

3. **Seed Conditioning (Cleaning).** Further cleaning may be required to eliminate all dirt, debris, weed, and other crop seeds. Commercial seed conditioning (91, 86, 139) utilizes various kinds of specialized equipment, such as screens of different sizes (Fig. 6–10, page 170), seed shape (Fig. 6–11, page 171), air lifters (Fig. 6–12a and b), and gravity separators (Fig. 6–12c and d). The basis for these types of separation is that there are differences in sizes, shapes, and densities between good seed, poor seed, and other debris.

(a)

(b)

Figure 6–6
Cole crop (*Brassica*) seed production is also from a dry dehiscent fruit. (a) Turnip at full harvest maturity. (b) Cabbage seed field mowed and windrowed. Windrowing is done before the fruit shatters, and windrowing allows additional maturation and drying before being combined with a windrow pickup unit. (c) Cole crop fruit is a silique, which is a dry, dehiscent fruit that opens along two suture lines, exposing the seeds attached to a papery septum.

Figure 6–7
Seeds with non-dehiscent and dehiscent fruits often require additional drying after harvest. (a) Portable field drying wagons alongside a permanent bin dryer used for drying prairie wildflower seeds. (b) Open wire screen racks used for air drying woody plant seeds. (c) Forced-air dryer.

Figure 6–8
Purple coneflower (*Echinacea*) seed production is an example of crop requiring the dry seed harvesting method. It has a fruit that shatters at maturity. (a) Seed production field in full bloom. (b) Field at harvest maturity before heads shatter and release seeds. (c) Combine for harvesting and threshing seeds. (d) The combine must be calibrated for cutting height and maximum seed retention. (e) The reel rotates and the paddles force plant stems into the (f) blades of the cutting bar.

Figure 6–9
Sandersonia seed removal from a dry dehiscent capsule. (a) Hand-cut fruiting stems are cut and windrowed under protective cover for additional drying. (b) Pods are passed through a threshing machine to remove seeds. (c) The threshing cylinder with a rasp-bar is the most common thresher. (d and e) Proper threshing captures up to 90 percent of the available seeds, but additional conditioning is usually needed to remove fruit debris.

Figure 6–10
Seed conditioning based on seed size and shape. (a) Hand screens manually sift seeds from plant debris; (b) Mechanical cleaner and seed sizing units use aspiration (air movement) combined with screens of various shapes and sizes to remove seed debris and separate seeds into various size classes. (c) Close up of screens in a scalper unit that separates good seed from plant debris and other unwanted material.

Figure 6–11
Seed conditioning based on seed shape. (a) An indent cylinder that separates seeds based on seed size (length). (b) A spiral separator uses gravity and centripetal force to separate round from flat seeds. Round seeds move faster down the separator. These are useful for cole crop seeds like cabbage and broccoli.

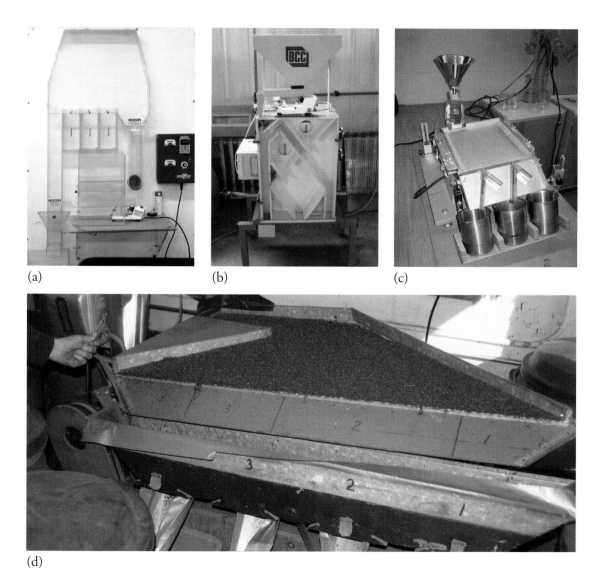

Figure 6–12
Seed conditioning based on seed density. (a) The wall mounted air separator uses a vacuum to lift seeds. Seeds are separated from lighter plant debris. (b) Standalone movable air separator. (c and d) Gravity tables have a tilted platform that uses vibration or air flow to separate seeds. Denser seeds walk toward the higher edge of the platform. Both types of units can be used to upgrade seed lots by directing seeds into bins based on density (weight).

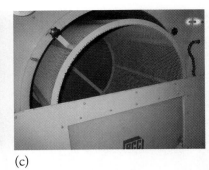

Figure 6–13
Seed extraction and conditioning in pines (a) Drying oven used to release seeds from pine cones. (b) Winged seeds extracted from the cones. (c) Seeds are tumbled to remove the wing attached to the seeds.

Conifer cones also fit in this category of dry dehiscent fruits, but their cones require special procedures (119):

1. **Drying.** Cones of some species will open if dried in open air for 2 to 12 weeks (Fig. 6–13). Others must be force-dried at higher temperatures in special heating kilns. Under such conditions, cones will open within several hours or, at most, 2 days. The temperature of artificial drying should be 46 to 60°C (115 to 140°F), depending upon the species, although a few require even higher temperatures. For example, Jack pine (*Pinus banksiana*) and red pine (*P. resinosa*) need high temperatures [77°C (170°F)] for 5 to 6 hours. Caution must be used with high temperatures, because overexposure will damage seeds. After the cones have been dried, the scales open, exposing the seeds.

2. **Extraction.** Seeds should be removed immediately upon drying, since cones may close without releasing the seeds. Cones can be shaken by tumbling or raking to dislodge seeds. A revolving wire tumbler or a metal drum is used when large numbers of seeds are to be extracted.

3. **Dewinging.** Conifer seeds have wings that are removed except in species whose seed coats are easily injured, such as incense cedar (*Calocedrus*). Fir (*Abies*) seeds are easily injured, but wings can be removed if the operation is done gently. Redwood (*Sequoia* and *Sequoiadendron*) seeds have wings that are inseparable from the seed. For small seed lots, dewinging can be done by rubbing the seeds between moistened hands or trampling or beating seeds packed loosely in sacks. For larger lots of seeds, special dewinging machines are used (Fig. 6–13c).

4. **Cleaning.** Seeds are cleaned after extraction to remove wings and other light chaff. As a final step, separation of heavy, filled seed from light seed is accomplished by gravity or pneumatic separators.

Type 3: Plants with Fleshy Fruits Plants with fleshy fruits include important fruit and vegetable species used for food such as berries, pomes (apples), and drupes (plums), as well as many related tree and shrub species used in landscaping or forestry. In general, fleshy fruits are easiest to handle if ripe or overripe. However, fruits in the wild are subject to predation by birds (45).

For **extraction** of small seed lots, fruits may be cut open and seeds scooped out, treaded in tubs, rubbed through screens, or washed with water from a high-pressure spray machine in a wire basket (Fig. 6–14). Another device that removes seeds from small-seeded fleshy fruits is an electric mixer or **blender** (Fig. 6–15) (122). To avoid injuring seeds, the metal blade of the blender can be replaced with a piece of rubber or Tygon tubing. It is fastened at right angles to the revolving axis of the machine (147). A mixture of fruits and water is placed in the mixer and stirred for about 2 minutes. When the pulp has separated from the seed, the pulp is removed by flotation. This procedure is satisfactory for fruits of serviceberry (*Amelanchier*), barberry (*Berberis*), hawthorn (*Crataegus*), strawberry (*Fragaria*), huckleberry (*Gaylussacia*), juniper (*Juniperus*), rose (*Rosa*), and others (122).

For larger lots, separation is by maceration, fermentation, mechanical means, or washing through screens. The basic procedures include:

1. **Maceration.** Vegetable crops such as tomato, pepper, eggplant, and various cucurbits are produced in commercial fields and may utilize special macerating machinery as a first step in seed extraction (126).

Figure 6–14
Small seed lots of small, fleshy seeds can have the fruit pulp removed by rubbing fruits against a screen and washing away the pulp.

Cucumber and other vine crops, for example, are handled with specially developed macerating machines (Fig. 6–16, page 174). Maceration crushes the fruits and mixes the pulverized mass with water that is diverted into a tank releasing the seeds, but additional handling is often required to separate seeds from the macerated pieces of fruit.

2. **Fermentation.** Macerated fruits can be placed in large barrels or vats and allowed to ferment for up to 4 days at about 21°C (70°F), with occasional stirring.

Figure 6–15
A method for small batch extraction of seeds from fleshy fruits uses a blender (a) or food processor retrofitted with a rubber or plastic impeller for maceration followed by floatation (b and c) to remove seeds from the pulp. (d) Commercial macerators (i.e., Dybvig) use the same principles of water and flailing impellers to extract seeds. They work well for fruit crops like cherry, peach, and plum.

Figure 6–16
Watermelon is an example of a crop that requires the wet seed harvesting method for seed extraction from a fleshy fruit. (a) Field ready for harvest. Withholding water knocks down the vines prior to harvest. (b) Custom seed harvester for large fleshy fruit. (c) Fruit is crushed and the pulp is separated from the seeds. (d) Seeds with a small amount of adhering pulp. (e) A washing unit provides final separation of pulp and seeds. (f) Large rotating dryers reduce seed moisture to its storage level.

If the process is continued too long, sprouting of the seeds may result. Higher temperature during fermentation shortens the required time. As the pulp separates from seeds, heavy, sound seeds sink to the bottom of the vat, and the pulp remains at the surface. Following extraction, the seeds are washed and dried either in the sun or in dehydrators. Additional cleaning is sometimes necessary to remove dried pieces of pulp and other materials. Extraction by fermentation is particularly desirable for tomato seed, because it can help control bacterial canker (35, 89).

3. **Chemical Treatment.** Alternatives to fermentation are various chemical treatments. The advantage of chemical treatments is that it takes less time (less than 24 hours) to separate seeds from macerated pulp. Like fermentation, overexposure to the chemical can reduce seed quality. Chemical treatments include acid treatment for tomato seed extraction (98), and digestive enzymes—like pectinase used in orange seed extraction—for understock production (12).

4. **Flotation.** Another alternative to separate seeds from fleshy fruits is floatation, which involves placing seeds and pulp in water so that heavy, sound seeds sink to the bottom and the lighter pulp, empty seeds, and other extraneous materials float to the top. This procedure can also be used to remove lightweight, unfilled seeds and other materials from dry fruits, such as acorn fruits infested with weevils, but sometimes both good and bad seeds will float. Small berries of some species, such as *Cotoneaster*, juniper (*Juniperus*), and *Viburnum*, are somewhat difficult to process because of small size and the difficulty in separating the seeds from the pulp. One way to handle such seeds is to crush the berries with a rolling pin, soak them in water for several days, and then remove the pulp by flotation.

Figure 6–17
Various drying units for seeds. (a) A spinning centripetal dryer. (b) A large rotating forced air dryer.

After seeds are thoroughly washed to remove fleshy remnants, they are dried (Fig. 6–17), except seeds of recalcitrant species that must not be allowed to dry out. If left in bulk for even a few hours, seeds that have more than 20 percent moisture will heat; this impairs viability. Drying may either occur naturally in open air if the humidity is low or artificially with heat or other devices. Drying temperatures should not exceed 43°C (110°F); if the seeds are quite wet, 32°C (90°F) is better. Drying too quickly can cause seeds to shrink and crack, and can sometimes produce hard seed coats. The minimum safe moisture content for storage of most orthodox seeds differs by species but is usually in the range of 4 percent to 15 percent.

SEED TESTING

In the United States, state laws regulate the shipment and sale of agricultural and vegetable seeds within each state. Seeds entering interstate commerce or those sent from abroad are subject to the **Federal Seed Act,** adopted in 1939. Such regulations require the shipper to use labeling (Fig. 6–18) of commercially produced seeds that includes:

1. Name and cultivar
2. Origin
3. Germination percentage
4. Percentage of pure seed, other crop seed, weed seed, and inert material

Regulations set minimum standards of quality, germination percentage, and freedom from weed seeds. Special attention must be paid to designated **noxious weeds** for a particular growing region. Laws in some states (117) and in most European countries regulate shipment and the sale of tree

> **noxious weeds**
> Weeds that vary from state to state, but that have been designated as weed species that must be identified in the seed lot and may cause the whole seed lot to be unsaleable.

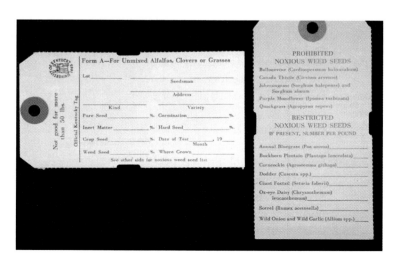

Figure 6–18
State and federal seed laws require testing seed lots prior to sale. Information for a seed lot includes standard germination percentage according to accepted seed-testing rules, purity of the seed lot (percentage of seeds that are the desired crop and its trueness to type), percentage of weed seeds, and the amount of noxious weed seeds in the seed lot. Noxious weeds are designated as weeds that are particularly undesirable, and tolerances may differ for a crop or region of the country.

seed, but there are no federal laws governing the tree seed trade.

Seed testing provides information in order to meet legal standards, determines seed quality (39), and establishes the rate of sowing for a given stand of seedlings (37). It is desirable to retest seeds that have been in storage for a prolonged period.

Procedures for testing agriculture and vegetable seeds in reference to the Federal Seed Act are given by the U.S. Department of Agriculture. The most current version of the Federal Seed Act can be found at the U.S. Electronic Code of Federal Regulations (53). The **Association of Official Seed Analysts, Inc.** (www.aosaseed.com), (5) publishes the "rules" for seed testing for the major edible food crops as well as many ornamental plant species. International rules for testing seeds are published by the **International Seed Testing Association** (www.seedtest.org) (73). The Western Forest Tree Seed Council also publishes testing procedures for tree seed and other useful information in their online woody plant seed manual (www.nsl.fs.fed.us/wpsm).

seed testing associations Organizations that set the standards for seed testing and can also train and certify seed analysts.

A high-quality seed lot is a function of the following characteristics that are routinely tested by seed companies or private and state seed labs (116):

1. Germination (viability)
2. Purity
3. Vigor
4. Seed health
5. Noxious weed seed contamination

Sampling for Seed Testing

The first step in seed testing is to obtain a uniform sample that represents the entire lot under consideration (Fig. 6–19). Equally sized (usually measured by weight) primary samples are taken from evenly distributed parts of the seed lot, such as a sample from each of several sacks in lots of less than five sacks or from every fifth sack with larger lots. The seed samples are thoroughly mixed to make a composite sample. A representative portion is used as a submitted sample for testing. This sample is further divided into smaller lots to produce a working sample (i.e., the sample upon which the test is actually to be run). The amount of seed required for the working sample varies with the kind of seed and is specified in the Rules for Seed Testing (5).

Figure 6–19
A sample from each seed lot must be tested prior to sale usually by a state-certified seed lab. The seed analyst uses a seed sorter to randomly select a seed sample for testing from the submitted seed lot. A portion of the seed lot will be tested for purity, while an additional subsample will be evaluated for standard germination.

Viability Determination

Viability can be determined by several tests, the **standard germination, excised embryo,** and **tetrazolium** tests being the most important.

Standard Germination Tests In the standard germination test, germination percentage is determined by the percent of **normal seedlings** produced by pure seeds (the kind under consideration). To produce a good test, it is desirable to use at least 400 seeds picked at random and divided into lots of 100 each. If any two of these lots differ by more than 10 percent, a retest should be carried out. Otherwise, the average of the four tests becomes the official germination percentage. Seeds are placed under optimum environmental conditions of light and temperature to induce germination. The conditions required to meet legal standards

standard germination The most common test for seed quality. It is performed according to standards set by seed-testing associations, often by certified seed analysts. It represents the percentage of seedlings in a seed lot that germinate normally. In a standard germination test, only seeds that are normal are counted as germinated.

normal seedlings Seedlings described for the major crops (often in pictures) in the rules for seed testing. In general, normal seedlings have elongated radicle and hypocotyl and at least one enlarged cotyledon.

are specified in the rules for seed testing, which may include type of test, environmental conditions, and length of test (5, 73).

Various techniques are used for germinating seeds in seed-testing laboratories (127). Small seeds are placed on plastic germination trays or in Petri dishes (Fig. 6–20). The most common substrate used by commercial seed technology labs for germination tests are blue blotter or washed paper towels, available from commercial suppliers. These products ensure uniformity and reproducibility in their tests. Containers are placed in germinators in which temperature, moisture, and light are controlled according to the established standard germination rules. To discourage the growth of microorganisms, all materials and equipment should be kept scrupulously clean, sterilized when possible, and the water amount carefully regulated.

The **rolled towel test** (Fig. 6–21a, b, and c, page 178) is commonly used for testing large seeds like cereal grains. Several layers of moist paper toweling, about 2.8 by 3.6 cm (11 by 14 in) in size, are folded over the seeds and then rolled into cylinders and placed vertically in a germinator.

A germination test usually runs from 1 to 4 weeks but could continue up to 3 months for some slow-germinating tree seeds with dormancy. Usually a first count is taken at 1 week and germinated seeds are discarded with a final count taken later. At the end of the test, seeds are divided into (a) normal seedlings, (b) hard seeds, (c) dormant seeds, (d) abnormal seedlings, and (e) dead or decaying seeds. A normal seedling should have a well-developed root and shoot, although the criterion for a "normal seedling" varies with different kinds of seeds (Fig. 6–21d). "Abnormal seedlings" can be the result of age of seed or poor storage conditions; insect, disease, or mechanical injury; frost damage; or mineral deficiencies. Any non-germinated seeds should be examined to determine the possible reason. "Hard seeds" have not absorbed water. Dormant seeds are those that are firm, swollen, and free from molds but do not germinate.

Under seed-testing rules, certain environmental requirements to overcome dormancy may be specified routinely for many kinds of seeds (5, 73). These may include chilling stratification or hormone treatment with gibberellins or potassium nitrate.

Excised-Embryo Test The excised-embryo test is used to test seed viability of woody shrubs and trees whose dormant embryos require long treatment periods to relieve dormancy before true germination will take

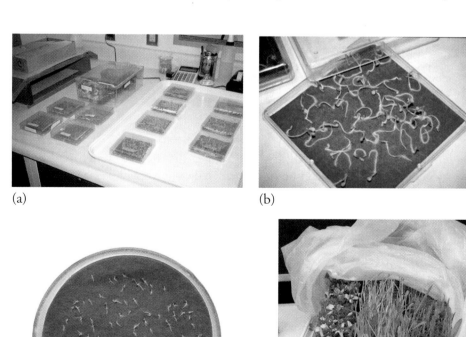

Figure 6–20
A standard germination test is required for seed lots prior to sale. The two most common test procedures include the (a, b, and c) Petri dish and (d) rolled towel tests. The tests and the procedures used for standard germination are detailed in accepted publications like the rules for testing seeds from the Association of Official Seed Analysts (4, 5). Included in these rules will be the preferred test (i.e., Petri dish or rolled towel); the environment for the test (i.e., 20/30, this indicates daily cycles of 16 hours at 20°C followed by 30°C for 8 hours); whether light is required during the test; any seed pretreatments for dormant seeds (e.g., treatment with gibberellin or potassium nitrate); and the number of days for the first and last evaluation (counts).

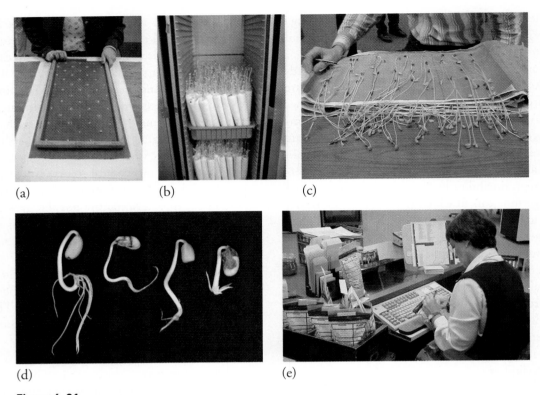

Figure 6–21
Commercial seed labs process a large number of seed samples. They must keep accurate records of each seed lot and must be efficient to process samples in a timely manner while maintaining high reproducibility from seed lot to seed lot. (a) A seed analyst uses a template board to place a standard number of seeds in precise locations on the germination paper for the rolled towel or Petri dish tests. (b) Rolled towels are held upright in the growth chamber. (c) After the number of days indicated in the testing rules, the seed analyst counts the number of normal seedlings. (d) The seed analyst must determine if a seedling is normal and can be counted as germinated. These seedlings are "abnormal" because either the shoot or root has not developed normally after the final count for this seed test. (e) Results are recorded in a computer database.

place (5, 44, 67). In this test, the embryo is excised from seeds that are soaked for 1 to 4 days and germinated following standard germination conditions (see Fig. 6–22).

The excision must be done carefully to avoid injury to the embryo. Any hard, stony seed coverings, such as the endocarp of stone fruit seeds, must be removed first. The moistened seed coats are cut with a sharp scalpel, razor blade, or knife, under clean but nonsterile conditions with sterilized instruments. The embryo is carefully removed. If a large endosperm is present, the seed coats may be slit and the seeds covered with water, and after about a half-hour, the embryo may float out or be easily removed.

Tetrazolium Test The **tetrazolium test** (6) is a biochemical test for viability determined by the red color appearing when seeds are soaked in a 2,3,5-triphenyl-tetrazolium chloride (TTC) solution (Fig. 6–23). Living tissue changes the TTC to an insoluble red compound (chemically known as formazan); in nonliving tissue the TTC remains uncolored. The test is positive in the presence of dehydrogenase enzymes involved in respiration. This test was developed in Germany by Lakon (87), who referred to it as a **topographical test** since loss in embryo viability begins to appear at the extremity of the radicle, epicotyl, and cotyledon tips. The reaction takes place equally well in dormant and nondormant seed. Results can usually be obtained within 24 hours (see Box 6.2, page 180). The TTC solution deteriorates with exposure to light but will remain in good condition for several months if stored in a dark bottle. The solution should be discarded if it becomes yellow. A 0.1 to 1.0 percent concentration is commonly used. The pH should be 6 or 7. In the hands of a skilled technologist, this test can be used for seed-quality evaluation and as a tool in seed research (101).

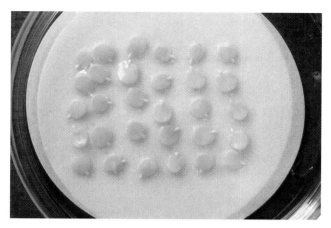

Figure 6–22
The excised-embryo test is a quick evaluation method used for dormant seed. Eastern redbud (Cercis) seeds require at least four months of moist chilling to satisfy dormancy and another 2 weeks for a standard germination test. In comparison, isolated embryos removed from the seed coverings will germinate in 5 days.

X-ray Analysis X-ray analysis of seeds (80) can be used as a rapid test for seed soundness (2). X-ray photographs do not normally measure seed viability but provide an examination of the inner structure for mechanical disturbance, absence of vital tissues, such as embryo or endosperm, insect infestation, cracked or broken seed coats, and shrinkage of interior tissues (Fig. 6–24).

Standard X-ray equipment is used to assess seeds. Dry seeds are exposed for 1/2 to 3 minutes at 15- to 20-kilovolt tube potential. Seed with dimensions less than 2 mm are too small to show details. Since X-rays do not injure the seed, further tests for viability can be conducted on the same batch (2). Prototype machines that provide fast, automatic, online sorting have been proposed (140). These procedures have the potential to remove nonviable seeds as well as seeds with morphological characteristics that are linked to poor vigor.

Figure 6–23
Tetrazolium chloride (TZ) is used to test seed viability. Portions of the embryo will stain red (an indication of respiration) if they are viable. The seed analyst must determine if vital portions of the embryo are living, which would indicate positive germination potential. (a and b) A positive TZ corn seed test showing that the embryo and scutellum are viable while the white endosperm is non-living at maturity. (c and d) A poor TZ test in gasplant (Dictamnus). White embryos are non-viable and the embryo (d) although generally red-stained would probably be abnormal because the shoot area (arrow) did not stain.

(a)

(b)

Figure 6–24
Examples of the X-ray tests for the 1999 (a) and 2005 (b) harvests of *Gaura biennis* capsules. Note the number of filled and empty (aborted) seeds in the capsules. Courtesy of the Ornamental Plant Germplasm Center, The Ohio State University.

> ### BOX 6.2 GETTING MORE IN DEPTH ON THE SUBJECT
> ### TETRAZOLIUM TESTING
>
> Details vary for different seeds, but general procedures include (6, 73, 127):
>
> 1. Any hard covering such as an endocarp, wing, or scale must be removed. Tips of dry seeds of some plants, such as *Cedrus*, should be clipped.
> 2. First, seeds should be soaked in water in the dark; moistening activates enzymes and facilitates the cutting or removal of seed coverings. Seeds with fragile coverings, such as snap beans or citrus, must be softened slowly on a moist medium to avoid fracturing.
> 3. Most seeds require preparation for TTC absorption. Embryos with large cotyledons, such as *Prunus*, apple, and pear, often comprise the entire seed, requiring only seed coat removal. Other kinds of seed are cut longitudinally to expose the embryo (corn and large-seeded grasses, larch, some conifers); or transversely one-fourth to one-third at the end away from the radicle (small-seeded grasses, juniper, *Carpinus*, *Cotoneaster*, *Crataegus*, *Rosa*, *Sorbus*, *Taxus*). Seed coats can be removed, leaving the large endosperm intact (some pines, *Tilia*). Some seeds (legumes, timothy) require no alteration prior to the tests.
> 4. Seeds are soaked in the TTC solution for 2 to 24 hours. Cut seeds require a shorter time; those with exposed embryos somewhat longer; intact seeds 24 hours or more.
> 5. Interpretation of results depends on the kind of seed and its morphological structure. Completely colored embryos indicate good seed. Conifers must have both the megagametophyte and embryo stained. In grass and grain seeds, only the embryo itself colors, not the endosperm. Seeds with declining viability may have uncolored spots or be unstained at the radicle tip and the extremities of the cotyledons. Nonviability depends on the amount and location of necrotic areas, and correct interpretation depends on standards worked out for specific seeds (127).
> 6. If the test continues too long, even tissues of known dead seeds become red due to respiration activities of infecting fungi and bacteria. The solution itself can become red because of such contamination.

Purity Determination

purity A determination assessed in a seed lot by a seed analyst who is certified for purity tests. It involves meticulous evaluation of a seed lot for any foreign material including seeds.

Purity is the percentage by weight of the "pure seed" present in a sample. Purity determination requires a trained seed analyst, usually from a state or private seed lab. In the United States, the **Society of Commercial Seed Technologists** provides training and testing to certify **Registered Seed Technologists** (116).

There are two aspects to pure seed: a **physical** and a **genetic** component (4, 116). Pure seed must be separated from other physical contaminants such as soil particles, plant debris, other inert material, and weed seeds (Fig. 6–25). Seed standards list tolerances for levels of pure seed in a sample. They usually are based on the seed type and seed class (i.e., Certified vs. Registered seed—see Chapter 5). References are available with detailed seed anatomy to help seed technologists to identify crop and weed seeds (18). Special care must be taken to document the occurrence of noxious weeds in a sample. **Noxious weeds** are identified as being particularly bad weeds for a region of the country and can vary by state. Occurrence of a single seed of some noxious weed species in a sample can render an entire seed lot unacceptable for public sale.

Purity testing also identifies the genetic purity of a seed lot. The seed analyst determines if the sample is the proper cultivar and identifies the percentage of seeds that are either other contaminating cultivars or inbreds in a hybrid seed lot (see Chapter 5). Genetic purity can be difficult to determine and relies on an assortment of tests that include field visits by regulatory personnel, seed color, seed and seedling morphology, chemical tests, isozyme (characteristic seed proteins) separation by electrophoresis (4, 116), and DNA fingerprinting (see Box 6.3) (99).

Vigor Testing

Although state and federal seed laws currently require only purity and standard germination tests for seed lots, seed companies and many crop producers are performing **vigor** tests prior to sale or use (95). The Association of Official

vigor (of a seed lot) An estimate of the seed's ability to germinate when the environmental conditions are not ideal for germination. Seed lots with high vigor show a high germination percentage and uniform seedling emergence.

Figure 6–25
Purity of seeds is determined by visual examination of individual seeds in a weighed seed sample taken from the larger lot in question. (a) Impurities may include other crop seed, weed seed, and inert, extraneous material. In this seed lot, several different types of impurities were discovered in the seed lot. (b) Each was placed in a small dish and will be weighed. (c) Purity is also evaluated in field or greenhouse trials. This petunia seed lot shows a percentage of white variants reducing its purity. White plants may be from self-pollinated plants from the female inbred parent that should have been removed during production.

BOX 6.3 GETTING MORE IN DEPTH ON THE SUBJECT
TESTS FOR GENETIC PURITY

Details for cultivar identification are published in the Association of Official Seed Analysts' handbook for purity testing (4). These can include:

Chemical Tests There are a number of chemical treatments used to separate cultivars of specific species (31). Examples include a fluorescence test for fescue and ryegrass (Fig. 6–26a), hydrochloric acid for oat, and peroxidase for soybean. The chemical reaction usually gives a characteristic color that identifies the seed. Chemical tests are usually used in association with other tests, like seed shape and color, to help determine purity.

Protein Electrophoresis A more sophisticated evaluation for cultivar identification uses differences that exist in seed proteins or enzymes. Some plant enzymes are present in different forms (isozymes) that can be separated by electrophoresis to give a pattern that is characteristic of a cultivar. Electrophoresis is a form of chromatography that uses an electrical current to separate proteins on a gel. Isozymes migrate to different locations on the gel to form a pattern that identifies the cultivar.

DNA Fingerprinting This technique also uses the basic principle of electrophoresis but separates fragments of DNA such as RAPDs (random amplified polymorphic DNA), RFLPs (random fragment length polymorphisms) and SCARs (sequence-characterized amplified region) rather than proteins (99). Since these techniques use amplified DNA, the test is very accurate and can identify a larger number of cultivars than can isozyme analysis. DNA fingerprinting is the same process being used by law enforcement to identify suspects in criminal cases.

Strip Tests for Genetically Modified Organisms (GMOs)
The presence of specific GMO seeds can be detected using commercially available strip tests that identify the presence of an antibody for the genetically modified trait (Fig. 6–26b). For example, Bt corn is genetically transformed to produce *Bacillus thuringiensis* proteins (Cry1Ab and Cry1Ac) that are toxic to caterpillars. The strip test contains antibodies to the Bt proteins. If the extract from the seed sample contains these proteins, they will react with the strip's antibodies and produce a double-lined color reaction.

(Continued)

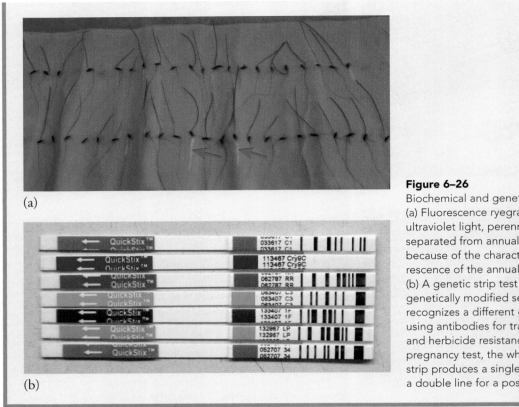

Figure 6–26
Biochemical and genetic tests for purity. (a) Fluorescence ryegrass tests. Under ultraviolet light, perennial ryegrass can be separated from annual ryegrass (red arrow) because of the characteristic white fluorescence of the annual ryegrass radicle. (b) A genetic strip test for the presence of genetically modified seeds. Each strip recognizes a different genetic modification using antibodies for traits such as insect and herbicide resistance. Similar to a pregnancy test, the white portion of the strip produces a single line for negative and a double line for a positive identification.

Seed Analysts (7) states that "seed vigor comprises those seed properties which determine the potential for rapid, uniform emergence, and development of normal seedlings under a wide range of field conditions." Standard germination tests do not always adequately predict seedling emergence under field conditions (Fig. 6–27). Seed vigor tests can provide a grower with additional information that can help predict germination where conditions may not be ideal (110). For many vegetable crops, there is a positive relationship between seed vigor and crop yields (38, 39, 85, 135). Various vigor tests have been developed and certain tests are applied to different species (49). Vigor tests include accelerated aging, controlled deterioration, cold test, cool test, electrolyte leakage, seedling growth rate, and seedling grow-out tests (see Box 6.4) (5, 58, 73, 116).

Seed Health (1)

Seed companies usually have the personnel and facilities to evaluate the health of a seed lot. Seed health comprises the occurrence of diseases, insects, or nematodes in the seed lot (70, 93). Detection of these organisms requires specialized equipment and trained personnel. Seed health is integral to the performance of the seed lot. It has also become increasingly important as international trading agreements (like the World Trade Organization and the North American Free

Figure 6–27
Here is a good example of the impact of seed vigor on stand establishment. In both cases, all pansy seeds have germinated in each plug flat, but seedlings on the left (a) are all at the same stage of growth, while the plug flat on the right (b) has numerous seedlings that are less developed than the majority of seedlings.

BOX 6.4 GETTING MORE IN DEPTH ON THE SUBJECT
SEED VIGOR TESTS

Details for procedures used to conduct vigor tests are found in the Association of Official Seed Analysts' handbook on seed vigor testing (7). The more commonly conducted vigor tests include (Fig. 6–28).

Aging Tests Controlled deterioration and accelerated aging (AA) are established vigor tests for agronomic, horticultural, and forestry species. Both tests are based on the premise that vigor is a measure of **seed deterioration.** Hampton and Coolbear (60) concluded that aging tests were the most promising vigor tests for most agronomic species. Both methods are described in detail in the AOSA vigor testing methods (59).

seed deterioration The loss of vigor and viability in a seed during storage.

Controlled deterioration (92) exposes seeds to high temperature (40 or 45°C) for a short duration (24 or 48 hours) after the moisture content has been raised to approximately 20 percent. Seed moisture is raised prior to exposure to high temperature and maintained by keeping seeds in sealed watertight packages. Germination is usually assessed as radicle emergence, but normal germination improves results in some cases.

Accelerated aging is similar to controlled deterioration but differs in the way seed moisture is increased and, therefore, modifies the duration of the test (133). It is a test commonly used for agronomic and vegetable seeds. Prior to a standard germination test, seeds are subjected to high temperatures (40 to 45°C) and high relative humidity (near 100 percent) for 2 to 5 days. This is done by suspending seeds on a stiff nylon frame suspended above water in specially designed boxes (Fig. 6–28a). This partially hydrates the seed without permitting radicle emergence. Higher-vigor seeds tolerate this stress better than

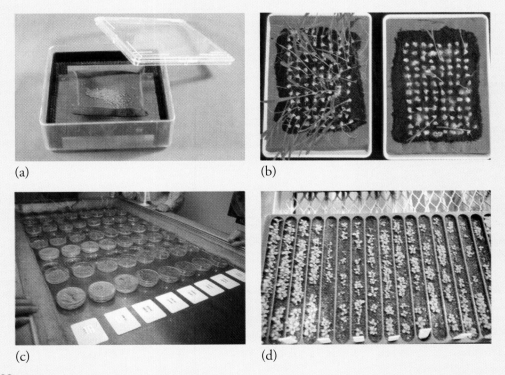

Figure 6–28
Various seed vigor tests. (a) Impatiens seeds in accelerated aging boxes. The frame inside the box keeps seeds suspended above water or a solution of saturated salts. (b) Sweet corn seeds sprouting in the cold test. Seeds are placed on moist towels or Kimpack and covered with field soil. It is easy to see that the seed lot on the *left* has higher vigor (seedling emergence) compared with the seed lot on the *right*. (c) A thermal gradient table provides numerous temperatures to simultaneously test germination of a single seed lot, which is useful for determining seed vigor by evaluating germination at minimal and maximal temperatures. Breeders also use thermal gradient tables to evaluate a genotype's tendency for producing seed susceptible to thermodormancy (like lettuce). (d) For many horticultural crops, standard germination and seedling vigor is evaluated in a seedling grow-out test. The environment for this test is standard greenhouse conditions where the crop will be commercially grown.

(Continued)

low-vigor seeds, as shown by higher normal germination percentages in the standard germination test conducted after the aging treatment. For smaller-seeded species, like flower seeds, lower relative humidity is used to reduce rapid seed hydration. This variation is called the saturated salt accelerated aging test, because it uses saturated salts rather than water to control humidity in the accelerated aging boxes (150).

Cold Test (59) This is the preferred vigor test for corn seed (Fig. 6–28b). Seeds are planted in boxes, trays, or rolled towels that contain field soil and held at 10°C for 7 days before being moved to 25°C. The number of normal seedlings that emerge are counted after 4 days.

Cool Test This is a vigor test that uses procedures identical to the standard germination test, except the temperature is lowered to 18°C. A similar tool being used to evaluate vegetable and flower seed vigor is the thermal gradient table (Fig. 6–28c). This provides a range of temperatures by circulating warm and cold water to the table. This determines the range of germination for a seed lot. Higher vigor seeds germinate better at the extreme temperatures on the table.

Electrolyte Leakage Seeds tend to "leak" electrolytes when imbibed, and the amount of electrolyte leakage usually increases as seeds deteriorate. Electrical conductivity can be measured by using a conductivity meter. Conductivity measurements have been correlated with field emergence, especially in large-seeded crops like peas and corn (94).

Seedling Growth Seedling grow-out tests can be conducted under greenhouse or growth-chamber conditions, and vigor calculated based on seedling emergence and uniformity (Fig. 6–28d). An alternative to plug and flat germination includes evaluations like the slant-board test that uses similar conditions as the standard germination test for percentage germination. After a period of time at a controlled temperature (this varies between species), shoot and root length or seedling weight is determined (Fig. 6–29a). This permits a determination of strong versus weak seedlings in a seed lot. Measuring individual seedlings can be tedious, but advances in computer-aided image analysis offer an alternative to hand measurements (Fig. 6–29b) (71, 105). Ball Seeds Inc. (West Chicago, IL) has introduced the Ball Vigor Index that employs computer analysis of video images of seedlings in plug trays after a predetermined number of days. The index is suggestive of seedling greenhouse performance.

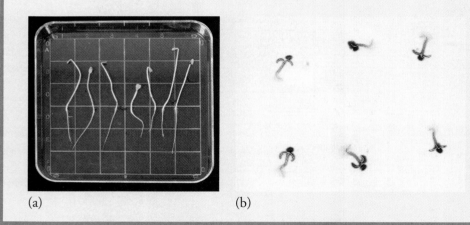

Figure 6–29
(a) A slant-board test for lettuce. Seedling must be grown in an upright orientation to get straight seedlings. Radicle length is then measured by hand. (b) Computer-aided measurements of digital images of petunia from Petri dish germination.

Trade Agreement) require clean seed be made available for international sale.

Specific procedures to standardize seed health tests are available (137). Three types of tests for seed health include:

1. **Visual evaluation** of a seed sample for characteristic structures like spores or sclerotia of pathogens, or the presence of insects.
2. **Incubation of seed** on moist germination paper or agar and inspection for disease growth.
3. **Biochemical tests,** such as ELISA tests, which detect the presence of specific disease organisms.

SEED TREATMENTS TO IMPROVE GERMINATION

Presowing seed treatments has become a common practice in the seed industry. Seed treatments may be applied by seed producers or on the farm. The objective of seed treatments is to either enhance the potential for germination and seedling emergence or to help mechanical seed sowing (75, 120, 132). Types of seed treatments include:

1. Seed protectants
2. Germination enhancement

3. Inoculation with microorganisms (nitrogen-fixing bacteria)
4. Coatings to help mechanical sowing

Facilities that treat seeds must consider the following aspects for quality seed treatment (56):

1. Seeds must be treated **uniformly.**
2. The material must continue to **adhere to the surface** of the seed during sowing.
3. The treatment should **not reduce seed quality.** Any physical damage due to high temperature or mechanical injury must be minimized and monitored by seed testing.
4. The treatment should be **safely applied** and allow for safe handling by the seed consumer.
5. Treatments to **help mechanical sowing** must produce a **uniform size** and **shape** for each seed.
6. All seeds treated with a pesticide must be **colored** to avoid accidental ingestion by humans or animals. Color can also enhance the appearance of the seed.

Modern seed treatments require specialized equipment and facilities (30, 56, 57). The equipment varies depending on the type of seed treatment. Historically, the first seed treatment incorporated pesticides in simple powders (74). These are still used today, especially for on-site farm application, because they require the least specialized equipment. However, powders and the dust from them present a problem for safe handling. Most commercial treatment of seeds is from **liquid slurries.** These are preferred because they treat seeds more uniformly, are safer to apply and handle, and are relatively cheap.

Recently, **polymer film coatings** have become a popular seed treatment because the pesticide can be incorporated into the polymer that is applied in a thin, uniform coat or film (57). The advantages of film coatings are the ability to incorporate chemical or biological materials into the coating for safe handling (this material does not rub off when handled), uniform coating size, and an attractive appearance. The cost has been prohibitive for general use with many large-volume agronomic crops, but film-coated seeds have become more widely available on high-value flower and vegetable seed.

Seed Protectants

Seed protectants can be grouped as

1. **Chemical treatments** against pathogens, insects, and animals.
2. **Heat treatment** against pathogens and insects.
3. **Inoculation** with **beneficial microbes** against harmful fungi.
4. **Safners,** to reduce herbicide injury (19, 120).

Chemical Treatment A seed stores food reserves to provide energy and carbon for seedling growth, which makes seeds a primary food source for humankind. However, insects, pathogens, and animals also target seeds as a food source. Strategies to protect seeds probably date to man's earliest use of seeds as a food crop (74). Chemical treatments for seeds can be seen in the 1800s with the use of copper sulfate against a variety of cereal diseases (120). In the 1900s, mercury compounds were very effective against seed and seedling pathogens. These were banned in most parts of the world in the 1980s because of health risks. The 1940s and 1950s saw the introduction of the first broad-spectrum fungicides (like captan and thiram), starting the modern use of seed protectants for diseases.

The most common and important seed treatments are the chemical and physical treatments against seed-borne pathogens (20) and insects (79). It is important to understand that these treatments will not improve germination in seeds with a genetically low potential for germination or in mechanically injured seeds. These treatments are especially beneficial where germination is delayed due to poor environmental conditions such as excessive water in the field, or cool soils. Under these conditions, seed leakage stimulates fungal spore germination and growth. A chemical seed treatment can protect the seed until the seedling emerges.

Seed treatment may be designed to protect seed from soil-borne pathogens, disinfest the seed from pathogens on the seed surface, or eliminate pathogens inside the seed (20). Chemical seed protectants can be applied as **powders, liquids, slurries,** or incorporated into a pellet or **film coating** (57, 75).

Biocontrol Although chemical treatments dominate industry seed treatments, the novel use of treating seeds with **beneficial microbes** presents an interesting alternative to chemical treatments (100, 112, 118). Various

> **seed protectants**
> Seed treatments used most often for field-seeded crops that are prone to insect and disease attack. Early season plantings that are slower to emerge because of cool soils benefit from seed protectants.
>
> **geneficial microbes**
> An alternative to chemicals for seed protection against soil-borne diseases. These microbes compete with pathogenic microbes to help seedlings emerge before they are attacked.

biocontrol agents provide protection to seeds by producing antibiotic substances; decreasing competition for space and nutrients; and reducing parasitism (63, 100). Common biocontrol agents include bacterial strains like *Eterobacter, Pseudomonas, Serratia,* and fungal strains like *Gliocladium* and *Trichoderma.* Several studies show disease prevention with biologicals to be as effective as chemical treatment with fungicides (27, 129). A second approach is to treat seeds with materials extracted from fungi or bacteria that activate the plant's natural defense system (145).

Heat Treatment (Thermotherapy) High temperature to control seed-borne diseases has been in use since 1907 (74). Dry seeds are immersed in hot water (49 to 57°C; 120 to 135°F) for 15 to 30 minutes, depending on the species (10, 11). After treatment, the seeds are cooled and spread out in a thin layer to dry. To prevent injury to the seeds, temperature and timing must be regulated precisely; a seed protectant should subsequently be used, and old, weak seeds should not be treated. Hot water is effective for specific seed-borne diseases of vegetables and cereals, such as *Alternaria* blight in broccoli and onion, and loose smut of wheat and barley.

Microwave and UV radiation also can be used to disinfest seeds (121). Aerated steam (see Chapter 3) is an alternate method that is less expensive, easier to manage, and less likely to injure seeds than hot water. Seeds are treated in special machines in which steam and air are mixed and drawn through the seed mass to rapidly (in about two minutes) raise the temperature of the seeds to the desired temperature. The treatment temperature and time vary with the organism to be controlled and the kind of seed. Usually the treatment is 30 minutes, but it may be as little as 10 or 15 minutes. Temperatures range from 46 to 57°C (105 to 143°F). At the end of treatment, temperatures must be lowered rapidly to 32°C (88°F) by evaporative cooling until dry. Holding seeds in moisture-saturated air at room temperature for 1 to 3 days prior to the steam-air treatment will improve effectiveness.

Hot water is also used to kill insects in seeds. For example, oak (*Quercus*) seed is soaked in water at 49°C (120°F) for 30 minutes to eliminate weevils commonly found in acorns (149). As with heat treatments to eliminate disease, precise temperature and timing must be maintained or seeds will be damaged.

Seed Coating

Seed coating uses the same technology and equipment used by the pharmaceutical industry to make medical pills (82, 131). Seed coatings include **pelleted** and **film-coated** seeds (26).

Pelleted Seeds The objective of coating seeds as a pellet is to provide a **round, uniform shape** and **size** to small or unevenly shaped seeds in order to aid precision mechanical sowing (Fig. 6–30). Pelletized seeds are tumbled in a pan while inert powders (like clay or diatomaceous earth) and binders form around seeds to provide a uniform, round shape (Fig. 6–31). Recent advances in coating materials and processing using rotary coaters has allowed seed producers to produce thinner pellets (Fig. 6–30b). These are usually termed **encrusted** seeds for very thin coatings (1 to 5 times the seed size) or mini-pellets (10 to 25 times the seed size). Compare this with a traditional pellet that may be 50 to 100 times the seed size (Fig. 6–30c and d). Encrusted seeds are similar to film-coated seeds but are less expensive to produce. Pellets can be distinguished by either "splitting" or "melting" when the coating is wetted, with many growers preferring the split-type pellets (Fig. 6–30e). Many ornamental flower seeds are commonly pelletized for precision sowing one seed per cell in a plug flat (see Chapter 7). An increasing number of direct-seeded vegetable crops are also being pelletized. It is common for lettuce seed sown in Florida and California to be pelletized to provide uniform spacing and sowing depth that reduces the need to hand-thin the crop.

> **pelleted seeds** Seeds that have a round, uniform shape that make it easier for machine sowing. Pelleted seeds are most commonly used in greenhouse bedding plant production and precision-sown vegetable crops.

Polymer Film-Coated Seeds Film coating (Fig. 6–32, page 188) uses a thin polymer film to cover the seed (82, 114). Film coating only adds 1 to 5 percent to the weight of a seed compared with more than 1,000 percent for pelletized seed, but this can still aid in precision sowing by improving flowability. Fungicides and beneficial microbes can be added to both pellets and film coatings (see seed treatments, page 184) and is the major benefit to film coating (57). Novel films are being employed that allow seeds to imbibe only when the soil temperature has warmed to prevent imbibitional chilling injury in sensitive plants (103).

Germination Enhancement

Commercial practices that provide germination enhancement are **seed sizing, priming,** and **pregermination** (48, 57).

Seed Sizing Seed lots sold as "elite" seeds have been sized to provide larger seed. In addition, seed sizing eliminates lightweight and cracked seeds (Fig. 6–33,

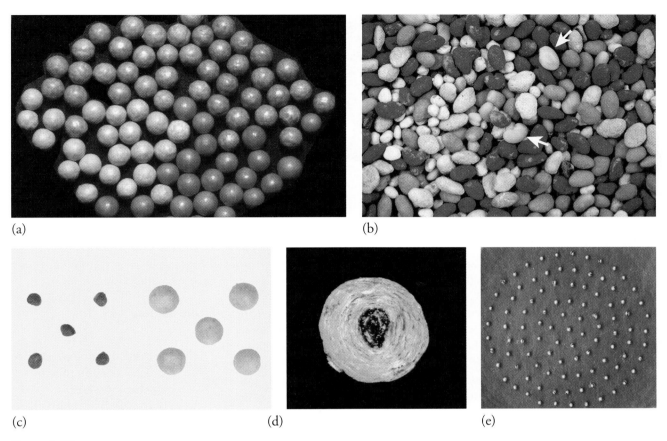

Figure 6–30
Seed Pellets (a) Pelleted seeds showing the uniformly round shape to help in mechanical sowing. Colors may indicate seed differences (primed vs. untreated) or just be cosmetic. (b) A collection of encrusted pasture legume seeds. Notice how the seed shape is still evident with the lighter pelleti coating; the arrows indicate non-encrusted seeds. (c and d) Seed pelleting adds considerable size to a seed as well as a uniform, round shape. (c) On the *left* are raw seeds versus pelleted seeds on the *right*. (d) A cross-section of a pelleted seed showing how the coating (light blue) adds significant volume to the seed. (e) Pellets showing the split-coat habit as it hydrates. Splitting allows easy penetration by the radicle of the germinating seed.

Figure 6–31
Pan type seed coater for pelletizing seeds. Seeds tumble in this seed coating machine while layers of a bulking material and binder build the pellet around the seed.

page 188). This can provide seeds with a higher potential for germination viability and vigor. Elite seeds also may be the seeds selected by seed companies to be further enhanced by seed priming.

seed priming A controlled hydration seed treatment that induces faster, more uniform germination. This effect is most noticeable when seeds are sown in less-than-favorable environments. Primed seeds are most often used in greenhouse bedding plant production to shorten the time to produce seedling plugs, and in crops like pansy and lettuce to avoid reduced germination due to high temperature.

Seed Priming Seed **priming** is a controlled seed-hydration treatment that can reduce the time it takes for seedlings to emerge. It uses basic principles of water potential to hold seeds in an imbibed condition, but prevent germination

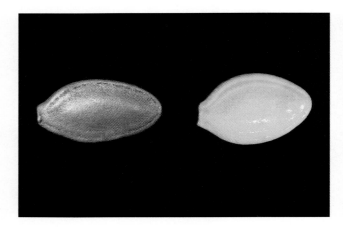

Figure 6–32
Film coating is used to improve flowability of seeds during planting and as a carrier for pesticides. Several examples of film coating on corn seed. Seeds on the left are untreated.

Pregermination The goal of each grower is to establish a "stand" (seedling emergence) of 100 percent (54), which means a plant at each appropriate field spacing or greenhouse plug cell (see Chapter 8). This can be accomplished by using transplants or sowing more seeds than are required and thinning seedlings to the appropriate spacing. An additional treatment to improve stand establishment is pregermination of seeds. In concept, pregermination can take place under optimum conditions and any seeds showing radicle emergence are sown, providing near 100 percent stand. Two types of pregermination sowing techniques have been used:

1. **Fluid drilling** to sow germinated seeds in a gel to protect emerged radicles.
2. **Pregerminated** seeds that use a technique to dry seeds after the radicle emerges prior to sowing.

(radicle emergence) (24, 97). After being hydrated for an extended time, seeds are dried back to near the original dry weight. These seeds can be handled as normal raw seeds or pelleted prior to sowing (82). Growth substances (28) or biologicals (termed *biopriming*; 20, 27) also can be included in the priming solution for added seed enhancement.

Primed seeds will usually show higher seed vigor compared with raw seeds (97). The physiological basis for seed priming is discussed in Chapter 7. Priming can provide faster, more uniform seedling emergence for field and greenhouse crops, especially when environmental conditions for germination are not ideal. The grower must weigh the additional cost of primed seed with this potential for improved seedling emergence. It is common to prime crops like lettuce (106) and pansy (29) to overcome problems of reduced germination due to conditions of high temperature (thermodormacy, see Chapter 7).

Fluid Drilling. **Fluid drilling** (55, 107) is a system involving the treatment and pregermination of seeds followed by their sowing suspended in a gel. Seeds are pregerminated under conditions of aeration, light, and optimum temperatures for the species (Fig. 6–22). Among the procedures that can be used are (a) germinating seeds in trays on absorbent blotters covered with paper, or (b) placing seeds in water in glass jars or plastic columns through which air

fluid drilling A technique to sow pregerminated seeds where the radicle has emerged and is vulnerable to damage. Germinated seeds are mixed in a gel for sowing. Fluid drilling has not been used extensively because of the expense and difficulty in timing. It has its greatest utility in high-value vegetable crops sown in cool soils for early harvest.

(a)

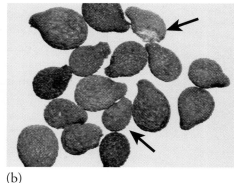
(b)

Figure 6–33
(a) "Elite" or enhanced seeds have additional seed conditioning to remove any broken seeds and have been sized to give larger, more uniform seeds. (b) Notice the broken and small seeds (arrow) in the seed lot on the *right*.

is continuously bubbled and fresh water continuously supplied. Growth regulators, fungicides, and other chemicals (51) can potentially be incorporated into the system. Chilling (10°C, 50°F) of thermodormant celery seeds for 14 days has produced short, uniform radicle emergence without injury (47). Pregerminated seeds of various vegetables have been stored for 7 to 15 days at temperatures of 1 to 5°C (34 to 41°F) in air or aerated water. Separating out germinated seeds by density separation has improved the uniformity and increased overall stand (130).

Various kinds of gels are commercially available. Among the materials used are sodium alginate, hydrolyzed starch-polyacrylonitrile, guar gum, synthetic clay, and others. Special machines are needed to deposit the seeds and gel into the seed bed.

Pregerminated Seeds. **Pregerminated seeds** were introduced commercially in 1995 for bedding plant species (impatiens), but L. H. Bailey introduced the concept as early as 1897. A quote from Bailey's *The Nursery Book* (8) demonstrates that "new" is truly a relative term as he describes **"regermination."** "It is a common statement that seeds can never revive if allowed to become thoroughly dry after they have begun to sprout. This is an error. Wheat, oats, buckwheat, maize, pea, onion, radish, and other seeds have been experimented upon in this direction, and they are found to regerminate readily, even if allowed to become thoroughly dry and brittle after sprouting is well progressed. They will even regerminate several times."

pregerminated seeds A technique for bedding plants that uses a special process to synchronize radicle emergence and then slowly dry seeds prior to sowing. Under the right conditions, this treatment ensures near 100 percent germination.

Pregermination involves germination of seeds under controlled conditions to synchronize germination in order to induce the radicle to emerge about one-sixteenth of an inch. Germinated seeds are separated from nongerminated seeds, and then seeds are dried slowly to near their original dry weight (26). The advantages of using pregerminated seeds include production of 95 percent or better usable seedlings; fast, uniform germination; and because the seeds are dry, mechanical seeders can be used to sow them. The disadvantages of using pregerminated seeds are increased cost (up to 25 percent), seeds have a shorter shelf life (around 35 days at 5°C or 40°F), and growers must have optimized seedling growing conditions to take advantage of the benefits of pregermination.

SEED STORAGE

Seeds are usually stored for varying lengths of time after harvest. **Viability** at the end of storage depends on (a) the initial viability at harvest, as determined by factors of production and methods of handling; and (b) the rate at which deterioration takes place. This rate of physiological change, or aging (96, 111), varies with the kind of seed and the environmental conditions of storage, primarily temperature, and humidity.

viability A measure of whether the seed is alive and can germinate.

Seed Longevity

Plant species can be separated as **recalcitrant** or **orthodox** seeds based on their genetic potential to tolerate storage.

Recalcitrant or Short-Lived Seeds **Recalcitrant** seeds do not tolerate significant drying after seed development. Most recalcitrant seeds cannot tolerate seed moistures below 25 percent, and some species are also sensitive to chilling temperatures. This group is represented by species whose seeds normally retain viability for as little as a few days, months, or at most a year following harvest (see Chapter 5). However, with proper handling and storage, seed longevity may be maintained for significant periods. A list of species with short-lived seeds has been compiled by King and Roberts (83). The group includes:

orthodox seeds Seeds that tolerate drying after seed development; can usually be stored for years in this dry state. The majority of crop plants have orthodox seeds.

recalcitrant seeds Seeds that do not tolerate drying after seed development. They offer special challenges in storage because they are short-lived.

1. Certain spring-ripening, temperate-zone trees such as poplar (*Populus*), maple (*Acer*) species, willow (*Salix*), and elm (*Ulmus*). Their seeds drop to the ground and normally germinate immediately.

2. Many tropical plants grown under conditions of high temperature and humidity; these include such plants as sugarcane, rubber, jackfruit, macadamia, avocado, loquat, citrus, many palms, litchi, mango, tea, choyote, cocoa, coffee, tung, and kola.
3. Many aquatic plants of the temperate zones, such as wild rice (*Zizania*), pondweeds, arrowheads, and rushes.
4. Many tree nut and similar species with large fleshy cotyledons, such as hickories and pecan (*Carya*), birch (*Betula*), hornbeam (*Carpinus*), hazel and filbert (*Corylus*), chestnut (*Castanea*), beech (*Fagus*), oak (*Quercus*), walnut (*Juglans*), and buckeye (*Aesculus*).

Orthodox Seeds

The majority of important crop plants are species with **orthodox seeds.** Orthodox seeds tolerate drying after seed development and can be stored in a dry state (usually 4 percent to 10 percent moisture) for extended periods of time. Species with orthodox seed behavior vary in the length of time they tolerate storage.

Medium-Lived Seeds. Medium-lived seeds remain viable for periods of 2 or 3 up to perhaps 15 years, providing that seeds are stored at low humidity and, preferably, at low temperatures. Seeds of most conifers, fruit trees, and commercially grown vegetables, flowers, and grains fall into this group. Crop species can be grouped according to the ability of seeds to survive under favorable ambient storage conditions (Table 6–2). The Relative Storability Index (78) indicates the storage time where 50 percent or more of seeds can be expected to germinate. Seed longevity will be *considerably longer* under controlled low temperature and humidity storage.

Long-Lived Seeds. Many of the longest-lived seeds have hard seed coats that are impermeable to water. Plant families that produce seeds with hard seed coats include the legume, geranium, and morning glory families. If the hard seed coat remains undamaged, such seeds can remain viable for at least 15 to 20 years. The maximum life can be as long as 75 to 100 years and perhaps more. Records exist of seeds being kept in museum cupboards for 150 to 200 years while still

Table 6–2
RELATIVE STORABILITY INDEX[a]

Crop	Category 1 (1 to 2 yr)	Category 2 (3 to 5 yr)	Category 3 (>5 yr)
Agronomic			
	Bermuda grass	Barley	Alfalfa
	Cotton	KY Bluegrass	Clover
	Field corn	Fescue	Sugar beet
	Millet	Oats	Vetch
	Peanut	Rape seed	
	Soybean	Rice	
	Sunflower	Wheat	
Vegetable			
	Green bean	Broccoli, cabbage, cauliflower	Beet
	Lettuce	Cucumber	Tomato
	Onion	Melon	
	Pepper	Pea	
		Spinach	
		Sweet corn	
Flower			
	Begonia	Alyssum	Hollyhock
	Coreopsis	Carnation	Morning glory
	Pansy	Coleus	Salpiglossis
	Primrose	Cyclamen	Shasta daisy
	Statice	Marigold	Stocks
	Vinca	Petunia	Zinnia

[a] The relative storability index is the expected 50 percent germination in a seed lot stored under favorable ambient conditions. Storage life would be longer under controlled low temperature conditions.

Source: Adapted from Justice and Bass, 1979.

retaining viability (115). There are a number of claims of seeds from ancient tombs germinating after thousands of years. However, these lack definitive scientific support (115). Indian lotus (*Nelumbo nucifera*) seeds that had been buried in a Manchurian peat bog were originally estimated to be more than 1,000 years old and germinated perfectly when the impermeable seed coats were cracked (21). However, recent carbon-14 dating of these and other lotus seeds estimate the age of these seeds to be only 100 to 430 years old (115)!

A systematic study was initiated by Beal in 1879 at Michigan State University to study long-term survival of buried seed. This study is still ongoing, and in 1981 (84) and 2001 (136), three species continued to show germination after 100 and 120 years, respectively. These species were *Malva rotundifolia*, *Verbascum blattaria*, and *Verbascum thapsus*. Some weed seeds retain viability for many years (50 to 70 years or more) while buried in the soil, even though they have imbibed moisture (113). Longevity seems related to dormancy induced in the seeds by environmental conditions deep in the soil.

Storage Factors Affecting Seed Germination

As seeds **deteriorate,** they:

1. first **lose vigor,**
2. then the **capacity for normal germination,**
3. and finally **viability.**

Storage conditions that reduce seed deterioration are those that slow respiration and other metabolic processes without injuring the embryo. The most important conditions are low moisture content of the seed, low storage temperature, and modification of the storage atmosphere. Of these, the moisture-temperature relationships have the most practical significance. Harrington (64) introduced a "rule of thumb" that indicated that seeds lose half their storage life for every 1 percent increase in seed moisture between 5 percent and 14 percent. Also, seeds lose half their storage life for every 5°C increase in storage temperature between 0 and 50°C. This is, of course, a generalized theory that varies between species. More accurate mathematical models have been developed to predict seed longevity at various temperature and moisture contents (43).

The most important factors impacting extended seed longevity in storage are seed moisture content and storage temperature.

Moisture Content Control of seed moisture content is probably the most important factor in seed longevity and storage. Most crop species have orthodox seeds where dehydration is their natural state at maturity. These seeds are best stored at a non-fluctuating low moisture content (43).

Seeds of orthodox species are desiccation-tolerant and, for most, 4 percent to 6 percent moisture content is favorable for prolonged storage (33), although a somewhat higher moisture level is allowable if the temperature is reduced (138). For example, for tomato seed stored at 4.5 to 10°C (40 to 50° F), the percent moisture content should be no more than 13 percent; if 21°C (70°F), 11 percent; and if 26.5°C (80°F), 9 percent.

Various storage problems arise with increasing seed moisture (64). At 8 percent or 9 percent or more, insects are active and reproduce; above 12 percent to 14 percent, fungi are active; above 18 percent to 20 percent, heating may occur due to seed respiration; and above 40 percent to 60 percent, germination occurs.

If the moisture content of the seed is too low (1 percent to 2 percent), loss in viability and reduced germination rate can occur in some kinds of seeds (17). For seeds stored at these low moisture levels, it would be best to rehydrate with saturated water vapor to avoid injury to seed (104). Moisture in seeds is in equilibrium with the relative humidity of the air in storage containers, and increases if the relative humidity increases and decreases if it is reduced (64). Thus, moisture percentage varies with the kind of storage reserves within the seed (13, 14). Longevity of seed is best if stored at 20 percent to 25 percent relative humidity (115).

Since fluctuations in seed moisture during storage reduce seed longevity (15), the ability to store seeds exposed to the open atmosphere varies greatly in different climatic areas. Dry climates are conducive to increased longevity; areas with high relative humidity result in shorter seed life. Seed viability is particularly difficult to maintain in open storage in tropical areas.

Storage in hermetically sealed, moisture-resistant containers is advantageous for long storage, but seed moisture content must be low at the time of sealing (16). Seed moisture content of 10 percent to 12 percent (in contrast to 4 percent to 6 percent) in a sealed container is worse than storage in an unsealed container (33, 115).

Recalcitrant seeds owe their short life primarily to their sensitivity to low moisture content. For instance, in silver maple (*Acer saccharinum*), seed moisture content was 58 percent in the spring when fruits were released from the tree. Viability was lost when moisture content dropped below 30 percent to 34 percent (76). Citrus seeds can withstand only slight drying (15) without loss of viability. The same is true for seeds of some water plants, such as wild rice, which can be stored directly in water at low temperature (102). The large fleshy seeds of oaks (*Quercus*), hickories (*Carya*), and walnut (*Juglans*) lose viability if allowed to dry after ripening (119).

Viability of recalcitrant seeds of the temperate zone can be preserved for a period of time if kept in a moist environment and the temperature is lowered (21). Under these conditions many kinds of seeds can be kept for a year or more. Seeds of some tropical species (e.g., cacao, coffee), however, show chilling injury below 10°C (50°F).

Temperature Reduced temperature invariably lengthens the storage life of seeds and, in general, can offset the adverse effect of a high moisture content. Subfreezing temperatures, at least down to −18°C (0°F), will increase storage life of most kinds of seeds, but moisture content should not be high enough to allow the free water in the seeds to freeze and cause injury (115). Refrigerated storage should be combined with dehumidification or with sealing dried seeds in moisture-proof containers.

Cryopreservation. Survival of seeds exposed to ultralow temperatures (**cryopreservation**) has been known since 1879 (25). There is renewed interest in storage of seeds by cryopreservation because it is potentially a cost-effective way to preserve germplasm for long periods of time with minimal loss of genetic information due to chromosomal mutations that accompany seed deterioration (124). Seeds are cryopreserved by immersion and storage in liquid nitrogen at −196°C (Fig. 6–34). Seed moisture must be low for survival, and gradual cooling and warming rates limit damage to the seed like cracks in the seed coat (115).

Cryopreservation of seeds has not replaced standard long-term storage at −18°C because long-term effects on seed survival have yet to be determined (142). However, numerous species have been stored for short periods of time in liquid nitrogen with promising results (123, 125). Research is continuing, especially at the National Seed Storage Lab (see Getting More In Depth on the Subject box on conserving genetic resources) to make cryopreservation an important tool for seed preservation. Cryopreservation technology is also being applied to other tissue like pollen and dormant buds for possible preservation of germplasm (9, 81).

cryopreservation The storage of seeds or vegetative organs at an ultralow temperature. This is usually in liquid nitrogen at −196°C.

Types of Seed Storage

Although optimal seed storage conditions are cold temperature and low relative humidity, it is not always possible to maintain these conditions for commercial seed lots because of economic reasons. Typical conditions for commercial storage listed from least to most expensive include: (Fig. 6–35)

Figure 6–34
Germplasm storage. (a) Movable storage cabinets for seed storage. (b and c) Seed storage in liquid-nitrogen–filled dewers.

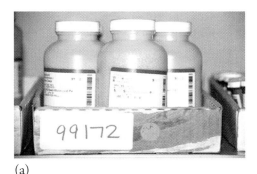
(a)

(b)

(c)

(d)

(e)

Figure 6–35
Various seed storage methods. (a) Small, high value seeds in plastic containers. (b) Vegetable seeds stored in sealed cans. (c) Large-seeded vegetables in bulk storage in waxed boxes. (d) Conditioned storage for crop seeds. (e) Refrigerator storage for flower seeds.

1. **Open storage** without humidity or temperature control
2. Storage in **sealed containers** with or without temperature control
3. **Conditioned storage** with humidity and temperature control

Open Storage without Humidity or Temperature Control Many kinds of orthodox seeds need to be stored only from harvest until the next planting season. Under these conditions, seed longevity depends on the relative humidity and temperature of the storage atmosphere, the kind of seed, and its condition at the beginning

---

BOX 6.5 GETTING MORE IN DEPTH ON THE SUBJECT
CONSERVING GENETIC RESOURCES

Crop cultivars produced for food, fiber, and ornamentals represent only a small proportion of the worldwide gene pool that could have economic benefit in the future. This is a genetic resource that is most easily and economically preserved by storing seed from diverse populations of crop plants. Facilities that provide long-term storage of seeds or other plant parts are called "gene banks" (108).

The International Board for Plant Genetic Resources (72) was established in 1974 to promote an international network of gene banks to conserve genetic resources mainly by storing seeds for the long term. (62). This organization provides handbooks and describes the criteria for facilities that store seed germplasm (41, 42, 62). Facilities are described for either long-term or medium-term storage. Long-term storage facilities provide an environment and testing regime to maintain seed viability and plant recovery for from 10 to more than 20 years. Medium-term storage facilities are designed to preserve seeds for 5 to 10 years before having to regrow the crop to produce fresh seed. In 1984, more than 100 storage facilities (55 with long-term storage) had been established worldwide (62).

The major facility in the United States for preserving germplasm resources is the National Seed Storage Laboratory, established in 1958 on the Colorado State University campus (115, 141). Seeds are actively acquired from public agencies, seed companies, and individuals engaged in plant breeding or seed research. Descriptive material is recorded for each new accession on the Germplasm Resources Information Network. Seed samples are tested for viability, dried to approximately 6 percent moisture, and stored at –18°C (0°F) in moisture-proof bags. Seed lot sizes vary for storage from between 3,000 to 4,000 seeds for cross-pollinated species and 1,500 to 3,000 seeds for pure lines. Seed lots are tested every 5 or 10 years for germination. Seeds can be made available to breeders and researchers on request. This facility also conducts seed storage research and is one of the leading centers for research on cryopreservation of seeds. Information on germplasm can be obtained online at http://www.ars-grin.gov.

of storage. Basic features (78) of the storage structures include (a) protection from water, (b) avoidance of mixture with other seeds or exposure to herbicides, and (c) protection from rodents, insects, fungi, and fire. Retention of viability varies with the climatic factors of the area in which storage occurs. Poorest conditions are found in warm, humid climates; best storage conditions occur in dry, cold regions. Fumigation or insecticidal treatments may be necessary to control insect infestations.

Open storage can be used for many kinds of commercial seeds for at least a year (i.e., to hold seeds from one season to the next). Seeds of many species, including most agricultural, vegetable, and flower seeds, will retain viability for longer periods up to 4 to 5 years (17, 78), except under the most adverse conditions.

Sealed Containers Packaging dry seeds in hermetically sealed, moisture-proof containers is an important method of handling and/or merchandising seeds. Containers made of different materials vary in durability and strength, cost, protective capacity against rodents and insects, and ability to retain or transmit moisture. Those completely resistant to moisture transmission include tin or aluminum cans (if properly sealed), hermetically sealed glass jars, and aluminum pouches. Those almost as good (80 percent to 90 percent effective) are polyethylene (3 mil or thicker) and various types of aluminum-laminated paper bags. Somewhat less desirable, in regard to moisture transmission, are asphalt and polyethylene-laminated paper bags and friction-top tin cans. Paper and cloth bags give no protection against moisture change (46). Small quantities of seeds can be stored satisfactorily in small moisture-proof containers like mason jars or plastic food containers.

Seed may be protected against moisture uptake by mixing with a desiccant (32, 78). A useful desiccant is silica gel treated with cobalt chloride. Silica gel (one part to ten parts seed, by weight) can absorb water up to 40 percent of its weight. Cobalt chloride turns from blue to pink at 45 percent RH and can act as a useful indicator of excess moisture. Seeds should not be stored in contact with the desiccant. Seeds in sealed containers are more sensitive to excess moisture than when subjected to fluctuating moisture content in open storage. Seed moisture content of 5 percent to 8 percent or less is desirable, depending on the species.

Conditioned Storage Conditioned storage includes use of dehumidified and/or refrigerated facilities to reduce temperature and relative humidity (115). Such facilities are expensive but are justified where particularly valuable commercial seeds are stored. It is also justified for research, breeding stocks, and germplasm.

Also in some climatic areas, such as in the highly humid tropics, orthodox seeds cannot be maintained from one harvest season to the next planting season.

Cold storage of tree and shrub seed used in nursery production is generally advisable if the seeds are to be held for longer than 1 year (68, 119). Seed storage is useful in forestry because of the uncertainty of good seed-crop years. Seeds of many species are best stored under cold, dry conditions (149). Ambient relative humidity in conditioned storage should not be higher than 65 percent to 75 percent RH (for fungus control) and no lower than 20 percent to 25 percent.

It is important to control humidity in refrigerated storage since the relative humidity increases with a decrease in temperature and moisture will condense on the seed. At 15°C (59°F), this equilibrium moisture may be too high for proper seed storage. Although the seed moisture content may not be harmful at those low temperatures, rapid deterioration will occur when the seeds are removed from storage and returned to ambient uncontrolled temperatures. Consequently, refrigeration should be combined with dehumidification or sealing in moisture-proof containers (64).

Low humidity in storage can be obtained by judicious ventilation, moisture proofing, and dehumidification as well as by the use of sealed moisture containers, or the use of desiccants, as described previously. Dehumidifiers utilize desiccants (silica gel) or saturated salt solutions. The most effective storage is to dry seeds to 3 percent to 8 percent moisture, place in sealed containers, and store at temperatures of 1 to 5°C (41°F). Below-freezing temperatures can be even more effective if the value of the seed justifies the cost.

Moist, Cool Storage for Recalcitrant Seeds. Many recalcitrant seeds that cannot be dried can be mixed with a moisture-retaining medium, placed in a polyethylene bag or other container, and refrigerated at 0 to 10°C (32 to 50°F). The relative humidity in storage should be 80 percent to 90 percent. Examples of species whose seeds require this storage treatment are silver maple (*Acer saccharinum*), buckeye (*Aesculus* spp.), American hornbeam (*Carpinus caroliniana*), hickory (*Carya* spp.), chestnut (*Castanea* spp.), filbert (*Corylus* spp.), citrus (*Citrus* spp.), loquat (*Eriobotrya japonica*), beech (*Fagus* spp.), walnut (*Juglans* spp.), litchi, tupelo (*Nyssa sylvatica*), avocado (*Persea* spp.), pawpaw (*Asimina triloba*), and oak (*Quercus* spp.). The procedure is similar to moist-chilling (stratification). Acorns and large nuts may be dipped in paraffin or sprayed with latex paint before storage to preserve their moisture content (69).

DISCUSSION ITEMS

By far, more plants are propagated from seed for the production of food, fiber, and for ornamental use than any other propagation method. There are more recent advancements in techniques related to seed germination than any other area of plant propagation. It has become standard to purchase seeds treated with a pre-sowing treatment for vegetable and flower production. As examples, most pansy seed are primed to avoid thermodormancy for summer sowing. Lettuce seed is commonly pelleted to facilitate mechanical sowing, as are many flower seeds. Newer techniques (like pregermination) also must be evaluated by growers and may become important in the future. Seed quality and handling makes a large contribution to the production practices discussed in Chapter 8.

1. Contrast seed viability vs. vigor. How do these characteristics of seeds affect different horticulture crop production?
2. Standard germination is the number of normal seedlings produced in a seed lot. How does this compare to radicle emergence as a measure of viability?
3. Discuss disease protection of seeds by chemical vs. biological materials such as using the fungus *Trichoderma*.
4. What are the advantages of pelleted and film-coated seed?
5. Compare seed storage of orthodox vs. recalcitrant seeds.
6. Discuss strategies to conserve genetic resources.

REFERENCES

1. Agarwal, V. K. 2006. *Seed health*. Lucknow: International Book.
2. Allison, C. J. 1980. X-ray determination of horticultural seed quality. *Comb. Proc. Intl. Plant Prop. Soc.* 30:78–86.
3. Asgrow. 1959. A study of mechanical injury to seed beans. *Asgrow Monograph* 1. New Haven, CT: Associated Seed Growers.
4. Association of Official Seed Analysts. 1991. Cultivar purity testing handbook, #33. *Assn. Offic. Seed Anal.*
5. Association of Official Seed Analysts. 1993. Rules for testing seeds. *J. Seed Tech.* 16:1–113.
6. Association of Official Seed Analysts. 1999. Tetrazolium testing handbook, #29. *Assn. Offic. Seed Anal.*
7. Association of Official Seed Analysts. 2002. Seed vigor testing handbook, #32. *Assn. Offic. Seed Anal.*
8. Bailey, L. H. 1897. *The nursery book*. New York: The MacMillan Co.
9. Bajaj, Y. P. S. 1979. Establishment of germplasm banks through freeze storage of plant tissue culture and their implications in agriculture. In W. R. Sharp et al., eds. *Plant cell and tissue culture principles and applications*. Columbus: Ohio State Univ. Press. pp. 745–74.
10. Baker, K. F. 1972. Seed pathology. In T. T. Kozlowski, ed. *Seed biology*, Vol. 2. New York: Academic Press.
11. Baker, K. F. 1980. Pathology of flower seeds. *Seed Sci. Tech.* 8:575–89.
12. Barmore, C. R., and W. S. Castle. 1979. Separation of citrus seed from fruit pulp for rootstock propagation using a pectolytic enzyme. *HortScience* 14:526–27.
13. Barton, L. V. 1941. Relation of certain air temperatures and humidities to viability of seeds. *Contrib. Boyce Thomp. Inst.* 12:85–102.
14. Barton, L. V. 1943. Effect of moisture fluctuations on the viability of seeds in storage. *Contrib. Boyce Thomp. Inst.* 13:35–45.
15. Bass, L. N. 1943. The storage of some citrus seeds. *Contrib. Boyce Thomp. Inst.* 13:4–55.
16. Bass, L. N. 1953. Seed storage and viability. *Contrib. Boyce Thomp. Inst.* 17:87–103.
17. Bass, L. N. 1980. Flower seed storage. *Seed Sci. Tech.* 8:591–99.
18. Baxter, D., and L. O. Copeland. 2008. Seed purity and taxonomy. *Application of purity testing techniques to specific taxonomical groups of seeds*. East Lansing: Michigan State University Press.
19. Bazin, M., J. F. Morin, and J. P. Vergneau. 1989. New technologies in seed protection. *Acta Hort.* 253:268–69.
20. Bennett, M. A., V. A. Fritz, and N. W. Callan. 1992. Impact of seed treatments on crop stand establishment. *HortTechnology* 2:345–49.
21. Bewley, J. D., and M. Black. 1994. *Seeds: Physiology of development and germination*. New York: Plenum Press.
22. Bonner, F. T. 1974. Seed testing. In C. S. Schopmeyer, ed. *Seeds of woody plants in the United States*.

U.S. Dept. Agr. Handbook 450, Washington, DC: U.S. Govt. Printing Office. pp. 136–52.

23. Bonnet-Masimbert, M., and J. E. Webber. 1995. From flower induction to seed production in forest tree orchards. *Tree Physiology* 15:419–26.

24. Bradford, K. J. 1986. Manipulations of seed water relations via osmotic priming to improve germination under stress conditions. *HortScience* 21:1105–12.

25. Brown, H. T., and F. Escombe. 1879. Note on the influence of very low temperatures on the germinative power of seeds. *Proc. Roy. Soc. London* 62:160–65.

26. Bruggink, G. T. 2005. Flower seed priming, pregermination, pelleting and coating. In M. B. McDonald and F. Y. Kwong, eds. *Flower seeds: biology and technology*. Wallingford, UK: CAB International. pp. 249–62.

27. Callan, N. W., D. E. Mathre, and J. B. Miller. 1990. Biopriming seed treatment for biological control of *Pythium ultimum* preemergence damping-off in *sh2* sweet corn. *Plant Dis.* 74:368–72.

28. Cantliffe, D. J. 1991. Benzyladenine in the priming solution reduces thermodormancy of lettuce seeds. *HortTechnology* 1:95–9.

29. Carpenter, W. J., and J. F. Boucher. 1991. Priming improves high-temperature germination of pansy seed. *HortScience* 26:541–44.

30. Clayton, P. B. 1993. Seed treatment. In G. A. Matthews and E. C. Hislop, eds. *Application technology for crop protection*. Wallingford, UK: CAB International. pp. 329–49.

31. Cooke, R. J. 1995. Variety identification: Modern techniques and applications. In A. S. Basra, ed. *Seed quality: Basic mechanisms and agricultural implications*. New York: Food Products Press. pp. 279–318.

32. Copeland, L. O., and M. B. McDonald. 2001. *Principles of seed science and technology*, 4th ed. New York: Chapman and Hall.

33. Crocker, W., and L. V. Barton. 1953. *Physiology of seeds*. Waltham, MA: *Chronica Botanica*.

34. Delouche, J. C. 1980. Environmental effects on seed development and seed quality. *HortScience* 15:775–80.

35. Desai, B. B., P. M. Kotecha, and D. K. Salunkhe. 1997. *Seeds handbook: Biology, production, processing, and storage*. New York: Marcel Dekker.

36. Dirr, M. A., and C. W. Heuser, Jr. 2009. *The reference manual of woody plant propagation*, 2nd ed. Athens, GA: Varsity Press.

37. Doijode, S. D. 2006. Seed quality in vegetable crops. In A. S Basra, ed. *Handbook of seed science and technology*. Binghamton, NY: Food Products Press, The Harworth Press.

38. Egli, D. B. 1998. *Seed biology and the yield of grain crops*. Wallingford, UK: CAB International.

39. Elias, S. 2006. Seed quality testing. In A. S. Basra, ed. *Handbook of seed science and technology*. Binghamton, NY: Food Products Press, The Harworth Press.

40. Elias, S., R. Baalbaki, and M. McDonald. 2007. Seed moisture testing handbook. *Assn. Offic. Seed Anal.*

41. Ellis, R. H., and E. H. Roberts. 1980. Improved equations for the prediction of seed longevity. *Ann. Bot.* 45:13–30.

42. Ellis, R. H., T. D. Hong, and E. H. Roberts. 1985. *Handbooks for seed technology for genebanks. Vol. I. Principles and methodology.* Handbooks for genebanks: No. 2. Intern. Board for Plant Genetic Resources, IBPGR Secretariat, Rome.

43. Ellis, R. H., T. D. Hong, and E. H. Roberts. 1985. *Handbooks for seed technology for genebanks. Vol. II. Principles and methodology.* Handbooks for genebanks: No. 3. Intern. Board for Plant Genetic Resources, IBPGR Secretariat, Rome.

44. Flemion, F. 1938. A rapid method for determining the viability of dormant seeds. *Contrib. Boyce Thomp. Inst.* 9:339–51.

45. Fordham, A. J. 1984. Seed dispersal as it concerns the propagator. *Comb. Proc. Intl. Plant Prop. Soc.* 34:531–34.

46. Freire, M. S., and P. M. Mumford. 1989. The efficiency of a range of containers in maintaining seed viability during storage. *Seed Sci. Tech.* 14:371–81.

47. Furatani, S. C., B. H. Zandstra, and H. C. Price. 1985. Low temperature germination of celery seeds for fluid drilling. *J. Amer. Soc. Hort. Sci.* 110:149–53.

48. Geneve, R. L. 1996. New developments in seed germination. *Comb. Proc. Intl. Plant Prop. Soc.* 46:546–49.

49. Geneve, R. L. 2005. Vigor testing in flower seeds. In M. B. McDonald and F. Y. Kwong, eds. *Flower seeds: Biology and technology*. Wallingford, UK: CAB International. pp. 311–32.

50. George, A. T. 2000. *Vegetable seed production*. New York: CAB International.

51. Ghate, S. R., S. C. Phatak, and K. M. Batal. 1984. Pepper yields from fluid drilling with additives and transplanting. *HortScience* 19:281–83.

52. Globerson, D., A. Sharir, and R. Eliasi. 1981. The nature of flowering and seed maturation of onions as a basis for mechanical harvesting of seeds. *Acta Hort.* 111:99–101.

53. GPO. 2009. http://ecfr.gpoaccess.gov/cgi/t/text/text-idx?c=ecfr&tpl=/ecfrbrowse/Title07/7cfr201_main_02.tpl

54. Gray, D. 1978. The role of seedling establishment in precision cropping. *Acta Hort.* 83:309–15.

55. Gray, D. 1981. Fluid drilling of vegetable seeds. *Hort. Rev.* 3:1–27.

56. Halmer, P. 1994. The development of quality seed treatments in commercial practice—objectives and achievements. In *Seed treatment: Progress and prospects.* BCPC Monograph No. 57. pp. 363–74.

57. Halmer, P. 2000. Commercial seed treatment technology. In M. Black and J. D. Bewley, eds. *Seed technology and its biological basis.* London, UK: Sheffield Academic Press. pp. 257–86.

58. Hampton, J. G. 1995. Methods of viability and vigor testing: A critical appraisal. In A. S. Basra, ed. *Seed quality: Basic mechanisms and agricultural implications.* New York: Food Products Press. pp. 81–118.

59. Hampton, J. G., and D. M. TeKrony. 1995. *Vigor testing methods,* 3rd ed. Zurich: International Seed Testing Association.

60. Hampton, J. G., and P. Coolbear. 1990. Potential versus actual seed performance—can vigour testing provide an answer? *Seed Sci. Tech.* 18:215–28.

61. Hamrick, D. 2005. Ornamental bedding plant industry and plug production. In M. B. McDonald and F. Y. Kwong, eds. *Flower seeds: Biology and technology.* Wallingford, UK: CAB International. pp. 27–38.

62. Hanson, J. 1985. *Procedures for handling seeds in genebanks.* Practical manuals for genebanks: No. 1. Intern. Board for Plant Genetic Resources, IBPGR Secretariat, Rome.

63. Harman, G. E., and E. B. Nelson. 1994. Mechanisms of protection of seed and seedlings by biological seed treatments: Implications for practical disease control. In *Seed treatment: Progress and prospects.* BCPC Monograph No. 57. pp. 283–92.

64. Harrington, J. F. 1972. Seed storage and longevity. In T. T. Kozlowski, ed. *Seed biology.* New York: Academic Press. pp. 145–245.

65. Hawthorn, L. R. 1961. Growing vegetable seeds for sale. In A. Stefferud, ed. *Seeds: Yearbook of agriculture.* Washington, DC: U.S. Govt. Printing Office. pp. 208–15.

66. Hawthorn, L. R., and L. H. Pollard. 1954. *Vegetable and flower seed production.* New York: Blakiston Co.

67. Heit, C. E. 1955. The excised embryo method for testing germination quality of dormant seed. *Proc. Assn. Off. Seed Anal.* 45:108–17.

68. Heit, C. E. 1967. Propagation from seed. 10. Storage methods for conifer seed. *Amer. Nurs.* 126(20):14–5.

69. Heit, C. E. 1967. Propagation from seed. 11. Storage of deciduous tree and shrub seed. *Amer. Nurs.* 126(21):12–3, 86–94.

70. Hewett, P. D., and W. J. Rennie. 1986. Biological tests for seeds. In K. A. Jeffs, ed. *Seed treatment.* Surrey, UK: BCPC Publications. pp. 51–82.

71. Hoffmaster, A. F., L. Xu, K. Fujimura, M. B. McDonald, M. A. Bennett, and A. F. Evans. 2005. The Ohio State University Seed Vigor Imaging System (SVIS) for soybean and corn seedlings. *J. Seed Tech.* 27:7–26.

72. International Board for Plant Genetic Resources. 1979. *A review of policies and activities 1974–1978 and of the prospects for the future.* IBPGR Secretariat, Rome.

73. ISTA. 1999. International Seed Testing Association: International rules for seed testing. *Seed Sci. Tech.* 27(Suppl.).

74. Jeffs, K. A. 1986. A brief history of seed treatment. In K. A. Jeffs, ed. *Seed treatment.* Surrey, UK: BCPC Publications. pp. 1–5.

75. Jeffs, K. A. and R. J. Tuppen. 1986. Requirements for efficient treatment of seeds. In K. A. Jeffs, ed. *Seed treatment.* Surrey, UK: BCPC Publications. pp. 17–50.

76. Jones, H. A. 1920. Physiological study of maple seeds. *Bot. Gaz.* 69:127–52.

77. Jorgensen, K. R., and R. Stevens. 2004. Seed collection, cleaning, and storage, Chap. 24. In S. B. Monsen, R. Stevens, and N. Shaw, eds. *Restoring western ranges and wildlands.* Ft. Collins, CO: USDA Forest Service Gen. Tech. Rep. RMRS-GTR-136.

78. Justice, O. L., and L. N. Bass. 1979. *Principles and practices of seed storage.* London: Castle House Pub.

79. Jyoti, J. L., A. M. Shelton, and A. G. Taylor. 2003. Film-coating seeds with chlorpyrifos for germination and control of cabbage maggot (*Diptera: Anthomyiidae*) on cabbage transplants *J. Entomol. Sci.* 38(4):553–65.

80. Kamra, S. K. 1964. The use of x-rays in seed testing. *Proc. Intl. Seed Testing Assn.* 29:71–9.

81. Kartha, K. K. 1985. *Cryopreservation of plant cells and organs.* Boca Raton, FL: CRC Press.

82. Kaufman, G. 1991. Seed coating: A tool for stand establishment; a stimulus to seed quality. *HortTechnology* 1:96–102.

83. King, M. W., and E. H. Roberts. 1980. Maintenance of recalcitrant seeds in storage. In H. F. Chin and E. H. Roberts, eds. *Recalcitrant crop seeds.* Kuala Lumpur, Malaysia: Tropical Press SDN. BHD. pp. 53–89.

84. Kivilaan, A., and R. S. Bandurski. 1981. The one hundred-year period for Dr. Beal's seed viability experiment. *Amer. J. Bot.* 68:1290–2.

85. Kolasinska, K., J. Szyrmer, and S. Dul. 2000. Relationship between laboratory seed quality tests and field emergence of common bean seed. *Crop Sci.* 40:470–75.

86. Kwong, F. Y., R. L. Sellman, H. Jalink, and R. van der Schoor. 2005. Flower seed cleaning and grading. In M. B. McDonald and F. Y. Kwong, eds. *Flower seeds: Biology and technology.* Wallingford, UK: CAB International. pp. 225–47.

87. Lakon, G. 1949. The topographical tetrazolium method for determining the germinating capacity of seeds. *Plant Physiol.* 24:389–94.

88. Landis, T. D., R. W. Tinus, and J. P. Barnett. 1998. The container tree nursery manual. Volume 6, Seedling propagation. *Agric. Handbk. 674.* Washington, DC: USDA Forest Service.

89. Liptay, A. 1989. Extraction procedures for optimal tomato seed quality. *Acta Hort.* 253:163–65.

90. Lovelace, R. 1993. Establishing and maintaining a seed orchard. *Comb. Proc. Intl. Plant Prop. Soc.* 495–96.

91. Macdonald, B. 1986. *Practical woody plant propagation for nursery growers,* Vol. 1. Portland, OR: Timber Press.

92. Mathews, S. 1980. Controlled deterioration: A new vigour test for crop seeds. In P. D. Hebblethwaite, ed. *Seed production.* London: Butterworths. pp. 647–60.

93. Maude, R. B. 1996. *Seedborne diseases and their control: Principles and practices.* Wallingford, UK: CAB International.

94. McDonald, M. B. 1980. Assessment of seed quality. *HortScience* 15:784–88.

95. McDonald, M. B. 1994. The history of seed vigor testing. *J. Seed Tech.* 17:93–101.

96. McDonald, M. B. 1999. Seed deterioration: Physiology, repair and assessment. *Seed Sci. Tech.* 27:177–237.

97. McDonald, M. B. 2000. Seed priming. In M. Black and J. D. Bewley, eds. *Seed technology and its biological basis.* New York: Plenum Press.

98. McDonald, M. B, and L. O. Copeland. 1997. *Seed production: Principles and practices.* New York: Chapman and Hall.

99. McDonald, M. B, L. O. Copeland, L. J. Elliot, and P. M. Sweeney. 1994. DNA extraction from dry seeds for RAPD analyses in varietal identification studies. *Seed Sci. Tech.* 22:171–76.

100. McQuilken, M. P., P. Halmer, and D. J. Rhodes. 1998. Application of microorganisms to seeds. In H. D. Burges, ed. *Formulation of microbial biopesticides, beneficial microorganisms and nematodes.* Dordrecht: Kluwer Acad. Pub. pp. 255–85.

101. Miller, A. 2005. Tetrazolium testing for flower seeds. In M. B. McDonald and F. Y. Kwong, ed. *Flower seeds: Biology and technology.* Wallingford, UK: CAB International. pp. 299–309.

102. Muenscher, W. C. 1936. Storage and germination of seeds of aquatic plants. New York (Cornell Univ.) *Agr. Exp. Sta. Bul.* 652, pp. 1–17.

103. Ni, B. R. 2001. Alleviation of seed imbitional chilling injury using polymer film coating. In A. Biddle, ed. *Seed treatments: Challenges and opportunities.* BCPC Monograph No. 76. pp. 73–80.

104. Nutile, G. E. 1964. Effect of desiccation on viability of seeds. *Crop Sci.* 4:325–28.

105. Oakley, K., S. T. Kester, and R. L. Geneve. 2004. Computer-aided digital image analysis of seedling size and growth rate for assessing seed vigour in impatiens. *Seed Sci. Tech.* 32:907–15.

106. Perkins-Veazie, P., and D. J. Cantiliffe. 1984. Need for high quality seed for priming to effectively overcome thermodormancy in lettuce. *J. Amer. Soc. Hort. Sci.* 109:368–72.

107. Pill, W. G. 1991. Advances in fluid drilling. *HortTechnology* 1:59–64.

108. Plucknett, D. L., N. J. H. Smith, J. T. Williams, and N. M. Anishetty. 1987. *Gene banks and the world's food.* Princeton, NJ: Princeton Univ. Press.

109. Pollock, B. M., and E. E. Roos. 1972. Seed and seedling vigor. In T. T. Kozlowski, ed. *Seed biology,* Vol. 1. New York: Academic Press.

110. Powell, A. A. 2006. Seed vigor and its assessment. In A. S. Basra, ed. *Handbook of seed science and technology.* Binghamton, NY: Food Products Press, The Harworth Press.

111. Priestley, D. A. 1986. *Seed aging.* Ithaca, NY: Cornell Univ. Press.

112. Rhodes, D. J., and K. A. Powell. 1994. Biological seed treatments—the development process. In *Seed treatment progress and prospects.* BCPC Monograph No. 57. pp. 303–10.

113. Rhodes, E. H. 1972. Dormancy: A factor affecting seed survival in the soil. In E. H. Roberts, ed. *Viability of seeds.* London: Chapman and Hall. pp. 32–59.

114. Robani, H. 1994. Film-coating horticultural seed. *HortTechnology* 4:104–5.

115. Roos, E. E. 1989. Long-term seed storage. *Plant Breeding Rev.* 7:129–58.

116. Roos, E. E., and L. E. Wiesner. 1991. Seed testing and quality assurance. *HortTechnology* 1:65–9.

117. Rudolf, P. O. 1965. State tree seed legislation. U.S. Forest Service, *Tree Planters' Notes* 72:1–2.

118. Scheffer, R. J. 1994. The seed industry's view on biological seed treatments. In *Seed treatment: progress and prospects.* BCPC Monograph No. 57. pp. 311–14.

119. Schopmeyer, C. S., ed. 1974. *Seeds of woody plants in the United States.* U.S. Dept. Agr. Handbook 450. Washington, DC: U.S. Govt. Printing Office.

120. Schwinn, F. J. 1994. Seed treatment—A panacea for plant protection? In *Seed treatment: Progress and prospects.* BCPC Monograph No. 57. pp. 3–15.

121. Sherf, A. F., and A. A. MacNab. 1986. *Vegetable diseases and their control*, 2nd ed. New York: John Wiley & Sons.

122. Smith, B. C. 1950. Cleaning and processing seeds. *Amer. Nurs.* 92(11):13–4, 33–5.

123. Stanwood, P. C. 1985. Cryopreservation of seed germplasm for genetic conservation. In K. K. Kartha, ed. *Cryopreservation of plant cells and organs.* Boca Raton, FL: CRC Press.

124. Stanwood, P. C., and L. N. Bass. 1981. Seed germplasm preservation using liquid nitrogen. *Seed Sci. and Tech.* 9:423–37.

125. Stanwood, P. C., and E. E. Roos. 1979. Seed storage of several horticultural species in liquid nitrogen (–196°C). *HortScience* 14:628–30.

126. Steiner, J. J., and B. F. Letizia. 1986. A seed-cleaning sluice for fleshy-fruited vegetables from small plots. *HortScience* 21:1066–7.

127. Stephenson, M., and J. Mari. 2005. Laboratory germination testing of flower seeds. In M. B. McDonald and F. Y. Kwong, eds. *Flower seeds: Biology and technology.* Wallingford, UK: CAB International. pp. 263–97.

128. Struve, D. K., J. B. Jett, and D. L. Bramlett. 1987. Production and harvest influences on woody plant seed germination. *Acta Hort.* 202:9–21.

129. Taylor, A. G., and G. E. Harman. 1990. Concepts and technologies of selected seed treatments. *Annu. Rev. Phytopath.* 28:321–39.

130. Taylor, A. G., and T. J. Kenny. 1985. Improvement of germinated seed quality by density separation. *J. Amer. Soc. Hort. Sci.* 110:347–49.

131. Taylor, A. G., D. H. Paine, N. Suzuki, B. A. Nault, and A. McFaul. 2004. Coating technologies for seed treatment applications. In C. S. Vavrina and G. E. Welbaum, eds. *Acta Hort.* 631:49–54.

132. Taylor, A. G., P. S. Allen, M. A. Bennett, K. J. Bradford, J. S. Burris, and M. K. Misra. 1998. Seed enhancements. *Seed Sci. Res.* 8:245–56.

133. TeKrony, D. M. 1993. Accelerated aging test. *J. Seed Tech.* 17:111–20.

134. TeKrony, D. M. 2006. Seeds: The delivery system for crop science. *Crop Sci.* 46:2263–9.

135. TeKrony, D. M., and D. B. Egli. 1991. Relationship of seed vigor to crop yield: A review. *Crop Sci.* 31:816–22.

136. Telewski, F. W., and J. A. D. Zeevaart. 2002. The 120-yr period for Dr. Beal's seed viability experiment. *Amer. J. Bot.* 89:1285–8.

137. Tempe, J. de, and J. Binnerts. 1979. Introduction to methods of seed health testing. *Seed Sci. Tech.* 7:601–36.

138. Toole, E. H. 1958. *Storage of vegetable seeds.* USDA Leaflet 220 (rev.).

139. Van der Berg, H. H., and R. Hendricks. 1980. Cleaning flower seeds. *Seed Sci. Tech.* 8:505–22.

140. Van der Burg, W. J., H. Jalink, R. A. van Zwol, J. W. Aartse, and R. J. Bino. 1994. Nondestructive seed evaluation with impact measurements and x-ray analysis. *Acta Hort.* 362:149–57.

141. Volk, G. M., and C. Walters. 2004. Preservation of genetic resources in the national plant germplasm clonal collections. *Plant Breeding Rev.* 23:291–344.

142. Walters, C., L. J. Wheeler, and P. C. Stanwood. 2004. Longevity of cryogenically-stored seeds. *Cryobiology* 48:229–44.

143. Watkins, J. T. 1992. The effect of environment and culture on vegetable seed quality. *HortTechnology* 2:333–34.

144. Watkins, J. T. 1998. Seeds quality problems commonly encountered during vegetable and flower seed production. *Seed Technology* 20:125–30.

145. Welbaum, G. E. 2006. Natural defense mechanisms in seeds. In A.S Basra, ed. *Handbook of Seed Science and Technology.* Binghamton, NY: Food Products Press, The Harworth Press.

146. Wells, J. S. 1985. *Plant propagation practices*, 2nd ed. Chicago: American Nurseryman Publ.

147. Wyman, D. 1953. Seeds of woody plants. *Arnoldia* 13:41–60.

148. Young, J. A., and C. G. Young. 1986. *Collecting, processing and germinating seeds of wildland plants.* Portland, OR: Timber Press.

149. Young, J. A. 1992. *Seeds of woody plants in North America*, rev. ed. Portland, OR: Dioscorides Press.

150. Zhang, J. H., and M. B. McDonald. 1997. The saturated salt accelerated aging test for small-seeded crops. *Seed Sci. Tech.* 25:123–31.

*7 Principles of Propagation from Seeds

learning objectives

- Describe the process of germination.
- Compare methods for measuring germination.
- Define the environmental and disease factors influencing germination.
- Describe the types of seed dormancy and how dormancy controls germination.

INTRODUCTION

Seed germination, from an ecological standpoint, is the beginning of the next sexual generation. It is the first adaptive step toward colonizing an environmental niche. Therefore, plant species have developed a variety of seed germination and dormancy strategies that make the study of seed germination one of the most fascinating areas of plant growth and development.

From a human ecology standpoint, humankind's recognition that seeds were highly nutritious and could be selected and used to propagate crop plants was pivotal to establishing communities that were self-sustaining for food. Seeds are the genetic repositories of thousands of years of selection for crop plants.

From the standpoint of modern commercial crop production, more plants are propagated from seeds for food, fiber, and ornamental use than any other method of propagation. In Chapter 7, we will summarize the important physiological mechanisms responsible for seed germination and dormancy. A command of these basic principles allows growers to take full advantage of cultural practices to optimize plant production.

THE GERMINATION PROCESS

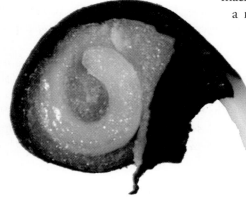

A **seed** is a ripened ovule. At the time of separation from the parent plant, it consists of an **embryo** and **stored food supply,** both of which are encased in a **protective covering** (Fig. 7–1). The activation of the seed's metabolic machinery leading to the emergence of a new seedling plant is known as **germination.** For germination to be initiated, three conditions must be fulfilled (51, 128):

seed The next sexual generation for a plant. It consists of an embryo, food storage tissue, and a protective covering.

germination The committed stage of plant development following radicle emergence from the seed coverings, which leads to a seedling.

1. The seed must be **viable;** that is, the embryo must be alive and capable of germination.
2. The seed must be subjected to the **appropriate environmental conditions:** available water, a proper temperature range, a supply of oxygen, and, sometimes, light.

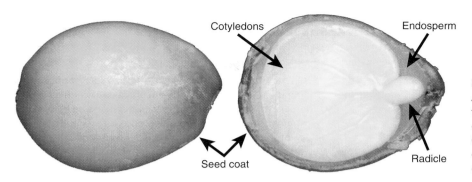

Figure 7–1
A seed consists of an embryo, a food supply (usually endosperm or cotyledon) and a protective covering (seed coat or pericarp). Intact seed on the left and half seed on the right, exposing the embryo.

3. Any **primary dormancy** condition present within the seed (19, 56) must be overcome. Processes leading to removal of primary dormancy result from the interaction of the seed with its environment. If the seeds are subjected to adverse environmental conditions, a **secondary dormancy** can develop and further delay the period when germination takes place (133, 142).

Transition from Seed Development to Germination

As reviewed in Chapter 4, many seeds lose water during the maturation drying stage of seed development. These seeds are either **dormant** or nondormant at the time they are shed from the plant. However, some seeds either do not enter the maturation drying stage of seed development and germinate prior to being shed from the plant (vivipary or precocious germination) or can tolerate only a small degree of desiccation (recalcitrant seeds). Figure 7–2 illustrates the fate of various seeds as they approach the end of seed development. Viviparous and recalcitrant seeds are discussed in detail in Chapters 4 and 6. The discussion of seed germination in this chapter will focus on the basic process of seed germination in orthodox seeds that complete maturation drying and are dormant or nondormant after separation from the mother plant.

dormancy The condition where seeds will not germinate even when the environment is suitable for germination.

Phases of Early Germination

Early seed germination begins with imbibition of water by the seed and follows a triphasic (three-stage) increase in seed fresh weight due to increasing water uptake (Fig. 7–3, page 202); the three phases are described as follows:

1. **Imbibition** is characterized by an initial rapid increase in water uptake.
2. The **lag** phase follows imbibition and is a period of time where there is active metabolic activity but little water uptake.
3. **Radicle protrusion** results from a second period of fresh weight gain driven by additional water uptake.

These processes rely on the **water potential** of the cells

water potential As it relates to seed germination, is a measure of the potential for a cell to take up water from its surrounding environment. Changes in the seed's water potential are the driving force behind germination.

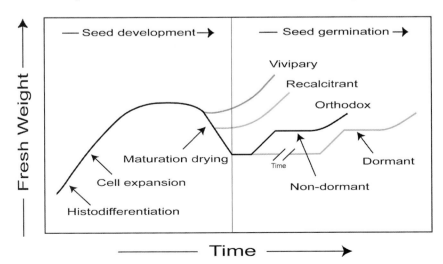

Figure 7–2
The transition from seed development to seed germination. Seeds may end seed development and display viviparous, recalcitrant, or orthodox seed behavior. Viviparous and recalcitrant seeds germinate before completing the maturation drying stage of development. Orthodox seeds continue to dry to about 10 percent moisture and can be either nondormant (sometimes termed quiescent) or dormant.

BOX 7.1 GETTING MORE IN DEPTH ON THE SUBJECT
WATER POTENTIAL AND SEED GERMINATION

Water potential as it impacts water movement in plants is described as

Water potential (ψ_{cell}) = Matric potential (ψ_m) + Osmotic potential (ψ_π) + Pressure potential (ψ_p)

Matric potential is the major force responsible for water uptake during imbibition. Matric forces are due to the hydration of dry components of the seed including cell walls and macromolecules like starch and proteins. Water uptake due to matric forces during imbibition is usually rapid, as might be expected because the seed is very dry (less than 10 percent moisture) at the end of seed development; see Chapter 4.

Osmotic potential and pressure potential determine water uptake during the radicle protrusion phase of seed germination. The initial stage of radicle protrusion is due to enlargement of the cells in the radicle corresponding to increased water uptake. Osmotic potential is a measure of the osmotically active solutes in a cell, including molecules like organic or amino acids, sugars, and inorganic ions. Osmotic potential is expressed as a negative value. As the number of osmotically active solutes increases in a cell, the osmotic potential becomes more negative (i.e., from −0.5 MPa to −1.0 MPa). This can result in more water moving into the cell. [Note: Water potential is expressed as either megapascals (MPa) or bars. One MPa is equal to 10 bars.]

On the other hand, **pressure potential** is an opposing force and is expressed as a positive value. The pressure potential is the **turgor force** due to water in the cell pressing against the cell wall. It is also an expression of the ability of the cell wall to expand. Cell wall loosening in the radicle is determined by the physical properties of the cell wall and the counterpressure exerted by the seed tissues covering the radicle (Fig. 7–4). A combination of increasing osmotic potential (more negative) and/or change in the pressure potential can result in cell enlargement and initiate radicle protrusion. This is termed **growth potential** (21). Thus, changes in osmotic potential of radicle cells, and cell wall loosening in radicle or seed covering cells, are essential components controlling radicle growth and germination. An understanding of this concept is essential to understanding aspects of seed dormancy, effects of hormones on germination, and treatments like seed priming.

growth potential The relative force generated by the radicle during germination. Conceptually, a seed germinates when the radicle force is sufficient to penetrate the seed coverings. This is accomplished by an increase in radicle growth potential and/or weakening of the seed coverings.

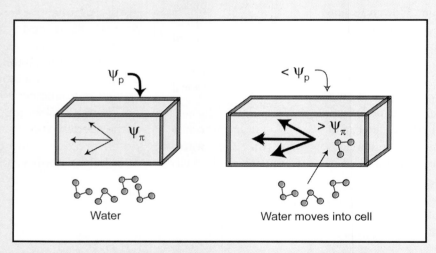

A change in the elastic properties of the cell wall or increase in the osmotic potential of the cell can lead to cell enlargement.

Figure 7–4
Schematic representation of water uptake in a cell. The opposing forces of osmotic potential (ψ_π) and pressure potential (ψ_p) determine water uptake by the cell and the cell's ability to expand.

in the seed and embryo (see text box on water potential and Chapter 3 for additional information).

Water Uptake by Imbibition (Phase I) Most seeds are dry (less than 10 percent moisture) after completing seed development. This results in a very low water potential in dry seeds of near −100 to −350 MPa (216, 217). **Imbibition** is a physical process related to matric forces that occurs in dry seeds with water-permeable seed coats whether they are alive or dead,

imbibition The initial stage of water uptake in dry seeds.

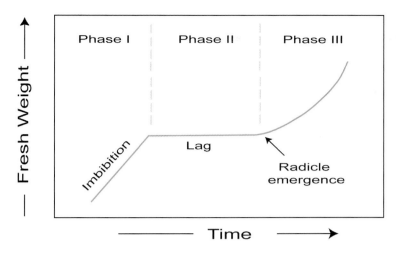

Figure 7–3
There are three phases to germination that can be described by the seed's increase in fresh weight (water uptake). These include the imbibition, lag, and radicle emergence phases.

dormant or nondormant. There are two stages to imbibition (Fig. 7–5) (186, 218). Initially, water uptake is very rapid over the first 10 to 30 minutes. This is followed by a slower wetting stage that is linear for up to an hour for small seeds or several hours (5 to 10) for large seeds. Water uptake eventually ends as the seed enters the lag phase of germination.

The seed does not wet uniformly during imbibition. There is a "wetting front" that develops as the outer portions of the seed hydrate while inner tissues are still dry. Seed parts may wet differentially depending on their contents. Starch is more hydrophobic than protein, and the starchy endosperm will hydrate more slowly compared to the protein-rich embryo. Another characteristic of seeds during imbibition is that they are **"leaky."** Several compounds, including amino acids, organic acids, inorganic ions, sugars, phenolics, and proteins can be detected as they leak from imbibing seeds (218). "Leakiness" is due to the inability of cellular membranes to function normally until they are fully hydrated (25, 168). However, there are some seeds, like members of the cucumber family, which have a perisperm envelope that surrounds the embryo and inhibits ion leakage (241).

The quantity of leaked solutes is diagnostic for seed quality and is the basis for the electrolyte leakage assay for seed vigor testing (see page 184). Solute leakage is also important because it influences detection of the seed by insects and fungi (both pathogenic and beneficial—like mycorrhiza) during germination. Seeds that are slow to germinate or which leak excessively due to poor seed quality are more susceptible to attack by insects and diseases.

Seeds can be physically damaged during imbibition. Seed coverings are usually very hydroscopic, thus slowing the influx of water that could damage internal tissues. Therefore, seeds with physical damage to the seed coverings may be injured by the inrush of water

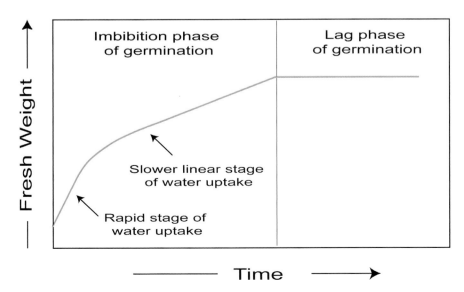

Figure 7–5
Water uptake during seed imbibition. A rapid initial stage followed by a slower linear stage of water uptake is typical for most seeds.

during imbition. Raising the moisture content of seeds (up to 20 percent) prior to sowing can reduce imbibitional injury in susceptible seed lots. In addition, some tropical and subtropical species (like cotton, corn, lima bean) are sensitive to chilling injury when imbibed in cold soil (117).

Lag Phase of Germination (Phase II)   Although the lag phase is characterized as a period of reduced or no water uptake following imbition, it is a highly active period physiologically (Fig. 7–3) (25, 218). Phase II is a period of metabolic activity that prepares the seed for germination. Cellular activities critical to normal germination during the lag phase include:

1. **Mitochondria "Maturation."** Mitochondria are present in the dry seed and these must be rehydrated, and membranes within the mitochondria must become enzymatically active. Within hours of imbition, mitochondria appear more normal when viewed by electron microscopy, and both respiration and ATP synthesis increase substantially.
2. **Protein Synthesis.** Although mRNA is present within the dry seed (see Box 7.2 below), protein synthesis does not occur until polysomes form after seed hydration. New proteins are formed within hours of the completion of imbition. New protein synthesis during the lag period is required for germination.
3. **Storage Reserve Metabolism.** This is the enzymatic breakdown of storage macromolecules to produce substrates for energy production and amino acids for new protein synthesis. Reserve metabolism also produces osmotically active solutes (like sucrose) that can lead to a change in water potential of cells within the embryo in preparation for radicle protrusion.
4. Specific **enzymes,** including those responsible for **cell wall loosening** in the embryo or tissues surrounding the embryo, can be produced.

Radicle Protrusion (Phase III)   The first visible evidence of germination is protrusion of the radicle. This is initially the result of cell enlargement rather than cell division (14, 106). However, soon after radicle elongation begins, cell division can be detected in the radicle tip (163, 201).

Radicle protrusion is controlled by the opposing forces between the growth potential in the embryo and the physical resistance presented by the seed coverings (Fig. 7–6). Radicle protrusion occurs when (a) the water potential of the cells in the radicle becomes more negative due to metabolism of storage reserves; (b) cell walls in the hypocotyl and radicle become more flexible to allow cell expansion; or (c) cells in the seed tissues surrounding the radicle weaken to allow cell expansion in the radicle (25, 179). A combination of these factors may be involved to control germination, depending on the species and the tissues covering the radicle.

In non-endospermic seeds like radish (*Brassica*) and lentil (*Lens*), the seed coat is thin and presents very little resistance to radicle protrusion (Fig. 7–7a). In these seeds, changes in the water potential of the cells in the radicle and cell wall flexibility are responsible for radicle elongation (211). In this case, the activity of gibberellin may be to promote germination by a change

---

BOX 7.2   GETTING MORE IN DEPTH ON THE SUBJECT
PROTEIN SYNTHESIS AND mRNA IN SEEDS

In the dry seed, there is a complement of mRNA made during the final stages of seed development (see Chapter 4). There are two types of stored mRNA in dry seeds: **residual** and **conserved mRNA** (25). Residual mRNA are messages left over from seed development. They persist in dry seeds but are rapidly degraded after imbition and are not involved in germination. Conserved or stored mRNA are produced (transcribed) during late seed development, stored in dry seed, and translated into proteins during the lag phase of germination. Translation of conserved mRNA is an important step in the germination process. In *Arabidopsis*, germination (radicle protrusion) can still occur even if transcription is inhibited (195). This suggests that all of the mRNA required for germination is pre-packaged in the seed as conserved mRNA.

Conserved mRNA have the genetic code for both "housekeeping" genes necessary for normal cellular activities and for germination-specific proteins like "germin" (149). Germin is an oxylate oxidase that may function to release calcium from calcium oxylate. Changes in cellular calcium have been shown to be important during germination (206). Most conserved mRNA are degraded within several hours of imbition and a new mRNA population must be made (transcribed) for germination to be completed. All of the components for new mRNA synthesis (DNA and RNA polymerases and ribonucleotide triphosphate precursors) are present in dry seed, and new mRNA can be detected in the lag phase. Using microarray analysis, it has been calculated that *Arabidopsis* seeds express over 6,500 genes during germination (196).

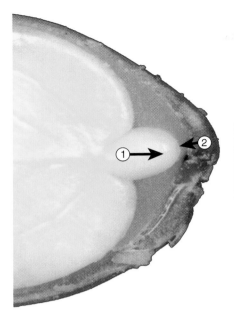

Gibberellin promotes, while ABA inhibits changes in growth potential of the radicle.

① Growth potential in radicle cells.
vs.
② Physical resistance of the seed coverings.

Gibberellins promote, while ABA inhibits enzymatic cell wall loosening in the seed coverings.

Figure 7–6
The balance of forces involved in germination. In many seeds, the seed coverings provide a physical resistance to radicle emergence. The ability of the radicle to penetrate the seed coverings determines the speed of germination and can be an important mechanism for controlling germination in dormant seeds. Adapted from Bradford and Ni, 1993.

in the embryo's water potential, allowing phase III water uptake (hydraulic growth), while the action of abscisic acid is to inhibit germination by preventing this water potential change.

In endospermic dicot seeds, the seed coverings (especially the endosperm cap) can be a significant barrier to germination in some species. Endosperm properties are especially important under conditions that reduce germination, like low temperature conditions in pepper (237) or dormancy as in iris (27), redbud (*Cercis*) (94), and lilac (*Syringa*) (132). In *Arabidopsis*, a single outer layer of endosperm is sufficient to impede germination (84). Partial control of germination by gibberellin and prevention of germination by abscisic acid may be mediated by the induction or inhibition of hydrolytic enzymes acting on the endosperm (179, 187). In solanaceous seeds (such as tomato and tobacco), hydrolytic cell wall enzymes (like endo-b-mannanase and extensins) soften endosperm cell walls (49, 184), and other cell wall enzymes (like β-1,3-glucanases) cause cell-to-cell separation (189, 248), permitting germination by reducing

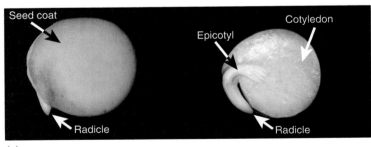

(a)

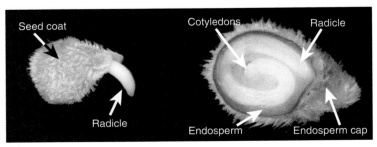

(b)

Figure 7–7
Seed morphology related to germination. (a) Lentil (lens) is non-endospermic and most of the seed cavity is filled with cotyledon tissue. The seed coat restricts radicle protrusion. (b) In tomato the embryo is embedded in endosperm, the endosperm cap covers the radicle, and is the restraint to radicle protrusion.

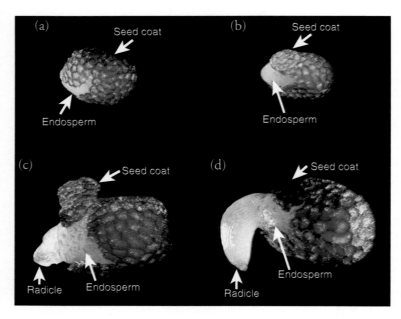

Figure 7–8
Petunia seed demonstrating two-step germination (a) Seed coat cracks. (b) Endosperm stretches over emerging radicle. (c) Radicle protrudes from endosperm. (d) Hypocotyl and radicle elongation.

the force of the seed coverings restricting radicle elongation and, finally, releasing the radicle for germination. In some seeds with an endosperm cap, germination proceeds in two stages (84). First, radicle elongation initiates seed coat cracking while the endosperm stretches over the radicle (Fig. 7–8). In the second step, the endosperm ruptures releasing the radicle. Hormones may act differently in each step. For example, in tobacco and *Arabidopsis*, abscisic acid does not inhibit initial radicle elongation and seed coat rupture, but does inhibit endosperm rupture (174).

The perisperm also can be a barrier for germination as observed in members of the cucumber family (Fig. 7–9). In cucumber and melon, the perisperm-endosperm forms an envelope surrounding the embryo. This envelope shows reduced and selected permeability to ions due to lipid and callose content of the envelope (197, 241). In dormant seeds or non-dormant seeds germinated at low temperature (74), removal of embryos from the seed coat and perisperm-endosperm envelop is sufficient to permit germination. Prior to radicle emergence, cell wall enzymes work to make the

(a)

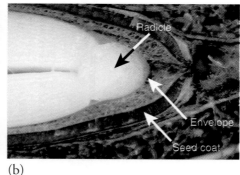

(b)

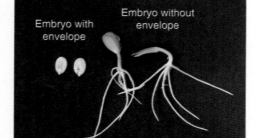

(c)

Figure 7–9
In the cucumber family, there is a perisperm (or perisperm + endosperm) envelope that surrounds the embryo. It is usually a semi-permeable membrane that limits the movement of ions into or away from the embryo. (a) Isolated seeds of prickly cucumber (Sycos) with and without the surrounding envelope. (b) A longitudinal section showing the location of the envelope between the embryo and seed coat. (c) Seeds isolated from the seed coat with the envelope intact do not germinate, but those with the envelope removed germinate readily.

envelope more ion-permeable and to weaken the envelope around the radicle tip (197).

Seedling Emergence

Seedling emergence begins with elongation of the root and shoot meristems in the embryo axis, followed by expansion of the seedling structures (Fig. 7–10). The embryo consists of a shoot **axis** bearing one or more **cotyledons** and a root axis **(radicle).** The seedling stem is divided into the **hypocotyl, cotyledonary node,** and the **epicotyl.** The hypocotyl is the stem section between the cotyledons and the radicle. In some seedlings, there is a noticeable swelling at the hypocotyl-radicle juncture called the collet or collar. The epicotyl is the section between the cotyledons and the first true leaves.

Once growth begins, fresh and dry weight of the new seedling plant increases, as storage tissue weight decreases. The respiration rate, as measured by oxygen uptake, increases steadily with advance in

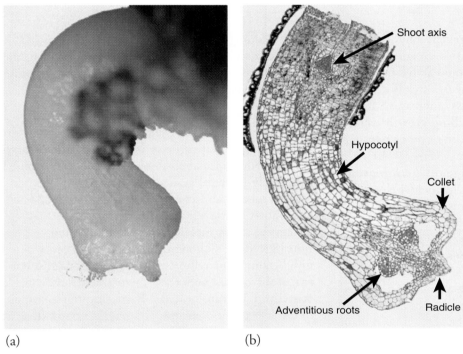

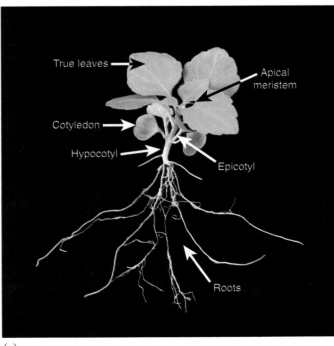

Figure 7–10
A seedling usually consists of radicle, hypocotyl, and shoot axis.

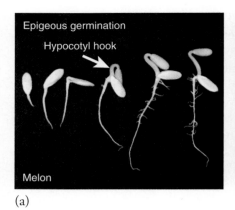
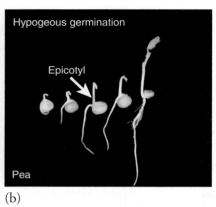

Figure 7–11
Typical patterns of germination include (a) epigeous where the hypocotyl hook raises the cotyledons above the soil and (b) hypogeous where the cotyledons remain below ground and the epicotyl and shoot emerge from the soil.

growth. Seed storage tissues eventually cease to be involved in metabolic activities except in plants where persistent cotyledons become active in photosynthesis. Water absorption increases steadily as root mass increases. Initial seedling growth usually follows one of two patterns (Fig. 7–11). In **epigeous germination,** the hypocotyl elongates, forms a **hypocotyl hook,** and raises the cotyledons above the ground. **Hypogeous germination** is the other pattern of germination and is characterized by a lack of hypocotyl expansion so only the epicotyl emerges above the ground, and the cotyledons remain within the seed coverings.

epigeous Germination when a seedling emerges from the soil using the hypocotyl hook to penetrate the soil first.

gypogeous Germination when a seedling emerges from the soil using the shoot tip to penetrate first.

Storage Reserve Utilization

Initially, new embryo growth is dependent on the storage reserves manufactured during seed development and stored in the endosperm, perisperm, or cotyledons. The major storage reserves are:

1. **proteins**
2. **carbohydrates** (starch)
3. **lipids** (oils)

These are converted to **amino acids** or **sugars** to fuel early embryo growth (Fig. 7–12). The embryo is dependent on the energy and structural materials from stored reserves until the seedling emerges into the light and can begin photosynthesis.

Use of Storage Proteins Storage proteins are stored in specialized structures called **protein bodies.** Protein bodies are located in cotyledons and endosperm. Enzymes (proteinases) are required to catabolize storage proteins into amino acids that, in turn, can be used by the developing embryo for new protein synthesis (25). These enzymes can be present in stored forms in the dry seed, but the majority of **proteinases** are synthesized as new enzymes following imbibition.

Use of Storage Carbohydrates (Starch) Starch is a major storage material in seeds and is mostly stored in the endosperm but can also be found in the embryo and cotyledons. Catabolism of starch has been studied extensively in cereal grains (barley, wheat, and corn), and shows a coordinated system for starch mobilization (82, 109). Following imbibition, gibberellin in the embryo axis and the **scutellum** is translocated to the cells of the **aleurone layer** (Fig. 7–13). The aleurone is a layer of secretory cells that surrounds the non-living starchy

aleurone A special layer of cells that surrounds the endosperm in monocot seeds. It is responsible for making the enzymes used to degrade storage materials in the endosperm to be used by the embryo for germination.

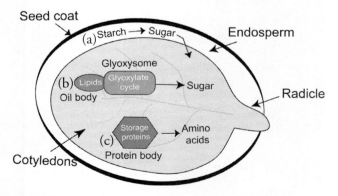

Figure 7–12
The general pattern of seed reserve mobilization leading to germination. These include the conversion of (a) starch to sugar, (b) lipids to sugar, and (c) storage protein to amino acids.

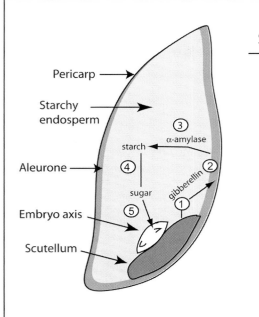

Figure 7–13
The cereal grain model for starch mobilization in seeds.

endosperm. Gibberellin initiates the de novo synthesis of numerous enzymes in the aleurone that are secreted into the endosperm, including amylases that hydrolyse starch to sugar.

The major starch-degrading enzyme is α-**amylase.** It hydrolyses starch in the **starch grains** of the endosperm to simple glucose and maltose sugar units that are eventually synthesized into sucrose for transport to the embryo axis. Enzymes break down the cell walls of the endosperm to allow movement of sucrose to the scutellum for transport to the growing axis. Gibberellin-initiated synthesis of α-amylase has been studied extensively in the cereal aleurone system and has significantly improved our understanding of the molecular mechanisms for hormone-regulated gene expression in plants (25).

Use of Storage Lipids (Oils) Lipids are stored in specialized structures called **oil bodies** located in the endosperm and cotyledons of seeds. Catabolism of lipids in seeds is a complex, unique interaction among the oil bodies, **glyoxysomes,** and mitochondria (Fig. 7–14, page 210). The main storage forms of lipids in the oil body are **triacylglycerides.** In the oil body, triacylglycerides are catabolized to glycerol and free fatty acids. Free fatty acids are moved to the glyoxysome. Glyoxysomes are specialized structures only present in oil-storing seeds. They function to convert free fatty acids to the organic acids, malate, and succinate using enzymes in the glyoxylate cycle. *Glyoxysomes and the glyoxylate cycle are unique to germinating seeds* and are not found in any other part of the plant. The end result of this biochemical process is the production of sucrose from storage lipids for use by the developing embryo.

Measures of Germination

A seed lot completes germination when either the radicle protrudes through the seed coverings or the seedling emerges from soil or media. In either case, the time required for individual seeds in a seed lot to complete germination usually produces a sigmoidal germination curve (Fig. 7–15, page 210). This sigmoid curve is indicative of the way a seed population behaves. There is an initial delay in the start of germination and then a rapid increase in the number of seeds that germinate, followed by a decrease in their appearance over time. In addition, when a population of seeds is graphed as the number of seeds that germinate per unit of time, the curve roughly follows a near normal distribution (Fig. 7–16, page 210). These two germination curves are the basis for describing characteristics of a seed lot and for models that predict the time to seedling emergence under a variety of environments (31, 33).

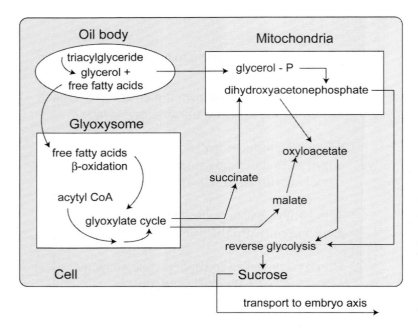

Figure 7–14
Lipid conversion to starch is a complex set of enzymatic reactions coordinated between the oil body, mitochondria, and glyoxysome. Key steps include the conversion of triacylglycerides (the storage form for oils in the seed) to glycerol and free fatty acids, the production of succinate and malate in the glyoxylate cycle in the glyoxysome, and then reverse glycolysis (sometimes termed gluconeogenesis) to produce sucrose for use by the embryo.

Important aspects of seed germination can be measured by three parameters:

1. **Percentage**
2. **Speed (Rate)**
3. **Uniformity**

Seed lots, even within a species, can vary in their germination patterns related to these three parameters. The ideal seed lot germinates at nearly 100 percent and has a fast germination rate that produces uniform seedling emergence.

Germination percentage is the number of seeds that produce a seedling from a seed population expressed as a percentage. For example, if 75 seeds germinate from a seed lot of 100 seeds, the germination percentage would be 75 percent (75 germinated seeds divided by 100 seeds in the seed lot multiplied by 100 to give a percentage).

Germination speed or rate is a measure of how rapid a seed lot germinates. It is the time required for a seed lot to reach a predetermined germination percentage; for example, the time required for a seed lot to reach 50 percent germination based on the final germination

> **germination percentage** Not the same as **germination rate**. Percentage is a measure of the number of seeds that germinate, while rate is a measure of how fast the seeds germinate.

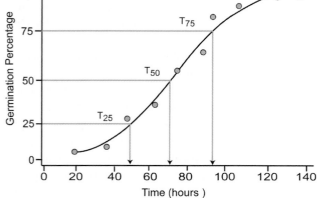

Figure 7–15
Typical sigmoidal germination curve for a sample of germinating seeds. After an initial delay, the number of seeds germinating increases then decreases.

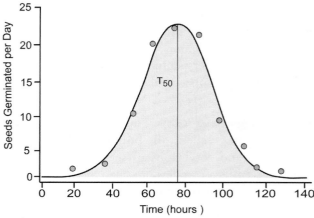

Figure 7–16
The number of seeds that germinate per unit of time can usually be represented as a normally distributed curve.

> **BOX 7.3 GETTING MORE IN DEPTH ON THE SUBJECT**
> **MEASURES OF GERMINATION RATE**
>
> Germination curves are the best representation of germination patterns. The mathematical equations used to describe a particular germination pattern can vary for seed lots and germination environment. In some cases, a simple sigmoid equation explains the germination pattern. In other cases, the pattern is better represented by other curve-fitting equations like Richard's or Weibull functions (23, 37, 38) because they provide better estimates of early and late aspects of germination.
>
> Although germination curves adequately describe germination data, researchers have attempted to represent cumulative germination data as a single germination value. Obviously, germination percentage is a single value that can be used to compare seed lots for superior germination characteristics. Standard germination tests are basic for describing seed lots as prescribed by Federal and International seed laws (see Chapter 4). But germination patterns can be very different, while final germination percentages can be nearly identical (Fig. 7–14). In addition to germination percentage, germination speed can be adequately represented as the T_{50}, and germination uniformity can be expressed as the standard deviation of the population mean.
>
> Numerous single values for germination have been developed to describe germination, including Kotowski's coefficient of velocity, Czabator's germination value, Maguire's speed of germination, Diavanshir and Poubiek's germination value, and Timson's cumulative germination. These were nicely compared by Brown and Mayer (37), who concluded that there were inherent problems in using a single value to describe germination. Therefore, it is most descriptive to use three independent properties of germination—percentage, speed, and uniformity—when comparing seed lots.

percentage. This value is the T_{50} and can be seen on the sigmoidal and normal distribution curves (Figs. 7–15 and 7–16). Since this value is calculated as 50 percent germination based on the final germination, it is a more meaningful descriptor for high-germinating seed lots.

Germination uniformity measures how close in time seeds germinate or seedlings emerge. In some seed lots, the time between the first and last seedling emergence is clustered closely around the mean time to 50 percent emergence, while in others this time is spread out. One way to express germination uniformity is as the standard deviation around the mean. This can be reported as the time to 75 percent germination (T_{75}) minus the time to 25 percent germination (T_{25}).

These properties are nicely illustrated in the germination curves for seed lots in Figure 7–17. All three seed lots have 100 percent germination. In the higher-vigor seed lot (seed lot #1), the T_{50} is reached sooner than in the other two seed lots. However, even though seeds in seed lot #2 germinate much later than those in seed lot #1, they both have the same germination uniformity as indicated by the width of the curve. Also, notice how the T_{50} for seed lots #1 and #3 are the same, but the germination is more uniform for seed lot #1.

Environmental Factors Influencing Germination

Factors in the environment that impact germination properties include:

1. **Temperature**
2. **Water**
3. **Gases**
4. **Light**

Temperature Temperature is a most important environmental factor regulating the timing of germination, partly due to dormancy release and partly due to climate adaptation. Temperature control is also essential in subsequent seedling growth. Dry, non-imbibed seeds can withstand extremes of temperature. For disease

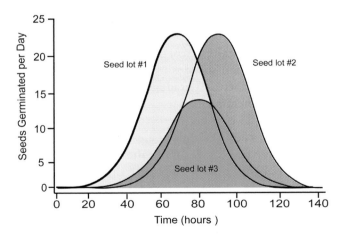

Figure 7–17
Germination curves for three seed lots of tomato illustrate how germination can be described. Seed lot #1 has high vigor. This is shown with high germination, reduced T_{50}, and a small standard deviation around the T_{50}. Seed lot #2 has high germination but requires more time to germinate. Seed lot #3 has reduced overall germination and although the time to 50 percent germination is not different from seed lot #1, this seed lot does not germinate uniformly.

Figure 7–18
A thermogradient table allows for the simultaneous evaluation of germination over a range of temperatures. The table is an aluminum plate with a differential heating source at each end of the table, which establishes a linear temperature gradient across the table.

control, seeds can be placed in hot water for short periods without killing them. In nature, brush fires are often effective in overcoming dormancy without damaging seeds. Seeds show prolonged storage life when stored at low temperatures, even below freezing for dry seeds (see Chapter 8).

Temperature effects on germination. Temperature affects both germination percentage and germination speed (75). Three temperature points (**minimum, optimum,** and **maximum**), varying with the species, are usually designated for seed germination. **Optimum** temperature for seed germination produces the largest percentage of seedlings in the shortest period of time. The optimum temperature for non-dormant seeds of most commercially produced plants is between 25 and 30°C (77 and 86°F) but can be as low as 15°C (59°F). **Minimum** is the lowest temperature for effective germination, while **maximum** is the highest temperature at which germination occurs. Above the maximum temperature, seeds are either injured or go into secondary dormancy.

Germination speed is usually slower at low temperatures but increases gradually as temperatures rise, similar to a chemical rate-reaction curve (145). Above an optimum level, a decline occurs as the temperature approaches a lethal limit where the seed is injured. Germination percentage, unlike the germination speed, may not change dramatically over the middle part of the temperature range, if sufficient time is allowed for germination to occur.

Thermoinhibition is the inhibition of germination by high or low temperature. It is commercially important in vegetable (lettuce and celery) and flower (pansy) crops whose crop cycles can require germination when soil or greenhouse substrate temperatures exceed approximately 30°C (86°F) or in direct-seeded warm season vegetables (sweet corn and cucumber) sown into cold soils (less than 15°C, 59°F). Thermoinhibition can be impacted by environmental (light and temperature), physiological (hormones), and genetic factors (44). A thermogradient table (Fig. 7–18) can be used to screen for thermotolerant genotypes or the effectiveness of seed treatments. Seed priming generally has been an effective treatment to circumvent thermoinhibition (169). Seeds of different species, whether cultivated or native, can be categorized into temperature-requirement groups. These are related to their climatic origin.

Cool-Temperature Tolerant. Seeds of many kinds of plants, mostly native to temperate zones, will germinate over a wide temperature range from about 4°C (39°F) (or sometimes near freezing) up to the lethal limit—from 30°C (86°F) to about 40°C (104°F). The optimum germination temperature for many cool-tolerant seeds—including broccoli, cabbage, carrot, alyssum, and others—is usually about 25 to 30°C (77 to 86°F).

Cool-Temperature Requiring. Seeds of some cool-season species adapted to a "Mediterranean" climate require low temperatures and fail to germinate at temperatures higher than about 25°C (77°F). Species of this group tend to be winter annuals in which germination is prevented in the hot summer but takes place in the cool fall when winter rains commence. Seeds that require cool temperatures include various vegetables, such as celery, lettuce, and onion, as well as some flower seed—coleus, cyclamen, freesia, primrose (*Primula*), delphinium, and others (7).

Warm-Temperature Requiring. Seeds of another broad group fail to germinate below about 10°C (50°F) (asparagus, sweet corn, and tomato) or 15°C (59°F) (beans, eggplant, pepper, and cucurbits). These species primarily originated in subtropical or tropical regions. Other species, such as lima bean, cotton, soybean, and sorghum, are also susceptible to "chilling injury" when exposed to temperatures of 10 to 15°C (50 to 59°F) during initial imbibition. Planting in a cold soil can injure the embryo axis and result in abnormal seedlings (117, 190).

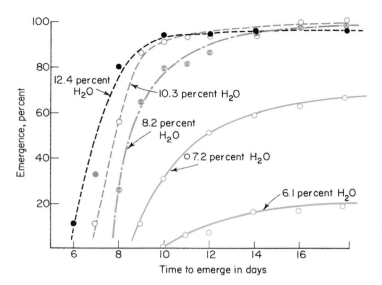

Figure 7–19
Effect of different amounts of available soil moisture on the germination (emergence) of 'Sweet Spanish' onion seed in Pachappa fine sandy loam. From Ayers, 1952.

Alternating Temperatures. Fluctuating day/night temperatures, rather than providing a constant temperature, often gives better results for both seed germination and seedling growth. Use of fluctuating temperatures is a standard practice in seed testing laboratories. The alternation should be a 10°C (18°F) difference (231). This requirement is particularly important with some freshly harvested seeds (6). Seeds of a few species will not germinate at all at constant temperatures. It has been suggested that one of the reasons imbibed seeds deep in the soil do not germinate is that soil temperature fluctuations disappear with increasing soil depth (19).

Water For many non-dormant seeds, water availability is the only factor limiting germination at suitable temperatures. The mechanism for water uptake by seeds has been discussed in detail as it relates to the phases of germination (see page 201).

The rate of water movement into the seed is dependent on the water relations between the seed and its germination medium. Water moves from areas of high (more positive value) water potential to areas of low (more negative value) water potential (see page 201). The water potential of the seed is more negative than moist germination substrates, so water moves into the seed. Rate of water movement within the soil or germination substrate depends on (a) pore structure (texture), (b) compaction, and (c) the closeness and distribution of soil-seed contact. As moisture is removed by the imbibing seed, the area nearest the seed becomes dry and must be replenished by water from adjacent soil. Consequently, a firm, fine-textured seed bed in close contact with the seed is important in maintaining a uniform moisture supply.

Osmotic potential in the soil solution depends on the presence of solutes (salts). Excess soluble salts (high salinity) may exert strong negative pressure (exosmosis) and counterbalance the water potential in seeds. Salts may also produce specific toxic effects. These may inhibit germination and reduce seedling stands (8, 116). Such salts originate in the soil or may come from the irrigation water or excessive fertilization. Since the effects of salinity become more acute when the moisture supply is low and, therefore, the concentration of salts is increased, it is particularly important to maintain a high moisture supply in the seed bed where the possibility of high salinity exists. Surface evaporation from subirrigated beds can result in the accumulation of salts at the soil surface even under conditions in which salinity would not be expected. Planting seeds several inches below the top edge of a sloping seed bed can minimize this problem (24).

Water stress can reduce germination percentage (69, 111). Germination of some seeds, particularly those that can be difficult to germinate (e.g., beet, lettuce, endive, or celery) are reduced as moisture levels are decreased. Such seeds may contain inhibitors that require leaching. Seeds of other species (e.g., spinach), when exposed to excess water, produce extensive mucilage that restricts oxygen supply to the embryo, reducing germination (5). In these cases, germination improves with less moisture. Substrate moisture content can also impact germination percentages during plug production of flower crops (46, 47).

Moisture stress strongly reduces seedling emergence rate from a seed bed. This decline in emergence rate occurs as the available moisture decreases to a level approximately halfway through the range from field capacity to permanent wilting point (Fig. 7–19) (8, 69, 111). Once the seed germinates and the radicle emerges, the seedling water supply depends on the ability of the root system to grow into the surrounding soil and the new roots' ability to absorb water.

Seed Priming. **Seed priming** is a form of controlled seed hydration that can improve the germination properties of a seed lot, particularly germination rate and uniformity (Fig. 7–20) (169, 222). Controlled seed hydration has a long tradition as a seed treatment. Theophrastus (4th Century BC) observed that cucumber seeds soaked in water prior to sowing would induce faster emergence (80). In 1600, Oliver de Serres described the "clever trick" of soaking grains (wheat, rye, or barley) for two days in manure water followed by drying in the shade before planting the seeds. He noted that soaked seeds emerged more quickly, avoiding "the danger of being eaten away by soil pests" (222). In experiments conducted in 1855, Charles Darwin hinted at the possibilities for osmotic seed priming (3). Darwin submerged seeds in salt water to show that they could move across the sea between land masses as a means to explain geographic distribution of plant species. Not only did seeds survive immersion in cold salt water for several weeks, but some species, like cress and lettuce, showed accelerated germination.

The potential significance of this observation to agriculture was not recognized in Darwin's time. However, in 1963, Ells (77) treated tomato seed with a nutrient solution and observed improved germination. At the same time, it was observed that seeds dried following various times of imbibition showed quicker germination after subsequent rehydration (167). This was termed "imbibitional drying" (112). Heydecker et al. (120) used polyethylene glycol to treat seeds, and this prompted interest in "priming seeds" (121) that has led to a commercially significant practice for the seed industry.

Seed priming is a seed presowing treatment that can significantly enhance germination efficiency in a diverse group of plants including agronomic, vegetable, and ornamental crops (240). It is a treatment for controlled seed hydration. Priming permits the early metabolic events of germination to proceed while the seeds remain in the lag phase of germination (Fig. 7–21). Radicle emergence is prevented by the water potential of the imbibitional medium. After priming is complete, the seed is dried to nearly its original water content. Various techniques have been used to control seed hydration while not permitting radicle emergence (32, 143). These treatments provide conditions for priming that have an imbibitional medium that (a) has a water potential usually between –1.0 and –2.0 MPa (–10 and –20 bars); (b) temperatures between 15 and 25°C (59 and 78°F); and (c) keeps seeds in the lag phase of germination for an extended time (up to 20 days, but usually less than 2 weeks) (39).

Techniques used for seed priming include:

1. **Osmotic priming** by imbibing seeds in osmotic solutions.
2. **Matrix priming** using solid carriers with appropriate matric potential.
3. **Drum priming** that hydrates seeds with water in a tumbling drum.

> **priming**
> A pregermination treatment that enhances germination. It is a controlled hydration treatment that allows seeds to begin the germination process, but prevents radicle emergence.

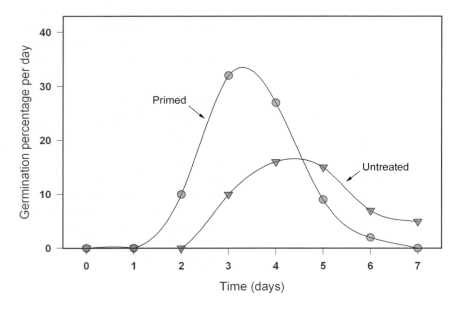

Figure 7–20
A major advantage for primed seeds is faster, more uniform germination. This is illustrated for a primed and controlled seed lot of purple coneflower (*Echinacea purpurea*). Both sets of seeds germinated at the same percentage, but primed seeds germinated faster and more uniformly Geneve, et al. 1991.

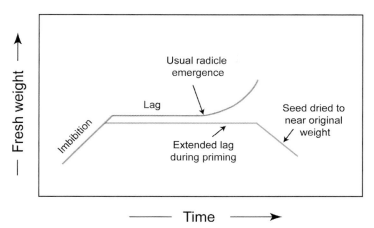

Figure 7–21
Phases of germination related to water uptake modified to describe seed behavior during seed priming. Seed priming extends the time the seed remains in the lag phase of germination. Primed seeds are dried to near their original weight prior to radicle emergence.

In osmotic seed priming (also termed **osmoconditioning** or **osmopriming**), osmotic solutions are made using various inorganic salts, or more commonly polyethylene glycol (PEG) (118, 143). The osmotic potential of the solution, temperature during priming, and duration of priming vary for different species but must hydrate seeds in an aerated solution without allowing the radicle to emerge (30). Seeds are primed in aerated solutions (60) using either a bubble column or stirred bioreactor (101, 102, 180). Following priming, seeds must be dried using forced air, fluidized beds (166), or centripetal dryers.

Problems with aeration, large solution volumes, and disposal of PEG has prompted the use of matrix seed priming (also termed **matriconditioning** or **solid matrix priming**) as an alternative to osmotic priming, especially in large-seeded species like beans (114, 223). Matrix priming uses similar water potential, temperature, and treatment duration as osmotic priming, but uses materials like moistened vermiculite, Leonardite shale, diatomaceous silica, or calcined clay to prime seeds. Materials are mixed with seeds at a ratio of 0.2 to 1.5 g of material to 1 g seed and 60 to 300 percent water (based on dry weight of solid material), depending on the matrix material. The material is usually removed prior to sowing but may be left on the seed.

Drum priming is simple in concept, but sophisticated in practice (207, 236). The amount of water

---

BOX 7.4 GETTING MORE IN DEPTH ON THE SUBJECT
PHYSIOLOGY OF SEED PRIMING

Several biochemical changes occur during priming (36, 141, 143). There is very little increase in DNA synthesis during priming. This is expected because during priming seeds remain in the lag phase of germination prior to the onset of cell division. In contrast to DNA synthesis, RNA synthesis increases during priming. However, seeds primed in the presence of RNA synthesis inhibitors indicate that RNA synthesis is not required for the observed priming effect on seeds. One characteristic of primed seeds is that they resume RNA synthesis quicker than non-primed seeds during germination. It is not clear if this is a cause or an effect of the priming process.

Protein synthesis increases substantially during and following priming (143, 169). This includes both the quantity and the type of proteins being made. Inhibiting protein synthesis during priming prevents enhanced germination, indicating that protein synthesis is an important part of the priming process. Metabolic enzymes involved in storage reserve mobilization have been shown to increase, including α-amylase, malate dehydrogenase, and isocitrate lyase, which implies that one mechanism for priming is a change in the osmotic potential of cells in the embryo due to the increase in osmotically active solutes like sugars and amino acids mobilized from starch and proteins. However, there is also some evidence to suggest that cell wall properties of the seed coverings also change during priming.

Transcriptome and microarray experiments showed that approximately 20 percent of genes were expressed differentially during priming compared to untreated seeds (155). Of the priming-specific genes, subsets of genes for signal transduction and energy production were down-regulated, while subsets for cellular stress tolerance and transcription were up-regulated. However, there were a significant number of genes showing differential regulation that remain with unknown function.

Primed seeds tend to have a shorter life in storage than nonprimed seeds, and the benefits of priming can be lost during storage (169). Primed seed storage life may be most impacted by conditions during drying after seed hydration. Temperature should be cool and the drying rate rapid enough to prevent germination processes to proceed any further. However, rapid drying can cause damage in some seeds. Also, in primed pepper seeds, a brief heat shock for 3 hours at 40°C reduced subsequent seed deterioration in storage (40).

required to obtain seed hydration that allows priming but prevents radicle emergence is determined for a quantity of seeds. This amount of water is applied to seeds in a fine spray as seeds slowly rotate in a drum to provide uniform seed hydration. The drum is positioned on a scale that continually weighs the seeds, signaling a computer to add additional water as necessary to maintain the predetermined hydrated seed weight. Otherwise, the parameters of hydration, temperature, and duration are similar to other priming treatments.

Seed priming has become a commercially important seed treatment, especially for high-value seeds where uniform germination is required; for example, plug production of bedding plants (see Chapter 8). The major benefit found in primed seeds is more rapid and uniform germination compared to untreated seeds. It has also become important for crops that experience thermoinhibition, including summer seeded lettuce (43, 232) and summer greenhouse-sown pansy (46).

Water and Temperature Models for Germination Germination is primarily a function of temperature and

germination models
Mathematical equations based on a seed's response to available water and temperature. They are useful for determining the time required for germination to occur under variable environmental conditions.

available water (85, 239). Mathematical equations have been developed to predict the time required to complete germination. **Germination models** are based on temperature (**thermal time**), moisture availability (**hydrotime**), or a combination of moisture and temperature (**hydrothermal time**).

Models can be useful to predict germination, but they are also useful to help conceptualize environmental effects on germination. Faster-germinating seeds require less accumulated thermal time to germinate than slower-germinating seeds in the seed lot. At the same time, seeds vary in their base water potential that permits radicle emergence. If the temperature or water potential in the soil or germination substrate falls below the base values for that seed, then germination is delayed. For example, if adequate temperature and moisture is available following sowing, quicker-germinating seeds (those that require less thermal or hydrotime) are able to germinate and emerge (83). As the seed bed dries out (water potential falls below base water potential), or temperature is reduced, the remaining seeds in the population are unable to germinate. These seeds germinate only after warmer conditions or subsequent irrigation. This helps visualize how environmental changes could lead to erratic seedling emergence under field conditions. Using this concept, Finch-Savage et al. (87) was able to schedule irrigation at critical stages of germination predicted by thermal time to optimize seedling emergence for various vegetable crops.

Aeration Effects on Germination Exchange of gases between the germination substrate and the embryo is essential for rapid and uniform germination. **Oxygen (O_2)** is essential for the respiratory processes in germinating seeds. Oxygen uptake can be measured shortly

---

BOX 7.5 GETTING MORE IN DEPTH ON THE SUBJECT
GERMINATION MODELS

Thermal Time
At constant moisture levels (water potential), germination has been described by a thermal-time model (93, 229). **Thermal time** is the accumulated hours above a predetermined base temperature that is required for germination. The **base temperature** must be determined for each species and is defined as the minimum temperature where germination occurs. The thermal time required for seeds within a seed lot to germinate can vary, but the base temperature for a species or seed population is relatively stable. Thermal time is a good predictor of germination under conditions that are not limiting water availability to the seed (113).

Hydrotime
Hydrotime is an analogous calculation to thermal time, where the temperature does not vary, and the germination rate is a function of the time above a **base water potential**. Seeds within a seed lot or between seed lots can differ in their ability to germinate at given water potentials because the base water potential is not the same for all seeds in a seed lot. The base water potential is defined as the minimum water potential needed to initiate radicle emergence (31, 34, 105). Therefore, models to describe the time to radicle emergence based on water potential are population-based models. There is evidence that the base water potential in seeds can change due to seed priming and treatments to relieve dormancy (34, 239), and that this change can account for more rapid germination in treated seeds because they now require less hydrotime to germinate.

Hydrothermal Time
Under field conditions, where moisture and temperature vary, the time to radicle emergence can be predicted using a hydrothermal time model (4, 59). This model uses both base values for temperature and water potential to predict the time to radicle emergence.

after water imbibition. Rate of oxygen uptake is an indicator of germination progress and has been suggested as a measure of seed vigor. In general, O_2 uptake is proportional to the amount of metabolic activity taking place. Oxygen supply is limited where there is excessive water in the germinating medium. Poorly drained outdoor seed beds, particularly after heavy rains or irrigation, can have soil pore spaces so filled with water that little oxygen is available to seeds. The amount of oxygen in the germination medium is affected by its low solubility in water and its slow ability to diffuse. Thus, gaseous exchange between the soil and the atmosphere, where the O_2 concentration is 20 percent, is reduced significantly by soil depth and, in particular, by a hard crust on the surface, which can limit oxygen diffusion (15, 111). Seeds of different species vary in their ability to germinate at very low oxygen levels, as occurs under water (172). Seeds of some water plants germinate readily under water, but their germination is inhibited in air. In some species, such as white mustard, basil, and spinach, mucilaginous layers in seed coats or fruit tissue are produced (particularly under high moisture conditions), which can restrict gaseous exchange (Fig. 7–22). The mucilage may provide contact between the soil environment and the seed for better water uptake, but under wet conditions the mucilage can restrict oxygen diffusion to the seed and inhibit germination (119).

Carbon dioxide (CO_2) is a product of respiration and, under conditions of poor aeration, can accumulate in the soil. At lower soil depths, increased CO_2 may inhibit germination to some extent but probably plays a minor role, if any, in maintaining dormancy. In fact, high levels of CO_2 can be effective in overcoming dormancy in some seeds (145).

Light Effects on Germination Light has been recognized since the mid-20th Century as a germination-controlling factor (57). Recent research demonstrates that light acts in both dormancy induction and release and is a mechanism that adapts plants to specific niches in the environment, often interacting with temperature. Light effects on germination can involve both **quality** (wavelength) and **photoperiod** (duration). See Chapter 3 for a detailed description of light.

Light-sensitive seeds are characterized by being small in size, and a shallow depth of planting is an important factor favoring survival (171). If covered too deeply, the epicotyl may not penetrate the soil. Some important flower crops requiring light for germination include alyssum, begonia, *Calceolaria*, coleus, *Kalanchoe*, primrose, and *Saintpaulia* (13). Germination can also be inhibited by light in species, such as *Phacelia*, *Nigella*, *Allium*, *Amaranthus*, and *Phlox*. Some of these are desert plants where survival is enhanced if the seeds are located at greater depths where adequate moisture might be assured. Certain epiphytic plants, such as mistletoe (*Viscum album*) and strangling fig (*Ficus aurea*), have an absolute requirement for light and lose viability in a few weeks without it. Additional aspects of light on germination are discussed under photodormancy (see page 226).

Disease Control during Seed Germination

Control of disease during seed germination is one of the most important tasks of the propagator. The most universally destructive pathogens are those resulting in "damping-off," which may cause serious loss of seeds, seedlings, and young plants. In addition, there are a number of fungal, viral, and bacterial diseases that are seed-borne and may infect certain plants (11). In such cases, specific methods of control are required during propagation. (See the discussion of sanitation in Chapter 3.)

Damping-Off Damping-off is a term long used to describe the death of small seedlings resulting from attacks by certain fungi, primarily *Pythium ultimum* and *Rhizoctonia solani*, although other fungi—for example, *Botrytis cinerea* and *Phytophthora* spp.—may also be involved (Fig. 7–23). Mycelia and spores from these organisms occur in soil,

damping-off The collective term for various disease organisms that can cause early seedling death.

Figure 7–22
When basil seeds are imbibed, the outer cells in the seed coat exude a mucilage that encompasses the seed.

Figure 7–23
Damping-off in a seedling tray and plug flat. (a) In a community seeded flat, damping-off can move from seedling to seedling, killing whole areas of the flat. (b) One symptom of damping-off is a constricted hypocotyl and severe wilting.

in infected plant tissues, or on seeds, from which they contaminate clean soil and infect clean plants. *Pythium* and *Phytophthora* produce spores that are moved about in water.

The environmental conditions prevailing during the germination period will affect the growth rate of both the attacking fungi and the seedling. For instance, the optimum temperature for the growth of *Pythium ultimum* and *Rhizoctonia solani* is between approximately 20 and 30°C (68 and 86°F), with a decrease in activity at both higher and lower temperatures. Seeds that have a high minimum temperature for germination (warm-season plants) are particularly susceptible to damping-off, because at lower or intermediate temperatures (less than 23°C or 75°F), their growth rate is low at a time when the activity of the fungi is high. At high temperatures, not only do the seeds germinate faster, but the activity of the fungi is less. Field planting of such seeds should be delayed until the soil is warm. On the other hand, seeds of cool-season plants germinate (although slowly) at temperatures of less than 13°C (55°F), but since there is little or no activity of the fungi, they can escape the effects of damping-off. As the temperature increases, their susceptibility increases because the activity of the fungi is relatively greater than that of the seedling.

The control of damping-off involves two separate procedures: (a) the complete elimination of the pathogens during propagation, and (b) the control of plant growth and environmental conditions, which will minimize the effects of damping-off or give temporary control until the seedlings have passed their initial vulnerable stages of growth.

If damping-off begins after seedlings are growing, treatment with a fungicide may sometimes control its spread. The ability to control attacks depends on their severity and on the modifying environmental conditions (see Chapter 3).

Symptoms resembling damping-off are also produced by certain unfavorable environmental conditions in the seed bed. Drying, high soil temperatures, or high concentrations of salts in the upper layers of the germination medium can cause injuries to the tender stems of the seedlings near the ground level. The collapsed stem tissues have the appearance of being "burned off." These symptoms may be confused with those caused by pathogens. Damping-off fungi can grow in concentrations of soil solutes high enough to inhibit the growth of seedlings. Where salts accumulate in the germination medium, damping-off can be particularly serious.

DORMANCY: REGULATION OF GERMINATION

In some cases, seeds may be non-dormant when they are separated from the plant. **Non-dormant** seeds need only be imbibed at permissive temperatures to initiate germination. In other cases, seeds display **primary dormancy**.

Dormancy is a condition in which seeds will not

primary dormancy
A common condition of seeds when they are shed from the plant. Seeds with primary dormancy will not germinate even under normally permissive conditions for germination.

germinate even when the environmental conditions (water, temperature, and aeration) are permissive for germination.

Seed dormancy prevents immediate germination but also regulates the time, conditions, and place that germination will occur. In nature, different kinds of primary dormancy have evolved to aid the survival of the species (19, 146, 147, 181, 182, 225, 228) by programming germination for particularly favorable times in the annual seasonal cycle.

Secondary dormancy is a further survival mechanism that can be induced under unfavorable environmental conditions and may further delay the time germination occurs. Some seeds will cycle between dormant and non-dormant states numerous times before germinating. Knowledge of the ecological characteristics of a species' natural habitat can aid in establishing treatments to induce germination (200, 246).

Domestication of seed-propagated cultivars of many crop plants, such as grains and vegetables, undoubtedly has included selection for sufficient primary dormancy to prevent immediate germination of freshly harvested seed, but not enough to cause problems in propagation. Dormancy facilitates seed storage, transport, and handling. Changes take place with normal dry storage handling of many agricultural, vegetable, and flower seeds to allow germination to proceed whenever the seeds are subjected to normal germinating conditions. Problems can occur when seed testing is attempted on freshly harvested seeds. Seeds of some species are sensitive to high temperature and light conditions related to seed dormancy. Many weed seeds persist in soil due to either primary or secondary dormancy, and provide "seed banks" that produce extensive weed seed germination whenever the soil is disturbed (19). Practical problems occur with nursery propagation of seeds of many tree and

---

BOX 7.6 GETTING MORE IN DEPTH ON THE SUBJECT
ECOLOGICAL ADVANTAGES OF SEED DORMANCY

Seed dormancy is an evolutionary adaptation to delay germination after the seed has been shed from the plant. There are numerous advantages to germination delay:

1. Permitting germination only when environmental conditions favor **seedling survival.** For example, temperate species require a period of moist, chilling conditions (i.e., winter conditions) before germination in the spring; desert species germinate only after rainfall; small-seeded species require light; and even species that require extremely high temperatures prior to germination to become the primary species in an area following a forest fire.

2. Creation of a **"seed bank."** In nature, a seed bank ensures that not all seeds of a species germinate in a single year. This is insurance against years when flowering or fruiting may not occur due to some catastrophic environmental reason. Some seeds remain dormant in a seed bank for decades. Although this is a wonderful ecological adaptation, it is also the basis for persistent weed problems in agricultural fields. Some species take this concept one step further and produce **polymorphic seeds.** In this case, seeds produced on the same plant or different plants in a population have different degrees of dormancy. Often these seeds have a different physical appearance. A classic example is found in cocklebur (*Xanthium pennsylvanicum*). Each cocklebur fruit contains two seeds of different sizes (Fig. 7–24). One seed is non-dormant, while the other seed is dormant, and is for the seed bank and future germination.

3. Dormancy can also **synchronize germination** to a particular time of the year, which ensures that spring-germinating seedlings have the entire growing season to grow and develop or that summer-germinating seedlings are at a proper stage of development entering the winter. Although environmental cues signal flowering for most crops, synchronizing germination also ensures a population of plants at the same stage of development to facilitate genetic outcrossing when all plants flower at the same time.

Figure 7–24
Cocklebur (*Xanthium*) fruits have two seeds. The smaller of the two seeds is dormant (red arrow). This is an example of polymorphic seed production.

shrub species, which require specific treatments to overcome dormancy in order to satisfy the requirements needed to bring about germination (see Chapters 19 and 20).

KINDS OF PRIMARY SEED DORMANCY

Propagators of cultivated plants have long recognized germination-delaying phenomena and have learned to manipulate different kinds of seed dormancy. The first recorded discussion of seed dormancy was by Theophrastus around 300 BC (80). He recognized that most seeds germinated less after time in storage (seed deterioration), while other seeds germinated at a higher percentage (dormancy release). Much scientific thought has gone into defining a uniform terminology for different kinds of seed dormancy. A historically early system for dormancy categories was formulated by Crocker in 1916 (56, 58), who described seven kinds of seed dormancy based primarily on treatments to overcome them. Subsequently, Nikolaeva (182) defined a system based predominantly upon physiological controls of dormancy. Atwater (7) has shown that morphological characteristics, including both seed morphology and types of seed covering characteristic of taxonomic plant families, could be associated with dormancy categories particularly significant in seed testing. More recently, a universal terminology for seed and bud dormancy was proposed (150). It uses the terms *eco-*, *para-*, and *endo*-dormancy to refer to dormancy factors related to the environment (eco), physical or biochemical signals originating external to the affected structure (para), and physiological factors inside the affected structure (endo). These terms are better at describing bud dormancy than the many different seed dormancy conditions.

Dormancy will be discussed in this chapter (Table 7–1) using a system adapted from Crocker (56) and Nikolaeva (182), and further modified by Baskin and Baskin (19). Major categories include

I. Primary dormancy
   a. exogenous
   b. endogenous
   c. combinational
II. Secondary dormancy
   a. thermodormancy
   b. conditional

Primary dormancy is a condition that exists in the seed as it is shed from the plant. In contrast, secondary dormancy occurs in seeds that were previously nondormant but reenter dormancy because the environment was unfavorable for germination.

Primary Exogenous Dormancy

Exogenous dormancy is imposed upon the seed from factors outside the embryo, including the seed coat and/or fruit parts. The tissues enclosing the embryo can impact germination by:

1. inhibiting water uptake,
2. modifying gas exchange (i.e., limit oxygen to the embryo),
3. preventing inhibitor leaching, and
4. supplying inhibitors to the embryo.

Physical Dormancy Seeds with physical dormancy fail to germinate because seeds are impermeable to water. Physical dormancy is most often caused by a modification of the seed coverings (seed coat or pericarp) becoming hard, fibrous, or mucilaginous during dehydration and ripening. For most seeds with physical dormancy, the outer integument layer of the seed coat hardens and becomes impervious to water. Cells of the outer integument coalesce and deposit water-repellant materials within the cells and on their surface. These materials include lignin, suberin, cutin, and waxes (76, 202). These cells are **macrosclereids** but can also be referred to as Malpighian or palisade cells (Fig. 7–25, page 222). Seeds with this condition are often termed "hard" seeds.

> **macrosclereid cells** Cells in the seed coat that are responsible for preventing water uptake in seeds with exogenous, physical dormancy. An older term for these cells was Malpighian cells in honor of the early 17th-century plant anatomist Marcello Malpighi of Italy.

Physical dormancy is a genetic characteristic found in species from at least 15 plant families, including Fabaceae, Malvaceae, Cannaceae, Geraniaceae, and Convolvulaceae. Physically dormant seeds in the Anacardiaceae have impermeable fruit coats. Among cultivated crops, hard seeds are chiefly found in the herbaceous legumes, including clover and alfalfa, as well as many woody legumes (*Robinia, Acacia, Sophora,* etc.). The degree to which seeds are impervious to water is also increased by environmental (dry) conditions during seed maturation and environmental conditions during seed storage. Drying at high temperatures during ripening will increase hardseededness. Harvesting slightly immature seeds and preventing them from drying can reduce or overcome this condition in some cases.

Seeds with physical dormancy become impermeable to water late in seed development when they fall below 20 to 15 percent moisture (19). In papilionoid

Table 7–1
CATEGORIES OF SEED DORMANCY

Types of dormancy	Causes of dormancy	Conditions to break dormancy	Representative genera
I. Primary dormancy	Dormancy condition at the end of seed development.		
a. Exogenous dormancy	Imposed by factors outside the embryo.		
i. Physical	Impermeable seed coat.	Scarification	*Baptisia, Convolvulus, Gleditsia, Lupinus*
ii. Chemical	Inhibitors in seed coverings.	Removal of seed coverings (fruits). Leaching seeds.	*Beta, Iris*
b. Endogenous dormancy	Imposed by factors in the embryo		
i. Physiological	Factors within embryo inhibit germination.		
1. Nondeep	Embryo growth potential inadequate to escape seed coverings. Can be light sensitive.	Short periods of moist chilling. After-ripening (dry storage).	Most common form of dormancy. *Lactuca, Primula, Cucumis, Impatiens.*
2. Intermediate	Embryo growth potential inadequate to escape seed. Embryo germinates if separated from the seed coat.	Moderate periods (up to 8 weeks) of moist chilling (stratification).	Common in temperate woody plants. *Cercis, Cornus, Pinus.*
3. Deep	Embryo does not germinate when removed from seed coat or will form a physiological dwarf.	Long periods (>8 weeks) of moist chilling (stratification).	*Dictamnus, Euonymus, Prunus, Rhodotypos.*
ii. Morphological	The embryo is not fully developed at the time the seed sheds from the plant.	Warm or cold stratification.	*Anemone, Daucus, Cyclamen, Viburnum*
iii. Morphophysiological	Combination of an underdeveloped embryo and physiological dormancy.	Cycles of warm and cold stratification.	*Asimina, Helleborus, Ilex, Magnolia, Asarum, Paeonia, Trillium.*
c. Combinational dormancy	Combinations of exogenous and endogenous dormancy conditions. Example: physical (hard seed coat) plus physiological dormancy.	Sequential combinations of dormancy-releasing treatments. Example: scarification followed by cold stratification.	*Cercis, Tilia*
II. Secondary dormancy			
a. Thermodormancy	After primary dormancy is relieved, high temperature induces dormancy.	Growth regulators or cold stratification.	*Apium, Lactuca, Viola*
b. Conditional dormancy	Change in ability to germinate related to time of the year.	Chilling stratification.	Many species with endogeneous dormancy display conditional dormancy.

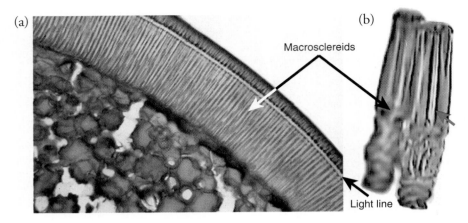

Figure 7–25
(a) Cross-section of a redbud (*Cercis canadensis*) seed showing the typical macrosclereid layer in the seed coat. Notice the light line that is the top half of each macrosclereid cell.
(b) Individual macrosclereid cells from a chemically digested seed coat. These cells show the interior lumen (red arrow) surrounded by the non-living thickened cell walls.

legume seeds, the point of seed attachment (hilum) acts as a valve during late stages of development; it opens to allow water vapor to escape in a dry atmosphere, and closes in a moist atmosphere to prevent water uptake (Fig. 7–26) (127). This valve action allows the last bit of water to leave the seed as the seed coat becomes impermeable.

Seed coat impermeability is maintained by a layer of palisade-like macrosclereid cells. There is usually a single area of the seed coat that acts as a **water gap** to initiate imbibition (22). For many legumes, the area is the lens (strophiole) or hilum (Fig. 7–27). For example, in *Albizia lophantha* (65), a small opening at the lens near the hilum is sealed with a corklike plug that can be dislodged with vigorous shaking or impact (110) or by exposure to dry heat as in a fire (65). For members of the Malvaceae, it is a chalazal plug that must be dislodged to allow imbibition. In the Convolvulaceae, there are two bulges (bumps) on either side of the hilar rim that raise up to initiate imbibition after exposure to dry heat (Fig. 7–28) (129).

These water gap structures act as environmental sensors to detect appropriate times for germination (22). For many seeds, it is temperature that is the environmental cue to relieve physical dormancy. Some seeds require relatively high temperatures (greater than 35°C, 95°F) and either moist or dry conditions to relieve dormancy. For others, daily fluctuations (greater than 15°C change; i.e., 50°C down to 25°C) in temperature allow imbibition. Temperature is postulated to be a way for seeds to detect differences in the seasonal year or whether they are in an open or protected area—that is, detecting a gap in the forest canopy after tree fall or fire. The higher temperature or temperature fluctuation would occur in the open area, ensuring less competition due to the shade of other plants.

In cultivation, any method to break, soften, abrade, or remove the seed coverings is called scarification and is immediately effective for inducing imbibition and germination (see Chapter 8 for specific methods for scarification). Physical abrasion breaks through the impermeable outer cell layer to admit water to the permeable cells below. Acid scarification removes the water-repelling materials on the surface of the macroscleries, exposing the inner lumen of the

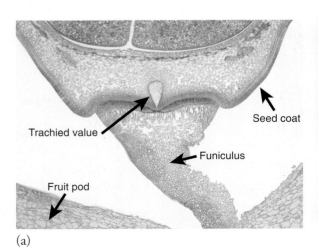

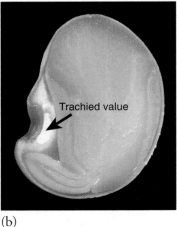

Figure 7–26
A characteristic of papillionoid legume seeds is the presence of the trachied valve under the hilum. This valve opens or closes during the final stages of maturation drying to allow water to leave the seed. The hilum is also the location of initial water entry following dormancy release. (a) Immature bean (*Phaseolus*) seed. (b) Nearly mature scholar tree (*Sophora japonica*) seed.

PRINCIPLES OF PROPAGATION FROM SEEDS CHAPTER SEVEN 223

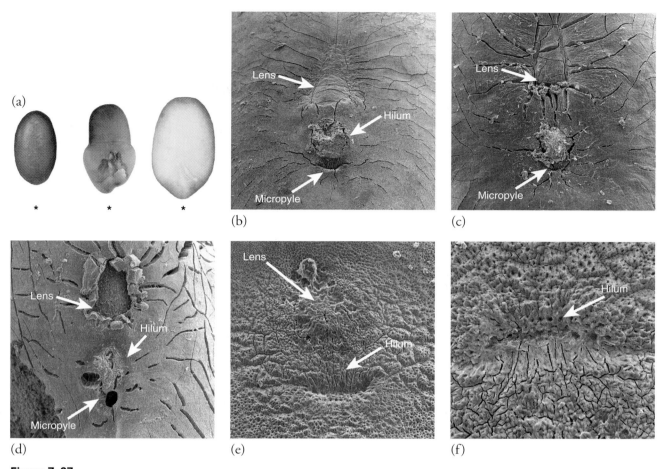

Figure 7–27
Alleviation of physical dormancy in honeylocust (*Gleditsia triacanthos*) seeds. (a) Seeds treated with moist heat showing imbibition at the hilar seed end (*). Electron micrographs for (b) untreated, (c) heat-treated, (d) initial imbibition in heat-treated seed, (e) surface etching in acid-treated seeds, and (f) close-up of hilum in acid-treated seeds showing open tops on the macrosclereids.

cell for water transport (Fig. 7–27e and f) (35, 158). Heat treatments (like hot water) tend to target the water gap structures for permeability. Figure 7–27 clearly shows that the hilar region of honeylocust (*Gleditsia*) seeds is the initial entry point for water in heat-treated seeds.

Chemical Dormancy Chemicals that accumulate in fruit and seed-covering tissues during development and remain with the seed after harvest may act as **germination inhibitors** (79).

Germination inhibitors have been extracted from the fruits and seeds of a number of species (19); however,

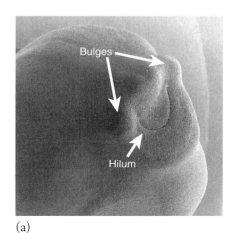
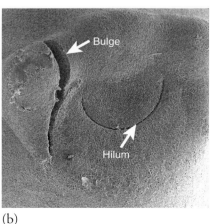

Figure 7–28
Dormancy release in seeds of some members of the Convolvulaceae involves the two bulges on either side of the hilum that raise up to permit imbibition. (a) Dormant seeds and (b) non-dormant seeds beginning to imbibe water. Electron micrographs by Gehan Jayasuriya.

proving their function as causal agents for dormancy does not necessarily follow. Nevertheless, germination can sometimes be improved by prolonged leaching with water, removing the seed coverings, or both (73, 182). Some examples include:

1. Fleshy fruits, or juices from them, can strongly inhibit seed germination. This occurs in citrus, cucurbits, stone fruits, apples, pears, grapes, and tomatoes. Likewise, dry fruits and fruit coverings, such as the hulls of guayule, *Pennisetum ciliare*, wheat, as well as the capsules of mustard (*Brassica*), can inhibit germination. Some of the substances associated with inhibition are various phenols, coumarin, and abscisic acid.
2. Specific seed germination inhibitors play a role in the ecology of certain desert plants (145, 242, 243). Inhibitors are leached out of the seeds by heavy soaking rains that also provide sufficient soil moisture to ensure survival of the seedlings. Since a light rain shower is insufficient to cause leaching, such inhibiting substances have been referred to as "chemical rain gauges."
3. Dormancy in iris seeds is due to a water and ether-soluble germination inhibitor in the endosperm, which can be leached from seeds with water or avoided by embryo excision (5).

Inhibitors have been found in the seeds of such families as Polygonaceae, Chenopodiaceae (*Atriplex*), Portulaceae (*Portulaca*), and other species in which the embryo is peripherally located. Likewise, seeds of a group of such families as Brassicaceae (mustard), Linaceae (flax), Violaceae (violet), and Lamiaceae (*Lavendula*) have a thin seed coat with a mucilaginous inner layer that contains inhibitors (7).

In many seeds, the inner seed coat becomes membranous but remains alive and semipermeable. In the Asteraceae, for instance, this layer coalesces with the remnant layers of the endosperm. These layers of integument and remnants of the endosperm and nucellus remain physiologically active during ripening and for a period of time after the seed is separated from the plant (Fig. 7–29). Such physiologically active layers play a role in maintaining primary dormancy, mainly because this semipermeable nature restricts aeration and inhibitor movement.

Primary Endogenous Dormancy

Seeds with endogenous dormancy fail to germinate primarily because of factors within the embryo. These factors can be either **physiological** or **morphological.**

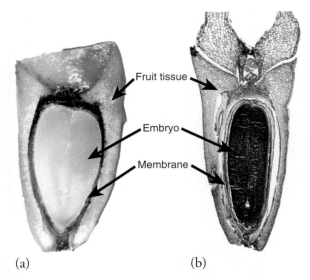

Figure 7–29
The seed (achene) of purple coneflower (*Echinacea*) showing the location of the semipermeable layer that is involved with seed dormancy. (a) Transverse section through the seed and its (b) photomicrograph.

Endogenous Physiological Dormancy The most common mechanism for delaying germination is **physiological dormancy.** The basic model for maintenance of physiological dormancy is that the embryo lacks the **growth potential** to allow the radicle to escape the restraint of the seed coverings (84). Growth potential is the force used by the radicle to penetrate seed coverings (21). Many species with physiological dormancy have seeds that germinate normally if the seed coverings over the radicle are cut or the embryo is removed from the coverings (the exception is deep physiological dormancy). The physical strength of the endosperm and seed coverings has been shown to restrict germination in both herbaceous (lettuce, pepper, and tomato) and woody (redbud and lilac) plants. Dormancy in these species is overcome by weakening seed coverings, by increasing growth potential in the embryo (see Fig. 7–6, page 205), or by a combination of seed covering and embryo effects. This interaction between the embryo and the seed coverings has been clearly demonstrated by the genetic control of dormancy in wheat (91). Dormancy in wheat is a

> physiological dormancy A condition mainly controlled by factors within the embryo that must change before the seed can germinate.
>
> morphological dormancy Seeds that have an embryo that is less than one-quarter of the size of the seed when it is shed from the plant.

multigenic trait. Genes associated with red color in seed coverings restrain embryo expansion, while a separate set of genes control internal embryo conditions impacting growth potential. This combination is to be expected, because the seed coverings are maternal tissue while the embryo is the result of sexual reproduction.

Endogenous physiological dormancy can be separated into three types based on their "depth" of dormancy. These include **nondeep, intermediate,** and **deep** physiological dormancy, but it should be recognized that the delineation between types may not always be clear cut.

Nondeep Physiological Dormancy. By far, endogenous, nondeep physiological dormancy is the most common form of dormancy found in seeds (19) and the most intensely studied because this is the form of dormancy found in the model plant, *Arabidopsis*. This type of dormancy includes species respond to short periods of **chilling stratification** (see Box 7.6), that require light or darkness to germinate **(photodormancy)**, and species that can undergo an **"after-ripening"** period for dormancy release. After-ripening is the time required for seeds in dry storage to lose dormancy.

> **after-ripening** Technique used historically to indicate any change that occurs in seeds leading to release from endogenous physiological dormancy. However, it is more appropriately used to describe changes that occur in seeds during dry storage that lead to dormancy release.

BOX 7.6   GETTING MORE IN DEPTH ON THE SUBJECT
CHILLING STRATIFICATION

Moist-chilling is the environmental signal alleviating physiological dormancy. A typical response for seeds that require chilling stratification is shown in Figure 7–30. Nursery propagators have known since early times that such seeds required moist-chilling (25, 235, 250). This requirement led to the horticultural practice of **stratification,** in which seeds are placed between layers of moist sand or soil in boxes (or in the ground) and exposed to chilling temperatures, either out-of-doors or in refrigerators (see Chapter 8). **Successful stratification requires seeds to be stored in a moist, aerated medium at chilling temperatures for a certain period of time.**

Moisture
Dry dormant seeds absorb moisture by imbibition to around 50 percent (25). Seed moisture should remain relatively constant during stratification. Dehydration stops the stratification process (115), and seeds may revert to secondary dormancy. When the end of the chilling period is reached, seed coverings "crack," and the radicle eventually emerges, sometimes even at low temperatures.

Aeration
The amount of oxygen needed during stratification is related to temperature (52). At high temperature, moist seed coverings of dormant, imbibed seeds can restrict

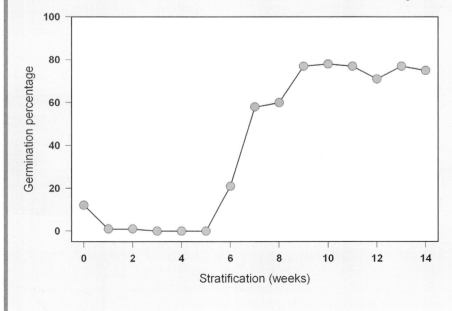

Figure 7–30
Pawpaw (*Asimina triloba*) is typical of species that require chilling stratification (89). It shows the typical population effect, where some seeds in a seed lot require only a few weeks of chilling, while others require longer times to be released from dormancy (89).

(Continued)

oxygen uptake because of (a) low oxygen solubility in water and (b) oxygen fixation by phenolic substances in the seed coats. At chilling temperatures, however, the embryo's oxygen requirement is low and oxygen is generally adequate.

Temperature
Temperature is the single most important factor controlling stratification. The most effective temperature regimes for moist-chilling are similar to those during the winter and early spring of the natural environment of the species. Temperatures somewhat above freezing [1 to 7°C (33 to 45°F)] are generally most effective, with more time required at higher and lower temperatures with a minimum at −5°C (23°F) (213). There is a particular maximum temperature, known as the compensation temperature, where no progress is made toward dormancy release (1, 214, 235). For apple, this point has been determined to be 17°C (62°F) (1), but it apparently varies with individual species (215) and different stages of stratification (221). Toward the end of the stratification period, the maximum temperature for germination gradually increases and the minimum temperature gradually decreases. This period has been called conditional dormancy (170).

Time
The time required to stratify seeds depends on the interaction of (a) the genetic characteristics of the seed population (137, 138, 213, 246), (b) conditions during seed development (235), (c) environment of the seed bed, and (d) management of seed handling.

Mechanism for Action
Stratification appears to relieve dormancy through a combination of physiological changes to the embryo and tissues surrounding the embryo. The embryo can be shown to increase in growth potential while seed coverings (especially the endosperm in angiosperms and the megagametophyte in gymnosperms) become weaker. These active changes occur through gene activation (173) and increased enzyme activity (198), and the result is an embryo that can produce more radicle force to escape the seed coverings, and seed coverings that are weaker, presenting less of a barrier to germination. Examples of these changes are discussed in more detail under physiological dormancy.

Photodormancy. Seeds that require either light or dark conditions to germinate have historically been termed **photodormant**, skotodormant, or photoblastic. It should be recognized that photodormancy may not completely fit the definition of dormancy if you consider light as a required environmental parameter for germination, similar to temperature and water. However, it is clear that light impacts germination timing in many species. Seeds from species with non-deep physiological dormancy (especially small-seeded species) often display a requirement for light or darkness to germinate (Fig. 7–31). The basic mechanism of light sensitivity in seeds involves a photochemically reactive pigment called **phytochrome**, widely present in plants (25, 63, 224, 233).

phytochrome A photoreceptor pigment used by plants to perceive light.

Exposure of the imbibed seed to red light causes the phytochrome to change from the biologically inactive red (P_r) to the active far-red form of phytochrome (P_{fr}), which stimulates germination. Exposing the seed to far-red light or darkness causes a change back to the inactive P_r form, which inhibits germination. These changes are reversible and can be repeated many times, the last treatment being the one that determines germination (Fig. 7–32). Borthwick

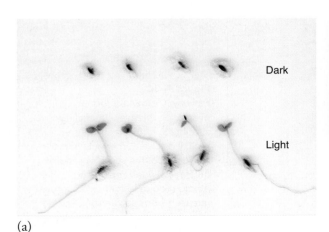

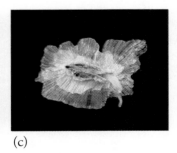

Figure 7–31
(a) Empress tree (*Paulownia*) is a light-sensitive seed that requires light to germinate. Examples of seeds germinated (b) without or (c) with light.

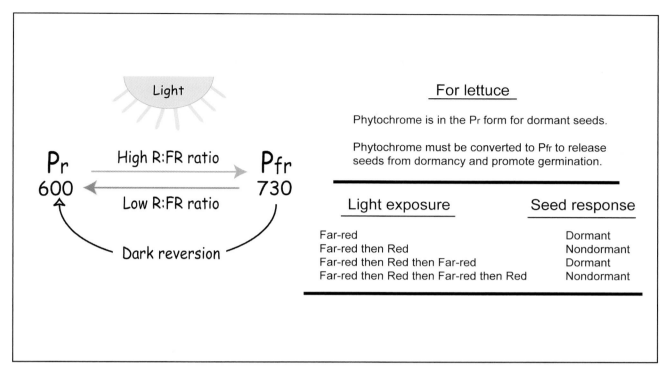

Figure 7–32
Phytochrome controls the dormancy condition of photodormant seeds. Lettuce seeds are the model to study the photoreversibility of phytochrome. The last quality of light the seeds are exposed to determines the dormancy state. Far-red light (730 nm) or darkness keeps seeds dormant, while red light (600 nm) will relieve dormancy.

and co-workers at the USDA in Beltsville, MD, used lettuce seeds to demonstrate this in their classic studies that established the concept of photoreversibility and, eventually, the discovery of the two forms of phytochrome (see Box 7.7).

In natural sunlight, red (R) wavelengths dominate over far-red (FR) at a ratio of 2:1, so that phytochrome tends to remain in the active P_{fr} form. Under a foliage canopy, far-red is dominant and the R:FR ratio may be as low as 0.12:1.00 to 0.70:1.00, which can inhibit seed germination (201). This inhibited germination explains why in agricultural settings, weed seeds show reduced germination as a crop canopy covers the soil. Also, in natural ecosystems, seedling survival would not be favored if the seed germinates in close proximity to other plants, where there would be intense competition for light, nutrients, and water by the established plant population. Red light penetrates less deeply into the soil than far-red, so that the R:FR ratio becomes lower with soil depth, until eventually darkness is complete. Imbibed light-sensitive seeds buried in the soil will remain dormant until such time as the soil is cultivated or disturbed, thereby exposing them to light. Light sensitivity can be induced in some seeds by exposing imbibed non–light-sensitive seeds to conditions inhibiting germination, such as high temperature or high osmotic pressure (244).

For some seeds, there is a distinct light and temperature interaction regarding dormancy and germination. A light requirement can be offset by cool germination temperatures and, sometimes, by alternating temperatures. Lettuce seeds generally require light to germinate; however, they lose their light requirement and can germinate in darkness if the temperature is below 25°C (77°F). Seeds may also lose their requirement for light after a period of dry storage. For years, birch (*Betula*) seeds were thought to require chilling stratification to permit germination. However, there is no chilling required if seeds are germinated in light at warm temperatures (250).

The light quality seen by the mother plant can subsequently impact the light requirement for seed germination. For example, lettuce seeds produced from plants grown in a high R:FR ratio germinated at 100 percent at 23°C (73°F) and over percent at 30°C (86°F) in the dark, while seeds from plants grown with a low R:FR ratio germinated approximately 35 percent at 23°C (73°F) and less than 5 percent at 30°C (86°F) (53).

Likewise, seeds of some plants (*Chenopodium album*) are dormant if plants are exposed to long days and nondormant if exposed to short days (25).

A seed is a composite of maternal-only genetics (seed coat) and a combination of maternal/paternal genetics (endosperm and embryo). Each can influence dormancy and germination potential. This maternal vs. paternal inheritance factor can be illustrated in reciprocal crosses of petunia (98). In petunia (*Petunia* ×*hybrida*), the requirement for light was maternally inherited, while endogenous dormancy within the embryo was under paternal control.

After-Ripening. Nondeep physiological dormancy is the general type of primary dormancy that exists in many, if not most, freshly harvested seeds of herbaceous plants (19, 182, 224). For most cultivated cereals, grasses, vegetables, and flower crops, nondeep physiological dormancy may last for 1 to 6 months and disappears with dry storage during normal handling procedures (95). Cucumber displays nondeep physiological dormancy and is typical of many crops. Cultivated cucumber (*Cucumis sativus* var. *sativus*) has been selected over many years of cultivation for a short dormancy period. It loses dormancy in dry storage at room temperature after several weeks (15 to 30 days). The hardwickii cucumber (*Cucumis sativus* var. *hardwickii*) is considered a wild progenitor species of the cultivated cucumber, and it can remain dormant for up to 270 days (245). The release from dormancy for hardwickii cucumber seeds in dry storage at various temperatures is presented in Figure 7–34. The shorter storage time required to satisfy dormancy at warmer temperatures is typical of seeds with nondeep physiological dormancy. For most seeds, there is a negative log-linear relationship between after-ripening time and temperature to reach 50 percent germination (199). After-ripening is also impacted by seed moisture. In general, there is a reduction in after-ripening time as the seed moisture constant rises to approximately 25 percent. After-ripening slows or stops at greater seed moisture

---

BOX 7.7 GETTING MORE IN DEPTH ON THE SUBJECT
PHYTOCHROME AND SEED GERMINATION

Seeds sense their environment to schedule germination. The two major environmental signals perceived by seeds are temperature and light. From an ecological standpoint, light perception by the seed acts as an indicator of the light available for seedling growth. In general, small seeds require light to germinate including many herbaceous plants and pioneering tree species. They perceive light to indicate:

1. how deeply the seed is **buried** in the soil,
2. **gaps** in the forest canopy, and
3. **soil disturbance** that might indicate an opportunity for growth—like animal grazing or agricultural tillage.

Light is perceived in plants by light receptors called phytochrome. Phytochrome is a chromoprotein that undergoes photoconversion to exist in a red (P_r) or far-red (P_{fr}) form (Fig. 7–32). Exposure of plants to sunlight (which has a high R:FR spectral ratio) or red light (maximum absorption at 660 nm) causes phytochrome to convert to the P_{fr} form. Conversely, exposure to darkness or far-red light (maximum absorption at 730 nm) causes phytochrome to be in the P_r form.

Discovery of phytochrome mutants and subsequent isolation of phytochrome genes shows that phytochrome is encoded as a multigene family with at least five genes coding for different phytochromes (called PHY A-E) (48). Interestingly, separate phytochromes can have different functions and can act differently in seedlings compared with seeds. The two important phytochromes for germination are PHYA and PHYB (215). PHYB is responsible for the **low fluence response** (LFR) and PHYA is responsible for the **very low fluence response** (VLFR). The PHYB low fluence response is seen in seeds with the classic, photoreversible R:FR ratio that was initially thought to control all phytochrome responses. Seeds that are not exposed to red light after the initial hours of imbibition eventually may employ a PHYA response. PHYA accumulates in dark imbibed seeds until the seed will respond to a relatively wide light spectral range (even FR) to initiate germination.

Studies mostly involving *Arabidopsis* and lettuce provide strong evidence that light dramatically alters the gibberellin/abscisic acid interaction controlling germination (215). Red light promotes gibberellin biosynthesis (66, 227, 230) and reduces enzymes that inactivate gibberellin, while decreasing abscisic acid levels (215). In addition, there is evidence that cytokinin may also participate in light-activated germination. For example, in Scots pine (*Pinus sylvestris*) red light can reduce abscisic acid levels and increase cytokinin content in a manner that could cause dormancy release (194). Treatments with hormones can offset the light effect, as illustrated in Figure 7–33.

These hormone interactions most likely control germination by initiating changes in embryo growth potential as well as decreasing the strength of the seed coverings (209). In lettuce, endosperm cells covering the radicle tip change in response to light that contributes to release from dormancy (193). In radish seeds, far-red light inhibits germination even in seeds without seed coats. This response is reversed in red light by increasing the growth potential of the embryo (212).

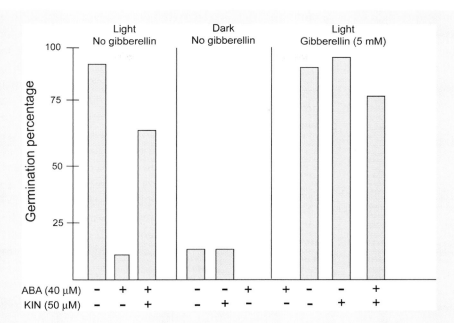

Figure 7–33
Interaction of light and three applied hormones on the germination of 'Grand Rapids,' a light- and temperature-sensitive cultivar of lettuce with physiological dormancy. In the light, untreated seeds germinate and ABA inhibits germination in the light. Kinetin partially overcomes the ABA inhibition. Germination is inhibited in the dark and kinetin does not overcome the dark inhibition in lettuce seeds. ABA completely inhibits germination in the dark. Gibberellic acid overcomes dark-imposed dormancy with or without kinetin. ABA negates the promotive effect of gibberellic acid on germination in the dark and kinetin counteracts this ABA effect and permits gibberellic acid to act. Redrawn from Khan et al., 1971.

levels. In nature, temperature and seed moisture content are changing on a continual basis, but the relationship among after-ripening time, temperature, and seed moisture remains consistent for a particular plant type, and time to dormancy release can be predicted using a hydrothermal time model (10).

Nondeep physiological dormancy in commercial flower and vegetable seeds is often transitory and disappears during dry storage (after-ripening) so that it is generally gone before the grower sows the seeds. Consequently, it is primarily a problem with seed-testing laboratories that need immediate germination.

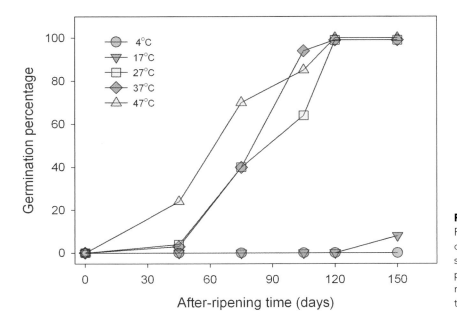

Figure 7–34
Release from dormancy in hardwickii cucumber (*Cucumis sativus* var. *hardwickii*) stored dry at various temperatures. The period required to after-ripen seeds and relieve dormancy is shorter at higher temperatures (245).

In seed-testing laboratories such seeds respond to various short-term treatments, including short periods of chilling, alternating temperatures, and treatment with potassium nitrate and gibberellic acid (see page 175).

Intermediate and Deep Physiological Dormancy. Seeds with intermediate and deep physiological dormancy are characterized by a requirement for a period of 1 to 3 (sometimes more) months of chilling while in an imbibed and aerated state. This type of dormancy is most common in seeds of trees and shrubs and some herbaceous plants of the temperate zone (56, 250). Seeds of this type ripen in the fall, overwinter in the moist leaf litter on the ground, and germinate in the spring.

Seeds displaying **intermediate physiological dormancy** usually require chilling stratification to release the seeds from dormancy (181, 182). These seeds are distinguished from those with deep physiological dormancy by three key factors:

1. Embryos isolated from the surrounding seed coverings of seeds with intermediate physiological dormancy germinate readily.
2. The length of time required at chilling temperatures to satisfy dormancy is considerably shorter compared to seeds with deep physiological dormancy.
3. Intact seeds with intermediate physiological dormancy often respond to gibberellic acid as a substitute for chilling, while seeds with deep physiological dormancy do not.

There is a correlation between the seed-chilling requirements and the bud-chilling requirements of the plants from which the seeds were taken (192). In studies with almond, a high quantitative correlation was observed between the mean time for dormancy release for seeds and buds in seedling populations, and the mean for both the seed and pollen parents (137). However, there was a low correlation between the time required to release dormancy in each individual seed compared to the buds of the new plant coming from that embryo (138). This difference suggests that dormancy involves both a genetic component within the embryo and a maternal component from the seed parent (interaction between the embryo and seed coverings as discussed previously). As a result, a great deal of variability in individual seed germination time can occur within a given seed lot and between different seed lots of the same species collected in different years and different locations.

For seeds with intermediate physiological dormancy, there is an interaction between temperature and seed moisture content. Chilling stratification is not effective unless seeds are hydrated. In nature, the degree of seed hydration varies depending on the environment. Therefore, there is a critical moisture content below which seeds would not be positively affected by chilling for dormancy release. In several conifer species, the critical moisture content appears to be approximately 25 percent moisture (99). About 33 percent seed moisture allows dormancy release to proceed without allowing germination during prolonged storage (131). Downie, et al. (71) also observed that dormancy release in spruce (*Picea glauca*) seeds was achieved at a moisture content starting at approximately 25 percent. In this condition, cellular components are hydrated, but not enough to support turgor-driven cell expansion.

Seeds exhibiting **deep physiological dormancy** usually require a relatively long (8 to 20 weeks) period of moist-chilling stratification to relieve dormancy. Excised embryos from seeds displaying deep physiological dormancy usually will not germinate, or the seedlings produced may be abnormal. Typically, non-chilled excised embryos develop into physiological dwarfs (Fig. 7–35) (56, 90).

Physiological dwarfing in excised embryos from non-chilled seeds has been shown to result from exposure of the apical meristem to warm germination temperature before chilling stratification is complete (191). In peaches, temperatures of 23 to 27°C (73 to 80°F) and higher produced symptoms of physiological dwarfing, but at lower temperatures the seedlings grew relatively normally. In almonds, exposing incompletely stratified seed to high temperatures subsequently induced physiological dwarfing in the seedling.

Figure 7–35
Physiological dwarfing of seedlings from almond. Seedlings on the left have been exposed to chilling stratification, while seedlings on the right were grown from embryos isolated from dormant seeds that were never exposed to chilling temperatures.

Pinching out the apex can circumvent dwarfing by forcing lateral growth from non-dwarfed lower nodes. Exposing seedlings to long photoperiods or continuous light (90, 148), provided that this action is taken before the apical meristem becomes fully dormant, has also offset dwarfing. Repeated application of gibberellic acid has also overcome dwarfing (17, 90). Some experiments have shown that systematic removal of the cotyledons from the dormant embryo can induce germination and overcome physiological dwarfing, suggesting the existence of endogenous inhibitors present within the cotyledons (25).

Mechanisms of Dormancy Release

A competing two-component system maintains seed dormancy in seeds with nondeep and intermediate physiological dormancy. There is an embryo and a seed covering component that interact to maintain dormancy. The seed coverings present a significant barrier to germination because embryos can germinate and grow if isolated from seed coverings. Therefore, dormancy release involves changes in the restraint of the seed coverings and an increase in embryo growth potential. **Growth potential** is the force used by the radicle to penetrate the seed coverings (see Fig. 7–6, page 205). One way to observe changes in growth potential is to germinate isolated embryos on solutions containing increasing amounts of an osmoticum-like polyethylene glycol, which provides a gradient of more negative water potentials restricting water availability to the seed. This gradient can be illustrated using embryos isolated from cucumber as they after-ripen (245) and eastern redbud (*Cercis canadensis*) seeds during moist chilling stratification (94) where embryos develop a higher growth potential, as measured by radicle length, as seeds come out of dormancy (Table 7–2).

Mechanisms for after-ripening are not well understood partly because they take place at low embryo hydration levels where there is little enzyme activity. They may involve non-enzymatic mechanisms that alter membrane properties (108), remove inhibitors, interact with stress reactions via antioxidants (9), and degrade certain proteins. Molecular studies suggest that after-ripened embryos have switched at the transcriptional level to be able to express important dormancy-related genes previously silenced in dormant embryos (28, 42).

Possible mechanisms for changes in embryo growth potential during stratification include changes in membrane fluidity at chilling temperatures (less than 15°C, 59°F) and differential enzyme activity for storage reserves (25). Protease and lipase enzymes have been shown to increase during chilling stratification, and one lipase shows a temperature optimum of 4°C (39°F) for activity (154). In general, there is a decrease in storage lipids and an increase in sugars and amino acids from storage reserves during chilling stratification. This increase in osmotically active solutes could, in part, explain the increase in growth potential seen in embryos following chilling stratification and the subsequent release from dormancy.

The seed coverings also participate in physiological dormancy. For many seeds, the endosperm surrounding the radicle forms an endosperm cap that provides

Table 7–2
Isolated Embryo Growth on Polyethylene Glycol (PEG) Solutions as an Indication of Embryo Growth Potential during Dormancy Release in Cucumber by After-Ripening (245), and Eastern Redbud by Moist Chilling Stratification (94)

	Water potential MPa	Radicle length (cm) after 4-days			
		Time (days)			
		0	60	120	180
Cucumber	0	1.4	2.8	4.3	5.4
	−1.0	0	0	0.8	2.9
		Time (days)			
	Water potential MPa	0	30	60	
Redbud	0	1.2	1.5	1.7	
	−0.6	0.7	0.8	1.0	
	−1.0	0.36	0.45	0.6	

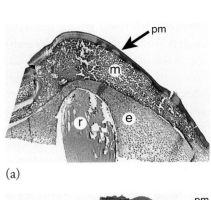
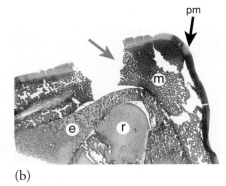
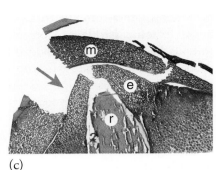
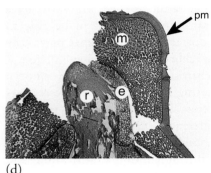

Figure 7–36
Longitudinal section of the hilar end of a germinating redbud (*Cercis canadensis*) seed showing the radicle elongating to rupture the seed coat. (a) Hand section showing endosperm enclosing radicle tip. (b) Photomicrograph of germinating seed shows the seed coat rupturing (red arrow) and the endosperm stretching as the radicle grows. (c) Finally the endosperm also ruptures (green arrow). (d) The radicle emerges. Abbreviations are radicle (r), endosperm (e), Mesophyll layer of the seed coat (m), and palisade layer of the seed coat (pm) (130).

sufficient restraint to prevent germination in dormant seeds (Fig. 7–36). Some seed-enclosing fruit structures, such as walnut shells (58), stone fruit pits (182), and olive stones (55), are very rigid and restrict embryo expansion. In addition, layers of the fleshy fruit may dry and become part of the seed covering, as in *Cotoneaster* or hawthorn (*Crataegus*). In the caryopsis or achenes of grains or grasses, the fruit covering becomes fibrous and coalesces with the seed. Water may be absorbed through these hard seed coverings, but the difficulty arises in the cementing material that holds the dehiscent layers together, as shown in walnut. Originally, Nikolaeva (182) placed these types of species in a separate exogenous mechanical dormancy category that is still referred to as "coat-imposed" dormancy (84). It seems more appropriate to discuss them here because although the seed coverings are a barrier to seed germination, these seeds still require chilling stratification (and a change in embryo growth potential) to be released from dormancy.

Endosperm weakening by cell-wall–degrading enzymes is required to initiate germination in a number of species including tomato (183), pepper (238), and *Datura* (210). The puncture force required for the radicle to penetrate the endosperm layer in ash (*Fraxinus*) seeds is reduced during stratification presumably by cell wall enzymes that provide localized weakening of the surrounding tissues (84). For eastern redbud seeds (*Cercis canadensis*), puncture force was slightly reduced during chilling stratification but was considered secondary to the greater change in embryo growth potential, which was better correlated with germination potential in intact seeds (94).

Several conifer species show this interaction between covering materials and embryo growth potential for release from dormancy. The megagametophyte (seed storage endosperm tissue in conifer seeds) that surrounds the conifer embryo can be a considerable barrier to germination and may be the primary mechanism maintaining dormancy. Cell-wall–altering enzymes are associated with weakening the megagametophyte, especially in the area covering the radicle that contributes to release from dormancy. In white spruce (*Picea glauca*) endo-β-mannanase (70) and yellow cedar (*Chamaecyparis nootkatensis*) pectin methyl esterase (198) enzyme activity increase during chilling stratification.

In addition to the endosperm cap, the seed coat or pericarp can also contribute to the restraint to germination in dormant seeds (64). These tissues are entirely maternal in origin, and, therefore, differences in dormancy related to the seed coat can be maternally inherited. Seed coat mutations for pigmentation in *Arabidopsis* (146) and tomato (72) show the importance of the seed coat in controlling germination. Seed coats with reduced pigmentation tend to decrease the time to radicle emergence, while those with increased pigmentation tend to delay germination. Each of these conditions is related to the physical restraint of the coverings. There is also a strong correlation between the pericarp coloration in cereal grasses (rice and wheat) and dormancy. Those seeds with red pigmentation in the pericarp tend to have deeper dormancy than those without pigmentation (104).

Several studies using global genomic approaches are beginning to elucidate those genes important to dormancy imposition and release, especially in seeds with non-deep physiological dormancy (88, 123, 124).

Morphological Dormancy Dormancy occurs in some seeds where the embryo is not fully developed at the time of seed dissemination. Seeds are considered to have morphological dormancy if they require more than 30 days to germinate, have an embryo that fills less than 1/2; of the mature seed, and have an embryo that must grow inside the seed before the radicle can emerge (Fig. 7–37) (19). The process of embryo enlargement is usually favored by a period of warm temperature, but can also take place during chilling temperatures.

It is generally felt that seeds with morphological dormancy (a high ratio of endosperm to embryo) are more primitive than seeds where the embryo fills the seed cavity and consequently contain little or no endosperm (20, 161). The types of embryos observed in seeds with morphological dormancy include **rudimentary, linear, spatulate,** and **undifferentiated** embryo types (Fig. 7–37) (7, 20).

Rudimentary embryos are small, have about the same width as length, and do not have readily identifiable seedling parts. These are found in various families, such as Ranunculaceae (anemone, *Ranunculus*), Papaveraceae (poppy, *Romneya*), and Araliaceae (ginseng, *Fatsia*). Effective aids for inducing germination include (a) exposure to temperatures of 15°C (59°F) or below, (b) exposure to alternating temperatures, and (c) treatment with chemical additives such as potassium nitrate or gibberellic acid.

Linear (torpedo-shaped) and **spatulate** (spoon-shaped) embryos are longer than they are wide. Each can be up to one-half the size of the seed cavity and have easily observed cotyledons and radicles (Fig. 7–37a and b). Important families and species in this category include Apiaceae (carrot), Ericaceae (rhododendron, heather), Primulaceae (cyclamen, primula), and Gentianaceae (gentian). Other conditions, such as semipermeability of the inner seed coats and internal germination inhibitors may be involved. A warm temperature of at least 20°C (68°F) favors germination, as does gibberellic acid treatment.

Morphological dormancy occurs in gymnosperms (ginkgo, cycads), dicots, and monocots from both temperate and tropical ecosystems. Various tropical species have seeds with embryos that require an extended period at warm temperatures for germination to take place. For example, seeds of various palm species require 3 months of warm temperatures at 38 to 40°C (100 to 104°F) before visible signs of germination (175). Other examples include *Actinidia* and *Annona squamosa,* whose seeds require 2 or 3 months of warmth, respectively, to complete germination (182).

Seeds with **undifferentiated** embryos are very small (often from only a few to 100 cells in size) and have not reached the stage of cotyledon or radicle organization. They also may lack substantial seed storage materials (Fig. 7–37c). Families with undifferentiated embryos include the orchids (Orchidaceae), non-chlorophytic plants that rely on fungal support (Ericaceae, Monotropaceae, Pyrolaceae) and parasitic

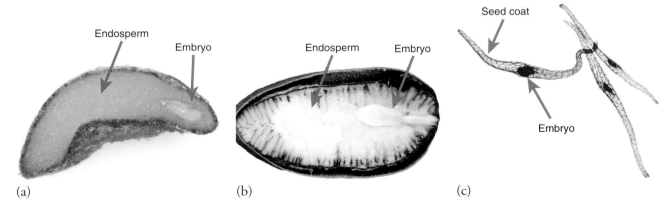

Figure 7–37
Seeds with morphological dormancy. (a) Linear embryo type in heavenly bamboo (*Nandina*) and (b) spatulate type in pawpaw (*Asimina triloba*). (c) Orchid seeds are in the unclassified seed category with an undifferentiated embryo. The outer seed coat is composed of a single papery layer that facilitates wind dissemination.

plants (Orobanchaceae, Rafflesiaceae). Orchids have undifferentiated embryos when the seed is shed from the mother plant and require a mycorrhizal fungus association for germination. Orchids are germinated commercially by special tissue culture methods, as discussed in Chapter 18.

Morphophysiological Dormancy Seeds with morphophysiological dormancy have a underdeveloped embryo that also displays physiological dormancy. In some cases, the morphological dormancy must be satisfied before physiological dormancy release. For example, warm stratification to permit the embryo to grow to a critical size, followed by moist chilling for physiological dormancy. In others, the physiological dormancy precedes morphological dormancy. For example, moist chilling to relieve physiological dormancy, followed by warm temperature for embryo growth prior to germination. There are at least eight types of morphophysiological dormancy that are recognized based on different combinations of physiological and morphological dormancy release conditions (19). Two groups that are relatively important for horticultural crops include simple and epicotyl types.

Most seeds with **simple morphophysiological** dormancy usually require warm (at least 15°C) followed by chilling (1 to 10°C) conditions, during which time the embryo develops and then breaks physiological dormancy. Various temperate zone herbaceous plants and trees fall into this category, including windflower (*Anemone*), twinleaf (*Jeffersonia*), ash (*Fraxinus*), yew (*Taxus*), and holly (*Ilex*) (182). In nature, these seeds are usually shed from the plant with an underdeveloped embryo that must have a warm period for growth to initiate inside the seed coverings. Once the embryo reaches a certain size, it can then respond to chilling temperature to release the seed from physiological dormancy. Therefore, these seeds require warm followed by cold stratification to satisfy dormancy. In some species, there is a difference between cultivated and wild forms with respect to morphophysiological dormancy. For example, in *Anemone,* cultivated 'de Caen' seeds showed only morphological dormancy (required only warm treatment), while wild populations of *Anemone coronaria* displayed morphophysiological dormancy and required warm followed by moist chilling stratification (125).

Seeds with **epicotyl dormancy** display the most fascinating dormancy patterns found in seeds. These seeds have separate dormancy conditions for the radicle and epicotyl (18, 58, 182). These species fall into two subgroups. In one group, seeds initially germinate during a warm period of 1 to 3 months to produce root and hypocotyl growth beyond the seed coverings, but then require 1 to 3 months of subsequent chilling to enable the epicotyl to grow. This group includes various lily (*Lilium*) species, *Viburnum* spp., peony (*Paeonia*), black cohosh (*Cimicifuga racemosa*), and liverwort (*Hepatica acutiloba*). The dormancy-breaking response of the epicotyl to chilling is sensitive to the stage of radicle growth (17). For peony, 85 percent of the epicotyls exposed to 7 weeks of chilling grew if the radicle had reached 4 cm in length. In contrast, only 40 percent of the epicotyls were released from dormancy under the same conditions with smaller 2 to 3 cm radicles.

In the second group, both the epicotyl and the radicle require chilling to relieve dormancy, but each is released from dormancy at different times. Seeds in this group require a chilling period to relieve radicle dormancy, followed by a warm period to allow the radicle to grow, and then a second cold period to release the epicotyl from dormancy. In nature, such seeds require at least two full growing seasons to complete germination. These are the seeds for which the term **double dormancy** was first coined. Examples include bloodroot (*Sanguinaria*), *Trillium,* and lily-of-the-valley (*Convallaria*). There are also seed population differences in this group.

> **double dormancy** One of the original terms used to describe morphophysiological dormancy. It was used to describe seeds that took 2 years to germinate.

Barton (16) showed that in both bloodroot and Solomon's seal (*Polygonatum*), about half of the seeds showed simple epicotyl dormancy, while the other half showed the epicotyl and radicle required chilling.

Primary Combinational Dormancy

Combinational dormancy refers to seeds that have both physical and physiological dormancy. There are two types of combinational dormancy based on the sequence of environmental cues required for complete dormancy release (19). One type requires an initial period of warm temperature to relieve nondeep physiological dormancy prior to alleviation of physical dormancy and imbibition. The second requires loss of physical dormancy to allow imbibition, followed by a cold stratification period to relieve physiological dormancy. To induce germination, all blocking conditions must be eliminated in the proper sequence.

In the most typical form of combinational dormancy, physical dormancy must be relieved followed by conditions that relieve endogenous physiological dormancy. Therefore, the seed coat must be modified to allow water to penetrate to the embryo, and then chilling stratification can release the seed from physiological dormancy. This is not a common form of dormancy. It is found in redbud (*Cercis*), buttonbush (*Ceanothus*), golden raintree (*Koelreuteria*), sumac (*Rhus*), and linden (*Tilia*) (19).

SECONDARY DORMANCY

In nature, primary dormancy is an adaptation to control the time and conditions for seed germination. If for some reason seeds fail to germinate after primary dormancy is broken, seeds of many species can reenter dormancy. This re-entry is called **secondary dormancy.** It is a further adaptation to prevent germination of an imbibed seed if other environmental conditions are not favorable (25, 56, 133, 143). These conditions can include unfavorably high temperature, prolonged light or darkness (skotodormancy), water stress, and anoxia. These conditions are particularly involved in the seasonal rhythms (**conditional dormancy**) and prolonged survival of weed seeds in soil (25).

conditional dormancy
A continuum seen in many seeds in nature as they cycle through periods of dormancy and nondormancy; it is detected as the seed's ability to germinate over a range of temperature.

Secondary Dormancy and Light

Induction of secondary dormancy is illustrated by classical experiments with freshly harvested seeds of lettuce (142). If germinated at 25°C (77°F), seeds require light, but if imbibed with water for 2 days in the dark, excised embryos germinate immediately, illustrating that only primary dormancy was present. If imbibition in the dark continues for as long as 8 days, however, excised embryos will not germinate, because they have developed secondary dormancy. Release from this type of secondary dormancy can be induced by chilling, sometimes by light, and, in various cases, treatment with germination-stimulating hormones, particularly gibberellic acid.

Baby blue eyes (*Nemophila*) seeds require darkness to germinate. If these seeds are exposed to light for a period of time, they enter secondary dormancy and will no longer germinate in the dark without a chilling treatment (50).

Thermodormancy

For some species like lettuce (*Lactuca*), celery (*Apium*), *Schizanthus,* and pansy (*Viola*), germination at high temperatures (at least 30°C, 86°F) can induce **thermodormancy.** Thermodormancy should not be confused with the thermal inhibition most seeds experience when the temperature exceeds the maximum temperature for germination. Seeds experiencing thermodormancy will not germinate when the temperature returns to near optimum temperatures, while thermal-inhibited seeds will germinate when temperatures are lowered. Lettuce (139) and celery (185) seeds become thermodormant at 35°C (95°F), and can be relieved by exogenous application for combinations of GA_3, cytokinin (kinetin), and ethylene. It is most probable that impairment for ethylene production or action has the greatest endogenous influence on thermodormancy in lettuce (177).

thermodormancy
A type of secondary dormancy that prevents seeds from germinating at high temperature.

Conditional Dormancy

As seeds come out of dormancy, or begin to enter secondary dormancy, they go through a transition stage where they will germinate, but only over a narrow range of temperatures (Fig. 7–38, page 236). This transition stage is termed **conditional dormancy** (19, 234). In this way, seeds of many species cycle through years of dormancy and non-dormancy based on germination temperature. A common **dormancy cycle** for seeds would follow this basic sequence:

1. Seeds shed from the plant have primary dormancy and fail to germinate regardless of temperature.
2. Seeds are exposed to dormancy-releasing environmental conditions and gradually lose dormancy. These conditionally dormant seeds germinate only over a narrow range of temperatures.
3. Fully non-dormant seeds germinate over a wide range of temperatures.
4. If non-dormant seeds fail to germinate because the environment is unfavorable, they again become conditionally dormant and will germinate only over a narrow range of temperatures.
5. Eventually, conditionally dormant seeds enter secondary dormancy, where they fail to germinate regardless of temperature.

This type of dormancy cycle can be repeated over many years (Fig. 7–39, page 236). Dormancy cycles ensure that seeds germinate when the environment is most suitable for seedling survival (42). It is also the basis for persistent weed problems in field-grown crops.

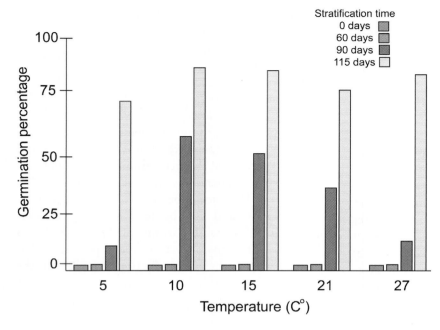

Figure 7–38
Conditional dormancy in *Cotoneaster divaricatus*. After 115 days of stratification, seeds are fully non-dormant and germinate well across all temperatures. After 90 days, seeds are conditionally dormant and germinate better at 10 and 15°C compared to other temperatures. Seeds not receiving stratification or those only stratified for 60 days are dormant and fail to germinate at any temperature. Adapted from Meyer M. M. Jr. 1988. *HortScience* 23:1046–7.

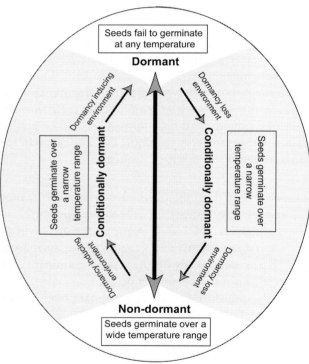

Figure 7–39
Dormancy cycling in seeds showing dormant, non-dormant, and conditionally dormant states.

DORMANCY CONTROL BY PLANT HORMONES

Much experimental evidence supports the concept that specific endogenous growth-promoting and growth-inhibiting compounds are involved directly in the control of seed development, dormancy, and germination (26, 147, 153). Evidence for hormone involvement comes from correlations between hormone concentrations with specific developmental stages, effects of applied hormones, mutants for hormone production or perception, and genome-wide microarray analysis (88). The two most important hormones controlling seed dormancy and dormancy release are abscisic acid and gibberellin, and their interaction (Fig. 7–40). ABA controls the establishment and maintenance for dormancy, while GA appears to control initiation and completion of germination. The ratio of ABA to GA-induced signal transduction is as important as the active hormone levels for dormancy release (84, 147). Other hormones have a modifying impact on this relationship.

Abscisic Acid (ABA)

ABA plays a major role in preventing "precocious germination" of the developing embryo in the ovule. ABA increases during late stages of seed development and is a major factor in the induction of primary dormancy (136, 152). ABA-deficient mutants show reduced primary dormancy, while transgenic plants overexpressing ABA show increased primary dormancy (176). ABA-deficient and ABA response mutants in *Arabidopsis* (134), sunflower (81), and tomato (103) indicate that ABA must be present during seed development to induce dormancy.

However, endogenous ABA levels may not show a strong correlation with seed dormancy. For several woody plants, including peach (67, 157) walnut (162), plum (156), apple (15), and hazelnut (247), ABA concentrations are high in both the seed coat and a lesser

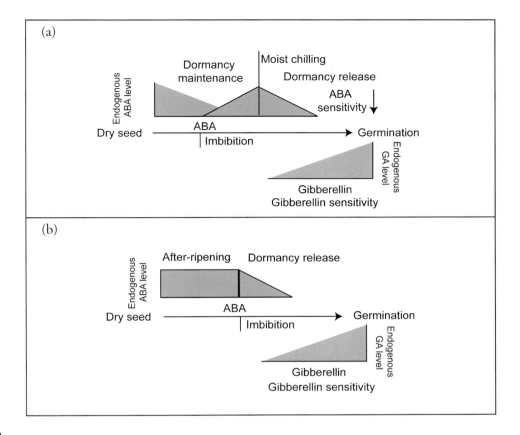

Figure 7–40
A model for the relationship between ABA and gibberellin during dormancy release by moist chilling stratification and dry seed after-ripening. (a) After seed development, ABA levels can be high in seeds. However, imbibition reduces ABA levels from stored sources, but new synthesis of ABA maintains seed dormancy. Moist chilling tends to reduce ABA levels. Non-dormant seeds show reduced ABA sensitivity, reduced ABA synthesis, and increased ABA catabolism, all resulting in lower ABA levels. This is coupled with an increase in gibberellin synthesis and increased gibberellin sensitivity. (b) After-ripening occurs in dry seeds, where there is little change in ABA levels due to low metabolism in the dry state. Upon imbibition of after-ripened seeds, ABA levels are reduced through increased catabolism and inhibition of new ABA synthesis. Again, this is coupled with increased gibberellin synthesis and sensitivity. Modified from Finkelstein et al. 2008 (88).

amount in the cotyledons in freshly harvested, dormant seeds. In peach, ABA concentration drops to near zero after 30 days stratification, but seeds do not fully come out of dormancy for an additional 8 weeks of stratification (97, 157). In apple, ABA levels can remain high during stratification even as the seeds become non-dormant (12, 208). Endogenous ABA may be reduced during treatments to relieve dormancy, but this does not appear to be a strict requirement. It is becoming apparent that continued ABA synthesis following imbibition is the major factor required to maintain dormancy (2, 100). In *Arabidopsis*, ABA levels drop following imbibition regardless of whether the seeds are dormant or nondormant (2). However, after 4 days of imbibition, dormant seeds resume ABA synthesis, while non-dormant seeds do not. Thus, non-dormant seeds show up-regulation for genes involved in ABA catabolism and down-regulation of those for ABA synthesis. The opposite is found for maintenance of the dormancy state. In addition, treatments to relieve dormancy can induce a reduction in ABA sensitivity—that is, it takes more exogenous ABA to inhibit germination in stratified seeds compared to untreated seeds (Fig. 7–41, page 238).

An important aspect of ABA action includes the negative regulation of gibberellin levels. ABA inhibits gibberellin-biosynthesis enzymes and promotes gibberellin-degradation enzymes that impact endogenous gibberellin accumulation (215). ABA regulation of active gibberellin levels directly impacts dormancy and the seed's ability to germinate.

Gibberellins

Gibberellins (GA) are important for both the control and promotion of seed germination (188). Several mutants in tomato and *Arabidopsis* that are impaired for gibberellin biosynthesis fail to germinate without

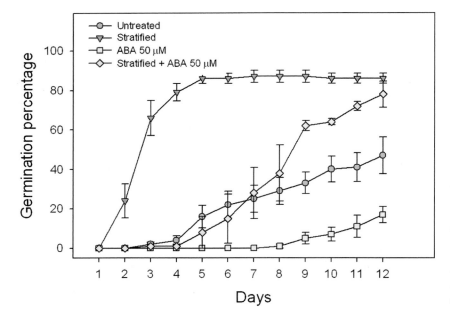

Figure 7–41
Chilling stratification changes the seeds' sensitivity to abscisic acid (ABA). Untreated seeds of purple coneflower (*Echinacea tennesseensis*) germinate slowly with a germination percentage below 50 percent. Stratified seeds germinate quickly at about 85 percent germination. ABA dramatically inhibits germination in untreated seeds, but only slows germination in stratified seeds.

application of exogenous gibberellin (122). These seeds act like dormant seeds because of a failure to make gibberellin. Gibberellins stimulate germination by inducing enzymes that weaken the seed coverings (endosperm or seed coat) surrounding the radicle, inducing mobilization of seed storage reserves, and stimulating cell expansion in the embryo (84). Gibberellin synthesis and perception are affected by numerous environmental signals that also influence release from dormancy. These include light, temperature (including stratification), and nitrate levels. Applied gibberellins [commercially as gibberellic acid (GA_3) or (GA_{4+7})] can relieve certain types of dormancy, including nondeep and intermediate physiological dormancy, photodormancy, and thermodormancy.

Gibberellins occur at relatively high concentrations in developing seeds but usually drop to a lower level in mature dormant seeds, particularly in dicotyledonous plants. Dormancy release treatments increase gibberellin biosynthesis as well as gibberellin sensitivity (147). During stratification, gibberellins are either synthesized at the chilling temperatures or are converted to an available (or unbound) form (41, 107, 164). In *Arabidopsis*, dormant seeds show high expression of an enzyme that deactivates gibberellin, while non-dormant after-ripened or stratified seeds show increases in multiple gibberellin biosynthesis genes that increase endogenous gibberellin levels (86, 249).

There is an interaction between ABA and gibberellin during dormancy release; ABA must be reduced before gibberellins can promote germination. Dormancy induction and release in filbert (*Corylus avellana*) seeds illustrates this point. At the time of ripening, a significant amount of abscisic acid can be detected in the seed covering (247) as well as a detectable amount of gibberellin in the embryo (203, 204). When the seed is dried following harvest, the embryo becomes dormant, and gibberellin levels decrease significantly (205). Stratification for several months is required for germination. The gibberellin level remains low during this chilling period but increases after the seeds are placed at warm temperatures when germination begins (Fig. 7–42). Gibberellic acid applied to the dormant seed (29) can replace the chilling requirement (Fig. 7–42). However, ABA applied with gibberellin offsets the gibberellin effect and prevents germination (204).

A major mode-of-action for gibberellin is the deactivation of gene-expression repressors called DELLA proteins (88). At least 360 genes are repressed by DELLA proteins prior to seed germination (45). RGL2 is a major DELLA protein target for gibberellin. Prior to germination, gibberellin initiates a signal transduction pathway that deactivates RGL2. The result is the expression of a number of genes associated with germination, including genes for important cell wall enzymes involved in endosperm weakening.

Ethylene

Ethylene gas is an important naturally occurring hormone involved in many aspects of plant growth. Inhibitor studies and mutant seeds impaired for ethylene production or perception indicate that ethylene is not required for germination, because these seeds germinate at fairly high percentages (165). However, it is becoming increasingly evident that ethylene production may be linked

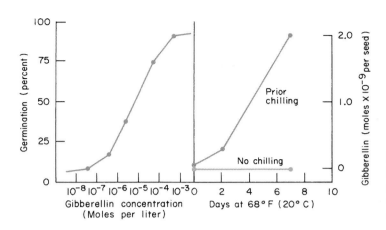

Figure 7–42
Interaction of gibberellin, stratification, and germination in filbert seeds. Reproduced by permission from A.W. Galston and P.S. Davies, Control mechanisms in plant development, Prentice-Hall, Englewood Cliffs, N.J. 1970.

with aspects of germination rate and seed vigor. For a majority of seeds, there is a burst of ethylene production that occurs simultaneously with radicle emergence.

Ethylene is involved with dormancy release in some seeds (135). For some species, a strong correlation has been shown between treatments that overcome dormancy and the ability of the seed to produce ethylene. There are also examples where exogenous ethylene application as either the gas or ethephon (an ethylene-releasing compound) alleviates seed dormancy without additional dormancy-breaking treatments (135). One dramatic example is the response of seeds of a hemiparasitic weed called witchweed (*Striga*) to ethylene exposure (78, 159). Witchweed can be a devastating plant parasite on grain (corn) crops, especially in tropical areas. Germinating grain seeds provide a signal for witchweed seeds to germinate and subsequently infest the host plant. Ethylene triggers the dormant seeds to germinate without the required host being present. This practice has been proposed as an eradication practice for infested fields.

Beech (*Fagus sylvatica*) requires chilling stratification to release seeds from dormancy. In addition, gibberellic acid or ethylene application relieves dormancy. Several genes expressed during dormancy release in beech seeds were found to be related to ethylene receptor genes (160). Mutant screens in *Arabidopsis* for reduced seed dormancy or reduced response to ABA also uncovered ethylene receptor genes (96). It has been shown that there is a significant antagonism between ethylene and seed sensitivity to ABA, and it is assumed that at least one mechanism for ethylene-induced dormancy release is reducing the embryo's sensitivity to ABA. In addition, *Arabidopsis* seeds become progressively more dormant as the number of receptor genes are knocked out for ethylene (219).

Ethylene production and application has also been implicated as a mechanism to alleviate thermodormancy for chickpea (92), lettuce (126), and sunflower (54), possibly because of an interaction with polyamines.

Cytokinin

Cytokinin activity tends to be important in early developing fruits and seeds but decreases and becomes difficult to detect as seeds mature. Cytokinin does not appear essential for germination. However, exogenous application of cytokinin can offset ABA effects and rescue seeds from thermodormancy (220). The antagonistic interaction between cytokinin and ABA may involve cytokinin-enhanced ethylene production, which, in turn, reduces the seed's sensitivity to ABA (165). It has also been suggested that cytokinin plays a "permissive" role in germination by allowing gibberellins to function (140, 144).

Auxin

Auxin does not appear to play a major role in seed dormancy (147). Some auxin-response mutants show increased seed dormancy, but exogenously applied auxin does not substitute for dormancy release treatments such as after-ripening or stratification. Auxin and stored forms of auxin are present in the seed at maturity and are important for post-germinative growth rather than in initial germination or dormancy release. Auxin has a much greater role in embryo formation during seed development, and it is possible that it is important in morphophysiological dormancy, but this has not been investigated.

Brassinosteroids

Brassinosteroids are naturally occurring steroid-based plant hormones. They can induce similar behavior in plants as gibberellins. Brassinosteroids can induce germination in gibberellin mutants, but this stimulation in germination is apparently in a gibberellin-independent manner (151). Brassinosteroid mutants germinate normally, suggesting that they play a modifying role in dormancy release, possibly by reducing ABA sensitivity.

Nitrogenous Compounds

Nitrogenous compounds are known to stimulate seed germination, but their role is not clear compared to traditional plant hormones. Nitrogenous compounds implicated in germination and dormancy release include nitrate, nitrite, thiourea, nitric oxide, ammonium and cyanide. Use of potassium nitrate has been an important seed treatment in seed-testing laboratories. One suggested role for these compounds is as a possible means of sensing soil nitrogen availability (88). A second possibility is that these compounds interact with enzymes in the pentose phosphate pathway (86), which involves the production of NADPH and oxygen that are required for the catabolism of ABA. Thiourea overcomes certain types of dormancy, such as dormancy in *Prunus* seeds, as well as the high-temperature inhibition of lettuce seeds (226). The effect of thiourea may be due to its cytokinin activity.

Butenolides

Numerous species from Mediterranean climates show increased germination following fire. Butenolides have been shown to be the active components in plant-derived smoke that stimulates germination (61). This discovery has led to the proposal for a new group of plant growth regulators called karrikins (178), of which KAR_1 has been shown to enhance germination in approximately 1,200 species in more than 80 genera worldwide (68). It appears that KAR_1 action requires gibberellin biosynthesis and stimulates germination through an interaction with ABA and gibberellin. Interestingly, it is also effective in stimulating germination in parasitic weed species (62).

DISCUSSION ITEMS

The physical, physiological, and biochemical concepts of seed germination provide important background understanding for many practical germination practices and the ecological implications for seed germination. Newer practices like seed priming and pre-germination and older practices to satisfy dormancy are easier to understand after the basic principles are presented.

1. When does DNA synthesis take place during seed germination? When does it take place during seed priming?
2. Compare the effects of water and temperature on seed germination.
3. Contrast thermal time and hydrotime models for germination.
4. How do components of water potential affect seed germination?
5. Compare quiescent with dormant seeds.
6. Compare primary and secondary dormancy.
7. What are the advantages and disadvantages of seed dormancy?
8. What are the ecological and agronomic implications of a seed bank?
9. Compare types of physiological dormancy.
10. Contrast thermodormancy with thermal inhibition on seed germination.
11. Discuss how hormone mutants are adding to our understanding of seed germination.

REFERENCES

1. Abbott, D. L. 1955. Temperature and the dormancy of apple seeds. *Rpt. 14th Intern. Hort. Cong.* 1:746–53.

2. Ali-Rachedi, S., D. Bouinot, M. H. Wagner, M. Bonnet, and B. Sotta. 2004. Changes in endogenous abscisic acid levels during dormancy release and maintenance of mature seeds: studies with the Cape Verde Islands ecotype, the dormant model of *Arabidopsis thaliana*. *Planta* 219:479–88.

3. Allan, M. 1977. *Darwin and his flowers: The key to natural selection.* New York: Taplinger Pub.

4. Allen, P. S., S. E. Meyer, and M. A. Khan. 2000. Hydrothermal time as a tool in comparative germination studies. In M. Black, K. J. Bradford, and J. Vázquez-Ramos, eds. *Seed Biology: Advances and Applications.* Wallingford, UK: CAB International pp. 401–10.

5. Arditti, J., and P. R. Pray. 1969. Dormancy factors in iris (Iridaceae) seeds. *Amer. J. Bot.* 56(3): 254–59.

6. Association of Official Seed Analysts. 1993. Rules for testing seeds. *J. Seed Tech.* 16:1–113.

7. Atwater, B. R. 1980. Germination, dormancy and morphology of the seeds of herbaceous ornamental plants. *Seed Sci. Tech.* 8:523–73.

8. Ayers, A. D. 1952. Seed germination as affected by soil moisture and salinity. *Agron. J.* 44:82–4.

9. Bailly, C. 2004. Active oxygen species and antioxidants in seed biology. *Seed Sci. Res.* 14:93–107.

10. Bair, N. B., S. E. Meyer, and P. S. Allen. 2006. A hydrothermal after-ripening time model for seed dormancy loss in *Bromus tectorum* L. *Seed Sci. Res.* 16:17–28.

11. Baker, K. F. 1972. Seed pathology. In T. T. Kozlowski, ed. *Seed biology,* Vol. 2. New York: Academic Press.

12. Balboa-Zavala, O., and F. G. Dennis. 1977. Abscisic acid and apple seed dormancy. *J. Amer. Soc. Hort. Sci.* 102:633–37.

13. Ball, V., ed. 1998. *Ball red book*, 16th ed. Batavia, IL: Ball Publ. Co.

14. Barroco, R. M., K. Van Poucke, J. H. W. Bergervoet, L. De Veylder, S. P. C. Groot, D. Inze, and G. Engler. 2005. The role of the cell cycle machinery in resumption of postembryonic development. *Plant Physiol.* 137:127–40.

15. Barthe, P., and C. Bulard. 1983. Anaerobiosis and release from dormancy in apple embryos. *Plant Physiol.* 72:1005–10.

16. Barton, L. V. 1944. Some seeds showing special dormancy. *Contrib. Boyce Thomp. Inst.* 13:259–71.

17. Barton, L. V., and C. Chandler. 1957. Physiological and morphological effects of gibberellic acid on epicotyl dormancy of tree peony. *Contrib. Boyce Thomp. Inst.* 19:201–14.

18. Baskin, C. C., and J. M. Baskin. 1985. Epicotyl dormancy in seeds of *Cimicifuga racemosa* and *Hepatica acutiloba*. *Bull. Torrey Bot. Club* 112:253–57.

19. Baskin, C. C., and J. M. Baskin. 1998. *Seeds: Ecology, biogeography, and evolution of dormancy and germination.* New York: Academic Press.

20. Baskin, C. C., and J. M. Baskin. 2007. A revision of Martin's seed classification system, with particular reference to his dwarf-seed type. *Seed Sci. Res.* 17:11–20.

21. Baskin, J. M., and C. C. Baskin. 1971. Effect of chilling and gibberellic acid on growth potential of excised embryos of *Ruellia humilis*. *Planta* 100: 365–69.

22. Baskin, J. M., C. C. Baskin, and X. Li. 2000. Taxonomy, anatomy and evolution of physical dormancy in seeds. *Plant Species Biol.* 15:139–52.

23. Berry, G. J., R. J. Cawood, and R. G. Flood. 1988. Curve fitting of germination data using the Richard's function. *Plant Cell Environ.* 11:183–88.

24. Berstein, L., A. J. MacKenzie, and B. A. Krantz. 1955. The interaction of salinity and planting practice on the germination of irrigated row crops. *Proc. Soil Sci. Soc. Amer.* 19:240–43.

25. Bewley, J. D., and M. Black. 1994. *Seeds: Physiology of development and germination.* New York: Plenum Press.

26. Black, M. 1980/1981. The role of endogenous hormones in germination and dormancy. *Israel J. Bot.* 29:181–92.

27. Blumenthal, A., H. R. Lerner, E. Werker, and A. Poljakoff-Mayber. 1986. Germination preventing mechanisms in *Iris* seeds. *Ann. Bot.* 58:551–61.

28. Bove, J., P. Lucas, B. Godin, L. Ogé, M. Jullien, and P. Grappin. 2005. Gene expression analysis by cDNA-AFLP highlights a set of new signaling networks and translational control during seed dormancy breaking in *Nicotiana plumbaginifolia*. *Plant Mol. Biol.* 57:593–612.

29. Bradbeer, J. W., and N. J. Pinfield. 1967. Studies in seed dormancy. III. The effects of gibberellin on dormant seeds of *Corylus avellana* L. *New Phytol.* 66:515–23.

30. Bradford, K. J. 1986. Manipulations of seed water relations via osmotic priming to improve germination under stress conditions. *HortScience* 21: 1105–12.

31. Bradford, K. J. 1990. A water relations analysis of seed germination rates. *Plant Physiol.* 94:840–49.

32. Bradford, K. J. 1995. Water relations in seed germination. In J. Kigel and G. Galili, eds. *Seed development and germination.* New York: Marcel Dekker. pp. 351–96.

33. Bradford, K. J. 1997. The hydrotime concept in seed germination and dormancy. In R. H. Ellis, M. Black, A. J. Murdoch, and T. D. Hong, eds. *Basic and applied aspects of seed biology.* Boston: Kluwer Acad. Pub. pp. 349–60.

34. Bradford, K. J., and O. A. Somasco. 1994. Water relations of lettuce seed thermoinhibition. 1. Priming and endosperm effects on base water potential. *Seed Sci. Res.* 4:1–10.

35. Brant, R. E., G. W. McKee, and R. W. Cleveland. 1971. Effect of chemical and physical treatment on hard seed of Penngift crown vetch. *Crop Sci.* 11:1–6.

36. Bray, C. M. 1995. Biochemical processes during the osmoconditioning of seeds. In J. Kigel and G. Galili, eds. *Seed development and germination.* New York: Marcel Dekker. pp. 767–89.

37. Brown, R. F., and D. G. Mayer. 1988. Representing cumulative germination. 1. A critical analysis of single-value germination indices. *Ann. Bot.* 61:117–25.

38. Brown, R. F., and D. G. Mayer 1988. Representing cumulative germination. 2. The use of the

Weibull function and other empirically derived curves. *Ann. Bot.* 61:127–38.

39. Bruggink, G .T. 2005. Flower seed priming, pregermination, pelleting and coating. In M. B. McDonald and F. Y. Kwong, eds. *Flower seeds: Biology and technology.* Wallingford, UK: CAB International. pp. 249–62.

40. Bruggink, G. T., J. J. J. Ooms, and P. van der Toorn. 1999. Induction of longevity in primed seeds. *Seed Sci. Res.* 9:49–53.

41. Bulard, C. 1985. Intervention by gibberellin and cytokinin in the release of apple embryos from dormancy: A reappraisal. *New Phytol.* 101:241–49.

42. Cadman, C. S. C., P. E. Toorop, H. W. M. Hilhorst, and W. E. Finch-Savage. 2006. Gene expression profiles of *Arabidopsis* Cvi seeds during dormancy cycling indicate a common underlying dormancy mechanism. *Plant Journal* 46:805–22.

43. Cantliffe, D. J., 1991. Benzyladenine in the priming solution reduces thermodormancy of lettuce seeds. *HortTechnology* 1:95–9.

44. Cantliffe, D. J., Y. Sung, and W. M. Nascimento. 2000. Lettuce seed germination. *Hort. Rev.* 224:229–75.

45. Cao, D., H. Cheng, W. Wu, H. M. Soo, and J. Peng. 2006. Gibberellin mobilizes distinct DELLA-dependent transcriptomes to regulate seed germination and floral development in *Arabidopsis. Plant Physiol.* 142:509–25.

46. Carpenter, W. J., and S. W. Williams. 1993. Keys to successful seeding. *Grower Talks* 11:34–43.

47. Carpenter, W. J., and S. Maekawa. 1991. Substrate moisture level governs the germination of verbena seed. *HortScience* 26:786–88.

48. Casal, J. J., and R. A. Sánchez. 1998. Phytochromes and seed germination. *Seed Sci. Res.* 8:317–29.

49. Chen, F., and K. J. Bradford. 2000. Expression of an expansin is associated with endosperm weakening during tomato seed germination. *Plant Physiol.* 124:1265–74.

50. Chen, S. S. C. 1968. Germination of light-inhibited seed of *Nemophila insignis. Amer. J. Bot.* 55:1177–83.

51. Ching, T. M. 1972. Metabolism of germinating seeds. In T. T. Kozlowski, ed. *Seed biology*, Vol. 2. New York: Academic Press.

52. Comé, D., and T. Tissaoui. 1973. Interrelated effects of imbibition, temperature, and oxygen on seed germination. In W. Heydecker, ed. *Seed ecology.* University Park, PA: Pennsylvania State Univ. Press.

53. Contreras, S., M. A. Bennett, J. D. Metzger, D. Tay, and H. Nerson. 2009. Red to far-red ratio during seed development affects lettuce seed germinability and longevity. *HortScience* 44:130–34.

54. Corbineau, F., R. M. Runicki, and D. Come. 1989. ACC conversion to ethylene by sunflower seeds in relation to maturation, germination and thermodormancy. *Plant Growth Reg.* 8:105–15.

55. Crisosto, C., and E. G. Sutter. 1985. Role of the endocarp in 'Manzanillo' olive seed germination. *J. Amer. Soc. Hort. Sci.* 110(1):50–2.

56. Crocker, W. 1916. Mechanics of dormancy in seeds. *Amer. J. Bot.* 3:99–120.

57. Crocker, W. 1930. Effect of the visible spectrum upon the germination of seeds and fruits. In *Biological effects of radiation.* New York: McGraw-Hill. pp. 791–828.

58. Crocker, W. 1948. *Growth of plants.* New York: Reinhold.

59. Dahal, P., and K. J. Bradford. 1994. Hydrothermal time analysis of tomato seed germination at suboptimal temperature and reduced water potential. *Seed Sci. Res.* 4:71–80.

60. Darby, R. J., and P. J. Salter. 1976. A technique for osmotically pre-treating and germinating small quantities of seeds. *Ann. Applied Biol.* 83:313–15.

61. Daws, M. I., H. W. Pritchard, and J. Van Staden. 2008. Butenolide from plant-derived smoke functions as a strigolactone analogue: evidence from parasitic weed seed germination. *South African J. Bot.* 74:116–20.

62. Daws, M. I., J. Davies, H.W. Pritchard, N. A. C. Brown, and J. Van Staden. 2007. Butenolide from plant-derived smoke enhances germination and seedling growth of arable weed species. *Plant Growth Regul.* 51:73–82.

63. De Greef, J. A., H. Fredricq, R. Rethy, A. Dedonder, E. D. Petter, and L. Van Wiemeersch. 1989. Factors eliciting germination of photobalstic kalanchoe seeds. In R. B. Taylorson, ed. *Recent advances in the development and germination of seeds.* New York: Plenum Press. pp. 241–60.

64. Debeaujon, I., L. Lepiniec, L. Pourcel, and J. Routaboul. 2008. Seed coat development and dormancy. In H. Nonogaki and K. J. Bradford, eds. *Seed development, dormancy and germination.* Wallingford, UK: CAB International.

65. Dell, B. 1980. Structure and function of the strophiolar plug in seed of *Albizia lophantha. Amer. J. Bot.* 67(4):556–61.

ed. *Recent advances in the development and germination of seeds.* New York: Plenum Press. pp. 181–90.

234. Vegis, A. 1964. Dormancy in higher plants. *Annu. Rev. Plant Physiol.* 15:185–224.

235. Villiers, T. A. 1972. Seed dormancy. In T. T. Kozlowski, ed. *Seed biology,* Vol. 2. New York: Academic Press. pp. 220–82.

236. Warren, J. E., and M. A. Bennett. 1997. Seed hydration using the drum priming system. *HortScience* 32:1220–1.

237. Watkins, J. T., and D. J. Cantliffe. 1983. Mechanical resistance of the seed coat and endosperm during germination of *Capsicum annuum* at low temperature. *Plant Physiol.* 72:146–50.

238. Watkins, J. T., D. J. Cantliffe, D. J. Huber, and T. A. Nell. 1985. Gibberellic acid stimulated degradation of endosperm in pepper. *J. Amer. Soc. Hort. Sci.* 1101:61–5.

239. Welbaum, G. E., K. J. Bradford, K. O. Yim, D. T. Booth, and M. O. Oluoch. 1998. Biophysical, physiological and biochemical processes regulating seed germination. *Seed Sci. Res.* 8:161–72.

240. Welbaum, G. E., Z. Shen, M. O. Oluoch, and L. W. Jett. 1998. The evolution and effects of priming vegetable seeds. *Seed Technol.* 20:209–35.

241. Welbaum, G. E., W. J. Muthui, J. H. Wilson, R. I. Grayson, and R. D. Fell. 1995. Weakening of muskmelon perisperm envelope tissue during germination. *J. Exp. Bot.* 46:391–400.

242. Went, F. W. 1949. Ecology of desert plants. 11. The effect of rain and temperature on germination and growth. *Ecology* 30:1–13.

243. Went, F. W., and M. Westergaard. 1949. Ecology of desert plants. III. Development of plants in the Death Valley National Monument, California. *Ecology* 30:26–38.

244. Wesson, G., and P. F. Wareing. 1969. The induction of light sensitivity in weed seeds by burial. *J. Exp. Bot.* 20(63):414–25.

245. Weston, L. A., R. L. Geneve, and J. E. Staub. 1992. Seed dormancy in *Cucumis sativus* var. *hardwickii* (Royle) Alef. *Scientia Hort.* 50:35–46.

246. Westwood, M. N., and H. O. Bjornstad. 1948. Chilling requirement of dormant seeds of fourteen pear species as related to their climatic adaptation. *Proc. Amer. Soc. Hort. Sci.* 92:141–49.

247. Williams, P. M., J. D. Ross, and J. W. Bradbeer. 1973. Studies in seed dormancy. VII. The abscisic acid content of the seeds and fruits of *Corylus avellana* L. *Planta* (Berl.) 110:303–10.

248. Wu, C. T., G. Leubner-Metzger, F. Meins, Jr., and K. J. Bradford. 2001. Class I β-1,3-glucanase and chitinase are expressed in the micropylar endosperm of tomato seeds prior to radicle emergence. *Plant Physiol.* 126:1299–313.

249. Yamauchi, Y., M. Ogawa, A. Kuwahara, A. Hanada, Y. Kamiya, and S. Yamaguchi. 2004. Activation of gibberellin biosynthesis and response pathways by low temperature during imbibition of *Arabidopsis thaliana* seeds. *Plant Cell* 16:367–78.

250. Young, J. A., and C. G. Young. 1992. *Seeds of woody plants in North America.* rev. ed. Portland, OR: Dioscorides Press.

*8
Techniques of Propagation by Seed

learning objectives
- Define the major systems for seedling production.
- Describe the procedures for direct-seeding crops.
- Describe the procedures for seedling production in temporary nursery beds.
- Describe the procedures for producing transplants under protected culture.
- Define the procedures for transplanting to permanent locations.

INTRODUCTION

Producing plants from seeds is the most important propagation method for agronomic, forestry, vegetable, and flowering bedding plants. These methods vary from field seeding operations to very sophisticated greenhouse transplant production systems. For example, vegetables may be started by direct field seeding or from transplants. Bedding plants and herbaceous perennials are started primarily as transplants grown in small transplant containers called plugs. Woody seedlings are usually started from field transplant beds at close spacing to produce bare-root liners for nursery production or understocks for grafting.

Seedling propagation involves careful management of germination conditions and knowledge of the requirements of individual kinds of seeds. Success depends on fulfilling the conditions detailed in the previous chapters on seeds. These include:

1. Using seeds of **proper genetic characteristics** to produce the cultivar, species, or provenance desired. This can be accomplished by obtaining seeds from a reliable dealer, buying certified seed, or—if producing one's own—following the principles of seed selection described in Chapter 5.
2. Using **good-quality** seeds. Seeds should germinate rapidly to withstand possible adverse conditions in the seed bed and provide a high percentage of usable seedlings.
3. **Manipulating seed dormancy.** This is accomplished by applying pregermination treatments or by properly timing planting.
4. Supplying **proper environment** to the seeds and resulting seedlings, including supplying sufficient water, proper temperature, adequate oxygen, and either light or darkness (depending on the kind of seed) to the seeds and resulting seedlings until they are well established. A proper environment also includes control of diseases and insects and prevention of excess salinity.

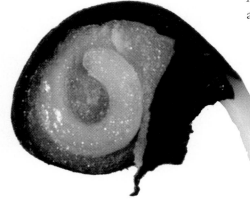

SEEDLING PRODUCTION SYSTEMS

Although many horticultural production systems rely on seedling propagation, these systems can vary depending on the crop being produced. Therefore, there are three basic systems that are relied upon to produce seedlings:

1. **Field seeding** at relatively low density in the location where the plant is to remain during the production cycle.

2. Seeding in **field nurseries** at relatively high density to produce seedlings that will be transplanted to a permanent location.
3. Seeding in **protected conditions,** as in a greenhouse, cold frame, or similar structure, and then transplanting to the permanent location.

Field Seeding

Direct **field seeding** is used for commercial field planting of agronomic crops (grains, legumes, forages, fiber crops, oil crops), lawn grasses, many vegetable crops, and some woody perennials (Fig. 8–1). The method may also be used by hobbyists for home vegetable and flower gardens. Compared to transplants, directly seeded plants are less expensive and can grow continuously without the check in growth often seen by transplanting (53). Frequently, direct-field-seeded vegetables and other crops are precontracted for processing, whereas the more expensive transplants are targeted as a fresh market crop. On the other hand, there are many potential field problems that must be overcome to provide the proper environmental conditions for good uniform germination. Likewise, cold weather may decrease growth. Seeding rates are critical to providing proper plant spacing for optimum development of the crop. If the final plant density is too low, yields will be reduced because the number of plants per unit area is low; if too high, the size and quality of the finished plants may be reduced by competition among plants for available space, sunlight, water, and nutrients.

field seeding
A common propagation method for agronomic, forestry, and vegetable crops and woody plants for liner production.

The following factors maximize direct-seeding success:

1. Good site selection and seed bed preparation
2. Using high-quality seed
3. Planting at the correct time
4. Seed treatments to facilitate sowing or to relieve dormancy
5. Selecting an appropriate mechanical seeder
6. Using the correct sowing depth
7. Sowing seed at an appropriate rate
8. Applying proper postsowing care

Field Seeding for Vegetable Crops   Direct field seeding is a common propagation method for many vegetable crops including corn, peas, beans, and spinach.

Site Selection and Seed Bed Preparation.   The ideal site for vegetable production is relatively flat with good soil water drainage. High production areas such as California and Texas often use laser-assisted grading to produce a level seed bed (Fig. 8–2a, page 252). A good **seed bed** should have a loose but fine physical texture that produces close contact between seed and soil so that moisture can be supplied continuously to the seed. Such a soil should provide good aeration, but not too much or it dries too rapidly. The surface soil should be free of clods and of a texture that will not form a crust (Fig. 8–2b). Soil impedance due to crusting from an improperly prepared seed bed or adverse environmental conditions during seeding can substantially

seed bed   The area where seeds will be planted and seedlings will emerge. It may be an entire field for field-seeded crops or a smaller area for future transplants.

(a)

(b)

Figure 8–1
Examples of a direct-seeded (a) vegetable (spinach) and (b) nursery crop (oak).

Figure 8–2
Seed bed preparation. (a) Some vegetable production areas are leveled with laser guides. (b) Crusting can occur following site preparation if it is flooded for prolonged periods. (c) Tillage equipment. (d) Seed bed following sowing.

reduce seedling emergence (61). Several materials, including organic polymers and phosphorus-containing compounds, have been developed to reduce soil crusting and to aid in seedling emergence (53, 55). The subsoil should be permeable to air and water with good drainage and aeration. Adequate soil moisture should be available to carry the seeds through germination and early seedling growth stages, but the soil should not be waterlogged or anaerobic (without oxygen). A medium loam texture, not too sandy and not too fine, is best. A good seed bed is one in which three-fourths of the soil particles (aggregates) range from 1 to 12 mm in diameter (39).

Seed bed preparation requires special machinery for field operations, and spading and raking or rototilling equipment for small plots (Fig. 8–2). Adding organic or soil amendments may be helpful, but these should be thoroughly incorporated and have time to decompose. Seed bed preparation may include soil treatments to control harmful insects, nematodes, disease organisms, and weed seeds. Weed control can be facilitated by careful seed bed preparation, cultivation, and may include chemical herbicide application (see Chapter 3). Three types of chemical controls are available:

1. Preplant fumigation is effective and also kills disease organisms and nematodes.
2. Pre-emergence herbicides can be applied before the weed seeds emerge but can reduce germination of the desired species.
3. Post-emergence herbicides can be applied as soon as the weed seedlings emerge.

Wide ranges of selective and nonselective commercial products are available. Such materials should be used with caution, however, since improper use can cause injury to the young nursery plants. Not only should the manufacturers' directions be followed, but also preliminary trials should be made before large-scale use.

Select High-Quality Seed. Quality is based on seed-testing data, as discussed in Chapter 7. A low sowing rate requires high-quality seeds that produce not only high germination percentages but also vigorous, uniform, healthy seedlings.

Choose Correct Planting Times. Planting time is determined by the germination temperature requirements of the seed, available soil moisture, and the need to meet production schedules. These are determined according to the individual crop and vary with the particular kind of seed. Early season sowing of seeds that require warm soil temperatures can result in slow and uneven germination, disease problems, and "chilling" injury to seedlings of some species, causing growth abnormalities. High soil temperatures can result in excessive drying, injury, or death to seedlings, or induction of thermodormancy in the case of heat-sensitive seeds such as lettuce, celery, and various flower seeds (see Chapter 21).

Seed Treatments to Facilitate Sowing. It is often desirable to use seeds that have been pretreated for protection with a pesticide (fungicide and/or insecticide) or enhanced for germination by a seed coating or priming treatment (see Chapter 6). These treatments can speed up germination, increase uniformity, and offset some environmental hazards in the seed bed. **Coated seeds** have improved flowability, and uniform size can improve the seeding precision of mechanical planters.

> **coated seeds** Seeds with an altered shape that makes them easier to sow with precision seeding machines. The coating may also improve flowability, which is the ability for seeds to flow out of the seeder.

Choose the Proper Mechanical Seeder for Outdoor Planting. The first mechanical seed drill was developed by Henry Smith in 1850 (37). Today, most field-sown crops are seeded mechanically. Selection of a seeder is determined by the following:

1. Size and shape of the seed
2. Soil characteristics
3. Total acreage to be planted
4. Need for precision placement of the seed in the row

Mechanical seeders contain three basic components: a **seed hopper** for holding seeds and a **metering system** to deliver seeds to the **drill**. A drill opens the furrow for planting the seed. The drill controls seeding depth and must provide good seed-to-soil contact while minimizing soil compaction that might impede seedling emergence. The most common type of drill is a simple "Coulter" drill that places seeds into an open furrow. "Dibber" drills that punch individual holes to place seed have also shown good seeding performance (21, 27). In some cases, a press wheel may be used to help cover the seed, and attachments to the seeder may supply fertilizer, pesticides, or anticrusting agents before or after depositing the seed.

Mechanical seeders (Fig. 8–3) are available as either **random** or **precision** seeders (9, 53).

Random seeders meter seed in the row without exact spacing. They are less complex than precision seeders and are useful when spacing between plants in the row is not critical, and thinning is not applied to achieve final plant stand as in many agronomic crops. Random seeders use gravity to drop seeds through holes located at the bottom of the hopper. The size of these holes and tractor's speed determine the seeding rate.

> **random seeders** Seeders that use gravity and tractor speed to place seeds in the ground.

Precision seeders selectively meter seed from the hopper to maintain a preset spacing in the row, and can greatly reduce the number of seeds required to seed an acre compared to random seeding. For example, to achieve the same stand for California lettuce, seeding rates were reduced by

> **precision seeders** Seeders that use belts, plates, or vacuum to place single seeds at a selected seed spacing.

(a)

(b)

(c)

(d)

Figure 8–3
Examples of filed seeding machinery. (a) A single row Planet Junior. (b) A row crop seeder used for crops such as corn. (c) A multiple row seeder for drilling grasses. (d) A precision row seeder for vegetable crops.

84 percent using precision, compared with random seeders (40). Precision seeders use a separate power take-off on the tractor drive to power the planter and control seeding rate. Several types of precision seeders are available. These include **belt, plate, wheel,** and **vacuum** seeders (9).

The **belt seeder** uses a continuously cycling belt that moves under the seed supply. Holes in the belt at specified intervals determine seed spacing. When operating correctly, one seed will move by gravity to occupy one hole on the belt and be released as it passes over the furrow.

The **plate seeder** also uses gravity to fill holes in a metal plate rotating horizontally through the seed hopper. The number of holes in the plate and the speed of plate rotation determine seed spacing.

The **wheel seeder** employs a rotating wheel oriented in a vertical position at a right angle to the bottom of the seed hopper. Seed fills the opening at the top of the wheel (bottom of the hopper) by gravity and is carried 180 degrees where it is deposited into the furrow opening.

Vacuum seeders (Fig. 8–4) are replacing gravity seeders in the vegetable industry because they can more precisely deliver single seeds at a specified row spacing, especially small seeds (like tomato), irregularly shaped seeds (like lettuce), or uncoated seeds (41). In a comparison of vacuum and belt seeders using several different vegetable seeds (56), no difference was observed for seed placement for carrot or onion seeds. The belt seeder performed better for cabbage seed, while the vacuum seeder was more precise for cucumber seed placement. The vacuum seeder utilizes a vertical rotating plate in the hopper with cells under vacuum that pick up a single seed. Seeds are released into the planting furrow by removing the vacuum in the cell as it rotates above the seed drop tube or planting shoe (54). A **"singulator device"** helps displace extra seeds prior to planting. Some vacuum seeders use a burst of air to clean the cell after the seed has been dropped to avoid skips from a clogged seed hole.

For all precision seeders, different sized holes in belts or plates can be used for seeds of different species that are different sizes. In many cases, uniform seed size or pelleted seeds improve the precision of in-row spacing. However, because of seed quality, environmental factors, insect, disease, or animal predation, seeds are usually spaced at a higher density than is optimum for a final stand, and the grower must physically thin seedlings to the desired plant density following emergence. Most direct-seeded vegetable crops are planted with precision seeders.

Seeders have also been adapted for direct-seeding vegetable crops in no-till production systems (52, 76). The challenge in reduced tillage systems is planting through existing crop residue or covercrops to establish an adequate plant stand.

In addition to conventional seeders, **gel seeders** used for fluid drilling have been developed to deliver pregerminated seeds (see pregermination, page 188). Pregerminated seeds are incorporated into a gel and extruded or fed into the furrow via a pumping system or by having the seed tank under pressure using compressed air (53). Although this method can improve seedling

(a)

(b)

Figure 8–4
The inside of a vacuum seeder showing the rotating plate that picks up and delivers single seeds. This plate is for sowing spinach.

> **BOX 8.1 GETTING MORE IN DEPTH ON THE SUBJECT**
> **CALCULATING SOWING RATE**
>
> The following formula is useful in calculating the rate of seed sowing (23, 44):
>
> $$\text{Weight of seeds to sow per unit area} = \frac{\text{Density (plants/units area) desired}}{\text{*Purity percentage} \times \text{*Germination percentage} \times \text{*Field factor} \times \text{Seed count (number of seeds per unit weight)}}$$
>
> *Expressed as a decimal.
>
> Field factor is a correction term that is applied based on the expected losses that experience at that nursery indicates will occur with that species. It is a percentage expressed as a decimal.

emergence (especially under adverse environmental conditions), gel seeding is still only a minor planting system compared with conventional seeding of dry seeds because of the cost and complexity of the operation (41).

Use Correct Sowing Depth. Depth of planting is a critical factor that determines the rate of emergence and stand density. If too shallow, the seed may be in the upper surface that dries out rapidly; if too deep, emergence of the seedling is delayed. Depth varies with the kind and size of seed and, to some extent, the condition of the seed bed and the environment at the time of planting. When exposure to light is necessary, seeds should be planted shallowly. A rule of thumb is to plant seeds to a depth that approximates three to four times their diameter.

Determine Proper Sowing Rate. The sowing rate is critical in direct sowing in order to produce a desired plant density (see Box 8.1). This rate is a minimum and should be adjusted to account for expected losses in the seed bed, determined by previous experience at that site. Many seed companies will help producers set up spacing requirements for direct-seeding precision planting equipment.

Rates will vary with the spacing pattern. Field crops or lawn seeds may be broadcast (i.e., spaced randomly over the entire area) or drilled at given spaces. Other field crops, particularly vegetables, are row planted, so that the rate per linear distance in the row must be determined. Crops may be grown in rows on raised beds, particularly in areas of low rainfall where irrigation is practiced and excess soluble salts may accumulate to toxic levels through evaporation. Overhead sprinkling and planting seed below the crest of sloping seed beds may eliminate or reduce this problem.

Supply Postplanting Care. Adequate moisture must be supplied to the seed once the germination process has begun. In many areas, there is adequate natural rainfall to support seed germination. In areas with irregular rainfall, supplemental irrigation is usually supplied by overhead sprinklers, subsurface furrow flooding to raised seed beds, or by trickle irrigation (Fig. 8–5, page 256). The soil should also be kept from drying out and developing a crust. This is primarily a function of seed bed preparation but may be avoided by light sprinkling, shading, and covering with light mulch. With row planting, excess seed is planted, and then the plants are thinned to the desired spacing. Thinning is expensive and time-consuming and can be reduced by precision planting. Competition from weeds must also be controlled by herbicide, tillage, or mulching to ensure a vigorous seedling stand.

Field Nurseries for Transplant Production

Outdoor **field nurseries** where seeds are planted closely together in beds are used extensively for growing transplants of conifers and deciduous plants for forestry (62), for ornamentals (19, 44), to provide understock liners for some fruit and nut tree species (Fig. 8–6, page 256) (28, 43, 60), and vegetable transplants (Fig. 8–7, page 257) (24). The conditions for optimum seed germination and seedling emergence are very similar to those previously described for field-seeding vegetables. However, field transplant nurseries produce seedlings at a close spacing using smaller acreage and more controlled management. It is more common to produce woody plant seedlings in transplant nurseries than direct-sowing them to a permanent location. Practices for successful production in a transplant nursery include:

> **field nurseries**
> Nurseries that contain seeds sown at high density in the field for future transplanting to a wider spacing.

1. Site selection and seed bed preparation
2. Time of the year for sowing

Figure 8–5
Irrigation examples. (a) Spinach crop on central pivot irrigation. The elevated pipe and irrigation heads travel through the field. (b) A lettuce crop being irrigated after sowing with movable pipes. These will be located temporarily in one field and moved to other fields as necessary. (c) A sunflower crop being furrow irrigated. Water is siphoned out of the main canal to temporarily flood each row.

3. Sowing rates
4. Plant after-care
5. Harvesting field-grown transplants

Site Selection and Seed Bed Preparation Nursery production requires a fertile, well-drained soil of medium to light texture. Site selection and preparation for planting may include rotation with other crops and incorporation of a green manure crop or animal manure (65). Preplant measures for weed control are essential aspects of most nursery operations.

A common size of seed bed is 1.1 to 1.2 m (3.5 to 4 ft) wide with the length varying according to the size of the operation. Beds may be raised to ensure good

Figure 8–6
Field seeding for woody plant liner production. (a) Seed bed prepared for sowing. Each bed is approximately 4 ft wide. (b) These beds are covered with burlap to help retain moisture. (c and d) These pine and barberry seedlings were sown with a five-row drill to permit cultivation for weed control.

(a)

(b)

Figure 8–7
Vegetable field transplant nurseries. (a) Cabbage seedlings planted at high density will be pulled as transplants. (b) A floating row cover provides protection for early seeded transplants.

drainage, and, in some cases, sideboards are added after sowing to maintain the shape of the bed and to provide support for glass frames or lath shade. Beds are separated by walkways 0.45 to 0.6 m (1.5 to 2 ft) wide. North-south orientation gives more even exposure to light than east-west orientation.

Seeds may be either broadcast over the surface of the bed or drilled into closely spaced rows with seed planters. For economy, seeds should be planted as closely together as feasible without overcrowding, which increases damping-off and reduces vigor and size of the seedling (35), resulting in thin, spindly plants and small root systems. Seedlings with these characteristics do not transplant well (34).

Time of the Year for Sowing Several vegetable species, including tomato, pepper, cabbage, broccoli, and onion, can be produced from transplants produced in field nurseries. This is an alternative to direct seeding and is less expensive than container-grown transplants produced in greenhouses. Warm-season crops are usually seeded in spring and may be covered with plastic or fabric (floating) row covers to prevent frost injury (Fig. 8–7b). Cool-season crops are seeded in early spring or summer for a fall harvest.

For many species (especially woody and herbaceous native plants), seeds must be treated to overcome seed dormancy conditions (see Chapter 7). The two most common treatments used by commercial propagators include scarification for species with hard seed coats, and stratification for species that require periods of warm or chilling conditions to alleviate dormancy.

Scarification. **Scarification** is the process of physically or chemically altering the seed coverings to improve germination in dormant seeds. It is a horticultural necessity for species with physical dormancy (hard, impermeable seed coats) to permit water uptake. Such seeds include members of the legume, geranium, morning glory, and linden families. Scarification (usually in the form of brushing) is also commonly applied to cereals and grasses to remove the structures covering the caryopsis (glumes, palea, and lemma) that can reduce germination. Three types of treatments are commonly used as scarification treatments. These include **mechanical, chemical,** and **heat** treatments.

Mechanical Scarification. Mechanical scarification is simple and effective with seeds of many species, and commercial equipment is available that tumbles seeds in drums against an abrasive material (Fig. 8–8). These seeds are dry after such treatment and may be stored or planted immediately by mechanical seeders. Scarified seeds are more susceptible to injury from pathogenic organisms, however, and may not store as well as comparable non-scarified seeds.

Small amounts of relatively large seeds can be scarified by rubbing with sandpaper, abraded with a file, or cutting with clippers (Fig. 8–8). For large-scale mechanical operations, commercial scarifiers are used. Small seeds of legumes, such as alfalfa and clover, are

scarification A treatment that allows water to penetrate seeds with a hard seed coat. Scarification may be physical, chemical (acid), or involve high temperature.

Figure 8–8
Seed scarifier used to abrade hard seeds with physical dormancy, such as legumes.

often treated in this manner to increase germination (14). Seeds may be tumbled in drums lined with sandpaper or in concrete mixers containing coarse sand or gravel (7, 62). The sand or gravel should be of a different size than the seed to facilitate subsequent separation of the sand from the seed prior to sowing.

Scarification should not proceed to the point at which the seeds are injured. The seed coats generally should be dull but not so deeply pitted or cracked as to expose the inner parts of the seed. To determine the optimum time a test lot can be germinated, the seeds may be soaked to observe swelling, or the seed coats may be examined with a hand lens.

Chemical (Acid) Scarification. Dry seeds are placed in containers and covered with concentrated sulfuric acid in a ratio of about one part seed to two parts acid (see Box 8.2). The amount of seed treated at any one time should be restricted to no more than about 10 kg (22 lbs) to avoid uncontrollable heating. Containers should be glass, earthenware, or wood—not metal or plastic. The mixture should be stirred cautiously at intervals during the treatment to produce uniform results and to prevent accumulation of the dark, resinous material from the seed coats, which is sometimes present. Since stirring tends to raise the temperature, vigorous agitation of the mixture should be avoided in order to prevent injury to the seeds. The time of treatment may vary from as little as 10 minutes for some species to 6 hours or more for other species. Since treatment time may vary with different seed lots, making a preliminary test on a small lot is recommended prior to treating large lots (36, 47).

At the end of the treatment period, the acid is poured off, and the seeds are quickly washed to remove any acid residue. Glass funnels are useful for removing the acid from small lots of seed. Placing seeds in a large amount of water with a small amount of baking soda (sodium bicarbonate) will neutralize any adhering acid, or the seeds can be washed for 10 minutes in running water. The acid-treated seeds can either be planted immediately when wet or dried and stored for later planting.

Large seeds of most legume species respond to the simple sulfuric acid treatment, but variations are required for some species (47). Some roseaceaous seeds (*Cotoneaster, Rosa*) have hard pericarps that are best treated partially with acid followed by warm stratification. A third group, such as *Hamamelis* and *Tilia*, have very "tough" pericarps that may first need to be treated with nitric acid and then with sulfuric acid.

High Temperature Scarification. In nature, physical dormancy appears to be relieved most often by high temperature exposure. This process can be mimicked by placing seeds on moist or dry sand at temperatures above 35°C (95°F). The requirement for moist or dry heat, as well as the temperature and duration of the treatment, varies between species (4, 51).

Hot water scarification is a common alternative to acid and mechanical scarification (64), but it usually yields more variable results. Drop the seeds into 4 to 5 times their volume of hot water 77 to 100°C (170 to 212°F). Seeds can be treated for several minutes, but prolonged exposure to heat will kill them. Start by removing the seeds immediately after exposure and allow them to soak in the gradually cooling water for 12 to 24 hours. Microwave energy has also been reported to be an effective heat treatment (73). Following heat treatment and imbibition, non-swollen seeds can be separated from the swollen ones by suitable screens and either re-treated or subjected to some other treatment. Usually the seeds should be planted immediately after the hot water treatment; some kinds of seed have been dried and stored for later planting without impairing the germination percentage, although the germination may be reduced.

Stratification. **Stratification** is a method of handling dormant seeds in which imbibed seeds are subjected to a period of chilling or warm temperatures to alleviate dormancy conditions in the embryo.

> **stratification** A period of moist-warm or moist-chilling conditions that satisfies dormancy in seeds with endogenous, physiological dormancy.

The term originated because nurseries placed seeds in

BOX 8.2 GETTING MORE IN DEPTH ON THE SUBJECT
USING ACIDS SAFELY

Always use proper safety precautions while using acids for scarification, including personal safety equipment like gloves, face shield, eye protection, and lab coat. An eye wash and a source of running water must be available in case of an accident. Request the MSDS safety sheet from your chemical supplier for additional safety precautions. There are several web sites that offer this information including www.msdssearch.com and www.msdssolutions.com.

stratified layers interspersed with a moist medium, such as soil or sand, in out-of-doors pits during winter (Fig. 8–8). The term *moist-chilling* has been used as a synonym for stratification. However, with temperate species displaying epicotyl dormancy (like *Chionanthus*—fringetree) or underdeveloped embryos (like *Ilex*—hollies), a warm-moist stratification of several months followed by a moist-chilling stratification is required to satisfy dormancy conditions, though it may require more than one season to achieve under natural conditions. Several tropical and semitropical species (like palms) require a period of warm stratification prior to germination to allow the embryo to continue development after fruit drop.

Outdoor Planting for Stratification. Seeds requiring a cold treatment may be planted out-of-doors directly in the seed bed, cold frame, or nursery row at a time of the year when the natural environment provides the necessary conditions to relieve dormancy (Figs. 8–9 and 8–10). This is the most common treatment for seeds with endogenous physiological dormancy. Several different categories of seeds can be handled in this way with good germination in the spring following planting.

Seeds must be planted early enough in the fall to allow them to become imbibed with water and to get the full benefit of the winter chilling period. Seeds need to be protected against freezing, drying, and rodents (Fig. 8–11, page 260). The seeds generally germinate promptly in the spring when the soil begins to warm up but while the soil temperature is still low enough to inhibit damping-off organisms and to avoid high-temperature inhibition.

Seeds with a hard endocarp, such as *Prunus* species (the stone fruits, including cherries, plums, and peaches), show increased germination if planted early enough in the summer or fall to provide 1 to 2 months of warm temperatures prior to the onset of chilling (43). Thus, seeds that require high temperatures followed by chilling can be planted in late summer to fulfill their warm-temperature requirements followed by the subsequent winter period that satisfies the chilling requirement.

Refrigerated Stratification. An alternative to outdoor field planting is refrigerated stratification (Fig. 8–12, page 260). This is a useful technique for small seed lots or valuable seeds that require special handling. Dry seeds should be fully imbibed with water prior to refrigerated stratification. Soaking at a warm temperature for 12 to 24 hours may be sufficient for seeds without hard seed coats or coverings.

After soaking, seeds are usually mixed with a moisture-retaining medium for the stratification

(a)

(b)

Figure 8–9
(a) The term stratification comes from the old practice of layering seeds and sand. (b) On old-style outdoor stratification box for yew (*Taxus*) seeds.

(a)

(b)

Figure 8–10
Raised beds for outdoor seeding. (a) Wildflower seeds sown outdoors with inverted flat holders used to protect seeds from predation. (b) These conifer seedlings will spend the first year in raised seedbeds.

Figure 8–11
(a) Wire screen used to protect acorns from rodent and squirrel predation. (b) Nursery fabric used to protect outdoor seed beds.

period. Almost any medium that holds moisture, provides aeration, and contains no toxic substances is suitable. These include well-washed sand, peat moss, chopped or screened sphagnum moss [0.6 to 1.0 cm (1/4 to 3/8 in.)], vermiculite, and composted sawdust. Fresh sawdust may contain toxic substances. A good medium is a mixture of one part coarse sand to one part peat, or one part perlite to one part peat, moistened and allowed to stand 24 hours before use. Any medium used should be moist but not so wet that water can be squeezed out.

Seeds are mixed with 1 to 3 times their volume of the medium or they may be stratified in layers, alternating with similarly sized layers of the medium. Suitable containers are boxes, cans, glass jars with perforated lids, or other containers that provide aeration, prevent drying, and protect against rodents. Polyethylene bags are excellent containers either with or without media. Stratification of seeds in a plastic bag without a surrounding medium has been called naked chilling (18). A fungicide may be added as a seed protectant. Seeds may also benefit from surface disinfection prior to

Figure 8–12
Examples of refrigerated stratification. (a) Small batches of seeds can be mixed with moist vermiculite and placed in polyethylene bags. (b) Conifer (pine) seeds are hydrated and placed in polyethylene bags without any substrate. (c) Hazelnut (*Corylus*) seeds mixed with a bark substrate in large plastic tubs were placed into large refrigerated storage units. (d) A technician removing seeds that had germinated while being stratified.

imbibition and stratification with a 10 percent bleach solution for 10 to 15 minutes followed by multiple rinses with water to remove the bleach.

The usual chilling stratification temperature is 1 to 10°C (33 to 50°F). At higher temperatures, seeds often sprout prematurely. Lower temperatures (just above freezing) may delay sprouting. No progress toward dormancy release occurs above 15°C (60°F) (26). Warm stratification temperatures are usually above 25°C (77°F) and can be quite high in tropical species, like palms (Fig. 8–13) at 30 to 35°C (85 to 95°F).

The time required for stratification depends on the kind of seed and, sometimes, on the individual lot of seed as well (see Chapter 7). For seeds of most species, 1 to 4 months is sufficient. During this time, the seeds should be examined periodically; if they are dry, the medium should be remoistened. The seeds to be planted are removed from the containers and separated from the medium, using care to prevent injury to the moist seeds. A good method is to use a screen that allows the medium to pass through while retaining the seeds. The seeds are usually planted without drying to avoid injury and reversion to secondary dormancy. Some success has been reported for partially drying previously stratified seeds, holding them for a time at low temperatures, then planting them "dry" without injury or loss of dormancy release. Beech (*Fagus*) and mahaleb cherry seeds were successfully dried to 10 percent and then held near freezing (72). Similarly, stratified fir (*Abies*) seed has been dried to 20 to 35 percent and then stored for a year at low temperatures after stratification (20).

Sowing Rates for Outdoor Seeding   The optimum seed density primarily depends on the species but also on the nursery objectives. If a high percentage of the seedlings is to reach a desired size for field planting, low

Figure 8–13
Palm seed (a) has morphological dormancy in which the embryo is small (b) and must develop within the seed at moist, warm temperatures before germination can occur. Several seeds are planted in each container that are placed in racks (c), watered, and covered with plastic (d) for several months to relieve morphological dormancy. Containers are moved to the greenhouse where (e) several seedlings emerge per container.

(a) (b)

Figure 8–14
Planting density depends on the ultimate use of the seedlings. (a) Oak seedlings were planted at a high density and will be sold as seedling liners. (b) These ginkgo seedlings were drilled at a lower density and may be used as seedling liners or could be field budded.

densities might be desired; but if the seedlings are to be transplanted into other beds for additional growth, higher densities (with smaller seedlings) might be more practical (Fig. 8–14). Once the actual density is determined, the necessary rate of sowing can be calculated from data obtained from a germination test and from experience at that particular nursery (see Box 8.3).

Seeds can be planted by (a) broadcasting by hand or seeders, (b) hand spacing (larger seeds), or (c) drilling by hand with push drills, or drilling with tractor-drawn precision drills. Seeds of a particular lot should be thoroughly mixed before planting to ensure that the density in the seed bed will be uniform. Treatment with a fungicide for control of damping-off is often desirable. Small conifer seeds may be pelleted for protection against disease, insects, birds, and rodents. Depth of planting varies with the kind and size of seed. In general, a depth of three to four times the diameter of the seed is satisfactory. Seeds can be covered by soil, coarse sand, or by various mulches.

BOX 8.3 GETTING MORE IN DEPTH ON THE SUBJECT
SEEDING TIMES FOR HERBACEOUS AND WOODY PERENNIAL SEEDLING PRODUCTION

Seeds are planted in the nursery in the summer, fall, or spring depending on the dormancy conditions of the seed, the temperature requirements for germination, the management practices at the nursery, and the location of the nursery (in a cold-winter or a mild-winter area). Planting time varies for several general categories of seed (44, 62).

Summer Seeding
Seeds of some species, such as maple (*Acer*), poplar (*Populus*), elm (*Ulmus*), and willow (*Salix*), ripen in spring or early summer. Such seeds should be planted immediately after they ripen, as they do not tolerate drying and their viability declines rapidly (see recalcitrant seeds, page 189). Other species with morphological and morphophysiological dormancy, like *Clematis*, holly (*Ilex*), ash (*Fraxinus*), windflower (*Anemone*) and twinleaf (*Jeffersonia*), should be planted in summer or early fall to allow 6 to 8 weeks of warm stratification in the seed bed prior to the winter chilling (4).

Fall Seeding
Seeds of species with physiological dormancy that require moist-chilling can be fall seeded where winter temperatures have appropriate periods of cold temperature to satisfy dormancy. Certain species [apple (*Malus*), pear (*Pyrus*), Cherry (*Prunus*), and yew (*Taxus*)] are adversely affected by high germination temperatures, which produce secondary

dormancy. Germination temperatures of 10 to 17°C (50 to 62°F) are optimum. Seeds of these species should be planted in the fall, and germination will take place in late winter or early spring.

Spring Seeding

Many kinds of seeds—including most conifers (pine, fir, spruce) and many deciduous hardwood species—benefit from moist-chilling stratification but do not germinate until soil temperatures have warmed up, and are not inhibited by high soil temperatures. Optimum germination temperatures are 20 to 30°C (68 to 86°F). Such seeds can be fall planted, but spring planting following refrigerated stratification often results in superior germination and seedling emergence. Non-dormant seeds or those with only physical dormancy (black locust (*Robinia*), yellowwood (*Cladrastis*) and Kentucky coffeetree (*Gymnocladus*)) are planted in the spring either outdoors or under protected cultivation (greenhouse or coldframes) to take advantage of the long growing season. Soil firming may be done to increase the contact of seed and soil. It is used for California lettuce, for example, and carried out with a tamper, hand roller, or tractor-drawn roller either before sowing or immediately afterward. Rodent and bird protection may be necessary.

Plant After-Care During the first year in the seed bed, the seedlings should be kept growing continuously without any check in development. A continuous moisture supply, cultivation or herbicides to control weeds, and proper disease and insect control contribute to successful seedling growth. Fertilization (especially nitrogen) is usually necessary, particularly when mulch has been applied, since decomposition of organic material can reduce nitrogen availability. In the case of tender plants, glass frames can be placed over the beds, although for most species a lath shade is sufficient. With some species, shade is necessary throughout the first season; with others, shade is necessary only during the first part of the season (Fig. 8–15).

Harvesting Field-Grown Transplants Vegetable transplants can be harvested after 6 to 10 weeks in the seed bed. These are usually "pulled," bundled, and used as bare-root transplants. In the United States, vegetable transplant beds are either located on the producer's farm or shipped to northern growing areas from southern transplant nurseries. A large number of vegetable and tobacco transplants are being produced in plug systems and "float beds" (see page 272), which are replacing the more traditional field-nursery-produced transplants.

In contrast, woody plants can remain in the "liner" bed for a year or more before being transplanted to a permanent location (see Chapter 3). For some species, the plants may be shifted to a transplant bed after 1 year and then grown for a period of time at wider spacing (Fig. 8–14). This basic procedure is used to propagate millions of forest tree seedlings, both conifer and deciduous species.

(a)

(b)

(c)

Figure 8–15
Several examples of seedling shading. The shading is usually temporary for the first few months of seedling growth. (a) Burlap over a wire covering. (b) Snow fence is commonly used for conifer seedling shading. (c) Conventional shade cloth for ornamental liners.

(a) (b)

(c)

(d)

Figure 8–16
Transplanting liners at higher spacing. (a) A tractor pulls the transplanting unit with several workers (b) placing seedlings into the (c) planting wheel. (d) Soil is mounded around the seedlings to complete the planting operation.

Liners produced in a seed-bed nursery are often designated by numbers to indicate the length of time in a seed bed and the length of time in a transplant bed. For instance, a designation of 1–2 means a seedling grown 1 year in a seed bed and 2 years in a transplant bed or field. Similarly, a designation of 2–0 means a seedling produced in 2 years in a seed bed and no time in a transplant bed (Fig. 8–16).

Seedling liners are lifted mechanically by undercutting the plants and shaking off the soil around the roots (Fig. 8–17). Bareroot liners are graded into size classes prior to being overwintered in large refrigerated coolers for spring sales.

Specialty Systems for Direct-Seeded Crops

Direct-Seeded Nursery Row Production Planting directly in separate nursery rows is one of the primary methods used to propagate rootstocks of many fruit and nut tree species (28, 60). Cultivars are budded or grafted to the seedlings in place (see Chapters 12 and 13). The method is also used to propagate shade trees and ornamental shrubs, either as seedlings or on rootstocks as budded selected cultivars (Fig. 8–18).

Deciduous fruit, nut, and shade tree propagation usually begins by planting seeds or liners in nursery rows. Where plants are to be budded or grafted in place, the width between rows is about 1.2 m (4 ft) and the seeds are planted 7.6 to 10 cm (3 to 4 in) apart in the row (see Fig. 8–17). Seeds known to have low germination must be planted closer together to get the desired stand of seedlings. Large seed (walnut) can be planted 10 to 15 cm (4 to 6 in) deep, medium-sized seed (apricot, almond, peach, and pecan) about 7.6 cm (3 in), and small seed (myrobalan plum), about 3.8 cm (1.5 in). Spacing may vary with soil type. If germination percentage is low and a poor stand results, the surviving trees, because of the wide spacing, may grow too large to be suitable for budding. Plants to be grown to a salable size as seedlings without budding could be spaced at shorter intervals and in rows closer together.

Fall planting of fruit and nut tree seeds is commonly used in mild-winter areas such as California (28). Seeds are planted 2.5 to 3.6 cm (1 to 1.5 in) deep and 10 to 15 cm (4 to 6 in) apart, depending on size, and then covered with a ridge of soil 15 to 20 cm (6 to 8 in) deep, in which the seeds remain to stratify during winter. The soil ridge is removed in the spring just before seedling emergence. Herbicide control of weeds and protection of the seeds from rodents become important considerations during these procedures.

Field Seeding for Reforestation or Naturalizing
Field seeding of forest trees is accomplished in reforestation either through natural seed dissemination or planting. Costs and labor requirements of direct seeding are lower than those for transplanting seedlings,

Figure 8–17
Harvesting bareroot liners. (a, b, and c) The liners are mechanically undercut and lifted. The lifting tines vibrate to shake off as much soil as possible. (d) Workers collect the plants and group them in bundles. (e) Liners are graded into size classes before being placed into (f) cold storage.

provided soil and site conditions favor the operation (17). The major difficulty is the very heavy losses of seeds and young plants that result from predation by insects, birds, and animals; drying, hot weather; and disease (62). A proper seed bed is essential, and an open mineral soil with competing vegetation removed is best. The soil may be prepared by burning, disking, or furrowing. Seeds may be broadcast by hand or by special planters, or drilled with special seeders. Seeds should be coated with a bird and rodent repellent.

Wildflower seed mixtures can be naturalized to provide landscape color for public or private lands at a low cost. In many locations, wildflower establishment has become an alternative to mowing on highway

Figure 8–18
Some fruit and ornamental trees (like these dogwoods) are direct seeded with wide spacing so the plants can be budded in the nursery row.

right-of-ways. Seed germination and seedling establishment are improved by tillage for seed bed preparation and a straw mulch covering for seeds (15). For highly erodible sites, a "nurse" grass crop plus wildflower seed mixture can improve wildflower establishment (16). Weed competition is a serious problem for wildflower plantings, which must be managed to ensure a successful stand. Successful strategies include the use of herbicides, tillage, fumigation, and solarization (covering soil with plastic to trap solar radiation and allow heat to pasteurize soil).

Production of Transplants Under Protected Conditions

Seedling production is used extensively to produce flowers and vegetables for outdoor transplanting. Historically, this method has been used to extend the growing season by producing seedlings under protection for transplanting to the field as soon as the danger of spring frosts is over, or by placing seedlings under individual protectors to avoid freezing. This procedure also avoids some of the environmental hazards of germination and allows plants to be placed directly into a final spacing. Optimum germination conditions are provided in greenhouses, cold frames, or other structures to ensure good seedling survival and uniformity of plants.

Seedling growing has become an extensive bedding plant industry to produce small ornamental plants for home, park, and building landscaping, as well as vegetable plants for home gardening (2, 38). Commercial vegetable growing also relies heavily on the production of transplants, involving highly mechanized operations beginning with seed germination and ending with transplanting machines that place individual plants into the field.

Production Systems for Transplants Traditionally, bedding plants and vegetables have been produced by germinating seeds in flats and transplanting seedlings to larger containers prior to field or landscape planting. However, modern greenhouse producers have adopted plug production as the preferred method for transplant production (1, 38, 69, 75). Plug production provides numerous advantages over conventional flat seeding, and specialized plug growers produce acres of plugs under glass each spring. Many bedding plant growers find they can purchase plugs from specialized plug producers more economically than producing seedlings themselves. In either case, seedlings are moved to larger cell packs by the bedding plant grower for "finishing" prior to sale to the consumer. The advantages of plug production include:

1. Optimization of the number of plants produced per unit of greenhouse space.
2. Specialization in plug production allows growers to invest in equipment to control environmental conditions during germination.
3. Fast production (most plugs are sold within four to six weeks of seeding) allows growers to seed multiple crops per season, permits accurate crop scheduling, and allows plugs to be shipped easily to the end user.
4. Because plugs are transplanted to larger-size containers with the roots and original medium intact, plugs transplant easily with a high degree of uniformity. Plugs do not experience the same "transplant shock" and check in growth as seedlings removed from seedling flats.

Flat Production. Traditional bedding plant production relied on flat production of seedlings. Seeds were planted in a germination flat or container, and later germinated seedlings were "pricked out" and transplanted to develop either in a transplant flat at a wider spacing or in individual containers where they remained until transplanted out-of-doors (Fig. 8–19). This method is still utilized by small bedding plant producers but has largely been replaced by mechanized plug production.

Plug Production. The first crops to be produced in plugs were vegetable transplants in the 1960s by the Florida-based Speedling Corporation (6). Today, millions of vegetable and flower transplants are produced annually in greenhouses under carefully controlled environmental conditions for optimizing germination and plant growth. This has become possible mainly through the development of the plug system (2, 13, 69, 75). A **plug** is a seedling produced in a small volume of medium contained in a small cell, of which between 72 to 800 are contained on a single sheet of polystyrene, Styrofoam, or other suitable material (Fig. 8–20). Plug flats are filled mechanically with a growing substrate, and seeds are sown mechanically into each cell. Standard plug trays are 55 × 28 cm (21.5 × 11 in) or 25 × 51 cm (10 × 20 in), and individual cell sizes may range down to 1 × 1 cm (3/8 × 3/8 in). Cell size dictates the length of time a crop of plugs takes to produce and the time required for the bedding plant grower to finish the crop. Generally, the larger the cell, the longer it takes the plug grower to produce the plug. For the bedding plant grower, the larger the cell (plug), the less time it takes to finish the crop (32). Considerations for the bedding plant grower include crop scheduling, economics between purchasing

> **plugs** Small-celled transplant flats used to produce many seedlings in a small greenhouse area.

Figure 8–19
Community flats were a common sowing technique before the development on single unit plugs. (a) Several vegetable varieties sown into a single flat. (b) Commercial vegetable transplants being sown in community flats and covered with expanded clay to help reduce moisture loss. (c) Sowing density is important to prevent crowding in community flats. (d) Once the seedlings have unfolded their true leaves, they are pricked out of the flat and moved to a larger spacing. Dibble boards were commonly used to make transplant holes at equal spacing in a transplant flat.

larger plugs and greenhouse production costs, number of greenhouse turns (using the same space for multiple crops), and mechanical transplanting equipment requirements.

High seed germination and seedling uniformity are critical for good plug production (67, 71). Seed germination may be on the greenhouse bench in sophisticated computer-controlled environments or in specialized germination rooms that provide optimum temperature and moisture conditions, and light, if necessary. It is important to have high-quality, high vigor seed to maximize germination rate, seedling uniformity, and mechanical handling (8). Pelleting and seed priming (see Chapter 6) are common seed enhancements for plug production.

Figure 8–20
Plug production. (a and b) Seedlings germinated and grown on movable benches. (c) Good uniform germination with one usable seedling per cell. Plants are in stage 2 of production with fully expanded cotyledons. (d) A finished begonia plug in stage 4.

Plug Growth Stages. The four morphological stages of seedling growth are: (29, 69, 75)

Stage 1: sowing to germination (radicle emergence)

Stage 2: germination to full cotyledon spread and root system establishment

Stage 3: seedling plug growth (unfolding of three or four leaves; root growth)

Stage 4: seedling plug getting ready to transplant or ship (more than four leaves)

Providing precise environmental control for each of the stages is essential in plug production. Warm temperature and consistent moisture are essential for stage 1 but usually are reduced in stage 2 and in later stages (see Table 8–1). Light may be required for germination in stage 1 for some crops and relative humidity is held at at least 95 percent, often provided by fog in growth rooms. A starter fertilizer charge may be applied to the substrate in stage 1. A moderate light level and low fertilization is typical for stage 2 growth. Substrate water content is reduced compared to stage 1 and varies depending on the crop. High light and a complete fertilization (N, K, P) is particularly important in stage 3 (74) but must be monitored carefully (63). Plant growth regulators may be applied in stage 3 to control seedling height. As seedlings enter stage 4, they are usually "toned" in preparation for shipping and transplanting. Therefore, substrate moisture and temperature are usually reduced, compared to stage 3. Nitrogen fertilization is reduced or may be withheld in stage 4.

Production of Woody Plant Seedlings in Containers. Production of seedling trees and shrubs in containers is an intensive alternative to field production (Figs. 8–21 and 8–22). Seeds may be sown in germination flats or direct-seeded into plug-trays (48, 57). Later they are moved to slightly larger containers or transplanted directly into the containers where they will remain until transplanted out-of-doors.

Container-grown tree seedlings are grown in deep containers, and root pruning is essential to induce a desirable, well-branched root system (45). Root pruning can be done physically prior to the first transplanting, soon after the roots reach the bottom of the flat (30, 31). More commonly, plants can be grown in open-bottom containers where air-pruning removes roots that protrude from the bottom of the container (Fig. 8–21). Metal or plastic screen-bottomed flats (25) can also stimulate formation of branch roots. Seedlings may be produced in plastic containers from which the seedling plug is removed prior to planting, or they may be containers made of substances such as peat or fiber blocks that are planted with the seedling (Fig. 8–22).

Control Methods to Maximize Transplant Production

Efficient indoor transplant production can be a very sophisticated operation with a substantial monetary investment in greenhouse facilities. In many cases, profit is determined by producing a high density of seedlings in as short a time as possible. Factors to consider include:

1. Germination facilities
2. Substrate
3. Mechanical seed sowing

Table 8–1
REQUIREMENTS FOR SEED GERMINATION DURING PLUG PROPAGATION OF THREE POPULAR BEDDING PLANTS

	Petunia	Pansy	Impatiens
Stage 1			
Temperature	75–78°F (24–26°C)	62–68°F (17–20°C)	75–80°F (21–27°C)
Moisture	100% RH	100% RH	100% RH
Light	90 µmol · sec^{-1} · m^{-2}	80 µmol · sec^{-1} · m^{-2}	90 µmol · sec^{-1} · m^{-2}
Fertilizer	25–75 ppm KNO$_3$ 1 application (1–3 days)	25–50 ppm KNO$_3$ (1–7 days)	None
Stage 2			
Temperature	75°F (24°C)	66°F (18°C)	72–75°F (22–24°C)
Moisture	85% RH	75% RH	75% RH
Light	90 µmol · sec^{-1} · m^{-2}	80 µmol · sec^{-1} · m^{-2}	90 µmol · sec^{-1} · m^{-2}
Fertilizer	50 ppm 20–10–20 (3–7 days)	None	None

Source: Ball, 1998.

(a)

(b)

(c)

(d)

Figure 8–21
An alternative to field production in conifers is container production. (a) Conifer production in an open roof greenhouse. (b) Conifer plug production on movable benches. (c and d) Pine seedlings in deep, narrow containers held in trays that permit air circulation beneath the container for air pruning.

4. Watering systems
5. Temperature control
6. Seedling growth
7. Transplanting

Germination Facilities. Indoor seedling production occurs in several types of structures including greenhouses, **cold frames,** and hotbeds, as described in Chapter 3. Some bedding plant operations have special **germination growth rooms** (Fig. 8–23, page 270) where seed flats are placed on carts or shelves in an enclosed area and subjected to controlled environments for germination prior to being moved to the greenhouse (68).

cold frame Structure that uses passive solar heating to protect transplants. Cold frames are often used as a transition environment for transplants between greenhouse and field conditions to "harden" transplants and reduce transplant shock.

growth rooms Structure used by large bedding-plant producers that control germination conditions to optimize seedling emergence in plug flats.

Growth rooms need controlled lighting (daylength and irradiance), temperature, and relative humidity (67). Flats are irrigated prior to moving to the growth room, and the high humidity (at least 95 percent) keeps the substrate moisture optimal for germination. Flats remain in the growth room until the end of

(a)

(b)

(c)

Figure 8–22
Conifer production in alternative substrates. (a and b) Spruce seedlings in expanded peat and (c) foam-like peat product.

Figure 8–23
Growth rooms designed for seed germination in plug trays. (a) Plug trays are brought into the germination room on movable racks. These rooms are well insulated for temperature control and fitted with fluorescent lamps. (b) The germination substrate stays moist because the rooms are held at greater than 95 percent humidity with high-pressure fog.

plug growth stage 1—radical emergence—and are then moved to the greenhouse.

Substrates (Media). Substrates and fertilizer used for seedling propagation are discussed in detail in Chapter 3. Germination substrate for herbaceous bedding plants must retain moisture, supply nutrients, permit gas exchange, and provide support for the seedling (22). Common mixes are combinations of peat moss, perlite, ground or shredded bark, coconut coir, and vermiculite, and they may be fortified by mineral nutrients or slow-release fertilizers. These mixes are available commercially, but may be made on-site for custom blends. Air and water content should be maintained for good germination and seedling growth (49). Small seeds should have a finer and more compact medium than is used for larger seeds. Plug flats are usually filled with substrate mechanically.

Mechanical Seed Sowing. Seeds may be broadcast over the surface of the transplant flats or planted in rows (Fig. 8–19, page 267). Advantages of row planting are reduced damping-off, better aeration, easier transplanting, and less drying out. Planting at too high a density encourages damping-off, makes transplanting more difficult, and produces weaker, non-uniform seedlings. Suggested rates are 1,000 to 1,200 seeds per 29 × 54 cm (11 × 22 in) flat for small-seeded species (e.g., petunia) and 750 to 1,000 for larger seeds. Small seeds are dusted on the surface; medium seeds are covered lightly to about the diameter of the seed. Larger seeds may be planted at a depth of two to three times their minimum diameter.

mechanical seed sowing
A method required for large-scale production of vegetable and flower seeds for greenhouse production because seed size is small and large quantities of seeds must be precisely sown in each small plug in the flat.

Efficient plug production requires the use of a mechanical seeder (3). The objective of plug production is to get a usable seedling in each cell. The choice of seeder depends on several factors including cost, seeding speed, number of flats to be seeded, and the need for flexibility to sow a variety of seed shapes and sizes. When evaluating a seeder, growers must consider the machine's ability to deliver seeds at the desired speed without skipping cells due to poor seed pickup or delivery, sowing multiple seeds per cell, and sowing seeds without seed "bounce" that can reduce the precise location of the seed in each plug cell.

Three types of seeders are commonly available to plug growers. These are **template, needle,** and **cylinder (drum)** seeders (Figs. 8–24 and 8–25, page 272).

The **template seeder** (Fig. 8–24a) is the least expensive type of seeder. It uses a template with holes that match the location of cells in the plug flat. Template seeders use a vacuum to attach seeds to the template. Releasing the vacuum drops the seeds either directly into the plug flat or into a drop tube to precisely locate seeds in each cell of the plug flat. Templates with different size holes are available to handle different size and shape seeds. A differently sized template is also required for each plug flat size. It is a relatively fast seeder because it sows an entire flat at once. However, this is the least mechanized of the commercially available seeders. It requires the operator to fill the template with seeds, remove the excess, and then move the template to the flat

(a)

(b)

(c)

(d)

Figure 8–24
Mechanical seeders for plug production. (a) Template seeders sow an entire flat in one operation. (b, c, and d) Needle seeders use a vacuum to remove a line of seeds from a tray and drop them into the flat. Seeds may be placed directly onto each cell or drop tubes (d) may help place the seeds.

for sowing. Template seeders work best for round, semi-round, or pelleted seeds.

The **needle seeder** (Fig. 8–24b, c, and d) is an efficient and moderately priced seeder. It is fully mechanical, requiring little input from the operator. Individual needles or pickup tips, under vacuum pressure, lift single seeds from a seed tray and deposit one seed directly in each plug cell or into drop tubes for more accurate seeding. A burst of air can be used to deposit seeds and clean tips of unwanted debris. The needle seeder can sow a variety of seed sizes and shapes including odd-shaped seeds like marigold, dahlia, and zinnia. Although slower than the cylinder seeder, it is still relatively fast, sowing up to 100,000 seeds per hour. Small- and moderate-sized plug growers choose needle seeders because of the flexibility in seeding and cost.

The **cylinder** or **drum seeders** (Fig. 8–25, page 272) have a rotating cylinder or drum that picks up seeds using vacuum from a seed tray and drops one seed per plug cell. This is the fastest, most precise, and most costly of the commercial seeders. It is fully mechanical. Most drum seeders require a different drum for each plug flat, but newer models of cylinder seeders have several hole sizes per cylinder that can be selectively put under vacuum pressure and can be computer-adjusted for different flat types. These can sow single or multiple seeds per cell at a time. Sophisticated seeders "eject" seeds from the drum using an air or water stream for precise seeding location in the flat.

These seeders work best with round, semi-round, or pelleted seeds. Large plug growers must have the capacity to sow millions of plugs per year of over 100 different types of bedding plants (66). They choose cylinder seeders because they sow a high volume of seeds quickly—up to 800,000 seeds per hour.

Watering Systems. The moisture content of the growing medium can be critical to germination success (5, 12). Species like *coleus, begonia,* and *alyssum* require a wet medium (saturated); *impatiens, petunia,* and *pansy* require a moist medium (wet but not saturated); while *asters, verbena,* and *zinnia* prefer a drier medium (watered only prior to sowing) for good germination (67).

For smaller growers, seed flats may be held under polyethylene tents (see Chapter 3) or, in small operations, covered with spun fabric or vermiculite to keep the surface from drying out (Fig. 8–26a and b, page 273). Covered flats should not be exposed directly to sunlight, as excessive heat buildup injures the seedling.

Several systems for delivering water to seed flats are available including **automated watering systems** (Fig. 8–26). These include overhead and subirrigation systems (42). Overhead irrigation can be as simple as a hose with a fine-holed "rose"

automated watering systems A system that reduces labor costs and can provide more even moisture to plug trays.

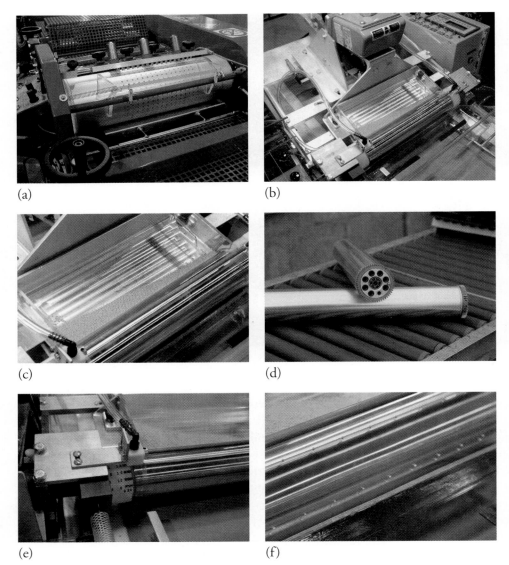

Figure 8–25
The fastest seed-sowing machines are cylinder and drum seeders. (a) Drum seeders have a large rotating drum that uses a vacuum to pick up a line of seeds and deposit them onto the plug tray when that line rotates into position above the tray. (b, c, d, e, and f) Cylinder seeders are becoming more common than drum seeders because they offer more flexibility. (b and c) High end cylinder seeders have computer controls for precise adjustment of seed placement depending on the plug tray size. (d) A cylinder removed from the machine to show the vacuum tubes. (e) One advantage of the cylinder seeder is the ability to make changes to the sowing rate and placement without removing the cylinder. (f) A seeder set to sow two seeds per cell.

irrigation nozzle or a timed mist system. Automated boom sprayers provide fine control of overhead irrigation. The boom travels the length of the greenhouse, providing a spray of water to the flats. The speed of the boom and irrigation timing can be computer-controlled.

Subirrigation systems have the advantage of providing even moisture while reducing water runoff. **Capillary mat systems** (Fig. 8–26e) deliver water from a reservoir to the mat where the growing medium "pulls" water into the flat or plug cell by capillary action. **Ebb and flood** systems use a sealed bench that is flooded periodically, and then the nutrient solution drains passively back into a holding tank (58). A variation on these systems is **"float bed"** production (Fig. 8–26f), in which a Styrofoam flat is floated in a water bed containing a nutrient solution (46). Regardless of the system used, water quality must be monitored during production (42).

In most cases, seeds are sown on the surface of the medium by mechanical seeders. Seeds can be covered with vermiculite or porous fabric or plastic sheets to maintain even moisture until seedlings emerge. In modern palletized greenhouse operations, germination occurs in specialized germination rooms or in greenhouse sections designed to optimize germination conditions, then the entire movable bench is transferred on special rails to additional greenhouse sections designed for seedling growth.

Temperature Control. Temperature requirements for germination vary depending on the plant species being grown. In general, most bedding plants can be germinated in one of three temperature regimes: 26 to 30°C (78 to 80°F), 21 to 22°C (70 to 72°F), or 18 to 19°C (64 to 66°F). Non-optimal temperatures can lead to erratic or poor germination and emergence. In many cases, **bottom**

Figure 8–26
Methods to maintain even moisture for seed germination. (a) A spun-woven fabric covering seed flats. (b) A top-coating of vermiculite is commonly applied to plug-seeded flats. (c) A solid set irrigation system with irrigation nozzles on PVC risers. (d) A traveling boom system moves down the greenhouse and can selectively irrigate seed flats. (e) A capillary mat system subirrigates the flats for very even moisture control. (f) Float bed production of transplants. The float bed is lined with a plastic pool liner and the seedlings float on the nutrient solution in Styrofoam trays.

bottom heat Heat provided by recirculating hot-water systems is common for spring-seeded crops.

heat is used to warm the germination medium to the appropriate temperature. However, high temperature during germination may also lead to thermoihibition or thermodormancy in some crops. See Chapter 21 for specific germination temperatures for different plant species.

Seedling Growth. The principal objective of seedling production is to develop healthy, stocky, vigorous plants capable of further transplanting with little check in growth. The usual procedure in production is to move the flats to lower temperatures (10°C or less) compared to germination temperatures, and expose them to full sunlight. High temperatures and low light tend to produce spindly, elongated plants that will not survive transplanting. Such growth is termed "stretching."

Height control for quality plant production may require the use of growth regulators (10) or strict environmental control like water management (70) or temperature differentials (50). For example, plant height can be reduced by growing plants with a cooler day time temperature compared to the night time temperature. This is referred to as a negative DIF.

Once root systems grow into the medium, irrigation can be scheduled to keep the medium somewhat dry on the surface but moist underneath. Such irrigation helps prevent disease and produces sturdy seedlings. Fertilization should provide a good root-to-shoot ratio in the plug without excessive shoot growth, because a good root system is as important as above-ground shoot growth for plugs. Poor root systems will negatively impact the vigor of the transplant and hamper mechanical transplanting, which relies on a firm plug for the robot's "fingers" to lift. Plugs are hardened or toned in stage 4 by reducing the frequency of irrigation and fertilization in preparation for transplanting (11).

Transplanting. For seedlings grown in community flats, transplanting should begin when the first true leaves have fully expanded. Holes are made in the medium at the correct spacing with a small dibble. The roots of each small seedling are inserted into a hole, and the medium is pressed around them to provide good contact. Dibble boards are often used to punch holes for an entire flat at once (Fig. 8–19, page 267). As soon as the flat is filled, it is thoroughly watered.

Plug-grown seedlings are transplanted at stage 4 (see page 268). These can be transplanted by hand, but as seasonal labor has become relatively more costly and difficult to acquire, even smaller bedding plant growers are increasingly using mechanical transplanters in place of hand labor to transplant plugs. These transplanters lift or push seedlings from the plug flat into six or four packs for growing on before sale. To be efficient, every cell in the plug flat must have a usable seedling, otherwise there will be skips in the transplant containers. Plug growers must backfill flats with missing plants. This may be done by hand or with machine vision robots (Fig. 8–27).

Mechanical transplanters may be as simple as a mechanical press that pushes seedlings into dibbled cell packs or as sophisticated as robots that lift tightly spaced plugs and expand to transplant them to larger spaced cell packs (Fig. 8–28). These machines are a substantial investment for the grower, but there is often a long-term cost saving due to the increase in transplanting rate and the reduction in the temporary labor force required for spring transplanting.

Transplanting Seedling Material to Permanent Locations

The final step in seedling production is transplanting to a permanent location (59). Seedlings may be transplanted bare-root (vegetable transplants or deciduous fruit, nut, and shade trees), in cells or modular containers (bedding plants, vegetables, forest trees), balled and burlapped (evergreen trees), or containerized (ornamental shrubs and trees).

Bare-root transplanting invariably results in some root damage and **transplant shock,** both of which check growth. Some transplant shock can be observed even in container transplants. With vegetable plants these may result in premature seed-stalk formation, increased susceptibility to disease, and reduced yield potential. Handling prior to transplanting should involve hardening-off, achieved by temporarily

> **transplant shock**
> A check in plant growth that is observed following transplanting, compared to direct-seeded crops. It is more severe if the transplants are too large and have confined root systems in the plug tray, or if the weather is unfavorable in the field.

(a)

(b)

(c)

(d)

Figure 8–27
It is essential to fill as many cells as possible in a plug tray with usable seedlings. (a) A worker backfilling a plug flat that had poor or non-uniform seedling production. (b) A machine vision robot that automates the backfilling process. (c) Flats enter the machine and a computer determines cells to be removed and refilled. (d) The robot literally blows out the cell with an air jet, making it easy to mechanically refill with a usable seedling.

(a) (b) (c) (d)

Figure 8–28
Transplanting plugs into larger containers. (a) Hand transplanting. (b) A semi-automatic machine that pushes the plugs out the bottom of specially designed cell flats into the larger container below the punch. (c and d) A transplanting robot lifts the seedling out of the plug flat with mechanical fingers, then expands along a track to place the seedlings into larger containers at a wider spacing.

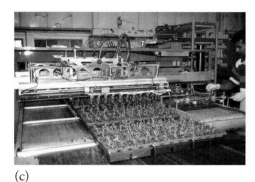

Figure 8–29
Bedding plants being hardened-off prior to shipment and sale. These flats are rolled out on to pipe benches, but could easily be moved back under greenhouse protection if the weather becomes severe.

withholding moisture, reducing temperature, and gradually shifting from protected to outdoor conditions over a period of 1 week to 10 days (Fig. 8–29). Hardening-off can cause carbohydrates to accumulate, making the plant better able to withstand adverse environmental conditions.

Ornamental and Vegetable Bedding Plants During the transition to the new site, deterioration must be prevented if the plants are bare-root. Following planting, conditions must be provided for rapid root regeneration. Planting should be done as soon as possible. If not, transplants can be kept (no more than 7 to 10 days) in moist, cool (10°C, 50°F) storage. Longer-term (several weeks) plug storage is possible (33) by maintaining high humidity but avoiding direct watering in order to prevent disease.

Field beds should be moderately well pulverized, although not necessarily finely prepared, and well watered but not saturated (59). Transplanting is done in the field by hand or by machine. Afterward, a good amount of irrigation should be applied to increase moisture to the roots and settle the soil, but not saturate it. A starter solution containing fertilizers that are high in phosphorus can be applied, but if the soil is dry, it should be diluted. Temporary shade may be used for the first few days.

Trees and Shrubs Transplanting of bare-root evergreen forest trees follows principles similar to those described. Seedling plants should be dug in the nursery in the fall after proper physiological "hardening-off." Seedlings are packed into moisture-retaining material (vermiculite, peat moss, sawdust, shingletoe) and kept in low-temperature (2°C, 35°F), humid (at least 90 percent RH) storage. Polyethylene bags without moisture-holding material are satisfactory. Some kinds of sawdust can be toxic, particularly if fresh. Bare-root nursery stock of deciduous plants and container-grown stock are handled as described for rooted cuttings in Chapter 10.

DISCUSSION ITEMS

This chapter brings together all the concepts presented in previous seed chapters into production techniques for horticultural and forestry crops, including treatments to overcome dormancy, use of pretreated seeds, and optimizing seedling emergence. The major production schemes to propagate nursery, vegetable, and flower crops from seed are included in this chapter:

1. Compare field, greenhouse flat, and greenhouse plug systems for transplant production.
2. Compare plug production with float-bed production for transplants.
3. Compare mechanical seeders used for field vs. greenhouse sowing.

REFERENCES

1. Armitage, A. M. 1994. Ornamental bedding plants. In *Ornamental production science in horticulture 2.* Wallingford, UK: CAB International.

2. Ball, V. 1998. *Ball red book,* 16th ed. Batavia, IL: Ball Pub.

3. Bartok, J., Jr. 1994. Facilities planning and mechanization. In J. Holcomb, ed. *Bedding plants IV.* Batavia, IL: Ball Pub. pp. 233–44.

4. Baskin, C. C., and J. M. Baskin. 1998. *Seeds: Ecology, biogeography, and evolution of dormancy and germination.* San Diego, CA: Academic Press.

5. Biernbaum, J. A., and N. B. Versluys. 1998. Water management. *HortTechnology* 8:504–9.

6. Thomas, B. M. 1993. Overview of the speedling, incorporated, transplant industry operation. *HortTechnology* 3:406–8.

7. Bonner, F. T., B. F. McLemore, and J. P. Barnett. 1974. Presowing treatment of seed to speed germination. *Seeds of woody plants in the United States. Agric. Handbook No. 450,* Washington, DC: U.S. Forest Service. pp. 126–35.

8. Boyle, T. H. 2003. Influence of seed germination percentage and number of seeds sown per cell on expected numbers of seedlings in plug trays. *HortTechnology* 13:689–92.

9. Bracy, R. P., and R. L. Parish. 1998. Seeding uniformity of precision seeders. *HortTechnology* 8:182–85.

10. Britten, A. 2000. PGRs at seeding reduce early stretch. In J. VanderVelde, ed. *GrowerTalks on plugs 3,* Batavia, IL: Ball Pub. pp. 44–6.

11. Cantliffe, D. J. 1993. Pre- and postharvest practices for improved vegetable transplant quality. *HortTechnology* 3:415–18.

12. Carpenter, W. J., and S. Maekawa. 1991. Substrate moisture level governs the germination of verbena seed. *HortScience* 26:786–88.

13. Cooley, J. 1985. Vegetable plant raising using Speedling transplants. *Comb. Proc. Intl. Plant Prop. Soc.* 35:468–71.

14. Copeland, L. O., and M. B. McDonald. 2001. *Principles of seed science and technology,* 4th ed. New York: Chapman and Hall.

15. Corley, W. L. 1991. Seedbed preparation alternatives for establishment of wildflower meadows and beauty spots. *Southern Nurserymen's Assoc. Res. Conf.* 36:278–79.

16. Corley, W. L., and J. E. Dean. 1991. Establishment and maintenance of wildflowers on erodible sites. *Southern Nurserymen's Assoc. Res. Conf.* 36:280–81.

17. Deer, H. J., and W. F. Mann, Jr. 1971. Direct seeding pines in the South. *U.S. Dept. Agr. Handbook 391.* Washington, DC: U.S. Govt. Printing Office.

18. Delong, S. K. 1985. Custom seed preparation for optimum conifer production. *Comb. Proc. Intl. Plant Prop. Soc.* 35:259–63.

19. Dirr, M. A., and C. W. Heuser, Jr. 2009. *The reference manual of woody plant propagation: From seed to tissue culture,* 2nd ed. Athens, GA: Varsity Press.

20. Edwards, D. G. W. 1986. Special prechilling techniques for tree seeds. *J. Seed Tech.* 10:151–71.

21. Finch-Savage, W. E., M. Rayment, and F. R. Brown. 1991. The combined effects of a newly designed dibber drill, irrigation and seed covering treatments on lettuce and calabrese establishment. *Ann. Applied Biol.* 118:453–60.

22. Fonteno, W. C. 1994. Growing media. In J. Holcomb, ed. *Bedding plants IV.* Batavia, IL: Ball Pub. pp. 127–38.

23. Fordham, D. 1976. Production of plants from seed. *Comb. Proc. Intl. Plant Prop. Soc.* 26:139–45.

24. Frantz, J. M., and G. E. Welbaum. 1995. A comparison of four cabbage transplant production systems. In *Proceedings of the fourth national symposium on stand establishment.* Department of Vegetable Crops, University of California, Davis. pp. 169–74.

25. Frolich, E. F. 1971. The use of screen bottom flats for seedling production. *Comb. Proc. Intl. Plant Prop. Soc.* 21:79–80.

26. Geneve, R. L. 2003. Impact of temperature on seed dormancy. *HortScience* 38:336–41.

27. Gray, D., and J. Reed. 1995. Use of a dibber drill and coulter drill with press wheel to improve seedling emergence in onion and lettuce. In *Proceedings of the fourth national symposium on stand establishment.* Department of Vegetable Crops, University of California, Davis. pp. 125–32.

28. Hall, T. 1975. Propagation of walnuts, almonds and pistachios in California. *Comb. Proc. Intl. Plant Prop. Soc.* 25:53–7.

29. Hamrick, D. 2005. Ornamental bedding plant industry and plug production. In M. B. McDonald and F. Y. Kwong, eds. *Flower seeds: Biology and technology.* Wallingford, UK: CABI. pp. 27–38.

30. Harris, R. W., W. B. Davis, N. W. Stice, and D. Long. 1971. Root pruning improves nursery tree quality. *J. Amer. Soc. Hort. Sci.* 96:105–9.

31. Harris, R. W., W. B. Davis, N. W. Stice, and D. Long. 1971. Influence of transplanting time in nursery production. *J. Amer. Soc. Hort. Sci.* 96:109–10.

32. Healy, W. 2000. Fast cropping works. In J. VanderVelde, ed. *GrowerTalks on plugs 3.* Batavia, IL: Ball Pub. pp. 49–55.

33. Heins, R., N. Lange, T. F. Wallace, Jr., and W. Carlson. 1994. Plug storage. *Greenhouse Grower.* Willoughby, OH: Meister Pub.

34. Heit, C. E. 1964. The importance of quality, germinative characteristics and source for successful seed propagation and plant production. *Comb. Proc. Intl. Plant Prop. Soc.* 14:74–85.

35. Heit, C. E. 1967. Propagation from seed. 5. Control of seedling density. *Amer. Nurs.* 125(8):14–15, 56–59.

36. Heit, C. E. 1967. Propagation from seed. 6. Hardseededness, a critical factor. *Amer. Nurs.* 125(10):10–2, 88–96.

37. Hendrick, U. P. 1933. *A history of agriculture in the state of New York.* New York: Hill and Wang.

38. Holcomb, E. J. 1995. *Bedding plants IV.* Batavia, IL: Ball Pub.

39. Hoyle, B. J., H. Yamada, and T. D. Hoyle. 1972. Aggresizing—to eliminate objectionable soil clods. *Calif. Agr.* 26(11):3–5.

40. Inman, J. W. 1967. Precision planting—a reality for vegetables. *Paper No. PC–67–12. Amer. Soc. Ag. Eng.* Paper No. PC-67-12.

41. Inman, J. W. 1995. New developments in planting and transplanting equipment. In *Proceedings of the fourth national symposium on stand establishment.* Department of Vegetable Crops, University of California, Davis. pp. 19–22.

42. Langhans, R. W., and E. T. Paparozzi. 1994. Irrigation. In J. Holcomb, ed. *Bedding plants IV.* Batavia, IL: Ball Pub. pp. 139–50.

43. Lawyer, E. M. 1978. Seed germination of stone fruits. *Comb. Proc. Intl. Plant Prop. Soc.* 28:106–9.

44. MacDonald, B. 1986. *Practical woody plant propagation for nursery growers,* Vol. 1. Portland, OR: Timber Press.

45. Maclean, N. M. 1968. Propagation of trees by tube technique. *Comb. Proc. Intl. Plant Prop. Soc.* 18:303–9.

46. Maglianti, C. G. 1987. Speedling float growing tobacco transplants on water. *Amer. Soc. Ag. Eng.* Fiche no. 87–1573, 3p.

47. McMillan-Browse, P. D. A. 1978. Scarification—a detail of technique. *Comb. Proc. Intl. Plant Prop. Soc.* 28:191–92.

48. Menzies, M. I., and J. T. Arnott. 1992. Comparisons of different plant production methods for forest trees. In K. Kurata and T. Kozai, eds. *Transplant production systems.* Dordrecht, Kluwer Acad. Pub. pp. 21–44.

49. Milks, R. R., W. C. Fonteno, and R. A. Larson. 1989. Hydrology of horticultural substrates: III. Predicting air and water content in limited-volume plug cells. *J. Amer. Soc. Hort. Sci.* 114:57–61.

50. Moe, R., K. Willumsen, I. H. Ihlebekk, A. I. Stup, N. M. Glomsrud, and L. M. Mortensen. 1995. DIF and temperature DROP responses in SDP and LDP, a comparison. *Acta Hort.* 378:27–33.

51. Morrison, D. A., K. McClay, C. Porter, and S. Rish. 1998. The role of the lens in controlling heat-induced breakdown of testa-imposed dormancy in native Australian legumes. *Ann. Bot.* 82:35–40.

52. Morse, R. D. 1999. No-till vegetable production—its time is now. *HortTechnology* 9:373–79.

53. Orzolek, M. D., and D. R. Daum. 1984. Effect of planting equipment and techniques on seed germination and emergence: A review. *J. Seed Tech.* 9:99–113.

54. Ozmerzi, A., D. Karayel, and M. Topakci. 2002. Effect of sowing depth on precision seeder uniformity. *Biosystems Eng.* 82:227–30.

55. Page, F. R., and M. J. Quick. 1979. A comparison of the effectiveness of organic polymers as soil anti-crusting agents. *J. Sci. Food Agric.* 30:112–18.

56. Parish, R. L., P. E. Bergeron, and R. P. Bracy. 1991. Comparison of vacuum and belt seeders for vegetable planting. *Applied Eng. Agric.* 7:537–40.

57. Pinney, T. S., Jr. 1986. Update of GROPLUG® system. *Comb. Proc. Intl. Plant Prop. Soc.* 36:577–81.

58. Poole, R. T., and C. A. Conover. 1992. Fertilizer levels and medium affect foliage plant growth in an ebb and flow irrigation system. *J. Environ. Hort.* 10:81–86.

59. Price, H. C., and B. H. Zandstra. 1988. Maximize transplant performance. *Amer. Veg. Grower* 36(4):10–6.

60. Rom, R. C., and R. F. Carlson, eds. 1987. *Rootstocks for fruit crops.* New York: John Wiley & Sons.

61. Royle, S. M., and T. M. Hegarty. 1978. Soil impedance and its effect on calabrese emergence. *Acta Hort.* 72:259–66.

62. Schopmeyer, C. S., ed. 1974. Seeds of woody plants in the United States. *U.S. For. Ser. Agr. Handbook 450.* Washington, DC: U.S. Govt. Printing Office.

63. Scoggins, H. L., P. V. Nelson, and D. A. Bailey. 2000. Development of the press extraction method for plug substrate analysis: Effects of variable extraction force on pH, electrical conductivity, and nutrient analysis. *HortTechnology* 10:367–69.

64. Singh, D. P., M. S. Hooda, and F. T. Bonner. 1991. An evaluation of scarification methods for seeds of two leguminous trees. *New For.* 5:135–49.

65. Steavenson, H. 1979. Maximizing seedling growth under midwest conditions. *Comb. Proc. Intl. Plant Prop. Soc.* 29:66–71.

66. Stelk, B. 1993. Seed sowing success starts with the right equipment. *GrowerTalks* 67:33–7.

67. Styer, R. C. 2000a. Improve your germination. In J. VanderVelde, ed. *GrowerTalks on plugs 3*, Batavia, IL: Ball Pub. pp. 56–60.

68. Styer, R. C. 2000b. The ideal germination chamber. In J. VanderVelde, ed. *GrowerTalks on plugs 3*. Batavia, IL: Ball Pub. pp. 60–6.

69. Styer, R. C., and D. S. Koranski. 1997. *Plug and transplant production: A grower's guide.* Batavia, IL: Ball Pub.

70. Styer, R. C., and D. Koranski. 2000a. Controlling the root-to-shoot ratio. In J. VanderVelde, ed. *GrowerTalks on plugs 3*. Batavia, IL: Ball Pub. pp. 74–86.

71. Styer, R. C., and D. Koranski. 2000b. Key tips for bench-top germination. In J. VanderVelde, ed. *GrowerTalks on plugs 3*. Batavia, IL: Ball Pub. pp. 86–8.

72. Suszka, B. 1978. Germination of tree seed stored in a partially after-ripened condition. *Acta Hort.* 83:181–8.

73. Tran, V. N. 1979. Effects of microwave energy on the strophiole, seed coat and germination of *Acacia* seeds. *Aust. J. Plant Physiol.* 6:277–87.

74. Van Iersel, M. W., R. B. Beverly, P. A. Thomas, J. G. Latimer, and H. A. Mills. 1998. Fertilizer effects on the growth of impatiens, petunia, salvia, and vinca plug seedlings. *HortScience* 33:678–82.

75. VanderVelde, J., ed. 2000. *GrowerTalks on plugs 3*. Batavia, IL: Ball Pub.

76. Wilkins, D. E., F. Bolton, and K. Saxton. 1992. Evaluating seeders for conservation tillage production of peas. *Applied Eng. Agric.* 8:165–70.

part three
Vegetative Propagation

CHAPTER 9 **Principles of Propagation by Cuttings**
CHAPTER 10 **Techniques of Propagation by Cuttings**
CHAPTER 11 **Principles of Grafting and Budding**
CHAPTER 12 **Techniques of Grafting**
CHAPTER 13 **Techniques of Budding**
CHAPTER 14 **Layering and Its Natural Modifications**
CHAPTER 15 **Propagation by Specialized Stems and Roots**
CHAPTER 16 **Principles and Practices of Clonal Selection**

The chapters in Part III deal with vegetative procedures to propagate clones. The importance of clones as a category of cultivar cannot be overestimated. Essentially, all fruit and nut tree cultivars, as well as many rootstocks, are clones. Many major floriculture crops (e.g., roses, chrysanthemum, and carnation) are clones. Some important vegetable crops are clones, including potatoes and sweet potato. Some major plantation crops (i.e., sugar cane, banana, and pineapple) are also clones. Although forest species have traditionally been produced as seedlings, bamboo, poplar, and willow have been propagated vegetatively for centuries. Methods of vegetative propagation discussed include cutting, grafting, budding, layering, separation, and division.

9
Principles of Propagation by Cuttings

learning objectives

The first section of this chapter explores the biological approaches utilized to understand the regenerative process of adventitious root and bud (and shoot) formation. After reading the first section, you should be able to:

- Describe the observations made of adventitious root and bud (and shoot) formation.
- Explain how hormonal control affects root and bud (and shoot) formation.
- Explain the biochemical basis for adventitious root formation.
- Discuss the biotechnological advances in asexual propagation.

The second section of the chapter deals with the management and manipulation of adventitious root and bud (and shoot) formation. After reading the second section, you should be able to:

- Discuss the management of stock plants to maximize cutting propagation.
- Describe the factors involved in the treatment of cuttings.
- List the environmental conditions necessary in the manipulation of cuttings.

INTRODUCTION

The main focus of this chapter is on **adventitious root formation**, since it is the primary regenerative process required in most cutting propagation. **Adventitious bud and shoot development,** events important in the regeneration of leaf and root cuttings, are also discussed. **Adventitious organs** include new roots and buds that are formed from cells and tissue of previously developed shoots and roots.

Cutting propagation is the most important means for **clonal regeneration** of many horticultural crops: ornamentals, fruits, nuts, and vegetables. Adventitious root formation is a prerequisite to successful cutting propagation. In forestry, cutting propagation has been around for hundreds of years. Vegetative propagation of forest planting stock through adventitious rooting is one of the most exciting emerging technologies in forestry. Yet, many economically important woody plants have a low genetic and physiological capacity for adventitious root formation, which limits their commercial production. Furthermore, rooting and acclimatization of tissue-culture–produced plants will need to be improved if biotechnology (manipulating genes for new flower color, disease resistance, fruit yield, etc.) is to be incorporated into the propagation and production of genetically transformed woody plant species. Labor costs contribute more than 50 percent of

adventitious roots Roots that arise on aerial plant parts, underground stems and old root parts.

adventitious buds (and shoots) Arise from any plant part other than terminal, lateral, or latent buds on stems. Adventitious buds form irregularly on older portions of a plant and not at the stem tips or in the leaf axils. Unlike dormant buds, adventitious buds do not have a bud trace all the way to the pith. An adventitious bud is an embryonic shoot.

adventitious organs Organs that rise from the dedifferentiation of parenchyma cells; when they originate from callus (also composed of parenchyma cells) their **organogenesis** is termed **indirect.**

cutting propagation The clonal multiplication of plants with propagules of stems, leaves, or roots.

clonal regeneration or reproduction The asexual reproduction of genetically uniform copies (**clones**) of plants using propagules such as stem, leaf, and root cuttings.

propagation costs, so there is considerable financial incentive to streamline propagation techniques and improve rooting success.

Commercial propagators have developed technologies that successfully manipulate environmental conditions to maximize rooting (i.e., *intermittent mist* and *fog systems,* temperature, and light manipulation). What has lagged behind is the knowledge of the biochemistry, genetic and molecular manipulation of rooting. While we know a lot about the biology and manipulation of cuttings, the fundamental events of what triggers adventitious root formation remain largely unknown. The new tools of biotechnology offer exciting opportunities to understand the molecular keys to rooting and to enable propagators to develop new cultivars that can be commercially rooted.

Figure 9–1
The ultimate in adventitious root production is shown on this screwpine (*Pandanus utilis*). Prop roots (arrow) arise from the shoots, grow into the soil, and support the tree.

DESCRIPTIVE OBSERVATIONS OF ADVENTITIOUS ROOT AND BUD (AND SHOOT) FORMATION

Propagation by **stem** and **leaf-bud cuttings (single-eye cuttings)** requires only that a new adventitious root system be formed, because a potential shoot system (a bud) is already present. **Root cuttings** and **leaf cuttings** must initiate both a new shoot system—from an adventitious bud—as well as new adventitious roots.

The formation of adventitious roots and buds is dependent on plant cells to **dedifferentiate** and develop into either a root or shoot system. The process of **dedifferentiation** is the capability of previously developed, differentiated cells to initiate cell divisions and form a new meristematic growing point. Since this characteristic is more pronounced in some cells and plant parts than in others, the propagator must do some manipulation to provide the proper conditions for plant regeneration. A sound understanding of the underlying biology of regeneration is very helpful in this regard.

dedifferentiation
The early stage of adventitious root or bud formation when differentiated cells are triggered to form new meristematic regions.

grow into the ground and support the tree (Fig. 9–1). Plants that are regenerated from rhizomes, bulbs, and other such structures also develop adventitious roots (see Chapter 15).

Adventitious roots are of two types:

- **preformed roots** (Figs. 9–2 and 9–3)
- **wound-induced roots** (Figs. 9–3 and 9–4)

preformed root initials and primordia
Develop naturally on stems while they are still attached to the parent plant and roots may or may not emerge prior to severing the stem piece.

Adventitious Root Formation

Adventitious roots form naturally on various plants. Corn, screwpine (*Pandanus utilis*), and other monocots develop "brace" roots, which arise from the intercalary regions at the base of internodes. Screwpine produces long, aerial, prop roots from their shoots that

Figure 9–2
Preformed aerial roots at node of *Ficus pumila*.

Adventitious Roots

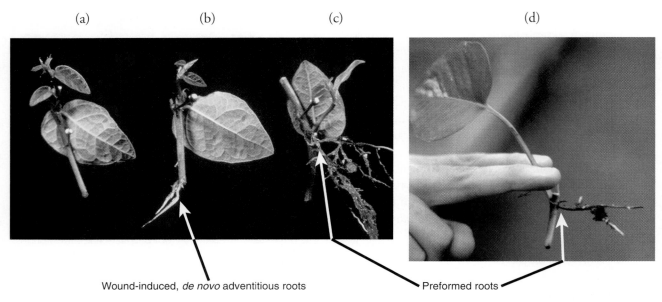

Figure 9–3
Leaf bud cuttings of *Ficus pumila* with (a) unrooted cutting, (b) wound induced, de novo and (c) preformed adventitious roots. (d) Philodendron cutting with preformed adventitious root from node.

Preformed or Latent Root Initials Preformed or latent root initials generally lie dormant until the stems are made into cuttings and placed under environmental conditions favorable for further development and emergence of the primordia as adventitious roots. In poplar (*Populus xrobusta*), root initials form in stems in midsummer and then emerge from cuttings made the following spring (257).

Figure 9–4
Emergence of adventitious roots in mung bean (*Vigna*) stem cuttings. Observe the tendency of the roots to form in longitudinal rows.

In some spices, primordia develop into aerial roots on the intact plant and become quite prominent (Figs. 9–1 and 9–2). Such preformed root initials occur in a number of easily rooted genera, such as willow (*Salix*), hydrangea (*Hydrangea*), poplar (*Populus*), coleus, jasmine (*Jasminum*), currant (*Ribes*), citron (*Citrus medica*), and others. The position of origin of these preformed root initials is similar to **de novo adventitious root formation** (Table 9–1) (185). In some of the clonal apple rootstocks and in old trees of some apple and quince cultivars, these preformed latent roots cause swellings, called **burr knots.** Species with preformed root initials generally root rapidly and easily, but cuttings of many species without such root initials root just as easily.

In willow, latent root primordia can remain dormant, embedded in the inner bark for years if the stems remain on the tree (2, 43). Their location can be observed by peeling off the bark and noting the protuberances on the wood, with

de novo adventitious roots Roots that are formed "anew" (from scratch) from stem or leaf cells that experience a stimulus, such as wounding, to dedifferentiate into roots.

burr knots Preformed roots that are not desirable and are selected against in modern apple rootstock breeding programs. Though rooting of cuttings is easier, clusters of burr knots can later girdle the stem.

Table 9–1
ORIGIN OF PREFORMED ROOT INITIALS (PRIMORDIA, BURR KNOTS, AND/OR ROOTGERMS) IN STEMS OF WOODY PLANTS

Origin	Genera
Rays	
Wide rays	*Populus*
Medullary rays, associated with buds	*Ribes*
Nodal and connected with wide radial bands of parenchyma	*Salix*
Internodal medullary rays	*Salix*
Medullary ray	*Citrus*
Phloem ray parenchyma	*Hydrangea*
Cambium	
Cambial ring in branch and leaf gap; 1 and 2° medullary rays	*Malus*
Cambial region of an abnormally broad ray	*Acer, Chamaecyparis, Fagus, Fraxinus, Juniperus, Populus, Salix, Taxus, Thuja, Ulmus*
Leaf and bud gaps	
Bud gap	*Cotoneaster*
Median and lateral leaf trace gaps at node	*Lonicera*
Parenchymatous cells in divided bud gap	*Cotoneaster*

Source: M. B. Jackson (154).

corresponding indentations on the inside of the bark that was removed.

Wound-Induced Roots On the other hand, wound-induced roots develop only after the cutting is made, in response to wounding in preparing the cutting. In effect, they are considered to be formed ***de novo*** (anew) (59, 154). Any time living cells at the cut surfaces are injured and exposed, a **response to wounding** begins (48).

Wounding Response. The subsequent wound response and root regeneration process includes three steps:

1. The outer injured cells die, a necrotic plate forms, the wound is sealed with a corky material (suberin), and the xylem may plug with gum. This plate protects the cut surfaces from desiccation and pathogens.
2. Living cells behind this plate begin to divide after a few days and a layer of **parenchyma cells** form *callus* which develops into a **wound periderm.**
3. Certain cells in the vicinity of the vascular cambium and phloem begin to divide and initiate *de novo* adventitious roots.

Stages of De Novo Adventitious Root Formation. The developmental changes that occur in *de novo* adventitious root formation of wounded roots can generally be divided into four stages:

Stage I: Dedifferentiation of specific differentiated cells.

Stage II: Formation of root initials from certain cells near vascular bundles, or vascular tissue, which have become meristematic by dedifferentiation.

Stage III: Subsequent development of root initials into organized **root primordia.**

Stage IV: Growth and **emergence of the root primordia** outward through other stem tissue plus the formation of vascular (conducting) tissue between the root primordia and the vascular tissues of the cutting.

While most scientists divide the process of adventitious root formation into four stages, rooting of Monterrey pine hypocotyl cuttings are divided (*Pinus radiata*) into three stages: preinitiative, initiative, and postinitiative with continuous division of derivatives to form **meristemoids** (255, 256).

Parenchyma cells
The basic cells from which all other differentiated cells and tissues are derived, including adventitious organs.

Wound periderm
A mass of callus cells that forms a protective layer behind the wounded surface of a cutting.

meristemoid A cell or group of cells constituting an active locus of meristematic activity in a tissue composed of somewhat older, differentiated cells; they can develop into root primordia or adventitious buds.

Table 9–2
TIME OF ADVENTITIOUS ROOT FORMATION IN JUVENILE AND MATURE LEAF-BUD CUTTINGS OF *FICUS PUMILA* TREATED WITH IBA

	Juvenile	Mature
Anticlinal cell divisions of ray parenchyma	Day 4	Day 6
Primordia	Day 6	Day 10
First rooting[a]	Day 7	Day 20
Maximum rooting[b]	Day 14	Day 28

[a]Based on 25 percent or more cuttings with roots protruding from stem.
[b]Based on 100 percent rooting and maximum root number.
Source: Davies et al. (59).

Time to Form Adventitious Roots The time for root initials to develop after cuttings are placed in the propagating bed varies widely. In one study (260), they were first observed microscopically after 3 days in chrysanthemum, 5 days in carnation (*Dianthus caryophyllus*), and 7 days in rose (*Rosa*). Visible roots emerged from the cuttings after 10 days for the chrysanthemum, but 3 weeks were required for the carnation and rose.

Phloem ray parenchyma cells in juvenile (easy-to-root) cuttings of creeping fig (*Ficus pumila*) undergo early anticlinal cell division and root primordia formation more quickly than mature (difficult-to-root) plants under optimal auxin treatments (Table 9–2). Once primordia are formed, there is a comparable time period (7 to 8 days) between root primordia elongation (emergence) and maximum rooting in both the easy-to-root and difficult-to-root plants (59). This delay was also reported with *Agathis australis*, where primordia formation was variable in cuttings from different-aged stock plants—but once root primordia formed, root emergence consistently occurred within a three-to-four-week period (185, 294, 295).

The Anatomical Origin of Wound-Induced Adventitious Roots The precise location inside the stem where adventitious roots originate has intrigued plant anatomists for centuries. Probably the first study of this phenomenon was made in 1758 by a French dendrologist, Duhamel du Monceau (72). A great many subsequent studies have covered a wide range of plant species (10, 185).

Adventitious roots usually originate on **herbaceous plants** just outside and between the vascular bundles (224), but the tissues involved at the site of origin can vary widely depending upon plant species and propagation technique (1). In tomato, pumpkin, and mung bean (22), adventitious roots arise in the phloem parenchyma; in *Crassula* they arise in the epidermis, while in coleus they originate from the pericycle (42). Root initials in carnation cuttings arise in a layer of parenchymatous cells inside a fiber sheath; the developing root tips, upon reaching this band of impenetrable fiber cells, do not push through it but turn downward, emerging from the base of the cutting (260).

Adventitious roots in stem cuttings of **woody perennial plants** usually originate from living parenchyma cells, in the young, secondary phloem (Figs. 9–6 and 9–7, page 286), but sometimes in vascular rays, cambium, phloem, callus, or lenticels (Table 9–3, page 286) (101, 126, 185).

Generally, the origin and development of *de novo* adventitious roots takes place next to and just outside the central core of vascular tissue. Many easy-to-root woody plant species develop adventitious roots from phloem ray parenchyma cells. Figure 9–7, page 286, depicts the first **anticlinal division** of a phloem ray cell during dedifferentiation (Stage I). Further cell divisions occur and the meristematic area becomes more organized with the formation of a root initial (Stage II) (Fig. 9–8, page 287). Ultimately a fully developed root primordia forms in the phloem and cortex (Fig. 9–9, page 287). Upon emergence from the stem (Fig. 9–10, page 287), the adventitious roots have already developed a root cap as well as a complete vascular connection with the originating stem.

> **anticlinal division**
> Cell division that occurs when the cell wall plate is formed perpendicular to the circumference of the stem.

The Relationship of Stem Structure and Rooting Ability There have been attempts to correlate stem structure with the rooting ability of cuttings. A continuous **sclerenchyma**

> **sclerenchyma ring**
> Composed of sclereid cells that are highly lignified and used for structural support of the stem. In some rare occasions these cells may impede the rooting process.

PRINCIPLES OF PROPAGATION BY CUTTINGS CHAPTER NINE 285

BOX 9.1 GETTING MORE IN DEPTH ON THE SUBJECT
DEVELOPMENTAL PHASES IN ADVENTITIOUS ROOT AND SHOOT FORMATION

Figure 9–5 depicts the developmental phases in the organogenesis of adventitious root and shoot formation. Cells in potential sites must become competent to respond to chemical/metabolic signals that trigger induction, which enables subsequent dedifferentiation and adventitious organ development. See page 283 for a discussion of developmental stages of wound-induced roots, page 299 for biochemical and page 303 for molecular implications on cell competency to root.

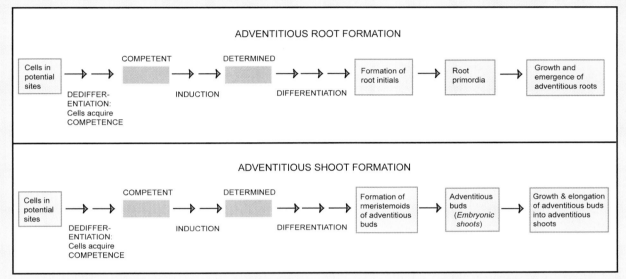

Figure 9–5
Developmental phases in the organogenesis of adventitious root and shoot formation. Modified from Christianson and Warnick (46); Davies et al. (57, 59).

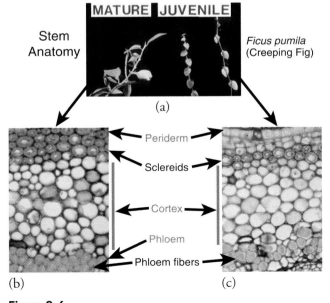

Figure 9–6
Shoot morphology (a) in juvenile and mature *Ficus pumila*. Cross section from (b) mature and (c) juvenile stems from the outside periderm to phloem fibers. Rarely are sclereids or phloem fibers a barrier that prevents adventitious rooting.

ring (Fig. 9–6) between the phloem and cortex, exterior to the point of origin of adventitious roots, occurs as the stem matures and gets older. Sclereids and fibers are impregnated with **lignin,** which provides structural support and mechanical barriers for pest resistance.

> **lignin** An abundant plant polymer in cell walls that provides structural support and mechanical barriers for pest resistance.

Sclereids occur in difficult-to-root species such as olive stem cuttings, mature English ivy (*Hedera helix*) (102), and creeping fig (*Ficus pumila*) (59), while easy-to-root types are characterized by discontinuity or fewer cell layers of this sclerenchyma ring (Fig. 9–6) (15).

Easily rooted carnation cultivars have a band of sclerenchyma present in the stems, yet the developing root primordia emerge from the cuttings by growing downward and out through the base (260). In other plants, in which an impenetrable ring of sclerenchyma could block root emergence, this same rooting pattern can occur. Rooting is related to the genetic potential

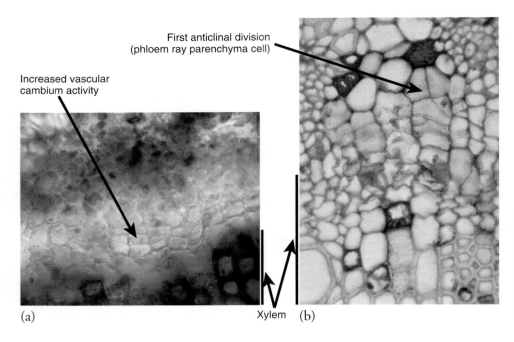

Figure 9–7
Early events of rooting with (a) increased vascular cambium activity and (b) first anticlinal division of phloem ray parenchyma cell during stage I—dedifferentiation in *Ficus pumila* (59).

Table 9–3
ORIGIN OF WOUND-INDUCED *DE NOVO* ADVENTITIOUS ROOTS IN STEMS OF WOODY PLANTS

Origin	Genera
Cambial and ray	
Cambial and phloem portions of ray tissues	*Acanthopanax, Chamaecyparis, Cryptomeria, Cunninghamia, Cupressus, Metasequoia*
Medullary rays	*Vitis*
Cambium	*Acanthus, Lonicera*
Fascicular cambium	*Clematis*
Phloem ray parenchyma	*Ficus, Hedera*
Secondary phloem in association with a ray	*Malus* (Malling stocks), *Camellia,* 'Brompton' plum
Phloem area close to the cambium	*Pistacia*
Cambium and inner phloem ray also in leaf gap	*Griselinia*
Bud and leaf gaps	
Outside the cambium in small groups	*Rosa, Cotoneaster, Pinus, Cephalotaxus, Larix, Sciadopitys, Malus, Acanthus*
Pericycle Callus, internal	
Irregularly arranged parenchymatous tissues	*Abies, Juniperus, Picea, Sequoia*
Callus, external	
Callus tissues (external)	*Abies, Cedrus, Cryptomeria, Ginkgo, Larix, Pinus, Podocarpus, Sequoia, Sciadopitys, Taxodium, Pinus*
Bark and basal callus	*Citrus*
Within callus at base of cutting	*Pseudotsuga*
Other	
Hyperhydric outgrowth of the lenticels	*Tamarix*
Margin of differentiating resin duct or parenchyma within the inner cortex	*Pinus*

Source: M. B. Jackson (154).

and physiological conditions for root initials to form, rather than to the mechanical restriction of a sclerenchyma ring barring root emergence (59, 245, 293).

Thus, two patterns of adventitious root formation emerge: **direct root formation** of cells in close proximity to the vascular system (i.e., generally more easy-to-root species); and **indirect root formation**, where nondirected cell divisions, including callus formation, occur for an interim period before cells divide in an organized pattern to initiate adventitious root primordia (i.e., generally more difficult-to-root species). See the flow diagram of adventitious root formation (Fig. 9–11, page 288) (98, 185).

Callus Formation: Rooting and Bud (and Shoot) Organogenesis

Root Organogenesis Callus is an irregular mass of parenchyma cells in various stages of

callus An irregular mass of parenchyma cells in various stages of lignification.

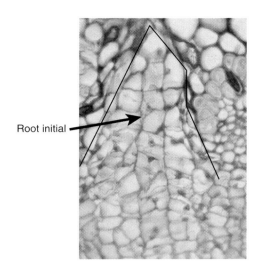

Figure 9–8
Root initial development in *Ficus pumila* with the meristematic zone in the phloem ray becoming more organized during stage II of adventitious root formation—root initial formation.

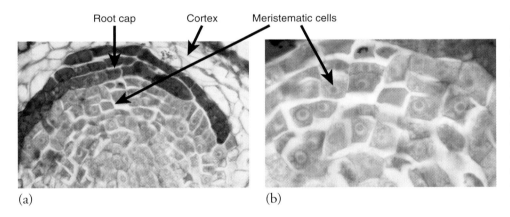

Figure 9–9
Development of a fully organized meristem during stage III of adventitious root formation—root primordia formation. (a) The root cap of the adventitious root has become organized, and (b) meristematic cells are characterized with isodymetric cell walls, deeply staining cytoplasm, and large nuclei in a *Ficus pumila* cutting.

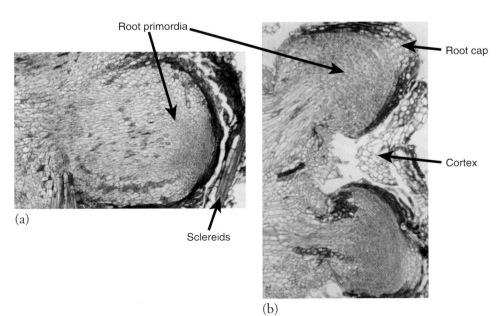

Figure 9–10
Elongation of root primordia during stage IV of adventitious root formation—root elongation. (a) Longitudinal section with root primordia elongating through the cortex, pushing out sclereids in the exterior of the cortex.
(b) Cross-section of two adventitious primordia elongating through the cortex and periderm in a *Ficus pumila* cutting.

BOX 9.2 GETTING MORE IN DEPTH ON THE SUBJECT
STEM STRUCTURE AND ROOTING

With most difficult-to-root species, stem structure does not influence rooting potential. While a sheath of lignified tissue in stems may in some cases act as a *mechanical barrier* to root emergence, there are so many exceptions that this **is not** the primary cause of rooting difficulty (Fig. 9–10). Moreover, auxin treatments and rooting under mist (15, 59) cause considerable cell expansion and proliferation in the cortex, phloem, and cambium, resulting in breaks in continuous sclerenchyma rings—yet in some difficult-to-root cultivars, even with wounding, there is still no formation of root initials.

lignification that commonly develops at the basal end of a cutting placed under environmental conditions favorable for rooting. Callus growth proliferates from cells at the base of the cutting, primarily from the vascular cambium, although cells of the cortex and pith may also contribute to its formation (Table 9–3).

Roots frequently emerge through the callus, leading to the belief that callus formation is essential

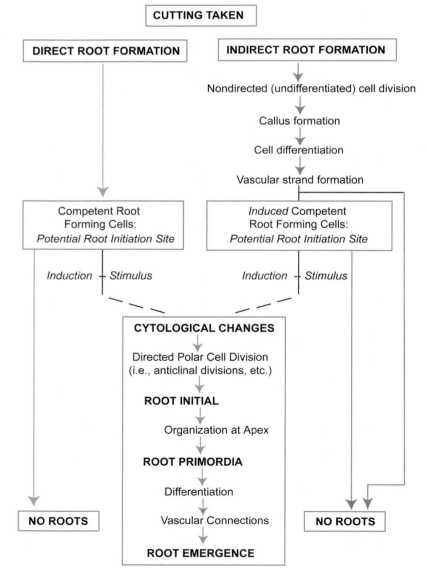

Figure 9–11

Flow diagram of adventitious root formation through direct (cells in close proximity to vascular system—i.e., generally more easy-to-root species) and indirect model (interim period of undifferentiated cell division—i.e., generally more difficult-to-root species). When a potential root initiation site is already present the initial cell divisions lead to root production *in situ*. When a site is not present, alternative routes leading to the creation of a site are shown. Rooting does not always occur. Modified from Lovell and White (185) and Geneve (98).

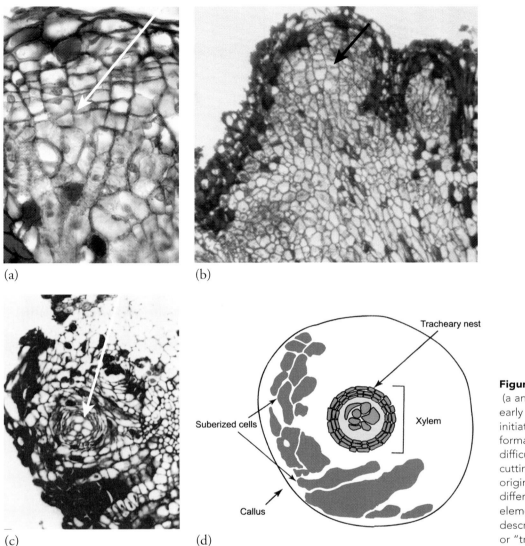

Figure 9–12
(a and b) Cell divisions in early *de novo* root primordia initiation from callus formation at base of mature, difficult-to-root *Ficus pumila* cutting. (c and d) Primordia originating in the vicinity of differentiating tracheary elements that have been described as "callus xylem" or "tracheary nests" (59).

for rooting. In easy-to-root species, the formation of callus and the formation of roots are independent of each other, even though both involve cell division. Their simultaneous occurrence is due to their dependence upon similar internal and environmental conditions.

In some species, callus formation is a precursor of adventitious root formation, while in other species excess callusing may hinder rooting. Origin of adventitious roots from callus tissue has been associated with difficult-to-root species (Table 9–3) (59, 142), such as pine (*Pinus radiata*) (41), *Sedum* (310), and the mature phase of English ivy (*Hedera helix*) (98). Adventitious roots originate in the callus tissue formed at the base of the cutting and from "tracheary nests," such as in callus of creeping fig (*Ficus pumila*) (Fig. 9–12). It is possible to have adventitious roots originating from different tissues on the same cutting—epicotyl stem cuttings of pine (*Pinus sylvestris*) can form roots from resin duct wound (callus) tissue, central and basal wound (callus) tissue, and vascular tissue (Fig. 9–13, page 290) (93).

Shoot Organogenesis Adventitious bud differentiation and subsequent adventitious shoot formation may also be obtained by direct organogenesis or via secondary organogenesis from disorganized calli (95). Shoot formation occurs by direct morphogenesis when the apical ends of epicotyl microcuttings of Troyer citrange are inserted vertically in a solid medium (204); conversely at the basal end, shoot formation occurs by indirect organogenesis through callus formation. When epicotyl explants are placed horizontally on the medium, shoot regeneration at both ends occurs by indirect organogenesis through callus formation.

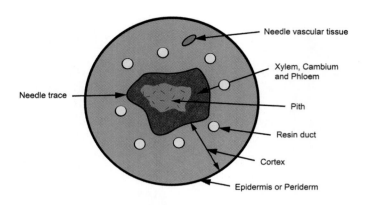

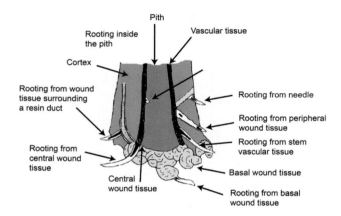

Figure 9–13
It is possible to have adventitious rooting originating from different tissues on the same cutting. *Top*: Tissue map of transverse section of epicotyl stem of one-year-old *Pinus sylvestris*. *Bottom*: Schematic longitudinal section showing examples of rooting occurring from resin duct wound (callus) tissue. No single cutting developed roots from all potential tissues.
Redrawn from Flygh et al. (93).

Leaf Cuttings—Adventitious Bud (and Shoot) and Root Formation

Many plant species, including both monocots and dicots, can be propagated by leaf cuttings (113). The origin of new shoots and new roots in leaf cuttings is quite varied and develops from primary or secondary meristems:

- **Preformed, primary meristems** are groups of cells directly descended from embryonic cells that have never ceased to be involved in meristematic activity.
- **Wound-induced, secondary meristems** are groups of cells that have differentiated and functioned in some previously differentiated tissue system and then dedifferentiate into new meristematic zones (*de novo*), resulting in the regeneration of new plant organs. This is the most common type of **meristem** in leaf cuttings.

meristem tissue
Tissue composed of undifferentiated cells that can continue to synthesize protoplasm and produce new cells by division.

Leaf Cuttings with Preformed, Primary Meristems
Detached leaves of *Bryophyllum* produce small plantlets from notches around the leaf margin (see Fig. 10–19, page 301). These small plants originate from so-called *foliar "embryos,"* formed in the early stages of leaf development from small groups of vegetative cells at the edges of the leaf. As the leaf expands, a foliar embryo develops until it consists of two rudimentary leaves with a stem tip between them, two root primordia, and a "foot" that extends toward a vein (134, 309). As the leaf matures, cell division in the foliar embryo ceases, and it remains dormant. If the leaf is detached and placed in close contact with a moist rooting medium, the young plants rapidly break through the leaf epidermis and become visible in a few days. Roots extend downward, and after several weeks many new independent plants form while the original leaf dies. Thus the new plants develop from ***latent primary meristems***—from cells that have not fully differentiated. Production of new plants from leaf cuttings by the renewed activity of primary meristems is found in species such as the piggyback plant (*Tolmiea*) (see Fig. 10–19) and walking fern (*Camptosorus*).

Leaf Cuttings with Wound-Induced, Secondary Meristems
In leaf cuttings of *Begonia rex*, *Sedum*, African violet (*Saintpaulia*), snake plant (*Sansevieria*) (see Fig. 10–18, page 297), *Crassula*, and lily, new plants may develop from secondary meristems arising from differentiated cells at the base of the leaf blade or petiole as a result of wounding.

meristematic cells Cells that synthesize protoplasm and produce new cells by division. They vary in form, size, wall thickness, and degree of vacuolation, but have only a primary cell wall.

In African violet, new roots and shoots arise *de novo* by the formation of **meristematic cells** from previously differentiated cells in the leaves. The roots are produced from thin-walled cells lying between the vascular bundles. The new shoots arise from cells of the subepidermis and the cortex immediately below the epidermis. Adventitious roots first emerge, form branch roots, and continue to grow for several weeks before adventitious buds and their subsequent development into adventitious shoots occurs. Root initiation and development are independent of adventitious bud and shoot formation (284). The same process occurs with many begonia species (Figs. 9–14 and 9–15). Although the original leaf supplies metabolites to the young plant, it does not become a part of the new plant.

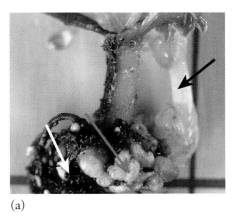
(a)

(b)

Figure 9–14
(a) Adventitious shoot (upper black arrow), adventitious buds (blue arrow) and roots (white arrow) from a leaf cutting of Rieger begonia. An adventitious bud is an embryonic shoot. (b) At high cytokinin concentration, only buds and budlike tissue are visible (arrow) with poor shoot development; roots formed but were removed before the photograph was taken (57).

(a)

(b)

(c)

Figure 9–15
(a) Adventitious shoots and adventitious roots arise at the base of the petiole (arrow) of a leaf cutting of Rieger begonia. (b) Application of a cytokinin mixed with talc to leaf cutting petiole base. (c) For sufficient, normal-appearing adventitious shoot production from a leaf cutting, without excessive adventitious bud formation, the 0.01 percent (100 ppm) treatment was optimal (arrow) (57). The original leaf blade was removed prior to taking the photo.

In lily (*Lilium longiflorum*) and *L. candidum*, the bud primordium originates in parenchyma cells in the upper side of the bulb scale (see Figs. 15–3 and 15–15), whereas the root primordium arises from parenchyma cells just below the bud primordium. Although the original scale serves as a source of food for the developing plant, the vascular system of the young bulblet is independent of that of the parent scale, which eventually shrivels and disappears (287).

In several species (e.g., sweet potato, *Peperomia*, and *Sedum*), new roots and new shoots on leaf cuttings arise in callus tissue that develops over the cut surface through the activity of secondary meristems. The petiole of *Sedum* leaf cuttings forms a considerable pad of callus within a few days after the cuttings are made. Root primordia are organized within the callus tissue, and shortly thereafter four or five roots develop from the parent leaf. Following this, bud primordia arise on a lateral surface of the callus pad and develop into new shoots (310).

Root Cuttings—Adventitious Bud (and Shoot) and Root Formation

Development of adventitious shoots, and in many cases adventitious roots, must take place if new plants are to be regenerated from root pieces (root cuttings) (251). Regeneration of new plants from root cuttings takes place in different ways, depending upon the species. Commonly, the root cutting first produces an adventitious shoot, and later produces roots, often from the base of the new shoot rather than from the original root piece itself. With root cutting propagation of apples, and the storage roots of sweet potato, these adventitious shoots can be removed and rooted as stem cuttings when treated with auxin (239). In other plants, a well-developed root system has formed by the time the first shoots appear.

In some species, adventitious buds form readily on roots of intact plants, producing **suckers.** When roots of such species are dug, removed, and cut into pieces, buds are even more likely to form. In young roots, such buds may arise in the pericycle near the vascular cambium (248). The developing buds first appear as groups of thin-walled cells having a prominent nucleus and a dense cytoplasm (80). In old roots, buds may arise in a callus-like growth from the phellogen; or they may appear in a callus-like proliferation from vascular ray tissue. Bud primordia may also develop from wound callus tissue that proliferates from the cut ends of injured surfaces of the roots (224), or they may arise at random from cortex parenchyma (239).

Sometimes regeneration of new root meristems on root cuttings is more difficult than the production of adventitious buds (2, 33). New roots may not always be adventitious and can develop from latent lateral root initials contained in the root piece or attached lateral roots.

BOX 9.3 GETTING MORE IN DEPTH ON THE SUBJECT
IMPORTANCE OF ADVENTITIOUS BUD FORMATION IN LEAF CUTTINGS

The limiting factor in leaf cutting propagation is generally the formation of adventitious buds, **not** *adventitious roots.* Adventitious roots form on leaves much more readily than do adventitious buds. In some plants, such as the India rubber fig (*Ficus elastica*), the cutting must include a portion of the old stem containing an axillary bud (a leaf-bud cutting) because although adventitious roots may develop at the base of the leaf, an adventitious shoot is not likely to form. In fact, rooted leaf cuttings of some species will survive for years without producing an adventitious shoot.

BOX 9.4 GETTING MORE IN DEPTH ON THE SUBJECT
PROPAGATION OF CHIMERAL PLANTS FROM LEAF AND ROOT CUTTINGS

One of the chief advantages claimed for asexual propagation is the exact reproduction of all characteristics of the parent plant. With root and leaf cuttings, however, this generalization does not always hold true. In periclinal chimeras, in which the cells of the outer layer are of a different genetic makeup from those of the inner tissues, the production of a new plant by root cuttings (derived from nonmutated, "wild type" inner tissues) results in a plant that is different in appearance from the parent. This is well illustrated in the thornless boysenberry and the 'Thornless Evergreen' trailing blackberry, in which stem or leaf-bud cuttings produce plants that retain the (mutated) thornless condition, but root cuttings develop into (normal, nonmutated) thorny plants. This is because the tissues forming the root cutting originate from normal, nonmutated cells. Likewise, with leaf cuttings, adventitious buds would have to originate from both mutated and normal cells for the chimera to be expressed. See Chapter 16 for more information on chimeras.

Generally, such branch roots arise from differentiated cells of the pericycle adjacent to the central vascular cylinder (21). Adventitious root initials have been observed to arise in the region of the vascular cambium in roots.

A list of plants commonly propagated by root cuttings can be found in Chapter 10, Table 10–2.

Polarity and Organ Formation in Cuttings

The polarity inherent in shoots and roots is shown dramatically in the rooting of cuttings (Fig. 9–16). Polarity is the quality or condition inherent in a cutting that exhibits different properties in opposite parts; that is, stem cuttings form shoots at the distal end (nearest to the shoot tip), and roots form at the proximal end (nearest to the crown, which is the junction of the shoot and root system). Root cuttings of many species form roots at the **distal** end and shoots at the **proximal** end. Changing the position of a stem cutting with respect to gravity does not alter this tendency (Fig. 9–16) (28). Polarity is also observed in leaf cuttings even though roots and shoots arise at the same position, usually the base of the cutting (see Fig. 9–14).

In 1878, Vöchting (286) advanced the theory that polarity could be attributed to individual cellular components, since no matter how small the piece, regeneration was consistently polar. A general explanation of polarity is that when tissue segments are cut, the physiological unity is disturbed. This must cause a redistribution of some substance, probably auxin, thus accounting for the different growth responses. The correlation of polarity of root initiation with auxin movement has been noted in several instances (115, 188, 240, 251, 289). It is also known that the polarity in auxin transport varies in intensity among different tissues. The polar movement of auxins is an active transport process, mediated by a membrane transport carrier, which occurs in phloem parenchyma cells (154, 176, 307). See page 39 in Chapter 2 for discussion on auxin transport.

CORRELATIVE EFFECTS: HOW HORMONAL CONTROL AFFECTS ADVENTITIOUS ROOT AND BUD (AND SHOOT) FORMATION

The Effects of Buds and Leaves

In 1758, Duhamel du Monceau (72) explained the formation of adventitious roots in stems on the basis of the downward movement of sap. Sachs, the noted German plant physiologist (1882), postulated the existence of a specific root-

> **correlative effect** The control of one organ over the development of another, which is mediated by phytohormones. Auxin produced from axillary buds is transported basipetally down the shoot and is important in subsequent root formation at the base of a cutting.

forming substance manufactured in leaves, which moves downward to the stem base where it promotes adventitious root formation (244). It was shown by van der Lek (1925) that sprouting buds promoted root initiation just below the buds in cuttings of such plants as willow, poplar, currant, and grape (175). It was assumed that hormone-like substances formed in the developing buds and these were transported through the phloem to the cutting base where they stimulated root initiation.

The existence of a specific root-forming factor was first determined by Went in 1929 when he discovered that leaf extracts from chenille (*Acalypha*) plants applied back to chenille or papaya (*Carica*) tissue induce root formation (292). Bouillenne and Went found substances in cotyledons, leaves, and buds that stimulated the rooting of cuttings; they called this material **"rhizocaline"** (35, 292).

> **rhizocaline** A hypothetical chemical complex, that was considered important in the biochemical events leading to root initiation.

Figure 9–16
Polarity of root regeneration in grape hardwood cuttings. Cuttings at left were placed for rooting in an inverted position, but roots still developed from the morphologically basal (proximal) end. Cuttings at right were placed for rooting in the normal, upright orientation with roots forming at the basal end.

Bud Effects on Rooting In Went's 1934 pea test for root-forming activity of various substances, it is significant that the presence of at least one bud on the pea cutting was essential for root production (292). After auxins were discovered, it was shown that a budless cutting would not form roots even when treated with an auxin-rich preparation. This finding indicated again that a factor other than auxin, presumably one produced by the bud, was needed for root formation. In 1938, Went postulated that specific factors other than auxin were manufactured in the leaves and were necessary for root formation. Thus, rhizocaline was more than just auxin. Later studies (83, 198) with pea cuttings confirmed this theory.

For root initiation, the presence of a metabolically active shoot tip (or a lateral bud) is necessary during the first three or four days after the cuttings are made (115). But after the fourth day the shoot terminal and axillary buds can be removed without interfering with subsequent root formation.

"rest period"
A physiological condition of the buds of many woody perennial species beginning shortly after the buds are formed. While in this condition, they will not expand into flowers or leafy shoots even under suitable growing conditions. After exposure to sufficient chilling hours (1 to 6°C (33 to 43°F), however, the "rest" influence is broken, and the buds will develop normally with the advent of favorable growing temperatures.

Bud removal from cuttings in certain species will stop root formation, especially in species without preformed root initials (175). In some plants, if the tissues exterior to the xylem are removed, just below a bud, root formation is reduced, indicating that some root-promoting compound(s) travels through the phloem from the bud to the base of the cutting. If hardwood, deciduous cuttings are taken in midwinter when the buds are in the **rest period,** they have either no effect or can inhibit rooting (88, 145). But if the cuttings are made in early fall or in the spring when the buds are active and not at rest, they show a strong root-promoting effect.

Conversely, with cuttings of apple and plum rootstocks, the capacity of shoots to regenerate roots increases during the winter, reaching a high point just before budbreak in the spring; this root regeneration is believed to be associated with a decreasing level of bud dormancy following winter chilling (144). Studies with Douglas-fir cuttings showed a pronounced relationship between bud activity and the rooting of cuttings—cuttings taken in early fall (September to October in the United States) root the poorest (238).

Leaf Effects on Rooting It has long been known that the presence of leaves on cuttings exerts a strong stimulating influence on rooting (Fig. 9–17). The stimulatory effect of leaves on rooting in stem cuttings is nicely shown by studies (234) with avocado. Cuttings of difficult-to-root cultivars under mist soon shed their leaves and die, whereas leaves on the cuttings of cultivars that have rooted are retained as long as nine months. While the presence of leaves can be important in rooting, leaf retention is more a consequence of rooting than a direct cause of rooting. After five weeks in the rooting bed, there was five times more starch in the base of the easy-to-root avocado cuttings than there was at the beginning of the tests. In hibiscus, rooting is also enhanced when leaves are retained on the cuttings (279).

Carbohydrates translocated from the leaves are important for root development. However, the strong root-promoting effects of leaves and buds are probably due to other, more direct factors (38). Leaves and buds produce auxin, and the effects of the polar apex-to-basal

Figure 9–17
Effect of leaves, buds, and applied auxin on adventitious root formation in leafy 'Old Home' pear cuttings. *Top:* Cuttings treated with auxin (indolebutyric acid at 4,000 ppm for five seconds). *Bottom:* Untreated cuttings. *Left to right:* with leaves; leaves removed; buds removed; one-fourth natural leaf area. Courtesy W. Chantarotwong.

(basipetal) transport of auxins enhances rooting at the base of the cutting.

Plant Growth Substances

All classes of growth regulators—auxins, cytokinins, gibberellins, ethylene, and abscisic acid, as well as ancillary compounds such as growth retardants/inhibitors, polyamines, and phenolics—influence root initiation either directly or indirectly (64). However, auxins have the greatest effect on root formation in stem cuttings, while cytokinins are used to stimulate adventitious bud formation in leaf cuttings. The other plant growth regulators and ancillary compounds can influence organogenesis, but not consistently enough to merit their commercial use in propagation. See Table 9–4 for a synopsis on plant growth regulator effects on adventitious bud and shoot formation. Chapter 2 has a more complete description of plant hormones, plant growth regulators, and how they function.

Auxins In the mid-1930s and later, studies of the physiology of auxin action showed that auxin was involved in such varied plant activities as stem growth, adventitious root formation (115, 275, 276, 292), lateral bud inhibition, abscission of leaves and fruits, and activation of cambial cells. Auxins can induce gene activity and are also signaling molecules in developmental events of adventitious root formation (39, 307).

Indole-3-acetic acid (IAA) was identified as a naturally occurring compound having considerable auxin activity (115). Indole-3-acetic acid (see Fig. 2–25) was subsequently tested for its activity in promoting roots on stem segments, and in 1935 investigators demonstrated the practical use of this material in stimulating root formation on cuttings (276). About the same time it was shown that two synthetic materials, **indole-3-butyric acid (IBA)** and a-**naphthalene acetic acid (NAA)** (see Fig. 2–25), were even more effective than the naturally occurring or synthetic IAA for rooting (29). Today, IBA and NAA are still the most widely used auxins for rooting stem cuttings and for rooting tissue-culture–produced microcuttings. It has been repeatedly confirmed that auxin is required for initiation of adventitious roots on stems, and indeed, it has been shown that divisions of the first root initial cells are dependent upon either applied or endogenous auxin (96, 116, 188, 266).

Indole-3-butyric acid, although less abundant than IAA, is also a naturally occurring substance in plants (11, 82, 186). In *Arabidopsis*, endogenously

Table 9–4
Plant Growth Regulator Effects on Adventitious Root and Bud (and Shoot) Formation

Plant growth regulator	Adventitious root formation	Adventitious bud and shoot formation
Auxins	Promote	Inhibit; low auxin: high cytokinin ratio promote
Cytokinins	Inhibit; high auxin: low cytokinin ratio promote	Promote
Gibberellins	Inhibit	Inhibit; can enhance shoot elongation *after* organ formation
Ethylene	Can promote with auxin-induced rooting of some herbaceous plants; with woody plants generally not directly involved in rooting—but in small concentrations and for short durations may enhance competency to root (68)	Not promotive
ABA	Inhibit; however, used in combination with auxin can promote rooting in some species	Inhibits; however was reported to stimulate adventitious bud formation of a herbaceous species
Other potential hormones and ancillary compounds Retardants/inhibitors, polyamines, jasmonate, brassinosteroids, phenolics polyamines, salicylate, flavoinds, peroxidases	Used in combination with auxin can promote or inhibit rooting in some species	Not promotive; may depress shoot development

conjugation of plant hormones Plant hormones that are important in the regulation of physiologically active phytohormone levels, and are deactivated ("bound") hormones attached to other molecules via ester, glycoside, or amide bonds. The conjugated hormones may later be liberated via enzymatic hydrolysis and regain their activity, for example, IAA-aspartate is an auxin conjugate.

formed IAA is more readily transported than endogenously formed IBA (11). IAA also **conjugates** via amide bonds, while IBA conjugates from ester bonds.

In apple (*Malus*), when IBA is applied to stem cuttings or microcuttings to stimulate rooting, it is, in part, converted to IAA (282, 307). IBA may also enhance rooting via increased internal-free IBA or may synergistically modify the action of IAA or endogenous synthesis of IAA; IBA can enhance tissue sensitivity for IAA and increase rooting (282). In avocado microcuttings, IBA increased endogenous IAA and indole-3-acetyl-aspartic acid (IAA-asp) before root differentiation occurred, and as root formation proceeded (94). The same IBA response occurred in juvenile and mature phase microcuttings of chestnut (9); however, more endogenous IAA was detected in mature (recalcitrant) than juvenile (easy-to-root) tissue, indicating that endogenous IAA was not limiting rooting capacity.

In mung bean cuttings, IBA applied to the cutting base was transported to the upper part of the cuttings to a greater extent than IAA, and rapidly metabolized into IBA conjugates. These IBA conjugates were reported to be superior to free IBA in serving as an auxin source during later stages of rooting (297).

Cytokinins Cytokinins have the greatest effect on initiating buds and shoots from leaf cuttings and in tissue culture systems (31, 57, 79, 241, 281). Natural and synthetic cytokinins include **zeatin, zeatin riboside, kinetin, isopentenyladenine (2iP), thidiazuron (TDZ),** and **benzyladenine (BA or BAP)** (See Chapter 2). Generally, a high auxin/low cytokinin ratio favors adventitious root formation and a low auxin/high cytokinin ratio favors adventitious bud formation (36, 133) (Figs. 9–15 and 9–18). Cuttings of species with high natural cytokinin levels have been more difficult to root than those with low cytokinin levels (212). Applied synthetic cytokinins normally inhibit root initiation in stem cuttings (217). However, cytokinins at very low concentrations, when applied to decapitated pea cuttings at an early developmental stage (84), or to begonia leaf cuttings (133), promote root initiation, while higher concentrations inhibit root initiation. Application to pea cuttings at a later stage in root initiation does not show such inhibition; the influence of cytokinins in root initiation may thus depend on the particular stage of initiation and the concentration (32, 58, 256). To date, the quantitative determination of endogenous cytokinins at various stages of rooting has yet to be determined (281).

It has been suggested that the few cases of rooting success using exogenous applications indicate that cytokinins have an indirect rather than a direct role on rooting (281). Cytokinins may also be indirectly involved in rooting through effects on rejuvenation and

---

BOX 9.5 GETTING MORE IN DEPTH ON THE SUBJECT
CHANGES IN AUXIN REQUIREMENTS DURING ADVENTITIOUS ROOT FORMATION

With pea cuttings, the role of auxins in the intricate developmental processes of rooting occurred in two basic stages (83, 85, 197):

- A **root initiation stage** in which root meristems were formed (including dedifferentiation, root-initial, and root-primordia formation). This stage could be further divided into:

    a. An **auxin-active stage**, lasting about 4 days, during which auxin had to be supplied continuously for roots to form, coming either from terminal or lateral buds, or from applied auxin (if the cutting has been decapitated) (85, 197).

    b. An **auxin-inactive stage** occurred next. Withholding auxin during this stage (which lasts about 4 days) did not adversely affect root formation.

- **Elongation of root primordia stage,** during which the root tip grows outward through the cortex, finally emerging from the epidermis of the stem (see Fig. 9–10). A vascular system develops in the new root and becomes connected to adjacent vascular bundles of the stem. At this stage there was no further response to applied auxin.

> **BOX 9.6 GETTING MORE IN DEPTH ON THE SUBJECT**
> **DIFFERENCES IN ROOTING RESPONSES OF IBA AND IAA**
>
> Variability in forming adventitious roots has been attributed to differences in auxin metabolism (27). However, the endogenous auxin concentration or type of auxin applied, (i.e., IBA compared to IAA), do not always explain rooting differences. Response to type of auxin is also species dependent (67, 225). While the more difficult-to-root *Grevillea* (Proteaceae) species had a reduced rooting response to IBA application when compared to the easy-to-root species, there were no differences in endogenous levels of IAA (170). Both IAA and IBA transport is mediated by different transport protein complexes (228). Difficult-to-root *Prunus avium* conjugated IBA more rapidly than the easy-to-root cultivar (82). Only free IBA was observed in the easy-to-root cultivar, suggesting that the difficult-to-root cultivar could not hydrolyze (de-conjugate) IBA during the appropriate developmental points of ARF. In young (easy-to-root) *Sequoia sempervirens* explant cuttings, higher levels of IAA were found after IBA treatment, whereas the mature (more difficult-to-root clone) had higher free IBA and conjugated IBA (27). Rooting was attributed to differences in auxin metabolism, and not to cell competency or sensitivity to form adventitious roots. In summary, the enhanced rooting of IBA compared to IAA has been attributed to differences in receptor binding, compartmentalization, greater stability and differences in tissue sensitivity between the two auxins (67, 82, 307).

> **BOX 9.7 GETTING MORE IN DEPTH ON THE SUBJECT**
> **AUXIN: ADVENTITIOUS ROOTING AND MOLECULAR STUDIES**
>
> One explanation for auxin activity of IBA is that it is a "slow-release" form of IAA (82, 282). IBA may supply plants with a continuous IAA source when it is required for root initiation. Biochemical studies in numerous plants and genetic studies of *Arabidopsis* with IBA-responsive mutants indicate that IBA acts primarily via its conversion to IAA through peroxisomal fatty beta-oxidation (11). Mutants and genes of *Arabidopsis* involved in auxin biosynthesis, conjugation (inactivation of auxin), conjugate hydrolysis (activation of auxin), and degradation are being used to determine the complex mechanisms by which auxins are controlled (307). While we know the gross effects of auxin on rooting, we don't fully know the molecular basis, that is, the function of auxins as signaling molecules during root induction, initiation, and development (11, 39, 258). Molecular biology can help determine upstream and downstream regulators of IAA. Identifying genes involved in converting IBA to IAA is important to understanding auxin regulation and the contribution of IBA to active auxin pools (including *de novo* synthesis and conjugate hydrolysis of IAA.

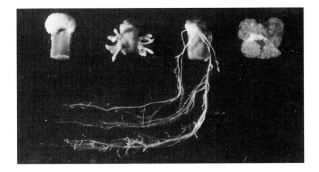

Figure 9–18
Effects of adenine sulfate (a cytokinin precursor) and indoleacetic acid (auxin) on growth and organ formation in tobacco stem segments. *Far left:* Control. *Central left:* Adenine sulfate, 40 mg per liter. Bud formation with decrease in root formation. *Central right:* Indoleacetic acid, 0.02 mg per liter. Root formation with prevention of bud formation. *Far right:* Adenine sulfate, 40 mg per liter plus indoleacetic acid, 0.02 mg per liter. Growth stimulation but without organ formation. Courtesy Folke Skoog.

accumulation of carbohydrates at the cutting base (i.e., carbohydrate loading) (281).

Leaf cuttings provide good test material for studying auxin-cytokinin relationships since such cuttings must initiate both roots and shoots. Cytokinin application at relatively high concentrations promoted bud formation and inhibited root formation of *Begonia* and *Bryophyllum* (134) leaf cuttings, while auxins, at high concentrations, stimulated roots and inhibited buds. Too high a cytokinin concentration applied to leaf cuttings maximizes adventitious bud formation but reduces the quality of new shoots (Figs. 9–14 and 9–15); from a horticultural standpoint, adventitious shoot quality, not just adventitious bud formation, is an important criterion in regenerating new plants from leaf cuttings (57). The considerable seasonal changes in the regenerative ability of *Begonia* leaf cuttings are due to a complex interaction of **environmental cues**: temperature, photoperiod, and irradiance, which affect the

> **BOX 9.8** **GETTING MORE IN DEPTH ON THE SUBJECT**
> **CHANGES IN CYTOKININ REQUIREMENT DURING SHOOT ORGANOGENESIS**
>
> As with auxin and rooting, there are also changes in cytokinin requirement during shoot organogenesis (46). Three phases of shoot organogenesis can be distinguished: (a) formation of cell competence, (b) shoot induction, and (c) shoot development (Fig. 9–5). During induction, the leaf cutting or explant perceives exogenous cytokinin and auxin compounds and becomes committed to the development of shoots. With a highly shoot organogenic *Petunia hybrida* line, there was an 1.7-fold increase in endogenous cytokinins during shoot induction and 2.6-fold cytokinin increase during the shift from the induction to shoot development phase; conversely, isoprenoid cytokinins did not accumulate with mutant explants, incapable of shoot induction (6). Hence, the early stages of shoot development are influenced by cytokinin uptake and metabolism, which subsequently affects accumulation of isoprenoid cytokinins and the activity of cytokinin oxidase (6).

levels of endogenous cytokinins, auxins, and other growth regulators (137).

Gibberellins (GA) The gibberellins (see Fig. 2–28) are a group of closely related, naturally occurring compounds first isolated in Japan in 1939 and known principally for their effects in promoting stem elongation. At relatively high concentrations (i.e., 10^{-3} M), they have consistently inhibited adventitious root formation (250). This inhibition is a direct local effect that prevents the early cell divisions involved in transformation of differentiated stem tissues to a meristematic condition. Gibberellins have a function in regulating nucleic acid and protein synthesis and may suppress root initiation by interfering with these processes, particularly transcription (125). At lower concentrations (10^{-11} to 10^{-7} M), however, gibberellin has promoted root initiation in pea cuttings, especially when the stock plants were grown at low light levels (125).

In *Begonia* leaf cuttings, gibberellic acid (138) inhibited both adventitious bud and root formation, probably by blocking the organized cell divisions that initiate formation of bud and root primordia. Inhibition of root formation by gibberellin depends on the developmental stage of rooting. With herbaceous materials, inhibition is usually greatest when GA is applied 3 to 4 days after cutting excision (125). However, woody plant species such as willow (*Salix*) (116) and fig (*Ficus*) (59) were not adversely affected by GA during root initiation but were inhibited if GA was applied after root primordia were initiated. GA caused the reduction in cell numbers in older established primordia, which was deleterious to root formation. *The biochemical and physiological mechanisms by which applied gibberellins inhibit adventitious rooting remains unknown* (115).

Ethylene (C_2H_4) Ethylene can enhance, reduce, or have no effect on adventitious root formation (64). In 1933, Zimmerman and Hitchcock (311) showed that applied ethylene at about 10 mg/liter (ppm) causes root formation on stem and leaf tissue as well as the development of preexisting latent roots on stems. They and other scientists (312) also showed that auxin applications can regulate ethylene production and suggested that auxin-induced ethylene may account for the ability of auxin to cause root initiation. Centrifuging *Salix* cuttings in water, or just soaking them in hot or cold water, stimulates ethylene production in the tissues as well as root development, suggesting a possible causal relationship between ethylene production and subsequent root development (161, 162, 206). High auxin concentrations will also trigger ethylene evolution.

Ethylene promotion of rooting occurs more frequently in intact plants than cuttings, herbaceous rather than woody plants, and plants having preformed root initials. Rooting cuttings of ethylene-insensitive tomato mutants has shown that the promotive effect of auxin on adventitious rooting is enhanced in plants that are responsive to and sensitive to ethylene (47). The commercial ethylene receptor blockers, STS and 1-MCP, also inhibit rooting. However, the effects of ethylene on rooting are not as predictable or consistent as those of auxin (115). While a large body of evidence suggests that endogenous ethylene is not directly involved in auxin-induced rooting of cuttings (206), ethylene may be necessary in minute quantities for initiating cell division as a prerequisite for root initiation in cuttings (34). Ethylene effects are of very short duration, whereas higher concentrations and longer time exposure to ethylene inhibits rooting. It is possible that ethylene changes the competency of cells for receiving auxin signals (68).

Abscisic Acid (ABA) Reports on the effect of abscisic acid (ABA) on adventitious root formation are contradictory (14, 136, 230)—apparently depending upon the concentration, environmental, and nutritional status of the stock plants from which the cuttings are taken. ABA is important to rooting, since it (a) antagonizes the effects of gibberellins and cytokinins, both of which can inhibit rooting, and (b) influences the ability of cuttings to withstand water stress during propagation. If the role of ABA in rooting is to be understood, then endogenous ABA levels will need to be determined at the site of root initiation, during the developmental stages of rooting (64).

Other Potential Hormones and Ancillary Compounds

There are ancillary compounds that modify main hormone effects on rooting, and adventitious bud and shoot formation. These compounds include growth retardants/inhibitors, flavonoids, peroxidases, and phenolics. Other potential phytohormones include jasmonic acid (jasmonate), polyamines, brassionosteroids and salicylic acid (salicylate). Salicylate has been reported to enhance rooting in combination with auxin (64, 229).

Growth Retardants/Inhibitors. Growth retardants, generally applied to reduce shoot growth, have been used to enhance rooting based on the rationale that they (a) antagonize GA biosynthesis or activity (GA is normally inhibitory to rooting) or (b) reduce shoot growth, resulting in less competition and consequently more assimilates are available for rooting at cutting bases (66). Synthetic anti-gibberellins and inhibitors of GA biosynthesis include chlormequat chloride (CCC), paclobutrazol (PP333, Bonzi), uniconazole (a triazole growth retardant related to PP333), morphactins, ancymidol (Arest), gonadotropins, and daminozide (SADH, Alar) (64, 231). Growth retardants frequently promote rooting (generally in combination with exogenous auxin) (66, 128). However, the mode of action of how these compounds enhance rooting is not well understood. Hence, rooting enhancement by GA biosynthesis inhibitors has been inconsistent, and none are commercially used for rooting (64).

The Polyamines. The effect of polyamines on rooting of woody plant species is quite variable. Putrescine, spermidine, and spermine in combination with IBA improved rooting of hazel microshoots (235). Conversely, higher levels of endogenous putrescine, spermidine, and spermine were found in mature phase (recalcitrant) than juvenile (easy-to-root) microshoots of chestnut (9). The rooting of olive microshoots increased by using polyamines along with NAA, but rooting of almond, pistachio, chestnut, jojoba, apricot, and walnut did not increase (243). In NAA-treated English ivy (*Hedera helix*) cuttings, there were increases in endogenous polyamines, particularly putrescine (99). *Polyamines may serve as secondary messengers for rooting.* To date, polyamine enhancement of rooting occurs only in the presence of auxin.

Classification of Plant Rooting Response to Growth Regulators

Plants can be divided into three classes with regard to growth regulator effects on rooting:

- **Easy-to-Root**—plants that have all the essential endogenous substances (**root morphogens**) plus auxin. When cuttings are made and placed under proper environmental conditions, rapid root formation occurs. Auxin may further enhance rooting, but is generally not required.

  > **root morphogen**
  > An endogenous substance(s) that stimulates rooting. It may be auxin or a combination of substance(s) with auxin that promote rooting.

- **Moderately Easy-to-Root**— plants in which the naturally occurring root morphogen(s) are present in ample amounts, but auxin is limited. Auxin is needed to enhance rooting.

- **Difficult-to-Root (Recalcitrant)**—plants that lack a rooting morphogen(s) and/or lack the cell sensitivity to respond to the morphogen(s), even though natural auxin may or may not be present in abundance. External application of auxin gives little or no rooting enhancement.

  > **recalcitrant plants**
  > Plants that are difficult to root from cuttings. They lack a rooting morphogen(s) and/or lack the cell sensitivity to respond to the morphogen(s), even though natural auxin may or may not be present in abundance. External application of auxin gives little or no rooting response.

THE BIOCHEMICAL BASIS FOR ADVENTITIOUS ROOT FORMATION

The biochemical basis for root formation implies that there are root-promoting and root-inhibiting substances produced in plants and their interaction is thought to be involved in rooting. Therefore, this theory considers that difficult-to-root cuttings either lack the appropriate root-promoting substances or are high in root-inhibiting substances.

While we know much about the biology and manipulation of cuttings, the **primary chemical stimulus** for dedifferentiation and root initial formation (the critical steps of adventitious root formation) and the subsequent organization of root primordia **remains unknown** (65, 115). The following is a brief history of post–World War II research on the biochemistry of rooting.

Endogenous Rooting Inhibitors

In the early 1950s, endogenous chemical inhibitors were reported to retard rooting in selected plant species, as indicated in the following section. This was found to be the case with selected grape cultivars; leaching cuttings with water enhanced the quantity and quality of roots. Difficult-to-root hardwood cuttings of wax flower (*Chamaelaucium uncinatum*) have a cinnamic acid derivative that inhibits rooting, while no detectable levels of this phenolic compound were found in easy-to-root softwood cuttings (50). Cuttings of difficult-to-root mature eucalyptus (49, 215), chestnut (285), and dahlia cultivars (18, 19) also had higher rooting inhibitors than easy-to-root forms.

Rooting Co-Factors (Auxin Synergists)

rooting bioassay The use of a plant organ or tissue to respond morphologically to chemical stimulation, such as the rooting response of mung bean hypocotyl cuttings to various chemicals.

Various **model rooting bioassay systems** have been used to test adventitious root formation. The easy-to-root mung bean (*Vigna*) was used by Hess (140, 141) as a rooting bioassay to screen biochemical effects on rooting (Fig. 9–4). Hess was not able to demonstrate any difference in rooting inhibitors between the juvenile easy-to-root, and mature difficult-to-root forms of English ivy (*Hedera helix*). Instead, he determined that the juvenile, easy-to-root forms of English ivy, and easy-to-root cultivars of chrysanthemum and *Hibiscus rosa-sinensis* contained greater nonauxin rooting stimuli than their difficult-to-root forms (140, 141). He termed these nonauxin rooting stimuli *rooting co-factors,* which was a *modification of the rhizocaline theory* that biochemical factors, other than just auxin, were controlling rooting. These rooting co-factors were naturally occurring substances that appeared to act synergistically with indoleacetic acid in promoting rooting.

Rooting co-factors have since been found in maple (*Acer*) species (168). Fadl and Hartmann (87, 88) isolated an endogenous root-promoting factor from basal sections of hardwood cuttings of an easily rooted pear cultivar ('Old Home'). Extracts from basal segments of similar cuttings of a difficult-to-root cultivar ('Bartlett'), treated with IBA, did not show this root-promoting factor. The action of these phenolic compounds in root promotion was theorized to be in protecting the root-inducing, naturally occurring auxin—indoleacetic acid—from destruction by the enzyme indoleacetic acid oxidase (109).

Jarvis (157) attempted to *integrate the biochemical with developmental anatomy of adventitious root formation by examining the four developmental stages of rooting* (Fig. 9–19). His premise was that (a) the initial high concentrations of auxin needed in early rooting events are later inhibitory to organization of the primordium and its subsequent growth—hence the importance of regulating endogenous auxin concentration with the IAA oxidase/peroxidase enzyme complex playing a central role (i.e., IAA oxidase metabolizes or breaks down auxin); and (b) IAA oxidase activity is controlled by phenolics (*o*-diphenols are inhibitory to IAA oxidase), while borate complexes with *o*-diphenols result in greater IAA oxidase activity—and hence a reduction of IAA to levels that are optimal for the later organizational stages of rooting.

With *in vitro* rooting of poplar (*Populus*) shoots, endogenous free IAA activity is highest during root induction, followed by a peak of soluble peroxidase activity and a subsequent decrease in free IAA preceding root emergence (132). These events correspond to the initiative phase of rooting suggested by Jarvis (157).

Biochemical Changes During the Development of Adventitious Roots

Once adventitious roots have been initiated in cuttings, considerable metabolic activity occurs as new root tissues are developed and the roots grow through and out of the surrounding stem tissue. Protein synthesis and RNA production were both shown to be indirectly involved in adventitious root development in etiolated stem segments of willow (*Salix tetrasperma*) (155) and in seasonal rooting of *Ficus* (see Fig. 9–31, page 317) (51). To date, it is not clear to what extent RNA metabolism is altered within that small pool of cells actually involved in root initiation (156). More definitive studies need to include microautoradiographic and histochemical approaches.

During the rooting of hydrangea cuttings, enzymatic changes were identified during the development

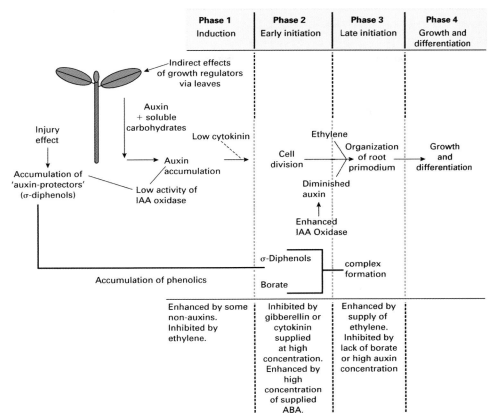

Figure 9–19
Hypothesized scheme of Jarvis (157) which proposes the role of phenolics, IAA oxidase/peroxidase, borate, and phytohormones in the four developmental stages of adventitious root production.

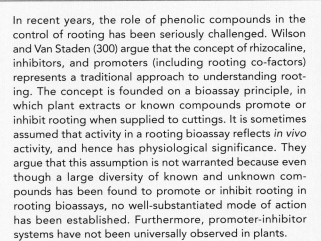

BOX 9.9 GETTING MORE IN DEPTH ON THE SUBJECT
A SYNOPSIS OF RHIZOCALINE, INHIBITORS, AND ROOTING CO-FACTORS IN ROOTING

In recent years, the role of phenolic compounds in the control of rooting has been seriously challenged. Wilson and Van Staden (300) argue that the concept of rhizocaline, inhibitors, and promoters (including rooting co-factors) represents a traditional approach to understanding rooting. The concept is founded on a bioassay principle, in which plant extracts or known compounds promote or inhibit rooting when supplied to cuttings. It is sometimes assumed that activity in a rooting bioassay reflects *in vivo* activity, and hence has physiological significance. They argue that this assumption is not warranted because even though a large diversity of known and unknown compounds has been found to promote or inhibit rooting in rooting bioassays, no well-substantiated mode of action has been established. Furthermore, promoter-inhibitor systems have not been universally observed in plants.

There is no good existing evidence that hypothesized rhizocaline consists of an auxin-phenolic conjugate, and other explanations for the actions of phenolics are not well substantiated. *Possibly the action of rooting promoters and inhibitors is mediated by chemical injury (see the later discussion on wounding in this chapter). Irrespective of their chemical identity, low concentrations promote rooting, while higher concentrations are inhibitory* (300).

Wilson (301) further proposed that a **rooting morphogen** can be assumed to induce roots in woody stem cuttings. Whereas auxins promote rooting of most herbaceous cuttings, they may have little effect on more difficult-to-root woody cuttings. The interaction between a rooting morphogen(s) of vascular origin and potential sites for root initiation are likely to be dynamic and variable. Potential rooting sites are not equally sensitive to the rooting morphogen, since each cell has a unique lineage, ontogeny, and position (i.e., the competency of cells varies, which affects their ability to respond to the morphogen and root). Hence, he concluded that no simply defined morphogen can be said to limit rooting.

PART THREE VEGETATIVE PROPAGATION

> **BOX 9.10 GETTING MORE IN DEPTH ON THE SUBJECT**
> **TYING IT ALL TOGETHER—INTEGRATING THE MORPHOLOGY: HORMONAL, PHYSIOLOGICAL, AND BIOCHEMICAL RELATIONSHIPS OF ADVENTITIOUS ROOTING**
>
> Much of the research dealing with hormones and rooting has been based on exogenous treatments (115). In contrast, little work has critically tested the roles of endogenous hormones (9) and their interactions with applied hormones. Particularly lacking is research aimed at determining how hormones might regulate gene expression and thereby influence rooting, directly and indirectly. Hence, it is difficult to distinguish between possible controlling roles of hormones on rooting and indirect hormonal effects on other physiological processes of cuttings (115).
>
> Likewise, physiological and biochemical studies have largely addressed the influences of plant growth regulators on the biochemistry of rooting without focusing on changes in gene expression (Fig. 9–20) (121). Essentially, these studies are post-translational and are geared on finding the missing chemical component(s) of rooting.
>
> Figure 9–21 attempts to synthesize the early morphological, physiological, and biochemical events of adventitious root formation—commencing with the severing of the stem cutting from the stock plant, wounding, perceived dehydration, decline in photosynthesis, the signaling cascade of chemicals and phytohormones, and gene expression.
>
> Using the tools of molecular biology with auxin and ethylene mutants, microarray analysis and proteomics, more is being learned about gene expression and the primary control of rooting (11, 39, 258, 307). See Figures 9–22 and 9–23 on microarray analysis of gene expression **during the synchronized development of different stages of adventitious root formation** of *Pinus contorta* hypocotyl cuttings (39).

INVESTIGATION OF ROOTING BY PROCESS		
EXPERIMENTAL TREATMENTS	MEDIATING PROCESSES	OBSERVATION OF EFFECTS
Light Temperature Oxygen Carbon Dioxide Water Auxins	TRANSCRIPTIONAL	LIMITED ASSESSMENTS AVAILABLE
Gibberellins Cytokinins Ethylene Minerals Organic Nitrogen Phenolics Carbohydrates Amino acids Polyamines	TRANSLATIONAL	LIMITED ASSESSMENTS AVAILABLE
Histones Acid / Bases Nucleotides Nucelosides Nucleic Acids Root Symbionts Various Inhibitors Many Others	POST-TRANSLATIONAL	MOST PAST ASSESSMENTS HERE

INVESTIGATION OF ROOTING BY DISCIPLINE				
Morphology	Cytology Anatomy	Physiolgogical Biochemistry (Most past assessments here)		Molecular Genetics

Figure 9–20
Some environmental and chemical factors (in the left column) that have been implicated in rooting. Investigation of rooting research is by process (upper section) and investigation by discipline (lower section). In past research, effects of experimental treatments may have been at any or all process levels, but were usually assessed only post-translationally, in physiological and/or biochemical studies. From Haissig et al. (121).

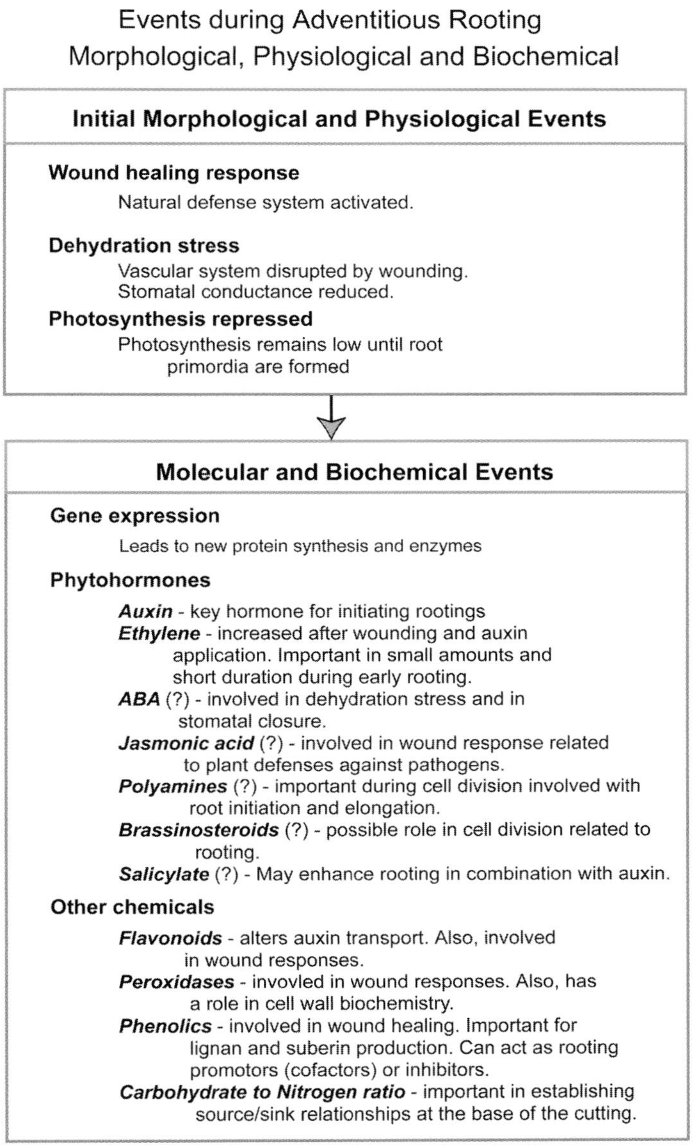

Events during Adventitious Rooting
Morphological, Physiological and Biochemical
Initial Morphological and Physiological Events
Wound healing response
Natural defense system activated.
Dehydration stress
Vascular system disrupted by wounding.
Stomatal conductance reduced.
Photosynthesis repressed
Photosynthesis remains low until root primordia are formed
Molecular and Biochemical Events
Gene expression
Leads to new protein synthesis and enzymes
Phytohormones
Auxin - key hormone for initiating rootings
Ethylene - increased after wounding and auxin application. Important in small amounts and short duration during early rooting.
ABA (?) - involved in dehydration stress and in stomatal closure.
Jasmonic acid (?) - involved in wound response related to plant defenses against pathogens.
Polyamines (?) - important during cell division involved with root initiation and elongation.
Brassinosteroids (?) - possible role in cell division related to rooting.
Salicylate (?) - May enhance rooting in combination with auxin.
Other chemicals
Flavonoids - alters auxin transport. Also, involved in wound responses.
Peroxidases - invovled in wound responses. Also, has a role in cell wall biochemistry.
Phenolics - involved in wound healing. Important for lignan and suberin production. Can act as rooting promotors (cofactors) or inhibitors.
Carbohydrate to Nitrogen ratio - important in establishing source/sink relationships at the base of the cutting.

Figure 9–21
Early morphological, physiological and biochemical events in rooting a cutting. See Fig. 9–23 for detailed description of gene expression during discrete rooting stages.

Stages of Adventitious Root Development
Dedifferentiation
Root initial
Root primordia
Root formation
Root elongation
Percentage of genes up or down-regulated
100
75
50
25
0
33%
67%
82%
18%
87%
13%
42%
58%
23%
77%
Up-regulated
Down-regulated
0 to 3 days
3 to 6 days
6 to 9 days
9 to 12 days
12 to 33 days

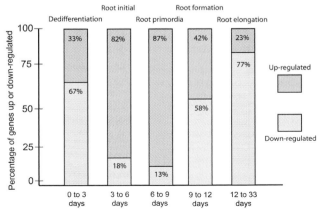

Figure 9–22
Some 220 genes are differentially expressed during the five phases (time period–days) of adventitious root development in *Pinus contorta*. The histogram shows the percentage of genes up-regulated (increased gene expression) or down-regulated (decreased) during rooting (39).

Days	Phase of Development	Up ↑ or Down ↓ Gene Regulation
0 to 3	Dedifferentiation	↑ Cell replication ↑ Cell wall weakening ↑ Water stress ↓ Cell wall synthesis ↓ Auxin transport ↓ Photosynthesis
3 to 6	Root Intitial	↑ Flavonoid pathway enzymes
6 to 9	Root Primordia	↑ Auxin transport ↑ Auxin responsive ↑ Cell wall synthesis ↑ Hypersensitive response proteins ↑ Pathogenesis proteins ↓ Cell wall weakening ↓ Cell wall modification ↓ Water stress
9 to 12	Root Formation	↑ Auxin transport
12 to 33	Root Elongation	↓ Water stress ↓ Cell replication

Figure 9–23
Microarray analysis of gene expression during the synchronized development of different stages of adventitious root formation of *Pinus contorta* hypocotyl cuttings. Transcript levels of 220 genes and their encoding proteins were up-regulated (↑increased expression) or down-regulated (↓decreased expression) (39).

of preformed root initials into emerging roots (201, 202). Initially, the enzymes peroxidase, cytochrome oxidase, succinic dehydrogenase, and starch-hydrolyzing enzymes increased in the phloem and xylem ray cells of the vascular bundles. During subsequent root development, enzyme activity shifted from the vascular tissues to the periphery of the vascular bundles. These increases in enzyme activity occurred 2 to 3 days after the cuttings were made. Peroxidase activity has been used as a predictive marker of the inductive phase of rooting (97).

During rooting, starch is converted to soluble carbohydrate. In hydrangea, starch disappeared from the endodermis, phloem and xylem rays, and pith—in tissues adjacent to the developing root primordia—and was converted to soluble carbohydrate. Similarly, in the development of adventitious roots on IBA-treated plum cuttings, as soon as callus and roots started forming, pronounced carbohydrate increases of sucrose, glucose, fructose, and sorbitol—and starch losses—occurred at the base of the cuttings where rooting occurs (37). While soluble carbohydrates are not the cause of rooting, the developing callus and roots at the cutting base act as a "sink" for the movement of soluble carbohydrates from the top of the cutting.

MOLECULAR/BIOTECHNO-LOGICAL ADVANCES IN ASEXUAL PROPAGATION

Biotechnological Advances In Asexual Propagation

While the physiology of adventitious root formation is better known than the genetic and molecular events of rooting, researchers are identifying specific genes affecting rooting in model systems (i.e., using plants such as *Arabidopsis,* tobacco, loblolly pine, lodgepole pine, and English ivy). They are trying to discover the regulatory sequencing of genes in the rooting process. Artificially inducing roots by nonpathogenic *Agrobacterium,* and the potential transformation of cells using a disarmed **plasmid** from a root-inducing bacterium or from an auxin-inducing fragment of the **T-DNA** may play important roles in the vegetative propagation of plants (see Chapter 2). Applying biotechnology studies at the earlier transcriptional and translational

plasmids Small molecules of extra-chromosomal DNA that carry only a few genes and occur in the cytoplasm of a bacterium.

T-DNA The portion of the root-inducing (Ri) plasmid (e.g., from *Agrobacterium rhizogenes*) that is inserted into the plant genome (e.g., of a difficult-to-root species) and stabilized; hence this normally difficult-to-root species is potentially "transformed" to an easy-to-root clone. See Chapter 2 for molecular biology terminology.

periods to determine gene expression can reveal the controls of rooting, adventitious bud formation, tuberization, and other developmental processes important to vegetative propagation. Once the regulatory sequences between genes and the rooting process of a species are known, plants may be genetically transformed with a higher rooting potential. As an example, if an enzyme negatively affected rooting, then antisense DNA or RNA could be used to turn off the gene that produced the enzyme. Initially, the genetically transformed plant would be micropropagated, and then once established *ex vitro* (outside the test tube), conventional cutting propagation techniques would be used to mass-produce the genetically transformed plant (54).

It has not yet been fully determined which genes or gene groups affect rooting. Changes in gene expression were observed during the formation of adventitious root primordia of sunflower (*Helianthus annus*) hypocotyl cuttings (213), rooting of *Arabidopsis* (67), and rooting of juvenile and mature English ivy (246).

Today, difficulties in rooting *in vitro* and *ex vitro*, developing successful tissue culture multiplication systems, and transformation systems for rooting limit the production of transgenic woody plants (e.g., commercially important plants for the production of fruits, nuts, wood, paper, and landscape ornamentals) (60, 236, 249). Some difficult-to-root woody species have been genetically "transformed" to easy-rooters. Rooting of kiwi (*Actinidia deliciosa*) cuttings was improved by introducing genes from the root-inducing bacterium *Agrobacterium rhizogenes* (242, 243) (see Chapter 2).

Progress is being made by using this root-inducing bacterium to enhance root regeneration of bare-root almond stock (265) and *in vitro* rooting of difficult-to-root apple (214). *Agrobacterium rhizogenes* have been used as an effective rooting agent in hazelnut (*Corylus avellana*) cuttings (12), and with *in vitro* and *ex vitro* rooting of pine (*Pinus*) and larch (*Larix*). How the bacterium enhances rooting is not well understood. It may be modifying the root environment by secreting hormones or other compounds, or by transforming plant cells (194).

MANAGEMENT AND MANIPULATION OF ADVENTITIOUS ROOT AND SHOOT FORMATION

Great differences in the rooting ability of cuttings exist among species and cultivars. Stem cuttings of some cultivars root so readily that the simplest facilities and care give high rooting percentages. On the other hand, cuttings of many cultivars or species have yet to be rooted. Cuttings of some "difficult" cultivars can be

---

BOX 9.11 GETTING MORE IN DEPTH ON THE SUBJECT
ADVANCES IN THE BIOTECHNOLOGY OF ROOTING

Because rooting potential is complex and likely controlled by many genes with unknown modes of action and inheritance, molecular studies are essential to revealing the basic mechanism of rooting. While few results have been obtained to date, there are rootless mutants (89, 112) and some differences in genes and gene products that have been identified in physiologically mature and juvenile materials (67, 107, 254, 306).

In studies of tobacco plants transformed with root-inducing (Ri T-DNA) of *Agrobacterium rhizogenes*, rooting of the transformed tobacco explants was due to *genes that increased auxin sensitivity* of the tissue. Rooting of transformed plants was not due to genes that regulated auxin production, or to a substantially altered balance of auxin to cytokinin ratio (259). In other studies with nonrooting tobacco mutants, sensitivity to auxin was due to general alteration of the cellular response to auxin and was not due to the increased rate of conjugation of auxins by these tissues, or by disruption of auxin transport (40). *Thus, there are implications that the lack of cell competency in difficult-to-root species may be due to a lack of cell sensitivity to auxin rather than to a suboptimal level of endogenous auxin.*

Just as in biochemical studies, understanding the molecular events of rooting is difficult because only a very few cells in an explant or cutting are directly involved in regeneration—the specific features of these cells are swamped by those of the other cells. Therefore, validation by microscopic studies is needed to determine characteristics specifically in the cells involved in the regeneration event.

BOX 9.12 GETTING MORE IN DEPTH ON THE SUBJECT
GENE REGULATION IN ROOTING OF *PINUS CONTORTA* HYPOCOTYL CUTTINGS

In a very challenging study, the histological events of adventitious root formation of *Pinus contorta* hypocotyl cuttings were correlated to gene expression during five rooting stages using microarray analysis (39). Essentially RNA was harvested at discrete stages of rooting and hybridized to microarrays. The transcript levels of 220 genes and their encoding proteins were either up-regulated (↑increased expression) or down-regulated (↓ decreased expression) (39). Not surprisingly, the highest number of genes were differentially expressed (either up- or down-regulated) during days 0 to 3 (response to: severing the cutting, wounding, exogenous auxin treatment, perceived water stress, decreased photosynthesis, and decreased auxin transport) (Figs. 9–22 and 9–23). The highest up-regulation occurred between days 3 (root initial) to 9 (more defined root meristem—root primordia), which included increased auxin transport, auxin-responsive transcription, cell wall synthesis, and pathogenesis- and hypersensitive-induced response proteins—the latter suggesting further development of a defense barrier—as part of the "wound-healing response." Highest down-regulation occurred during days 0 to 3, and days 12 to 33 (fully developed roots were elongating) roots were fully functional in water uptake, so genes affiliated with water stress and cell replication had reduced expression.

BOX 9.13 GETTING MORE IN DEPTH ON THE SUBJECT
PROTEOMICS AND ROOTING

Since adventitious rooting is known to be a quantitative genetic trait, research is being done with **proteomic analysis** Using different mutant genotypes of *Arabidopsis* has led to the identification of eleven proteins whose abundance was either positively or negatively correlated with endogenous auxin, number of adventitious root primordia, and/or number of mature adventitious roots (258). The identification of regulatory pathways associated with adventitious rooting could lead to valuable markers for future identification of genotypes with better rooting ability.

proteomics The large-scale study of proteins, particularly their structures and functions. The complement of proteins and modifications made to a particular set of proteins will vary with time and distinct requirements during the various stages of adventitious root formation.

BOX 9.14 GETTING MORE IN DEPTH ON THE SUBJECT
CELL COMPETENCY-TO-ROOT

The formation of new centers of cell divisions—called *de novo* meristems, that differentiate into adventitious roots—requires that a cell or group of cells (e.g., phloem ray parenchyma cells or callus cells) embark upon a new developmental program (199). What is the molecular mechanism that controls adventitious organ formation? What is the molecular basis for the plasticity that allows differentiated cells (phloem ray parenchyma) to start new developmental programs? How many different signals are needed for root induction? Why is there a decline or loss of competence for the formation of adventitious roots in physiologically mature-phase shoot tissue, compared with physiologically juvenile-phase tissue? Competence-to-root can be assessed by determining whether tissue is capable of responding in a specific way to inductive treatments (208). A model of the events in the organogenic process of rooting is given in Figure 9–5. Our understanding of cell competency-to-root will be enhanced via the molecular tools, such as microarray analysis of gene regulation during the five discrete stages of rooting in *Pinus contorta* hypoctyl cuttings (Figs. 9–22 and 9–23) (39), and proteomics (258).

rooted only if specific influencing factors are taken into consideration and if the cultivars are maintained at the optimum condition. With most species, the careful selection of cutting material from stock plants or containerized plants, management of cuttings, and control of environmental conditions during rooting are the difference between commercial success or failure. The remainder of the chapter discusses these influencing factors that include:

1. Management of stock plants to maximize cutting propagation
2. Treatment of cuttings
3. Environmental manipulation of cuttings

> **BOX 9.15 GETTING MORE IN DEPTH ON THE SUBJECT**
> **CURRENT STATUS OF ADVENTITIOUS ROOT BIOLOGY**
>
> Significant new biotechnology has not emerged in commercial rooting operations (60, 221). Cuttings are still rooted by a brief exposure (quick-dipped) in a solution containing a moderate to high auxin concentration or via a rooting powder formulation—techniques developed 60 years ago. Where improvements have been made is in the selection and manipulation of stock plants, maximization of environmental controls, and media manipulation during the propagation and transplanting of rooted liner plants.
>
> Much research has focused on finding the **Primary Causes of Rooting**: *genetic potential, metabolic factors, and physiological condition*. Generally, cuttings that do not root are considered deficient in rooting promoters, including hormones. The search for the primary chemical stimulus to root initiation (60, 115, 121) is merely one way of looking at the mechanism of adventitious rooting (e.g., the concept of rooting promoters and inhibitors may have led to undue emphasis on the "*ultimate mechanism of adventitious rooting*").
>
> Hopes for genetic engineering techniques reside in their potential power, which is easily manifest in traits under simple genetic control. However, many genes with unknown modes of action and inheritance control rooting potential. Some 220 genes are either up- or down-regulated during the five discrete development stages of *Pinus contorta* hypocotyl cuttings (39).
>
> Often overlooked are the **Secondary Causes of Poor Rooting**: many leafy woody (and herbaceous!) cuttings have major limitations affecting their *survival* (i.e., they are quite susceptible to stress prior to developing roots) and require good management to avoid mortality (302). Among the secondary causes of poor rooting are low photosynthetic and transpirational capacity of cuttings, loss of plant inertia (abscission of leaves; failure of recently rooted cuttings to put on an initial growth flush prior to fall dormancy, thus incurring high winter losses), environmental stress—inadequate water regimes, desiccation, anaerobic conditions—adverse effects of high auxins on cutting buds and shoots, and so on (305). These problems are discussed in greater detail in the remainder of this chapter and in Chapter 10 where the technology to enhance cutting rootability and survival is addressed.

MANAGEMENT OF STOCK PLANTS TO MAXIMIZE CUTTING PROPAGATION

Selection and Maintenance of Stock Plants for Cutting Propagation

Management of stock plants (or containerized plants) to maximize rooting begins with the *selection* of source material that is easy-to-root (juvenile), *maintenance* of stock plants in the juvenile/transition phase to maximize rooting, and *rejuvenation* of stock plant material (reversal from the mature to a juvenile/transition phase) to reestablish high rooting potential (Table 9–5).

The remainder of this chapter addresses these factors in detail.

Since many containerized ornamental nurseries no longer use stock plants, it is essential to maintain quality control of all production container plants from which propagules are taken. Propagules should be collected from stock plants free of viruses, bacteria,

> **Table 9–5**
> **STOCK PLANT MANAGEMENT. SELECTION AND MAINTENANCE OF STOCK PLANTS FOR CUTTING PROPAGATION REJUVENATION OF STOCK PLANTS: HEDGING, PRUNING, GRAFTAGE, MICROPROPAGATION MANIPULATION OF ENVIRONMENTAL CONDITIONS AND PHYSIOLOGICAL STATUS**
>
> - Water status
> - Temperature
> - Light: duration (photoperiod), irradiance, spectral quality (wavelength)
> - Stock plant etiolation: banding, blanching, shading
> - Girdling
> - Carbon dioxide enrichment
> - Carbohydrates
> - Managing carbohydrate/nitrogen levels of stock plants
>
> **Selection of Cuttings from Stock Plants**
> - Type of wood selected
> - Seasonal timing
> - Predictive indices of rooting

fungi, and other pathogenic organisms. See Chapters 10 and 16 for the discussions on stock plant maintenance including nutritional status, disease control, and the importance of assuring trueness-to-type.

For new cultivars to be commercially successful, they must be relatively easy to propagate and suitable for existing propagation and production systems. New cultivars are, in part, selected for their ease of rooting. Despite how desirable the form, flower color, ornamental characteristics, or yield (fruit crops), it is not economically feasible to use cutting propagation with a new cultivar that has less than 50 percent rooting. Nurseries continually select for plants that are easy to root through the annual harvesting and rooting of cuttings from previously rooted containerized plants in production blocks or stock plants. This **serial propagation** of new generations of rooted cuttings helps *maintain* easy-to-root characteristics of a cultivar.

serial propagation
The annual harvesting and rooting of cuttings from previously rooted, containerized plants to help maintain a high rooting potential from generation to new cutting generation.

There are other horticultural and forestry practices that can maintain stock plants in a physiologically juvenile or transition phase and improve rooting success (53, 146, 151, 167). The development of systems for obtaining whole populations of juvenile and partially juvenile/transition cuttings has revolutionized clonal forestry. For example, seedling and clonal populations of elite germplasm of Monterey pine, loblolly pine, and Douglas-fir are grown as stock plants. They are then subjected to hedging and pruning systems and serial-cutting practices to maintain a high rooting potential. This has exciting opportunities for clonally multiplying elite germplasm and increasing timber yield. The hedging or shearing treatments given Monterey pine (*Pinus radiata*) trees (see Fig. 14–16), stooling of apples (see Fig. 14–1), and pecans (see Fig. 14–12) are quite effective in maintaining juvenility and increasing the rooting potential of cuttings taken from them, compared with nonhedged trees (177, 195).

Rejuvenation of Stock Plants

In difficult-to-root woody plant species, the ease of adventitious root formation declines with the age of parent stock, resulting in a propagation enigma, since desirable characteristics are frequently not expressed until after a plant has reached maturity. The transition from the *juvenile* to the *mature* phase has been referred to as *phase change, ontogenetic aging,* or *meristem aging.*

There are progressive changes in such morphological and developmental characteristics as leaf shape, branching pattern, shoot growth, vigor, and the ability to form adventitious buds and roots (106, 110, 111, 205) (see Chapter 16). Experiments with apple, pear, eucalyptus, live oak, and Douglas-fir have shown that the ability of cuttings to form adventitious roots decreased with increasing age of the plants from seed; in other words, when the stock plant changed from the juvenile to the mature phase. With many woody species, it is the **physiological** or **ontogenetic age,** not **chronological age,** of the cutting that is most important in rooting success (see Chapter 16).

In some species, such as apple, English ivy, olive, eucalyptus, and Koa tree (*Acacia koa*), differences in certain morphological characteristics, such as leaf size and shape, make it easy to distinguish between the mature and the lower, juvenile portions of the plant. In some kinds of deciduous trees, such as oak and beech, leaf retention late into the fall occurs on the basal parts of the tree and indicates the part **(cone of juvenility)** still in the physiologically juvenile stage (see Fig. 16–22). Ideally, cuttings should be taken from juvenile wood.

Inducing Rejuvenation In rooting cuttings of difficult species it would be useful to be able to **induce rejuvenation** to the easily rooted juvenile or transition stage from plants in the mature form. This has been done in several instances by the following methods:

- **Rejuvenation of apple** can be done with mature trees by causing **adventitious buds/shoots to develop from root pieces,** which are then made into softwood stem cuttings, and rooted.
- **Forcing epicormic sprouts** of 2- to 10-cm (1- to 4-in) wide × 24 cm (9.5 in) long branch segments of adult hardwoods is done to produce softwood cuttings with higher rooting success in red and white oaks, white ash, maple, honeylocust, and other species (Fig. 9–24) (91, 223, 280).
- By **removing terminal and lateral buds and spraying stock plants** of *Pinus sylvestris* with a mixture of cytokinin, tri-iodobenzoic acid, and Alar (daminozide), many fascicular buds can be forced out. With proper subsequent treatment, high percentages of these shoots can be rooted (296).
- **Chemical manipulation with gibberellin sprays** on English ivy stock plants can stimulate growth and reversion of some of the branches to the juvenile stage, and improve rooting of cuttings (264).
- In some plants juvenile wood can be obtained from mature plants by **forcing juvenile growth from sphaeroblasts,** wartlike protuberances containing meristematic and conductive tissues sometimes

Figure 9–24
Forcing softwood cuttings from woody stem segments to propagate hardwood species. (a) River birch shoot forcing under intermittent mist, (b) shoot forcing of white ash and silver maple, and (c) epicormic shoots from forced silver maple—will later be harvested as softwood cutting and rooted under mist (223). Courtesy J. E. Preece.

found on trunks or branches. These are induced to develop by disbudding and heavily cutting back stock plants. Using the mound-layering (stooling) method on these rooted sphaeroblast cuttings produces rooted shoots that continue to possess juvenile characteristics (see Fig. 16–23).

- **Grafting mature forms onto juvenile forms** has induced a change of the mature to the juvenile stage, provided that the plants are held at fairly high temperatures (264); such transmission of the juvenile rooting ability from seedlings to mature forms by grafting has also been accomplished in rubber trees (*Hevea brasiliensis*) (209), and with serial graftage of mature difficult-to-root scions onto seedling rootstock of eucalyptus (*Eucalyptus xtrabutii*) (Fig. 9–25, page 310).
- Ready-rooting cuttings can be produced from **stock plants that are produced via micropropagation.** Epigenetic (non-permanent) changes that occur with rejuvenation of tissue *in vitro* has tremendous potential to enhance rooting ability. Stock plants derived from micropropagation exhibit certain juvenile/transition characteristics and produce an increased number of higher-rooting, thin-stemmed cuttings than conventionally produced stock plants (4, 108, 147, 167, 218, 222, 269). The tissue culture effect can be long-lasting depending on the plant species and proper maintenance via severe hedge pruning of stock plants (147, 148). However, without proper stock plant maintenance, the rejuvenation effect may last only one to two generations of cuttings (219). To

> **epigenetic change**
> The heritable changes in gene expression, resulting in changes in phenotype (appearance) or physiology (adventitious rooting potential). There is modification of the activation of certain genes, but no changes in basic DNA structure. These changes may remain through cell division and may last for multiple generations.

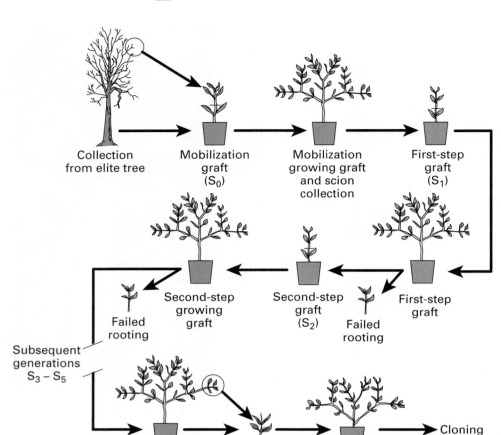

Figure 9–25
Scheme for rejuvenation techniques used in serial graftage of ten-year-old *Eucalyptus xtrabutii* onto juvenile seedling understock. Six serial grafts (S_1 to S_6) were needed before mature grafted scions could be used as cuttings and rooted.

maintain high rooting potential and avoid clonal variation (i.e., habitation and irregular growth), there are advantages of periodically replenishing tissue culture systems with new explant sources and producing new tissue-culture–derived stock plants from which cuttings are selected (see Chapter 16). Stock plants derived from transgenic plants with higher rooting potential or from somatic embryogenesis (synthetic seed technology) may also be used to restore high rooting potential (74, 210), see Figure 17–45.

Manipulating the Environmental Conditions and Physiological Status of the Stock Plant

The physiological condition of stock plants is a function of genotype (species, cultivar) and environmental conditions (water, temperature, light, CO_2, and nutrition).

Water Status There may be *advantages of periodic, controlled drought stress to stock plants*. Controlled water stress of eucalyptus (*Eucalyptus globus*) stock plants enhanced the survivability and rooting of cuttings (303). However, there is experimental evidence to support the view that extreme drought stress of stock plants is not desirable. Studies with cacao and pea (226) cuttings showed reduced rooting when the cuttings were taken from stock plants having a water deficit. Plant propagators often emphasize the desirability of taking cuttings early in the morning when the plant material is in a turgid condition. Unrooted cuttings are particularly vulnerable to water stress, since rehydration of the tissue is very difficult without a root system. Furthermore, droughted cuttings are more prone to disease and pest problems.

Temperature Information on temperature interactions with stock plant water relations, irradiance, and CO_2 is limited. Research has shown that there is a complex interaction of temperature and stock plant photoperiod on the level of endogenous auxins and other hormones (137). With deciduous woody species (apple, plum), higher air temperatures can produce more rapid growth of stock plants and the production of higher-rooting, thinner-stemmed cuttings (148). *In general*, the air temperature of stock plants (12 to 27°C, 54 to 81°F) appears to play only a minor role in the ease of rooting of cuttings (196).

BOX 9.16 GETTING MORE IN DEPTH ON THE SUBJECT
UTILIZING STOCK PLANTS FOR CUTTINGS

Stock plants for cuttings of selected fruit tree rootstocks and woody ornamentals are maintained as hedges rather than allowed to grow to a tree form. Proper hedge management (pruning) of permanent stock plants can maintain large numbers of cuttings in an apparent juvenile stage of development. See Chapter 10 for stock plant pruning and girdling systems.

Severe or hard pruning gives rise to many shoots suitable for cuttings, but their higher rooting potential is not necessarily due to greater vigor (as has long been supposed), but rather to the less vigorous (thinner-diameter), subordinate shoots that root better than the more vigorous ones (e.g., *Prunus, Rhododendron, Syringa*, etc.) (149, 151, 152).

Enhanced rooting potential, with relatively thin shoots of both hardwood and softwood cuttings, is only achieved with an improved propagation environment. Thin cuttings are more susceptible to basal rotting, so good drainage of media and mist management are critical; when planting hardwood cuttings directly in the field, there must be a compromise between the thinner shoots that root more quickly and larger-stem-diameter cuttings that survive longer in the poor conditions often present in field soil during winter (149).

Competence-to-root appears to be controlled independently in individual shoots and is indirectly related to shoot thickness, which favors the subordinate (subdominant) shoots that develop in the shoot hierarchy of the severely pruned hedges (Fig. 9–26). Rooting potential among shoots in a hedge is then more dependent on their relative position, rather than their proximity to the ground (Fig. 9–27). The most vigorous shoots are the poorest rooters but make better hardwood cuttings (151). Thin-stemmed shoots are better propagated as softwood cuttings. They have a higher leaf-to-stem ratio, and greater accumulation of dry matter at the cutting base before the first roots emerge (150).

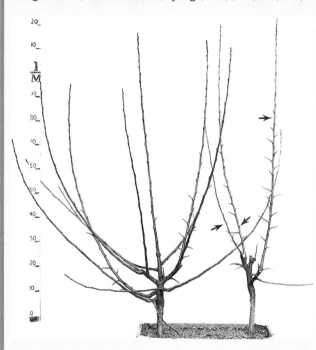

Figure 9–26
Optimum rooting of hardwood cuttings for 'M-26' apple rootstock occurs from the subordinate, thinner shoots that develop in the shoot hierarchy (framework) of severely pruned stock plant hedges. *Left:* An unpruned stock plant with subordinate and dominant (spiny) shoots. *Right:* The thinner, subdominant cuttings have been collected, while only the dominant, spiny shoots (arrows) remain to provide the framework for next year's generation of shoots (cuttings). Courtesy B. H. Howard, Horticulture Research International, East Malling, England.

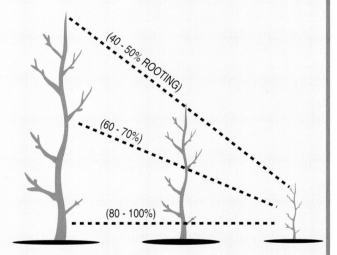

Figure 9–27
Rooting potential (typical values in brackets) of hardwood cuttings in a hardpruned (severely cut back) hedge is more influenced by the relative position of the shoots than by their absolute position in terms of distance between themselves or from the root system (149).

Light *Light duration* (photoperiod), *irradiance* [(W × m^{-2}) or photon flux (μmol · m^{-2} · s^{-1})], and *spectral quality* (wavelength) influence the stock plant condition and subsequent rooting of cuttings (196). (See Chapter 3 for discussion.) For instance, sufficient irradiance to stock plants is needed to maintain minimal endogenous auxin for rooting chrysanthemum cuttings. Conversely, too high an irradiance can cause photo destruction of auxin or adversely affect stock plant water relations.

There is some evidence that the **photoperiod** under which the stock plants are grown exerts an influence on the rooting of cuttings taken from them (45, 159, 196). This could be a photosynthetic or morphogenic effect. If the photoperiod influences photosynthesis, it may be related to carbohydrate accumulation, with best rooting obtained under photoperiods promoting increased carbohydrates. If manipulation of photoperiod favors vegetative growth (rooting) and suppresses reproductive growth (flowering), then the effect is photomorphogenic (124, 262). Long-day conditions (sufficient hours of light to satisfy the critical photoperiod) have been used with some short-day flowering cultivars of chrysanthemum; where flowering is antagonistic to rooting, the long-day conditions promote vegetative growth and enhance rooting (90). Likewise, with some woody perennials where the onset of dormancy shuts down vegetative growth and/or reduces rooting, propagators can manipulate stock plants by extending the photoperiod with low irradiance from an artificial light source (see Fig. 3–14).

Controlling photoperiod and the daily light integral is not always sufficient to maintain vegetative growth. For many crops, ethephon (Florel) is applied once every two to three weeks at rates ranging from 200 to 750 parts per million or higher. Ethephon releases the gas ethylene, which can abort open flowers and flower beds. Ethephon can also increase the cutting yield of annual stock plants by increasing the laters' branching.

Conflicting reports on the influence of **light quality** on stock plants and subsequent rooting of cuttings is attributed to the effect of red and far-red light on rooting (196). *In vitro* rooting of pear cultivars was enhanced under red light and inhibited under far-red light and darkness, which indicates involvement of the phytochrome system in rooting (17). Using light emitting diodes (LED), rooting of *in vitro Tripterospermum* was inhibited by blue and promoted by red light (203). Red shade cloth, which increases the red, while reducing the blue and green spectra, is used in commercial propagation to enhance root initiation and development of cuttings (see Fig. 3–11).

etiolation The development of plants or plant parts in the absence of light, resulting in such characteristics as small unexpanded leaves, elongated shoots, and lack of chlorophyll, which yields a yellowish or whitish color.

Stock Plant Etiolation Reducing irradiance levels of stock plants can sometimes enhance rooting of difficult-to-root species. By definition, **etiolation** is the total exclusion of light; however, plant propagators also use this term when forcing new stock plant shoot growth under conditions of heavy shade. Softwood cuttings are then taken from new growth and often root more readily. **Banding** is a localized light exclusion pretreatment which excludes light from that portion of a stem that will be used as the cutting base (13, 191). Banding can be applied to etiolated shoots or applied to light-grown shoots which are still in the softwood stage. In the latter case, a band of Velcro or black adhesive tape is said to blanch the underlying tissues, since the stock plant shoot accomplished its initial growth in light prior to banding. Shading refers to any stock-plant growth under reduced light conditions (159). For techniques of banding, blanching, etiolation, and shading, see Chapter 10, Table 10–3 and see Figures 10–25 and 10–26.

Anatomical and physiological changes can occur in etiolated stem tissue that enhance rooting. Etiolation of chestnut (*Castanea*) (i.e., covering the stool bed with soil) caused a greater accumulation of starch grains, but no significant change in stem anatomy; however, girdling and then etiolating the shoots increased parenchyma and storage cells above the girdle, reduced sclerenchyma formation, and was the only treatment that rooted (20). Exclusion of light by etiolation, stem banding, or shading greatly enhances a stem's sensitivity to auxin (192, 193). Translocatable factors produced distal to (above) an etiolated segment also enhanced the etiolation effect (13). Etiolation may also reduce the production of lignin [for structural support cells (sclereids, fibers)]; thus, instead of forming lignin, phenolic metabolites may be channeled to enhance root initiation (53, 81).

Rooting in cuttings of *Syringa vulgaris* 'Madame Lemoine' was enhanced when stock plants were grown in the dark for a short period after bud break. Cuttings grown initially in the dark were found to have relatively thin stems, resulting in a higher leaf-to-stem ratio than normal light-grown ones. This was associated with a net accumulation of dry matter at the cutting base before the first roots emerged (Fig. 9–28) (150). A cutting must produce and/or rely on stored carbohydrates in excess of its maintenance requirement for successful rooting to occur, which is why stock plant manipulation (etiolation, hedging) and the rooting environment of the cutting are so critical for successful rooting.

Girdling Girdling, or otherwise constricting the stem, blocks the downward translocation of carbohydrates, hormones, and other possible root-promoting factors and can result in an increase in root initiation. Girdling

Figure 9–28
(a) Etiolation frames (arrow) in place over stock plant hedges of *Syringa vulgaris*. (b) Improved rooting following etiolation of *S. vulgaris* 'Madame Lemoine' (far left) and *S. vulgaris* 'Charles Joly' (second from right). Cuttings from nonetiolated stock plants have poor rooting [second from left and far right (arrows)]. Courtesy B. H. Howard, Horticulture Research International, East Malling, England (149).

shoots prior to their removal for use as cuttings can improve rooting. This practice has been remarkably successful in some instances. For example, rooting of citrus and hibiscus cuttings was stimulated by girdling or binding the base of the shoots with wire several weeks before taking the cuttings (263).

In cuttings from mature trees of the water oak (*Quercus nigra*), a threefold improvement in rooting was obtained when cuttings were taken from shoots that had been girdled 6 weeks previously, especially if a talc powder combined with a mixture of auxin, growth retardant, carbohydrate, and a fungicide was rubbed into the girdling cuts (127). Enhanced rooting of cuttings taken from girdled stock plants has also been obtained with sweet gum, slash pine, and sycamore. Girdling just below a previously etiolated stem section was particularly effective in promoting rooting in apple cuttings (69).

Carbon Dioxide Enrichment With many species, carbon dioxide enrichment of the stock plant environment has increased the number of cuttings that can be harvested from a given stock plant, but there is considerable variation of rooting response among species. Principal reasons for increased cutting yields are increased photosynthesis, higher relative growth rate, and greater lateral branching of stock plants (196). Any benefits of CO_2 enrichment have been limited to greenhouse-grown stock plants and cuttings during conditions when propagation house vents are closed and ambient CO_2 becomes a limiting factor to photosynthesis (i.e., October–March in central Europe). Without adequate light (supplementary greenhouse lighting during low-light-irradiance months), CO_2 enrichment is of minimal benefit (200) (see Chapter 3).

Carbohydrates The relationship between carbohydrates and adventitious root formation remains controversial. Since Krause and Kraybill (169) hypothesized the importance of the carbohydrate-to-nitrogen (C/N) ratio in plant growth and development, rooting ability of cuttings has been discussed in relation to carbohydrate content. The carbohydrate pools of sugars (soluble carbohydrates) and storage carbohydrates (starches or insoluble carbohydrates) are important to rooting as building blocks of complex macromolecules, structural elements, and energy sources (105, 119, 120, 267).

Although stock plant carbohydrate content and rooting may sometimes be positively correlated (122, 139), *carbohydrates do not have a regulatory role in rooting.* A positive correlation between carbohydrate content and rooting may reveal that the supply of current photosynthate is insufficient for supporting optimal rooting (283). High C/N ratios in tissue of cuttings promote rooting but do not accurately predict the degree of rooting response (267). Cuttings use stored carbohydrates in root regeneration, but only in small amounts. Differences in C/N ratios are due mainly to nitrogen rather than carbohydrate content. Nitrogen has been negatively correlated to rooting (122), which suggests that the correlation between high C/N ratios and rooting may be due to low N levels.

Managing Carbohydrate/Nitrogen Levels of Stock Plants Rooting can be enhanced by controlling nitrogen fertility of stock plants such that cutting shoot development is not stimulated by high N levels (233, 291). This avoids the disadvantage of adventitious rooting competing with rapidly developing shoots for carbohydrates, mineral nutrients, and hormones (119).

Generally, maintaining stock plants under a high carbohydrate/high nitrogen level is optimal for rooting cuttings under mist, and a high carbohydrate/low-to-moderate nitrogen ratio is optimal for rooting dormant hardwood cuttings. Cuttings of *Hypericium, Ilex, Rosa,* and *Rhododendron* rooted best when stock plants were suboptimally fertilized, resulting in less-than-maximal shoot growth (233). Very low nitrogen leads to reduced vigor, whereas high nitrogen caused excess vigor; either extreme is unfavorable for rooting. Adequate nitrogen is necessary for nucleic acid and protein synthesis.

It is important to distinguish between the role of carbohydrates in enabling a cutting to survive (*until it roots*) and the role of carbohydrates in rooting itself. In species where unrooted hardwood cuttings were propagated directly in the field without mist, survival is necessary before rooting occurs, hence the need to compromise between thin rooting cuttings (which root better, but have poorer field survival) and larger diameter, carbohydrate-rich cuttings that survive better in the field, but have lower rooting capacity.

To maintain high carbohydrate/low-to-moderate nitrogen ratios of stock plants for optimal rooting of hardwood cuttings, producers can manipulate stock plants as follows:

- Reducing nitrogen fertilization, thus reducing shoot growth and allowing for carbohydrate accumulation.
- Selecting cutting material from lateral shoots, which have slower growth rates and higher carbohydrate storage than fast-growing terminal shoots. [But for plants showing a plagiotropic growth pattern (see Fig. 9–29), use of lateral shoots should be avoided.]
- For maintenance of adequate carbohydrate levels, photosynthate production of greenhouse-grown stock plants can be controlled by increasing light irradiance of supplementary high-pressure sodium-vapor lights (see Fig. 3–14).

Selection of Cuttings from Stock Plants

Type of Wood Selected from Stock Plants In woody perennials, types of materials to use range from softwood terminal shoots of current growth to dormant hardwood cuttings. No one type of cutting material is best for all plants. What may be ideal for one species would be a failure for another. See Table 10–1 for a synopsis of propagation systems with different cutting types. Procedures for certain species or cultivars, however, often may be extended to related species or cultivars (see Chapters 19, 20, and 21).

Differences Between Lateral and Terminal Shoots. In general, with exceptions, softwood cuttings root better from terminal shoots, and the more lignified, semi-

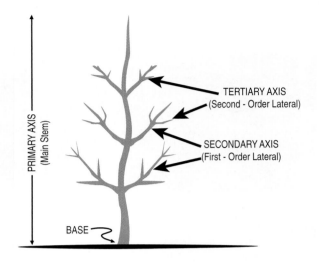

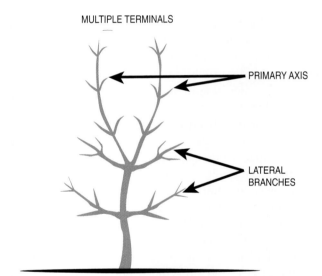

Figure 9–29
The line drawing shows the location where cuttings were taken on stock plants of Frasier fir (*Abies fraseri*). *Top:* A schematic of the branch order. *Bottom:* Demonstrates multiple terminals used as cuttings. Cuttings from lateral branches root readily, but have an undesirable horizontal growth habit (plagiotropic) after rooting. Cuttings taken from the tips of primary axes (main stem) produce symmetrical, upright (orthotropic) trees. Redrawn from Blazich and Hinesley (26).

hardwood cuttings root better from lateral shoots. In rooting different types of softwood plum cuttings taken in the spring, there was a marked superiority in rooting of lateral shoots, compared with terminal shoots. Similarly, lateral branches of Fraser fir (*Abies*) (Fig. 9–29), white pine, and Norway and Sitka spruce gave consistently higher percentages of rooted cuttings than did terminal shoots (26, 261). In rhododendrons, too, thin cuttings made from lateral shoots consistently

BOX 9.17 GETTING MORE IN DEPTH ON THE SUBJECT
SELECTION OF DIFFERENT TYPES OF CUTTINGS

Leafy, softwood cuttings may be the best way to propagate certain difficult-to-root species [e.g., maple (*Acer*), crabapple (*Malus*), redbud (*Cercis*)]. Softwood cuttings tend to have higher auxin and lower endogenous carbohydrate. They have a moderate light requirement, since some photosynthesis enhances their rooting. Their propagation requires more intensive (critical) water management, using mist or fog. Whereas **dormant, hardwood cuttings** have low auxin and high carbohydrate storage, photosynthesis is initially not needed for rooting, and they can be propagated under lower light, without mist or under less critical mist regimes.

---

give higher rooting percentages than those taken from vigorous, strong terminal shoots. In certain species, however, plants propagated from cuttings taken from lateral branches may have an undesirable growth habit: They tend to become **plagiotropic** and have a horizontal branchlike growth after rooting, whereas cuttings taken from primary axes grow upright **(orthotropically)** and produce symmetrical trees, for example, yew (*Taxus cuspidata*), coffee, Norfolk Island pine, and *Podocarpus* (see Figs. 9–29 and 16–27). This effect on growth is referred to as *topophysis*.

plagiotropic
A horizontal branchlike growth habit that is generally not horticulturally desirable.

orthotropic
A desirable, upright growth allowing production of symmetrical plants.

Differences Between Various Parts of the Shoot. With some woody plants, hardwood cuttings are made by sectioning shoots a meter long and obtaining 4 to 8 cuttings from a single shoot. Marked differences are known to exist in the chemical composition of such shoots from base to tip (277). Variations in root production on cuttings taken from different portions of the shoot are often observed, with the highest rooting, in many cases, found in cuttings taken from the basal portions of the shoot. Cuttings prepared from shoots of three cultivars of the highbush blueberry (*Vaccinium corymbosum*) have greater rooting if taken from the basal portions of the shoot rather than from terminal portions (211). Exceptions are found in rose (122) and other species. The number of preformed root initials in woody stems (in some species at least) distinctly decreases from the base to the tip of the shoot (116). Consequently, the rooting capacity of basal portions of such shoots would be considerably higher than that of the apical parts. This factor is of little importance, however, in cuttings of easily rooted species, which root readily regardless of the position of the cutting on the shoot.

Flowering or Vegetative Wood. With most plants, cuttings can be made from shoots that are in either a flowering or a vegetative condition. Again, with easily rooted species it makes little difference which is used, but in difficult-to-root species the state of the plant can be an important factor. For example, in blueberry (*Vaccinium atrococcum*), hardwood cuttings from shoots bearing flower buds do not root as well as cuttings with only vegetative buds. Herbaceous dahlia cuttings bearing flower buds are more difficult to root than cuttings having only vegetative buds (19).

Flowering is a complex phenomenon and can serve as a **competing sink** to the detriment of rooting. Removal of flower buds increased rooting in rhododendron by eliminating the strong competing sink of flower buds for metabolites necessary for rooting (158). With many ornamental species (e.g., *Abelia, Ligustrum, Ilex*, etc.) it is commercially desirable to remove flower buds from cuttings for more rapid root development, earlier vegetative growth, and more efficient liner production (164).

BOX 9.18 GETTING MORE IN DEPTH ON THE SUBJECT
SEASONAL TIMING AND TYPE OF CUTTING WOOD

In propagating **deciduous** species, *hardwood cuttings* can be taken during the dormant season (from leaf fall, when buds are dormant, and before buds start to force out in the spring). Leafy *softwood* or *semihardwood cuttings* could be prepared during the growing season, using succulent and partially matured wood, respectively. The narrow- and broad-leaved evergreen species have one or more flushes of growth during the year, and cuttings can be obtained year-round in relation to these flushes of growth.

Seasonal Timing Seasonal timing, or the period of the year in which cuttings are taken, can play an important role in rooting (51). With many species there is an optimal period of the year for rooting (3). Propagators strive to *maintain the plants momentum* by rooting during these optimal periods to maximize the rooting process and speed up the production of liners. Climate permitting, it is possible to make cuttings of easy-to-root species throughout the year.

Certain species, such as privet, can be rooted readily if cuttings are taken almost any time during the year; on the other hand, excellent rooting of leafy olive cuttings under mist can be obtained during late spring and summer, whereas rooting drops almost to zero with similar cuttings taken in midwinter. Seasonal changes influenced rooting of both juvenile and mature (difficult-to-root) creeping fig (*Ficus pumila*) cuttings; however, treating juvenile (easy-to-root) cuttings with IBA overcame the seasonal fluctuation in rooting (Fig. 9–30). Shoot RNA was found to be an index of bud activity and subsequent seasonal rooting differences (Fig. 9–31). Highest shoot RNA levels and increased vascular cambial activity occurred during peak rooting periods in both the easy-to-root and difficult-to-root forms (51). As previously discussed, micorray analysis of gene expression is being used to better understand rooting events (39). RNA is harvested at distinct developmental periods of rooting and then hybridized to microarrays (Fig. 9–23).

Softwood cuttings of many deciduous woody species [e.g., cherries, lilac (*Syringa*)] taken during spring or summer usually root more readily than hardwood cuttings procured in the winter. The Chinese fringe tree (*Chionanthus retusus*) is notoriously difficult to root, but by taking cuttings during a short period in midspring, high rooting percentages can be obtained.

The effect of timing is also strikingly shown by difficult-to-root deciduous azalea cuttings. These root readily if the cuttings are taken from succulent growth in early spring; by late spring, however, the rooting percentages decline rapidly. For any given species, small experiments are required to determine the optimum time to take cuttings, which is more related to the physiological condition of the plant than to any given calendar date.

Often the effects of timing are merely a reflection of the response of the cuttings to environmental conditions at different times of the year. When hardwood cuttings of deciduous species are taken and planted in the nursery in early spring, after the rest period of the buds has been broken by winter chilling, the results are quite often a complete failure, since the buds quickly

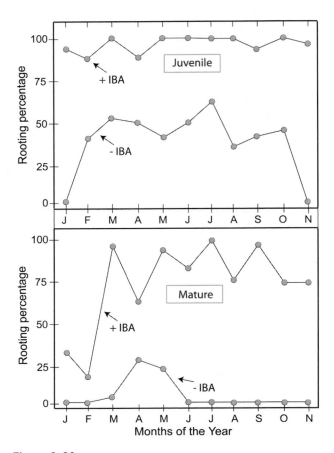

Figure 9–30
Seasonal fluctuation in percent rooting of juvenile (easy-to-root) and mature (difficult-to-root) *Ficus pumila* cuttings. Mature control (–IBA) plants root only from March to May, while juvenile control roots poorly from November to January. When treated with IBA (+IBA), the juvenile cuttings overcome the seasonal fluctuation in rooting, whereas mature cuttings still show poor rooting in January and February (51).

open with the onset of warm days. The newly developing leaves will start transpiring and remove the moisture from the cuttings before they have the opportunity to form roots, and they soon die. Newly expanding buds and shoots are also competing sinks for metabolites and phytohormones, to the detriment of rooting. This competition has been shown with rose (*Rosa multiflora*) under an intermittent mist system where water stress was not a factor (122). If cuttings can be taken and planted in the fall while the buds are still in the rest period, roots may form and be well established by the time the buds open in the spring.

Broad-leaved evergreens usually root most readily if the cuttings are taken after a flush of growth has been completed and the wood is partially hardened-off or lignifed. This occurs, depending upon the species, from spring to late fall. In rooting cuttings of *narrow-leaved evergreens,* best results may be expected if the cuttings

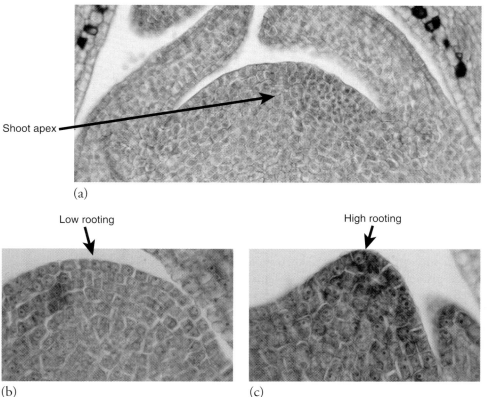

Figure 9–31
Shoot RNA and seasonal effects. (a) Shoot apex of *Ficus pumila* during (b) low seasonal rooting and (c) high rooting, with greater shoot RNA staining during the latter (51).

are taken during the period from late fall to late winter (171). With junipers and yew (*Taxus*), rooting was lowest during the season of active vegetative growth and highest during the dormant period. Furthermore, the low temperatures occurring at the time when such coniferous evergreens root best apparently is not a requirement, since juniper stock plants held in a warm greenhouse from early fall to midwinter produced cuttings that rooted better than outdoor-grown stock plants exposed to seasonal conditions (172).

In many containerized ornamental nurseries, cuttings from difficult-to-root species are taken early in the propagation season, whereas cuttings of easy-to-root species are taken later in the summer. This seasonal scheduling of propagation also more efficiently utilizes propagation facilities and personnel.

Predictive Indices of Rooting Predictive indices of rooting could facilitate clonal selection for rooting traits and reduce rooting variability within a clone. Rooting can be enhanced through well-directed research for predicting and improving propagation potential, and developing more efficient propagation systems, that is, stock plant selection and manipulation, environmental manipulation of cuttings, etc. (304).

Methods to Document the Most Advantageous Time to Collect Cuttings Various predictors of optimum rooting have been used, including calendar days, days from bud-break, use of plants as phenological indicators (plant growth characteristics), number of hours of sunlight, degree-day chilling units of *Juniper* stock plants (189), and the morphological condition of the

---

BOX 9.19 GETTING MORE IN DEPTH ON THE SUBJECT
ELECTRICAL IMPEDANCE AND SEASONAL ROOTING

Seasonal changes in electrical impedance of shoots and leaves of olive (*Olea europaea*) cuttings is related to rooting ability (190). Impedance measurements reveal information about extra- and intracellular fluids and the condition of cell membranes. They have also been used to estimate general plant health, nutrient status, and tissue stress damage. Seasonal rooting ability has been correlated to intracellar and extracellular resistances of shoots and leaves of olive cuttings.

> **BOX 9.20 GETTING MORE IN DEPTH ON THE SUBJECT**
> **PREDICTIVE INDICES OF ROOTING**
>
> Even with relatively easy-to-root clones, rooting ability can vary unpredictably—after 51 harvests of cuttings from 1 clone of eucalyptus (*Eucalyptus globulus*) rooting varied from 14 percent to 100 percent, largely due to environmental variation (304). There are excellent opportunities for developing indices based on *stock plant morphology*, which are more practical if they are sufficiently accurate and easy to measure. For instance, with tropical pine cuttings, shoot attributes of stock plants, including primary needle length, are highly correlated to rooting of cuttings (114). Selection of *Eucalyptus* clones and families has been used to minimize variation in propagation potential, including correlating rooting ability with growth rates of stock plants, leaf thickness, and speculatively, with the frequency of stem sclerenchyma or rays. With *Eucalyptus globulus*, rooting was positively correlated to preharvest extension rates (cm growth per shoot per week) and cutting productivity of stock plants (weekly harvest number per plant) (304).

stock plant. In a novel approach, a degree-day (heat-unit) system was utilized to predict successful rooting in difficult-to-root adult Chinese pistache (*Pistacia*) (73). Maximum rooting occurred when cuttings were collected from stock plants with green softwood stems, which had 380-degree days (using a threshold temperature of 7.2°C/45°F) after bud-break.

TREATMENT OF CUTTINGS

Only high-quality cuttings should be collected for propagation. As the wise instructions to employees in a commercial propagation department go—*"A cutting that is barely good enough is never good enough, so don't put it in the bunch!"* Quality control of cuttings begins with stock plant quality control. Propagation is the foundation on which production horticulture hinges. Marginal quality propagules delay product turnover and create cultural and quality problems throughout the production cycle (7).

Storage of Cutting Material

Propagators prefer to collect propagules from stock plants early in the day when cuttings are still turgid. If the cuttings cannot be stuck immediately, they are misted to reduce transpiration and held overnight in refrigeration facilities (see Fig. 10–55) at 4 to 8°C (40 to 48°F) and generally stuck the next day.

Cuttings of some temperate-zone woody species have been stored at low temperatures for extended periods without any deleterious effects on subsequent root formation and leaf retention. Storage of rhododendron (*Rhododendron catawbiense*) cuttings in moist burlap bags at either 21 or 2°C (70 or 36°F) for 21 days did not reduce rooting (62), although carbohydrate concentrations in the bases of cuttings changed with time and storage temperature. However, cuttings of Foster's holly root poorly even after the shortest cold storage.

Cuttings of many tropical foliage, greenhouse, and nursery crops are imported from Central and South America, the Middle East and other international locations to be rooted and finished in the United States or Europe. It can take 3 to 10 days to deliver cuttings from Central America to U.S. nurseries. Duration in transit can affect cutting quality due to excess respiration, light exclusion, moisture loss, pathogen invasion, and/or ethylene buildup. Croton (*Codiaeum variegatum*) cuttings had excellent quality when stored 5 to 10 days at 15 to 30°C (59 to 86°F) or 15 days at 15 to 20°C (59 to 68°F) (288).

Unrooted cuttings (URCs) of chrysanthemum, poinsettia, and carnation are routinely shipped by air transportation. See Figure 10–11, page 357 which illustrates URCs being precooled to lower temperatures prior to shipping from a vegetative cutting facility in Central America to the United States. Storage life of URCs of geranium (*Pelargonium xhortorum* Bailey) was improved by high-humidity storage in polyethylene bags at 4°C (39°F) and low-irradiance illumination. Prestorage application of antitranspirants was detrimental, but soaking cutting bases in 2 to 5 percent sucrose for 24 hours prior to storage improved rooting (216). The ethylene inhibitor, silver nitrate, was more effective in maintaining storage life than silver thiosulfate, which reduced rooting (216). Abscisic acid will reduce transpiration in geranium cuttings, which may be of practical value in the shipment and storage of geranium cuttings (5).

In general, successful storage of unrooted cuttings depends on storage conditions, state of the cuttings, and species. It is important that dry matter losses and pathogens be minimized. Within the storage unit, it is best to maintain nearly 100 percent humidity, and the temperature should be as low as the hardiness of the given species can tolerate (16). Reduced oxygen and ethylene levels and high CO_2 [controlled atmospheric storage (CA)] help to maintain rooting capacity (16). Storage duration can vary from days to several months, depending on cutting carbohydrate reserves, cold hardiness, and degree of lignification (woodiness of the material) (see the discussion in Chapter 10).

Auxins

Before the use of root-promoting growth regulators (auxins) in rooting stem cuttings, many chemicals were tried with limited success (165). The discovery that auxins, such as **indoleacetic acid (IAA), indole-3-butyric acid (IBA)**, and α-**naphthalene acetic acid (NAA),** stimulated the production of adventitious roots in cuttings was a milestone in propagation history (29, 30, 312). The response, however, is not universal. As discussed earlier, cuttings of some difficult-to-root species still root poorly after treatment with auxin, so auxin is not always the limiting chemical component in rooting, as discussed earlier in this chapter.

An ancient practice of some Middle Eastern and European gardeners in early days was to embed grain seeds into the split ends of cuttings to promote rooting. This seemingly odd procedure had a sound physiological basis, for it is now known that germinating seeds are good sources of auxin, which aids root formation in cuttings.

Mixtures of IBA and NAA Mixtures of root-promoting substances are sometimes more effective than either component alone. For example, equal parts of indole-3-butyric acid (IBA) and α-naphthalene acetic acid (NAA), when used on a number of widely diverse species, were found to induce a higher percentage of cuttings to root and more roots per cutting than either auxin alone (64). Species are also known to react differently when treated with equal amounts of NAA or IBA; NAA was more effective than IBA in stimulating rooting of Douglas-fir (225).

Adding a small percentage of certain phenoxy compounds to either IBA or NAA increased rooting and produced root systems better than those obtained with phenoxy compounds alone (64, 143). Amino acid conjugates of IAA sometimes stimulate better rooting than IAA alone. It has been suggested that the activity of IAA in rooting may depend on its covalent bonding to low molecular weight phenolic compounds (i.e., chemical linkage with sugars, sugar alcohols, etc.) (64, 117).

The acid form of auxin is relatively insoluble in water but can be dissolved in a few drops of alcohol or ammonium hydroxide before adding to water. Salts of some auxins may be more desirable than the acid form in some instances because of their comparable activity and greater solubility in water (313). Also, solvents used to dissolve the acid formulations at higher concentrations—alcohol, NaOH, and others—may be toxic to cuttings.

The aryl esters of both IAA and IBA, and amides of IBA [Phenyl-IAA **(P-IAA)**, Phenyl-IBA **(P-IBA),** phenyl thioester **(P-ITB),** and phenyl amide **(NP-IBA)** (see Fig. 2–25)], have been reported to be more effective than the acid forms in promoting root initiation (64, 118, 268). The physiologically active phenyl-modified auxins are probably enzymatically hydrolyzed after cellular uptake, yielding the free parent acid (i.e., IAA or IBA) and phenolic moiety or portion (64). Again, this is species-dependent. It may also be that these formulations are less toxic to plant material than the acid form.

Auxins are commercially applied as a 1- to 5-second basal, quick-dip, or talc application (30). However foliar sprays of auxin on cuttings are gaining in popularity to reduce worker exposure and the amount of auxin used in the propagation industry. The auxin IBA has an LD_{50} (lethal dosage) of 200 and is considered a pesticide, so there are concerns about worker safety and future restrictions. While foliar sprays of auxin may inhibit shoot growth, for most species there has been good rooting success (29, 30). There are advantages of using water soluble IBA salt formulations as foliar sprays, for example, Hortus IBA http://www.rooting-hormones.com/IBAsalts.htm (71).

For general use in rooting stem cuttings of the majority of plant species, IBA and/or NAA are recommended (64). To determine the best auxin and optimum concentration for rooting any particular species under a given set of conditions, small trials are necessary and should be repeated over several occasions, since repeated experiments can give conflicting results. Also, see Chapters 19, 20, and 21 for specific species recommendations.

Auxin Suppression of Bud-Break of Cuttings
Application of auxins to stem cuttings at high concentrations can inhibit bud development, sometimes to the point at which no shoot growth will take place even though root formation has been adequate. Application of auxins to root cuttings may also inhibit the initiation and development of shoots from such root pieces. Basally applied IBA increased rooting but inhibited bud-break of single-node rose stem cuttings. IBA was translocated to the upper part of the cutting, where it inhibited bud-break and increased ethylene synthesis of the cuttings (272).

Early bud-break and shoot growth of newly rooted cuttings are important in the overwinter survival of *Acer, Cornus, Hamamelis, Magnolia, Prunus,* and *Rhododendron* (305). These species need to put on a growth flush (after rooting but prior to winter dormancy) so that sufficient levels of carbohydrates are stored in the root system to ensure winter survival. Hence, there is concern about auxin-suppressing budbreak and growth of rooted cuttings—and reduced winter survival. See Chapter 10 for discussion on post-propagation care of rooted cuttings.

Shelf-Life of Auxins There is often a question of how long the various root-promoting preparations will keep without losing their activity. Bacterial destruction of IAA occurs readily in unsterilized solutions. A widely distributed species of *Acetobacter* destroys IAA, but the same organism has no effect on IBA. Uncontaminated solutions of NAA and 2,4-D maintained their strength for as long as a year. Of course, alcohol solutions of auxin will depress microbial activity.

IAA is sensitive to light and is readily inactivated. Concentrated IBA solutions in 50 percent isopropyl alcohol are quite stable and can be stored up to 6 months at room temperature in clear glass bottles under low-light conditions without loss in activity (237). Both NAA and 2,4-D seem to be light-stable. Indoleacetic acid oxidase in plant tissue will break down IAA but has no apparent effect on IBA or NAA.

Movement of Auxins in Cuttings In stem tissue, auxin generally moves in a basipetal direction (apex to base). The naturally occurring auxins, IAA and IBA, and the synthetic auxin, NAA, are translocated via polar transport, while the synthetic auxin, 2,4-D, has little polar transport (29). Synthetic auxins were originally applied to cuttings at the apical end to conform to the natural downward flow. As a practical matter, it was soon found that basal applications gave better results. Sufficient movement carried the applied auxin into parts of the cutting where it stimulated root production. In tests using radioactive IAA for rooting leafy plum cuttings, IAA was absorbed and distributed throughout leafy cuttings in 24 hours, whether application was at the apex or base (270). However, with basal application, *most* of the radioactivity remained in the basal portion of the cuttings. Leafless cuttings absorbed the same amount of IAA as leafy cuttings, indicating that transpiration "pull" was not the chief cause of absorption and translocation.

In a study comparing auxin uptake of radioactive NAA in cuttings treated by dilute soak, quick-dip, or talc methods, auxin movement occurred in the vascular system of the stem with the talc and aqueous auxin applications (100). In contrast, auxin in a 50 percent alcohol solution entered the stem from the cut surface and the epidermis throughout the area of the stem quick-dipped in solution. Hence, the solvent used for the quick-dip application facilitated auxin movement through the epidermis and the cut surface of the stem.

Mineral Nutrition of Cuttings During Rooting

Optimal nutrition is needed for adventitious rooting and to assure that root development and production of rooted liner plants precedes smoothly. While it is important to maintain stock plants under optimum nutrition prior to the collection of cuttings, it is difficult to quantify the effect of nutrition on root primordia initiation versus root primordia elongation (25, 273). Mobilization studies have been conducted to examine the movement of mineral ions into the base of cuttings during root initiation. The redistribution of nitrogen in stem cuttings during rooting was accelerated by auxin treatment of plum (271). However, N was not mobilized, nor was any redistribution of P, K, Ca, and Mg detected during root initiation in stem cuttings of Japanese holly (23, 24).

There are conflicting reports on mobilization. During root initiation in chrysanthemum cuttings, P, but not N, K, or Ca, was mobilized. Although considered immobile, redistribution of Ca was reported during rooting of Japanese holly. Apparently, Ca was redistributed to support tissue development in the upper cutting sections and not for root growth and development.

The importance of N in root initiation is supported by nutrition studies on rooting of cuttings and the importance of N in nucleic acid and protein synthesis (25). The influence of N on root initiation and development also relates to such factors as carbohydrate availability, C/N ratio, and hormonal interactions.

Zinc can promote the formation of the auxin precursor, tryptophan, and the subsequent formation of auxin (IAA) from tryptophan. Conversely, Mn acts as an activator of the IAA-oxidase enzyme system and B may enhance IAA-oxidase activity, thus regulating endogenous auxin levels (Fig. 9–19) (157, 290). Higher endogenous auxin levels are required for early root initiation than for later root development (see Fig. 9–19 and discussion). If root initiation is related to the relative activity of IAA and IAA-oxidase, then rooting may be correlated with changes in relative Zn, Mn, and B concentration at the site of root initiation during the developmental stages of *de novo* rooting.

In a study with poinsettia where mineral element concentration was analyzed during the developmental

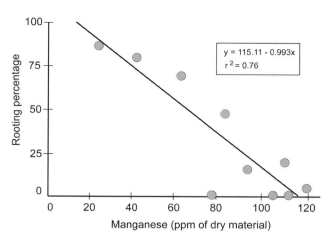

Figure 9–32
Correlation between manganese content of leaves of different avocado clones and rooting percentage. From data of Reuveni and Raviv (234).

stages of rooting, Fe, Cu, and Mo increased in the basal portion of stem cuttings during early root initiation, while P, K, Ca, and Mg decreased (273). During root primordia elongation and root emergence, Fe, Cu, Mo, Mg, Mn, B, and Zn concentration continued to increase at the cutting bases, but P and K concentrations remained low compared to when cuttings were initially inserted into the propagation medium.

High levels of Mn were found (234) in leaves of cuttings taken from difficult-to-root avocado cultivars, whereas cuttings from easy-to-root cultivars had a much lower manganese level (Fig. 9–32). The negative correlation with rooting may be linked to manganese's activating the IAA-oxidase system and lowering endogenous IAA levels (157).

Leaching of Nutrients

The development of intermittent mist revolutionized propagation, but mist can severely leach cuttings of nutrients. This is a particular problem with cuttings of difficult-to-root species, which take a longer time to root under mist. Mineral nutrients such as N, P, K, Ca, and Mg are leached from cuttings while under mist (24).

Nitrogen and Mn are easily leached; Ca, Mg, S, and K are moderately leached; and Fe, Zn, P, and Cl are leached with difficulty (278). Both leaching and mineral nutrient mobilization contribute to foliar deficiencies of cuttings (25). The amount of leaching depends on the growth stage of the cutting material: leafy hardwood cuttings are reported to be more susceptible than softwood or herbaceous cuttings. Apparently, young, growing tissues more quickly tie up nutrients by using them in the synthesis of cell walls and other cell components. Greater leaching occurs with leafy hardwood cuttings, since a greater portion of nutrients is in exchangeable forms.

High leaching rates are avoided by reducing the misting frequency and using mist nozzles that supply smaller volumes of water (247). Foliar nutrition of poinsettia cuttings was significantly reduced during the first week on the mist bench (273, 298)

As a whole, mist application of nutrients has not been a viable technique to maintain cutting nutrition. Nutrient mist application can inhibit rooting (163) and stimulate algae growth, which causes sanitation and media aeration problems (308).

A commercial technique is to apply moderate levels of controlled slow-release macro- and microelements to the propagation media either preincorporated into the media prior to sticking cuttings or by top-dressing (broadcast) during propagation. These supplementary nutrients do not promote root initiation (160) but rather improve root development after root primordia initiation has occurred. Hence, turnover of rooted cuttings occurs more quickly and plant growth is maintained by producing rooted liners that are more nutritionally fit. Optimum levels of fertilization for rooting need to be determined on a species-specific basis (see Chapter 10).

Wounding

Cuttings are naturally wounded when excised from stock plants. Additional basal wounding is beneficial in rooting cuttings of certain species, such as

---

BOX 9.21 GETTING MORE IN DEPTH ON THE SUBJECT
WOUNDING-RELATED COMPOUNDS (WRCS)

- Wounding of cuttings results in destruction of cell compartments (vacuoles, vesicles, per- oxisomes, plastids), which leads to synthesis and/or release of catabolic enzymes (glucanases, peroxidases, phospholipases, lipoxygenases) present in cell organelles.

- Breakdown products of these cell structures are called wounding-related compounds (WRCs).
- WRCs play an important role in rooting and enhance rooting when applied with low auxin concentration (67).

rhododendrons and junipers, especially cuttings with older wood at the base. Following wounding, callus production and root development frequently are heavier along the margins of the wound. Wounded tissues are stimulated into cell division and production of root primordia (Figs. 9–33 and 9–34) (187), due to a natural accumulation of auxins and carbohydrates in the wounded area and to an increase in the respiration rate in the creation of a new "sink area." In addition, injured tissues from wounding produce ethylene, which can indirectly promote adventitious root formation (67, 68, 206, 311). See the schematic on the physiological and biochemical events in severing a cutting for rooting (Fig. 9–21).

It has been proposed that wounding a cutting initiates a chemical signal that induces changes in the metabolism of affected cells (300). A listing of metabolic responses to wounding is given in Table 9–6. Potentially, cells at the base of the cutting influenced by wounding have enhanced receptivity to respond to auxin and other morphogens (nonauxin endogenous compounds) essential to rooting (Fig. 9–21) (300, 301).

Wounding cuttings may also permit greater absorption of applied growth regulators by the tissues at the base of the cuttings. In stem tissue of some species, there is a sclerenchymatic ring of tough fiber cells in the cortex external to the point of origin of adventitious roots. There is evidence in a few species (15) that newly formed roots may have difficulty

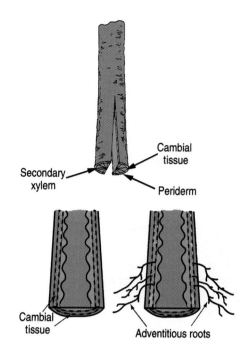

Figure 9–33
(a) Split-base treatment to enhance root initiation in leafless, dormant apple rootstock cuttings. (b) The inner surface of a hardwood cutting wounded by splitting. When split longitudinally, much more cambium is exposed than with a normal cut across the stem base; cambial cells are able to regenerate in response to auxin treatment and produce cambial callus. (c) Roots emerging from cambial callus. Redrawn from MacKenzie et al. (187) and Howard, Horticulture Research International, East Malling, England (150).

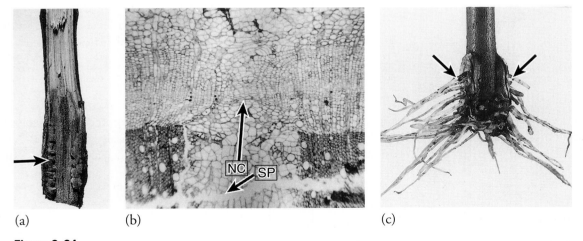

Figure 9–34
(a) Split-base treatment. One-half of the split-stem base has been removed to show the nodular callus (arrow) in the split. (b) Split base—transverse section near the apex of the wound. The split (Sp) here is narrow and, consequently, the new cambium (NC) (tip of arrow) reforms across the split instead of forming callus protrusion and roots as in split base. (c) Note roots emerging in a double rank from the same side of the wound (187).

Table 9–6
SOME PLANT METABOLIC RESPONSES TO WOUNDING

Increase in ascorbic acid	Increase in phenolics
Increase in fatty acids	Evolution of ethylene
Increase in lipids	Increase in terpenoids
Systemic chemical signal	Systemic electrical signal
New membrane synthesis	Peroxidation of membranes
Weakened cell membranes	Induction of cyanide-insensitive pathway
Ion influx into cells	Increased capacity for protein synthesis

Source: Wilson and van Staden (300).

penetrating this band of cells. In those species, a shallow wound would cut through these cells and enhance the emergence of the developing roots.

ENVIRONMENTAL MANIPULATION OF CUTTINGS

Water Relations—Humidity Control

The loss of water from leaves may reduce the water content of the cuttings to such a low level that they do not survive. Propagation systems are designed to maintain:

- **An atmosphere** with low evaporative demand, minimizing transpirational water losses from cuttings and, thereby, avoiding substantial tissue water deficits (cuttings without roots lack effective organs to replace transpired water lost), and cells must maintain adequate *turgor* for the initiation and development of roots (55);
- **Acceptable temperatures** for the regeneration processes occurring at the cutting base, while avoiding the heat stress of leaves; and
- **Light levels** suitable for photosynthesis and carbohydrate production for the maintenance of the cuttings and for use, once root initiation has occurred, without causing water stress (181). See Chapter 3 for the discussion on environmental management and the water relations of propagules.

The water status of cuttings is a balance between transpirational losses and uptake of water. Water absorption through the leaves is *not* the major contributor to water balance in most species. Rather, the cutting base and any foliage immersed in the propagation media are main entry points for water (182). Relative water content is lowest during the first days of sticking poinsettia cuttings and increases with primordia development and root elongation (274, 298). Water uptake of cuttings is directly proportional to volumetric water content of the propagation media, with wetter media improving water uptake (Fig. 9–35) (103, 232). However, excess water reduces media aeration (86) and can lead to anaerobic conditions and the death of cuttings.

Water uptake in cuttings declines after they are initially inserted into propagation media. This decline

Figure 9–35
Water uptake by cuttings is directly proportional to the volumetric water content of the rooting medium. Here, softwood cuttings of *Escallonia xexoniensis* are inserted in a peat-pumice mix containing 15, 20, 40, and 60 percent by volume of water, and in water (left to right). The degree of wilting relates to the water content. While the cutting in 100 percent water is turgid, most species will not tolerate such an anaerobic environment. Courtesy K. Loach.

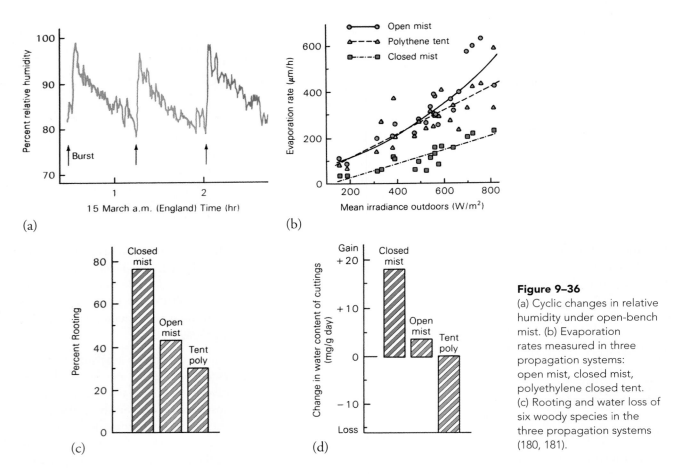

Figure 9–36
(a) Cyclic changes in relative humidity under open-bench mist. (b) Evaporation rates measured in three propagation systems: open mist, closed mist, polyethylene closed tent. (c) Rooting and water loss of six woody species in the three propagation systems (180, 181).

in hydraulic conductivity of cuttings is apparently caused by blockage of xylem vessels and/or collapse of tracheids, which is similar to post-harvest problems observed with cut flowers (153). Another advantage in wounding cuttings is to increase the contact area between the cutting base and propagation medium, thus improving water uptake of cuttings (55, 103, 182). With the formation of functional adventitious roots, new vascular connections occur between the roots and stem. Thus, the hydraulic contact between the propagation medium water and the cutting is maintained.

Degree of stomatal opening can be a useful indicator to determine if a given propagation system is maintaining adequate turgor of cutting leaves. Simpler and more useful systems are to measure evaporation rates directly with evaporimeters (Fig. 9–36) (131, 179, 182) or measuring transpirational losses (298). When water deficits cause stomata to close, CO_2 diffusion into the leaf is restricted, limiting photosynthesis and any subsequent carbohydrate gain in the cuttings. Carbon gain due to photosynthesis is probably more important *after* root initiation has occurred to promote rapid development of roots. It has been reported that translocation of photosynthate from leaves of intact plants continues under moderate or severe stress.

Vapor Pressure Deficit (VPD) Water loss from cuttings is the difference between vapor pressure between the cutting leaf and surrounding air of the mist bed (Figs. 9–37 and 9–38). Water potential of unrooted loblolly pine cuttings has been correlated with VPD, mist application, and rooting percentage (173, 174). Ambient VPD (measurement of general propagation house area) is not dynamic enough to be used as a controlling mechanism. However, VPD determined at the stem-cutting level with temperature and relative humidity probes being misted along with the stem cuttings to provide real-time data of the cuttings is sufficiently sensitive as a dynamic controlling mechanism for misting. If VPD between the cutting and air is high, misting occurs more frequently, and misting is less frequent when there is low VPD (174). Cuttings can tolerate a certain amount of water stress, and moderate stress [-1.0 MPa (-10 bars)] enhances rooting of

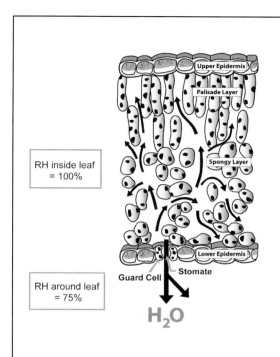

Figure 9–37
Controlling vapor pressure deficit (VPD) during cutting propagation. Leaf cross section with high 100 percent internal relative humidity (RH). Water vapor exits the leaf stomata into the lower RH (lower water potential) of the outside surrounding air.

loblolly pine cuttings (173). While leafy cuttings would not tolerate as low a stress, moderate stress is beneficial (178, 180, 182).

The driving force that determines the rate at which cuttings lose water is the difference in pressure between water vapor in the leaves (V_{leaf}) and that in the surrounding air (V_{air}). Commercial propagation systems aim to minimize this difference either by decreasing V_{leaf} through reducing leaf temperature (e.g., with intermittent mist) and/or by increasing the V_{air} by preventing the escape of water vapor (i.e., with an enclosed polytent). Enclosed systems use humidification since

(a)

(b)

Figure 9–38
(a and b) Water loss from cuttings during mist propagation is the difference between the vapor pressure between the leaf and the surrounding environment. This is vapor pressure deficit (VPD). For vapor pressure deficit models, sensors (light, temperature, humidity) send data to a computer that calculates the VPD for the greenhouse environment. Crop models use VPD to estimate water loss from cuttings to initiate misting. Also see Figure 10–42, page 389.

only the V_{air} is increased. Intermittent mist affects primarily V_{leaf} but also provides a modest increase in V_{air}. Methods used to control water loss of leaves (181) are:

- **intermittent mist**—open and enclosed mist systems;
- **nonmisted enclosures**—outdoor propagation under low tunnels or cold frames, or nonmisted enclosures in a glasshouse or polyhouse (shading, tent, and contact polyethylene systems, wet tents); and
- **fogging systems.**

Intermittent Mist. Intermittent mist has been used in propagation since the 1940s and 1950s (178). Mist systems minimize V_{leaf}, which lowers the leaf-to-air vapor pressure gradient and slows down leaf transpiration.

> **intermittent mist** The periodic application of small amounts of water or "mist" to the leaves and shoots of cuttings during propagation.

Mist also lowers ambient air temperature, and the cooler air consequently lowers leaf temperature by **advection**, in addition to cooling occurring through evaporation of the applied film of water (181). Advective cooling occurs only minimally in enclosed nonmist systems. Since intermittent mist lowers medium temperature, suboptimal temperatures can occur, which reduce rooting. A common commercial technique to control the rooting medium temperature is to use bottom (basal) heat both with indoor and outdoor mist systems (see Figs. 3–9 and 10–37). **Enclosed mist** (see Fig. 3–18) utilizes polyethylene-covered structures in glasshouses that reduce the fluctuation in ambient humidity that is common to open-bench mist (Figs. 3–18 and 10–36). Enclosed mist also ensures more uniform wetting of foliage since air currents are reduced. There are advantages to using enclosed mist with difficult-to-root species, compared to **open mist** (Fig. 9–38) or a polytent system without mist (see Fig. 10–36, page 384).

> **advection** The horizontal movement of a mass of air that causes changes in temperature or in other physical properties of air (i.e., movement of cool air mass).

Major advantages of **enclosed systems** are their simplicity and low cost. The main disadvantage is that they trap heat if light irradiance is high. The trapped heat reduces the relative humidity of the air, and leaf temperature rises to increase the leaf-to-air vapor pressure gradient and, consequently, leaves lose water. Shading must be used with these systems. Polyethylene films have a low permeability to water vapor loss but allow gas exchange. They are used to cover outdoor propagation structures as well as for closed mist systems in greenhouses. Modified polyethylene films are now available with additives of vinyl acetate, aluminum, or magnesium silicates, which increase their opacity to long-wave radiation (i.e., reduce heat buildup). Polyethylene-covered structures have replaced many of the traditional glass-covered cold frames (see Fig. 3–17).

Nonmisted Enclosures. Nonmisted enclosures in a glasshouse or polyhouse can be used for difficult-to-root species and have the advantage of avoiding nutrient-leaching problems of intermittent mist, yet affording greater environmental control than outdoor propagation. The shading system entails applying shading compounds to greenhouse roofs and/or utilizing automatically operating light-regulated shading curtains (see Fig. 3–18). Shading systems are integrated with temperature control by ventilated fogging or pad-and-fan cooling and heating.

Contact Systems. Contact systems entail laying polyethylene, spun-bound polyester, or polypropylene sheets *directly onto cuttings* that are watered-in (see Fig. 10–36). When the irradiance and air temperature can be controlled, leaves tend to be cooler under contact polyethylene because they are in direct contact with the polyethylene and are moistened by condensation forming under the cover. Thus, there is the dual benefit that some evaporative cooling can occur and that water loss from the foliage is reduced since the condensation contributes to the relative humidity of the air rather than solely internal water from the leaf tissue. Hence, there is less internal water stress than with drier leaves in an **indoor polytent system**. Well-managed, enclosed nonmist systems offer a low-tech, cost-effective alternative to mist and fog systems, and may be superior to mist when irradiance and temperature levels are relatively low (207).

Fog Systems. **Fog systems** maximize V_{air} by raising the ambient humidity. Fog generators produce very fine water droplets that average 15 µm in diameter and remain suspended in the air for long periods to maximize evaporation (see Figs. 10–44 and 10–45). Their surface/volume ratio is high (compared with larger water droplets produced from a deflector-type mist nozzle) so that the finely divided mist particle has a larger surface, which also increases evaporation (181). With fog, water passes into the air as a vapor rather than condensing and

> **fog systems** Similar to intermittent mist, except the particle size of the water applied is much finer and water does not condense on the surfaces of the cutting.

wetting leaf surfaces, as mist does. Thus, fog systems avoid foliar leaching and oversaturation of media. Since the propagation media are not saturated and cooled to the same degree as with mist, suboptimal rooting temperatures are avoided and less basal heat is needed. A disadvantage of the fog system is the higher initial cost and maintenance.

There are advantages to using fog over either open or enclosed mist systems, particularly with difficult-to-root plants (129, 130) and with the acclimation and *ex vitro* rooting needs of tissue-culture-produced liners (see Chapter 18). Propagators must decide which is the most cost-effective system for their particular needs.

Temperature

Temperature of the propagation medium can be suboptimal for rooting due to the cooling effect of mist or seasonally related ambient air temperature. It is more satisfactory and cost-effective to manipulate temperature by heating at the propagation bench level rather than by heating the entire propagation house. See Chapters 3 and 10 for heating equipment systems.

The consensus regarding the optimum medium temperature for propagation is 18 to 25°C (65 to 77°F) for temperate-climate species and 7°C (13°F) higher for warm-climate species (75, 166). Daytime air temperatures of about 21 to 27°C (70 to 80°F) with nighttime temperatures about 15°C (60°F) are satisfactory for rooting cuttings of most temperate species, although some root better at lower temperatures. High air temperatures tend to promote bud elongation in advance of root initiation and to increase water loss from the leaves. It is important that adequate moisture status be maintained by the propagation system so that cuttings gain the potential benefit of the higher basal temperature.

Root initiation in cuttings is temperature-driven, but subsequent root growth is strongly dependent on available carbohydrates. This is particularly evident in leafless hardwood cuttings in which excessive root initiation and growth can so deplete stored reserves that there are insufficient available carbohydrates for satisfactory bud growth. The same principle holds true for leafy cuttings (semihardwood, softwood, herbaceous), where shoot growth can divert carbohydrates away from developing root initials and thereby slow root growth (Fig. 9–39).

The optimum air temperature for growing a crop is probably the best for rooting cuttings (220). Bottom heat should be manipulated in two phases, with a higher beginning temperature for **root initiation** and a lower temperature for **root development** and growth (75, 166). Optimum temperature for root initiation in *Forsythia* and *Chrysanthemum* was 30°C (86°F), whereas root development (elongation of primordia and protruding of roots from the stem cutting) was optimum at lower temperatures of 22 to 25°C (72 to 77°F). Respiration is reduced at the lower temperature, which allows more optimum photosynthate accumulation for root development.

A system for predicting rooting stages in poinsettia cuttings was developed using root-zone temperature-based models (299). Optimum rooting for root initiation and elongation stages were 28°C (82°F) and 26°C (79°F), respectively. Rooting did not occur at 20°C (68°F) or less, and was reduced at 32°C (90°F) or more (Fig. 9–40, page 328). See Chapter 3 for the discussion on temperature in propagation.

Figure 9–39
Cuttings of *Forsythia Xintermedia* 'Lynwood' rooted under open-bench mist (left) and in a misted, polyethylene enclosure (right). Note that cuttings rooted in the warmer, more humid enclosure break bud and grow faster than those under open mist. However, for some species and circumstances, too much top growth can divert carbohydrates away from developing root initials and slow root growth. Courtesy K. Loach.

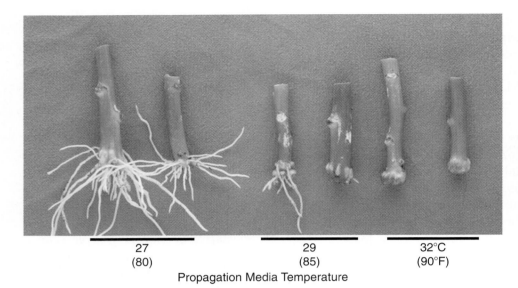

Figure 9–40
Effect of temperature on rooting poinsettia cuttings at 27, 29, and 32°C (80, 85, and 90°F). A temperature of 27°C was optimal. Root induction and initiation temperature is higher than during the later stages of root elongation (300).

Light

As discussed earlier in the stock plant manipulation section and in Chapter 3, light is a contributing factor in the adventitious root and bud formation of cuttings (70, 76).

Irradiance. Cuttings of some woody plant species root best under relatively low irradiance (159, 181). However, cuttings of certain herbaceous plants, such as chrysanthemum, geranium, and poinsettia, root better when the irradiance increased to 116 W/m^2 during trials in winter months. Very high irradiance (174 W/m^2) damaged leaves on the cuttings, delayed rooting, and reduced root growth. With selected temperate species under an English propagation system, acceptable light ranges were 20 to 100 W/m^2 (181). Propagators need to determine irradiance levels to fit their particular production systems. See Chapter 3, page 52 for an explanation of light units.

Most vegetative annuals used for greenhouse and nursery production root within 2 to 3 weeks. Managing light intensity is a key component for successful rooting (183, 184). When light levels are too high, cuttings experience stress and wilt, which delays rooting. When light levels are too low, root formation is delayed, increasing propagation time (Fig. 9–41). Desirable levels of light vary with the stage of root development (77).

Photoperiod. In some species, the photoperiod under which the cuttings are rooted may affect root initiation; long days or continuous illumination are generally more effective than short days (44), although in other species photoperiod has no influence (253).

The relationship of photoperiod and organogenesis is complex, since photoperiod can affect shoot development as well as root initiation. For example, in propagation by leaf cuttings, there must be development of adventitious buds and roots. Using *Begonia* leaf cuttings (135), where the light irradiance was adjusted so that the total light energy was about the same under both long days and short days, it was found that short days and relatively low temperatures promoted adventitious bud formation on the leaf pieces, whereas short days suppressed adventitious root formation. Roots formed best under long days with relatively high temperatures.

In rooting cuttings of 'Andorra' juniper, pronounced variations in rooting occurred during the year, but the same variations took place whether the cuttings were maintained under long days, short days, or natural daylength (171). A number of tests have been made of the effect of photoperiod on root formation in cuttings, but the results are conflicting; hence, it is often difficult to generalize (8, 76, 78, 171).

Herbaceous short-day flowering crops such as poinsettia and chrysanthemum are routinely rooted under long-day conditions to stimulate rooting and inhibit the competing sink of flowering. Once rooted, the plants are switched to short-days to encourage flowering.

In some plants, photoperiod will control growth after the cuttings have been rooted. Certain plants cease active shoot growth in response to natural decreases in daylength. This is the case with spring cuttings of deciduous azaleas and dwarf rhododendrons, which had rooted and were potted in late summer or early fall.

Figure 9–41
Effect of photosynthetic daily light interval (DLI) on producing marketable rooted liners of petunia. The higher the DLI, with sufficient environmental controls to minimize desiccation, the more rapidly rooted liners are produced (183, 184). Photo courtesy R. Lopez & E. Runkel.

Improved growth of such plants was obtained during the winter in the greenhouse if they were placed under continuous supplementary light, in comparison with similar plants subjected only to the normal short winter days. The latter plants, without added daylength, remained in a dormant state until the following spring.

High carbohydrate reserves are important for rooted cuttings (liners), since spring growth in a deciduous plant depends on reserves accumulated during the previous growing season. With red maple (*Acer rubrum*), a night interruption lighting period to extend the natural photoperiod in order to maintain high carbohydrate reserves enhanced growth of rooted liners; however, this was not economically justified, since growth of natural daylength liners was comparable after 2 years of field culture (252).

Light Quality. Lighting that provides more red than far-red light increases rooting in many greenhouse crops (196). It is conceivable with certain plant species that root initiation is regulated by red and far-red light through the phytochrome system. Radiation in the orange-red end of the spectrum seems to favor rooting of cuttings more than that in the blue region, but there are conflicting reports. Using light emitting diodes (LEDs) red light enhance and blue light inhibited *in vitro* rooting of *Tripterospermum* (203). Red shade cloth (e.g., ChromatiNet Red http://www.polysack.com/) that enhances the red and far-red, while reducing the blue, green, and yellow spectra, is being used in mist propagation and tissue culture production to enhance rooting (see Fig. 3–11).

Photosynthesis of Cuttings. Photosynthesis by cuttings is not an absolute requirement for root formation, as has been observed in leafy cuttings forming roots when placed in the dark (61) and with leafless hardwood cuttings that root. Increasing light irradiance has not always promoted rooting, and net photosynthesis of unrooted cuttings is saturated at relatively low PAR (irradiance measured as photosynthetically active radiation) levels (61); hence, high PAR does not enhance photosynthesis and could potentially lead to desiccation of cuttings. Unrooted *Acer rubrum* cuttings are much more prone to drought stress,

> **BOX 9.22 GETTING MORE IN DEPTH ON THE SUBJECT**
> **OPTIMIZING ENVIRONMENTAL CONTROLS IN THE ROOTING OF LEAFY UNROOTED CUTTINGS (URCs)**
>
> Over one billion unrooted cuttings (URCs) are produced offshore and sent to greenhouse and nursery operations in the United States. Most vegetative annual URCs can be fully rooted within 2 to 3 weeks—if proper environmental conditions are maintained. While growers have little influence on the stockplant management techniques and the methods employed to harvest, store, and ship these URCs, they can improve how they propagate URCs to reduce rooting time and increase profitability. The critical environmental factors to manage during rooting are:
>
> - controlling light intensity;
> - providing adequate mist;
> - maintaining high relative humidity;
> - maintaining desirable air and media temperatures; and
> - limiting air flow around leaves (to minimize desiccation and maintain a low vapor pressure deficit between leaves and surrounding air) (183, 184).
>
> **Managing Light Intensity** Desirable levels of light vary with the stage of root development.
>
> **Stage 1: sticking to callus formation.** During the early stages of propagation, maximum recommended light intensity is between 120 to 200 $\mu mol \cdot m^{-2} \cdot s^{-1}$ (600 to 1,000 foot-candles) to provide photosynthate for callus formation and root initiation without causing desiccation. In addition, light transmission through the propagation house should be indirect or diffuse via exterior shade or retractable shade curtains.
>
> **Stage 2: after initial rooting.** Once roots have initiated (generally 5 to 12 days after sticking), maximum light intensity can be increased to 200 to 500 $\mu mol \cdot m^{-2} \cdot s^{-1}$ (1,000 to 2,500 foot-candles). Light should be diffuse.
>
> **Stage 3: after roots fill half the plug.** Once cuttings are moderately well rooted into the plug tray or liner (generally 10 to 16 days after sticking), light levels should be increased to near production levels of 500 to 800 $\mu mol \cdot m^{-2} \cdot s^{-1}$ (2,500 to 4,000 foot-candles) to acclimate plants to the post-propagation environment.
>
> *Ideal propagation conditions for rooting and growth of rooted liners* for New Guinea impatiens (*Impatiens hawkeri*) and petunia (*Petunia* × *hybrid*) include:
>
> - 8.5 to 5.4 daily light integral (DLI) $\mu mol /m^{-2}/day$ (see Fig. 9–41)
> - 12- to 13-hour photoperiod *to keep cuttings in vegetative condition*
> - maintaining air temperature [20 to 23°C (68 to 73°F)] cooler than media temperature [20 to 23°C (68 to 73°F)], *which retards shoot growth and promotes root development*
> - 89 to 85 percent relative humidity (0.3 kPa)
> - humid, still air to minimize the vapor pressure deficit and mist frequency
> - mist applied minimally to prevent wilting and just long enough so water coats leaf surface, but does not fall off (183, 184).

which lowers photosynthetic rates and stomatal conductance (253).

It has long been thought that the carbohydrate content of cuttings is important to rooting, and carbohydrates do accumulate in the base of cuttings during rooting (119). The amount of carbohydrates accumulated at cutting bases has been correlated with photosynthetic activity (61), but carbohydrates can also accumulate in the upper portion of leafy cuttings until after roots have formed (38). With leafy stem cuttings, the leaf-derived influx of carbohydrates determines the intensity of adventitious root formation (227).

In poinsettia cuttings, stomatal conductance and photosynthetic levels were initially low and remained low until root primordia were first microscopically observed (274); stomatal conductance and photosynthesis increased rapidly as root primordia began to elongate and emerge from the cuttings (Fig. 9–42). Most likely, root primordia were producing phytohormones such as cytokinins, which increased stomatal conductance and subsequently affected photosynthetic rates. In the rooting of cuttings, initial lower light irradiance could be used to hasten root initiation by reducing water stress (181), and light irradiance increased during root primordia emergence to support rapid primordia elongation and root system development.

If a generalization can be made, photosynthesis in cuttings is probably more important *after* root initiation has occurred, and helps aid root development and the more rapid growth of a rooted liner (63) (see Chapter 10).

Accelerated Growth Techniques

Accelerated growth techniques (AGT) were developed by the forestry industry to speed up the production of liners from vegetative propagules and from seed propagation (123). Woody perennial plants undergo cyclic growth, and many tree species experience dormancy. Liners are grown in protective culture facilities where photoperiod is extended and water, temperature, carbon dioxide, nutrition, mycorrhizal fungi (56), and growing

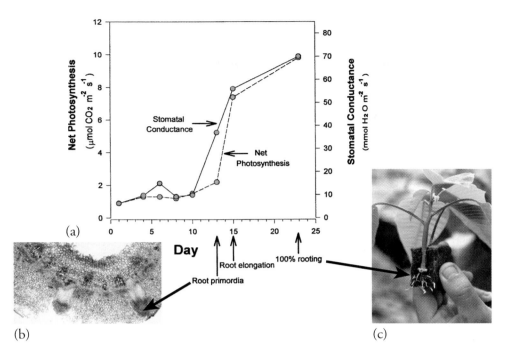

Figure 9–42
(a) Influence of adventitious root formation on gas exchange of poinsettia (*Euphorbia pulcherrima* cv. Lilo) cuttings. (b) Root primordia were microscopically observed at day 13, when photosynthesis began to increase. (c) Maximum photosynthesis was at 100 percent rooting (274).

media are optimized for each woody species and for each different phase of growth. See Figure 3–30 and the discussion on AGT in Chapter 3.

This concept is also being used in the propagation of horticultural crops where supplementary lighting with high-pressure sodium vapor lamps and injection of CO_2 gas into mist water are used to enhance rooting of holly (*Ilex aquifolium*) (92). The promotive effects on cuttings have been attributed to enhanced photosynthesis. In another study, CO_2 injection into enclosed fog tunnels enhanced root formation of *Chamelaucium* and Australian fuschia (*Correa*). This was attributed, in part, to decreased leaf transpiration and increased water potential of cuttings, implying that the higher CO_2 reduced stomatal conductance and improved water relations of the cuttings (104).

There is a growing trend for **modeling** propagation environments to determine optimal light, temperature, water, CO_2, and nutritional regimes (52, 174, 298, 299). See the discussion in Chapter 10 and earlier discussion on dynamic system models using vapor pressure deficit (VPD), transpiration, and temperature. Computers can be programmed to monitor the propagation environment and adjust environmental conditions as needed through automated environmental control systems (see Figs. 3–13, 3–14, 3–15, and 9–38).

DISCUSSION ITEMS

1. What are the developmental stages of wound-induced *de novo* adventitious roots?
2. What is callus, and how does it contribute to the formation of adventitious roots?
3. What organs must be formed adventitiously in both leaf and root cuttings?
4. How are correlative effects important in the control of adventitious root and bud formation?
5. What is the historical importance of rhizocaline in studies of adventitious root formation?
6. Discuss the most important phytohormones controlling adventitious root and bud formation.
7. What are some advantages of integrating molecular, biochemical, physiological, and anatomical developmental approaches to rooting studies?
8. What are some of the proposed roles of root inhibitors and rooting cofactors in adventitious root formation?
9. How are stock plants manipulated to maximize the rooting of cuttings?
10. How does the physiological age of a stock plant influence the rooting process?
11. How does the type of wood (hardwood, softwood, semihardwood) selected from stock plants influence the rooting process?
12. What is meant by seasonal timing, and why can it be advantageous to collect cuttings of selected plant species during specific times of the year?
13. What are the most effective compounds for stimulating adventitious root formation, and

what hormone (phytohormone) groups are they from?
14. What are the most effective compounds for stimulating adventitious bud and shoot formation, and what hormone (phytohormone) group are they from?
15. How does mineral nutrition affect the rooting of cuttings, and why can leaching of nutrients be a problem during propagation under intermittent mist?
16. What are some of the anatomical and physiological effects of wounding on the rooting of cuttings?
17. How do propagators manipulate the water relations and humidity control of cuttings with intermittent mist, fog, and enclosed propagation systems? In your discussion, include the terminology water potential, turgor potential, leaf and air water vapor pressure.
18. What is the influence of temperature on root initiation and development—how can a propagator manipulate temperature to maximize rooting of stem cuttings?
19. What is the influence of temperature on bud initiation and shoot development, and how can a propagator manipulate temperature to maximize leaf cutting propagation? (When answering the question, remember what organs are formed adventitiously from leaf cuttings.)
20. Give examples of environmental parameters that are manipulated with accelerated growth techniques (AGT) to enhance rooting of cuttings.
21. What is the influence of photosynthesis on the rooting of cuttings, and how does rooting influence photosynthetic rates of cuttings? Based on photosynthetic rates of unrooted and rooted cuttings, how can light irradiance be manipulated to maximize rooting?

REFERENCES

1. Altamura, M. M., F. Capitani, D. Serafini-Fracassini, P. Torigiani, and G. Falasca. 1991. Root histogenesis from tobacco thin cell layers. *Protoplasma* 161:31–42.

2. Altman, A. Y., and Y. Waisel, eds. 1997. *Biology of root formation and development.* New York: Plenum Publishing.

3. Anand, V. K., and G. T. Heberlein. 1975. Seasonal changes in the effects of auxin on rooting in stem cuttings of *Ficus infectoria. Physiol. Plant.* 34:330–34.

4. Arnaud, Y., A Franclet, H. Tranvan, and M. Jacques. 1993. Micropropagation and rejuvenation of *Sequoia sempervirens* (Lamb) Endl: A review. *Ann Sci. For.* 50:273–95.

5. Arteca, R. N., D. S. Tsai, and C. Schlagnhaufer. 1985. Abscisic acid effects on photosynthesis and transpiration in geranium cuttings. *HortScience* 20:370–72.

6. Auer, C. A., V. Motyka, A. Březinová, and M. Kaminek. 1999. Endogenous cytokinin accumulation and cytokinin oxidase activity during shoot organogenesis of *Petunia hybrida. Physiol. Plant.* 105:141–37.

7. Baker, K. F. 1984. The obligation of the plant propagator. *Comb. Proc. Intl. Plant Prop. Soc.* 34:195–203.

8. Baker, R. L., and C. B. Link. 1963. The influence of photoperiod on the rooting of cuttings of some woody ornamental plants. *Proc. Amer. Soc. Hort. Sci.* 82:596–601.

9. Ballester, A., M. C. San-José, N. Vidal, J. L. Fernández-Lorenzo, and A. M. Vieitez. 1999. Anatomical and biochemical events during *in vitro* rooting of microcuttings from juvenile and mature phases of chestnut. *Ann. Bot.* 83:619–29.

10. Barlow, P. W. 1994. The origin, diversity and biology of shoot-borne roots. In T. D. Davis and B. E. Haissig, eds. *Biology of adventitious root formation.* New York: Plenum Press.

11. Bartel, B., S. LeClere, M. Magidin, and B. K. Zolman. 2001. Inputs to the active indole-3-acetic acid pool: *de novo* synthesis, conjugate hydrolysis, and indole-3-butyric acid β-oxidation. *J. Plant Growth Regul.* 20:198–216.

12. Bassil, N. V., W. M. Proebsting, L. W. Moore, and D. A. Lightfoot. 1991. Propagation of hazelnut stem cuttings using *Agrobacterium rhizogenes. HortScience* 26:1058–60.

13. Bassuk, N., and B. Maynard. 1987. Stock plant etiolation. *HortScience* 22:749–50.

14. Basu, R. N., B. N. Roy, and T. K. Bose. 1970. Interaction of abscisic acid and auxins in rooting of cuttings. *Plant Cell Physiol.* 11:681–84.

15. Beakbane, A. B. 1969. Relationships between structure and adventitious rooting. *Comb. Proc. Intl. Plant Prop. Soc.* 19:192–201.

16. Behrens, V. 1988. Storage of unrooted cuttings. In T. D. Davis, B. E. Haissig, and N. Sankhla, eds. *Adventitious root formation in cuttings.* Portland, OR: Dioscorides Press.

17. Bertazza, G., R. Baraldi, and S. Predieri. 2005. Light effects on *in vitro* rooting of pear cultivars of different rhizogenic ability. *Plant Cell. Tiss. Organ Cult.* 41:139–43.

18. Biran, I., and A. H. Halevy. 1973. Endogenous levels of growth regulators and their relationship to the rooting of dahlia cuttings. *Physiol. Plant.* 28:436–42.

19. Biran, I., and A. H. Halevy. 1973. The relationship between rooting of dahlia cuttings and the presence and type of bud. *Physiol. Plant.* 28:244–47.

20. Biricolti, S., A. Fabbri, F. Ferrini, and P. L. Pisani. 1994. Adventitious rooting in chestnut: An anatomical investigation. *Scientia Hort.* 59:197–205.

21. Blakely, L. M., S. J. Rodaway, L. B. Hollen, and S. G. Croker. 1972. Control and kinetics of branch root formation in cultured root segments of *Haplopappus ravenii*. *Plant Physiol.* 50:35–49.

22. Blazich, F. A., and C. W. Heuser. 1979. A histological study of adventitious root initiation in mung bean cuttings. *J. Amer. Soc. Hort. Sci.* 104:63–7.

23. Blazich, F. A., and R. D. Wright. 1979. Nonmobilization of nutrients during rooting of *Ilex crenata* cv. Convexa stem cuttings. *HortScience* 14:242.

24. Blazich, F. A., R. D. Wright, and H. E. Schaffer. 1983. Mineral nutrient status of 'Convexa' holly cuttings during intermittent mist propagation as influenced by exogenous auxin application. *J. Amer. Soc. Hort. Sci.* 108:425–29.

25. Blazich, F. A. 1988. Mineral nutrition and adventitious rooting. In T. D. Davis, B. E. Haissig, and N. Sankhla, eds. *Adventitious root formation in cuttings.* Portland, OR: Dioscorides Press.

26. Blazich, F. A., and L. E. Hinesley. 1995. Fraser fir. *Amer. Nurs.* 181:54–67.

27. Blazkova, A., B. Sotta, H. Tranvan, R. Maldiney, M. Bonnet, J. Einhorn, L. Kerhoas, and E. Miginiac. 1997. Auxin metabolism and rooting in young and mature clones of *Sequoia sempervirens*. *Physiol. Plant.* 99:73–80.

28. Bloch, R. 1943. Polarity in plants. *Bot. Rev.* 9:261–310.

29. Blythe, E. K., J. L Sibley, K. M. Tilt, and J. M. Ruter. 2008a. Methods of auxin application in cutting propagation: A review of 70 years of scientific discovery and commercial practice: Part I *North Amer. Plant Prop.* 20:13–25.

30. Blythe, E. K., J. L. Sibley, K. M. Tilt, and J. M. Ruter. 2008b. Methods of auxin application in cutting propagation: A review of 70 years of scientific discovery and commercial practice: Part 2. *North Amer. Plant Prop.* 20:19–27.

31. Boe, A. A., R. B. Steward, and T. J. Banko. 1972. Effects of growth regulators on root and shoot development of Sedum leaf cuttings. *HortScience* 74:404–5.

32. Bollmark, M., and L. Eliasson. 1986. Effects of exogenous cytokinins on root formation in pea cuttings. *Physiol. Plant.* 68:662–66.

33. Bonnett, H. T., Jr., and J. G. Torrey. 1965. Chemical control of organ formation in root segments of *Convolvulus* cultured *in vitro*. *Plant Physiol.* 40:1228–36.

34. Boot, K. J. M., L. Goosen-deRoo, H. P. Spain, and J. W. Kijne. 2000. Adventitious root formation in stem segments of tobacco. In *Adventitious Root Formation: Third International Symposium on Adventitious Root Formation.* Veldhoven, The Netherlands.

35. Bouillenne, R., and M. Bouillenne-Walrand. 1955. Auxines et bouturage. *Rpt. 14th Inter. Hort. Cong.* 1:231–38.

36. Bouza, L., M. Jacques, B. Sota, and E. Miginiac. 1994. Relations between auxin and cytokinin contents and *in vitro* rooting of tree Peony (*Paeonia suffruticosa* Andr.). *J. Plant Growth Regul.* 15:69–73.

37. Breen, P. J., and T. Muraoka. 1973. Effect of indolebutyric acid on distribution of ^{14}C photosynthate in softwood cuttings of Marianna 2624 plum. *J. Amer. Soc. Hort. Sci.* 98:436–39.

38. Breen, P. J., and T. Muraoka. 1974. Effect of leaves and carbohydrate content and movement of ^{14}C-assimilate in plum cuttings. *J. Amer. Soc. Hort. Sci.* 99:326–32.

39. Brinker, M., L. van Zyl, W. Liu, D. Craig, R. R. Sederoff, D. H. Clapham, and S. von Arnold. 2004. Microarray analyses of gene expression during adventitious root development in *Pinus contorta*. *Plant Physiol.* 135:1526–39.

40. Caboche, M., J. F. Muller, F. Chanut, G. Aranda, and S. Cirakoglu. 1987. Comparison of the growth-promoting activities and toxicities of various auxin analogs on cells derived from wild type and a nonrooting mutant tobacco. *Plant Physiol.* 83:795–800.

41. Cameron, R. J., and G. V. Thomson. 1969. The vegetative propagation of *Pinus radiata*: Root initiation in cuttings. *Bot. Gaz.* 130:242–51.

42. Carlson, M. C. 1929. Origin of adventitious roots in *Coleus* cuttings. *Bot. Gaz.* 87:119–26.

43. Carlson, M. C. 1950. Nodal adventitious roots in willow stems of different ages. *Amer. J. Bot.* 37:555–67.

44. Carpenter, W. J., G. R. Beck, and G. A. Anderson. 1973. High intensity supplementary lighting during rooting of herbaceous cuttings. *HortScience* 8:338–40.

45. Christiansen, M. V., E. N. Eriksen, and A. S. Andersen. 1980. Interaction of stock plant irradiance

and auxin in the propagation of apple rootstocks by cuttings. *Scientia Hort.* 12:11–7.

46. Christianson, M. L., and D. A. Warnick. 1985. Temporal requirement for phytohormone balance in the control of organogenesis *in vitro*. *In Vitro Cell. Dev. Bio.—Plant.* 112:494–97.

47. Clark, D. G., E. G. Gubrium, J. E. Barrett, T. A. Nell, and H. J. Klee. 1999. Root formation in ethylene-insensitive plants. *Plant Physiol.* 121:53–60.

48. Cline, M. N., and D. Neely. 1983. The histology and histochemistry of the wound healing process in geranium cuttings. *J. Amer. Soc. Hort. Sci.* 108:450–96.

49. Crow, W. D., W. Nicholls, and M. Sterns. 1971. Root inhibitors in *Eucalyptus grandis*: Naturally occurring derivatives of the 2,3-dioxabicyclo (4,4,0) decane system. Tetrahedron Letters 18. London: Pergamon Press. pp. 1353–6.

50. Cuir, P., S. Sulis, F. Mariani, C. F. Van Sumere, A. Marchesini, and M. Dolci. 1993. Influence of endogenous phenols on rootability of *Chamaelaucium uncinatum* Schauer stem cuttings. *Scientia Hort.* 55:303–14.

51. Davies, F. T. 1984. Shoot RNA, cambial activity and indolebutyric acid effectively in seasonal rooting of juvenile and mature *Ficus pumila* cuttings. *Physiol. Plant.* 62:571–75.

52. Davies, F. T. 1985. Plant modeling: Developing an approach. *Comb. Proc. Intl. Plant Prop. Soc.* 35:770–76.

53. Davies, F. T., and H. T. Hartmann. 1988. The physiological basis of adventitious root formation. *Acta Hort.* 227:113–20.

54. Davies, F. T. 1993. What's new in the biology of adventitious root formation. *Comb. Proc. Intl. Plant Prop. Soc.* 43:382–84.

55. Davies, F. T. 2005. Optimizing the water relations of cuttings during propagation. *Comb. Proc. Intl. Plant Prop. Soc.* 55:585–92.

56. Davies, F. T. 2008. How mycorrhizal fungi can benefit nursery propagation and production systems. *Comb. Proc. Intl. Plant Prop. Soc.* 58:539–48.

57. Davies, F. T., and B. C. Moser. 1980. Stimulation of bud and shoot development of Rieger begonia leaf cuttings with cytokinins. *J. Amer. Soc. Hort. Sci.* 105:27–30.

58. Davies, F. T., and J. N. Joiner. 1980. Growth regulator effects on adventitious root formation in leaf bud cuttings of juvenile and mature *Ficus pumila*. *J. Amer. Soc. Hort. Sci.* 100:643–46.

59. Davies, F. T., J. E. Lazarte, and J. N. Joiner. 1982. Initiation and development of roots in juvenile and mature leafbud cuttings of *Ficus pumila* L. *Amer. J. Bot.* 69:804–11.

60. Davies, F. T., T. D. Davis, and D. E. Kester. 1994. Commercial importance of adventitious rooting to horticulture. In T. D. Davis and B. E. Haissig, eds. *Biology of adventitious root formation.* New York: Plenum Press.

61. Davis, T. D., and J. R. Potter. 1981. Current photosynthate as a limiting factor in adventitious root formation in leafy pea cuttings. *J. Amer. Soc. Hort. Sci.* 106:278–82.

62. Davis, T. D., and J. R. Potter. 1985. Carbohydrates, water potential and subsequent rooting of stored rhododendron cuttings. *HortScience* 20:292–93.

63. Davis, T. D. 1988. Photosynthesis during adventitious rooting. In T. D. Davis, B. E. Haissig, and N. Sankhla, eds. *Adventitious root formation in cuttings.* Portland, OR: Dioscorides Press.

64. Davis, T. D., and B. E. Haissig. 1990. Chemical control of adventitious root formation in cuttings. *Plant Growth Regul. Soc. Amer. Quart.* 18:1–17.

65. Davis, T. D., and B. E. Haissig. 1994. *Biology of adventitious root formation.* New York: Plenum Press.

66. Davis, T. D., and N. Sankhla. 1989. Effect of shoot growth retardants and inhibitors on adventitious rooting. In T. D. Davis, B. E. Haissig, and N. Sankhla, eds. *Adventitious root formation in cuttings.* Portland, OR: Dioscorides Press.

67. De Klerk, G. J., W. Van der Krieken, and J. C. de Jong. 1999. The formation of adventitious roots: New concepts, new possibilities. *In Vitro Cell. Dev. Bio.—Plant.* 35:189–99.

68. De Klerk, G. J. 2000. Multiple effects of ethylene during rooting of micropropagules. In *Adventitious root formation: Third international symposium on adventitious root formation.* Veldhoven, The Netherlands.

69. Delargy, J. A., and C. E. Wright. 1978. Root formation in cuttings of apple (cv. Bramley's Seedling) in relation to ringbarking and to etiolation. *New Phytol.* 81:117–27.

70. Delargy, J. A., and C. E. Wright. 1979. Root formation in cuttings of apple in relation to auxin application and to etiolation. *New Phytol.* 82:341–47.

71. Drahn, S. R. 2007. Auxin application via foliar sprays. *Comb. Proc. Intl. Plant Prop. Soc.* 57:274–77.

72. Duhamel du Monceau, H. L. 1758. *La physique des arbres.* Paris: Guerin and Delatour.

73. Dunn, D. E., J. C. Cole, and M. W. Smith. 1996. Calendar date, degree days, and morphology

influence rooting of *Pistacia chinensis*. *J. Amer. Soc. Hort. Sci.* 121:269–73.

74. Durzan, D. J. 1988. Rooting in woody perennials: Problems and opportunities with somatic embryos and artificial seeds. *Acta Hort.* 227:121–25.

75. Dykeman, B. 1976. Temperature relationship in root initiation and development of cuttings. *Comb. Proc. Intl. Plant Prop. Soc.* 26:201–7.

76. Economou, A. S., and P. E. Read. 1987. Light treatments to improve efficiency of *in vitro* propagation systems. *HortScience* 22:751–54.

77. Eliasson, L. 1980. Interaction of light and auxin in regulation of rooting in pea stem cuttings. *Physiol. Plant.* 48:78–82.

78. Eliasson, L., and L. Brunes. 1980. Light effects on root formation in aspen and willow cuttings. *Physiol. Plant.* 48:261–65.

79. Ellis, D. D., H. Barczynsha, B. H. McCown, and N. Nelson. 1991. A comparison of BA, zeatin and thidiazuron for adventitious bud formation from *Picea glauca* embryos and epicotyl explants. *Plant Cell Tiss. Organ Cult.* 27:281–87.

80. Emery, A. E. H. 1955. The formation of buds on roots of *Chamaenerion angustifolium* (L.). *Scop. Phytomorphology* 5:139–45.

81. Englert, J. M., B. K. Maynard, and N. L. Bassuk. 1991. Correlation of phenolics with etiolated and light-grown shoots of *Carpinus betulus* stock plants. *Comb. Proc. Intl. Plant Prop. Soc.* 41:290–95.

82. Epstein, E., and J. Ludwig-Müller. 1993. Indole-3-butyric acid in plants: Occurrence, synthesis, metabolism and transport. *Physiol. Plant.* 88:382–89.

83. Eriksen, E. N. 1973. Root formation in pea cuttings. I. Effects of decapitation and disbud-ding at different development stages. *Physiol. Plant.* 28:503–6.

84. Eriksen, E. N. 1974. Root formation in pea cuttings. III. The influence of cytokinin at different development stages. *Physiol. Plant.* 30:163–67.

85. Eriksen, E. N. 1974. Root formation in pea cuttings. II. The influence of indole-3-acetic acid at different development stages. *Physiol. Plant.* 30:158–62.

86. Erstad, J. L. F., and H. R. Gislerød. 1994. Water uptake of cuttings and stem pieces as affected by different anaerobic conditions in the rooting medium. *Scientia Hort.* 58:151–60.

87. Fadl, M. S., and H. T. Hartmann. 1967. Isolation, purification, and characterization of an endogenous root-promoting factor obtained from the basal sections of pear hardwood cuttings. *Physiol. Plant.* 42:541–49.

88. Fadl, M. S., and H. T. Hartmann. 1967. Relationship between seasonal changes in endogenous promoters and inhibitors in pear buds and cutting bases and the rooting of pear hardwood cuttings. *Proc. Amer. Soc. Hort. Sci.* 91:96–112.

89. Faivre-Rampant, O., C. Kevers, and T. Gasper. 2000. IAA-oxidase activity and auxin protectors in nonrooting, rac, mutant shoots of tobacco *in vitro*. *Plant Sci.* 153:73–80.

90. Fischer, P., and J. Hansen. 1977. Rooting of chrysanthemum cuttings: Influence of irradiance during stock plant growth and of decapitation and disbudding of cuttings. *Scientia Hort.* 7:171–78.

91. Fishel, D. W., J. Zaczek, and J. E. Preece. 2003. Positional influence on rooting of shoots forced from the main bole in swamp white oak and northern red oak. *Can. J. Forest. Res.* 33:705–11.

92. French, C. J., and W. C. Lin. 1984. Seasonal variations in the effect of CO_2 mist and supplementary lighting from high pressure sodium lamps on rooting of English holly cuttings. *HortScience* 19:519–21.

93. Flygh, G., R. Grönroos, L. Gulin, and S. Von Arnold. 1993. Early and late root formation in epicotyl cuttings of *Pinus sylvestris* after auxin treatment. *Tree Physiol.* 12:81–92.

94. García-Gomez, C., A. Sanchez-Romero, A. Barcelo-Munoz, A. Heredia, and F. Pliego-Alfaro. 1994. Levels of endogenous indole-3-acetic acid and indole-3-acetyl-aspartic acid during adventitious rooting in avocado microcuttings. *J. Expt. Bot.* 45:865–70.

95. García-Luis, A., Y. Bordón, J. M. Moreira-Dias, R. V. Molina, and J. L. Guardiola. 1999. Explant orientation and polarity determine the morphogenic response of epicotyl segments of Troyer citrange. *Ann. Bot.* 87:715–23.

96. Gasper, T., and M. Hofinger. 1989. Auxin metabolism during rooting. In T. D. Davis, B. E. Haissig, and N. Sankhla, eds. *Adventitious root formation in cuttings.* Portland, OR: Dioscorides Press.

97. Gasper, T., C. Kevers, and J. F. Hausman. 1997. Indissociable chief factors in the inductive phase of adventitious rooting. In A. Altman and Y. Waisel, eds. *Biology of root formation and development.* New York: Plenum Publishing.

98. Geneve, R. L. 1991. Patterns of adventitious root formation in English Ivy. *J. Plant Growth Regul.* 10:215–20.

99. Geneve, R. L., and S. T. Kester. 1991. Polyamines and adventitious root formation in the juvenile and mature phase of English ivy. *J. Exp. Bot.* 42:71–5.

100. Geneve, R. L. 2000. Root formation in relationship to auxin uptake in cuttings treated by the

dilute soak, quick dip and talc methods. *Comb. Proc. Intl. Plant Prop. Soc.* 50:409–12.

101. Ginzburg, C. 1967. Organization of the adventitious root apex in *Tamarix aphylla*. *Amer. J. Bot.* 54:4–8.

102. Girouard, R. M. 1969. Physiological and biochemical studies of adventitious root formation. Extractable rooting co-factors from *Hedera helix Can. J. Bot.* 47(5):287–99.

103. Grange, R. I., and K. Loach. 1983. The water economy of unrooted leafy cuttings. *J. Hort. Sci.* 58:9–17.

104. Grant, W. J. R., H. M. Fan, W. J. S. Downton, and B. R. Loveys. 1992. Effects of CO_2 enrichment on the physiology and propagation of two Australian ornamental plants, *Chamelaucium unicanatum* (Schauer) × *Chamelaucium floriferum* (MS) and *Correa schlechtendalii* (Behr). *Scientia Hort.* 52:337–42.

105. Greenwood, M. S., and G. P. Berlyn. 1973. Sucrose: Indoleacetic acid interactions on root regeneration by *Pinus lambertiana* embryo cuttings. *Amer. J. Bot.* 60:42–7.

106. Greenwood, M. S. 1987. Rejuvenation in forest trees. *J. Plant Growth Regul.* 6:1–12.

107. Greenwood, M. S., C. Diaz-Sala, P. B. Singer, A. Decker, and K. W. Hutchinson. 1997. Differential gene expression during maturation-caused decline in adventitious rooting ability in loblolly pine (*Pinus taeda* L.). In A. Altman and Y. Waisel, eds. *Biology of root formation and development.* New York: Plenum Press.

108. Gupta, P. K., and D. J. Durzan. 1987. Micropropagation and phase specificity in mature, elite Douglas fir. *J. Amer. Soc. Hort. Sci.* 112:969–71.

109. Hackett, W. P. 1970. The influence of auxin, catechol, and methanolic tissue extracts on root initiation in aseptically cultured shoot apices of the juvenile and adult forms of *Hedera helix*. *J. Amer. Soc. Hort. Sci.* 95:398–402.

110. Hackett, W. P. 1985. Juvenility, maturation and rejuvenation in woody plants. *Hort. Rev.* 7:109–15.

111. Hackett, W. P. 1989. Donor plant maturation and adventitious root formation. In T. D. Davis, B. E. Haissig, and N. Sankhla, eds. *Adventitious root formation in cuttings.* Portland, OR: Dioscorides Press.

112. Hackett, W. P. 1997. The use of mutants to understand competence for shoot-borne root initiation. In A. Altman and Y. Waisel, eds. *Biology of root formation and development.* New York: Plenum Publishing.

113. Hagemann, A. 1932. Untersuchungen an Blattstecklingen. *Gartenbauwiss.* 6:69–202.

114. Haines, R. J., T. R. Copley, J. R. Huth, and M. R. Nester. 1992. Shoot selection and the rooting and field performance of tropical pine cuttings. *Forest Sci.* 38:95–101.

115. Haissig, B. E., and T. D. Davis. 1994. A historical evaluation of adventitious rooting research to 1993. In T. D. Davis and B. E. Haissig, eds. *Biology of adventitious root formation.* New York: Plenum Press.

116. Haissig, B. E. 1972. Meristematic activity during adventitious root primordium development. Influences of endogenous auxin and applied gibberellic acid. *Plant Physiol.* 49:886–92.

117. Haissig, B. E. 1974. Influence of auxins and auxin synergists on adventitious root primordium initiation and development. *New Zealand For. Sci.* 4:299–310.

118. Haissig, B. E. 1983. N-phenyl indolyl-3-butyramide and phenyl indole-3-thiolobutyrate enhance adventitious root primordium development. *Physiol. Plant.* 57:435–40.

119. Haissig, B. E. 1984. Carbohydrate accumulation and partitioning in *Pinus banksiana* seedlings and seedling cuttings. *Physiol. Plant.* 61:13–19.

120. Haissig, B. E. 1986. Metabolic processes in adventitious rooting of cuttings. In M. B. Jackson, ed. *New root formation in plants and cuttings.* Dordrecht: Martinus Nijhoff Publishers.

121. Haissig, B. E., T. D. Davis, and D. E. Riemenschneider. 1992. Researching the controls of adventitious root formation. *Physiol. Plant.* 84:310–17.

122. Hambrick, C. E., F. T. Davies, Jr., and H. B. Pemberton. 1991. Seasonal changes in carbohydrate/nitrogen levels during field rooting of *Rosa multiflora* 'Brooks 56' hardwood cuttings. *Scientia Hort.* 46:137–46.

123. Hanover, J. W. 1976. Accelerated-optimal-growth: A new concept in tree production. *Amer. Nurs.* 144(10):11–2, 58, 60, 64–5.

124. Hansen, J., L. H. Strömquist, and A. Ericsson. 1978. Influence of the irradiance on carbohydrate content and rooting of cuttings of pine seedlings (*Pinus sylvestris* L.). *Physiol. Plant.* 61:975–79.

125. Hansen, J. 1988. Influence of gibberellins on adventitious root formation. In T. D. Davis, B. E. Haissig, and N. Sankhla, eds. *Adventitious root formation in cuttings.* Portland, OR: Dioscorides Press.

126. Harbage, J. F., D. P. Stimart, and R. F. Evert. 1994. Anatomy of adventitious root formation in microcuttings of *Malus domestica* Borkh. 'Gala'. *J. Amer. Soc. Hort. Sci.* 118:680–88.

127. Hare, R. C. 1977. Rooting of cuttings from mature water oak (*Quercus nigra*). *Southern J. Appl. For.* 1:24–5.

128. Hare, R. C. 1981. Improved rooting powder for chrysanthemums. *HortScience* 16:90–1.

129. Harrison-Murray, R. S., B. H. Howard, and R. Thompson. 1988. Potential for improved propagation of cuttings through the use of fog. *Acta Hort.* 227:205–10.

130. Harrison-Murray, R. S., and R. Thompson. 1988. In pursuit of a minimum stress environment for rooting leafy cuttings: Comparison of mist and fog. *Acta Hort.* 227:211–16.

131. Harrison-Murray, R. S. 1991. A leaf-model evaporimeter for estimating potential transpiration in propagation environments. *J. Hort. Sci.* 66:131–39.

132. Hausman, J. F. 1993. Changes in peroxidase activity, auxin level and ethylene production during root formation by poplar shoots raised *in vitro*. *J. Plant Growth Regul.* 13:263–68.

133. Heide, O. M. 1965. Interaction of temperature, auxin, and kinins in the regeneration ability of begonia leaf cuttings. *Physiol. Plant.* 18:891–920.

134. Heide, O. M. 1965. Effects of 6-benzylamino-purine and 1-naphthaleneacetic acid on the epiphyllous bud formation in *Bryophyllum*. *Planta* 67:281–96.

135. Heide, O. M. 1965. Photoperiodic effects on the regeneration ability of begonia leaf cuttings. *Physiol. Plant.* 18:185–90.

136. Heide, O. M. 1968. Stimulation of adventitious bud formation in begonia leaves by abscisic acid. *Nature* 219:960–61.

137. Heide, O. M. 1968. Auxin level and regeneration of begonia leaves. *Planta* 81:153–59.

138. Heide, O. M. 1969. Non-reversibility of gibberellin-induced inhibition of regeneration in begonia leaves. *Physiol. Plant.* 22:671–79.

139. Henry, P. H., F. A. Blazich, and L. E. Hinesley. 1992. Nitrogen nutrition of containerized eastern red cedar. II. Influence of stock plant fertility on adventitious rooting of stem cuttings. *J. Amer. Soc. Hort. Sci.* 117:568–70.

140. Hess, C. E. 1962. Characterization of the rooting co-factors extracted from *Hedera helix* L. and *Hibiscus rosa-sinensis*. L. *Proc. 16th Inter. Hort. Cong.*: 382–88.

141. Hess, C. E. 1968. Internal and external factors regulating root initiation. In *Root growth: Proc. 15th Easter School in Agricultural Science.* University of Nottingham. London: Butterworth.

142. Hiller, C. 1951. *A study of the origin and development of callus and root primordia of Taxus cuspidata with reference to the effects of growth regulators.* Master's thesis. Ithaca, NY: Cornell Univ.

143. Hitchcock, A. E., and P. W. Zimmerman. 1942. Root inducing activity of phenoxy compounds in relation to their structure. *Contrib. Boyce. Thomp. Inst.* 12:497–507.

144. Howard, B. H. 1965. Increase during winter in capacity for root regeneration in detached shoots of fruit tree rootstocks. *Nature* 208:912–13.

145. Howard, B. H. 1968. Effects of bud removal and wounding on rooting of hardwood cuttings. *Nature* 220:262–64.

146. Howard, B. H., R. S. Harrison-Murray, J. Vesek, and O. P. Jones. 1988. Techniques to enhance rooting potential before cutting collection. *Acta Hort.* 227:1976–86.

147. Howard, B. H., O. P. Jones, and J. Vasek. 1989. Growth characteristics of apparently rejuvenated plum shoots. *J. Hort. Sci.* 64:157–62.

148. Howard, B. H., O. P. Jones, and J. Vasek. 1989. Long-term improvement in the rooting of plum cuttings following apparent rejuvenation. *J. Hort. Sci.* 64:147–56.

149. Howard, B. H. 1991. Stock plant manipulation for better rooting and growth from cuttings. *Comb. Proc. Intl. Plant Prop. Soc.* 41:127–30.

150. Howard, B. H., and M. S. Ridout. 1992. A mechanism to explain increased rooting in leafy cuttings of *Syringa vulgaris* 'Madame Lemoine' following dark-treatment of the stock plant. *J. Hort. Sci.* 59:131–39.

151. Howard, B. H. 1994. Manipulating rooting potential in stock plants before collecting cuttings. In T. D. Davis and B. E. Haissig, eds. *Biology of adventitious root formation.* New York: Plenum Press.

152. Howard, B. H., and R. S. Harrison-Murray. 1997. Relationships between stock plant management and rooting environments for difficult-to-propagate cuttings. *Comb. Proc. Intl. Plant Prop. Soc.* 47:322–27.

153. Ikeda, T., and T. Suzaki. 1985. Influence of hydraulic conductance of xylem on water status in cuttings. *Can. J. For. Res.* 16:98–102.

154. Jackson, M. B., ed. 1986. *New root formation in plants and cuttings.* Dordrecht: Martinus Nijhoff Publishers.

155. Jain, M. K., and K. K. Nanda. 1972. Effect of temperature and some antimetabolites on the interaction effects of auxin and nutrition in rooting etiolated stem segments of *Salix tetrasperma*. *Physiol. Plant.* 27:169–72.

156. Jarvis, B. C., S. Yasmin, and M. T. Coleman. 1985. RNA and protein metabolism during

adventitious root formation in stem cuttings of *Phaseolus aureus*. *Physiol. Plant.* 64:53–9.

157. Jarvis, B. C. 1986. Endogenous control of adventitious rooting in non-woody species. In M. B. Jackson, ed. *New root formation in plants and cuttings.* Dordrecht: Martinus Nijhoff Publishers.

158. Johnson, C. R. 1970. *The nature of flower bud influence on root regeneration in the Rhododendron shoot.* Ph.D. Dissertation. Oreg. State Univ., Corvallis, OR.

159. Johnson, C. R., and A. N. Roberts. 1971. The effect of shading rhododendron stock plants on flowering and rooting. *J. Amer. Soc. Hort. Sci.* 96:166–68.

160. Johnson, C. R., and D. F. Hamilton. 1977. Effects of media and controlled-release fertilizers on rooting and leaf nutrient composition of *Juniperus conferta* and *Ligustrum japonicum* cuttings. *J. Amer. Soc. Hort. Sci.* 102:320–22.

161. Kawase, M. 1964. Centrifugation, rhizocaline, and rooting in *Salix alba*. *Physiol. Plant.* 17:855–65.

162. Kawase, M. 1971. Causes of centrifugal root promotion. *Physiol. Plant.* 25:64–70.

163. Keever, G. J., and J. H. B. Tukey. 1979. Effect of nutrient mist on the propagation of azaleas. *HortScience* 14:755–56.

164. Keever, G. J., G. S. Cobb, and D. R. Mills. 1987. Propagation of four woody ornamentals from vegetative and reproductive stem cuttings. *Ornamentals Res. Rep. 5,* Alabama Agr. Exp. Sta., Auburn University, Montgomery.

165. Kefford, N. P. 1973. Effect of a hormone antagonist on the rooting of shoot cuttings. *Plant Physiol.* 51:214–16.

166. Kester, D. E. 1970. Temperature and plant propagation. *Comb. Proc. Intl. Plant Prop. Soc.* 20:153–63.

167. Kester, D. E. 1982. The clone in horticulture. *HortScience* 18:831–37.

168. Kling, G. J., J. M. M. Meyer, and D. Seigler. 1988. Rooting co-factors in five *Acer* species. *J. Amer. Soc. Hort. Sci.* 113:252–57.

169. Kraus, E. J., and H. R. Kraybill. 1918. *Vegetation and reproduction with special reference to the tomato.* Oreg. Agr. Exp. Sta. Bul. 149.

170. Krisantini, S., M. Johnston, R. R. Williams, and C. Beveridge. 2006. Adventitious root formation in Grevillea (*Proteaceae*), an Australian native species. *Scientia Hort.* 107:171–75.

171. Lanphear, F. O., and R. P. Meahl. 1961. The effect of various photoperiods on rooting and subsequent growth of selected woody ornamental plants. *Proc. Amer. Soc. Hort. Sci.* 77:620–34.

172. Lanphear, F. O., and R. P. Meahl. 1966. Influence of the stock plant environment on the rooting of *Juniperus horizontalis* 'Plumosa.' *Proc. Amer. Soc. Hort. Sci.* 89:666–71.

173. LeBude, A. V., B. Goldfarb, F. A. Blazich, F. C. Wise, and J. Frampton. 2004. Mist, substrate water potential and cutting water potential influence rooting of stem cuttings of loblolly pine. *Tree Physiol.* 24:823–31.

174. LeBude, A. V., B. Goldfarb, F. A. Blazich, J. Frampton, and F. C. Wise. 2005. Mist level influences vapor pressure deficit and gas exchange during rooting of juvenile stem cuttings of loblolly pine. *HortScience* 40:1448–56.

175. Lek, H. A., and A. van der. 1925. Root development in woody cuttings. *Meded. Landbouwhoogesch. Wageningen.* 38(1).

176. Leopold, A. C. 1964. *The polarity of auxin transport in meristems and differentiation.* Brookhaven Symposia in Biology. Rpt. 16. Upton, NY: Brookhaven Natl. Lab., pp. 218–34.

177. Libby, W. J., A. G. Brown, and J. M. Fielding. 1972. Effect of hedging *radiata pine* on production, rooting, and early growth of cuttings. *New Zealand J. For. Sci.* 2:263–83.

178. Loach, K. 1979. Mist propagation: Past, present, future. *Comb. Proc. Intl. Plant Prop. Soc.* 29:216–29.

179. Loach, K. 1983. Propagation systems in New Zealand: A means of comparing their effectiveness. *Comb. Proc. Intl. Plant Prop. Soc.* 33:291–94.

180. Loach, K. 1987. Mist and fruitfulness. *Horticulture Week.* April 10, 1987:28–9.

181. Loach, K. 1988. Controlling environmental conditions to improve adventitious rooting. In T. D. Davis, B. E. Haissig, and N. Sankhla, eds. *Adventitious root formation in cuttings.* Portland, OR: Dioscorides Press.

182. Loach, K. 1988. Water relations and adventitious rooting. In T. D. Davis, B. E. Haissig, and N. Sankhla, eds. *Adventitious root formation in cuttings.* Portland, OR: Dioscorides Press.

183. Lopez, R. G., and E. S. Runkle. 2008. Photosynthetic daily light integral during propagation influences rooting and growth of cuttings and subsequent development of New Guinea impatiens and petunia. *HortScience* 43:2052–9.

184. Lopez, R. G., and E. S. Runkle. 2005. Managing light during propagation. *Greenhouse Product News* 15(6):1.

185. Lovell, P. H., and J. White. 1986. Anatomical changes during adventitious root formation. In M. B. Jackson, ed. *New root formation in plants and cuttings.* Dordrecht: Martinus Nijhoff Publishers.

186. Ludwig-Müller, J., and E. Epstein. 1994. Indole-3-butyric acid in *Arabidopsis thaliana* III. *In vivo* biosynthesis. *J. Plant Growth Regul.* 14:7–14.

187. MacKenzie, K. A. D., B. H. Howard, and R. S. Harrison-Murray. 1986. The anatomical relationship between cambial regeneration and root initiation in wounded winter cuttings of the apple rootstock M.26. *Ann. Bot.* 58:649–61.

188. Maini, J. S. 1968. The relationship between the origin of adventitious buds and the orientation of *Populus tremuloides* root cuttings. *Bul. Ecol. Soc. Amer.* 49:81–2.

189. Major, J. E., and S. C. Grossnickle. 1990. Chilling units used to determine rooting of stem cuttings of junipers. *J. Environ. Hort.* 8:32–5.

190. Mancuso, S. 1998. Seasonal dynamics of electrical impedance parameters in shoots and leaves relate to rooting ability of olive (*Olea europaea*) cuttings. *Tree Physiol.* 19:95–101.

191. Maynard, B. K., and N. L. Bassuk. 1987. Stock plant etiolation and blanching of woody plants prior to cutting propagation. *J. Amer. Soc. Hort. Sci.* 112:273–76.

192. Maynard, B. K., and N. L. Bassuk. 1991. Stock plant etiolation and stem banding effect on the auxin dose-response of rooting in stem cuttings of *Carpinus betulus* L. 'Fastigiata.' *J. Plant Growth Regul.* 10:305–11.

193. Maynard, B. K., and N. L. Bassuk. 1992. Stock plant etiolation, shading, and banding effects on cutting propagation of *Carpinus betulus*. *J. Amer. Soc. Hort. Sci.* 117:740–44.

194. McAfee, B. J., E. E. White, L. E. Pelcher, and M. S. Lapp. 1993. Root induction in pine (*Pinus*) and larch (*Larix*) spp. using *Agrobacterium rhizogenes*. *Plant Cell Tiss. Organ. Cult.* 34:53–62.

195. Medina, J. P. 1981. Studies of clonal propagation on pecans at Ica, Peru. *Plant Propagator* 26:11–3.

196. Moe, R., and A. S. Andersen. 1988. Stock plant environment and subsequent adventitious rooting. In T. D. Davis, B. E. Haissig, and N. Sankhla, eds. *Adventitious root formation in cuttings.* Portland, OR: Dioscorides Press.

197. Mohammed, S., and E. N. Eriksen. 1974. Root formation in pea cuttings. IV. Further studies on the influence of indole-3-acetic acid at different development stages. *Physiol. Plant.* 32:94–6.

198. Mohammed, S. 1975. Further investigations on the effects of decapitation and disbudding at different development stages of rooting in pea cuttings. *J. Hort. Sci.* 50:271–73.

199. Mohnen, D. 1994. Novel experimental systems for determining cellular competence and determination. In T. D. Davis and B. E. Haissig, eds. *Biology of adventitious root formation.* New York: Plenum Press.

200. Molitor, H. D., and W. U. von Hentig. 1987. Effect of carbon dioxide enrichment during stock plant cultivation. *HortScience* 22:741–46.

201. Molnar, J. M., and L. J. LaCroix. 1972. Studies of the rooting of cuttings of *Hydrangea macrophylla*: Enzyme changes. *Can. J. Bot.* 50:315–22.

202. Molnar, J. M., and L. J. LaCroix. 1972. Studies of the rooting of cuttings of *Hydrangea macrophylla*: DNA and protein changes. *Can. J. Bot.* 50:387–92.

203. Moon, H. K. 2006. Growth of Tsururindo (*Tripterospermum japonicum*) cultured *in vitro* under various sources of light-emitting diode (LED) irradiation. *J. Plant Biol.* 49:174–79.

204. Moreira-Dias, J. M., R. V. Molina, Y. Bordón, J. L. Guardiola, and A. García-Luis. 2000. Direct and indirect shoot organogenic pathways in epicotyl cuttings of Troyer citrange differ in hormone requirements and in their response to light. *Ann. Bot.* 85:103–10.

205. Morgan, D. L., E. L. McWilliams, and W. C. Parr. 1980. Maintaining juvenility in live oak. *HortScience* 15:493–94.

206. Mudge, K. W. 1988. Effect of ethylene on rooting. In T. D. Davis, B. E. Haissig, and N. Sankhla, eds. *Adventitious root formation in cuttings.* Portland, OR: Dioscorides Press.

207. Mudge, K. W., V. N. Mwaja, F. M. Itulya, and J. Ochieng. 1995. Comparison of four moisture management systems for cutting propagation of bougainvillea, hibiscus and kei apple. *J. Amer. Soc. Hort. Sci.* 120:366–73.

208. Murray, J. R., M. C. Sanchez, A. G. Smith, and W. P. Hackett. 1993. Differential competence for adventitious root formation in histologically similar cell types. In T. D. Davis and B. E. Haissig, eds. *Biology of adventitious root formation.* New York: Plenum Press.

209. Muzik, T. J., and H. J. Cruzado. 1958. Transmission of juvenile rooting ability from seedlings to adults of *Hevea brasiliensis*. *Nature* 181:1288.

210. Nórgaard, J. V. 1992. Artificial seeds in micropropagation. *Comb. Proc. Intl. Plant Prop. Soc.* 42:182–84.

211. O'Rourke, F. L. 1944. Wood type and original position on shoot with reference to rooting in

hardwood cuttings of blueberry. *Proc. Amer. Soc. Hort. Sci.* 45:195–97.

212. Okoro, O. O., and J. Grace. 1978. The physiology of rooting *Populus* cuttings. II. Cytokinin activity in leafless hardwood cuttings. *Physiol. Plant.* 44:167–70.

213. Oliver, M. J., I. Mukherjee, and D. M. Reid. 1994. Alteration in gene expression in hypocotyls of sunflower (*Helianthus annuus*) seedlings associated with derooting and formation of adventitious root primordia. *Physiol. Plant.* 90:481–89.

214. Patena, L., E. G. Sutter, and A. M. Dandekar. 1988. Root induction by *Agrobacterium rhizogenes* in a difficult-to-root woody species. *Acta Hort.* 227:324–29.

215. Paton, D. M., R. R. Willing, W. Nichols, and L. D. Pryor. 1970. Rooting of stem cuttings of eucalyptus: A rooting inhibitor in adult tissue. *Austral. J. Bot.* 18:175–83.

216. Paton, F., and W. W. Schwabe. 1987. Storage of cuttings of *Pelargonium xhortorum*. *J. Hort. Sci.* 62:79–87.

217. Pierik, R. L. M., and H. H. M. Steegmans. 1975. Analysis of adventitious root formation in isolated stem explants of Rhododendron. *Scientia Hort.* 3:1–20.

218. Pliego-Alfaro, F., and T. Murashige. 1987. Possible rejuvenation of adult avocado by graftage onto juvenile rootstocks *in vitro*. HortScience 22:1321–4.

219. Plietzsch, A., and H. H. Jesch. 1998. Using *in vitro* propagation to rejuvenate diffficult-to-root woody plants. *Comb. Proc. Intl. Plant Prop. Soc.* 48:171–76.

220. Preece, J. E. 1993. Basics of propagation by cuttings—temperature. *Comb. Proc. Intl. Plant Prop. Soc.* 43:441–43.

221. Preece, J. E. 2003. A century of progress with vegetative plant propagation. HortScience 38:1015–25.

222. Preece, J. E. 2008. Stock plant physiological factors affecting growth and morphogenesis. In E. F. George, M. A. Hall, and G. J. De Klerk, eds. *Plant propagation by tissue culture.* Dordrecht: Springer. pp. 403–22.

223. Preece, J. E., and P. Read. 2007. Forcing leafy explants and cuttings from woody species. *Propagation Ornamental Plants* 7:138–44.

224. Priestley, J. H., and C. F. Swingle. 1929. *Vegetative propagation from the standpoint of plant anatomy,* USDA Tech. Bul.

225. Proebsting, W. M. 1984. Rooting of Douglas-fir stem cuttings: Relative activity of IBA and NAA. HortScience 19:854–56.

226. Rajagopal, V., and A. S. Andersen. 1980. Water stress and root formation in pea cuttings. *Physiol. Plant.* 48:114–49.

227. Rapaka, V. K., B. Bessler, M. Schreiner, and U. Druge. 2005. Interplay between initial carbohydrate availability, current photosynthesis, and adventitious root formation in *Pelargonium* cuttings. *Plant Sci.* 168:1547–60.

228. Rashotte, A. M., J. Poupart, C. S. Waddell, and G. K. Muday. 2003. Transport of the two natural auxins, indole-3-butyric acid and indole-3-acetic acid, in *Arabidopsis. Plant Physiol.* 133:761–72.

229. Raskin, I. 1992. Salicylate—a new plant hormone. *Plant Physiol.* 99:799–803.

230. Rasmussen, S., and A. S. Andersen. 1980. Water stress and root formation in pea cuttings. II. Effect of abscisic acid treatment of cuttings from stock plants grown under two levels of irradiance. *Physiol. Plant.* 48:150–54.

231. Read, P. E., and V. C. Hoysler. 1969. Stimulation and retardation of adventitious root formation by application of B-Nine and Cycocel. *J. Amer. Soc. Hort. Sci.* 94:314–16.

232. Rein, W. H., R. D. Wright, and J. R. Seiler. 1991. Propagation medium moisture level influences adventitious rooting of woody stem cuttings. *J. Amer. Soc. Hort. Sci.* 116:632–36.

233. Rein, W. H., R. D. Wright, and D. D. Wolf. 1991. Stock plant nutrition influences the adventitious rooting of 'Rotundifolia' holly stem cuttings. *J. Environ. Hort.* 9:83–5.

234. Reuveni, O., and M. Raviv. 1981. Importance of leaf retention to rooting avocado cuttings. *J. Amer. Soc. Hort. Sci.* 106:127–30.

235. Rey, M., C. Díaz-Sala, and R. Rodríguez. 1994. Exogenous polyamines improve rooting of hazel microshoots. *Plant Cell Tiss. Organ. Cult.* 36:303–8.

236. Ritchie, G. A. 1994. Commercial applications of adventitious rooting to forestry. In T. D. Davis and B. E. Haissig, eds. *Biology of adventitious root formation.* New York: Plenum Press.

237. Robbins, J. A., M. J. Campidonica, and D. W. Burger. 1988. Chemical and biological stability of indole-3-butyric acid (IBA) after long-term storage at selected temperatures and light regimes. *J. Environ. Hort.* 6:33–8.

238. Roberts, A. N., and L. H. Fuchigami. 1973. Seasonal changes in auxin effect on rooting of Douglas-fir stem cuttings as related to bud activity. *Physiol. Plant.* 28:215–21.

239. Robinson, J. C., and W. W. Schwabe. 1977. Studies on the regeneration of apple cultivars from

root cuttings. I. Propagation aspects. *J. Hort. Sci.* 52:205–20.

240. Robinson, J. C., and W. W. Schwabe. 1977. Studies on the regeneration of apple cultivars from root cuttings. II. Carbohydrate and auxin relations. *J. Hort. Sci.* 52:221–33.

241. Rowland, L. J., and E. L. Ogden. 1992. Use of a cytokinin conjugate for efficient shoot regeneration from leaf sections of highbush blueberry. *HortScience* 27:1127–9.

242. Rugini, E., A. Pellegrineschi, M. Mencuccini, and D. Mariotti. 1991. Increase of rooting ability in the woody species kiwi (*Actinidia deliciosa* A. Chev.) by transformation with *Agrobacterium rhizogenes* rol genes. *Plant Cell Rept.* 10:291–95.

243. Rugini, E. 1992. Involvement of polyamines in auxin and *Agrobacterium rhizogenes*-induced rooting of fruit trees *in vitro*. *J. Amer. Soc. Hort. Sci.* 117:532–36.

244. Sachs, J. 1880 and 1882. *Stoff und Form der Pflanzenorgane.* I and II. *Arb. Bot. Inst. Würzburg* 2:450–88; 4:689–718.

245. Sachs, R. M., F. Loreti, and J. DeBie. 1964. Plant rooting studies indicate sclerenchyma tissue is not a restricting factor. *Calif. Agr.* 18:4–5.

246. Sanchez, M. C., A. G. Smith, and W. P. Hackett. 1995. Localized expression of a proline-rich protein gene in juvenile and mature ivy petioles in relation to rooting competence. *Physiol. Plant.* 93:207–16.

247. Santos, K. M., P. R. Fisher, and W. R. Argo. 2008. A survey of water and fertilization management during cutting propagation. *HortTechnology* 18:597–604.

248. Schier, G. A. 1973. Origin and development of aspen root suckers. *Can. J. For. Res.* 3:39–44.

249. Schuerman, P. L., and A. M. Dandekar. 1993. Transformation of temperate woody crops: Progress and potentials. *Scientia Hort.* 55:101–24.

250. Sircar, P. K., and S. K. Chatterjee. 1974. Physiological and biochemical changes associated with adventitious root formation in *Vigna* hypocotyl cuttings. II. Gibberellin effects. *Plant Propagator* 20:15–22.

251. Skoog, F., and C. Tsui. 1948. Chemical control of growth and bud formation in tobacco stem and callus. *Amer. J. Bot.* 35:782–87.

252. Smally, T. J., and M. A. Dirr. 1988. Effect of night interruption photoperiod treatment on subsequent growth of *Acer rubrum* cuttings. *HortScience* 23:172–74.

253. Smally, T. J., M. A. Dirr, A. M. Armitage, B. W. Wood, R. O. Teskey, and R. F. Severson. 1991. Photosynthesis, leaf water, carbohydrate, and hormone status during rooting of stem cuttings of *Acer rubrum*. *J. Amer. Soc. Hort. Sci.* 116:1052–7.

254. Smith, D. L., and N. V. Fedoroff. 1995. LRP1, a gene expressed in lateral and adventitious root primordia of *Arabidopsis*. *Plant Cell* 7:735–45.

255. Smith, D. R., and T. A. Thorpe. 1975a. Root initiation in cuttings of *Pinus radiata* seedlings. I. Developmental sequence. *J. Expt. Bot.* 26:184–92.

256. Smith, D. R., and T. A. Thorpe. 1975b. Root initiation in cuttings of *Pinus radiata* seedlings. II. Growth regulator interactions. *J. Expt. Bot.* 26:193–202.

257. Smith, N. G., and P. F. Wareing. 1972. The distribution of latent root primordia in stems of *Populus xrobusta* and factors affecting emergence of preformed roots from cuttings. *J. Forestry* 45:197–210.

258. Sorin, C., L. Negroni, T. Balliau, H. Corti, M. P. Jacquemot, M. Davanture, G. Sandberg, M. Zivy, and C. Bellini. 2006. Proteomic analysis of different mutant genotypes of *Arabidopsis* led to the identification of 11 proteins correlating with adventitious root development. *Plant Physiol.* 140:349–64.

259. Spano, L., D. Mariotti, M. Cardarelli, C. Branra, and P. Costantino. 1988. Morphogenesis and auxin sensitivity of transgenic tobacco with different complements of Ri T-DNA. *Plant Physiol.* 87:476–83.

260. Stangler, B. B. 1949. An anatomical study of the origin and development of adventitious roots in stem cuttings of *Chrysanthemum morifolium* Bailey, *Dianthus caryophyllus* L., and *Rosa dilecta* Rehd. Ph.D. dissertation. Cornell Univ., Ithaca NY.

261. Steele, M. J., M. M. Yeoman, and M. P. Coutts. 1990. Developmental changes in Sitka spruce as indices of physiological age. II. Rooting of cuttings and callusing of needle explants. *New Phytol.* 114:11–120.

262. Steponkus, P. L., and L. Hogan. 1967. Some effects of photoperiod on the rooting of *Abelia grandiflora* Rehd. 'Prostrata' cuttings. *Proc. Amer. Soc. Hort. Sci.* 91:706–15.

263. Stoltz, L. P., and C. E. Hess. 1966. The effect of girdling upon root initiation carbohydrates and amino acids. *Proc. Amer. Soc. Hort. Sci.* 89:734–43.

264. Stoutemyer, V. T., O. K. Britt, and J. R. Goodin. 1961. The influence of chemical treatments, understocks, and environment on growth phase changes and propagation of *Hedera canariensis*. *Proc. Amer. Soc. Hort. Sci.* 77:552–57.

265. Strobel, G. A., and A. Nachmias. 1988. *Agrobacterium rhizogenes:* A root inducing bacterium. In T. D. Davis, B. E. Hassig, and N. Sankhla, eds.

Adventitious root formation in cuttings. Portland, OR: Dioscorides Press.

266. Strömquist, L., and J. Hansen. 1980. Effects of auxin and irradiance on the rooting of cuttings of *Pinus sylvestris*. *Physiol. Plant.* 49:346–50.

267. Struve, D. K. 1981. The relationship between carbohydrates, nitrogen and rooting of stem cuttings. *Plant Propagator* 27:6–7.

268. Struve, D. K., and M. A. Arnold. 1986. Aryl esters of IBA increase rooted cutting quality of red maple 'Red Sunset' softwood cuttings. *HortScience* 21:1392–3.

269. Struve, D. K., and R. D. Lineberger. 1988. Restoration of high adventitious root regeneration potential in mature *Betula papyrifera* Marsh. softwood stem cuttings. *Can. J. For. Res.* 18:265–69.

270. Strydom, D. K., and H. T. Hartmann. 1960. Absorption, distribution, and destruction of indoleacetic acid in plum stem cuttings. *Plant Physiol.* 35:435–42.

271. Strydom, D. K., and H. T. Hartmann. 1960. Effect of indolebutyric acid on respiration and nitrogen metabolism in Marianna 2624 plum softwood stem cuttings. *Proc. Amer. Soc. Hort. Sci.* 76:124–33.

272. Sun, W. Q., and N. L. Bassuk. 1993. Auxin-induced ethylene synthesis during rooting and inhibition of bud-break of 'Royalty' rose cuttings. *J. Amer. Soc. Hort. Sci.* 118:638–43.

273. Svenson, S. E., and F. T. Davies, Jr. 1995. Change in tissue elemental concentration during root initiation and development of poinsettia cuttings. *HortScience* 30:617–19.

274. Svenson, S. E., F. T. Davies, Jr., and S. A. Duray. 1995. Gas exchange, water relations, and dry weight partitioning during root initiation and development of poinsettia cuttings. *J. Amer. Soc. Hort. Sci.* 120:454–59.

275. Thimann, K. V., and F. W. Went. 1934. On the chemical nature of the root-forming hormone. *Proc. Kon. Ned. Akad. Wet.* 37:456–59.

276. Thimann, K. V., and J. B. Koepfli. 1935. Identity of the growth-promoting and root-forming substances of plants. *Nature* 135:101–2.

277. Tukey, H. B., and E. L. Green. 1934. Gradient composition of rose shoots from tip to base. *Plant Physiol.* 9:157–63.

278. Tukey, H. B., Jr., H. B. Tukey, and S. H. Wittwer. 1958. Loss of nutrients by foliar leaching as determined by radioisotopes. *Proc. Amer. Soc. Hort. Sci.* 71:496–506.

279. Van Overbeek, J., S. A. Gordon, and L. E. Gregory. 1946. An analysis of the function of the leaf in the process of root formation in cuttings. *Amer. J. Bot.* 33:100–7.

280. Van Sambeek, J. W., J. E. Preece, and M. V. Coggeshall. 2002. Forcing epicormic sprouts on branch segments of adult hardwoods for softwood cuttings. *Comb. Proc. Intl. Plant Prop. Soc.* 52:417–24.

281. Van Staden, J., and A. R. Harty. 1988. Cytokinins and adventitious root formation. In T. D. Davis, B. E. Haissig, and N. Sankhla, eds. *Adventitious root formation in cuttings*. Portland, OR: Dioscorides Press.

282. Vander Krieken, W. M., H. Breteler, M. H. M. Visser, and D. Mavridou. 1993. The role of the conversion of IBA into IAA on root regeneration in apple: Introduction of a test system. *Plant Cell Rpt.* 12:203–6.

283. Veierskov, B. 1988. Relations between carbohydrates and adventitious root formation. In T. D. Davis, B. E. Haissig, and N. Sankhla, eds. *Adventitious root formation in cuttings*. Portland, OR: Dioscorides Press.

284. Venverloo, G. J. 1976. The formation of adventitious organs. III. A comparison of root and shoot formation on *Nautilocalyx* explants. *Z. Pflanzenphysiol.* 80:310–22.

285. Vieitez, J., D. G. I. Kingston, A. Ballester, and E. Vieitez. 1987. Identification of two compounds correlated with lack of rooting capacity of chestnut cuttings. *Tree Physiol.* 3:247–55.

286. Vöchting, H. 1878. *Uber Organbildung in Pflanzenreich*. Bonn: Verlag Max Cohen.

287. Walker, R. I. 1940. Regeneration in the scale leaf of *Lilium candidum* and *L. longiflorum*. *Amer. J. Bot.* 27:114–17.

288. Wang, Y. T. 1987. Effect of temperature, duration and light during simulated shipping on quality and rooting of croton cuttings. *HortScience* 22:1301–2.

289. Warmke, H. E., and G. L. Warmke. 1950. The role of auxin in the differentiation of root and shoot primordia from root cuttings of *Taraxacum* and *Cichorium*. *Amer. J. Bot.* 37:272–80.

290. Weiser, C. J., and L. T. Blaney. 1967. The nature of boron stimulation to root initiation and development in beans. *Proc. Amer. Soc. Hort. Sci.* 90:191–99.

291. Welander, M. 1995. Influence of environment, fertilizer and genotype on shoot morphology and subsequent rooting of birch cuttings. *Tree Physiol.* 15:11–8.

292. Went, F. W. 1934. A test method for rhizocaline, the root-forming substance. *Proc. Kon. Ned. Akad. Wet.* 37:445–55.

293. White, J., and P. H. Lovell. 1984a. The anatomy of root initiation in cuttings of *Griselinia littoralis* and *Griselinia lucida*. *Ann. Bot.* 54:7–20.

294. White, J., and P. H. Lovell. 1984b. Anatomical changes which occur in cuttings of *Agathis australis* (D. Don) Lindl. 1. Wounding responses. *Ann. Bot.* 54:621–32.

295. White, J., and P. H. Lovell. 1984c. Anatomical changes which occur in cuttings of *Agathis australis* (D. Don) Lindl. 2. The initiation of root primordia and early root development. *Ann. Bot.* 54: 633–45.

296. Whitehill, S. J., and W. W. Schwabe. 1975. Vegetative propagation of *Pinus sylvestris*. *Physiol. Plant.* 35:66–71.

297. Wiesman, Z., J. Riov, and E. Epstein. 1989. Characterization and rooting ability of indole-3-butyric acid conjugates formed during rooting of mung bean cuttings. *Plant Physiol.* 91:1080–4.

298. Wilkerson, E. G., R. S. Gates, S. Zolnier, S. T. Kester, and R. L. Geneve. 2005a. Transpiration capacity in poinsettia cuttings at different stages and the development of a cutting coefficient for scheduling mist. *J. Amer. Soc. Hort. Sci.* 130:295–301.

299. Wilkerson, E. G., R. S. Gates, S. Zolnier, S. T. Kester, and R. L. Geneve. 2005b. Predicting rooting stages in poinsettia cuttings using root zone temperature-based models. *J. Amer. Soc. Hort. Sci.* 130:302–7.

300. Wilson, P. J., and J. Van Staden. 1990. Rhizocaline, rooting co-factors, and the concept of promoters and inhibitors of adventitious rooting—a review. *Ann. Bot.* 66:476–90.

301. Wilson, P. J. 1994. The concept of a limiting rooting morphogen in woody stem cuttings. *J. Hort. Sci.* 69:591–600.

302. Wilson, P. J. 1998. The discipline of forest tree propagation. *South. African For. J.* 183:47–52.

303. Wilson, P. J. 1998. Environmental preferences of *Eucalyptus globulus* stem cuttings in one nursery. *New Zealand J. For. Sci.* 28:293–303.

304. Wilson, P. J. 1999. The growth and form of potted mother plants of *Eucalyptus globulus* Labill. ssp. *globulus* in relation to the rooting ability of stem cuttings. *J. Hort Sci. Biotech.* 74:645–50.

305. Wilson, P. J., and D. K. Struve. 2004. Overwinter mortality in stem cuttings. *J. Hort. Sci. Biotech.* 79:842–49.

306. Woo, H. H., and W. P. Hackett. 1994. Differential expression of a chlorophyll a/b binding protein gene and a proline rich protein gene in juvenile and mature phase English ivy (*Hedera helix*). *Physiol. Plant.* 92:69–78.

307. Woodward, A., and B. Bartel. 2005. Auxin: Regulation, action and interaction. *Ann. Bot.* 95:707–35.

308. Wott, J. A., and J. H. B. Tukey. 1967. Influence of nutrient mist on the propagation of cuttings. *Proc. Amer. Soc. Hort. Sci.* 90:454–61.

309. Yarborough, J. A. 1932. Anatomical and developmental studies of the foliar embryos of *Bryophyllum calycinum*. *Amer. J. Bot.* 19:443–53.

310. Yarborough, J. A. 1936. Regeneration in the foliage leaf of Sedum. *Amer. J. Bot.* 23:303–7.

311. Zimmerman, P. W. 1933. Initiation and stimulation of adventitious roots caused by unsaturated hydrocarbon gases. *Contrib. Boyce Thomp. Inst.* 5:351–69.

312. Zimmerman, P. W., and F. Wilcoxon. 1935. Several chemical growth substances which cause initiation of roots and other responses in plants. *Contrib. Boyce Thomp. Inst.* 7:209–29.

313. Zimmerman, P. W. 1937. Comparative effectiveness of acids, esters, and salts as growth substances and methods of evaluating them. *Contrib. Boyce Thomp. Inst.* 8:337–50.

10
Techniques of Propagation by Cuttings

learning objectives

- Describe the different types of cuttings.
- Explain how stock plants can be manipulated to maximize adventitious root formation.
- Explain how cuttings are prepared for propagation.
- Describe how the propagation environment is managed.
- Identify the management practices used in propagation.
- Describe how cuttings are managed after rooting.

INTRODUCTION

Cutting propagation utilizes a portion of stem, root, or leaf that is cut from the parent or stock plant and induced to form roots and shoots by chemical, mechanical, and/or environmental manipulation. In most cases the new independent plant produced is a **clone,** which is identical to the parent plant.

Cuttings are the most important means of propagating ornamental shrubs—deciduous species as well as the broad- and narrow-leaved types of evergreens. Cuttings are extensively used in commercial greenhouse propagation of many florists' crops—poinsettias, chrysanthemums, geraniums—and in propagating foliage crops, bedding plants, certain fruit crops, and some vegetables (horseradish, chicory, artichoke, sweet potato), and forestry species (37, 124).

Generally, vegetative propagation is more costly (per unit propagule) than sexual (seedling) propagation. Propagating leafy cuttings requires the use of *protected culture* (glass-, polycarbonate-, or polyethylene-covered structures), bottom-heated rooting systems, and *intermittent mist* and/or *fog systems,* which increase production costs. However, for many species, the superiority of clonally produced cultivars justifies the higher cash value that is necessary to offset the added propagation costs associated with this process.

TYPES OF CUTTINGS

Cuttings are made from the vegetative portions of the plant, such as **stems, modified stems** (rhizomes, tubers, corms, and bulbs), **leaves,** or **roots.** Cuttings can be classified according to the part of the plant from which they are obtained:

cuttings Portions of stems, roots, or leaves that are detached from a plant and used to clonally multiply new plants.

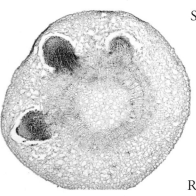

    Stem cuttings
      Hardwood
        Deciduous
          Narrow-leaved evergreen
        Semi-hardwood
        Softwood
        Herbaceous
      Leaf cuttings
    Leaf-bud cuttings (single-eye or single-node cuttings)
Root cuttings

The preferred type of cutting depends on individual circumstances. The least expensive and easiest method is usually selected. For easy-to-root woody perennial plants, hardwood stem cuttings in an outdoor nursery are frequently used because of the simplicity and low cost. For more tender herbaceous species, or for those more difficult to propagate, it is necessary to resort to the more expensive and elaborate facilities required for rooting the leafy types of cuttings. In today's containerized nurseries, a larger portion of easy- and difficult-to-root species are propagated with intermittent mist (see Fig. 10–39, page 387), fog (see Fig. 10–46, page 393), or contact polyethylene sheet systems (see Fig. 10–36, page 384). Root cuttings of some species are also satisfactory, but sufficient cutting material may be difficult to obtain.

In selecting cutting material, stock plants must be disease-free, moderately vigorous, and of known identity. Propagators should avoid stock plants that have been injured by frost or drought, defoliated by insects, or that have overly vigorous growth. Likewise, stock plants stunted by excessive flowering or fruiting, or by lack of soil moisture or proper nutrition, are not acceptable. A poor-quality cutting slows down the whole production process, creates cultural problems, and produces an inferior plant.

A desirable practice for the propagator is the establishment of stock blocks as a source of propagating material, where uniform, true-to-type, pathogen-free mother plants can be maintained and held under optimal nutrition for the best rooting of cuttings taken from them. Another advantage of stock plants is that they can be manipulated (via layering, stooling, hedging-severe pruning) to enhance rooting. However, in many container nurseries no stock plants are maintained. Rather, propagules are collected from the container plants in production—hence the need for good cultural controls, and the maintenance of cultivar records/identity of all plants in production.

Propagators need to keep thorough records of procedures and the seasonal condition of plant materials, and conduct small tests to achieve optimum success for their particular propagation system (32, 42). The type of cutting utilized and cultural manipulation during propagation is dependent on the market for which the plant is being targeted (e.g., ornamental trees with multistemmed flowering stems versus high-branched trees with strong, straight central leaders). Other considerations are the plant species, environmental conditions, propagation system utilized, propagation facilities, available personnel, and ultimately what is cost-effective for the producer.

Stem Cuttings

Stem cuttings can be divided into four groups, according to the nature of the stem tissue used: **hardwood, semi-hardwood, softwood,** and **herbaceous.** In propagation by stem cuttings, segments of shoots containing lateral or terminal buds are obtained with the expectation that under the proper conditions **adventitious roots** will develop, and thus produce independent plants. See Table 10–1 for a comparison of different types of cuttings.

> **adventitious roots**
> Roots that arise on aerial plant parts, underground stems, and old root parts.

The type of wood, the stage of growth used in making the cuttings, and the time of year when the cuttings are taken are some of the important factors in the satisfactory rooting of plants. Information concerning these factors is given in Chapters 9, 19, 20, and 21.

Hardwood Cuttings (Deciduous Species) Hardwood cuttings are those made of matured, dormant, firm wood after leaves have abscised. The use of hardwood cuttings is one of the least expensive and easiest methods of vegetative propagation. Hardwood cuttings are easy to prepare, are not readily perishable, may be shipped safely over long distances if necessary, and require little or no special equipment during rooting. Hardwood cuttings are easily transplanted after rooting, and some producers report that liners produced from hardwood cuttings are larger than those from softwood cuttings (60).

The low cost of hardwood cutting propagation makes feasible high-density meadow orchards, consisting of precocious dwarfed fruit trees. Some peach cultivars, for example, can be propagated on a large scale from rooted hardwood cuttings (54, 74).

Hardwood cuttings are prepared during the dormant season—late fall, winter, or early spring—usually from wood of the previous season's growth, although with a few species—fig, olive, and certain plum cultivars—2-year-old or older wood can be used. Hardwood cuttings are most often used in propagation of deciduous woody plants, although some broad-leaved evergreens, such as the olive, can be propagated by leafless hardwood cuttings. Many deciduous ornamental shrubs are started readily by this type of cutting. Some common ones are privet, forsythia, wisteria, honeysuckle, willow, poplar (*Populus*), dogwood, *Potentilla, Sambucus,* crape myrtle, and *Spiraea.* Rose rootstocks such as *Rosa multiflora* are propagated in great quantities by hardwood cuttings (Figs. 10–1, 10–2, and 10–3, pages 348–49). A few fruit species are propagated commercially by this method, for example, fig, quince,

Table 10-1
PROPAGATION SYSTEMS WITH DIFFERENT TYPES OF CUTTINGS

Cutting type	Hardwood (Deciduous)	Hardwood (Evergreen)	Semi-hardwood	Softwood	Herbaceous	Leaf	Root	Leaf-bud (Single eye or node)
Description	Mature, dormant, or quiescent hardwood stems; woody species.	Mature Hardwood stems; woody species.	Partially mature wood on current season's growth; woody species.	New, soft succulent growth; woody species.	Succulent stems from nonwoody plants.	Leaf blade or leaf blade and petiole; generally from non-woody, herbaceous plants.	Root pieces from thin to fleshy roots; woody and herbaceous plants.	Leaf blade + petiole + short piece of stem with attached axillary bud; woody and herbaceous plants.
Season propagated	Dormant season: late fall to early spring.	Dormant season: late fall to late winter.	Late spring to late summer.	Spring to early summer.	Year-round—with greenhouse-forced, and/or tropical field production.	Year-round—as long as leaves are available.	Take in late winter or early spring when roots contains stored carbohydrates—but before new shoot growth.	Generally during growing season; year-round for tropical plants.
Propagation system	Field propagated; also greenhouse propagated with light intermittent mist, fog, humidity tent, or contact polyethylene.	Light intermittent mist, fog, humidity tent, or contact polyethylene.	Intermittent mist, fog, humidity tent, or contact polyethylene.	Intermittent mist, fog, or humidity tent.	Intermittent mist, fog, or humidity tent.	Intermittent mist, fog, or humidity tent.	Depending on species, directly planted into field or direct planted in flats, and covered with contact polyethylene, or stuck in containers and held in dormant storage.	Intermittent mist, fog, humidity tent, or contact polyethylene.
Cutting length	10–76 cm (4–30 in); normally at least two nodes with basal cut just below the node; mallet, heel, and straight (most common) cuttings used; longer cuttings for rootstocks.	10–20 cm (4–8 in)	7.5–15 cm (3–6 in)	7.5–12.5 cm (3–5 in)	7.5–12.5 cm (3–5 in)	Varies with species, and leaf size, e.g., Sansevieria 7.5–10 cm (3–4 in); other species just use section of leaf.	Small, delicate roots—2.4–5.0 cm (1–2 in); somewhat fleshy—5.0–7.5 cm (2–3 in); large roots—5–15 cm (2–6 in)	2.0–7.5 cm (1–3 in); bud may sometimes be placed 1.3–2.5 cm (0.5–1.0 in) below the surface.

Cutting type	Hardwood (Deciduous)	Hardwood (Evergreen)	Semi-hardwood	Softwood	Herbaceous	Leaf	Root	Leaf-bud (Single eye or node)
Chemical treatment	IBA or NAA at 2,500–5,000 ppm; generally, 10,000 ppm maximum with difficult-to-root.	IBA or NAA at 2,000 ppm or slightly higher. Difficult-to-root; 5,000–10,000 ppm.	IBA or NAA at 1,000–3,000 ppm, generally, 5,000 ppm maximum.	IBA or NAA at 500 to 1,250 ppm; generally, 3,000 ppm maximum.	Auxin not needed, or may apply IBA or NAA from 500–1,250 ppm.	Apply cytokinin for adventitious buds, e.g., BA at 100 ppm for begonia, African violets (spray application).	Usually no hormones applied. Can apply cytokinins for adventitious buds.	IBA or NAA at 1,000 to 3,000 ppm with Ficus.
Examples of plants	Privet, Forsythia, landscape Roses, Willow, Sycamore, Crape Myrtle, Euonymus, Dogwood, Fig, Quince, Apple, Pear, Plum rootstock.	Juniper, Yew, Spruce, Firs (Abies), Chamaecyparis, Cryptomeria, Thuja, Tsuga.	Holly, Pittosporum, Rhododendron, Citrus, Olive, Euonymus.	Lilac, Forsythia, Maple, Magnolia, Weigela, Apple, Peach, Pear, Plum, Chionanthus, Crape Myrtle, Spirea.	Geraniums, Poinsettia, Dieffenbachia, Chrysanthemum, most Floral Crops; Pineapple (slips and suckers); Sweet Potato (slips).	Begonia, African violet, Sansevieria, Sedums.	Poppy, Aralia, Geranium, Viburnum, Horse Radish, Sumac (Rhus), Stokesia.	Blackberry, Raspberry, Boysenberry, Maple, Ficus, Rhododendron, Pothos, Philodendron, Clematis, English Ivy.
Other observations	One of the least expensive; easy to ship over long distances; little equipment needed, i.e., field propagation without mist; start when you can remove leaves without tearing bark; central and basal parts generally better cuttings than apical; cuttings can be pretreated with bottom heat, then field planted; or direct rooted in field during spring or fall.	Very slow to root; bottom heat 23–27°C (75–80°F) helpful; basal wounding may be helpful; important to maintain adequate moisture levels during rooting.	Leaf may be trimmed to control transpiration; wounding may be helpful	Roots relatively quickly (2–5 weeks); bottom heat helps; 23–27°C (75–80°F); requires more attention and equipment (mist/fog); do not use weak, thin, or thick, heavy cuttings; remove all flower buds; use lateral or side branches for cuttings. Very susceptible to desiccation.	Probably the fastest to root; can be susceptible to decay; bottom heat is helpful.	Periclinal chimeras will not reproduce true-to-type; with fibrous-rooted begonias—cut large leaves into triangular sections containing a piece of large vein; entire leaf blade plus petiole used with African violet.	To maintain polarity and avoid planting upside down, make a straight cut near the crown (proximal) end, and a slanting cut at the distal end; should be planted with the proximal end up or cutting planted horizontally; periclinal chimeras cannot be propagated by root cuttings.	Useful when propagating materials are scarce—since twice as many plants can be produced as with stem cuttings; each node can be used as a cutting; high humidity essential and bottom heat is helpful.

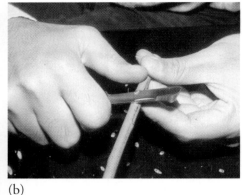

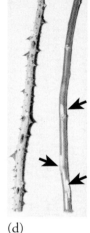

Figure 10–1
Preparing hardwood cuttings of deciduous *Rosa multiflora* rootstock. (a) Bandsaw used to cut 15 cm (6 in) cuttings from rose canes. (b) "De-eying" all lower axillary buds to prevent suckering of rootstock. (c) De-eyed cuttings in bundles of 50. (d) Dormant rose budwood (left) and thornless, de-eyed (arrows) rootstock (right) prior to sticking dormant, unrooted cuttings in an open field in November in Texas.

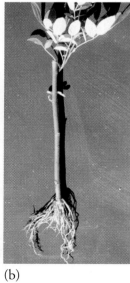

Figure 10–2
(a) Raised, shaped beds, recently propagated with *Rosa multiflora* hardwood cuttings. Some 12,000 to 15,000 cuttings per acre are stuck in November in Texas. (b) Rooted, leafed-out cutting removed in April for photograph. The rooted rootstocks will be T-budded during the spring, grown for a second year, harvested bare-root with a U-blade in the fall of the second year, and processed as landscape roses.

olive, mulberry, grape, currant, pear, gooseberry, pomegranate, and some plums.

The propagating material for hardwood cuttings should be taken from healthy, moderately vigorous stock plants growing in full sunlight. The wood selected should not have abnormally long internodes or be from small, weakly growing interior shoots. Generally, hardwood cutting material is ready when you can remove the leaves without tearing the bark. Wood of moderate size and vigor is the most desirable.

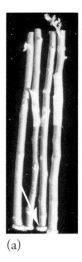
(a)

(b)

(c)

(d)

Figure 10–3
(a) *Rosa multiflora* hardwood cuttings that were simultaneously chip-budded and callused for rooting. Too high a temperature—27°C (80°F)—caused profuse callusing at the base (arrow), but poor rooting and field survival. (b) Winter callusing (arrow) of hardwood cuttings in cooler storage. (c) Deciduous hardwood cuttings of pear rooted under protective cover. (d) Narrow-leaved, evergreen standard and heel (arrow) hardwood cuttings of Juniper.

The cuttings should have an ample supply of stored carbohydrates to nourish the developing roots and shoots until the new plant becomes self-sustaining. Tip portions of shoots are usually discarded because they are often low in stored carbohydrates and commonly contain unwanted flower buds. Central and basal parts generally make the best cuttings, but there are exceptions (73).

Hardwood cuttings vary in length from 10 to 76 cm (4 to 30 in). The diameter of the cuttings may range from 0.6 to 2.5 cm (1/4 in or 0.25 in to 1 in), depending upon the species. Long cuttings, when they are to be used as rootstocks for fruit trees, permit the insertion of the cultivar bud (scion) into the original cutting following rooting, rather than into a smaller new shoot arising from the original cutting.

At least two nodes are included in the cutting: the basal (bottom) cut is usually just below a node and the top cut 1.3 to 2.5 cm (1/2 in or 0.5 in to 1 in) above a node. When hardwood cuttings of *Rosa multiflora* are to be field-planted in raised soil beds, it is common procedure to "de-eye" or remove all lower basal axillary buds prior to sticking cuttings to prevent suckering from the base of the cutting (Fig. 10–1).

Where it is difficult to distinguish between the top and base of the cuttings, it is advisable to make one of the cuts at a slant rather than at a right angle. In large-scale operations, bundles of material are cut to the desired lengths by band saws or other types of mechanical cutters rather than individually, by hand (Fig. 10–1).

Hardwood cuttings will desiccate, so it is important that they not dry out during handling and storage. After cuttings are cut with a band saw, some producers will take the bundled cuttings and dip the tops (apex) in wax. The wax helps reduce desiccation and indicates the orientation of the cuttings for late fall or spring field propagation. For large-scale commercial operations, the planting of cuttings is mechanized.

Deciduous hardwood cuttings are dormant and leafless when propagated. Three propagation systems of deciduous hardwood cuttings are described: (a) direct fall planting, (b) initiation of rooting with bottom heat, and (c) direct spring planting.

Direct Fall Planting. In regions with mild winters or reliable snow cover, cuttings can be made in the autumn and planted immediately in the nursery. Rooting may take place during the dormant season, or the formation of roots and shoots may occur simultaneously the following spring. Hardwood cuttings will take longer to root in the field at the relatively low and declining soil temperatures of fall. Hardwood cuttings of peach and peach × almond hybrids have been successfully rooted in the nursery, provided they were treated prior to planting with the **auxin**,

> auxin A phytohormone or plant growth regulator that can stimulate cuttings to root. Common auxins include indole-3-butyric acid (**IBA**) and α-naphthalene-acetic acid (**NAA**).

indole-3-butyric acid **(IBA)** (74). Hardwood cuttings for landscape and cut rose rootstock are fall-planted in production fields in Spain, California, and Texas (Fig. 10–2).

Field-propagated hardwood cuttings are dug after a growing season as rooted liners using an apparatus such as a modified potato digger, or with a U-blade attached to a tractor. However, rooted, hardwood cuttings of field-propagated landscape roses are budded (grafted) and left in the field for an additional season and sold as finished plants.

Deciduous hardwood cuttings are leafless, cannot photosynthesize, and survive on stored carbohydrate reserves. Thus, it is important to avoid temperatures that cause excessive callusing and the loss of stored reserves—otherwise, rooting and field survival are poor (Fig. 10–3).

Initiating Roots with Bottom Heat. This method has been successful for difficult-to-root species such as some apple, pear, and plum rootstocks. Cuttings are collected in either the fall or late winter, the basal ends treated with IBA at 2,500 to 5,000 ppm. Cuttings are then bundled and placed upright for about 4 weeks on a 20-cm (8-in) sand base with heating mats or circulating hot-water tubing just below the sand surface. The bundled cuttings are packed between 6 cm (3 in) of moist peat and maintained with bottom heat at 18 to 21°C (65 to 70°F). The top portion of the cuttings is left exposed to the cool or ambient outdoor temperatures. Covered, unheated sheds are used for protection against excessive moisture from rains. The East Malling Research Station in England (currently known as Horticulture Research International, East Malling) developed commercial procedures for propagating difficult species by this method (80, 82, 83). Cuttings must be transplanted before buds begin growth; this is usually done as roots first emerge. This procedure is probably best suited for regions that experience relatively mild winters (78, 79). When soil or weather conditions are not suitable for planting after roots become visible, cuttings are left undisturbed under **protected culture** in the rooting bed, with the bottom heat off (Fig. 10–3). They are transplanted in the nursery when conditions become suitable (20).

Direct Spring Planting. Cutting material of easily rooted species is gathered during the dormant season, wrapped in newspaper or slightly damp peat moss in a polyethylene bag, and stored at 0 to 4.5°C (32 to 40°F) until spring. The cutting material should not be allowed to dry out or to become excessively wet during storage. At planting time, the cuttings are made into proper lengths and planted into a field nursery propagation bed without intermittent mist, or in propagation flats with a very light intermittent mist.

Stored cutting material should be examined frequently. If signs of bud development appear, lower storage temperatures should be used, or the cuttings should be made and planted without delay. If buds are forcing out when the cuttings are planted, leaves will form and the cuttings will die due to water loss from the leaves and depletion of stored carbohydrate reserves prior to rooting.

Hardwood Cuttings (Narrow-Leaved Evergreen Species) Hardwood cuttings of narrow-leaved evergreens are also dormant. However, unlike deciduous plants, their foliage is retained when propagated. Narrow-leaved evergreen cuttings must be rooted under moisture conditions that will prevent excessive drying as they usually are slow to root, sometimes taking several months. Some species root much more readily than others. In general, *Chamaecyparis, Thuja,* and the low-growing *Juniperus* species root easily and the yews (*Taxus* spp.) fairly well, whereas the upright junipers, the spruces (*Picea* spp.), hemlocks (*Tsuga* spp.), firs (*Abies* spp.), and pines (*Pinus* spp.) are more difficult. In addition, there is considerable variability among the different species in these genera regarding the ease of rooting of cuttings. Cuttings taken from young seedling stock plants root much more readily than those taken from older trees because of the juvenility factor. Auxins such as IBA will enhance rooting.

> **protected culture** The use of temperature-controlled or unheated propagation structures or greenhouses to minimize environmental fluctuations of cuttings or rooted liner plants.

BOX 10.1 GETTING MORE IN DEPTH ON THE SUBJECT
CALLUSING AND ROOTING

Callusing and rooting can occur simultaneously. Callusing can sometimes interfere with rooting and—except for some difficult-to-root species that initiate roots from callus—is not a prerequisite to rooting.

Figure 10–4
(a and b) Hardwood cuttings of narrow-leaved evergreens being prepared for sticking. (c) Bundled cuttings ready for propagating. (d) Quick-dipping (arrow) cuttings in auxin rooting solution. Photos (a), (b), and (d) courtesy V. Priapi.

Narrow-leaved evergreen cuttings ordinarily are best taken between late fall and late winter (Fig. 10–4). Cuttings taken from stock plants should be processed rapidly. Cuttings are usually best rooted in a greenhouse or polyhouses with relatively high light irradiance and under conditions of high humidity or very light misting, but without heavy wetting of the leaves. However, cuttings can also be outdoor propagated in mist beds (Fig. 10–5). A bottom heat temperature of 24 to 26.5°C (75 to 80°F) has given good results. Dipping the cuttings into a fungicide helps prevent fungal diseases. Sand alone is a satisfactory rooting medium, as is a 1:1 mixture of perlite and peat moss. Some individual cuttings take longer to root than others. The slower-rooting ones can be inserted again in the rooting medium and, eventually, root.

The type of wood to use in making the cuttings varies considerably with the particular species being rooted. As shown in Figure 10–4, the cuttings are made 10 to 20 cm (4 to 8 in) long with all the leaves removed from the lower half. Mature terminal shoots of the previous season's growth are usually used. In some instances, as in with juniper (*Juniperus chinensis* 'Pfitzeriana,') older and heavier wood also can be used, thus resulting in a larger plant when it is rooted. On the other hand, some propagators use small tip cuttings, 5 to 8 cm (2 to 3 in) long, placed very close together in a flat for rooting. In some species, such as *Juniperus excelsa*, older growth taken from the sides and lower portion of the stock plant roots better than the more succulent tips. Basal wounding benefits rooting of some narrow-leaved evergreen species (see Fig. 10–28, page 373).

Semi-Hardwood Cuttings Semi-hardwood cuttings are those made from woody, broad-leaved evergreen species, and leafy summer and early fall cuttings of deciduous plants with partially matured wood. Cuttings of broad-leaved evergreen species are generally taken during the summer (or late spring through early fall in warmer climates) from new shoots just after a flush of growth has taken place and the wood is partially matured. Many broad-leaved evergreen shrubs, such as *Camellia, Pittosporum, Rhododendron, Euonymus,* evergreen azaleas, and holly, are commonly propagated by semi-hardwood cuttings. A few fruit species, such as citrus and olive, can also be propagated in this manner.

Semi-hardwood cuttings are made 7.5 to 15 cm (3 to 6 in) long with leaves retained at the upper end (Fig. 10–6). If the leaves are very large, they can be trimmed one-third to one-half their size to reduce the leaf surface area, which lowers transpirational water loss and allows closer spacing in the cutting bed (see Fig. 10–28). The shoot terminals are often used in making cuttings, but the basal parts of the stem will

Figure 10–5
(a and c) Outdoor propagation of narrow-leaved evergreen cuttings. (b) Propagation in mistbed under protected culture. (c) Cuttings of *Thuja, Taxus,* and *Juniperus* cultivars, are struck in the concrete sand of the beds between the railroad ties (117). (d) Rooted cuttings being harvested.
Photos (c) and (d) courtesy V. Priapi.

Figure 10–6
Semi-hardwood cuttings of (a and b) Magnolia and (c and d) hibiscus.

BOX 10.2 GETTING MORE IN DEPTH ON THE SUBJECT
USE OF LONG CUTTINGS

While conventional cuttings are typically 5 to 20 cm (2 to 8 in) long, there has been success with propagating long cuttings of 50 to 152 cm (20 to 60 in) for rooting more difficult-to-root shade and forestry species, as well as rootstocks of fruit trees and standard roses (Fig. 10–7) (137, 138). Some of the difficult-to-root species include *Acer plantanoides, Carpinus betulus, Pyrus* sp, *Quercus robur, Tilia cordata* and *Ulmus* 'Regal,' as well as apple, cherry, and pear. Long cuttings have increased rooting success, better over-winter survival, and subsequently faster production time as rooted liners. What is important for success is: (a) using high-humidity, high-pressure fog systems, (b) propagating semi-hardwood cuttings in late June and July (Hanover, Germany), (c) use of current-year shoot growth and taking cuttings from low on the stock plants with species such as sycamore (*Acer pseudoplantnus*), and (d) the over-wintering method (unheated greenhouse) for survival and growth. Cuttings are propagated in ground beds or 11-cm (4-in) liner pots with a peat-sand media and controlled release fertilizer (138). Only leaves of the basal 10 to 20 cm (4 to 8 in) of the cutting are removed. Cuttings are treated with 0.5 percent (5,000 ppm) IBA.

Another benefit of using long-cuttings is that by cutting back the stock plant to promote long shoot growth, there is a *rejuvenation effect that enhances the competence to root*. Another feature is that the fog system *enhances environmental conditions during rooting* with better control of vapor pressure deficit (avoiding moisture loss from the leaf, without over-wetting). Hence a larger-sized propagule can be supported, which has more leaf surface area and vegetative buds. This can lead to *greater carbohydrate and basal auxin accumulation.*

Figure 10–7
(a) A majority of cuttings are 5–20 cm (2–8 in) long. However, long cuttings of 50–152 cm (20–60 in) are used to propagate ornamental and fruit crops. (b) Long, rooted semi-hardwood cuttings of rose (*Rosa* 'Pfaenders' rootstock for standard roses) in a greenhouse propagation bed. (c) Nine-month-old rooted liners of elm (*Ulmus* 'Regal'), sycamore maple (*Acer pseudoplatanus*), pear (*Pyrus* 'Williams Christ'), (Linden) *Tilia cordata,* and English oak (*Quercus robur*) propagated from long cuttings. Part of the advantage of long cuttings may be that the pruning management of the stock plants enhances rejuvenation and rooting. Photos courtesy of W. Spethmann.

often root, too. The basal cut is usually just below a node. The cutting wood should be obtained in the cool, early morning hours when leaves and stems are turgid. Cuttings should be placed in large containers, which are covered with clean moist burlap to maintain a high humidity, or put in large polyethylene bags (see Fig. 10–55, page 399). Cuttings should be kept out of the sun until they can be stuck and propagation is initiated.

Leafy cuttings are rooted under conditions that will keep water loss from the leaves at a minimum. They are commercially rooted under intermittent mist or fog. Rooting is also done in cool, temperate climates or during fall in the southern United States under polyethylene sheets laid over the cuttings (see Fig. 10–36, page 384). Bottom heat, auxin treatment, and sometimes wounding are also beneficial. High shade levels (93 percent reduction of ambient sun) with fog systems can increase rooting performance of oak and maple cuttings (164). In Denmark, semi-hardwood cuttings of dogwood, *Deutzia,* forsythia, ligustrum, and *Spiraea* are watered in, covered with white, opaque polyethylene sheets, and rooted in the field (75). High rooting accelerates axillary bud growth and winter survival of rooted liners.

Softwood Cuttings Cuttings prepared from the soft, succulent, new spring growth of deciduous or evergreen species are **softwood cuttings.** The softwood condition for most woody plants ranges from 2 to 8 weeks. Softwoods are produced during growth flushes and may occur just once per year, as with the fringe tree (*Chionanthus virginicus*), elm (*Ulmus parvifolia*), and Euonymus (*Euonymus alatus*) or several times during the year [e.g., April to August in Texas with crape myrtle (*Lagerstroemia*) and spirea]. August softwood cuttings in the southeastern United States are not physiologically the same as June softwood cuttings (e.g., June softwood cuttings generally survive the winter better than August cuttings). Many ornamental woody plants can be started by softwood cuttings (Fig. 10–8). Typical examples are the hybrid French lilacs, *Forsythia, Magnolia, Weigela, Spiraea,* maples, and flowering dogwood. Various crab apple cultivars can also be started in this manner (24). Although not commonly done, apple, peach, pear, plum, apricot, and cherry can be propagated by softwood cuttings.

For some difficult-to-root species, softwood cuttings may be the only commercial method to clonally regenerate cultivars. Softwood cuttings generally root easier and quicker (2 to 5 weeks) than other types but require more attention and sophisticated equipment. This type of cutting is always made with leaves attached. They must, consequently, be handled carefully to prevent desiccation and be rooted under conditions that will avoid excessive water loss from the leaves. Temperature should be maintained during rooting at 23 to 27°C (75 to 80°F) at the base of cuttings. As long as light is adequate (but not excessive), ambient air temperature of the mist or fog system can rise to 30 to

(a)

(b)

Figure 10–8
Hydrangea quercifolia 'Snow Queen' propagated by softwood cuttings. (a) Mist propagation bed. (b) Rooted liner.

32°C (86 to 90°F) without detriment to rooting. High shade levels (91 percent reduction of ambient sun) with fog systems can increase IBA effectiveness and the rooting performance of many woody plant taxa (163).

The proper type of cutting material must be obtained for making softwood cuttings. Extremely fast-growing, soft, tender shoots are not desirable because they are likely to deteriorate before rooting. The best cutting material has some degree of flexibility but is mature enough to break when bent sharply. Weak, thin, interior shoots should be avoided as well as vigorous, abnormally thick, or heavy ones. Average growth from portions of the plant in full light is the most desirable to use. (With more difficult-to-root species, shading and etiolation of the stock plants are sometimes done prior to taking cuttings. See the later section on stock plant manipulation.) Some of the best cutting materials are the lateral or side branches of the stock plant. Heading back the main shoots will usually force out numerous lateral shoots from which cuttings can be made. Softwood cuttings are 7.5 to 12.5 cm (3 to 5 in) long with two or more nodes. The basal cut is usually made just below a node.

The leaves on the lower portion of the cutting are removed, but those on the upper part are retained. Large leaves can be trimmed to minimize transpirational loss and to occupy less space in the propagating bed. But drastically cutting back leaves, and the mutual shading of leaves of crowded cuttings in a flat, can reduce rooting and encourage diseases such as *Botrytis*. For difficult-to-root plants, factors that favor a *high leaf-to-stem ratio*—such as *not* trimming leaves, and selecting cuttings with relatively thin stems—favor rooting; whereas leaf-trimming and selecting thick, fleshy-stemmed cuttings may result in stem rotting (Fig. 10–9). Flowers and flower buds should be removed.

Softwood cuttings stress easily, so it is important to collect cutting material early in the day. Softwood cuttings should be kept moist, cool, and turgid at all times. Laying the cutting material or prepared cuttings in the sun for even a few minutes will cause serious damage. Soaking the cutting material or cuttings in water for prolonged periods to keep them fresh is undesirable. Refrigerated storage (4 to 8°C, 40 to 47°F) for 1 to 2 days is another option. Cuttings of some species, such as forsythia (*Forsythia xintermedia*), can be safely stored for a month.

Some producers of ornamental shrubs will fall-propagate hardwood cuttings in polyhouses, with heated benches under intermittent mist, and then harvest softwood cuttings from the developing hardwood shoots as the various species flush in the spring (3). Before the end of summer, both the original crop of hardwood cuttings and the bonus crop of softwood cuttings can be lined out.

There has been some very innovative research for forcing softwood cuttings (116, 152). Large branch segments of adult hardwoods from 2 to 10 cm (1 to 4 in) wide × 24 cm (9.5 in) long are placed under intermittent mist (see Fig. 9–24). This process encourages latent (dormant) axillary buds to force out and produce epicormic sprouts, which are harvested later as softwood cuttings and rooted under mist.

Herbaceous Cuttings Herbaceous cuttings are made from succulent with little woody tissue like geraniums, chrysanthemums, poinsettia, coleus, carnations, many

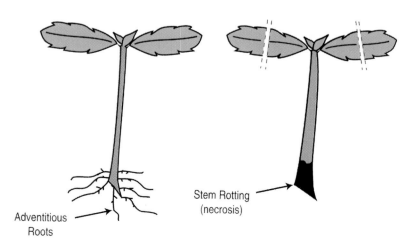

Figure 10–9

For difficult-to-root species, factors that favor a high leaf:stem ratio, such as not trimming leaves and selecting cuttings with relatively thin stems (*left*), favor rooting; whereas leaf-trimming and selecting thick, fleshy-stemmed cuttings may result in stem rot (*right*). Redrawn from B. H. Howard (83).

BOX 10.3   GETTING MORE IN DEPTH ON THE SUBJECT
FROM A PROPAGATOR'S NOTEBOOK: CHARLOTTE LEBLANC (90)

Propagating Difficult-to-Root Plants
With recalcitrant, deciduous plants such as *Acer palmatum*, *Stewartia*, *Styrax*, *Betula nigra*:

- There is a short window when cuttings will root—collect cuttings just after the first growth flush has finished and when the wood is just beginning to firm up.
- Carefully select cutting wood from only the best container or stock plants.
- Take larger cuttings of 13 to 15 cm (5 to 6 in) that have a lot of foliage, because they'll have better growth after rooting. This will later produce a larger rooted liner with a more extensive root system.
- Harvest cuttings early in the morning. Baskets containing cuttings are covered with burlap soaked with a disinfectant and algaecide (quaternary ammonium) to help cuttings remain cool and moist.
- Cuttings are misted before going into the cooler and periodically sprayed with a mist bottle during cutting preparation and auxin application to prevent desiccation.
- Cuttings are quick-dipped with the auxins Chloromone K+, K-IBA or Dip'N Grow (IBA+NAA) diluted with sterile water and then mixed 50 percent (1:1) with Celluwet (water-thickening additive from Griffin Labs).
- Use tall tree tube containers, which allow sufficient mist to be applied without over-wetting the media.
- Use well-drained media: 3 parts well composted pine bark to 1 part perlite.
- Incorporate control release fertilizer (Osmocote 18-6-12) in propagation media.
- Rooted liners are very sensitive to cold the first winter; once plants are dormant, maintain just above freezing in a protected structure (quonset house) during first winter (Florida).
- Apply soluble fertilizer just as the buds start to break in the spring.

---

bedding plants, ground covers, and foliage crops. Unlike softwood cuttings, herbaceous cuttings can be taken anytime of the year, weather permitting. They are 8 to 13 cm (3 to 5 in) long with leaves retained at the upper end (Figs. 10–10, 10–11, and 10–12, page 358). Many florists' crops are propagated by easily rooted herbaceous cuttings. They are rooted under the same conditions as softwood cuttings. Bottom heat is also helpful. Under proper conditions, rooting is rapid and in high percentages. Although auxins are usually not required, they are often used to gain uniformity in rooting and the development of heavier root systems.

Types of cuttings (apical versus basal) and nodal position of herbaceous cuttings can influence shoot growth and finished plant quality of rooted liners. For example, basal cuttings of *Hedera helix* and *Schefflera arboricola* develop longer shoots and more roots than apical cuttings. With pothos (*Epipremnum aureum*), a 3-cm (1-in) or longer internode section below the node and a fraction of

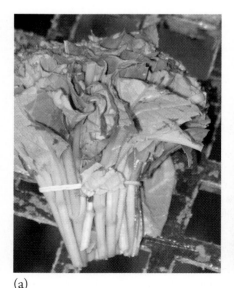

(a)   (b)

Figure 10–10
(a and b) Holding herbaceous cuttings under mist (arrow) in bundles of 50 cuttings until they can be processed and stuck. The cuttings are kept under low light, cool temperatures, and high relative humidity—all of which reduces vapor pressure deficit. This is one important aspect of "maintaining the plant's momentum" for propagation success.

Figure 10–11
With millions of unrooted cuttings (URCs) being produced and shipped internationally, propagators need to be diligent in their handling, storing, processing, and sticking. (a) Stock plants in Costa Rica, (b and c) taking cuttings—URCs, (d) sealing cuttings into plastic bags, (e) cuttings move through a chilling unit that rapidly takes the temperature from approximately 27°C (80°F) down to 13°C (55°F); labeling to alert shippers and propagators. (g) Sticking herbaceous poinsettia URCs produced in Guatemala for propagation and production in Florida. (h) URCs of poinsettia direct stuck into small liner cells containing paper pot sleeves (tube) for propagation under mist.

the old aerial root should be retained on cuttings for most rapid axillary shoot development (154). With foliage plants such as *Dracena* spp., branching of herbaceous canes during propagation is done by cutting one-third to one-half of the way through the canes (Fig. 10–13).

suberized (suberization)
The process of forming a protective, semiimpermeable layer that occurs under the wounded surface area of a cutting.

Cuttings of succulent plants are normally callused and **suberized** for a week or more before being inserted into rooting media; that is, basal ends of pineapple slips and suckers, and cactus cuttings (cladodes) (Fig. 10–14, page 359)(89). These practices tend to prevent the entrance of decay organisms.

Leaf-Bud Cuttings

A **leaf-bud cutting** (**single-eye** or **single-node cutting**) consists of a leaf blade, petiole, and a short piece of the stem with the attached axillary bud (Fig. 10–15, page 359). Leaf-bud cuttings differ from leaf cuttings in that only adventitious roots need form. The axillary bud at the nodal area of the stem provides the new shoot. A number of plant species, such as the black raspberry (*Rubus occidentalis*), blackberry, boysenberry, lemon, camellia, maple,

Figure 10–12
Herbaceous carnation (*Dianthus caryophyllus*) cuttings. (a) Greenhouse stock plants. (b and c) Harvesting and preparing cuttings. (d) Carnation cuttings ready for sticking. (e) Sticking cuttings. (f) Propagating under mist.

Figure 10–13
(a) Branching of herbaceous canes of *Dracaena* during propagation is done by cutting one-third to one-half way through the canes (28). (b) Sansiveria cane cutting with new axillary shoot (black arrow) and adventitious roots (white arrow).

Figure 10–14
(*Opuntia ficus-indica*) cactus pear. (a) For forage and fruit (arrow) consumption. (b) Harvested cladode. (c) Caldodes are processed in $CuSO_4$, then allowed to air-dry and suberize for several days prior to field propagating. (d) Rooted cladode with new axillary shoot (arrow).

rhododendron, and vines are readily started by leaf-bud cuttings; many tropical shrubs and most herbaceous greenhouse plants are usually started by stem cuttings.

Leaf-bud cuttings are particularly useful when propagating material is scarce, because they will produce at least twice as many new plants from the same amount of stock material as stem cuttings. Each node can be used as a cutting.

Basal treatment of the cutting with auxin will stimulate root production (Table 10–1). The size of cuttings ranges from 2 to 7 cm (1 to 3 in). High humidity is essential, and bottom heat is desirable for rapid rooting.

Figure 10–15
Leaf bud cutting or single eye cuttings—more propagules can be produced than with larger shoot cuttings. (a) *Ficus pumila* and (b) several leaf-bud cuttings of pothos directly stuck for rooting.

Figure 10–16
(a) Leaf cuttings of Rieger begonias are taken from greenhouse stockplants. (b) New adventitious shoots and roots have formed from a leaf cutting, which was removed prior to the photo. (c) Adventitious roots and buds will form at the base of the petiole (arrow). Adventitious bud formation is the limiting factor in leaf and root cutting propagation.

Leaf Cuttings

In **leaf cuttings,** the leaf blade, or leaf blade and petiole, is utilized in starting new plants. Adventitious buds, shoots, and roots form at the base of the leaf and develop into the new plant; the original leaf does not become a part of the new plant. Only a limited number of plant species can be propagated by leaf cuttings.

African violets (*Saintpaulia*), begonias, and peperomia are routinely propagated by leaf cuttings. Leaf cuttings of begonias can be made of an entire leaf (leaf blade plus petiole), the leaf blade only, or just a portion of the leaf blade. The new plant forms at the base of the petiole or midrib of the leaf blade (Figs. 10–16 and 10–17).

Figure 10–17
Various types of leaf cuttings with new plantlets developing (arrows): (a) African violet, (b) Cape primrose (*Streptocarpus*), (c) Begonia leaf pieces, (d) Begonia whole leaf cut on main veins, and (e) Sedum.

Figure 10–18
Leaf cuttings of *Sansevieria*. (a) Leaf blade with adventitious roots. (b and c) New plant with shoots and adventitious shoots and roots. The original leaf cutting does not become a part of the new plant.

In starting plants with fleshy leaves, such as *Begonia rex,* by leaf cuttings, the large veins are cut on the undersurface of the mature leaf, which is then laid flat on the surface of the propagating medium (Fig. 10–17). The leaf is pinned or held down in some manner, with the natural upper surface of the leaf exposed. After a period of time under humid conditions, new plants form at the point where each vein was cut. The old leaf blade gradually disintegrates.

Another method, sometimes used with fibrous-rooted begonias, is to cut large, well-developed leaves into triangular sections, each containing a piece of a large vein. The thin outer edge of the leaf is discarded. These leaf pieces are then inserted upright in sand or on a peat-perlite medium, with the pointed end down. The new plant develops from the large vein at the base of the leaf piece (Fig. 10–17).

One type of propagation by leaf cuttings is illustrated by *Sansevieria*. The long tapering leaves are cut into sections 8 to 10 cm (3 to 4 in) long, as shown in Figure 10–18. These leaf pieces are inserted into the rooting medium, and, after a period of time, a new plant forms at the base of the leaf piece. The variegated form of *Sansevieria, S. trifasciata laurenti* is an example of a periclinal chimera that will not reproduce true-to-type from leaf cuttings: to retain its characteristics, it must be propagated by division of the original plant. (See the discussion in Chapter 9 on propagule types and maintaining chimeras.)

New plants arise from leaves in a variety of ways. The piggy-back plant (*Tolmiea*) develops a new plantlet at the junction of its leaf blade and petiole, even while the leaf is still growing on the mother plant (Fig. 10–19). With *Bryophyllum,* many new plantlets or *foliar embryos* arise at the margins of the leaf (Fig. 10–19). If used for propagation, the leaf itself eventually deteriorates.

Leaf cuttings should be rooted under the same conditions of high humidity as those used for softwood or herbaceous cuttings. Cuttings are commercially rooted under mist or high-humidity tents. Most leaf cuttings root readily, but the **limitation to propagation is adventitious bud and shoot development.** Hence, cytokinins can be used to induce buds to form (Fig. 9–15) (35). Methods of applying cytokinins are the same as those discussed for auxins later in the chapter.

adventitious buds (and shoots) Buds and shoots that rise from any plant part other than terminal, lateral, or latent buds on stems. Adventitious buds form irregularly on older portions of a plant and not at the stem tips or in the leaf axils. Unlike dormant buds, adventitious buds do not have a bud trace all the way to the pith. An adventitious bud is an embryonic shoot.

Root Cuttings

Best results with root cuttings are likely to be attained if the root pieces are taken from young stock plants in late winter or early spring when the roots are still supplied with stored carbohydrates but before new growth starts. Taking the cuttings during the spring when the parent plant is rapidly making new shoot growth should be avoided. Root cuttings of the Oriental poppy (*Papaver orientale*) should be taken in midsummer, the dormant period for this species. Securing cutting material can be quite labor-intensive, so it is more cost-effective to collect root cuttings by trimming roots from nursery plants as they are dug.

The correct polarity should be maintained when planting root cuttings. To avoid planting upside-down, the proximal end (nearest the crown of the plant) may be

Figure 10–19
(a) The piggy-back plant (*Tolmiea menziesii*) forms plantlets (arrows) on its leaves. The new plants arise at the junction of the leaf blade and petiole. The plant is propagated by leaf cuttings. (b) Leaf cuttings of *Bryophyllum crenatodaigremontianum* and (c) *Bryophyllum daigremontianum*. New plantlets develop from foliar "embryos" in the notches at the margin of the leaf (arrows). Leaves are partially covered or pegged down to hold the leaf margin in close contact with the rooting medium.

made with a straight cut and the distal end (away from the crown) with a slanting cut. The proximal end of the root piece should always be up. Insert the cutting vertically so that the top is at about soil level (Figs. 10–20 and 10–21, page 364). Cuttings may also be planted horizontally 2.5 to 5 cm (1 to 2 in) deep, to avoid the possibility of planting upside-down.

Not all species should be propagated by root cuttings. If root cuttings are used to propagate chimeras of aralia and geranium with variegated foliage, the new plants produced lose their variegated form. Propagation by root cuttings is very simple, but the root size of the plant being propagated may determine the best procedure to follow.

Root Cuttings of Plants with Small, Delicate Roots
Root cuttings of plants with small, delicate roots should be started in propagation flats or cell packs in the greenhouse, hotbed, or heated polyhouse. The roots are cut into short lengths, 2.5 to 5 cm (1 to 2 in) long, and scattered horizontally over the surface of the medium. Then they are covered with a layer of medium measuring 1 to 2 cm [1/2; (or 0.5) in]. After watering, a polyethylene cover is placed over the flat to prevent drying until the plants are started. Figure 10–21 shows a rooting chamber with small root cuttings of gerarium rooting in cells of propagation flats.

The flats are set in a shaded place. After the plants become well formed, they can be transplanted to other flats or lined-out in nursery rows for further growth. See Table 10–2 for a list of selected species that can be propagated by root cuttings.

Root Cuttings of Plants with Somewhat Fleshy Roots
Cuttings of plants with fleshy roots are best started in a flat in the greenhouse or hotbed (e.g., the lilacs in Fig. 10–21). The root pieces should be 5 to 7.5 cm (2 to 3 in) long and planted vertically, observing correct polarity. New adventitious shoots should form rapidly, and as soon as the plants become well established with good root development, they can be transplanted. These root pieces can also be stuck directly in containers and held in dormant storage in cool greenhouses during the winter season, then undergo a period of active spring growth, followed by midsummer planting in the field.

Root Cuttings of Plants with Large Roots, Propagated Out-of-Doors
Large root cuttings are made 5 to 15 cm (2 to 6 in) long (Fig. 10–22, page 365). They are tied in bundles, care being taken to keep the same ends together in order to avoid planting upside-down later. The cuttings are packed in boxes of damp sand, bark, or peat moss for about 3 weeks and held at about 4.5°C (40°F). After this, they should be planted 5 to 7.5 cm (2 to 3 in)

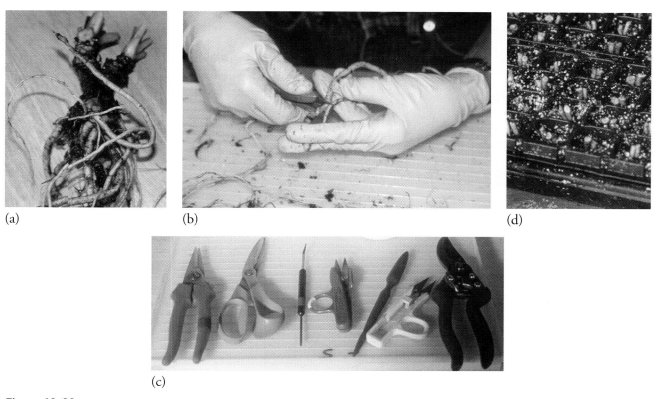

Figure 10–20
Root cuttings of the herbaceous perennial sea kale (*Crambe maritima*). (a) Plant ready for harvesting root cuttings. (b and c) prepping root cuttings and tools used. (d) Root cuttings in propagation flats.

apart in well-prepared nursery soil with the tops of the cuttings level with, or just below, the top of the soil.

Some deciduous shrubs grown from root pieces are converted to softwood summer cuttings by taking elongating shoots from the root pieces and rooting these softwood cuttings under mist. This conversion technique is reported to produce heavier, faster-growing plants and reduce production time by 1 year with *Aronia, Clethra, Comptonia, Euonymus, Spiraea,* and *Viburnum* species (113). However, advances in softwood stem cutting technology have reduced the usage of root cuttings with many taxa (39).

SOURCES OF CUTTING MATERIAL

In cutting propagation, the source of the cutting material is very important (8). The stock plants and other sources from which the cutting material is obtained should be:

- Free of disease and insect pests
- True-to-name and type
- In the proper physiological state so that cuttings root successfully.

Acquiring Sources of Cutting Material

Several sources are possible for obtaining cutting material:

a. *Stock plants specially maintained as a source of cutting material* (Fig. 10–23, page 366). Although such plants may occupy valuable land space, this is probably the ideal source of cutting material. There is an accurate history and identity of each stock plant. To maintain high rooting potential, it is much easier to use techniques such as hedging-back, mounding, stooling, and banding on stock plants than on non-permanently maintained container-produced plants. By culturally maintaining uniformity of growth in stock plants, the propagator ensures that evenly graded batches of cutting material are available during a given period. Consequently, the success rate and uniformity of rooting is that much greater.

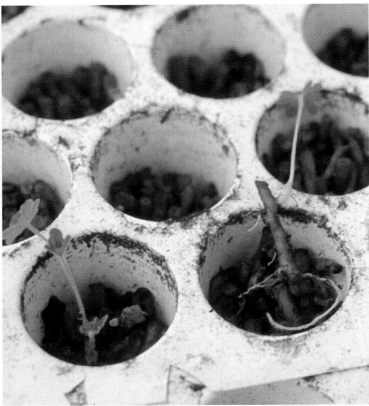

Figure 10–21
Root cuttings: (a) rooting chamber for propagating root cuttings of geranium and lilac, (b) root cuttings of geranium are maintained in the cells of propagation flats and finished off in pots, (c) rooted lilac cuttings are transplanted to containers or lined-out in the field.

Field-grown stock plants are insurance for future propagules, particularly during unusually cold winters that might wipe out an unprotected, container-grown crop.

b. *Prunings from nursery plants as they are trimmed and shaped.* Many nurseries use prunings as the primary source of their cutting material. Sometimes, however, the trimming is not done at the optimum time to root the cuttings, and the unrooted cuttings must be stored. Most ornamental nurseries in the southern United States take cuttings from containerized plants rather than stock blocks. As a general rule of thumb, cuttings of easy-to-root species are taken during normal production pruning cycles. Conversely, softwood cuttings of more difficult-to-root species are taken during a brief window–of time when rooting is optimum.

c. *Tissue-culture-produced liners.* It is becoming more common to use tissue-culture-produced liners as sources of stock plants in the development of new cultivars and disease-indexed plants. Conventional macropropagation techniques can then be used after establishment of micropropagated stock plants.

d. *Buying in small, rooted liner plants or unrooted cuttings (URCs).* See Figure 10–11 and the discussion on URCs (page 330), and Box 10.4: "To Propagate or Not to Propagate."

e. *From plants growing in the landscape in parks, around houses or buildings, or in the wild.* For nursery production, care must be taken to ensure proper identification of the species and cultivar prior to propagation.

Stock Plant Manipulation

Stock plants are manipulated to maximize the rooting potential prior to taking cuttings. Various techniques to manipulate stock plants are described in the following sections.

Table 10–2
SOME SPECIES THAT CAN BE PROPAGATED BY ROOT CUTTINGS

Actinidia deliciosa (kiwifruit)	*Malus* spp. (apple, flowering crab apple)
Aesculus parviflora (bottle-brush buckeye)	*Myrica pennsylvanica* (bayberry)
Ailanthus altissima (tree-of-heaven)	*Papaver orientale* (oriental poppy)
Albizia julibrissin (silk tree)	*Phlox* spp. (phlox)
Anemone japonica (Japanese anemone)	*Plumbago* spp. (leadwort)
Aralia spinosa (devil's walking stick)	*Populus alba* (white poplar)
Artocarpus altilis (breadfruit)	*Populus tremula* (European aspen)
Broussonetia papyrifera (paper mulberry)	*Populus tremuloides* (quaking aspen)
Campsis radicans (trumpet vine)	*Prunus glandulosa* (flowering almond)
Celastrus scandens (American bittersweet)	*Pyrus calleryana* (oriental pear)
Chaenomeles japonica (Japanese flowering quince)	*Rhus copallina* (shining sumac)
Chaenomeles speciosa (flowering quince)	*Rhus glabra* (smooth sumac)
Chlerodendrum trichotomum (glory-bower)	*Rhus typhina* (staghorn sumac)
Comptonia peregrina (sweet fern)	*Robinia pseudoacacia* (black locust)
Daphne genkwa (daphne)	*Robina hispida* (rose acacia)
Dicentra spp. (bleeding heart)	*Rosa blanda* (rose)
Eleutherococcus sieboldianus (fiveleaf aralia)	*Rosa nitida* (rose)
Eschscholzia californica (California poppy)	*Rosa virginiana* (rose)
Ficus carica (fig)	*Rubus* spp. (blackberry, raspberry)
Forsythia xintermedia (forsythia)	*Sassafras albidum* (sassafras)
Geranium spp. (geranium)	*Stokesia laevis* (Stokes aster)
Hypericum calycinum (St. Johnswort)	*Styphnolobium japonica* (Japanese pagoda tree)
Koelreuteria paniculata (golden-rain tree)	*Symphoricarpos xchenaultii* (snowberry)
Liriope spp. (liriope)	*Syringa vulgaris* (lilac)
Liquidambar styraciflua (American sweet gum)	*Ulmus carpinifolia* (smooth-leaved elm)

Pruning and Girdling Annual pruning is an important aspect of stock plant management in relation to (a) maintenance of juvenility to improve rooting, (b) plant shaping for easier and faster collection of propagules (see Figs. 9–26, page 369, and 9–27, page 372), (c) increased cutting production, (d) timing of flushes, and (e) reducing reproductive shoots (132).

Types of pruning (132) include:

Modified Stooling. In modified stooling, plants are severely cut back to their base but not mounded with soil as with traditional stooling; this eliminates reproductive shoots and is beneficial for *Hydrangea* and *Senecio*.

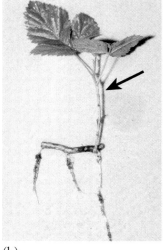

Figure 10–22
Propagation by root cuttings with (a) sassafras and (b) blackberry. New adventitious shoots (arrows) and roots form from the root cutting.

(a)   (b)

(a)

(b)

(c)

Figure 10–23
Stock plants maintained as a source of cutting material: (a) stock plants of *Pittosporum tobira*, (b) *Liriope* spp., and (c) hydrangea for greenhouse and nursery propagation.

Hard Pruning. With hard pruning, stock plants are cut back to half their size annually. This avoids irregular growth that can occur from modified stooling, and eliminates reproductive growth (e.g., *Forsythia*, *Weigela*, and heather). The advantage of hard-pruned hedges is not to increase the vigor of shoots or to mimic juvenile material, as has been long assumed. Thick cuttings from vigorous shoots may survive better than thinner shoots, but the thinner shoots root faster as long as propagation conditions are designed to rapidly drain

---

BOX 10.4 GETTING MORE IN DEPTH ON THE SUBJECT
TO PROPAGATE OR NOT TO PROPAGATE

It may be more profitable <u>NOT</u> to propagate your entire plant inventory. It is very important that every item produced in a greenhouse or nursery wholesale business be profitable. Sometimes this means that it is cheaper for the grower to buy-in liner plants propagated by a company specializing in custom propagation liners. This is part of a *business-to-business (B2B) niche*. Custom propagators sell seedling plugs, rooted cuttings, tissue-culture produced liners, grafted or budded plants, tree whips, etc. to wholesalers, which are then shifted up into larger containers or field-planted, and finished-off.

There is also huge international business of unrooted cuttings (URCs) that are produced offshore and shipped to wholesalers to be propagated and finished off. The advantage to the wholesaler is that they do not have to maintain stock plants or stock blocks, which take up valuable production space. URCs can lower production costs by 30 percent compared to buying rooted liners, and vendors can offer new varieties more quickly than the wholesaler can produce (71).

Conversely, it may be more profitable for a nursery or greenhouse company to specialize as a custom propagation business—propagating and vending their rooted liner cuttings (plants) to other wholesalers to be finished off. Propagating nursery liner plants can increase revenue per unit of land by five-fold, compared to finished container production. There is also a faster turnover of liner plants, more frequent sales, higher sales dollars generated by propagation employees, and cheaper shipping costs (18).

away excess water; for instance, thinner cuttings are particularly prone to rotting (Fig. 10–9) (83). The faster rooting of thinner hardwood cuttings suggests that the rooting potential among shoots in a hedge is more influenced by the relative positions of shoots than by their absolute position in terms of the distance between each other or from the root system (Figs. 9–26 and 9–27). In general, hedges should be grown to produce the maximum number of relatively thin-stemmed cuttings with a high leaf-to-stem ratio if the species is difficult-to-root, such as syringa (*Syringa vulgaris*). Conversely, large, fleshy-stemmed cuttings are perfectly acceptable for the easy-to-root *Forsythia xintermedia*.

Moderate Pruning. Plants are cut back by one-third to one-half of the previous annual shoot each year, and there is less die-back than with the foregoing two methods. This type of pruning is used with *Viburnum* and deciduous azaleas.

Light Pruning. Light pruning implies tipping back or just normal removal of cuttings from the stock.

Hedging. The severity of pruning to maintain the hedge form is generally determined by the ease with which cuttings can be collected from the stock plant [e.g., *Berberis* and *Pyracantha* are heavily pruned, while *Eleagnus* and dogwood (*Cornus*) are lightly trimmed].

Double Pruning. Spring pruning produces a flush of cuttings for summer softwood cuttings or semi-hardwood fall cuttings. In England, a second trimming in June delays the softwood cutting collection period until fall; cutting production is increased, but growth is weaker than during the normal summer flush.

With *Dracaena* stock plant production, **incisions** are made above axillary buds by cutting one-third to one-half through the cane (Fig. 10–13). This breaks apical dominance and induces additional buds to develop in the plant without sacrificing the apical heads. This technique further promotes greater branching during field propagation (28).

Girdling shoots of stock plants prior to taking cuttings has been used successfully to root slash pine, sweetgum, sycamore, and 19- to 57-year-old water oak (76, 77). The treatment consists of girdling shoots by removing 2.5 cm (1 in) of bark, applying IBA talc, and wrapping the shoot with polyethylene film and aluminum. Once primordia become visible as small bumps in the callus, the cutting is removed from the stock plant and rooted under mist (Fig. 10–24).

girdling The constriction of the phloem of a stem by wounding, bending, or tying with a band. Girdling can be used to enhance the rooting potential of cuttings or layers, before they are removed from the stock plant.

Etiolation, Shading, Blanching, and Banding A modification of the traditional technique of etiolation and blanching, using Velcro adhesive fabric strips as the blanching material, is shown in Figs. 10–25 and 10–26 (101). This etiolation technique for *softwood cutting propagation* has improved the rooting success of a wide range of difficult-to-root woody species (Table 10–3) (102, 103). (See Chapter 9 for a discussion of the physiological implications of these techniques.) Rooting, subsequent bud-break, and growth of difficult-to-root apple cultivars have been promoted by a short period of banding light-grown shoots of stock plants with Velcro, before taking cuttings (144). Stock plant etiolation and stem banding are estimated to be 30 percent more expensive than traditional cutting propagation but 50 percent less expensive than grafted plants and 30 percent less or equal to that of micropropagated plants (102). The trade-off in using these light exclusion techniques is extending the range of plant species that can be propagated on their own roots, the increased success of plant establishment, and extending the production season (i.e., propagation occurs earlier by forcing containerized stock plants in the greenhouse, which can allow for additional top-growth of rooted liners—shortening production times and reducing costs).

ROOTING MEDIA

There is no universal or ideal rooting mix for cuttings. An appropriate propagation medium depends on the species, cutting type, season, and propagation system (e.g., with fog, a high water-holding medium is less of a problem than with intermittent mist). The cost and availability of the medium components are other considerations.

The rooting medium has four functions:

- To hold the cutting in place during the rooting period
- To provide moisture for the cutting
- To permit exchange of air at the base of the cutting
- To create a dark or opaque environment by reducing light penetration to the cutting base

Media Substrates

Propagation media includes an **organic component:** peat, sphagnum moss, or softwood and hardwood barks. The **coarse mineral component** is used to increase the proportion of large, air-filled pores and drainage and includes perlite, vermiculite, expanded shale, coarse sand or grit, pumice, scoria, polystyrene,

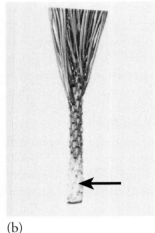

Figure 10–24
Girdling systems to enhance rooting. (a) A girdled shoot of slash pine (*Pinus caribaea*) with 2.5 cm (1 in) of bark removed (arrow), treated with an IBA talc slurry, then wrapped with plastic film and aluminum foil. (b) Root primordia visible as small bumps on the callus (arrow). (c) The cutting is later removed from the stock plant and rooted under mist. Only girdled cuttings will root when removed from the stock plant. Courtesy R. C. Hare.

and rockwool (6, 21, 111). Rarely is mineral soil used as a propagation medium component, except in field propagation of hardwood or semi-hardwood cuttings. Most propagators use a combination of organic and mineral components [e.g., peat-perlite, peat-expanded shale, peat-vermiculite-perlite, bark-haydite (clay and shale), peat-rockwool, etc.] (see Fig. 10–27, page 372) (125). Sometimes the mineral component is used alone (e.g., sand, rockwool, Oasis cubes, perlite) or in combination (e.g., vermiculite-perlite, sand-polystyrene). Sufficient coarse mineral component should be added to improve aeration.

A trend in U. S. nursery propagation is for partial replacement of expensive peat with softwood bark. In container production, barks have generally replaced peat as the dominant media component. Coconut coir (mixed

---

BOX 10.5   GETTING MORE IN DEPTH ON THE SUBJECT
CUTTING PRODUCTION IS AFFECTED BY SCAFFOLD DEVELOPMENT OF STOCK PLANTS

The vegetatively propagated annuals markets continue to grow rapidly, with greenhouse-produced crops such as geranium, poinsettia, chrysanthemum, petunia, *Verbena*, *Nemesia*, etc. Scaffold management during the early stages of stock plant production can significantly impact cutting production. Increasing the number of pinches performed on stock plants during scaffold development can increase the weekly production rate and the cumulative yield of cuttings harvested (55). It is also a useful tool for stock plant growers to manipulate a crop's timing. See Figures 9–26 and 9–27 for the effect of shoot hierarchy—whether the cutting was produced from directed pruning of the stock plant's frame work or scaffold—on rooting success of cuttings from severely pruned stock plants hedges of woody plant species.

 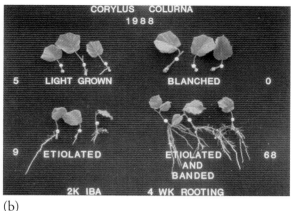

Figure 10–25
(a) Etiolation of *Kalmia latifolia* 'Obtusa red.' Black velcro (arrows) with an auxin talc is wrapped around the base of the etiolated shoots, which are gradually exposed to higher light irradiance. After the shoots green up, they are removed and rooted under mist as softwood cuttings. (b) Highest rooting occurred with *Corylus colurna* shoots that were etiolated and banded with Velcro, then made into cuttings and treated with 2,000 mg per L IBA. Courtesy B. K. Maynard.

with perlite) is also being utilized as a peat substitute in propagation (140). In Europe, Australia, and Israel, rockwool sheets and cubes are used for direct sticking of cuttings (Fig. 10–27). Rockwool can be handled efficiently (no containers or trays need be filled), and since cuttings are directly stuck into cubes or sheets, there is minimal disturbance of roots, which avoids transplant shock when rooted cuttings are shifted up to **liner production** stages (Fig. 10–27).

There are myriad premixed commercial propagation media available, which can help reduce production steps, save time and labor, and assure better media standards [e.g., Promix BX (sphagnum peat, perlite, vermiculite)], W. R. Grace Co; see Chapter 3 for other examples]. There is generally no

> **liners** Small plants that are produced from rooted cuttings, seedlings (plugs), or tissue culture.

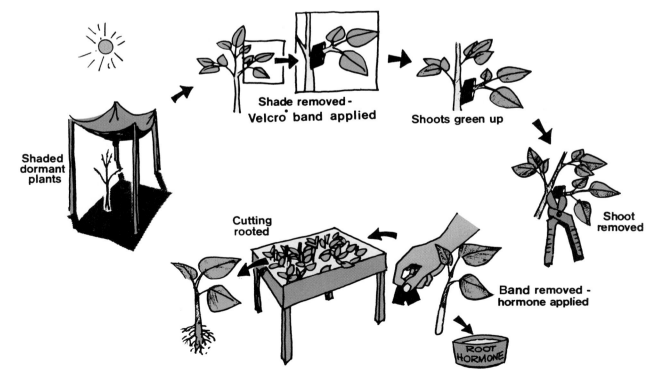

Figure 10–26
Scheme for etiolated softwood cutting propagation using Velcro fabric strips as the blanching material. Courtesy B. K. Maynard and N. L. Bassuk.

Table 10–3
ENHANCED ROOTING OF CUTTINGS FROM STOCK PLANT ETIOLATION, SHADING, AND BANDING TREATMENTS

Treatment	Species	
Banding/blanching	Acer platanoides	Platanus occidentalis
	Tilia cordata	Rhododendron cvs.
	Pinus elliottii	Rubus idaeus
Etiolation	Artocarpus heterophyllus	Syringa vulgaris cvs.
	Bryophyllum tubiflorum	Malus sylvestris
	Cinnamomum camphora	Mangifera indica
	Clematis spp.	Persea americana
	Corylus maxima	Prunus domestica
	Cotinus coggygria	Rubus idaeus 'Meeker'
	Fallopia baldschuanicum	Tilia tomentosa
Etiolation plus banding	Acer spp.	Pinus strobus
	Betula papyrifera	Malus xdomestica
	Carpinus betulus	Persea americana
	Castanea mollissima	Pistachia vera
	Corylus americana	Syringa vulgaris
	Pinus spp.	Carpinus betulus
	Quercus spp.	Tilia spp.
	Hibiscus rosa-sinensis cvs.	
Shading	Crassula argentea	Rhododendron spp.
	Schefflera arboricola	Rosa spp.
	Hibiscus rosa-sinensis	Euonymus japonicus
	Picea sitchensis	

Source: Maynard and Bassuk (101–103).

advantage in adding a **wetting agent (surfactant)** to propagation media, and with some species a wetting agent may decrease rooting (13). However, many commercial mixes come with wetting agents, and some propagators will incorporate a wetting agent with peat-perlite-based propagation mixes.

wetting agent (surfactant) Substance that reduces the surface tension of water, which allows better absorption of a chemical into plant tissue and can enhance water absorption of propagation media.

An ideal propagation medium provides sufficient porosity to allow good aeration and has a high water-holding capacity yet is well drained and free from pathogens. Pathogens can occur in peat (*Pythium, Penicillium,* etc.) and other organic components of media. A good commercial practice is to pasteurize with aerated steam. Some mineral components, such as vermiculite, perlite, calcined clay, and rockwool, are relatively sterile due to their high-temperature exposure during manufacturing, and need not be pretreated for pathogens. Integrated pest management (IPM) is used during rooting to control damping-off organisms (*Pythium, Phytophthora, Rhizoctonia, Pestalotiopsis, Glomerella, Botrytis, Peronospora*). IPM includes the selective use of fungicides and pesticides on cuttings, as well as biological and cultural controls (see Chapter 3).

The key to successful propagation medium is **good water management**. Rarely can a rooting response be attributed to differences in aeration due to the physical properties of the various media (147). Gaseous diffusion proceeds relatively freely through the propagation media. Most of the aerobic requirements for rooting (94) are supplied by diffusion of oxygen through the aerial portion of the cutting to its base.

Water films, both within and around the base of the cutting, can obstruct the free passage of oxygen to developing root initials. Thus, it is very difficult to pinpoint the relationship of rooting with physical characteristics such as volumetric air and water contents of the medium (94). During cool winter months and/or with a closed mist system, the medium should be sufficiently loose to allow adequate aeration even when water utilization is lower.

Chemical and physical standards of a propagation media used at a successful commercial nursery are listed in Table 10–4, page 372.

BOX 10.6 GETTING MORE IN DEPTH ON THE SUBJECT
TECHNIQUES OF SHADING, ETIOLATION, BANDING, AND BLANCHING (101)

- Field-grown or containerized dormant stock plants are ready for treatment when all chilling or rest requirements have been fulfilled and the buds begin to swell, usually in early spring in the field or midwinter in a greenhouse.

- During the **shading** process, entire plants or several branches of a plant are then covered with an opaque material, usually black cloth or plastic. Sufficient space is needed to allow for the new growth to extend. A wire or wooden frame may be used to support the covering.

    shading The partial reduction of light to near 100 percent light exclusion that can occur during stock plant manipulation and/or propagation.

- Cuts should be made in the covering material, or corners left slightly open near the top of the structure to allow for ventilation. A small heat buildup under the structure is desirable, but enough ventilation should be supplied so that plants are not scorched. It is neither necessary nor desirable to exclude 100 percent of the light. Between 95 and 98 percent light exclusion is preferable.

- Initial growth is allowed to progress in the dark (*etiolation*) until new shoots are between 5 and 7 cm (2 and 3 in) long, after which time the shade is gradually removed over the period of 1 week so as not to scorch the very tender shoots.

- On the first day of shade removal, **banding** is initiated with the placement of self-adhesive black bands at the base of each new shoot (the future cutting base). Black plastic electrical tape works well. To speed later removal, the end of the tape should be folded over to form a small loop. These bands keep the base of the shoot in an etiolated condition while the tops of the shoots are allowed to turn green in the light. (*Banding mimics mound stooling and is applied to actively developing tissue of shoots above ground.*)

- The bands are approximately 2.5 cm wide × 2.5 cm (1 in) long but can vary in length. Strips of Velcro are made up of two pieces—one a "woolly" side and the other a material with "hooks" or "nubs" that adhere to the wool when they are pressed together. The Velcro band "sandwiches" the shoot so that the hooks gently pierce the surface of the shoot when they are pressed together. In addition, auxin in a talc base is added to both parts of the Velcro before banding the shoot, thereby delivering a rooting stimulus to the shoot prior to its being made into a cutting. The hooks of the Velcro aid in wounding the stem so there is better penetration of the stem with auxin. Typically, up to 8,000 ppm IBA in talc is used. Both pieces of Velcro are dipped in the talc preparation, the excess powder is tapped off, and then the Velcro band is firmly pressed onto the shoot base.

- Velcro bands are generally left on the shoots for 4 weeks, although periods as short as 2 weeks and as long as 12 weeks have also been successful.

- After 4 weeks, the cuttings are removed from the stock plant (cut just below the banded area). The bands are removed and the cutting bases are treated again with hormone before sticking them in the rooting medium. This second IBA treatment is generally a quick dip of 4,000 ppm in 50 percent ethanol.

- The area underneath the band is typically yellowish-white and swollen, and occasionally root primordia can be seen already emerging.

- Cuttings are stripped of lower leaves and placed in a rooting medium consisting of peat and perlite (1:1 by volume). Bottom heat (25°C, 77°F) is supplied during the winter months. Mist is applied intermittently, beginning at 7 seconds of mist every 2 minutes, becoming less frequent with time and amount of cloud cover. Fifty percent Saran shading is also applied over the rooting bench to keep down air temperatures around the cuttings. Greenhouse ambient temperature is maintained at 20°C (68°F), and daylength is regulated to 16 hours of daylight using 60-W incandescent bulbs spaced 1 meter (1.1 yd) apart and hung 1 meter (1.1 yd) above the bench.

- The cuttings are left in the rooting bench from 2 to 5 weeks for deciduous plants and up to 12 weeks for pines.

- After rooting, the cuttings are potted up, fertilized, and kept under long days (16 hours) to encourage growth. Depending on the time of year, they may be placed outside to continue growing during the late spring and summer (for winter-propagated plants) or they may be hardened-off and placed in a protective overwintering structure (for summer-propagated plants).

BOX 10.7 GETTING MORE IN DEPTH ON THE SUBJECT
BLANCHING

All of the steps listed in Box 10.6 on blanching are followed except that stock plants are not initially covered, so new growth occurs in the light. When the soft, green shoots are 5 to 7 cm long, banding with Velcro or electrical tape (as previously described) proceeds in exactly the same way as etiolation and banding.

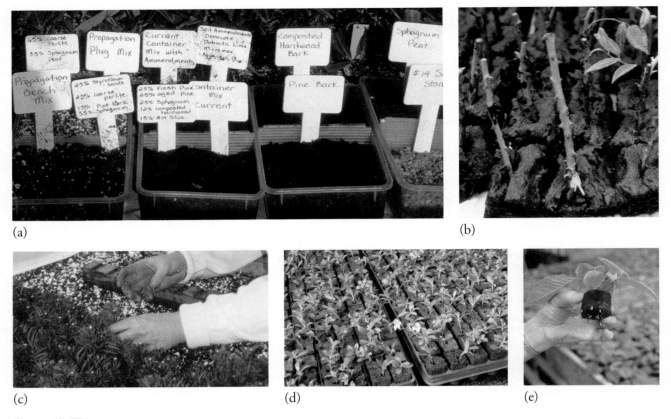

Figure 10–27
(a) Various rooting media used by propagators. (b) Rockwool. (c) Peat-perlite mix. Stabilized peat (d) and compressed peat pellet (e) for direct-sticking.

Current Propagation Systems Besides rockwool rooting cubes and blocks, and the standard plastic propagation flat, there is a variety of *propagation unit systems* to work with (100). These include Jiffy-7s (compressed peat blocks), synthetic foam media Rootcubes, Japanese paper pot containers with an expandable honeycomb form, bottomless polystyrene or plastic trays with wedge or cylindrical cells for air pruning of roots (Speedling, Gro-Plug, etc.), and various modifications of plastic flats (see Chapter 3). A number of propagators are using

Table 10–4
SUGGESTED CHEMICAL AND PHYSICAL STANDARDS FOR ROOTING MEDIUM

Property	Comments
Chemical	
pH	4.5–6.5; 5.5–6.5 preferred
Buffer capacity	As high as possible
Soluble salts	400–1,000 ppm (1 media: 2 water by volume)
Cation exchange capacity	25 to 100 meg/liter
Physical	
Bulk density	0.3–0.80 g/cm^3 (dry) or 0.60–1.15 g/cm^3 (wet)
Air-filled porosity	15–40% by volume, ideally 20–25% range
Water-holding capacity	20–60% volume after drainage
Particle stability	Materials should resist decomposing quickly; decomposition can alter other media components

Source: Maronek, Studebaker, and Oberly (99).

a Danish (Ellegaard) paper tube-sleeve insert that fits into plastic flat cells for direct rooting of cuttings (see Figs. 3–19 and 3–20).

WOUNDING

Root production on stem cuttings can be promoted by wounding the base of the cutting. This has proven useful in a number of species, such as juniper, arborvitae, rhododendron, maple, magnolia, and holly (156). Wounds may be produced in cuttings of narrow-leaved evergreen species, such as arborvitae or Thuja, by stripping off the lower side branches of the cuttings (Fig. 10–28). The benefits of **stripping** basal leaves is species-dependent; *Berberis* and *Juniperus* cuttings benefit, whereas *Spiraea, Forsythia,* and *Weigela* do not. Stripping basal leaves of cuttings reduces the propagation bench space required for some species, allows the propagator more flexibility to work with different size propagules, may serve to improve the contact area between the cutting and media (see Chapter 9), and can potentially improve absorption of auxins. Also, auxin is predominately absorbed via the cut (wounded) surface of a cutting and not through the epidermis or periderm (63, 72).

A common method of wounding: making one to four vertical cuts with the tip of a sharp grafting knife or pruning shears down each side of the cutting for 2.5 to 5 cm (1 to 2 in), penetrating through the bark and into the wood (Fig. 10–29, page 374). Wounding hardwood cuttings by splitting the stem base can be an effective technique to allow auxin to reach the cambium where it can stimulate rooting (see Figs. 9–33 and 9–34).

Larger cuttings, such as magnolias and rhododendrons, may be more effectively wounded by removing a thin slice of bark for about 2.5 cm (1 in) from the base on two sides of the cutting, exposing the cambium but not cutting deeply into the wood (156). A simple carrot peeler works just as well for wounding rhododendron cuttings (Fig. 10–28). For the greatest benefit, the cuttings should be treated with auxins after wounding, either in a talc or liquid quick-dip. The device shown in Figure 10–29 can be used to make rapid and uniform wounded cuttings (Fig. 10–30).

One type of wounding to avoid is inadvertent crushing and damaging of basal cutting tissue with dull shears. It is from this basal tissue that both the movement of water from the propagation media must occur for cutting survival, as well as root initiation for regenerating new plants. Always use sharp, periodically disinfected pruning shears, such as Felco No. 7 (Fig. 10–29). A pruning shear system with tubing supplying disinfectant has been developed to reduce pathogen contamination when making cuttings.

TREATING CUTTINGS WITH AUXINS

Treating cuttings with auxins, which are plant growth regulators, increases the percentage of cuttings that form roots, hastens root initiation, and increases uniformity of rooting. However, some difficult-to-root species will not respond to auxin treatment. Plants whose cuttings root easily may not justify the additional expense and effort of using these materials. Best use of auxins is with

(a)

(b)

(c)

(d)

Figure 10–28
Wounding and stripping of cuttings: (a) A potato peeler (arrow) is used for wounding rhododendron cuttings—part of the leaf surface area has been removed by the pruning shears. (b) Wounding a cutting with a knife—notice the thumb protection (arrow) for the propagator. (c and d) Preparing a *Thuja* cutting by stripping off the lower needles.

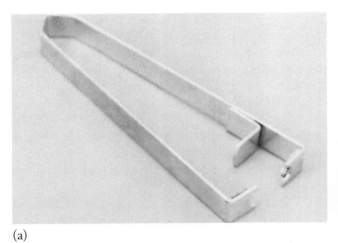
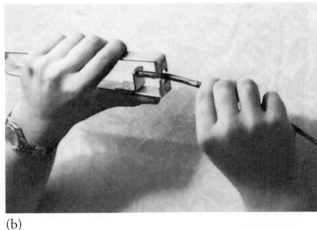

Figure 10–29
(a and b) Tool designed for making wounding cuts in the base of cuttings to stimulate rooting. Four sharp prongs make the actual cuts as the cutting is pulled through the opening, as shown in the photo on the right. (c) Wounding the basal side of a magnolia cutting with pruners.

moderately-difficult- to difficult-to-root species. Although treatment of cuttings with auxins is useful in propagating plants—and can increase efficiency by reducing production time from propagule to rooted liner—the ultimate size and vigor of such treated plants is no greater than that obtained with untreated plants.

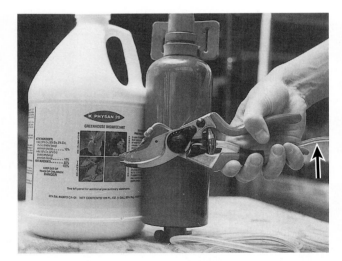

Figure 10–30
A pruning shear system with tubing (arrow) supplying disinfectant to the blades of the shear. This reduces potential pathogen contamination when making cuttings.

Rooting Chemicals, Formulations, and Carriers

The plant growth regulator most reliable in stimulating adventitious root production in cuttings are the auxins: **indole-3-butyric acid (IBA)** and α-**naphthaleneacetic acid (NAA),** although others can be used (14, 15, 115). IBA and NAA are often used in combination. IBA is the best auxin for general use because it is nontoxic to plants over a wide concentration range, and is effective in promoting rooting of a large number of plant species. IBA is a relatively stable compound, and its shelf-life can be extended by darkness and refrigeration (Fig. 10–31). NAA is quite stable both when mixed as a powder or a liquid. Generally, if a cutting does not respond to IBA, other root-promoting compounds will not compensate. IBA may be toxic to softwood cuttings of certain species, which leads to poor cutting regrowth and overwintering losses. These chemicals are available in commercial preparations and dispersed in talc or in concentrated liquid formulations that can be diluted with water (aqueous solution) to the proper strength (see Table 10–5).

The **potassium salt formulations** enable IBA (e.g., **K-IBA**) and NAA (e.g., **K-NAA**) to be dissolved in water. Otherwise, the acid formulations of these auxins need to be dissolved initially in alcohol (isopropyl,

Figure 10–31
Auxin formulations: (a) Dip 'N Grow liquid rooting compound. (b) Woods rooting compound. (c) Seradix rooting powder. (d) Hortus IBA water-soluble salts. (e) Refrigeration of liquid auxin formulations to extend its shelf life.

ethanol, or methanol), acetone, or another solvent or carrier before water can be added. It is not advisable to use any carriers that have not been registered with the EPA. There may be some benefit in mixing auxins with carriers, such as polyethylene glycol (Carbowax, laboratory-grade PEG) or propylene glycol (potable water antifreeze)—particularly with extremely difficult-to-root species that require higher auxin concentrations and are sensitive to alcohol (26, 27, 42, 44). Windshield washer fluid (methanol based) is also an effective carrier of auxin (27). At auxin concentrations above 5,000 ppm there is less burning with propylene glycol than alcohol (9). The auxins NAA and IBA do not readily dissolve in 50 percent polyethylene glycol (available as antifreeze), so it is necessary to heat the solvent to 72°C (160°F), although upon cooling after mixing, auxins will not precipitate out (9).

Ethyl alcohol (95 percent/190 proof ethanol or grain natural spirits that can be purchased at liquor stores and later diluted to a 50 percent concentration) causes less burning to plant tissue than isopropyl alcohol (rubbing alcohol). Some U. S. nurseries will use K-IBA (water-based solution) with cuttings during active growth stages (softwood, semi-hardwood) and use IBA with ethanol during dormant periods (hardwood) to avoid burning and dehydrating the tissue of the cuttings. See Table 10–5 for a listing of commercial rooting compounds containing K-salt formulations of auxin.

Table 10–5
PARTIAL LIST OF COMMERCIAL ROOTING COMPOUNDS, SOURCES, FORMULATIONS, AND INGREDIENTS

Trade name	Source	Formulation	Ingredient
C-mone, C-mone K (Chloromone K) C-mone K+	Coor Farm Supply Services, Inc., Smithfield, NC	Liquid (isopropyl alcohol)	1 and 2% IBA, 1% K-IBA 1% K-IBA. 0.5% NAA
Dip'N Grow	Dip'N Grow, Inc., (Astoria-Pacific, Inc.) Clackamas, OR (www.dipngrow.com)	Liquid (alcohol—ethanol and isopropyl)	1% IBA + 0.5% NAA + boron
Hormex	Brooker Chemical Corp., North Hollywood, CA	Powder (talc) Liquid	Rooting Powder—0.1 to 4.5% IBA Hormex Concentrate—0.13% IBA + 0.24% NAA + Vitamin B-1
Hormodin	MSD-Agvet (Merck & Co.) Rahway, NJ	Powder (talc)	0.1, 0.3, 0.8% IBA
Hormo-Root	Rockland Chemical Co. Newfoundland, NJ	Powder (talc)	1, 2, 3, 4.5% IBA
Hortus IBA	Hortus USA Corp. Inc., New York, NY (www.rooting-hormones.com)	IBA water-soluble salts Powder and water-soluble tablet forms	Up to 1.0% IBA 0.5 to 1.0% IAA 0.1, 0.3, 0.8% IBA
Rootone	Dragon Chemical Corp. Roanoke, VA (www.dragoncorp.com)	Powder (talc)	0.2% 1-1-Napthaleneacetamide, 4% Thiram (fungicide)
Roots	Sure-Gro IP Inc., Brantford, Ontario, Canada	Liquid	0.4% IBA + ethazol (fungicide)
Synergol	Certis, Amesbury, Wiltshire, Great Britain (www.certiseurope.co.uk)	Liquid	0.5% K-IBA + 0.5% K-NAA + fungicides and other additives
Woods Rooting Compound	Earth Science Products Corp., Wilsonville, OR	Liquid (ethanol)	1.03% IBA + 0.56% NAA

Note: The EPA registration number is on the finished formulation container. If no number is present, the product has not been registered and/or is being sold illegally (45).

What is needed are not new solvents or carriers for auxins, but rather new, more effective auxin formulations. With some woody plant species, the **aryl esters** of IAA and IBA, such as Phenyl indole-thiolobutyrate (P-ITB), and the **aryl amid** of IBA are equal or more effective than the acid formulation in promoting root initiation (43, 45, 141).

With the interest in organic production, which prohibits use of synthetic auxins, there are organic rooting products with known auxin effect such as algae extract, brewer's yeast, and seaweed dry extract. An extract of macerated seeds (Terrabal Organico) was as effective as IBA in the propagation of olive cuttings for organic production (23).

BOX 10.8 GETTING MORE IN DEPTH ON THE SUBJECT
FINISHED (END-USE) FORMULATIONS

Until recently, it was possible for propagators to prepare their own solutions and talc formulations with technical-grade auxins. Technical-grade IBA has never undergone registration with the United States Environmental Protection Agency (EPA). It contains a trace amount of dioxin, to which some people are very sensitive. Always wear gloves during propagation when handling formulations containing auxin and other chemicals.

In the United States, nurserymen/propagators can only use **finished (end-use) formulations** containing auxin(s) that are registered with the EPA. Some of the EPA-approved formulations include the liquid formulations Dip'N Grow and Wood's Rooting Compound, the water-soluble salt formulation Hortus IBA Water Soluble Salts, and the commercial talc (powder) auxin formulations Hormodin and Hormo-Root (Table 10–5 and Fig. 10–31). Information is presented in Chapters 19, 20, and 21 for rooting with technical-grade auxins since recommended auxin concentrations can still be legally made by diluting end-use formulation products. *Note the **EPA registration number** on the container.* If the number is not present, the product has not been registered or is not being sold legally (45). See Box 10.10, page 377 for examples of making recommended auxin concentrations with end-use formulation products for quick-dip applications.

BOX 10.9 GETTING MORE IN DEPTH ON THE SUBJECT
MOST COMMON AUXIN CONCENTRATIONS

Generally, auxin concentrations of *500 to 1,250 ppm* are used to root the majority of **softwood** and **herbaceous cuttings**; *1,000 to 3,000 ppm with a maximum of 5,000 ppm* for **semi-hardwood cuttings**; and *1,000 to 3,000 ppm with a maximum of 10,000 ppm* for **hardwood cuttings** (see Table 10–1). One Texas nursery that produces some 3,000 cultivars and species propagates over 70 percent of their cuttings with 3- to 5-second quick-dips in aqueous dilutions of Dip'N Grow at either 1,250 ppm IBA + 625 ppm NAA (1 Dip'N Grow:7 water, v/v), or 2,000 ppm IBA + 1,000 ppm NAA. No product endorsement is intended.

Methods of Applying Auxins

Commercial Powder Preparations Regardless of application method, woody, difficult-to-root species are treated with higher auxin concentrations than tender, succulent, and easily rooted species. Fresh cuts can be made at the base of the cuttings shortly before they are dipped into the powder. The operation is faster if a bundle of cuttings is dipped at once rather than dipping individual cuttings. The inner cuttings in the bundle need to receive as much powder as those on the outside. The powder adhering to the cuttings after they are lightly tapped is sufficient. It may be beneficial to pre-wet cutting bases with water so that powder adheres better.

Enhanced rooting has occurred by predipping cuttings in 50 percent aqueous solutions of acetone, ethanol, or methanol prior to applying IBA talc (81). Some propagators have used *combined treatments of auxin quick-dips* (solution concentrate) followed by *auxin mixed in talc* applied to cutting bases.

In using powder and liquid preparations, place a small portion of the chemical stock material into a temporary container, sufficient for the work at hand, and discard any remaining portion after use. This is a better procedure than dipping the cuttings into the entire prepared chemical stock, which can lead to early deterioration via moisture, fungal, or bacterial contamination.

Talc preparations have the advantage of being easy to use (Figs. 10–31 and 10–32). However, uniform rooting may be difficult to obtain due to variability in the amount of the talc adhering to the base of cuttings, the amount of moisture at the base of the cutting, the texture of the stem (i.e., coarse or smooth), and loss of the talc during insertion of the cutting into the propagation medium.

Talc formulations are generally less effective than IBA in solution at comparable concentrations (26). In talc formulation, auxins (which have low solubility in the acid forms) must first go into solution *after* cuttings are stuck in the propagation media. Hence, there is a time delay before auxins are absorbed through the cutting base. Only in a few limited cases does talc outperform the quick-dip method (12); for example, certain species of *Elaeagnus, Rhododendron,* and holly (*Ilex*).

BOX 10.10 GETTING MORE IN DEPTH ON THE SUBJECT
PREPARING QUICK-DIPS WITH TECHNICAL-GRADE IBA OR NAA IN 50 PERCENT ALCOHOL FOR EXPERIMENTAL ROOTING TRIALS

Concentration: 500 to 10,000 ppm
Duration of basal dip: 3 to 5 seconds

Final concentration	Auxin (per liter of solution)	
(ppm)	(mg)	(g)
500	500	0.5
1,000	1,000	1.0
5,000	5,000	5.0
10,000	10,000	10.0

- To make a **10,000 ppm** (10,000 mg/liter or **1 percent stock solution**) of auxin, dissolve 10 g of auxin in 15 to 20 ml of alcohol (ethyl, isopropyl, or methyl), then top to 1,000 ml (1 liter) with 50 percent alcohol. To make 1 liter (1,000 ml) of a **1,000 ppm** auxin solution from the **1 percent stock solution,** add 100 ml of the 1 percent stock solution and 900 ml of 50 percent alcohol.
- With K-IBA or K-NAA follow the same procedures, except that water is used as the solvent and no alcohol is needed.
- Avoid precipitation problems with auxin solutions by using distilled or deionized water, not tap water.
- Label solutions and color-code different solution concentrations with food dyes, which can be purchased at supermarkets.

Figure 10–32
(a and b) Liquid auxin quick-dips of 1 to 5 seconds. (c) Application of auxin by talc. (d) Spray application at end of day reduces exposure of the propagators to auxin.

Dilute Solution Soaking Method The basal part—2.5 cm (1 in)—of the cuttings is soaked in a dilute solution of the material for up to 24 hours just before the cuttings are inserted into the rooting medium. The concentrations used vary from about 20 ppm for easily rooted species to about 200 ppm for the more difficult species. During the soaking period, the cuttings should be held at about 20°C (68°F) but not placed in the sun.

In general, this is a slow, cumbersome technique that is not commercially popular. Equipment is needed for soaking cuttings, and with the long time duration, there can be variability of results, with environmental changes occurring during the soaking period (95). There has been success with soaking basal portions of hard-to-root cuttings (*Prunus*, conifers, evergreen, and deciduous shrubs) for a maximum of 4 hours at 50 to 150 ppm IBA. In some dilute solution soaking studies from 5 to 50 minutes, improved uniformity of rooting roses was reported with a "hormonal time" of 1000 µM (203 ppm) IBA per minute (84). Longer dilution soaks increased root number, but inhibited bud break. After soaking, cuttings are then propagated under mist (97).

Quick-Dip (Concentrated Solution Dip) In the quick-dip method, a concentrated solution varying from 500 to 10,000 ppm (0.05 to 1.0 percent) of auxin in aqueous solution or 50 percent alcohol is prepared, and the basal 0.5 to 1 cm (1/5 to 2/5 in) of the cuttings are dipped in it for a short time (usually 3 to 5 seconds, sometimes longer). Then the cuttings are inserted into the rooting medium. Cuttings are most efficiently dipped as a bundle, not one by one (Figs. 10–32 and 10–33). There is no absolute ideal dipping depth; however, the majority of auxin is absorbed at the cut surface of the cutting base (63, 72). Hence, consistency in dipping time and maintaining the correct concentration are more important criteria.

Many propagators prefer the quick-dip compared to a talc application because of the consistency of results and application ease (17, 40, 81). Greater rooting and more consistent rooting response have been reported with quick-dips than with talc, due to more

Figure 10–33
Innovations in propagation. (a) Rather than counting individual cuttings to be placed in a bundle of 50, the average weight of the total cuttings in bundles is measured on a scale (black arrow) and the number of cuttings estimated. (b, c, and e) Bundles of cuttings are placed in a large bin that is flooded with preformulated auxin concentrations for a given time. (c) The vacuum (below the bin, arrow) drains the tray and the auxin is recycled for other cuttings; this takes the guesswork out of what constitutes a 1-, 3-, or 5-second quick-dip. The auxin solution is discarded at the end of the day. (d) Auxin preparations are stained with food dye to denote different concentrations and (a) stored in color-coded containers (white arrows).

BOX 10.11 GETTING MORE IN DEPTH ON THE SUBJECT
PREPARING AN IBA: NAA QUICK-DIP WITH AN EPA-APPROVED END-USE FORMULATION PRODUCT

Concentration: 500 to 10,000 ppm

Duration of basal dip: 3 to 5 seconds

Use an EPA-approved end-use formulation (e.g., Dip'N Grow (1% IBA + 0.5% NAA)

The Dip'N Grow is the stock solution (1% IBA + 10,000 ppm: 0.5% NAA + 5,000 ppm). To make 1 liter (1,000 ml) solution of 250 ppm IBA: 125 ppm NAA, use the formula:

Conc D'NG × VolD'NG = Con sol × Vol sol
10,000 ppm IBA (D'NG concentrate) × Vol.
= 250 ppm × 1,000 ml;
25 ml of D'NG concentrate stock solution
+ 975 ml 50% alcohol = 250 ppm IBA: 125 ppm NAA in one liter.

Final concentration	IBA (per liter of solution)			NAA (per liter of solution)			Stock concentrate and dilution		
	(ppm)	(mg)	(g)	(ppm)	(mg)	(g)	Dip'N Grow concentrate	50% alcohol	Final solution
250 IBA: 125 NAA	250	250	0.25	125	125	0.125	25 ml	975 ml	1,000 ml
500 IBA: 250 NAA	500	500	0.5	250	250	0.25	50 ml	950 ml	1,000 ml
1,000 IBA: 500 NAA	1,000	1,000	1.0	500	500	0.5	100 ml	900 ml	1,000 ml
5,000 IBA: 2,500 NAA	5,000	5,000	5.0	2,500	2,500	2.50	500 ml	500 ml	1,000 ml
10,000 IBA: 5,000 NAA	10,000	10,000	10.0	5,000	5,000	5.0	1,000 ml	0 ml	1,000 ml

uniform coverage and reduced environmental influence on chemical uptake (41, 95). Interestingly, the solvent used for quick-dip application facilitates auxin movement through the epidermis as well as the cut surface of the cutting (62, 95). With talc (powder) or aqueous (water) solutions, auxin moves into the stem in the vascular system. In contrast, auxin in ethanol can enter the cut surface as well as the epidermis throughout the area of the stem dipped in solution (62).

There can be advantages of mixing solution thickening additives, such as carboxymethyl cellulose (CMC), which is also used as a thickener (viscosity modifier) and water retention agent in the food industry for ice cream and salad dressing. It gives the quick-dip solution the consistency of thick gravy or heavy motor oil. The polymers adhere to the base of the cutting, allowing auxin to remain in contact with the tissue longer. Some commercial examples include Celluwet (Griffin Labs), Dip-Gel™ (Dip'N Grow, Inc.), and Horta-Sorb (Whitfield Forestry) (15).

Quick-dip stock solutions must be tightly sealed when not in use because the evaporation of the alcohol will increase the auxin concentration. Use only a portion of the material at a time, just sufficient for immediate needs, discarding it after use at the end of the day rather than pouring it back into the stock solution. On extremely hot days, in open areas where evaporation is high, it is best to discard the old and add fresh solution several times during the day. Stock solutions that contain a high percentage of alcohol will retain their activity almost indefinitely if kept clean. No matter what the product, protect yourself and workers from undue exposure. Work in a well-ventilated room. Use rubber or plastic gloves when working with any of these rooting compounds.

Auxins used in excessive concentrations can inhibit bud development, cause yellowing and dropping of leaves (abscission), blackening of the stem (basal necrosis), and eventual death of the cuttings. An effective, nontoxic concentration has been used if the basal portion of the stem shows some swelling, callusing, and profuse root production just above the base of the cutting.

Alternative Auxin Application Methods Rather than quick-dipping, some propagators will spray auxin on cuttings that are stuck in propagation flats. The auxin is applied to the point of runoff, when beads of liquid just start to initially roll off the foliage into the rooting media (Fig. 10–32). A 50 to 250 ppm IBA spray solution is used for chrysanthemum, begonia, *Dieffenbachia,* heath, and hibiscus (87). There has been a renewed interest in auxin spray application. See the Getting More in Depth on the Subject box.

Excellent rooting has occurred with the **total immersion of whole cuttings** of herbaceous cuttings of plumbago, ivy, clematis, delphinium, *Ficus,* and others for a few seconds—at 50 to 250 ppm IBA.

BOX 10.12 GETTING MORE IN DEPTH ON THE SUBJECT
SPRAY APPLICATION OF AUXINS

In recent years, there has been renewed interest in applying auxins as aqueous sprays on cuttings (16, 47). One nursery reports that many taxa respond equally well to water soluble IBA applied as a spray after sticking (within the first 24 hours), compared to traditional hand, quick-dip methods (48). Some formulations of soluble IBA include Hortus IBA water soluble salts and K-IBA, which is a potassium salt formulation that is readily dissolved in water. See Table 10–5 for a partial list of commercial auxins. Deionized water is most desirable (but may not be practical); mix only the quantity of auxin needed for the day's application.

Spray applications between 200 to 2,000 ppm (mg liter^{-1}) are used with a small backpack sprayer or with hose and reel-type sprayers with or without a boom-style irrigator (48) for larger areas. Applications should be sprayed evenly over the cuttings "until run-off." IBA is a pesticide and should be treated as such. Applications should be made (after the propagation crew has left) by a trained, licensed pesticide applicator, wearing appropriate protective equipment. The IBA is applied at the end of the day or very early in the morning when light levels are low and mist requirements are minimal. Some propagators wait several days after sticking to apply water-based aqueous sprays of auxin with good rooting results. It is best to apply auxin sprays to cuttings within the first week to 10 days of propagating. After treating cuttings with auxin at the time of sticking, some growers will apply an additional auxin spray application within the first two weeks, which enhances rooting of slower-to-root cuttings (87).

Advantages of using spray application include a) minimizing employee exposure (since only the applicator applies the chemical), b) developing a more streamlined and sanitary approach to propagation so cuttings spend less time in storage and the cutting prep room—where problems associated with lengthened exposure to temperature, humidity, and handling can occur, c) reduced labor costs, d) reduced chemical usage, and e) better cost effectiveness.

Rooting was enhanced when *Berberis, Cotoneaster, Lavandula, Prunus, Pyracantha,* and *Viburnum* cuttings were totally immersed for 2 minutes at 1,000 ppm—compared with powder formulations. Immersing the entire cuttings into the concentrated solution dip containing a wetting agent (surfactant) has been more effective in promoting rooting in some cases than just dipping the base alone (149). There is an initial retardation of shoot growth, but this does not seem to be a disadvantage.

Auxins can also be applied in low concentrations to stabilized, rooting media substrate plugs, such as compressed peat pellets and cuttings stuck later (Fig. 10–27) (16). In a comparable situation for air-layering *Mahonia,* the sphagnum peat used for air-layering was soaked with low IBA concentrations (60 ppm), which enhanced rooting (157).

Pretreatment of stock plants with foliar application of auxins prior to removing cuttings has also been used to promote rooting (114, 115).

PREVENTATIVE DISEASE CONTROL

Disease-Free Stock Plants

As part of a preventative disease program, cuttings should be harvested from disease-free stock plants under nonstress conditions. Collect turgid cuttings early in the day to assure optimum water conditions (90). The pruning shears used to collect cuttings should be disinfected periodically (Figs. 10–30 and 10–34). Physan 20 (benzyl chloride), isopropyl alcohol, and monochloramine are better disinfectants than sodium hypochlorite (Clorox), which is quickly inactivated when it comes in contact with organic matter (stem material, media components). Monochloramine was found to be equal in efficacy to alcohol—less corrosive and costly—and with excellent stability under high organic contamination (133). See the referenced article on how a propagator can make monochloramine from local materials (133). Also, refer back to Chapter 3 for

(a)

(b)

(c)

(d)

Figure 10–34

Preventative disease control measures: (a and b) Collecting cuttings in buckets containing cups for periodically disinfecting knives and shears. (c and d) Soaking cuttings in a broad-spectrum fungicide and bactericide prior to treating with rooting hormones and sticking. (d) Cuttings put in wire basket and soaked in chemical bucket. (With the current Worker Protection regulations, individuals utilizing chemicals with cuttings are considered to be pesticide handlers and need to be properly trained. Any chemical usage needs to comply with the manufacturer's recommendation; see the OSHA web site, www.osha.gov).

> **BOX 10.13 GETTING MORE IN DEPTH ON THE SUBJECT**
> **AVOID ROOTING INCONSISTENCY**
>
> To avoid rooting inconsistency (with *any* application method!), it is critical that propagators strictly adhere to established procedures.
>
> It is also widely held that the most critical factor affecting the response of cuttings to hormone treatment is the concentration of auxin in the liquid or powder preparation—this is important. However, **it is the total dose of auxin received by those tissues capable of responding that determines rooting.** For quick-dip preparations, factors such as duration and depth of dipping, and the position in which the cutting is dried, affect the amount of solution taken in through the cut end of the stem. Both factors are as important as the auxin concentration when treating cuttings (82). With powder carriers, the surface moisture at the cut end influences the transfer of auxin. Hence, more uniform results occur by predipping stem cutting bases in organic solvents, controlling the set quantity of powder adhering to the cutting bases, and care in retaining talc during the sticking (planting) phase of propagation. **Systematically following standardized procedures of a technique (method of application)** is as important as the concentration of auxin applied. Propagators must set up simple **standard operational procedures (SOPs)**. Procedures that will improve overall survival and rooting of cuttings, and increase the uniformity of response (83).
>
> Standard operational procedures (SOPs) The step-wise tasks and methods needed for efficiently performing a propagation process, such as the collection, pretreatment, and posttreatment handling of cuttings for rooting.

preventative measures discussed, including integrated pest management **(IPM),** current best management practices **(BMP),** and post-propagation care of rooted liner plants.

Chemical Treatment of Cuttings

Once cuttings are collected, they should be selectively treated with broad-spectrum fungicidal dips prior to sticking and/or chemical drenches during propagation. Cuttings can be dipped in solutions of Agribrom, which is an oxidizing biocide that controls pathogens (86). One Texas nursery immerses unrooted cuttings in a 25 ppm Agribrom bath. The nursery reports that it is more effective for disease control than dipping cuttings in baths containing fungicides for control of damping-off organisms, and agricultural streptomycin for bacterial control. The cutting bases are allowed to dry (keeping the leaves wet) and then are trimmed and treated with auxin. Cuttings can also be dipped and disinfected in biodegradable, quaternary ammonium products such as Physan 20 (www.physan.com) and Consan (www.consan.net), which also have fungicidal and algaecidal properties (Fig. 10–34). ZeroTol (Biosafe Systems), hydrogen dioxide, is a strong oxidizing agent that is used as an algaecide and fungicide. Cuttings are immersed in dilute solutions and then prepped, quick-dipped with rooting hormone solution and propagated. (See Chapter 3.)

Benomyl was the most widely used fungicide for propagation and ornamental use in the United States. It is no longer labeled for use on ornamentals or as a soil drench, but is marketed as a general-purpose fungicide. Future trends are for fewer chemicals labeled for horticultural usage. Some commercial substitutes for Benomyl include Topsin M, Domain, Cleary 3336, and SysTec 1998, all of which have the systemic activity of thiophanate methyl (32).

Beneficial Microbes for Enhancing Rooting and Pathogen Control

As plants have evolved, so have rhizosphere organisms, some of which show great promise for propagation systems. The use of **biocontrol agents** (beneficial bacteria, actinomycetes, mycorrhizal fungi, and other beneficial fungi living and functioning on or near roots in the **rhizosphere** soil) to control pathogens and enhance rooting in propagation is still in its infancy (33, 36, 93, 98). Although industry still relies on chemical application of auxins to stimulate rooting of cuttings, and application of pesticides to control pathogens and pests during propagation, utilizing beneficial microbes is a novel approach to reduce chemical treatments, control soil pathogens (93), and enhance rooting and cutting survival (105). The beneficial fungus, *Gliocladium virens* (SoilGard 12G, W. R. Grace & Co.), may be an alternative to Benomyl. It has been cleared by the EPA for biological control of *Rhizoctonia colani* and *Phythium ultimum*,

rhizosphere The zone of soil immediately adjacent to plant roots in which the kinds, numbers, or activities of microorganisms differ from that of the bulk soil.

BOX 10.14 GETTING MORE IN DEPTH ON THE SUBJECT
PRECAUTION AND USAGE OF PESTICIDES

Always follow directions and conduct small tests to check for phytotoxicity, and protect yourself and workers from chemical exposure. See the discussion of fungicides/pesticides in Chapter 3. **If the chemical being used in propagation has *not been labeled*** by the manufacturer for a particular function (e.g., drenching cuttings, immersing cuttings, etc.) **then it is being used <u>illegally</u>**.

Methyl bromide (MB)—used in fumigating and sterilizing propagation beds and media for very effective control of diseases, insects, nematodes, and weeds—will be completely phased out in the United States by the year 2015. The USDA has a special web site on MB alternatives for agriculture (http://www.ars.usda.gov/is/np/mba/mebrhp.htm).

---

which are two of the principal pathogens causing damping-off diseases. With hardwood cutting propagation of roses and cherry, crown gall (*Agrobacterium tumefaciens*) is controlled by dipping cutting bases in a special *Agrobacterium* isolate that is antagonistic to the virulent form.

Beneficial bacterium (*Agrobacterium rhizogenes*) in combination with auxin can increase rooting of recalcitrant clones of elm (*Ulmus*) and pine (*Pinus*), probably by producing root-inducing compounds, but without genetic transformation of host cells (122, 126).

Until recently, it was thought that mycorrhizal fungi enhanced root development and cutting survival *after* colonization of adventitious roots (153), but new evidence indicates that mycorrhizae can enhance root initiation prior to root colonization (33, 46, 51, 112, 130, 131, 139). Some outstanding nurseries in the United States incorporate mycorrhiza during propagation as a value-added product for improved stress- and disease-resistance (33). *Trichoderma harzianum,* a fungus that controls soil-borne pathogens, enhanced root and shoot growth of chrysanthemum cuttings during propagation, possibly by the production of growth-regulating substances or by chemically antagonizing or competing with pathogens. (See Chapter 3 for the discussion of biocontrol agents and IPM methods, including cultural controls.)

ENVIRONMENTAL CONDITIONS FOR ROOTING LEAFY CUTTINGS

For successful rooting of leafy cuttings, some essential environmental requirements are:

- Rooting media temperature of 18 to 25°C (65 to 77°F) for temperate species and 7°C (12°F) higher for most tropical species
- Atmosphere conducive to low water loss and maintenance of turgor in leaves
- Ample, but not excessive, light—100 W/m^2 with selected temperate woody species (exceptions are with species propagated under full sun irradiance in outdoor mist beds)
- Clean, moist, well-aerated, and well-drained rooting medium

A wide range of equipment is satisfactory for providing these conditions—ranging from "low-tech" systems of rooting *Dracena* cane cuttings in a small polyethylene bag filled with sphagnum (Fig. 10–35) to contact polyethylene systems where semi-hardwood and hardwood cutting are stuck in propagation flats, watered-in, and covered with a poly sheet before being taken to very elaborate controlled environment propagation facilities with raised benches, automatic mist and fog systems, and computerized environmental control of relative humidity, temperature, photoperiod,

(a)

(b)

Figure 10–35
(a and b) "Low-tech" system of commercially rooting *Dracaena* cane cuttings in a polyethylene bag filled with sphagnum.

light irradiance, and CO_2 enrichment (see Figs. 3–13 and 10–36) (109).

Intermittent Mist System

Intermittent mist systems are widely used and have given propagators great flexibility in rooting softwood, semi-hardwood, hardwood, and herbaceous cuttings. The small water droplets of the mist provide a film of water over the cuttings and media. An important function of the film of water on the leaf surface is to intercept the irradiation of light so that water is *evaporated from the leaf surface* rather than

> **intermittent mist**
> A thin film of water produced through a pressurized irrigation system, which cools the atmosphere and leaf surface of cuttings.

Figure 10–36
Enclosed-case systems. (a and b) Rooting cuttings in enclosed polyethylene tents under mist. (c) Nonmisted polytent. (d and f) Nonmisted contact polyethylene sheet system. (d) Note condensation on the underside of the polysheet. (e) A modified contact poly system for rooting rose cuttings—greatly reduced levels of water are applied on top and seep through the holes (arrow) in the plastic to keep cuttings moist. (f) The shade cloth shown can be readily pulled (arrow) if light irradiance becomes too high.
Photo (e) courtesy Bill Barr.

from internal leaf tissue. Intermittent mist controls water loss from cuttings by reducing both leaf and surrounding air temperature via evaporative cooling and by raising relative humidity. To counteract the lower media temperatures caused by mist, bottom heat is frequently used in outdoor and indoor rooting structures (see Fig. 10–37).

Open Mist Systems **Open mist** systems are used in outdoor propagation in cold frames, polyethylene tunnels, and lath and shade houses, and under full sun (Fig. 10–38). The open mist system is also used in glass and poly-covered greenhouses and set up on the floor area or on, or above, the propagation bench (Figs. 10–38 and 10–39). A very short duration (3 to 15 seconds) is used for misting. Unless mist actually wets the leaves, rooting is likely to be unsatisfactory. Besides using fixed risers containing mist nozzles (Fig. 10–39), mechanized **traveling boom systems** are used to deliver mist to cuttings.

Enclosed Mist Systems **Enclosed mist** systems are covered polyethylene structures inside greenhouses to reduce the fluctuation in ambient humidity and ensure more uniform coverage of mist, since air currents that disturb mist patterns are avoided (Fig. 10–36). This system has been very effective in propagating difficult-to-root species, softwood cuttings of large-leaved species (e.g., *Corylus maxima*), and broad-leaved evergreens; it is not effective for conifers. The enclosed mist system has fewer disease problems than the open mist system since there is less mist required, less media saturation, and *fewer foliar leaching problems* (96).

> **foliar leaching** The rapid depletion of essential nutrient reserves from a cutting, caused by intermittent mist.

Mist Nozzles

The choice of mist nozzles is based on (a) cost, (b) maintenance, (c) convenience in operation, (d) availability from suppliers, (e) size of mist droplet [ideally, 50 to 100 μm (0.002 to 0.004 in)], (f) amount of water used (fine orifice mist nozzles use less water but clog up more readily), and (g) mist pattern (sufficient coverage while avoiding overwetting media).

The two main types of nozzles are **pressure jet** or **whirl-type nozzle** and the **deflection** or **anvil nozzle.** In the whirl-type, water is forced under pressure through small grooves set on angles to each other, which produces a mist when water exits the orifice. There are

(a)

(b)

(c)

(d)

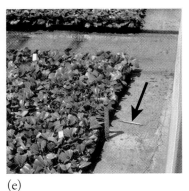
(e)

Figure 10–37
Bottom heating: (a, b, and c) Ground bed heating in a glass propagation house with hot water solar panels. (c) Cross section of solar panel with larger tube feeding hot water into smaller capillaries (arrow). (d) Outdoor propagation facility relying on bottom heat by circulating hot water through PVC pipe embedded in scoria. (e) Outdoor hot-water-heated concrete bed; the temperature probe (arrow) is normally inserted into a propagation flat.

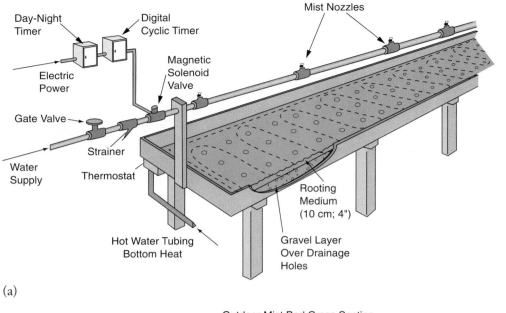

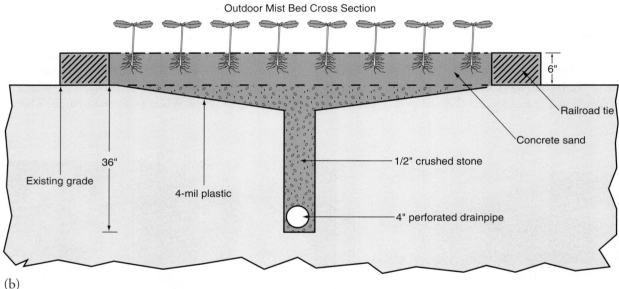

Figure 10–38
(a) Basic component parts of an open intermittent mist propagating installation with bottom heat supplied by hot water tubing. A 24-hour (day-night) timer turns the mist system on in the morning and off at night. The second is a digital, short interval timer to provide the intermittent mist cycles. (b) Cross section of an outdoor mist bed. Cuttings of *Thuja*, *Taxus*, and *Juniperus* cultivars, etc., are stuck in the concrete sand of the beds between the railroad ties, which has a crushed stone base with a drainpipe for better drainage. See Figure 10–5.

improved designs that operate under lower water pressures, which are nondripping and self-cleaning. Many of these nozzles have a low water output of 9 to 20 liters (2 to 5 gallons) per hour. Pressure jet nozzles have curved internal grooves, and when pressurized water is forced through the grooves, the impact at the orifice of the nozzle breaks up the water flow into mist. The Spray Systems Parasol is one such nozzle used in U. S. nurseries that has a larger, more maintenance-free orifice (Fig. 10–40).

The deflection nozzle develops a mist when pressurized water passes through the orifice and strikes a flat surface or anvil. The larger aperture in this type of nozzle reduces clogging but uses more water. Again, there are many variations in orifice size and water efficiency between the two principal types of nozzle. Excellent hard plastic nozzles that are less expensive and more durable than metal ones are also available (Figs. 10–39 and 10–40) (143).

Figure 10–39
Versatility of mist systems hung from the propagation roof allowing more efficient propagation bench utilization per unit area. (a, b, and c) Netafim plastic impact nozzle system. (b) Netafim sprinkler with a check valve to prevent dripping between misting intervals. (c) Red shade cloth shifts light quality to the red and far-red, which can enhance rooting of cuttings. (d and e) Boom mist propagation system for large propagation areas.

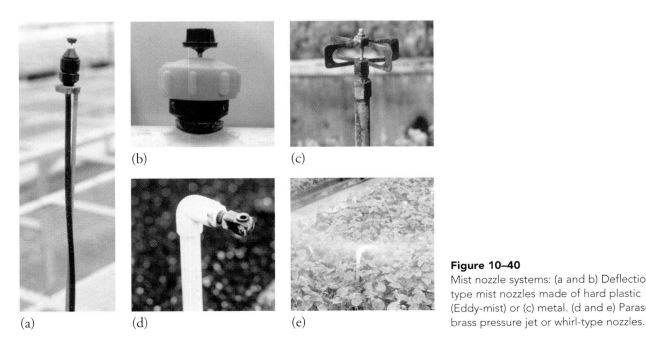

Figure 10–40
Mist nozzle systems: (a and b) Deflection-type mist nozzles made of hard plastic (Eddy-mist) or (c) metal. (d and e) Parasol, brass pressure jet or whirl-type nozzles.

Mist Controls

Applying mist intermittently during the daylight hours frequently enough to keep a film of water on the leaves, but no more, gives better results than continuous mist. Since it would be impractical to turn the mist on and off by hand at short intervals throughout the day, automatic-control devices are necessary. Several types are available, all operating to control a solenoid (magnetic) valve in the water line to the nozzles.

In a mist installation, the cuttings will be damaged if the leaves are allowed to become dry for very long. Even 10 minutes without water on a hot, sunny day can be disastrous. In setting up the control system to provide an intermittent mist, every precaution should be taken to guard against accidental failure of the mist applications. This includes the use of a "normally open" solenoid valve; that is, one constructed so that if electric power is lost, the valve is open and water passes through it. Application of electricity closes the valve and shuts off the water. If an accidental power failure occurs or any failure in the electrical control mechanism takes place, the mist remains on continuously, avoiding desiccation damage to the cuttings.

There are two types of **control systems** for scheduling the intervals between misting events: *static* and *dynamic*.

static control systems Control systems that rely on clocks and timers to manage intermittent mist and fog systems.

Static Control Systems Static control systems (Fig. 10–41) can lead to an inefficient usage of water, causing cuttings to wilt (too little water) or stress due to excessive water—causing foliar leaching, media saturation, anaerobic conditions, and poor rooting. Static systems are operated by timers that turn the mist on at preset intervals. These are described below.

Timers. Electrically operated timer mechanisms operate the mist as desired. A successful type uses two timers acting together in series—one turns the entire system on in the morning and off at night; the second, an interval timer, operates the system during the daylight hours to produce an intermittent mist at any desired combination of timing intervals, such as 6 seconds **ON** and 2 minutes **OFF**. Time clocks for regulating the application of water are preferred by many propagators because they are easily installed, inexpensive, and dependable. Some electronic timers are very versatile and can operate many banks of mist nozzles in sequence (Fig. 10–41). Timers have the disadvantage of not responding to daily fluctuation in light irradiance, cloud cover, relative humidity, or temperature. Although mechanically and electronically reliable, the propagator must make daily adjustments to this equipment.

Dynamic Control Systems For controlling mist or fog application, **dynamic control systems** rely on plant or environmental parameters to determine the water status of cuttings (Fig. 10–42), including electronic and mechanical leaves, light

dynamic control systems Control systems that rely on environmental parameters to determine water status of cuttings. They are more precise in regulating water management than static control systems.

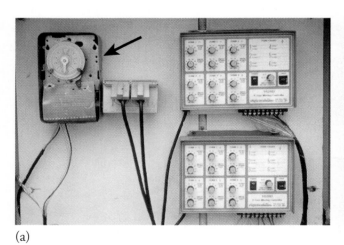

(a)

(b)

Figure 10–41
Static control systems rely on clocks and timers to manage intermittent mist and fog systems. (a) A 24-hour clock (arrow) turns the system on in the morning and off around dusk, or can be adjusted manually. (b) Time clock controlling the minutes between mist interval "on" time and the seconds of actual mist duration is wired to the 24-hour clock.

   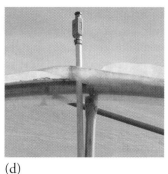

(a) (b) (c) (d)

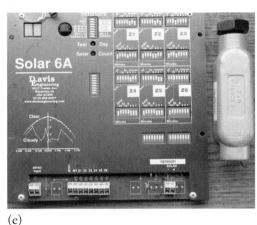 

(e) (f)

Figure 10–42

Dynamic control systems rely on environmental parameters to determine water status of cuttings. They are more precise in regulating water management than static control systems. Examples of dynamic mist control include: (a) Artificial leaf (screen balance), (b, c, and d) quantum light sensors. (e) The computer triggers a misting event after a certain number of accumulated light units. (f) Crop models use vapor pressure deficit (VPD) to estimate water loss from cuttings to initiate misting.

sensors, and humidistats described in this section and in Chapter 3. Predicting and controlling misting frequency can be vastly improved with the aid of environmental sensors linked to computers programmed for evapotranspiration models (e.g., Penman-Monteith equation) (61, 63). These dynamic monitoring techniques allow for more efficient water usage, less stressed cuttings, and better and faster rooting. See the discussion in Chapter 9, page 324, on dynamic controls and using vapor pressure deficit (VPD) at the stem cutting level in providing real-time data for dynamic control of mist application (91, 92). Transpiration capacity in poinsettia cuttings at different stages and development has been used when scheduling mist (61, 159).

Screen Balance. Another type of control is based on the weight of water. A small stainless-steel screen is attached to a lever that actuates a switch. When the mist is on, water collects on the screen until its weight trips the switch, shutting off the solenoid. When the water evaporates from the screen it raises, closing the switch connection, which opens the solenoid, again turning on the mist. This type of control is best adapted to regions where considerable fluctuation in weather patterns may occur throughout the day, from warm and sunny to overcast, cool, and rainy; the unit compensates for changes in leaf evaporation. The Mist-A-Matic is a common screen balance unit (Fig. 10–42). These units have greater maintenance requirements than time clocks, and are prone to salt deposits, algae growth, and wind currents, which distort balance accuracy.

Photoelectric Cell. Controls based on the relationship between light irradiance and transpiration contain a photoelectric cell that conducts current in proportion to light irradiance. In essence, these systems are

---

BOX 10.15 GETTING MORE IN DEPTH ON THE SUBJECT
SAFETY ISSUES

The *danger of electrical shock* should always be kept in mind when installing and using any electrical control unit in a mist bed where considerable water is present. Low-voltage systems are safer. The complete electrical installation should be done by a competent electrician.

controlled by light irradiance, and convert light energy into electrical energy. The photoelectric cell activates a magnetic counter, or charges a condenser, so that after a certain period of time the solenoid valve is opened and the mist is applied. The higher the light irradiance, the more frequently the mist is applied. Between dusk and dawn, very little water is transpired. During cloudy days, less mist is used than during bright, sunny days. Such a control system would not be well suited for outdoor mist beds, where transpiration is affected by wind movement and light irradiance. Solar-activated mist-control devices can be quite suitable for greenhouse propagation, where wind velocities are negligible, and light is the most important environmental parameter contributing to evapotranspiration of the cuttings (Fig. 10–42). The Weather Watcher solar-powered mist controller (Jeffery Electronics, New South Wales, Australia) makes sole use of solar energy to control mist systems. It also uses 70 percent less water than intermittent mist benches controlled by conventional time clocks, which conserves water and reduces potential leaching of nutrients from cuttings during rooting (19).

Computerized Controllers. There are computerized propagation controllers that can be programmed to monitor air, media, leaf cutting temperature, light irradiance, and vapor pressure differences between air and leaf; the environmental information can then be coupled with the frequency and duration of mist or fog needed.

These systems are common in Holland, which has more than 2,000 hectares (5,000 acres) of glasshouses with computer-controlled climates that regulate temperature, supplementary light, shade, and CO_2 enrichment. Likewise, larger North American and English propagators are using computer-controlled systems to record environmental conditions of temperature, light, and mist in order to **model** optimum conditions for rooting cuttings (Figs. 10–42 and 10–43) (32, 61).

modeling The use of environmental inputs (temperature, light, humidity) that are recorded and then analyzed with mathematical equations. **Models** are then developed to program computers linked to environmental sensors for more precise control of the propagation house environment.

Enclosure Systems (Closed-Case Propagation)

Rooting of cuttings can be done with simple **enclosure systems (closed-case propagation)** outdoors with **low polyethylene tunnels** (sun tunnels), or **cold** or **hot frames** covered with glass or polyethylene. Enclosed systems are

polyethylene (also known as poly) A plastic covering used to cover propagation greenhouses.

(a)

(b)

(c)

(d)

Figure 10–43
Using plant modeling for determining optimal root propagation temperatures. (a) Temperature probe. (b) Data logger for compiling propagation media temperature data. (c and d) Assessing rooting response to various temperature regimes.

cold frames Propagation structures that are covered with poly, lath, or other covering material and generally not heated.

hot frames Propagation structures that are covered with poly and heated during cold weather.

contact polyethylene systems An enclosed propagation process where watered-in hardwood or semi-hardwood cuttings are tightly covered by a sheet of poly in a propagation house under shade control.

also used inside a greenhouse with **contact polyethylene systems,** where thin, 1- to 3-mil polyethylene sheets are laid in direct contact with watered-in cuttings on a raised bench or on propagation flats placed on the floor (Fig. 10–36). **Indoor polytents,** which are nonmisted polyethylene tents supported by wire or wooden frames, are another low-cost way to propagate (Fig. 10–36). Nonmisted enclosures in a greenhouse can be used to propagate difficult-to-root species, and they have the advantage of avoiding the nutrient leaching problems of mist propagation—yet afford greater environmental controls than outdoor propagation. See the discussion in Chapter 9.

With enclosed systems, the water loss from leaves is reduced by an increase in relative humidity and reduction in vapor pressure deficit (VPD), but enclosures also tend to trap heat. Leaf tissue is not readily cooled since there is minimal air movement and, consequently, evaporative cooling. To help reduce the heat load, light irradiance reaching the enclosed poly system is regulated by shading, and the greenhouse temperature is controlled by fan and pad cooling. Another variation of the contact poly system is to use **rooting beds on the ground, out-of-doors** in full sun, and lay **Microfoam** sheets [0.63 cm (1/4 in) or 0.25 in thick] directly on cuttings in the fall, covering them with white 4-mil co-polymer film sealed to the ground by gravel or pieces of pipe (70). With propagation in temperate climates, the ideal cycle may be to root cuttings under contact polyethylene film in fall and winter, and utilize intermittent mist during spring and summer.

Fog Systems

In **fog systems,** true fog is made by fog generators and atomizers that produce very fine water droplets from 2 to 40 microns (μm). By comparison, human hair is about 100 μm in diameter (Fig. 10–44). High-pressure foggers produce both **fog** and **micromist.** In reality, the best atomizers used in micromist systems produce an array of droplets ranging in size from 2 to 100 μm in diameter (106). The volume median diameter of such a micromist is about 40 μm so that half of the volume of water is in droplets larger than 40 μm. Fog remains airborne sufficiently long for evaporative cooling, and for an increased relative humidity of 93 to 100 percent to occur. Manufacturers may claim that their atomizers produce 10 to 20 μm droplets, but what is important is the *average* micron size of the water droplet (Fig. 10–45). As a general rule of thumb in greenhouse propagation, all droplets smaller than 40 μm will stay airborne as fog, but droplets larger than 40 μm tend to settle and condense as water on leaf surfaces, especially in a high-humidity propagation environment (106). With true fog, water is suspended in the air as a vapor, whereas mist droplets (generally 50 to 100 μm) lose their suspension, fall onto the surface of leaves and media, and condense. This liquid from mist cools the leaf surface where it evaporates, but leaches nutrients from the leaf and can easily overwet the media. There are many advantages with fog systems; however, they are more costly to purchase, install, and maintain than conventional mist systems.

fog systems Similar to intermittent mist, except the particle size of the water applied is much finer and water does not condense on the surfaces of the cutting.

Fogging Equipment There are three major types of fogging equipment.

Centripetal Foggers (Direct-Pressure Swirl Jet Atomizers) for Ventilated High Humidity. These are self-contained units incorporating a large fan that forces a stream of air through water ejected from a rapidly rotating nozzle (Fig. 10–45). The water is atomized into an average 30+ μm droplet, which is then forced

Propagation Water System	Fog	Micromist	Mist	Sprinklers (coarse mist; rainsize drops)
Droplet size range	2–40 μm[1]	2–100 μm	50–100+ μm	100+ μm
Average droplet size	15 μm	40 μm	>50 μm	>100+ μm

[1]Human hair has an average diameter of 100 μm.

Figure 10–44
A comparison of fog, micromist, and mist systems used for propagation (106).

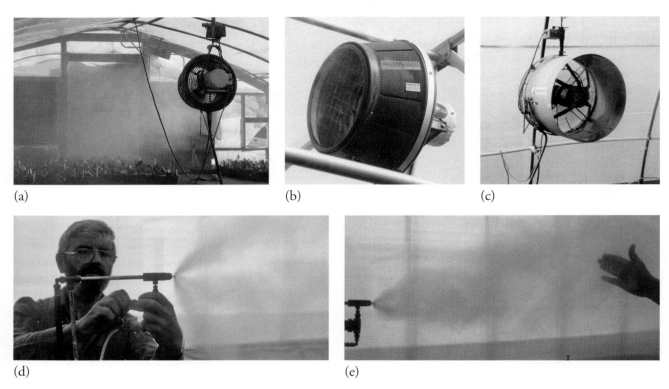

Figure 10–45
Fog systems: (a, b, and c) Centripetal foggers for ventilated high humidity. (a and b) AquaFog Humidifier (AgriTech). (d and e) Pneumatic or ultrasonic humidifier nozzle systems (Sonicore Ultrasonic Humidifier) have many advantages over intermittent mist, even though they are more expensive. Photos (d) and (e) courtesy of K. Loach.

into a cooling air stream through the propagation house by a fan attached to the rear of the unit. The water droplet size can be two times larger than those generated by impaction-pin atomizers. Centripetal foggers produce "wet fog," since they wet the leaf and humidify the air. They combine the advantages of both mist and dry fog (2 to 40 µm) and are very suitable for rooting large-leaved cuttings during the spring and summer. Larger mist droplets tend to fall out closer to the fan, making that area wetter, while smaller-size mist is dispersed at greater distances. High-humidity propagation is successful only when incoming air passes through the fogger while operating for effective ventilation (108). Greenhouses must be shaded, and good fan ventilation is essential. Best results are obtained with an oscillating humidifier that produces a large volume [10 to 50 gal (38 to 190 liters) per hour] of fog with 20- to 30-µm droplets. An example system is the AquaFog Humidifier (http://www.cloudtops.com/aquafog.htm) (Fig. 10–45). Another variation is the Humidifan (http://www.humidification.usgr.com/humidifan.php), which has a single motor and is operated without a nozzle, eliminating potential blockage problems.

High-Pressure Fogging (Impaction Pin Atomizers). In these systems, water is forced under high pressure (500 to 1,000 psi) through mist nozzles with very fine orifices. The water hits an impact pin attached to the nozzle, which atomizes droplets to less than 20 µm in size, subsequently forming a dense fog. Individual nozzles typically put out 5 to 8 liters (1 to 2 gal) per hour and are spaced 2 m (6 ft) apart. This is the "Mercedes" of fog systems. It is more expensive, but the most energy-efficient system for producing true fog droplets. One such system is produced by Mee (http://www.meefog.com/) (Fig. 10–46).

Pneumatic or Ultrasonic Humidifier Nozzles (Air Atomizers). These systems use compressed air and water. Water is disrupted by passage through a field of high-frequency sound waves generated by compressed air in a resonator located in front of the nozzle. In essence, water is being accelerated and atomized to fog. The nozzle orifice is much larger and less prone to blockage than are high-pressure fogging nozzles. Outputs range from 20 to 55 liters (5 to 14 gal) per hour. Air atomizing fog systems are more cost-effective for small propagation areas, but are not practical for large propagation ranges. Energy

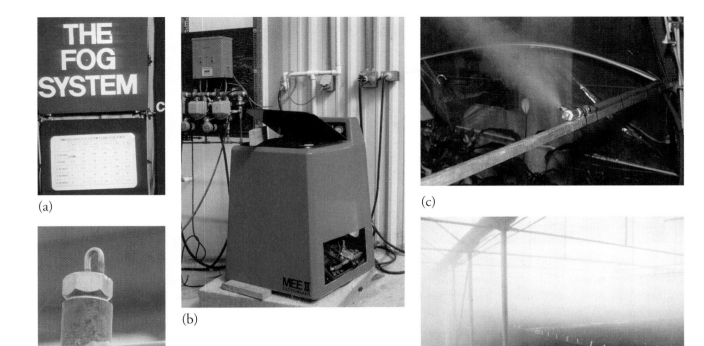

Figure 10–46
(a, b, c, d, and e) High pressure fog systems. (b, c, and d) Fog produced by a Mee system that uses a nozzle (d) with very small orifice that generates fog under extremely high pressure.

requirements for producing fog with this system are 20 times more than with impaction pin atomizers. One such unit is the Sonicore ultrasonic humidifier (Fig. 10–45). Some propagators are combining mist systems with low pressure fog systems for rooting softwood cuttings of difficult-to-root taxa (88). This combined system reduces the amount of mist required, enhances rooting, reduces the acclimation period following rooting, and improves the quality of the rooted liner.

Fogging Controllers The key to successful fogging hinges on a good ventilation system to avoid heat buildup from stagnant warm air. Fog systems must sense relative humidity and vapor pressure deficit (Fig. 10–42), but accurate control of high humidity is problematic, because time clocks are not satisfactory for controlling the rate of fogging. Most fog controllers operate to maintain a fixed relative humidity, which is the simplest and least expensive option, albeit a less–than-perfect one (96).

Alternative Systems: Subirrigation

A subirrigation system supplies water to the base of a cutting by capillary action through a coarse medium, which is immersed in a reservoir of water maintained at a low level below the base of the cuttings (68). It allows rooting of softwood, semi-hardwood cuttings and herbaceous perennial plants with little or no supplementary mist (Fig. 10–47). Subirrigation is more economical and maintenance-free than mist and fog systems, does not have the problem of excessive heat buildup characteristic of enclosure systems, and unlike mist systems, there is no foliar leaching of nutrients. Cuttings under subirrigation propagation usually develop strong, healthy root systems (120). This system is species-specific—it works with some species, while inferior rooting can occur with other taxa (2, 120).

PREPARING THE PROPAGATION BED, BENCH, ROOTING FLATS, AND CONTAINERS, AND INSERTING THE CUTTINGS

The rooting frames or benches should preferably be raised or, if on the ground, equipped with drainage tile, to assure adequate drainage of excess water (Figs. 10–37 and 10–38). It has become popular to propagate in flats and liner containers placed on the ground of greenhouses,

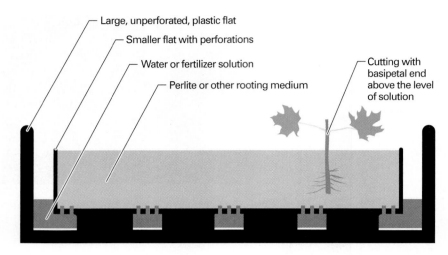

Figure 10–47
A subirrigation propagation system.

quonset-style houses, or outdoors on washed gravel or concrete-base mist beds that have been sloped for good water drainage (Fig. 10–48), all to avoid the high cost of propagation bench construction and utilize space more efficiently.

The frames or flats should be deep enough so that about 10 cm (4 in) of rooting medium can be used, and a cutting of average length—7.5 to 13 cm (3 to 5 in)—can be inserted up to half its total length, with the end of the cutting still 2.5 cm (1 in) or more above the bottom of the flat (Figs. 10–5 and 10–6). The rooting medium should be watered thoroughly before the cuttings are inserted, which should be as soon as possible after they are prepared. It is very important that the cuttings be protected from drying at all stages during their preparation and insertion.

After a section of the rooting bench or flat or small liner container has been filled with cuttings, it should be watered well to settle the rooting medium around the cuttings.

Direct Sticking (Rooting)

Direct sticking or direct rooting of cuttings into small liner plastic containers for rooting, as opposed to sticking in conventional flats or rooting trays, is an important technique for utilizing personnel and materials more efficiently (107). Over 50 percent of cutting propagation costs are due to labor. By direct sticking, the production step of transplanting rooted cuttings and potential transplant shock due to a disturbed root system is avoided (Figs. 3–21 and 10–48). The plant materials must be easy to root (greater than 80 percent rooting) to justify

Figure 10–48
Systems for sticking cuttings. (a) More cuttings can be rooted per unit area in a conventional plastic rooting flat, but additional labor is needed to initially transplant rooted cuttings into small liner pots, and then transplant into larger containers or produce as a field-grown crop. (b and c) Direct sticking (direct rooting) allows cuttings to be rooted directly into small liner pots which saves labor and avoids transplant shock to the root system. (d) Direct sticking into large 3.8-liter (1-gal) containers with no transplanting steps. Notice sloped incline (arrow) for better drainage.

the additional propagation space required, but the labor savings and versatility of this system are substantial.

PREVENTING OPERATION PROBLEMS WITH MIST AND FOG PROPAGATION SYSTEMS

Difficulties may arise in **operating mist and fog systems.** Low water pressure can be a problem. Many propagators like to operate with a minimum mist-line water pressure of 356 kPa (50 psi) to assure that fine mist (*not coarse mist*) is produced and that uniform coverage is maintained within the specification and the spacing of the nozzle used (97). A pressure regulator tank maintains sufficient pressure so the mist can be produced on demand (Fig. 10–49).

If there is much sand or debris in the water, filters or strainers should be installed in the supply line and cleaned periodically. Filters should always precede the solenoid valve in a mist line (Fig. 10–38). Fog systems utilize very elaborate filtration systems (particularly high-pressure fog systems with ultrafine nozzle orifices).

Cuttings close to mist nozzles frequently become overly wet and rot because of water leaking from the mist nozzle area. To prevent dripping from the nozzle between mist cycles, newer nozzles have pressurized cutoff valves that shut down as they go below 20 psi (Fig. 10–39). There are also pressure-release systems attached to mist lines that assure rapid cutoff to minimize drip between cycles.

Controlling Pathogens, Algal Growth, Mosses, and Liverworts via Irrigation Water

Pathogen control begins with clean propagation water for mist and fog systems. The process of disinfection, or destroying pathogenic microorganisms such as *Pythium* and *Rhizoctonia,* is most commonly done through chlorination of water (see Fig. 3–34), however, bromination and ozonation are other options (56). (See the discussion in Chapter 3.) Bromination is the least expensive and most effective disinfection method, and has less phytotoxicity problems than applied chlorine (56). Ozonation is highly effective as a disinfectant, and unlike chlorine and bromine, produces no by-products, but ozone generation is currently not cost-effective.

Algal growth often develops a gelatinous green coating on and around mist-propagation installations after an extended period of operation (Fig. 10–50). It is composed principally of blue-green (*Oscillatoria, Phormidium,* and *Arthrospira*) and green (*Stichococcus* and *Chlamydomonas*) algae (30). Algae will reduce aeration of propagation media, plug up nozzle orifices, and create other cultural problems. Chlorination is frequently used to control algae. The biocide Agribrom can be injected into high-pressure fog systems and conventional intermittent mist systems at 25 ppm to control algae, fungi, bacteria, viruses, and other microorganisms (104). Zerotol, hydrogen dioxide, is also an effective algaecide. Algaecides such as Algimine, Algofen (dichlorophen), Algae-Go 36-20, Cyprex (dodine acetate), and Agribrom (86) also can be used and are effective on mosses and liverworts (146). Diluted household vinegar and chlorox give good control of algae and moss along walkways.

Water Quality

The **quality of water** used in mist can influence the rooting response (see also Chapter 3 for the discussion on water management and treatment systems). A complete water analysis should be done (by

> **water quality** The amount of soluble salts (**salinity**) in irrigation water, measured with an electrical conductivity meter.

(a) (b)

Figure 10–49
Mist water quality can affect photosynthesis. (a) Carbonates in poor quality mist accumulate on leaves and can reduce photosynthetic rates (arrow). (b) Filtration, de-ionizing system (black arrow) for removing anions and cations from mist irrigation water. A pressure regulator tank (white arrow) is used to maintain sufficient water pressure for the mist system.

(a) (b)

Figure 10–50
Some correct and incorrect ways to propagate. (a) Poor sanitation with algae build-up (arrow)—can harbor disease and insects, and creates a poor propagation and work environment for personnel. (b) Good cultural and chemical practices: Sanitizing concrete pads before starting the next propagation crop.

the local municipal water department if using municipality water; otherwise, by university soil labs or commercial nutrient and water analyses laboratories) to determine pH, total soluble salts, SAR (sodium absorption ratio), total carbonates, electrical conductivity, and so on. High pH is commonly controlled by acidification (acid injection of sulfuric or phosphoric acid) into mist water. Water high in salts, such as sodium, or potassium carbonate, bicarbonates, or hydroxides can be detrimental, especially when coupled with low calcium levels. Bicarbonates can coat the leaves of cuttings and reduce photosynthetic levels during propagation (Fig. 10–49). Adding gypsum ($CaSO_4$) to the rooting media is one way to partially offset sodium problems. Sometimes a reverse osmosis system is used for removing salts from propagation mist water (see Figs. 3–28 and 10–49). Another option could be switching to a propagation system that requires less water on the foliage (e.g., a contact poly system, or closed mist versus a traditional open mist with greater water demands).

Chlorinated, chloraminated, or brominated (Agribrom, http://www.chemtura.com/) mist water (see Fig. 3–34) reduces algae growth and controls some damping-off organisms (104, 133, 134). This is particularly appropriate, since many Australian, European, and North American nursery and greenhouse propagators are forced to capture and reuse runoff water. (See the discussion in Chapter 3.)

MANAGEMENT PRACTICES

Record Keeping

There are many factors that go into making a good propagator: education, training, personal interest, a keen eye, and the ability to learn from success and failure. Good record keeping is essential in helping the propagator to hone skills and reduce failures (52, 128). With such a large number of different cultivars and species, it is difficult to remember the details of propagating a particular crop. Written records (electronic and printed copies) and pictures (color prints or digital photos) are important. Both show new propagation personnel how to propagate plants, and what optimum results look like. Videotapes can be effective in training personnel. Successful propagators rely on computerized databases for propagation scheduling and planning (64).

It is equally important to protect your records with printed hard copies and backup files on computer disks. One Texas nursery had an electrical fire that occurred at night and burned down their propagation facilities, destroying their computers and all propagation file records.

Propagation is the critical first key step in producing a finished crop (57). If problems occur in propagation, then scheduling and planning are for naught, and production, marketing, and sales are delayed. It is from good record keeping that the data and details can be used to develop lists of scheduling, and planning in propagating and producing a finished crop. Examples of record keeping forms are shown in Figures 10–51, 10–52, and 10–53.

In order to manage a propagation facility, one must collect data on critical activities. Record keeping and evaluation is the first management step. Record keeping compels the propagator to monitor cuttings for subtle changes in callusing and rooting, from which optimum environmental conditions can be determined. The propagator recognizes what type of cuttings can be rapidly produced, and those that cannot. The system also becomes an excellent method to track experimental results, leading to improved techniques and implementation of new crop

	Daily Propagation Record Sheet				
Date	Quantity Propagated	Cultivar	Description of Activity Performed	Number of Personnel	Total Personal Hours

Figure 10–51
A sample record sheet that a propagator could also computerize to chart daily progress of materials propagated, units produced, activity performed, and total personnel hours—in conjunction with Figure 10–52. Modified from (52, 121).

Record Card/File for Cutting Propagation
Cutting
Botanical Name: _____
Common Name: _____
Cultivar: _____
Date Propagated: _____
Date Rooted: _____
Cutting Type (i.e., semi-hardwood, terminal, basal, etc.): _____
Cutting Size (length or number of nodes): _____
Stock Plant Characteristics & Any Pretreatment (shading, banding, etc.): _____
Cutting Treatment (wounding, stripping cuttings, etc.): _____
Auxin(s): Formulation _____ Concentration _____ Method of Application _____
Rooting Medium: _____
Propagating System (mist, fog, contact polysheets, etc.): _____
Environmental Requirements (bottom heat, temperature, special mist conditions, light conditions, etc.): _____
Flat, Bed, or Container Size Planted & Location _____
No. of Cuttings per Flat _____ (or) No. of Direct-Stuck Liner Pots per Flat _____
Source of Cuttings _____
Propagator's Name and ID No. (to correspond with Label No. on propagation flat) _____
Date Rooted Cuttings Potted Up: _____ No. of Liners _____
Area to be Placed, Customer, or Department Shipped to: _____
Results: Total Rooted _____ % Rooted _____
Total Rooted Cuttings Shifted-up Liner Pots _____
% Rooted Cuttings Shifted-up to Liner Pots _____
Total Rooted Liner Pots Shifted-up to One-Gallon Containers _____
% Rooted Liner Pots Shifted-up to One-Gallon Containers _____
Observations & Comments _____

Figure 10–52
A sample record card charting the propagation history in a production cycle of a plant cultivar from propagation through linear production. This is easily computerized. Modified from (52, 121).

Genus	Species	Cultivar	Patent #(if appropriate)
Nandina	*domestica*	'Gulf Stream'	#5656
Company Catalog No.	**Propagation System**		**Date Propagated**
No. 4928	Direct stuck into 3P liner pots		14-Sept-2010
No. Liner Pots (Cuttings) Per Flat; Location Cuttings Taken from Stock Plants			
36 liner pots (cuttings) per flat;	Cuttings harvested from Section D, Area 1, from 1-gallon plants		
Propagator's ID # (to track who propagated the tray)			
No. 18			

Figure 10–53
Some sample propagation information to be printed on plastic labels and inserted in propagation flats or direct-stuck liner pots in liner trays.

(a) (b) (c) (d)

Figure 10–54
The importance of ergonomics and efficiency of movement. (a) Poor ergonomics with uncomfortable back posture, compared to (b and c) good ergonomics with correct posture, close proximity of materials and economy of movement, which enhance worker efficiency. (d) Ultimate of efficiency with propagators "floating" about the containers sitting in a trellis system for direct sticking in a flood floor system.

propagation systems. Data from records can be used by propagation managers to make lists of plants to be propagated each month, schedule critical propagation facilities, assist in budgeting, order supplies, schedule labor needs, and establish yearly, monthly, and daily propagation quotas (52, 128).

Costing Variables

Costing variables need to be closely monitored (through record keeping). For projecting costs, many nurseries and greenhouse operations have some form of costing labor (*loaded labor rates*) that includes salaries, wages, benefits, maintenance costs, grounds upkeep, and depreciation (121). *Reasonable Expectancy* (RE) programs are used in propagation for timing propagation procedures, figuring crew averages, and developing rates per worker hour, which has led to **piecework systems,** where propagation workers receive cash incentives when their units of propagules produced exceed the daily established quotas (110). Once a costing basis has been established, the cost of a procedure can be determined (121). Mechanization during propagation can enhance worker productivity and is readily integrated into piecework systems (119).

The value of producing a propagation crop is generally based on the price of selling the crop to commercial customers, or in buying the liner crop if it was not propagated in-house. Labor is the single most expensive item in a propagation budget, so nurseries tend to base the majority of their costing on labor hours. A way to reduce labor costs is to enhance the work efficiency of employees through improved ergonomics and efficiency of movement of people and materials in completing a propagation task (Fig. 10–54).

Addressing and changing *standard operational procedures (SOPs)* in propagation is done with the implementation of an **action plan/cost-benefit analysis** (121). The planning process used is called an action plan, and it states the goal and the stepwise procedure for producing a crop. Each step identifies the person responsible, due dates, and the final completion date. The action plan forces the individual doing the planning to outline the details. The action plan is also coupled with a cost-benefit analysis. Costs of propagating the crop can be figured by the *Reasonable Expectancy* (RE) established and by *loaded labor* rates, revenues can be figured, and the fixed assets required can be calculated, all of which allows identification of key weaknesses. Action plans are thus written to improve each shortfall. Using this analysis, nurseries have realized that the increased labor in carrying out an additional propagation procedure may be more costly than the gain resulting from increased rooting percentages (i.e., the extra labor and time in stripping or wounding a cutting to get marginally higher rooting percentages may not be cost-effective). By forcing propagators to evaluate each crop, some producers may decide that it is *more cost-effective not to propagate certain crops*.

Figure 10–55
Timing and Scheduling: "Maintaining the plant's momentum" to minimize stress. (a) Harvesting the right kind of cutting wood during the optimum season—that is, shoot tips of *Nandina* with no brown wood, trimmed to 4 cm. (b) Harvesting cuttings early in the day, when plants are stress free. (c and d) Storing cuttings in cool-moist refrigerated environments until they can be processed and stuck. (e) Processed cuttings covered with moist burlap until stuck.

A more profitable alternative would be to buy rooted liners of selected species from outside contractors.

There is no single correct way for costing cuttings (7), just as there is no one correct way to propagate a species, but the importance of accurate record keeping and cost analysis helps the propagator determine: (a) which crops are profitable, (b) which crops need changes in propagation/production procedures, and (c) which crops should be dropped from production. The bottom line is to realize an acceptable profit on *all* plants produced (121).

Timing and Scheduling

Commercial priorities determine scheduling in a nursery (11). When cuttings are stuck it may be decided by competing heavy labor demands in the spring to help with shipping nursery product to retailers and mass merchandisers, the availability of propagation space, and efficient use of personnel, rather than the optimum biological time to take cuttings. However, with some species, it is critical that cuttings be taken during a specific period if rooting is to occur (Figs. 10–55 and 10–56). For example, elm (*Ulmus parvifolia*) cuttings must be taken 6 to 8 weeks after bud-break. With some species, taking cuttings during the optimum time of the year is more crucial than using auxins. As a general rule of thumb, more difficult-to-root plants are stuck early in the propagation season. Easier-to-root cuttings have greater flexibility in propagation requirements; they root more quickly and tie up propagation space for shorter periods (32, 128).

Controlling Plant Wastage (Scrapage)

Efficiently run nursery and greenhouse companies are constantly striving to reduce *plant wastage* (**scrapage**); that is, to reduce the percentage of plants that are propagated and later discarded because of poor quality and/or poor market demand. Plant wastage is caused by poor propagation and production techniques, scheduling problems, and poor marketing strategies. Reducing the

> **scrapage**
> The discarding and economic loss of plants that occurs due to poor crop quality or low market sales.

Figure 10–56
Maintaining the Plant's Momentum. Collecting cuttings during the right time of the year can be more critical than applying auxin. The larger, 1-year-old Indian hawthorn (*Rhaphiolepis indica*) (*left*) was propagated during the spring with optimal rooting and growth, whereas the 2-year old, smaller plant (*right*) was propagated during the fall (in Texas) when growth had slowed and the momentum was lost.

plant residency period—so that the time from propagation to production to sale of *quality*, finished plants is condensed—is important in holding down production costs and maintaining profitability.

plant residency
The period of time that a plant occupies space in production, from propagation through point of sale.

Market-Led Propagation Systems

There have been some shifts in the nursery and greenhouse industries from production-led to market-led propagation systems (148). Problems with traditional *production-led propagation* include:

- The marketing strategy is constrained by the production process (i.e., plants for next year's spring sales are produced from late spring onwards and require additional space for overwintering).
- Cuttings for the year's production are taken at one time, requiring large amounts of propagation and production space in the greenhouse, which may be poorly utilized during other periods.
- The utilization of propagation and production facilities and systems is poor; bottlenecks occur at crucial stages [e.g., filling propagation flats with media, sticking cuttings, and shifting (transplanting) rooted liners up to larger containers].
- Mass factory production techniques, rather than more careful individual selection, can lead to variable quality and high failure rates.

With a *market-led propagation system*, the producer and retail outlets negotiate the quantity of plants required at particular times during the sales season so the producer can adjust the growing program accordingly to deliver plants in prime condition, as required. In addition, by careful selection of plant species, and improved propagation techniques and facilities, it is possible to enhance the sales appeal of many species by producing better-quality plants in flower (color!) for delivery to garden centers throughout the selling season (148). Figure 10–57 charts a market-driven propagation system for *Abelia* in England, where it is propagated seven different times during the year to meet market demands—instead

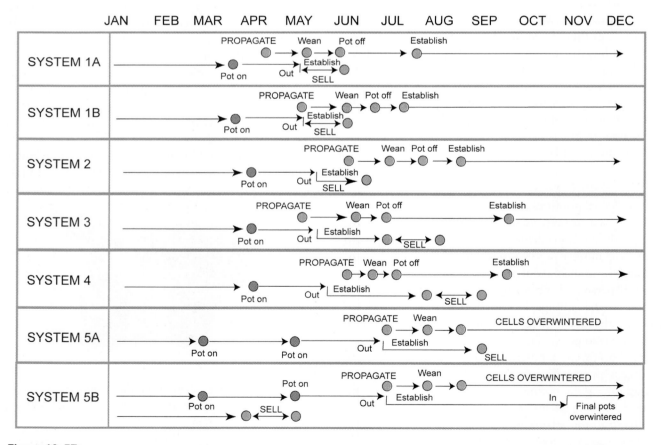

Figure 10–57

An example of a market-driven propagation system for *Abelia xgrandiflora* in England that more efficiently utilizes propagation facilities and delivers finished plants during designated selling periods. There are seven different propagation periods and seven targeted sales periods that extend the marketing period of the crop (148).

> **BOX 10.16** **GETTING MORE IN DEPTH ON THE SUBJECT**
> **QUALITY**
>
> In propagation, quality is the result of meticulous attention to detail (Ben Davis).

of mass propagating in more traditional production systems during only one period in July.

CARE OF CUTTINGS DURING ROOTING

Cutting Nutrition

Cuttings must be taken from nutritionally healthy stock plants (31). Most cuttings have sufficient tissue nutrients to allow root initiation to occur. Intermittent mist will rapidly deplete nutrients from cutting leaves, and until a cutting initiates roots, its ability to absorb nutrients from propagation media is limited.

As a whole, mist application of nutrients has not been a feasible technique for maintaining cutting nutrition (136, 162). The algae formed create sanitation and media aeration problems, which inhibit rooting.

A commercial technique that works with many plant species is to supply a low level of controlled-release fertilizer (CRF) either **top-dressed (broadcast)** on the media or **preincorporated** (e.g., Osmocote, Phycote, Nutricote, etc.). Top-dressing Osmocote 18N-6P-12K at 6.8 to 13.8 g/m² (2.6 to 5.3 oz/ft²) enhanced both root and shoot development of *Ligustrum japonicum* (too high a rate will delay the rooting of cuttings). These supplementary nutrients do not promote root initiation, but rather improve root development once root primordia formation and subsequent root elongation have occurred. Generally, CRF formulations with a slow initial release rate (e.g., 12 to 14 months, Osmocote 17N-3.1P-8.1K) are better suited than short-release (3-month formulations) for top-dressing on the surface of propagation media under mist systems (1). Biosolids (from sewage waste systems) added to rooting media as a fertilizer can improve root development of *Bougainvillea* (5).

Dilute liquid fertilizer can be applied to the propagation medium *after* roots have been initiated. In general, propagation turnover occurs more quickly and plant growth is maintained by producing rooted liners that are nutritionally balanced. However, with some species, such as *Thuja occidentalis,* there is no initial benefit of fertilizer application on root development and early growth of rooted liners (25). Other species, such as *Stewartia pseudocamillia,* are initially very sensitive to fertilizer in propagation media.

Environmental Conditions

The temperature of leafy softwood, herbaceous, or semi-hardwood stem cuttings and leaf-bud or leaf cuttings should be carefully controlled throughout the rooting period. Polyethylene-covered structures and glass-covered frames, exposed to a few hours of strong sunlight, will build up excessively high and injurious temperatures. Such structures should always be shaded by saran screens or whitewash on the poly or glass to reduce light irradiance.

If **bottom heat** is provided, thermometers or remote recording sensors should be inserted in the rooting medium, to the level of the base of the cutting, and checked at frequent intervals (Fig. 10–37). Excessively high temperatures in the rooting medium, even for a short time, are likely to result in the death of the cuttings.

As previously mentioned, it is important to *maintain humidity* as high as possible in rooting leafy cuttings to maintain a low vapor pressure deficit and keep water loss from the leaves to a minimum. If cuttings become droughted for any length of time, the cuttings will not root, even if they are rehydrated and high-humidity conditions are reestablished. However, most nursery and greenhouse operations use intermittent mist, contact polyethylene sheets, or fog to overcome such problems. Too much water can be as deleterious as too little, but there can be some advantages of mild drought stress that enhance rooting (91).

Adequate drainage must be provided so that excess water can escape and not cause the rooting medium to become soggy and waterlogged. When peat or sphagnum moss is used as a component of the rooting medium, it is especially important to see that it does not become excessively wet.

High light is not needed early in propagation, and in fact it can stress cuttings. Many nursery plants are propagated under 30 to 50 percent shade. Once cuttings initiate roots, light levels can be increased to

---

top-dressed (broadcast) The application of fertilizer to the propagation media surface of a flat or liner pot.

preincorporated fertilizer Fertilizer that is mixed into the propagation media prior to sticking cuttings.

> **BOX 10.17 GETTING MORE IN DEPTH ON THE SUBJECT**
> **CONTROLLED-RELEASE FERTILIZERS (CRF) AND MANAGING CUTTING NUTRITION**
>
> A number of propagators utilize CRF preincorporated into the propagation media, because it improves root development and more rapidly establishes rooted liner plants. Apex/Polyn 20N-8P-8K (12 to 14 month) and Scott's Osmocote 18N-5P-9K (5 to 6 month) have mini-prill formulations that allow five-fold more uniform distribution compared to traditional sized prills. This is important for incorporating CRF into the small cells of propagation trays (49). Using CRF allows a drier environment during propagation and rooted liner production (compared to liquid fertilizer application), helps improve liner growth, reduces liner production time, is more targeted (which facilitates nutrition of individual plants), and reduces algae growth on walkways. While CRF benefits rooted liner production with most evergreen and deciduous taxa, it does not work well with *Rhododendron* taxa during rooting (49).
>
> The management of water and nutrition are inextricably linked with more efficient fertilization usage and reducing leaching during propagation (129). The Young Plant Research Center at the University of Florida (http://hort.ifas.ufl.edu/yprc/) focuses on the production, propagation, and shipping for the young plant industry. Their web site includes technical information on water and nutrition management for producers of unrooted cuttings and rooted liner plants.

support increased photosynthesis, which is important for producing a rooted liner more quickly. Conversely in the rooting of herbaceous crops, such as petunia, which root readily within two weeks, the higher the daily light interval (provided there are sufficient environmental controls to minimize desiccation), the more rapidly rooted liners are produced (see Fig. 3–11). Over the past couple of years new types of colored shade cloth have come on the market to enhance rooting. Redshade cloth shifts wavelengths to red and far-red which can enhance rooting (Fig. 3–10).

Supplemental carbon dioxide (CO_2) can enhance the rooting of cuttings when CO_2 levels become too low to maintain photosynthesis (i.e., 75 ppm CO_2) (127). See Chapter 3 for the discussion on *supplemental carbon dioxide* (CO_2) (see Fig. 3–13) and accelerated growth techniques (AGT) (see Fig. 3–30).

Sanitation and IPM

It is also necessary to maintain sanitary conditions in the propagating frame. Leaves that drop should be removed promptly, as should any obviously dead cuttings. Pathogens find ideal conditions in humid, closed propagating structures with low light irradiance. If not controlled, pathogens can destroy thousands of cuttings in a short time. Disease control is done on a preventive and scheduled weekly basis, by selectively rotating fungicides. See the discussion on IPM in Chapter 3 and the preventative disease control measures outlined earlier in this chapter. Disease problems under mist or fog propagation conditions have generally not been serious. Frequent washing of leaves by aerated water can remove spores (e.g., powdery mildew) before they are able to germinate. With the higher allowable light irradiance conditions under mist, a newly rooted cutting can produce carbohydrates through photosynthesis in excess of its maintenance requirements for root development to proceed. Rapid rooting and later growth of the rooted liner along with good ventilation and movement of air can also decrease the incidence of disease (83). Cuttings propagated with a fog system experience less stress. The cuttings are not subjected to high irradiance, heat, or desiccation, and optimum water status is maintained without severe leaching. Hence, they are much less susceptible to disease problems.

Pests such as mites, aphids, and mealy bugs are controlled by miticides and insecticides, and by immediately roguing (discarding) infested cuttings and other IPM techniques.

Weed Control

Weeds can be a serious problem during propagation. Weeds should be removed to prevent their seeding and competing with cuttings. Weed control in propagation begins with (a) using weed-free, pasteurized, or gas-sterilized rooting media, (b) keeping the perimeters adjacent to the propagation area free of weeds, and (c) herbiciding the propagation area and occasionally doing spot weeding by hand (29, 66).

Herbicide Use in Propagation

In general, it is best not to use herbicides during propagation. Some preemergent herbicides, such as Ronstar and Rout, can be used for effective weed control in flats of unrooted cuttings of selected species of *Rhododendron, Euonymus, Ilex,* and *Cotoneaster* (85). However, granular dinitroaniline herbicides (Rout 3G, OH-2 3G,

Ronstar 2G, Snapshot 2.5TG, SWGC 2.68G, etc.) can also suppress root initiation and development in stem cuttings (65). Sensitivity of cuttings to herbicides is dependent on the herbicide, application rate, and plant species. The rooting and subsequent liner growth of hibiscus was adversely affected by preemergence herbicides (applied after cuttings were stuck), while Asian jasmine was unaffected (34). Preemergence herbicides can effectively be used in propagation when applied to propagation flats, prior to sticking cuttings, but propagators should:

- choose the correct herbicide to control the particular weed species,
- follow label directions,
- make sure that a herbicide is labeled for use in propagation, particularly if applied to a confined greenhouse/propagation house as opposed to outside open propagation beds,
- *always* conduct trials to evaluate specific herbicides and the depth of sticking cuttings of individual species to be propagated, and
- determine potential phytotoxicity, reduced rooting, and reduced growth of the cutting species prior to large-scale application of any herbicide (65).

Suspicions have persisted in the nursery industry that preemergence herbicides applied to container plants and stock plants cause reduced rooting of stem cuttings. Research shows that herbicides applied at normal-use rates generally have no effect on rooting of cuttings of most woody landscape species, even when stock plants are treated repeatedly over several years (22).

HARDENING-OFF AND POST-PROPAGATION CARE

Hardening-Off of Rooted and Unrooted Cuttings

Hardening-off is the process of gradually acclimating rooted cuttings from the high humidity of mist, fog, or a contact polysheet system in order to reduce humidity. This weaning process enables the rooted cutting to become more self-sufficient in absorbing nutrients and water through the root system, in photosynthesizing, and in conditioning new developing leaves and stems to better tolerate the stresses of lower relative humidity, coupled with higher temperature and light irradiance.

hardening-off The stress adaptation process or **acclimation** that occurs as a rooted cutting is gradually weaned from a higher to a lower relative humidity environment during propagation.

Cuttings *deteriorate* when they are left under mist too long after they have rooted. This reduces root quality, causes premature leaf drop, and slows down the plant's momentum, which can delay the production period and produce a poorer quality plant. Cutting deterioration is one reason why flats or propagation liner containers of easy-to-root and difficult-to-root species are not mixed together in the same propagation system or house. One species would need to have the mist reduced and the plant hardened-off and removed, while the slower-rooting species is still rooting. Hardening-off encourages better root development from rooted cuttings. The *key to plant survival* is to reduce mist once roots start to develop and allow secondary root growth but avoid excessive root development, which can be particularly detrimental to shoot development of leafless hardwood cuttings (158). Chemically, root pruning with copper-treated propagation containers can also help direct root development (see Fig. 3–23) (142).

There are several ways of successfully weaning rooted cuttings from the mist conditions:

- The cuttings may be left in place in the mist bed but with the duration of the misting periods gradually decreased, either by lessening the "on" periods and increasing the "off" periods or by leaving the misting intervals the same but gradually decreasing the time for which the mist is in operation each day.
- Another method is to root the cuttings in flats and move the flats after rooting to a lathhouse or cold frame, where they are "hardened-off" and then potted into containers as rooted liners. Cuttings may be left in the rooting medium until the dormant season, when they can be dug more safely, to be either lined out in the nursery row for further growth or potted and brought into the greenhouse. If the rooted cuttings are left in the rooting medium for a considerable time, it is advisable to fertilize them at intervals with a nutrient solution or top-dress with a controlled-release fertilizer.

Some propagators **direct stick** cuttings in small containers set up in flats or in modules preformed in plastic trays (Fig. 10–48). Then, after rooting, the plants may be easily moved for transplanting without disturbing the roots. Another direct sticking system is to root the cutting in a solid, block-type rooting medium, which, after rooting, permits transplanting without disturbing the roots. Such products are made from compressed peat and synthetic materials such as rockwool, polystryrene, and others (Fig. 10–27).

- Another method is to pot the rooted cuttings immediately after rooting and hold them for a time in a cool, humid, shaded location (e.g., a fog chamber, closed frame, or greenhouse).
- With contact poly, slits can be made in the polyethylene with a knife; these are gradually increased in size and number over time to lower relative humidity and increase light irradiance and ultraviolet exposure.

Avoiding Overwintering Problems of Rooted Liners
Poor winter survival of rooted cuttings can occur with certain deciduous woody plants, such as maple, beech, dogwood, *Hamamelis,* lilac, magnolia, *Prunus,* oak, rhododendron, *Stewartia,* and viburnum (160). It is generally attributable to poor cold hardiness or insufficient reserves to sustain the cuttings. Newly rooted cuttings go dormant in the fall, but die either during the winter or after bud-break in the spring (90, 95). With some species it is essential that after rooting a flush of growth occurs in mid- to late summer so that adequate carbohydrate reserves are produced, which assures winter survival.

Propagating early enough in the season to allow sufficient rooting, followed by shoot extension growth during the season of propagation—generally increases overwinter survival and can create greater storage capacity in the cutting (160). Late summer flushes of growth can be accomplished by extending the photoperiod (Fig. 3–14) and manipulating fertilizer regimes (31, 50, 69). However, there must be sufficient time after growth flush(es) for rooted liners to harden-off before the onset of winter. Rooted liners should be allowed to go dormant, while maintaining them above freezing in a protected structure (90). The rooted liners are then transplanted to containers or lined out in the field in the spring. See the discussion in Chapter 9.

Residual auxin applied to enhance rooting may suppress bud-break and growth flushes of rooted cuttings—and reduce winter survival. In rose cuttings, basally applied IBA increased rooting, but also increased ethylene synthesis, and subsequently inhibited bud-break of the rooted cuttings (145).

Cold Storage of Rooted and Unrooted Cuttings

Sometimes it may be convenient to collect cuttings when nursery plants are pruned and store them for later propagating. Most nurseries have refrigerated storage facilities (4 to 8°C, 40 to 47°F) for holding cuttings 1 to 2 days or longer before processing for propagation (Fig. 10–58). Cuttings of *Rhododendron catawbiense* can be stored for 21 days in moist burlap bags at 2 to

(a) (b) (c) (d)

Figure 10–58
Cutting storage: (a and b) Refrigerated storage for holding rooted and unrooted cuttings. (a) High humidity is maintained by overhead sprinkler systems (arrows). (c) Cold storage of rooted and unrooted cuttings. (d) Winter-lifted rooted deciduous cuttings on long strips of plastic with moist peat moss; the rooted liners, peat moss, and plastic are rolled up like a jelly roll and placed in cold storage. They will later be transplanted in the field in late spring (Maryland, USA) (117). Photo (d) Courtesy V. Priapi.

21 C (36 to 70°F) with no reduction in rooting (38). Softwood cuttings of Kurume-type azaleas were taken in spring and held for 10 weeks in polyethylene bags at −0.5 to 4.5°C (31 to 40°F) with no adverse effect on rooting (118). Likewise, unrooted cuttings of junipers (*Juniperus*), *Thuja*, and *Taxus* can be stored for several months at 0°C (32°F) in sealed polybags and still root well in the spring (10).

Many nurseries will overwinter rooted cuttings either in flats or in small liner containers protected by minimum-heat-maintained structures (e.g., quonset, polyhoop houses, greenhouses). It is possible to store rooted cuttings of certain species safely for up to 5 months at 1 to 4°C (34 to 39°F) in polyethylene bags (135). Cuttings of thirty-one woody ornamentals stored for 6 months had better survival at 0°C than at 4.5°C (32°F versus 40°F), although with some species there was no difference (58).

With rooted, deciduous hardwood cuttings that are lifted from propagation beds in late winter, it may be necessary to refrigerate them for several months until they can be transplanted. Figure 10–58 shows a plant roll system, where dormant, rooted deciduous cuttings are placed on long strips of plastic with moist peat moss. Then the plants, peat moss, and plastic are rolled up like a jelly roll and placed in cold storage for up to 2 months at 1°C (34°F) until field planted (117).

As previously mentioned, newly rooted softwood cuttings of selected species can be difficult to overwinter, and cannot be transplanted to the field from their rooting beds. Storing rooted cuttings in *cold storage* allows growers to commit valuable greenhouse overwintering space to the production of other crops. It also permits earlier deliveries to warmer regions because the plants are available for shipping all winter. The planting season can also be extended because the rooted cuttings are held in a state of dormancy into the spring (161). Late in the year (during the slow season) is probably the safest time to harvest crops for refrigerated storage. With rooted softwood cuttings, their natural growth cycle has been disrupted, so they need not be dormant before cold storage (161). Cuttings should be allowed to partially dry before pulling them from the rooting beds, and then put in polybags packed with materials to act as insulation and to absorb excess water. The polybags allow gas exchange, but retain cutting moisture to *prevent desiccation* during cold storage. Rooted softwood cuttings of seven species successfully survived up to 7 months when stored at either −2 or 2°C (28 and 35°F) (161).

Douglas-fir cuttings can be stored successfully at −1°C (30°F). This lower storage temperature also prevents mold. Cuttings are stored at nearly 100 percent humidity by placing them in clear polybags with a large block of water-saturated oasis material. The cold storage conditions may be satisfying the chilling requirement for bud dormancy, and therefore satisfying the requirement to have a sufficient cold period prior to taking cuttings in early winter (123, 124).

Unrooted chrysanthemum and carnation cuttings can be stored in sealed plastic bags for several weeks at −0.5°C (31°F) for subsequent rooting. In tests on the effects of storage on subsequent performance of plants, cuttings rooted after storage gave better results than those stored after rooting. Prestorage of chrysanthemum for 12 days at 10°C (50°F) enhanced rooting of cuttings compared with nonstored cuttings

---

BOX 10.18 GETTING MORE IN DEPTH ON THE SUBJECT
REFRIGERATED STORAGE SYSTEMS

The **direct-cooled refrigeration system** (similar to the common kitchen refrigerator) is the most popular for nursery stock. Fans circulate air over cooling coils to directly cool the cuttings and the system's interior. This system is relatively stable as long as the humidity is not too high. Excessive humidity in the storage chamber will condense and freeze the coils, requiring defrosting and subsequent temperature fluctuations. It is generally not advisable to install misting systems or sprinkle water on the floor to raise humidity, because it causes more condensation on the coils and increases potential mold development. Unrooted *Taxus* cutting wood is stored by a Michigan nursery in slotted pallet boxes to allow air circulation. Temperature is maintained at 1 to 2°C (34 to 36°F) and relative humidity is maintained with a humidifier (135). To avoid desiccation and retain moisture, cuttings are packed in an insulation material and placed in polyethylene bags.

The **jacketed cooler** is another popular system for storing plants, cutting wood, and understock. It is essentially a box within a box. Refrigerated air passes between the boxes, cooling the inner walls. It is a very stable cooling system, and it can maintain nearly 100 percent relative humidity, which means cuttings do not require packing in sealed polybags to maintain high relative humidity. Because the jacketed cooler is a closed system, growers can convert it into a controlled-atmosphere (C.A.) storage system. Researchers are studying the usefulness of controlled atmosphere storage on nursery crops.

Figure 10–59
Effect of length of time at a prestorage temperature of 10°C (50°F) on rooting of *Chrysanthemum morifolium* 'Pink Boston.' Cuttings were stored 0 days (left) and 12 days (right). Both were propagated at the same time and evaluated after 7 days (151). Courtesy P. A. Van de Pol.

(Fig. 10–59). Roots were initiated while cuttings were in storage (151). This knowledge could enable producers to store cuttings for a more convenient time to propagate, and also reduce the time needed for propagating under a mist or fog system.

Storage and subsequent rooting of carnations was better if auxin was applied after storage temperatures of *less than* 13°C (55°F), whereas auxin was more effective if applied *prior* to cuttings being maintained at storage temperatures above 13°C (150). Again, the *implications are for using storage for convenience of propagating, as well as for reducing the propagation period.*

Carnation cuttings, either rooted or unrooted, store well at −0.5°C (31°F) for at least 5 months if placed in polyethylene-lined boxes with a small amount of moist sphagnum or peat moss. The poly film should not be sealed.

The proper storage temperature is species-dependent. Some species survive cold temperature better than others. The storage unit for unrooted cuttings should be maintained at close to 100 percent humidity—pathogens must be controlled. Topsin (Atochem N.A.), Domain (Sierra-Grace), Cleary 3396 (W. A. Cleary), Alliette, and other systemic fungicides should be considered. Storage temperature should be as low as possible without impairing rooting of cuttings or survival of rooted liners.

HANDLING FIELD-PROPAGATED PLANTS

Bare-Root Species

Rooted deciduous, hardwood cuttings are dug in the nursery row during the dormant season after the leaves have dropped. With fast-growing species, the cuttings may be sufficiently large to dig after one season's growth. Slower-growing species may require 2 or even 3 years to become large enough to transplant.

Most deciduous trees and shrubs harvested bare-root in the late fall and early winter will lose more than 90 percent of their root system between the nursery digging process and the final transplanting site. Conversely, smaller-sized liners may have 90 percent of their root system still intact. Many species of herbaceous perennials are also handled bare-root, although harvest time often depends on the species. By properly performing five major steps when working with bare-root plants—(a) harvesting, (b) processing, (c) storing, (d) shipping, and (e) transplanting—most nurseries can avoid problems (Fig. 10–60). With some species it may be advantageous to root-prune plants 1 year prior to digging by pruning the roots with a sharp-shooter spade, or mechanically digging, slightly lifting, and placing the rootball in its original ground location (67, 155). This slows growth in the nursery, but promotes a more compact, fibrous root system and reduces transplant shock.

The digging should take place on cool, cloudy days when there is no wind. In large-scale nursery operations, some type of mechanical digger is generally used that "undercuts" the plants. A sharp U-shaped blade travels 30 to 60 cm (1 to 2 ft) below the soil surface under the

---

BOX 10.19 GETTING MORE IN DEPTH ON THE SUBJECT
HANDLING FIELD-PRODUCED PLANTS

The handling procedures described in this section also apply to nursery plants propagated as seedlings, as tissue-culture-produced liners, or as budded or grafted trees.

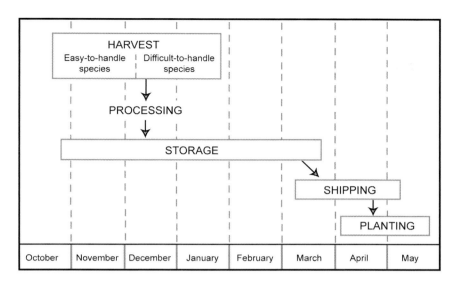

Figure 10–60
A general outline and schedule for handling bare-rooted trees and shrubs in Oregon (53).

nursery row, cutting through the roots. Sometimes a horizontal, vibrating, "lifting" blade is also attached, and travels behind the cutting blade. This blade lifts the plants out of the soil, and shakes the soil from the root system, making them easy to pull by hand.

Once a bare-root plant is dug, it is imperative that the roots not dry out. Excessive moisture loss or desiccation causes large transplant losses. Roots of bare-root plants will lose water five times faster than the stems (53). After the plants are dug, they should be quickly heeled-in in a convenient location, placed in cold storage, or replanted immediately in their permanent location. **Heeling-in** is placing dug, **bare-rooted** deciduous or coniferous nursery plants close together in trenches with the roots well covered. This is a temporary provision for holding the young plants until they can be set out in their permanent location.

Commercial nurseries often store quantities of deciduous plants for several months through the winter in cool, dark rooms with the roots protected by damp bark wood shavings, moist sawdust, or some similar material. Nursery stock to be kept for extended periods should be held in cold storage, ideally with jacketed cooler systems, with high relative humidity (+95 percent) and temperatures of 0 to 2°C (32 to 35°F). Bare-root plants are usually placed onto pallets, and the pallets are stacked on top of each other (59). If bare-root liners leaf out during storage, they should be stripped of their leaves prior to transplanting in the field (4).

Generally, *water stress* and the *root regeneration potential* are the chief factors limiting transplanting success. Waxing the stems of stored, dormant roses and other shrubs has been a common nursery practice.

Some of the waxes used in the fresh-fruit industry, such as Shield Brite (Shield-Brite Corp., Kirkland, WA), have been somewhat effective in reducing water loss and improving survival of bare-root plants (53). The use of anti-transpirants and film-forming compounds, or dipping roots in hydrogels is not nearly as effective as defoliating plants (4). The anti-transpirant, Moisturin (Burke's Protective Coatings, Washougal, WA), enhanced transplant survival and growth of bare-root Washington hawthorns and roses, which are difficult to transplant because of excessive desiccation during postharvest production. However, with selected oak species, where transplanting success is limited by poor root regeneration potential, antidesiccants are ineffective (53). The greatest amount of water loss during postharvest handling comes after storage—during transport and transplanting to the final site. In general, any treatment that reduces water loss and plant stress during handling and transplanting is beneficial.

Balled and Burlapped Stock

Unless very small, plants of broad- or narrow-leaved evergreen species usually are not handled successfully bare-root, as is done with dormant and leafless deciduous plants. The presence of leaves on evergreen plants means the roots must have continuous contact with soil. Therefore, large, salable plants of broad- or narrow-leaved evergreens, and occasionally deciduous plants, are either grown in containers or dug and sold "balled and burlapped" (B&B). By the latter method, the plants are removed from the soil by carefully digging a trench around each individual plant or using a digging machine such as a mechanized tree spade (Fig. 10–61). The ball is gently placed on a large

> **BOX 10.20 GETTING MORE IN DEPTH ON THE SUBJECT**
> **DIFFICULT-TO-HANDLE BARE-ROOT SPECIES (53)**
>
> Difficult-to-handle bare-root species include *Betula* (birch) and *Crataegus* (hawthorn). Generally, poor transplanting success and survival is due to *excessive desiccation during post-production handling*. However, some species of *Quercus* (oak) exhibit poor or slow *root regeneration potential* (RRP) after transplanting.
>
> There are physiological differences in dormancy and stress tolerance between easy-to-handle *Acer platanoides* (Norway maple) and difficult-to-handle *Crataegus phaenopyrum* (Washington hawthorn). Washington hawthorn does not achieve the same level of deep dormancy, cold-hardiness, and desiccation tolerance that Norway maple does. The desiccation tolerance of Norway maple increases substantially in November (Oregon), whereas desiccation tolerance in Washington hawthorn does not begin until late December. Thus, it is best to schedule difficult-to-transplant species for harvest as late in the season as possible to allow desiccation tolerance to increase (see Fig. 10–60) (53).

square of burlap, which is then pulled tightly around the ball, pinned with nails or hog wings, and wrapped with twine. When done properly, the burlap ensures that roots are kept in contact with the soil, and the plant can be moved safely for considerable distances and replanted successfully (Fig. 10–61). Larger specimen B&B plants are sometimes set in a wire cage to help keep the rootball intact during handling and shipping. Some field-grown B&B plants are *shifted* into larger, rigid-plastic containers or wooden boxes to ease handling and marketing, and to increase the sale value.

(a)

(b)

(c)

(d)

Figure 10–61
(a and b) Container-produced plants with (b) white rootskirt (arrow), an insulating sleeve to reflect light and reduce heat load. (c and d) Field-grown shade trees. (d) Mechanically dug "balled and burlapped" (B&B) tree being processed for shipping.

CONTAINER-GROWN PLANTS AND ALTERNATIVE FIELD PRODUCTION SYSTEMS

Production of Plants in Containers

Container production has replaced much of the traditional nursery field production methods, primarily because of handling ease, improved plant marketability, greater cultural control, and faster product turnover (Fig. 10–61). However, many plant species are still field-grown because of their growth habits and lower production costs. Field-grown plants do well in climates experiencing low temperatures, which limit survival of containerized crops without overwintering protection.

See Chapter 3 for the discussion on best management practices, post-propagation care of liners, handling *container-grown* plants (irrigation systems, winter protection, root development in containerized plants), alternatives to traditional production systems (**pot-in-a-pot system, in-ground plastic containers,** and **in-ground fabric containers** or **grow bags**) (see Fig. 3–25), and avoiding transplant problems.

DISCUSSION ITEMS

1. Why are cuttings the most important means of propagating ornamental shrubs and selected florist crops?
2. How does the nature of the wood affect the type of stem cutting that can be taken? What types of stem cuttings are there?
3. How does the propagation of leaf cuttings differ from leaf-bud cuttings? What growth regulator would you apply for each type of propagule and why?
4. What are some ways that stock plants can be manipulated to enhance the rooting process?
5. What are the differences among etiolation, shading, blanching, and banding?
6. Why has direct sticking (direct rooting) become more popular in the propagation of nursery crops in various production areas of the world?
7. Auxins are the most important growth regulators for stimulating rooting. What are the most important commercial auxins, and in what types of formulations (carriers) are they applied?
8. What are the functions of organic and coarse mineral components in propagation media?
9. What advantages are there for using quick-dip applications of auxin vs. talc applications?
10. What is the role of integrated pest management (IPM) and best management practices (BMP) in plant propagation?
11. What are the advantages and disadvantages in using intermittent mist systems?
12. In propagation management, what is the importance of Reasonable Expectancy programs, costing variables, piece-work systems, and the implementation of an action plan/cost-benefit analysis?
13. Why has timing and scheduling become that much more important in market-led propagation systems?
14. What are some important techniques to harden-off (acclimate) rooted liner cuttings?

REFERENCES

1. Ahmad, M., P. B. Lombard, and R. L. Ticknor. 1992. Effect of slow-release fertilizers on propagation medium and on rooting and growth of cuttings. *Comb. Proc. Intl. Plant Prop. Soc.* 42:238–41.

2. Aiello, A. S., and W. R. Graves. 1998. Success varies when using subirrigation instead of mist to root softwood cuttings of woody taxa. *J. Environ. Hort.* 16:42–7.

3. Alward, T. M. 1984. Softwood cuttings taken from developing hardwood cuttings. *Comb. Proc. Intl. Plant Prop. Soc.* 34:535–36.

4. Askew, J. C., C. H. Gilliam, H. G. Ponder, and G. J. Keever. 1985. Transplanting leafed-out bare root dogwood liners. *HortScience* 20:219–91.

5. Atzmon, N., Z. Wiesman, and P. Fine. 1997. Biosolids improve rooting of Bouganinvillea (*Bougainvillea glabra*) cuttings. *J. Environ. Hort.* 15:1–5.

6. Avery, J. D., and C. B. Beyl. 1991. Propagation of peach cuttings using foam cubes. *HortScience* 26:1152–4.

7. Badenhop, M. B. 1984. How much does it cost to produce a rooted cutting? *Amer. Nurs.* 160:104–11.

8. Baldwin, I., and J. Stanley. 1981. How to manage stock plants. *Amer. Nurs.* 153:16, 74–80.

9. Barnes, H. W. 1989. Propylene glycol quick-dips: Practical applications. *Comb. Proc. Intl. Plant Prop. Soc.* 39:427–32.

10. Barnes, H. W. 1993. Storage and production of selected conifers from hardwood cuttings. *Comb. Proc. Intl. Plant Prop. Soc.* 43:397–99.

11. Beeson, R. C., Jr. 1991. Scheduling woody plants for production and harvest. *HortTechnology* 1:30–6.

12. Berry, J. B. 1984. Rooting hormone formulations: A chance for advancement. *Comb. Proc. Intl. Plant Prop. Soc.* 34:486–91.

13. Bhat, N. R., T. L. Prince, H. K. Tayama, and S. A. Carver. 1992. Root cutting establishment in media containing a wetting agent. *HortScience* 27:78.

14. Blythe, E. K., J. L. Sibley, K. M. Tilt, and J. M. Ruter. 2008a. Methods of auxin application in cutting propagation: A review of 70 years of scientific discovery and commercial practice: Part I *North Amer. Plant Prop.* 20:13–25.

15. Blythe, E. K., J. L Sibley, K. M. Tilt, and J. M. Ruter. 2008b. Methods of auxin application in cutting propagation: A review of 70 years of scientific discovery and commercial practice: Part 2 *North Amer. Plant Prop.* 20:19–27.

16. Blythe, G., and J. L. Sibley. 2003. Novel methods of applying rooting hormones in cutting propagation. *Comb. Proc. Intl. Plant Prop. Soc.* 53:406–10.

17. Bonaminio, V. P. 1983. Comparison of quick-dips with talc for rooting cuttings. *Comb. Proc. Intl. Plant Prop. Soc.* 33:565–68.

18. Brun-Wibaux, F. X. 2002. Liner production: Asset or Liability? *Comb. Proc. Intl. Plant Prop. Soc.* 52:459–63.

19. Burger, D. W. 1994. Intermittent mist control via solar cells. *HortTechnology* 4:273–74.

20. Carlson, R. F. 1966. Factors influencing root formation in hardwood cuttings of fruit trees. *Mich. Quart. Bul.* 48:449–54.

21. Carter, A., and M. Slee. 1991. Propagation media and rooting cuttings of *Eucalyptus grandis*. *Comb. Proc. Intl. Plant Prop. Soc.* 41:36–9.

22. Catanzaro, C. J., W. A. Skroch, and P. H. Henry. 1993. Rooting performance of hardwood stem cuttings from herbicide-treated nursery stock plants. *J. Environ. Hort.* 11:128–30.

23. Centeno, A., and M. Gomez del Campo. 2008. Effect of root-promoting products in the propagation of organic olive (*Olea europaea* L. cv. Cornicabra) nursery plants. *HortScience* 43:2066–9.

24. Chapman, D. J. 1989. Consider softwood cuttings for tree propagation. *Amer. Nurs.* 169:45–50.

25. Chong, C. 1982. Rooting response of cuttings of two cotoneaster species to surface-applied Osmocote slow-release fertilizer. *Plant Propagator* 28:10–2.

26. Chong, C., O. B. Allen, and H. W. Barnes. 1992. Comparative rooting of stem cuttings of selected woody landscape shrub and tree taxa to varying concentrations of IBA in talc, ethanol and glycol carriers. *J. Environ. Hort.* 10:245–50.

27. Chong, C., and B. Hamersma. 1995. Automobile radiator antifreeze and windshield wiper fluid as IBA carriers for rooting woody cuttings. *HortScience* 30:363–65.

28. Cialone, J. 1984. Developments in dracaena production. *Comb. Proc. Intl. Plant Prop. Soc.* 34:491–94.

29. Cochran, D. R., C. H. Gilliam, D. J. Eakes, G. R. Wehtje, and P. R. Knight. 2006. Pre-emergent herbicide use in propagation of *Loropetalum chinesnse* 'Ruby'. *Comb. Proc. Intl. Plant Prop. Soc.* 56:560–65.

30. Coorts, G. D., and C. C. Sorenson. 1968. Organisms found growing under nutrient mist propagation. *HortScience* 3:189–90.

31. Davies, F. T. 1988. Influence of nutrition and carbohydrates on rooting of cuttings. *Comb. Proc. Intl. Plant Prop. Soc.* 38:432–37.

32. Davies, F. T. 1991. Back to the basics in propagation. *Comb. Proc. Intl. Plant Prop. Soc.* 41:338–42.

33. Davies, F. T. 2008. Opportunities from down under—How mycorrhizal fungi can benefit nursery propagation and production systems. *Comb. Proc. Intl. Plant Prop. Soc.* 58:539–48.

34. Davies, F. T., and S. A. Duray. 1992. Effect of preemergent herbicide application on rooting and subsequent liner growth of selected nursery crops. *J. Environ. Hort.* 10:181–86.

35. Davies, F. T., and B. C. Moser. 1980. Stimulation of bud and shoot development of Rieger begonia leaf cuttings with cytokinins. *J. Amer. Soc. Hort. Sci.* 105:27–30.

36. Davies, F. T., Jr. 2000. Benefits and opportunities with mycorrhizal fungi in nursery propagation and production systems. *Comb. Proc. Intl. Plant Prop. Soc.* 50:482–89.

37. Davies, F. T., T. D. Davis, and D. E. Kester. 1994. Commercial importance of adventitious rooting to horticulture. In T. D. Davis and B.E. Haissig, eds. *Biology of adventitious root formation*. New York: Plenum Press.

38. Davis, T. D., and T. R. Potter. 1985. Carbohydrates, water potential, and subsequent rooting of stored rhododendron cuttings. *HortScience* 20:292–93.

39. Del Tredici, P. 1996. Cutting through the confusion. *Amer. Nurs.* 184:22–8.

40. Dirr, M. A. 1982. What makes a good rooting compound? *Amer. Nurs.* 155:33–40.

41. Dirr, M. A. 1983. Comparative effects of selected rooting compounds on the rooting of *Photinia xfraseri*. *Comb. Proc. Intl. Plant Prop. Soc.* 33:536–40.

42. Dirr, M. A. 1989. Rooting response of *Photinia xfraseri* Dress 'Birmingham' to 25 carrier and carrier plus IBA formulations. *J. Environ. Hort.* 7:158–60.

43. Dirr, M. A. 1990. Effects of P-ITB and IBA on the rooting response of 19 landscape taxa. *J. Environ. Hort.* 8:83–5.

44. Dirr, M. A. 1992. Update on root-promoting chemicals and formulations. *Comb. Proc. Intl. Plant Prop. Soc.* 42:361–65.

45. Dirr, M. A. 1994. The latest status of IBA and other root-promoting chemicals. *Nurs. Manag.* 10:24–5.

46. Douds, D. D., Jr., G. Becard, P. E. Pfeffer, L. W. Doner, T. J. Dymant, and W. M. Kayser. 1995. Effect of vesicular-arbuscular mycorrhizal fungi on rooting of *Sciadopitys verticillata* Sib & Zucc. cuttings. *HortScience* 30:133–34.

47. Drahn, S. R. 2003. Replacing manual dips with water soluble IBA. *Comb. Proc. Intl. Plant Prop. Soc.* 53:373–77.

48. Drahn, S. R. 2007a. Auxin application via foliar sprays. *Comb. Proc. Intl. Plant Prop. Soc.* 57:274–77.

49. Drahn, S. R. 2007b. Propagating with controlled-release fertilizer. *Comb. Proc. Intl. Plant Prop. Soc.* 57:521–22.

50. Drew, J. J., M. A. Dirr, and A. M. Armitage. 1993. Effects of fertilizer and night interruption on overwinter survival of rooted cuttings of *Quercus* L. *J. Environ. Hort.* 11:97–101.

51. Druege, U., H. Baltuschat, and P. Franken. 2007. *Piriformospora indica* promotes adventitious root formation in cuttings. *Scientia Hort.* 112:422–26.

52. Elliott, F. A. 1989. Benefits of good record keeping in propagation. *Comb. Proc. Intl. Plant Prop. Soc.* 39:135–38.

53. Englert, J. M., L. H. Fuchigami, and T. H. H. Chen. 1993. Bare-root basics—how to handle bare-root trees and shrubs after harvesting. *Amer. Nurs.* 177:56–61.

54. Erez, A., and Z. Yablowitz. 1981. Rooting of peach hardwood cuttings for the meadow orchard. *Scientia Hort.* 15:137–44.

55. Faust, J. E., and L. W. Grimes. 2004. Cutting production is affected by pinch number during scaffold development of stock plants. *HortScience* 39:1691–4.

56. Ferraro, B. A., and M. J. Brenner. 1997. Disinfection of nursery irrigation water with chlorination, bromination, and ozonation. *Comb. Proc. Intl. Plant Prop. Soc.* 47:423–26.

57. Finnerty, T. L. 1994. Native woody plant propagation—three key steps. Part II. Planning, scheduling, and good record keeping. *North Amer. Plant Prop.* 6:16–17.

58. Flint, H. L., and J. J. McGuire. 1962. Response of rooted cuttings of several woody ornamental species to overwinter storage. *Proc. Amer. Soc. Hort. Sci.* 80:625–29.

59. Foster, S. 1994. Handling bare-root tree whips at Greenleaf Nursery. *Comb. Proc. Intl. Plant Prop. Soc.* 44:478–79.

60. Fourrier, B. 1984. Hardwood cutting propagation at McKay nursery. *Comb. Proc. Intl. Plant Prop. Soc.* 34:540–43.

61. Geneve, R., R. Gates, S. Zoinier, J. Owen, and S. Kester. 1999. A dynamic control system for scheduling mist propagation in poinsettia cuttings. *Comb. Proc. Intl. Plant Prop. Soc.* 49:300–3.

62. Geneve, R. L. 2000. Root formation in relationship to auxin uptake in cuttings treated by the dilute soak, quick dip, and talc methods. *Comb. Proc. Intl. Plant Prop. Soc.* 50:409–12.

63. Geneve, R. L., S. Zolnier, E. Wilkerson, and S.T. Kester. 2004. Environmental control systems for mist propagation of cuttings. *Acta Hort.* 630:297–303.

64. Gilbert, J. 1997. Computer propagation and production planning with the help of a database. *Comb. Proc. Intl. Plant Prop. Soc.* 47:407–13.

65. Gilliam, C. H., D. J. Eakes, and J. W. Olive. 1993. Herbicide use during propagation affects root initiation and development. *J. Environ. Hort.* 11:157–59.

66. Gilliam, C. H. 1996. Weed control in propagation. *Comb. Proc. Intl. Plant Prop. Soc.* 46:663–66.

67. Gilman, E. F., and T. H. Yeager. 1988. Root initiation in root-pruned hardwoods. *HortScience* 23:775.

68. Giroux, G. J., B. K. Maynard, and W. A. Johnson. 1999. Comparison of perlite and peat: Perlite rooting media for rooting softwood stem cuttings in a subirrigation system with minimal mist. *J. Environ. Hort.* 17:147–51.

69. Goodman, M. A., and D. P. Stimart. 1987. Factors regulating overwinter survival of newly propagated stem tip cuttings of *Acer palmatum* Thunb. 'Bloodgood' and *Cornus florida* L. var. rubra. *HortScience.* 22:1296–8.

70. Gouin, F. R. 1981. Vegetative propagation under thermoblankets. *Comb. Proc. Intl. Plant Prop. Soc.* 30:301–5.

71. Grove, R. 2003. The benefits of buying unrooted cuttings to the propagator. *Comb. Proc. Intl. Plant Prop. Soc.* 53:325–28.

72. Guan, H., and G. J. De Klerk. 2000. Stem segments of apple microcuttings take up auxin predominately via the cut surface and not via the epidermal surface. *Scientia Hort.* 86:23–32.

73. Hambrick, C. E., J. F. T. Davies, and H. B. Pemberton. 1991. Seasonal changes in carbohydrate/nitrogen levels during field rooting of *Rosa multiflora* 'Brooks 56' hardwood cuttings. *Scientia Hort.* 46:137–46.

74. Hansen, C. J., and H. T. Hartmann. 1968. The use of indolebutyric acid and captan in the propagation of clonal peach and peach-almond hybrid rootstocks by hardwood cuttings. *Proc. Amer. Soc. Hort. Sci.* 92:135–40.

75. Hansen, J., and K. Kristiansen. 2000. Root formation, bud growth and survival of ornamental shrubs propagated by cuttings of different planting dates. *J. Hort. Sci. Biotech.* 75:568–74.

76. Hare, R. C. 1976. Rooting of American and Formosan sweetgum cuttings taken from girdled and nongirdled cuttings. *Tree Planters Notes* 27:6–7.

77. Hare, R. C. 1977. Rooting of cuttings from mature water oak. *South J. Appl. Forest.* 1:24–5.

78. Hartmann, H. T., W. H. Griggs, and C. J. Hansen. 1963. Propagation of ownrooted Old Home and Bartlett pears to produce trees resistant to pear decline. *Proc. Amer. Soc. Hort. Sci.* 82:92–102.

79. Hartmann, H. T., C. J. Hansen, and F. Loreti. 1965. Propagation of apple rootstocks by hardwood cuttings. *Calif. Agr.* 19:4–5.

80. Howard, B. H. 1968. The influence of indolebutyric acid and basal temperature on the rooting of apple rootstock hardwood cuttings. *J. Hort. Sci.* 43:23–31.

81. Howard, B. H. 1985. Factors affecting the response of leafless winter cuttings of apple and plums to IBA applied in powder formulation. *J. Hort. Sci.* 60:161–68.

82. Howard, B. H. 1985. The contribution to rooting in leafless winter plum cuttings of IBA applied to the epidermis. *J. Hort. Sci.* 60:153–59.

83. Howard, B. H. 1993. Understanding vegetative propagation. *Comb. Proc. Intl. Plant Prop. Soc.* 43:157–62.

84. Jan van Telgen, H., B. Eveleens, and N. Garcia. 2007. Improving rooting uniformity in rose cuttings. *Propagation Ornamental Plants.* 7:190–94.

85. Johnson, J. R., and J. A. Meade. 1986. Preemergent herbicide effect on the rooting of cuttings. *Comb. Proc. Intl. Plant Prop. Soc.* 36:567–70.

86. Klupenger, D. 1999. Cleanliness in propagation and the use of agribrom. *Comb. Proc. Intl. Plant Prop. Soc.* 49:602–3.

87. Kroin, J. 1992. Advances using indole-3-butyric acid (IBA) dissolved in water for root cuttings, transplanting, and grafting. *Comb. Proc. Intl. Plant Prop. Soc.* 42:489–92.

88. Kuszmaul, R. 1999. Economical low pressure fog system. *Comb. Proc. Intl. Plant Prop. Soc.* 49:335–36.

89. Lazcano, C. A., F. T. Davies, Jr., V. Olalde-Portuga, J. C. Mondragon, S. A. Duray, and A. Estrada-Luna. 1999. Effects of auxin and wounding on adventitious root formation of prickly-pear cactus cladodes (*Opuntia amyclaea* T.). *HortTechnology* 9:99–102.

90. LeBlanc, C. 2005. A propagator's notebook. *Comb. Proc. Intl. Plant Prop. Soc.* 55:25:34–38.

91. LeBude, A. V., B. Goldfarb, F. A. Blazich, F. C. Wise, and J. Frampton. 2004. Mist, substrate water potential and cutting water potential influence rooting of stem cuttings of loblolly pine. *Tree Physiol.* 24:823–31.

92. LeBude, A. V., B. Goldfarb, F. A. Blazich, J. Frampton, and F. C. Wise. 2005. Mist level influences vapor pressure deficit and gas exchange during rooting of juvenile stem cuttings of loblolly pine. *HortScience* 40:1448–56.

93. Linderman, R. G. 1993. Effects of biocontrol agents on plant growth. *Comb. Proc. Intl. Plant Prop. Soc.* 43:249–52.

94. Loach, K. 1985. Rooting of cuttings in relation to the propagation medium. *Comb. Proc. Intl. Plant Prop. Soc.* 35:472–85.

95. Loach, K. 1988. Hormone applications and adventitious root formation in cuttings—a critical review. *Acta Hort.* 227:126–33.

96. Loach, K. 1989. Controlling environmental conditions to improve adventitious rooting. In T. D. Davis, B. E. Haissig, and N. Sankhla, eds. *Adventitious root formation in uttings*. Portland, OR: Dioscorides Press.

97. Macdonald, B. 1986. *Practical woody plant propagation for nursery growers*. Portland, OR: Timber Press.

98. MacKenzie, A. J., T. W, Starman, and M. T. Windham. 1995. Enhanced root and shoot growth of chrysanthemum cuttings propagated with the fungus *Trichoderma harzianum*. *HortScience* 30:496–98.

99. Maronek, D. M., D. Studebaker, and B. Oberly. 1985. Improving media aeration in liner and container production. *Comb. Proc. Intl. Plant Prop. Soc.* 35:591–97.

100. Maunder, C. 1983. A comparison of propagation unit systems. *Comb. Proc. Intl. Plant Prop. Soc.* 33:233–38.

101. Maynard, B. K., and N. L. Bassuk. 1987. Stockplant etiolation and blanching of woody plants prior to cutting propagation. *J. Amer. Soc. Hort. Sci.* 112:273–76.

102. Maynard, B. K., and N. L. Bassuk. 1990. Comparisons of stock plant etiolation with traditional propagation methods. *Comb. Proc. Intl. Plant Prop. Soc.* 40:517–23.

103. Maynard, B. K., and N. L. Bassuk. 1992. Stock plant etiolation, shading, and banding effects on cutting propagation of *Carpinus betulus*. *J. Amer. Soc. Hort. Sci.* 117:740–44.

104. Mazalewski, R. L. 1989. Practical aspects of high pressure fog systems. *Comb. Proc. Intl. Plant Prop. Soc.* 39:101–5.

105. McLean, C., A. C. Lawrie, and K. L. Blaze. 1994. The effect of soil microflora on the survival of cuttings of *Epacris impressa*. *Plant Soil* 166:295–97.

106. Mee, T. R. 1994. Understanding fog technology. *Comb. Proc. Intl. Plant Prop. Soc.* 44:250–53.

107. Merker, R. 1985. System of direct stick propagation. *Comb. Proc. Intl. Plant Prop. Soc.* 35:182–83.

108. Milbocker, D. C. 1987. The use of humidifan in propagation. *Comb. Proc. Intl. Plant Prop. Soc.* 37:513–18.

109. Molnar, J. M., and W. A. Cumming. 1968. Effect of carbon dioxide on propagation of softwood, conifer, and herbaceous cuttings. *Can. J. Plant Sci.* 48:595–99.

110. Motley, B. 1994. Incentive pay in propagation. *Comb. Proc. Intl. Plant Prop. Soc.* 44:449–53.

111. Noland, D. A., and D. J. Williams. 1980. The use of pumice and pumice-peat mixtures as propagation media. *Plant Propagator* 26:6–7.

112. Normand, L., H. Bärtsdhi, J. C. Debaud, and G. Gay. 1996. Rooting and acclimatization of micropropaged cuttings of *Pinus pinaster* and *Pinus sylvestris* are enhanced by the ectomycorrhizal fungus *Hebeloma cylindrosporum*. *Physiol. Plant.* 98:759–66.

113. Orndorff, C. 1987. Root pieces as a means of propagation. *Comb. Proc. Intl. Plant Prop. Soc.* 37:432–35.

114. Preece, J. E. 1987. Treatment of stock plant with plant growth regulators to improve propagation success. *HortScience* 22:754–59.

115. Preece, J. E. 2003. A century of progress with vegetative plant propagation. *HortScience.* 38:1015–25.

116. Preece, J. E., and P. Read. 2007. Forcing leafy explants and cuttings from woody species. *Propagation Ornamental Plants.* 7:138–44.

117. Priapi, V. M. 1993. Outdoor mist propagation. *Amer. Nurs.* 178:30–5.

118. Pryor, R. L., and R. N. Stewart. 1963. Storage of unrooted azalea cuttings. *Proc. Amer. Soc. Hort. Sci.* 82:483–84.

119. Rainey, M. 2003. Improving productivity and morale through mechanization *Comb. Proc. Intl. Plant Prop. Soc.* 53:321–24.

120. Regan, R., and A. Henderson. 1999. Using subirrigation to root stem cuttings: A project review. *Comb. Proc. Intl. Plant Prop. Soc.* 49:637–44.

121. Richey, M. L. 1989. Costing variables in propagation techniques. *Comb. Proc. Intl. Plant Prop. Soc.* 39:502–6.

122. Rinallo, C., L. Mittempergher, G. Frugis, and D. Mariotti. 1999. Clonal propagation in the genus *Ulmus*: Improvement of rooting ability by *Agrobacterium rhizogenes* T-DNA genes. *J. Hort. Sci. Biotech.* 74:502–6.

123. Ritchie, G. A., Y. Tanaka, and S. D. Duke. 1992. Physiology and morphology of Douglas-Fir rooted cuttings compared to seedlings and transplants. *Tree Physiol.* 10:179–94.

124. Ritchie, G. A. 1994. Commercial applications of adventitious rooting to forestry. In T. D. Davis and B. E. Haissig, eds. *Biology of adventitious root formation.* New York: Plenum Press.

125. Sabalka, D. 1986. Propagation media for flats and for direct sticking: What works? *Comb. Proc. Intl. Plant Prop. Soc.* 36:409–13.

126. Saborio, F., M. M. Moloney, P. Tung, and T. A. Thorpe. 1999. Root induction of *Pinus ayachauite* by co-culture with *Agrobacterium tumefacines* strains. *Tree Physiol.* 19:383–89.

127. Sanders, W. 1996. Experiences with carbon dioxide enrichment for production of rooted cuttings. *Comb. Proc. Intl. Plant Prop. Soc.* 46:175–77.

128. Santana, C. 1995. Production scheduling of cuttings. *Comb. Proc. Intl. Plant Prop. Soc.* 45:316–19.

129. Santos, K. M., P. R. Fisher, and W. R. Argo. 2008. A survey of water and fertilizer management during cutting propagation. *HortTechnology* 18:597–604.

130. Scagel, C. F. 2004. Changes in cutting composition during early stages of adventitious rooting of miniature rose altered by inoculation with arbuscular mycorrhizal fungi. *J. Amer. Soc. Hort. Sci.* 129:624–34.

131. Scagel, C. F., K. Reddy, and J. M. Armstrong. 2003. Mycorrhizal fungi in rooting substrate influences the quantity and quality of roots on stem cuttings of Hick's yew. *HortTechnology* 13:62–6.

132. Scott, M. A. 1987. Management of hardy nursery stock plants to achieve high yields of quality cuttings. *HortScience* 22:738–41.

133. Skimina, C. A. 1984. Use of monochloramine as a disinfectant for pruning shears. *Comb. Proc. Intl. Plant Prop. Soc.* 34:214–20.

134. Smith, I. 1993. Government regulations and nursery accreditation. *Comb. Proc. Intl. Plant Prop. Soc.* 43:117–21.

135. Snyder, W. E., and C. E. Hess. 1956. Low temperature storage of rooted cuttings of nursery crops. *Proc. Amer. Soc. Hort. Sci.* 67:545–48.

136. Sorenson, D. C., and G. D. Coorts. 1968. The effect of nutrient mist on propagation of selected woody ornamentals. *Proc. Amer. Soc. Hort. Sci.* 92:696–703.

137. Spethmann, W. 2004. Use of long cuttings to reduce propagation time of rose and fruit rootstocks and street trees. *Comb. Proc. Intl. Plant Prop. Soc.* 54:223–31.

138. Spethmann, W. 2007. Increase of rooting success and further shoot growth by long cuttings of woody plants. *Propagation Ornamental Plants.* 7:160–66.

139. Stein, A., J. A. Fortin, and G. Vallée. 1990. Enhanced rooting of *Picea mariana* cuttings by ectomycorrhizal fungi. *Can. J. Bot.* 68:468–70.

140. Stoven, J., and H. Kooima. 1999. Coconut-coir-based media versus peat-based media for propagation of woody ornamentals. *Comb. Proc. Intl. Plant Prop. Soc.* 49:373–74.

141. Struve, D. K., and M. A. Arnold. 1986. Aryl esters of IBA increase rooted cutting quality of red maple 'Red Sunset' softwood cuttings. *HortScience* 21:1392–3.

142. Struve, D. K., M. A. Arnold, R. Beeson, Jr., J. M. Ruter, S. Svenson, and W. T. Witte. 1994. The copper connection: The benefits of growing woody ornamentals in copper-treated containers. *Amer. Nurs.* 179:52–61.

143. Sumner, P. E. 1987. Uniformity analysis of various types of mist propagation nozzles. *Comb. Proc. Intl. Plant Prop. Soc.* 37:522–26.

144. Sun, W. Q., and N. L. Bassuk. 1991. Stem banding enhances rooting and subsequent growth of M.9 and MM.106 apple rootstock cuttings. *HortScience* 26:1368–70.

145. Sun, W. Q., and N. L. Bassuk. 1993. Auxin-induced ethylene synthesis during rooting and inhibition of budbreak of 'Royalty' rose cuttings. *J. Amer. Soc. Hort. Sci.* 118:638–43.

146. Svenson, S. E., B. Smith, and B. Briggs. 1997. Controlling liverworts and moss in nursery production. *Comb. Proc. Intl. Plant Prop. Soc.* 47:414–22.

147. Tilt, K. M., and T. E. Bilderback. 1987. Physical properties of propagation media and their effects on rooting of three woody ornamentals. *HortScience* 22:245–47.

148. Vallis, G. 1991. The development of a market-led propagation system. *Comb. Proc. Intl. Plant Prop. Soc.* 41:134–41.

149. Van Braght, J., H. van Gelder, and R. L. M. Pierik. 1976. Rooting of shoot cuttings of ornamental shrubs after immersion in auxin-containing solutions. *Scientia Hort.* 4:91–4.

150. Van de Pol, P. A., and J. V. M. Vogelezang. 1983. Accelerated rooting of carnation 'Red Baron' by temperature pretreatment. *Scientia Hort.* 20:287–94.

151. Van de Pol, P. A. 1988. Partial replacement of the rooting procedure of *Chrysanthemum morifolium* cuttings by pre-rooting storage in the dark. *Acta Hort.* 226:519–24.

152. Van Sambeek, J. W., J. E. Preece, and M. V. Coggeshall. 2002. Forcing epicormic sprouts on branch segments of adult hardwoods for softwood cuttings *Comb. Proc. Intl. Plant Prop. Soc.* 52:417–24.

153. Verkade, S. D. 1986. Mycorrhizal inoculation during plant propagation. *Comb. Proc. Intl. Plant Prop. Soc.* 36:613–18.

154. Wang, Y. T., and C. A. Boogher. 1988. Effect of nodal position, cutting length, and root retention on the propagation of golden pothos. *HortScience* 23:347–49.

155. Wells, J. S. 1952. Pointer on propagation. Propagation of *Taxus*. *Amer. Nurs.* 96:13, 37–8, 43.

156. Wells, J. S. 1962. Wounding cuttings as a commercial practice. *Comb. Proc. Intl. Plant Prop. Soc.* 12:47–55.

157. Wells, R. 1986. Air layering: an alternative method for the propagation of *Mahonia aquifolium* 'Compacta'. *Comb. Proc. Intl. Plant Prop. Soc.* 36:97–9.

158. Whalley, D. N., and K. Loach. 1981. Rooting of two genera of woody ornamentals from dormant, leafless (hardwood) cuttings and their subsequent establishment in containers. *J. Hort. Sci.* 56:131–38.

159. Wilkerson, E. G., R. S. Gates, S. Zolnier, S. T. Kester, and R. L. Geneve. 2005. Transpiration capacity in poinsettia cuttings at different stages and the development of a cutting coefficient for scheduling mist. *J. Amer. Soc. Hort. Sci.* 130:295–301.

160. Wilson, P. J., and D. K. Struve. 2004. Overwintering mortality in stem cuttings. *J. Hort. Sci. Biotech.* 79:842–44.

161. Wood, T., and A. C. Cameron. 1989. Cold storage—overwintering your softwood cuttings in the refrigerator releases valuable greenhouse space. *Amer. Nurs.* 170:49–54.

162. Wott, J. A., and J. H. B. Tukey. 1967. Influence of nutrient mist on the propagation of cuttings. *Proc. Amer. Soc. Hort. Sci.* 90:454–61.

163. Zaczek, J. J., C. W. Heuser, Jr., and K. C. Steiner. 1997. Effect of shade levels and IBA during the rooting of eight tree taxa. *J. Environ. Hort.* 17:130–33.

164. Zaczek, J. J., C. W. Heuser, Jr., and K. C. Steiner. 1999. Low irradiance during rooting improves propagation of oak and maple taxa. *J. Environ. Hort.* 17:130–33.

11
Principles of Grafting and Budding

INTRODUCTION

Since the beginning of civilization, fruit and nut trees have been grafted because of the difficulty in propagating by cuttings, and the superiority and high value of the grafted crop. *Grafting* is among the most expensive propagation techniques, surpassing even micropropagation. *Budding,* which is a form of grafting, is three times more costly than cuttings and fourteen times more expensive than seedling propagation (89). The horticulture and forestry industries have sought to develop clonal propagation systems that avoid labor-intensive graftage. Yet, traditional and highly efficient grafting and budding systems are essential for the propagation of many woody plant species. New markets continue to require grafted and budded plants for improved plant quality, fruit yield, superior forms, and better adaptation to greater ecological ranges. In the southeastern United States, where high temperatures and periodic flooding of soils (low soil oxygen) are the norm, cultivars of birch, fir, oak, and other species are grafted onto adapted rootstock (Fig. 11–1) (129). The propagator benefits via new markets, while the consumer gains a greater variety of better-adapted landscape plants. The acid-loving blueberry can be produced in more basic pH soils when grafted to pH-tolerant rootstock (Fig. 11–2).

With the greater reliance on integrated pest management and reduced availability of pesticides and soil fumigants, disease-tolerant rootstocks are playing a greater role not only with woody perennial fruit crops and ornamentals, but also with grafted vegetable crops (Figs. 11–3 and 11–4, page 417) (34, 39, 67, 82, 85, 86). Organic growers of high value heirloom tomatoes are using grafted plants as a management tool to reduce crop loss from soilborne diseases (131).

This chapter reviews the biology of grafting and budding. Chapters 12 and 13 describe the techniques of grafting and budding, respectively. Chapter 19 enumerates grafting and budding systems for selected fruit and nut trees, as Chapter 20 does for selected woody ornamental plants. A better understanding of the fundamental biology of grafting (and the causes of graft incompatibility) will enhance the development of superior cultivars and increase the ecological range of species for new markets in horticulture and forestry.

THE HISTORY OF GRAFTING

The origins of grafting can be traced to ancient times (110). There is evidence that the art of grafting was known to the Chinese at least as early as 1560 BC. Aristotle (384–322 BC) and Theophrastus (371–287 BC) discussed grafting in their

learning objectives

- Describe the role of grafting in human history.
- Distinguish between the use of seedling and clonal rootstocks.
- Describe how natural grafting can affect tree performance.
- Describe how the rootstock and scion heal together during grafting.
- Define how specific genetic, environmental, and management factors and polarity affect graft success.
- Determine what kinds of plants can be grafted.
- Define graft incompatibility—its symptoms, causes, and control.
- Describe important ways the rootstock (root system) influences the scion (shoot system) and vice versa.

Figure 11–1
Cleft-grafted-variegated English Holly on *Ilex* 'Nellie Stevens' rootstock adapted to the high temperature, periodic flooding, low oxygen soils of the southeastern United States.

Figure 11–2
Pushing the ecological envelope. Using an inlay bark graft of 'Tif Blue' blueberry (*Vaccinium ashei*) on a farkelberry (*Vaccinium arboreum*) rootstock, which tolerates a more basic soil pH, allows the acid-loving blueberry to be produced in a site with higher soil pH. (a) New scion growth with aluminum foil and poly bag protecting the graft area. (b) Healed graft union, and (c) 'Tif Blue' blueberry crop.

Figure 11–3
Grafting vegetables is a common practice in Japan, Korea, the Mediterranean basin, and Europe. It is used for managing soil-borne diseases, enhancing tolerance of low temperature and salinity, and for increasing plant vigor and yield. (a) Grafted melon scion on *curcurbita* rootstock with a grafting clip. (b) Melons grafted (white arrow) on *Fusarium*-resistant *Curcurbita* rootstock in Israel, (b) compared to susceptible, non-grafted melons (black arrows). Courtesy M. Edelstein.

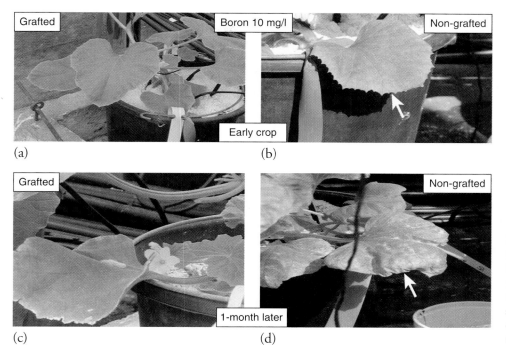

Figure 11–4
(a and c) Melon grafted onto boron-resistant *Cucurbita* rootstock. (b) Non-grafted melon showing boron susceptibility early in crop cycle and (d) 1 month later.
Photos courtesy M. Edelstein.

writings with considerable understanding. During the days of the Roman Empire, grafting was very popular, and methods were precisely described in the writings of that era. Paul the Apostle, in his Epistle to the Romans, discussed grafting between the "good" and the "wild" olive trees (Romans 11:17–24).

The Renaissance period (AD 1350–1600) saw a renewed interest in grafting practices. Large numbers of new plants from foreign countries were imported into European gardens and maintained by grafting. By the 16th Century, the cleft and whip grafts were widely used in England and it was realized that the cambium layers must be matched, although the nature of this tissue was not then understood or appreciated. Propagators were handicapped by a lack of a good grafting wax; mixtures of wet clay and dung were used to cover the graft unions. In the 17th Century, orchards in England were planted with budded and grafted trees.

Early in the 18th Century, Stephen Hales, in his studies on the "circulation of sap" in plants, approach-grafted three trees and found that the center tree stayed alive even when severed from its roots. Duhamel studied wound healing and the uniting of woody grafts. The graft union at that time was considered to act as a type of filter that changed the composition of the sap flowing through it. Thoüin (163), in 1821, described 119 methods of grafting and discussed changes in growth habit resulting from grafting. Vöchting (171), in the late 19th Century, continued Duhamel's earlier work on the anatomy of the graft union. Development of some of the early grafting techniques have been reviewed by Wells (178).

Liberty Hyde Bailey in *The Nursery Book* (8), published in 1891, described and illustrated the methods of grafting and budding commonly used in the United States and Europe at that time. The methods used today differ very little from those described by Bailey.

TERMINOLOGY

Grafting is the art of joining two pieces of living plant tissue together in such a manner that they will unite and subsequently grow and develop as one composite plant. As any technique that will accomplish this could be considered a method of grafting, it is not surprising that innumerable procedures for grafting are described in the literature. Through the years, several distinct methods have become established that enable the propagator to cope with almost any grafting problem. These are described in Chapter 12 with the realization that there are many variations of each, and that there are other forms that can give similar results. Figure 11–5 illustrates a grafted plant and the parts involved in the graft.

grafting The union of a root system (**understock**) with a shoot system (**scion**) in such a manner that they subsequently grow and develop as one **composite** (compound) plant.

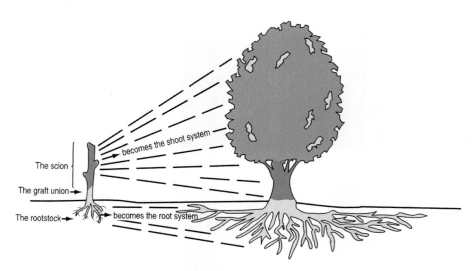

Figure 11–5
In grafted plants the shoot system consists of growth arising from one (or more) buds on the scion. The root system consists of an extension of the original rootstock. The graft union remains at the junction of the two parts throughout the life of the plant.

budding A form of grafting that uses a smaller scion piece—sometimes just a piece of the stem with an axillary bud.

Budding is a form of grafting. However, the scion is reduced in size and usually contains only one bud. An exception to this is patch budding of pecan, where secondary and tertiary buds are adjacent at the same node to the primary bud. The various budding methods are described in Chapter 13.

The **scion** becomes the new shoot system of the graft. It is composed of a short piece of detached shoot containing several dormant buds, which, when united with the rootstock, comprises the upper portion of the graft. The stem, or branches, or both, of the grafted plant will grow from the scion. The scion should be of the desired cultivar and free from disease.

The **rootstock (understock, stock)** is the lower portion of the graft, which develops into the root system of the grafted plant. It may be a seedling, a rooted cutting, a layered or micropropagated plant. If the grafting is done high in a tree, as in topworking, the rootstock may consist of the roots, trunk, and scaffold branches.

The **interstock (intermediate stock, interstem)** is a piece of stem inserted by means of two graft unions between the scion and the rootstock. Interstocks are used to avoid incompatibility between the rootstock and scion, to produce special tree forms, to control disease (e.g., fire-blight resistance), or to take advantage of their growth-controlling properties.

Vascular cambium is a thin tissue located between the bark (periderm, cortex, and phloem) and the wood (xylem) (see Fig. 11–6). Its cells are meristematic; that is, they are capable of dividing and forming new cells.

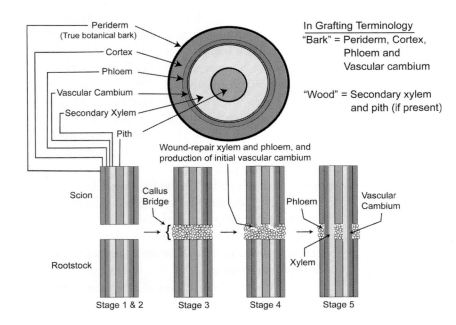

Figure 11–6
Top: Grafting terminology of the "bark" and "wood" and associated tissues with schematic drawing of a stem cross section of a young woody plant stem. *Bottom*: Schematic longitudinal section of the stages of graft union formation: (Stage 1) Lining up vascular cambiums of the rootstock and scion, and (Stage 2) subsequent wound healing response. (Stage 3) Callus bridge formation. (Stage 4) Wound-repair xylem and phloem occur in the callus bridge just prior to initial cambium formation. (Stage 5) The vascular cambium is completed across the callus bridge and is forming secondary xylem and phloem.

For a successful graft union, it is essential that the cambium of the scion be placed in close contact with the cambium of the rootstock.

callus Tissue composed of parenchyma cells, which is a response to wounding. Callus development is important in graft union formation.

Callus is a term applied to the mass of parenchyma cells that develop from and around wounded plant tissues. It occurs at the junction of a graft union, arising from the living cells of both the scion and rootstock. The production and interlocking of these parenchyma (or callus) cells constitute one of the important steps in callus bridge formation between the scion and rootstock in a successful graft.

SEEDLING AND CLONAL ROOTSTOCK SYSTEMS

Rootstocks can be divided into two groups: seedling and clonal.

Utilization and Propagation of Seedling Rootstock

Seedling rootstocks propagated from seed can be mass-produced relatively simply and economically. Viruses are transmitted from parent to progeny in very low percentages or not at all except in specific instances. Seedling plants tend to have deeper rooted and more firmly anchored plants than rootstocks grown from cuttings (e.g., plum and apple rootstock).

Seedling rootstock may show genetic variation leading to variability in growth and performance of the scion variety. The variation can arise from natural heterozygosity of the source or from cross-pollination—both are more likely if the rootstock is from an unknown, unselected source. Selection of special mother-tree (elite) seed source trees or a special clone can provide uniform, special seedling rootstocks for specific crops (see Chapter 5).

Uniformity of seedling variability can be controlled by managing production conditions in the nursery, including digging nursery trees of the same age, one row at a time, and discarding off-type or slow-growing seedlings or budded trees. In most nurseries, the young trees are graded by size, and all those of the same grade are sold together. Many fruit crops grown on uniform seedling rootstocks show no more variability resulting from the rootstock than from unavoidable environmental differences in the orchard—principally soil variability.

Utilization and Propagation of Clonal Rootstock

Clonal rootstocks are those vegetatively propagated by stool layering, rooted cuttings, or micropropagation. Micropropagation of clonal rootstocks makes possible the production of great numbers of such plants, upon which the scion cultivar can be grafted or budded (76, 77). Rootstock of citrus is produced from apomictic seed and is genetically uniform; this is a more cost-effective method of propagating clonal rootstock than traditional asexual techniques.

Clonal rootstocks are desirable not only to produce uniformity, but also to utilize special characteristics such as disease resistance. Clonal rootstock also influence the size and growth habit of the grafted plant and flowering and fruit development of the scion. Each particular scion-rootstock combination requires an extensive evaluation period in different environments before its future performance can be predicted.

Historically, clonal rootstocks for fruit crops received much attention in European and Middle Eastern countries, going back centuries. Today, much of the apple production around the world is on clonal rootstocks for size control and fruit yield. Other fruit crops, such as pear, quince, plum, cherries, grapes, citrus, and others are routinely propagated on clonal rootstock (179).

Only pathogen-free scions and rootstock material should be utilized in the nursery. To maintain rootstock influence, deep planting of the nursery tree or grafted vegetable—which may lead to **"scion rooting"**—must be avoided, as illustrated in Figure 11–7. The deeper the graft union below the soil surface, the higher the incidence of scion rooting is likely to be (31).

REASONS FOR GRAFTING AND BUDDING

Grafting and budding serve many different purposes:

- Perpetuating clones desired for their fruiting, flowering, or growth characteristics that cannot be readily maintained or economically propagated by other asexual means
- Combining different cultivars into a composite plant as scion, rootstock, and interstock—each part providing a special characteristic
- Changing cultivars of established plants (topworking), including combining more than one scion cultivar on the same plant
- Repairing graftage for injuries—including inarching and bridge graftage

Figure 11–7
An incompatible graft with the melon scion forming adventitious roots above the grafted *Cucurbita* rootstock. The melon will establish its own roots above graft, which is not desirable. Courtesy M. Edelstein.

- Disease indexing—testing for virus diseases
- Study of plant developmental and physiological processes

Each of these reasons is discussed in detail in the following pages.

Perpetuating Clones Desired for Their Fruiting, Flowering, or Growth Characteristics That Cannot Be Readily Maintained or Economically Propagated by Other Asexual Means

Cultivars of some groups of plants, including most fruit and nut species and many other woody plants, such as selected cultivars of fir, eucalyptus, beech, oak, and spruce, are not propagated commercially by cuttings because of poor rooting. Additional individual plants often can be started by the slow and labor-intensive techniques of layering or division. But for propagation in large quantities, it is necessary to resort to budding or grafting scions of the desired cultivar on compatible seedling rootstock plants.

In forestry, grafting is used almost exclusively for the clonal production of genetically improved seed orchards of Monterey pine (*Pinus radiata*), hoop pine (*Araucaria cunninghamii*), slash pine (*P. elliottii*), Caribbean pine (*P. caribaea*), eucalyptus (*Eucalyptus nitens*), Douglas-fir (*Pseudotsuga menziesii*), and others (120). The major advantage of using grafts is that superior germplasm from older, elite trees can be clonally regenerated as parent trees for seed orchards. Frequently, trees selected for breeding or seed orchard purposes are so old (often greater than 15 or 20 years) that clonal production by rooted cuttings is either impossible or far more costly than grafting. Where graft incompatibility is not a serious problem, grafting scions of elite trees onto established seedling rootstock is a quick, straightforward, and cost-effective way of developing seed orchards.

Combining Different Cultivars into a Composite Plant as Scion, Rootstock, and Interstock—Each Part Providing a Special Characteristic

Obtaining the Benefits of Certain Scions Grafting selected cultivars can enhance plant growth rates, fruit characteristics and yield, and plant form. "Weeping" forms of landscape plants can be obtained by grafting (Fig. 11–8). Cactus and succulents are easily grafted to produce unusual plant forms, as shown in Figure 11–9.

Obtaining the Benefits of Certain Rootstocks There are a number of benefits of grafting onto selected rootstock, including greater plant resistance to biotic and

Figure 11–8
"Weeping" plant forms may be obtained by grafting. Rootstock of an upright willow is grafted at the top by a side graft with another cultivar having a hanging growth pattern.

Figure 11–9
Grafted ornamental (a) cactus and (b) succulents. An easily rooted cultivar is used as the rootstock and an unusual attractive type is used as the scion. These grafts are made in large quantities in Japan and Korea, and shipped to wholesale nurseries in other countries for rooting, potting, and growing until ready for sale in retail outlets.

abiotic stress
A condition caused by environmental factors such as drought, low temperature, low oxygen, and salinity, which reduce growth and can sometimes kill plants.

biotic stress
A condition caused by living organisms such as insects, pathogens, and nematodes that reduce growth and can sometimes kill plants.

abiotic stress, size control, enhanced reproductive growth, reduction in nursery production time, and increased transplanting success.

Greater Resistance to Environmental Stress and Disease. For many kinds of plants, rootstocks are available that tolerate unfavorable **abiotic stress** conditions—such as heavy, wet soils, salinity, and drought (Figs. 11–1, 11–2, and 11–4) (47, 124–126, 129). Other rootstocks may resist **biotic stresses** such as soil-borne insect, nematodes, viruses, or pathogens (34, 86) better than the plant's own roots (Fig. 11–3). See Chapters 19 and 20 for detailed discussions of the rootstocks available for the various fruit and ornamental species. Special rootstocks for glasshouse, poly-covered high tunnel production and field production of vegetable crops are used in Europe, the Middle East, Asia, and North America to avoid root diseases such as *Monosporascus, Fusarium* and *Verticillium* wilt (34, 67, 131). In the Netherlands, greenhouse cucumbers are grafted onto *Cucurbita ficifolia*, and commercial tomato cultivars are grafted onto vigorous F_1 hybrid, disease-resistant rootstocks (21).

Controlling Size of Grafted Plant. For some species, size-controlling rootstocks are available that can cause the composite grafted plant to have exceptional vigor or to become dwarfed (Fig. 11–10). Scions grafted onto selected rootstock of some citrus, pear, and apple rootstocks produce larger size and/or better-quality fruit than when grafted onto other rootstock (179).

Hastening Reproductive Maturity. Scions of many fruit crops can be established more quickly in the orchard and come into bearing more rapidly when grafted onto dwarfing rootstock (169), as opposed to being grown as seedlings or as rooted cuttings. (An exception to this is peach production in Mexico, where very vigorous seedlings are selected for fruit production—seedling plants fruit as rapidly as grafted plants.) It is also possible to hasten the onset of maturity by grafting cultivars onto larger, established trees. Such grafting takes advantage of an existing large root system of the rootstock plant to speed up maturation of the scion.

Hastening Plant Growth Rate and Reducing Nursery Production Time. In nursery production of shade trees, budded or grafted trees grow more rapidly than seedling or cutting-produced trees; for example, *Acer platanoides* 'Crimson King' budded on a vigorous rootstock (see Fig. 13–4), and budded *Tilia cordata* or budded *Zelkova serrata* grow more in 1 year than rooted cuttings will in 3 or 4 years (53).

Improving Transplanting Success. Some plants rooted by cuttings make such poor root systems that they are difficult or impossible to transplant; for example, the Koster spruce (*Picea pungens*) can be rooted in commercial numbers, but cannot be successfully transplanted unless the root system is produced from grafted plants (53). Many Asiatic maples form poor root systems from cuttings and must be grafted (170).

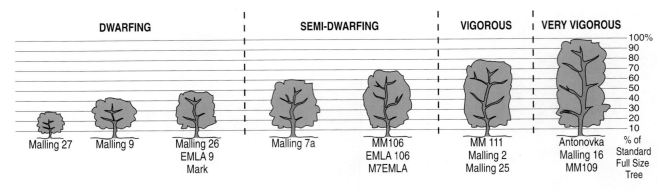

Figure 11–10
Relative size of apple trees on different rootstock. The reduction in tree size ranges from dwarfing (25 to 50 percent of a standard full-size tree) to semi-dwarfing (60 to 70 percent) to vigorous to very vigorous (same size as a seedling tree). With the exception of Antonovka, all listed are clonal rootstock. The *absolute size* of the mature, composite tree is determined by soil, climate, culture, and the vigor of the scion cultivar (e.g., the scions of the vigorous cultivar 'Mutsu' are twice as large as 'Golden Delicious' on 'Malling 9' dwarfing rootstock).

Obtaining the Benefits of Certain Interstocks (Double-Working) In addition to the rootstock and scion, one may insert a third plant system between them by grafting. Such a section is termed an **interstock, interstem, intermediate stock,** or **intermediate stem section.** This is done by making two grafts (see Fig. 12–50), or double budding. For example, a thin plate (minus the bud) of 'Old Home' pear interstock is budded on the quince rootstock, then a shield bud of the 'Bartlett' scion is inserted directly over the 'Old Home' plate and wrapped with a budding rubber (see Fig. 13–21).

double-working The grafting or budding of an interstock (interstem) between the rootstock and scion.

There are several reasons for using **double-working** in propagation:

- The interstock makes it possible to avoid certain kinds of incompatibility.
- The interstock may possess a particular characteristic (such as disease resistance or cold-hardiness) not possessed by either the rootstock or the scion.
- A certain scion cultivar may be required for disease resistance in cases where the interstock characteristics are the chief consideration, such as in the control of leaf blight on rubber trees (*Hevea*) (84).
- The interstock may reduce vegetative growth and enhance reproductive growth of the tree. For example, when a stem piece of the dwarfing 'Malling 9' apple rootstock is used as an interstock and inserted between a vigorous rootstock and a vigorous scion cultivar, it reduces growth of the composite tree and stimulates flowering and fruiting in comparison with a similar tree propagated without the interstock [Fig. 11–11 (132).]

- Obtaining special forms of plant growth. By grafting certain combinations together it is possible to produce unusual types of plant growth, such as "tree" roses (Fig. 11–12) or "weeping" cherry, birch, or willow cultivars (Fig. 11–8).

Nurseries supplying trees on seedling or clonal rootstocks, or with a clonal interstock, should identify such stocks on the label just as they do for the scion cultivar.

Changing Cultivars of Established Plants (Topworking)

A fruit tree, or an entire orchard, may be replaced with a more desirable cultivar. It could be unproductive, or an old cultivar whose fruits are no longer in demand; it could be one with poor growth habits, or possibly one that is susceptible to prevalent diseases or insects. **Topworking** has sometimes been done by California pro-

topworking The grafting of a new cultivar onto established trees in the orchard.

ducers of peach, plum, and nectarine every 2 to 3 years to take advantage of newer, more promising cultivars and thus remain competitive on the market. Examples of topworking are shown in Figure 11–13, page 424.

In an orchard of a single cultivar of a species requiring cross-pollination, provision for adequate cross-pollination can be obtained by topworking scattered trees throughout the orchard to a proper pollinating cultivar. A single pistillate (female) plant of a dioecious (pistillate and staminate flowers borne on separate individual plants) species, such as the hollies (*Ilex*), may be unfruitful because of the lack of a nearby staminate (male) plant to provide proper pollination. This problem can be

Figure 11–11
Effect of interstock on the size of six-year-old 'Cox's Orange Pippin' apple scion grafted on a vigorous 'MM 104' rootstock: (a) Cox/'M 9' dwarfing interstock/'MM 104', (b) Cox/'M 27' dwarfing interstock/'MM 104', (c) Cox/'MM 104' vigorous interstock/'MM 104', (d) Cox/'M 20' dwarfing interstock/'MM 104'.

Figure 11–12
Double-working. (a) Used in the production of specialty "tree" roses, where the interstock (arrow) of *Multiflore de la Grifferaie* forms the straight trunk of the tree rose. (b) Doubleworking citrus in Sicily with micrografted scion grafted on Troyer citrange interstock (arrow) grafted onto sour orange rootstock.

corrected by grafting a scion taken from a staminate plant onto one branch of the pistillate plant.

The home gardener may be interested in growing several cultivars of a fruit species together on a single tree of that species by topworking each primary scaffold branch to a different cultivar. In a few cases, different species can be worked on the same tree. For example, a single citrus tree would grow oranges, lemons, grapefruit, mandarins, and limes; or plum, almond, apricot, and nectarine can be grafted on peach rootstock. Some different cultivars (or species), however, grow at different degrees of vigor, so careful pruning is required to cut back the most vigorous cultivar on the tree to prevent it from becoming dominant over the others.

Walnut and pistachio are difficult to transplant. Producers will plant seedling rootstock in the orchard and then graft 2 years later.

Repair Graftage for Injuries

Occasionally, the roots, trunk, or large limbs of trees are severely damaged by winter injury, cultivation implements, diseases, or rodents. By the use of bridge

Figure 11–13
TopWorking. (a) Inlay bark graft in top working an orchard. (b) Top worked citrus grove in Sicily using an inlay bark graft. (c) Smaller citrus liner with inlay bark graft.

grafting, or inarching, such damage can be repaired and the tree saved. This is discussed in detail in Chapter 12.

Disease Indexing—Testing for Virus Diseases

Virus diseases can be transmitted from plant to plant by grafting. This characteristic makes possible testing for the presence of the virus in plants that may carry the pathogens but show few or no symptoms. By grafting scions or buds on a plant suspected of carrying the virus onto an indicator plant known to be highly susceptible, and which shows prominent symptoms, detection is easily accomplished. This procedure is known as indexing (see Chapter 16).

In order to detect the presence of a latent virus in an asymptomatic carrier, it is not necessary to use combinations that make a permanent, compatible graft union. For example, the 'Shirofugen' flowering cherry (*Prunus serrulata*) is used to detect viruses in peach, plum, almond, and apricot. Cherry does not make a compatible union with these species, but a temporary, incompatible union is a sufficient bridge for virus transfer (see Fig. 16–30).

Thermotherapy Thermotherapy is a heat treatment used to rid scion material of viruses (see Fig. 16–34). After the virus-free material is indexed, as indicated previously, or tested with serological techniques, it can be multiplied by traditional grafting/budding techniques. Micrografting under aseptic tissue culture conditions is another technique used to clean up viruses and bacterial problems with budwood (112).

Study of Plant Developmental and Physiological Processes

Grafting has enabled plant biologists to study unique physiological and developmental processes, beginning in the 1700s with Stephen Hale's studies on the circulation of plant sap. Grafting has been used successfully to study transmissible factors (98) in flowering (42), tuber initiation, the control of branching (20), and promotion of cold-hardiness between induced and noninduced organs. The use of multiple graft combinations, including reciprocal and autografting, has facilitated studies on promoters and inhibitors in adventitious rooting (57), root regeneration potential, and rejuvenation of mature phase plant material (119).

NATURAL GRAFTING

Occasionally, branches become naturally grafted together following a long period of being pressed together without disturbance. In commercial orchards, limbs of fruit trees are sometimes deliberately "braced" together and allowed to naturally graft, forming a stronger scaffold system to better support the fruit load of the tree (see Fig. 12–34).

Natural grafting of roots is not as obvious but is more significant and widespread, particularly in stands of forest species of pine, hemlock, oak, and Douglas-fir (59, 97). Such root grafts are common between roots of the same tree or between roots of trees of the same species. Grafts between roots of trees of different species are rare. In the forest, living stumps sometimes occur, kept alive because their roots have become grafted to those of nearby intact, living trees, allowing the exchange of nutrients, water, and metabolites (95, 97).

The anatomy of natural grafting of aerial roots has been studied (128). Natural root grafting also permits transmission of fungi, viruses, and phytoplasmas from infected trees to their neighbors (128). This problem can occur in orchard and nursery plantings of trees and in urban shade tree sites where numerous root grafts may result in the slow spread of pathogens throughout the planting. Natural root grafting is a potential source of error in virus-indexing procedures where virus-free and virus-infected trees are grown in close proximity (60). In addition, fungal pathogens causing oak wilt and Dutch elm disease can be spread by such natural root connections.

FORMATION OF THE GRAFT UNION

A number of detailed studies have been made of graft union formation, with woody (9, 11, 35, 49, 133, 156, 168) and herbaceous plants (52, 91, 101, 105, 123, 152, 159, 164, 188). Just as ***de novo* meristems** are necessary for adventitious bud and root formation, a *de novo*-formed meristematic area (new vascular cambium) must develop between the scion and rootstock if successful graft union formation is to occur (188). The parts of the graft that are originally prepared and placed in close contact do not themselves move about or grow together. Rather, the union is accomplished entirely by cells that develop after the actual grafting operation has been made. The graft union is initially formed by rapidly dividing callus cells, originating from the scion and rootstock, which later differentiate to form the vascular cambium (a lateral meristem) and the associated vascular system.

de novo meristems
New meristematic areas initiated from parenchyma cells such as the vascular cambium that must develop in the callus bridge of a grafted plant.

The development of a compatible graft is typically comprised of three major events: **adhesion** of the rootstock and scion, proliferation of callus cells at the graft interface or **callus bridge,** and **vascular differentiation** across the graft interface (106).

The scion will not resume its growth successfully unless a vascular connection has been established so that it may obtain water and mineral nutrients. Likewise, degeneration of the rootstock will occur if the phloem in the graft union is disrupted from sending carbohydrates and other metabolites from the scion to the root system. In addition, the scion must have a terminal meristematic region—a bud—to resume shoot growth and, eventually, to supply photosynthate to the root system.

Considering in more detail the **steps involved in graft union formation** (Figs. 11–6 and 11–14), the first one listed below is a preliminary step, but nevertheless, it is essential, and one over which the propagator has control.

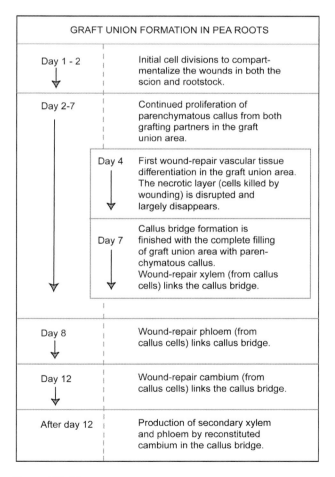

Figure 11–14
Graft union formation in grafted pea roots (91, 159). This sequence of grafting events is common to topgrafting and root grafting in many other woody and herbaceous plant species. What will vary is the time period in grafting events with different species.

1. **Lining Up Vascular Cambiums of the Rootstock and Scion**

   The statement is often made that successful grafting requires that the cambium layers of rootstock and scion must "match." Although desirable, it is unlikely that complete matching of the two cambium layers occurs since they are only one to several cell layers thick. In fact, it is only necessary that the cambial regions be close enough together so that the parenchyma cells from both rootstock and scion produced in this region can become interlocked. In a **mismatched rootstock and scion,** where one partner has a greater diameter than the other, lining up the periderm on at least one side of the rootstock and scion generally assures that their vascular cambia are close enough to interconnect through the callus bridge (see Figs. 12–5 and 13–6). The cambium is critical for maintaining vascular connections in the callus bridge.

   Two badly matched cambial layers may delay the union or, if extremely mismatched, prevent the graft union from taking place, leading to graft failure (152). With vanilla, which is a herbaceous, monocotyledonous plant, the cambium layer is not necessarily required for forming the graft union, since any parenchyma cells capable of dividing will produce callus tissue and lead to the formation of a union between the rootstock and scion (111). However, a continuous cambium layer in the graft union is necessary for successful graft union formation with woody perennial angiosperms and gymnosperms.

   It is essential that the two original graft components be held together firmly by some means, such as wrapping, tying, stapling, or nailing, or better yet, by wedging (as in the cleft graft, or machine-notched chip budding)—so that the parts will not move about and dislodge the interlocking parenchyma cells after proliferation has begun.

2. **Wounding Response**

   A **necrotic layer** or **isolation layer** forms from the cell contents and cell walls of the cut scion and rootstock cells. Cells are killed at the cutting of the scion and rootstock at least several cell layers deep. Much of the necrotic layer material later disappears, or it may remain in pockets between subsequently formed callus produced by actively dividing parenchyma cells. Undifferentiated **callus tissue** is produced from uninjured, rapidly dividing parenchyma cells (adjacent and internal to the necrotic layer). The callus tissue initially forms a **wound periderm.**

3. **Callus Bridge Formation**

   Callus formation is a prerequisite for successful graft union formation. New parenchymatous callus proliferates in 1 to 7 days from both the rootstock and scion (Figs. 11–6, 11–14, 11–15, and 11–16) (164, 168). The callus tissue continues to form by further cell divisions of the outer layers of undamaged parenchyma cells [in the *cambial region, cortex, pith* (159)—or *xylem ray parenchyma* (9)] in the scion and rootstock. The actual cambial tissue plays a lesser role in callus formation of the wound periderm and callus bridge formation than originally supposed (146, 159). New parenchyma cells produced are adjacent and internal to the necrotic layer; soon they intermingle and interlock, filling up the spaces between scion and stock (Fig. 11–17, page 428).

   In grafting scions on larger, established rootstocks (e.g., topworking in the field), the rootstock produces most of the callus. However, when the graft partners are of equal size, the scion forms much more callus than does the rootstock (35, 159, 164). This difference is explained by natural polarity, since the root-tip–facing end of the scion (proximal end) forms more callus than the shoot-tip–facing (distal end) (see Fig. 11–26, page 437) (24). In budding, the sizes of the cut surfaces are so different that it is difficult to distinguish which grafting partner contributes the most callus (28).

   Adhesion between cells of the scion and rootstock is aided by "cement" or binding material, which projects in a beadlike manner from the surface of the callus cells of both grafting partners. A general fusion of the cell walls then follows (9, 72, 159). The beadlike projections are a mix of pectins, carbohydrates, and proteins (96). The cells do not need to divide to produce the cement, and the cement can bond the graft partners, regardless of the absence or presence of the necrotic layer (159).

   It is not clear if a specific cell-to-cell recognition in grafting is required as part of adhesion and the events that follow successful graft union formation. The formation of superimposed sieve areas and sieve plates (in phloem sieve elements), pits and perforation plates (in xylem elements), and the **plasmodesmata** (in vascular parenchyma) may require cellular

   > plasmodesmata Minute cytoplasmic threads that extend through openings in cell walls and connect the protoplasts of adjacent living cells at the graft interface.

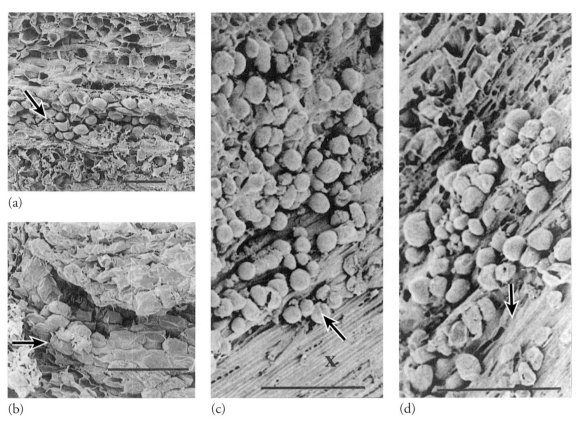

Figure 11–15
Early callus bridge formation in Sitka spruce (*Picea sitchensis*). (a) Scanning electron photos of a cross section of rootstock wound surface at seven days with a cluster of callus cells (arrow) formed in the cortical region. (b) Scion wound surface at seven days with callus cells (arrow) associated with the needle trace (nt) in the outer cortex. (c) Rootstock wound surface of a nine-day-old graft with well-established callus originating from ray cells in the xylem (x) close to the cambium (arrow). (d) Scion wound surface of a nine-day-old graft showing callus formation mainly in the cambium region. Callus is also produced from ray cells in the xylem (arrow), and from phloem parenchyma cells. Courtesy of J. R. Barnett (9).

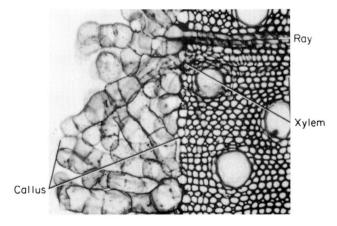

Figure 11–16
Callus production from incompletely differentiated xylem, exposed by excision of a strip of bark. x120. Photo courtesy K. Esau.

recognition or cellular communication (101). For cell recognition, the pectin fragments during the adhesion process may act as signaling molecules. Cell recognition is discussed later in the section on graft compatibility-incompatibility.

Underneath the necrotic layer, parenchyma cells show an increase in **cytoplasmic activity** with, in some plants at least, a very pronounced accumulation of **dictyosomes** along the graft interfaces (Fig. 11–18) (101,

dictyosomes A series of flattened plates or double lamellae that accumulate along the graft interface—one of the component parts of the Golgi apparatus. They secrete materials into the cell wall space between the graft components via vesicle migration to the plasmalemma.

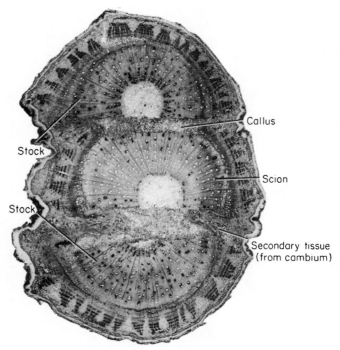

Figure 11–17
Cross section of a *Hibiscus* wedge graft showing the importance of callus development in the healing of a graft union. Cambial activity in the callus has resulted in the production of secondary tissues that have joined the vascular tissues of the stock and scion ×10. Photo courtesy K. Esau.

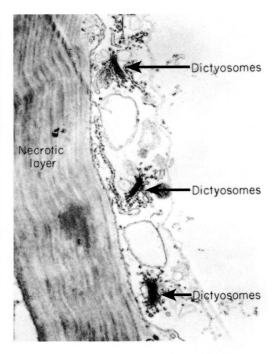

Figure 11–18
Accumulation of dictyosomes along the cell walls adjacent to the necrotic layer at six hours after grafting in the compatible autograft in *Sedum telephoides* ×17,500. Courtesy R. Moore and D. B. Walker (101).

105, 106). These dictyosomes appear to secrete materials into the cell wall space between the graft components via vesicle migration to the **plasmalemma,** resulting in a rapid adhesion between parenchymatous cells at the graft interface.

4. **Wound-Repair Xylem and Phloem: Differentiation of Vascular Cambium Across the Callus Bridge**
   In both woody and herbaceous plants, the initial xylem and phloem are generally differentiated prior to the bridging of vascular cambium across the callus bridge (Figs. 11–6 and 11–14) (35, 56, 159).

BOX 11.1 GETTING MORE IN DEPTH ON THE SUBJECT
WOUNDING RESPONSE

Some literature refers to a "wound healing response" (25), "wound healing process" (33), or "healing of the graft union." A wounded area of a plant is not healed per se by the replacement of injured tissues; rather, it is compartmentalized or walled-off from the rest of the plant as a defensive mechanism to eliminate invasion of pathogens, and so on (139, 142, 150, 151). This is all part of the *response* or *reaction* to wounding, which occurs in grafting, budding, or the propagation of a cutting. A *necrotic plate* or *isolation layer* at the graft interface is first formed, which helps adhere the grafted tissues together, especially near the vascular bundles (164). Wound repair occurs by meristematic activity, which results in the initial formation of a wound periderm between the necrotic layer and uninjured tissue—the wound periderm becomes suberized to further reduce pathogen entry (33). In grafting, the close physical contact of scion and stock cells, and pressure exerted on the graft union area from the scion and rootstock tied or wedged together prevents the necrotic layer from forming a barrier to graft union formation. Profuse callusing causes the majority of the necrotic layer to disappear (in most situations) (159, 164). Further meristematic activity occurs in graft union formation, culminating with the formation of a vascular cambium in the callus bridge area.

BOX 11.2 GETTING MORE IN DEPTH ON THE SUBJECT
SYMPLASTIC AND APOPLASTIC CONNECTIONS BETWEEN THE SCION AND ROOTSTOCK

In the callus bridge, parenchyma cells of the graft partners are interconnected by plasmodesmata (72, 104); these cytoplasmic strands form continuous, *symplastic cell connections,* linking cell membranes that form a potential pathway of communication among cells in the graft bridge. This pathway may be important in cell recognition and compatibility/incompatibility response, which is discussed later. *Apoplastic connections* occur during adhesion of the graft with cell walls of both graft partners coming together and adhering by means of their extracellular pectin-containing beads.

In a compatible graft, the wound response is followed by dissolution of the necrotic layer, perhaps as a prerequisite to the formation of secondary plasmodesmata between cells of the graft partners (164).

The secondary plasmodesmata are formed *de novo* across the fused callus walls, particularly near cut vascular strands (80). In the *de novo* formation of plasmodesmata, development of continuous cell connections starts with the thinning and loosening of local wall regions, opening the chance of fusion of *plasmalemma* (cell or protoplast membrane) and **endoplasmic reticulum** between the adjoining cells (80). Golgi vesicles bud off from individual dictyosomes and secrete cell wall material as part of this process (Fig. 11–19). Sieve elements in the connecting phloem of the grafting partners are also interconnected, further demonstrating *symplastic connections* between the graft partners (100).

endoplasmic reticulum (ER)   A membrane system that divides the cytoplasm into compartments and channels. Rough ER is densely coated with ribosomes, whereas smooth ER has fewer ribosomes.

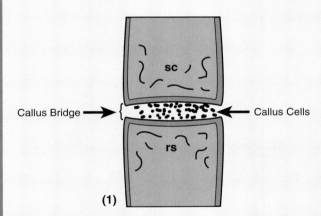
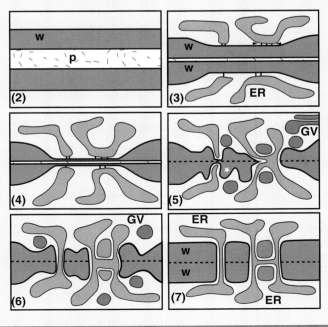

Figure 11–19
Schematic diagram of secondary (*de novo*) formation of plasmodesmata at the graft interface (callus bridge). (1) Approaching callus cells of scion (sc) and rootstock (rs). Pectic material (p) between adjoining callus cell walls. Region between arrows: wall parts where secondary plasmodesmata will be formed, as shown in detail. Formation of *continuous* cell connections (2 to 7) by plasmalemma and endoplasmic reticulum (ER) fusion of adjoining cells (5, 6) within wall parts that have been thinned synchronously with both cell partners. Elongation of the branched and single strands during rebuilding the modified wall parts (6, 7). W = cell wall, GV = golgi vesicles, *new deposited wall material.
Redrawn from Kollman and Glockmann (80).

The **wound-repair xylem** (wound-type vascular elements) is generally the first *differentiated* tissue to bridge the graft union, followed by **wound-repair phloem** (Fig. 11–20). Initial xylem tracheary elements and, frequently, initial phloem sieve tubes form directly by differentiation of callus into these vascular elements. A vascular cambium layer subsequently forms between the vascular systems of the scion and rootstock.

Exceptions to this developmental sequence are in bud graftage in citrus, apple, and rose where a **wound cambium** differentiates prior to the bridging of vascular tissue, and in autografts of *Sedum* (*Crassulaceae*) where procambial differentiation occurs before vascular differentiation (101). With budding, the scion is considerably smaller and normally limited to one bud and a short shoot piece; hence, any early vascular differentiation from callus cells is probably limited by lower phytohormone levels. The vascular cambium can form independent of any xylem or phloem (28), or the cambium may differentiate between the wound-bridging xylem and phloem (159). It is important that the vascular cambium unite so that the continuity of wound-bridging xylem and phloem can be maintained, and so that secondary vascular development occurs for successful graft union formation.

> wound-repair xylem (Wound-type vascular elements) generally the first *differentiated* tissue to bridge the graft union, followed by **wound-repair phloem**.

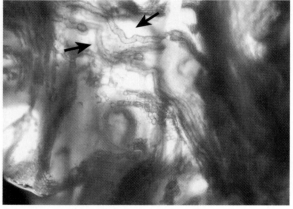
(a)

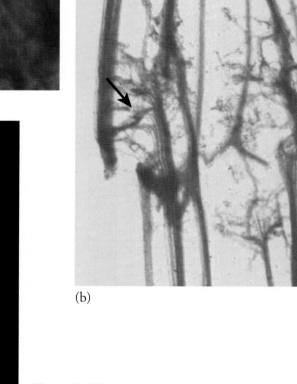
(b)

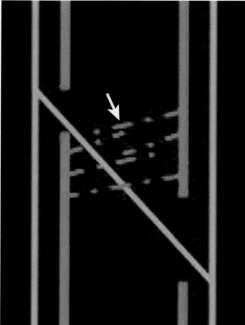
(c)

Figure 11–20
Vascular connections between melon and *Cucurbita* rootstock. (a) Early vascular strands in callus bridge area which are from wound-repair xylem and wound-repair phloem. (b) Vascular connections after 14 days. (c) Schematic of vascular connections (dotted red lines between scion and rootstock). Courtesy M. Edelstein.

BOX 11.3 GETTING MORE IN DEPTH ON THE SUBJECT
CORRELATIVE EFFECTS OF SCION BUDS AND LEAVES ON XYLEM AND PHLOEM FORMATION

The first vascular tissues produced in the callus bridge are wound-repair xylem and phloem. The new wound-repair xylem tissue originates from the activities of the scion tissues, rather than from that of the rootstock (147, 187). The amount of initial graft-bridging xylem is strongly influenced by the presence of leaves and branches on the scion, and not by the presence of the rootstock (159). The scion buds are effective in inducing differentiation of vascular elements in the tissues onto which they are grafted. Such bud influence has been shown by inserting a scion bud into a root piece of *Cichorium* rootstock. Under the influence of auxin produced by the bud, the old parenchyma cells differentiate into groups of conducting xylem elements (59).

Induction of vascular tissues in callus is under the control of phytohormones (principally auxins) and other metabolites originating from growing points of shoots (166). Auxins (IAA or NAA) will cause the induction of wound-repair xylem, while auxins and carbohydrates can induce wound-repair phloem in callus tissue (3, 115). Auxin can also induce cambial formation when applied to wounded vascular bundles of cactus rootstock (152). For successful graft union formation of *in vitro* grafted internodes, auxin is an absolute requirement, cytokinin stimulates graft development, but gibberellic acid is inhibitory (118). Auxins enhance grafting success in root-grafting pecan trees (186). In cactus grafts, auxin can also promote vascular connections (Fig. 11–21) (152).

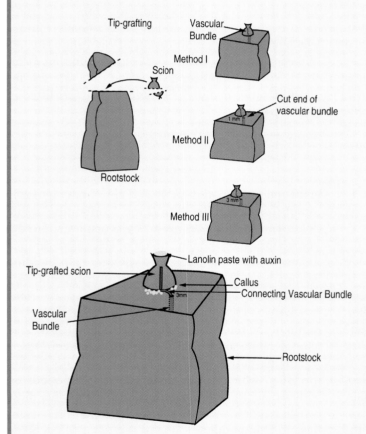

Figure 11–21
Schematic of tip grafting of cactus. *Top*: In Method I, the vascular bundles of the scion and rootstock were placed together, or 1 mm (Method II) or 3 mm (Method III) apart. *Bottom*: Auxin in lanolin paste promoted vascular connections between misaligned graft partners and increased the diameter of the connecting vascular bundle. Redrawn from Shimomura and Fuzihara (152).

At the edges of the newly formed callus mass, parenchyma cells touching the cambial cells of the rootstock and scion differentiate into new cambium cells within 2 to 3 weeks after grafting. This cambial formation in the callus mass proceeds farther and farther inward from the original rootstock and scion cambium, and on through the **callus bridge,** until a continuous cambial connection forms between rootstock and scion.

5. **Production of Secondary Xylem and Phloem from the New Vascular Cambium in the Callus Bridge**

The newly formed cambial layer in the callus bridge begins typical cambial activity, laying down new secondary xylem toward the inside and phloem toward the outside.

In the formation of new vascular tissues following cambial continuity, the type of cells formed by the cambium is influenced by the cells of the graft partners adjacent to the cambium. For example, xylem ray cells are formed where the cambium is in contact with xylem rays of the rootstock, and xylem elements where they are in contact with xylem elements (122).

Production of new xylem and phloem thus permits the vascular connection between the scion and the rootstock. It is essential that this stage be completed before much new leaf development arises from buds on the scion. Otherwise, the enlarging leaf surfaces on the scion shoots will have little or no water to offset that which is lost by transpiration, and the scion quickly will become desiccated and die. It is possible, however, even though vascular connections fail to occur, that enough translocation can take place through the parenchyma cells of the callus to permit survival of the scion. In grafts of vanilla orchid, a monocot, scions survived and grew for 2 years with only union of parenchyma cells; however, the grafted plants did not survive when subjected to transpirational stress (111).

GRAFT UNION FORMATION IN T- AND CHIP BUDDING

bark (In grafting) composed of tissues from the periderm, cortex, phloem, and vascular cambium.

wood (In grafting) composed of secondary xylem with some pith (in younger woody plants).

In T-budding, the bud piece usually consists of the **"bark"** (periderm, cortex, phloem, cambium), and often some **"wood"** (xylem tissue). Attached externally to this is a lateral bud subtended, perhaps, by a leaf petiole. In budding, this piece of tissue is laid against the exposed xylem and cambium of the rootstock, as shown diagrammatically in Figure 11–22.

Detailed studies of the grafting process in T-budding have been made for the rose (28), citrus (93, 94), and apple (108).

In the apple, when the flaps of bark on either side of the "T" incision on the rootstock are raised, separation occurs from the young xylem. The entire cambial zone remains attached to the inside of the bark flaps. Very shortly after the bud shield is inserted, a necrotic plate or layer of material develops from the cut cells. Next, after about two days, callus parenchyma cells start developing from the rootstock xylem rays and break through the necrotic plate. Some callus parenchyma from the bud scion ruptures through the necrotic area in a similar manner. As additional callus is produced, it surrounds the bud shield and holds it in place. The callus originates almost entirely from the rootstock tissue, mainly from the exposed surface of the xylem cylinder. Very little callus is produced from the sides of the bud shield (scion).

Cell proliferation continues rapidly for 2 to 3 weeks until all internal air pockets are filled with callus. Following this, a continuous cambium is established between the bud and the rootstock. The callus then begins to lignify, and isolated xylem tracheary elements appear. Lignification of the callus is completed between 5 to 12 weeks after budding (108, 172). The developmental stages and time intervals for graft union formation in T-budded citrus are listed in Box 11.4.

More Rapid Union Development in Chip Budding

Anatomical studies (155) have been made comparing graft union formation in T- and chip budding. Early union formation between 'Lord Lambourne' apple scion and 'Malling 26' dwarfing rootstock showed a more rapid and complete union of xylem and cambial tissues of the scion and rootstock after chip budding compared to T-budding. This is probably due to a much closer matching of the scion tissue to the rootstock stem (Fig. 11–22). Also in T-budding, the cambium of the rootstock is lifted in the flap of "bark," so considerable callus in-filling and development of new cambium must occur. There is more flexibility in chip budding, which can be done over longer periods on either an active or dormant rootstock, than T-budding, which requires an active rootstock. In part this advantage to chip budding is due to less callus filling being needed, and because there is no requirement for an active cambium to lift the flap of rootstock bark, as there is with T-budding.

The previously mentioned advantages of chip budding compared with T-budding have also been demonstrated with 'Crimson King' maple on *Acer platanoides* rootstock, 'Conference' pear on 'Quince A' rootstock, and 'Rubra' linden on *Tilia platyphyllos* rootstock.

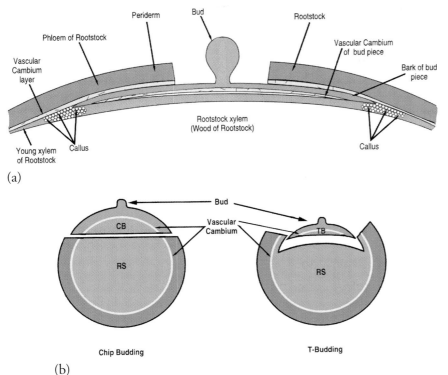

Figure 11–22
(a) Tissues involved in healing of an inserted T-bud as prepared with the "wood" (xylem) attached to the scion bud piece. Graft union formation occurs when callus cells developing from the young xylem of the rootstock intermingle with callus cells forming from exposed cambium and young xylem of the T-bud piece. As the bark is lifted on the rootstock for insertion of the bud piece it detaches by separation of the youngest xylem and cambial cells. (b) A cross section of a chip bud (CB), T-bud (TB), and rootstock (RS). Because the chip bud substitutes exactly for the part of the rootstock that is removed, the cambium of the roots and scion are placed close together, resulting in a rapid and strong union. When a T-bud (*right*) is slipped under the "bark," the cambium of the rootstock and scion are not adjacent, and the initial union formation can be weak and slow. *Redrawn from B. H. Howard (68).*

BOX 11.4   GETTING MORE IN DEPTH ON THE SUBJECT
STAGES AND TIME INTERVALS IN GRAFT UNION FORMATION OF T-BUDDED CITRUS (94)

Stage of development	Approximate time after budding
• First cell division	24 hours
• First callus bridge	5 days
• Differentiation of cambium	
a. In the callus of the bark flaps (rootstock)	10 days
b. In the callus of the shield bud (scion)	15 days
• First occurrence of xylem tracheids	
a. In the callus of the bark flaps	15 days
b. In the callus of the shield	20 days
• Lignification of the callus completed	
a. In the bark flaps	25 to 30 days
b. Under the shield	30 to 45 days

FACTORS INFLUENCING GRAFT UNION SUCCESS

As anyone experienced in grafting or budding knows, the results are often inconsistent. An excellent percentage of "takes" occur in some operations, but in others the results are disappointing. A number of factors can influence the healing of graft unions. Factors that influence graft union success include:

- Incompatibility
- Plant species and type of graft
- Environmental conditions during and following grafting

- Growth activity of the rootstock
- Polarity
- The craftsmanship of grafting
- Virus contamination, insects, and diseases
- Plant growth regulators and graft union formation
- Post-graftage—bud-forcing bethods

Incompatibility

One of the symptoms of incompatibility in grafts between distantly related plants is a complete lack, or a very low percentage, of successful unions. Incompatibility is discussed in greater detail starting on page 441. Grafts between some plants known to be incompatible, initially will make a satisfactory union, even though the combination eventually fails.

Plant Species and Type of Graft

Some plants—including hickories, oaks, and beeches—are much more difficult to graft than others even when no incompatibility is involved. Nevertheless, such plants, once successfully grafted, grow very well with a perfect graft union. In grafting apples, grapes, and pears (Figs. 11–23 and 11–24), even the simplest techniques usually give a good percentage of successful unions, but

Figure 11–23
Some species form profuse callusing (arrow), which helps increase graft union success. Pear is easily grafted by a whip-and-tongue graft.

grafting certain stone fruits, such as peaches and apricots, requires more care and attention to detail. Strangely enough, grafting peaches to some other compatible species, such as plums or almonds, is more successful than reworking them back to peaches. One method of grafting may give better results than another, or budding may be more successful than grafting, or vice versa. For example, gymnosperms are grafted, whereas many angiosperm cultivars tend to be budded,

(a)

(b)

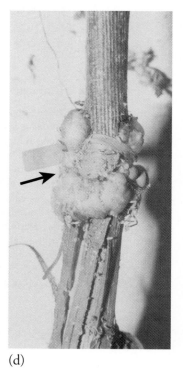

(c) (d)

Figure 11–24
A high take occurs when grapes are saddle grafted, but the same graft is unsuccessful with roses, which did not form sufficient callus. (a) Heitz saddle graft bench graft tool. (b) Unsuccessful saddle graft with rose. (c and d) Successful saddle-grafted grape with profuse callusing in the callus bridge area.

rather than grafted (19). In topworking native black walnut (*Juglans regia*) to the Persian walnut (*Juglans hindsii*) in California, the bark graft method is more successful than the cleft graft. In nursery propagation of pecans, patch budding in Texas is preferred to the whip graft, which does better in climates with higher humidity, such as Mississippi.

Some species, such as mango (*Mangifera indica*) and camellia (*Camellia reticulata*) are so difficult to propagate by the usual grafting and budding methods that they are **approach grafted** (see Fig. 12–27, page 437). Both graft partners are maintained for a time after grafting onto their own roots as containerized plants. This variation among plant species and cultivars in their grafting ability is probably related to their ability to produce callus parenchyma, and differentiate a vascular system across the callus bridge.

The genetic limits of grafting are discussed on page 439.

Environmental Conditions During and Following Grafting

Certain environmental requirements must be met for callus tissue to develop.

Temperature Compared to field grafting and budding, temperature levels for greenhouse containerized rootstock and **bench grafting** can be readily controlled, thereby permitting greater reliability of results and more flexibility of scheduling grafting and budding over a longer period of time. Temperature has a pronounced effect on the production of callus tissue (Fig. 11–25). In apple grafts, little, if any, callus is formed below 0°C (32°F) or above about 40°C (104°F). At 32°C (90°F) and higher, callus production is retarded and cell injury increases with higher temperatures. Cell death occurs around 40°C (104°F). In bench grafting, callusing may be allowed to proceed slowly for several months by storing the grafts at relatively low temperatures, 7 to 10°C (45 to 50°F), or, if rapid callusing is desired, they may be kept at higher temperatures for a shorter time. Maintaining too high a temperature in order to induce rapid callus development of bench-grafted plants can deplete needed carbohydrate reserves, which limits field survival (see Fig. 10–3) (38).

Following bench grafting of grapes, a temperature of 24 to 27°C (75 to 80°F) is about optimal; 29°C

bench grafting
A grafting procedure that is done on a bench in a protected environment with bare-root or containerized rootstock.

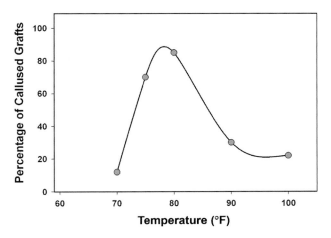

Figure 11–25
Influence of temperature on the callusing of walnut (*Juglans*) grafts. Callus formation is essential for the healing of the graft union. Maintaining an optimum temperature following grafting is very important for successful healing of walnut grafts. Adapted from data of Sitton (154).

(85°F) or higher results in profuse formation of a soft type of callus tissue that is easily injured during transplanting operations. At 20°C, callus formation is slow, and below 15°C (60°F) it almost ceases. Mango (*Mangifera indica*) is a tropical evergreen species that can be grafted year-round, provided the temperature is suitable for callusing. Optimum grafting temperature for mango is comparable to temperate-zone grape cultivars (24 to 28°C) (4). However, callusing of mango is somewhat more tolerant to high temperature than grapes [i.e., at 38°C (100°F) unions formed within 20 days, even though further high temperature exposure caused tissue injury and death of callus cells]. Conversely, mangoes are less tolerant of low temperatures—grafts failed to develop at 20°C or lower (4).

Outdoor grafting operations should thus take place at a time of year when favorable temperatures are expected and the vascular cambium is in an active state. These conditions generally occur during the spring months. Delay of outdoor grafting operations performed late in the spring (e.g., in the southern United States where excessively high temperatures may occur) often results in failure. For top-grafting walnut in California during high temperature conditions, whitewashing the area of the completed graft union promoted healing of the union. The whitewash reflected the radiant energy of the sun, which lowered the bark temperature to a more optimal level.

Moisture and Plant Water Relations The cambium of the graft partners and parenchyma cells comprising the important callus tissue are thin-walled and tender, with no provision for resisting desiccation. If exposed to drying air they will be killed. This was found to be the case in studies of the effect of humidity on the healing of apple grafts. Air moisture levels below the saturation point inhibited callus formation; desiccation of cells increased as the humidity dropped. *In vitro* studies (43) of stem pieces of ash (*Fraxinus excelsior*) have shown that callus production on the cut surfaces was markedly reduced as the water potential decreased.

Water is one of the driving forces for cell enlargement and is necessary for callus bridge formation between the stock and scion. Water must be utilized initially from scion tissue, and if below a certain water potential, insufficient water is available for callus formation. Failed grafts of well-hydrated Sitka spruce rootstocks produced no callus at the graft union, suggesting that callus formation at the cut surface is controlled or dependent on the formation of callus from the scion (10). Until vascular connections are formed between the rootstock and scion, the callus bridge provides the initial pathway for water, bypassing damaged xylem vessels and tracheids of the scion and rootstock. Within the first 3 to 4 days of callus bridge formation, there is a recovery of scion water potential (10); with maturation of the connecting tracheids, water potential and osmotic potential continue to increase (15, 16). Photosynthesis declines and does not increase until xylem connections become reestablished (18).

Unless the adjoining cut tissues of a completed graft union are kept at a very high humidity level, the chances of successful healing are poor. With most plants, thorough waxing of the graft union or sealing of the graft union with polyethylene grafting tape, Parafilm, or Buddy Tape (Aglis & Co. Ltd.) helps retain the natural moisture of the tissues, which is all that is necessary. Often root grafts are not waxed but stored in a moist (not overly wet) packing material during the callusing period. Slightly damp peat moss or wood shavings are good media for callusing, providing adequate moisture and aeration.

Growth Activity of the Rootstock

Some propagation methods, such as T-budding and bark grafting, depend on the bark "slipping," which means that the vascular cambium is actively dividing, producing young thin-walled cells on each side of the cambium. These newly formed cells separate easily from one another, so the bark "slips" (Fig. 11–22). Chip budding can be done on a dormant or active rootstock. Hence, there is much more flexibility in scheduling chip budding, because there is no requirement for an active cambium to lift the flap of rootstock bark, as with T-budding.

Initiation of cambial activity in the spring results from the onset of bud activity, because shortly after the buds start growth, cambial activity can be detected beneath each developing bud, with a wave of cambial activity progressing down the stems and trunk. This stimulus is due, in part, to production of auxin originating in the expanding buds (175). Callus proliferation—essential for a successful graft union—occurs most readily at the time of year just before and during "bud-break" in the spring, because auxin gradients diminish through the summer and into fall. Increasing callus proliferation takes place again in late winter, but this is not dependent upon the breaking of bud dormancy.

When T-budding seedlings in the nursery in late summer, it is important that they have an ample supply of soil moisture just before and during the budding operation. If they should lack water during this period, active growth is checked, cell division in the cambium stops, and it becomes difficult to lift the bark flaps to insert the bud. At certain periods of high growth activity in the spring, plants exhibiting strong root pressure (such as the walnut, maple, and grape) show excessive sap flow or "**bleeding**" when cuts are made preparatory to budding and grafting. Grafts made with such moisture exudation around the union will not heal properly. Such

> **bleeding** A process in which a plant has strong root pressure that causes excess sap flow that can reduce grafting success.

"bleeding" at the graft union can be overcome by making slanting knife cuts below the graft around the tree. Cuts should be made through the bark and into the xylem to permit such exudation to take place below the graft union. Containerized rootstock plants of *Fagus*, *Betula*, or *Acer* are relocated to a cool place with reduced watering until the "bleeding" stops. Then plants are grafted after the excessive root pressure subsides.

On the other hand, dormant containerized rootstocks of junipers or rhododendrons, when first brought into a warm greenhouse in winter for grafting, should be held for several weeks at 15 to 18°C (60 to 65°F) until new roots begin to form. Then the rootstocks are physiologically active enough to be successfully grafted.

When the rootstock is physiologically overactive (excessive root pressure and "bleeding"), or underactive (no root growth), some form of side graft can be used, in which the rootstock top is initially retained. On the other hand, **top-grafting,** in which

> **top-grafting** A form of grafting in which the shoot of the rootstock is completely removed at the time the graft is made (e.g., in-lay bark graft of pecan).

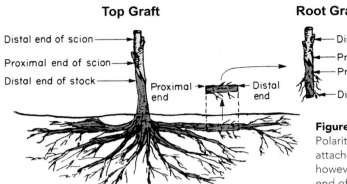

Figure 11–26
Polarity in grafting. In *topgrafting*, the proximal end of the scion is attached to the distal end of the rootstock. In *root grafting*, however, the proximal end of the scion is joined to the proximal end of the rootstock.

the top of the rootstock is completely removed at the time the graft is made, is likely to be successful in plants in which the rootstock is neither overactive nor underactive (44).

Polarity in Grafting

Distal and Proximal Ends Correct polarity is strictly observed in commercial grafting operations. As a general rule, (and as shown in Fig. 11–26), in *top-grafting*, the *proximal end* of the scion should be inserted into the *distal end* of the rootstock. But in normal *root grafting*, the *proximal end* of the scion should be inserted into the *proximal end* of the rootstock.

Should a scion be inserted with reversed polarity "upside-down," it is possible for the two graft unions to be successful and the scion to stay alive for a time (Fig. 11–27). But in bridge grafting, the reversed scion does not increase from its original size, whereas the scion with correct polarity enlarges normally (Fig. 11–28).

Nurse-Root Grafting **Nurse-root grafting** is a **temporary graft system** to allow a difficult-to-root plant to form its own adventitious roots. The rootstock may be turned upside-down, its polarity reversed, and then grafted to the desired scion. A temporary union will form, and the rootstock will supply water and mineral nutrients to the scion, but the scion is unable to supply necessary organic materials to the rootstock, which eventually dies.

In nurse-root grafting, the graft union is purposely set well below the ground level, and the scion itself produces adventitious roots, which ultimately become the entire root system of the plant. See Figure 12–26, page 487 for greater detail of nurse-root grafting systems.

In T-budding or patch budding, the rule for observance of correct polarity is not as exacting. The buds (scion) can be inserted with reversed polarity and

Figure 11–27
Inverse graft of grape with graft union forming between the distal end of the scion to the distal end of the rootstock. Notice that the shoot reorients itself via gravitational response.

BOX 11.5 GETTING MORE IN DEPTH ON THE SUBJECT
PROXIMAL AND DISTAL ENDS

The **proximal end** of either the shoot or the root is that which is nearest the stem-root junction (crown) of the plant. The **distal end** of either the shoot or the root is that which is farthest from the stem-root junction of the plant and nearest the tip of the shoot or root.

proximal end   The end closest to the crown of the plant, whereas the **distal end** is farthest away from the crown.

crown   The junction of the root and shoot system of a plant.

Figure 11–28
Bridge graft on a pear tree five months after grafting. Center scion was inserted with reversed polarity. Although the scion is alive it has not increased from its original size. The two scions on either side were inserted with normal polarity and have grown rapidly.

still make permanently successful unions. As shown in Figure 11–29, inverted T-buds start growing downward, then the shoots curve and grow upward. In the inverted bud piece, the cambium is capable of continued functioning and growth. There is a twisting configuration in the xylem, phloem, and fibers formed from cambial activity that apparently allows for normal

Figure 11–29
Two-year-old 'Stayman Winesap' apple budded on 'McIntosh' seedling by inverted T-bud (reversing the scion bud polarity). Note the development of stronger, wide-angle crotches. Courtesy Arnold Arboretum, Jamaica Plain, MA.

translocation and water conduction. However, it is still desirable to maintain polarity when budding.

The Craftsmanship of Grafting

The art and craftsmanship in grafting and budding is critical for successful grafting. This is particularly true with difficult-to-graft species, such as conifers (e.g., *Picea pungens*), which callus poorly, making alignment of the cambial layers of the rootstock and scion critical. Conversely, the grafting technique is less critical in grape or pear grafts, which callus profusely and have high grafting success (Figs. 11–23 and 11–24).

Sometimes the techniques used in grafting are so poor that only a small portion of the cambial regions of the rootstock and scion are properly aligned. Graft union formation may be initiated and growth from the scion may start; however, after a sizable leaf area develops, and if high temperatures and high transpiration occurs, water movement through the limited conducting area is insufficient, and the scion subsequently dies. Other errors in technique resulting in graft failure include insufficient or delayed waxing, uneven cuts, use of desiccated scions, and girdling that occurs when polyethylene wrapping tape is not removed expeditiously after graft "take" occurs.

Virus Contamination, Insects, and Diseases

Some delayed incompatibilities are caused by viruses and phytoplasma (mycoplasma-like organisms). The cherry leaf roll virus causes blackline in walnut when it is initially spread by virus-infected pollen of the symptomless English walnut (*Juglans regia*). The virus then travels down the scions of *J. regia* into the susceptible rootstocks—California black walnut (*J. hindsii*) or Paradox walnut (*J. hindsii* × *J. regia*). The black walnut rootstock (used for resistance to *Phytophthora* root-rot in the soil) has a hypersensitive reaction and puts down a chemical barrier to wall-off the virus, which causes the graft to fail, and a characteristic black line forms at the graft union. Apple union necrosis and decline (AUND) (37) and brownline of prune (99) is caused by the tomato ring-spot virus that is transmitted by soil-borne nematodes to the rootstock and then to the graft union. Graft unions appear to be normal until the virus has moved, either from the rootstock or the fruiting branches to the graft union. Because of tissue sensitivity and death of the scion cells (in prunes and apples) or rootstock cells (in walnut), the graft union deteriorates and graft failure occurs. Virus and phytoplasma-induced delayed incompatibility is probably more common than expected (142).

Using virus-infected propagating materials in nurseries can reduce bud "take," as well as the vigor of the resulting plant (121). In stone fruit propagation, bud-wood, free of ring-spot virus, has consistently greater "takes" than infected bud-wood.

Top-grafting olives in California is seriously hindered in some years by attacks of the American plum borer (*Euzophera semifuneralis*), which feeds on the soft callus tissue around the graft union, resulting in the death of the scion. In England, nurseries are often plagued with the red bud borer (*Thomasiniana oculiperda*), which feeds on the callus beneath the bud-shield in newly inserted T-buds, causing them to die.

Plant Growth Regulators and Graft Union Formation

Plant growth regulators, particularly auxin, applied to tree wounds or to graft unions give variable results in wounding response and graft union formation (93, 118, 152). Auxin (IBA, NAA) and cytokinin (BA) enhance graft success when applied to the base of side-grafted *Picea* scions, while the plant growth retardant, dikegulac, stimulated scion growth by retarding rootstock development (17). Cytokinins enhance patch budding of Persian walnut. The eloquent work of Shimomura (152) in tip grafting of cactus demonstrated how auxins enhanced vascular connections of deliberately misaligned scions (Fig. 11–21). TIBA, a well-known inhibitor of basipetal transport of auxin, inhibited vascular connections in the graft union; however, by subsequent reapplication of auxin, the inhibitory effect of TIBA was eliminated and vascular connections occurred.

However, unlike auxin usage in cutting propagation, no plant growth regulators are routinely used in commercial grafting and budding systems. In general, plant growth regulators do not uniformly enhance grafting, nor do they overcome graft incompatibility.

Post-Graftage—Bud-Forcing Methods

After graft union formation has occurred in grafting or budding, it is often necessary to force out the scion or the scion bud. In field budding of roses, 2 to 3 axillary buds of the rootstock remain distal to the scion bud. The axillary buds of the rootstock, which develop into photosynthesizing branches, are initially important for the growth of the composite plant. But they can inhibit growth of the scion through apical dominance, which is an auxin response. By **"crippling"** (cutting halfway through the rootstock shoot above the bud union and breaking the shoot over the rootstock stem), girdling, or totally removing the rootstock above the scion bud union, apical dominance is broken and the scion bud rapidly elongates (Fig. 11–30) (50).

With budded citrus, plants on which rootstock shoots remained attached (lopping, or bending the rootstock shoot to its base and tying it in position) had the greatest gains in scion growth. This was due to the greater transfer of photosynthate from the rootstock leaves to scion shoots during growth flushes, and to roots during periods between growth flushes (181, 182).

GENETIC LIMITS OF GRAFTING

Since one of the requirements for a successful graft union is the close matching of the callus-producing tissues near the cambium layers, grafting is generally confined to the dicotyledons in the angiosperms, and to gymnosperms. Both have a vascular cambium layer existing as a continuous tissue between the xylem and the phloem. Grafting is more difficult, with a low percentage of "takes" in monocotyledonous plants. Monocots have vascular bundles scattered throughout the stem, rather than the continuous vascular cambium of dicots. However, there are cases of successful graft unions between monocots. By making use of the meristematic properties found in the intercalary tissues (located at the base of internodes), successful grafts have been obtained with various grass species as well as the large tropical monocotyledonous vanilla orchid (111).

Before a grafting operation is started, it should be determined that the plants to be combined are capable of uniting and producing a permanently successful union. There is no definite rule that can exactly predict the ultimate outcome of a particular graft combination except that the more closely the plants are related botanically, the better the chances are for the graft union to be successful (71). However, there are numerous exceptions to this rule.

Grafting Within a Clone

A scion can be grafted back onto the plant from which it came, and a scion from a plant of a given clone can be grafted onto any other plant of the same clone. For example, a scion taken from an 'Elberta' peach tree could be grafted successfully to any other 'Elberta' peach tree in the world.

---

crippling The bending (restriction) or cutting halfway through the rootstock stem above the bud union to helps force out the bud and maintain growth of the grafted plant.

Figure 11–30
Forcing or "crippling" of (a and b) T-budded apples; (c and d) Chip budded roses. The rootstock is partially severed on the same side (arrows) that the rootstock was budded. This breaks the apical dominance of the rootstock shoot system on the scion, and helps force out the scion bud. By not totally severing the rootstock top, growth of the composite plant is maximized, since the shock of total severance to the composite plant is avoided, and photosynthate is still produced by the rootstock (182). The rootstock shoot system will be totally severed later, and the scion will fully develop into the shoot system of the composite plant.

Grafting Between Clones Within a Species

In tree fruit and nut crops, different clones within a species can almost always be grafted without difficulty and produce satisfactory trees. However, in some conifer species, notably Douglas-fir (*Pseudotsuga menziesii*), incompatibility problems have arisen in grafting together individuals of the same species, such as selected *P. menziesii* clones onto *P. menziesii* seedling rootstock (36). Incompatibility is also a problem in grafting clones of deciduous species, such as red maple (*Acer rubra*), Chinese chestnut (*Castanea mollissima*), and red oak (*Quercus rubra*).

Grafting Between Species Within a Genus

For plants in different species but in the same genus, grafting is successful in some cases but unsuccessful in others. Grafting between most species in the genus *Citrus*, for example, is successful and widely used commercially. Almond (*Prunus amygdalus*), apricot (*Prunus armeniaca*), European plum (*Prunus domestica*), and Japanese plum (*Prunus salicina*)—all different species—are grafted commercially on rootstock of peach (*Prunus persica*). But on the other hand, almond and apricot, both in the same genus, cannot be intergrafted successfully. The 'Beauty' cultivar of Japanese plum (*Prunus salicina*) makes a good union when grafted on almond, but another cultivar of *P. salicina*, 'Santa Rosa,' cannot be successfully grafted on almond. Thus, compatibility between species in the same genus depends on the particular genotype combination of rootstock and scion.

Reciprocal interspecies grafts are not always successful. For instance, 'Marianna' plum (*Prunus cerasifera* × *P. munsoniana*) on peach (*Prunus persica*) roots makes an excellent graft combination, but the reverse—grafts of the peach on 'Marianna' plum roots—either soon die or fail to develop normally (2, 90).

Grafting Between Genera Within a Family

When the plants to be grafted together are in the same family but in different genera, the chances of a successful union become more remote. Cases can be found in which such grafts are successful and used commercially, but in most instances such combinations are failures. Intergeneric grafts are rarely used in conifers. However, high success rates occur between Nootka cypress

(*Chamaecyparis nootkatensis*) grafted on Chinese arborvitae (*Platycladus orientalis*) rootstock (71).

Trifoliate orange (*Poncirus trifoliata*) is used commercially as a dwarfing rootstock for the orange (*Citrus sinensis*), which is a different genus. The quince (*Cydonia oblonga*) has long been used as a dwarfing rootstock for certain pear (*Pyrus communis* and *P. pyrifolia*) cultivars. The reverse combination, quince on pear, though, is unsuccessful. The evergreen loquat (*Eriobotrya japonica*) can be grafted on deciduous and dwarfing quince rootstock (*Cydonia oblonga*). See Westwood (179) for other examples of graft compatibility between related pome genera.

Intergeneric grafts in the nightshade family, Solanaceae, are quite common. Tomato (*Lycopersicon esculentum*) can be grafted successfully on Jimson weed (*Datura stramonium*), tobacco (*Nicotiana tabacum*), potato (*Solanum tuberosum*), and black nightshade (*Solanum nigrum*).

Grafting Between Families

Successful grafting between plants of different botanical families is usually considered to be impossible, but there are reported instances in which it has been accomplished. These are with short-lived, herbaceous plants, though, for which the time involved is relatively brief. Grafts, with vascular connections between the scion and rootstock, were successfully made (114) using white sweet clover, *Metilotus alba* (Leguminosae) as the scion, and sunflower, *Helianthus annuus* (Compositae) as the rootstock. Cleft grafting was used, with the scion inserted into the pith parenchyma of the stock. The scions continued growth with normal vigor for more than 5 months. To date, there are no reported instances in which woody perennial plants belonging to different families have been successfully and permanently grafted together.

GRAFT INCOMPATIBILITY

The ability of two different plants, grafted together, to produce a successful union and to develop satisfactorily into one composite plant is termed **graft compatibility** (142). **Graft failure** can be caused by anatomical mismatching, poor craftsmanship, adverse environmental conditions, disease, and graft incompatibility. **Graft incompatibility** occurs because of (a) adverse physiological responses between the grafting partners, (b) virus or phytoplasma transmission, and (c) anatomical abnormalities of vascular tissue in the callus bridge (Figs. 11–31 and 11–32).

Graft incombality is an interruption in cambial and vascular continuity leading to a smooth break at

graft compatibility The ability of two different plants, grafted together, to produce a successful union and to develop satisfactorily into one composite or compound plant.

graft failure An unsuccessful graft caused by anatomical mismatching, poor craftsmanship, adverse environmental conditions, disease, or graft incompatibility.

graft incompatibility An interruption in cambial and vascular continuity leading to a smooth break at the point of the graft union, causing graft failure. It is caused by adverse physiological responses between the grafting partners, disease, or anatomical abnormalities.

the point of the graft union. Normal vaccular tissue does not develop in the graft union (Figs. 11–31 and 11–32). Consequently, the gap formed is filled

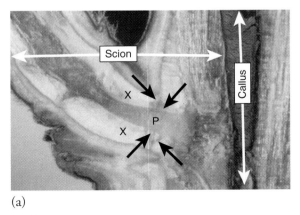

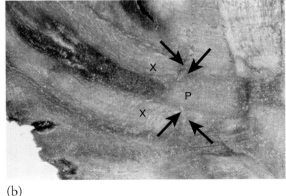

(a) (b)

Figure 11–31
Graft incompatibility in 'Jonagold' apple scions budded to dwarfing 'Mark' rootstock. (a) Unstained section, with callus tissue between the rootstock and scion. (b) Section stained with toluidine blue O. The xylem (x) in the graft union is interrupted by parenchyma tissue (arrows) which limits water flow and survival of the scion. Courtesy of M. R. Warmund (176).

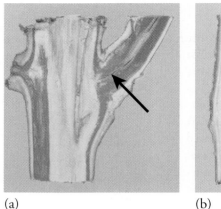 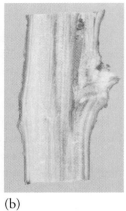

Figure 11–32
(a) Compatible apple chip bud with vascular continuity indicated by red dye, azosulfonate. (b) Unsuccessful chip bud with vascular discontinuity, as indicated by no visible dye. Courtesy M. R. Warmund.

Figure 11–34
Graft incompatibility occurring some 15-plus years after the Monterey pine (*Pinus radiata*) was grafted.

in by proliferating ray tissue that does not lignify normally (109).

Incompatiblity can occur within a period of days or years (Figs. 11–33 and 11–34). Delayed incompatibility can take as long as 20 years to occur with confiers and oaks. Some apricot cultivars grafted onto myrobalan plum rootstick will not break at the graft union until the trees are fully grown and bearing crops (46).

The distinction between a compatible and an incompatible graft union is not always clear-cut. Incompatible rootstock-scion combinations can completely fail to unite. Frequently they unite initially with apparent success (Figs. 11–33 and 11–34) (35) but gradually develop distress symptoms with time, due either to failure at the union or to the development of abnormal growth patterns. Incompatibility of citrus and Monterey pine (*Pinus radiata*) may occur 15 or more years after grafting (Fig. 11–34). Nelson (113) has developed an extensive survey of incompatibility in horticultural plants which should be consulted before attempting graft combinations between species whose graft reactions are unknown to the grafter. Other summaries of graft compatibility have been published (2).

External Symptoms of Incompatibility

Graft union malformations resulting from incompatibility can usually be correlated with certain external symptoms. The following symptoms have been associated with incompatible graft combinations:

- Failure to form a successful graft or bud union in a high percentage of cases.
- Yellowing foliage in the latter part of the growing season, followed by early defoliation. Decline in vegetative growth, appearance of shoot die-back, and general ill health of the tree, including drought stress (Fig. 11–35).
- Premature death of the trees, which may live for only a year or two in the nursery.
- Marked differences in growth rate or vigor of scion and rootstock.
- Differences between scion and rootstock in the time at which vegetative growth for the season begins or ends.
- Overgrowths at, above, or below the graft union (Fig. 11–36).

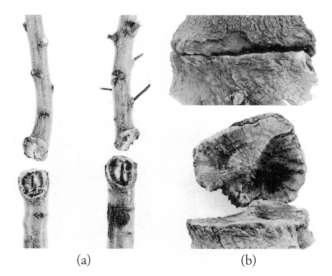

Figure 11–33
Breakage at the graft union resulting from incompatibility. (a) One-year-old nursery trees of apricot on almond seedling rootstock. (b) Fifteen-year-old 'Texas' almond tree on seedling apricot rootstock, which broke off cleanly at the graft union—a case of "delayed incompatibility" symptoms.

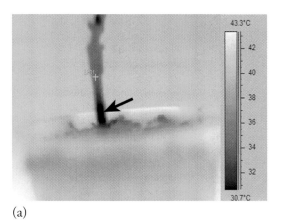
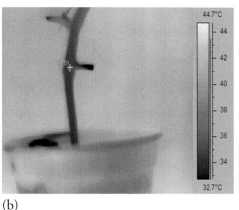

Figure 11–35
Graft compatibility affects water uptake. (a) Arava melon showing hotter scion and cooler temperature in *Cucurbita* rootstock (arrow) with noncompatible grafting combination. (b) Compatible graft showing uniform temperature between scion and rootstock. Differences in temperature gradients determined with a thermal camera. Courtesy M. Edelstein.

- Suckering of rootstock (Fig. 11–37).
- Graft components breaking apart cleanly at the graft union.

An isolated case of one or more of the preceding symptoms (except for the last) does not necessarily mean the combination is incompatible. Incompatibility is clearly indicated by trees breaking off at the point of union, particularly when they have been growing for some years and the break is clean and smooth, rather than rough or jagged. This break may occur within a year or two of the union, for instance, in the apricot on almond roots (see Fig. 11–33), or much later with conifers and oaks (Fig. 11–34). While the scion overgrowing the rootstock (or rootstock outgrowing the scion) at the graft union is not a reliable indicator, it is sometimes associated with incompatibility (Figs. 11–38 and 11–39) (2, 26).

Anatomical Flaws Leading to Incompatibility

With incompatible cherry (*Prunus*) grafts, the number of well-differentiated phloem sieve tubes is much lower at and below the union. There is a greater autolysis of cells, and generally a very low degree of phloem differentiation (149). Poor differentiation of the phloem below the union may be due to a lack of hormones, carbohydrates, and other factors—the size of the sieve tubes depends on auxin, cytokinin, and sucrose levels (149). With incompatible apricot/plum (*Prunus*) grafts, some callus differentiation into cambium and vascular tissue does occur; however, a large portion of the callus never differentiates (Fig. 11–40) (48). The union that occurs is mechanically weak.

With incompatible apple grafts, vascular discontinuity occurs with xylem interrupted by parenchyma tissue (Figs. 11–31 and 11–32) (176), which disrupts normal xylem function leading to death of the budded scion.

Nontranslocatable (Localized) Incompatibility

For lack of better terminology, physiological factors of graft incompatibility has been traditionally classified as **nontranslocatable (localized)** or **translocatable** (109). It is difficult to distinguish differences between the symptoms of nontranslocatable and translocatable incompatibility. Anatomical symptoms of incompatibility can include phloem degeneration or phloem compression, and cambial or vascular discontinuity in

Figure 11–36
Physiological incompatibility between scion and rootstock. Scion overgrowth caused by blockage of assimilates translocating from the scion to the rootstock, causing a weak root system. The melon scion grafted on *Cucurbita* rootstock later died as a result of insufficient support from the rootstock. Photo courtesy M. Edelstein.

Figure 11–37
Undesirable suckering of rootstocks. (a) *Hamamelis vernalis* 'Sandra' grafted on *Hamamelis vernalis* rootstock, and (b) rootstock suckers on recently grafted *Ulmus alata* 'Lace Parasol' grafted onto seedling *Ulmus alata*. The suckers will need to be removed. Photo courtesy B. Upchurch.

Figure 11–38
While rootstock outgrowth is not desirable, a large, strong tree can still develop. (a) Sweet orange rootstock used for dwarfing, overgrowing the grapefruit scion. (b) Rootstock overgrowing scion on *Morus alba* 'Platanifolia.' Photo (b) courtesy B. Upchurch.

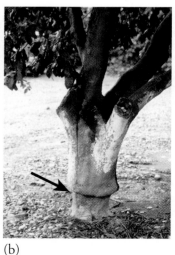

Figure 11–39
Scion or rootstock outgrowth can still lead to a large, strong tree. Such outgrowth (arrows) is more related to the genetic tendency for growth, than to incompatibility. (a) Scion overgrowing rootstock: *Acer pentaphyllum* on *A. pseudoplatanus* rootstock, and (b) grapefruit scion on sour orange rootstock, which tolerates alkaline, heavy soils, but can be susceptible to *Trestiza*. Photo (a) courtesy B. Upchurch.

> **BOX 11.6 GETTING MORE IN DEPTH ON THE SUBJECT**
> **TYPES OF GRAFT INCOMPATIBILITY**
>
> **Anatomical Flaws**
> - Incompatible cherry (*Prunus*) grafts with poor phloem development and/or weak unions
> - Incompatible apricot/plum (*Prunus*) grafts—mechanically weak unions
> - Some budded apple (*Malus*) combinations—vascular discontinuity
>
> **Nontranslocatable (Localized) Incompatibility**
> - 'Bartlett' pear on quince roots; incompatibility overcome with 'Old Home' interstock
>
> **Translocatable Incompatibility**
> - 'Hale's Early' peach on 'Myrobalan B' plum roots
> - 'Nonpareil' almond on 'Marianna 2624' plum roots
> - Peach cultivars on 'Marianna 2624' plum roots
>
> **Pathogen-Induced Incompatibility (Virus, Phytoplasma)**
> - Citrus quick decline or Tristeza
> - Pear decline
> - Walnut blackline
> - Apple union necrosis and decline (AUND)
> - Prune brownline

the union area, causing mechanical weakness and subsequent breakdown of the union. Nontranslocatable incompatibility includes graft combinations in which a **mutually compatible interstock** overcomes the incompatibility of the scion and rootstock. The interstock prevents physical contact of the rootstock and scion and affects the physiology of the normally incompatible scion and rootstock. In some innovative research, membrane filters placed between graft partners demonstrated that physical contact is not necessary to develop compatible grafts (104, 106). A good example of nontranslocatable incompatibility is 'Bartlett' ('Williams') pear grafted directly onto dwarfing quince rootstock. When mutually compatible 'Old Home' or ('Beurré Hardy') is used as an interstock, the three-graft combination is completely compatible, and satisfactory tree growth takes place (107, 122, 132).

Translocatable Incompatibility

Translocatable incompatibility includes certain graft/rootstock combinations in which the insertion of a mutually compatible interstock does not overcome

Figure 11–40
Callus bridge formation in graft union of compatible and incompatible *Prunus* spp. (a and b) Compatible 'Luizet' apricot grafted on 'Myrobalan' standard plum rootstock. (a) Callus in graft union from a compatible graft 21 days after grafting. The cells show an orderly disposition and are uniformly stained (160× magnification). (c) Callus from incompatible graft of 'Monique' apricot on 'Myrobalan' standard plum rootstock ten days after grafting. The cells show an irregular disposition and the cell walls are thick and irregular. Courtesy P. Errea (48).

incompatibility. Apparently, some biochemical influence moves across the interstock and causes phloem degeneration. This type of incompatibility can be recognized by the development of a brownline or necrotic area in the bark at the rootstock interface. Consequently, carbohydrate movement from the scion to the rootstock is restricted at the graft union.

'Hale's Early' peach grafted onto 'Myrobalan B' plum rootstock is an example of translocatable incompatibility. The tissues are distorted and a weak union forms. Abnormal quantities of starch accumulate at the base of the peach scion. If the mutually compatible 'Brompton' plum is used as an interstock between the 'Hale's Early' peach and the 'Myrobalan B' rootstock the incompatibility symptoms persist, with an accumulation of starch in the 'Brompton' interstock. 'Nonpareil' almond on 'Marianna 2624' plum rootstock shows complete phloem breakdown, although the xylem tissue connections are quite satisfactory. In contrast, 'Texas' almond, on 'Marianna 2624' plum rootstock produces a compatible combination. Inserting a 15-cm (6-in) piece of 'Texas' almond as an interstock between the 'Nonpareil' almond and the 'Marianna' plum rootstock fails to overcome the incompatibility between these two components. Bark disintegration occurs at the normally compatible 'Texas' almond/'Marianna' plum graft union (79).

Pathogen-Induced Incompatibility

Viruses and **phytoplasmas (mycoplasma-like organisms)** cause pathogen-induced incompatibility. Cases of this incompatibility are widespread, and more are continually being found. In certain cases abnormalities first attributed to rootstock-scion incompatibility were later found to be due to latent virus or phytoplasma introduced by grafting from a resistant, symptomless partner to a susceptible partner (32, 41, 95). Figure 11–41 shows such an occurrence in apple.

> **phytoplasmas (mycoplasma-like organisms)** Organisms that can cause pathogen-induced incompatibility in grafted plants.

Tristeza, which comes from the Spanish and Portuguese word *triste*, meaning "sad" or "wretched," is an important example of virus-induced incompatibility in citrus. Failure of sweet orange (*Citrus sinensis*) budded onto sour orange (*C. aurantium*) rootstock in South Africa (1910) and in Java (1928) was at one time blamed on incompatibility, even though this combination was a commercial success in other parts of the world. Incompatibility was believed due to production of a substance by the scion that was toxic to the rootstock (167). Subsequent studies involving Tristeza or "quick decline" of orange in Brazil and California made clear that the toxic substance from the sweet orange scions was instead a virus tolerated by the sweet orange, but lethal to sour orange rootstock (22, 177).

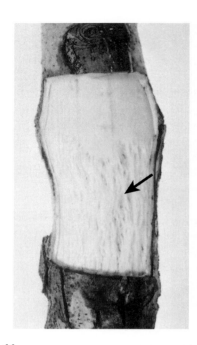

Figure 11–41
Latent viruses in the scion portion of graft combination may cause symptoms to appear in a susceptible rootstock following grafting. Here "stem pitting" virus symptoms (arrow) have developed in the sensitive 'Virginia Crab' apple rootstock. The wood of the scion cultivar—above the graft union—is unaffected. Courtesy H. F. Winter.

Other examples of virus-induced incompatibility include blackline in English walnut (*Juglans regia*), which infects susceptible walnut rootstock; apple union necrosis and decline (AUND) (37); and brownline of prune (99), which is caused by tomato mosaic virus that is transmitted by soil-borne nematodes to the rootstock, and then to the graft union. Pear decline is due to a phytoplasma, rather than a virus.

The major causes for graft incompatibility include (a) physiological and biochemical factors; (b) modification of cells and tissues at the graft union; and (c) cell recognition between grafting partners.

Causes and Mechanisms of Incompatibility

Physiological and Biochemical Mechanisms Tissue compatibility or incompatibility in plants can be regarded as a physiological tolerance or intolerance, respectively, between different cells (103, 105, 106). Although incompatibility is clearly related to genetic differences between rootstock and scion, the mechanisms by which incompatibility is expressed are not clear. The large number of different genotypes that can be combined by grafting produces a wide range of different physiological, biochemical, and anatomical interactions when grafted. Several hypotheses have been advanced in attempts to explain incompatibility.

One proposed physiological and biochemical mechanism concerns incompatible combinations of certain pear cultivars on quince rootstock (61). The incompatibility is caused by a **cyanogenic glucoside, prunasin,** normally found in quince but not in pear tissues. Prunasin is translocated from the quince into the phloem of the pear. The pear tissues break down the prunasin in the region of the graft union, with hydrocyanic acid (cyanide) as one of the decomposition products (Fig. 11–42). The presence of the hydrocyanic acid leads to a lack of cambial activity at the graft union, with pronounced anatomical disturbances in the phloem and xylem at the resulting union. The phloem tissues are gradually destroyed at and above the graft union. Conduction of water and materials is seriously reduced in both xylem and phloem. *The presence of cyanogenic glycosides in woody plants is restricted to a relatively few genera.* Hence, this reaction cannot be considered a universal cause of graft incompatibility.

Phenolic compounds have also been implicated in graft incompatibility (49). Phenolic compounds are widespread in plants and present in the biochemical responses to stress and wounding. They play a role in lignification (27), which occurs in graft union formation.

Modification of Cells and Tissue The **lignification** processes of cell walls are important in the formation of strong unions in pear-quince grafts. Inhibition of lignin formation and the establishment of a mutual middle lamella results in weak graft unions. In compatible pear-quince graft combinations, the lignin in cell walls at the graft union is comparable to adjacent cells outside the union (27). Conversely, adjoining cell walls in the graft union of incompatible combinations contain no lignin, and are interlocked only by cellulose fibers.

With incompatible apricot-plum (*Prunus*) grafts, some callus differentiation into cambium and vascular tissue does occur; however, a large portion of the callus never differentiates (Fig. 11–40) (48). The union that occurs is mechanically weak.

Cell Recognition of the Grafting Partners It has been postulated that the critical event deciding compatible and incompatible grafts may occur when the callus cells first

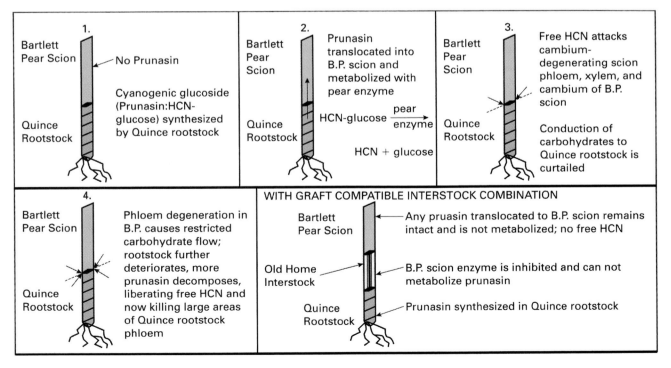

Figure 11–42
Nontranslocatable incompatibility of Bartlett pear scion overcome with 'old Home' interstock on quince rootstock (61).

cellular recognition
The union of specific cellular groups on the surfaces of the interacting cells that results in a specific defined response [e.g., pollen-stigma compatibility-incompatibility recognition responses with glycoprotein surface receptors in flowering plants (30)].

touch (189). There may be **cellular recognition** that must occur in successful graft union formation. Alternatively, the failure of procambial differentiation in incompatible grafts may be the result of a direct form of cellular communication between the graft partners (101).

In a compatible graft, the wound response is followed by a dissolution of the necrotic layer, perhaps as a prerequisite to the formation of secondary plasmodesmata between cells of the graft partners (165). *There is direct cellular contact* of plasmodesmata (minute cytoplamic threads that extend through openings in cell walls and connect the protoplasts of connecting cells) in the callus bridge that symplastically connects the grafting partners (Fig. 11–19) (81). This forms a potential communication pathway among cells in the graft bridge, which may be important in cell recognition and compatibility/incompatibility responses.

Conversely, cellular recognition may not be a factor in grafting compatibility/incompatibility. Partners of compatible and incompatible grafts adhere during the early stages of graft union formation; this passive event does not require mutual cell recognition [grafted *Sedum* will even adhere to inert wooden objects (101, 103)], nor is it related to compatibility (106). Adhesion of graft partners results from the deposition and subsequent polymerization of cell wall materials that occur in response to wounding. Callus proliferation is not related to graft compatibility-incompatibility systems, since it does not require a recognition event to occur; that is, callus proliferation occurs in wounded cuttings, as well as in incompatible and compatible graft systems (101, 103).

Vascular differentiation in the callus bridge, which typically occurs from the severed vascular strands of the scion and rootstock, can occur even when the scion and rootstock are physically separated by a porous membrane filter (inserted in order to prevent direct cellular contact without impeding the flow of diffusible substances between the graft partners) (102, 104); this was done with *autografts* of *Sedum* (a herbaceous species), which may not be representative of graftage in woody perennial plants. Nonetheless, it is evidence that successful graft union formation can occur in the absence of direct cellular contact, and does not require a positive recognition system.

autograft The scion and rootstock are from the same plant or species.

Tissue alignment [e.g., vascular cambium of woody plants, vascular bundles of cacti (152)] determines what cell types and tissue will be differentiated in the callus bridge. It has been proposed that phytohormones are released from wounded vascular bundles into the surrounding tissue where they function as morphogenic substances inducing and controlling the regeneration of cambium and vascular tissue (3). This hypothesis can be applied to graft union formation, with phytohormones such as auxin as potential morphogens needed for graft union formation. Auxin should not be considered as a specific recognition molecule per se because of its common occurrence and involvement in numerous other developmental processes (104, 106). Phytohormones (and carbohydrates, etc.), predominantly released from

---

BOX 11.7 GETTING MORE IN DEPTH ON THE SUBJECT
CELLULAR RECOGNITION

It is currently not known if some kind of cell-to-cell recognition in grafting must occur as part of adhesion and the events that follow in successful graft union formation. Possibly, the formation of superimposed sieve areas and sieve plates (in sieve elements), pits and perforation plates (in xylem elements), and the plasmodesmata (in vascular parenchyma) require some sort of cellular recognition or cellular communication (101). Evidence suggests that in the graftage of *Cucumis* and *Cucurbita*, changes in protein banding may be due to polypeptides migrating symplastically across the graft union via the connecting phloem (165). Translocation of signaling molecules, such as polypeptides in the phloem, could be significant in cell recognition and compatibility between the graft partners. (In graft incompatibility, phloem degeneration frequently takes place at the graft union.)

Pectin fragments formed during the adhesion process of grafting may act as signaling molecules—and influence cell recognition. In Sitka spruce, the beadlike projections from callus formed during graftage are in part composed of pectins, proteins, carbohydrates, and fatty acids. These beadlike projections, besides binding or cementing cells, may serve a more active role in cell recognition and the successful merging of tissues of the grafting partners (96).

the scion, enable vascular connections to develop and join as a functional unit in the graft union, without any cellular recognition required.

A model for graft compatibility-incompatibility is presented that suggests grafts will be incompatible only if naturally occurring morphogens that promote the formation of a successful graft (e.g., auxin) are overridden by toxins [e.g., hydrocyanic acid, benzaldehyde (62, 63)] that elicit graft incompatibility (Fig. 11–43) (106).

There is probably no universal cause of graft incompatibility in plants (145). Most likely, graft compatibility-incompatibility is a combination of the auxin-toxin interactions of Figure 11–43 and/or some chemical recognition response. To date, we have little understanding of the molecular chain of events that occurs during wounding (180) and graft union formation, or how those chains of events vary between compatible-incompatible graft partners. In Douglas-fir, graft incompatibility is apparently controlled by multiple genes with additive effects (36).

Predicting Incompatible Combinations

Accurately predicting whether or not the components of the proposed scion-stock combination are compatible would be tremendously valuable. An *electrophoresis test* was used for testing *cambial peroxidase* banding patterns of the scion and rootstock of chestnut, oak, and maple (138, 140–145). Peroxidases mediate lignin production. Increased peroxidase activity occurs in incompatible *heterografts,* compared with compatible autografts, and adjacent rootstock and scion cells must produce similar lignins and have identical peroxidase enzyme patterns to ensure the development of a functional vascular system across the graft union (40). With electrophoresis, if the peroxidase bands match, the combination may be compatible; if they do not, incompatibility may be predicted. Using electrophoresis is an important step in developing diagnostic tests for graft compatibility. Perhaps serological tests for graft compatibility may be developed in the future, to complement those currently used in disease diagnostic kits of plant pathogens.

The introduction of new *Prunus* rootstock can be difficult (and very costly!) because incompatibility can occur some years after grafting. The composite tree can grow "normally" for years, and then a breakdown occurs at the graft union area. It is now known that with incompatible apricot-plum (*Prunus*) grafts, some callus differentiation into cambium and vascular tissue does occur; however, a large portion of the callus never differentiates (Fig. 11–40) (48). Early detection of graft incompatibility in fruit trees is greatly facilitated since this process can be detected histologically within weeks after grafting (48).

Magnetic resonance imaging (MRI) can be used to

> **heterograft** The scion and rootstock are from a different cultivar or species.

> **Magnetic Resonance Imaging (MRI)**
> A diagnostic imaging technique that can be used for detecting vascular continuity in the callus bridge.

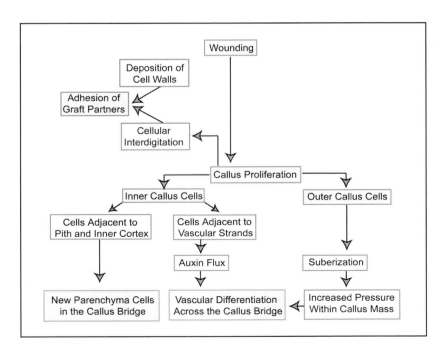

Figure 11–43
A model to explain the development of a compatible graft union. The stages are adhesion of the scion and rootstock, proliferation of callus cells to form the callus bridge, and vascular differentiation across the graft interface. The outer callus cells are from the periderm and outer cortex. The pressure exerted on the graft is from the physical contact of the scion to the rootstock—and the development of a suberized periderm. Auxin is a potential morphogen, enhancing vascular dedifferentiation. In this model, incompatibility is not caused by specific cellular recognition events between the graft partners. Rather, incompatibility may occur when a toxin, such as hydrocyanic acid (HCN) or benzaldehyde, counteracts naturally occurring morphogens (e.g., auxin), thus inhibiting or degenerating vascular tissues in the graft union (106).

BOX 11.8 GETTING MORE IN DEPTH ON THE SUBJECT
CELLULAR RESPONSES OF COMPATIBLE AND INCOMPATIBLE GRAFTS

At the cellular level, the initial stages of graft union formation were similar between the incompatible combination (heterographs) of *Sedum telephoides* (Crassulaceae) on *Solanum pennellii* (Solanaceae) and those occurring in a compatible autograft of *Sedum* on *Sedum*. However, after 48 hours, *Sedum* cells in the incompatible graft deposited an insulating layer of suberin along the cell wall. The cell walls later underwent lethal senescence and collapse and formed a necrotic layer of increasing thickness (Fig. 11–44) (101, 103). Associated with this cellular senescence in *Sedum* cells was a dramatic increase in a hydrolytic enzyme, acid phosphatase (102). Rather than callus cells interlocking, the thick necrotic layer prevented cellular connections, which led to scion desiccation and eventual death. Interestingly, the *Solanum* rootstock did not show the rejection response that the *Sedum* scion did.

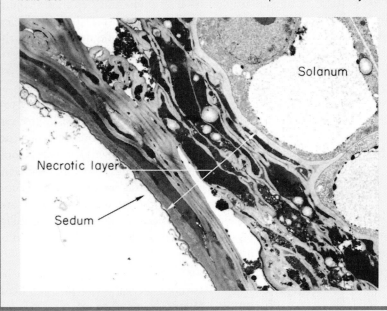

Figure 11–44
The graft interface of an incompatible graft between Sedum telephoides and *Solanum pennellii* at eight days after grafting. Lethal cellular senescence in *Sedum* has resulted in the formation of a necrotic layer of collapsed cells that separates the two graft partners. ×5,000. Courtesy R. Moore and D. B. Walker (101).

---

detect vascular discontinuity in bud unions of apple (176). A high MRI signal intensity is associated with bound water in live tissue and the establishment of vascular continuity between the rootstock and scion. MRI may be useful for detecting graft incompatibilities caused by poor vascular connections.

Correcting Incompatible Combinations

There is not a practical, cost-effective way to correct large-scale plantings of incompatible graft partners. Plants are normally rogued and discarded. With some isolated specimen trees of value, if the incompatibility were discovered before the tree died or broke off at the union, a bridge graft could be done with a mutually compatible rootstock. Another costly alternative is to inarch with seedlings of a compatible rootstock. The inarched seedlings would eventually become the main root system (see Chapter 12).

SCION-ROOTSTOCK (SHOOT-ROOT) RELATIONSHIPS

Combining two (or more, in the case of interstocks) different plants (genotypes) into one plant by grafting—one part producing the top and the other part the root system—can produce growth patterns that are different from those that would have occurred if each component part had been grown separately. Some of these effects are of major importance in horticulture and forestry, while others are detrimental and should be avoided. These altered characteristics may result from (a) specific characteristics of the graft partners not found in the other; for example, resistance to certain diseases, insects, or nematodes, or tolerance of certain adverse environmental or soil conditions; (b) interactions between the rootstock and the scion that alter size, growth, productivity, fruit quality, or other horticultural attributes; and (c) incompatibility reactions. In practice, it may be difficult to separate which influencing factor is dominant in any given graft combination growing in a particular environment.

Long-term results depend on the rootstock-scion combination, environment (climate, edaphic factors such as soil), propagation, and production management, which affects yield, quality, plant form, and ornamental characteristics (if applicable), and by extension, the economics of production.

Effects of the Rootstock on the Scion Cultivar

Size and Growth Habit Size control, sometimes accompanied by change in tree shape, is one of the most significant rootstock effects. Rootstock selection in apple has produced a complete range of tree sizes—from dwarfing to very vigorous—by grafting the same scion cultivar to different rootstocks (Fig. 11–10).

That specific rootstocks can be used to influence the size of trees has been known since ancient times. Theophrastus—and later the Roman horticulturists—used dwarfing apple rootstocks that could be easily propagated. The name "Paradise," which refers to a Persian (Iranian) park or garden—*pairidaeza*—was applied to dwarfing apple rootstocks about the end of the 15th Century.

A wide assortment of size-controlling rootstocks has now been developed for certain of the major tree fruit crops. Most notable is the series of clonally propagated apple rootstocks collected and developed at the East Malling Research Station in England, beginning in 1912. These apple rootstocks were classified into four groups, according primarily to the degree of vigor imparted to the scion cultivar: **dwarfing, semi-dwarfing, vigorous,** and **very vigorous**—same size as seedling rootstock (Fig. 11–10). Similarly, the size-controlling effects of the rootstock on sweet cherry (*Prunus avium*) scion cultivars has been known since the early part of the 18th Century. Mazzard (*P. avium*) seedling rootstocks produce large, vigorous, long-lived trees, whereas *P. mahaleb* seedlings, as a rootstock, tend to produce smaller trees that do not live as long. However, individual seedlings of these species, when propagated asexually and maintained as clones, can produce different, distinct rootstock effects. Rootstock effects on tree size and vigor are recognized also in citrus, pear, peach, olive and other species. A discussion of specific rootstocks for the various fruit and nut crops is given in Chapter 19.

Fruiting *Fruiting precocity, fruit bud formation, fruit set, and yield* of a tree can be influenced by the rootstock used. In general, fruiting precocity is associated with dwarfing rootstocks, and delay in fruiting with vigorous rootstocks. Apple rootstocks are used primarily for reducing tree size and for increasing precocity and yield efficiency.

Besides being more precocious, intensive plantings of small trees resulting from dwarfing rootstock intercept more light and have less internal shading, which is related to greater dry matter production and fruit yield. The higher ratio of fruit weight to trunk and branch weight (partitioning of photosynthate to fruit rather than wood formation) may also contribute to higher *yield efficiencies* for trees growing on dwarfing rootstock than more vigorous clonal and seedling rootstock (121, 160).

Vigorous, strongly growing rootstocks, in some cases, result in a larger plant that produces a bigger crop (per individual tree) over many years. On the other hand, trees on dwarfing rootstocks are more fruitful, and if closely planted, produce higher yields per hectare (acre). The producer's cash flow and return on investment are much improved because an apple crop on dwarfing rootstock produces more fruit earlier. Furthermore, the management costs of harvesting, pruning, spraying, and general maintenance are much greater on large trees.

Size, Quality, and Maturity of Fruit There is considerable variation among plant species in regard to the effect of the rootstock on fruit characteristics of the scion cultivar. However, *in a grafted tree there is no transmission of fruit traits characteristic of the rootstock to the fruit produced by the grafted scion*. For example, quince, commonly used as a dwarfing pear rootstock, has fruits with a pronounced tart and astringent flavor, yet this flavor does not appear in the pear fruits. The peach is often used as a rootstock for apricot, yet apricot fruits do not have any characteristics of peach fruits.

Although there is no transfer of fruit characteristics between the rootstock and the scion, certain rootstocks can affect fruit quality of the scion cultivar. A good example of this is the "black-end" defect of pears. 'Bartlett,' 'Anjou,' and some other pear cultivars on several different rootstocks often produce fruits that are abnormal at the calyx end. While the fruit quality and yield of tomatoes and cucurbits is generally enhanced with the correct stock-scion combination, sometimes melon fruit quality is impaired when grafted on disease resistant *Cucurbita* rootstock (39). Rootstocks of chili peppers (*Capsicum annuum*) can increase the level of capsaicin, which influences the "hotness" of peppers (185).

In citrus, striking effects of the rootstock appear in fruit characteristics of the scion cultivar (23). If sour

Figure 11–45
Stock-scion relations. (a) Seedless 'Marsh White' grapefruit scion on rough lemon rootstock compared to (b) 'Marsh White' on sour orange rootstock, which has a thinner peel (arrows)—and is also sweeter and juicer.

cucumber (*Cucumis sativus*) scions are grafted onto figleaf gourd rootstocks (*Cucurbita ficifolia*), there is greater low-root temperature resistance compared to **own-rooted** cucumber plants, a phenomenon is attributed to greater water absorption capacity of the figleaf gourd rootstock exposed to root temperatures of less than 20°C (68°F), which causes own-rooted cucumber plants to wilt, as the result of reduced water absorption (1).

Increased Nitrogen Efficiency. Grafted vegetable crops with very vigorous root systems can absorb more inorganic nutrients than own-rooted plants (82, 85). The organic nitrogen and fruit yield increased with grafted melon cultivars (*Cucumis melo*) on *Cucurbita maxima xmoschata* rootstock, compared with own-rooted plants (137). There was greater nitrogen utilization and assimilation in the grafted than own-rooted plants. Mini-watermelon plants (*Citrullus lanatus*) grafted on *Cucurbita* rootstock had a higher marketable yield, higher nutritional status (including nitrogen, potassium, and phosphorus), photosynthesis, and water uptake than non-grafted plants under limited-water supply (136).

Extending Scion Tolerance of Adverse Edaphic Conditions. For many kinds of plants, rootstocks are available that tolerate unfavorable conditions, such as heavy, wet soils (124–126, 129), or high soil pH (Fig. 11–2). In the southeastern United States, where high temperatures and periodic flooding of soils (low soil oxygen) are the norm, cultivars of birch (*Betula*), fir (*Abies*), and oak (*Quercus*) are grafted onto rootstock that tolerate these atmospheric and edaphic environments (Fig. 11–1) (125, 126, 129). 'Whitespire' Japanese birch (*Betula populifolia*) is an excellent landscape tree for the southeastern United States. It tolerates heat and drought but will not tolerate poorly drained soils. The ecological niche of 'Whitespire' may be expanded by grafting it onto flood-tolerant rootstock of river birch (*B. nigra*) (125). Compared with many other genera of temperate woody plants, trees in the genus *Prunus* are often intolerant of poor drainage conditions. Ornamental *Prunus* cultivars can be adapted to poorly drained landscape sites by grafting onto more flood-tolerant 'Newport' plum (*Prunus hybrida*) and 'F-12/1' Mazzard cherry (*P. avium*) (124). Japanese Momi fir (*Abies firma*) is one of the few firs that will tolerate the heavy clay, wet soil conditions, and heat of the southeastern United States. Consequently, it is being recommended as the rootstock for more desirable fir cultivars (129).

orange (*Citrus aurantium*) is used as the rootstock, fruits of sweet orange, tangerine, and grapefruit are smooth, thin-skinned, and juicy, with excellent quality, and they store well without deterioration (Fig. 11–45). Sweet orange (*C. sinensis*) rootstocks also result in thin-skinned, juicy, high-quality fruits. The larger fruit size of 'Valencia' oranges is associated with the dwarfing trifoliate orange rootstock, whereas sweet orange rootstocks produce smaller fruits. Semi-dwarfing clonal rootstock will enhance the fruit size of 'Red Delicious' and 'Granny Smith' apples, compared with seedling rootstock, while 'Gala' is unaffected by rootstocks (69).

Miscellaneous Effects of the Rootstock on the Scion Cultivar Rootstocks can also increase cold-hardiness, nitrogen efficiency, enhance tolerance of adverse edaphic conditions, and increase disease and insect resistance of the grafted scion.

Cold-Hardiness. In citrus, which rootstock is used can affect the winter-hardiness of the scion cultivar. Grapefruit cultivars on 'Rangpur' lime rootstock survive cold better than those on sour orange or rough lemon rootstock. The rootstock can affect the rate of maturity of the scion wood as it hardens-off in the fall (54). Greater low-root temperature resistance can occur in grafted, herbaceous vegetable crops. When

own-rooted The propagation of a plant by rooted cutting, as opposed to propagating the cultivar on a grafted rootstock.

Disease and Pest Resistance. Some rootstocks are more tolerant to adverse soil pests, such as nematodes (*Meloidogyne* spp.), than others; for example, 'Nemaguard' peach rootstock. The growth of the scion cultivar is subsequently enhanced by the rootstock's ability to withstand these soil pests. Grape cultivars susceptible to the insect pest, phylloxera (*Dacylosphaera vitifoliae*), are grafted onto resistant rootstocks. Many cucurbits and solonaceous crops are grafted for enhanced disease resistance and tolerance of abiotic stress (Figs. 11–3 and 11–4) (34, 39, 82, 86, 131). Grafting with disease-resistant rootstock also offers new IPM management strategies for organic vegetable production (131).

Effect of the Scion Cultivar on the Rootstock

Although there is a tendency to attribute all cases of dwarfing or invigoration of a grafted plant to the rootstock, the effect of the scion on the behavior of the composite plant may be as important as that of the rootstock.

Effect of the Scion on the Vigor and Development of the Rootstock Scion vigor can have a major effect on rootstock growth, just as rootstocks can affect scion growth. If a strongly growing scion cultivar is grafted on a weak rootstock, the growth of the rootstock will be stimulated so as to become larger than it would have been if left ungrafted. Conversely, if a weakly growing scion cultivar is grafted onto a vigorous rootstock, the growth of the rootstock will be lessened from what it might have been if left ungrafted. In citrus when the scion cultivar is less vigorous than the rootstock cultivar, it is the scion cultivar rather than the rootstock that determines the rate of growth and ultimate size of the tree (66).

Effect of Interstock on Scion and Rootstock

The ability of certain dwarfing rootstock clones, inserted as an interstock between a vigorous top and vigorous root, to produce dwarfed and early bearing fruit trees has been used for centuries to propagate dwarfed trees. The degree of size control induced in apples by various dwarfing rootstock is shown in Figure 11–10. Dwarfing of apple trees by the use of a 'Malling 9' as an interstock was a common commercial practice for many years (Fig. 11–11). This dwarfing method had the advantage of allowing the use of well-anchored, vigorous rootstock rather than a brittle, poorly anchored dwarfing clone. Sometimes excessive suckering from the roots occurred due to the dwarfing interstock, even in rootstock types that normally do not sucker freely. Today, apple interstocks are rarely used except in China (G. Fazio, personal communication).

Possible Mechanisms for the Effects of Rootstock on Scion and Scion on Rootstock

While many of the effects of rootstock-scion relations are known, the fundamental mechanisms of control, particularly on the molecular basis, are not well understood. A better understanding of the mechanisms controlling growth and development in grafted plants would speed up the design, development, and commercialization of new composite plant systems. By understanding these mechanisms, breeders could better predict the growth responses of new potential graft partners (while they are still on the "drawing board") and develop more efficient screening tests—rather than relying on cumbersome trial and error processes that may take up to 10 years or more in evaluating grafted, woody perennial plants.

Without question, the nature of the rootstock-scion relationship is very complex and differs among genetically different combinations. Furthermore, in a composite plant system, size control, plant form, flowering, fruiting, disease resistance, flood tolerance, etc. are not controlled by the same genes or physiological/morphological mechanisms. Theories advanced as possible explanations for the interaction between the rootstock and scion include: (a) **anatomical factors,** (b) **nutritional and carbohydrate levels,** (c) **absorption and translocation of nutrients and water,** (d) **phytohormones and correlative effects,** and (e) **other physiological factors.**

Anatomical Factors The roots and stems of dwarfing apple rootstocks, which can reduce vegetative growth and increase flowering, are characterized by several anatomical features. These include: (a) a high ratio of bark (periderm, cortex, and phloem tissue) to wood (xylem tissue); (b) a large proportionate volume occupied by living cells (axial parenchyma and ray parenchyma cells) relative to functionally dead xylem cells (vessels and fibers); and (c) fewer and smaller xylem vessels (13, 14, 92, 153).

Much of the functional wood tissue of roots of dwarfing apple stocks is composed of living cells, whereas in nondwarfing, vigorous rootstocks, the wood consists of a relatively large amount of lignified tissue without living cell contents (i.e., a larger vessel/tracheid system for more efficient water transport). At the graft interface between the scion bud and dwarfing apple rootstock, xylem vessels with smaller than normal diameter are formed, whereas semi-dwarfing rootstock produces normal xylem after a brief interruption (157). It has been proposed that failure of auxin to cross the bud-union interface in the case of the dwarfing rootstock leads to reduced rootstock xylem formation, and hence a reduced supply of water and minerals to the scion, thus causing the dwarfing effect (87). Defects in the graft union that cause a partial discontinuity of the vascular tissues may in part explain the marked depletion of solutes, nutrients, and cytokinins (produced from root apices) in the sap contents of dwarfing interstocks and rootstocks (73).

Conversely, with kiwifruit, the roots of flower-promoting rootstock tend to have more and larger xylem vessels, more crystalline idioblasts, and more starch grains (173, 174). Most likely the greater water supply from the rootstock to the scion in early spring determines the abundance of flower production of the kiwifruit scions.

Morphologically, dwarfing rootstock have fewer coarse roots (diameter greater than 2 mm) and fewer fine roots (diameter less than 2 mm) than more vigorous apple rootstocks (5, 6). There is not always a clear relationship between root length growth and size control characteristics of dwarfing versus vigorous rootstock. However, there are fewer active root tips in dwarfing than vigorous apple rootstock (51). The roots and shoots of vigorous apple rootstocks also have a longer growing season than dwarfing rootstock (78).

Nutritional and Carbohydrate Levels Dwarfing rootstock of apple tends to partition a greater proportion of carbon to reproductive areas (**spurs,** spur leaves, fruit) and less to the tree branch and frame dry weight, compared with nondwarfing rootstocks (160). The greater water and nutrient uptake of the vigorous rootstock contributes to the production of new vegetative growth, which is a **competing sink** with reproductive growth.

The rootstock affects the partitioning of the dry matter between above- and below-ground tree components. Vigorous rootstocks accumulate more dry matter in the shoot and root system than dwarfing stock (6, 161). At the time apples are being harvested, the insoluble root starch supply is greater, but soluble sucrose and sorbitol are less in vigorous rootstock compared with dwarfing rootstock.

It appears that apple rootstock does not influence mineral nutrition at the site of flower formation (65). Most likely, rootstock effects on flowering are due to internal control mechanisms that affect the proportion of spurs that become floral (64).

To summarize, dwarf apple rootstocks do affect precocity and flowering, in part, because of differences in carbohydrate metabolism and the greater carbon partitioning to the reproductive areas. The contributing influence of hormones, which also affects carbon partitioning and flowering, is discussed below.

Absorption and Translocation of Nutrients and Water
Apple rootstocks affect Ca, Mg, Mn, and B uptake, but there is no apparent direct relationship of mineral status with rootstock vigor, productivity, or spur characteristics (65).

Rootstocks do differ in their ability to absorb and translocate P (74), but a direct role of phosphorus at the site of flower formation induced by rootstock seems unlikely (65). In a study of the translocation of radioactive phosphorus (^{32}P) and calcium (^{45}Ca) from the roots to the tops of 1-year 'McIntosh' apple trees grown in solution culture, it was shown that more than three times as much of both elements was found in the scion top when vigorous rootstock was used in comparison with the dwarfing rootstock (29). This may indicate a superior ability of the vigorous rootstock to absorb and translocate mineral nutrients to the scion in comparison with the dwarfing rootstock. Or it may only mean that roots of the dwarfing rootstock, with their higher percentage of living tissue, formed a greater "sink" for these materials, retaining them in the roots.

Interstocks of such dwarfing apple clones as 'Malling 9' will cause a certain amount of dwarfing, suggesting reduced translocation due to partial blockage at the graft unions or to a reduction in movement of water or nutrient materials (or both) through the interstock piece. Differences among rootstocks in water translocation have been demonstrated with a steady-state, heat-balance technique that accurately measures xylem sap flow rate and sap flow accumulation over time. Under nonstress conditions, sap flow was greater in 'Granny Smith' apple scions grafted to very vigorous seedling (standard) rootstock, while sap flow was similar between the dwarfing and semi-dwarfing rootstock (70). Moisture stress affects the sap flow of the vigorous seedling rootstock the least and reduced sap flow on the dwarfing rootstock the most.

---

spurs The principal fruiting unit in apple, which may be classified as *short shoots.* The *terminal bud* of a spur may be either vegetative, containing only leaves, or reproductive. *Reproductive buds* of apple are *mixed buds* that produce both flowers and leaves.

competing sink The competition of two independent growth processes (such as flowering and adventitious root formation) for the same limited metabolic resources (e.g., carbohydrates, proteins).

Sweet cherries grafted on dwarfing rootstock have smaller and fewer xylem vessels in the scion and graft union, and irregular vessel orientations in the vascular tissue compared to non-dwarfing rootstock; this difference could contribute to greater hydraulic resistance in the graft union, resulting in reduced scion growth (dwarfing) (116). With peach trees grafted on rootstocks with differing size-controlling potentials, the higher root resistance (reduced sap flow) plays a central role in the dwarfing mechanism induced by size-controlling rootstock (semi-dwarfing). Interestingly, the root system accounted for the *majority* of resistance of water flow through the tree and had *no effect* on hydraulic conductance through the scion or graft union (12).

Conversely, with olive trees (*Olea europaea* L.), while there was lower hydraulic conductance in dwarfing than vigorous rootstock during the first several months, after 1 year hydraulic conductivity was the same between dwarfing or vigorous rootstocks (55).

In summary, as long as mineral elements are not limiting, the greater uptake of P and Ca by the more vigorous rootstock does not adequately account for dwarfing effects (29). While rootstocks can influence leaf mineral nutrition, results have been inconsistent (65). In general, sap flow is greatest in vigorous and least in dwarfing rootstock. While differences in sap flow may be attributed to differences in root characteristics, xylem anatomy or other features of the hydraulic architecture from the roots to the graft union, or the union itself, the primary influence probably lies more in the nature of the growth characteristics of such rootstocks.

Phytohormones and Correlative Effects Plants maintain a constant root/shoot ratio, and any attempt to alter this ratio results in the plant redirecting its growth pattern until the ratio is reestablished. This also applies to grafted plants and plants transplanted into a landscape site or orchard. Producing a composite plant by grafting onto a dwarfing rootstock is an alteration in the normal growth pattern (87). Growth in the composite plant will be redirected until equilibrium is reached between the rootstock-scion system. Intimately involved in redirecting plant growth are the **correlative effects** of root (rootstock)/shoot (scion) systems, mediated by phytohormones. Auxins, which are produced predominantly in the shoot system, are basipetally translocated through the phloem and into the root system, where they affect root growth. Cytokinins are produced predominantly from root apices, and are translocated primarily through the xylem, where they can influence physiological responses and growth in the scion.

correlative effects
The influence of one organ over another, due to phytohormones (e.g., high ABA produced in the root tips of dwarfing apple rootstock reduces the vegetative growth of the scion).

Of the phytohormones, auxin plays one of the most important roles in dwarfing rootstock control of apple scion growth (75). The dwarfing effect may be explained by reduced auxin transport into the graft union of the dwarf rootstock (87); this could alter the hormonal balance between shoots and roots, and account for the reduced vegetative growth and vigor of the scion. Auxin affects vascular differentiation, and is important for stimulating cambial activity and xylem development (1) in the graft union area and the vascular system of the grafting partners. Dwarfing yields greater reduction in cambial activity and xylem formation in the graft union than vigorous rootstock (158) because of the dwarfing's reduced capacity to support polar auxin transport (not auxin uptake into cells), and a reduced capacity for auxin efflux from transporting cells (158). Since auxin is known to stimulate its own transport (58), lower endogenous auxin levels in the dwarfing rootstock may limit its capacity to support polar auxin transport. A chain of events is set off with less auxin being transported, which leads to reduced cambial activity and subsequently reduced xylem formation. Reduced xylem formation limits conduction in the dwarf rootstock, which concurs with the reports on lower xylem sap flow (70).

There is evidence for greater auxin accumulation in the scion of dwarfing apple understock. With apple, hydraulic conductivity of the graft tissue was lower for grafted trees on dwarfing rootstocks, compared to semi-vigorous rootstocks. The amount of functional xylem tissue in the graft union and scion initially increased with rootstock vigor (7). However, as the grafted *tree aged, any differences in sap flow become marginal.* The dwarfing tree compensated for hydraulic limitations imposed by the graft tissue and abnormal xylem anatomy (compared to more vigorous rootstock) by initially reducing its transpiring leaf area, and producing a smaller canopy (smaller tree). As the dwarfed tree aged, the cross-sectional area of the graft union increased (7), brought about by greater auxin accumulation (reduced transport) in the graft tissue of the dwarfing rootstock, which led to *increased xylem development later, as the dwarf tree aged.*

Auxin can indirectly affect cytokinin production. Reduced auxin transport leads to a smaller root system in the dwarf rootstock that produces less cytokinin, and/or the root metabolism is sufficiently altered to affect cytokinin synthesis. Subsequently, there is less cytokinin translocated upward from the roots to the shoots and reduced top growth occurs; hence, the dwarfing effect. This correlative effect is mediated by auxin and cytokinin as growth in the composite plant is redirected and equilibrium is reached between the dwarf rootstock/scion system.

Abscisic acid (ABA) and **gibberellic acid (GA)** may also play a role in the correlative effects of dwarfing rootstock. Root apices are an important site of ABA synthesis. The dwarfing 'Malling 9' apple rootstock contains lower amounts of growth-promoting materials—but more growth inhibitors—than does the very vigorous 'Malling 16' rootstock (88). ABA levels are also reported to be higher in dwarfing rootstock (184), and in the stems of dwarfed apple trees, than in more vigorous ones (135).

There are higher ratios of ABA:IAA (auxin) in dwarfing than vigorous apple rootstock, a finding confirmed using gas chromatography-mass spectrophotometer selective ion monitoring techniques (78). Higher ABA:IAA ratios may lead to greater differentiation of phloem and related tissues in dwarfing rootstocks, which could explain why dwarfing rootstocks have higher bark (periderm, cortex, phloem, vascular cambium) to wood (xylem) ratios than vigorous rootstocks. The higher concentration of ABA in shoot bark of dwarfing compared with vigorous rootstock is a potentially useful **marker** in selecting for dwarfing apple rootstock (78).

marker A morphological, biochemical, genetic indication of a trait (e.g., higher ABA in shoot bark of dwarfing compared with a vigorous apple rootstock).

There are conflicting reports that higher GA is found in more vigorous rootstock. Earlier reports concluded there was little evidence to support a role for gibberellin in vigorous, compared to dwarfing rootstock (87, 134, 135). However, in other studies, dwarfing (M9) interstock labeled GA_3 was lower, and glycosyl conjugated GA_3 (inactive GA_3 form) was higher compared to nondwarfing (MM115) interstocks (130). However, a problem with hormonal studies is that the composition of the xylem sap often has very little resemblance to that flowing through the intact, transpiring trees. Hormone and ion concentrations in osmotically exuding sap do not always reflect the condition of the intact plant (5). For instance, slow-flowing sap concentrates solutes faster than fast-flowing sap diluted solutes. Apparently, xylem-borne substances are not delivered in proportion to sap flow, suggesting that differences in tree transpiration or leaf area have considerable influence on signal molecule concentration and delivery (5).

In summary, with apple, auxin is directly involved in dwarfing rootstock effects, and cytokinins (which are affected by auxin-mediated root growth and subsequent cytokinin biosynthesis) are either directly or indirectly involved in plant size control. There is a strong case for ABA-mediated dwarfing effects, while there are conflicting reports on the role of GA. Most likely, there is an interaction of factors affecting dwarfing phenomena such as phytohormones, anatomical factors, nutrition and carbohydrate levels, sap flow, and translocation of carbohydrates across the graft union.

Other Physiological Factors A wide range of physiological characteristics have been found to affect rootstocks, scions, and their resulting interactions (87, 127, 162). For example, rootstocks have been found to influence transpiration rate and crop water-use efficiency in peach; leaf conductance and osmotic potential in apple; and midday leaf water potential in citrus, peach, and

---

BOX 11.9 GETTING MORE IN DEPTH ON THE SUBJECT
MOLECULAR APPROACHES TO STOCK-SCION RELATIONS

There has been recent progress with homografts and heterografts of *Arabidopsis thaliana* as a model system for graft union development. Using mutants of *A. thaliana* could lead to a greater understanding of the fundamental genetic and molecular aspects of graft union formation and plants' stock-scion relations (52). In other developments, the phloem sap transports carbohydrates, amino acids, other nutrients, and specific RNA molecules [small regulatory RNA (183)]. In heterografs of potato (scion) and tomato (understock) graft, transmittable RNA from the leafless tomato rootstock caused changes in the leaf morphology of the potato scion (83). In grafting of transgenic tobacco, **gene silencing** was transmitted by a diffusible messenger that mediated the de novo, post-transcriptional silencing from silenced rootstock to non-silenced scions (117). Hence, grafting enables signaling in plants via RNA and protein movement. While plant yield, desirable dwarfing characteristics, and disease resistance are complex, multi-gene traits, there is future potential for genetic engineering to manipulate desirable RNA that could enhance or suppress scion phenotype characteristics (110). See Box 2.2, pages 35–6 for discussion of micro RNA and gene silencing.

apple trees. Rootstock-scion combinations can also influence net photosynthesis and growth characteristics of grafted *Prunus* species under droughted conditions (127). The greater tolerance to flooding found in selected rootstock of *Prunus* (124) and fir (*Abies*) is probably due to physiological and/or morphological mechanisms (45) that allow selected rootstock to handle anaerobic conditions better than other rootstock.

Net photosynthesis of leaves tends to be higher with apple scions on vigorous rootstock than on dwarfing rootstock (148). But photosynthetic rates cannot be used to explain differences in yield and yield efficiencies induced by the rootstock. Part of this complexity is because the presence of fruit increases leaf net photosynthesis by some unknown mechanism (148).

Cytokinins are known to promote photosynthesis, and root-produced ABA—translocated in xylem sap—can reduce stomatal conductance and photosynthetic rates in the shoot system.

More needs to be done with the molecular basis of rootstock-scion relations. It is possible that certain genes are being turned on and off and/or that genetic information may be transmitted between the graft partners of the composite plant (115). Epigenetic changes occur in grafting with the speeding up of maturation on grafted versus seedling-grown plants (see discussion on epigenetic changes in Chapter 16). Conversely, micropropagated dwarfing apple rootstocks that are grafted can have more juvenile-like characteristics, which delays bearing and fruit cropping of trees (76).

DISCUSSION ITEMS

1. What have been some historical reasons for grafting compared to other propagation methods?
2. Compare budding and grafting.
3. What are the differences between seedlings and clonal rootstock? What are the advantages of each system?
4. Using an interstock (double working) is expensive. Why is it still used as a propagation technique?
5. What are some of the ecological advantages of natural root grafting? How can it be a disadvantage in the dissemination of diseases, such as oak wilt and Dutch elm disease?
6. What are the stages of graft union formation?
7. Does cellular recognition take place in grafting, and, if so, how might that be important to graft compatibility/incompatibility?
8. Why is there potentially more rapid graft union development and frequently a higher percentage of "takes" in chip budding compared to T-budding?
9. What environmental conditions are desirable during and following grafting?
10. What are the genetic limits of grafting,(i.e., when is grafting most likely to be successful)?
11. What are the different types of graft incompatibility, and what causes them?
12. What are some techniques to help predict graft incompatibility?
13. What are some possible mechanisms for size control (dwarfing) in stock-scion relations?

REFERENCES

1. Ahn, S. J., Y. J. Im, G. C. Chung, B. H. Cho, and S. R. Suh. 1999. Physiological response of grafted-cucumber leaves and rootstock roots affected by low root temperature. *Scientia Hort.* 81:397–408.

2. Alexander, J. H. 1998. A summary of graft compatibility from the records of the Arnold Arboretum. *Comb. Proc. Intl. Plant Prop. Soc.* 48:371–83.

3. Aloni, R. 1987. Differentiation of vascular tissues. *Annu. Rev. Plant Physiol.* 38:179–204.

4. Asante, A. K., and J. R. Barnett. 1998. Effect of temperature on graft union formation in mango (*Mangifera indica* L.) *Trop. Agric. (Trinidad)* 75:401–4.

5. Atkinson, C. J., and M. Else. 2001. Understanding how rootstocks dwarf fruit trees. *Compact Fruit Tree* 34:46–9.

6. Atkinson, C. J., M. Policarpo, A. D. Webster, and A. M. Kuden. 1999. Drought tolerance of apple rootstocks: Production and partitioning of dry matter. *Plant Soil* 206:223–25.

7. Atkinson, C. J., M. A. Else, L. Taylor, and C. J. Dover. 2003. Root and stem hydraulic conductivity as determinants of growth potential in grafted trees of apple (*Malus pumila* Mill.). *J. Exp. Bot.* 54:1221–29.

8. Bailey, L. H. 1891. *The nursery book.* New York: Rural Publishing Company.

9. Barnett, J. R., and I. Weatherhead. 1988. Graft formation in Sitka spruce: A scanning electron microscope study. *Ann. Bot.* 61:581–87.

10. Barnett, J. R., and I. Weatherhead. 1989. The effect of scion water potential on graft success in Sitka spruce (*Picea sitchensis*). *Ann. Bot.* 64:9–12.

11. Barnett, J. R., and H. Miller. 1994. The effect of applied heat on graft union formation in dormant *Picea sitchensis* (Bong.). Carr. *J. Exp. Bot.* 45:135–43.

12. Basile, B., J. Marsal, L. I. Solari, M. T. Tyree, D. R. Brylan, and T. M. Dejong. 2003. Hydraulic conductane of peach trees grafted on rootstocks with differing size control potentials. *J. Hort. Sci. Biotech.* 78:768–74.

13. Beakbane, A. B. 1953. Anatomical structure in relation to rootstock behavior. *Rpt. 13th Inter. Hort. Cong.* Vol. I. pp. 152–58.

14. Beakbane, A. B., and W. S. Rogers. 1956. The relative importance of stem and root in determining rootstock influence in apples. *J. Hort. Sci.* 31:99–110.

15. Beeson, R. C., Jr., and W. M. Proebsting. 1988. Scion water relations during union development in Colorado blue spruce grafts. *J. Amer. Soc. Hort. Sci.* 113:427–31.

16. Beeson, R. C., Jr., and W. M. Proebsting. 1988. Relationship between transpiration and water potential in grafted scions of *Picea*. *Physiol. Plant.* 74:481–86.

17. Beeson, R. C., Jr., and W. M. Proebsting. 1989. *Picea* graft success: Effects of environment, rootstock disbudding, growth regulators, and antitranspirants. *HortScience* 24:253–54.

18. Beeson, R. C., Jr., and W. M. Proebsting. 1988. Photosynthate translocation during union development in *Picea* grafts. *Can. J. For. Res.* 18:986–90.

19. Beeson, R. C., Jr. 1991. Scheduling woody plants for production and harvest. *HortTechnology* 1:30–6.

20. Beveridge, C. A., J. J. Ross, and I. C. Murfet. 1994. Branching mutant rms-2 in *Pisum sativum*: Grafting studies and indole-3-acetic acid levels. *Plant Physiol.* 104:953–59.

21. Biggs, F., and T. Biggs. 1990. Tomato grafting. *Comb. Proc. Intl. Plant Prop. Soc.* 40:97–101.

22. Bitters, W. P., and E. R. Parker. 1953. Quick decline of citrus as influenced by top-root relationships. *Calif. Agr. Exp. Sta. Bul.* 733.

23. Bitters, W. P. 1961. Physical characters and chemical composition as affected by scions and rootstocks. In W. B. Sinclair, ed. *The orange: Its biochemistry and physiology.* Berkeley: Univ. Calif. Div. Agr. Sci.

24. Bloch, R. 1943. Polarity in plants. *Bot. Rev.* 9:261–310.

25. Bloch, R. 1952. Wound healing in higher plants. *Bot. Rev.* 18:655–79.

26. Bradford, F. C., and B. G. Sitton. 1929. Defective graft unions in the apple and pear. *Mich. Agr. Exp. Sta. Tech. Bul.* 99.

27. Buchloh, G. 1960. The lignification in stock-scion junctions and its relation to compatibility. In J. B. Pridham, ed. *Phenolics in plants in health and disease.* Long Island City, NY: Pergamon Press.

28. Buck, G. J. 1953. The histological development of the bud graft union in roses. *Proc. Amer. Soc. Hort. Sci.* 63:497–502.

29. Bukovac, M. J., S. H. Wittwer, and H. B. Tukey. 1958. Effect of stock-scion interrelationships on the transport of ^{32}P and ^{45}Ca in the apple. *J. Amer. Soc. Hort. Sci.* 33:145–52.

30. Burnet, F. M. 1971. Self-recognition in colonial marine forms and flowering plants in relation to the evolution of immunity. *Nature (London)* 232:230–35.

31. Carlson, R. F. 1967. The incidence of scion-rooting of apple cultivars planted at different soil depths. *Hort. Res.* 7:113–15.

32. Cation, D., and R. F. Carlson. 1962. Determination of virus entities in an apple scion/rootstock test orchard. *Quart. Bul. Mich. Agr. Exp. Sta. Rpt. I,* 45(2):435–43, 1960. *Rpt. II,* 45(17):159–66.

33. Cline, M. N., and D. Neely. 1983. The histology and histochemistry of the wound healing process in geranium cuttings. *J. Amer. Soc. Hort. Sci.* 108:452–96.

34. Cohen, R., Y. Burger, C. Horev, A. Porat, and M. Edelstein. 2005. Performance of Galia type melons grafted onto Cucurbita rootstock in *Monosporascus cannonballus*-infested and non-infested soils. *Ann. Appl. Biol.* 146:381–87.

35. Copes, D. L. 1969. Graft union formation in Douglas-fir. *Amer. J. Bot.* 56:285–89.

36. Copes, D. L. 1974. Genetics of graft rejection in Douglas-fir. *Can. J. For. Res.* 4:186–92.

37. Cummins, J. N., and D. Gonsalves. 1982. Recovery of tomato ring-spot virus from inoculated apple trees. *J. Amer. Soc. Hort. Sci.* 107:798–800.

38. Davies, F. T., Jr., Y. Fann, and J. E. Lazarte. 1980. Bench chip budding of field roses. *HortScience* 15:817–18.

39. Davis, A. R., P. Perkins-Veazie, R. Hassell, S. R. King, and X. Zhang. 2008. Grafting effects on vegetable quality. *HortScience* 43:1670–72.

40. Deloire, A., and C. Hebant. 1982. Peroxidase activity and lignification at the interface between stock and scion of compatible and incompatible grafts of *Capsicum* on *Lycopersicon*. *Ann. Bot.* 49:887–91.

proteins in *Cucumis sativus* grafted on two *Cucurbita* species. *J. Plant Physiol.* 143:189–94.

166. Torrey, J. G., D. E. Fosket, and P. K. Hepler. 1971. Xylem formation: A paradigm of cytodifferentiation in higher plants. *Amer. Sci.* 59:338–52.

167. Toxopeus, H. J. 1936. Stock-scion incompatibility in citrus and its cause. *J. Pom. and Hort. Sci.* 14:360–64.

168. Troncoso, A., J. Liñán, M. Cantos, M. M. Acebedo, and H. F. Rapoporta. 1999. Feasibility and anatomical development of an *in vitro* olive cleft-graft. *J. Hort. Sci. Biotech.* 74:584–87.

169. Tydeman, H. M., and F. H. Alston. 1965. The influence of dwarfing rootstocks in shortening the juvenile phase of apple seedlings. *Ann. Rpt. E. Malling Res. Sta. for 1964.* pp. 97–8.

170. Vertrees, J. D. 1991. Understock for rare *Acer* species. *Comb. Proc. Intl. Plant Prop. Soc.* 41:272–75.

171. Vöchting, H. 1892. *Veber transplantation am pflanzenköper.*

172. Wagner, D. F. 1969. *Ultrastructure of the bud graft union in* Malus. Ph.D. disseration.1969. Ames: Iowa State Univ.

173. Wang, Z. M., K. J. Patterson, K. S. Gould, and R. G. Lowe. 1994. Rootstock effects on budburst and flowering in kiwifruit. *Scientia Hort.* 57:187–99.

174. Wang, Z. M., K. S. Gould, and K. J. Patterson. 1994. Comparative root anatomy of five *Actinida* species in relation to rootstock effects on kiwifruit flowering. *Ann. Bot.* 73:403–14.

175. Wareing, P. F., C. E. A. Hanney, and J. Digby. 1964. The role of endogenous hormones in cambial activity and xylem differentiation. In M. H. Zimmerman, ed. *The formation of wood in forest trees.* New York: Academic Press.

176. Warmund, M. R., B. H. Barritt, J. M. Brown, K. L. Schaffer, and B. R. Jeong. 1993. Detection of vascular discontinuity in bud unions of 'Jonagold' apple on mark rootstock with magnetic resonance imaging. *J. Amer. Soc. Hort. Sci.* 118:92–6.

177. Webber, H. J. 1943. The "tristeza" disease of sour orange rootstock. *Proc. Amer. Soc. Hort. Sci.* 43:160–68.

178. Wells, R. B. 1986. A historical review of grafting techniques. *Comb. Proc. Intl. Plant Prop. Soc.* 35:96–101.

179. Westwood, M. N. 1993. *Temperate-zone pomology: Physiology and culture.* 3rd ed. Portland, OR: Timber Press.

180. Wildon, D. C., J. F. Thain, P. E. H. Minchin, I. R. Gubb, A. J. Riley, Y. D. Skipper, H. M. Doherty, P. J. O'Donnell, and D. J. Bowles. 1992. Electrical signalling and systemic proteinase inhibitor induction in the wounded plant. *Nature* 360:62–5.

181. Williamson, J. G., W. S. Castle, and K. E. Koch. 1992. Growth and 14C-photosynthate allocation in citrus nursery trees subjected to one or three bud-forcing methods. *J. Amer. Soc. Hort. Sci.* 117:37–40.

182. Williamson, J. G., and B. E. Maust. 1995. Growth of budded, containerized, citrus nursery plants when photosynthesis of rootstock shoots is limited. *HortScience* 30: 1363–5.

183. Wu, X., D. Weigel, and P.A. Wigge. 2006. Signaling in plants by intercellular RNA and protein movement. *Genes Dev.* 16:151–58.

184. Yadava, U. L., and D. F. Dayton. 1972. The relation of endogenous abscisic acid to the dwarfing capability of East Malling apple rootstocks. *J. Amer. Soc. Hort. Sci.* 97:701–5.

185. Yagishita, S., Y. Hirata, J. Okochi, K. Kimura, H. Miukami, and H. Ohashi. 1985. Characterization of graft-induced change in capsaicin contents of *Capsicum annuum* L. *Euphytica* 34:297–301.

186. Yates, I. E., and D. Sparks. 1992. Pecan cultivar conversion by grafting onto roots of 70-year-old trees. *HortScience* 27:3–7.

187. Yeager, A. F. 1944. Xylem formation from ring grafts. *Proc. Amer. Soc. Hort. Sci.* 44:221–22.

188. Yeoman, M. M., and R. Brown. 1976. Implications of the formation of the graft union for organization in the intact plant. *Ann. Bot.* 40: 1265–76.

189. Yeoman, M. M. 1984. Cellular recognition systems in grafting. *Encyclopedia of Plant Physiology.* Heidelberg: Springer–Verlag. pp. 453–72.

*12 Techniques of Grafting

learning objectives

- Explain the requirements for successful graftage.
- Describe the techniques of detached scion graftage, approach graftage, and repair graftage.
- Discuss the preparation for grafting—tools, accessories, machines, automation, and processing scionwood.
- Explain the craftsmanship of grafting—manual techniques, record keeping, and mechanization.
- Describe the aftercare of grafted plants—in bench grafting systems, and field and nursery grafting systems.
- Identify field, bench, and miscellaneous grafting systems.

INTRODUCTION

Since people first learned to graft plants, a myriad of grafting techniques have been developed. In *The Grafter's Handbook,* Garner (19) enumerates and describes some forty different grafts.

Here we describe the most important grafting methods. Among them, a person who can use a sharp knife can find one that meets any specific grafting need. However, success in grafting depends not only on a technically correct graft but in preparation of the scion and rootstock for graftage. Equally critical are the optimum time for grafting, and proper aftercare.

With high labor costs, only a few of the more efficient grafts are utilized in United States woody ornamental nurseries, including the side veneer, splice (whip graft), and whip-and-tongue graft; use of approach and repair graftage is limited. With fruit crops, depending on the species, a number of different apical, side, and root grafts are utilized around the world. Chip budding and T-budding, which are described in detail in Chapter 13, are two of the most common budding methods for woody ornamentals and fruit crops. Vegetable grafting has increased dramatically worldwide—and is commonly done in Asia and Europe where land is intensively used and crops are not rotated. Grafting onto rootstock resistant to soil pathogens and environmental stress helps increase yield and reduce chemical usage (12, 21, 34). For example, some of the most important grafts with cucurbit vegetables (melon, squash) include hole insertion grafting, tongued approach, and one cotyledon graft (also known as the splice, slant, or the Japanese tube graft), which are described in the chapter. Some robotic vegetable grafting machines can produce 800 grafts per hour.

This chapter is divided into three sections: (a) the **types of grafts,** (b) **production processes of graftage**—including the preparation, craftsmanship, and aftercare of grafted plants, and (c) **grafting systems,** including field grafting, bench grafting, and miscellaneous grafting systems—such as herbaceous graftage, cutting grafts, and micrografting.

REQUIREMENTS FOR SUCCESSFUL GRAFTING

For any successful grafting operation, producing a plant, as shown in Figure 12–1, requires five important elements:

1. *The rootstock and scion must be compatible.* They must be capable of uniting. Usually, but not always, closely related plants, such as two apple cultivars, can be grafted together. Distantly related plants, such as oak and apple, cannot make a successful graft combination (see Chapter 11 for a discussion of these factors).

Figure 12–1
Cultivar of Japanese maple (*Acer palmatum*) grafted on seedling rootstock. The characteristics of the genetically different scion and rootstock remain distinctly different after grafting, exactly to the junction (arrow) of the graft union. (a) Prepping seedling rootstock. (b) Attaching darker scion via side veneer graft. (c) Composite plant with grafted cultivar leafing out. *Courtesy B. Upchurch.*

2. *The **vascular cambium** of the scion must be placed in direct contact with that of the rootstock.* The cut surfaces should be held together tightly by wrapping, nailing, wedging, or some similar method. Rapid development of the graft union is necessary so that the scion may be supplied with water and nutrients from the rootstock by the time the buds start to open.

   > **vascular cambium** The tissue responsible for the formation of new xylem and phloem in the development of a successful graft union.

3. *The grafting operation must be done at a time when the rootstock and scion are in the proper physiological stage.* Usually, this means that the scion buds are dormant while at the same time, the cut tissues at the graft union are capable of producing the callus tissue necessary for healing of the graft. For deciduous plants, dormant scionwood is collected during the winter and kept inactive by storing at low temperatures. The rootstock plant may be dormant or in active growth, depending upon the grafting method used.

4. *Immediately after the grafting operation is completed, all cut surfaces must be protected from desiccation.* The graft union is covered with tape, grafting wax, Parafilm tape, Buddy Tape, or the grafts are placed in moist material or a covered grafting frame.

5. *Proper care must be given to the grafts for a period of time after grafting.* Shoots (suckers) coming from the rootstock below the graft will often choke out the desired growth from the scion. In some cases, shoots from the scion will grow so vigorously that they break off unless staked and tied or cut back.

TYPES OF GRAFTS

Grafting may be classified according to the part of the rootstock on which the scion is placed—a root, or various places in the top of the plant. Types of grafts can be categorized as (1) **detached scion graftage,** which includes apical, side, bark, and root graftage; (2) **approach graftage,** where the root system of the scion and the shoot system of the rootstock are not removed until after successful graft union formation occurs; and (3) **repair graftage of established trees.** The grafts that are categorized in Tables 12–1 and 12–2 are described in greater detail later in the chapter.

> **detached scion graftage** A type of graft used when a section of the shoot of the scion is removed and grafted to the apex or side of the rootstock. It is also used in grafting roots (**root graftage**).
>
> **approach graftage** The root system of the scion and shoot system of the rootstock are not removed until after successful graft union formation occurs.
>
> **repair graftage** Graft used in repairing or reinforcing injured or weak trees.

Table 12–1
TYPES OF GRAFTS

I. **Detached Scion Graftage**
   A. **Apical Graftage**
      Whip-and-tongue graft
      Splice graft [whip graft; with vegetables—One cotyledon graft (OCG) or Japanese tube graft]
      Cleft graft (split graft)
      Wedge graft (saw-kerf graft)
      Saddle graft
      Four-flap graft (banana graft)
      Hole Insertion Graft (HIG) or Terminal/Tip Insertion graft with vegetables
   B. **Side Graftage**
      Side-stub graft
      Side-tongue graft
      Side-veneer graft
      Side insertion graft (SIG) with vegetables
   C. **Bark Graftage**
      Bark graft (rind graft)
      Inlay bark graft
   D. **Root Graftage**
      Whole-root and piece-root grafting
      Nurse-root grafting
II. **Approach Graftage**
   Spliced approach graft
   Tongued approach graft (TAG)
   Inlay approach graft
III. **Repair Graft**
   Inarching
   Bridge graft
   Bracing

Table 12–2
UTILIZATION AND ROOTSTOCK CRITERIA OF SELECTED GRAFTS

Graft type	Diameter of rootstock	Rootstock condition	Uses
Whip-and-tongue graft	Small: 6 to 13 mm (1/4 to 1/2 in); same diameter as scions	Dormant; however, active with bench grafting of container rootstock	Bench grafting; container grafting; some topworking in field; root grafting; a popular graft
Whip graft (splice graft) —also called **One cotyledon graft (OCG) or Japanese tube graft** with vegetables.	Small: 6 to 13 mm (1/4 to 1/2 in); same diameter as scions; See Figure 12–46 for schedule.	Dormant; however, active with bench grafting of container rootstock, greenwood grafting, and vegetable crops	Bench grafting; container grafting; some topworking in field; grafting of vegetable liner plants; root grafting; a popular graft
Cleft graft (split graft)	Moderate: 2.5 to 10 cm (1 to 4 in)	Dormant—before active growth starts in spring	Topworking in field
Wedge graft (saw-kerf graft)	Moderate: 2.5 to 10 cm (1 to 4 in)	Dormant—before active growth starts in spring	Topworking in field
Saddle graft	Small: 6 to 19 mm (1/4 to 3/4 in); same diameter as scion	Dormant	Bench grafting via hand or machine; container grafting; root grafting
Four-flap graft (banana graft)	Small: up to 2.5 cm (1 in); same diameter as scions	Active; bark must be slipping	Topworking small caliper trees
Hole insertion graft (HIG) or Terminal/Top insertion graft			Bench grafting; container grafting of liner vegetable plants

Graft type	Diameter of rootstock	Rootstock condition	Uses
Side-stub graft	Small to moderate rootstock larger than scion: grafted on rootstock branches up to 2.5 cm (1 in) in diameter	Dormant	Topworking in field
Side-tongue graft	Small: 6 to 19 mm (1/4 to 3/4 in); diameter of scion slightly smaller than rootstock	Dormant	Bench grafting; container grafting of broad- and narrow-leaved evergreen species
Side-veneer graft	Small: 6 to 19 mm (1/4 to 3/4 in); same diameter as scion	Dormant	Bench grafting; container grafting of smaller liner potted plants; a popular graft for conifers, deciduous trees and shrubs, and fruit crops
Side insertion graft (SIG)	Works well using vegetable rootstock with wide hypocotyls.	Active	Bench grafting; container grafting of liner vegetable plants
Bark graft (rind graft)	Large: 2.5 to 30 cm (1 to 12 in).	Active; bark must be slipping	Topworking in field
Inlay bark graft	Large: 2.5 to 30 cm (1 to 12 in).	Active; bark must be slipping	Topworking in field (e.g., pecans)
Spliced approach graft	Small: 6 to 19 mm (1/4 to 3/4 in); same size as scion; exception is mango grafting in India on larger, established trees	Active	Container grafting with difficult-to-graft species; scion and stock grafted as two independent, self-sustaining plants; only limited topworking in field
Tongued approach graft (TAG)	Small: 6 to 19 mm (1/4 to 3/4 in); same size as scion	Active	Container grafting with difficult-to-graft species; scion and stock grafted as two independent, self-sustaining plants; also used with vegetables
Inlay approach graft	Small: 6 to 19 mm (1/4 to 3/4 in); bark of rootstock is thicker than scion	Active	Container grafting with difficult-to-graft species; scion and stock grafted as two independent, self-sustaining plants
Inarching	Large: 15 cm (6 in) and larger	Dormant	Used to replace a weak or damaged root system of an established tree
Bridge graft	Large: 15 cm (6 in) and larger	Active; bark must be slipping	Repair injury to trunk of tree
Bracing	Limbs of tree bound by pulling together two strong young lateral shoots from limbs to be braced	Active or dormant	Natural grafting used to strengthen scaffolding limbs of a tree

Detached Scion Graftage— Apical Graftage

There are many variations of apical graftage. As the name suggests, the scion is inserted into the top of the severed rootstock shoot.

Whip-and-Tongue Graft The whip-and-tongue graft, shown in Figures 12–2 and 12–3, is particularly useful for grafting relatively small material about 6 to 13 mm (1/4 to 1/2 in) in diameter. It is highly successful if done properly because there is considerable vascular cambium contact,

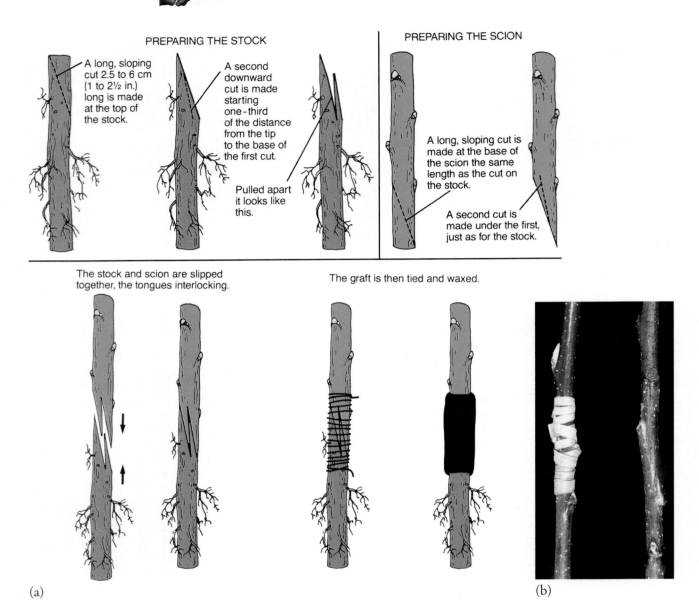

Figure 12–2
Whip-and-tongue graft. (a) This method is widely used in grafting small plant material and is especially valuable in making root grafts as illustrated here. (b) Whip-and-tongue of bench grafted pear.

plus it heals quickly and makes a strong union. Preferably, the scion and rootstock should be of equal diameter. The scion should contain two or three buds, and the graft made in the smooth internodes area below the lower bud.

The cuts made at the top of the rootstock should be the same as those made at the bottom of the scion. First, a smooth, sloping cut is made, 2.5 to 6 cm (1 to 2 1/2 in) long; longer cuts are made when working with large material. This first cut should preferably be made with one single stroke of the knife, in order to leave a smooth, flat surface. To do this, the knife must be razor sharp. Wavy, uneven cuts made with a dull knife will not result in a satisfactory union.

On each of these cut surfaces, a reverse cut is made. It is started downward at a point about one-third of the distance from the tip and should be about one-half the length of the first cut. To obtain a smooth-fitting graft, this second cut should not just split the grain of the wood but should follow along under the first cut, tending to parallel it.

The rootstock and scion are then inserted into each other, with the tongues interlocking. It is extremely important that the vascular cambium layers match along at least one side, preferably along both sides. The lower tip of the scion should not overhang the stock, because it increases the likelihood of the formation of

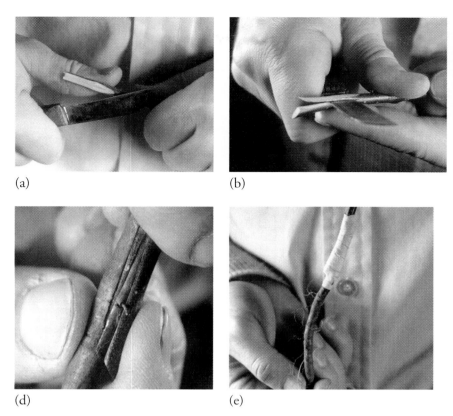

Figure 12–3
Procedures in making a whip-and-tongue graft: (a) Slice cut is made across both the rootstock and scion. (b) A second cut is made to the tongue; the grafter's hands are locked together to avoid injury. (c and d) Fitting and locking the tongues of the graft partners. (e) Wrapping the whole-root apple graft with grafting tape.

large callus knots. The use of scions larger than the rootstock should be avoided for the same reason.

After the scion and rootstock are fitted together, they are securely tied with budding rubber strips, plastic (poly) budding/grafting tape, or raffia. It is important that the tissues in the graft union area not dry out, so either sealing the graft union with grafter's wax, Parafilm, or Buddy Tape, or placing the plants under high relative humidity, is essential until the graft union has formed.

In **bench graftage** (page 502) the bare-root grafted plants can be stored in a grafting box (without sealing the graft union with grafter's wax) and packed with slightly moist peat or bark. Grafted plants in liner pots can be placed in a polytent in a temperature-controlled greenhouse (Fig. 12–4). If bare-root, bench-grafted plants are to be directly planted in a field nursery, the graft union is temporarily placed below the soil level. Any poly budding tape will need to be removed after graft union formation to prevent girdling the stem. Grafts wrapped with budding rubbers and temporarily covered with soil or media should be inspected later; the rubber decomposes very slowly below ground and may cause a constriction at the graft union.

If the whip-and-tongue graft is used in **field grafting**, the graft union of the **topworked** (page 422) plant must be tied and sealed with grafter's wax, Parafilm, or Buddy Tape. Aftercare of grafted plants is further described in the section "Production Processes of Graftage" (page 502).

Splice Graft (Whip Graft) The splice graft is simple and easy to make (Fig. 12–5). It is the same as the whip-and-tongue graft except that the second, or "tongue," cut is not made in either the rootstock or scion. A simple slanting cut of the same length and angle is made in both the rootstock and the scion. These are placed together and wrapped or tied as described for the whip graft. If the scion is smaller than the rootstock it should be set at one side of the rootstock so that the vascular cambium layers will match along that side (Fig. 12–5).

The splice graft is particularly useful in grafting plants that have a very pithy stem or that have wood that is not flexible enough to permit a tight fit when a tongue is made as in the whip-and-tongue graft. The splice graft is used in greenhouse production of vegetable crops for grafting disease-resistant rootstocks. For vegetable crops such as cucurbits or *Solanaceae*, this graft is sometimes referred to as One Cotyledon Grafting (OCG), the slant graft, or Japanese tube graft

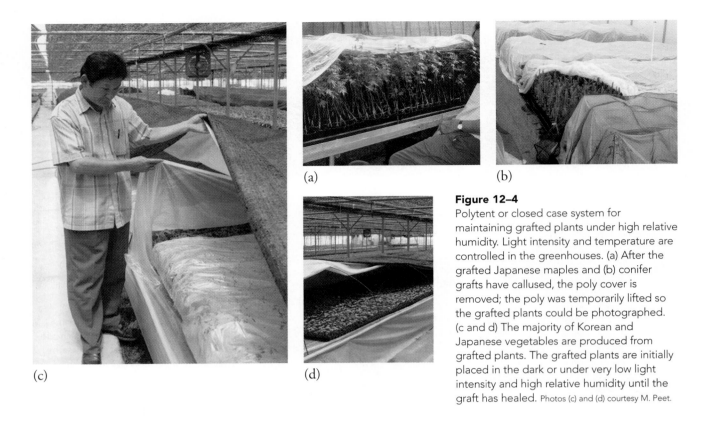

Figure 12–4
Polytent or closed case system for maintaining grafted plants under high relative humidity. Light intensity and temperature are controlled in the greenhouses. (a) After the grafted Japanese maples and (b) conifer grafts have callused, the poly cover is removed; the poly was temporarily lifted so the grafted plants could be photographed. (c and d) The majority of Korean and Japanese vegetables are produced from grafted plants. The grafted plants are initially placed in the dark or under very low light intensity and high relative humidity until the graft has healed. Photos (c) and (d) courtesy M. Peet.

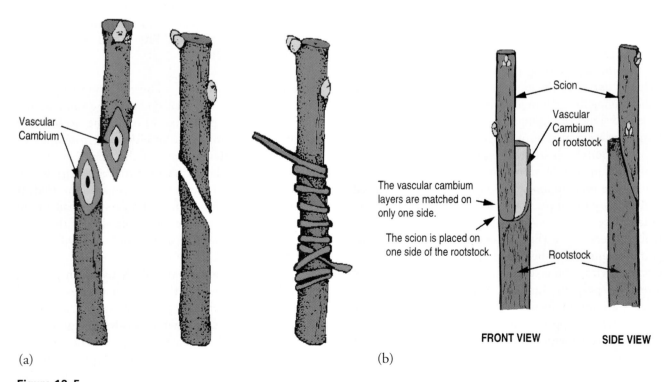

Figure 12–5
Splice graft (whip graft). (a) Procedures in making the splice graft with a slice cut that slants across the grafting partners. Ideally, the rootstock and scion are of the same caliber. (b) Method of making a splice graft when the scion is considerably smaller than the rootstock. It is important that the cambium layers be matched on one side.

BOX 12.1 GETTING MORE IN DEPTH ON THE SUBJECT
CORRECT INSERTION OF THE SCION

In all types of grafting, the scion must be inserted right side up. That is, the apical tip of the buds on the scion should be pointing upward and away from the rootstock. The graft will not be successful if this rule is not observed.

(Figs. 12–6 and 12–7) (11, 12, 21). The graft can be performed manually or with sophisticated, robotic grafting machines; see Figures 12–43, 12–44, and 12–45, pages 498–99. The rootstock and scion must be held together while tying the splice graft. In field grafting, it is not a convenient method to use at ground level, and must be performed higher up on the rootstock, where the grafter must do both the cutting and tying. The whip-and-tongue does not have this limitation, since the tongue holds the graft together, so that

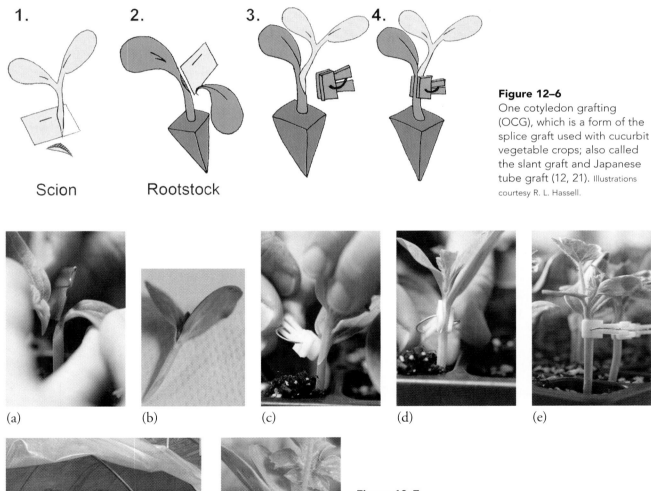

Figure 12–6
One cotyledon grafting (OCG), which is a form of the splice graft used with cucurbit vegetable crops; also called the slant graft and Japanese tube graft (12, 21). Illustrations courtesy R. L. Hassell.

Figure 12–7
One cotyledon grafting (OCG): A form of splice graft used with cucurbits. (a) Preparation of squash rootstock leaving a single cotyledon leaf. (b) Watermelon scion with slant cut. (c) Plastic clip used to hold scion and rootstock. (d and e) Plastic clips used to hold watermelon scion and squash rootstock. (f) Grafts are allowed to heal under very high humidity and dark to very low light conditions until graft union formation has occurred. (g) Successfully healed OCG. Photos courtesy of R. L. Hassell.

grafter The person cutting the stock and scion and inserting the scion piece.

Tier The person who completes the grafting process by tying, and sometimes waxing, the graft area.

the **grafter** has both hands free for tying, or the fitted graft can be left for a helper or **"tier"** to tie and seal. A splice graft is used in bench grafting and grafting of container plants.

Cleft Graft (Split Graft) The cleft or split graft is one of the oldest methods of field grafting. It is used to topwork trees, either in the trunk of a small tree or in the scaffold branches of a larger tree (Figs. 12–8 and 12–9). Cleft grafting is used for *crown grafting* (see the "Grafting Systems" section, page 504) or grafting smaller plants such as established grapevines or camellias. In topworking trees, this method should be limited to rootstock branches about 2.5 to 10 cm (1 to 4 in) in diameter, and to species with fairly straight-grained wood that will split evenly.

Although cleft grafting can be done any time during the dormant season, the chances for successful healing of the graft union are best if the work is done in early spring just when the buds of the rootstock are beginning to swell, but before active growth has started. If cleft grafting is done after the tree is in active growth, the bark of the rootstock may separate from the wood, making it difficult to obtain a good union. When this separation occurs, the loosened bark must be firmly nailed back in place. The scions should be made from dormant, 1-year-old wood. Unless the grafting is done early in the season (when the dormant scions can be collected and used immediately), the

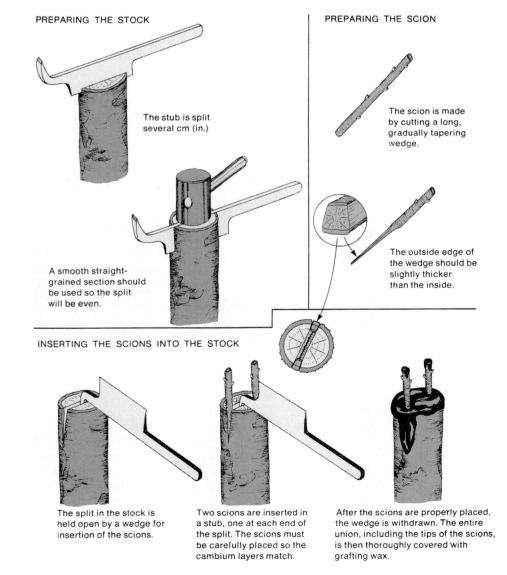

Figure 12–8
Steps in making the cleft graft (split graft). This method is very widely used and is quite successful if the scions are inserted so that the cambium layers of stock and scion match properly.

Figure 12–9

(a) Tools used in making a cleft (split) graft. (b) Making the cleft by splitting the rootstock top. (c) Scionwood with the outside wedge slightly thicker than the inside. (d) Inserting the first of two scions. The split rootstock is temporarily separated by the tool. (e) Cleft grafted yellow kiwi (scion) to replace the standard kiwi.

scionwood should be collected in advance and refrigerated. In sawing off the branch for this and other top-working methods, the cut should be made at right angles to the main axis of the branch.

In making the cleft graft, a heavy knife, such as a butcher knife, or one of several special cleft grafting tools, is used to make a vertical split for a distance of 5 to 8 cm (2 to 3 in) down the center of the stub to be grafted (Figs. 12–8 and 12–9). This split is made by pounding the knife in with a hammer or mallet. The branch is sawed off in such a position that the end of the stub that is left is smooth, straight-grained, and free of knots for at least 15 cm (6 in). Otherwise, the split may not be straight, or the wood may split one way and the bark another. The split should be in a tangential rather than radial direction in relation to the center of the tree to permit better placement of the scions for their subsequent growth. Sometimes the cleft is made by a longitudinal saw cut rather than by splitting. After a good, straight split is made, a screwdriver, chisel, or the wedge part of the cleft-grafting tool is driven into the top of the split to hold it open.

Two scions are inserted, one at each side of the stock where the vascular cambium layer is located. The scions should be 8 to 10 cm (3 to 4 in) long, about 10 to 13 mm (3/8 to 1/2 in) thick, and should have two or three buds. The basal end of each scion should be cut into a gently sloping wedge about 5 cm (2 in) long. It is not necessary that the end of the wedge come to a point. The side of the wedge which is to go to the outer side of the rootstock should be slightly wider than the inside edge. Thus, when the scion is inserted and the tool is removed, the full pressure of the split rootstock will come to bear on the scions at the position where the vascular cambium of the rootstock touches the vascular cambium layer on the outer edge of the scion. Since the bark of the rootstock is almost always thicker than the bark of the scion, it is usually necessary for the outer surface of the scion to set slightly in from the outer surface of the rootstock in order to match the vascular cambium layers.

The long, sloping wedge cuts at the base of the scion should be smooth, a single cut on each side made with a sharp knife. Both sides of the scion wedge should press firmly against the rootstock for their entire length. A common mistake in cutting scions for this type of graft is to make the cut on the scion too short and the slope too abrupt, so that the point of contact is only at the top. Slightly shaving the sides of the split in the stock will often permit a smoother contact.

After the scions are properly made and inserted, the tool is withdrawn, without disturbing the scions, which should be held tightly by the pressure of the rootstock so that they cannot be pulled loose by hand. No further tying or nailing is needed unless very small rootstock branches have been used. In this case, the top of the rootstock can be wrapped tightly with poly grafting tape or adhesive tape to hold the scions in place more securely.

Thorough waxing of the completed graft is essential. The top surface of the stub should be entirely covered, permitting the wax to work into the split in the stock. The sides of the grafted stub should be well covered with wax as far down the stub as the length of the split. The tops of the scions should be waxed but not necessarily the bark or buds of the scion. Two or three days later, all the grafts should be inspected and rewaxed where openings appear. Lack of thorough and complete waxing in this type of graft is a common cause of failure.

Wedge Graft (Saw-Kerf Graft) Wedge grafting is illustrated in Figure 12–10. Like the cleft graft, it can be made in late winter (in mild climates) or early spring before the bark begins to slip (separates easily from the wood).

The diameter of the stock to be grafted is the same as for the cleft graft—5 to 10 cm (2 to 4 in), and the scions are also the same size—10 to 13 cm (4 to 5 in) long and 10 to 13 mm (3/8 to 1/2 in) in thickness.

A sharp, heavy, short-bladed knife is used for making a V-wedge in the side of the stub, about 5 cm (2 in) long. Two cuts are made, coming together at the bottom and as far apart at the top as the width of the scion. These cuts extend about 2 cm (3/4 in) deep into the side of the stub. After these cuts are made, a screwdriver is pounded downward behind the wedge chip from the top of the stub to knock out the chip, leaving a V-shaped opening for insertion of the scion. The base of the scion is trimmed to a wedge shape exactly the same size and shape as the opening. With the two vascular cambium layers matching, the scion is tapped downward, firmly into place, and slanting outward slightly at the top so that the vascular cambium layers cross. If the cut is long enough and gently tapering, the scion should be so tightly held in place that it would be difficult to dislodge.

In a stub that is 5 cm (2 in) wide, 2 scions should be inserted 180 degrees apart; in a 10-cm (4 in) stub, 3 scions should be used, 120 degrees apart. After all scions are firmly tapped into place, all cut surfaces, including the tips of the scion, should be waxed thoroughly.

Saddle Graft The saddle graft can be bench grafted by hand or machine (see Fig. 12–41, page 497). The rootstock and scion should be the same size. The scion is prepared by cutting upward through the bark and into the wood on opposite sides of the scion (Fig. 12–11, page 476). The knife should penetrate more deeply into the wood as the cuts are lengthened. Before the knife is withdrawn, it is turned towards the middle of the scion piece, and the saddle shape is gradually formed by removing pieces of the wood. The rootstock is cut transversely and receives two upward cuts on either side to expose the vascular cambium of the rootstock, in order to match vascular cambium in the saddle of the scion. The apex of the rootstock is carved to fit the saddle. The graft needs to be tied, and all exposed cut surfaces sealed or stored in a grafting case until the graft union has formed. The saddle graft is used for bench grafting grape and *Rhododendron* cultivars (19).

Four-Flap Graft (Banana Graft) The four-flap graft is used in topworking small-caliper trees or tree limbs up to 2.5 cm (1 in) in diameter. This field graft is normally done manually [Figs. 12–12 (page 476) and 12–13 (page 477)], but there is a tool that aids in stripping the rootstock bark flaps from the wood (Fig. 12–13). Both the scion and rootstock should be of equal diameter, and the best fit is obtained when the scion is slightly larger than the rootstock. The four-flap graft is done with pecans in Texas from April to mid-May, when the rootstock bark is actively slipping (39). Scionwood, which is collected while dormant during the winter, is taken from cold storage and used immediately.

The rootstock with a primary stem or lateral limb is severed horizontally with sharp pruning shears. On the rootstock where the horizontal cut was made, 4 vertical, equally spaced cuts 4 cm (1.5 in) long are made with a grafting knife that penetrates from the bark down to the interior wood. A 15 cm (6 in) piece of scionwood with 3 axillary buds is cut on 4 sides with a knife. Cuts are made on the scion through the bark down to the wood—without removing much wood. There should be 4 thin slivers of bark, with the vascular cambium at the corners, which gives the prepared scion a square diameter appearance. The 4 flaps of bark are pulled down 4 cm (1.5 in) on the rootstock, and the inner wood is removed with pruning shears. The scion piece is inserted upright on the rootstock and the 4 flaps of the rootstock are pulled up to cover the 4 cut surfaces of the scion. A rubber band is rolled up onto the flaps to hold them in place. The cut flap areas are then tied with flagging tape, green floral tape, or white budding tape. The tip of the scion is painted with tree paint or sealed with white glue to prevent it from drying out.

Then the taped graft area is covered with aluminum foil to protect it from heat. A hole is made in the corner of a clear poly bag (freezer bag) and the poly

PREPARING THE ROOTSTOCK

A heavy sharp knife is pounded into the side of the stub to make two cuts to form a V.

A screwdriver is used to flip out the V-shaped chip, leaving a space for insertion of the scion.

PREPARING THE SCION

The scion should be about 10 to 13 cm (4 to 5 in.) long, 10 to 12 mm (3/8 to 1/2 in.) thick, and with 2 or 3 healthy vegetative buds. The basal ends should be cut to a V-shaped wedge, matching the opening in the stock.

(b)

(c)

INSERTING THE SCIONS INTO THE ROOTSTOCK

The scion is gently tapped into the V-shaped opening in the stock, matching the cambium layers at a slight angle so that the cambium of stock and scion cross.

Scion should be inserted at an angle so that the cambium layers of stock and scion are closely matched, barely crossing each other.

After scions are in place all cut surfaces are thoroughly covered with grafting wax.

(d)

(a)

(e)

(f)

(g)

Figure 12–10
(a) Wedge graft (saw-kerf graft). Sometimes called the saw-kerf because the cuts in the side of the rootstock can be made with a saw, rather than with the sharp tool depicted. (b, c, d, e, and f) Wedge graft of cherry whips in field using one scion piece. (b) Trimming scionwood with grafting knife, (c and d) inserting scionwood into rootstock, (e) wrapping graft with poly, (f) sealing with grafting wax—notice wax container (arrow) and (g) tied graft with poly and grafting wax covering bottom of scion and graft union area (arrow) used to fill in tissue separation.

475

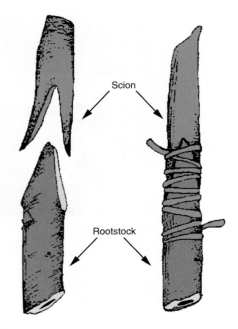

Figure 12–11
Steps in making the saddle graft. The scion is cut to have a saddle appearance and the understock to form a blunt point. The cambium layers are matched up and the graft tied.

slid down over the graft area so that it is just covering the aluminum foil (no poly should cover the apex of the scion, nor should it touch any exposed wood). Air is expelled so the poly fits snugly over the aluminum foil, and it is tied at both ends with stretchable plastic budding tape or rubber bands. The function of the poly bag is to maintain a high relative humidity in the graft area. In 4 to 6 weeks after the graft has taken, the ties, poly bag, and aluminum foil are removed.

The vegetative growth of the rootstock plant must be kept in check, since many new shoots will appear on the rootstock below the graft. Some of these shoots are needed for maintaining tree vigor, but the rootstock shoots should not become dominant or exceed the height of the scion—the growing tips of the rootstock shoots will have to be removed several times during the growing season. After 2 to 3 years, all rootstock branches are removed below the graft and the scion becomes the dominant shoot system.

Hole Insertion Graft (HIG) or Terminal/Top Insertion Graft This technique is used for grafting watermelon to squash rootstock (12, 21). This is most popular graft used in China because it is suitable for *Lagenaria* (Cucurbita) and interspecific squash as rootstocks, requires few materials, is highly efficient (1,500+ plants/day/worker), and allows simpler management techniques (34). When both cotyledons and first true leaf start to develop, the rootstock plant is ready to graft (7 to 10 days after sowing). Remove the growing point with a sharp probe, and then

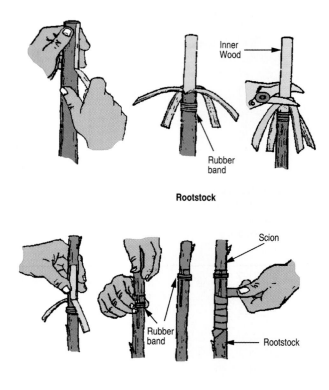

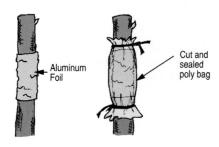

Figure 12–12
The four-flap or banana graft. *Top:* The top of the rootstock is cut horizontally, and the bark is cut vertically into four strips. The four bark flaps are peeled down and the inner wood removed. *Middle:* The scion bark is removed and the wood retained. The flaps of the rootstock cover the cut surfaces of the scionwood and are temporarily held by a rubber band. The graft is then tied with white grafting tape. *Bottom left:* Aluminum foil is wrapped around the graft to exclude heat from the graft. *Bottom right:* The grafted area covered with aluminum foil is wrapped with a cut poly bag, which is sealed to retain high relative humidity until the graft takes (39).

open a hole on the upper portion of the rootstock hypocotyl with a bamboo needle or 1.4-mm drill bit. The scion is then cut on a 35- to 45-degree angle, on both sides, on the hypocotyls and inserted into the hole made in the rootstock. The cut surfaces are matched together, held with or without a grafting clip and transferred to a humidity chamber or healing room. Grafted plants should not be older than 33 days before transplanting (Figs. 12–14 and 12–15, page 478) (21).

Figure 12–13
Steps in the four-flap or banana graft. (a) Preparing the rootstock by severing the rootstock top. (b) The four bark flaps of the rootstock with the "wood" of the rootstock removed. (c and d) Prepping the scion by removing the bark of the scion with wood left intact. (e and f) Flaps of the rootstock cover the cut surfaces of the scionwood and are temporarily held by grafting tape, then covered with aluminum foil and poly to prevent desiccation. (g and h) A tool for the four-flap graft, which is slid over the rootstock and used to cut and peal the four flaps.

Detached Scion Graftage—Side Grafting

There are many types of side graftage. As the name suggests, the scion is inserted into the side of the rootstock, which is generally larger in diameter than the scion. This method has proven useful for large-scale propagation of nursery trees (36). Generally, the rootstock shoot is removed after the graft takes, and the scion becomes the dominant shoot system.

Side-Stub Graft The side-stub graft is useful in grafting branches of trees that are too large for the whip-and-tongue graft, yet not large enough for other methods such

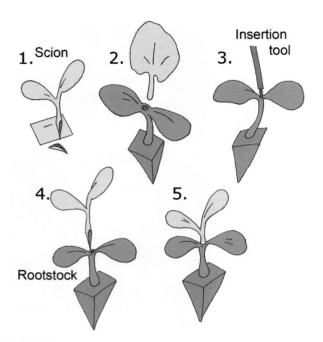

Figure 12–14
Hole insertion graft (HIG) or terminal/top insertion graft (12, 21).
Illustrations courtesy R. L. Hassell.

Figure 12–15
(a and b) Hole insertion graft (HIG) for grafting watermelon to squash rootstock. This is the most popular graft used in China because it is suitable for *Lagenaria* (*Cucurbita*) and interspecific squash as rootstocks, requires few materials, has high efficiency, 1500+ plants/day/worker, and simpler management techniques (34). Photos courtesy of R. L. Hassell.

as the cleft or bark graft. For this type of side graft, the best rootstocks are branches about 2.5 cm (1 in) in diameter. An oblique cut is made into the rootstock branch with a chisel or heavy knife at an angle of 20 to 30 degrees. The cut should be about 2.5 cm (1 in) deep and at such an angle and depth that when the branch is pulled back, the cut will open slightly but will close when the pull is released.

The scion should contain two or three buds and be about 7.5 cm (3 in) long and relatively thin. At the basal end of the scion, a wedge about 2.5 cm (1 in) long is made. The cuts on both sides of the scion should be very smooth, each made by one single cut with a sharp knife. The scion must be inserted into the rootstock at an angle, as shown in Figure 12–16, to obtain maximum contact of the vascular cambium layers. The grafter inserts the scion into the cut while the upper part of the rootstock is pulled backward, being careful to obtain the best cambium contact, then the rootstock is released. The pressure of the rootstock should grip the scion tightly. The scion can be further secured by driving two small flat-headed wire nails [20 gauge, 1.5 cm (5/8 in) long] into the stock through the scion. Wrapping the rootstock and scion at the point of union with nursery tape also may be helpful. After the graft is completed, the rootstock may be cut off just above the union. This must be done very carefully or the scion may become dislodged. The entire graft union must be thoroughly covered with grafting wax, sealing all openings. The tip of the scion also should be covered with wax or sealed with white glue (57).

Side-Tongue Graft The side-tongue graft, shown in Figure 12–17, page 480, is useful for small plants, especially some of the broad- and narrow-leaved evergreen species. The rootstock plant should have a smooth section in the stem just above the crown of the plant. The diameter of the scion should be slightly smaller than that of the rootstock. The cuts at the base of the scion are made in the same way as for the whip-and-tongue graft. Along a smooth portion of the stem of the rootstock a thin piece of bark and wood, the same length as the cut surface of the scion, is completely removed. Then a reverse cut is made downward in the cut on the rootstock starting one-third of the distance from the top of the cut. This second cut in the rootstock should be the same length as the reverse cut in the scion. The scion is then inserted into the cut in the rootstock, the two tongues interlocking, and the vascular cambia matching. The graft is wrapped tightly, using one of the methods described for the whip-and-tongue graft.

The top of the rootstock is left intact for several weeks until the graft union has started to heal. Then it may be cut back above the scion gradually or all at once to force the buds on the scion into active growth.

Side-Veneer Graft The side-veneer graft is widely used for grafting small potted liner plants such as seedling conifers, deciduous trees and shrubs, and fruit crops (Figs. 12–18, 12–19, and 12–20, pages 480–83). A shallow downward and inward cut from 25 to 38 mm (1 to 1 1/2 in) long is made in a smooth area just above the crown of the rootstock. At the base of this cut, a second

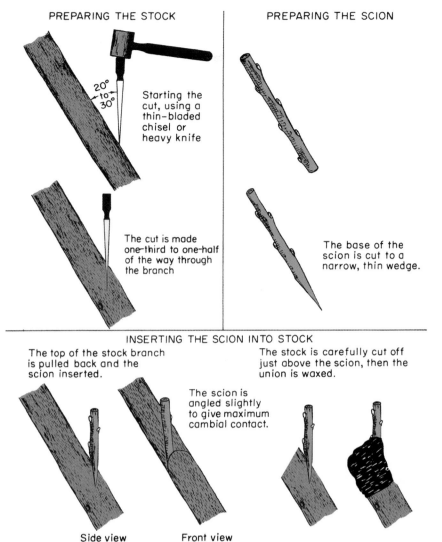

Figure 12–16
Steps in preparing the side-stub graft. A thin-bladed chisel, as illustrated here, is ideal for making the cut, but a heavy butcher knife could be used satisfactorily.

short inward and downward cut is made, intersecting the first cut, that removes the piece of wood and bark. The scion is prepared with a long cut along one side and a very short one at the base of the scion on the opposite side. These scion cuts should be the same length and width as those made in the rootstock so that the vascular cambium layers can be matched as closely as possible.

After inserting the scion, the graft is tightly wrapped with poly budding strips, budding rubbers, Buddy Tape, or with nursery adhesive tape. The graft may or may not be covered with wax, depending upon the species. A common practice in side grafting small potted plants of some woody ornamental species is to plunge the grafted plants into a slightly moist medium, such as peat moss, so that it just covers the graft union. Inserting the grafted liner plants in polytents in temperature-controlled greenhouses is another common practice (Fig. 12–4). To maintain high humidity, the newly grafted plants may also be placed for healing in a mist propagating house (but the grafts are not directly placed under mist), or set in grafting cases. The latter are closed boxes with a transparent cover, which permits retention of high humidity around the grafted plant until the union has healed. The grafting cases are kept closed for a week or so after the grafts are put in, and then gradually opened over a period of several weeks; finally, the cover is taken off completely.

After the union has healed, the rootstock can be cut back above the scion either in gradual steps or all at once.

Side Insertion Graft (SIG) The Side Insertion Graft (SIG) has been largely replaced by the OCG or Japanese tube graft, hole insertion graft (HIG), and tongue approach graft (TAG) (12, 21). The SIG is suitable for rootstocks with wide hypocotyls (Fig. 12–21, page 483). Production of rootstocks and scions is the

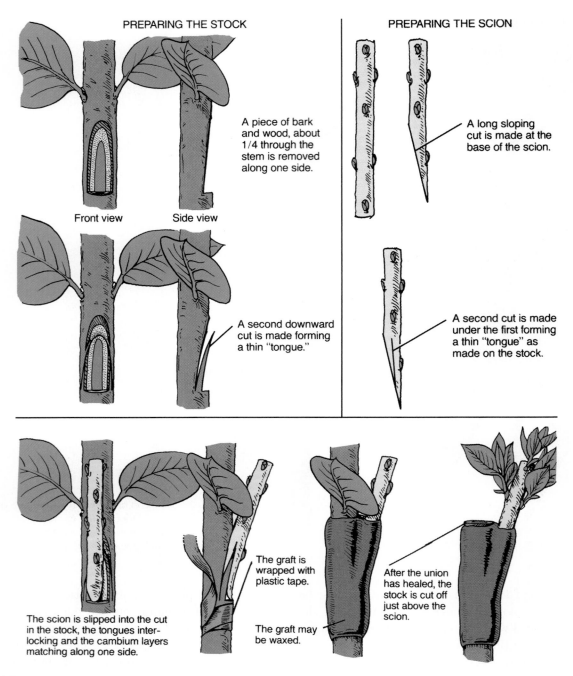

Figure 12–17
Sidetongue graft. This method is very useful for grafting broadleaved evergreen plants. Final tying may be done with budding rubbers, poly tape, or waxed string. The graft may be waxed, or wrapped with a sealing tape such as Parafilm or Buddy Tape.

same as that described for hole insertion grafting. A slit is cut on the hypocotyl of the rootstock with a razor blade and held open with a toothpick. A 35- to 45-degree-angle cut, on both sides is made on the hypocotyl of the scion. Then the scion is inserted into the slit in the hypocotyl of the rootstock and the toothpick is removed. Two cut surfaces are matched together and held with a grafting clip or silicone sleeve. The top of the rootstock is cut off 5 days after grafted plants are moved from the high-humidity growth chamber (21).

Detached Scion Graftage—Bark Grafting

Bark grafting is done in topworking established plants. The rootstock must be in an active stage of growth so that the bark will slip. The scion is inserted between the bark and wood of the rootstock. Bark grafting can be

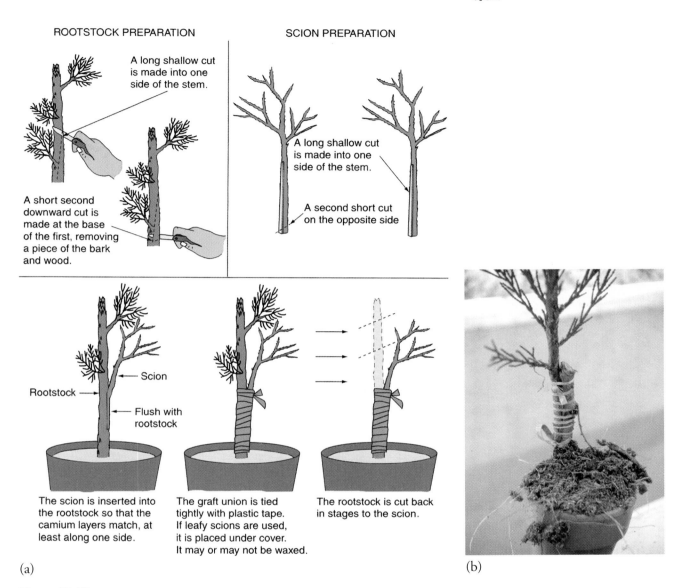

Figure 12–18
(a) Steps in making the side-veneer graft. This method is one of the most popular grafts for propagating conifers and shrubs. The graft is quite versatile and can be used on a larger number of species than other grafts, such as the whip-and-tongue.
(b) Side-veneer grafted conifer.

performed on branches ranging from 2.5 cm (1 in) up to 30 cm (1 ft) or more in diameter. The latter size is not recommended, because it is difficult to heal over such large stubs before decay-producing organisms attack.

Scions must be collected for deciduous species during the dormant season and held under refrigeration. For evergreen species, freshly collected scionwood can be used. In the bark graft, scions are not as securely attached to the rootstock as in some of the other methods and are more susceptible to wind breakage during the first year, even though healing has been satisfactory. Therefore, the new shoots arising from the scions probably should be staked during the first year, or cut back to about half their length, especially in windy areas.

After a few years' growth, the bark graft union is as strong as the unions formed by other methods. *Two modifications of the bark graft are described next.*

Bark Graft (Rind Graft) Several scions are inserted into each rootstock stub (Fig. 12–22, page 484). For each scion, a vertical knife cut 2.5 to 5 cm (1 to 2 in) long is made at the top end of the rootstock stub through the bark to the wood. The bark is then lifted slightly along both sides of this cut, in preparation for the insertion of the scion. The dormant scions should be 10 to 13 cm (4 to 5 in) long, contain 2 or 3 buds, and be 6 to 13 mm (1/4 to 1/2 in) thick. One cut—about 5 cm (2 in) long—is made along one side at the base of the scion. With large scions,

Figure 12-19
(a) Cultivars of the highly diverse Japanese maple (*Acer palmatum*) are grafted onto seedling rootstock. (b and c) A budding rubber (arrow) is used to wrap the dormant, leafless scion to the rootstock. (d) Side-veneer-grafted Japanese maples in potted liner pots which will be moved to a polytent area for callusing, and (e) post-allusing liner production.

this cut extends about one-third of the way into the scion, leaving a "shoulder" at the top. This shoulder reduces the thickness of the scion to minimize the separation of bark and wood after insertion in the rootstock. The scion should not be cut too thin, or it will be mechanically weak and break off at the point of attachment to the rootstock. If small scions are used, no shoulder is necessary. On the side of the scion opposite the first long cut, a second, shorter cut is made, as shown in Figure 12–22, bringing the basal end of the scion to a wedge shape. The scion is then inserted between the bark and the wood of the rootstock, centered directly under the vertical cut through the bark. The longer cut on the scion is placed against the wood, and the scion's shoulder is brought down until it rests on top of the stub. The scion is then ready to be fastened in place. The scion is nailed into the wood, using two nails per scion. Flat-headed nails 15 to 25 mm (5/8 to 1 in) long, of 19- or 20-gauge wire, depending on the size of the scions, are satisfactory. The bark on both sides of the scion should be nailed down securely or it will tend to peel back from the wood.

Another method commonly used with soft-barked trees, such as the avocado, is to insert all the scions in the stub and then hold them in place by wrapping waxed string, adhesive tape, or poly budding tape around the stub. This method is more effective than nailing for preventing the scions from blowing out, but probably does not give as tight a fit. A combination of nailing and wrapping are advisable for maximum strength. If a wrapping material is used, it must be checked to avoid constricting the rootstock. After the stub has been grafted and the scions fastened by nailing or tying, all cut surfaces, including the end of the scions, should be covered thoroughly with grafting wax.

Inlay Bark Graft Two knife cuts about 5 cm (2 in) long are made through the bark of the rootstock down to the wood, rather than just one (Fig. 12–23, page 485). The distance between these two cuts should be exactly the same as the width of the scion. The piece of bark between the cuts should be lifted and the terminal two-thirds cut off. The scion is prepared with a smooth

Figure 12–20
Side veneer graft of *Eugenia* (Myrtaceae). (a, b, and c) Graft wrapped with Parafilm tape. (d) Healed graft. Photos courtesy J. Griffis.

slanting cut along one side at the basal end, completely through the scion. This cut should be about 5 cm (2 in) long but without the shoulder, in contrast to the bark graft. On the opposite side of the scion, a cut about 13 mm (1/2 in) long is made, forming a wedge at the base of the scion. The scion should fit snugly into the opening in the bark with the longer cut inward and with the wedge at the base slipped under the flap of remaining bark.

The scion should be nailed into place with two nails, the lower nail going through the flap of bark covering the short cut on the back of the scion. If the bark along the sides of the scion should accidentally become disturbed, it must be nailed back into place. Flat-point staples in the vertical position, or budding or flagging tape have all been used to secure the graft (40). The inlay bark graft is well adapted for use with thick-barked trees, such as walnuts and pecans, on which it is not feasible to insert the scion under the bark; it is used when topworking an existing orchard (Fig. 12–24, page 486).

Detached Scion Graftage—Root Graftage

A number of plants are propagated commercially by root grafting—apples, pears, grapes, and selected woody ornamental shrubs and trees (17, 18).

Root Grafting (Whole-Root and Piece-Root Graftage)
In root grafting, the rootstock seedling, rooted cutting, or layered plant is dug up, and the roots are used as the rootstock for the graft. The entire root system may be used (**whole-root graft**—Figs. 12–3 and 12–25, page 487), or the roots may be cut up into small pieces and each piece used as a rootstock (**piece-root graft**—Fig. 12–25). Both methods give satisfactory results. Since the roots used are relatively small [0.6 to 1.3 cm (1/4 to 1/2 in) in diameter], the **whip-and-tongue graft** is frequently used. In England, Rhododendron cultivars are saddle-grafted on roots of *R. ponticum;* the root graft is then tied and placed in a propagation case (19). Tree peony and herbaceous peony are root-grafted with a cleft graft using the root of herbaceous peony. Root grafts are usually bench-grafted indoors during the late winter or early spring. The scionwood collected previously is held in storage, while the rootstock plants are also dug in the late fall and stored under cool [1.5 to 4.5°C (35 to 40°F)] and moist conditions until the grafting is done. The term *bench grafting* is given to this process, because it is performed indoors with dormant scions and rootstocks at benches by skilled grafters as part of a large-scale operation.

In making root grafts, the root pieces should be 7.5 to 15.0 cm (3 to 6 in) long and the scions about the same length, containing 2 to 4 buds. After the grafts are

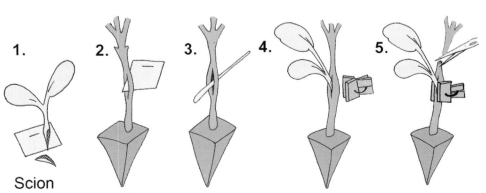

Figure 12–21
Steps in preparing the Side Insertion Graft (SIG). The SIG has been largely replaced by the OCG or Japanese tube graft hole insertion graft (HIG), and tongue approach graft (TAG). The SIG is suitable for rootstocks with wide hypocotyls (12, 21).
Illustrations courtesy R. L. Hassell.

PREPARING THE ROOTSTOCK

A vertical cut 2.5 to 5 cm (1 to 2 in.) long is made through the bark to the wood.

The bark on both sides of the cut is slightly separated from the wood.

PREPARING THE SCION

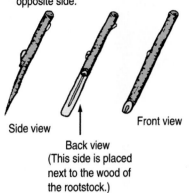

The scion is cut as shown below, a long cut with a shoulder on one side, and a shorter cut on the opposite side.

Side view

Back view (This side is placed next to the wood of the rootstock.)

Front view

INSERTING THE SCIONS INTO THE ROOTSTOCK

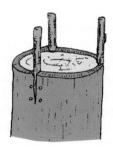 

Figure 12–22
Steps in preparing the bark graft (rind graft). In grafting some thick-barked plants, the vertical cut in the bark is unnecessary; the scion is inserted between the bark and wood of the stock.

made and properly tied, they are bundled together in groups of 50 to 100 and stored for callusing in damp sand, peat moss, or other packing material.

Nurse-Root Grafting Stem cuttings of a difficult-to-root species can sometimes be induced to develop adventitious roots by making a temporary "nurse-root" graft. The plant to be grown on its own roots is temporarily grafted as the scion. The scion may be made longer than usual and the graft planted deeply, with the major portion of the scion below ground. **Scion rooting** can be promoted by applying an auxin, such as indole-3-butyric acid, into several vertical cuts made through the bark at the base of the scion, above the graft union before planting. The grafts are set deeply, so that most of the scion is

scion rooting The development of adventitious roots from the grafted scion, desirable in **nurse-root grafting,** which is a temporary graft, but problematic in other grafts where the size control or disease-resistant characteristics of the rootstock may be lost.

covered (mound layered) with soil (28). After one season of growth the scions have roots, and the temporary nurse rootstock is cut off and discarded. The rooted scion is replanted to grow on its own roots; it can later be used as a rootstock and grafted to a scion fruit cultivar.

Methods of nurse-root grafting include *reversing the polarity of the nurse-root rootstock.* The rootstock piece will eventually die if it is grafted onto the scion in an inverted position (Fig. 12–26, page 487) (37). A graft union is formed—the inverted rootstock piece sustains the scion until it roots—but the rootstock fails to receive sufficient carbohydrates from the scion and eventually dies, leaving the scion on its own roots. Another method is *girdling the rootstock just above the graft union at the scion base.* The rootstock is girdled with budding rubber strips (0.016 gauge) (7). Budding rubbers disintegrate within a month when exposed to sun and air; however, when buried in the soil, they will last as long as 2 years, allowing sufficient time for the scion to become rooted. In a third method, an *incompatible rootstock is used.* When the graft is planted deeply, scion roots will gradually become more dominant in sustaining the plant. Examples of this are apple scions on pear rootstock, and lilac scions on ash rootstock.

PREPARING THE ROOTSTOCK

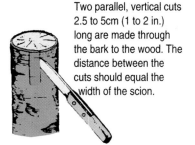

Two parallel, vertical cuts 2.5 to 5cm (1 to 2 in.) long are made through the bark to the wood. The distance between the cuts should equal the width of the scion.

A horizontal cut is made between the two vertical cuts and most of the piece of bark is removed. A small flap is left at the bottom.

PREPARING THE SCION

The scions are made with a long sloping cut on one side and a shorter cut on the opposite side.

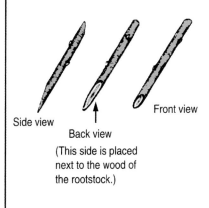

Side view

Back view
(This side is placed next to the wood of the rootstock.)

Front view

INSERTING THE SCION INTO THE ROOTSTOCK

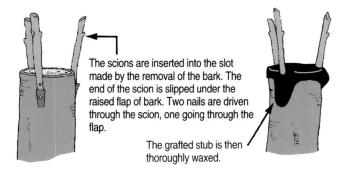

The scions are inserted into the slot made by the removal of the bark. The end of the scion is slipped under the raised flap of bark. Two nails are driven through the scion, one going through the flap.

The grafted stub is then thoroughly waxed.

Figure 12–23
Inlay bark graft. With Texas pecans, the scion of the inlay bark graft is stapled or nailed, and aluminum foil and a cut poly bag are used instead of grafting wax (40).

Approach Graftage

The distinguishing feature of approach grafting is that two independent, self-sustaining plants are grafted together. After a union has occurred, the top of the rootstock plant is removed above the graft, and the base of the scion plant is removed below the graft. Sometimes it is necessary to sever these parts gradually rather than all at once. Approach grafting provides a means of establishing a graft union between certain plants which are otherwise difficult to successfully graft. It is usually performed with one or both of the plants growing in a container. Rootstock plants in containers may also be placed adjoining an established plant that is to furnish the scion part of the new, grafted plant (Fig. 12–27, page 488).

This type of grafting should be done at times of the year when growth is active and rapid healing of the graft union will take place. Three useful methods of making approach grafts are described as follows, and illustrated in Figure 12–28.

Spliced Approach Graft In the spliced approach graft, the two stems should be approximately the same size (Fig. 12–28, page 489). An exception to this is the spliced approach graft of mango in India, where the scion is considerably smaller than the field-grown rootstock; the scion, in a pot, is hung from the branch of the larger rootstock (19). At the point where the union is to occur, a slice of bark and wood 2.5 to 5 cm (1 to 2 in) long is cut from both stems. This cut should be the same size on each so that identical cambium patterns are made. The cuts must be perfectly smooth and as nearly flat as possible so that when they are pressed together there will be close contact of the vascular cambium layers. The two cut surfaces are bound tightly together with raffia or poly grafting tape, then the

Figure 12–24
Topworking an existing orchard using the inlay bark graft for (a) citrus, (b) pecan and (c, d, and e) peaches. (b) For topworking pecans in Texas, the inlay bark graft is covered with aluminum foil to reduce the heat load and polyethylene to retain moisture; conditions are too hot for using grafting wax. (c, d, and e) Topworked peach orchard in Israel using an inlay bark graft. (c and d) The grafts have aluminum covers to reduce heat buildup.

whole union should be covered with grafting wax. After the parts are well united (which may require considerable time in some cases) the rootstock above the union, and the scion below the union are cut, and the graft is completed. It may be necessary to reduce the leaf area of the scion if it is more than the root system of the rootstock can initially sustain.

Tongued Approach Graft (TAG) The tongued approach graft is the same as the spliced approach graft, except that after the first cut is made in each stem to be joined, a second cut—downward on the stock and upward on the scion—is made, thus providing a thin tongue on each piece. By interlocking these tongues a very tight, closely fitting graft union can be obtained (Fig. 12–29, page 489).

For grafting vegetable crops, after the rootstock has fully developed cotyledons and scion has cotyledon and first true leaf, plants are pulled out from the tray (21). Make a cut at a 35- to 45-degree angle into the hypocotyl of the rootstock approximately halfway with a razor blade, and make a cut of the opposite angle on the hypocotyl of the scion. Cuts need to be made so that the scion will be on top of the rootstock when completed. Two cut hypocotyls are placed together and sealed with aluminum foil or Buddy Tape to help healing and prevent the graft from drying out. The two plants are transplanted into a bigger cell that will accommodate the two root balls. The top of the rootstock is cut off 5 days after grafting, and the bottom of the scion is cut off 7 days after the top of the rootstock is removed (Fig. 12–29).

two or more small, flat-headed wire nails. Then the entire union must be thoroughly covered with grafting wax. After the union has healed, the rootstock can be cut off above the graft and the scion below the graft.

Repair Graftage

Inarching Inarching is similar to approach grafting in that both rootstock and scion plants are on their own roots at the time of grafting. It differs in that the top of the new rootstock plant usually does not extend above the point of the graft union, as it does in approach grafting. Inarching is used to replace roots damaged by cultivation equipment, rodents, or disease. It can be used to very good advantage to save a valuable tree or improve its root system (Fig. 12–30, page 489).

Seedlings (or rooted cuttings) planted beside the damaged tree, or suckers arising near its base, are grafted into the trunk of the tree to provide a new root system to supplant the damaged roots. The seedlings to be inarched into the tree should be spaced about 13 to 15 cm (5 to 6 in) apart around the circumference of the tree if the damage is extensive. A damaged tree usually will stay alive for some time unless the injury is very severe. The procedure for inarching is to plant seedlings of a compatible species around the tree during the dormant season, and graft when active growth commences in early spring. Inarching may also enhance growth of uninjured, older trees (22).

As illustrated in Figures 12–30 and 12–31 (page 490), the graft is similar to an inlay bark graft. The upper end of the seedling, which should be 6 to 13 mm (1/4 to 1/2 in) thick, is cut shallowly along the side for 10 to 15 cm (4 to 6 in). This cut should be on the side next to the trunk of the tree and deep enough to remove some of the wood, exposing two strips of cambium tissue. Another, shorter cut, about 13 mm (1/2 in) long, is made on the side opposite the long cut, creating a sharp, wedge-shaped end on the seedling stem.

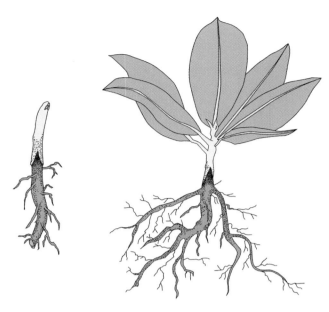

Figure 12–25
Root graftage with a saddle graft on a small root section or piece-root (*left*) and a whole root rootstock (*right*). See Figure 12–3 which is a whip and tongue graft using a whole root apple rootstock.

Inlay Approach Graft The inlay approach graft may be used if the bark of the rootstock plant is considerably thicker than that of the scion plant. A narrow slot, 7.5 to 10 cm (3 to 4 in) long, is made in the bark of the rootstock plant by cutting two parallel channels and removing the strip of bark between (Fig. 12–28); this can be done only when the rootstock plant is actively growing and the bark "slipping." The slot should be exactly as wide as the scion to be inserted. The stem of the scion plant, at the point of union, should be given a long, shallow cut along one side, of the same length as the slot in the rootstock plant and deep enough to go through the bark and slightly into the wood. This cut surface of the scion branch should be laid into the slot cut in the rootstock plant and held there by nailing with

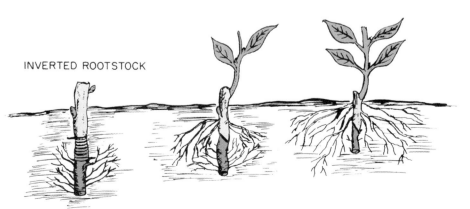

Figure 12–26
Reversing the polarity of the rootstock piece of the root graft is one method of "nurse-root" grafting. The nurseroot graft is a temporary graft used to induce the scion to develop its own roots. The nurse root sustains the plant until the scion roots form, then it dies. In the method shown, the rootstock piece is inverted, so the distal of the rootstock is temporarily joined to the proximal of the scion.

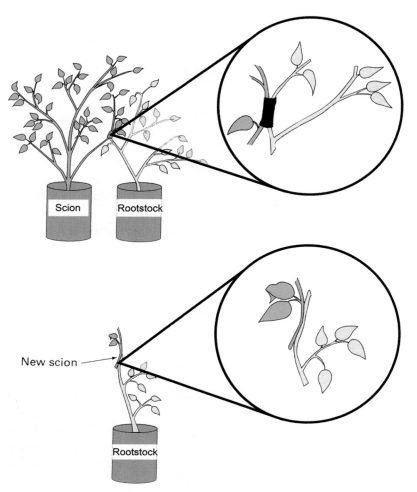

Figure 12–27

Approach grafting. *Top:* Initial grafting of the containerized scion and rootstock plants. *Below:* Completed graft union with scion severed from its own roots, and shoots of rootstock severed above the graft union.

A long slot is cut in the trunk of the tree by removing a piece of bark the width of the seedling and just as long as the cut surface made on the seedling. A small flap of bark is left at the upper end of the slot, under which the wedge end of the seedling is inserted. The seedling is nailed into the slot with four or five small, flat-headed wire nails. The nail at the top of the slot should go through the flap of bark and through the end of the seedling. If the bark of the tree along the sides of the seedling is accidentally pulled loose, it should be nailed back into place. The entire area of the graft union should then be thoroughly waxed.

Bridge Grafting Bridge grafting is another form of repair grafting—used when there is injury to the trunk, such as by cultivation equipment, rodents, disease, or winter injury. If the damage to the bark is extensive, the tree is almost certain to die, because the roots will be deprived of their carbohydrate supply from the top of the tree. Trees of some species, such as the elm, cherry, and pecan, can compartmentalize extensively injured areas by the development of a wound periderm of callus tissue. However, most woody species with severely damaged bark should be bridge grafted if they are to be saved, as illustrated in Figure 12–32, page 491.

An *interstock bridge graft system* has been used with mature apple trees for grafting M9 dwarfing rootstock (as the interstock) onto semi-dwarfing apple rootstock, leading to 20 percent reduced shoot growth, but a 30 percent increase in yield and increased soluble sugars and starch in the scion (53). A ring of bark 8 cm wide was removed from the trunk about 30 cm from ground level. Bridge grafts composed of 1-cm-wide split interstocks were inserted perpendicular around the ring, and then tightly wrapped with plastic during graft healing (New Zealand).

Bridge grafting is best performed in early spring as active growth of the tree is beginning and the bark is slipping easily. The scions should be obtained when dormant from 1-year-old growth, 6 to 13 mm (1/4 to 1/2 in) in diameter, of the same or compatible species, and refrigerated until grafted. In an emergency, one

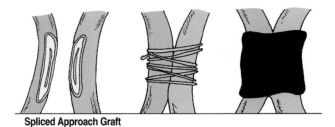

Figure 12–28
Three methods of making an approach graft: spliced-, tongued- and inlay-approach graft.

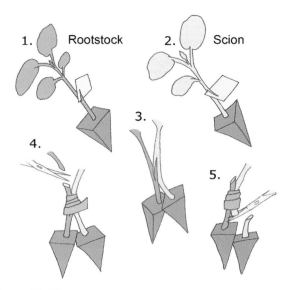

Figure 12–29
Tongued Approach Graft (TAG) with cucurbits. An easy graft that requires no special equipment or graft healing chambers, has a high success rate, but is labor intensive (12, 21). Illustrations courtesy R. L. Hassell.

Figure 12–30
(a) Inarches that have just been inserted (arrows). The one on the left has been waxed. The one on the right has been nailed into place and is ready for waxing. (b) Inarching can be used for invigorating established trees by replacing a weak rootstock with a more vigorous one. Here a Persian walnut tree has been inarched with Paradox hybrid seedlings (*Juglans hindsii* × *J. regia*).

may successfully perform bridge grafting late in the spring, using scionwood whose buds have already started to grow; the developing buds or new shoots are removed.

The first step in bridge grafting is to trim the wounded area back to healthy, undamaged tissue by removing dead or torn bark. A scion is inserted every 5 to 7.5 cm (2 to 3 in) around the injured section and attached at both the upper and lower ends into live, undamaged bark. It is important that the scions are right side up. If reversed, a union may form, but the scions will not enlarge in diameter as they would if inserted correctly. Figure 12–33, page 491 shows the details of making a bridge graft.

After all the scions have been inserted, the cut surfaces must be thoroughly covered with grafting wax; particular care should be taken to work the wax around the scions, especially at the graft unions.

Bracing Bracing is a form of natural branch grafting that is used by fruit producers to strengthen scaffolding limbs of a tree in order to support the weight of the fruit crop. Natural grafting of roots or shoot systems occurs in species such as fig (*Ficus*), rubber trees

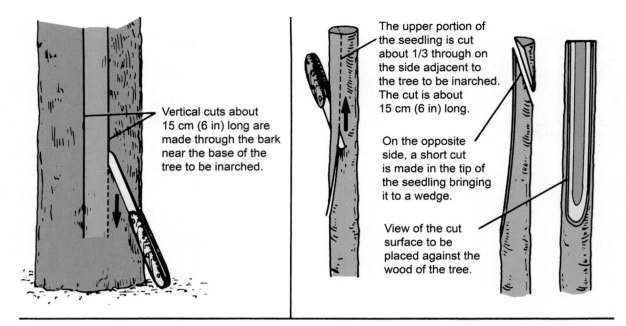
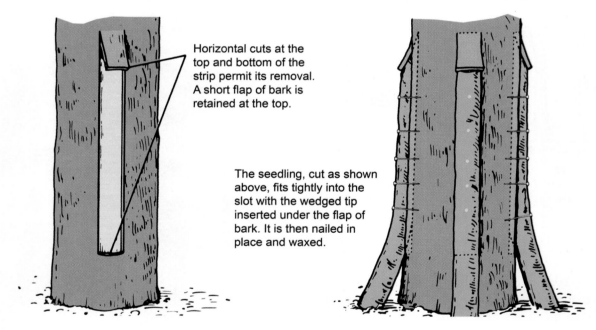

Figure 12–31
Steps in inarching a large plant, with smaller ones planted around its base.

(*Hevea*), birch (*Betula*), beech (*Fagus*), ash (*Fraxinus*), maple (*Acer*), pine (*Pinus*), and climbing species such as English ivy (*Hedera*) (see Chapter 11). Branches and trunks can naturally graft when they come in contact with each other during early development. The union begins with compression and constant and increasing pressure that ruptures the outer bark of the graft partners, followed by continued secondary growth, which leads to graft union formation and the joining of the independent vascular systems of the partners.

When bracing limbs, fruit producers will pull together two strong, young lateral shoots from the limbs to be braced. A rope or cord is used to temporarily brace the larger limbs. The weaved smaller shoots, which will naturally graft, are tied with waxed string or poly tape to keep them together (Fig. 12–34, page 492).

PRODUCTION PROCESSES OF GRAFTAGE

Success in grafting depends 45 percent on preparation, including the quality and preparation of the scion and rootstock material, 10 percent on craftsmanship, and 45 percent on the aftercare of the grafted plant (26, 38). The production goals of grafting are achieving a high success rate, or "take," and obtaining high speed and accuracy in performing the graft. Preparation for grafting begins with the proper tools and accessories, as well as the selection and handling of the scion and rootstock (Table 12–3). Since grafting is a repetitive, labor-intensive process, grafting machines and grafting automation, including robotics, continue to play a greater role.

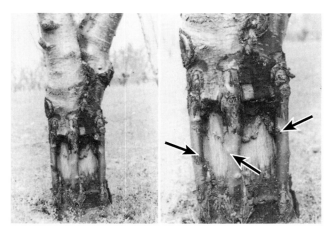

Figure 12–32
Injured trunk of a cherry tree successfully bridge grafted (arrows) by a modification of the bark graft.

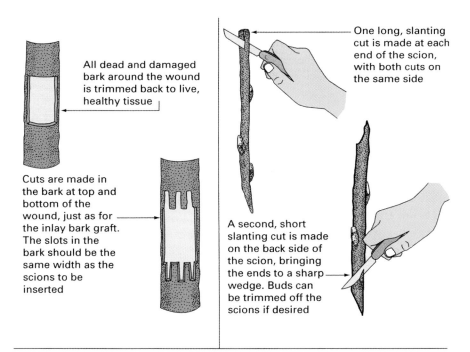

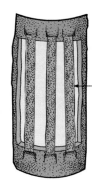

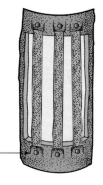

Figure 12–33
A satisfactory method of making a bridge graft, using a modification of the inlay bark graft.

> **BOX 12.2 GETTING MORE IN DEPTH ON THE SUBJECT**
> **LARGE-SCALE BRIDGE GRAFTING**
>
> After World War I, thousands of fruit trees were bridge grafted in France to repair damage that occurred during the war. Alternatively, trunks from mutilated trees less than 20 cm (8 in) in diameter were cut off and crown grafted (23).

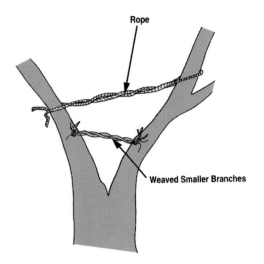

Figure 12–34
Bracing of fruit tree limbs by encouraging natural grafting. The tree limbs are braced with a twisted rope or electrical cord, and smaller shoots from the limbs are woven together and graft naturally as secondary growth occurs.

Preparation for Grafting

Tools and Accessories for Grafting Common tools and accessories used for grafting include grafting knives, tying and wrapping materials, and grafting waxes. Special equipment needed for any particular method of grafting is illustrated, along with the description of the method, in the remaining pages of this chapter. For example, grafting planes are sometimes used for more accurate fitting of scions with a hard and thick wood (Fig. 12–35) (61).

Knives. The two general types of knives used for propagation work are the budding knife and the grafting knife (Fig. 12–36). Where a limited amount of either budding or grafting is done, the budding knife can be used satisfactorily for both operations. The knives have either a folding or a fixed blade. The fixed-blade type is stronger, and if a holder of some kind is used to protect the cutting edge, it is probably the most desirable. A well-built, sturdy knife of high-carbon steel is essential. Grafting blades are flat on one side and have a tapered edge on the other to make a sharp, clean cut. Grafting knives are available for either right- or left-handed people.

Tying and Wrapping Materials. Grafting methods, such as the whip-and-tongue, splice (whip), and side-veneer graft, and budding methods, such as chip budding and T-budding, require that the graft union be held together by tying until the parts unite. A number of materials can be used for tying or wrapping grafts and budding. Some of these tying materials can also seal and help maintain a high relative humidity in the graft union area, which can help eliminate production steps for applying a hot wax sealant on top of the tying materials, or the need for maintaining the grafts in special poly chambers or "sweat boxes."

Table 12–3
PRODUCTION PROCESSES OF GRAFTAGE

Preparation for Graftage
- Tools and accessories for grafting
- Grafting machines and grafting automation/robotics
- Selection, handling, and storage of scionwood
- Handling of rootstock.

The Craftsmanship of Grafting
- Manual techniques: speed, accuracy, enhancing success rates
- Record keeping

Aftercare of Grafted Plants
- In bench grafting systems
- In field and nursery grafting systems

Figure 12–35
The grafting plane is a small woodworking plane equipped with a disposable heavy-duty razor blade. It allows more precise fitting of scions with hard and thick wood (61).

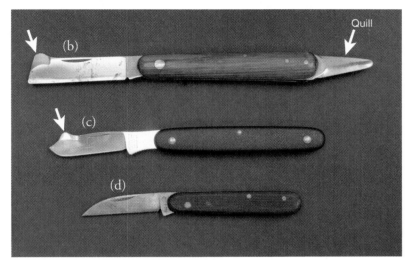

Figure 12–36
Folding, budding-grafting knives. (a) Tina budding-grafting knife. (a, b, and c) The metal flap (arrow) opposite the sharp end of the blade is used to separate the bark during the "T"–cut of the rootstock. (b) The quill is also used to separate the bark during T-budding. (d) Grafting knives are sharpened on one side, so there is a *flat side* of the blade for better control.

Some common tying materials for budding and grafting include:

- **Budding rubbers** (also used in grafting).
- Clear or colored polyethylene or polyvinyl chloride **(PVC) budding and grafting strips,** which are 0.5 to 1.3 cm (3/16 to 1/2 in) wide and slightly elastic, allowing for a more secure wrap; this is also called flagging tape, green floral tape, white budding tape, or orange grafting tape.
- **Plastic clips and silicon tubing** (for manual and robotic grafting).
- **Raffia** (strips of palm leafstalk fiber—an older wrapping material, but still used).

Since PVC budding and grafting strips are not self-adhesive, they must be tied with a half-hitch knot (Fig. 12–37), which is done at the final turn of the tape by slipping it under the previous turn. With the exception of budding rubber, which deteriorates in full sunlight (but not when buried with the graft below ground), or Buddy Tape, the wrapping materials must be removed later to prevent girdling the plant.

Figure 12–37
(a) Sometimes a loop is included in the final half-hitch knot of the rubber (arrow) or polyethylene budding and grafting strips. This allows easy unraveling of the strips after the graft has "taken" and avoids potential girdling problems in this side veneer graft of Japanese maple. Various tying materials: (b) raffia, (c) poly tape, (d) nursery tape, and (e and f) Buddy Tape (similar to Parafilm) used to cover magnolia scion pieces.

Self-Adhering Tying Materials. A time-tested tying material is **waxed string** or **twine**, which adheres to itself and to the plant parts without tying. It should be strong enough to hold the grafted parts together, yet weak enough to be broken by hand.

Nursery adhesive tape is similar to surgical adhesive tape but lighter in weight and not sterilized; it is more convenient to use than waxed cloth tape. Adhesive tape is useful for tying and sealing whip grafts. When using any kind of tape or string for wrapping grafts, it is important not to use too many layers or the material may eventually girdle the plant unless it is cut. When this type of wrapping is covered with soil, it usually rots and breaks before damage can occur. On a limited scale, adhesive tapes, such as **duct** and **electrical tape** can be used, while **masking tape** tends to unravel (54). Regardless of the wrapping material, it is best to remove or cut it *after* the graft has taken to avoid girdling.

Self-sealing tying materials include **Parafilm tape,** which has been used with successful results to wrap graft unions rapidly (5) and for chip-budding roses (Fig. 13–6). This material is a waterproof, flexible, stretchable, thermoplastic film with a paper backing. The film is removed from the paper, wrapped around the graft union, and pressed into place by hand. **Buddy Tape** (buddytape.com) is similar to Parafilm tape, but thinner and more economical. It seals and holds the graft or bud piece in place, and is thin enough for the bud to elongate and pass through it once the graft "takes" (Fig. 12–37). Sometimes budding rubbers are used to tie a graft, which is then sealed with Parafilm tape. Self-sealing cure **crepe rubber sheets** are used for herbaceous grafts and small woody plant material. Rubber patches up to 4 cm (1 1/2 in) are fixed with a staple and used for budding (Fig. 12–38).

Miscellaneous Fastenings and Wrapping Material. Miscellaneous fastenings and wrapping material include 18-gauge, 2-cm (3/4-in) **nails,** 1.6-cm (5/8-in) **flat-point staples,** as used in the inlay bark graft of pecans (40), and **plastic graft clips,** used in manual and machine splice grafting of vegetable crops (Figs. 12–7 and 12–39). The combination of **aluminum foil** and **polyethylene bags** wrapped around a four-flap or inlay bark graft replaces the need for waxing the graft (which would melt and be unsuitable in spring field grafting in Texas or other warm regions) (Figs. 12–13 and 12–24). **Metal shoot guide clips** (Fig. 13–4) are used for field budded, dormant rootstock to compel upright growth from the bud. **Silicon tubing** has been used to hold graft unions of single-node scions of oak and ash with high rooting success (Fig. 12–40, page 496) (14).

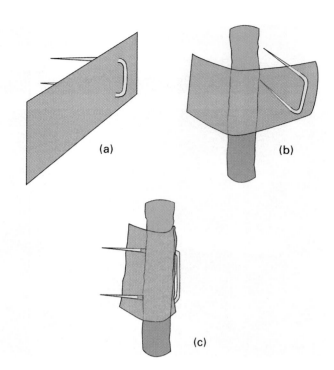

Figure 12–38
Rubber patch tying materials: (a) Rubber patches that come with a fixed stainless steel staple. (b) The rubber sheet is pulled over the bud and pulled securely around the rootstock. (c) The staple is pushed through the flap to secure the patch. The rubber patch normally deteriorates later with UV light from the sun or can easily be removed.

Splints made of *toothpicks, bamboo,* or *metal skewers* are used with bench grafting of herbaceous plants such as cacti. The splints are later removed after the graft has taken.

In a novel approach for developing robotic grafting systems for vegetable crops in Japan, Chinese cabbage seedling (scions) are horizontally grafted to turnip rootstocks. In grafting *Solanaceae* and *Cucurbitacae* vegetable crops, the graft partners are joined by a chemical adhesive, followed by spraying a chemical hardener to set up and solidify the adhesive around the graft (see Fig. 12–43, page 498) (29, 34, 47, 48).

Whether a sealant, such as grafting wax, may be applied depends on the type of graft, the grafting system, and type of material used. Sealants are generally not used with budding, since tying with budding rubbers, rubber patches, Buddy Tape, or Parafilm tape is sufficient to alleviate desiccation problems. If the bench-grafted plant is to be placed in a high relative humidity graft box or temperature-controlled polytent, or immediately outplanted in the field with the union below the soil surface, waxing may be omitted.

Figure 12–39
Grafting of heirloom tomatoes in high tunnels can lead to increased yield, earlier season extension (compared to field-grown), soil-borne disease resistance and increased nutrient uptake with a potential reduction in fertilizer inputs. (a, b, c, and d) OCG or Japanese tubegrafting system. (c and d) Plastic clip holding graft. (e and f) High tunnel production of grafted heirloom tomatoes. The system enhances IPM—integrated pest management—and organic production systems for vegetable crops (52).
Photos courtesy S. O'Connell, M. Peet, C. Rivard, F. Louws, and S. McDonough.

Grafting Waxes. Grafting wax has two chief purposes: (a) It seals over the graft union, thereby preventing the loss of moisture and death of the tender, exposed cells of the cut surfaces of the scion and rootstock. These cells are essential for callus production and healing of the graft union. (b) It prevents the entrance of various decay-producing organisms that rot wood.

An ideal grafting wax should adhere well to the plant surfaces, not be washed off by rains, not be so brittle as to crack and chip during cold weather or so soft that it will melt and run off during hot days, but still be pliable enough to allow for the swelling of the scion and the growth enlargement of the rootstock without cracking. **Hot waxes** require heating, while **cold waxes** contain volatile solvents that keep the wax liquid. The cold wax solidifies when the solvents evaporate. Most nurseries develop their own hot wax, which is low-melting, soft, and flexible, so that subsequent handling of the graft does not cause cracking and flaking. Thermostatically controlled wax heaters are available to provide instant liquid wax when needed. *The wax should be hot enough to flow easily, yet not be boiling, which damages plant tissue.*

Various recipes for making hot and cold waxes are listed by Garner (19). For hot wax, blocks of premixed grafting wax containing the necessary ingredients (e.g., TrowBridge's grafting wax, Walter E. Clark & Son, Orange, Conn., USA) can be purchased from nursery supply houses.

hot waxes Waxes that are paraffin based and must be heated to melt and apply.

cold waxes Waxes that contain a volatile solvent that preserves the wax liquid. After application the solvents evaporate and the wax hardens.

Grafting Machines Several *bench grafting* machines or devices have been developed to prepare graft and bud unions, and a few have been widely used, especially in propagating grapevines (2, 3).

Various bench grafting machines for the wedge graft or French-V are available, including a portable and

Figure 12–40
High grafting success has been obtained using silicone tubing to hold graft unions together of single-node scions of oak and ash: (a) Digital caliper to measure stem diameter. (b) Single node of oak grafted using silicone tube (arrow). (c) Oak graft after 8 weeks (arrow showing tubing). (d) Healed ash graft after 12 weeks. Photographs courtesy of G. Douglas (14).

bench-mountable device made in New Zealand by Raggett Industries, Ltd., Gisborne (www.raggettindustries.co.nz) (Fig. 12–41). This device makes a type of wedge graft, cutting out a long V-notch in the rootstock and a corresponding long, tapered cut at the base of the scion. By reversing the position of the rootstock and scion, it could also make a saddle graft (Fig. 12–11). Although the cuts fit together very well, the operation is slow because the graft union must be either tied with a budding rubber or poly tape or stapled together. This machine has been used successfully in propagating grapes and fruit trees.

There are machines for making omega grafts, which are hand-operated (Fig. 12–42) or foot-operated. One device for grape grafting is the Pfropf-Star grafting machine manufactured in Germany. It cuts through both the rootstock and scion, one laid on top of the other, making an omega-shaped cut and leaving the two parts interlocked. While these machines work fine for grape grafting, most ornamental nurseries that graft do not use machines. Instead, they rely on hand-grafting, which may be faster and more reliable given the larger number of genera and species grafted (35). Without question, finding skilled grafters is a severe production problem; hence the importance of developing mechanized and automated grafting systems.

Grafting Automation/Robotics Prototypes and commercial robotic machines for grafting vegetable seedlings have been developed (Figs. 12–43, 12–44, and 12–45, pages 498–99). The production of grafted vegetable crops is becoming more common in the United States. However, grafted vegetable seedlings are used extensively in heavily populated countries such as Japan, Korea, some other Asian countries, and in parts of Europe, where the land use is highly intensive, farming areas quite small, and crops are not rotated. Grafted seedlings account for 81 percent of the commercial outdoor and greenhouse vegetable production in Korea, and 54 percent and 81 percent, respectively, for Japan (12, 29, 30, 33). Vegetable rootstock used are resistant to soil-borne pathogens and nematodes, which build up under these intensive cultivation conditions. Some of the commercialized grafting robots can graft 800 or more *Solanaceae* vegetable seedlings per hour (Figs. 12–43 and 12–45) (49).

Selection, Handling, and Storage of Scionwood
Proper selection, handling, and storage of scionwood is important.

Kind of Wood. Since bench or field grafting of deciduous species takes place in winter or early spring, it is necessary to use the scionwood that grew the previous fall.

In selecting such scion material the following points should be observed:

- For most species, the wood should be 1 year old or less (current season's growth). Avoid including older growth, although with certain species, such as the fig or olive, 2-year-old wood is satisfactory, or even preferable, if it is of the proper size.
- Healthy, well-developed vegetative buds should be present. Avoid wood with flower buds. Usually, vegetative buds are narrow and pointed, whereas flower buds are round and plump (see Fig. 13–3).
- The best type of scion material is vigorous (but not overly succulent), well-matured, hardened shoots from the upper part of the tree, which have grown 60 to 90 cm (2 to 3 ft) the previous summer. Such growth develops on relatively young, well-grown, vigorous plants; high production of scion material can be promoted by pruning the plant back heavily the previous winter. Water sprouts from older trees sometimes make satisfactory scionwood, but suckers arising from the base of grafted trees should not be used, since they

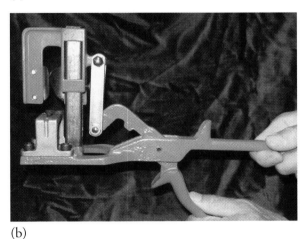

Figure 12–41
(a and b) Wedge grafting made by French-V grafting devices. (b) Raggett top grafter. (c) Grape graft union healing with profuse callusing (arrow) one month after bench grafting with a wedge graft or French-V. Grafts can be made with these devices much faster than by the whip-and-tongue graft method. These machines can also make a saddle graft by these machines, by reversing the cuts so the scion piece has the saddle shape (see Fig. 12–11).

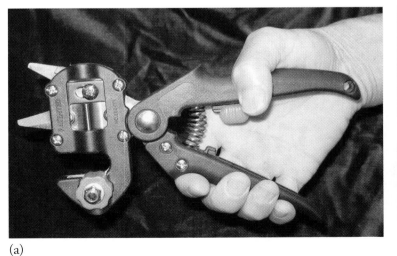

Figure 12–42
(a) Grafting tool for making an omega graft. (b and c) Omega graft locks in place and then is held together with grafting tape. The scion and stock must be the same diameter.

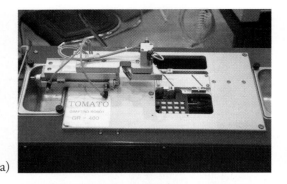

Figure 12–43
Commercial grafting robot for *Solanaceae* (tomatoes, melons, cucurbits) vegetables on seedling rootstock suitable for intensive planting and resistant to disease, insect and environmental stress. Plant vigor and yield can also be enhanced with superior rootstock. (a) Tomato and (b) melon grafting robots.

may consist of rootstock material. A satisfactory size is from 0.6 to 1.2 cm (1/4 to 1/2 in) in diameter.
- The best scions are obtained from the center portion or from the basal two-thirds of the shoots. The terminal sections, which are likely to be too succulent, pithy, and low in stored carbohydrates, should be discarded. Mature wood with short internodes should be selected.

Source of Material. Scionwood should be taken from source plants of the correct cultivar known to be pathogen-tested and genetically true-to-type (see Chapter 16). Virus-diseased, undesirable sports, chimeras, and virus-like genetic disorders must be avoided. Source plants may be of three basic types:

1. Plants produced in an orchard, vineyard, ornamental field, container nursery, or landscape are selected when the flowering, fruiting, and growth habits are known. It is best to take propagation material from bearing plants whose production history is known. Visual inspection, however, may not reveal the true condition of the proposed source plant and, appropriate indexing and progeny tests are required to be sure (see Chapter 16).
2. In commercial nurseries, special scion blocks, where plants are grown particularly for propagation, may be maintained. Such plants are handled differently than they would be for producing a crop. For example, fruit trees may be pruned back each year to produce a large annual supply of long, vigorous shoots well-suited for scionwood. Such special blocks would usually be handled to conform to registration and certification programs and would be subject to isolation, indexing, and inspection requirements. In addition, it is important to maintain source identity of scion material through the entire propagation sequence, so that over a period of time proper

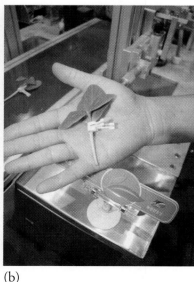

Figure 12–44
Finished grafts from a commercial melon grafting robot. (a) The grafted plants are moved on a conveyor belt system for processing. (b) Splice grafted scion and rootstock held together by a grafting clip.

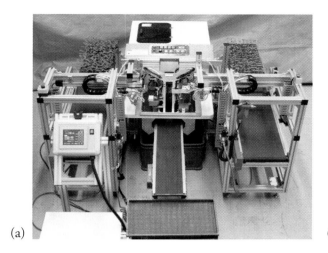 

Figure 12–45
Two techniques used in herbaceous grafting are manual and robotic grafting. (a) Fully automated grafting robot for cucurbits (BRAIN, Saitama, Japan), and (b) semiautomated Korean grafting machine. Plants are grafted at the 2 to 4 leaf stage. The advent of OCG, "tube-grafting" or "Japanese top-grafting" has become the most popular graft for tomato. Grafting robots can make up to 800 grafts/hr, whereas an individual can make 1,000 grafts/day. Photos (a) and (b) courtesy of C. Kubota and M. Peet, respectively.

sources of the various cultivars can be identified and maintained.

3. For vegetable grafting, commercial seed from selected rootstocks and scions are sown under protected cultural conditions. Selected rootstock for vegetable grafting is listed in Table 12–4. A timeline for grafting heirloom tomatoes, starting with sowing rootstock seed 2 to 5 days prior to sowing scion seed, is depicted in Figure 12–46.

Collection and Handling. For deciduous plants to be grafted in early spring, the scionwood can be collected almost any time during the winter season when the plants are fully dormant (6). In climates with severe winters, the wood should not be gathered when it is frozen, and any wood that shows freezing injury should not be used. Where considerable winter injury is likely, it is best to collect dormant scionwood and put it in cold storage after leaf fall but before the onset of winter.

Storage. Scionwood collected prior to grafting must be properly stored. It should be kept slightly moist and at a low enough temperature to prevent elongation of the buds. A common method is to wrap the wood, in bundles of 25 to 100 sticks, in heavy, waterproof paper or in polyethylene sheets or bags. *All bundles must be labeled accurately.*

Polyethylene bags are useful for storing small quantities of scionwood. They allow the passage of oxygen and carbon dioxide, which are exchanged during

Table 12–4
SELECTED SCION AND ROOTSTOCK COMBINATIONS FOR VEGETABLE GRAFTING (50)

Scion	Rootstock
Watermelon (*Citrullus vulgaris* syn. *C. lanatus*)	Bottle gourd (*Lagenaria siceraria*) White gourd (*Benincasa hispida*) *Cucurbita* spp.
Melon (*Cucumis melo*)	White gourd (*Benincasa hispida* Cogn.) *Cucumis* spp.; *Cucurbita* spp. *C. moschata* x*C. maxima*
Cucumber (*Cucumis sativus* L.)	Pumpkin (*Cucurbita* spp.) *Cucurbita ficifolia*
Aubergine (*Solanum melongena* L.)	*Solanum integrifolium*; *Solanum torvum*; *Solanum melongena*
Tomato (*Lycopersicon esculentum*)	Tomato (*Lycopersicon esculentum*)

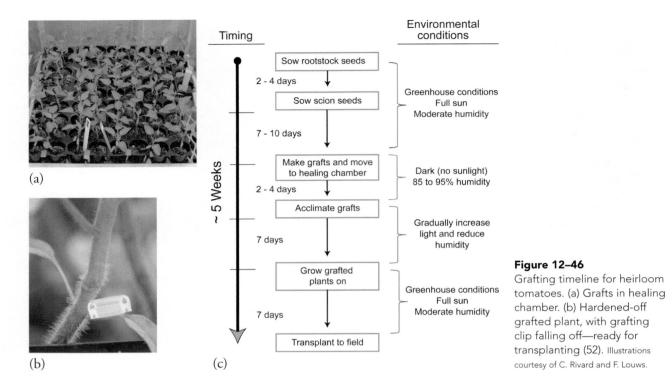

Figure 12–46
Grafting timeline for heirloom tomatoes. (a) Grafts in healing chamber. (b) Hardened-off grafted plant, with grafting clip falling off—ready for transplanting (52). Illustrations courtesy of C. Rivard and F. Louws.

the respiration process of the stored wood, but retard the passage of water vapor. Sometimes the natural moisture in the wood is sufficient, so slightly moist packing material is not needed in the sealed poly bags. In commercial field rose production, the scionwood (budwood) is harvested dormant, wrapped in slightly moist newspaper and sealed in poly bags, and maintained at –1.7 to –0.6°C (29 to 31°F) for up to 7 months.

The temperature at which the wood is stored is important. If it is to be kept only 2 or 3 weeks before grafting, the temperature of a home refrigerator—about 5°C (40°F)—is satisfactory. If stored for a period of 1 to 3 months, scionwood should be held at about 0°C (32°F) (6) to keep the buds dormant. However, buds of some species, such as the almond and sweet cherry, will start growth after about 3 months, even at such low temperatures. Do not store scionwood in a home freezer because the very low temperatures, about –18°C (0°F), can injure the buds.

Storage of scions should not be attempted if succulent, herbaceous plants are being grafted; such scions should be obtained at the time of grafting and used immediately. Certain broad-leaved evergreen species, such as camellias, olives, and citrus, can be grafted in the spring before much active growth starts, without previous collection and storage of the scionwood. Grafts are taken directly from the tree as needed, using the basal part of the shoots containing dormant, axillary buds. The leaves are removed at the time of collection.

Attempting to use scionwood in which the buds are starting active growth is almost certain to result in failure. In such cases, the buds quickly leaf out before the graft union has healed; consequently, the leaves withdraw water from the scions by transpiration, and cause the scions to die. In addition, the strong competing sink of a developing shoot can interfere with graft union formation.

In topworking pecans (4), good results are obtained by using precut scions; that is, scions cut in advance by skilled persons at a convenient time, that are then held in cold storage in polyethylene bags for up to 9 days before being inserted in the graft unions. Grafting success is reduced only slightly by the use of precut scions.

Handling of Rootstock for Bench Grafting. Seedling rootstock of maple (*Acer*) is established in liner pots for 1 year, brought into a greenhouse in the fall after leaf drop, and placed in bottom-heated benches at 13 to 16°C (55 to 60°F). Bench grafting of the container rootstock with a splice graft is done in January and February (Canada), when white roots appear along the perimeter of the rootballs (26).

In North Carolina, rootstock liners of woody ornamental plants are allowed to harden-off in minimum-temperature–controlled poly houses in the fall, and maintained just above freezing. When new roots emerge from the rootstock in late winter, plants are

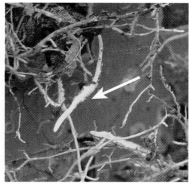

(a) (b) (c) (d)

Figure 12–47
Proper rootstock preparation. (a) Hardening-off *Acer palmatum* rootstock liners in late September (North Carolina). (b) New roots emerging from rootstock in January prior to grafting. (c) Proper after-care of grafted, labeled plants in poly covered, temperature-controlled hoop house. (d) Buds swelling on grafted *A. palmatum* 'Fireglow' in mid to late March (56).
Courtesy B. L. Upchurch.

ready for grafting (Fig. 12–47) (56). In general, bench grafting is best when new, white root tips of 6 mm (1/4 in) occur or buds start to swell on the rootstock of potted liner plants.

The Craftsmanship of Grafting

Grafting is both an art and a skill. Successful grafting is a **repetitive task** that requires a high degree of accuracy and speed; to become skilled, it is essential for the grafter to eliminate all unnecessary movements. To increase grafting efficiency, it is important to organize the workplace so that scion material, knives, and grafting tape are all within easy reach (42). Grafting is generally more efficient with a **team approach,** in which each worker performs a certain task, in order to reduce inefficient motion of materials and repetitive picking up and putting down of different tools.

Manual Techniques: Speed, Accuracy, and Enhancing Success Rates Tips on improving grafting techniques and ergonomics include the following (42):

- Concentrate on accuracy first, and allow grafting speed to build up—aim at initially completing at least 200 bench grafts a day.

- Use a graft method that is less time-consuming (yet still successful!) and that can be done with lesser skill and preparation of the rootstock [e.g., bench grafting with a splice (whip) graft, compared to a whip-and-tongue or side-veneer graft (43)]; this works well with *Betula, Cornus, Fagus, Ginkgo, Quercus,* and *Acer* (26).

- Grafting is best with two people: in bench grafting with the whip-and-tongue, one person does the graft, and the other moves the potted liner rootstocks and waxes the graft union. In T-budding field roses, the budder prepares the rootstock and inserts the shield bud of the scion, while the "tier" follows and ties the budding rubber around the budded graft (see Fig. 13–13).

- The grafting knife should always be held with a relaxed grip to improve accuracy and reduce repetitive strain injuries (e.g., **carpal tunnel syndrome**); it is necessary to restrict and control your arm movements.

  carpal tunnel syndrome Nerve damage in the wrist caused by the stress of repetitive hand-arm movements.

- There are two basic cuts in grafting: the **slice cut,** which is made using the arm and shoulder to pull the knife [e.g., in the making of a splice (**whip**) graft]; and the **cross cut,** in which the grafter's arm and knife are rotated using the thumb

as a pivot, with both hands joined to prevent the knife from cutting the grafter—a whip-and-tongue is created when the scion and rootstock are sliced and then cut across in **four** (economical) movements.
- Hang the grafting tape around your neck so that you know where it is and it stays free of contamination.
- In most cases, speed is more important than 100 percent accuracy; increase speed by developing a routine when grafting, and avoiding useless movements.
- Try to hold the grafting knife in your hand at all times (e.g., in your little finger).
- Make sure that everything is at a height that is easy to reach.

Record Keeping   Maintaining good records is important for successful grafting.

- *Keep records on the grafters* to determine daily quantities grafted; some operations pay on a **piecework** or bonus system (see Chapter 10). Records kept by the supervisor can be constructively used to help grafters improve their technique and efficiency (9).
- *Keep records on the plant material* to determine the optimum windows of time to graft (43) and to assure having the best available material to graft. Records should be kept on the conditions of the grafting material, grafting problems encountered, and aftercare of the grafts. Should a crop failure occur, records can generally help pinpoint the cause and help managers improve efficiency (9).
- *Importance of developing and sticking to a grafting time-line schedule.* A timeline schedule is critical for commercial success (Fig. 12–46).

piecework   A bonus system awarded to workers when they exceed daily graft production quotas.

AFTERCARE OF GRAFTED PLANTS

In Bench Grafting Systems

A common method of bench grafting is to wrap the union with budding rubbers, poly or plastic tape, plastic clips and silicone tubing, biodegradable cloth tape, or older materials such as raffia. Depending on the wrapping material, the entire union may be covered with grafting wax.

Root Grafting   The root grafts may be placed under refrigeration at 7°C (45°F) for about 2 months. For general callusing purposes, temperatures from 7 to 21°C (45 to 70°F) are the most satisfactory. The callusing period for apples can be shortened to around 30 days if the grafts are stored at a temperature of about 21°C (70°F) and at a high humidity. To use this higher callusing temperature, the material should be collected in the fall and the grafts made before any cold weather has overcome the rest period of the scion buds. After the unions are well healed, the grafts must be stored at cool temperatures—2 to 4°C (35 to 40°F)—to overcome the "rest period" of the buds and to hold them dormant until planting (24). The root grafts are lined-out in early spring in the nursery row directly from the low-temperature storage conditions.

Hot-Pipe Callusing System   With some plants, the graft union should be kept warm, 24 to 27°C (75 to 82°F), but the roots and the buds on the scion should be kept cool, about 5°C (41°F), to prevent premature growth before the graft union has callused and healed together. An ingenious system for regulating temperature was developed for whip-grafting hazelnut (*Corylus*), which are notoriously difficult to root graft (Fig. 12–48). This hot-pipe callusing system keeps the graft union warm by recirculating hot water in a PVC pipe onto which the graft is placed. The scion and roots protrude into areas of lower temperatures to keep the plant dormant until ready for transplanting. This hot-pipe callusing system, when used outdoors in late winter or early spring, has increased the root grafting "takes" of hazelnut and other difficult-to-graft species (31). Some aeration of the callusing grafts is required, so airtight containers should not

Figure 12–48
Hot-pipe callusing system for bench grafting difficult plants. The graft union is placed in a slot in a large plastic pipe. Inside the large pipe is a smaller pipe through which thermostatically controlled hot water circulates. Insulating material laid over this pipe retains the heat. The protected roots and scions protrude into areas of cooler temperatures, which retards their development.

be used. Virtually any graft used in bench grafting can be callused with the hot-pipe system, including apples, pears, peaches, and plums (32), and ornamentals such as *Acer, Cedrus, Corylus,* and *Fagus*. Not all species respond—the higher graft union temperature does not enhance graftage of spruce (*Picea*), for instance.

Closed Case Waxing may be omitted if the **bare-root grafts** are to be protected from drying by packing the grafts in boxes containing slightly moist peat. Some producers still dip the grafts in rose wax from the scion end to the taped union of the graft prior to boxing (51). Then the boxes are moved to a callusing room at 21°C (70°F) for about 12 days. Once the grafts have formed sufficient callus, the boxed grafts are held in cold storage at 2°C (35°F) until outplanted in the field (51). Another form of the closed case is the use of a **polytent** in a heated greenhouse for callusing the grafts of **potted rootstock liners** (Fig. 12–4). Provided light irradiance is controlled, glass mason jars can be used as a closed-case system for grafted plants in containers (Fig. 12–49).

(a)

(b)

Figure 12–49
(a and b) A closed case system of covering containerized grafted plants with glass Mason jars. (a) Uncovered plant. Condensate is visible inside the glass jars. (b) These grafted plants are maintained under shade to reduce the light irradiance and minimize heat buildup.

Open Case (Open Bench) Grafting is also done in a temperature-controlled greenhouse or unheated polyhouse (depending on the season). Waxing can be omitted in the bench grafting of **potted rootstock liners** by plunging the container and burying the graft in slightly moist peat moss or bark in a temperature-controlled greenhouse. The medium is bottom heated and kept at 18 to 21°C (65 to 70°F) for 3 to 6 weeks for callusing; ideally, the air temperature should be cooler to discourage any initial top growth. Wrapping a graft with poly grafting tape is also sufficient without waxing. Optimum periods for grafting selected ornamental species in a greenhouse are listed in Table 12–5.

Outplanting of Bare-Root Grafts. As soon as the ground can be prepared in the spring, the grafts are lined-out in the nursery row 10 to 15 cm (4 to 6 in) apart. Grafts should be planted before growth of the buds or roots begins. If growth starts before the grafts can be planted, they should be moved to lower temperatures (−1 to 2°C, 30 to 35°F). The grafts are usually planted deep enough so that the graft union is just below the ground level, but if the roots are to arise only from the rootstock, the graft should be planted with the union well above the soil level [i.e., 7 to 15 cm (3 to 6 in)]. It is very important to prevent *scion rooting* where certain definite influences, such as dwarfing or disease resistance, are expected from the rootstock.

After one summer's growth, the grafts should be large enough to transplant to their permanent location. If not, the scion may be cut back to one or two buds, or headed-back somewhat to force out scaffold branches and allowed to grow a second year. With the older root system—a strong, vigorous top is obtained the second year.

Aftercare in Field and Nursery Grafting Systems

Aftercare of field- and nursery-grafted plants is described in "Production Processes of Graftage" starting on page 492. In the section "Types of Grafts" (page 466), see the descriptions for grafting and aftercare using such grafts as the whip-and-tongue, four-flap, and inlay bark graft. Chapter 11 describes heading-back (**lopping**) as well as the **crippling** techniques that are used in field and container nurseries to encourage the

> **crippling or lopping**
> Bending (restriction) or cutting halfway through the rootstock stem above the bud union to help force out the bud and maintain growth of the grafted plant. The rootstock tops are later cut off.

Table 12–5
"Optimum Windows" of the Year When Selected Ornamental Species Can Be Grafted in the Greenhouse in Oregon, USA (43)

Crop	Jan.	Feb.	Mar.	Apr.	May	June	July	Aug.	Sept.	Oct.	Nov.	Dec.
Abies	X	X										X
Acer palmatum	X	X	X	X	X	X	X	X	X	X		
Aesculus	X	X	X				X	X	X			X
Carpinus	X	X	X	X						X	X	
Cedrus	X	X	X						X			
Cercis		X	X	X			X	X				
Cornus	X	X	X						X	X		
Fagus	X	X	X	X					X	X	X	
Ginkgo	X	X	X	X				X	X			
Hamamelis	X	X	X	X				X	X	X		
Larix	X	X										
Liquidambar		X	X					X	X			
Liriodendron		X	X						X	X		
Picea	X	X					X	X				X
Pinus spp.	X	X									X	X
Wisteria	X	X	X	X								

scions of grafted and budded plants to overcome the apical dominance of the rootstock and begin final production development.

Aftercare in Topworking Systems (Top-Grafting and Top-Budding) After the actual top-grafting (or top-budding) operation is finished, much important work needs to be done before the topworking is successfully completed. A good grafting job can be ruined by improper care of the grafted trees.

The trees should be carefully inspected 3 to 5 days after grafting, and the graft unions rewaxed if cracks or holes appear in the wax. However, using wax in top-grafting pecan trees is not feasible, given the high-temperature conditions of Texas, so inlay bark grafts are covered with aluminum foil and polybags to control desiccation and heat stress during the grafting process (Figs. 12–13 and 12–24). In Israel the grafts of topworked peaches have aluminum covers to reduce heat buildup (Fig. 12–24).

FIELD, BENCH, AND MISCELLANEOUS GRAFTING SYSTEMS

Some of the different grafting systems have been described in the sections on **types of grafting** and the **production processes of graftage. Grafting systems** are categorized as field, bench, and miscellaneous grafting systems (Table 12–6).

Field Grafting Systems

Crown Grafting The **crown graft** originally referred to scions grafted onto larger rootstock. The large rootstock stem was grafted with a number of scions, which sometimes were in a crown-like circle (19). Today, the term includes grafting onto an established rootstock with single or multiple scions, using the whip-and-tongue, cleft, wedge, side-veneer, inlay bark graft, and others. The choice of the graft depends on the species and size of the rootstock. In California, seedling walnut trees are planted in the nursery and then grafted at the crown—close to the junction of the root and shoot—of the rootstock.

> **crown grafting** Grafting that is done at the **crown** of the rootstock, which is the junction of the root and shoot system. In earlier times it referred to grafting several scions in a crown-like circle onto an established larger rootstock.

Crown grafting of deciduous plants is done from late winter to late spring. In each species, grafting should take place shortly before new growth starts. The scions should be prepared from mature, dormant wood of the previous season's growth.

If the graft is above the soil level, the union must be well tied (or nailed) and sealed to firmly hold the graft and prevent desiccation. However, when the operation is performed just below, at, or just above the soil

> **Table 12–6**
> **FIELD, BENCH, AND MISCELLANEOUS GRAFTING SYSTEMS**
>
> Field Grafting Systems
> - Crown graftage
> - Topworking
> - Frame working
> - Repair graftage—see page 487
> - Double-working (by grafting and budding)—sometimes bench grafted
>
> Bench Grafting Systems—see page 502
> - Root graftage
> - Nurse-root graftage
> - Grafting of plants in liner pots under protected culture
> - Herbaceous grafting—including grafting of vegetable crops
> - Approach graftage—sometimes field grafted
>
> Miscellaneous Grafting Systems
> - Cutting grafts—simultaneous rooting and grafting (stenting) of roses, citrus, etc.
> - Micrografting

level, it is possible to cover the graft union, or even the entire scion, with soil and thus eliminate the necessity for waxing or ceiling. In all cases, the union should be tied securely with tape to hold the grafted parts together until the graft takes.

Topworking Topworking is used primarily to change the cultivar of an established plant—tree, shrub, or vine—by grafting (see Fig. 12–24). Topworking can be done with any of the apical or side graftage grafts described earlier in this chapter, depending on the plant species. This procedure may be preferred to removal and replacement of the entire plant, since a return to flowering and fruiting is faster with topworking an established plant than with transplanting a new nursery plant—particularly if the topworked plant is young, healthy, and well cared for. Plants that are old, diseased, or of a short-lived species are not satisfactory candidates for topworking.

topworking Using either top-grafting or top-budding when a scion is grafted or budded onto an established plant in the orchard.

Preparation for Topworking. Top-grafting is usually done in the spring, shortly before new growth starts. The exact time depends on the method to be used. The cleft, side, whip, and wedge grafts can be done before the bark is slipping, but the bark graft must be done when the bark is slipping, preferably just as the buds of the stock tree are starting to grow.

It is usually advisable to obtain an ample amount of good-quality scionwood prior to grafting and store it under the proper conditions, although for broad-leaved evergreens, such as avocado or citrus, scionwood can be collected at the time of the grafting operation. See the earlier section on the selection and handling of scionwood (page 496).

In preparing for topworking, one must decide for each individual rootstock tree how many scaffold branches, if any, should be used (usually 3 to 5). However, no scaffold branches are retained when topworking pecans in Texas with the inlay bark method.

Double-Working (Grafting or Budding) A double-worked plant has three genetically distinct parts: the rootstock, the interstock, and the scion (Fig. 12–50). Such a plant has two unions, one between the rootstock and interstock and one between the interstock and the scion. The interstock may be less than 25 mm (1 in) in length or extensive enough to include the trunk and secondary scaffold branches of a tree.

Double-working is used for various purposes (see Chapter 11 and Chapter 13). Examples of double-working are (a) the propagation of 'Bartlett' pears on quince as a dwarfing rootstock by using a compatible interstock such as 'Old Home' or 'Hardy' pear (Fig. 12–50, page 506), and (b) the propagation of dwarfed apple trees consisting of the scion cultivar grafted onto a dwarfing 'M 9' or 'M 27' interstock that is grafted onto a more vigorous rootstock such as 'MM 106,' 'M 111,' or apple seedlings (Fig. 11–11) (10). Another form of double working is using a bridge-graft with a dwarfing rootstock (as the interstock) on older, mature apple rootstock in an established orchard, as previously described (53).

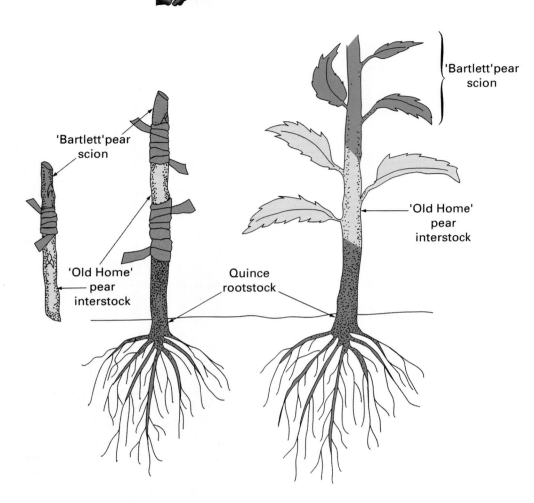

Figure 12–50
There are three genetically distinct parts and two graft unions in a doubleworked plant. The 'Bartlett' pear scion is grafted on an 'Old Home' pear interstock—then grafted to quince rootstock.

Several methods are used for developing double-worked nursery trees. The grafting in these techniques can be done with the whip graft:

- Rootstock "liners"—seedlings, clonal rooted cuttings, or rooted layers—are set out in the nursery row in early spring. These are then fall-budded with the interstock buds, growth from which, a year later, is fall-budded with the scion cultivar buds. Generally, 3 years are required to produce a nursery tree by this method.
- The interstock piece is bench grafted onto the rooted rootstock—either a seedling or a clonal stock—in late winter. After callusing, the grafts are lined-out in the nursery row in the spring, and fall-budded to the scion cultivar. By this method, the nursery tree is propagated in 2 years.
- A variation of the previous method is to prepare, by bench grafting, two graft unions—the scion grafted to the interstock and the interstock grafted to the rooted rootstock. After callusing, the completed graft, with two unions, is lined-out in the nursery row. Depending on growth rate, a nursery tree can be obtained in 1 or 2 years.
- Double-shield budding (T-budding) is used for double-working in one operation by budding (Fig. 13–21). A nursery tree is produced in 1 year or, if growth is slow, 2 years after budding.
- The interstock shoots still on the plant can be T-budded in late summer with the scion buds inserted about 15 cm (6 in) apart. During late winter, the budded interstock shoots are removed with the budded scion at the terminal end of each piece and bench grafted with a whip graft onto seedling rootstocks. After callusing, the completed graft—now consisting of rootstock interstock and a budded scion—is ready for planting in the nursery row (16).

Bench Grafting Systems

The term **bench grafting (bench working)** traditionally refers to any graft procedure performed on a rootstock and scion that are not initially planted, including root graftage, nurse-root graftage, or any graftage performed on bare-root rootstock. Bench grafting also applies to potted liner rootstock that is grafted on a bench or table, as is commonly done with selected

(a)

(b)

(c)

(d)

Figure 12–51
(a and b) Bench grafting of vegetables in Canada and Japan. (c) Machine grafted cucumber cuttings. (d) Growing grafted seedlings indoors under artificial lights (29). Courtesy C. Kubota.

woody ornamental species (43) or selected vegetable crops (Fig. 12–51) (29). Certain approach grafts are bench grafted, while others, such as the spliced approach graft of mangos in India, are field grafted, with the potted rootstock grafted to the established scion in the field.

Herbaceous types of plants are grafted for various purposes, such as studying virus transmission, stock-scion physiology, and grafting compatibility, as well as for the commercial greenhouse and field production of selected vegetable crops, particularly in Japan, Korea, and Europe (Fig. 12–51). The rootstock is grafted shortly after seed germination, while the plants are quite small. Such material is generally very soft, succulent, and susceptible to injury. The one cotylendon graft (OCG) or Japanese tube graft, is described in detail (see Figs. 12–6, 12–7, and 12–46) (21). Automated procedures of vegetable grafting using robotics and plastic grafting clips were described earlier (Figs. 12–43, 12–44, and 12–45).

Miscellaneous Grafting Systems

Cutting-Grafts In the cutting-graft, a leafy scion is grafted onto a leafy, unrooted stem piece (which is to become the rootstock), and the combination is placed in a rooting medium under intermittent mist for simultaneous grafting and rooting of the rootstock. Leaves must be retained on the rootstock piece in order for it to root. This procedure was utilized many years ago in studying stock-scion physiology in citrus (20), and has been used in commercial propagation of various types of citrus on clonal dwarfing rootstocks (13). It is also of value in propagating certain difficult-to-root conifers (55), rhododendrons (15, 41), and macadamias (1), as well as a number of apple, plum, and pear cultivars (44). It is used in the Netherlands and Israel in propagating greenhouse roses, where it is called **stenting** (Figs. 12–52 and 12–53) (58, 59).

> **stenting** A Dutch term for simultaneously grafting and rooting the rootstock.

For citrus, a simple splice graft is used. The slope of the cut is at a 30-degree angle 1.3 to 2 cm (1/2 to 3/4 in) long; the union is tied with a rubber band. The base of the rootstock is dipped into an auxin, such as IBA, and then the grafts are placed under mist, or in a closed case, in flats of the rooting medium over bottom heat. After healing of the union and rooting of the stock, the grafts are allowed to harden by discontinuing the mist and bottom heat for about 2 weeks. Then the grafts are ready to be planted in 3.8-liter (1-gal) containers.

Micrografting Grafting of tiny plant parts can be done aseptically using tissue culture techniques described in Chapters 17 and 18, in which the small grafts are grown in closed containers until they are large enough to be transferred to open conditions. Micrografting has been used mostly with citrus, apple, and some *Prunus* species to develop virus-free plants, where a virus-free shoot tip can be obtained but cannot be rooted. The shoot tip is grafted aseptically onto a virus-free seedling, thus providing a complete virus-free plant from which other

Figure 12–52
Simultaneously grafting and rooting roses for cut flower production in Israel. (a and b) Rooting in Rockwool rooting blocks. (b and c) Grafted rose with plastic graft clip for simultaneous rooting and successful callus bridge formation. (d) Rooted, stented rose.

Figure 12–53
Roses in the Netherlands being propagated by simultaneous rooting and grafting. (a) *Left:* Shoot cut apart to be used for rootstocks. Only internodes are used. *Right:* Sections cut for scions. One leaf is used per scion. (b) Saddle graft made with an Omega grafting machine. (c) Graft wrapped with tape for healing. (d) Completed graft with union healed and stock well rooted, ready for planting. In the Netherlands this process is called "stenting," a contraction of the Dutch words stekken ("to strike a cutting") and enten ("to graft"). Courtesy P. A. Van de Pol (58, 59).

"clean" plants can then be propagated (8, 25, 27, 46). Tests conducted some years after the shoot-tip grafts were made and the resultant trees were fruiting showed that the fruits were normal for the cultivar, disease-free, and with no variations appearing (45). The procedures for micrografting are described in Chapter 17.

Micrografting can also be done without tissue culture techniques. The non-aseptic propagation of very small nursery trees by grafting tiny seedlings with match-like scions, then growing the minute grafts long enough to have a viable plant, is a promising procedure (60). This is useful particularly in the tropics, where nursery plants sometimes must be shipped long distances, often by air, into inaccessible regions. Quantities of such tiny plants can be transported much more readily than full-sized nursery trees.

DISCUSSION ITEMS

1. What are five important requirements necessary for successfully producing a grafted plant?
2. What are the three major types (classifications) of grafts, and what criteria are used to categorize a specific graft into one of these three types?
3. What are the three types of nurse-root grafts? Why use an expensive process such as nurse-root grafting?
4. How does approach graftage differ from repair graftage? Give examples of grafts used.
5. Why are grafting waxes not used as much today as in the past? What kinds of substitutes are being used in their place?
6. Why is there increased interest in grafting automation/robotics?
7. What are some important considerations in the collection, handling, and storage of scionwood for grafting?
8. The craftsmanship of grafting includes both art and skill. How can an individual improve techniques and ergonomics to become a better grafter?
9. What are some good cultural techniques to enhance aftercare in topworking (topgrafting) an orchard with a new cultivar?
10. How do field grafting systems differ from bench grafting systems? Include examples of the various types of grafts in your answer.
11. Why the interest in herbaceous grafting, and with what crops is this technique of commercial importance? Why does this type of grafting lend itself to automation/robotics?
12. Contrast cutting grafts with micrografting, and give examples of their commercial use.

REFERENCES

1. Ahlswede, J. 1985. Twig grafting of macadamia. *Comb. Proc. Intl. Plant Prop. Soc.* 34:211–14.
2. Alley, C. J. 1957. Mechanized grape grafting. *Calif. Agr.* 11:3,12.
3. Alley, C. J. 1970. Can grafting be mechanized? *Comb. Proc. Intl. Plant Prop. Soc.* 20:244–48.
4. Anonymous. 1968. Pre-cut scions. *Agr. Res.* 17:11.
5. Beineke, W. F., 1978. Parafilm: A new way to wrap grafts. *HortScience* 13:284.
6. Bhar, D. S., R. J. Hilton, and G. C. Ashton. 1966. Effect of time of cutting and storage treatment on growth and vigor of scions of *Malus pumila* cv. McIntosh. *Can. J. Plant Sci.* 46:69–72.
7. Brase, K. D. 1951. The nurse-root graft, an aid in rootstock research. *Farm Res.* 17:16.
8. Burger, D. W. 1985. Micrografting: A tool for the plant propagator. *Comb. Proc. Intl. Plant Prop. Soc.* 34:244–48.
9. Carpenter, E. L. 1989. How records can improve grafting. *Comb. Proc. Intl. Plant Prop. Soc.* 39:413–15.
10. Cummins, J. N. 1973. Systems for producing multiple-stock fruit trees in the nursery. *Plant Propagator* 19:7–9.
11. Davis, A. R., P. Perkins-Veazie, R. Hassell, S. R. King, and X. Zhang. 2008a. Grafting effects on vegetable quality. *HortScience* 43:1670–2.
12. Davis, A. R., P. Perkins-Veazie, Y. Sakata, S. Lopez-Galarza, J. V. Maroto, S. G. Lee, Y. C. Huh, Z. Sun, A. Miguel, S. R. King, R. Cohen, and J. M. Lee. 2008b. Cucurbit grafting. *Crit. Rev. Plant Sci.* 27:50–74.

13. Dillon, D. 1967. Simultaneous grafting and rooting of citrus under mist. *Comb. Proc. Intl. Plant Prop. Soc.* 17:114–17.

14. Douglas, G. C., and J. McNamara. 1996. A tube method for grafting small diameter scions of the hardwoods *Quercus, Fraxinus, Betula* and *Sorbus*. *Comb. Proc. Intl. Plant Prop. Soc.* 46:221–26.

15. Eichelser, J. 1967. Simultaneous grafting and rooting techniques as applied to Rhododendrons. *Comb. Proc. Intl. Plant Prop. Soc.* 17:112.

16. Fisher, E. 1977. The pre-budded interstem: A new technique. *Fruit Var. J.* 31:14–5.

17. Flemer, W., III. 1986. New advances in bench grafting. *Comb. Proc. Intl. Plant Prop. Soc.* 36: 545–49.

18. Gaggini, J. B. 1985. Bench grafting of trees under polyethylene. *Comb. Proc. Intl. Plant Prop. Soc.* 34:646–47.

19. Garner, R. J. 1988. *The grafter's handbook.* 5th ed. New York: Oxford Univ. Press.

20. Halma, F. F., and E. R. Eggers. 1936. Propagating citrus by twig-grafting. *Proc. Amer. Soc. Hort. Sci.* 34:289–90.

21. Hassell, R. L., F. Memmott, and D. G. Liere. 2008. Grafting methods for watermelon production. *HortScience* 43:1677–9.

22. Hearman, J., A. B. Beakbane, R. G. Hatton, and W. A. Roach. 1936. The reinvigoration of apple trees by the inarching of vigorous rootstocks. *J. Pom. Hort. Sci.* 14:376–90.

23. Hottes, A. C. 1937. *Plant propagation: 999 questions answered.* New York: De La Mare Co.

24. Howard, G. S., and A. C. Hildreth. 1963. Induction of callus tissue on apple grafts prior to field planting and its growth effects. *Proc. Amer. Soc. Hort. Sci.* 82:11–5.

25. Huang, S., and D. F. Millikan. 1980. *In vitro* micrografting of apple shoot tips. *HortScience* 15:741–43.

26. Intven, W. J., and T. J. Intven. 1989. Apical grafting of *Acer palmatum* and other deciduous plants. *Comb. Proc. Intl. Plant Prop. Soc.* 39:409–12.

27. Jonard, R., J. Hugard, J. Macheix, J. Martinez, L. Mosella-Chancel, J. Luc Poessel, and P. Villemur. 1983. *In vitro* micrografting and its application to fruit science. *Scientia Hort.* 20:147–59.

28. Kerr, W. L. 1935. A simple method of obtaining fruit trees on their own roots. *Proc. Amer. Soc. Hort. Sci.* 33:355–57.

29. Kubota, C., M. A. McClure, N. Kokalis-Burelle, M. G. Bausher, and E. N. Rosskopf. 2008. Vegetable grafting: History, use, and current technology status in North America. *HortScience* 43:1664–9.

30. Kurata, K. 1994. Cultivation of grafted vegetables II. Development of grafting robots in Japan. *HortScience* 29:240–44.

31. Lagerstedt, H. B. 1982. A device for hot callusing graft unions of fruit and nut trees. *Comb. Proc. Intl. Plant Prop. Soc.* 31:151–59.

32. Lagerstedt, H. B. 1984. Hot callusing pipe speeds up grafting. *Amer. Nurs.* 160:113–17.

33. Lee, J. M. 1994. Cultivation of grafted vegetables 1. Current status, grafting methods, and benefits. *HortScience* 29:235–39.

34. Lee, J. M., and M. Oda. 2003. Grafting of herbaceous vegetable and ornamental crops. *Hort. Rev.* 28:61–124.

35. Legare, M. 2007. The future of grafting. *Comb. Proc. Intl. Plant Proc. Soc.* 57:380–84.

36. Leiss, J. 1987. Modified side graft for nursery trees. *Comb. Proc. Intl. Plant Prop. Soc.* 36:543–44.

37. Lincoln, F. B. 1938. Layering of root grafts—a ready method for obtaining self-rooted apple trees. *Proc. Amer. Soc. Hort. Sci.* 35:419–22.

38. MacDonald, B. 1986. *Practical woody plant propagation for nursery growers.* Portland, OR: Timber Press.

39. McEachern, G. R., and A. Stockton. 1993. The four-flap graft. In G. R. McEachern and L. A. Stein, eds. *Texas pecan handbook.* TAEX Hort Handbook 105. College Station, TX: Texas Agricultural Extension Service, Texas A&M University.

40. McEachern, G. R., S. Helmers, L. Stein, J. Lipe, and L. Shreve. 1993. Texas inlay bark graft. In G. R. McEachern and L. A. Stein, eds. *Texas pecan handbook.* TAEX Hort Handbook 105. College Station, TX: Texas Agricultural Extension Service, Texas A&M University.

41. McGuire, J. J., W. Johnson, and C. Dawson. 1987. Leaf-bud or side graft nurse grafts for difficult-to-root Rhododendron cultivars. *Comb. Proc. Intl. Plant Prop. Soc.* 37:447–49.

42. McPhee, G. R. 1992. Grafting techniques. *Comb. Proc. Intl. Plant Prop. Soc.* 42:51–3.

43. Meacham, G. E. 1995. Bench grafting, when is the best time? *Comb. Proc. Intl. Plant Prop. Soc.* 45:301–4.

44. Morini, S. 1984. The propagation of fruit trees by grafted cuttings. *J. Hort. Sci.* 59:287–94.

45. Nauer, E. M., C. N. Roistacher, T. L. Carson, and T. Murashige. 1983. *In vitro* shoot-tip grafting to eliminate citrus viruses and virus-like pathogens produces uniform bud-lines. *HortScience* 18:308–9.

46. Navarro, L., C. N. Roistacher, and T. Murashige. 1975. Improvement of shoot tip grafting

in vitro for virus-free citrus. *J. Amer. Soc. Hort. Sci.* 100:471–79.

47. Oda, M., and T. Nakajima. 1992. Adhesive grafting of Chinese cabbage on turnip. *HortScience* 27:1136.

48. Oda, M., K. Okada, H. Sasaki, S. Akazawa, and M. Sei. 1997. Growth and yield of eggplants grafted by a newly developed robot. *HortScience* 32:848–49.

49. Oda, M., T. Nagaoka, T. Mori, and M. Sei. 1994. Simultaneous grafting of young tomato plants using grafting plates. *Scientia Hort.* 58:259–64.

50. Passam, H. C. 2003. Use of grafting makes a comeback. *Fruit Veg. Tech.* 3:7–9.

51. Patrick, B. 1992. Budding and grafting of fruit and nut trees at Stark Brothers. *Comb. Proc. Intl. Plant Prop. Soc.* 42:354–56.

52. Rivard, C., and F. Louws. 2006. *Grafting for disease resistance in heirloom tomatoes*. Raleigh, NC: North Carolina Cooperative Extension Service, AG-675. E07:45829.

53. Samad, A., D. L. McNeil, and Z. U. Khan. 1999. Effect of interstock bridge grafting (M9 dwarfing rootstock and same cultivar cutting) on vegetative growth, reproductive growth and carbohydrate composition of mature apple trees. *Scientia Hort.* 79:23–8.

54. Singha, S. 1990. Effectiveness of readily available adhesive tapes as grafting wraps. *HortScience* 25:579.

55. Teuscher, H. 1962. Speeding production of hard-to-root conifers. *Amer. Nurs.* 116:16.

56. Upchurch, B. L. 2006. Grafting with care. *Amer. Nurs.* 203:18–22.

57. Upshall, W. H. 1946. The stub graft as a supplement to budding in nursery practice. *Proc. Amer. Soc. Hort. Sci.* 47:187–89.

58. Van de Pol, M. H., A. J. Joosten, and H. Keizer. 1986. Stenting of roses, starch depletion and accumulation during the early development. *Acta Hort.* 189:51–9.

59. Van de Pol, P. A., and A. Breukelaar. 1982. Stenting of roses: A method for quick propagation by simultaneously cutting and grafting. *Scientia Hort.* 7:187–96.

60. Verhey, E. W. M. 1982. Minute nursery trees, a breakthrough for the tropics? *Chronica Hort.* 22:1–2.

61. Westergaard, L. 1997. Improved grafting techniques for nursery stock 1997. *Comb. Proc. Intl. Plant Prop. Soc.* 42:354–56.

*13 Techniques of Budding

learning objectives

- Discuss the importance and utilization of budding.
- Describe the different types of rootstocks utilized for budding.
- Explain the management practices of summer, spring, and June budding.
- Identify the different types of budding.
- Describe the processes of top-budding (topworking), double-working by budding, and microbudding.

INTRODUCTION

Budding is a form of grafting in which the **scion** consists of a single bud and a small section of bark with or without the wood. In other forms of **grafting** the scion has several buds.

Budding accounts for the vast majority of grafted nursery stock. Millions of fruit and shade trees and roses are budded annually. *In turn, nearly all of this production is field-grown,* rather than container-grown. **Rootstocks** are lined-out in the field and grown until suitable for budding. After a season of scion growth, the trees are harvested when dormant and sold for orchard production or as landscape plants.

budding A form of grafting that uses a smaller scion piece—sometimes just a portion of the stem with an axillary bud.

rootstock, understock, or stock The root system of the budded plant.

Chip budding and **T-budding** are the two most important types of budding for woody ornamentals and fruit trees (see Table 13–1, page 522). Chip and T-budding are much simpler and, therefore, much faster than manual grafting techniques. Single **budders,** working with one or more persons tying the buds **(tiers)**, can bud 2,000 to 4,000 buds in a day—sometimes more. With well-grown rootstocks, healthy buds, and skillful budders, nurseries expect to achieve 90 to 100 percent successful unions or **"bud takes."** Managing production of budded liner plants is one of the more complex logistical problems in nursery production.

tier The person who completes the grafting process by tying and sometimes waxing the graft area.

topworking (top-budding) The process of budding onto an existing rootstock with new scion material.

In certain cases, budding is useful for **topworking (top-budding),** but because budding is confined to shoots less than 2.5 cm (1 in) diameter, it can only be used to topwork young trees or smaller shoots of older trees.

IMPORTANCE AND UTILIZATION OF BUDDING

Advantages of Budding Compared with Grafting

Budding makes very efficient use of scionwood, because only a single bud is needed to propagate a new tree. This efficiency reduces both the number of trees required to supply scionwood and the labor to maintain the trees and collect wood. Budding also makes good use of plant material in cases when scionwood of a particular clone is limited. Budding may also result in a stronger union, particularly during the first few years, than is obtained by some grafting techniques, which reduces the likelihood that wind will damage the trees. The simplicity and speed of budding, especially the T-budding and chip budding techniques, makes these techniques useful for amateur horticulturists. A single, well-learned method can be used in a wide variety of applications.

Combining Budding and Grafting Techniques

A strategy of some ornamental and fruit tree nurseries in England is to chip bud or T-bud initially in the nursery row. Grafting with a whip-and-tongue graft is used as a backup for budded plants that do not take. There are other variations of this method where different budding and grafting systems are combined in nursery field production.

Conditions for Budding

Budding methods, such as T-budding and patch budding, depend on the **bark's "slipping."** This term is used to define the condition in which the "bark" (periderm, cortex, phloem, and vascular cambium) can be easily separated from the **"wood"** (xylem)—see Chapter 11 and Figures 11–6 and 11–22. Bark slippage denotes the period of the year when the plant is in active growth, when the cambium cells are actively dividing, and when newly formed tissues are easily separated as the bark is lifted from the wood. In Oregon, the bark slips from late June through August (19). However, adverse growing conditions, such as lack of water, insect or disease problems, defoliation, or low temperatures, may reduce growth, lead to a tightening of the bark, which will seriously interfere with the budding operation. If the bark has only the slightest adherence to the wood, the percentage of bud take will be severely limited. Irrigating nursery rootstock prior to and after budding enhances bud take. Of the budding types described here, only one—the chip bud—can be done when the bark is not slipping.

> bark Composed of tissue from the periderm to the phloem. For most budding systems the bark must be *slipping*, which occurs with active rootstock growth.
>
> wood The secondary xylem that makes up the major mass of the rootstock tissues.

Budding height varies on nursery rootstock. Budding is done higher on dwarfing rootstock for fruit crops to prevent scion rooting. Higher budding also increases the dwarfing effect of the rootstock. Peaches tend to be budded low, about 2.5 to 5 cm (1 to 2 in) above the ground, while citrus is budded higher (see Fig. 13–12, page 527). Budding is done low on ornamental shade trees (see Fig. 13–4, page 517), since tree shape and appearance are important.

ROOTSTOCKS FOR BUDDING

Rootstock should have the desired characteristics of vigor, proper growth habit, and resistance to soilborne pests, as well as being easily propagated. Rootstock may be a rooted cutting, a rooted layer, a seedling, or a micropropagated plant (see Chapter 12). The length of time before budding depends on rootstock vigor, length of growing season, and climate. As little as six months to one year's growth in the nursery row is needed to produce a rootstock plant large enough to be budded, but seedlings of slow-growing species, such as pecan, and those grown under unfavorable conditions may require two or three seasons.

To produce nursery trees free of harmful pathogens (such as viruses, mycoplasmas, fungi, or bacteria) the rootstock plant, as well as the budwood, must be free of such organisms, see discussion in Chapter 16.

TIME OF BUDDING—SUMMER, SPRING, OR JUNE

Traditionally, budding is done when rootstocks are actively growing, so that the cambium divides and the bark separates readily from the wood (e.g., using the T-budding method). In the 1980s, however, chip budding replaced T-budding at many nurseries. This was based on research in England, which showed that chip buds of some species form a better graft union in cooler growing regions (7). Present consensus in the United States is to use chip budding for harder-to-graft varieties, and T-budding on easier budding material.

For any budding technique, however, well-developed vegetative buds of the scion variety must be available. For spring budding, the scion buds are used from dormant, stored budwood, whereas **quiescent** (nonelongating) buds are used from the current season's wood

> quiescent (quiescence) Buds that are inhibited from growing and elongating via apical dominance of more distal buds produced during the current season on the same shoot.

spring budding
Budding that is done as soon as new seasonal growth occurs, in late March to early May, depending on location. A 1-year scion/2-year rootstock is generally produced.

June budding
Budding that is done from May to early June, which produces a smaller budded plant with a 1-year scion/ 1-year rootstock.

summer budding
Sometimes referred to as "fall budding," which is a misnomer since the budding occurs from mid-July to early September, not in autumn. A 1-year scion/2-year rootstock is generally produced.

in June and summer budding.

Budding is normally done during three periods of the year. In the Northern Hemisphere, these periods are March to early May (**spring budding**), May to early June (**June budding**), and mid-July to early September (**summer budding**) (Figs. 13–1 and 13–2, page 516). In the Southern Hemisphere, similar periods would be September to early November (*spring budding*), early November to early December (*June budding*), and mid-January to early March (*summer budding*).

The budding schedules listed (spring, June, summer) are for *production of nursery plants* that will later be dug from the nursery row and transplanted into an orchard or landscape site. Budding schedules can also be applied to container-produced nursery plants such as citrus, pecans, and selected woody ornamentals. **Top-budding (topworking)** *is done with established trees and shrubs in the field* and is scheduled during the year based on the plant species and budding method to be utilized. **Bench budding** (e.g., chip budding) can be done with dormant rootstock during wintertime, *working on a table*. Bench budding also includes budding onto *liner rootstocks in containers under protected culture conditions* (greenhouse, polyhouse). The budding could extend throughout the year, depending on the species and budding method. See Table 12–5 for optimum production windows when grafting/budding in the greenhouse (12).

bench budding (bench working) Budding under protected culture onto a dormant rootstock using a bench or onto a containerized rootstock that may be active or dormant.

Summer Budding

Summer is still the most important time for budding many species of fruit tree nursery stock and shade trees. Summer budding is sometimes referred to as "fall budding," which is a misnomer since the budding occurs from mid-July to early September, not in autumn. Summer budding is particularly important in northern areas where the growing season is short. The rootstock plants are large enough by midsummer to accommodate the bud, and the plants are still actively growing.

During the spring following lining-out, soil is mounded around the rootstock stem to protect the bark from sunburn and physical damage. During the spring and early summer, vigorous growth increases stem diameter; continued rapid growth is required for optimal budding. Poor growing conditions caused by drought, temperature extremes, insects, or disease may prevent or impair budding. During the period up to budding, suckers are removed from the roots, and sprouts are removed from the trunk area at or below where the bud will be inserted. Just prior to budding, the soil around the trunk is removed to expose the smooth, straight stem where the bud will be placed.

Budding starts when rootstock liners have sufficient growth, and vegetative buds on scion varieties are mature and well-developed. Buds are usually ready by late July (in Oregon). Budding peaks in August and finishes in September, particularly in areas with longer seasons. Nurseries maintain scion blocks to produce large amounts of high-quality scionwood. For tree fruits, scionwood is collected from trees provided by certification programs, depending on the state and nursery. These trees are certified as true-to-type; that is, they produce the accepted phenotype of a given fruit variety, and are free of known viruses and related pathogens (see Chapter 16). To produce a certified fruit tree, all parts of the tree must originate from certified blocks—scion, seed, stool, and cutting.

Scionwood (budwood, budsticks) is collected from the current season's growth and should be used promptly after cutting. Although budwood can be refrigerated for a short time, it is best to collect the budsticks as they are being used, a day's supply at a time. As the budsticks are selected, the leaves should be cut off immediately, leaving only a short piece of the leaf petiole attached to the bud to aid in handling the bud later on. Budsticks should be kept from drying by wrapping in material, such as clean, moist burlap, and placed in a cool, shady location until they are needed.

scionwood, budsticks, or budwood The scion pieces from which buds are collected for budding.

The best buds to use are usually on the middle and basal portions of the budwood. These buds are *quiescent*, and their growth is checked by apical dominance of more terminal buds. Buds on the succulent terminal portion of the shoot should be discarded. In

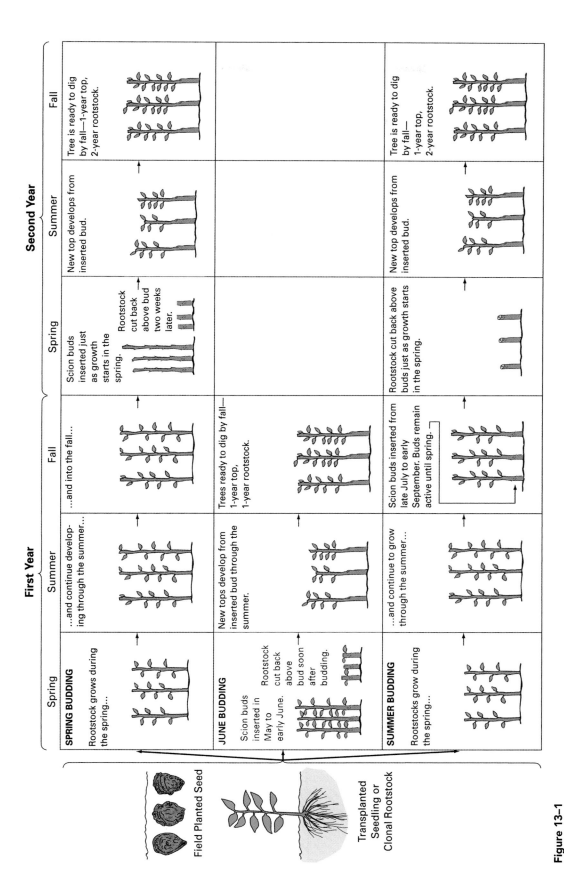

Figure 13–1
Comparison of the steps in spring, June, and summer budding for nursery production. The actual techniques in budding are not difficult, but it is very important that the various operations be done at the proper time. Rootstocks are field propagated from fall-planted or early spring-planted stratified seeds, seedling transplants, rooted cuttings or layers, or micropropagated plantlets.

Figure 13–2
It is important in June budding that the rootstock be cut back to the bud properly. *Far left:* The shield bud is inserted high enough on the rootstock so that there are several leaves below the bud. *Center left:* Three or four days after budding, the stock is partially cut back about 9 cm (3 1/2 in) above the bud. *Center:* Ten days to two weeks after T-budding, the rootstock is completely removed just above the bud. *Center right:* This forces the shield bud and other buds on the rootstock into growth; the latter must subsequently be removed. *Far right:* Appearance of the budded tree after the new shoot has made considerable growth.

certain species, such as the sweet cherry, buds on the basal portion of the shoots are flower buds which, of course, should not be used (Fig. 13–3).

After the buds have been inserted and tied, nothing is done to the budded area until the following spring. However, it may be necessary to remove budding strips after bud take to prevent girdling. *Although the rootstock is eventually cut off above the bud, in no case should this be done immediately after the bud has been inserted.* The union of the bud piece to the rootstock is greatly facilitated by the normal movement of water, nutrients, and photosynthate in the stem of the rootstock. This flow is stopped if the top of the rootstock is cut off above the bud.

If the budding operation is done properly, the bud piece should unite with the rootstock in 2 to 3 weeks, depending on growing conditions. Abscission of the leaf petiole next to the bud is a good indication that the bud has united, especially if the bark piece retains its normal light brown or green color, and the bud stays plump. On the other hand, if the leaf petiole does not drop off cleanly but adheres tightly and starts to shrivel and darken, and the bark piece turns black—the operation has failed. If the bark of the rootstock is still slipping and budwood is still available, budding may be repeated.

Even though the union has formed, buds on most deciduous species usually do not grow or "push out" in the fall, since they are either quiescent or in a physiological **dormancy (rest)**. By spring, the chilling winter temperatures have satisfied the rest (dormancy) requirement, and the bud is ready to grow. Exceptions to the need for dormancy occur in maples, roses, honey locust, and certain other plants, where some of the buds start growth in the fall. In northern areas, such fall-forced buds usually fail to mature before cold weather starts and are likely to be winter-killed.

dormancy (rest) Buds that are inhibited from growing and elongating until sufficient fall and winter chilling needs are met to overcome an internal physiological requirement.

In the spring, the rootstock is cut off immediately above the bud, just before new growth begins. Cutting

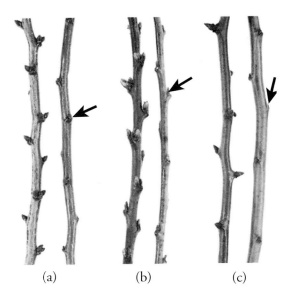

Figure 13–3
In budding it is important to use vegetative rather than flower buds. Vegetative buds (arrows) are usually small and pointed, while flower buds are larger and more plump. Differences between vegetative and flower buds in three fruit species are illustrated here. (a) Almond: the shoot on the left has primarily flower buds and should not be selected for budding. The shoot on the right has vegetative buds, which are more suitable. (b) Peach: the shoot on the right has excellent vegetative buds, while those on the left shoot are mostly flower buds. (c) Pear: all the buds on the shoot at the left are flower buds, while buds on the shoot at the right are good vegetative buds, suitable for budding.

back the rootstock breaks apical dominance of the upper axillary buds and forces the inserted bud into growth. In citrus, the rootstock is partially cut above the bud and the top is bent or lopped over, away from the bud. This procedure is referred to as **"crippling"** or **"lopping"** (see Fig. 11–30). Leaves of the rootstock still supply the roots with photosynthate, but the partial cutting forces the bud into growth more rapidly than severely damaging the rootstock top. After the new shoot

crippling or lopping
Bending (constriction) or cutting halfway through the rootstock stem above the bud union. This technique helps to force out the bud and maintain growth of the budded plant. At a later date, the rootstock stem is completely removed.

from the bud has started to grow, the rootstock top is completely removed. In colder northern regions, summer-inserted buds are sometimes covered with soil during the winter until danger of frost has passed, and are then uncovered and topped-back in late spring.

Where strong winds occur, support for the newly developing shoot may be necessary in species where new shoots grow vigorously. Sometimes the rootstock is cut off several centimeters above the bud, and the projecting stub is used as a support to tie the tender young shoot arising from the bud. Pecan patch buds are forced in the spring by girdling the rootstock a few centimeters above the inserted bud and stripping the phloem away; the tender scion shoot growing from the patch bud can then be tied to the rootstock. This stub is removed after the shoot has become well-established. With pistachio, stakes may be driven into the ground next to the rootstock, and developing scion shoots tied at intervals during their growth.

Metal shoot guide clips such as "Grow Straights" are used to obtain upright scion growth as the bud elongates (see Fig. 13–4). Since they are attached to the rootstock or inserted into the soil, they also give some mechanical support to the elongating scion shoot.

Figure 13–4
(a) Spring budded *Acer platanoides* using a chip-bud system. (b) A chip-budded crab apple is being wrapped with poly tape (arrow), which will be removed after the graft has taken to prevent girdling the plant. (c) Grow Straight metal shield (arrow) to produce straight, upright growth from the scion bud. The top of the rootstock has been cut off to force out the scion bud. (d) The metal shield system with 'Crimson King' maple T-budded to a seedling rootstock. Courtesy of K. Warren (19).

Cutting back to force the scion bud also forces many latent buds on the rootstock. These must be rubbed off as soon as they appear, or they will soon choke out the inserted bud. It may be necessary to go over the budded plants several times before the developing shoot inhibits these "sprouts" from appearing. Nursery people refer to this procedure as **"suckering."**

suckering Sprouts from the rootstock that can crowd out and inhibit growth of the scion; hence, they must be removed.

The shoot arising from the inserted bud becomes the scion or top portion of the plant (Figs. 13–2 and 13–4). After one season's growth in the nursery, this shoot will have developed sufficiently to enable the plant to be dug and moved to its permanent location during the following dormant season. Such a tree would have a *1-year-old top and a 2- or perhaps 3-year-old rootstock, but it is still considered a "yearling" tree*. If the top makes insufficient growth the first year, it can be allowed to grow a second year, and is then known as a 2-year-old tree. Abnormal, slow-growing, stunted trees should be discarded.

Spring Budding

Spring budding is similar to summer budding, except budding takes place the following spring as soon as active growth of the rootstock begins and the bark separates easily from the wood. The period for successful spring budding is shorter, and budding must be completed before the rootstocks make any significant growth. As with summer budding, most spring-budded, yearling trees will have a *1-year-old top and a 2-year-old rootstock, when dug in the late fall or winter* (Fig. 13–1).

Budsticks are chosen from the same type of shoots—with regard to vigor of growth and type of buds (Fig. 13–3)—that would have been used in fall budding. However, they are collected when dormant, in late fall or winter, and placed in cold storage. Budwood collected in late winter must still be dormant, before there is any evidence of the buds swelling. Budwood is stored at about –2 to 0°C (29 to 32°F) to hold the buds dormant.

The budsticks should be wrapped in bundles with slightly damp sphagnum, wood shavings, peat moss, moistened newspapers, or paper towels to prevent drying out.

With chip budding, the bark does not need to slip, and dormant rootstock can be worked with. In Oregon, spring chip budding of shade trees is done just before bud-break of the rootstock.

Other types of budding are done in the spring as soon as the bark on the rootstock slips easily. When the bud unions have formed, about 2 to 4 weeks after budding, the top of the rootstock must be cut off above the bud to force the inserted bud into active growth. At the same time, axillary buds on the rootstock begin to grow and should be removed. In California, spring budding is used to propagate plums, apricots, and almonds.

Sometimes shoots from the rootstock are allowed to develop in order to prevent sunburn and help nourish the plant. The shoots must be held in check, however, and eventually removed. In 2-year field rose bush production in California, Texas, and Spain, spring T-budded rootstock plants are not cut back until the late winter of the second and final growing season. In Texas, corn choppers, or machines that can cut and mulch the rootstock tops, are used to top the rootstock about 8 to 15 cm (3 to 6 in) above the union. In a subsequent operation, pneumatic pruning shears are used to prune back the scion shoot so lateral breaks can occur, in order to produce a #1 grade rose bush.

Although the new shoot from the inserted bud gets a later start in spring budding than in summer budding, spring buds will usually develop rapidly enough, if growing conditions are favorable, to make a satisfactory top by fall.

In California and Oregon, some nurseries use spring budding extensively to shorten the production time needed to sell the crop (budded and harvested the same year). As an example, 'Marianna 2624' rootstock are propagated as hardwood cuttings in the nursery row in the fall. When the rooted rootstock cuttings begin growth, they are spring-budded in April with apricot cultivars. After the bud takes, the rootstock top is cut

---

BOX 13.1 GETTING MORE IN DEPTH ON THE SUBJECT
ADVANTAGES OF SUMMER BUDDING COMPARED WITH SPRING BUDDING

In general, summer budding is preferred to spring budding for several reasons:

- The higher temperatures of summer promote more rapid, extensive bud union formation.
- The budding season is longer.

- There is no need to store the budsticks.
- One can rebud spring-budded plants that did not take.
- The demands of other nursery activities are usually not so great for propagators in summer as they are in the spring.

back and a nursery plant is produced in one growing season. This smaller, yearling tree will have a *1-year-old top and 1-year-old rootstock, when dug in the fall or late winter* so the nursery can get new varieties and rootstocks into production sooner. The nursery can also respond more quickly to changing market conditions, fill orders more quickly, and improve their cash flow.

June Budding

In June budding, both rootstocks and tops of the budded tree develop during a single growing season, producing a *1-year-old top and 1-year-old rootstock*. Budding is done in the early part of the growing season as soon as the budsticks reach their proper maturity, typically early May to early June in California. The inserted bud (from the current season's growth) is forced into growth immediately. As a method of nursery propagation, June budding is confined to regions that have a long growing season. In the United States, this includes the central valley of California and some of the southern states—such as Tennessee and Arkansas. June budding is not done in Texas because budding failure is too high when temperatures exceed 35°C (95°F).

June budding is used mostly to produce stone fruits—peaches, nectarines, apricots, almonds, and plums. Peach seedlings are generally used as the rootstock because they have the necessary vigor to produce a large enough seedling rootstock for budding. Budding is done by the T-bud method. If seeds are planted in the fall, or stratified seeds planted as early as possible in the spring, the seedling rootstock usually attain sufficient size—30 cm (12 in) high and at least 30 mm (1/8 in) in diameter—to start budding in early May to early June. Preferably, June budding should not be done after late June, or a nursery tree of satisfactory size will not be obtained by fall. June-budded trees are not as large by the end of the growing season as those propagated by summer or spring budding, but they are of sufficient size—10 to 16 mm (3/8 to 5/8 in) caliper and 90 to 150 cm (3 to 5 ft) tall—to produce entirely satisfactory trees (13).

Budwood used in June budding consists of the current season's growth; that is, new shoots that have developed since growth started in the spring. By early May or early June, these shoots will usually have grown sufficiently to reach the proper state of maturity and have a well-developed bud in the axil of each leaf. Buds are quiescent at the time of budding, but do not enter a rest period until fall. Consequently, the shoot continues to grow all summer to produce the top portion of the budded tree.

For June-budded trees, handling subsequent to the actual operation of budding is somewhat more exacting than for fall- or spring-budded trees. The rootstocks are smaller and have less stored carbohydrates than those used in fall or spring budding. The object behind the following procedures—shown in Figure 13–2—is to keep the rootstock (and later the budded top) actively and continuously growing so as to allow no check in growth, while at the same time converting the seedling rootstock shoot to a budded top. The bud should be inserted high enough (about 14 cm; 5 1/2 in) on the stem so that a number of leaves—at least three or four—can be retained below the bud. Some nurseries will also remove all leaves to a height slightly above the budding site on the rootstock to facilitate budding and typing.

The method of T-budding with the "wood out," also known as a "flipped bud," should be used (see Fig. 13–11, page 526). The bud union forms quickly, since temperatures are relatively high, and rapidly growing, succulent plant parts are used. By 4 days after budding, the bud union has started, and the top of the rootstock can be partially cut back—about 9 cm (3 1/2 in) above the bud—leaving at least one leaf above the bud and several below it. This operation will force the inserted bud into growth and will check terminal growth of the rootstock, but it will also stimulate shoot growth from basal buds of the rootstock, which will produce additional leaf area. This continuous leaf area is necessary because it ensures that there will always be enough leaves to manufacture photosynthate for the small plant. The rootstock can be cut back to the bud 10 days to 2 weeks after budding. If the budding rubber has not broken, it should be cut at this time. Poly budding strips would need to be removed or cut on the opposite side of the bud. Other shoots arising from the rootstock should be headed back to retard their growth. After the inserted bud grows and develops a substantial leaf area, it can supply the plant with the necessary photosynthates. By the time the shoot from the inserted bud has grown about 25 cm (10 in) high, it should have enough leaves so that all other shoots and leaves of the rootstock can be removed. Later inspections should be made to remove any rootstock shoots below the budded shoot. The steps in summer, spring, and June budding are compared in Figure 13–1.

TYPES OF BUDDING

Chip Budding

Chip budding works well in regions with shorter growing seasons (e.g., Northern Unites States, England). Chip budding has gradually replaced T-budding as the primary budding method for many woody ornamental trees, shrubs, and fruit trees in many parts of the world (10, 14, 18). Commercial nurseries have switched to chip budding because of better takes and straighter, more uniform tree growth (15). Chip budding in late summer gives excellent results in budding grape cultivars on phylloxera or nematode-resistant rootstocks (Fig. 13–5, page 520) (5, 11).

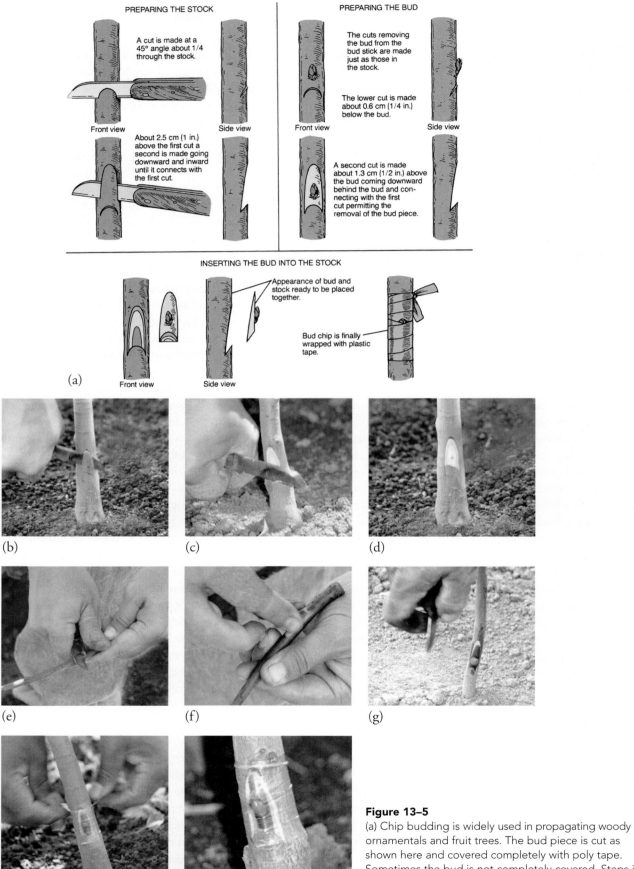

Figure 13–5
(a) Chip budding is widely used in propagating woody ornamentals and fruit trees. The bud piece is cut as shown here and covered completely with poly tape. Sometimes the bud is not completely covered. Steps in chip budding. (b) First downward cut in understock, (c) second downward cut, (d) finished cut, (e and f) removal of scion chip (g), insertion of bud, (h) wrapping bud, and (i) finished chip bud.

> **production "window"**
> The time during the season when a budding procedure can successfully be done (i.e., T-budding can only be done when the bark of the rootstock is slipping).

The **production "window"** for chip budding—or for bench grafting dormant rose rootstock (2, 3)—is greater than with T-budding, since chip budding can be done with dormant or active rootstock (Table 13–1). Plant water status and temperature extremes are less of a problem with chip budding. More vigorous initial growth has been reported with chip buds than with T-buds (6, 7). Studies in England have shown that a better union is obtained with chip budding than with T-budding (8, 14). See Chapter 11 for further discussion on the anatomical and physiological advantages of chip budding compared to T-budding (16). Chip budding is easily mechanized and performed with budding machines (Fig. 13–6, page 523).

Chip budding is generally used on rootstock with small diameter stems, about 13 to 25 mm (1/2 to 1 in). It works well with late winter (Texas) grafting of small pistachio rootstocks that have too thin a bark for spring T-budding. As illustrated in Figures 13–4, 13–5, 13–6, 13–7 (page 523), and 13–8 (page 524), a chip of bark is removed from a smooth place between nodes near the base of the rootstock and replaced by another chip of the same size and shape from the budstick, which contains a bud of the desired cultivar. The chips in both rootstock and budstick are cut out in the same manner. In the budstick, the first cut is made just below the bud and down into the wood at an angle of 30 to 45 degrees. The second cut is started about 25 mm (1 in) above the bud and goes inward and downward behind the bud until it intersects the first cut. (The order of making these two cuts may be reversed.) The chip is removed from the rootstock and replaced by the one from the budstick. The bark on the rootstock is generally thicker than that of the scionwood. Therefore, the chip removed from the rootstock is slightly larger than that removed from the scion. The cambium layer of the bud piece must be placed to coincide with that of the stock, preferably on both sides of the stem, but at least on one side (Fig. 13–6).

The chip bud must be wrapped to seal the cut edges and to hold the bud piece tightly into the rootstock since there are no protective flaps of bark to prevent the bud piece from drying out. Nursery adhesive tape, Parafilm (2) and Buddy Tape works well for this purpose, although white or transparent plastic tape is more often used, covering the bud. Wrapping must be done immediately to prevent drying out (1). When the bud starts growth, the tape must be cut, except with Parafilm or Buddy Tape that is easily penetrated by the expanding bud (Fig. 13–6).

The rootstock is not cut back above the bud until the union is complete. If the chip bud is inserted in the fall, the rootstock is cut back just as growth starts the next spring. If the budding is done in the spring, the rootstock is cut back about 10 days after the bud has been inserted.

T-Budding (Shield Budding)

This method of budding is known by both names—the "T-bud" designation arises from the T-like appearance of the cut in the rootstock, whereas the "shield bud" is derived from the shield-like appearance of the bud piece when it is ready for insertion in the rootstock.

T-budding is widely used by nurserymen in propagating nursery stock of many fruit trees, shade trees, roses, and some ornamental shrubs. Its use is generally limited to rootstocks that range from 6 to 25 mm (1/4 to 1 in) in diameter and are actively growing so that the bark will separate readily from the wood (Table 13–1).

The bud is inserted into the rootstock 5 to 25 cm (2 to 10 in) above the soil level, where the bark is smooth (Fig. 13–9). Opinions differ on the proper side of the rootstock in which to insert the bud. If extreme weather conditions are likely to occur during the critical graft union period following budding, the bud is placed on the side of the rootstock in order to give as much protection from prevailing winds as possible. Some believe that placing the bud on the windward side gives less chance for the young shoot to break off. Otherwise, it probably makes little difference where the bud is inserted, and the convenience of the operator and the location of the smoothest bark are controlling factors. When rows of closely planted rootstocks are budded, it is more convenient to have all the buds on the same side for later inspection and manipulation.

The cuts to be made in the rootstock plant are illustrated in Figures 13–9, 13–10, 13–11; and 13–12, (pages 525–27). Most budders prefer to make the vertical cut first and then the horizontal crosscut at the top of the T. As the horizontal cut is made, the knife is given a twist to open the flaps of bark for insertion of the bud. Neither the vertical nor horizontal cut should be made longer than necessary, because additional tying would be required later to close the cuts.

After the proper cuts are made in the rootstock and the incision is ready to receive the bud, the shield piece or shield bud is cut out of the budstick.

To remove the bark shield with the bud, an upward slicing cut is started at a point on the stem about 13 mm (1/2 in) below the bud, continuing under the bud to

Table 13–1
UTILIZATION AND ROOTSTOCK CRITERIA OF SELECTED TYPES OF BUDDING

Type of budding	Diameter of rootstock	Rootstock condition	Uses	Plant species
Chip-Budding	Small: 13 to 25 mm (1/2 to 1 in)	Only budding system that can be done on either dormant or active rootstock	Nursery budding; bench budding; container budding; top-working in orchard; one of the two most popular budding systems	Wide variety of fruit crops and ornamental plants; thin-barked species such as pistachio that do not T-bud well
T-Budding (Shield Budding)	Small: 6 to 25 mm (1/4 to 1 in)	Active; bark must be slipping	Nursery budding; container budding; top-working in orchard; one of the two most popular budding systems; used in "June Budding" with the "wood out" or flipped bud method	Wide variety of fruit crops and orna-mental trees and shrubs, including apples, peach, roses, citrus
Inverted T-Incision of Rootstock	Small: 6 to 25 mm (1/4 to 1 in)	Active; bark must be slipping	Nursery budding; container budding; used in high rainfall areas or with species that have excess sap flow "bleeding"; the horizontal portion of the T-cut is made at the bottom rather than the top to allow the water or sap to drain; the shield bud is inserted with normal polarity	Citrus, chestnuts
Patch Budding	Small: 13 to 25 mm (1/2 to 1 in); same diameter as scion (budstick)	Active; bark must be slipping on both rootstock and scion piece (budstick)	Nursery budding; container budding; slower and more difficult graft than T-budding; most important pecan grafting system in Texas nurseries	Pecans, rubber tree (*Hevea brasiliensis*), walnuts
Flute and Ring (Annular) Budding	Small: 13 to 25 mm (1/2 to 1 in); same diameter as scion (budstick)	Active; bark must be slipping on both rootstock and scion piece (budstick)	Modification of patch budding; bud patch of flute bud has greater circumference than conventional patch bud; bud patch of ring (annular) budding completely wraps around the rootstock	Citrus, other tree and shrub species
I-Budding	Small: 13 to 25 mm (1/2 to 1 in); bark of rootstock is thicker than budstick	Active; bark must be slipping on both rootstock and scion piece (budstick)	Nursery budding; container budding; utilized when bark of rootstock is thicker than budstick; bud patch is cut in the form of a rectangle or square, just as for patch budding	
Microbudding	Very small; less than 6 mm (1/4 in)	Active	Similar to T-budding except bud piece reduced to very small size using only the bud and a small piece of wood under it. An inverted T-cut is made and the microbud is slipped into it, right side up. A modification of micro-budding is using 2 mm diameter rootstock, with the bud inserted as in a wedge graft; personal communication M. Skaria, http://aghs.tamuk.edu/uploads/media/ Microbudding_overview.pdf	

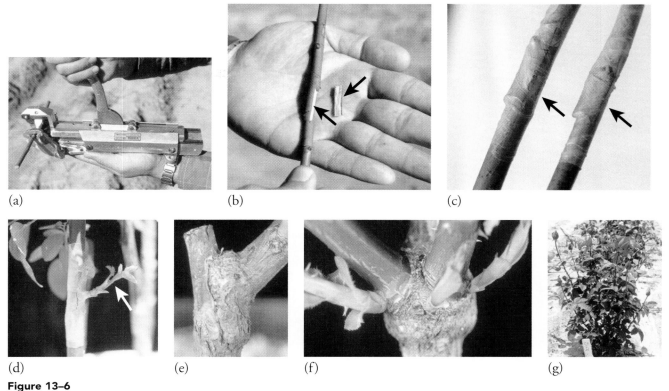

Figure 13–6
Chip budding of field rose bushes. (a) A tool for chip budding. (b) Machine-cut chip bud—the scion bud with stem section (*right*, arrow) is slid into the grooves of the machine-notched rootstock (*left*). (c) Ideally, the scion and rootstock should be of equal diameter. When the diameters are different, it is important that the cambiums of the graft partners be matched on one side. (d) The scion bud (arrow) penetrates through the Parafilm tape. (e) Chip-budded plants form a strong graft union (arrow) (see Chapter 11 for the discussion on chip budding). The rootstock top has been cut off and the scion shoot system established. (f) Multiple shoot breaks from the chip budded scion. (g) Chip-budded 'Mirandy' on *Rosa multiflora* rootstock.

about 2.5 cm (1 in) above. The shield piece should be thin, but thick enough to have some rigidity. A second horizontal cut is then made 1.3 to 1.9 cm (1/2 to 3/4 in) above the bud, permitting the removal of the shield piece. On many cultivars of shade trees in Oregon, professional budders use only a single cut to remove the bud. Two cuts are used on "wood-out" cultivars (see Box 13.2, page 524).

The next step is the insertion of the shield piece containing the bud into the incision in the rootstock. The shield is pushed downward under the two raised flaps of bark until its upper, horizontal cut matches the same cut on the stock. The shield should fit snugly in place, well covered by the two flaps of bark but with the bud itself exposed (see Fig. 11–22). A **"budder"** and **"tier"** in T-budding field roses are shown in Figures 13–13 and 13–14, pages 527–28.

Waxing is not necessary, but the bud union must be wrapped with poly tape, budding rubbers, buddy tape, or parafilm to hold the two components firmly together until healing is completed. Parafilm and tape works well for tying and sealing the bud, and the elongating shoot easily penetrates the tape (2, 3). Rubber budding strips, especially made for wrapping, are widely used for this purpose (Fig. 13–13), because their elasticity provides sufficient pressure to hold the bud securely in place. The rubber, being exposed to the sun and air, usually deteriorates, breaks, and drops off after

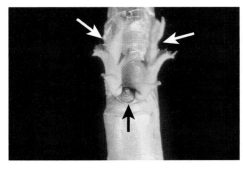

Figure 13–7
Chip budded rose. Not all chip buds have just one bud. With this chip budded rose, the primary axillary bud did not develop (black arrow), so the two secondary axillary buds (on the same chip bud) have elongated to form shoots. One of the shoots will be removed to allow the other to become the dominant shoot system.

(a)

(b)

(c)

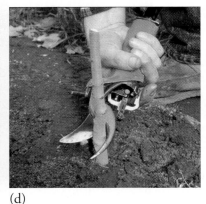
(d)

Figure 13–8
Understock removal prior to bud break of chipbud liners in the spring. (a) Healed chipbud. (b) Understock has been precut to remove most of the top; (c and d) pneumatic pruners make clean cuts and are ergonomic.

BOX 13.2 GETTING MORE IN DEPTH ON THE SUBJECT
WRAPPING MATERIALS FOR DIFFICULT-TO-BUD FIELD-GROWN TREES

Degradable budding rubbers that do not cover the actual bud work best when chip-budding on very rapidly growing rootstock, such as birch (*Betula pedula* 'Darecarlica') (8, 9). However, polyethylene budding strips work best for chip-budding black locust (*Robinia pseudoacacia* 'Frisia'), which has small sunken buds that are not subjected to physical damage and pressure of the poly strips. The pressure of the poly strips prevents the formation of large pads of undifferentiated rootstock callus, which hold the scion in place, but do not form a successful union (9).

---

several weeks, at which time the bud should be healed in place. If the budding rubber is covered with soil, the rate of deterioration will be much slower. Use of this material eliminates the need to cut the wrapping ties, which can be costly if many thousands of plants have been budded. The rubber will expand as the rootstock grows, reducing the danger of constriction.

In tying the bud, the ends of the budding rubbers are held in place by inserting them under the adjacent turn. The bud itself should not be covered. The amount of tension given the budding rubber is quite important: it should not be too loose, or there will be too little pressure holding the bud in place; on the other hand, if the rubber is stretched too tightly, it may be so thin that it will deteriorate rapidly and break too soon, before the bud union has taken place. Often the tying is done from the top down to avoid forcing the bud out through the horizontal cut. Novice budders should avoid the tendency to over-tie, as if the bud were a mummy. The inner layers of the tie will not deteriorate, so girdling can occur. See Chapter 12 for the description of different tying and sealing materials for budding and grafting. See Figure 13–15 for T-budding in a 2-year landscape rose production system and Figure 13–16 for T-budding dogwoods (*Cornus*) in Tennessee.

Inverted T-Incision of the Rootstock

In areas that experience a lot of rainfall during the budding season, water running down the stem of the rootstock may enter the T-cut, soak under the bark of the rootstock, and

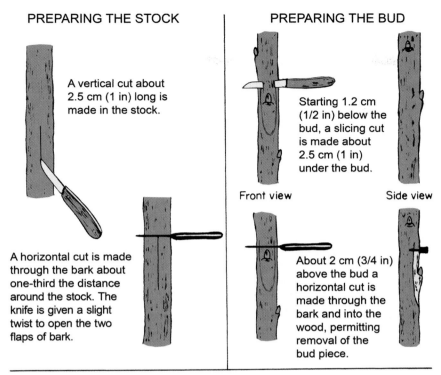

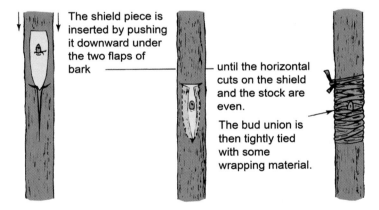

Figure 13–9
Basic steps in making the T-bud (shield bud).

bleeding Excessive sap flow that can occur from the rootstock during budding, such as with chestnuts; an inverted T-incision allows better drainage and better healing.

prevent the shield piece from healing into place (Table 13–1). Under such conditions, an inverted T-incision in the rootstock gives better results, since excess water is shed—*the scion bud is inserted with normal polarity*. The inverted T-incision method is widely used in citrus budding, even though the conventional met-hod also gives good results. In species that **bleed** (excessive sap flow) during budding, such as chestnuts, the inverted T-incision allows better drainage and better healing. Proponents of both conventional and inverted T-budding can be found, and in a given locality the usage of either with a given species tends to become traditional.

In the inverted T-incision method, the rootstock has the transverse cut at the bottom rather than at the top of the vertical cut. In removing the shield piece from the budstick, the knife starts above the bud and cuts downward below it. The shield is removed by making the transverse cut 13 to 19 mm (1/2 to 3/4 in) below the bud. The shield piece containing the bud is inserted with normal polarity into the lower part of the incision and pushed upward until the transverse cut of the shield meets that made in the rootstock.

Patch Budding

The distinguishing feature of patch budding and related methods is that a rectangular patch of bark is completely removed from the rootstock and replaced with a patch of bark of the same size containing a bud of the cultivar to be propagated (Table 13–1).

Figure 13-10
T-budding crab apples. (a) Making "T" insertion in rootstock. (b) Shield bud from scion. (c and d) Inserting the shield bud and taping with poly tape that will later need to be removed to prevent girdling. (e) Staked, 1-year-old crabapples. Courtesy K. Warren.

Figure 13-11
T-budding sequence. (a) Scion cuts. (b) Scion with "wood in" or "wood out" (arrow). (c) Scion shield inserted in t-cut. (d) Wrapped with bud exposed. (e) Budding band removal after bud heals.

Figure 13–12

(a and b) Containerized nursery production of citrus in Sicily. (a) Black poly on trunks is for herbicide and sun scald protection. These citrus were T-budded on sour orange and Troyer citrange rootstocks.

Figure 13–13

T-Budding field roses in a 2-year rose production cycle for landscape roses. (a) Budder is using the quill end of the budding knife to insert the shield bud into the "T" cut of the rootstock. (b) The 2nd person, the "tier," wraps the budding rubber around the T-bud, being careful not to cover the bud tip. (c) The budder and tier working in conjunction.

Patch budding is somewhat slower and more difficult to perform than T-budding. It is widely and successfully used on thick-barked species, such as walnuts and pecans, in which T-budding sometimes gives poor results, presumably owing to the poor fit around the margins of the bud—particularly the top and bottom. Patch budding, or one of its modifications, is also extensively used in propagating tropical species, such as the rubber tree (*Hevea brasiliensis*).

Patch budding requires that the bark of both the rootstock and budstick be slipping easily. In propagating nursery stock, the diameter of the rootstock and the budstick should preferably be about the same, about 13 to 25 mm (1/2 to 1 in) (17).

Special knives (see Fig. 13–17, page 530) have been devised to remove the bark pieces from the rootstock and the budstick. Some type of double-bladed knife that makes two transverse parallel cuts 2.5 to 3.5 cm (1 to 1 3/8 in) apart is necessary. These cuts, about 25 mm (1 in) in length, are made through the bark to the wood in a smooth area of the rootstock about 10 cm (4 in) above the ground. Then the two

Figure 13–14
Improving the ergonomics of field budding. (a) To enhance the efficiency of T-budding field roses in California, the "budder" (left, arrow) and "tier" (b) are supported above the rootstock plants with a body harness, which minimizes muscle strain. The technicians work in the shade and propel the carts down the field row with their legs. (b) The budder inserts the shield bud, which will be wrapped with a budding rubber by the tier. The box next to the budder's head contains the scionwood (budwood). Both the budder and tier work as a team and are paid on piecework, with an incentive bonus based on successful "takes." (c) Two teams of budders and tiers.

transverse cuts are connected at each side by vertical cuts made with a single-bladed knife.

The patch of bark containing the bud is cut from the budstick in the same manner as the bark patch is removed from the rootstock. Using the same two-bladed knife, the budder makes two transverse cuts through the bark, one above and one below the bud. Then two vertical cuts are made on each side of the bud so that the bark piece will be about 25 mm (1 in) wide. The cut on the right side of the bud should form a 90-degree angle with the horizontal cut. Now the bark piece containing the bud is ready to be removed. It is important that it be slid off sideways rather than being lifted or pulled off. There is a small core of wood, the bud trace, which must remain inside the bud if a successful **"take"** is to be obtained. By sliding the bark patch to one side, this core is broken off, and it stays in the bud. If the bud patch is lifted off, this core of wood is likely to remain attached to the wood of the budstick, and the bud will fail.

After the bud patch is removed from the budstick, it must be inserted immediately on the rootstock, which should already be prepared, needing only to have the bark piece removed. The patch from the budstick should fit snugly at the top and bottom into the opening in the rootstock, since both transverse cuts were made with the same knife. It is more important that the bark piece fits tightly at top and bottom than along the sides. These procedures are illustrated in Figures 13–18 and 13–19 (pages 531–32).

Now the inserted patch is ready to be wrapped. Often the bark of the rootstock will be thicker than the bark of the inserted bud patch so that upon wrapping, it is impossible for the wrapping material to hold the bud patch tightly against the rootstock. In this case, the bark of the rootstock is pared down around the bud patch so

---

BOX 13.3 GETTING MORE IN DEPTH ON THE SUBJECT
WOOD-IN AND WOOD-OUT WITH T-BUDDING

In T-budding, there are two methods of preparing the shield—with the **"wood-in"** or with the **"wood-out."** (Fig. 13–11). The terms refer to the sliver of wood just under the bark of the shield piece, which remains attached if the second, or horizontal, cut is deep and goes through the bark and wood, joining the first slicing cut. Some professional budders prefer to remove this sliver of wood, but others retain it. In budding certain species, however, such as maples and walnuts, success is usually obtained with "wood-out" buds. To prepare the shield with the wood-out, the second horizontal cut, above the bud, should be just deep enough to go through the bark and not the wood. If the bark is slipping, the bark shield can be snapped (slipped) loose from the wood (which still remains attached to the budstick) by pressing it against the budstick and sliding it sideways. The small core of wood comprising the vascular tissues remains with the wood. If the shield is pulled outward rather than slid sideways from the wood, the core pulls out of the bud—leaving a hole in the shield and eliminating any chances of success. **June-budded** fruit trees require that the shield piece be prepared with the wood-out. In most other instances, however, the wood is left in. In spring budding, using dormant budwood, this sliver of wood is tightly attached to the bark and cannot be removed (Fig. 13–11).

Figure 13–15
(a) Treating field with methyl bromide prior to sticking rootstock hardwood cuttings to root. (b) Roses are spring budded during the first production year, so scion (budwood) is collected dormant during early winter and stored at −1°C (31°F). (c and d) Maintaining budwood in moist burlap in (d) a movable field work station with budwood kept under shade. (e) Rose field in bloom during 2nd year production prior to digging roses in fall. (f) Digging roses bare-root with a U-blade and shaker; the spring-budded roses are dug as a 1-year scion–2-year rootstock. (g) Reducing the shoot and root system during processing and (h) packaged, dormant landscape rose ready for spring sale and planting into the landscape.

Figure 13–16
(a and b) Tennessee liner nursery of budded dogwoods. The frame on wheels will be tarped on top for shade during budding process for improved ergonomics and comfort of the budders. (c, d, and e) Field–budding dogwood on seedling rootstock.

that it will be of the same thickness, or preferably slightly thinner than the bark of the bud patch, until the wrapping material will hold the bud patch tightly in place.

The patch bud should be covered by a material that not only holds the bark tightly in place but covers all the cut surfaces. Air must not be able to enter under the patch in order to avoid drying and subsequent death of the tissues, but the bud itself must not be covered during wrapping. The most satisfactory material is nursery adhesive tape or ploy budding strips.

Patch buds, especially with walnuts, should not be wrapped so that they cause a constriction at the bud union. When the rootstock is growing rapidly, the tape is cut about 10 days after budding. A single vertical knife cut on the side opposite the bud is sufficient, but care should be taken not to cut into the bark.

Summer Patch Budding In California, Texas, and other areas with hot summers, patch budding is best performed in late summer when both the seedling rootstock and the source of budwood are growing rapidly and their barks slip easily. The budsticks for patch budding done at this time should have the leaf blades cut off 2 to 3 weeks before the budsticks are taken from the tree. The petiole or leaf stalk is left attached to the base of the bud, but by the time the budstick is removed, this petiole has dropped off or is easily pulled off.

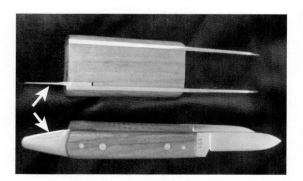

Figure 13–17
Double-bladed knives used for patch budding. The quill (arrow) is used to help separate and lift up cut bark during the budding process.

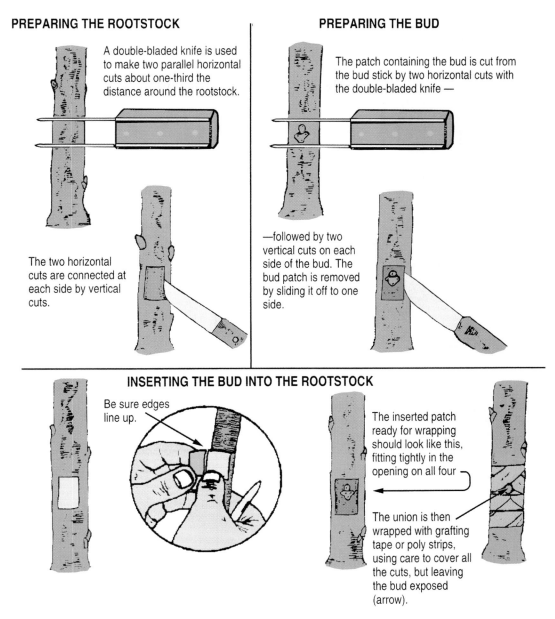

Figure 13–18
Steps in making the patch bud. This method is widely used for propagating thick-barked plants.

Spring Patch Budding Patch budding can be done in the spring after new growth has started on the rootstocks and the bark is slipping. The problem, however, in obtaining satisfactory buds to use at this time of year is that the bark of the budstick must separate readily from the wood. At the same time, the buds should not be starting to swell. There are two methods by which satisfactory buds can be obtained for patch budding in the spring.

One, used in Texas for pecans, uses budsticks selected during the dormant winter period and stored at low temperatures (about 27°C, 36°F) wrapped in *slightly moist* sphagnum or wood shavings to prevent drying out. Then, about 2 or 3 weeks before the spring budding is to be done, they are brought into a warm room. The budsticks may be left in the moist sphagnum or set with their bases in a container of water. The increased temperature will cause the cambium layer to become active, and soon the bark will slip sufficiently for the buds to be used. Although a few of the more terminal buds on each stick may start swelling in this time and cannot be used, there should be a number of buds in satisfactory condition.

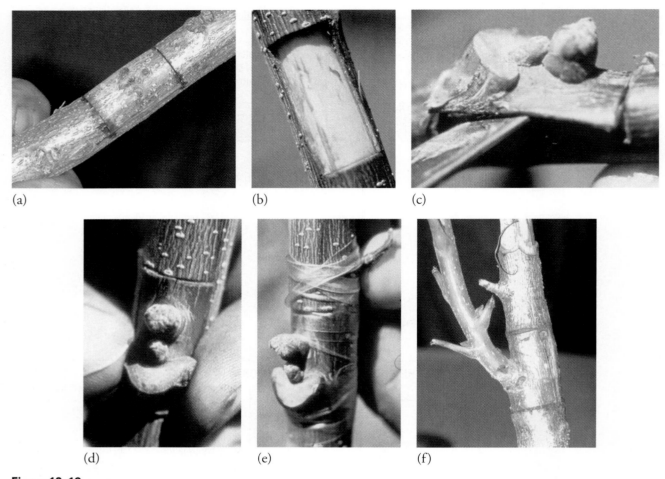

Figure 13–19
Patch budding process. (a and b) Making cuts on pecan rootstock. (c) Removing patch bud from scion. (d and e) Inserting and wrapping the patch bud without covering the axillary bud. (f) Elongated shoot from patch bud. Photos courtesy L. Lombardini.

The second method of obtaining buds for spring patch budding is to take them directly from the tree that is the source of the budwood, in early spring, before the buds begin to force. The scions are refrigerated and stored as dormant scions, later brought to room temperature for a few hours to stimulate cambial activity, and then budded. Pecans should not be budded until at least 2 weeks after the rootstock begins to grow.

I-Budding

In I-budding, the bud patch is cut in the form of a rectangle or square, just as for patch budding (Table 13–1). With the same parallel-bladed knife, two transverse cuts are made through the bark of the rootstock. These are joined at their centers by a single vertical cut to produce the shape of the letter I. Then the two flaps of bark can be raised to insert the bud patch beneath them. A better fit may occur if the side edges of the bud patch are slanted. In tying the I-bud, be sure that the bud patch does not buckle outward and leave a space between itself and the rootstock (see Fig. 13–20).

I-budding is most appropriate when the bark of the rootstock is much thicker than that of the budstick. If the patch bud were used in such cases, considerable paring down of the bark of the rootstock around the patch would be necessary. No paring is necessary with the I-bud method (Fig. 13–20).

Flute Budding and Ring Budding

See details in Figure 13–20 and Table 13–1.

TOP-BUDDING (TOPWORKING)

Young trees with an ample supply of vigorous shoots at a height of 1.2 to 1.8 m (4 to 6 ft) can be top-budded (topworked) rapidly with a high degree of success. Older trees can be top budded by severely cutting back the prior year to provide a quantity of

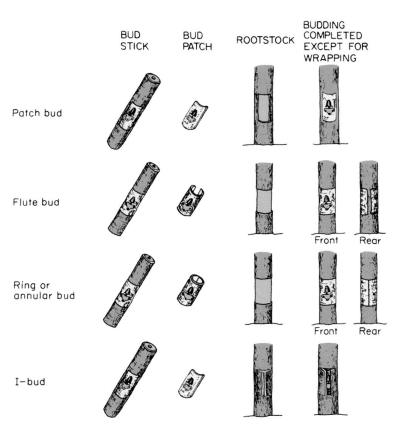

Figure 13–20
There are many variations of the patch bud, some of which are shown here. The naming of these types is somewhat confusing; the most generally accepted names are given.

vigorous water-sprout shoots fairly close to the ground.

Depending on the size of the tree, 10 to 15 buds are placed in vigorously growing branches 6 to 19 mm (1/4 to 3/4 in) in diameter in the upper portion of the tree—about shoulder height. A number of buds can be placed in a single branch, but usually only one will be saved to develop into secondary branches, to form the permanent new top of the tree. The T-bud or chip-bud method is used on thin-barked species, and the patch bud on those with thick bark.

Top-budding is usually done in midsummer, as soon as well-matured budwood can be obtained and while the rootstock is actively growing and the bark slipping. Orchard trees generally stop growth earlier in the season than young nursery trees; therefore, the budding must be done earlier. When top-budding is done at this time of year, the buds usually remain inactive until the following spring. As vegetative growth starts, the rootstock branches are cut back just above the buds to force the buds into active growth. They should develop into good-sized branches by the end of the summer. At the time the shoots are cut back to the buds, all nonbudded branches should be removed at the trunk. Inspect the trees carefully through the summer and remove all shoots that arise from any but the inserted buds. Top-budding is very labor intensive, so it is not commonly used.

frameworking A form of top-budding (topworking) where a few scaffold branches are retained on an established rootstock for multiple budding of a new scion.

DOUBLE-WORKING BY BUDDING

In propagating nursery trees, budding methods can be used to develop double-budded (**double-worked**) trees. **Interstocks** can be budded to the rootstocks; the following year the cultivar is budded on the interstock. Although effective, this process takes 3 years. Development of a double-worked tree—in one operation in 1 year—is possible by using the double-shield bud method (Fig. 13–21) (4, 7). First, a thin, budless shield piece of the desired interstock is inserted in the

double-working The budding of an interstock (interstem) between the rootstock and scion.

interstock or interstem The bud piece or stem section between the rootstock and the scion.

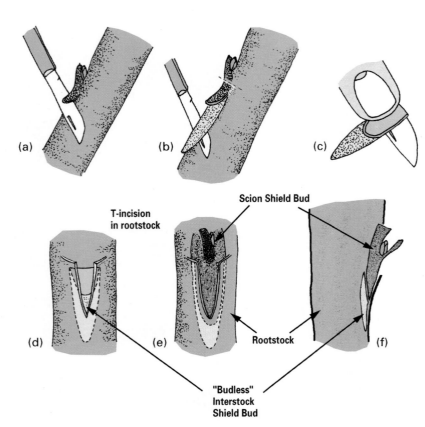

Figure 13–21
Double-working by budding. (a) A shallow incision is made in the interstock. (b) A second incision is made to debud the interstock shield bud (c), which is detached. (d) A T-cut incision has been made in the rootstock and the budless interstock shield bud inserted—as depicted by the shadowed area. (e) The scion shield bud is inserted in the rootstock T-incision on top of the budless interstock shield bud. (f) Side-view of the budding. The budless interstock will later grow and form a complete bridge between the scion and rootstock.

T-incision of the rootstock, then the scion shield bud is inserted on top of the interstock. If the scion shield bud is lined up properly with the interstock shield piece, the budless interstock will grow and form a complete bridge between the scion and rootstock.

Double-working is also done by bench-grafting on the lower union (interstock to rootstock) and transplanting the tree in the nursery. The scion is budded on the interstock later.

MICROBUDDING

Microbudding is used successfully in propagating citrus trees and can be utilized with other selected tree and shrub species. It has been of commercial importance in the citrus districts of southeastern Australia (20). Microbudding is similar to T-budding, except that the bud piece is reduced to a very small size. The leaf petiole is cut off just above the bud, and the bud is removed from the budstick with a razor-sharp knife. A flat cut is made just underneath the bud. Only the bud and a small piece of wood under it are used. In the rootstock an inverted T-cut is made, and the microbud is slipped into it, right side up. The entire T-cut, including the bud, is covered with thin, plastic budding tape. The tape remains for 10 to 14 days for spring budding and 3 weeks for fall budding, after which it is removed by cutting with a knife. By this time, the buds should have healed in place; subsequent handling is the same as for conventional T-budding.

A variation of micro-budding is described in Figure 13–22. Citrus greening, also called Huanglongbing or yellow dragon disease, is one of the more serious diseases of citrus. It is caused by bacteria in the phloem that is primarily spread by psyllid insects, and is devastating citrus in Brazil, Florida, and other major citrus-producing regions in the world. While the long-term solution is breeding genetic resistance, micro-budding is being adapted by commercial growers because it is an economical propagation system that allows a grove to be replanted and come into bearing within 1 to 2 years (see http://www.aphis.usda.gov/plant_health/plant_pest_info/citrus_greening/index.shtml).

Figure 13–22
"Microbudding" system of budding citrus to smaller rootstock. (a) Smaller 2 mm diameter rootstocks are used instead of pencil-size rootstocks. The bud size must match the rootstock. The rootstock is decapitated and the bud is inserted, as in a wedge graft, and capped. The wound calluses in 2 weeks and buds sprout in another 2 to 3 weeks. (b) Young, micro-budded citrus closely spaced at 100 plants per square foot, ready for shipment and field planting. (c) The smaller, micro-budded, field planted liners by-pass the 2-year conventional nursery, T-budding system and come into bearing within 14 to 18 months. (d) One-year old, container-grown 'Rio Red' grapefruit, bearing fruit. (e) Two-year old, field-grown 'Rio Red' grapefruit, bearing fruit. (f) Fruit load of 2.5-year old micro-budded 'Valencia' orange. Besides grapefruit and orange, micro budding has been used with lemons and kumquat. Courtesy M. Skaria.

DISCUSSION ITEMS

1. Why are chip budding and T-budding the two most important types of budding for woody ornamentals and fruit trees?
2. Why are terms such as bark, wood, and the slipping of the bark important for a grafter?
3. What are the advantages and disadvantages of different budding schedules: spring, June, and summer budding?
4. Why has chip budding become one of the most dominant budding systems for ornamental and fruit crops that are grown under relatively short production seasons?
5. Describe double-working by budding. Give some horticulture examples.
6. What are advantages and disadvantages of budding compared with grafting?

REFERENCES

1. Bremer, A. H. 1977. Chip budding on a commercial scale. *Comb. Proc. Intl. Plant Prop. Soc.* 27:366–67.

2. Davies, F. T., Y. Fann, and J. E. Lazarte. 1980. Bench chip budding of field roses. *HortScience* 15:817–18.

3. Fann, Y., F. T. Davies, Jr., and D. R. Paterson. 1983. Correlative effects of bench chipbudded 'Mirandy' roses. *J. Amer. Soc. Hort. Sci.* 108:180–83.

4. Garner, R. J. 1953. Double-working pears at budding time. *Ann. Rpt. E. Malling Res. Sta. for 1952.* pp. 174–75.

5. Harmon, F. N., and J. H. Weinberger. 1969. *The chip-bud method of propagating vinifera grape varieties on rootstocks*, USDA Leaflet 513.

6. Howard, B. H., D. S. Skene, and J. S. Coles. 1974. The effect of different grafting methods upon the development of one-year-old nursery apple trees. *J. Hort. Sci.* 49:287–95.

7. Howard, B. H. 1977. Chip budding fruit and ornamental trees. *Comb. Proc. Intl. Plant Prop. Soc.* 27:357–64.

8. Howard, B. H. 1993. Understanding vegetative propagation. *Comb. Proc. Intl. Plant Prop. Soc.* 43:157–62.

9. Howard, B. H., and W. Oakley. 1997. Budgrafting difficult field-grown trees. *Comb. Proc. Intl. Plant Prop. Soc.* 47:328–33.

10. Kidd, E. L., Jr. 1987. Asexual propagation of fruit and nut trees at Stark Brothers Nurseries. *Comb. Proc. Intl. Plant Prop. Soc.* 36:427–30.

11. Lider, L. A. 1963. *Field budding and the care of the budded grapevine*. Calif. Agr. Ext. Ser. Leaflet 153.

12. Meacham, G. E. 1995. Bench grafting, when is the best time? *Comb. Proc. Intl. Plant Prop. Soc.* 45:301–4.

13. Mertz, W. 1964. Deciduous June-bud fruit trees. *Comb. Proc. Intl. Plant Prop. Soc.* 14:255–59.

14. Osborne, R. H. 1987. Chip budding techniques in the nursery. *Comb. Proc. Intl. Plant Prop. Soc.* 36:550–55.

15. Patrick, B. 1992. Budding and grafting of fruit and nut trees at Stark Brothers. *Comb. Proc. Intl. Plant Prop. Soc.* 42:354–56.

16. Skene, D. S., H. R. Shepard, and B. H. Howard. 1983. Characteristic anatomy of union formation in T- and chip-budded fruit and ornamental trees. *J. Hort. Sci.* 58:295–99.

17. Taylor, R. M. 1972. Influence of gibberellic acid on early patch budding of pecan seedling. *J. Amer. Soc. Hort. Sci.* 97:677–79.

18. Tubesing, C. E. 1988. Chip budding of magnolias. *Comb. Proc. Intl. Plant Prop. Soc.* 87:377–79.

19. Warren, K. 1989. A crab apple system. *Amer. Nurs.* 170:31–5.

20. Wishart, R. D. A. 1961. *Microbudding of citrus*. S. Austral. Dept. Agr. Leaflet.

14
Layering and Its Natural Modifications

INTRODUCTION

Layering is a way of rooting cuttings in which adventitious roots are initiated on a stem while still attached to the plant. The rooted stem **(layer)** is then detached, transplanted, and becomes a separate plant on its own roots. Some plant species utilize layering as a natural means of reproduction, as *tip layers* of black raspberries and trailing blackberries (*Rubus*), runners of strawberry (*Fragaria*), and stolons of Bermuda grass (*Cynodon* spp.). Structures such as *offsets*, *suckers*, and *crowns* are handled essentially as rooted layers.

layer   A system of vegetative propagation where stems are rooted while still attached to the source plant.

Layering is an ancient nursery technique that was used extensively by European nurseries from the 18th to early 20th Centuries for propagating woody shrub and tree species (27, 51). Mound and trench layering were developed to mass produce hard-to-root clonal rootstocks for apple and other fruit species in England in the early 1900s (41, 46). To a large extent, layering as a nursery technique has been replaced by more modern methods of rooting and container production. Nevertheless, the procedure is highly reliable for hard-to-root clones and continues to be used in the production of horticulturally important rootstocks and other plants that are sufficiently valuable to justify the higher costs and labor requirement associated with layering. Fruit crops propagated by layering include hazelnut (filbert) (*Corylus* sp.), muscadine grape (*Vitis rotundifolia*), size-controlling apple rootstocks, and some tropical fruit plants such as mango (*Mangifera indica*) and litchi (*Nephelium*) (see Chapter 19). Layering is valuable in enabling an amateur or professional horticulturist to produce a relatively small number of large-sized plants of a special cultivar in an outdoor environment with a minimum of propagation facilities.

REASONS FOR LAYERING SUCCESS

Layering is a simple technique that induces adventitious roots on a stem while it is attached to the mother plant. It has proven to be a successful way to propagate a variety of plant species that are difficult or impossible to root from cuttings. There are several possible explanations for this increase in regeneration capacity:

1. Maintaining physical attachment of the stem to the mother plant.
2. Increased accumulation of photosynthates and hormones in the rooting area of the stem.
3. Excluding light to the stem in the rooting zone.

learning objectives

- Discuss the uses of layering in propagation.
- Understand the physiological characteristics of layering.
- Describe soil conditions for field layering.
- Describe the procedures for doing different layering techniques including simple, compound, serpentine, air, mound, trench, drop, and tip.
- Describe the different methods for natural layering including tip layering, runners, stolons, offsets, suckers, and crowns.

4. Invigoration and rejuvenation.
5. Utilizing seasonal effects on rooting.

Maintaining a Physical Attachment of Stem to the Mother Plant

This characteristic allows for a continual supply of water, minerals, carbohydrates, and hormones through the intact xylem and phloem to the rooting area. Layering eliminates both the water stress and leaching of nutrients and metabolites problems associated with rooting stem cuttings of hard-to-root genotypes for prolonged periods under intermittent mist.

Accumulation of Photosynthates and Hormones in the Rooting Area

During all types of layering, an important factor in rooting success is the ability of the stem to accumulate carbohydrates and auxin from the leaves and shoot tip. However, **girdling, incision,** or **bending** of the stem may be required to maximize this effect. Rooting may be further enhanced by adding auxins, such as indolebutyric acid (IBA), to the girdled cuts of layers, as with cutting propagation (42, 48).

girdling The practice of cutting through or removing the bark completely around the stem to interrupt the downward movement in the phloem, in order to cause the accumulation of carbohydrates and other substances just above the girdle without interfering with upward water conduction.

incision A cut made partially through the stem during layering to produce the same effect as girdling.

bending A practice used during layering to bend a one-year-old stem in a U-shape to duplicate the effect of girdling.

blanching Exclusion of light from the intact stem *after* it has grown by bending the Velcro or electrical tape or covering with a rooting media as in air layer.

Light Exclusion in the Rooting Zone

Exclusion of light from the area of the stem that will eventually form roots is common to all types of layering and is important for the success of propagating difficult-to-root plants from layers. A distinction must be made between **blanching,** which is the covering of an intact stem *after* it has grown, and **etiolation,** which is the effect produced as the shoot *elongates* in the absence of light (12, 23). (See Chapters 10 and 11 for discussions on *etiolation* and *blanching,* along with banding and shading on root initiation). Although some plants are able to produce roots on intact stems after blanching, many require phloem interruption by girdling as well. However, the greatest stimulus to root induction results when the initially developing shoots are continuously covered by the rooting media—as in trench layering—so that approximately 2.5 cm (1 in) of the base of the layered shoot is never exposed to light (12). A large measure of the success with which difficult-to-root plants are rooted by layering results from etiolation and blanching. Using the PVC black tape to cover the base of shoots from severely pruned stock plants has enhanced the rooting of apple (23).

etiolation The development of a plant stem or part in the absence of light.

Invigoration and Rejuvenation

A common procedure used in layering is to cut back stock plants, cover the base of the emerging shoot to exclude light, and provide a moist rooting environment. The production of rapid, new shoot development **(invigoration)** from the base of the plants that are in close proximity to the root system is similar to the hedging methods used to **rejuvenate** stock plants for improved rooting of cuttings. The cluster of new shoots is referred to as **stool shoots** and the process as **stooling.** Rejuvenation may be a key element in inducing roots on layered stems of difficult-to-root plants.

invigoration Rapid shoot growth in response to pruning.

rejuvenation A reversal of mature growth traits to those of a transition or juvenile phase.

stool shoots Clusters of shoots that emerge when a stem is cut to its base.

stooling The practice of cutting shoots back to the base, as described for mound layering.

Seasonal Patterns

Seasonal patterns are important for root initiation and development on stems during layering. In most cases, layering is started in the spring with the attached, dormant, previous season's shoots. Rooting on the shoot (layer) may not begin until later in the season and is associated with the accumulation of carbohydrates and other substances toward the latter part of the growth cycle.

MANAGEMENT OF PLANTS DURING LAYERING

Most layering methods are field nursery operations, which may continue in beds for 10 to 20 years, once established. The proper choice of the site for good soil, drainage, and climate is essential (21). Only virus-tested, true-to-type propagation sources should be used.

During the rooting period, layers are covered with soil or other rooting media not only to exclude light, but also to provide continuous moisture, allow for good aeration, and help insulate the layer from temperature extremes (33, 44, 45). Special attention needs to be paid to weed, disease, and insect control—see Chapter 3. Without good management, layering beds tend to decline over time.

In general, layering requires much hand labor and attention, which makes it an expensive procedure. Nevertheless, modern nurseries have developed specific kinds of mechanical devices and machinery to facilitate the management of large layering beds (Fig. 14–1) (8, 10, 24).

PROCEDURES IN LAYERING

The most commonly used systems to layer plants include:

- simple layering
- compound layering
- serpentine layering
- air layering
- mound layering or stooling
- trench layering
- drop layering

Of these, the most commercially important systems are mound layering for fruit understocks, air layering for some tropical fruits and ornamentals, and modified trench layering systems for clonal propagation of forestry trees such as the Monterey pine (*Pinus radiata*). Table 14–1 (page 541) shows comparison of different layering techniques.

Simple Layering

Simple layering is the bending of an intact shoot to the ground and covering a single portion of the stem between the base and shoot tip with soil or rooting medium so that adventitious roots form (Fig. 14–2, page 541). The method can be used to propagate a wide range of plants, indoor or outdoor, on woody shrubs that produce numerous new shoots annually (27, 41, 49, 52), or on trees that tend to produce suckers, such as filberts (*Corylus* sp.). Historically, European nurseries have established permanent layering beds by planting shrubs 2.4 to 3 m (8 to 10 ft) apart (51) and growing them in place for several years prior to use. Once the layering beds are established, the new shoots are bent to the ground annually for layering while new shoots develop for the following year. All available shoots are worked with, thus utilizing all of the area surrounding the plant.

simple layer A type of layering in which single one-year-old shoots are bent to the ground, covered with soil, and then (sometimes) girdled to stimulate root initiation of the stem.

Layering (Fig. 14–3, page 542) is usually done in the early spring using flexible, dormant, 1-year-old stems, which can be bent easily. These shoots are bent and "pegged down" at a location 15 to 20 cm (6 to 9 in) from the tip, forming a "U." Bending, twisting, cutting, or girdling at the bottom of the "U" stimulates rooting at that location. The base of the layer is covered, leaving the tip exposed.

Shoots layered in the spring usually will be rooted by the end of the growing season and removed either in the fall or in the next spring before growth starts. Mature shoots layered in summer should be left through the winter and either removed the next spring before growth begins or left until the end of the second growing season. When the rooted layer is removed from the parent plant, it is treated essentially as a rooted cutting.

New shoots growing from the base of the plant during the rooting year are used for layering during the next season. With this system a supply of rooted layers can be produced over a period of years by establishing a layering bed composed of stock plants far enough apart to allow room for all shoots to be layered.

Compound Layering

Compound layering is similar to simple layering except that the branch to be layered is laid horizontally to the ground and numerous shoots for rooting develop from various nodes rather than just one. This method was once used extensively in Europe but has been replaced by more modern methods of rooting cuttings and growing plants in containers.

compound layer A type of layering in which the entire horizontal shoot is covered with rooting media.

Permanent layering beds are established with plants spaced 1.8 to 3 m (6 to 10 ft) apart and grown for several years to establish a good root system. Then the vegetative top is cut back to 2.5 cm (1 in) from the ground and shoots are allowed to grow for the following season. Before the beginning of the season, long shoots

Figure 14–1
Apple rootstock production. (a) Apple stoolbeds of 8- to 10-year-old M9 EMLA rootstock, (b) stoolbeds of MM 111 planted at a 45 degree angle prior to layering, (c) layering with temporary electric cable clips to tie-down and train the layers flat in the planting trench, (d) sawdust applied on new growth in late May in England to etiolate the base of rootstock, (e) harvesting stoolbeds with a tractor-mounted rotary saw, (f) nursery-designed saw with replaceable teeth, (g) one-year-old, rooted apple layers with soil removed, and (h) brushing-off the winter covering of sawdust and soil in late March to begin the next stool crop cycle. Courtesy Nick Dunn.

Table 14–1
COMPARISON OF DIFFERENT LAYERING TECHNIQUES

Layering technique	Description	Plants propagated by these methods
Simple	An intact shoot is bent to the ground and a single portion of the stem between the base and shoot tip is covered with soil or rooting medium so that adventitious roots form.	Numerous tropical and temperate shrubs and some trees. Examples include: hazelnut (filbert), viburnum, forsythia.
Compound	A branch with numerous nodes is laid horizontally and covered with soil. Shoots develop from each node. Roots can form on the new shoots or the buried node.	Several woody shrubs as well as woody and herbaceous vines. Examples include: grape, wisteria, pathos.
Serpentine	Similar to compound layering except that each alternating node is covered with soil, leaving one node to root and the other node to develop a new shoot above ground.	Several woody shrubs as well as woody and herbaceous vines. Examples include: grape, wisteria, clematis, philodendron, pathos.
Air	A portion of the above ground stem is girdled and covered with a rooting substrate (sphagnum moss in a polyethylene covering). The girdled portion of the stem roots.	Several woody and herbaceous plants, especially tropical plants. Examples include: Citrus, Croton, Dracaena, Ficus.
Mound	Shoots are cut back to the ground and soil or rooting substrate (sawdust) is mounded around them to stimulate roots to develop at their bases.	Woody trees and shrubs, especially fruit tree rootstocks. Examples include: apple, cherry, hazelnut, oak, pecans.
Trench	The initial stem used to establish the layering system is laid horizontally in a trench. Shoots develop from nodes along the stem that are then covered with mounded rooting substrate (sawdust) similar to mound layering.	Woody trees and shrubs, that are difficult to establish from mound layers. Examples include: apple, cherry, quince, mulberry, walnut.
Drop	A modification of mound layering where the plants are grown in double-stacked containers rather than in the field.	Some woody shrubs. Examples include: barberry, boxwood, rhododendron.

(a)

(b)

Figure 14–2
(a and b) Simple layering of rose rootstock with adventitious roots (arrows) forming at nodal area.

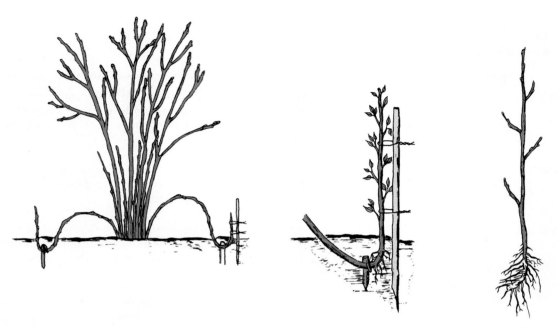

Figure 14–3
Steps in propagation by simple layering.

are bent horizontally and held to the ground with wire pegs. Once new shoots grow about 10 cm (4 in), the pegs are removed, a shallow trench is dug adjacent to the stem, and the shoots are laid in the bottom of the trench with additional pegs applied to hold them in place. Soil or other media is filled in as the shoots grow.

serpentine layer
A type of layering in which a horizontal shoot (or vine) is covered at individual nodes for rooting.

A variation of this method (sometimes called **serpentine layering**) is used for propagating plants that have long, flexible shoots; for example, the muscadine grape (*Vitis rotundifolia*) and ornamental vines such as *Wisteria* and *Clematis* (Fig. 14–4). The horizontal shoots are alternately covered and uncovered to produce roots at different nodes.

Air Layering (Pot Layerage, Circumposition, Marcottage, Gootee)

air layering A type of layering in which an aerial stem is girdled and enclosed with rooting media to produce rooted layers in the upper part of the plant.

Air layering includes wrapping an aerial stem with rooting medium and causing adventitious roots to form. An ancient method used to propagate a number of tropical and subtropical trees and shrubs (5, 25, 35, 38), including the litchi (*Nephelium*) (13), longan (*Euphoria*) (53), and the Persian lime (*Citrus aurantifolia*) (43). Today the method is useful for producing a few plants of relatively large size for special purposes. For instance, greenhouse and field production (in subtropical/tropical regions) of *Ficus* species, *Croton*, *Monstera*, and philodendron will result in rapid production of large plants (25, 36, 39, 43). Air layering has been used to root mature pines to obtain clones for research or to produce seed orchards (2, 14, 31). By using polyethylene film and aluminum foil to wrap the layers, outdoor air layering is possible with many woody plant species (Fig. 14–5) (13, 52). A container system that opens and closes around an air-layered shoot has been developed (Fig. 14–6, page 544).

Air layers are made in the spring or summer on stems of the previous season's growth or, in some cases, in the late summer with partially hardened shoots. Stems older than 1 year can be used in some cases, but rooting is less satisfactory and the larger plants produced are somewhat more difficult to handle after rooting. The presence of active leaves on the layered shoot speeds root formation. With tropical greenhouse plants, layering should be done after several leaves have developed during a period of growth.

First, girdle the stem by removing a strip of bark 1.8 to 2.5 cm (1/2 to 1 in) wide completely around the stem, depending on the kind of plant (Fig. 14–6). Width is generally three to four times wider than the branch diameter. Scrape the exposed surface to ensure complete removal of the phloem and cambium to avoid premature healing. A second method is to make a slanting upward cut (see Fig. 14–7, page 545) on one or both sides of the stem about 3 cm (1 1/2 in) long (referred to as a "double-slit"). Keep the two surfaces apart with some sphagnum or a piece of wood (e.g., toothpick). Girdling reduces water conductivity more than a double-slit technique, but need not impede rooting (5). Growing plants in 50 percent shade reduces water stress (4).

Application of IBA to the exposed wound can be beneficial. One method described for the commercial air layering of *Mahonia aquifolium* 'Compacta' is to insert a small amount of sphagnum moss soaked with 60 ppm IBA under the wounded flap of tissue (50). Increased concentrations up to 2 percent IBA in talc has increased rooting and survival in pecan air layers (42).

The cut area around the stem is enclosed in a medium that holds moisture and is well aerated. A suitable material is two handfuls of *slightly moistened* sphagnum moss, which has excess moisture squeezed out. If the moisture content of the sphagnum moss is too high, the stem will decay. The size of the rootball can be important: too large a rootball holds excess moisture that inhibits root growth—12 × 8 cm (4.7 × 3.1 in) is considered ideal by some propagators (52). Placement is also important. The top of the girdle should be in the top one-third of the rooting medium.

Polyethylene film, 20 to 25 cm (8 to 10 in) square, wrapped carefully about the branch so that the sphagnum moss is completely covered, is an excellent covering. The ends of the sheet should be folded (as when wrapping meat) with the fold placed on the lower

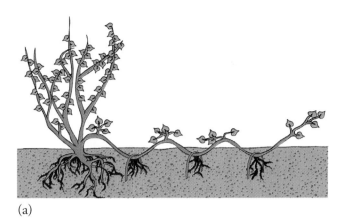

(a)

(b)

Figure 14–4
(a) Serpentine layering is similar to simple layering except more than one portion of the stem is alternately covered and exposed. It works well with "viney" plants such as grapes, *Wisteria*, *Clematis*, *Philodendron*, and climbing roses. (b) Serpentine layering of *Philodendron oxycardium* (*scandens*) 'Medio Picta' in smaller pots with shoot system still attached to stockplant (arrow).

(a)

(b)

(c)

(d)

(e)

Figure 14–5
(a) Air layering of *Ficus elastica* (stakes for supporting the layer), (b) air layering of *Ficus benjamina*, and (c) *Ficus macrophylla* (problem of incision area callusing over before rooting has occurred). (d) Rooted layers of *Ficus benjamina* and (e) *Ficus carica*.

Figure 14–6
(a) Air layering with a commercial pot system (RooterPot©) that encloses the air-layered shoot so roots can form. (b) Schematic of girdled shoot and open pot. (c-g) Air layering *Magnolia grandiflora* with auxin and Oasis blocks. (c) Auxin talc brushed onto the girdled area of the layer. (d) Moist Oasis blocks are wrapped around the layer and then (e) covered with aluminum foil to reduce heat and retain moisture. (f) Rooted layers in the shape of the Oasis blocks, prior to removing from the stockplant. (g) Detached rooted layers ready for planting. Photos (c), (d), (e), (f), and (g) courtesy R. L. Byrnes.

side. The two ends must be twisted to make sure that no water can seep inside. Aluminum foil is useful as an additional wrap and helps to maintain moderate temperatures by reflecting sunlight (6, 43). Foil is wrapped around the polyethylene, or may be used as the sole wrapping material in climates with high relative humidity. Adhesive tape, such as electricians' waterproof tape, serves well to wrap the ends; the winding should be started above the edge of the cover to enclose the ends, particularly the upper one, securely. Budding rubbers, twist ties, and florists' ties are other materials that can be used for this purpose. Prefilled plastic pouches and/or bags as well as Jiffy-7 pots enclosed with aluminum have been described as faster to apply (51). To avoid breakage during air layering, some operators will attach short canes as a "splint" across the girdled or incised section (Fig. 14–5).

The layer is removed from the parent plant when roots are observed through the transparent film. The earliest adventitious roots are generally thick and corky, and the propagator should wait until production of fibrous secondary roots occurs. In some plants, rooting occurs in two to three months or less. Layers made in spring or early summer are best left until the shoots become dormant in the fall, and are removed at that time. Removal of the layer for transplanting is best when growth is not active. In Costa Rica, air-layered Schefflera (*Brassaia*) can be pruned hard and harvested before roots emerge; the pre-rooted layers are then planted in small containers and sold as bonsai-like specimens (Fig. 14–8, page 545).

Pruning to reduce the top in proportion to the roots is usually advisable. Pot the rooted layer into a suitable container and place it under very reduced mist,

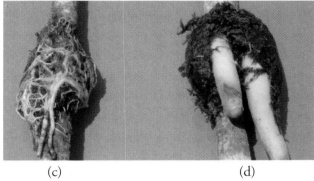

Figure 14–7
Air layering of *Ficus benjamina* with (a) double-slit and (b) complete girdle method produces, respectively, (c) fine roots and (d) coarse roots, as shown in *Dracaena marginata*.
Courtesy T. K. Broschat and H. Donselman (5).

Figure 14–8
(a) Air layers of *Schefflera* (*Brassaia*) in Costa Rica that are pruned hard and harvested *before* roots emerge. (a) The red solution (arrow) is a fungicide/wax dip for the cut surfaces. (b and c) These pre-rooted layers are then planted in small containers to root and sold as bonsai-like specimens.

fog, or under low-light conditions to allow the plant to acclimate. If dry, initially place the rooted layer in water within 5 to 10 minutes of removal. If potted in the fall, a sufficiently large root system usually develops by spring to permit successful growth under more optimal high-light conditions. Placing the rooted layers under mist for several weeks, followed by gradual hardening-off, is probably the most satisfactory procedure (37).

Mound (Stool) Layering or Stooling

Mound layering is a method where the shoots are cut back to the ground, and soil or rooting medium is mounded around them to stimulate roots to develop at their bases. This old nursery propagation method was

mound layering
A type of layering in which shoots on established plants are cut back to the base annually and mounded over with rooting media at intervals during growth to stimulate rooting.

BOX 14.1 GETTING MORE IN DEPTH ON THE SUBJECT
USE OF BIOTECHNOLOGY TO INCREASE ADVENTITIOUS ROOTS IN AIR LAYERS OF HARD-TO-ROOT CLONES OF ELM (*ULMUS*) RESISTANT TO DUTCH ELM DISEASE (40)

Dutch elm disease, caused by the fungus *Ophiostoma ulmi* (Buism), is killing elm (*Ulmus* sp.) around the world. Some traditional breeding programs have resulted in selection of clones resistant to the problem. Unfortunately, these clones are difficult to root for commercial use. Italian scientists (40) have reported that high rooting of air layers of resistant clones have been produced by the combination of (a) treatment with root-inducing genes from **Agrobacterium rhizogenes** plus auxin application as compared to (b) the bacteria alone or (c) the hormone alone. Previous research had shown positive results with treatment of *A. rhizogenes* with cuttings of other hard-to-root plants of chestnut (*Castanea*) and almond (*Prunus dulcis*). Similar favorable results have been obtained on *Fagus sylvatica* by these scientists with the same technique.

agrobacterium rhizogenes   A modified form of the crown gall bacterium which can transfer root-inducing genes into cells of another species.

standardized and improved to mass-propagate specific apple clonal rootstocks in England in the early 1900s. The procedure continues to be used commercially to propagate millions of apple, pear, and some other fruit tree roots each year. It is also useful for quince, currants, gooseberries, and oaks (3, 7, 9, 12, 15). **Stooling** produces **stool shoots,** whereas other layering techniques produce "layers" (Figs. 14–1 and 14–9).

Establishing the Stool Bed Plant healthy stock plants of suitable size (8 to 10 mm diameter) in loose, fertile, well-drained soil 1 year before propagation is to begin. The plants are set 30 to 38 cm (12 to 15 in) apart in a row, but the spacing between rows may vary with the different requirements of the nursery. Width between rows should allow for cultivation and hilling operations during spring and summer. A minimum of

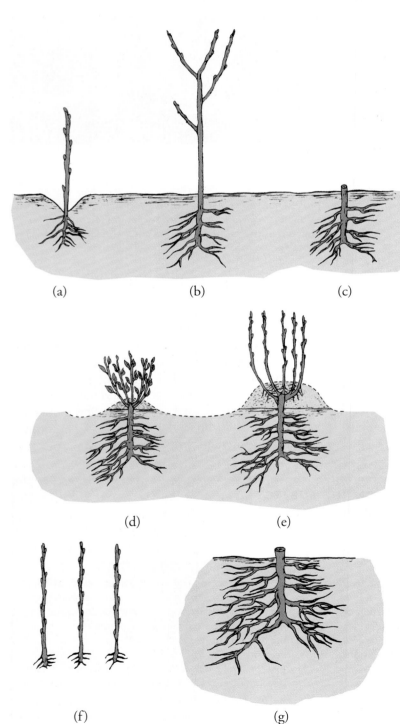

Figure 14–9
Steps in propagation by mound (stool) layering. (a) Stool bed started by planting rooted stool shoot in a small trench. (b) Mother plant is allowed to grow for one season. (c) Top is removed 2.5 cm (1 in) above ground just before growth starts. (d) When new shoots are 8 to 13 cm (1 to 5 in) high, soil or sawdust is added to half their height, and at later intervals until it is 6 to 20 cm (6 to 8 in) deep. (e) At end of season, roots have formed at base of covered stools. (f) Rooted stool shoots are cut off as closely as possible to the base and are lined out in the nursery row. (g) Mother stool with stool shoots previously harvested at the beginning of the next season. Additional new stool shoots will produce the next crop.

2.5 m (8 ft) row spacing is usually required to accommodate tractors (3). Plants are cut back to 45 to 60 cm (18 to 24 in) and allowed to grow for 1 year.

Managing the Bed Before new growth starts in the spring, plants are cut back to 2.5 cm (1 in) above ground level. Two to five new shoots usually develop from the crown the first year, more in later years. When these shoots have grown 7.6 to 12.7 cm (3 to 5 in), loose soil, bark, sawdust, or a soil-sawdust mixture is used to cover each shoot to one-half of its height (Fig. 14–10). When shoots have grown 19 to 25 cm (8 to 10 in), a second hilling operation takes place. Additional rooting medium is added around the bases of the shoots to about half the total height. A third and final hilling operation is made in midsummer when the shoots have developed approximately 45 cm (18 in). The bases of the shoots will then have been covered to a depth of 15 to 20 cm (6 to 8 in).

Harvest Stool shoots should have rooted sufficiently by the end of the growing season to be separated from the parent plant. Rooted shoots are cut close to their bases (see Figs. 14–1 and 14–11), handled as rooted liners (including grading, packing, and storing), and delivered to customers to be transplanted directly into the nursery row as "rooted liners."

Third Year After the shoots have been cut away, the stool beds remain exposed until new shoots have grown 7.6 to 12.7 cm (3 to 5 in). At this time, the **"hilling up"**

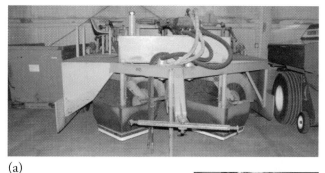

Figure 14–11
Equipment used in Oregon nurseries for mechanized field layering. (a) Machine for shaping soil and media around layers. (b) Machine for sweeping away media prior to cutting layers. (c) Machine for cutting and harvesting rooted stool layers; note sickle bar cutting system (arrow).

Figure 14–10
Mound layering of cherry understocks. (a) First year planting. (b and c) Mature field with mid-season mounding of sawdust. (d) Sawdust removed to show developing adventitious roots.

or mounding begins for the next year.

A stool bed can be used for 15 to 20 years with proper handling, providing it is maintained in a vigorous condition and diseases, insects, and weeds are controlled. Selective and biennial harvesting has been used to invigorate declining stool beds and produce large shoots for high budding, although the method has led to increased apple mildew infection (47). Sprays, containing the auxin NAA, have been used to eliminate small nonproductive shoots from apple stool beds (22).

Cutting back whole plants, then mound layering the vigorous juvenile shoots, has been described as a method of rooting 6- to 7-year-old seedling cashew (35), seedling pecan (30), and other difficult-to-root plants. Mounding (stooling) pecan rootstock has been successfully done in Peru (Fig. 14–12).

Girdling the bases of the shoots by wiring about 6 weeks after growth begins may stimulate rooting (29). The size of the root system in apple has been increased on shoots growing through the spaces of a galvanized

> "hilling up"
> A horticultural term that refers to the mounding up of soil or other media around the base during layering.

Figure 14–12
Clonal propagation of pecan by mound (stool) layering in Peru. (a) Five-year-old 'Stuart' pecan trees in the background and smaller-sized five-year-old pecan stooling bed in the foreground (arrow). (b and c) Stool plants. (b) Soil was removed for photo prior to rooting. (d and e) Large rooted pecan stools removed from the stool beds. (f) Rooted stools lined out in the field. From Medina (29, 30).

screen 0.5 cm (3/16 in) square laid in a 45-cm (18-in) strip down the row over the top of the cut-back stumps. New shoots grow through this screen and gradually become girdled as the season progresses (20).

Budded plants of apple (29) and citrus (9) have been produced by budding the shoot in place in the stool bed in the middle of the growing season. The budded shoots are transplanted to the nursery in the fall for an additional season's growth. Budding rootstocks in place is generally not recommended unless virus-free scions are used because the rootstocks may become permanently infected. A containerized stooling system for limited quantities of clonal apple rootstock is illustrated in Figure 14–13 (34).

Trench Layering

Trench layering is a layering method in which the mother plants are established in a sloping position such that shoots can be layered horizontally in

> **trench layering** A type of layering in which shoots of established plants are placed horizontally at the base of a trench, and the new shoots are covered at intervals during growth to induce etiolation.

BOX 14.2 GETTING MORE IN DEPTH ON THE SUBJECT
STORY OF MALLING APPLE ROOTSTOCKS

The Malling series of apple rootstocks represents a prime example of horticultural research at its best. For centuries, European and West Asian gardeners had been using clonal dwarfing rootstocks for apples in their gardens. Beginning in 1912, the E. Malling Research Station in England searched through European nurseries to collect sixteen such clones showing a range of size control and evaluated them over a number of years. At the same time, standard procedures for mound layering were being developed. The result was a range of size-controlling rootstocks ranging from *very dwarfing* to *very vigorous* (see Fig. 11–10), some of which have become standard rootstocks throughout the world. The second step was a new rootstock breeding program beginning in 1928 between the E. Malling Station and John Innes Horticultural Institute using some of these clones as parents. Another series of rootstock clones resulted, some of which have since become standard. The concept of clonal rootstocks with their special characteristics and propagation needs was later extended to cherries and plums, and resulted in the development of trench layering procedures. Although layering played an integral role in the development of clonal rootstocks, many of these plants are now being propagated by additional modern techniques including micropropagation.

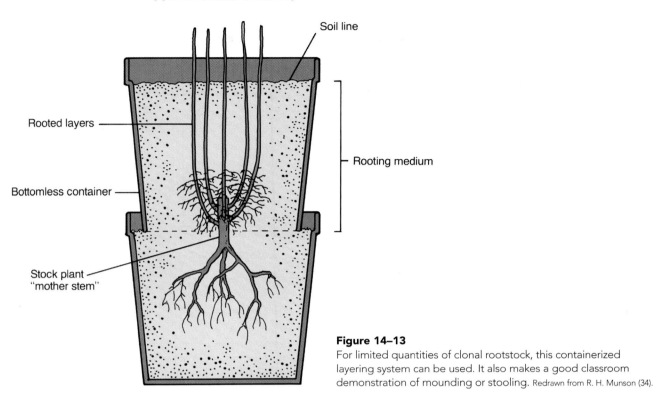

Figure 14–13
For limited quantities of clonal rootstock, this containerized layering system can be used. It also makes a good classroom demonstration of mounding or stooling. Redrawn from R. H. Munson (34).

the base of a trench (Figs. 14–14 and 14–15). Soil, bark, sawdust, or other rooting material is filled in around the new shoots as they develop to bring about etiolation. The procedure was developed to propagate the bacterial canker-resistant cherry rootstock Mazzard 'F12-1,' which could not be propagated by stooling. The method's success is attributed to horizontal positioning of the mother plant in the trench which breaks apical dominance—and careful etiolation of the emerging shoots. The method can be used for clones of other woody species that are difficult to root by mound layering (stooling), including quince (*Cydonia*), apple (*Malus*), mulberry (*Morus*), and walnut (*Juglans*) (26, 27). In New Zealand, a system similar to trench layering is used to produce rooted fascicle cuttings of Monterey pine (*Pinus radiata*) (Fig. 14–16, page 551).

Establishing the Layer Bed One-year-old nursery plants or well-rooted layers are planted 65 to 70 cm (20 to 30 in) apart in a straight line down the row (28). The distance between rows depends upon equipment. Double rows may also be used. The trees are planted at an angle of 30 to 45 degrees in order to get the required amount of growth and still allow layering. The plants are allowed to grow during the first year to establish a good root system. By the end of the growing season or during that winter, a shallow trench 5 × 23 cm (2 × 9 in) is dug down the row. The plants are brought down to a horizontal level and **"pegged"** carefully so that they are flat on the bottom of the trench along with strong lateral branches (Figs. 14–14 and 14–15). All of the plant must be level to produce even shoot growth.

> "pegged down"
> A horticultural term referring to the practice of holding the stem horizontally in place with any type of wire, metal, or wooden fastener.

Second Year Successful layering depends on etiolation. Buds are covered with about a 2.5 cm (1 in) of soil before they emerge. Subsequent applications of rooting medium such as sawdust are added periodically to etiolate 5 to 7.5 cm of the developing shoots. Final depth should be 15 to 19 cm. Rooting should take place by the end of the season.

Harvesting At the end of the season, the media (soil, pinebark, mulch, sawdust) is removed and rooted layers cut off close to the original branch, leaving a small stub for next year's growth. The process is repeated in subsequent years. A well-cared-for stock bed should last 15 to 20 years (28).

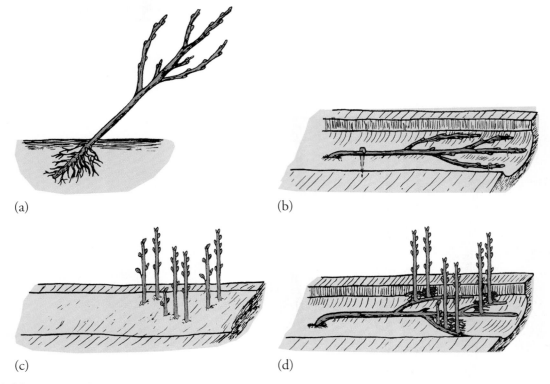

Figure 14–14
Steps in propagation by trench layering. (a) Mother plant after one year's growth in nursery. The trees were planted in the row at an angle of 30 to 45 degrees. The trees are 46 to 78 cm (18 to 30 in) apart down the row. (b) Just before growth begins, the plant is laid flat on the bottom of a trench about 5 cm (2 in) deep. Shoots are cut back slightly and weak branches removed. Tree must be kept completely flat with wooden pegs or wire fasteners. (c) Rooting medium, such as fine soil, peat moss, or sawdust, is added at intervals to produce etiolation on 5 to 7.5 cm (2 to 3 in) of the base of the developing shoots. Apply first 2.5- to 5-cm (1- to 2-in) layer before the buds swell. Repeat as shoots emerge and before they expand. Later coverings are less frequent and only cover half of shoot. Final medium depth is 15 to 19 cm (6 to 8 in). (d) At the end of the season the medium is removed and the rooted layers are cut off close to the parent plant. Shoots left at internodes can be layered the following season.

Figure 14–15
Trench layering in dwarf apple understock. (a) Initial plants are tipped at an angle in the row. (b and c) Tipped plants are tied down and mounded with sawdust. (d) Established stool bed.

Figure 14–16
Similar to trench layering. (a) Twenty-one-month-old seedlings of Monterey Pine (*Pinus radiata*) topped and pinned down to produce fascicle cuttings, (b) one-year-old stoolbed before topping in February (summer, New Zealand), and (c) three-year-old seed orchard established from rooted stools. Courtesy Jacqui Holiday.

Trench layering is used primarily for woody species that are difficult to propagate by other methods of stooling. In some cases (walnuts, for instance), plants were planted horizontally in the trench and the developing shoots layered the first year.

Drop Layering

Drop layering is a combination of crown division and layering that has been used for a limited number of shrubby species, such as dwarf ericaceous plants like Rhododendrons, barberry (*Berberis*), boxwood (*Buxus*), and some dwarf conifers, such as spruce (27). Well-grown and well-branched plants are planted deeply in a hole or trench and covered almost completely with only the tips of the branches exposed. New growth comes from the branch tips, but the older bases of the branches are blanched and roots form. At the end of the season the entire plant is dug and divided into all of its rooted parts, each used as rooted liners.

drop layering A type of layering in which the bases of shoots in plants established in containers are covered, as in mound layering.

PLANT MODIFICATIONS RESULTING IN NATURAL LAYERING

A number of plant species have specially modified stems and roots that allow them to naturally layer. Table 14–2 shows comparisons among structures used in natural layering.

Tip Layering

Tip layering occurs naturally in trailing blackberries, dewberries, and black and purple raspberries (*Rubus*). Biennial canes arise from the crown each spring. The canes are vegetative during the first year, fruitful the second, and then die. New canes are produced annually to produce the so-called "bramble bush." Propagation consists of removing the rooted tip layers at the end of the season and transplanting them into a permanent location. In the nursery, healthy young

tip layering A form of natural layering in which the stem tip of some species of *Rubus* form roots when inserted into the soil.

Table 14–2
COMPARISONS AMONG STRUCTURES USED IN NATURAL LAYERING

Type of modified stem structure	Growth habit	Plants propagated by these structures
Stolons	A trailing or arched stem that grows horizontally above or below the soil to form new plants at the nodes.	Dogwood (*Cornus stolonifera*), bugleweed (*Ajuga*), mint (*Mentha*)
Runners	A specialized type of stolon (usually without leaves) that arises from the axil of a leaf at the crown and grows horizontally above ground. New plants arise at the tip as daughter plants.	Strawberry (*Fragaria*), spider plant (*Chlorophytum*), strawberry geranium (*Saxifraga*)
Rhizomes	A horizontal stem distinguished from a stolon because it is also modified as a storage organ. Usually found in ferns and monocots (see Chapter 15).	Iris, Solomon's seal (*Polygonatum*)
Crowns	The growing point of a plant at the soil surface where new shoots are formed.	Many herbaceous perennials and ornamental grasses
Offsets	Short horizontal shoot at the base of the main stem that forms an independent crown.	Many bulbs, daylily, *Hosta*, palms
Suckers	Shoots that develop from underground roots or shoots. In most cases, these arise from roots.	Raspberry (*Rubus*), pawpaw (*Asimina*)

plants should be set aside as stock plants solely for propagation. The original plants are set 3.6 m (12 ft) apart to give room for subsequent layering. The plants are cut to within 23 cm (9 in) of the ground as soon as they are planted. Vigorous new canes are "summer topped" by pinching back 7.6 to 9.2 cm (3 to 4 in) of the tip after growth of 45 to 76 cm (18 to 30 in) to encourage lateral shoot production. "Summer topping" will increase the number of potential tip layers, and also can increase next season's fruit crop.

Canes begin to arch over in late summer. Their tips assume a characteristic appearance in that the terminal ends become elongated and the leaves small and curled to give a "rat-tail" appearance. The best time for layering is when only part of the lateral tips have attained this appearance. If the operation is done too soon, the shoots may continue to grow instead of forming a terminal bud. If done too late, the root system will be small.

Tips can be layered by hand, using a spade or trowel to make a hole, with one side vertical and one sloping slightly toward the parent plant. The tip is placed in the hole along the sloping side, and soil is pressed firmly against it. Placed thus, the tip stops growing and becomes "telescoped," forming an abundant root system and a vigorous vertical shoot. Rooting takes place below the tip of the current season's shoot. The shoot tip recurves upward to produce a sharp bend in the stem from which roots develop, as with the raspberry tip cutting (Fig. 14–17).

The plants are ready for digging at the end of the season. The rooted tip consists of a terminal bud, a large mass of roots, and 15 to 20 cm (6 to 8 in) of the old cane that serves as a "handle" and marks the location of the new plant. Since tip layers are tender, easily injured, and subject to drying out, digging just prior to replanting is

(a)

(b)

Figure 14–17
(a and b) Natural layering of raspberry tips; notice the white etiolated shoot tissue (red arrow) and roots occurring at the nodal area.

preferable. The remainder of the layered shoot attached to the parent plant is cut back to 23 cm (9 in) as in the first year. Economical quantities of shoots can be produced annually for as long as ten years. Rooted tip layers are planted in the late fall or early spring. New canes develop rapidly during the first season.

Runners

A **runner** is a specialized stem that develops from the axil of a leaf at the crown of a plant, grows horizontally along the ground, and forms a new plant at one of the nodes. The strawberry is a typical plant with this growth habit (Fig. 14–18). Other plants that produce runners used in propagation include bugle (*Ajuga*), strawberry geranium (*Saxifraga sarmentosa*), spider plant (*Chlorophytum comosum*) and the ground cover *Duchesnea indica*. Plants of these species grow as a rosette or crown. Some ferns, such as Boston fern (*Nephrolepis*), produce runner-like branches, as do certain orchid species (*Dendrobium*), which form small plants known as "keikis" (Fig. 14–19).

runner A natural form of layering in which a specialized stem grows laterally from the crown of the plant and takes root at alternate (or other) nodes to produce a new plant that provides for vegetative expansion of the original plant.

Most strawberry cultivars form runners in response to length of day and temperature. Runners are produced in long days of 12 to 14 hours or more with high midsummer temperatures. New plants produced at alternate nodes produce roots, but remain attached to the mother plant. New runners are, in turn, produced by daughter plants. The connecting stems die in the late fall and winter, separating each daughter plant from the others. In propagating by runners, daughter plants are dug when they have become well rooted, and then transplanted to the desired locations.

Figure 14–19
Phalaenopsis lueddemanniana–offsets called "Keikis," which produce new plantlets.

Stolons

Stolons are modified stems that grow horizontally to the ground and produce a prostrate or sprawling mass of stems growing along the ground; for example, in dogwood (*Cornus stolonifera*). The term also describes the horizontal stem structure occurring in Bermuda grass (*Cynodon dactylon*), *Ajuga*, mint (*Mentha*), and *Stachys*. Stolon-like underground stems are involved in tuberization of the potato tuber (see Fig. 15–22, page 580).

stolon A specialized underground stem that grows laterally from the crown of the plant to produce either another plant or a tuber.

A stolon can be treated as a naturally occurring rooted layer and cut from the plant and planted.

Offsets

An **offset** is a characteristic type of lateral shoot or branch that develops from the base of the main stem in certain plants. This term is applied generally to a shortened, thickened stem of rosette-like appearance. Many bulbs reproduce by producing typical offset bulblets from their base (see Chapter 15 for details). The term *offset* (or **offshoot**) also

offset A specialized leafy plant stem that develops from the base of many monocotyledonous plants and is used for propagation.
offshoot A synonymous term for offset.

Figure 14–18
Runners (arrow) arising from the crown of a strawberry (*Fragaria*) plant. New plants are produced at every second node. The daughter plants, in turn, produce additional runners and runner plants. Courtesy Marvin Pritts.

applies to lateral branches arising on stems of monocotyledons as in date palm (Fig. 14–20), pineapple, or banana (see Chapter 19).

Offsets are cut close to the main stem with a sharp knife. If well rooted, the offset can be potted and established like a rooted cutting. If insufficient roots are present, the shoot can be placed in a favorable rooting medium and treated as a leafy stem cutting. **Slips** and **suckers**, which are two types of pineapple offsets, readily form aerial rootlets that facilitate commercial field propagation (Fig. 14–21).

> **slip** An offshoot of pineapple that forms from the peduncle and is used in propagation.

If offset development is meager, cutting back the main rosette may stimulate the development of more offsets from the old stem in a manner similar to the removal of the terminal bud in the stimulation of lateral shoots in any other type of plant.

Natural multiplication by offsets tends to be slow and not well-suited for commercial propagation. The natural reproduction rate of many monocotyledonous plants can be increased greatly by micropropagation (see Chapters 17 and 18). Offsets can be induced in *Hosta* by cytokinins such as BA (11).

Offshoots of the date palm must be of sufficient size, and do not root readily if separated from the parent plant. They are usually layered for a year prior to removal (18, 19). Micropropagation is an alternate method.

Suckers

A **sucker** is a shoot that arises on a plant from below ground, as shown in Figure 14–22 (page 556). The most precise use of this term is to designate a shoot that arises from an adventitious bud on a root. In practice, shoots that arise from the vicinity of the crown are often called "suckers," even though they originate from stem tissue. Propagators generally designate any shoot produced from the rootstock below the bud union of a budded tree as a "sucker" and refer to the operation of removing them as **"suckering."** In contrast, a shoot arising from a latent bud of a stem several years old as, for instance, on the trunk or main branches, should be termed a **watersprout.**

Suckers are dug out and cut from the parent plant. In some cases, part

> **suckers** Adventitious shoots that emerge from a root or from the vicinity of the crown (often synonymous with watersprouts).

> **suckering** A term given to the practice of removing suckers and/or watersprouts.

> **watersprout** A term given to a shoot emerging from a latent bud on the crown or trunk of a tree.

(a)

(b)

(c)

(d)

Figure 14–20
Date palm propagation. (a) Tied new lateral trunks (offshoots) that develop from "mother" trunk or near its base, prior to their severance, (b) removing offshoots with a chisel and sledge hammer, (c) an inverted offshoot about 3 to 5 years old, with tied branches after removal from "mother" trunk, (d) a two-week-old planting of date palm offshoots in California; some will be thinned for final spacing and replanted at another site. Courtesy D. R. Hodel (18, 19).

  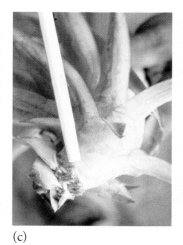 
(a) (b) (c) (d)

(e)

Figure 14–21
Slips, suckers, and crowns are three types of pineapple offsets that can be used in pineapple vegetative propagation. (a) Cross section of a pineapple showing the crown on top of the fruit and slip attached to the peduncle supporting the fruit, (b) slips (*left*) from the peduncle are more commonly used propagules than suckers (*right*), (c) preformed root primordia enable (d and e) slips to root readily when mass-propagated in black polyethylene-mulched field beds.

of the old root may be retained, although most new roots arise from the base of the sucker. It is best to dig out the sucker rather than pull it, to avoid injury to its base. Suckers are treated as a rooted layer or cutting, and are usually dug during the dormant season. Since suckers come from root tissue, they would have the same problems of reversion as shown by root cuttings and leaves in the case of a periclinal chimera.

Crown Division

The term **crown,** as generally used in horticulture, designates that part of a plant at the surface of the ground from which new shoots are produced. In trees or shrubs with a single trunk, the crown is principally a point of location near the ground surface marking the general transition zone between stem and root. In herbaceous perennials, the crown is the part of the plant from which new shoots arise annually. The crown of herbaceous perennials consists of many branches, each being the base of the current season's stem, which originated from the base of the preceding year's branch. These lateral shoots are stimulated to grow from the base of the old stem as it dies back after blooming. Adventitious roots develop along the base of the new shoots. These new shoots flower either the same year they are produced or the following year. As a result of annual production of new shoots and the dying back of old shoots, the crown may become extensive within a period of relatively few years, and may need to be divided to prevent overcrowding.

crown The root-stem juncture of a plant (also used to indicate the top of the tree in forestry).

Multibranched woody shrubs may develop extensive crowns. Although an individual woody stem may persist for a number of years, new vigorous shoots are continuously produced from the crown, and they eventually crowd out the older shoots. Crowns of such shrubs can be divided in the dormant season and treated as a large-rooted cutting.

Crown division is an important method of propagation for herbaceous perennials, and to some extent for woody shrubs, because of its simplicity and reliability. Such characteristics make this method particularly useful to the amateur or professional gardener who is generally interested in only a modest increase of a particular plant.

crown division A method of propagation in which the crown of a plant (usually an herbaceous perennial) is separated into parts with stem and root material attached.

Crowns of outdoor herbaceous perennials are usually divided in the spring just before growth begins, or in

Figure 14–22
Suckers arising as adventitious shoots from the roots of (a) breadfruit and (b) sassafras. After they are well rooted, the suckers may be cut from the parent plant and transplanted to their permanent location. Photograph (a) courtesy R. A. Criley.

late summer or autumn at the end of the growing season. As a general rule, plants that bloom in the spring and summer and produce new growth after blooming should be divided in the fall. Those that bloom in summer and fall and make little or no new growth until spring should be divided in early spring. Potted plants are divided when they become too large for the particular container in which they are growing. Division is necessary to maintain the variegated form in some plants usually propagated by leaf cuttings, such as *Sanseviera* (16).

In crown division, plants are dug and cut into sections with a knife, hand ax, handsaw, or other sharp instrument (Figs. 14–23 and 14–24). In herbaceous perennials, such as the Shasta daisy (*Chrysanthemum superbum*), daylily (*Hemerocallis*), *Hosta* (Fig. 14–25) or Liriope (Fig. 14–26), where an abundance of new rooted offshoots is produced from the crown, each may be broken from the old crown and planted separately. The older part of the plant clump is discarded.

For commercial cultivar production, division can be very slow. Shoot production can be increased by cutting back to the crown in the early spring after new growth starts, and treating with a cytokinin (1). The cytokinin, BA, enhances offset formation in a number of *Hosta* cultivars (11). While daylilies can be propagated by division, rapid multiplication is achieved by micropropagation (17, 32) (see Chapters 17 and 18). Asparagus is commercially propagated by seed, tissue culture, and crown division (Fig. 14–27, page 558).

Figure 14–23
Crown division. (a) Dividing blue fescue (*Festuca ovina* 'Glauca') with a knife, (b) using a hand ax with feather reed grass (*Calamagrostis acutiflora* 'Stricta'), and (c) cutting giant reed (*Arundo donax*) with a handsaw.
Photographs courtesy R. A. Simon.

Figure 14–24
Commercial herbaceous perennial division. (a, b, and c) Field dug perennials being divided. (d and e) Divided plants in cold storage.

Figure 14–25
Crown division of *Hosta*. (a) Dividing up crowns with a knife, and (b) processing and bagging divisions for liner production.

Figure 14–26
(a) Stock beds of field-grown *Liriope*. (b) Removing crowns, and (c and d) dividing into rooted liners or "bibs."

(a)

(b)

Figure 14–27
Asparagus is propagated by seed, tissue culture, and crown division. (a) Asparagus crown (arrow) with attached storage roots. (b) Male (♂) plants of asparagus, which is a cross-pollinated dioecious crop, are more productive in producing vegetative spears than female ones.

Many other herbaceous perennials propagated by division generally can also be propagated by micropropagation (54). Tissue culture enhances axillary branching, and subsequent multiplication can be made at high rates with fewer stock plants.

DISCUSSION ITEMS

1. Why would layering have more importance in nursery production 100 years ago than today?
2. Why does layering continue to have important uses in some modern horticultural and forestry nurseries?
3. Physiologically, why is layering a very successful method of vegetative propagation?
4. What are differences among blanching, etiolation, and banding during layering?
5. Why is layering a particularly useful tool for an amateur propagator?
6. Explain why some type of "bending" operation is highly beneficial for layering results.
7. Explain why girdling may be an important part of layering operations.
8. Compare mound and trench layering from the standpoint of physiology.
9. What are slips and suckers, and where do they originate?
10. Distinguish between "runners" and "stolons."

REFERENCES

1. Apps, D. A., and C. W. Heuser. 1975. Vegetative propagation of *Hemerocallis*—including tissue culture. *Comb. Proc. Intl. Plant Prop. Soc.* 25:362–67.
2. Barnes, R. D. 1974. Air-layering of grafts to overcome incompatibility problems in propagating old pine trees. *New Zealand J. For. Sci.* 4:120–26.
3. Brase, K. D., and R. D. Way. 1959. Rootstocks and methods used for dwarfing fruit trees. *N.Y. Agr. Exp. Sta. Bul. 783*.
4. Broschat, T. K., and H. M. Donselman. 1981. Effects of light intensity, air layering, and water stress on leaf diffusive resistance and incidence of leaf spotting in *Ficus elastica*. *HortScience* 16:211–12.
5. Broschat, T. K., and H. M. Donselman. 1983. Effect of wounding method on rooting and water conductivity in four woody species of air-layered foliage plants. *HortScience* 18:445–47.
6. Cameron, R. J. 1968. The leaching of auxin from air layers. *New Zealand J. Bot.* 6:237–39.
7. Carlson, R. F., and H. B. Tukey. 1955. Cultural practices in propagating dwarfing rootstocks in Michigan. *Mich. Agr. Exp. Sta. Quart. Bul.* 37:492–97.
8. Chase, H. H. 1964. Propagation of oriental magnolias by layering. *Comb. Proc. Intl. Plant Prop. Soc.* 14:67–9.
9. Duarte, O., and C. Medina. 1971. Propagation of citrus by improved mound layering. *HortScience* 6:567.
10. Dunn, N. D. 1979. Commercial propagation of fruit tree rootstocks. *Comb. Proc. Intl. Plant Prop. Soc.* 29:187–90.

11. Garner, J. M., G. J. Keever, D. J. Eakes, and J. R. Kessler. 1995. BA-induced offset formation in hosta dependent on cultivar. *Comb. Proc. Intl. Plant Prop. Soc.* 45:605–10.

12. Garner, R. J. 1988. *The grafter's handbook.* 5th ed. New York: Oxford Univ. Press.

13. Grove, W. R. 1947. Wrapping air layers with rubber plastic. *Proc. Fla. State Hort. Sci.* 60:184–89.

14. Hare, R. C. 1979. Modular air-layering and chemical treatments improve rooting of loblolly pine. *Comb. Proc. Intl. Plant Prop. Soc.* 29:446–54.

15. Hawver, G., and N. Bassuk. 2000. Improved adventitious rooting in *Quercus* through the use of modified stoolbed technique. *Comb. Proc Intl. Plant Prop. Soc.* 50:307–13.

16. Henley, R. W. 1979. Tropical foliage plants for propagation. *Comb. Proc. Intl. Plant Prop. Soc.* 29:454–67.

17. Heuser, C. W., and J. Harker. 1976. Tissue culture propagation of daylilies. *Comb. Proc. Intl. Plant Prop. Soc.* 25:269–72.

18. Hodel, D. R., and D. R. Pittenger. 2003a. Studies on the establishment of date palm (*Phoenix dactylifera* 'Deglet Noor') offshoots. Part I. Observations on root development and leaf growth. *Palms* 47:191–200.

19. Hodel, D. R., and D. R. Pittenger. 2003b. Studies on the establishment of date palm (*Phoenix dactylifera* 'Deglet Noor') offshoots. Part II. Size of offshoot. *Palms* 47:201–5.

20. Hogue, E. J., and R. L. Granger. 1969. A new method of stool bed layering. *HortScience* 4:29–30.

21. Howard, B. H. 1977. Effects of initial establishment practice on the subsequent productivity of apple stoolbeds. *J. Hort. Sci.* 52:437–46.

22. Howard, B. H. 1984. The effects of NAA-based Tipoff sprays on apple shoot production in MM.106 stoolbeds. *J. Hort. Sci.* 59:303–11.

23. Howard, B. H., R. S. Harrison-Murray, and S. B. Arjyal. 1985. Responses of apple summer cuttings to severity of stockplant pruning and to stem blanching. *J. Hort. Sci.* 60:145–52.

24. Howard, B. H. 1987. Propagation. In R. C. Rom and R. F. Carlson, eds. *Rootstocks for fruit crops.* New York: John Wiley & Sons.

25. Joiner, J. N., ed. 1981. *Foliage plant production.* Englewood Cliffs, NJ: Prentice Hall.

26. Maurer, K. J. 1950. Möglichkeiten der vegetativen Vermehrung der Walnuss. *Schweiz. Z. Obst. V. Weinb.* 59:136–37.

27. McDonald, B. 1986. *Practical woody plant propagation for nursery growers.* Portland, OR: Timber Press.

28. McKensie, R. 1993. Propagation of rootstocks by trench layering. *Comb. Proc. Intl. Plant Prop. Soc.* 43:345–47.

29. Medina, C., and O. Duarte. 1971. Propagating apples in Peru by an improved mound layering method. *J. Amer. Soc. Hort. Sci.* 96:450–51.

30. Medina, C. 1981. Studies of clonal propagation on pecans at Ica, Peru. *Plant Propagator* 27:10–1.

31. Mergen, F. 1955. Air layering of slash pine. *J. For.* 53:265–70.

32. Meyer, M. M. 1976. Propagation of daylilies by tissue culture. *HortScience* 11:485–87.

33. Modlibowska, I., and C. P. Field. 1942. Winter injury to fruit trees by frost in England (1939–1940). *J. Pom. Hort. Sci.* 19:197–207.

34. Munson, R. H. 1982. Containerized layering of *Malus* rootstocks. *Plant Propagator* 28:12–4.

35. Nagabhushanam, S., and M. A. Menon. 1980. Propagation of cashew (*Anacardium occidentale* L.) by etiolation, girdling and stooling. *Plant Propagator* 26:11–3.

36. Neel, P. L. 1979. Macropropagation of tropical plants as practiced in Florida. *Comb. Proc. Intl. Plant Prop. Soc.* 29:468–80.

37. Nelson, R. 1953. High humidity treatment for air layers of lychee. *Proc. Fla. State Hort. Soc.* 66:198–99.

38. Nelson, W. L. 1987. Innovations in air layering. *Comb. Proc. Intl. Plant Prop. Soc.* 37:88–9.

39. Poole, R. T., and C. A. Conover. 1988. Vegetative propagation of foliage plants. *Comb. Proc. Intl. Plant Prop. Soc.* 37:503–7.

40. Rinallo, C., L. Mittempergher, G. Frugis, and D. Marriott. 1999. Clonal propagation in the genus *Ulmus*: Improvement of rooting ability by *Agrobacterium rhizogenes* T-DNA genes. *J. Hort. Sci. and BioTech.* 7:500–6.

41. Rom, R. C., and R. F. Carlson. 1987. *Rootstocks for fruit crops.* New York: John Wiley & Sons.

42. Sparks, D., and J. W. Chapman. 1970. The effect of indole-3-butyric acid on rooting and survival of air-layered branches of the pecan, *Carya illinoinensis* Koch, cv. 'Stuart'. *HortScience* 5:445–46.

43. Sutton, N. E. 1954. Marcotting of Persian limes. *Proc. Fla. State Hort. Soc.* 67:219–20.

44. Thomas, L. A. 1938. Stock and scion investigations. II. The propagation of own-rooted apple trees. *J. Counc. Sci. Industr. Res. Org., Austral.* 11:175–79.

45. Tukey, H. B., and K. Brase. 1930. Granulated peat moss in field propagation of apple and quince stocks. *Proc. Amer. Soc. Hort. Sci.* 27:106–13.

46. Tukey, H. B. 1964. *Dwarfed fruit trees.* New York: Macmillan.

47. Vasek, J., and B. H. Howard. 1984. Effects of selective and biennial harvesting on the production of apple stoolbeds. *J. Hort. Sci.* 59:477–85.

48. Vieitez, E. 1974. Vegetative propagation of chestnut. *New Zealand For. Sci.* 4:242–52.

49. Wells, J. S. 1985. *Plant propagation practices.* Chicago, IL: Amer. Nurs. Publ.

50. Wells, R. 1986. Air layering: An alternative method for the propagation of *Mahonia aquifolia* 'Compacta.' *Comb. Proc. Intl. Plant Prop. Soc.* 36:97–9.

51. Wyman, D. 1952. Layering plants in Holland. *Amer. Nurs.* XCV (10).

52. Wyman, D. 1952. Air layering with polyethylene films. *J. Roy. Hort. Soc.* 77:135–40.

53. Young, P. J. 1994. Commercial marcotting of fruit trees. *Comb. Proc. Intl. Plant Prop. Soc.* 44:86–9.

54. Zilis, M., D. Zwagerman, D. Lamberts, and L. Kurtz. 1979. Commercial propagation of herbaceous perennials by tissue culture. *Comb. Proc. Intl. Plant Prop. Soc.* 29:404–13.

15
Propagation by Specialized Stems and Roots

INTRODUCTION

Bulbs, corms, tubers, tuberous roots and stems, rhizomes, and pseudobulbs are specialized vegetative structures that function primarily in the storage of food, nutrients, and water during adverse environmental conditions. Plants possessing these modified parts are generally herbaceous perennials whose shoots die down at the end of a growing season. The plant survives in the ground as a dormant, fleshy organ that bears buds to produce new shoots the following season. Collectively, plants that survive as underground storage organs are called **geophytes** (17, 70). Such plants are well suited to withstand periods of adverse growing conditions in their yearly growth cycle (68). However, the storage organs of these plants are never physiologically dormant even after their above-ground shoot system have died off. These highly specialized structures serve as "bio-processors," sensing and responding to the changing environment. The two principal climatic cycles for which such performance is adapted are the warm-cold cycle of the temperate zones and the wet-dry cycle of tropical and subtropical regions (22).

geophytes Types of plants that survive part of their annual life cycle as a dormant, fleshy, underground structure.

learning objectives
- Define structure.
- Characterize growth and development patterns.
- Describe propagation systems for each of the main classes of geophytes: bulbs, corms, tubers, tuberous roots and stems, rhizomes, and pseudobulbs.

Besides survival during adverse environmental conditions, these organs are equally important in the clonal regeneration of the species; their unique reproductive characteristics allow the plants to use both sexual and asexual systems for regeneration and adaptation.

Many horticulturists refer to all geophytes as bulbs, regardless of their morphology. Ornamental geophytes used extensively for cut flowers, pot plants, and landscaping (4, 17, 70) are usually listed as flowering bulbs. However, for propagation purposes, these specialized organs function in vegetative reproduction, and it is very important to distinguish among the various structures (e.g., bulbs vs. corms). The propagation procedure that utilizes naturally detachable structures, such as the bulb and corm, is generally called **separation.** In cases in which the plant is cut into sections, as is done with the rhizome, stem tuber, and tuberous root, the process is called **division.** A summary of propagation techniques used for species with specialized organs is presented in Table 15–1. Many

separation A type of clonal propagation that utilizes detachable structures on the plant as propagules.

division A type of clonal propagation that involves cutting or dividing the plant into sections with stems and roots.

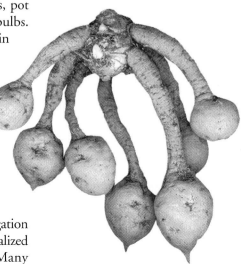

Table 15–1
PROPAGATION OF REPRESENTATIVE SPECIES WITH SPECIALIZED STRUCTURES

Specialized structure	Distinguishing characteristics	Plant species	Vegetative propagation			
			Separation	Stem cutting	Division	Micropropagation
Bulb	Underground structure produced mainly by monocots. Has a short modified stem enclosed in fleshy leaves (scales) modified for food storage.	Tulip Daffodil Onion Lily Hyacinth	X	X	X	X
Corm	Underground rounded stem consisting of compacted nodes with lateral buds. Corms are replaced each growth cycle by new corm on top or to the side of the old.	Crocus Gladiolus Liatris Freesia, Dasheen (Taro)	X		X	X
Tuber	Swollen underground stem modified for food storage with easily distinguished nodes and buds. Similar to corms except for lateral orientation.	Potato Caladium Anemone		X	X	X
Tuberous stem	Flattened swollen stem produced by enlargement of the hypocotyl at root-shoot junction. Perennial structure that can become large.	Tuberous begonia Gloxinia Cyclamen		X	X	X
Tuberous root	Enlarged fleshy root with shoots produced at one end and roots at the other. Biennial structure.	Dahlia Iris Sweet potato Daylily		X	X	X
Rhizome	Specialized stem that grows horizontally at or just below the ground.	Iris Bamboo Lily-of-the-valley Ginger root			X	X
Pseudobulb	Above-ground, enlarged stem with several nodes. Produced by orchids.	Epiphytic orchids			X	X

commercial vegetable crops are produced clonally by specialized structures (13).

BULBS

Definition and Structure

A **bulb** is a specialized underground organ consisting of a short, fleshy, stem axis (**basal plate**), bearing at its apex a growing point or a flower primordium enclosed by thick, fleshy scales (Figs. 15–1 and 15–2). Bulbs are mostly produced by monocotyledonous plants in which the usual plant structure is modified for storage and reproduction. Sorrel (*Oxalis* sp.) is the one dicot genus that produces bulbs (70).

bulb   A specialized underground storage structure that consists of a short, fleshy axis and roots or root primordia, axillary buds, and flower apices enclosed in thickened fleshy scales.

basal plate   The short, thickened stem of a bulb.

bulb scale   The expanded fleshy leaf base of a bulb that contains stored food.

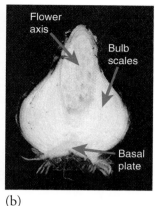

Figure 15–2
(a) Bulb with tunicate covering and adventitious roots (arrow). (b) Cross-section of bulb showing basal plate, bulb scales, and flower axis.

Most of the bulb consists of **bulb scales.** The outer bulb scales are generally fleshy and contain reserve food materials, whereas the bulb scales toward the center function less as storage organs and are more leaflike. In the center of the bulb, there is either a vegetative meristem or an unexpanded flowering shoot. Meristems develop in the axil of these scales to produce underground miniature bulbs, known as **bulblets,** which are known as **offsets,** since they can develop into a new plant (Fig. 15–3). In various species of lilies, bulblets may form in the leaf axils either on the underground portion or the aerial portion of the stem. Aerial bulblets are called

bulblet   A miniature bulb that forms in the axil of a bulb scale and provides a method of vegetative propagation.

offset   A latent shoot that develops from the base of the main stem in certain plants, such as an offset bulblet that is produced from a larger mother bulb.

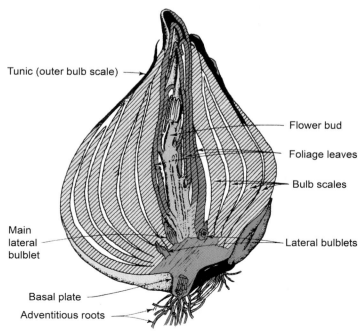

Figure 15–1
The structure of a tulip bulb—an example of a tunicate laminate bulb. Longitudinal section representing stage of development shortly after the bulb is planted in the fall.

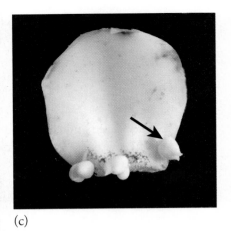

(a)　(b)　(c)

Figure 15–3
Propagation by offset bulbs. Aerial bulblets or bulbils of (a) lily (*Lilium*) and (b) sword lily (*Watsonia*). (c) Bulblets (arrow) from bulb scale cutting of *Lilium*.

bulbil   A type of bulblet produced in the aerial portion of the plant enclosed within a dry, membranous scale.

tunicate   A type of bulb scale characterized by concentric layers of fleshy tissue.

nontunicate   A type of bulb structure in which scales are fleshy, separate, and not enclosed in a membranous layer.

scaly   A synonym for nontunicate.

bulbils (Fig. 15–3). There are two types of bulbs: **tunicate** and **nontunicate (scaly).**

Tunicate (Laminate) Bulbs

Tunicate (laminate) bulbs are produced by the onion and garlic (*Allium*), daffodil (*Narcissus*), and tulip (*Tulipa*) as well as many genera in the family Amarylladaceae (60, 84). These bulbs have outer bulb scales that are dry and membranous. This covering, or tunic, provides protection from drying and mechanical injury to the bulb. The fleshy scales are in continuous, concentric layers, or lamina, so that the structure is more or less solid (Fig. 15–4).

There are three basic bulb structures, defined by type of scales and growth pattern. The amaryllis (*Hippeastrum*) is an example of one type, in which the expanded bases of leaves are used for food storage; there are no scale leaves in this type of bulb (Fig. 15–5). Tulip (*Tulipa*) is an example of the second type of bulb, a type that has only true scales, with leaves produced on the flowering or vegetative shoot (Fig. 15–6). A third type, which includes the daffodil (*Narcissus*), has both expanded leaf bases and true scales (69).

Adventitious root primordia are present on the dormant stored bulb. They do not elongate until planted under the right conditions and at the proper time, and they occur in a narrow band around the outside edge on the bottom of the basal plate.

Nontunicate Bulbs

Nontunicate (scaly) bulbs are represented by the lily (Figs. 15–6 and 15–7, page 566). These bulbs do not possess the enveloping dry covering. The scales are separate and attached to the basal plate. In general, nontunicate bulbs are easily damaged and must be handled more carefully than the tunicate bulbs, and must be kept continuously moist because they are injured by drying. In the nontunicate lily bulb, new roots are produced in midsummer or later, and persist through the following year (22, 74). In most lily species, roots also form on the stem above the bulb.

In many species, thickened **contractile roots** (Fig. 15–8) shorten and pull the bulb to a given level in the ground (69, 94). Tulips do not produce contractile roots,

contractile roots   The thickened, fleshy roots that pull the bulb to a deeper layer in the soil.

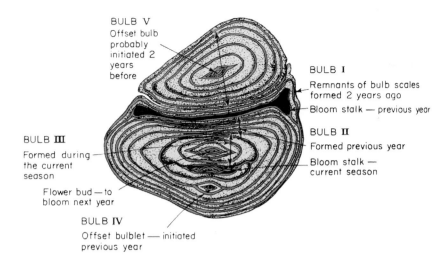

Figure 15–4
Cross section of a daffodil (*Narcissus*) bulb. The continuous, concentric leaf scales found in the laminate bulb are shown. Also shown is the perennial nature of the daffodil bulb, which continues to grow by producing a new bulb annually at the main meristem. Lateral offset bulbs are also produced; parts of five individual, differently aged bulbs are shown here. Redrawn from Huisman and Hartsema (36).

dropper A special kind of bulblike structure occurring in tulips, which grows to a deeper level to produce a new bulb.

but rather stolon-like structures called **droppers** (17, 69, 70), which are new bulbs that form at the end of a stolon (Fig. 15–8). Droppers are produced in seedling bulbs, and push their way from the soil surface down to the appropriate level, where the new bulb is formed.

Growth Pattern

An individual bulb goes through a characteristic cycle of development, beginning with its initiation as a meristem and terminating in flowering and seed production.

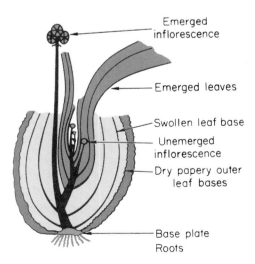

Figure 15–5
Diagram showing morphology and growth cycle of amaryllis (*Hippeastrum*) bulb. New bulbs continually develop from the center in a cycle of four leaves and an inflorescence. Bases of these leaves enlarge to become the scales that contain stored food. Oldest scales disintegrate. From Rees (69).

(a) (b)

(c)

Figure 15–6
(a) Tunicate bulbs of tulip (*Tulipa*). (b and c) Nontunicate, scaly bulb of lily (*Lilium hollandicum*) with adventitious roots.

This general developmental cycle is composed of two stages: (a) *vegetative* and (b) *reproductive*. In the vegetative stage, the bulblet grows to flowering size and attains its maximum weight. The subsequent reproductive stage includes the *induction* and *initiation* of flowering, *differentiation* of the floral parts, *elongation*

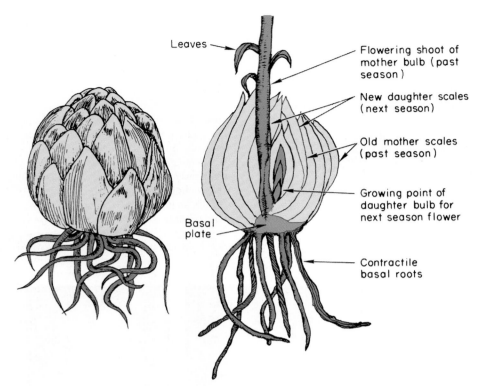

Figure 15–7
Left: Outer appearance of a scaly bulb of lily (*Lilium hollandicum*). Right: Longitudinal section of a bulb of *L. longiflorum* 'Ace,' after flowering stage, showing old mother bulb scales and new daughter bulb scales. Bulb obtained in fall near digging time (18).

of the flowering shoot, and finally *anthesis* (flowering). Sometimes seed production occurs. Various bulb species have specific environmental requirements for individual phases of this cycle that determine their seasonal behavior, environmental adaptations, and methods of handling for bulb forcing (84). Bulbs can be grouped into classes according to their time of bloom and method of handling.

Spring-Flowering Bulbs

Important commercial crops included in the spring-flowering group are the tulip (*Tulipa*), daffodil (*Narcissus*), hyacinth (*Hyacinthus*), and bulbous iris (*Iris*), although other kinds are grown in gardens (17, 70).

Bulb Formation The vegetative stage begins with the initiation of the bulblet on the basal plate in the axil of a bulb scale. In this initial period, which usually occupies a single growing season, the bulblet is insignificant in size, since it is present within another growing bulb and can be observed only if the bulb is dissected. Its subsequent pattern of development and the time required for the bulblet to attain flowering size differ for different species. The bulbs of tulip (*Tulipa*) and the iris (*Iris*) disintegrate upon flowering and are replaced by a cluster of new bulbs and bulblets initiated the previous season. The largest of these may have attained flowering size at this time, but smaller ones require additional years of growth (see Fig. 15–1). The flowering

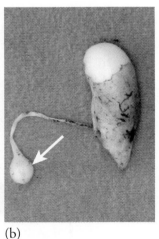

(a)   (b)

Figure 15–8
(a) Contractile roots (arrow) attached to a lily bulb, and (b) a dropper (arrow) of trout lily (*Erythronium*), which is a new bulb that forms at the end of a stolon.

bulb of the daffodil (*Narcissus*), on the other hand, continues to grow from the center, year by year, producing new offsets, which may remain attached for several years (see Fig. 15–4). The hyacinth (*Hyacinthus*) bulb also continues to grow year by year, but because the number of offsets produced is limited, artificial methods of propagation are usually used.

The size and quality of the flower are directly related to the size of the bulb. In fact, a bulb must reach a certain minimum size to be capable of initiating flower primordia. Commercial value is largely based on bulb size (2), although the condition of the bulb and freedom from disease are also important quality factors.

Increase in size and weight of the developing bulb takes place in the period during and (mostly) after flowering, as long as the foliage remains in good condition (14). Cultural operations that include irrigation; weed, disease, and insect control; and fertilization encourage vegetative growth. The benefit, however, is to the next year's flower production, because larger bulbs are produced. Conversely, adverse growing conditions, removal of foliage, and premature digging of the bulb results in smaller bulbs and reduced flower production.

Moderately low temperatures tend to prolong the vegetative period, whereas higher temperatures may cause the vegetative stage to cease and the reproductive stage to begin. Thus, a shift from cool to warm conditions early in the spring, as occurs in mild climates, will shorten the vegetative period, resulting in smaller bulbs, and consequently inferior blooms the following year (69). Commercial producing areas for hardy spring-flowering bulbs are largely in regions of cool springs and summers, such as the Netherlands and the Pacific Northwest of the United States.

The relative length of photoperiod apparently is not an important factor affecting bulb formation in most species. It has been shown, however, to be significant in some *Allium* species, such as onion and garlic (35, 49).

Flower Bud Formation and Flowering The beginning of the reproductive stage and the end of the vegetative stage is indicated by drying of the foliage and the maturation of the bulb. From this point on, there is no additional increase in size or weight of the bulb. The roots disintegrate, and the bulb enters a seemingly "dormant" period. However, important internal changes take place, and in some species the vegetative growing point undergoes transition to a flowering shoot. In nature, all bulb activity takes place underground during this period; in horticultural practice, the bulbs are dug, stored, and distributed during this three- to four-month period.

Temperature controls the progression from the vegetative stage to flowering (14, 34, 35). Differentiation of flower primordia for the spring flowering group occurs at moderately warm temperatures in late summer or early fall either in the ground or in storage. Subsequent exposure to lower but above-freezing temperatures is required to promote flower stalk elongation.

As temperatures increase in the spring, the flower stalk elongates and the bulb plant subsequently flowers. Bulbous iris is an exception in that flower induction is induced by low temperatures in fall and early winter. For species in groups II (i.e., *Narcissus*) and III (i.e., *Tulipa*), induction occurs in spring after the storage period.

The optimum temperature—determined by the shortest time in which the bulb would flower—has been established for these developmental phases of important bulb species, such as tulip (*Tulipa*) (14), hyacinth (*Hyacinthus*) (12, 64), lily (*Lilium*) (91), and daffodil (*Narcissus*), depending on cultivar. Such information is important in establishing schedules for the greenhouse bulb forcing market (10) (Figs. 15–9, 15–10). Holding the bulbs continuously at high temperatures [30 to 32°C (86 to 90°F) or more], or at temperatures near freezing, will inhibit or retard floral development and can be used to lengthen the period required for flowering. Flower bud development will continue with a shift to favorable temperatures. This treatment (below or above optimum) can be used when shipping bulbs from the Northern to Southern Hemisphere.

Summer-Flowering Bulbs

Lilies are important plants with a growth cycle geared to the seasonal pattern of the summer-winter patterns of the temperate zone. Their nontunicate bulbs do not go "dormant" in late summer and fall as does the tunicate type of bulb, but have unique characteristics that must be understood for proper handling. Although different lily (*Lilium*) species have somewhat different methods of reproduction (74, 95), the pattern, as determined for the Easter lily (*Lilium longiflorum*), can serve as a model (15, 18, 57, 90).

Lilies flower in the late spring or early summer at the apex of the leaf-bearing stem axis. The flower-producing bulb is known as the **mother bulb** and is made up of the basal plate, fleshy scales, and the flowering axis (Figs. 15–6, 15–10, and 15–11). Prior to flowering, a new daughter bulb(s) is developing within the mother bulb. It was initiated the previous fall and winter from a growing point in the axil of a scale at the base of the stem axis. During spring the daughter bulb initiates new scales and leaf primordia at the growing point. Natural chemical inhibitors in the daughter scales prevent elongation of the daughter axis, which remains dormant—but it can be promoted to grow by exposure to high temperature (37.5°C, 100°F), to low temperature (4.5°C, 40°F), or by treatment with gibberellic acid (89).

> **mother bulb** A cluster of bulbs still attached at the basal plate.

(a) (b) (c) (d)

Figure 15–9
Commercial rooting room for the greenhouse bulb forcing market. (a) Rooting room initially warm enough [9°C (48°F)] to allow bulb rooting, (b) later reduced to 5°C (41°F) to promote breaking dormancy of bulbs, reduced growth, and slowed stem elongation, (c and d) later moved to greenhouse for forcing.

(a) (b)

Figure 15–10
In the production of a lily crop, (a) producers count leaf emergence from potted bulbs (pot and soil removed) and use this as a gauge to speed up or slow down the production cycle, in (b) timing flowering plants for certain market periods. Photograph (a) courtesy A. E. Nightingale.

(a)

(b)

Figure 15–11
Easter lily propagation is exclusively done in the U.S. Pacific Northwest. Individual scales are directly planted into raised beds. Each scale (arrow) produces multiple plants in (a and b) dense rows.

After flowering of the mother bulb, no more scales are produced by the daughter bulb, but it increases in size (circumference) and weight until it equals the weight of the mother bulb surrounding it. Inhibitory effects of the daughter scales decrease, as does the response to dormancy-breaking treatments (89). Fleshy basal roots persist on the mother bulb through the fall and winter, and new adventitious roots develop in late summer or fall from the basal plate of the new bulb that has formed above the mother bulb. Warm temperatures promote root formation (15), so bulbs should be dug for transplanting after they "mature" in the fall; the top may or may not have died down. Bulbs should be handled carefully to avoid injury and to prevent drying. The commercial value of the bulb depends on size (transverse circumference) and weight at the time of digging (47), and on the condition of the fleshy roots and the bulb's freedom from disease.

Transition of the meristem to a flowering shoot does not take place until the stem axis has protruded through the "nose" of the bulb and the shoot is *induced* with chilling temperatures (44). The critical temperatures are 15.5 to 18.5°C (60 to 65°F) or less, and the chilling effect becomes most effective at 2 to 4.5°C (35 to 40°F). The cold period requirement to induce flowering is known as **vernalization.** The shoot emerges 10 to 15 cm (4 to 6 in) above ground after chilling and flower primordia start to develop. Storing bulbs at warm (21°C, 70°F or more) or low (−0.5°C, 31°F) temperatures will keep bulbs dormant and delay blooming (83). Moisture content of the storage medium is important: if too dry, the bulb will deteriorate; too wet, and it will decay.

vernalization The biological process in which flower primordia are induced by exposure to a period of chilling.

Following the flower induction stage, and with the onset of higher temperatures, the stem elongates, initiating first leaves, and then flowers. The outer scales of the "old" mother bulb rapidly disintegrate early in the spring as the new mother bulb produces the flower.

Stem bulblets may develop in the axils of the leaves underground or, in some species, bulbils may develop above ground. They appear at about the time of flowering.

Flowering time and the size and quality of the Easter lily (*Lilium longiflorum*) bloom can be closely regulated by manipulating the temperature at various stages following the digging of bulbs in the fall (18). Daylength also influences development, but to a lesser extent (91).

Tender, Winter-Flowering Bulbs

A number of flowering bulb species originating from geographical areas such as South Africa have growth cycles related to a wet-dry, rather than a cold-warm climatic cycle. The amaryllis (*Hippeastrum*) is an example (Fig. 15–5) (34, 65). This bulb is a perennial, growing continuously from the center with the outer scales disintegrating. New leaves are produced continuously from the center during the vegetative period, which extends from late winter to the following summer. A meristem is initiated in the axil of every fourth leaf (or scale) that develops. Thus, throughout the vegetative period, a series of vegetative offsets is produced. By fall, the leaves mature and the bulb becomes dormant, during which time the bulb should be dry. In this period, the *fourth* growing point from the center and any external to it differentiate into flower buds, and the shoot begins to elongate slowly. After 2 or 3 months of dry storage, the bulbs can be watered, which will cause the flowering shoots to elongate rapidly, and flowering will take place in midwinter. Maximum foliage development and bulb growth are essential to produce a bulb large enough to form a flowering shoot.

Commercial production of flower bulbs continues to expand into regions with warm climates and cheaper labor (39). While species such as *Hippeastrum, Narcissus, Anenmone, Ranunculus,* and *Ornithogalum* are adapted to relatively high temperatures, production of thermo-periodic bulbs with chilling requirements (*Tulipa, Allium, Eremurus*) is commercially done in warm-climate areas, such as Israel.

Propagation

Offsets Offsets, which are daughter bulbs still attached to the main bulb, are used to propagate many kinds of bulbs (Fig. 15–12). This method is sufficiently

(a)

(b)

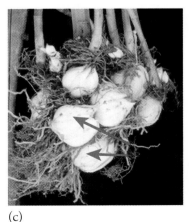
(c)

Figure 15–12
Propagation by separation of smaller, offset bulbs (arrows) in (a) daffodil (*Narcissus*), (b) tulip, and (c) lily.

rapid for the commercial production of tulip (*Tulipa*), daffodil (*Narcissus*), bulbous iris (*Iris*), and grape hyacinth (*Muscari*), but is too slow for the lily (*Lilium*), hyacinth (*Hyacinthus*), and *Amaryllis*.

If undisturbed, the offsets may remain attached to the mother bulb for several years. They can also be removed at the time the bulbs are dug and replanted into beds or nursery rows to grow into flowering-sized bulbs, although it may require several growing seasons, depending on the kind of bulb and size of the offset.

Tulip Tulip bulb planting takes place in the fall (14, 45). Two systems of planting are used: the bed system, used extensively in the Netherlands; and the row or field system, used mostly in the United States and England (Figs. 15–13 and 15–14). Beds are usually 1 m (3 ft) wide and separated by 31- to 45-cm (12- to 18-in) paths. The soil is removed to a depth of 9 cm (4 in), the bulbs set in rows 15 cm (6 in) apart, and the soil replaced. In the other system, single or double rows are placed wide enough apart to permit the use of machines. To improve drainage, two or three adjoining rows may be planted on a ridge. Bulbs are spaced 1 to 2 diameters apart, with small bulbs scattered along the row. A mulch may be applied after planting, but it should be removed the following spring before growth.

Planting stock consists principally of bulbs of the minimum size for flowering, 9 to 10 cm (3.6 to 4 in) or smaller in circumference. Since the time required to produce flowering size varies with the size of the bulb, the planting stock is graded so that all those of one size can be planted together. For instance, an 8-cm (3.2-in) or larger bulb normally requires a single season to become flowering size; a 5- to 7-cm (2- to 3-in) bulb, two seasons; and those 5 cm or less, three years (14).

During the flowering and subsequent bulb growing period of the next spring, good growing conditions should be provided so that the size and weight of the new bulbs will be at a maximum. Foliage should not be removed until it dries or matures. Important cultural operations include removal of competing weed growth, irrigation, fungicidal sprays to control *Botrytis* blight (8, 23), and fertilization. Beds should be inspected for disease early in the season and for trueness-to-cultivar at the time of blossoming. All diseased or off-type plants should be rogued out (32). It is desirable to remove the flower heads at blooming time, because they may serve as a source of *Botrytis* infection and can lower bulb weight.

Bulbs are dug in early to midsummer when the leaves have turned yellow or the outer tunic of the bulb has become dark brown in color. In the Pacific Northwest of the United States, where summer temperatures are cool and the leaves remain green for a longer period, digging may take place before the leaves dry. If the bulbs are dug too early or if warm weather causes early maturation, the bulbs may be small in size. In the United States and Holland, the bulbs are dug by machine (Figs. 15–13 and

(a)

(c)

(b)

Figure 15–13
Mechanical digging and harvesting tunicate bulbs in U.S. Pacific Northwest. (a) Mechanized digger harvesting bulbs. (b) Dug bulbs are then transferred up to a truck to be hauled to the processing area. (c) Harvesting process with workers trailing to collect missed bulbs.

Figure 15–14
Commercial propagation of tunicate bulbs in the Netherlands. (a) Flowers are cut to eliminate disease and competition to the bulbs. (b) Bulbs are dug and left to dry in the field. (c) Bulbs are further dried in open storage racks. (d and e) Grading bulbs into salable grades and offset grades for next season's production. (f) Storage of graded bulbs.

15–14). After the loose soil is shaken from the bulbs, they are placed in trays in well-ventilated storage houses for drying, cleaning, sorting, and grading. General storage temperatures are 18 to 20°C (65 to 68°F). To force early flowering, the bulbs should be held at 20°C for 3 to 5 weeks and then placed at 9°C (48°F) for 8 weeks. Later flowering can be produced by holding bulbs at 22°C (72°F) for 10 weeks. For shipment from the Northern to the Southern Hemisphere, the bulbs can be held at −1°C (31°F) until late December, when they are shifted to a higher temperature (25.5°C, 78°F) (12).

Daffodil Daffodil (*Narcissus*) bulbs are perennial and produce a new meristem growing point at the center every year (25, 30). Offsets are produced that grow in size for several years until they break away from the original bulb, although they are still attached at the basal plate. An offset bulb, when it first separates from the mother bulb, is known as a **split, spoon,** or **slab,** and can be separated from the mother bulb and planted. Within a year it becomes a **round,** or **single-nose,** bulb containing a single flower bud. One year later a new offset should be visible, enclosed within the scales of the original bulb, indicating the presence of two flower buds. At this stage the bulb is known as a **double-nose.** By the next year the offsets split

split An offset bulb when it first separates within the mother bulb.

spoon A synonym of split bulb.

slab A synonym of split bulb.

round A one-year-old mother bulb with a round shape.

single-nose A round bulb with a single flower bud.

double-nose A two-year-old bulb with two flower buds.

away and the bulb becomes a mother bulb. Grading of daffodil bulbs is principally by age, that is, as splits, round, double-nose, and mother bulbs. The grades marketed commercially are the round and the double-nose bulb. The mother bulbs are used as planting stock to produce additional offsets, and only the surplus is marketed. Offsets, or splits, are replanted for additional growth.

Storage should be at 13 to 16°C (55 to 60°F) with a relative humidity of 75 percent. To force earlier flowering, bulbs can be stored at 9°C (48°F) for 8 weeks. To delay flowering, store at 22°C (72°F) for 13 to 15 weeks. For shipment from the Northern to the Southern hemisphere, the bulbs can be held at 30°C (86°F) until October, then stored at −1°C (31°F) until late December, and then at 25°C (77°F) (12).

Hot-water treatment plus a fungicide for stem and bulb disease and nematode control is important (31). A three- or four-hour treatment at 43°C (110°F) is used, but that temperature must be carefully maintained or the bulbs may be damaged.

Lilies Lilies increase naturally, but except for a few species, this increase is slow and of limited propagation value except in home gardens (74, 95). Several methods of bulb increase are found among the different species.

bulb splitting
A unique bulb structure of certain *Lilium* species in which two to four bulbs are initiated at the base of the bulb.

budding-off Another unique bulb structure of certain *Lilium* species in which new bulbs form at the end of a rhizome-like structure.

For instance, *Lilium concolor, L. hansonii, L. henryi,* and *L. regale* increase by **bulb splitting** (Fig. 15–12). Two to four lateral bulblets are initiated about the base of the mother bulb, which disintegrates during the process, leaving a tight cluster of new bulbs. *Lilium bulbiferum, L. canadense, L. pardalinum, L. parryi, L. superbum,* and *L. tigrinum* multiply from lateral bulblets produced from the rhizome-like bulb. This process is sometimes called **budding-off.**

Bulblet Formation on Stems

Underground stem bulblets are used to propagate the Easter lily (*Lilium longiflorum*) and some other lily species (Fig. 15–12). In the field, flowering of the Easter lily (*Lilium longiflorum*) occurs in early summer. Bulblets form and increase in size from spring throughout summer (73). Between mid-August and mid-September in the Northern Hemisphere the stems are pulled from the bulbs and stacked upright in the field. Sprinkling periodically keeps the stems and bulblets from drying out. Similarly, the base of the stem can be **"heeled in"** the ground at an angle of 30 to 45 degrees or laid horizontally in trays at high humidity.

About mid-October the bulblets are planted in the field 10 cm (4 in) deep and an inch apart in double rows spaced 91 cm (36 in) apart. Here they remain for the following season. They are dug in September as yearling bulbs and again replanted, this time 12.5 cm (6 in) deep and 10 to 12.5 cm (4 to 6 in) apart in single rows. At the end of the second year, they are dug and sold as commercial bulbs.

Digging is done in September after the stem is pulled. The bulbs are graded, packed in peat moss, and shipped. Commercial bulbs range in size from 17.5 to 20 cm (7 to 10 in) in circumference. Lily bulbs must be handled carefully so they will not be injured, and must be kept from drying out. The fleshy roots should also be kept in good condition. For long-term storage that will prevent flowering, the bulbs should be packed in polyethylene-lined cases with peat moss at 30 to 50 percent moisture and stored at −1°C (31°F) (83).

Control of viruses, fungal diseases, and nematodes during propagation are important factors in bulb production. Methods of control include using pathogen-free stocks for propagation (3, 8, 73), micropropagation (41), growing plants in pathogen-free locations with good sanitary procedures, and treating the bulbs with fungicides.

Aerial stem bulblets (bulbils) are formed in the axil of the leaves of some species, such as *Lilium bulbiferum, L. sargentiae, L. sulphureum,* and *L. tigrinum* (Fig. 15–3). Bulbils develop in the early part of the season and fall to the ground several weeks after the plant flowers. Bulbils are harvested shortly before they fall naturally and are handled in essentially the same manner as underground stem bulblets. Increased bulbil production can be induced by disbudding as soon as the flower buds form. Likewise, some lily species that do not form bulbils naturally can be induced to do so by pinching out the flower buds and a week later cutting off the upper half of the stems. Species that respond to the latter procedure include *Lilium candidum, L. chalcedonicum, L. hollandicum, L. maculatum,* and *L. testaceum* (74).

Bulbil formation rarely occurs in Easter lily (*L. longiflorum*). Exogenous cytokinin application to foliage induced large numbers of bulbils in leaf axils on above-ground stems (62), which might be used for Easter lily propagation.

Stem Cuttings Lilies may be propagated by stem cuttings if the cutting is made shortly after flowering. Instead of roots and shoots forming on the cutting, as would occur in other plants, bulblets form at the axils of the leaves and then produce roots and small shoots while still on the cutting.

Leaf-bud cuttings, made with a single leaf and a small heel of the old stem, may be used to propagate a number of lily species. A small bulblet will develop in the axil of the leaf. It is handled in the same manner as the other methods described here.

Bulblet Formation on Scales (Scaling)

Scaling is a method where individual bulb scales are removed from the mother bulb, placed in appropriate growing conditions, and adventitious bulblets form at the base of each scale (Figs. 15–3c, 15–15). Since three to five bulblets will develop from each scale, the method is particularly useful to rapidly build up stocks of a new cultivar or to establish pathogen-free stocks. Almost any lily species can be propagated by scaling (50, 51, 88). Commercial-sized Easter lily bulbs are produced in two to three growing seasons. The usual sequence of development during scale propagation is: (a) callus initiation, (b) organized meristem differentiation, (c) leaf primordia formation, (d) development and enlargement of the bulblet, and (e) leaf emergence from the primordia (50).

> **scaling** A propagation procedure in which individual scales are removed and placed under appropriate conditions to cause the formation of adventitious bulblets.

Scaling is done soon after flowering in midsummer, although it might be done in late fall or even in midwinter. The bulbs are dug, the outer two layers of scales removed, and the mother bulb replanted for continued growth. Removing all of the scales to the core will reduce subsequent growth of the mother bulb. The scales should be kept from drying, and handled to avoid injury. Scales with evidence of decay should be discarded and the remaining dusted with or dipped into a fungicide. Naphthaleneacetic acid (1 ppm) will stimulate bulblet formation.

Scales are handled by the following methods:

- **Field planting.** Scales are planted in outdoor beds or frames no more than 6.25 cm (2 1/2 in) deep. Bulblets form on the scale during the first year to produce yearlings. The bulblets are replanted two more times, the third year producing commercials.
- **Incubation in trays or flats.** Scales are inserted vertically to about half their length in moist sand, peat moss, sphagnum moss, or vermiculite for six weeks at 18 to 21°C (65 to 70°F). Small bulblets and roots should form at the base within 3 to 6 weeks. The scales are transplanted either into the open ground or into pots or flats of soil, and then planted in the field the following spring. Subsequent treatment is the same as described for underground bulblets.
- **Plastic-lined boxes.** Scales may be packed in layers in moist vermiculite and incubated for 6 to 12 weeks at 15 to 26°C (60 to 80°F). The lower temperature encourages rooting. The boxes are placed in cold temperature over winter and planted in rows in the spring. Two years is required to reach planting size (54).
- **Polyethylene bags.** A simple method of propagating lilies by scales is to remove scales from the bulb, dust them with a fungicide, and place them in the bag in damp vermiculite so they do not touch. The bag is closed, tied, and maintained for 6 to 8 weeks where the temperature is about 21°C (70°F). After bulblets are well developed at the base of the scales, the bag with the scales still inside is refrigerated at 2 to 4.5°C (35 to 40°F) for at least eight weeks to overcome dormancy. The small bulblets can then be potted and placed in the greenhouse or out-of-doors for further growth.
- **Detached scale propagation.** This procedure carried out in the Netherlands involves scaling with preformed bulblets attached (88). Commercial bulbs for forcing are produced after only one growing season. Four types of plant development follow but only the **epigeous-type bulblet** is acceptable (Fig. 15–16).

> **detached scale propagation** A special propagation procedure in which the bulblet is produced before the scale is detached.
>
> **epigeous-type bulblet** A type of development that directly produces flower buds from a detached scale.

Several factors can affect scaling success. Longer bulb storage prior to scaling will decrease the harvest

Figure 15–15
Propagation of lily by scaling. (a) Adventitious bulblets form at base of scale (arrow), and (b) plantlet formed from bulb scale. Photograph courtesy R. A. Criley.

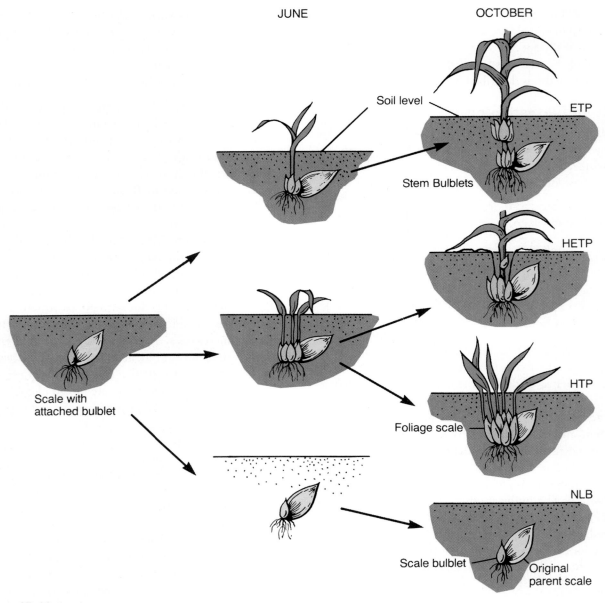

Figure 15–16
Types of development during scale propagation with attached bulblets. ETP, epigeous-type plant (bulblet flowers and produces the most desirable form); HETP, hypo-epigeous-type plant (rosettes first, then flowers); HTP, hypogeous-type plant (only a rosette with foliage leaves forms—not commercially desirable); NLB, non-greenleaf bulblet (bulblet remains dormant and does not flower—not desirable). Redrawn from Van Tuyl (88).

weight of newly generated bulbs (50). Outer or middle scales result in increased bulb weight and number of forcible commercial bulbs; innermost scales produce lower weights and few forcible commercial bulbs.

Tissue culture utilizing aseptic techniques has been used as a means of scaling, particularly to obtain, maintain, and multiply virus-free stock (see Chapters 17 and 18 and Fig. 15–17).

Since a bulb is a compressed shoot system, apical dominance limits axillary bud growth. Removing apical dominance, as in scaling, allows bulblets to form based on the same principle as in pruning an above-ground shoot structure. The bulbous iris cultivar 'Ideal' is an exception in that, although apical dominance does not limit numbers of bulblets formed, it does prevent lateral bud sprouting (19).

Basal Cuttage

The **basal cuttage** technique was developed for hyacinth, which has a slow natural increase, but other bulb species, such as *Scilla*, can also

basal cuttage The practice of cutting into the base of a bulb to stimulate adventitious bulblet formation on the base of scales.

Figure 15–17
Propagation of *Crinum*. (a) Mother bulb and single offset daughter bulblet (arrow). (b) Tri-scales with attached basal plate and elongating bulblets (arrow). (c) New bulblets will be (b) divided, subcultured, and multiplied via tissue culture (85).

scooping The basal cuttage carried out by cutting away the basal plant with a special scoop-like device.

be handled in this way. Specific methods are used to eliminate apical dominance and stimulate bulblet formation (12, 26). In **scooping,** the basal plate of a mature bulb is scooped out (Fig. 15–18) with a special curve-bladed scalpel, a round-bowled spoon, or a small-bladed knife; the cut should be deep enough to destroy the main shoot. Adventitious bulblets develop from the base of the exposed bulb scales. In **scoring,** three

scoring The basal cuttage carried out by cutting at right angles across the base of the bulb.

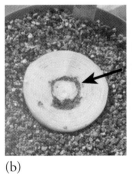
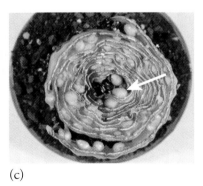
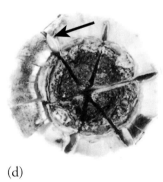
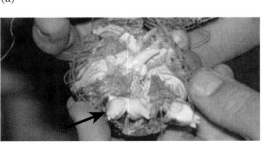

Figure 15–18
Propagation of bulbs by (a) splitting *Eucomis* bulb in half. (b) Scooping of *Eucomis* basal plate (arrow) and (c) later formation of multiple bulblets (arrow). (d) Scoring through the basal plate of hyacinth led to bulblet formation (arrow), and (e and f) multiple bulblets forming. Photographs (a), (b), and (c) courtesy R. A. Criley; photographs (e) and (f) courtesy W. B. Miller.

straight knife cuts are made across the base of the bulb (Fig. 15–18), each deep enough to go through the basal plate and the growing point. Growing points in the axils of the bulb scales grow into bulblets. Mature bulbs, 17 to 18 cm or more in circumference, which had been dug after the foliage had died down, are used.

Infected bulbs should be discarded to combat later decay, the tools disinfected frequently (with alcohol or mild carbolic acid solution, for instance), and the cut bulbs dusted with a fungicide. The bulbs are callused by placing them in dry sand or soil or in open trays, cut side down, at about 21°C (70°F) for a few days to a few weeks. After callusing, the bulbs are incubated in trays or flats, in dark or diffuse light, at 21°C (70°F), which is increased to 29.5 to 32°C (85 to 90°F) for 2 weeks. Humidity should be high (85 percent) for 2 1/2 to 3 months.

The mother bulbs are planted about 10 cm (4 in) deep in nursery beds in the fall. The next spring, bulblets produce leaves profusely. Normally, the mother bulb disintegrates during the first summer. The size-graded bulblets must be dug and replanted annually until they reach flowering sizes. Bulbs for greenhouse forcing should be 17 cm (6 3/4 in) or more in circumference; bulbs for bedding should be 14 to 17 cm (5 1/2 to 6 3/4 in) in circumference (2). On the average, a scooped bulb will produce 60 bulblets, but 4 to 5 years will be required to produce flowering sizes. A scored bulb will produce 24, requiring 3 to 4 years (12).

Hot-water treatment of hyacinth bulbs used for controlling *Xanthomonas hyacinthi* has been reported to induce bulblet formation and could substitute for basal cuttage (1). Bulbs must be at least 1.5 cm in circumference and treated from mid-July to early September. Treatment is 43°C (110°F) for 4 days or 38°C (100°F) for 30 days at relative humidity of 60 to 70 percent.

Leaf Cuttings

Leaf cuttings (also see Chapter 10) are successful for blood lily (*Haemanthus*), grape hyacinth (*Muscari*), hyacinth, and cape cowslip (*Lachenalia*) (20), although the range of species is probably wider than these few examples.

Leaves are taken when they are well developed and green. An entire leaf cut from the top of the bulb may in turn be cut into 2 or 3 pieces. Each section is placed in a rooting medium with the basal end several inches below the surface. The leaves should not be allowed to dry out, and bottom heat is desirable. Within 2 to 4 weeks small bulblets form on the base of the leaf, roots develop, and the bulblets are planted in soil.

Bulb Cuttings

Plants that respond to the **bulb-cutting** method of propagation include *Albuca, Chasmanthe, Cooperia, Haemanthus, Hippeastrum, Hymenocallis, Lycoris, Narcissus, Nerine, Pancratium, Scilla, Sprekelia,* and *Urceolina* (20).

> **bulb cutting** A method of propagation in which a bulb is cut into fragments of 3 or 4 bulb scales attached at the basal plate.

A mature bulb is cut into a series of 8 to 10 vertical sections, each containing a part of the basal plate. These sections are further divided by sliding a knife between each third or fourth pair of concentric scale rings and cutting through the basal plate. Each of these fractions makes a bulb cutting, and consists of a piece of basal plate and segments of 3 or 4 scales. This technique is also referred to as **bulb chipping,** or **fractional scale-stem cottage** (Fig. 15–18) (10).

> **bulb chipping** A synonym of bulb cutting.
>
> **fractional scale-stem cuttage** A synonym of bulb cutting.

The bulb cuttings are planted vertically in a rooting medium, such as peat moss and sand, with just their tips showing above the surface. Subsequent handling is the same as for ordinary leaf cuttings. A moderately warm temperature, slightly higher than for mature bulbs of that kind, is required. New bulblets develop from the basal plate between the bulb scales within a few weeks, along with new roots. At this time they are transferred to flats of soil to continue development.

A variation of this method, called **twin-scaling** (71, 79), involves dividing bulbs into segments, each containing a pair of bulb scales and a piece of basal plate. The segments are kept in plastic bags and incubated with damp vermiculite 3 to 4 weeks at 21°C (70°F) or planted in compost. Bulbils then develop at the edge of the basal plate (Fig. 15–19) (31).

> **twin-scaling** A variation of bulb cutting using a segment of two scales with a portion of the basal plate.

Bulb types most likely to produce a bulbil by twin-scaling consist of a third-year leaf base plus a second-year bulb scale, or scales and/or leaf bases. Bulb types with first-year organs or a flower stalk are less productive but can still form bulbils (29). Twin-scaling is important in the propagation of *Narcissus*.

BOX 15.1 **GETTING MORE IN DEPTH ON THE SUBJECT**
MICROPROPAGATION

Although the plant species covered in this chapter have special reproductive features that may make them relatively easy to propagate by traditional methods, most suffer from several important propagation problems. *First*, essentially all cultivars are clones, very vulnerable to systemic viruses and special selection procedures must be in place to obtain pathogen-free planting stock. *Second*, multiplication rates by natural division tend to be low. *Third*, seedlings have long juvenile periods before flowering. The slow rate of bulbs, corms, and other geophytes can be increased by micropropagation, utilizing enhanced axillary shoot proliferation, adventitious shoot formation, bulblet induction on scales, and sometimes, offshoots on flowerscapes (Table 15–2 and Fig. 15–17, page 575). Micropropagation is especially valuable to multiply new cultivars rapidly and to develop, maintain, and produce pathogen-free and virus-tested propagation sources. For example, millions of pathogen-free banana (*Musa*) plants are being produced around the world in micropropagation laboratories, particularly in developing countries (see Chapter 19). Micropropagation methods are described in Chapters 17 and 18.

Table 15–2
PARTIAL LIST OF GEOPHYTES CAPABLE OF TISSUE CULTURE PROPAGATION AND THEIR EXPLANT SOURCE

Family	Genus	Explant source				
		Bulb scale	Leaf	Stem	Bud	Flower petal
Liliaceae	*Lilium*	x	x	x	x	x
	Tulipa	x		x	x	x
	Hyacinthus	x	x	x	x	x
	Ornithogalum	x				x
	Muscari	x	x			x
	Fritillaria	x				
	Lachenalia	x				
	Alstroemeria			(Rhizome explants)		
Iridaceae	*Iris*	x		x	x	
	Gladiolus			x	x	x
	Freesia			x		
	Crocus				x	x
Amaryllidaceae	*Narcissus*	x		x	x	
	Hippeastrum	x	x			
	Nerine	x	x			

Source: Compiled from J. Van Aartrijk and P. C. G. Van der Linde (86); and DeHertogh and Le Nard (16).

Figure 15–19
Twin-scale propagation of *Narcissus* (daffodil) cut from whole bulbs and kept at 22°C in moist vermiculite for two months.
From A. R. Rees (71).

CORMS

Definition and Structure

A **corm** is a unique geophytic structure characteristic of certain important ornamentals, such as *Gladiolus* and *Crocus*. Here, the swollen base of the stem axis is enclosed by dry, scale-like leaves. A corm has a solid stem structure with distinct nodes and internodes. The bulk of corm consists of storage tissue composed of parenchyma cells. Dry leaf bases persist on the mature corm

corm A unique geophyte structure in which the base of the stem axis is swollen, has nodes and internodes, and is enclosed in a dry membranous tunic.

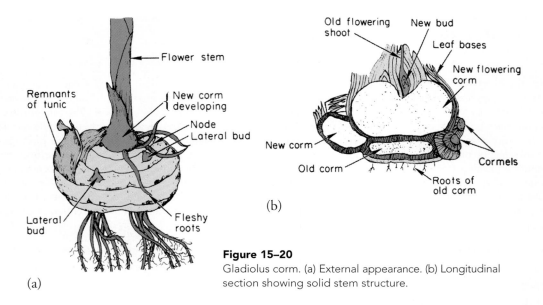

Figure 15–20
Gladiolus corm. (a) External appearance. (b) Longitudinal section showing solid stem structure.

attached to each node, enclosing the corm with a covering known as the **tunic,** which protects against injury and water loss. At the apex of the corm is a terminal shoot that will develop into the leaves and the flowering shoot. Axillary buds are produced at each of the nodes. In a large corm, several of the upper buds may develop into flowering shoots, but those nearer the base of the corm are generally inhibited from growing. Should something prevent the main buds from growing, lateral buds are capable of producing a shoot (see Figs. 15–20 and 15–21).

Two types of roots are produced from the corm: a *fibrous* root system developing from the base of the mother corm, and enlarged, fleshy *contractile* roots developing from the base of the new corm. The latter roots apparently develop in response to fluctuating temperatures near the soil surface, as well as exposure of the leaves to light (38). At lower soil depths, temperature fluctuations decrease and contraction ceases once the corm is at a given depth.

Growth Pattern

Gladiolus is semi-hardy to tender and, in areas with severe winters, the corm must be stored over winter and replanted in the spring (5, 11). At the time of planting, the corm is a vegetative structure (33, 66). New roots develop from its base, and one or more of the buds begin to develop leaves. Floral initiation takes place within a few weeks after the shoot begins to grow. At the same time the base of the shoot axis thickens, and a new corm for the succeeding year begins to form above the old corm. Stolon-like structures bearing miniature corms (**cormels**) on their tip develop from the base of the new corm.

> **cormel** A miniature corm produced on a short stolon from the base of a corm.

Gladiolus shows competition for assimilates between flower for the current year and corm development for the following year. Corm development is controlled by photoperiod (76). Short-day (SD) conditions stimulate while long-day (LD) conditions inhibit corm development. Final corm size and weight is a function of plant size, and larger plants are produced in LD

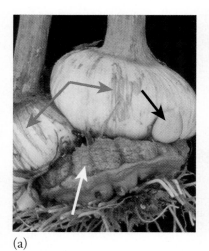
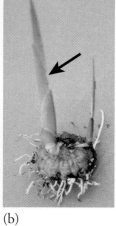

Figure 15–21
(a) Stage of gladiolus corm development during the latter part of the growing season. The remnants of the originally planted corm (white arrow) are evident just below the newly formed corms (blue arrow). A small white cormel has formed (black arrow). (b) Emerging flower stem of gladiolus (arrow).
Photograph (b) courtesy R. A. Criley.

conditions. Larger corm size and weight is apparently due to more total assimilates being available in the larger LD plants for corm production.

As new corms continue to enlarge, the old corm begins to shrivel and disintegrate as its contents are utilized in flower production. After flowering, the foliage continues to manufacture food materials, which are stored in the new corm. At the end of the summer, when the foliage dries, there are one or more new corms, and perhaps a great number of cormels.

Propagation

New Corms Propagation of cormous plants is principally by the natural increase of new corms. Flower production in corms, as in bulbs, depends on food materials stored in the corm the previous season, particularly during the period following bloom. In *Gladiolus,* cool nights and long growing periods are favorable for production of very large corms. Fertilization and other good management practices during bloom have their greatest effect on the next year's flowers. Plants are left in the ground for 2 months following blooming, or until frost kills the tops. After digging, the plants are placed in trays with a screen or slat bottom arranged to allow air to circulate between them, and cured at about 32°C (90°F) at 80 to 85 percent relative humidity. A few hours at 35°C (95°F) may be helpful. Then the new corms, old corms, cormels, and tops can be easily separated. The corms are graded according to size, sorted to remove any diseased ones, treated with a fungicide, and returned to a 35°C (95°F) temperature for an additional week. This curing process suberizes the wounds and helps combat *Fusarium* infection. The corms are then stored at 5°C (40°F) with a relative humidity of 70 to 80 percent in well-aerated rooms to prevent excessive drying. It may also be desirable to treat them with a suitable fungicide (48) immediately before planting.

Cormels Cormels are miniature corms that develop between the old and the new corms. One or two years' growth is required for them to reach flowering size. Shallow planting of the corms, only a few inches deep, results in greater production of cormels; increasing the depth of planting reduces cormel production.

Cormels are separated from the mother corms and stored over winter for planting in the spring. Dry cormels become very hard and may be slow to begin growth the following spring, but if they are stored at about 5°C (40°F) in slightly moist peat moss, they will stay plump and in good condition. Soaking dry cormels in cool running water for a day or two and keeping them moist until planting at the first sign of root development will hasten the onset of growth.

Pathogen-free cormels can be obtained by hot-water treatments, which should be done 2 to 4 months after digging (8). Holding cormels at room temperature to keep them dormant will increase tolerance to hot water treatment. The cormels are soaked in water at air temperature for 2 days, then placed in a 1:200 dilution of commercial 37 percent formaldehyde for 4 hours, and then immersed in a water bath at 57°C (135°F) for 30 minutes. At the end of the treatment, the cormels are cooled quickly, dried immediately, and stored at 5°C (40°F) in a clean area with good air circulation.

The cormels are planted in the field in furrows about 5 cm (2 in) deep in the manner of planting large seeds. Only grasslike foliage is produced the first season. The cormel does not increase in size but produces a new corm from the base of the stem axis, in the manner described for full-sized corms. At the end of the first growing season, the beds are dug and the corms separated by size. A few of the corms may attain flowering size, but most require an additional year of growth. The seven size grades in gladiolus are determined by diameter: the smallest, 0.9 to 1.2 cm (1/8 to 1/2 in) in diameter; the largest, 5 cm (2 in) or more (2).

Division of the Corm Large corms can be cut into sections, retaining a bud with each section; each of these should develop a new corm. Segments should be dusted with a fungicide because of the great likelihood of decay of the exposed surfaces (53).

TUBERS

Definition and Structure

A **tuber** is a special kind of swollen, modified stem structure that functions as an underground storage organ (Fig. 15–22). The potato (*Solanum tuberosum*) is a notable example of a tuber-producing plant, as is the *Caladium,* grown for its striking foliage, and the Jerusalem artichoke (*Helianthus tuberosus*).

tuber A swollen modified stem with nodes and internodes, which functions as a storage structure as well as an organ of vegetative propagation.
"eyes" The clusters of buds at the nodes of the potato tuber.

A tuber has all the parts of a typical stem but is very much enlarged. Externally, **eyes** are present as nodes, each containing one or more small buds subtended by a leaf scar. The arrangement of the eyes is a

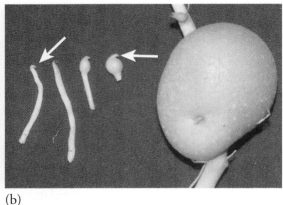

(a) (b)

Figure 15–22
Tubers of potato (*Solanum tuberosum*) showing their development from (a) white stolons arising from stem tissue; roots are darker, thinner; the tuber is attached to the stolon the tuber's morphological basal (proximal) end (arrow). (b) Tuberization (tuber formation) is characterized by the hook "gancho" at the subapical portion of the stolon (arrows) and subsequent tuber enlargement.

spiral, beginning with the terminal bud on the end opposite the scar resulting from the attachment to the stolon. The terminal bud is at the apical (distal) end of the tuber, oriented farthest from the crown of the plant. Consequently, tubers show apical dominance at any stem. Apical dominance also occurs with tuber pieces of yam (*Dioscorea alta*) where shoot production is much greater in locations at the proximal end than at the distal end of the tuber (67).

A tuber is highly nutritious and composed of enlarged parenchyma-type cells containing large amounts of starch, nitrogen, and small amounts of protein. It has the same internal structure as any stem with a pith, vascular system, and cortex.

Growth Pattern

The tuber is a storage structure that is produced in one growing season, remains dormant during the winter, and then functions to regenerate new shoots the following spring. After a new seasonal cycle begins, the shoots utilize the stored food in the old tuber, which then disintegrates (6). As the main shoot develops, adventitious roots are initiated at its base, and lateral buds grow out horizontally into the soil to produce elongated, etiolated stems (stolons) as shown in Figures 15–22 and 15–23. Continued elongation of the stolon takes place during long photoperiods and is associated with the presence of auxin and a high gibberellin level.

tuberization The biological process that leads to tuber formation.

Tuberization begins with inhibition of terminal growth and the initiation of cell enlargement and division in the subapical region of the stolons (Fig. 15–22). This process is associated with short or intermediate daylengths, reduced temperatures (particularly at night), high-light intensity, low mineral content, increased cytokinins and inhibitors (ABA), and reduction in gibberellin levels in the plant (9, 24, 56).

Tuberization is caused by the production of a tuber-inducing substance, which is linked to a tuberization regulatory protein that is produced in the leaves and the mother tuber (21, 63, 80). It seems to be necessary for the stolon tip to have attained a particular physiological age. Continued tuber enlargement is

(a)

(b)

Figure 15–23
Stolon (black arrow) and tuber (red arrow) production in *Cucurma* (ginger, Thai tulip) and *Phlomis* (Jerusalem sage).

dependent on a continuing adequate supply of photosynthate, but conditions that favor rapid and luxurious plant growth above ground, such as an abundance of nitrogen, or high temperatures, are not conducive to tuber production. In the fall, the tops of the plants die down and the tubers are dug. The buds of potato tubers are dormant for 6 to 8 weeks, a condition that must disappear before sprouting will take place.

Propagation

Division Traditionally, potatoes are propagated by planting tubers either whole or in pieces resulting from cutting into sections, each containing one or more axillary buds or eyes. These small pieces of tuber used for propagation are commonly referred to as **seed potatoes** (Fig. 15–24). The weight of the tuber piece should be 28 to 56 g (1 to 2 oz) to provide sufficient stored food for the new plant to become well established. In some areas only the eyes with a very small piece of tuber are sold and planted.

> **seed potato** The horticultural term applied to potato tubers when used for propagation.

Tubers are divided by machine or manually with a sharp knife shortly before planting. The cut pieces should be stored at warm (20°C, 68°F) temperatures and relatively high humidity (90 percent) for 2 to 3 days prior to planting. During this time the cut surfaces heal and become **suberized,** which protects the seed piece against drying and decay. Treatment of potato tubers prior to cutting for the control of *Rhizoctonia* and scab may be desirable.

> **suberization** The formation of suberin on the cut surface of a potato tuber as a wound-healing process.

Caladium tubers (75, 93) are cut into sections, usually 2 buds per piece. These sections are planted 7.6 to 9 cm (3 to 4 in) deep, 9 to 15 cm (4 to 6 in) apart in rows 45 to 60 cm (18 to 24 in) apart. After harvest, which begins in November, the tubers are dried in open sheds for 6 weeks or artificially dried for 48 hours. Further storage should be at temperatures above 16°C (60°F).

Tubercles *Begonia evansiana* and the cinnamon vine (*Dioscorea batatas*) produce small aerial tubers **(tubercles)** in the axils of the leaves. These tubercles are removed in the fall, stored over winter, and planted in the spring (20). Short days induce tuberization (63).

> **tubercles** The small aerial tubers produced in leaf axils of certain plant species.

TUBEROUS ROOTS AND STEMS

Definition and Structure

The tuberous root and stem class includes several types of structures with thickened tuberous growth that function as storage organs. Botanically, these differ from true tubers, although common horticultural usage sometimes utilizes the term *tuber* for all of them.

Fleshy and Tuberous Roots

Various herbaceous perennial species show massive enlargement of secondary roots. Typical examples are sweet potato (*Ipomoea batatus*) (Figs. 15–25 and 15–26, pages 583–84), cassava (*Manihot esculenta*), and *Dahlia* (Figs. 15–25 and 15–26). Sweet potato has a **fleshy root** from which both adventitious buds and roots are produced, while dahlia has a **tuberous root** with a

> **fleshy roots** Massive enlargement of a secondary root for carbohydrate storage and propagation exhibited by specific herbaceous perennial species.
>
> **tuberous root** The special swollen root system attached to the crown in specific herbaceous perennials.

(a) (b)

Figure 15–24
Tubers of potato (*Solanum tuberosum*). (a) "Seed potato," which is a vegetative propagule made of a diced, suberized tuber. The axillary buds ("eyes") later form shoots (arrow) and the root system arises from the newly forming shoot. (b) Minituber (arrow) produced from tissue culture, which facilitates rapid multiplication and exportation of disease-free propagules.

> **BOX 15.2 GETTING MORE IN DEPTH ON THE SUBJECT**
> **SPECIAL PROPAGATION SYSTEMS FOR POTATO (77, 82)**
>
> **Cuttings**
> Cuttings can be made of emerging sprouts of tubers, leaf-bud cuttings, and single-node cuttings of stems. These cuttings are handled as herbaceous cuttings. The technique is used for multiplication in "seed" production (see Chapter 16). Rooted cuttings are planted into the field to produce tubers about 50 to 70 g (0.18 to 0.24 oz) in size.
>
> **Micropropagation**
> Potato shoots can be propagated *in vitro* by standard shoot-tip culture procedures (Chapters 17 and 18). This procedure is used to multiply shoots as part of a system to produce "seed" potatoes. Their maintenance requires a specific medium.
>
> **Microtubers**
> **Microtubers** are very small tubers measuring 24 to 273 mg; 3 to 10 mm in diameter (1 to 10 mg; 0.12 to 0.4 in). They are produced on short stolons in aseptic culture. There are three stages in their production:
>
> > microtuber   The term that is applied to very small tubers (3 to 10 mm in diameter) that are produced directly on shoot tips in aseptic culture.
>
> > Stage I. Single-node cuttings are grown *in vitro* on a standard medium to produce rooted cuttings.
>
> > Stage II. New cuttings are made by removing tips and roots and placing them into a proliferation liquid media on a shaker or a bioreactor. Proliferation of axillary shoots occurs, and after about a month clusters with many buds form. These clusters can be divided and further multiplied in another cycle of the same conditions.
>
> > Stage III. Individual shoots are divided, placed on solid agar with a special tuberization media, and incubated for 8 weeks. Small microtubers result, ranging from 300 to 800 mg. These microtubers are used in seed potato multiplication systems.
>
> **Minitubers**
> **Minitubers** are small tubers ranging from 400 mg to 4.4 g (0.03 to 0.18 oz) and 5 to 25 mm (0.2 to 1 in) in size. They develop on *in vitro* plantlets *after* transplanting to growth media in the greenhouse. Following are three stages in their production:
>
> > minituber   The small tubers (5 to 25 mm in size) that are produced on tissue-cultured shoots after transplanting to the greenhouse.
>
> > Stage I. Multiplication of shoot-tip plantlets *in vitro* produces a given number of rooted shoots and developed stolons, which should remain intact with each multiplication.
>
> > Stage II. Rooted plantlets are transplanted to the greenhouse into a soil medium. Small tubers begin to develop within 2 to 3 weeks.
>
> > Stage III. Plants are lifted carefully after a month and tubers harvested, carefully removing them from stolons. A second harvest takes place after another 3 weeks. Final harvest of all minitubers takes place after an additional 3 weeks. These minitubers are then used in field plantings as the source material of "seed" potato production (Fig. 15–24).

section of the attached crown containing a preformed bud for shoot development. With dahlia, fibrous roots are commonly produced on the opposite (*distal*) end, and tuberous roots closer to the crown or stem (*proximal*) end.

Tuberous Stems **Tuberous stems** are produced by the enlargement of the hypocotyl section of the seedling plant, but may include the first nodes of the epicotyl and the upper section of the primary root (27, 37, 72). Typical plants with this structure are the tuberous begonia (*Begonia* × *tuberhybrida*) (28) and cyclamen (*Cyclamen persicum*) (Fig. 15–27, page 584). Tuberous stems have a vertical orientation with one or more vegetative buds produced on the upper end or crown; fibrous roots are produced on the basal part of the structure.

tuberous stem   The swollen stem structure that is produced by enlargement of the hypocotyl.

Growth Pattern

Tuberous roots are biennial. They are produced in one season, after which they go dormant as the herbaceous shoots die. The tuberous roots function as storage organs to allow the plant to survive the dormant period. In the following spring, buds from the crown produce new shoots, which utilize the food materials from the old root during their initial growth. The old root then disintegrates, and new tuberous roots are produced, which in turn maintain the plant through the following dormant period (46, 59).

Photoperiod, not temperature, is the dominant controlling factor of tuberization in dahlia (16). Tuberized roots are formed under short-day conditions (5 inductive cycles of an 11- to 12-hour critical photoperiod) or when a growth retardant is applied (59). Fibrous roots form under long-day conditions or when gibberellic acid is applied. Apparently, conditions favoring tuberous root growth are antagonistic to vegetative (shoot)

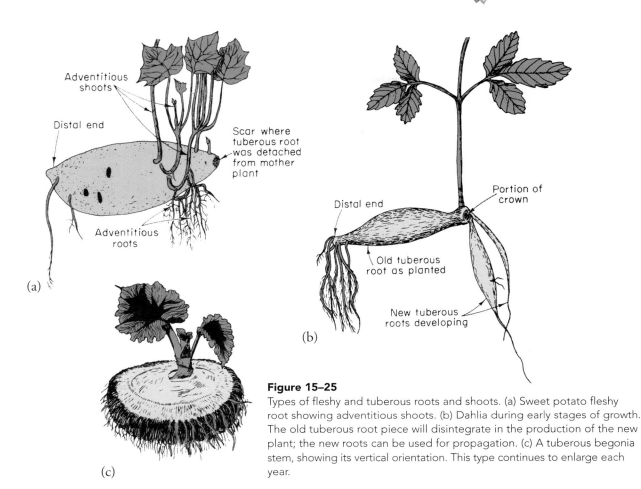

Figure 15–25
Types of fleshy and tuberous roots and shoots. (a) Sweet potato fleshy root showing adventitious shoots. (b) Dahlia during early stages of growth. The old tuberous root piece will disintegrate in the production of the new plant; the new roots can be used for propagation. (c) A tuberous begonia stem, showing its vertical orientation. This type continues to enlarge each year.

growth (59). Tuberous root formation in cassava is attributed to cytokinin control of meristematic activity (55).

Tuberous stems of tuberous begonia and cyclamen, on the other hand, are perennial and continue to enlarge laterally every year (17, 27). Normally, these species are commercially propagated by seed, but the tuber can be dug, stored, and used for annual propagation over a period of years. The commercial center for tuberous begonia is Belgium, while cyclamen is a common seasonal greenhouse crop.

Propagation

Division The usual method for propagating tuberous roots is by dividing the crown so that each section bears a shoot bud. *Dahlia,* for example, is dug with its cluster of roots intact, dried for a few days, and stored at 4 to 10°C (40 to 50°F) in sawdust or vermiculite. Open storage may result in shriveling. The root cluster is divided in the late winter or spring shortly before planting. In warm, moist conditions the buds begin to grow, and the tubers can be divided with assurance that each section will have a bud.

The tuberous stem of the tuberous begonia can be divided shortly after growth starts in the spring as long as each section has a bud. To combat decay, the cut surface should be dusted with a fungicide and each section dried for several days after cutting and before placing in a moist medium.

Adventitious Shoots The fleshy roots of a few species of plants such as sweet potato have the capacity to produce adventitious shoots if subjected to the proper conditions. The roots are laid in sand so that they do not touch one another, and covered to a depth of about 5 cm (2 in). The bed is kept moist. The temperature should be about 27°C (80°F) at the beginning and about 21 to 24°C (70 to 75°F) after sprouting has started. As the new shoots **(slips)** (Figs. 15–25 and 15–26)

> **slip** The term used to identify an adventitious shoot produced on a fleshy root.

come through the covering, more sand is added so that eventually the stems will be covered to a depth of 10 to 12.5 cm (4 to 5 in). Adventitious roots develop from the base of these adventitious shoots. After the slips are

nematodes and fungal diseases (92). This procedure can be modified in certain cultivars of the sweet potato by dividing the fleshy root into 20- to 25-g (7 to 8 oz) pieces, treating with a fungicide, then giving a pre-sprouting treatment for 4 weeks of 26.5°C (80°F) and 90 percent relative humidity before planting (7).

Cyclamen can be multiplied vegetatively by cutting off the upper one-third of the tuberous stem and notching the surface into 1-cm (0.16 in) squares. Adventitious shoots develop (12 to 13 per tuber), and can be used for propagation (61).

Stem Cuttings Vegetative propagation in plants of this group, such as *Dahlia* or tuberous begonia (*Begonia*), is often more satisfactory with stem, leaf, or leaf-bud cuttings. The cuttings will develop tuberous roots at their base. This process can be stimulated if the stem cutting initially includes a small piece of the fleshy root or stem. Vine cuttings from established beds can also be used in sweet potato (*Ipomoea batatus*) propagation.

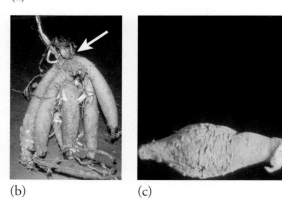

Figure 15–26
(a) Fleshy root of sweet potato with adventitious shoots or "slips" (arrow). (b) Tuberous roots of dahlia attached to crown (arrow). (c) Each separate tuberous dahlia root must have a section of the crown (arrow) bearing a bud that elongates into a shoot.

well rooted, they are pulled from the parent plant and transplanted into the field. If sweet potato roots are cut in half and the pieces are subjected to 43°C (110°F) for about 26 hours, slip production increases. This heat treatment overcomes the apical dominance and controls

Figure 15–27
A tuberous stem in cyclamen. (a) Tuber ready for transplanting. (b and c) Tuberous stems.

RHIZOMES

Definition and Structure

A **rhizome** is a specialized stem structure in which the main axis of the plant grows horizontally at, or just below, the ground surface. A number of economically important plants, such as bamboo, sugar cane, banana, and many grasses, as well as a number of ornamentals, such as rhizomatous *Iris* and lily-of-the-valley (*Convallaria*), have rhizome structures. Most are monocotyledons, although a few dicotyledons—for example, lowbush blueberry (*Vaccinium angustifolium*) and Nandina (*Nandina domestica*)—have analogous underground stems classed as rhizomes. Many ferns and lower plant groups have rhizomes or rhizome-like structures.

> **rhizome** A horizontally growing stem of specific plant species at, or near, the surface of the ground.

Figure 15–28 shows structural features of a rhizome (78). The lateral stem appears segmented because it is composed of nodes and internodes. A leaf-like sheath is attached at each node, which encloses the stem and, in an expanded form, becomes the foliage leaves. When the leaves and sheaths disintegrate, a scar is left at the point of attachment identifying the node and giving a segmented appearance. Adventitious roots and lateral growing points develop in the vicinity of the node. Upright-growing, above-ground shoots and flowering stems (**culms**) are produced

> **culm** The upright flowering stems produced on a rhizome.

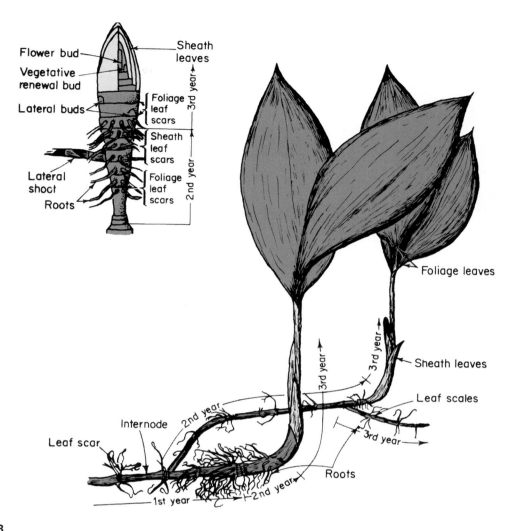

Figure 15–28
Structure and growth cycle of lily-of-the-valley (*Convallaria Mamalis*). *Right:* Section of rhizome as it appears in late spring or early summer with one-, two-, or three-year-old branches. A new rhizome branch begins to elongate in early spring and terminates in a vegetative shoot bud by the fall. The following spring, leaves of the bud unfold; food materials manufactured in the leaves by photosynthesis are accumulated in the rhizome. Growth the second season is again vegetative. Early in the third season a flower bud begins to form, and at the same time a vegetative growing point forms in the axil of the last leaf. *Top left:* Section of the three-year-old branch showing terminal flower bud and lateral shoot bud enclosed in leaf sheaths. Such a section is sometimes known as a *pip* or crown and is forced for spring bloom. In the early spring the flowering shoot expands, blooms, and then dies down, and the shoot bud begins a new cycle of development. Redrawn from Zweede (96).

either terminally from the rhizome tip or from lateral branches.

There are two general types of rhizomes (52). The first (**pachymorph**) is illustrated by *Sansevieria* (Fig. 15–29), rhizomatous *Iris* and by ginger (*Zingiber*) (Fig. 15–30). The rhizome is thick, fleshy, and shortened in relation to length. It appears as a many-branched clump made up of short, individual sections. It is **determinate;** that is, each clump terminates in a flowering stalk, growth continuing only from lateral branches. The rhizome tends to be oriented horizontally with roots arising from the lower side.

The second type of rhizome (the **leptomorph**) is illustrated by the lily-of-the-valley (Fig. 15–28). The rhizome is slender with long internodes. It

pachymorph The rhizome growth that exhibits determinate type of growth.

determinate A stem structure that terminates in a flowering stalk and then dies back, with new growth coming from lateral basal buds.

leptomorph The rhizome growth that exhibits indeterminate type of growth.

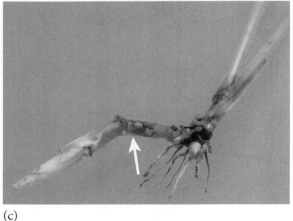
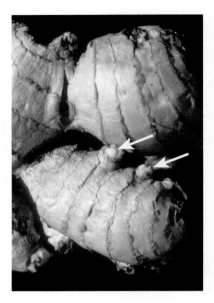

Figure 15–30
Rhizome of the tropical ginger plant (*Zingiber officinale*). This is easily propagated by division of the thickened rhizome, which is the source of commercial ginger. The axillary buds (arrow) will develop into the shoot system.

Figure 15–29
(a and b) Mother-in-law's tongue (*Sansevieria* spp.) and (c) Johnson grass (*Sorghum halepense*) are rhizomous plants that regenerate themselves clonally through rhizomes (arrows) from which shoots arise. Photo (c) courtesy R. A. Criley.

indeterminate A terminal bud that grows continuously in length from both terminal and lateral branches.

mesomorph The rhizome growth that is intermediate between determinate and indeterminate.

is **indeterminate;** that is, it grows continuously in length from the terminal apex and from lateral branch rhizomes. The stem is symmetrical and has lateral buds at most nodes, nearly all remaining dormant. This type does not produce a clump, but spreads extensively over an area.

Intermediate forms between these two types also exist. These are called **mesomorphs** (52).

Growth Pattern

Rhizomes grow by elongation of the growing points produced at the end of terminal and lateral branches. Length also increases by growth in the intercalary meristems in the lower part of the internodes. As the plant continues to grow and the older part dies, the several branches arising from one plant may eventually become separated to form individual plants of a single clone.

Rhizomes exhibit consecutive vegetative and reproductive stages, but the growth cycle differs somewhat in the two types described. In the pachymorph rhizome of *Iris*, a growth cycle begins with the initiation and growth of a lateral branch on a flowering section. The flowering stalk dies, but these new lateral branches produce leaves and grow vegetatively during the remainder of that season. Continued growth of the underground stem, storage of food, and the production of a flower bud at the conclusion of the vegetative period are dependent upon photosynthesis; consequently, foliage should not be removed during this period. A flowering stalk is produced the following spring and no further terminal growth can take place. In general, plants with this structure flower in the spring and grow vegetatively during the summer and fall.

Plants with a leptomorph habit as a general rule (with exceptions) grow vegetatively during the beginning of the growth period and flower later in the same period. The length of time during which an individual rhizome section remains vegetative varies with different kinds of plants. An individual branch of the lily-of-the-valley in Figure 15–28, for instance, is vegetative 3 years before a flower bud forms.

Bamboo is divided into clump growers (pachymorphs), which have constricted rhizomes, and running bamboos (leptomorphs), which spread rapidly by vigorous rhizomes that grow several feet or more (78). Generally speaking, the pachymorphs are more desirable for ornamental use. Some bamboo species remain in a vegetative juvenile phase for many years, but shift abruptly to a reproductive mature phase, and the entire plant produces flowers and often dies.

In some rhizomatous plants, such as blueberry (*Vaccinium*), rhizome development is increased by higher temperatures and a long photoperiod, and is correlated with vigorous above-ground growth and high-nitrogen status (40, 81).

Propagation

Division of Clumps and Rhizomes Division is the usual procedure for propagating plants with a rhizome structure, but the procedure may vary somewhat with the two types. In pachymorph rhizomes, individual culms are cut off at the point of attachment to the rhizome, the top is cut back, and the piece is transplanted to the new location. Leptomorph rhizomes can be handled in essentially the same way by removing a single lateral offshoot from the rhizome and transplanting it. The tip of the lily-of-the-valley (*Convallaria*) rhizome bearing a flower bud (Fig. 15–28), called a **pip**, is removed along with the rooted section below, and transplanted.

"pip" A piece of rhizome with a terminal flower bud produced by lily-of-the-valley (*Convallaria majalis*).

The bird of paradise (*Strelitzia reginae*) has a slow rate of multiplication when propagated by rhizome division. Mechanical induction of branching to eliminate apical dominance of a branch leaf sheath (fan) attached to the rhizome encourages lateral shoot formation and rapid multiplication (Fig. 15–31) (87). Division is usually carried out at the beginning of a growth period (as in early spring) or at or near the end of a growth period (in late summer or fall).

Propagation involves cutting the rhizome into sections, essentially stem cuttings, being sure that each piece has at least one lateral bud. Bananas (*Musa*) are also propagated in a similar way. Traditional clonal propagation of bananas is done with **suckers** or **pups** that contain a portion of the corm, rhizome, and roots from the mother plant (Fig. 15–32, page 589). However, tissue culture is the preferred method for large-scale, commercial banana production with disease-free, sterile triploid plantlets for new plantings.

Rhizome cuttings work well for the leptomorph rhizomes as long as a dormant lateral growing point is present at most nodes. The rhizomes are cut or broken into pieces, and adventitious roots and new shoots develop from the nodes. Rhizome-producing turf grasses are cut up into sections and the individual sprigs transplanted. New plants can be established readily by this method. The noxious weed, Johnson grass (*Sorghum halepense*), spreads by rhizomes (Fig. 15–29) and is difficult to eradicate.

Culm Cuttings In large rhizome-bearing plants, such as bamboos, the aerial shoot, or culm, may be used as a cutting. Whole culm cuttings are those in which the entire aerial shoot is laid horizontally in a trench. New branches arise at the nodes. Otherwise, a stem cutting of three- or four-node sections may be planted vertically in the ground.

PSEUDOBULBS

Definition and Structure

A **pseudobulb** (literally "false bulb") is a specialized storage structure, produced by many orchid species, consisting of an enlarged, fleshy section of the stem made up of one to several nodes (Fig. 15–33, page 589). In general, the appearance of the pseudobulb varies with the orchid species. The differences can be used to identify species.

pseudobulb An enlarged, special storage structure produced by many orchid species.

Growth Pattern

These pseudobulbs arise during the growing season on upright growths that develop laterally or terminally from the horizontal rhizome. Leaves and flowers form either at the terminal end or at the base of the pseudobulb, depending on the species. During the growth period, they accumulate stored food materials and water and assist the plants in surviving the subsequent dormant period.

Propagation

Offshoots In a few orchids, such as the *Dendrobium* species, the pseudobulb is long and jointed, and made up of many nodes. Offshoots develop at the nodes, and roots develop from the bases of the offshoots. The rooted offshoots are then cut from the parent plant and potted.

Division The most important commercial species of orchids, including the *Cattleya, Laelia, Miltonia,* and *Odontoglossum,* may be propagated by dividing the rhizome into sections, the exact procedure used being dependent on the particular kind of orchid. Division is

Figure 15–31

(a and b) Propagation of *Strelitzia reginae* (bird of paradise) by mechanical induction of branching to eliminate apical dominance and increase multiplication rate. To remove the apex, a transverse lateral incision is made above the basal plate through the basal leaf sheath of a branch, keeping the leaves in contact with the roots. (c) Lateral shoot formation occurs 4.5 months after excision of the apex. (d) After one year, multiple clusters with roots can be separated into (e) individual plantlets. Photos courtesy P. A. Van de Pol (87).

done during the dormant season and preferably just before the beginning of a new growth period The rhizome is cut back far enough from the terminal end to include 4 to 5 pseudobulbs in the new section. The old rhizome section is left with a number of old pseudobulbs, or **"back bulbs"** from which the leaves have dehisced. The section is then potted and new growth begins from the bases of the pseudobulbs and at the nodes. The removal of the new section of the rhizome from the old part stimulates new growth, or **"back breaks,"** to occur from the old parts of the rhizome.

back bulb The pseudobulbs that do not have foliage.

These new growths grow for a season and can be removed the following year.

An alternate procedure is to cut partly through the rhizome and leave it for 1 year. New back breaks will develop, which can be removed and potted.

back break The new growth that develops on a pseudobulb after separation from the plant.

Back Bulbs and Green Bulbs

Back Bulbs. These pseudobulbs, which do not have foliage, are commonly used to propagate clones of *Cymbidium* (Fig. 15–34). These are removed from the

Figure 15–32
Banana (*Musa* spp.) can be propagated by division of the rhizome (locally called corms) and suckers. (a) Underground stem structures consisting of a rhizome forming a sucker and nonrhizomatous sucker (*left*) (b) Propagation is done with suckers or pups that contain a portion of the rhizome and roots from the mother plant. (c) Banana inflorescence with fruit, (d and e) pseudostem of mother plant (arrow) which is cut after bananas are harvested, allowing for a new primary daughter shoot to become the dominant leader from which the next banana crop forms. Tissue culture is commonly used for large-scale, commercial banana production of disease-free, sterile triploid plantlets for new plantings. Photo (b) courtesy R. A. Criley.

Figure 15–33
Pseudobulbs (arrows) facilitate survival of orchids during adverse environmental conditions and can be divided-up and used as propagules with species such as (a) *Gongora quinquenervis*, (b) *Oncidium longifolium*, and (c) *Laelia anceps*.

Figure 15–34
Commercial propagation of cymbidium orchids from pseudobulbs. (a and b) Back and (c) green pseudobulbs.

> **BOX 15.3 GETTING MORE IN DEPTH ON THE SUBJECT**
> **MICROPROPAGATION**
>
> Commercial propagation of the major genera of orchids has been revolutionized by the biotechnological procedures of tissue culture. Aseptic seed germination was one of the first major steps introduced by Knudson (see Fig. 17–27, page 668) (42). Two problems accompanied this procedure: variability among seedlings and the very long juvenile period, up to 7 years. Commercial clonal production was limited because of very slow rates of increase and the widespread distribution of systemic viruses. Morel's discovery of vegetative proliferation of protocorms (58), combined with virus control and aseptic production, has revolutionized orchid production and led to mass-propagation of orchid cultivars (see Fig. 17–19, page 662).

plant, the cut end covered with grafting compound, and placed in a rooting medium for new shoots to develop. The shoot can be removed from the bulb and potted. The remaining back bulb can be repropagated and a second shoot will develop.

Green Bulbs. Green bulbs are pseudobulbs with leaves, which can also be used in *Cymbidium* propagation. Treatment with indolebutyric acid, either by soaking or by painting with a paste, has been beneficial (43).

green bulb Pseudobulbs that have foliage.

DISCUSSION ITEMS

1. Compare the morphology and life cycle of a corm vs. a bulb.
2. Compare the life cycle for a spring- vs. a fall-flowering bulb.
3. Contrast scaling and basal cuttage as propagation methods for bulbs.
4. What is the difference between a corm and a tuber?
5. Compare the morphology and life cycle in a tuber (potato) and a tuberous root (dahlia).

REFERENCES

1. Amano, N., and K. Tsutsui. 1980. Propagation of hyacinth by hot water treatment. *Acta Hort.* 109:279–87.

2. Amer. Assoc. Nurserymen, Inc., Commission on Horticulture Standards 1986. *American standard for nursery stock.* Washington, DC: Amer. Assoc. Nurs.

3. Baker, K. F., and P. A. Chandler. 1957. Development and maintenance of healthy planting stock. Sec. 13 in *Calif. Agr. Exp. Sta. Man. 23.*

4. Beytes, C., ed. 2003. *The Ball red book: Crop production.* Vol. 2. 17th ed. Chicago, IL: Ball Pub.

5. Benschop, M. 1993. *Crocus.* In A. DeHertogh and M. Le Nard, eds. *The physiology of flower bulbs.* Amsterdam: Elsevier. pp. 257–72.

6. Booth, A. 1963. The role of growth substances in the development of stolons. In J. D. Ivins and F. L. Milthorpe, eds. *The growth of the potato.* London: Butterworth. pp. 99–113.

7. Bouwkamp, J. C., and L. D. Scott. 1972. Production of sweet potatoes from root pieces. *HortScience* 7(3):271–72.

8. Byther, R. S., and G. A. Chastangner. 1993. Diseases. In A. DeHertogh and M. Le Nard, eds. *The physiology of flower bulbs.* Amsterdam: Elsevier. pp. 71–100.

9. Consultative Group on Intern. Agr. Res. 2001. CGIAR Research. Areas of research: *Potato (Solanum tuberosum).* www.cgiar.prg/areas.Potato.htm.

10. Christie, C. B. 1985. Propagation of amaryllids: A brief review. *Comb. Proc. Intl. Plant Prop. Soc.* 35:351–57.

11. Cohat, J. 1993. *Gladiolus.* In A. DeHertogh and M. Le Nard, eds. *The physiology of flower bulbs.* Amsterdam: Elsevier. pp. 297–320.

12. Crossley, J. H. 1957. Hyacinth culture; narcissus culture; tulip culture. *Handbook on bulb growing and forcing.* Northwest Bulb Growers Assoc., pp. 79–84, 99–104, 139–44.

13. Davies, F. T., Jr., D. E. Kester, and T. D. Davis. 1994. Commercial importance of adventitious rooting to horticulture. In Davis, T. D., B. E. Haissig, and N. Sankhla, eds. *Adventitious root formation in cuttings.* Portland, OR: Dioscorides Press.

14. DeHertogh, A. A., L. H. Aung, and M. Benschop. 1983. The tulip: Botany, usage, growth and development. *Hort. Rev.* 5:45–125.

15. DeHertogh, A. A., and N. Blakely. 1972. The influence of temperature and storage time on growth of basal roots of nonprecooled and precooled bulbs of *Lilium longiflorum* Thunb. cv. 'Ace'. *HortScience* 74:409–10.

16. DeHertogh, A. A., and M. Le Nard. 1993. *Dahlia*. In A. DeHertogh and M. Le Nard, eds. *The physiology of flower bulbs*. Amsterdam: Elsevier. pp. 273–84.

17. DeHertogh, A. A., and M. Le Nard. 1993. *The physiology of flower bulbs*. Amsterdam: Elsevier.

18. DeHertogh, A. A., A. N. Roberts, N. W. Stuart, R. W. Langhans, R. G. Linderman, R. H. Lawson, H. F. Wilkins, and D. C. Kiplinger. 1971. A guide to terminology for the Easter lily. *HortScience* 6:121–23.

19. Doss, R. P. 1979. Some aspects of daughter bulb growth and development and apical dominance in bulbous iris. *Plant Cell Physiol.* 20:387–94.

20. Everett, T. H. 1954. *The American gardener's book of bulbs*. New York: Random House.

21. Ewing, E. E. 1985. Cuttings as simplified models of the potato plant. In P. H. Li, ed. *Potato physiology*. New York: Academic Press.

22. Genders, R. 1973. *Bulbs*. New York: Bobbs-Merrill.

23. Gould, C. J. 1953. Blights of lilies and tulips. *Plant diseases: USDA yearbook of agriculture*. Washington, DC: U.S. Govt. Printing Office. pp. 611–16.

24. Gregory, L. E. 1965. Physiology of tuberization in plants (tubers and tuberous roots). In *Encyclopedia of plant physiology*, Vol. 15. Berlin: Springer-Verlag. pp. 1328–54.

25. Griffiths, D. 1930. Daffodils. *USDA Circ. 122*.

26. Griffith, D. 1930. The production of hyacinth bulbs. *USDA Circ. 112*.

27. Haegeman, J. 1979. *Tuberous begonias*. Vaduz: A. R. Gantner Verlag KG.

28. Haegeman, J. 1993. Begonia—Tuberous hybrids. In A. DeHertogh and M. Le Nard, eds. *The physiology of flower bulbs*. Amsterdam: Elsevier. pp. 227–39.

29. Hanks, G. R. 1985. Factors affecting yields of adventitious bulbils during propagation of *Narcissus* by the twin-scaling technique. *J. Hort. Sci.* 60(4):531–43.

30. Hanks, G. R. 1993. *Narcissus*. In A. DeHertogh and M. Le Nard, eds. *The physiology of flower bulbs*. Amsterdam: Elsevier. pp. 462–558.

31. Hanks, G. R., and A. R. Rees. 1979. Twin-scale propagation of *Narcissus*: A review. *Scientia Hort.* 10:1–14.

32. Harrison, A. D. 1964. *Bulb and corm production*. Bul. 62. London: Minist. Agr., Fish. and Foods. pp. 1–84.

33. Hartsema, A. M. 1937. Periodieke ontwikkeling *van Gladious hybridum* var. Vesuvius. *Verh. Koninkl. Ned. Akad. van Wet.* 36(3):1–34.

34. Hartsema, A. M. 1961. Influence of temperatures on flower formation and flowering of bulbous and tuberous plants. In *Encyclopedia of plant physiology*, Vol. 16. Berlin: Springer-Verlag. pp. 123–67.

35. Heath, O. V., and M. Holdsworth. 1948. Morphologenic factors as exemplified by the onion plant. *Symposia for the Soc. Exp. Biol.* II:326–50.

36. Huisman, E., and A. M. Hartsema. 1933. De periodieke ontwikkeling van *Narcissus pseudonarcissus L. Meded. Landbouwhoogesch., Wageningen*, DL. 37 (Meded. No. 38, Lab v. Plantenphys. onderz., Wageningen).

37. Jacobi, E. F. 1950. *Plantikunde voor tuinbouwscholen*. Zwolle, The Netherlands: W. E. J. Tjeenk Willink.

38. Jacoby, B., and A. H. Halevy. 1970. Participation of light and temperature fluctuations in the induction of contractile roots of gladiolus. *Bot. Gaz.* 131(1):74–7.

39. Kamenetsky, R. 2005. Production of flower bulbs in regions with warm climates. *Acta Hort.* 673:59–66.

40. Kender, W. J. 1967. Rhizome development in the lowbush blueberry as influenced by temperature and photoperiod. *Proc. Amer. Soc. Hort. Sci.* 90:144–48.

41. Kim, K., and A. A. DeHartogh. 1997. Tissue culture of ornamental flowering bulbs. *Hort. Rev.* 14:87–109.

42. Knudson, L. 1951. Nutrient solution for orchids. *Bot. Gaz.* 112:528–32.

43. Kofranek, A. M., and G. Barstow. 1955. The use of rooting substances in *Cymbidium* green bulb propagation. *Amer. Orch. Soc. Bul.* 24(11):751–53.

44. Langhans, R. W., and T. C. Weiler. 1968. Vernalization in Easter lilies. *HortScience* 3:280–82.

45. Le Nard, M., and A. A. DeHertogh. 1993. *Tulipa*. In A. DeHertogh and M. Le Nard, eds. *The physiology of flower bulbs*. Amsterdam: Elsevier. pp. 617–82.

46. Lewis, C. A. 1951. Some effects of daylength on tuberization, flowering, and vegetative growth of tuberous-rooted begonias. *Proc. Amer. Soc. Hort. Sci.* 57:376–78.

47. Lin, P. C., and A. N. Roberts. 1970. Scale function in growth and flowering of *Lilium longiflorum*, Thunb. 'Nellie White'. *J. Amer. Soc. Hort. Sci.* 95(5):559–61.

48. Magie, R. O. 1953. Some fungi that attack gladioli. In *Plant diseases: USDA yearbook of agriculture*. Washington, DC: U.S. Govt. Printing Office. pp. 601–7.

49. Mann, L. K. 1952. Anatomy of the garlic bulb and factors affecting bulb development. *Hilgardia* 21:195–251.

50. Matsuo, E., T. Ohkurano, K. Arisumi, and Y. Sakata. 1986. Scale bulblet malformations seen in *Lilium longiflorum* during scale propagation. *HortScience* 21:150.

51. Matsuo, E., and J. M. van Tuyl. 1986. Early scale propagation results in forcible bulbs of Easter lily. *HortScience* 21:1006–7.

52. McClure, F. A. 1966. *The bamboos: A fresh perspective*. Cambridge, MA: Harvard Univ. Press.

53. Mckay, M. E., D. E. Bythe, and J. A. Tommerup. 1981. The effects of corm size and division of the mother corm in gladioli. *Aust. Jour. Exp. Agric. Animal Husbandry* 21:343–48.

54. McRae, E. A. 1978. Commercial propagation of lilies. *Comb. Proc. Intl. Plant Prop. Soc.* 28:166–69.

55. Melis, R. J. M., and J. van Staden. 1985. Tuberization in cassava (*Manihot esculenta*). Cytokinin and abscisic acid activity in tuberous roots. *J. Plant Physiol.* 118:357–66.

56. Menzel, C. M. 1985. Tuberization in potato at high temperatures: Responses to exogenous gibberellin, cytokinin and ethylene. *Potato Res.* 28:263–66.

57. Miller, W. B. 1993. *Lilium longiflorum*. In A. DeHertogh and M. Le Nard, eds. *The physiology of flower bulbs*. Amsterdam: Elsevier. pp. 391–422.

58. Morel, G. M. 1964. Tissue culture: A new means of clonal propagation of orchids. *Amer. Orch. Soc. Bull.* 53:473–78.

59. Moser, B. C., and C. E. Hess. 1968. The physiology of tuberous root development in dahlia. *Proc. Amer. Soc. Hort. Sci.* 93:595–603.

60. Mulder, R., and I. Luyten. 1928. De periodieke ontwikkeling van der *Darwin* tulip. *Verh. Koninkl. Ned. Akad. van Wet.* 26:1–64.

61. Nakayama, M. 1980. Vegetative propagation of cyclamen by notching of tuber. II. Effect of scooping size and notching size on the regeneration of cyclamen tuber. *J. Japan. Soc. Hort. Sci.* 49(2):228–34.

62. Nightingale, A. E. 1979. Bulbil formation on *Lilium longiflorum* Thunb. cv. Nellie White by foliar applications of PBA. *HortScience* 14:67–8.

63. Nitsch, J. P. 1971. Perennation through seeds and other structures. In F. C. Steward, ed. *Plant physiology*, Vol. 6A. New York: Academic Press. pp. 413–79.

64. Nowak, J. 1993. *Hyacinthus*. In A. DeHertogh and M. Le Nard, eds. *The physiology of flower bulbs*. Amsterdam: Elsevier. pp. 335–48.

65. Okubo, H. 1993. *Hippeastrum* (Amaryllis). In A. DeHertogh and M. Le Nard, eds. *The physiology of flower bulbs*. Amsterdam: Elsevier. pp. 321–34.

66. Pfieffer, N. E. 1931. A morphological study of *Gladiolus*. *Contrib. Boyce Thomp. Inst.* 3:173–95.

67. Quamina, J. E., B. R. Phills, and W. A. Hill. 1982. Vine production from tuber pieces of various sizes and sections of yam (*Dioscorea alata* L.). *HortScience* 17:73.

68. Raunkiaer, C. 1934. *Life forms of plants and statistical plant geography*. Oxford: Clarendon Press.

69. Rees, A. R. 1972. *The growth of bulbs*. New York: Academic Press.

70. Rees, A. R. 1992. *Ornamental bulbs, corms and tubers*. Wallingford, UK: CAB International.

71. Rees, A. R., and G. R. Hanks. 1980. The twin-scaling technique for narcissus propagation. *Acta Hort.* 109: 211–16.

72. Reinders, E., and R. Prakken. 1964. *Leerboek der Plantkunde*. Amsterdam: Scheltema & Holkema N.V.

73. Roberts, A. N., and L. T. Blaney. 1957. Easter lilies: Culture. In *Handbook on bulb growing and forcing*. Northwest Bulb Growers Assoc. pp. 35–43.

74. Rockwell, F. F., E. C. Grayson, and J. de Graaf. 1961. *The complete book of lilies*. Garden City, NY: Doubleday.

75. Sheehan, T. J. 1955. Caladium production in Florida. *Fla. Agr. Ext. Circ. 128*.

76. Shillo, R., and A. H. Halevy. 1981. Flower and corm development in gladiolus as affected by photoperiod. *Scientia Hort.* 15:187–96.

77. Sieczka, J. B., and R. E. Thornton. 2001. Commercial potato production in North America. www.css.orst.edu/classes/css322/cppins.htm.

78. Simon, R. A. 1986. A survey of bamboos: Their care, culture and propagation. *Comb. Proc. Intl. Plant Prop. Soc.* 36:528–31.

79. Skelmersdale, L. 1978. Propagation of bulbous and bulbous-like plants. *Comb. Proc. Intl. Plant Prop. Soc.* 28:209–15.

80. Slater, J. W. 1963. Mechanisms of tuber initiation. In J. D. Ivins and F. L. Milthorpe, eds. *The growth of the potato*. London: Butterworth. pp. 114–20.

81. Smagula, J. M., and P. R. Hepler. 1980. Effect of nitrogen status of dormant rooted lowbush blueberry cuttings on rhizome production. *J. Amer. Soc. Hort. Sci.* 105:283–85.

82. Struik, P. C., and S. G. Wiersema. 1999. *Seed potato technology.* Wageningen: Wageningen Pers. pp. 1–383.

83. Stuart, N. W. 1954. Moisture content of packing medium, temperature and duration of storage as factors in forcing lily bulbs. *Proc. Amer. Soc. Hort. Sci.* 63:488–94.

84. Theron, K. I., and A. A. DeHertogh. 2001. Amarylladaceae: Geotropic growth, development and flowering. In J. Janick, ed. *Hort. Rev.* 25:1–70.

85. Ulrich, M. R., F. T. Davies, Y. C. Koh, S. A. Duray, and J. N. Egilla. (1999). "Micropropagation of *Crinum* sp. 'Ellen Bosanquet' by tri-scales" *Scientia Hort.* 82: 95–102.

86. Van Aartrijk, J., and P. C. G. Van der Linde. 1986. *In vitro* propagation of flower-bulb crops. In R. H. Zimmerman et al., eds. *Tissue culture as a plant production system for horticultural crops.* Dordrecht: Martinus Nijhoff Publishers.

87. Van de Pol, P. A., and T. F. Van Hell. 1988. Vegetative propagation of *Streilitzia reginae. Acta Hort.* 226:581–86.

88. Van Tuyl, J. M. 1983. Effect of temperature treatments on the scale propagation of *Lilium longiflorum* 'White Europe' and *Lilium* 'Enchantment.' *HortScience* 18:754–56.

89. Wang, S. Y., and A. N. Roberts. 1970. Physiology of dormancy in *Lilium longiflorum* Thunb. 'Ace.' *J. Amer. Soc. Hort. Sci.* 95:554–58.

90. Wang, Y. T., and A. N. Roberts. 1983. Influence of air and soil temperatures on the growth and development of *Lilium longiflorum* Thunb. during different growth phases. *J. Amer. Soc. Hort. Sci.* 108:810–15.

91. Weiler, T. C., and R. W. Langhans. 1972. Growth and flowering responses of *Lilium longiflorum* Thunb. 'Ace' to different day lengths. *J. Amer. Soc. Hort. Sci.* 97:176–77.

92. Welch, N. C., and T. M. Little. 1967. Heat treatment and cutting for increased sweet potato slip production. *Calif. Agr.* 21(5):4–5.

93. Wilfret, G. J. 1995. *Caladium.* In A. DeHertogh and M. Le Nard, eds. *The physiology of flower bulbs.* Amsterdam: Elsevier. pp. 239–48.

94. Wilson, K., and J. N. Honey. 1966. Root contraction in *Hyacinthus orientalis. Ann. Bot.* 30: 47–61.

95. Woodcock, H. B. D., and H. T. Stearn. 1950. *Lilies of the world.* New York: Scribner's.

96. Zweede, A. K. 1930. De periodieke ontwikkeling van *Convallaria majalis. Verh. Koninkl. Ned. Akad. van Wet.* 27:1–72.

*16
Principles and Practices of Clonal Selection

learning objectives

- Characterize clones as cultivars.
- Compare advantages and disadvantages for using clones in propagation.
- Describe how clones become cultivars.
- Describe causes and patterns of variation within clones.
- Characterize typical clonal sources used by nurseries.
- Describe the clonal selection and pedigree distribution system.

INTRODUCTION

The goal of vegetative propagation is to select a single source plant of superior characteristics and to reproduce populations of progeny plants with identical genotypes that are its direct descendants. In Chapter 2, this biological process is described as **cloning,** and the resulting population of plants as **clones.** A clone can also represent a taxonomic category of **cultivar** defined by the International Code of Nomenclature of Cultivated Plants (147) paralleling the categories of the various seedling cultivars described in Chapter 5.

Individual plants of a clone have a life cycle that differs fundamentally from an individual plant within a seedling population (as described in Chapter 2). Seedlings originate with different genotypes and exhibit all four phases of ontogenetic development: *embryonic, juvenile, transitional,* and *mature* (Fig. 2–1). A clonal plant, on the other hand, originates as a vegetative **propagule** (i.e., cutting, scion, bud, explant, layer, bulb segment) from some single plant source. Although all plants of the progeny population have an identical genotype and are therefore expected to have the same phenotype, some individuals may vary. The first part of this chapter describes the nature of clonal propagation, sources of variation, and principles for their control. The second part describes propagation systems for maximizing clonal advantages and managing potential clonal variability problems.

> cloning The process of vegetatively propagating a clone.
>
> clone The vegetative progeny of a single genotype such as an individual seedling, a mutant branch, a single plant of a clonal population, or a recombinant DNA segment.
>
> propagule Any plant part used as the starting point of a propagation process.

HISTORY

Clones can exist as a species adaptation in nature where vegetative reproduction occurs by special vegetative structures, such as bulbs, tip layers, rhizomes, runners, and other specialized structures (see Chapters 14 and 15). Although these structures provide a special advantage for colonizing a specific site, cloning alone as a reproductive strategy is not generally favored in nature because the process does not provide an opportunity for the genetic variation and evolutionary advancement that results from sexual reproduction. However, a combination of seedling and cloning strategies are characteristic of many species (34).

Clonal selection followed by vegetative propagation, on the other hand, provides a powerful tool for improvement of perennial crops in cultivation (1, 2). A single superior plant from a variable population of less desirable plants can be instantly selected and its genotype multiplied more or less indefinitely. Early domestication of vegetatively propagated species, such as potato (*Solanum tuberosum*), yam (*Dioscorea* sp.), sweet potato (*Ipomoea batatas*), bamboo (various genera), sugarcane (*Saccharum* sp.), and banana (*Musa* sp.), was due to their natural reproduction through vegetative structures with which they reproduced (56, 122). Clones of some woody species, such as grapes (*Vitis*) and figs (*Ficus*), could be easily multiplied simply by inserting cuttings into the ground to produce roots and a new plant (27, 167). Similarly, poplar (*Populus*) and willow (*Salix*) trees were readily propagated and came into use as hedges and fences to enclose fields and mark boundaries. Tree fruit plants, such as apple, pear, cherry, plum, peaches, plums, apricots, and citrus fruits, on the other hand, were less easily propagated by cuttings, and the propagation of selected individuals was accomplished by the discovery (or invention) of grafting methods or by layering. In the tropics, particularly where humidity is high, layering (Chapter 15) enabled cloning of tropical fruit species, including litchi (*Nephelium*), mango (*Mangifera*), and longan (*Euphoria*).

> **clonal selection** The process of selecting an individual plant or plant part to create a clone.

Great strides were made during the previous century to improve the technology for rooting cuttings (27) and to increase the clones available for human use (108). This technology included the utilization of structures (greenhouses, cold frames, hotbeds), environmental controls (misting, bottom heat), media development, hormone application, and other aids (Chapters 10 and 11). The application to ornamental horticulture has produced spectacular advances, including cultivar selection, propagation of floricultural crops and houseplants such as rose, carnation (*Dianthus*), chrysanthemum, poinsettia, and many foliage plants, as well as landscaping materials (deciduous and evergreen shrubs, conifers, landscape trees) (35, 84).

Traditionally, forestry (115) has relied upon seedling populations. In a few unique exceptions, Chinese fir (*Cunninghamia lanceolata*) and sugi (*Cryptomeria japonica*) have been propagated vegetatively for 1,000 years in China and Japan (88, 115). During the past 25 years, the concepts and practices of **clonal forestry** have become a major strategy for production of specific tree crops (2, 77), including *Eucalyptus* in South America, North Africa, and Europe, Monterey pine (*Pinus radiata*) in New Zealand (Fig. 16–1), and poplars (*Populus*) and willow (*Salix*) in the United States and Europe. Likewise, the technology of producing seed by vegetatively multiplying embryos in culture ("synthetic seeds") promises to support their direct use in propagating various agronomic, vegetable, and ornamental crops (see Chapters 17 and 18).

> **clonal forestry** A system of forestry management that utilizes clones in planting.

USING CLONES AS CULTIVARS

The following are the advantages and disadvantages of using clonal cultivars.

Advantages of clones as cultivars:	Disadvantages of clones as cultivars:
• Genetic improvement and selections	• Monoculture
• Uniformity of populations	• Slow reproduction rate
• Control of phases of plant development, such as earlier bearing of fruit trees	• Lack of genetic variation for breeding and selection
• Combine more than one genotype into a single plant, as in grafting	• Potential for propagating systemic pathogens from clonal sources
• Greater commercial value	• Insufficient genetic diversity for resistance against unforeseen pests or pathogens
• Facilitating propagation: sometimes only means to propagate	• Potential for latent genetic mutation

Advantages

Genetic Improvement and Selection Clonal selection followed by vegetative propagation is a major strategy for plant improvement of perennial crops. Cloning can select the best individuals for a specific purpose from within an entire population of seedlings and, by vegetative propagation, multiply them into a population of the same identical genotype without limit. Thus, a large genetic advance can be made in a single step without the multiple generations of seed propagation required of seed-propagated cultivars (1, 2, 3, 93,

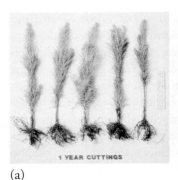

Figure 16–1
(a) Monterey Pine (*Pinus radiata*) 0-1-rooted layer cuttings after 1 year in stoolbed, and (b) 11-year-old, uniform stand of clonal *P. radiata* for timber production in New Zealand (see circled individual in photo for scale reference). Courtesy J. Holiday.

clonal "fixing" The concept of a clone as stabilizing a genotype for propagation.

136). From a plant breeding standpoint, the process of **clonal "fixing"** is equivalent to "fixing" genetic variation required for a self-pollinated population or an F_1 population in hybrid seed production (Chapter 5).

Most cultivars selected as clones are genetically heterozygous and their uniformity and unique characteristics would immediately be lost in the next seed-propagated generation. For example, 'Cherokee Princess' flowering dogwood (*Cornus florida*) was selected as a cultivar because of its large white bracts. Seedlings from this cultivar would be expected to have wide variation in bract size compared with the original cultivar.

Some important cultivars are seedless, and vegetative propagation is necessary to reproduce them. These include such important crops as banana (*Musa*), some fig (*Ficus*) cultivars, seedless grapes (*Vitis*), persimmons (*Diospyros*), and citrus.

Uniformity of Populations Commercial production of most perennials is based on the uniformity of individual plant size, growth rate, time of flowering, time of harvesting, type of product, and other phenotypic characteristics (Fig. 16–1). This characteristic makes economic industrial production of fruit and nut crops possible (159). Uniformity and the elimination of inferior individuals in the population have a major effect on increasing yield. Economic studies of clonal forests show a major genetic gain of one-third increased wood yield, due to increased uniformity and the elimination of less productive individuals found in a seedling mixture (2).

Control of Phases of Plant Development Juvenile, transitional, and mature (adult) phases can affect important traits, including age of flowering and seed production, phenotypic structure of the plant, and regeneration competence, particularly in the ability to initiate roots (Chapter 2). Selection of source can be used to maintain, enhance, or reverse specific phases.

Cultivars grown by vegetative propagation invariably come into flowering at an earlier age than comparable plants grown from seed, a result of propagating clones in their mature phase of development.

Cultivars vary greatly in their natural potential for initiating adventitious roots (Chapter 9). Herbaceous species and cultivars are invariably easy to root, accounting for the widespread use of carnations (*Dianthus*), chrysanthemums, foliage plants, and many florist crops. Selection of woody perennials is largely dependent on ease of rooting and is important when determining the range of species and cultivars grown. Some root readily by hardwood cuttings, but others respond to the many rooting technologies of hormone application, mist propagation, and so forth. For those difficult to root, success may depend on methods to manipulate the juvenile phase to increase rooting potential.

Combining More Than One Genotype into a Single Plant Grafting allows the combination of more than one genotype in the same plant. Separate genotypes can be chosen for the root, the interstem, and the fruiting part of the plant (Chapters 12 and 13). Tree roses require a specific interstock for the main stem with the flowering cultivar grafted on top (see Fig. 11–12). A weeping cultivar can be grafted on top of a larger upright plant (see Fig. 11–8). Other examples are given in Chapter 13.

In other cases, different cultivars can be grafted to different branches of the same plant. This practice can achieve a range of effects, such as combining early to late ripening in a backyard fruit tree, or adding a pollinating cultivar as a single limb in a self-incompatible cultivar.

Disadvantages

Monocultures A **monoculture** is the mass production of a single, genetically uniform crop within a single planting. Since all of the plants of the nursery, field, orchard,

monoculture The planting of a commercial cultivar that contains only a single genotype.

or plantation have the same genotype, each is equally vulnerable to specific environmental hazards, pests, and diseases, which may lead to an overall loss of genetic diversity for that crop (2, 38, 166). Much effort has been expended to control environmental hazards, insect pests, and plant diseases in commercial agriculture and horticulture, and the uniformity of clonal populations facilitates that control. However, such programs increase production costs and sometimes produce unacceptable environmental risks. More desirable and environmentally sound integrated pest and disease control methods (IPM) are described in Chapter 2.

To avoid monocultures in clonal forestry, a recommended policy is to plant a mixture of clones, with a minimum of 25 different genotypes in a single forest planting (77). This practice is also referred to as a *mosaic of clonal blocks,* and can have as much genetic diversity, if not greater, than sexually produced seedling populations derived from **mother block** plantings (88, 115). The pressure to utilize only the most productive clones in commercial agriculture may need to be offset by preservation of genetic diversity in germplasm collections, arboreta, natural areas, seed storage, and other systems.

mother block   A group of plants maintained by a nursery as a source planting.

Slow and Costly Reproduction   As a general statement, vegetative propagation is more expensive per plant than seedling production depending on the species and method of propagation. Consequently, the primary economic benefit is to produce cultivars with high individual value. In contrast, the individual value of most agronomic or vegetable crops, where seeds are the method of choice, is usually relatively low.

The natural method of increase for perennials, such as described in Chapters 14 and 15, is often slow and does not always lend itself to commercial production. For many of these species, micropropagation has produced a revolution in mass propagation and made many cultivars available to the public.

Potential for Latent Genetic Mutation   The unexpected appearance of off-type plants can produce serious economic problems. Propagation of these unwanted plants can sometimes occur before they can be identified and corrected; a situation that occurred in the early history of some micropropagated cultivars. A similar problem can occur if the wrong cultivar is selected at the beginning. These hazards are discussed in detail later in this chapter and in Chapter 17.

Potential for Systemic Infection and Insufficient Genetic Diversity for Resistance Against Unforeseen Pests or Pathogens   A serious hazard of clonal propagation is the high probability of systemic infection with viruses, viroids, and other transmissible pathogens. Once the pathogen infects, it spreads through the entire plant and is perpetuated into the next vegetative generation. Most of these pathogens are not transmitted to seedling progeny, although there are exceptions. In some, such as *Prunus necrotic ring spot virus,* the virus is seed-transmitted but in low percentages. A monoculture is also more susceptible to unforeseen pests or pathogens than a genetically diverse seedling population.

ORIGIN OF CLONES AS CULTIVARS

Selection of superior clones is important to many areas of horticulture including nursery, greenhouse, vegetable, and fruit crops. There are three basic ways that a new clone is developed: (1) *seedling selection*, (2) *mutation*, and (3) *biotechnology* including recombinant DNA technology (molecular biology).

Seedling Selection

Most cultivars grown as clones originated from a single plant in a seedling population. Historically, some well-known fruit cultivars were selected hundreds, sometimes thousands, of years ago and have been vegetatively propagated ever since (87, 92, 94, 108). 'Cabernet Sauvignon' and 'Sultana' (now known as 'Thompson Seedless') grapes have been grown horticulturally for about 2,000 years, as have some fig cultivars. The 'Bartlett' pear (also known as 'Williams Bon Chretien') originated as a seedling in England in 1770. 'Delicious' apple originated about 1870 in Jesse Hiatt's orchard near Peru, Iowa. 'Cavendish' banana, which is seedless, is the major banana of commerce through much of the world.

seedling selection   Selecting a single plant of a seedling population to be the start of a new clone cultivar.

More recently, fruit and nut cultivars are the product of commercial breeding programs for both scion cultivars and rootstocks (1, 92, 93, 136, 152). Many ornamentals, including rose, chrysanthemum, and carnation not only are the product of breeding but require specific nursery programs to deliver quality products to the end user. Woody ornamental shrub and tree clones have become sought-after cultivars for landscaping purposes (35, 84).

BOX 16.1 GETTING MORE IN DEPTH ON THE SUBJECT
PROPAGATION GENERATIONS

The designation of vegetative **propagation generations** can be important in defining variation patterns involving the phenotypic expression of individual plants of the clone (19, 67, 152). Clonal multiplication from a single seedling plant takes place in two separate patterns: (a) multiple propagation from the same plant (*horizontal*) and (b) consecutive generations of progeny from the same plant (*vertical*) (Fig. 16–2). It is convenient to refer to the seedling plant as the S_0 (originating as a seedling) generation. This plant exhibits a typical seedling life cycle as described in Chapter 2 and exhibits a "seedling" phenotype as will be described in this chapter.

The S_0 plant has been called the **ortet,** and the first vegetatively propagated generation as a population of **ramets** (137). This terminology originated as foresters began to utilize clonal forestry. The system can be useful in managing source plants and their progeny in relation to *rooting potential* and *true-to-type* phenotypes and in analyzing clonal variability in specific kinds of **inherited disorders** in fruit and nut cultivars (Box. 16.5, page 604).

ortet The original seedling tree in a vegetative propagation sequence.

ramet The vegetative progeny of a single seedling tree.

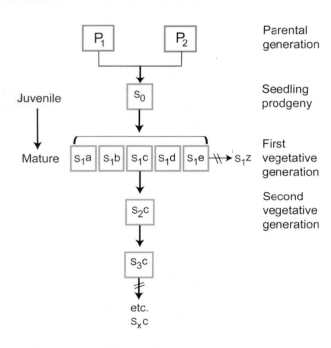

Figure 16–2
Model illustrating the pedigree pattern following initial selection and subsequent cloning of a single plant from a seedling population. The original seedling plant is designated as S_0. The vegetative progeny population propagated from this plant is designated as S_1 (first generation from the seedling). Individuals of this population are designated by lowercase letters, s_{1a}, s_{1b}, and so on. Subsequent consecutive generations are designated as s_2, s_3 to s_x.

Mutations and "Bud-Sports"

mutation A genetic change in genotype.

mutant A plant part with a changed genotype resulting from a mutation.

bud-sport A sudden change in a single branch due to a mutation apparently arising from a single bud.

A **mutation** is a permanent *genetic change involving some part of the DNA molecule*. The changed plant part is referred to as a **mutant.** When a mutant suddenly appears as a phenotypic change in the branch of a plant, horticulturalists commonly refer to it as a **bud-sport** or **bud-mutation** because it appears to have originated from a single bud (Fig. 16–3).

Spontaneous Origin Mutations may occur spontaneously as (a) chance rearrangement of the four bases in the DNA structure **(point mutations),** (b) rearrangements of different parts of the chromosome (deletions, duplications, **translocations,** and **inversions),** (c) addition or subtraction of individual chromosomes **(aneuploidy),** or (d) the multiplication of entire sets of chromosomes **(polyploidy).** See the spontaneous origin of mutations and bud-sports (Fig. 16–4, page 600). Most mutations are deleterious, but occasionally a bud-sport appears that has some horticultural advantage. For example, giant, vigorous "sports" have been observed in grapes, (*Vitis*) which turned out to be a result of polyploidy (39). These mutants were undesirable because they were nonproductive. In other crops where the goal

Figure 16–3
One branch of this crabapple shows a conspicuously columnar, upright form. This probably represents a bud sport where a mutation has occurred.

is not fruit production, this characteristic might be useful. Other reported undesirable mutations have produced misshapen fruit, low production, and susceptibility to disease (120). In the past, citrus orchards were notorious for the presence of inferior limb sports (127).

On the other hand, mutations within established clones have been and still are a potential source of variation within clones that can result in desirable new cultivars (151). Most modern apple (*Malus*) and pear (*Pyrus*) cultivars have red forms resulting from bud-sports. Many peach (*Prunus*) cultivars have produced bud-sports with a different time of maturity, which allows orchardists to extend their harvesting season. Compact, "spur-type" structures have been discovered in apple. Many, if not most, citrus cultivars have originated as bud-sports, including the 'Washington Navel' orange (118). The many cultivars of lemon (*Citrus limon*) have been shown through fingerprinting methods to have originated as bud-sports (48). Such a discovery made among hundreds of trees in an orchard can immediately result in a new cultivar. Many ornamental plants from chrysanthemum to shrubs and trees have also arisen from bud sports.

Induced Mutations The rate of mutation can be increased by treatment with specific **mutagenic agents** such as X-rays, gamma rays, neutrons, and specific chemicals. **Mutation breeding** has resulted in new cultivars (17, 19, 93, 151, 152). For example, 'Ruby Red' grapefruit (*Citrus paradisi*) originated as a bud-sport, and when subjected to radiation treatment, resulted in various mutants of which 'Star Ruby' and 'Rio Red' were introduced as cultivars (151). Colchicine treatment has also been used to induce mutations, in ornamentals, such as tetraploid daylilies and Heuchera 'Midnight Burgundy'.

mutagenic agent A chemical or radiation treatment that creates mutations.

mutation breeding A system of plant breeding that creates new cultivars by mutations.

Somaclonal Variation (61, 89) Somaclonal variation, which is a mutation that occurs in tissue culture (see Chapter 17), can sometimes be beneficial in the development of new clones, such as new *Spathiphyllum* and *Syngonium* cultivars. This type of variation was discovered when plants cells of certain species, for example, tomato (*Lycopersicon*) or tobacco (*Tobacum*), were grown in tissue culture to produce new plants (40). Regenerated plants sometimes differed from each other, as illustrated by the micropropagated hosta plants in Figure 17–44 (40). These variations may originate in cells in the deeper areas of the growing point below the meristem (24). Similarly, potato plants regenerated from protoplasts grown in culture can show variation, and potentially new culitvars are possible (121). Although these somaclonal variants can be useful to plant breeders, most are usually undesirable in propagation.

Biotechnology

Cell and Tissue Culture Technology The ability to grow cells, tissues, and protoplasts in culture provides procedures to identify and propagate new clones with potentially desirable characteristics (46). These procedures are described in Chapter 17.

cell and tissue culture technology Laboratory procedures by which cells and tissue are grown with in vitro culture, including clonal selection at the cell level.

Recombinant DNA Technology The process of **recombinant DNA technology** (commonly referred to as "molecular biology") is a method of genetic manipulation in which a novel gene is inserted into a plant's genome to produce a new genotype such as a new clone with insect or disease resistance. The concept was introduced in Chapter 2 (53, 107). Chapter 5 described how transgenic seedling cultivars are produced. The technology can be particularly effective in seed-propagated cultivars because once a gene becomes part of the genome, it can be manipulated through regular breeding strategies utilizing sexual reproduction.

recombinant DNA technology The process by which DNA from different origins are combined and cloned to be inserted into a separate genome.

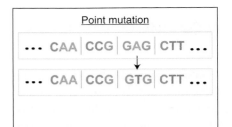

(a) A point mutation is a common result from chemical mutagenesis. It results in the change of a nucleic base within the DNA sequence. In this example, GAG codes for the amino acid glutamine. The base change from adenine to thymine changes the code to GTG, which now codes for valine.

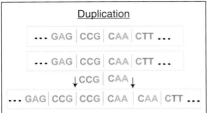

(b) During a duplication event a portion of the chromosome is duplicated and then inserted into the chromosome to form a new DNA sequence. In this example, the CCG CAA sequence was duplicated and inserted.

(c) A deletion is the removal of a segment of the chromosome. In this example, the portion of the DNA containing the CCG CAA sequence has been deleted from the chromosome.

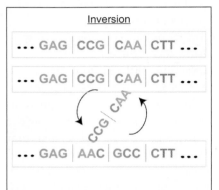

(d) During an inversion, a portion of the chromosome is removed and inverted. Then it is re-inserted into the same location from which it was removed.

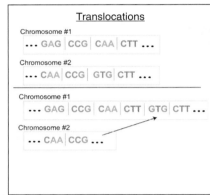

(e) Translocations involve the movement of a piece of chromosome from one non-homologous chromosome to another.

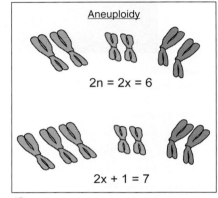

(f) Aneuploidy results in a cell with a different chromosome number compared to the original cell. This may be caused by the addition of a chromosome.

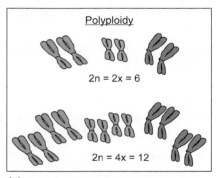

(g) Polyploidy is the result of a multiplication of the entire set of chromosomes.

Figure 16–4
A mutation is a permanent change in the DNA molecule, which can occur spontaneously via (a) point mutation, (b) duplication, (c) deletion, (d) inversion, (e) translocation, (f) aneuploidy, or (g) polyploidy.

BOX 16.2   GETTING MORE IN DEPTH ON THE SUBJECT
SOME EXAMPLES OF TRANSGENIC CLONES IN HORTICULTURE

Inserting genes to increase rooting of a specific cultivar could make it possible to propagate the plant from cuttings (108, 119). The insertion of an insect or disease-resistant gene(s) could improve a cultivar for a specific advantage while retaining its other commercial qualities (124). For instance, in potato (*Solanum tuberosum*), cultivars are available in either the original version or a transgenic version, which includes the Bt insect-resistance gene, and is distributed by Monsanto under the brand name NewLeaf http://www.naturemark.com/. Likewise, separate genes for resistance to VirusA and to Colorado potato beetle have been inserted into specific potato cultivars (75).

transgenic clone   A clone that is created by the introduction of recombinant DNA into its genome.

With clonal propagation, recombinant DNA technology has the potential to utilize the cloning principle, first, to change the genotype of a clonally propagated cultivar and, second, to multiply the new clone with established systems of vegetative propagation. The concept particularly applies to herbaceous and woody perennial crops, which would include fruits, nuts, ornamental trees, shrubs, and some vegetables such as potatoes. Many perennial plants have long seedling cycles, require many years to produce new cultivars by conventional breeding, and have highly heterozygous and complex genotypes. A major horticultural advantage claimed for recombinant DNA technology in perennial crops is that an established cultivar that lacks a major trait can be modified by inserting the gene(s) directly into the genome of the clone with no further breeding required (123, 124, 151). New genotypes of these plants, however, would be subject to the same kinds of clonal variability that are described in this chapter (82).

PHENOTYPIC VARIATIONS WITHIN CLONES

There are four fundamental sources of phenotypic variation that can develop within clones: (a) environment by genotype interactions (phenological changes), (b) ontogenetic aging (phase changes), (c) permanent genetic variation, and (d) infection by systemic pathogens, particularly viruses and similar organisms.

BOX 16.3 GETTING MORE IN DEPTH ON THE SUBJECT
ENVIRONMENTAL VARIATION

Some kinds of environmental variation are subtle. For example, 'Bartlett' pear (*Pyrus*) fruits produced in Washington and Oregon tend to be longer and more "pear-shaped" than those produced in California, apparently due to climatic differences (Fig. 16–5) (149). Fruit size, shape, and appearance may differ depending on the vigor of the plant, the size of the crop, and the age of the plant. Continued exposure to inadequate winter chilling in strawberries (*Fragaria*) can lead to deterioration in plant quality in subsequent years (16). An unfavorable post-bloom vegetative period can lead to deterioration of plants of many bulb species in a warm winter environment (see Chapter 15). A propagator needs to be familiar with not only what the plants look like when propagated, but also with how the plants perform when planted and grown in their ultimate site.

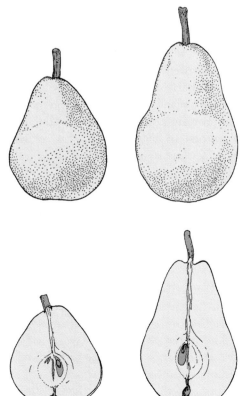

(a)   (b)

Figure 16–5
'Bartlett' pear fruit from different production sites. (a) Typical pear fruit from California grown under higher light and temperature, and lower relative humidity conditions than (b) pear fruit produced from Pacific Northwest (149). This is an example of differences in epigenetic expression due to environmental differences affecting phenotypic expression.

Genotype by Environmental Interactions

Phenotypes result from an interaction of the genotype with the environment. In some clones, differences among growing sites may be important. Some examples are given in Box 16.3 that discusses Environmental Variation. The introduction of a clonal selection as a future cultivar is invariably preceded by a series of field trials in which plots of vegetatively propagated individuals are grown at different environmental sites. The objective is to determine how much the positive phenotypic characteristics of the original seedling plant are due to its genetics and how much are due to the effects of the environment in which it was originally growing. Another reason for the field test is the practical objective of determining the range of adaptability of the cultivar. Traits that vary in different locations are said to have high **genotype by environmental interaction.**

> **genotype by environmental interaction** The variation in phenotype due to the effect of different environments on the same genotype.

Ontogenetic Aging (Phase Change)

Ontogenetic aging is a second cause of phenotypic variation among individuals within a clone. This phenomenon occurs when the phenotype of the individual plant changes with increasing age. This concept was introduced in Chapter 2 where the variations, referred to as **phase changes** (embryonic, juvenile, transitional, and mature), occur during seedling cycles. The significance of variations in propagation results from the fact that the **epigenetic control** for a specific phase is in the cells of individual growing points. Epigenetic control is perpetuated during vegetative propagation as part of the clonal cycle (see Fig. 2–1). This subject is so important that the relationship of phase change to propagation is discussed in a separate section. (See pages 613–19 as well as Chapter 9.)

> **ontogenetic aging (ontogeny)** The normal course of development during a plant's life cycle from a fertilized egg to a mature form.
>
> **phase change** The changes in phenotype associated with shifts from embryonic to juvenile to mature.
>
> **epigenetic control** A nonpermanent regulation of gene expression.

Genetic Variation

Detection of Mutants Mutations are considered to be normal and spontaneous, and they occur naturally in most organisms. Regardless of whether a mutation is by chance or induced, the sequence of discovery is the same (Fig. 16–6). The plant within the clone to which a mutanizing treatment is given is designated as V_0 with succeeding vegetative generations as V_1, V_2, and so on. Although a specific genetic change is usually permanent, the ability of a given mutation to produce a major change in the phenotype of the plant depends on a number of conditions:

- The allele for the new mutant must be dominant in order for the trait to be immediately expressed (see Chapter 2). If recessive, the allele appears after segregating in the next seedling generation, as described in Chapter 5.
- The cell in which the mutant appeared must not only survive but must divide sufficiently to occupy a significant sector of the growing point of the shoot in which it is initiated.
- The trait must be sufficiently conspicuous so that the new trait can be visually identified in the plant.
- Because of the unique arrangement of the meristematic cells within the apex (growing point), the new mutant initially develops as a **chimera** (see page 605), a plant that is a mixture of two genotypes, each expressed independently in separate layers of the shoot. Chimeras are so important in clones that the subject is covered in a separate section.
- Detection of a new mutant within a clone may require a series of vegetatively propagated (vertical) generations and multiple (horizontal) propagations from many buds of the same plant. Severe pruning can increase the number of growing points that are available to show mutated sectors. In Figure 16–6, this sequence is shown as V_1, V_2, and so on.

> **chimera** A plant that is composed of a mixture of tissue, with different genotypes that originates within meristematic tissue.

Somaclonal Decline. *Noninfectious bud-failure* is an inherited disorder in almond (*Prunus dulcis*) in which exposure to specific patterns of high summer temperature damage vegetative buds, which creates a "witches' broom" symptom. The potential for the disorder, which increases during annual growth cycles, is perpetuated

> **somaclonal decline** A unique kind of genetic variation in which an incremental genetic change occurs during annual shoot growth cycles that produces a progressive undesirable change in phenotype with age and consecutive cycles of propagation.

BOX 16.4 GETTING MORE IN DEPTH ON THE SUBJECT
EPIGENETIC VARIATION AND PHENOTYPIC EXPRESSION

The control of **epigenetic variation** and **phenotypic expression** is a central focus (and challenge!) of plant propagation. Plant propagators manipulate gene expression by hedging and stooling stock plants to slow down maturation in order to enhance rooting of cuttings, or conversely, speed-up maturation through budding and grafting to bring a fruit orchard into earlier bearing.

Epigenetic regulation of gene expression is a non-permanent change in gene expression that is retained during mitosis of a given plant's life cycle; it cannot be explained by changes in gene sequence (i.e., there are no mutations of the DNA) (60, 153). Gene activation means genes are "turned on," whereas the repression of gene expression is called **"gene silencing."** Modifications of DNA and protein can lead to **transcriptional** repression or activation, which are classified as *epigenetic regulation*. Although most genes use RNA in the form of *messenger RNA* (**mRNA**) as a **coding** intermediate for protein production (which can also produce enzymes) (see Fig. 2–19), there are many genes whose final products are RNA. These **noncoding RNA** include transfer and ribosomal RNA, as well as newly discovered "regulatory RNA" (11). Some of the noncoding, regulatory RNA include microRNAs (**miRNAs**). Thus, the non-permanent, *epigenetic regulation* caused by regulatory RNA is called **post-transcriptional gene silencing (PTGS)**, sometimes referred to as "*RNA silencing*," and is an RNA-degradation mechanism important in normal plant growth and development and as a defense against viruses (163). RNA silencing can also affect a plant's "competency" to form adventitious roots from cuttings.

As a woody perennial plant ages chronologically, its genome remains the same (provided there is not a somatic mutation) during its normal ontogeny (growth cycle). The plant goes through phase change from juvenile to the transition to the mature phase (see Fig. 2–1). These changes in phenotypic expression are further discussed and illustrated by (1) morphological differences in leaf shape and plant form (see Fig. 16–20, page 614), and (2) physiological changes, which can result in reduced rooting of cuttings taken from physiologically mature plants (see Fig. 16–22, page 615). Poor rooting of cutting from physiologically mature woody plants limits the clonal regeneration of important commercial species. Epigenetic regulation (control) of gene expression occurs as a non-permanent, phenotypic expression from the juvenile to mature phase; yet, the plant's genome remains the same. While poor rooting may be perpetuated through the life of a physiologically mature plant, it is "non-permanent" in that the next sexual generation reverts back to a seedling with an initial juvenile phase and high rooting potential—or if it is the next generation of a clonally produced plant, the rooted liner (clone) will generally go through one or more stages before reaching the "physiologically mature" phase.

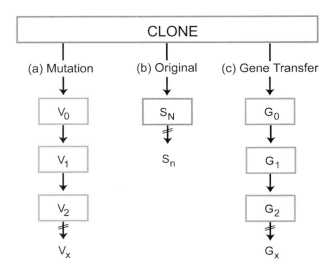

Figure 16–6
Model illustrating the pedigree pattern following genetic modification to single plants within the clone. (a) A mutant clonal line that can occur either spontaneously (see text) or as a result of a mutanizing agent. Both will probably develop into chimeras and may be initially unstable. Consecutive generations are designated as V_0, V_1, V_2, and so on. (b) The original clone, which remains unchanged. (c) Plants transformed by a foreign gene introduced by recombinant DNA technology. These plants would perform as (a) (i.e., those resulting from mutation).

in consecutive growth cycles and after vegetative propagation. Control is achieved by source selection and proper maintenance of source blocks (65, 68). *Crinkle, deep suture* in cherry (*Prunus avium*) (128, 158), and *June yellows* in strawberry (*Fragaria*) appear to be similar.

PATTERNS OF GENETIC CHIMERAS WITHIN CLONES

A *chimera* is a plant that is composed of a mixture of tissues with different genotypes, which are also sometimes referred to as **mosaics** (Fig. 16–7) (82). Most of the spontaneous "bud-sports" described previously originated as chimeras, as have variations produced by artificially induced mutations (72). Chimeras develop because of the unique architecture of the apical meristem. The strategic location of the mutated cell relative to the apex of the apical meristem (133, 144) determines the subsequent cellular distribution within the stem.

BOX 16.5 GETTING MORE IN DEPTH ON THE SUBJECT
EXAMPLES OF GENETIC VARIATION WITHIN CLONES

Variegation
Many types of mutation lead to production of defective plastids, which result in loss of chlorophyll, and are expressed as *albino* or **variegated** plants with sectors of albino and green areas (Fig. 16–7) (132, 133). Some of these mutations have occurred in chloroplasts and others in the mitochondria. In either case, their pattern in the clone follows that of a chimera, which is described further in this chapter.

variegation   The cellular-based contrasting genetic expression in different parts of the same organ (as a leaf), which is usually caused by a chimera.

Patterned Genes
Some genotypes produce variation because the variable expression of individual genes produces specific patterns within the plant (80, 81, 154). Their typical pattern is displayed throughout the clone and expressed by each plant. Some color variations result from patterns produced by single genes; for example, some variegated *coleus* and bicolored *petunia*. This type of variation is inherited in seedlings.

patterned genes   A differential gene expression (as a color pattern) that is not due to cell arrangement.

Transposons
Transposons (97) are unique sources of genetic variation, although often limited to certain species. The variation typically is observed as patterns of colored streaks and patches of different sizes and arrangements. Examples include variegated seeds of Indian corn (*Zea*) and color instabilities in flowers of some *Antirrhinum* (snapdragon) and *petunia* cultivars (Fig. 16–7) (80, 81). Studies initiated by Dr. Barbara McClintock, who received the Nobel Prize for this work in 1983 (83), and others have shown that these patterns are caused by unique mobile genetic factors. **Transposons (also called transposable elements)** have the capacity to move to different positions of the chromosome DNA and in doing so turn genes on or off. For this reason, they have often been referred to as **"jumping genes."** These resultant genetic changes are inherited and the characteristics are transmitted to seedling progeny.

transposons or transposable elements   A unique kind of genetic element in cells of some organisms that have the capacity to change position within the chromosome.

Somaclonal Variation
Infection by Viruses and Virus-Like Pathogens
The fourth major reason for phenotypic variation among individual plants of a clone is the presence of **systemic pathogens** (Fig. 16–7). This aspect is so important in the practical handling of clones that it is considered in a separate section.

systemic pathogens   Pathogens that infect the cells of a plant and spread throughout the plant, where they remain infective.

(a)

(b)

(c)

(d)

Figure 16–7
(a) Variegation in maple caused by a chimera. (b) Pattered variegation in petunia due to genetic selection. (c) Variation in corn caused by transposons (jumping genes) on chromosomes may result in colored, non-colored, and variegated grains of corn. (d) Variegation in camellia caused by a virus.

Origin of Chimeras Within Clones (69, 80, 144)

Layered Meristem The apical meristem of a plant shoot consists of cells that are arranged in 3 to 4 independent layers in most flowering plants. These layers of cells are defined as **histogens.** Figure 16–8 shows diagrammatically a typical layered arrangement found in a dicot plant, such as a peach (*Prunus*). The outer cell layers cover a "core" of inner cells to produce the arrangement known to botanists as the **tunica-corpus meristem.** For most dicots, there is a two-layered corpus, while monocots may only have a single corpus layer. In the typical dicot, the tunica layers are usually designated as L-I and L-II, while the corpus is comprised of the LIII and inner tissues. The outer L-I and L-II divide primarily by **anticlinal** divisions (perpendicular to the surface) and eventually become the epidermis and outer layers of the plant (Fig. 16–9). The L-III cells, on the other hand, first divide anticlinally but may later undergo **periclinal** divisions (parallel to the surface) to expand the shoot, particularly in the locations farther from the tip. The L-III and subsequent inner layers constitute the **corpus** portion of the meristem. The inner cell portions of the corpus can divide on any plane.

histogen The structured layers of cells in the plant growing points.

tunica The outside layers of cells in the meristem, referred to as the "cap."

anticlinal The direction of cell division in an end-to-end position to produce a continuous layer.

periclinal The direction of cell division in a side-to-side position to increase numbers of cell layers.

corpus The randomly distributed cells in the "body" of the meristem, underneath the tunica.

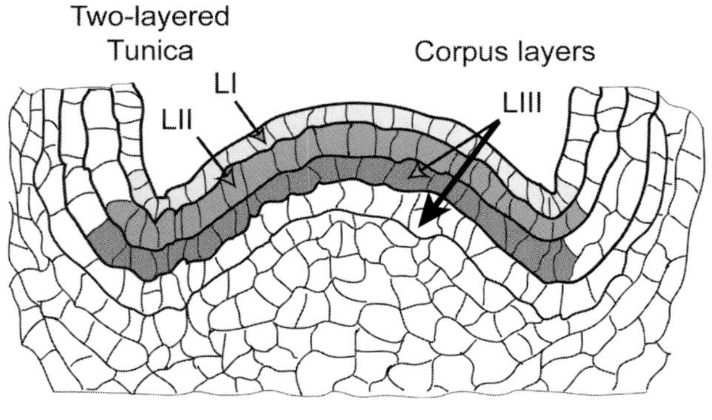

A chimera usually forms from a mutation in a single cell in any of the three meristem layers and can result in either a stable or unstable chimera. Figures 16–10 and 16–11 (page 607) illustrate the layered meristem development of the three chimera types — **mericlinal, periclinal,** and **sectorial: periclinal chimera** A stable chimera in which one or more of the cell layers is completely formed by mutated cells.

mericlinal chimera An unstable chimera in which the mutated tissue occupies only a part of the outside layer.

- **Mericlinal:** In this combination, cells carrying the mutant gene occupy only a part of the cell layer in the meristem. The fruit of 'Washington Navel' oranges is sometimes formed from a mericlinal chimera, where the mutant sector is only on the surface and only partly around the stem (Fig. 16–12, page 608). This type of chimera is unstable and tends to change into a periclinal chimera, revert to the nonmutated form, or continue to produce mericlinal shoots (Figs. 16–10 and 16–11).
- **Periclinal:** The mutated tissue occupies one or more outer layer of cells that completely surrounds an inner core of nonmutated tissue (Figs. 16–10 and 16–11). The most common form of dicot leaf variegation is termed the sandwich chimera. In this case, the L-II has a mutation and cannot make chlorophyll, while the L-I and L-III on either side of the L-II is made up of green chlorophyll producing cells. This type of variegation has a white or yellow leaf margin, because the green producing L-I and L-III cells do not extend to the edge of the leaf exposing the non-pigmented L-II cells (see Fig. 16–15, page 610). With a periclinal chimeral, a red-colored apple will likely have the red pigment only in the epidermal layer, whereas the cells of the inner tissue have alleles for green or yellow color. Similarly, blackberry (*Rubus*) cultivars may be *thornless* because the cells making up the epidermis do not have the *thorny* allele in their genome. Inner cells in the stem layer below the epidermis are likely to have a *thorny* allele but cannot express it.
- **Sectorial:** This is a third basic type of

sectorial chimera An unstable chimera in which the mutated tissue occupies a sector that encompasses all layers.

Figure 16–8
Chimeras: the dicot shoot meristem is usually organized into three distinct layers–LI, LII, LIII. Typically, LI gives rise to epidermal cells. LII provides the next inner layer of cells and also the gametes. LIII cells become the inner most cells and the vascular system. Cells in the tunica (L1 and L2) divide anticlinally, whereas cells in the corpus (below L3) divide anticlinally and periclinally.

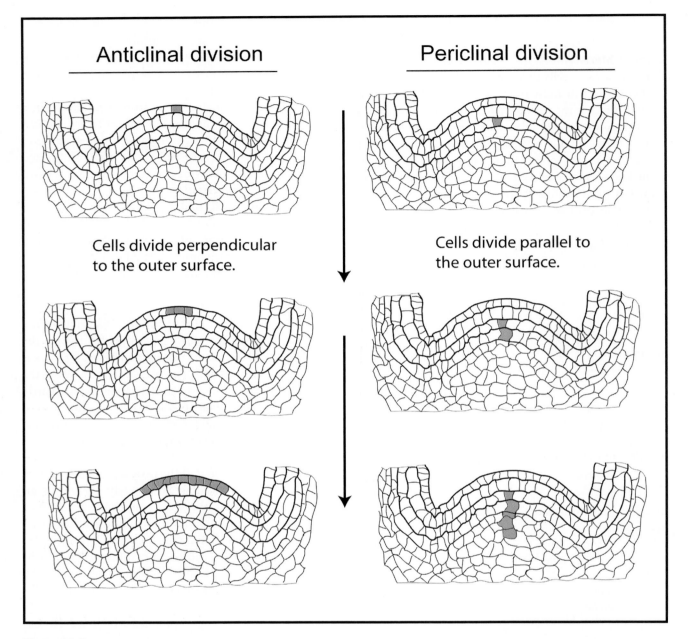

Figure 16–9
In anticlinal divisions the new cell wall plates of actively dividing cells form perpendicular to the shoot apex in Layer I, whereas in periclinal division, new cell wall plates form parallel to the shoot apex in cells dividing in Layer III and the corpus. The arrows indicate progressive division and growth of the shoot apex.

chimera in which the mutated tissue involves a sector of the stem but extends all the way from the surface to the center (Fig. 16–10). In actuality this type is rare in a plant with a layered apex, and would originate only in a very young stage of an embryo or in a root tip. If one occurred, it would be unstable and quickly revert either to a periclinal or mericlinal chimera.

By using larger-sized tetraploid (4x) cells as markers, Dermen demonstrated (32) how different layers of the apex tend to produce specific areas of the stem (Fig. 16–13, page 608). Cells in L-I give rise to the epidermal and outer layers of the stem. Cells in the L-II layer give rise to the cortex, some of the vascular tissues, and the reproductive structures. L-III gives rise to the pith and some of the vascular tissue. However, there is much fluidity in the makeup of the different parts. For example, periclinal division can shift individual cells from the L-II layer into the L-I layer, a phenomenon known as **displacement.** The reverse of this process, a

> **displacement** Cells in the outer layer shift into inner layers.

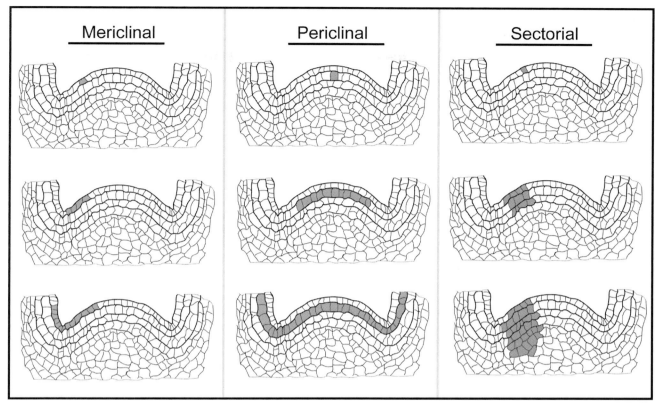

Figure 16–10
Mericlinal, periclinal, and sectorial chimera development. Only the periclinal chimera is stable and of horticultural importance. A segment of one or more apical layers is genetically different with mericlinal chimeras, while periclinal chimeras have one or more genetically distinct apical layers. In sectorial chimeras a segment of all apical cell layers is genetically distinct.

replacement Cells in the inner layer shift into outer layers.

shift from L-I to L-II, is known as **replacement** (32).

Unlayered Meristems Unlayered meristems are found in ferns and many gymnosperm species as well as most roots. The sequence of chimera formation as illustrated in Figure 16–14 (page 609) is as follows: (1) A mutation occurs in one of the few main cells of the meristem which serve as **initials;** (2) with cell division undirected, a small island of mutated cells develops; (3) the mutated area increases and occupies a solid part of the meristem; (4) when viewed as a cross section, the mutant area extends into the center of the stem and produces a sector of mutated tissue. This kind of sectorial chimeral arrangement is rare in angiosperms because of their layered meristem.

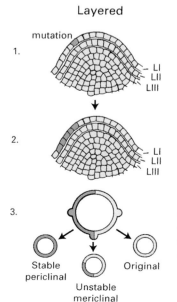

Figure 16–11
Origin of chimeras. Stable periclinal and unstable mericlinal chimeras develop because the mutant cell appears in an individual layer in the apical meristem. A mutation has occurred in a single cell of the L-I layer, which eventually develops into a chimera or reverts back to normal (original), non-mutated cells and tissue 'wild type' (see text).

Seedling Chimeras

Some variegated chimeral patterns are inherited in seedlings. Variegation in geranium was due to the presence of defective plastids, and part of the pattern resulted from the distribution of the chlorophyll-deficient plastids within cells (81). Sectorial chimeras sometimes arose very early in embryo development but degenerated rapidly into branches of nonmutated tissue, mutated tissue, more sectorial chimeras, mericlinal chimeras, and periclinal chimeras.

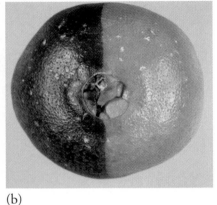
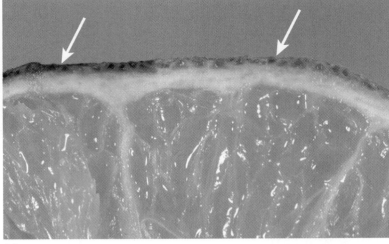

Figure 16–12
Vegetables and fruits can also form chimeras. (a) The periderm of the potato has mixed characteristics of a Russet and white potato, which is likely an unstable mericlinal chimera. (b) Chimeral orange. (c) An unstable mericlinal chimera of orange with differences in the epidermis of the pericarp (peel).

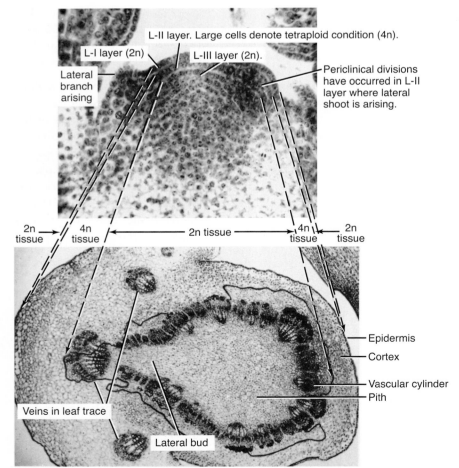

Figure 16–13
Relationship between cells of the histogenic layers of the apical meristem of a peach (*Prunus persica*) and cells of the stem that develop from them. L-I is diploid and produces the epidermis of the stem. L-II is identified in this example because the cells are tetraploid (4x) and larger in size. Cell progeny from this layer makes up the cortex and some of the vascular bundles. The L-III layer is diploid and becomes the inner part of the stem, including the pith and most of the vascular cylinder. From Dermen (32).

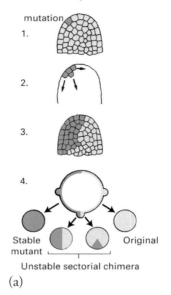

Figure 16–14
(a) Chimera development when a mutation occurs in an unlayered apical meristem. The result is a sectorial chimera, which affects a sector of tissue into the center of the stem. Unlayered growing points occur in ferns, gymnosperms, and in roots or young embryos of other plants (see text). (b) Sectorial chimera in a gymnosperm.

Graft Chimeras

One of the early botanical mysteries was the nature of unusual plants that developed after certain graft combinations were made. In 1644, a strange plant called a "bizzaria" orange was produced following the grafting of a sour orange (*Citrus aurantium*) onto a citron (*C. medica*) rootstock (142, 144). Similarly, unique **graft chimeras** resulted from medlar (*Mespilus germanica*) grafted onto hawthorn (*Crataegus monogyna*) and was known as a hawmedlar (+*Crataegomespilus durdarii*) (49); also *Laburnum* on *Cytisus* formed +*Laburnocytissus* (134). Early in the 20th century, a dispute raged as

> **graft chimera** A chimera sometimes created when an adventitious shoot emerges from callus at a graft that has tissues of both the stock and the scion.

BOX 16.6   GETTING MORE IN DEPTH ON THE SUBJECT
EXAMPLES OF CHIMERAL MUTATIONS

Chimeras in Potato
Some potato cultivars have been found to be chimeras by forcing adventitious buds from "debudded" tubers (modified stem structure). 'Norton Beauty,' which had mottled tubers, gave rise to 'Triumph,' which had red tubers (7). Similarly, 'Golden Wonder,' which had tubers with a thick brown russet skin, when propagated from the inside tissue, yielded plants characteristic of 'Langworthy', whose tubers have thin, white, smooth skin (23). Sometimes unstable, chimeras of tubers form, such as a mericlinal chimera with characteristics of a Russet and white potato (Fig. 16–12).

Fruit Chimeras
A mutation in the L-I or L-II of a structured or layered growing point normally results in a mericlinal chimera (Fig. 16–12). For example, if the mutation produces red pigment on a yellow apple, it would initially appear as a red streak on the fruit but only skin-deep. The stem on which the fruit is produced should have corresponding longitudinal streaks of cells with the yellow or red gene. Lateral buds produced on the stem could produce one of three kinds of fruit: those that produce all yellow fruit (nonmutated), those that produce all red fruit (periclinal chimera), and those that produce streaked fruit (mericlinal chimera).

Leaf Variegation
(Figs. 16–15 and 16–16)
The patterns of variegation in leaves are produced in response to the same basic principle shown by fruits or stems (132). Typically, dicot leaves are formed from three layers of cells L-I, L-II, and L-III. The epidermal layer is derived from the L-I, the middle part of the leaf blade from the L-II, and the central part including the midrib is derived from L-III. The various patterns of leaf variegation are often designated with a letter for each layer. For instance, a leaf with a white marginal variegation would be GWG (green-white-green). A GGG or WGG would be all green and WGW would show a green outer layer with an irregular central white sector.

Deviations from this basic pattern in different species can occur. The relative proportion of the leaf

(Continued)

that originates from the separate layers is not constant, and may vary in the pattern and the number of cells produced by each layer, thus the pattern may differ, and various shades of green may be present. Monocots, such as *Hosta*, only have L-I and L-II layers. So a typical variegation would be GW (green-white) or WG [(white-green), WG more typical in grasses]. GW would have an outer layer of green and inner layer of white, while WG would have a white margin and green inner section.

Figure 16–15

(a) Pinwheel-flowering African violets are chimeras and cannot be propagated true-to-type from leaf cuttings. (b) Adventitious shoots that originate from leaf cuttings flower as either single color or irregularly mottled bicolors. The cultivar 'Valencia' (*upper left* arrow) produced mostly solid-colored violet flowering plants (*lower right*) when leaves were tissue cultured. (c) Anthocyanin fails to develop in the abaxial epidermis of chimeral 'Valencia' African violet (*left* arrow) but does in the leaves of the solid violet flowering off type (*right*), with mottled bicolor-flowering plants having mottled coloration under their leaves (*center*). (d) Periclinal chimeral 'Valencia' African violet (*front* arrow) with its component phenotypes; plants that are violet flowering with dark leaves (*back right*) originating from epidermal tissue, and white-flowering lighter colored leaves (*back left*) originating from subepidermal tissue. (e) Periclinal chimeras with altered layering in the meristem originate occasionally. The periclinal chimera 'Silver Summit' (*top center* arrow) produced two reverse pinwheel flowering types (*bottom left and right*) as well as non-chimeral violet flowering and white-flowering plants (*center left and right*) (78). Courtesy of Dan Lineberger.

Figure 16–16
(a) Not all variegated plants are chimeras. The characteristic variegation of the African violet cultivar 'Tommie Lou' is faithfully reproduced from leaf cuttings and from leaf, petal, and subepidermal tissues *in vitro*. A clone with the same leaf variegation, but with pinwheel flowers ('Candy Lou'), did segregate during micropropagation, proving that the 'Tommie Lou' variegation was not chimeral. (b) Variegated periclinal chimeras in African violet have the typical dicot marginal variegation and segregate during propagation as do pinwheel flowering plants. (c) Plantlets regenerating from leaf of African violet plant grown *in vitro* because it appeared to develop no chlorophyll (albino). Epidermal peels revealed the stomatal guard cells to contain chloroplasts indicating the epidermis to be genetically green. The plant was likely a GWW (green white white) periclinal chimera. Their origin was likely adventitious from a single cell from the epidermal layer.
(d) The variegated clone *Rhododendron xlimbatum* 'President Roosevelt' (*right* white arrow) underwent chimeral rearrangement *in vitro* giving rise to the cultivar 'Carolina Jewel' (*left*). (e) *Rhododendron x limbatum* 'Carolina Jewel' was propagated faithfully *in vitro* for several years, indicating that chimeral rearrangements can produce stable periclinal chimeras (106).
Courtesy Dan Lineberger.

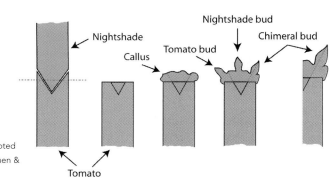

Figure 16–17
Method of producing a graft chimera between nightshade (*Datura*) and tomato (*Lycopersicon*). Adapted from W. N. Jones, *Plant chimeras and graft hybrids*. Courtesy Methuen & Co., London.

to whether or not these represented genetic hybrids produced by grafting. The question was settled by Winkler (160, 161) who was able to synthesize a true graft chimera of tomato (*Lycopersicon esculentum*) on black nightshade (*Solanum nigrum*) and vice versa. In the procedure, the scion of a young grafted plant was cut severely near to the callus area that arose during the healing of the wound (Fig. 16–17). Adventitious shoots from this callus turned out to be a chimera of tomato and nightshade tissue growing together, and not a hybrid. Winkler gave such mixed shoots the name "chimera" after the mythological monster that had a lion's head, a goat's body, and a dragon's tail. These combinations are rare, but a graft chimera of *Camellia* has been discovered (134).

Genetic Variability and Stability in Vegetative Tissue

Maintaining Stability of Chimeras The value of many cultivars in horticulture depends upon their chimeral structure. Maintaining the chimera during propagation is sometimes a problem, depending upon the nature of the chimera and the method of propagation. Periclinal chimeras are relatively stable and maintain their integrity as long as the growing point of the propagule has continuity with the shoot system of the source plant, as in stem cuttings, scions, layers, or division. Propagation that depends upon adventitious shoot formation (root cuttings, some kinds of leaf cuttings, and some types of tissue culture), results in **reversions** to the genotype of the inner (L-II and/or L-III) tissues. Figure 16–16c shows green plantlets regenerating from the albino (white) leaf of an African violet grown *in vitro* (103). The adventitious plantlets evidently arose because the adventitious shoots originated from single cells of the epidermis. Epidermal peels revealed the stomatal guard cells contained chloroplasts indicating the epidermis to be genetically green. The plant was likely a GWW periclinal chimera (103). A solid mutant plant can be reconstituted this way from a chimera—a technique that was used to produce thornless blackberries (see Box 16.7).

> **reversion** A shoot which emerges from inner tissue of a chimera and shows the genotype of the inner tissue.

Reversions of periclinal chimeras may also occur if a cell is displaced into apical cell layers, for example, a shift of cells from L-II into L-I. The visual result would be the random appearance of individual shoots or branches of the reverted phenotype. This problem of reversion can be acute for particular color "sports" of apple and pear. Sometimes reversion can result in a completely solid green shoot in an otherwise variegated plant.

Mericlinal and sectorial chimeras of plants are unstable, resulting in separate branches, some of which are completely periclinal, some completely nonmutant, and others mericlinal (mixtures). Continuous selection toward the mutant type during consecutive propagations would be necessary to **"fix"** the mutated (or non-mutated) form, whichever is desired.

Micropropagation (see Chapters 17 and 18) of chimeras follows the same rule. *Chimeras can only be reliably propagated from axillary buds. Shoot tip and axillary shoot cultures can be expected to reproduce the chimera.* Adventitious shoot initiation can result in **"chimeral breakdown."** However, micropropagation can be used to regenerate a nonchimeral plant from a chimeral plant (52).

> **chimeral breakdown** A reversion toward the inner tissue genotype, particularly during tissue culture propagation.

Detection of Undesirable Mutations Spontaneous mutations occur at regular, but very low rates in vegetatively propagated material. Nevertheless, where propagation of commercial cultivars involves millions of plants, the probability for a mutation to occur somewhere in the clone is high. Figures 16–16d and 16–16e illustrate the development and successful propagation of a stable, periclinal chimera rhododendron. In commercial propagation of chimeras, sometimes there can be a reversion back to the normal "wild type," or mutations to an undesirable chimera form (Fig. 16–18). The propagator vigorously rogues-out any off-types to perpetuate the desired form. Our knowledge of chimeras indicates that a mutation is initially latent and survives first as an unstable chimera, which is usually not detected until after the mutated cells occupy a significant area of the stem. Subsequently, the propagator will usually not

---

BOX 16.7 GETTING MORE IN DEPTH ON THE SUBJECT
MAINTAINING "THORNLESS" CHARACTERISTIC IN BLACKBERRY

Periclinal chimeras of thornless blackberry (*Rubus* sp.) exist in which the epidermis has a gene for the "thornless" mutation that overlays an inner core with the normal, non-mutated gene for "thorns". Roots develop naturally from genetically "thorny" internal tissue. Plants grown as root cuttings or that develop as suckers on a plant show reversion to the thorny, nonmutated genotype, because new shoots on roots are "suckers," which develop spontaneously from adventitious shoots originating from inner tissues. Similarly, breeding for thornlessness is difficult due to the gametes that originate from L-II tissue. Recovery of a completely *nonchimeral, thornless blackberry* has been attained by tissue culture in which new plants were regenerated from cells derived only from the outer layer (52, 98).

(a)

(b)

Figure 16–18
Chimeral reversion in (a) dogwood and (b) fuschia from the desirable chimeral variegation back to a non-mutated, green form (arrow), or other mutation. The propagator needs to vigorously rouge-out these "off-types" and be sure to propagate cuttings with the desired chimeral variegation.

observe the mutation in the plant in which it actually occurs; identification comes in the next generation of vegetative propagation, often after the customer receives the plant.

One method mutation breeders use to increase the probability of detecting a mutation is to prune plants severely to increase the number of shoots—such as those of fruit and nut trees—which speeds up the process that uncovers a mutation (72). Although the probability for any single mutation to occur may be low, the probability of finding it is high because of the single bud propagation and the volume of plants involved. This practice explains why off-type plants are usually detected in a commercial orchard rather than in the source orchard where the propagation material was obtained. If the character affects flowering or fruiting, the mutant may go undetected for a number of years after the plant is planted into its permanent location.

MANAGEMENT OF PHASE VARIATION DURING VEGETATIVE PROPAGATION

Differences Between Ontogenetic and Chronological Aging in Plants

Ontogenetic Aging "Aging" has two separate meanings in plant development. Ontogenetic aging refers to the phases of development that the seedling plant undergoes from embryonic to juvenile to intermediate to mature (adult) as was described in Chapter 2 (Fig. 2–1). The transition from seedling to flowering plant can be very rapid in annual plants (within days), but can take over 20 years for some trees. Although it is convenient to express the length of time required to make this transition in days or years, phase change may best be associated with the number of nodes that have been developed in the apical meristem (Fig. 16–19). Buds produced at a particular node retain the epigenetic potential of that node with

> "aging" Phenotypic changes with age, due either to ontogenetic or to chronological change, or both.

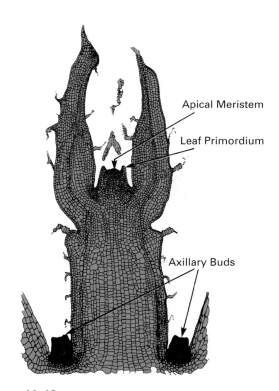

Figure 16–19
Apical buds on a seedling plant undergo shifts in their "ontogenetic age" as cells in their apical meristem divide and form new tissue. Axillary buds typically are dormant due to apical dominance, but retain the ontogenetic age at origin. Once lateral buds start to grow they undergo change in maturation in the same manner as the apical bud, but in their own pattern.

Figure 16–20
Foliage variation gradients within seedling trees and shrubs can be markers of juvenile and mature phenotypes. (a) Mature leaves of *Eucalyptus robusta* are alternate and lancelate in shape, whereas juvenile (arrow) are opposite and have a silver dollar form. (b) *Hedera helix* (English ivy) with juvenile and mature leaves and attached floral structures (arrow). (c) Needles of the basal juvenile part of many conifers tend to be acicular (sharp pointed), whereas needles on the (d) upper, more mature part of the juniper tend to be scale-like.

respect to its ontogenetic age. For example, buds produced at the base of the plant tend to be more juvenile because they were formed first during the juvenile phase of development.

Phase change is best identified by morphological characters for a specific phase, such as leaf shape (Figs. 16–20 and 16–21) or the ability to flower. In some plant species, there is very little obvious change in the plant's appearance as it grows from a seedling to its mature condition, a behavior called **homoblastic.** For other plant species, there

homoblastic Phase changes that involve little obvious change in phenotype with development and age.

Figure 16–21
The climbing vine, *Ficus pumila*, is heteroblastic with dimorphic differences, i.e., juvenile and mature forms. (a) Juvenile leaves are cordate-ovate, with aerial roots present (arrow) and the plant growth habit is horizontal (plagiotrophic). (b) Mature leaves (*left*) are elliptic to oblong-elliptic, and plants have an upright (orthotropic) growth habit, with shoots (c) forming fruit. (d) Plant in transition phase where the leaves (arrow) are small, like the juvenile, but have characteristics of the mature form.

heteroblastic Phase changes that show distinct change in phenotype during development.

is a distinct variation in specific traits that occur with age, which is called **heteroblastic**, such as physiologically juvenile and mature creeping fig (*Ficus pumila*) (Fig. 16–21) (29). Several genes and their gene products have been identified that are differentially expressed in the juvenile or mature phases on the same plant. These can be used as molecular or biochemical markers of phase development (51).

Chronological aging continues through the life of an individual plant whether seedling or vegetatively propagated. This characteristic may best be designated by the number of years that the plant has grown either from seed or vegetative propagule. When young and

---

BOX 16.8 GETTING MORE IN DEPTH ON THE SUBJECT
EQUIVALENCE OF PHASES IN ANIMALS AND PLANTS

Phase changes can be considered to be biologically equivalent in animals, humans, and plants in that embryonic, juvenile, adolescent, and mature phases are present in each. However, animals and plants differ in the way they grow and age. Cells in all parts of animals (and humans) more or less age together. Plants, on the other hand, grow from consecutive cell divisions in the meristem of apical and lateral shoots laid down in sequence over time. Juvenile to mature change occurs in the apical meristem as the shoot grows, but cells remaining behind retain their *initial ontogenetic age*; in other words, there are physiologically juvenile cells in a tree that chronologically is 100-years old. In perennials, ontogenetic aging continues in alternating cycles of growth and dormancy. In annuals and biennials, ontogenic aging terminates in reproductive structures, and the plant dies.

The phase of maturation in different locations of the plant is determined by the pattern of bud development in the plant. Juvenile and mature characteristics, such as flowering, leaf shape, or thorniness, can be used as markers of maturation, their location depending on the chronological age at which *ontogenetic maturity* is attained in particular parts of the plant.

BOX 16.9 GETTING MORE IN DEPTH ON THE SUBJECT
PARADOX IN TERMINOLOGY

The comparison of the two types of "aging" produces a paradox in horticultural terminology. That part of a seedling plant nearest to its base is the "oldest" in terms of chronology, but actually "youngest" (i.e., more physiologically juvenile) in terms of ontogenetic age (see Fig. 16–22).

Likewise, the outer peripheries of the stems and branches are the "oldest" in physiological maturity but "youngest" in chronology (13, 31, 66). The physiological age of a vegetatively propagated plant depends on the ontogenetic age of the plant part used for propagation.

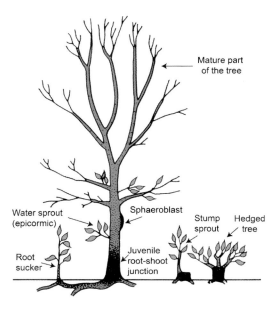

Figure 16–22
Juvenile-mature gradients occur in seedling trees from the base of the tree to the top. The Juvenile root-shoot junction, which is ontogenetically 'physiologically juvenile' with high rooting potential—even though chronologically it may be decades old. Flowering occurs in the ontogenetically 'physiologically mature' part of tree at the apical part, even though some of the flowering shoots may be chronologically only several months old; shoots taken from this region generally have low rooting potential. Juvenile structures arising from the 'cone of juvenility' (dark area) near the base (crown) of the tree include: adventitious root 'sucker,' watersprout (epicormic shoot), and sphaeroblast. Stump sprout from severe pruning, and shoots from heavily pruned or hedged bush. Rooting potential is highest from these structures close to the cone of juvenility. Redrawn with permission from Bonga (14).

growing under favorable conditions, the plant should be vigorous, healthy, and may become very floriferous. Eventually, as the plant gets older, vigor declines and, although flowering and fruiting may remain abundant, new growth will tend to be reduced. The plant eventually becomes senescent and may eventually die. Older plants can be returned to a vigorous and productive state by repropagation or by applying various horticultural practices, such as pruning, nitrogen fertilization, irrigation, controlling pests and diseases, and, in general, applying good management practices.

Phase Variation and Vegetative Propagation

Ontogenetic aging is important in vegetative propagation because buds (both terminal and lateral), when removed for vegetative propagation, perpetuate their ontogenetic age in the progeny plant. For instance, if the source plant was juvenile (or the propagule was taken from a juvenile part of the plant), the new progeny plant should initially express the **juvenile phase** of the source plant but shift toward the **mature phase** in subsequent growth. Eventually the plant stabilizes at the mature reproductive phase of development.

Vegetative propagation from mature branches of a seedling plant consequently may not reproduce the seedling phenotype in the S_1 generation (Fig. 16–2, page 598). Propagules taken from the base of the plant tend to produce biologically juvenile plants; those taken from the top or at the periphery of the plant tend to produce progeny that are biologically mature; the entire range may be produced if plants from intermediate locations are included. The recognition of this variability phenomenon by foresters led to the coining of the terms *ortet* (the S_0 plant) and *ramet* (S_1 plants) (Fig. 16–2) (138). These variations are not due to genetic change but to *epigenetic variation* in maturation that persists during vegetative propagation.

Juvenile Versus Mature Phenotypes Continuous growth and consecutive generations of vegetative propagation lead to the stabilization (i.e., "fixing") of the mature phase in individual plants of the clone and the disappearance of seedling phase characteristics. Most clonal cultivars grown for their fruits, seeds, or flowers exhibit the mature growth phase of the clonal cycle as the typical **"true-to-type"** form. Most individual plants come into flower and fruit reasonably quickly after a vegetative phase of development and undesirable traits associated with plants in the juvenile phase disappear. It is not appropriate to refer to the vegetative phase of the stabilized mature clone as "juvenile."

There are many examples of the application of phase changes in horticulture:

> **Mature phenotype**
> The characteristics that are associated with the adult phase, typically bushy growth, reduced vigor, no thorns, and profuse flowering.

> **true-to-type**
> Conforming to the phenotypic expectations of the specific cultivar.

> **juvenile phenotype**
> The characteristics that are associated with the juvenile phase in some species, usually upright growth, vigor, sometimes thorniness, and lack of flowering.

- Seedlings of many fruit and nut species, such as apple (*Malus*), pear (*Pyrus*), citrus, and walnut (*Juglans*), may take a long time to come into flowering (see Table 2–2). Cultivars grafted with scions from mature phase stock plants flower and produce fruit or nuts at an earlier chronological age (85) and exhibit a more uniform, desirable tree form. Similarly, a flowering ornamental plant grown from seed may require many years to flower. For example, seedling *Wisteria* and *Magnolia* take 7 and 15 years, respectively, to flower, whereas vegetatively propagated plants flower at a younger age (84). The value of seed orchards of specific genetic sources in forestry results from their production at an early age, compared to their natural behavior as seedlings (115, 166).
- Orchids, bulb species, and other herbaceous perennials experience phase change and may require 5 to 10 years of annual development from seed before flowering occurs (59). Once the mature (adult) stage is reached, annual flowering will occur with alternate phases of vegetative growth (bulbs, corms, pseudobulbs) and reproduction (flowering) (see Chapter 15).
- Some ornamental vines, such as *Hedera* sp. or creeping fig (*Ficus pumila*), are primarily vines in the juvenile phase, whereas the mature flowering phase is shrub-like (Figs. 2–3, 2–4 and 16–21). Both forms may be useful but for different purposes. *Eucalyptus* and *Pseudopanax* have strikingly unique juvenile foliage (Fig. 2–2). Some conifer cultivars have juvenile foliage (needle- or awl-like) distinctly different from the adult foliage (flat, rounded, or scalelike) (90). Some cultivars (*Chamaecyparis pisifera* 'Boulevard' and *Cryptomeria japonica* 'Elegans') are more or less "fixed" in the juvenile leaf phase. Others (*Pinus sylvestris*, *Cryptomeria japonica*), junipers (*Juniperus*), and Lawson cypress (*Chamaecyparis lawsonia*) may develop both types of foliage (see Fig. 2–4).

- The juvenile phase of most plants inherently has a higher rooting potential than the mature phase. The application of juvenility to cutting propagation was described in depth in Chapters 9 and 10. The easy-to-root juvenile vine form of ivy (*Hedera*) has been a model system for studying root initiation (50, 51). The juvenile phases of the conifers listed in the previous paragraph invariably root easily and can be maintained as cultivars. The rooting potential of many if not most tree species is correlated closely with seedling age and the retention of the juvenile phase (Fig. 16–23). A reason previously cited for cultivars of fruit and nut crops to be propagated by grafting was the difficulty of rooting associated with their mature phase. For clonal rootstocks, selection and maintenance of the juvenile phase are important strategies for propagation in order to improve rooting potential. See Box 16.10 for a description of a unique forestry breeding and propagation system that involves both seedling selection and clonal propagation.
- The juvenile apomictic seedlings of *Citrus* are very different from the typical mature phase of the cultivar propagated for commercial production. The seedlings are very vigorous, thorny, often have different shaped leaves, show low production, and have fruits of very inferior quality. On the other hand, the traits of uniformity and vigor make the apomictic seedlings highly desirable rootstocks (18, 45, 146), for which they are used.
- "Clonal forestry" depends on the ability to mass-propagate populations of juvenile propagules that not only exhibit easy rooting (Fig. 16–23), but also produce the desirable juvenile "seedling" phenotype growth characteristics (2). The desirable "true-to-type" form of forest trees grown for lumber and wood is straight-tapered trunk, reduced branching, lateness to flower, and other juvenile characteristics. Vegetatively propagated plants of Monterey pine (*Pinus radiata*), loblolly pine (*Pinus*), giant sequoia (*Sequoiadendron*), and eastern larch (*Larix*) taken from more mature locations in the tree showed reduced height and diameter, and more branching (47). Success in "clonal forestry" depends on the ability to control maturation and to select the proper "true-to-type" stage (2, 115). See Box 16.10.

Promoting Shifts from Juvenile to Mature Phases The long period of time required for seedling plants of many woody species to come into flower (see Table 2–1) has important applications in horticulture. Breeding programs of woody plants primarily use seedlings, but progress can be hampered by their long juvenile period (54). Breeding can result in shorter breeding and production cycles, since the juvenile period is genetically controlled (156). Growth and development can also be managed by environmental control. The most important concept is to keep the plant continuously growing in

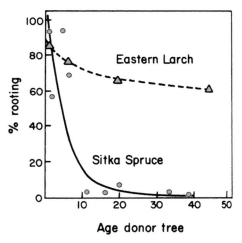

Figure 16–23
Graph showing the effect of age on donor tree relating to key tree characteristics that are important in clonal forestry. Rooting percentages of cuttings from donor trees of different ages. From Greenwood et al., (47) in *Clonal Forestry* by permission.

BOX 16.10 GETTING MORE IN DEPTH ON THE SUBJECT
IMPROVEMENT PROGRAMS IN FORESTRY SPECIES (27, 114, 115)

(a) Seed is taken from plants of seedling families that have been previously selected for superior growth rates, form, disease resistance, and so on. (b) These seedling blocks (referred to as ortets) are kept in a juvenile/transition phase by hedging and shearing. (c) Cuttings (ramets) are obtained and can be rooted because of their juvenile condition. (d) Rooted plants are transplanted to timber production sites for evaluation over time. (e) Superior ortet families (sourceplants, stockplants) are consequently identified. (f) Cuttings, taken from the seedling plants of these selected elite families which have been maintained by hedging and shearing are multiplied into commercial volumes ("bulking up"). (g) Clonally generated ramets are established in the forest planting in **mosaic blocks** of different clones to avoid monoculture production and to reinforce genetic diversity (115).

"bulking up" (in forestry propagation) Using vegetative propagation to multiply the supply of selected genotypes.

order to grow through the juvenile phases as quickly as possible. For example, birch seedlings grown continuously under long days in the greenhouse during winter passed through the juvenile into the mature phase much faster than under normal short-day winter conditions (116). Crab apple (*Malus hupehensis*) seedlings grown continuously in a greenhouse to 2.5 to 3.0 m (8 to 10 ft) started to flower 13 months from germination instead of 4 years (165).

Restricting growth, girdling, bending, growing on dwarfing rootstocks, and other horticultural practices are often cited to induce flowering (159). However, these are not useful until the plant approaches the transitional or mature phase. Prior to that phase, growth should be encouraged to allow the plant to grow through the juvenile stage. Micro-budding of citrus uses very young, containerized seedling rootstock, which avoids the nursery stage since the young grafted material are directly planted into the orchard (Fig. 13–22). Oranges and grapefruit are more precocious, producing fruit within 14 to 18 months after micro-budding (71).

Selection and Maintenance of the Juvenile Phase
Retention of the juvenile phase in a propagation source block may be important to increase rooting potential (see Chapter 9). Following are methods to select, delay, or preserve propagation sources in the juvenile phase:

- Collect material and propagate from plant parts showing juvenile characteristics. This may mean collecting from the base of a plant in the case of a seedling. Some cultivars are more or less "fixed" as juvenile (e.g., various conifer cultivars with juvenile foliage).
- Collect material from root sprouts (**suckers**) or from watersprouts (**epicormic** shoots) arising from near the base of the plant. These arise from latent lateral buds that are located in the "**cone of juvenility.**" The shoots may be forced by cutting back the plant to the ground to produce "**stump sprouts**" (Fig. 16–22) (43).
- Large wood stem segments of white ash and silver maple placed under intermittent mist: epicormic shoots can be forced and are later harvested as softwood cutting and rooted under mist (Fig. 16–24)(109, 110).

epicormic A shoot emerging from a latent bud on the base of a tree; synonymous with watersprouts.

cone of juvenility A cone-shaped area comprising the trunk and lower branches of a seedling tree, which tends to remain juvenile.

stump sprouts The vigorous shoots that are produced from the stump when a tree is pruned back severely.

Figure 16–24
Forcing softwood cuttings from large woody stem segments to propagate hardwood species. Epicormic shoot forcing of white ash and silver maple under intermittent mist. The forced epicormic shoots (arrows) from latent axillary buds will later be harvested as softwood cutting and rooted under mist. These axillary buds may have a chronological age of several years, but high rooting success because they retain juvenile traits of good rooting. Photo courtesy J.E. Preece (110).

- Some plant species produce **sphaeroblasts** (i.e., masses of adventitious buds located on the lower portion of the trunk) (Fig. 16–22). Examples of species with such characteristics include sequoia (*Sequoiadendron*), redwood (*Sequoia*) (15), some quince cultivars (*Cydonia oblonga*), and some apple (*Malus*) cultivars (138). Shoots developing from these structures invariably are juvenile, show characteristic morphology, and are easy to root.

sphaeroblasts The masses of adventitious buds that are produced on the lower trunk of some tree species.

- Establish **hedge rows** of stock plants produced by propagating juvenile material (see Chapters 10 and 11, and Figs. 9–27 and 10–23). The hedgerow method is perhaps the most important for providing large populations of juvenile cutting material (76). Plants are severely pruned annually to a constant height so they will produce a continuous supply of new shoots. The original material

hedge row A row of trees or bushes pruned back to a hedge in order to stimulate shoots for propagation.

(a)   (b)

Figure 16–25
A mounded and stooled callery pear. (a) The white, etiolated epicormic shoots originated from latent buds that were forced when the trunk was severed and covered with soil-media. These epicormic shoots have high rooting potential since they are close to the crown of the tree. (b) The epicormic shoots have acclimated to the light, greened-up, and have normal-appearing shoots and leaves. They can be severed from the stump, treated with auxin, and rooted under mist as stem cuttings.

may be seedling populations (114), rooted cuttings, or grafted plants on rootstocks. For the latter, care must be taken to eliminate rootstock shoots, or to choose a rootstock with morphologically distinct leaf characteristics.

- Juvenile material can also be maintained by consecutive ("serial") propagations of several generations from seedling material (2). **Serial propagation** is practiced in the annual propagation of new generations of cuttings collected from asexually propagated container-produced woody and herbaceous ornamentals.

  serial propagation Annual collection and rooting of cuttings from previously rooted plants.

- Establish "stoolshoot" beds by cutting shoots nearly to the ground (84). This is perhaps one of the reasons for rooting success in mounding/stooling and trench layering (see Chapter 15 and Figs. 14–1, 14–14, and 16–25).

Reversions from Mature to Juvenile Phase (Rejuvenation) The mature phase naturally reverts to the juvenile phase during seed reproduction, whether sexual or apomictic (67). However, reversions have also been induced from the mature to the juvenile phase in vegetative material. Examples include the following:

rejuvenation The shift from the adult phase to the juvenile or transitional phase.

- Grafting consecutive generations of scions in the mature phase to seedling rootstocks (15, 43) has produced reversion from mature to juvenile phases (Figs. 9–25 and 16–26). Successful results have been achieved with *Sequoia*, *Eucalyptus* (125), and other forest tree species (13, 14).

- Micropropagation has resulted in the reversion of plant shoots to the juvenile phase (see Chapter 17). For instance, Mullins (95) found that cultured somatic embryos from the ancient clone 'Sauvignon Blanc' produced "seedlings" with typical juvenile foliage. Micropropagated cultivars used as stock plants have resulted in increased rooting both *in vitro* as well as from *in vitro*-produced plantlets established as stock plants (57, 105) (see Chapter 17). Repetitive subculturing of mature walnut (*Juglans*) shoot tip culture led to rejuvenation of mature clones and enhanced rooting of micro-cuttings (36).

- Some hormone treatments have resulted in reversions under experimental conditions. For example, applications of gibberellic acid have induced juvenile growth in ivy (*Hedera*), which could be reversed with abscisic acid (50).

PATHOGENS AND PLANT PROPAGATION

The presence of disease can be the most important limiting factor in both plant production and plant propagation. In this section, the unique relationships between specific plant pathogens and vegetative propagation will be examined. In nature, plants have coexisted with many other organisms, including insects, mites, fungi, bacteria,

(a)   (b)   (c)

Figure 16–26
Rejuvenation produced by consecutive (serial) grafting from a ten-year-old mature *eucalyptus* tree. (a) Plant produced by cleft graft of scions from the ten-year-old tree onto a seedling rootstock showed mature leaves. (b) Plant produced from scions after three serial grafts. One of the shoots shows mature characteristics, the other juvenile. (c) Plant produced after grafting with scions after six serial grafts. All of the shoots produced juvenile (large) leaves. Cuttings from this plant rooted successfully, whereas rooting from the 10-year-old original mature tree was not successful (125). Photos courtesy Siniscalco and Pavolettoni.

and nematodes, to evolve more or less compatible interdependent communities. In many cases, mycorrhizal fungi, plants and microorganisms form mutually beneficial associations (26, 28, 129, 130, 155). When humans began to cultivate plants the balances changed. Humankind began to propagate specific kinds of crop plants and adopt monocultural systems, which were more vulnerable to attack by specific pests. As the civilized world expanded, individual plant species were transported to new areas of the world (see Chapter 1) where they were introduced to new environments and different pathogens and parasites. Pests and pathogens existing in one area "hitchhiked" along with plants to new areas and became established there. A classic example of this problem was the introduction of the "root louse" (*Phylloxera*) from the United States to Europe in infested grapevines. The high susceptibility of European grape cultivars, which were grown largely as rooted cuttings, devastated the European wine industry. The industry was saved by the subsequent introduction of resistant grape rootstocks from America to which the European cultivars could be grafted. Another example was the introduction of infected potato tubers to Ireland where local cultivars succumbed to *Phytophthora* fungus (86, 117).

BOX 16.11   GETTING MORE IN DEPTH ON THE SUBJECT
TOPOPHYSIS

Topophysis is defined as how the propagule's position on the plant affects the type of vegetative growth subsequently shown by the vegetative progeny (90, 102). Plants of certain species produced by cuttings taken from upright shoots (**orthotropic**) will produce plants in which the shoots grow vertically. Plants produced from cuttings taken from lateral shoots (**plagiotropic**) will grow horizontally, as occurs with *Podocarpus* (Fig.16–27). The phenomenon can occur in many conifer species primarily as a function of the adult form (2). With Fraser fir (*Abies fraseri*), lateral cuttings close to the base of the stock plant can lead to undesirable plagiotropic growth of rooted liners (Fig. 9–29). Horizontal and prostrate plant forms, which are desirable as ornamentals, may be obtained by taking cuttings from horizontally growing branches. Norfolk Island pine (*Araucaria*) is an example of topophysis that produces flat growing plants for pot culture. In some species, such as *Sequoia*, pruning can restore the orthotropic growth habit (15).

   topophysis   The effect of the source plant propagule's position on the phenotype of the progeny plants.

   orthotropic   The upright growth of the progeny plant, which comes from using the source's upright growing branches for cuttings.

   plagiotropic   The lateral growth of the progeny plant, which comes from using the source's laterally growing branches for cuttings.

Figure 16–27
Topophysis in *podocarpus*. (a) Plant propagated from a lateral shoot produces a horizontal growth habit (plagiotropic). (b) Podocarpus propagated from a vertically oriented cutting maintains an upright growth pattern (orthotropic).

small, but can be observed with the aid of an electron microscope. The etiology of many plant virus diseases is now known, and control procedures are a major aspect of propagation systems.

> **viruses** Microscopic organisms that replicate as obligate parasites within plant cells or the bodies of insect vectors, multiplying and growing in collaboration with chromosomes.

Cultivars and species differ in their tolerance to viruses (22, 96, 150). The effect may be sufficiently severe to kill the plant or make it stunted and unproductive. Others affect fruit and flower quality or produce symptoms of chlorosis, albinism, mosaic, and/or variegation patterns. Others are essentially symptomless and have little obvious effect on plant appearance, production, or growth (86). Opinions differ as to the seriousness of such latent viruses (22, 44). However, if tolerant themselves, infected plants are reservoirs of infection, and a threat to other potentially susceptible cultivars.

Systemic Pathogens and Clones

Propagation is a vulnerable phase of plant production since an environment of high moisture and warm temperatures is also optimal for pathogens. Some propagation practices were designed to suppress disease rather than to control the causes of disease.

Viruses These small submicroscopic organisms have a specific structure composed of nucleic acids (RNA or DNA) encased in an outer sheath of a protein (63, 74). **Viruses** replicate within plant cells or in the bodies of insect vectors as obligate parasites. Viruses are very

Phytoplasma (Mycoplasma-like Organisms) Phytoplasmas are specialized bacteria that are obligate parasites of phloem tissue and of specific phloem-feeding leafhoppers (*Homoptera*). These organisms multiply rapidly, utilizing host cells for their development and survival. They disrupt translocation in the phloem, causing wilting and yellowing. Hormonal disturbances can occur that result in proliferation of shoots (commonly named "witches' brooms") and other malformations, such as conversion of flower parts to leafy structures.

> **phytoplasma** A class of extremely small parasitic organisms that cause disease in plants.

BOX 16.12 GETTING MORE IN DEPTH ON THE SUBJECT
PLANT PATHOGEN–PLANT CULTIVAR COMPLEXES

Two major developments occurred during the middle 1800s, which, among others, have dominated plant culture ever since. One was the demonstration that plant diseases were caused by pathogens, as Louis Pasteur and others had shown earlier with human diseases. Since then, generations of plant pathologists have identified the **causal agents**, determined their **etiology**, and devised *control strategies* for numerous diseases that caused short-term disasters and long-term declines. A second major discovery was that plant diseases and pathogens could be controlled by *chemical* methods, first demonstrated by the famous Bordeaux Mixture (copper sulfate and lime), which controlled downy mildew on grapes (86, 117). Since that time, the "chemical revolution" has extended across the agricultural spectrum. Recent concerns about the environmental impact of chemicals as well as the destabilization of the natural ecosystem of plant and pathogen have led to revised concepts of control including Integrated Pest Management (IPM) (see Chapter 3), biological control, and genetically modified pest-protected plants.

> **causal agent** A pathogen (or other agent) that produces a specific disease.
>
> **etiology** Identifying the characteristics of a causal agent.

Plant diseases caused by *phytoplasma* include *aster yellows* in strawberry, some vegetables, and ornamentals, blueberry *stunt disease,* *X-disease* in stone fruit trees, and lethal yellowing of some palm species. Differential susceptibility to *phytoplasmas* (between scion cultivar and rootstock) causes *pear decline. Phytoplasmas* move among plants by insect vectors, usually leafhoppers or *psylla,* and are carried in propagating materials. A similar organism, known as *Spiroplasma citri,* causes *stubborn disease* in citrus.

Fastidious Bacteria

fastidious bacteria
The single-celled bacteria-like organisms that lack a nucleus and cause specific diseases in plants.

Fastidious bacteria are single-celled bacteria-like organisms that do not have a nucleus, are difficult to culture, and occur in the xylem and phloem of susceptible plants and in the bodies of leafhopper vectors (42). Fastidious bacteria are carried in vegetative propagating material. *Club leaf* in clover and greening in citrus are examples of phloem-limited fastidious bacteria. Citrus Greening is a highly destructive disease, caused by a fastidious bacteria, spread by psyllid insect vectors, that has yet to be cultured. As of this writing, there are major quarantines in Florida and Brazil, isolating citrus nurseries to psyllid and Citrus Greening-free areas, restricted movement of all citrus propagules, and vigorously rouging-out of infected plants (79) http://www.aphis.usda.gov/plant_health/plant_pest_info/citrus_greening/index.shtml. *Pierce's disease* in grape, *almond leaf scorch, phony peach, plum leaf scald,* and *leaf scorch* of various trees are caused by xylem-limited fastidious bacteria. Symptoms include leaf necrosis, wilting, small fruit, and decline. Control can be achieved by removing diseased orchard trees and wild hosts and using tolerant cultivars and rootstocks. To free infected bud-wood, dormant cuttings can be treated with hot water (see page 629).

Viroids Viroids are extremely small agents composed mostly of RNA about one-tenth the size of the smallest virus. Viroids cause such infectious diseases as *potato spindle-tuber, citrus exochortis, chrysanthemum mottle, chrysanthemum stunt, avocado sunblotch,* and *cadang-cadang* in coconut (33, 141).

Viroid structure is a circular, single-stranded RNA molecule of low molecular weight. Some, such as *citrus exocortis,* are transmitted mechanically on pruning shears and during grafting. Viroids are carried in vegetative propagules as well as in some seeds. Some multiply more rapidly at high temperatures.

Symptoms are sometimes slow to develop and difficult to identify. While detection of viroids through biological indexing on sensitive indicator plants is one of the most reliable indicators, molecular techniques of RT-PCR (real-time polymerase chain reaction) are used in virus-free citrus budwood certification programs (71). Other methods of detection involve the extraction and purification of nucleic acids, followed by electrophoresis to detect the unique RNA viroid (42, 74). Dot-blot hybridization assays use nucleic acid probes.

BOX 16.13 GETTING MORE IN DEPTH ON THE SUBJECT
VEGETATIVE PROPAGATION AND SYSTEMIC DISEASES

During the early history of horticulture, vegetatively propagated clones were associated with many production problems. Horticulturists and orchardists of the 1600s and 1700s (70, 90, 145) believed that fruit cultivars tended to "run out" as a result of aging and repeated propagation; hence, vegetatively propagated cultivars were thought to need replacement periodically with new seedling-produced cultivars (70). Some of these decline problems were caused by chronic diseases. The concept of viruses emerged as Adolf Mayer in 1886 demonstrated the transmission of *tobacco mosaic* with juice from a diseased plant to a healthy plant. Erwin Smith in 1888, a USDA scientist, showed that *peach yellows,* a disease that was devastating peach orchards of the eastern United States, could be transmitted from a diseased to a healthy plant by budding. Since that time, plant virologists have demonstrated that very small, transmissible organisms were the cause of numerous plant disorders and the primary reason for clonal degeneration (145).

Three aspects of the plant-virus complex should be emphasized: (1) Once a specific virus infects the plant it is systemic within that plant and all the plants vegetatively propagated from it; (2) Most viruses also become systemic within specific insects that feed on the host plant, particularly various aphids, which then spread the virus from plant to plant; and (3) The propagator can also spread these pathogens by collecting scion, buds, cuttings, and other propagules from infected plants (131). Today, controlling systemic pathogens in propagation sources is a major aspect of vegetative propagation systems.

In a few instances propagators purposely propagate some plant species with systemic diseases, such as phytoplasma infection of some commercial poinsettias, which induces free-branching of the crop. Variegation or mottling of the leaf of the ornament flowering maple (*Abutilon striatum*) is caused by the Abutilon mosaic viruses. Here the plants suffer not loss in health, but the enhanced ornamental features from the systemic diseases are desirable.

SELECTION AND MANAGEMENT OF PROPAGATION SOURCES

Characteristics of Sources

A source is the specific plant or group of plants from which propagules (cuttings, buds, scions, layers, etc.) are collected. The selection of a proper source is essential for successful vegetative propagation. In the previous chapters, the influence of source plants on the ability to make cuttings, to successfully graft or bud, and to do other procedures was considered. In the remainder of this chapter, we will consider three fundamental questions: How can we be sure that the plants produced are the correct cultivar (**true-to-name**), have not deviated from the standard for that cultivar (**true-to-type**), and are not **infected** with serious pathogens (pathogen-free)? The existence of any of the three problems can frustrate propagators, enrage customers, and sometimes cause large economic losses.

> **true-to-name** The characteristic of being the specified cultivar.

Trueness-to-Name Plant cultivars have names, and a successful nursery industry requires that plants being produced and sold be correctly labeled by scientific name, cultivar name, or both (157). The process begins by identifying the source of plant material at the time of collection, continues by maintaining records during propagation, and is complete when labels are applied to the final product. Nevertheless, mistakes can happen, plus a switch in labels or plants and/or change in name can be deliberately made.

Visual Inspection. Historically, identification of species and cultivar has been based on personal knowledge or written descriptions, but some problems occur. First, the plant may need to be grown under appropriate conditions to counteract any genotype x environment interaction. Plants must also be examined at the proper stage of development to ascertain their essential phenotypic characteristics. For example, fruiting or flowering cultivars should have their fruit or flowers inspected. Nevertheless, personal expertise of the propagator in identification continues to be essential.

Isozymes (91, 126, 143). Isozymes are different allelic forms of the same enzyme produced by a plant. They can be visualized by their different mobilities on an electrophoretic gel (see Fig. 2–20). The pattern of different **isozymes** is characteristic of specific genotypes and can be used to identify genetic differences and often specific cultivars.

> **isozyme** The genetic variants of specific enzymes, which can be identified by biochemical tests and used as genetic markers.

DNA-Based Marker Technology (73, 104, 112). Fundamentally, if two cultivars differ in appearance, they should have different genotypes. Technology has developed specific methods to identify cultivars and the novel patterns of their DNA. Several systems have been developed including **RFLP** (Restriction Fragment Length Polymorphism), **RAPD** (Randomly Amplified Polymorphic DNA), and SSRs (Simple Sequence Repeats). For more detailed information on these procedures see Box 16.15 and the information in Chapter 2.

Trueness-to-Type The propagation source must not only be the correct cultivar, but the characteristics of the progeny plants must conform to an accepted norm for that cultivar when grown in its ultimate location (i.e., be true-to-type). Trueness-to-type deviations among clones can result from several causes that have been described in this chapter including (a) juvenile to mature phases, (b) genetic variation, and (c) inherited

> **trueness-to-type** Corresponding to phenotypic characteristics of the source plant.

> ### BOX 16.14   GETTING MORE IN DEPTH ON THE SUBJECT
> ### PLANT PATENTS AND TRADEMARKS
>
>
>
> Also of concern to the propagator is the knowledge of any patents and trademarks associated with a plant cultivar. A **plant patent** is granted by the government to an inventor, who has invented or discovered and asexually reproduced a distinct and new variety of plant, other than a tuber propagated plant or a plant found in an uncultivated state. A plant propagator would pay a license fee to propagate the inventor's patented plant. A plant **trademark name** is intended to be used only to designate product origin or brands. There are issues with naming some plants which may violate the International Code of Nomenclature for Cultivated Plants (ICNCP), U. S. Trademark Law, and occasionally the U. S. Federal Trade Commission (FTC). Unfortunately, there can be confusion between a cultivar name and a company's marketing name. Further information can be found at the U. S. Patent and Trademark Office website http://www.uspto.gov/web/offices/pac/plant.

disorders and other conditions where the cause is unknown. Many of these problems are unique to specific plant species, cultivars, or individual propagation sources, and the procedures used for evaluation are often species- or cultivar-based.

Evaluating for trueness-to-type depends on **visual inspection** by individuals knowledgeable about a particular plant species or cultivar. True-to-type evaluation is primarily a process of **phenotypic selection** (visual inspection and evaluation of the **source** plants) versus **genotypic selection** (visual inspection and evaluation of the **progeny** plants). Phenotypic selection may fail because a plant judged "normal" may not show its latent potential for problems. Also, environmental or developmental interactions may provide a "false" reading by indicating that a plant is off-type when it is merely expressing environmental effects. For example, stock plants that are severely pruned for cuttings or scions may produce vigorous vegetative shoots that produce atypical fruit or nut phenotypes (68).

Genotypic selection requires that a **vegetative progeny test** be conducted. This procedure means that a specified sample of progeny plants be propagated and grown in a specified environment for a specified length of time. This procedure is analogous to the seedling progeny test described in Chapter 5.

phenotypic selection Source selection based on phenotypic appearance of the source plant.

genotypic selection Source selection based on phenotypic appearance of the vegetative progeny.

vegetative progeny test The practice of clonally propagating progeny to test their ability to reproduce the source plant.

Freedom from Pathogens Viruses and related organisms such as phytoplasmas, viroids, and fastidious bacteria have been among the most serious problems of clonally propagated cultivars. Once a plant is infected, the pathogen becomes systemic, spreads to unaffected parts of the plant, crosses graft unions, and becomes a source of inoculum in bud-wood, seed, or pollen. The infected source plant serves as a reservoir for disease dissemination by insect vectors or pollen transfer. The most important virus vectors are various species of aphid.

BOX 16.15 GETTING MORE IN DEPTH ON THE SUBJECT
TESTING FOR SSR FINGERPRINTS

The information described in Chapter 2 included the concept of **base** (nucleotide) **pairing** to characterize the pattern of genes on chromosomes, such as thymine: adenine (TA) and guanine: cytosine (GC). In characterizing base-pair patterns in many organisms, including humans, fruit flies, mice, and others, a discovery was made that chromosomes have two kinds of base-pair (bp) patterns. In one type, the bp pattern is variable, unique, and associated with specific genes. For instance, these could be CCGTCTCAACT on one strand with the complementary bases on the other. These are called **nonrepeats.** In the second type of pattern, the area is composed of consecutive **repeats** of the same base-pair. For instance, these could be CTCTCTCTCT with the complementary bases as GAGAGAGAGA on the other strand. The areas of repeating bp are called **microsatellites** or **simple sequence repeats (SSRs).** These unique areas are widely distributed in all organisms, and their patterns among individual genotypes show differences among clones (Fig. 16–28) (12, 148, 164) and can be used to analyze relationships among species (41, 48, 73).

base-pairs (bps)   The combinations of bases, such as thymine:adenine (TA) and guanine:cytosine (GC), which make up base sequences of genes.

nonrepeats   The areas of the chromosome in which base-pairs are not repeated, apparently marking specific areas where genes are present.

repeats   The areas of the chromosome in which the same base-pairs are repeated. The length of the repeated chromosomes determines specific genetic markers and is characteristic of specific genotypes.

simple sequence repeats (SSRs) or microsatellites   The areas of repeating base-pairs which act as markers in DNA fingerprinting.

Following is the general procedure for an SSR test: (a) DNA is extracted from the test organism and broken into fragments with restriction enzymes. (b) These extracts are heated to separate the two complementary DNA strands and then hybridized with two reference primers. (c) The extract is treated with DNA polymerase and heated, whereupon the fragments of the single DNA strand will replicate (up to 1 million times in a few hours) to provide workable amounts of material. (d) These amplified extracts are loaded onto an agarose gel in an electrophoresis plate. The microsatellite fragments will migrate to different positions on the gel corresponding to the number of repeating bps in the fragment. (e) These segments are stored in a genome library (laboratory) for future use as primers. (f) By repeating this process, the entire genome can become characterized and used as a reference for conducting future tests.

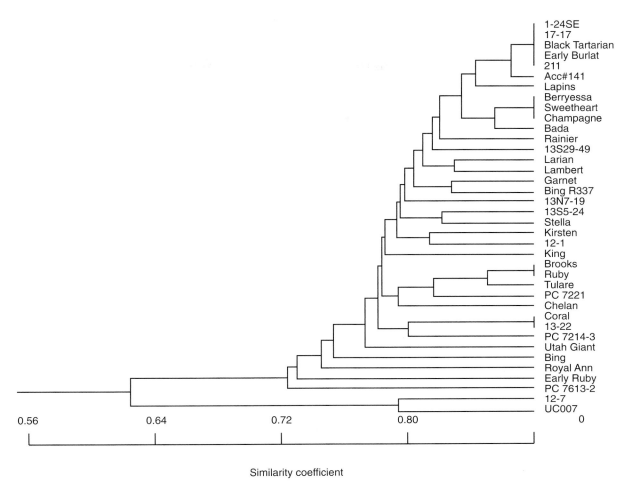

Figure 16–28
Analysis of marker data is used to develop a dendrogram showing genetic differences among cherry genotypes. See Figure 2–20 for an example of gel electrophoresis showing the migration of proteins, which is what is used for determining differences of simple sequence repeats (SSR) markers among cultivars of a species. Courtesy of S. Southwick and D. Struss.

During the 1950s, the concept of pathogen-tested source trees used for propagation was developed (10). Precautions taken included the elimination or absence of targeted pathogens that could affect the source plants, and the prevention of reinfections during propagation (9, 10, 135). Part of this work dealt with control of fungi and bacteria during greenhouse propagation. Similarly, nematodes were serious problems but were controlled by eliminating these pathogens in the nursery soil, mostly by fumigation (see Chapter 3). The most serious problems in clonal propagation are generally virus related (44, 55).

Methods to Detect Pathogens in Propagation Sources

Visual Inspection Pathogens alter the biochemical processes in susceptible plants, which may express characteristic symptoms. In highly susceptible plants, the pathogen may be self-eliminating by producing such drastic effects that the host plant does not survive. The absence of disease symptoms, however, does not indicate absence of a pathogen. Some, such as *Prunus necrotic ring spot virus*, may produce an initial shock, after which the plant recovers and shows no further visual symptoms (100). Virus symptoms are commonly expressed more at cool temperatures [less than 22°C (70°F)]; higher temperatures apparently inhibit their activity (44). Multiple infections by different viruses may produce a severe reaction, whereas each alone may produce little effect.

Visual inspection for specific symptoms is an important skill for the propagator to develop in order to detect problems among stock plants. It is not a reliable indicator, however, in the source plant itself. Rather, visual inspection becomes more important in the inspection of grafted sensitive indicator plants.

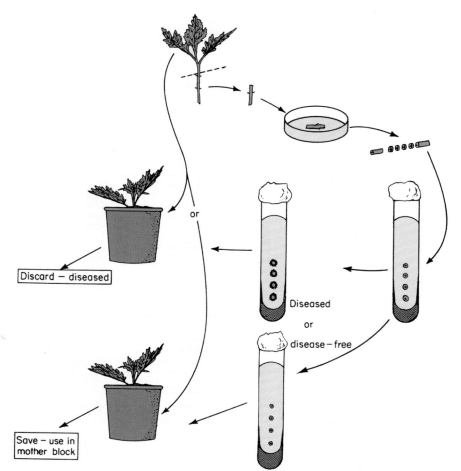

Figure 16–29
Culture indexing of chrysanthemum for *Verticillium* wilt pathogen. Stem sections are removed from surface-sterilized cuttings and placed on culture media. Only healthy plant sources are used to mass-propagate the clone or cultivar (10).

Culture Indexing Culture indexing is used to detect specific fungi and bacteria in the production of ornamentals, such as chrysanthemum, carnations (*Dianthus*), and geranium (Fig. 16–29). Pieces of surface-sterilized plant parts are placed on a medium favoring the growth of the pathogens (10, 62, 101, 113). Pathogens are identified by morphological features or biochemical tests.

culture indexing Growing vegetative progeny in culture to determine the presence of pathogens.

Virus Indexing The primary method for detecting viruses has been **indexing,** in which the virus is transmitted by grafting or budding to sensitive **indicator plants,** which then develop identifiable symptoms (Fig. 16–30) (74, 96, 150). One widely used procedure for indexing woody plants is to place a bud from the plant to be tested beneath a bud of the virus indicator plant onto the same rootstock growing in a container. The indicator bud is forced to grow. If the test bud contains a virus, it moves into the rootstock and up into the growing indicator shoot, where symptoms appear (Fig. 16–31, page 628) (44).

virus indexing The practice of inserting a bud into an indicator plant to test the presence and transmissibility of unknown viruses.

Leaf grafting is used in strawberry. Among *Prunus* spp., the necrotic reaction of highly sensitive tissue of 'Shirofugen' flowering cherry to the *Prunus* necrotic ring spot virus (Fig. 16–30) (96, 150) is a widely used test that requires about a month's incubation to complete. In contrast, 2 years is required to detect many viral diseases. Certain viruses can be detected in herbaceous hosts by sap transmission.

indicator plant A plant that is hypersensitive to specific viruses.

Serology Serology identifies unique proteins associated with a

serology An indexing procedure that identifies specific proteins by the production of antibodies in the blood of a test animal, such as a rabbit.

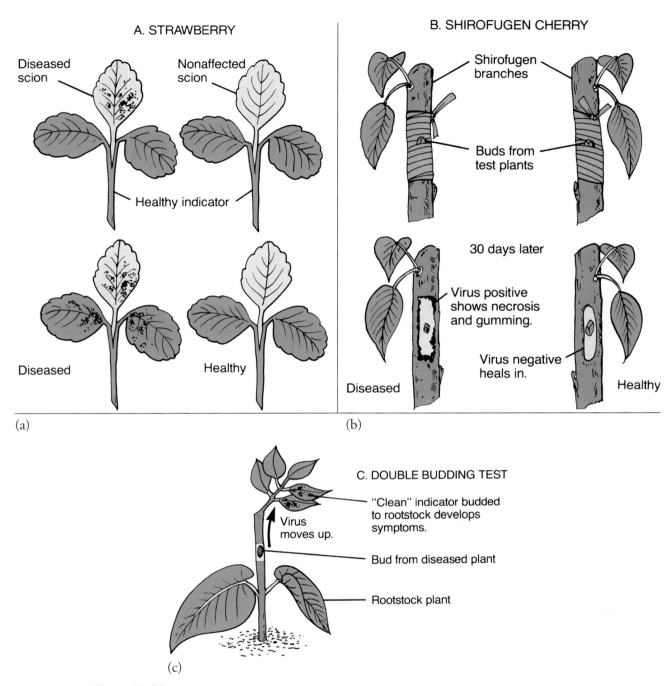

Figure 16–30

Indexing for virus transmission. (a) Excised leaf method for transmitting a virus from a symptomless plant to a sensitive indicator. A terminal leaf from the test plant is removed, a petiole of the indicator plant is split, and the excised test leaf is inserted. The graft union is wrapped with latex tape. Leaves on the indicator stock plant show symptoms. (b) 'Shirofugen' cherry indexing. Buds are removed from a test plant and inserted in sequence (usually three per test) in a 'Shirofugen' flowering cherry. This clone is very sensitive to certain fruit tree viruses, such as *Prunus* ring spot. If the test plant is infected, the bud will die; gum and necrotic tissue will appear around the bud shield within a month. (c) Double budding test for fruit trees. A rootstock plant is grown in a container in the greenhouse. A bud from the test plant is placed into the base of the plant and allowed to heal. At the same time a bud from a sensitive indicator plant is budded into the rootstock plant. If the test plant is diseased, the virus will move across the graft union, through the conducting system of the rootstock, and infect the indicator, which should develop specific symptoms.

(a) (b)

Figure 16–31
Virus indexing. (a) Buds are removed from the test plant and inserted (arrows) in sequence in the virus-sensitive rootstock; the virus-positive bud (above arrow) did not take, whereas the virus-negative bud (bottom arrow) has leafed out. (b) The test plant has become infected with a virus, leading to the defoliation of the foliage.

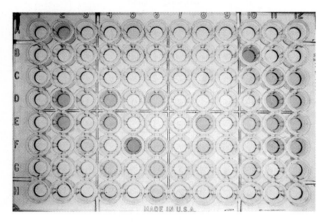

Figure 16–32
Typical serological results produced by the ELISA procedure. Each cell is a separate test whose results are shown by the intensity of the color staining. Colored cells are virus-positive, whereas clear cells are from virus-negative samples. ELISA strips are also used with seed technology.

ELISA A specific serological test to determine the antibody reaction to specific viruses.

particular pathogen. The **ELISA (enzyme-linked immunosorbent assay)** test (20) can be used to detect essentially any pathogen or virus (Fig. 16–32). Antibodies against specific proteins characteristic of particular virus or bacterial strains are produced by specialty laboratories using mammals such as rabbits and mice. After a series of antigen injections, blood is drawn and the serum fraction containing antibodies is separated from red blood cells. Whole serum or purified antibodies are employed in the serology tests with ELISA. Commercial kits can be obtained to screen hundreds of samples to produce immediate reactions.

Biochemical and Molecular Methods Methods to identify viral nucleic acids, such as **polyacrylamide gel electrophoresis,** are used to detect viroids and the replicated form (double-stranded RNA) of RNA plant viruses (63, 74). Molecular techniques of RT-PCR (real-time polymerase chain reaction) are used in virus-free citrus budwood certification programs (71).

polyacrylamide A gel used in electrophoresis tests.

Procedures to Eliminate Pathogens from Plant Parts

Successful mass-propagation of vegetatively propagated cultivars requires that infections of pathogens be controlled in the greenhouse and nursery. The most effective method of control is to eliminate pathogens from the source propagules that are to be used for propagation. The most important techniques for this purpose include selection from uninfected parts, shoot apex culture (meristemming), tip grafting, heat treatment, and thermotherapy.

Selection of Uninfected Parts Some parts of a plant may be infected, while others are not. Some soil-borne pathogens, such as *Phytophthora* spp., can be avoided by taking tip cuttings from uninfected portions of the plant (*Fusarium* and *Verticillium*) (10).

Shoot Apex Culture The small apical dome of a growing point (Fig. 16–33), including the meristem and a few subtending leaf primordia, is often free of virus and other pathogens even in systemically infected plants. Excision and aseptic culture of this small segment (**shoot apex culture**) can be the start of a new pathogen-free foundation clone (see Chapter 18 for culture procedures). The smaller the explant, the more effective the elimination of pathogens. For example, meristem tips

shoot apex culture The micropropagation of the shoot apex primarily for virus elimination.

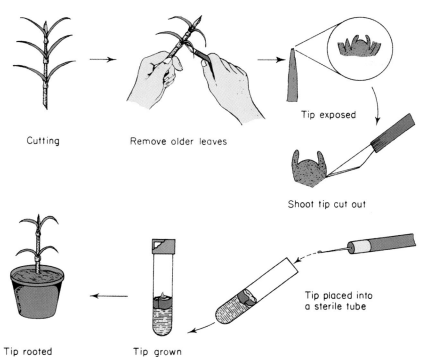

Figure 16–33
Meristem-tip culture of carnation (*Dianthus*). Following the arrows: the carnation cutting is obtained, the larger leaves are stripped away, and then the small, enclosing leaves at the extreme tip are removed to expose the growing point. The tip and the next subtending leaf primordia are removed with a scalpel and placed on the surface of a paper wick in a test tube with nutrient media. The shoot tip is grown in the tube until it is large enough to be transplanted to a container. A full-size plant is shown at bottom left.
Redrawn from W. O. Holley and R. Baker. 1963. *Carnation production*. Dubuque, IA: William C. Brown.

of 0.10 to 0.15 mm have given 100 percent virus elimination, but the percentage decreases as the size is increased to 1 mm. On the other hand, the smaller the explant, the more difficult it is to establish and to survive. A compromise is to use pieces 0.25 to 1.00 mm (0.01 to 0.04 in) long and confirm virus status by indexing. A combination of thermotherapy and meristem excision can increase the probability of virus exclusion.

Some examples where shoot apex culture has been used to eliminate viruses (111, 140) include the flower crops (amaryllis, carnation, chrysanthemum, dahlia, freesia, geranium, gladiolus, iris, lily, narcissus, nerine, orchid), fruit crops (apple, banana, citrus, gooseberry, grape, pineapple, raspberry, stone fruits, strawberry), vegetables (brussel sprouts, cauliflower, garlic, potato, rhubarb, sweet potato), and miscellaneous crops (cassava, hops, sugarcane, taro).

Micrografting   **Micrografting** is used for woody plants where tip cultures will not survive in aseptic culture. **Shoot tip grafts** can be produced by placing them on a shoot apex of an already-rooted seedling plant in culture. In *Citrus*, the procedure is carried out with buds of mature clones to avoid selection of the juvenile phenotype of nucellar seedlings (Chapter 17).

micrografting   Same as shoot tip grafting.

shoot tip grafting   *In vitro* grafting of the shoot apex to an *in vitro*-established rootstock plant, when roots do not readily form from the scion to be multiplied.

Heat Treatments   **Heat treatment** involves subjecting plant material to relatively high temperatures for short durations. Methods include hot water soaking, and/or exposure to hot air or aerated steam. The correct heat kills the pathogen but does not injure the plant. This treatment can be applied to plants, plant parts, bulbs, and seeds to control fungi, bacteria, and nematodes (8, 10) (See Chapter 3). Treatments may vary with different plant species, from 43.5 to 57°C (110 to 135°F) for 1 1/2; to 4 hours. Preconditioning vegetative material or reducing moisture content in seeds prior to heat treatment generally improves plant survival. Damage to heat-treated seeds can be alleviated by submersion in polyethylene glycol 6000 (8).

heat treatment   Subjecting a plant part to high temperatures of short duration to kill specific pathogens.

Thermotherapy   **Thermotherapy** involves exposure of plants to a lower temperature range than in heat treatments but at a longer duration (Fig. 16–34). Subjecting shoot tips to specific ranges of high temperature will

thermotherapy   The exposure of a plant to long periods of moderately high temperatures in order to inactivate viruses.

Figure 16–34
(a and b) Heat chambers for conducting thermotherapy of plant material to eliminate viruses. Plants are grown at approximately 38°C (100°F) for 4 to 6 weeks. At the conclusion of this period, buds are removed to be indexed for the continued presence of viruses. (b and c) Thermotherapy chambers can be individually controlled for the desired temperature.

eliminate or inactivate viruses within plants (64, 74, 99, 111, 140). Plants in containers are first preconditioned at higher temperatures and then placed at 37 to 38°C (98 to 100°F) for 2 to 4 weeks or longer (64, 99, 111, 140). After treatment, individual growing points are removed and budded to virus-free rootstocks. Each budded progeny must then be indexed to determine if the plant is indeed **"virus-free,"** though a more accurate term may be **"virus-tested."** The combination of thermotherapy and shoot apex culture may sometimes be used to improve the results when using only one of the procedures is not adequate. The plant is first heat-treated and then apical shoot tips about 0.33 mm long are removed and grown *in vitro* in test tubes. For plants that do not root easily, shoot tip grafting may be used to maintain the plant.

virus-free The concept of complete absence of viruses; it actually means only the absence of specific viruses which were tested for.

virus-tested The concept that only tests for specific harmful viruses were made.

Growing Seedlings Most viruses either are not transmitted through seeds, or the percentage of infected seedlings is low. This characteristic is apparently a major biological advantage of seed reproduction in nature, since most viruses are screened out in the sexual reproductive process. Horticulturally, this procedure has important applications. A new cultivar originating as a seedling generally starts its vegetative sequence (Fig. 16–2) as a "clean" individual. If protected from infection during the early testing process, the cultivar can often be introduced as a "virus-free" cultivar without further treatment.

PROPAGATION SOURCES AND THEIR MANAGEMENT

Propagation of cultivars in specific species requires that a supply of propagules be available either continually (some florists' crops, such as carnations, chrysanthemum, foliage plants) or at specific times of the year (fruit and nut crops, woody ornamentals). To do so requires that **source plants for propagation** be available to meet these needs. How to select, maintain, and manage propagation sources is one of the most basic decisions that a propagator or nursery has to make. As previously emphasized, the source material must be the correct cultivar, produce true-to-type plants, and generally be free of serious pathogens. Management of the material must be such that sufficient propagating material is available in an optimum condition at a time when it is to be used. The system must be economical and fit into the overall nursery operation. The situation of the individual nursery and requirements of the species and cultivar determine the method of source maintenance.

Traditionally, nurseries manage their own source procurement systems as part of their overall operations. As technological requirements to control disease and maintain quality have increased, specialized nursery operations, both private and governmental, have appeared that maintain the source material, supervise its distribution, and sometimes provide material, such as "liners," either rooted or unrooted, to produce commercial plants directly or to provide propagating material to establish source plants.

Commercial Plantings as Nursery Sources

This option is economical, requires no special stock blocks, and depends on commercial performance as the criterion of selection. Individual plants may be visually inspected prior to use and any "suspicious" off-type plant can be rejected whether or not the cause is known. For example, commercial orchards have been the traditional source of bud-wood for fruit and nut cultivars in California. The source trees are grown for crop and not bud-wood. Propagators use a limited amount of propagating material from each of many trees. Maintaining the identity of individual source plants is virtually impossible. Bud-wood "lines" may evolve from consecutive generations from different nursery sources. As growth in older trees declines, fewer new shoots are available and the source must be replaced by younger trees with greater vigor. Problems associated with any one tree may be diluted in the total population, but any variant will continue to persist and the proportion will, sometimes, increase (68).

European vineyardists, who have grown specific wine grape cultivars for centuries, practice selection from specific vines in their commercial plantings to establish new vineyards (162). Differences within cultivars have apparently evolved over time to produce "strains" or "clones" of established cultivars, which may not differ phenotypically but are claimed to have different product uses.

Sometimes field observations can result in the discovery of useful mutants, and many new cultivars of apples, pears, and peaches have been discovered in commercial orchards this way. Sometimes, undesirable mutants may have been avoided. Historically, however, use of commercial orchards as nursery sources has resulted in the increasing contamination by viruses and other pathogens of major cultivars of many, if not most, intensively clonally propagated crops including major fruit and nut crops, grapes, strawberries, and small fruit species. In almonds grown in California, the incidence of latent inherited disorders *noninfectious bud-failure* and *nonproductive syndrome* are associated with commercial orchard selection (68).

Production Material Within the Nursery

Collecting cuttings or buds from nursery material is comparable to field selection, except that the commercial nursery crop is the source of propagation material for the next cycle of propagation. The prunings of ornamental shrubs being grown in containers may be used as a source of cutting material, plants may be set aside and pruned severely to force growth, and surplus material may be utilized (84). Unsold budded fruit and nut trees may be maintained in the nursery row for several years by severe annual pruning.

The advantages are efficiency, convenience, and economy. However, the same problems of latent viruses, mutations, and genetic disorders described for field selection may develop.

Stock Blocks

A **stock block** includes source plants maintained in a more or less permanent location separate from commercial propagation blocks, and managed to produce cuttings, bud-wood, scionwood, or divisions (see Chapter 10). These blocks may be planted either in the ground in a permanent site, or in containers, in the open or under protection (84). Hedge rows can be maintained to keep tree source plants in a juvenile state (2, 76) and, similarly, to produce blocks for the production of fruit tree rootstocks (58). Management of these blocks consists of severe annual pruning geared to the production of scions, buds, and cutting material. A few plants might be maintained for observation of "trueness-to-type."

Fruit tree nurseries establish scion or bud-wood orchards where trees are planted relatively close

> stock blocks Plants maintained specifically for source production.

Figure 16–35

(a) Nursery stock plants grown in a scion block to provide scions and buds for fruit tree production. (b) Tag is attached to trees in a stock block to indicate that the trees have been tested as "virus-free."

together and pruned for the production of propagation material rather than for fruit (Fig. 16–35). Nurseries sometimes maintain some trees or a branch of each tree as a check for verification of cultivar and type.

Individual plants in stock blocks need to be examined for trueness-to-cultivar and trueness-to-type prior to their use. The plants may be allowed to fruit or flower, or a vegetative progeny test may be conducted. Severe pruning in the stock block may change the growth habit, general appearance, and fruit or nut quality such that either a false reading of off-type may be made, or specific variants may fail to be observed.

Maintaining stock blocks is expensive, time consuming, and may require much space. Once identity and trueness-to-type are verified, however, this system provides a safeguard for good quality and a convenient supply of propagation material. Unless these blocks are protected against reinfection and/or tested for disease periodically, the same disease contamination problems described for field and nursery commercial material may develop.

Clonal Selection and Pedigreed Production Programs

Clonal selection and pedigreed production programs originated as **"clean stock programs"** to control pathogens in vegetatively propagated crops. The **clonal selection, pedigreed production systems,** and **certification** of disease-tested propagation sources were first studied in potato (*Solanum tuberosum*), because pathogens on tuber propagules were a limiting factor in production (139). A Wisconsin (Potato) Seed Improvement organization was first formed in 1905, which established a Seed Potato Certification organization in 1913 (6) which expanded to 12 states and Canada. Crop Improvement organizations exist in most states combined within the Association of Official Certifying Agencies (AOCA) (see Chapter 1).

clean stock programs Programs that originated to eliminate pathogens from source blocks.

pedigreed production A program of maintaining, multiplying, and distributing plant material in consecutively controlled generations.

certification The practice carried out by a legal authority to certify that the commercial propagating material has been reproduced under a specified program to ensure pathogenic purity.

A landmark event in pathogen-free production was the publication in 1957 of Manual 23, *The UC System for Producing Healthy Container-Grown Plants* (10). Another significant event was the 1951 publication of the USDA Handbook No. 10, *Virus and Other Disorders with Virus-like Symptoms of Stone Fruits in North America*, and its revision in 1976 (150). Since then, many commercial crops have developed Registration and Certification programs (25, 30, 37, 113) similar to those used by seed producers, and described in Chapter 5. Most are subject to governmental regulations, but at the same time may be voluntary for individual nurseries.

Details of procedures for clean stock programs vary among different crops but all have the same three steps. Table 16–1 compares programs for three different commercial crops.

Step I. *Identify individual plants within the clone that are genetically true and free of serious pathogens.* In tree fruits, nuts, and vines (4, 5), this process begins with clonal selection among individual plants within the cultivar, although more than one source may be selected. There are two main scenarios

Table 16-1
COMPARISON OF CLEAN STOCK PROGRAMS IN REPRESENTATIVE CROPS

Phase	Potato (139)	Prunus (4)	Geranium (101)
I. Selection of sources Visual inspections, indexing for virus or other pathogens, meristem culture, thermotherapy, hot water	Clonal selection of "mother" plants from commercial plantings or as new cultivars. These are indexed and/or shoot-tip cultured to produce a *nuclear source*. Proc. I. *in vitro* cultured plantlets are multiplied to the desired number. Proc. II. Multiplied by sprout cuttings, which produces tubers for field production.	"Clonal selection" among individual trees: I. "Short index" either Shirofugen or ELISA or both. II. "Long index" 6 to 8 indicators = 2 years. Provides *nuclear source*.	Stage I. Cuttings selected from source plants are visually inspected, culture indexed, and heat-treated for 4 weeks. Stage II. Meristem-tip cultured S-I plants are indexed and multiplied for 4 weeks (*elite mother* block). Stage III. S-II plants reindexed, grown for visual observation of performance.
II. Maintenance Visual inspections, roguing, indexing	Plants are called *prebasic* stock. Proc. I. Plantlets are transplanted to greenhouses where they develop three monthly harvests of minitubers. Proc. II. Tubers planted in field for 5 years of consecutive generations.	Two trees planted in a *foundation block*. Isolated by distance or in a screenhouse. Annual Shirofugen index or ELISA; long index at other intervals. Trees are registered by location.	Stage IV. Plants from true-to-type and reindexed S-III plants are used to provide a *nuclear block* (3 months).
III. Multiplication and distribution	Field plantings: two steps I. *Basic seed* which involves 3 annual generations of propagation. II. *Certified seed*, 1 to 3 generations. Sold to potato producers.	I. To commercial nurseries; *Mother blocks; nursery increase blocks*; trees are registered by location. II. Commercial plants from mother blocks or increase blocks are certified.	Stage V. *Increase block* where plants are multiplied and used to produce stock plants. Stage VI. *Stock block* from which commercial cuttings are sold.

of selection: one is to screen populations of individual plants to identify those that are virus-free followed by tests for genetic identity; the second is to select first for genetic characteristics, and then to remove the offending pathogen by appropriate techniques, such as heat treatment, meristem-tip culture, and/or thermotherapy. Visual inspections of both the source plant (*phenotypic selection*) and the vegetative progeny plants (*genotypic selection*) may be required. Other crops, such as potato (*Solanum tuberosum*), geranium (*Pelargonium*), carnation (*Dianthus*), and chrysanthemum are selected as populations within the clone (see Table 16–1).

The selected plant(s) to be utilized as the initial sources in future propagation are usually called **nuclear plants,** depending upon the crop and program. The source may be identified for propagation purposes by a specific name and/or number.

> **nuclear plants** The initial source of a cultivar following its clonal selection in a Registration and Certification program.

Step II. *Maintain source plants in a protected **Foundation block** located to prevent reinfection.* Isolation may depend on distance from other sources, location, and method (greenhouses, screenhouses, or field). Standards of isolation, inspection, and reindexing specified in regulations must be followed. Since the cost and effort of maintaining such standards is high, only a few foundation plants may be maintained for each clone depending on crop requirements. These plants are identified as *foundation plants, elite stock plants, mother plants,* or other, depending on the crop. Individual plants may be **registered** with a certifying agency by clone identity and by location.

foundation block A group of plants that serves as the primary propagation source maintained under appropriate standards of isolation, indexing tests and inspection.

registration The practice of registering individual source plants from which collection is to be made, with a Registration agency such as a State Department of Agriculture.

Step III. *Multiply source and distribute to the public.* Multiplication takes place through a series of consecutive vegetative generations **(increase blocks)** under conditions and standards to lessen chances of reinfection until delivered to the ultimate user (buyer), and at the same time reduce the costs of production. Consecutive generations are always in a vertical sequence and/or expanded by horizontal propagations to maintain the original disease and genetic potential. The principles of **"nuclear stock"** and pedigreed production have as their outcome the possibility to obtain plants that can be **certified** as to the origin of the propagation source and the system by which they are produced. They do not certify the plants to be virus-free, although that is the goal.

increase block A temporary block produced to multiply the quantity of propagules.

Cultivar identity and trueness-to-type are particularly important in "clean stock" programs. High multiplication rates are inherent in the system in order to go from initial selection to commercial quantities. Consequently, the existence of the wrong cultivar or a genetic abnormality early in the system can build up to large numbers before it is detected. Some rapid methods of propagation, such as micropropagation, have resulted in a certain proportion of abnormalities in certain cultivars, particularly with chimeras (Chapter 17). Random mutation may occur, as discussed in this chapter.

Following are some guidelines for selection:

- Obtain nuclear stock from plants with a known history of production and from those that have been previously visually inspected.
- For some problems, such as genetic disorders (page 602), a vegetative progeny test must be conducted.
- Conduct visual inspections regularly throughout the propagation cycles and rogue out individuals as needed.
- Follow up with test plantings to determine if progeny plants are meeting expectations.

Fingerprint procedures are becoming available for routine cultivar identification (see Box 16.15, page 624). However, trueness-to-type problems primarily require visual inspections.

Repositories, Botanical Gardens, and Plant Collections

Plant collections exist around the world and can be a source of clonally propagated cultivars and species. Usually, these sources provide only enough material to establish a source block. The standards of quality already described should be followed when obtaining the material, but this may depend on the type of collection. Botanical gardens maintained by private, public, or governmental agencies in general emphasize the genetic aspects of species and cultivars, and may be part of a breeding and clonal selection program where propagation material may be controlled by patents and other protective protocols (see Chapter 2).

Clonal **repositories** are of two types: one maintains clonal germplasm, as in the United States Clonal Repository System (http://www.ars-grin.gov/); the second includes repositories to maintain and distribute Foundation clones for virus control programs. The USDA IR-2 Repository at Prosser, Washington, is of this second type (44). Many states and countries in the world maintain governmental repositories that provide virus-tested stocks of various crops.

repository A collection of plants maintained either to preserve genetic diversity or to maintain their virus status.

Quarantines and Movement of Vegetatively Propagated Material (46)

Vegetatively propagated material may be shipped from country to country or from state to state in the United States. Because of the potential risk of transporting

part four
Tissue Culture Propagation

CHAPTER 17 **Principles of Tissue Culture and Micropropagation**
CHAPTER 18 **Techniques for Micropropagation**

Plant tissue culture is the growth of plant organs or tissues in aseptic culture where the environment as well as the nutrient and hormone levels for growth are tightly controlled. Tissue culture has found application in a number of areas of plant science, including basic physiology, production of natural and pharmaceutical compounds, plant pathology, germplasm preservation, breeding, recovery of transgenic plants, and propagation. Micropropagation includes those tissue culture procedures used to propagate plants. It has proven successful for mass propagation, reducing the time for new cultivars to be introduced to the market, and has extended the range of plant genotypes that can be propagated. Micropropagation has become a standard nursery practice for certain crops, especially those that are difficult or slow to propagate by conventional methods. This section describes specific aspects of tissue culture with special emphasis on micropropagation.

*17 Principles of Tissue Culture and Micropropagation

learning objectives

- Explain the history of tissue culture and micropropagation.
- Describe the types of tissue culture systems.
- Describe micropropagation of plantlets.
- Determine the systems used for micropropagation.
- Describe reproduction of seedlings in tissue culture.
- Define culture systems for callus, cells, and protoplasts.
- Describe somatic embryogenesis and synthetic seed production.
- Define control of the tissue culture environment.
- Describe special problems encountered in tissue culture.
- Determine practices that lead to variation in micropropagated plants.

INTRODUCTION

One of the most interesting areas of biotechnology is tissue culture and micropropagation. Tissue culture is the ability to establish and maintain plant organs (embryos, shoots, roots, and flowers) and plant tissues (cells, callus, and protoplasts) in aseptic culture (Fig. 17–1). Micropropagation is a form of tissue culture used to regenerate (propagate) new plants.

Micropropagation exerts a high degree of control over each aspect of regeneration in tissue culture. Each step of the process can be manipulated (or programmed) by control of the tissue culture environment. Micropropagation is used as an accelerated form of clonal propagation, and is the method of choice to propagate plants that are slow to multiply or those that cannot be clonally propagated any other way. It is also used to regenerate plants that have been genetically modified (transformed) through biotechnology.

This chapter describes the biological nature of the tissue culture systems listed in Table 17–1 (pages 646–47) and defines the morphological, physiological, and genetic basis for the growth and development of plants in tissue culture.

A BRIEF HISTORY OF TISSUE CULTURE AND MICROPROPAGATION

Tissue culture is an inclusive term for the range of procedures used to maintain and grow plant cells (callus and protoplasts) and organs (stems, roots, and embryos) in **aseptic** (or ***in vitro***) culture. The technique is used for propagation, genotype modification (i.e., plant breeding), biomass production of biochemical secondary products, plant pathology, germplasm preservation, and scientific investigations. These procedures are a result of basic and applied research in scientific laboratories (botany, plant pathology, and genetics) since before the turn of the 20th Century (67, 270). The progression of scientific discoveries that has led to the modern practices of tissue culture is intertwined with many important discoveries in plant science.

aseptic The growth of plant tissues under conditions that are free of microbial contamination.

in vitro The culture of plant cells or organs in culture vessels (like test tubes) under controlled environment and nutritive growth medium.

Figure 17–1
Typical structures formed in tissue culture include (a) shoots, (b) roots, (c) flowers, (d) bulbs, (e) callus, (f) somatic embryos.

Plant tissue culture starts with the "cell theory" postulated in 1838 by Matthias Schleiden and Theodor Schwann (66, 67). They theorized that any plant cell could be **"totipotent"** and develop into a complete plant. The need for experimental evidence for totipotency drove the early development of tissue culture.

However, it was not until 1902 that Gottlieb Haberlandt attempted to grow isolated plant cells (86). He used a nutrient solution developed by Johann Knop in 1865 for hydroponic plant growth. Haberlandt supplemented Knop's medium with sucrose and asparagine, but his cells would only live for 20 days. They increased in size but failed to divide and grow. He did, however, observe that slices of potato tuber containing a vascular bundle would show some cell division. He concluded that these contained a substance that induced cell division and called this substance the "wound hormone."

Similarly, in 1934 Roger Gautheret from France grew callus from cambial tissue isolated from several woody plants including willow, sycamore, and elder (64). These were the first sustained dividing callus cultures, but they would only grow for about 6 months on an agar-based nutrient medium. He rightfully concluded that something was missing from the culture medium that would sustain unlimited cell growth in tissue culture.

totipotent The concept that a single cell has the genetic program to grow into an entire plant.

Table 17–1
TECHNIQUES USED TO REGENERATE PLANTS THROUGH TISSUE CULTURE

Structures formed	Regeneration method	Explant source	Uses
Plantlet formation	Axillary shoot formation		
	1. Meristem culture	Shoot tip less than 1 mm in size.	Initially developed as a micropropagation system but now mostly used for virus elimination.
	2. Shoot culture	Stem with 1 to 4 nodes.	Shoot cultures are the most often used micropropagation systems.
	Axillary branching	May include leaves and shoot tip.	Cultures are multiplied by cutting the clump into sections and subculturing each in separate containers.
	Nodal cultures	Long shoots are cut into single nodes and planted vertically in the medium.	Axillary buds at each node elongate and grow in length. The pattern is repeated by again cutting into nodal segments at each subculture.
	Stool shoots	A shoot of several nodes is laid horizontally on the medium. Lateral growing points form a thicket of small vertical growing shoots.	These "layers" may be subdivided at each subculture or the entire unit transferred to a new culture vessel when the medium is exhausted.
	Pseudocorms	Growing points of orchids in culture form pseudocorms.	Pseudocorms can be subcultured to propagate orchids.
	Minitubers	Potato plants in culture can form miniature tubers at the end of small stolons.	These storage organs can be removed and used in the production of virus-free planting stock.
	Micrografting	Small scion shoot tip usually grafted to a seedling understock.	Useful for virus elimination, propagation method, studying grafting problems, and rejuvenation.
	Adventitious shoot formation		
	1. Diploid plant regeneration (full complement of chromosomes)	Leaf pieces, petioles, bulb scales, stem internodes, roots, and callus.	Often used for micropropagation, especially in monocots. Adventitious shoot regeneration is one of the key steps in obtaining plants that have been genetically transformed.
	2. Regeneration of plants with different ploidy levels (haploid or triploid)	Anther or endosperm culture.	Used in breeding to obtain haploids or triploids. Shoots or somatic embryos may be obtained.
Seedling formation	Seed culture	Seeds	Primarily used to produce orchids. Orchid seeds lack the typical storage reserves found in other seeds and respond well to tissue culture.

(Continued)

Table 17–1 (*Continued*)

Structures formed	Regeneration method	Explant source	Uses
	Embryo culture	Embryos are isolated from the fruit and seed coverings.	Mature embryos germinate easily in tissue culture to form seedlings. Used for research, understocks for micrografting, and occasionally for propagation.
	Embryo rescue	Isolation of immature embryos.	Primarily used for breeding interspecific crosses. These crosses usually fail to set seed, but early embryo development can occur. Embryos complete their development in tissue culture.
	Ovule and ovary culture	Immature ovules or ovaries cultured.	Unfertilized ovules are excised, grown in culture, supplied with pollen, and subsequently fertilized *in vitro*. Used for plant breeding.
Callus formation	Callus cultures (stationary)	Any vegetative tissue.	Callus cultures are used for research, breeding, and genetic transformation studies. Callus cells can be used to produce enzymes, medicines, natural flavors, and colors.
	Callus suspension cultures	Callus subcultured from stationary cultures.	Suspension cultures are shaken constantly to perpetuate callus formation. Uses are the same as stationary callus cultures.
	Protoplast cultures	Protoplasts are isolated single cells without a cell wall. The cell wall has been digested by fungal enzymes.	Protoplasts are used in plant research to study basic cell function. Protoplasts can also be used in breeding. Under the right conditions, two protoplasts can fuse to form a single cell. The nuclei in these cells can merge, combining genetic information, even in species that are not sexually compatible. New cell walls form and the resultant callus can be induced to form adventitious shoots.
Somatic embryo formation	Direct somatic embryogenesis	Embryo, seedling, or leaf. Somatic embryos form directly from cells in the original explant.	Can be used to regenerate copies of the original mother plant or recovery of plants that have been genetically transformed.
	Indirect or induced somatic embryogenesis	Any plant part. Somatic embryos form after explant is induced to form callus.	Used to regenerate clonal copies of the mother plant or genetically transformed plants. This pattern of somatic embryogenesis has the greatest potential for mass propagation though synthetic seeds.

Several years later, in 1939, three researchers independently used the newly isolated growth hormone, auxin, to establish callus cultures with the potential for indefinite growth. These researchers were R. Gautheret (65) and P. Nobecourt (184) from France, working with carrot cells, and P. R. White (269) from the United States, working with tobacco cultures. In 1941, White and A. Braun (271) found that cells isolated from tobacco infected with crown gall (*Agrobacterium*) would display cell division and growth in culture without the addition of auxin.

In 1948, Folke Skoog and his colleagues discovered a breakdown compound from old herring sperm DNA that could induce organ formation in callus cultures of tobacco. Initially, they called this substance adenine and then kinetin because of its cell division activity. It was an artificially derived cytokinin in which they were able to demonstrate that the ratio of cytokinin and auxin could determine whether tobacco callus would make roots or shoots (Fig. 17–2) (228). Later in 1962, this system would be used to develop the universally used Murashige and Skoog medium (174).

In 1954, A. Hildebrandt and his colleagues were able to produce tissue cultures from single cells (170). Independently in 1959, Steward (234) and Reinert (213) reported the successful regeneration of somatic embryos from carrot cells (Fig. 17–3). Finally, in 1965, Hildebrandt was able to demonstrate that whole plants could be derived from a single tobacco cell (262). Thus, totipotency was shown to exist in plant cells.

In 1880, Julius von Sachs postulated that there were organ-forming substances in plants. In 1922, W. Kotte in Germany (135) and William Robbins in the United States (214) independently developed the first organ cultures. They grew isolated root tips of pea and corn in nutrient cultures. Yeast extracts added to the medium helped these roots to grow. By 1934, P. R. White, also from the United States, developed a medium that contained amino acids and vitamins that permitted continuous growth of tomato roots in culture (270). E. Ball in 1946 developed the first shoot cultures and recovered whole plants from nasturtium and lupine (9); this was the first demonstration of micropropagation. In 1950, G. Morel and C. Martin used meristem culture to eliminate virus in dahlia (168). Morel (167) also used meristem culture to micropropagate orchids in 1963. Toshio Murashige, of the University of California, Riverside, expanded this work and developed the concept of developmental stages of micropropagation leading to plantlet establishment (173).

Application of micropropagation began in the 1960s and early 1970s (87, 106, 168, 172, 219), extended to the development of commercial

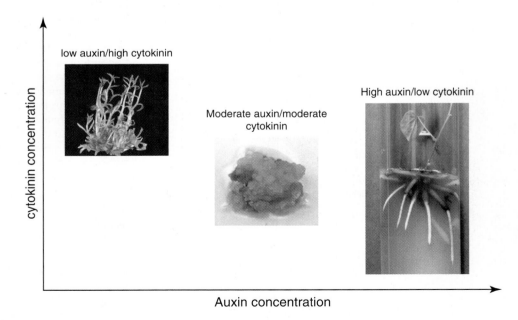

Figure 17–2
The relationship between the ratio of cytokinin to auxin on organogenesis in culture. A high cytokinin to auxin ratio tends to promote shoot organogenesis, while a low cytokinin to auxin ratio promotes rooting. More equal concentrations tend to promote both shoots and roots from the same culture or undifferentiated callus.

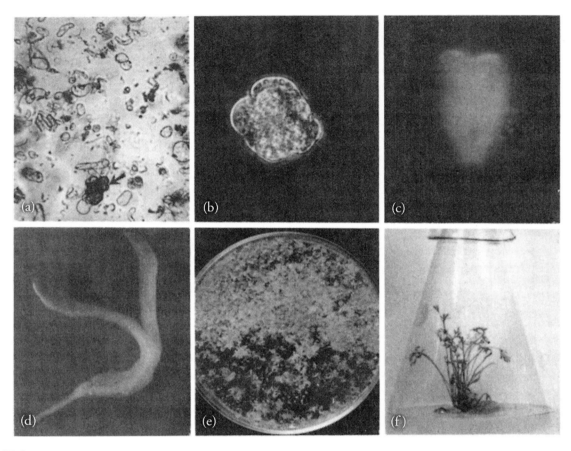

Figure 17–3
Totipotency of cells in carrot. (a) Highly magnified view of suspended cells and cell clumps derived from tissue cultures of carrot growing in a liquid medium. Arrow points to clump of cells—the beginning stage for a new plant. (b) Single-cell clump (higher magnification) illustrating the initial stages of somatic embryo development. (c) A more advanced heart-shaped stage of somatic embryo development. (d) Mature germinating somatic embryo. (e) Thousands of carrot plantlets that developed as somatic embryos. (f) Single carrot plant that grew from a single somatic embryo. Courtesy F. C. Steward and M. O. Mapes.

laboratory–nurseries in the 1970s in the United States, Europe, Australia, and Asia (20, 98, 103, 185, 235, 280), and expanded in the 1980s (58, 73, 283). In the 21st Century, China and India have become very active in tissue culture and micropropagation. Today, protocols for thousands of plant species have been developed, and many of these are in commercial propagation.

TYPES OF TISSUE CULTURE SYSTEMS

Table 17–1 lists the procedures used for tissue culture that are associated with propagation. The table is organized by the structures formed in tissue culture. These include the formation of:

1. Plantlets
2. Seedlings
3. Callus
4. Somatic embryos

Of these techniques, the process of plantlet formation via micropropagation has the most direct application for plant propagation and will be emphasized in this chapter.

Micropropagation of Plantlets from Tissue Culture

Cultural systems described under plantlet formation (Table 17–1) are based on the maintenance and multiplication of **microshoots** in culture to produce rooted **microcuttings**. A distinction is made between

microshoots Small shoots grown *in vitro*.

microcuttings Microshoots used as small cuttings induced to regenerate roots and eventually a plantlet.

microshoots originating from axillary buds and those arising adventitiously directly from tissue without axillary buds (i.e., leaf or petiole) or indirectly from callus developed from the original explant. In either case, the stages of micropropagation are very similar.

Developmental Stages in Micropropagation

Success of micropropagation is largely due to separating different developmental aspects of culture into stages, each of which is manipulated by media modification and environmental control (142, 74). Four distinct stages are recognized for most plants. The specific procedures for these stages are illustrated in Chapter 18 (see Fig. 18–14) and include:

Stage I: **Establishment**—placing tissue into culture and having it initiate microshoots.

Stage II: **Shoot multiplication**—inducing multiple shoot production.

Stage III: **Root formation**—initiating roots on microcuttings.

Stage IV: **Acclimatization**—gradually moving plants to open-air conditions.

Stage I: Establishment and Stabilization of Explants in Culture

The objectives for Stage I are to successfully place an **explant** into aseptic culture by avoiding contamination and then to provide an *in vitro* environment that promotes stable shoot production (151). Some references suggest that there is a Stage 0 where the source plants are manipulated prior to severance of the explant to start Stage I (27). Important aspects of Establishment Stage include:

explant The piece of the plant (propagule) used to initiate the micropropagation or tissue culture process.

1. Explant source selection
2. Explant disinfestation
3. Culture medium
4. Stabilization

Explant Source Selection. The selection and management of the source plant is an important aspect of successful micropropagation. (See Chapter 16 for specific information on clonal source plant selection.) Four aspects require particular attention:

1. Explant type
2. Genetic and epigenetic characteristics of the source plants
3. Control of pathogens
4. Physiological conditioning of the source plant to optimize its ability to establish in a culture

Explant Type. Many studies have demonstrated that the explant type can influence the success of a tissue culture system. For example, petiole explants may produce many more adventitious shoots than root explants under the same hormone treatments (71). Often the same explant type taken from different locations on the donor plant can impact regeneration ability, a phenomenon demonstrated nicely in European beech (*Fagus sylvatica*), where there was a different capacity for shoot regeneration shown between the upper and lower parts of the same leaf treated with the same cytokinin concentration (263). Also, depending on the species, location on the donor plant, and the growth regulator used, thin cell layer explants can initiate callus, shoots, roots, flowers, or somatic embryos (180).

Some plants can initiate cultures any time of the year, but for others, the time of the year an explant is collected can impact micropropagation. The differences may be between dormant and non dormant tissue, or some other subtle differences such as when the explant was taken within the rhythmic growth cycle shown in *Citrus* (51). Time of the year can also affect how difficult the explant is to clean and disinfest.

Genetic and Epigenetic Characteristics of the Source Plants. The correct genotype identification and trueness-to-type of the source plants are essential because a mistake at this stage can multiply the problem many times and cause much economic loss before it is eventually discovered.

Regeneration capacity can often vary depending on the genotype of the source tissue (99). Sunflower is a well-studied system that demonstrates the expected heritability of tissue culture shoot regeneration (178). In sunflower, the frequency of regeneration and the number of the shoots per explant varied among genetic inbred lines. Genetic crosses between these lines showed heritability estimates indicating that regeneration capacity can be selected and improved by conventional breeding.

Many thousands of different plant types have been successfully micropropagated. In general, those plants that are easy to propagate conventionally are also easy to micropropagate. For example, many herbaceous plants can be vegetatively propagated relatively easily by either conventional propagation or micropropagation (73, 74, 173, 191, 235). Some woody perennials have been proven difficult to conventionally propagate (i.e., from cuttings), and these usually prove to be more difficult to micropropagate as well. However, because of the high level of control presented through tissue culture, many of these recalcitrant plants have also been successfully micropropagated (8, 151, 154, 114).

Success in micropropagation of woody plant species is to a large extent a function of the juvenility status of the source plant (see Chapters 2 and 16). Seedling plants are invariably easier to micropropagate than mature cultivars of the same species (88). Manipulations to obtain juvenile explants are the same as those described for stock plants used to take cuttings and include **selection of explants** from the basal juvenile parts of stock plants, **induction of adventitious shoots on roots,** and **consecutive grafting** to seedling plants (described in Chapter 16).

adventitious shoots Shoots that arise from places where buds do not normally form, such as roots, leaves, flowers, and stem internodes.

Control of Pathogens (19, 27, 197). Pathogen problems include (a) contamination by fungi, molds, yeasts, and bacteria on the surface of stems or lodged in cracks, bud scales, and elsewhere; (b) systemic viruses and virus-like organisms; and (c) internal pathogens.

External contaminants are literally present everywhere—in the air and on the surface of plants, tables, hands, and so on. Spores move on dust particles carried by air currents. One must disinfest the explants, the tools, and the working area to remove contaminants from the surface. All work must be done in special transfer areas where contaminants have been eliminated and precautions taken to prevent recontamination (see Chapter 18 for specific procedures to disinfest explants).

Reduction in surface contaminants begins with the stock plants used as the source of explants. Often, contaminants are present only on the surface of plant parts. Internal structures, such as growing points of buds or inside seeds and fruits, tend to be relatively free of pathogens. However, if the plant is growing in a humid atmosphere, mycelia may invade plant interiors and become a persistent problem.

Reducing the surface contaminants on source plants prior to explant excision can aid in later disinfestation attempts (98). In general, stock plants growing in containers in protected environments like a greenhouse are "cleaner" sources of explants than are those growing out-of-doors. Overhead watering, sprinkling, or any activity that increases humidity around the plant should be avoided. Dormant plant parts can be taken and forced into growth under laboratory conditions to minimize contamination (209). For outdoor plants, keep plant parts off the ground and avoid the use of roots or underground portions as explant sources, if possible. Insect and mite populations should be controlled. In one case, withholding water and keeping plants cool and dry for 3 weeks was necessary to reduce contaminants on *Dieffenbachia*, even inside a greenhouse (132). If plants are growing out-of-doors, covering shoots with plastic bags or spraying them with fungicides may be useful.

Stock plants should be evaluated for pathogens prior to their selection as donor plants, as described in Chapter 16. Virus-contaminated plants may be exposed to elevated temperatures as a form of thermotherapy (27). Isolated plant parts are treated at approximately 50°C (122°F) for minutes to hours while growing plants are held at about 40°C (104°F) for several weeks. Alternatively, meristem culture (155) or a combination of meristem culture and thermotherapy can be employed to eliminate pathogens (see Chapter 16, thermotherapy section, page 629).

Physiological Conditioning. Source stock plants can be treated to provide explants that perform better in tissue culture. This point is sometimes referred to as **Stage 0** because it occurs on the source plant before any explant is taken (38). It may include (a) growing plants under controlled environments to achieve the proper stage of shoot development in relation to seasonal patterns, (b) producing healthy plants, and (c) treatment with growth regulators.

Stage 0 Stage that includes preconditioning procedures applied to the source plants prior to taking explants for Stage I of micropropagation that can improve establishments of explants in culture.

The light and temperature under which source plants are grown can influence tissue culture performance. Begonia explants produced more shoots in a shorter period of time from source plants grown under 25°C (77°F) and 160 $\mu mol \cdot m^{-2} \cdot sec^{-1}$ of light compared to 18°C (64°F) and 70 $\mu mol \cdot m^{-2} \cdot sec^{-1}$ (265). Light quality (red versus far-red) and photoperiod may also play a role in regeneration capacity, possibly by impacting the endogenous auxin and cytokinin levels (209).

Applying plant growth regulators to stock plants prior to explant removal can also impact regeneration capacity. Gibberellin inhibitors have increased shoot regeneration in tomato explants (42), and cytokinin sprays to petunia stock plants carried over to enhance shoot formation in tissue culture (50).

Explant Disinfestation. **Disinfestation** is the process of removing contaminants from the surface of the organ rather than from within the organ (27). Because the explant is being placed on a tissue culture medium containing nutrients and sugar, the explant must be completely free of contaminants; otherwise these will actively overgrow and consume the explant within a few days. Disinfestation requires the use of chemicals that are toxic to the microorganism but relatively nontoxic to plant material. Primary disinfestants include alcohol (ethyl, methyl, or isopropyl, usually about 80 percent) and bleach (calcium or sodium hypochlorite, usually with 5.25 to 6.0 percent active ingredient). Bleach is usually diluted at 10 to 20 percent before use. A few drops of a surfactant or detergent should be added to the bleach to improve surface coverage. The effectiveness of the bleach treatment is a time-dosage response; the disinfestation effectiveness increases with an increase in both time and concentration, but damage to living tissues also increases. Therefore, a compromise must be developed through testing for the kind of explant being treated.

Additional materials used to clean explants include hydrogen peroxide, silver nitrate, benzalkonium chloride, and mercuric chloride. Care should be taken, especially with mercuric chloride, to handle these materials safely and to properly dispose of them. Request the safety sheet (called an MSDS sheet) for these chemicals from your dealer for additional information about care in using these compounds.

A typical procedure for explant preparation would be to cut the tissue into short pieces several centimeters long, and wash them in tap water with a detergent. For woody pieces, a quick dip in alcohol following washing may be helpful. Plant parts are placed into the disinfesting solution (with a surfactant or detergent added) for 5 to 15 minutes. The solution is then poured off and the material rinsed two or three times with sterile water to remove all of the remaining solution. Modifications of this basic procedure that may improve effectiveness include prewashing, mechanical agitation, vacuum infiltration, and multiple treatments.

Microbial contaminants, consisting of yeast and various species of fungi and bacteria, usually appear on the agar surface within a few days to a week as white or opaque slime, or as variously colored colonies, sometimes with black spores and mycelium (59). When these contaminants can be identified visually (Fig. 17–4), the culture vessels containing them should be discarded quickly by autoclaving. Some bacterial contaminants, such as *Bacillus subtilis, Erwinia,* or *Pseudomonas,* sometimes remain inside the plant material without being detected for several months after the initial culture or incubation. Antibiotics are sometimes used in the medium to improve control of bacterial growth (27).

One should not take for granted that explants are free of viruses and any internal pathogen even if the extreme meristem tip of the shoot is utilized. **Culture**

disinfestation Chemical removal of surface contaminants that would otherwise grow in the tissue culture environment and kill the explant.

(a)

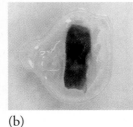
(b)

Figure 17–4
Typical microbial growth in a tissue culture from (a) fungal and yeast and (b) internal bacterial contamination.

indexing can be used to identify explants showing positive results for the presence of pathogens (see Chapter 16). Indexing at the initial explant stage, however, does not always identify all potential contaminants (98).

Culture Medium. (see Chapter 18 for specific information on media components). The culture medium usually includes a semisolid support (**agar** or other commercial product like Gelrite), a basal medium of inorganic nutrient elements (both major and minor), an energy source (primarily sucrose), and often some vitamin supplements. Several commercial media formulations are available (such as Murashige and Skoog – MS; Woody Plant – WPM; Anderson – AND), or media can be made from the component chemicals (see Chapter 18, Table 18–1). These available media are usually suitable for most tissue culture applications. However, a medium's components can be developed to specifically meet a particular plant's needs, especially those that show mineral deficiencies on standard media (215). The basic approach is to measure the leaf's chemical composition of a growing seedling plant, and then customize the medium's components to meet those requirements. Software is available to help formulate these media to ensure the mineral combinations are compatible and do not precipitate when being prepared (166). A similar approach has been developed by analyzing the mineral seed components of hazelnut (*Corylus*) seeds to develop a balanced nutrient medium for micropropagation of hazelnut (175).

> **agar** An algal extract that solidifies after heating. It is used as a support for explants in tissue culture.

Most cultures require auxin and cytokinin at particular ratios for establishment. Typically, the culture medium has lower amounts of hormones during establishment compared to the multiplication stage of micropropagation.

Stabilization. The explant initially grows by elongation of the main terminal shoot, with limited proliferation of axillary shoots. If the explant is successfully established, several microshoots are produced within a few weeks, the number depending upon the apical dominance of the particular kind of plant (38, 191). Established and stabilized cultures are ready for subculturing and can be moved along to Stage II of micropropagation. However, in some instances, explants may require repeated subculturing to produce a uniform, well-growing culture. Most annuals and herbaceous plants stabilize quickly within a few subcultures. Woody plants invariably require longer periods and sometimes stabilized cultures may never be achieved (Fig. 17–5). Growth characteristics change from unpredictable, often abnormal, shoot development, at first, to a uniform predictable growth pattern (Fig. 17–6) (151, 154). Difficulty in stabilization appears to be

(a)

(b)

Figure 17–5
Kentucky coffee tree (*Gymnocladus dioicus*) fails to stabilize when cultured from explants collected from mature trees. (a) Short shoots are produced that fail to elongate. (b) However, explants from juvenile explants (69) and explants from stump sprouts of mature trees will stabilize and form elongated shoots for micropropagation.

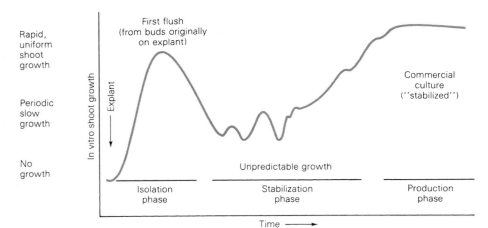

Figure 17–6
Generalized scheme of the three important phases in the microculture period through which a shoot must progress to be successfully microcultured. The second period is rooting and acclimatization of shoots. Reprinted by permission from D. D. McCown and B. H. McCown (154).

associated with the phase state of the explant: the more juvenile, the more easily stabilized. The characteristic seasonal growth pattern of the species being cultured has been associated with ease of stabilization.

Stage II: Shoot Multiplication The purpose of Stage II is to maintain the culture in a stabilized state and multiply the microshoots to the number required for rooting. The basic medium of Stage II is similar to Stage I and specific media formulations are provided in Chapter 18. Intensive commercial micropropagation may require the optimization of not only the inorganic nutrients but also other factors in the medium, which can be done by systematically testing a range of concentrations of individual media elements combined in various ways with other elements. Suggested methods have been a broad-spectrum approach using all possible combinations (38), or a factorial approach (2, 38, 235), where a range of each ingredient (others held constant) is tested in consecutive subcultures. Most micropropagation systems use a solid (agar solidified) medium, but liquid cultures may provide increased shoot multiplication and can be automated (108).

Growth Regulators. Cytokinin and auxin are used to support a basic level of growth but are equally important to directing the developmental response of the tissues in culture. Shoot initiation is strongly supported by cytokinin concentration (Fig. 17–7). In general, the minimum concentration of cytokinin that stimulates multiple shoot initiation is selected during the multiplication stage. Increased cytokinin levels may promote additional shoot proliferation, but can inhibit shoot elongation (Fig. 17–8). Requirements may vary at different stages of culture such that variable growth responses may occur during consecutive subcultures (Fig. 17–9). Adjustment of cytokinin concentrations may be necessary (235). Auxin is usually low or absent in Stage II cultures, but there are cases where auxin alone is used to induce shoot multiplication.

Figure 17–7
Illustration of shoot induction and morphology in tobacco leaf discs with increasing cytokinin concentration. (a) An insufficient concentration of cytokinin leads to little growth. (b) Suboptimal amounts induce few shoots and some callus. (c) An optimal concentration induces numerous well-formed elongating shoots. (d) At superoptimal concentrations, shoots are initiated but become malformed and fail to elongate.

Figure 17–8
Although cytokinin promotes shoot formation, high concentrations also inhibit shoot elongation. The culture of gas plant (*Dictamnus albus*) on the left was treated with 5 µM benzyladenine (BA), while the culture on the right was treated with 20 µM BA (123).

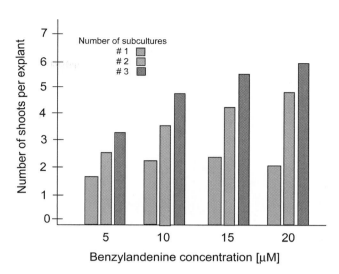

Figure 17–9
Relationship between cytokinin (BA) concentration and shoot initiation in shoot cultures of eastern redbud (*Cercis canadensis*) (275). The number of shoots per explant increases as the cytokinin level increases. This response is consistent between subcultures, but the overall number of shoots per explant increases until the cultures stabilize.

Subculturing. After microshoots have reached an appropriate length, they are harvested and either used to start new cultures or transferred to Stage III for rooting. The original culture can be trim-med or separated and **subcultured** to new multiplication media. The **propagation ratio** (PR) is the number of new microshoots that can be produced per subculture. In general, the PR is greater from adventitious shoots than from axillary shoot cultures. However, the proportion of off-type and aberrant shoots is more apt to be greater with the adventitious origin. Consequently, the highest PR may not be the most desirable to maintain quality and trueness-to-type.

> **subculture** The process of dividing stems or tissue from a tissue culture into smaller pieces and then transferring them to a fresh medium to multiply the tissue culture.

Subculturing frequency varies from 2 to 8 weeks, depending on the speed of shoot development. Promptness is essential to maintain shoot proliferation and to establish optimum shoot multiplication rates. Variation in routine may adversely affect multiplication rate, yield and quality of shoots, rooting, and subsequent growth. If subculturing is delayed too long, leaf yellowing and necrosis can develop.

Subculture technique needs to be optimized for the particular kind of culture, including size of new explant, method of cutting, and sometimes orientation of planting. Seasonal rhythmic patterns in proliferation occasionally occur (235), even though subculturing is done under uniformly controlled conditions. For cultures that show rhythmic patterns of proliferation, shoot numbers tend to be greater in summer than in winter.

BOX 17.1 GETTING MORE IN DEPTH ON THE SUBJECT
LIQUID CULTURES AND AUTOMATED MICROPROPAGATION SYSTEMS (108)

Liquid cultures contain the same nutritive and hormonal substances found in stationary cultures but lack a solidifying agent like agar. These usually require a mechanism to aerate the medium, or a method to keep the shoots from being submerged. Systems designed to automate micropropagation usually employ a liquid bioreactor culture system where the medium can be pumped into and away from the cultures (225, 253, 286). One automated method is the temporary immersion bioreactor system (54, 249), which acts to ebb and flood the explants with medium at regular intervals, and can be used for callus growth or shoot proliferation (Fig. 17–10). Bioreactors have also been developed for geophytic crops including potato where bud clusters proliferate in culture (286).

(Continued)

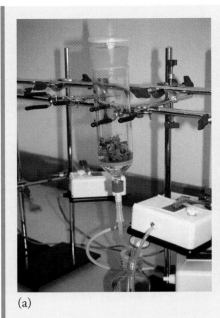 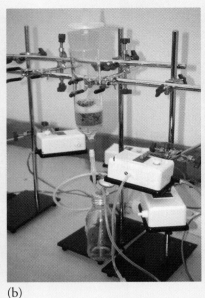 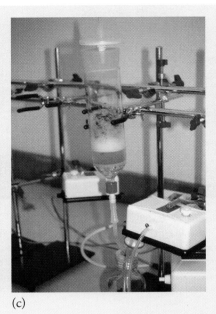

Figure 17–10
The temporary immersion bioreactor system uses pumps and filters to immerse tissue in nutrient medium for short durations before draining away. (a) Culture prior to immersion. (b) Liquid being pumped over the culture. (c) Liquid being reversed back to the reservoir.

An active area of research is the production of somatic embryos in bioreactors (109), especially with conifer species for forestation (247). Bioreactors coupled with image analysis to sort somatic embryos (276) have the potential to mass produce clonal conifer seedlings in quantities required for reforest plantations.

One approach to improve shoot elongation prior to taking microcuttings is using a liquid overlay or double phase culture (149), wherein short shoots remain on the agar medium and are covered with nutritive liquid medium that may contain hormones or activated charcoal (Fig. 17–11). This liquid overlay technique has been used to improve shoot multiplication as well as elongation, and it has been employed to improve somatic embryo initiation in conifers (196).

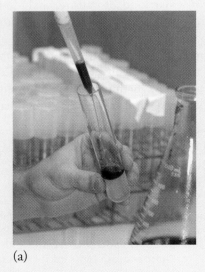  

Figure 17–11
Tissue overlay (also called double phase) culture. (a) Liquid medium being added over the shoot. (b) Explant submerged in liquid medium. (c) Overlay containing activated charcoal.

Stage III: Root Formation The function of Stage III is to root microcuttings and, in some cases, to prepare them for transplanting out of the aseptic, protected environment of the test tube to the outdoor conditions of the greenhouse or transplant area (47, 191, 284). Rooting can take place in the *in vitro* or *ex vitro* environment (Fig. 17–12).

In Vitro *Rooting*. Microcuttings that are rooted in the test tube are rooted *in vitro*. Microcuttings are moved to a root-inducing medium in which the growth regulator balance of the medium is changed to reduced (or no) cytokinin and increased auxin. The concentration of basal salts in the rooting medium is usually cut in half. Individual microcuttings may be placed into a root-inducing (often liquid) medium for a few days and then transferred to an auxin-free medium for rooting. Cultures may be rooted in light or darkness, and for some species, including additional factors in the medium such as an auxin synergist like phloroglucinol can be helpful (117, 122).

Ex Vitro *Rooting*. Many commercial as well as experimental micropropagation systems avoid *in vitro* rooting by treating microcuttings with auxin, inserting them directly in soilless greenhouse rooting medium, and placing them under mist or high humidity conditions for rooting. The procedure not only provides an excellent transition from the culture environment to the open air, but also saves labor requirements for handling plants in micropropagation. Protocols include either:

1. Treating microcuttings with auxin (usually IBA) as a quick-dip or talk application and stick in greenhouse medium under high humidity (see Chapter 18); or
2. Treating microcuttings with auxin for 5 to 15 days as *in vitro* treatment in agar or liquid culture, then washing off the agar and placing microcuttings under an *ex vitro* environment (275).

There is a distinct morphological difference between roots formed *in vitro* compared to *ex vitro*

(a)

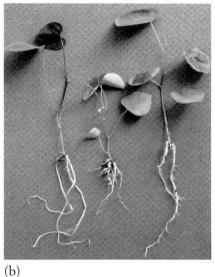
(b)

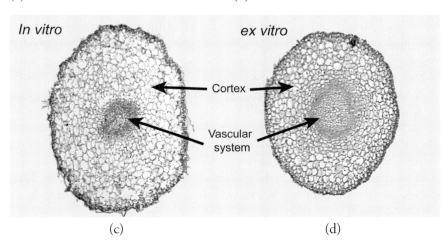
(c)   (d)

Figure 17–12
Root formation in microcuttings of eastern redbud (*Cercis canadensis*).
(a) Root formation *in vitro*. It is common that roots on NAA-treated microcuttings (*left*) are shorter and thicker than roots on IBA-treated microcuttings (*right*). IBA is used most often to root microcuttings in a wide variety of species. (b) Three cultivars of redbud microcuttings rooted *ex vitro*.
(c) Anatomy of *in vitro* and (d) *ex vitro* developed roots. Observe the swollen cortical cells and the less developed vascular system on *in vitro* roots.

roots (Fig. 17–12). *Ex vitro*–rooted microcuttings tend to form roots that are more "normal" than *in vitro*–rooted microcuttings, including less-swollen cortical cells and a more well-formed vascular system. After transplanting to a greenhouse medium, roots that developed *in vitro* may not survive. In other cases, *in vitro* roots must adopt a "normal" morphology to function properly (82, 150).

One positive aspect of *in vitro* culture is that many plants that are normally difficult to root from cuttings can be rooted as microcuttings (208, 242). This ability has been shown to be related to the number of subcultures (e.g., in apple) (231) and probably reflects the maturation state of the microcuttings. It is generally understood that microcuttings are in a more juvenile stage of development while in culture. Interestingly, this juvenile stage appears to be a short-term rejuvenation, because micropropagated plants can flower 1 or 2 years after transplanting to the field, while seedlings of the same species can take up to 15 years to flower.

Stage IV: Acclimatization This stage involves the shift from a heterotrophic (sugar-requiring) to an autotrophic (free-living) condition and the **acclimatization** of the microplant to the outdoor environment (195). Keeping the shoot actively growing is important because acclimatization and development of autotrophic conditions depend on new growth after transfer from the test-tube environment. Immediately upon transplanting, rooted microplants should be kept in very high humidity, gradually exposing them to outdoor conditions to prevent dehydration (47).

acclimatization The preferred term for the process of gradually moving tissue culture-grown plantlets to open-air conditions.

Microshoots developed *in vitro* have a unique leaf morphology (21). Leaf size and the number of cell layers that comprise the leaf are much reduced, compared to leaves developed outdoors. Because of the high humidity *in vitro*, leaves do not develop epicuticular waxes on the leaf surface in normal amounts or with the same chemical properties (Fig. 17–13) (195). As a result, leaves on microshoots desiccate quickly after they are removed from *in vitro* culture.

Stomata are also morphologically different on leaves of *in vitro* grown plants. For example, *in vitro* sweet gum (*Liquidambar*) leaves had stomata with guard cells that were rounded and raised on the lower leaf surface, while greenhouse plants typically had elliptical and sunken guard cells (268). Stomata on plants

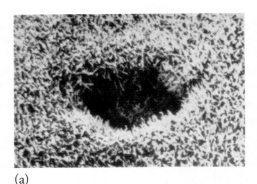

(a)

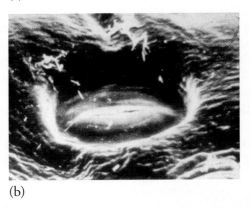

(b)

Figure 17–13
Scanning electron micrograph of stomata from the lower leaf surface of carnation plants grown in the greenhouse (*top*) or *in vitro* (*bottom*). Note lack of wax on *in vitro*–grown plants.
Courtesy of E. Sutter, University of California at Davis.

grown *in vitro* usually remain open and fail to close after initially being removed from the *in vitro* environment, again leading to rapid wilting and desiccation of these leaves (96).

Treatments to Enhance Acclimatization. Several methods have been developed to help rooted microcuttings acclimatize. Any of these methods may be carried out prior to or after transplanting (96).

The most common acclimatization treatment is to gradually reduce humidity around the rooted microcuttings. Humidity may be maintained by using intermittent mist, fog, or enclosing plantlets in a polyethylene tent. This is a critical step in the acclimatization process. Antitranspirants have been used with variable results and can be phytotoxic; they are not commonly used commercially.

Plantlets can also be "hardened" while they are still *in vitro* and prior to transplanting. Hardening involves various treatments to reduce humidity in the culture vessel. It may be as simple as removing the cap on the tissue culture vessel for 5 to 7 days prior to transplanting or using desiccants to reduce humidity. Cooling the bottom of the vessel will also reduce humidity.

Increasing the osmotic potential of the medium (usually with additional sucrose) can also be effective in hardening rooted microcuttings because it reduces available water to the shoots. Growth regulators (like paclobutrazol) have been used with variable success as an *in vitro* pretreatment before transplanting.

Inoculation of microcuttings with mycorrhizal fungi has also been shown to enhance rooting and/or acclimatization in numerous commercial crops including fruit, ornamental, and conifer (6, 248, 260). Mycorrhiza-treated plantlets show reduced transplant shock, greater control of stomatal conductance, and enhanced nutrient uptake and plant growth (52). In addition to mycorrhiza, beneficial microorganisms such as *Trichoderma* or *Bacillus* may be added to enhance plantlet growth by protecting plantlets from pathogenic organisms in the transplant substrate.

Systems Used to Regenerate Plantlets by Micropropagation

There are basically two developmental patterns for plantlet formation in tissue culture (Fig. 17–14), described by the way shoots originate from the initial explant in Stages I and II. Once shoots have formed, the rooting and acclimatization stages are the same for the various developmental patterns. These patterns are (a) **axillary shoot formation** and (b) **adventitious shoot formation.** Systems designed to produce predominately axillary shoots are meristem and shoot cultures, while systems designed to produce adventitious shoots use tissue for an initial explant source that does not contain an axillary bud like roots, stem internodes, leaves, and bulb scales. Shoots generated from callus also form adventitiously.

axillary shoots Shoots that form from existing buds at each node on the stem.

Axillary Shoot Formation Systems for axillary shoot formation are listed in Table 17-1, page 646. These can be separated into meristem culture and various forms of shoot culture.

Meristem Culture. **Meristem culture** utilizes the smallest part of the shoot tip as the explant (see Fig. 17–15, page 660), including the meristem dome and a few subtending leaf primordia. The number of additional structures depends on the length of the excised stem. The primary reason for this procedure is to produce a plantlet that is free of systemic viruses, virus-like organisms, and superficial fungi and bacteria (see Fig. 16–34). The meristem is usually free of disease organisms; therefore, the smaller the explant, the more effective the elimination of

meristem culture A procedure to eliminate diseases from plants, using a very small piece of tissue from the shoot tip as the initial explant.

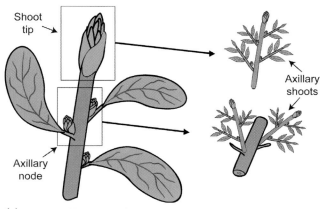

(a) Direct axillary shoot formation

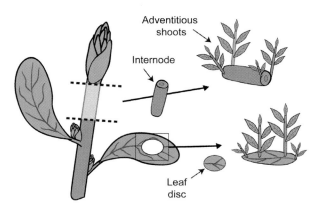

(b) Direct adventitious shoot formation

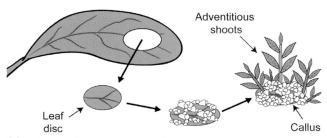

(c) Indirect adventitious shoot formation

Figure 17–14
Patterns of plantlet development by micropropagation: (a) Axillary shoot formation directly from existing buds from nodal explants. (b) Adventitious shoots formed directly from tissue without pre-existing buds. (c) Adventitious shoots formed indirectly after callus develops from the initial explant.

> **BOX 17.2** **GETTING MORE IN DEPTH ON THE SUBJECT**
> **CONCEPTUAL BASIS FOR ADVENTITIOUS PLANT ORGANOGENESIS**
>
> Since the classic experiments by Skoog and Miller (228), it has been recognized that the ratio of auxin to cytokinin determines whether an explant will form a root or shoot. This balance between auxin and cytokinin establishes regulatory networks that coordinate different steps in organ formation (201). Important hormones in the regulatory network include ethylene, gibberellin, abscisic acid, brassinosteroids, and polyamines.
>
> There are two general phases of organogenesis: dedifferentiation and redifferentiation. During the dedifferentiation phase, the plant cell must reverse its cell state and become **"competent"** to express its organogenic potential. There are two patterns for organ formation (Fig. 17–14). In the direct pattern, explant cells directly become competent for organ formation, while in the indirect pattern, there is an intervening cell division (callus) stage in which these new cells become competent. Competent cells can respond to auxin/cytokinin to form a root or shoot. Once these competent cells receive the inductive signal, they become **"determined"** to form the new adventitious organ (see Fig. 9–5).
>
> The concepts of competency and determination for plant organogenesis were first developed from a series of transfer experiments using field bindweed (*Convolvulus arvensis*) leaf explants (32). They transferred these leaf disks among shoot-inducing (SIM), root-inducing (RIM), or callus-inducing (CIM) media at different times to see when they were competent to respond to a hormone signal and when they were determined to form that organ. Cells are determined when they proceed to organ formation independent of the induction signal type. Using this method, it was discovered that an explant became competent after 3 days, and determined to form a shoot after 10 days. The actual shoot might not appear until 15 to 20 days. Similar experiments using *Arabidopsis* as a model system for organogenesis has dramatically advanced our insight into the gene activity during these phases of organogenesis (277, 278).

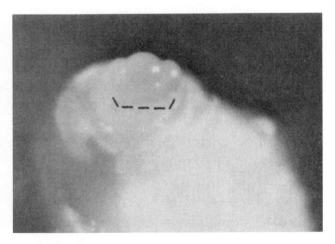

Figure 17–15
Shoot tip of carnation stem with outer leaves removed, showing the apical and lateral meristems (growing points). Part of the shoot tip to be excised for culturing is indicated in lines. Courtesy W. P. Hackett.

pathogens (125, 169). Meristem tips of 0.10 to 0.15 mm have given 100 percent virus elimination with the percentage decreasing as the size increased to 1 mm. On the other hand, the smaller the explant, the more difficult it is to establish and the lower the survival rates. A general compromise is to use explants 0.25 to 1.0 mm long. Virus status must be confirmed by indexing (144).

Growing the stock plant at a warm temperature (**thermotherapy;** see Chapter 16) prior to meristem excision can increase the probability of virus removal (43, 222). Some antiviral agents, such as Ribavirin (73, 191), added to the medium have increased the recovery of virus-free plants. Once a rooted plant is produced, it can be used as the start of a foundation clone in a "clean stock" program for that particular species and cultivar (see Chapter 16) (200).

Meristem-tip culture has been successful with many important herbaceous crops, and has become an essential production aspect for such commercial crops as carnation, chrysanthemum, orchid, geranium, potato, sweet potato, cassava, banana, and others (235). The procedure is more complex with woody plants, and micrografting is an alternative procedure. Improvements in the micropropagation of woody plants have extended its usefulness (19).

Axillary Shoot Cultures. Shoot cultures are like miniaturized stems, divisions, or layers (see Chapter 14) that grow from lateral and terminal meristems. Plantlets developed in this way tend to be reliable in the reproduction of the genotype of the source plant because the system involves extension of existing apical meristems.

Typical of biological material, different species and different clones have unique growth patterns and requirements that may differ in culture. Several different growth patterns have been described (74):

1. **Axillary branching.** In this system, the apical shoot tip is repressed and lateral shoots are stimulated to form a dense clump of shoots (Fig. 17–16). Cytokinin usually suppresses apical dominance, but in some cases, removal of the shoot tip increases axillary shoot development. The base of this clump may be thickened and callus-like. Cultures are multiplied by subculturing individual shoots or cutting the clump into sections and each in separate containers. This can often lead to a mixed culture with shoots originating from axillary stem buds and adventitious buds from the basal callus.
2. **Nodal cultures.** This pattern involves plants whose shoots have rather strong apical dominance. Long shoots are cut into single nodes and planted vertically in the medium (Fig. 17–17a). Axillary buds at each node elongate and grow in length. The pattern is repeated by again cutting into nodal segments at each subculture. Cotyledonary node cultures have been very effective in micropropagating a variety of plants, especially legumes (Fig. 17–17b).
3. **Stool shoots.** This pattern is analogous to that of plants grown by layering. A shoot of several nodes is

(a)

(b)

Figure 17–16
Multiple shoots formed by an axillary branching type of culture. (a) An explant with several nodes. (b) Shoots proliferating from axillary buds.

(a)

(b)

Figure 17–17
Shoots formed by a nodal type of culture. (a) Explants show strong apical dominance with very little axillary branching. The elongated shoot will be subcultured by being cut into several single-node explants. (b) A cotyledonary node explant showing elongation of shoot buds in the axils of the pair of cotyledons.

(a)

(b)

Figure 17–18
Stool shoots formed by horizontally placed explants. (a) Multiple node explants without leaves or an apical meristem are placed horizontally on the medium. (b) Single shoots arise from existing buds at each node.

laid horizontally on the surface of the medium, sometimes on a slant. Lateral growing points form a thicket of small vertical growing shoots (Fig. 17–18). These "layers" may be either subdivided at each subculture or restarted as individual horizontal shoots. Horizontal explant cultures often produce more shoots than upright explants. The stool shoots approach has been used to increase shoot formation in birch, maple, apple, and forsythia (151).

4. **Proliferation of pseudocorms.** Growing points of orchids in culture produce clusters of small protuberances that resemble the pseudocorms initiated when an orchid seed germinates (Fig. 17–19). Protocorms often form short rhizomes that can be subcultured to produce numerous plantlets. The structure is subcultured by cutting into sections.

5. **Minitubers.** Potato plants in culture can form miniature tubers at the end of small stolons extending from lateral meristems located at each node of potato plants when treated with high cytokinin levels, particularly in darkness (44). These storage organs can be removed and used in the production of virus-free planting stock. Similarly, yam (*Dioscorea*) produces root tubers at the base of stem node cuttings (118, 179). Minituber formation is also amenable to large-scale automated bioreactor production (152).

6. **Micrografting.** Micrografting uses procedures comparable to conventional grafting and budding of woody plants. Micrografting has several unique uses, including:

   a. Creation of disease-free plants by grafting small meristem tips (121).
   b. Virus indexing by micrografting to susceptible understocks (281).
   c. Early detection of graft incompatibility relationships (120).
   d. Propagation of novel plants created in tissue culture (Fig. 17–20).
   e. Serial micrografting mature scions on seedling understocks can be effective for rejuvenating scion (1, 4), which has been successful for avocado

(a)

(b)

Figure 17–19
Shoot formation via pseudocorms produced by orchids. (a) Proliferating pseudocorms on a charcoal medium. (b) Numerous pseudocorms with the medium washed away from the roots, ready for transplanting to the greenhouse.

Figure 17–20
A specialty cactus being micrografted.

Adventitious Shoot Formation Adventitious shoots are initiated either directly on the explant or indirectly in the callus that is produced from the explant (Fig. 17–14)(258). Direct formation of adventitious shoots is analogous to leaf and root cuttings (see Chapters 9 and 10), bulb scales, and bulbils (see Chapter 16). The selection of explant and the growth regulator regime determines the success of adventitious shoot initiation.

Indirect development of adventitious shoots first involves the initiation of basal callus from excised shoots in culture. Shoots arise from the periphery of the callus and are not initially connected to the vascular tissue of the explant. Adventitious shoot formation can generally result in high rates of multiplication, higher than rates from axillary shoot cultures. On the other hand, adventitious shoots can also result in increased numbers of aberrant, off-type plants resulting from conditions described in a later section.

Adventitious shoot cultures are most often used to regenerate diploid plants, but regenerants with different ploidy levels can also be obtained for plant breeding purposes.

(*Persea americana*) (1), Redwood (*Sequoia*) (4), apple (105), cashew (164), and pistachio (*Pistacia vera*) (186).

f. Recovery of poorly developed somatic embryos that lack a root meristem (198).

g. Small micrografted plants are a convenient way to send germplasm between countries.

Plants that have been successfully micrografted include many fruit and nut trees such as citrus, grape, pear, apple, cherry, granadilla (*Passiflora*), kiwifruit, walnut, cashew, pistachio, and almond, as well as forest tree species such as *Acacia,* rubber tree (*Hevea*), Douglas-fir (*Pseudotsuga*), *Sequoia,* and spruce (*Picea*).

Diploid Plant Regeneration Systems. Diploid plantlets are used for propagation or to regenerate plants that have been genetically modified *in vitro*. Diploid regeneration systems are listed in Table 17-1, page 646, and employ explants from almost any part of the plant that does not include pre-existing meristems.

1. **Leaf pieces.** Regeneration of shoots can occur on either pieces of the leaf blade (Fig. 17–21) or

---

> **BOX 17.3 GETTING MORE IN DEPTH ON THE SUBJECT**
> **SHOOT-TIP MICROGRAFTING *IN VITRO***
>
> Grafting very small meristem tips comparable to those described in the preceding section can be used as an alternative method to produce virus-free materials for various woody plants, such as citrus (176), apple (105), and *Prunus* (19, 177). For example, this procedure is important in citrus not only because it is successful, but also because explants can be used from ontogenetically mature trees (see Chapter 16), which avoids the juvenile phenotype of nucellar seedlings also used in virus cleanup in citrus (see Chapter 4).
>
> The technique for micrografting can be illustrated with citrus. Citrus embryos are excised from rootstock seeds, surface disinfested, and planted in standard inorganic salt medium with 1 percent agar (176). Embryos germinate in the dark in 2 weeks. Seedlings are then removed and decapitated to a 1 to 1.5 cm length; cotyledons and lateral buds are excised with a mounted razor blade. A 0.14- to 0.18-mm tip with 3-leaf primordia is used as the scion—this gives reasonable success, and the shoot tip eliminates viruses.
>
> An inverted T-bud cut is made in the seedling rootstock, cutting 1 mm down the stem, followed by a horizontal cut on the bottom. The excised shoot tip is placed inside the flap next to the cambium. Grafted plants are placed in a liquid medium. A filter paper bridge with a center hole supports the stem. Cultures are kept in the light for 3 to 5 weeks to heal. When two expanded leaves appear on the scion, the grafted plant is transplanted.
>
> A similar procedure has been used for apple (105) and plum (177). Rootstocks are either seedling plants or rooted stems. Shoot-tip scions taken from cultured plants reduce contamination problems.

Figure 17–21
A leaf explant showing direct formation of adventitious shoots.

petiole. Plants such as African violet (*Saintpaulia*) (14), *Salpiglossis* (145), and horseradish (163) easily regenerate in this manner.

2. **Root pieces** (Fig. 17–22). Roots can serve as initial explants and the subsequent shoots moved to a standard shoot-based micropropagation system.

Figure 17–22
Isolated root explant from Kentucky coffee tree (*Gymnocladus dioicus*) showing both (a) direct and (b) indirect shoot formation.

Crops that naturally produce root suckers can usually produce adventitious shoots on root explants.

3. **Thin layer explants.** These explants are derived from epidermal peels or thin transverse sections that are only a few cell layers thick (180, 250). They were originally derived from flower stem tissue of tobacco (257), but they can also be made from leaf and stem tissue (Fig. 17–23). There are thin layer systems for at least 15 different kinds of plants and depending on the explant type, thin layers can develop into shoots, roots, flowers, or somatic embryos. It is a useful tissue for genetic transformation because of the limited number of cells in the explant.

4. **Homogenized tissue.** A limited number of plants can have their tissue mechanically homogenized in a blender as initial explants. The homogenized tissue is mixed with liquid medium and spread out over an agar-based tissue culture medium for shoot formation, which results in a tremendous number of regenerated shoots. Plants amenable to this technique include several ferns (35), African violet (165), and begonia (265).

5. **Fragmented shoot apices** (10, 11). Grape shoot apices 0.1 mm in length are cut into 2 to 5 segments and cultured in drops of medium. These segments proliferate into small leafy structures that can develop into whole plants.

6. **Cotyledons and hypocotyls.** Excised cotyledon and hypocotyl segments are useful as starting explants for many types of plants. They can be important explants for plants that are difficult to regenerate from other plant parts, like conifers (16).

7. **Young needle fascicles.** These have been used to regenerate shoots from older conifer trees (16). Explants from rejuvenated plant tissues are particularly responsive (15).

8. **Immature inflorescences on flower stems** (Fig. 17–24). Segments from these structures are highly regenerative in some monocots, such as *Gladiolus* (287), *Hemerocallis* (101), *Iris* (162), *Hosta* (161), orchids (128), and *Freesia* (192), as well as some dicotyledons such as *Gerbera* (193) and chrysanthemum (216). Many geophyte (bulb) species appear to respond to culture from flower-stem tissue, including flowering onion (*Allium*), *Nerine*, daffodil (*Narcissus*), and *Ornithogalum* (129, 288).

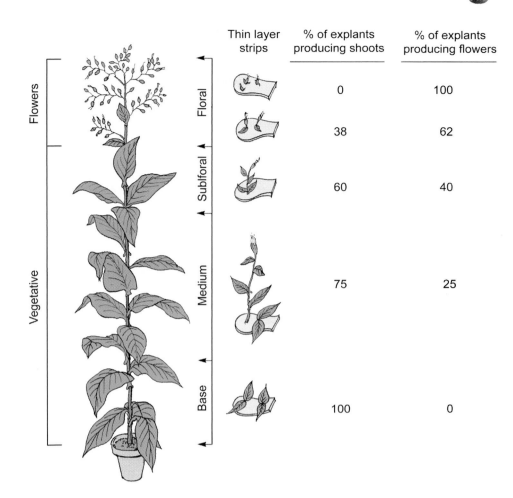

Figure 17–23
Thin layer explants of epidermal tissue showing organ initiation potential relative to location on the mother plant where the explant was taken (257). Redrawn from M. Tran Thanh Van, 1973.

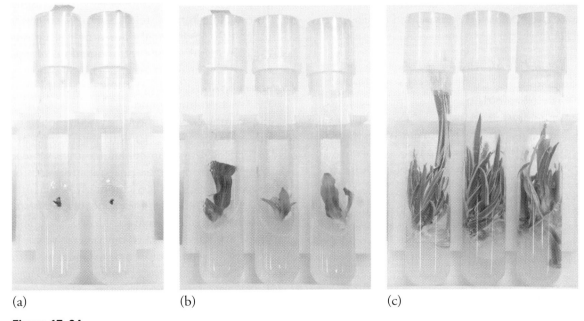

Figure 17–24
Hosta showing a progression from (a) initial flower stem explant to (b and c) multiple adventitious shoot formation.

(a) (b)

Figure 17–25
The formation of lily bulbs in tissue culture. (a) Bulbs initiating from leaf scales. (b) Vegetative leaf growth from regenerated bulbs.

9. ***Bulb scales and other storage structures*** (Fig. 17–25). Bulbous-type monocots (92, 106, 107) characteristically have rings of meristematic tissue at the base of bulb scales near the basal plate (see Chapter 16). Excised scale explants in culture develop adventitious shoots readily. Initiation of adventitious shoots begins in single parenchyma cells located either in the epidermis or just below the surface of the stem; some of these cells become meristematic and develop into pockets of small, densely staining cells (102).

Development of Plants with Altered Ploidy. There are some breeding strategies that require the production of plants with a haploid, triploid, or polyploid chromosome number rather than the usual diploid number found in most plants. The two major strategies for developing plants with an altered ploidy number include anther culture for production of haploids and endosperm culture to develop triploids.

1. ***Anther and pollen (microspore) culture.*** Guha and Maheshwari (83) accidentally discovered that pollen grains could develop into haploid embryos while working with *Datura*. Development of microspores into plants was later realized in tobacco by Nitsch and co-workers (18, 183). Since then, whole plants have been produced in a number of species either directly from the immature pollen grain (183, 241) or from callus that develops from the microspore (227, 241). Examples of plant species regenerated from anther culture include potato, tobacco, *Brassica*, corn, wheat, rice, rye, flax, apple, and lily. These may be regenerated from shoot cultures or through somatic embryogenesis. One reason so many important crop plants have been investigated for anther culture is that it can be used to create **dihaploids** (Box 17.4) after treating haploid plants with a chromosome doubling agent like colchicine (112). Dihaploids reduce the time for the release of new cultivars compared to traditional breeding programs because of reduced genetic recombination (dihaploids have the same genes on both chromosomes).

2. ***Endosperm culture.*** Endosperm cells can be used to initiate tissue cultures, and in some cases they become competent to form triploid shoots or somatic embryos (252). Triploid regenerants are seedless. Triploids can be important for improvement of seedless fruit crops like banana and citrus and for the development of seedless clones of ornamental crops that are or could become exotic invasive pests. Triploids usually show vigorous growth and have been used with crops where the vegetative parts of the plant have economic importance, as with mulberry leaves for silk worm production (251).

Reproduction of Seedling Plants in Tissue Culture

In vitro culture can be effectively used in a number of germination and reproduction procedures when protection from contaminating organisms is needed and

BOX 17.4 GETTING MORE IN DEPTH ON THE SUBJECT
ANTHER CULTURE

Anther culture is used for plant breeding to produce haploid plants (112). Subsequent doubling of the chromosomes results in an isogenic, homozygous line called a *dihaploid* (24). Two basic procedures have been used for tobacco (241). Figure 17–26 describes a simple technique that is effective with this plant. A second technique utilizes a "stress" period of the excised buds prior to culture. Flower buds are placed into a sealed container and stored in the dark for a period of time based on a time-temperature pattern, such as 7°C for 2 to 3 weeks. The temperature may be higher, but the time must be shorter. After this treatment the anthers are floated on liquid medium in a Petri dish. Embryos become discernible in about 14 days. It is necessary to transfer the embryos to new medium, or new medium must be added. Procedures have also been described for potato (266), *Brassica* spp. (126), cereals, and grasses (267).

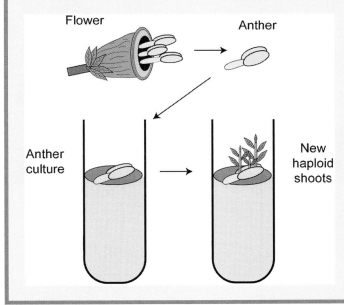

Figure 17–26
Anther culture is a procedure for obtaining haploid plants from normally diploid plants. A flower bud from *Nicotiana tabacum* is excised just as the petals are emerging from the bud (*upper left*). The immature anther is removed aseptically and planted on agar with a standard nutrient medium without hormones. Pollen should be at the uninucleate, microspore stage of development. Somatic embryos develop from callus derived from the haploid microspores. Haploid (1n) embryos germinate to form plantlets.

when specific genetic barriers need to be overcome. These procedures include:

1. Aseptic seed culture
2. Embryo culture and embryo rescue
3. Ovule culture and ovary culture

Aseptic Seed Culture Orchid seeds are extremely tiny, lack stored-food reserves, and in nature depend on a symbiotic relationship with specific fungi to obtain nutrients. It is estimated that 30,000 seeds are produced in a single *Cattleya* pod. In 1922, Lewis Knudson reported the successful culture of orchids in a medium supplied with minerals and sucrose (66). The basis of his success was his discovery that calcium hypochlorite (bleach) could surface-sterilize seeds but was not toxic to the germinating embryo. This **seed culture** procedure revolutionized orchid propagation at the time but was eventually replaced by micropropagation (see Chapter 18), though it is still commonly used to germinate hybrid seeds (see Box 17.5).

Embryo Culture and Embryo Rescue (212) **Embryo culture** involves the excision of an embryo from a seed and germinating it in aseptic culture (34, 104, 274). As early as 1904, the first attempts were made to grow immature embryos in nutrient solutions (191). These immature crucifer embryos germinated, but growth was weak. Hannig called this "precocious germination." A principal use of embryo culture is to **"rescue"** embryos that would have

seed culture
A procedure primarily used for orchid seed germination because the seed is so small and contains no seed storage reserves.

embryo culture
A procedure involving tissue culture of immature embryos that require controlled conditions to complete development. It is most commonly used by plant breeders for **embryo rescue** of genetic crosses that would not form seed on the plant.

BOX 17.5 GETTING MORE IN DEPTH ON THE SUBJECT
ASEPTIC ORCHID SEED GERMINATION

Aseptic **seed culture** for orchids is currently used in breeding programs to produce seedlings (Fig. 17–27). Approximately 7 years is required from germination to flowering. Seeds can be extracted from green pods that are about 60 percent mature after externally disinfesting the pod with alcohol. The usual procedure is to remove seeds from a mature fruit when it naturally dehisces and disinfest it with calcium or sodium hypochlorite. A small amount of seed is placed in a vial or flask and covered with five to ten times its volume with disinfestant plus a drop or two of a wetting agent. Seeds are kept immersed for 5 minutes, shaking periodically, after which the seeds will sink. The disinfestant is poured off and the seeds washed several times with sterile water. The seeds are poured over a sterile medium containing 1 to 1.5 percent agar, 2 percent sucrose, and basic mineral solution.

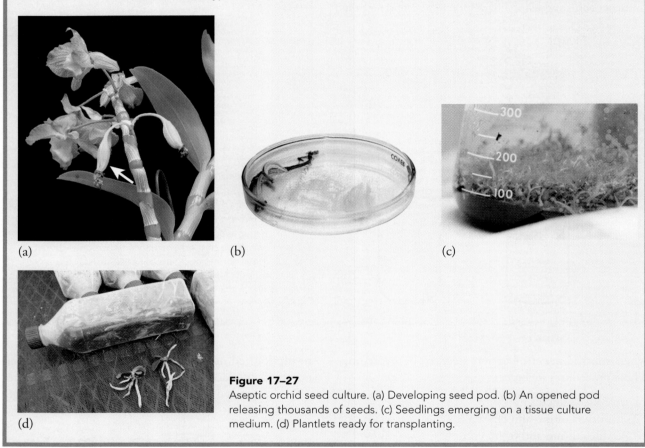

Figure 17–27
Aseptic orchid seed culture. (a) Developing seed pod. (b) An opened pod releasing thousands of seeds. (c) Seedlings emerging on a tissue culture medium. (d) Plantlets ready for transplanting.

aborted within the seed before the fruit was mature. Many interspecific and intergeneric hybridizations are initially successful but the embryo aborts during development. Laibach (191) first demonstrated embryo rescue in 1925 by growing embryos from interspecific crosses in flax (*Linum*). One reason wide-genetic hybrid crosses fail is that the endosperm does not develop properly (somatoplastic sterility) (36). The nutritive tissue culture medium serves as a substitute for the natural endosperm, allowing the embryo to grow to a stage where it will germinate or can be used to initiate shoot cultures.

Early ripening cultivars of many fruit tree species tend to mature the fruit before their embryos are fully developed. Under natural conditions these embryos can abort, but they can be excised and grown in an appropriate sterile culture (100, 127, 202). Hybrid production of seedless grapes also requires the rescue of the hybrid embryo early in development before it aborts.

A second use of embryo culture is to induce prompt germination of mature dormant seeds and thus shorten breeding cycles (143). Excising the embryo to eliminate dormancy-inducing restraints in the surrounding endosperm has been described for *Iris* (238),

Maranta (97), peach (127, 143), and others. Appropriate environmental sequences may be needed for embryos with internal epicotyl dormancy, as in *Paeonia* (160).

One protocol for the development of clover (*Trifolium*) embryo cultures (34) calls for the following sequence of stages:

1. High sucrose, moderate auxin, and low cytokinin for 1 to 2 weeks. At this time the embryo stops growing and must be transferred.
2. Normal sucrose, low auxin, and moderate cytokinin, which allows for resumption of embryo growth and, sometimes, shoot development.
3. Embryos with shoots are shifted to a medium with low auxin and high cytokinin to stimulate shoot proliferation. Shoots are rooted and transplanted.
4. Embryos with renewed, but disorganized, growth are transferred to a somatic embryogenesis induction medium.

Developing embryos up to the heart-shaped stage are difficult to handle physically and are sensitive to the medium used. Apparently, it is essential that the suspensor tissue be retained when the proembryo is excised at this particular stage. Embryos at this stage could best be handled by ovule culture.

Ovary and Ovule Culture Ovary and ovule culture include the aseptic culture of the excised ovary, ovule (fertilized or unfertilized), and placenta attached within the ovule. Although this technique was first utilized to investigate problems of fruit and seed development (12, 203), adaptations of the procedure have uses in propagation, particularly in genetic improvement.

> **ovary and ovule culture**
> A procedure that allows plant breeders to pollinate and fertilize plants in tissue culture. Pollen is added to the ovary while it is forming in tissue culture, resulting in fertilized embryos.

Unfertilized ovules have been excised, grown in culture, supplied with pollen, and subsequently fertilized *in vitro* (72, 205). Such procedures are most successful with plants that have multiple ovules. Pollen can be placed directly on the placenta inside the ovule, where pollen tubes can develop and grow immediately into the ovules without passing down the style.

Cultured ovules are useful for rescuing embryos that abort at a very young stage if not separated from the plant (77, 202, 204). This technique is simpler than culturing isolated embryos. The dissection of such small embryos is difficult and the medium required is more complex than culturing the entire ovule. Ovules can easily be disinfested and generally will grow in a basal medium of inorganic salts, sucrose, and sometimes growth regulators.

The technique of ovary (fruit) culture is less common and varies in nutritional requirements (203). For most species, a basic medium of inorganic salts and sucrose is essential, but auxin may be highly stimulatory.

BOX 17.6 GETTING MORE IN DEPTH ON THE SUBJECT
EMBRYO CULTURE

Culture procedures are based on the stage of embryo development (see Chapter 4). Methods of handling are different for the three broad categories of embryo development: (a) embryos from proembryo to approximately the heart stage; (b) immature embryos that are at approximately the heart stage up to about one-half full size; and (c) mature embryos whose cotyledons have reached nearly full size and are desiccation-tolerant. Fruits and seeds can be easily disinfested, opened, and the embryo extracted aseptically. Agar is the preferred support, but the ingredients of the basal medium vary greatly and individual species may require experimentation. Immature embryos separated from the ovule cease embryogenesis and show precocious abnormal germination. A major breakthrough in embryo nutrition came with the discovery by van Overbeek (259) in 1942 that 1 to 15 percent coconut milk (a liquid endosperm) would prevent precocious germination and allow embryo development to proceed. Since then, it has been found that success with embryos excised at this stage required high osmotic pressure and high nitrogen, either ammonium or nitrate, depending on the species. Important media constituents include sucrose (8 to 18 percent), casein hydrolysate (5 to 30 percent), a moderate or low level of auxin and cytokinin, and/or gibberellin, depending on the species. GA may, however, induce precocious germination, which can be overcome by abscisic acid. ABA has been shown to promote embryo development. A shift from embryonic to germination potential may be promoted by exposing the embryos to controlled desiccation or by chilling at 2°C (36°F) (99). Mature or nearly mature embryos can be germinated on a basal medium of only inorganic salts and agar. Sucrose (1 or 2 percent or none) may be helpful if the embryo is immature but otherwise may be inhibitory.

Callus, Cell, and Protoplast Culture Systems

callus Results from cell division in nondifferentiated parenchyma cells. Eventually, a callus culture does form stratified cell layers with outer meristematic layers and inner cells that can form vascular tissue.

Callus provides an important tissue culture system that can be subcultured and maintained more or less indefinitely. Several important pathways of development may follow:

1. **Callus cultures** may be used as a form of micropropagation. New roots, shoots, or complete plantlets may develop from individual cells or clumps of cells **(organogenesis).**

   organogenesis The process of developing adventitious shoots and/or roots. Changes take place in the cells that lead to the development of a unipolar structure (stem or root primordium) whose vascular system is often connected to the parent tissue.

2. **Cell suspensions** develop when callus cells disassociate in liquid culture medium on a rotary shaker. One interesting type of suspension culture is the **nodule culture** observed in several woody plants (153). Nodules are round aggregates of cells with a distinct morphology that can become organogenic when moved to a stationary culture. Cell suspensions are also used for the industrial production of secondary compounds, such as pharmaceuticals (230) and food coloring (22).
3. Cells may be treated to produce a **protoplast culture** by removing the cell wall around the cytoplasm and thus enhancing the ability to introduce specific materials into the cell or to join two cells together in a **somatic hybrid** or **cybrid.** Somatic hybrids have been important in citrus breeding (81).
4. **Somatic embryogenesis** may take place among individual or clumps of cells in the callus or cell suspension.

Callus and (particularly) cell cultures are potentially useful as methods of commercial propagation because of the high rates of multiplication and the possibility of industrialization. In practice, these methods have not been used directly, however, because significant amounts of genetic and/or epigenetic aberrations tend to develop during cell multiplication (157).

This class of culture system has been used extensively in the production of new and novel genotypes. Genetic variation induced in plants produced in culture is known as **somaclonal variation** (see Chapter 16). Variation may either be natural (57, 221), or may be induced by mutagenic treatments, or through specific genetic engineering techniques.

Callus Culture Callus is produced on explants *in vitro* as a response to wounding and growth substances, either within the tissue or supplied in the medium. Explants from almost any plant structure or part—seeds, stems, roots, leaves, storage organs, or fruits—can be excised, disinfested, and induced to form callus (74). Continued subculture at three- to four-week intervals of small cell clusters taken from these callus masses can maintain the callus culture for long periods (Fig. 17–28).

The most common cell culture medium is Murashige-Skoog (MS) (174), which established optimum rates for inorganic compounds, plus the Linsmaier and Skoog (147), which optimized organic supplements. Both media are rich in macroelements, particularly nitrogen, including nitrate (NO_3) and ammonium ions (NH_4), sucrose, and certain vitamins. Initiation of cell division and subsequent callus

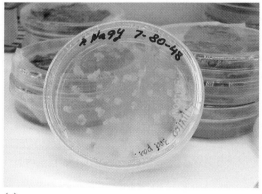

(a)

(b)

Figure 17–28
Callus developed on a stationary medium showing nicely that different parts of the callus have different cellular characteristics.

production requires a supply of cytokinin and auxin in the medium at the proper proportion (228). Auxin at a moderate to high concentration is the primary growth substance used to produce callus. The principal auxins include indoleacetic acid (IAA), naphthaleneacetic acid (NAA), and 2,4-dichlorophenoxyacetic acid (2,4-D), in increasing order of effectiveness. Cytokinin (such as kinetin, or benzyladenine) is supplied in a lesser amount if not adequate within the explant.

Although callus tissue cultures may outwardly appear to be uniform masses of cells, in reality their structure is relatively complex with considerable morphological, physiological, and genetic variation within the callus mass. Growth follows a typical logarithmic pattern. There is (a) a slow initial cell division induction period requiring auxin, (b) a rapid cell division phase involving active synthesis of DNA, RNA, and protein, followed by (c) a gradual cessation of cell division along with (d) differentiation into larger parenchyma and vascular-type cells. Cell division does not take place throughout the culture mass but is located primarily in a meristematic layer on the outer periphery of cells. The inner parts of the callus remain as an undivided mass of older tissue and, in time, may differ physiologically and genetically from cells of the outer layer. Division in the exterior layer decreases and the callus may appear "knobby" as cell division becomes restricted to specific islands of cells. Thus, variations in cell age and type may occur within the tissue culture mass. The inner cells are older and the exterior cells are younger, because a meristematic region persists around the periphery of the callus mass.

Organogenesis begins with dedifferentiation of parenchyma cells to produce centers of meristematic activity (meristemoids) (56, 220). In early studies by Skoog, tobacco callus produced shoots (228) if a relatively high cytokinin/auxin ratio was supplied. If the ratio was reversed, roots tended to form. An intermediate ratio produced both. Adenine was synergistic with cytokinin and increased inorganic phosphate (PO_4) was useful. Although the same basic pattern tends to follow with most other plants, an exact formula for optimizing conditions for regeneration is needed for each species or cultivar.

Cell Suspensions (108) A **suspension culture** is started by placing a piece of friable callus or homogenized tissue in liquid medium so that the cells disassociate from each other (Fig. 17–29). In **batch cultures,** cells are grown in a flask placed on a shaking device that allows air and liquid to mix. Rotating devices that result in continuous bathing of the tissue are available. Another device, called a **chemostat** or turbidostat, continuously cycles the media through the cell culture essentially in the same manner as in the culture of microorganisms. In a third method, cells on a filter paper layer are placed on a shallow liquid medium in a Petri dish with no agitation.

> **suspension cultures** Callus cells grown in liquid culture that is constantly agitated. Agitation breaks cells apart, preventing them from forming large callus clumps.

Growth of cells follows a typical pattern based on changes in rates of cell division. Cells first divide slowly (lag phase), then more rapidly (exponential), increasing to a steady state (linear), followed by a declining rate (deceleration) until a stationary state is reached. When cells are transferred to a new liquid medium, the process

(a)

(b)

Figure 17–29
Suspension culture of soybean (*Glycine*) containing proembryogenic masses.

will be repeated. Under proper environmental conditions with media control, the process can go on indefinitely.

Devices known as **bioreactors** are used to grow cell suspensions on a large commercial scale (286). Bioreactors were originally developed to grow microorganisms or other living cells for fermentation or to produce various secondary products for industrial use. These devices include provision for the introduction of fresh medium and removal of spent medium, along with proper environmental controls.

Culture media are usually similar to those for stationary callus culture and include a complete range of ingredients: inorganic salts, sucrose, vitamins, and a proper balance of growth substances.

Protoplast Culture **Protoplasts** are the living parts of plant cells, containing the nucleus, cytoplasm, vacuole, and various cellular structures surrounded by a semipermeable membrane (plasmalemma) but with the cell wall removed (134). The plant cell, in contrast to the animal cell, is surrounded by a firm, nonliving cell wall composed of cellulose and hemicellulose, and held together by pectin materials. Microbial enzymes digest the cell wall to produce a protoplast. Protoplasts can be obtained from cells in suspension or derived directly from mesophyll leaf cells.

> protoplasts Plant cells without a cell wall. The cell wall is removed by microbial enzymes.

The major advance that permitted protoplast cultures to be made (33) was the discovery that plant cell walls could be removed by enzymes that digest pectin and allow the protoplast surrounded by its cellular membrane to survive (Fig. 17–30). Commercial enzyme preparations are available. Maintaining an adequate osmotic pressure to prevent disruption of membranes is necessary; mannitol (0.45 to 0.8 M) has been used for this purpose. Protoplasts are cultured in media similar to those for cells except for the presence of the osmoticum. During subsequent culture, regeneration of the cell walls takes place rapidly within several days. When new cell walls are produced, cells resume division and can be used to start tissue cultures or cell suspensions.

Protoplasts are significant because many manipulations with the cells are possible (224) once freed of the enclosing cell walls; for example, viruses can be more easily incorporated into protoplasts. In plant breeding, the fusion of protoplasts from two different genotypes, such as two species, which combine two nuclei and two cytoplasms, has been accomplished in a process called **somatic** (or **parasexual**) **hybridization** or **cybrids** (26, 39, 81). Protoplast culture is particularly important in genetic engineering, since protoplasts can absorb DNA, proteins, and other large macromolecules. New genetic material can be incorporated directly into cells of an organism in a procedure that is a form of genetic transformation. Isolated protoplasts are also capable of taking up nuclei and chloroplasts (organelle transfer). Plants that have been regenerated from protoplasts include carrot, *Brassica*, potato, tobacco, citrus, onion, and sugar cane.

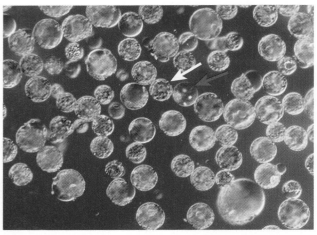

(a)

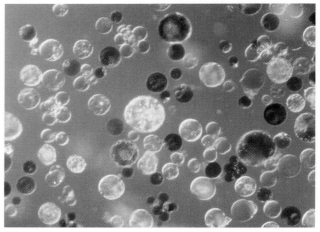

(b)

Figure 17–30
Protoplasts from tulip leaves and flower petals. (a) In leaf protoplasts, epidermal cells lack chloroplasts (red arrow) while those from palisade and mesophyll cells have chloroplasts (white arrow). (b) In flower petal protoplasts, the red and blue cells contain anthocyanins that give the flower petals their color. Courtesy Dr. George Wagner.

Somatic Embryogenesis and Synthetic Seed Production

Somatic embryogenesis is the development of embryos with distinct shoot and root meristems from vegetative cells and tissues within *in vitro* systems (223). Early in the study of cell suspension systems, F. C. Steward et al. (234) discovered that carrot cells, when treated with coconut milk, stopped multiplying and differentiated into miniature embryo-like structures that were called **embryoids**. At about the same time, Reinert (213) independently discovered the same phenomenon in carrot cells grown on agar using high auxin concentrations as the inducing agent. Since then, specific tissues in various species have been found to either have a capacity (**competence**) for somatic embryogenesis in culture systems or can be induced to develop competency in culture by specific treatments to the medium. Somatic embryos develop through stages similar to zygotic embryos as described in Chapter 4 on seeds (Fig. 17–31). However, the final size of the cotyledons is usually reduced, and there is no development of endosperm or seed coat (80). The genes involved in the competence to form somatic embryos and the regulation of genes common to both somatic and zygotic embryo development are an active area of basic research (45, 282). For instance, embryogenic callus could be identified by the expression of two embryo-related genes in barley (237). Elucidation of such genes will lead to a better understanding of how isolated plant cells gain the capacity to form somatic embryos.

> **somatic embryogenesis** The development of embryos from vegetative cells rather than from the union of male and female gametes to produce a zygote. In this process, a bipolar structure is produced with a root-shoot axis and a closed independent vascular system.

Applications of Somatic Embryogenesis to Propagation
Somatic embryogenesis can impact plant propagation in a number of unique ways including the production of synthetic seeds, as a tool for the plant breeder to enhance genetic improvement, and a mechanism to rejuvenate mature plants to make them easier to propagate from cuttings.

Mass Propagation—Synthetic Seeds. Unlike micropropagation from shoot cultures, mass propagation via somatic embryogenesis has not reached commercial application. However, somatic embryos are a clonal form of propagation and somatic embryogenesis has the potential for mass propagation under industrialized conditions at low cost per unit. If that happened, it would allow the clonal propagation of normally seed-propagated crops analogous to the production of apomictic seedlings for crops such as field crops (rice,

(a)

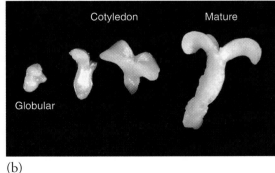
(b)

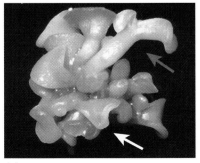
(c)

Figure 17–31
Somatic embryo formation. (a) Somatic embryos at various stages of development on a solid agar medium. (b) Somatic embryos follow a similar progression of development compared to zygotic embryos. There are recognizable globular, cotyledon, and mature stages. (c) A group of somatic embryos showing asymmetric development. Some have yet to reach beyond the heart-shaped stage (white arrow) while others are fully developed and ready for germination (red arrow).

BOX 17.7 GETTING MORE IN DEPTH ON THE SUBJECT
SYNTHETIC SEEDS

A synthetic seed (also called **synseed**) contains an embryo produced by somatic embryogenesis enclosed within an artificial medium that supplies nutrients and is encased in an artificial seed covering (210). Technology to carry out each of these steps is available for some crop seeds, although it is not available as a commercial operation (80, 130, 187). Plant patents protect many of these operations.

The proposed uses for synthetic seeds include:

1. Clonal propagation to replace traditional seed propagation.
2. A replacement to hand-pollinated hybrid plants.
3. Carriers for beneficial microorganisms, pesticides, and growth regulators (see Chapter 4).

The process of developing synthetic seeds starts with the development of somatic embryos, as previously described. Synchronous development is important to get a large number of somatic embryos at the same stage of development for further treatment. Considerable progress has been made in the development and utilization of large reactor vessels (187) to generate embryos similar to those used in fermenters or mass propagation of microbes and cells for industrial production of pharmaceuticals and other products. Somatic embryos are then either directly encapsulated or partially dehydrated before encapsulation. Materials that have been used for encapsulation include sodium alginate (a soluble hydrogel), carrageenan gum, Gelrite, and polyoxyethylene (polyox wafers) (80, 130, 211). Synthetic seeds can then be sown for germination like traditional seeds. Fluid drilling (80) has some potential as a planting procedure for seeds that are not encapsulated (Chapter 6). One major difference between synthetic and most natural crop seeds is the short storage life of synthetic seeds.

Vegetative plant parts, often regenerated in tissue culture, can also be encapsulated in a fashion similar to somatic embryos (Fig. 17–32), including shoot tips, axillary buds, and nodal segments. Encapsulated vegetative parts can be used as propagation units similar to synthetic seeds, but they have received the greatest attention as a mode of cold storage germplasm preservation and germplasm exchange (199).

Figure 17–32
Vegetative parts of African violet encapsulated in alginate beads.

---

alfalfa, orchard grass, soybean), vegetables (carrot, celery, lettuce), plantation crops (oil and date palm, coffee), and forest trees (conifers). To make this potential a reality, technical problems of engineering the entire process of embryogenesis to seedling production and planting need to be addressed. Also, the price needs to be competitive with conventional seed production. In concept, these are **synthetic seeds.** A synthetic seed is a somatic embryo enclosed in a synthetic seed coat suitable for sowing (see text box above on synthetic seeds). Problems with inherent genetic variability within the system need to be better understood and there must be careful field testing and monitoring for variability. The concerns of monoculture may also need to be addressed.

synthetic seeds
Somatic embryos enclosed in an artificial seed coat, which may be a way of sowing somatic embryos for mass propagation.

Nevertheless, the procedure holds considerable promise for having major applications to various species.

Genetic Improvement via Somatic Embryogenesis.
Researchers have used somatic embryogenesis as a method to recover plants that have unique genetic properties, including:

1. Variants recovered through **somaclonal variation.** Somatic embryogenesis could be useful for isolating somaclonal genetic variation within populations of cells. Somaclonal variability results because (a) variation is inherently present within the source plant, (b) the callusing system results in variation, or (c) it can be induced by mutagenic agents.
2. Plants modified through **genetic transformation.** Somatic embryos develop from a few cells (often single cells) (272), which makes them attractive targets for genetic transformation. Figure 17–33 shows embryogenic tissue transformed with a marker gene (GUS). Blue spots indicate integration of the new

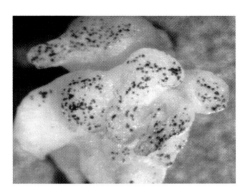

Figure 17–33
GUS expression (dark spots) after particle bombardment of embryonic tissue using gene gun (biolistics).

gene into the embryogenic cells. Transformation occurred by using the **gene gun** (see Chapter 2). Regeneration of somatic embryos from these cells would yield a genetically transformed plant.

3. **Haploids** from anther culture. As discussed previously (page 667), it is common to recover plants from anther culture by production of somatic embryos.
4. **Cybrids** that result from protoplast fusion (page 670) often regenerate from somatic embryos.
5. **Germplasm preservation** is always a concern for the plant breeder. Somatic embryos could be a convenient method to store clonal germplasm through low-temperature storage. For example, in tree improvement programs, 10 or more years might be required to evaluate individual plants for ornamental or timber quality. By that time, trees have entered a mature phase that makes them difficult to clonally propagate. Long-term storage to preserve the original embryogenic cell cultures by **cryopreservation** in liquid nitrogen (−196°C) has been suggested (85) to solve the problem, because it would allow the breeder to go back to original cultures to mass-produce these select trees even after 10 years. Similarly, shoots from tissue culture can also be used for cryopreservation (7).

Rejuvenation Through Somatic Embryogenesis. Somatic embryogenesis reestablishes the juvenile phase of a plant's life cycle in a way similar to that of zygotic and apomictic seed production (see Chapter 2). One advantage to rejuvenation is that it could restore regeneration capacity to mature plants that might make them amenable to micropropagation or conventional cutting propagation (88). For example, mature oak (76) and sweet gum (*Liquidambar*) (159) have been regenerated through somatic embryogenesis from developing flower tissue. These somatic seedlings could serve as a source of juvenile tissue to start micropropagation, cutting stock blocks, or stool beds.

Protocols for Somatic Embryogenesis

Protocols for somatic embryogenesis need to be established for each genotype, but certain generalizations can be made. Consequently, the system has the following stages (Fig. 17–34, page 676):

1. Selection of an appropriate explant material.
2. Induction stage—conditioning cells to express their embryonic potential.
3. Maintenance stage—cells aggregate into proembryogenic masses (PEM).
4. Development stage—removal of growth regulators permits embryo development.
5. Maturation stage—ABA and partial desiccation promote normal embryo formation.
6. Germination and conversion stage—somatic embryo germinates and initiates normal root and shoot growth to form a seedling.
7. Transplanting.

Selection of an Appropriate Explant Material The selection of explant source material is a critical decision and may require a systematic analysis of the embryogenic potential of different explant sources within the plant. Explants differ in their genetic background and include:

1. Embryo or seedling tissue
2. Seed tissues other than the embryo
3. Mature vegetative tissue such as leaf, flower, and root tissue

Somatic Embryogenesis from Embryo or Seedling Tissue. The developing zygotic embryo (at various stages of development) or seedling tissue are initial explant sources with high potential to form somatic embryos. This pattern of somatic embryogenesis is common in a wide variety of plants, but since the explant comes from the zygotic embryo, it has gone through sexual recombination and the resultant somatic embryos are not duplicate copies of the mother plant. This process makes a good explant for genetic transformation studies or for crops normally propagated by seeds, because the zygote has already gone through extensive genetic selection.

The suspensor (especially in gymnosperms) also shows somatic embryogenic potential. The **embryonal-suspensor mass** (ESM) appears at the very

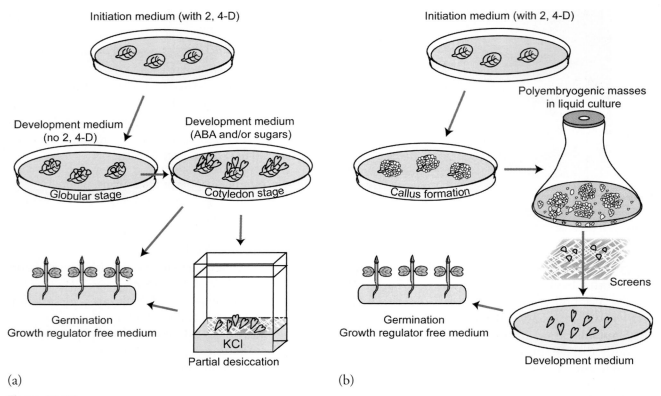

Figure 17–34
Two common patterns of somatic embryo formation. (a) Direct somatic embryogenesis. Immature zygotic embryos or embryo parts are placed on a 2,4-D induction medium. These are moved to a development medium without 2,4-D but supplemented with ABA. A gentle dehydration step may help somatic embryos to germinate. (b) Indirect somatic embryogenesis. Seedling tissue is induced to form polyembryogenic masses in a liquid culture containing 2,4-D. Polyembryogenic masses can be sized using a screen mesh for uniformity before being placed on a development medium without 2,4-D.

earliest stage that precedes embryo development (49, 84, 85). ESM tissue is prominent in conifers and can be used as an explant source to develop suspension cultures. It appears white and mucilaginous, stains red with acetocarmine (0.10 percent w/v), and fluoresces weakly under UV light as compared to green for embryonic cells and yellow for nonviable cells (49).

Somatic Embryogenesis from Seed Tissues Other Than the Embryo. Explants in this category come from the nucellus or integuments of young ovules of some polyembryonic species including citrus, mangosteen, and mango, as well as some monoembryonic species (loquat, rubber, apple, pear, cacao, and grape) (94). Since these explants are derived from maternal tissue in the ovule, somatic embryos derived from this tissue reproduce the genotype of the mother plant and are similar to certain types of apomixis (see Chapter 4).

Somatic Embryogenesis from Mature Vegetative Tissue. Leaf tissue has been used to initiate somatic embryo production in a variety of crops including carnation (273), grape (94), and sweet potato (279). In some cases, the leaf explant is from plants already established in tissue culture. Other vegetative organs such as roots (264) and flower tissue can be used as explant sources. For example, flower (ray) petals were used to initiate cultures in chrysanthemum that resulted in somatic embryo regeneration (244). Immature flower parts have also been used to establish somatic embryo cultures of several woody plants such as oak (76) and sweet gum (159), as well as geophyte (bulb) crops such as *Iris* (158) and *Gladiolus* (124). Somatic embryos from vegetative organs are copies of the donor plant and are a possible way to regenerate mature plants that are difficult to propagate clonally.

Induction Stage The first step in any somatic embryogenesis protocol is to treat the explant in a way that expresses its potential for somatic embryo formation. The most effective auxin for induction of somatic embryo potential is 2,4-dichlorophenoxyacetic acid (2,4-D), although other synthetic auxins

somatic embryos Embryos that form either by direct or indirect patterns. **Direct somatic embryogenesis** embryos form from cells of the explant; **Indirect somatic embryogenesis** embryos arise from callus induced to form from the initial explant.

[naphthaleneacetic acid (NAA) and Picloram] and cytokinins (BA) can also be effective. **Somatic embryos** can arise from two different pathways: (a) **Direct somatic embryogenesis** and (b) **indirect (induced) somatic embryogenesis** (158).

Direct Somatic Embryogenesis. Somatic embryos can develop directly from explants without an intervening callus stage (Fig. 17–34a). In these cases, the explant tissue already possesses the potential to form somatic embryos. Clumps of embryos form from embryogenic centers on the explant, usually from epidermal or subepidermal cells (Fig. 17–35). Developing zygotic embryo and seedling tissue explants most commonly regenerate somatic embryos by direct somatic embryogenesis, but this pathway has also been observed in leaf explants (273, 279). Somatic embryos produced by direct somatic embryogenesis usually are attached to the explant in clumps that make this pattern of embryogenesis less amenable to automation compared to indirect somatic embryogenesis.

Indirect (Induced) Somatic Embryogenesis. In many cases, the initial explant does not contain cells that can form somatic embryos directly. These explants must go through an initial callus induction stage and form somatic embryos in an indirect pattern from this intervening callus (Fig. 17–34b). Explants that do not contain non-predetermined cells directly from somatic embryos must be induced. This indirect method involves an induction period using callus or, more commonly, suspension cultures. Induction is achieved through the transfer of cells to a basal medium with a high concentration of auxin (2,4-D) and sometimes cytokinin. The result is cells with a high capacity to form somatic embryos. The phenomenon seems to be biologically widespread (139, 191, 272) but may require systematic analysis of the embryonic potential of different explant sources and the required culture conditions, as has been done for carrot (148), grasses (261), coffee (53), palm species (254), soybean (31), celery (256), and numerous woody species (113, 116).

Maintenance Stage Following induction, callus cells can be moved to a suspension culture with lower 2,4-D and cytokinin levels to increase and maintain cell proliferation. Although embryogenic cell suspension

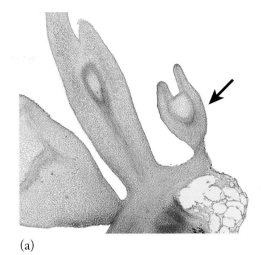

(a)

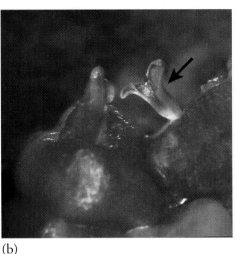

(b)

Figure 17–35
Somatic embryo forming directly from cotyledon tissue in eastern redbud (*Cercis canadensis*) (68). (a and b) Somatic embryo showing attachment to original explant at the radicle tip.

cultures outwardly appear to be uniform groups of cells, on closer inspection portions of these cell masses can be very different. For suspension cultures of banana (75), five different cell aggregate types were identified that differed in their potential to form somatic embryos. Often, cell aggregates can be identified that are well-organized into **proembryogenic masses (PEMs)** (Fig. 17–36, page 678). PEMs were first described by Halperin in 1966 (89) working with suspension cultures of carrot. Structurally distinct from callus and amenable for mass production, PEMs continue to develop in suspension cultures until they are moved to a stationary medium (agar-based development medium) to develop into mature somatic embryos. Three stages of PEM development have been recognized in spruce with stage 3 PEMs forming the

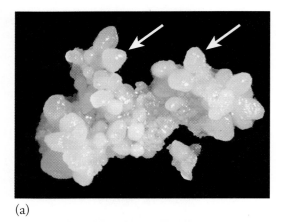

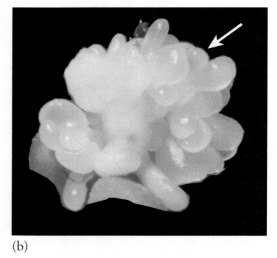

Figure 17–36
Somatic embryo induction. (a) Polyembryogenic masses are aggregates of organized cells that will continue to be produced on a 2,4-D medium in an agitated liquid culture. (b) These will form somatic embryos when moved to a stationary development medium where the 2,4-D has been removed.

most uniform embryos (232). PEMs are often passed through sizing screens to get uniformity and synchrony of development.

Development Stage In most cases, somatic embryos will not continue to develop until they are moved to a growth-regulator-free medium without 2,4-D or cytokinin. When somatic embryos form directly from the explant without callus formation, the entire original explant is moved to a growth-regulator-free medium to permit further embryo development (Fig. 17–37). Cells from suspension culture, or proembryo masses (PEMs), are shifted to a growth-regulator-free medium, sometimes high in ammonium nitrogen. Somatic embryos arise from clumps or small masses, develop polarity, and follow a pattern mimicking normal embryo development within the ovule (80). Development may be variable in rate and abnormal in appearance with secondary embryos developing from primary embryos (Fig. 17–37).

Maturation Stage Somatic embryos that are not exposed to a maturation step often develop abnormally (Fig. 17–38). In some cases, these embryos lack a functional shoot or root meristem. Abscisic acid (ABA) is added to the maturation medium to promote normal somatic embryo development and inhibit precocious germination (5, 232). Somatic embryos express many (most) of the genes for storage protein accumulation. However, they are generally expressed at lower levels and may be expressed at other times in development compared to zygotic embryos.

Although somatic embryos look relatively normal, they still may have trouble germinating. In these

Figure 17–37
Photos showing developmental stages during somatic embryogenesis in interior spruce [*Picea glauca* (Moench) Voss x*engelmanni* Parry ex Emgelm]. (a) Embryogenic tissue, (b) An immature somatic embryo at its early developmental stage. (c) Immature somatic embryos at pre-cotyledonary stage. (d) A mature somatic embryo with well-developed cotyledons. (e) A somatic embryo-derived seedling. Photos provided by Patrick von Aderkas and Lisheng Kong, University of Victoria, Victoria, BC Canada.

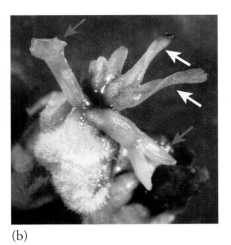

Figure 17–38
Malformed somatic embryo formation in cultures of eastern redbud (*Cercis canadensis*). (a) Photomicrograph of a somatic embryo showing malformed cotyledons and lack of an apical meristem. (b) Typical malformations include straplike fasciated cotyledons (white arrow), fused cotyledons (red arrow) and poor or missing apical meristem formation.

cases, a dehydration step can promote further development and provide the switch from development to germination, similar to the conditions observed for zygotic embryos (78). Dehydration of embryos can be accomplished slowly under a moderate humidity or by placing them on a medium with PEG or other osmotically active substances like alcohol sugars (manitol or sorbitol) to reduce embryo water content (37). Somatic embryos usually do not have the same capacity to withstand low moisture percentages as zygotic embryos, and dehydration should stop when somatic embryos have lost approximately 20 percent of their initial moisture. In some cases, somatic embryos may exhibit dormancy that is broken by chilling stratification (23), gibberellic acid (30), or cytokinin (78, 79).

Germination and Conversion Stage Once a mature somatic embryo has reached an appropriate size, it is plated onto a growth-regulator-free germination medium. Conversion is the production of functional shoot and root meristems following germination. Some somatic embryos show a poor ability to convert to seedlings (80). Ascorbic acid is sometimes included in the germination to aid conversion. Ethylene has also been implicated in poor somatic embryo formation and low conversion rates because development and conversion are improved by venting culture vessels or including ethylene inhibitors in the medium (140).

Transplanting Once an adequate shoot and root have developed, the seedling can be transplanted to a greenhouse medium as with any seedling plant or treated with a "synthetic seed coat" and handled as a synthetic seed.

CONTROL OF THE TISSUE CULTURE ENVIRONMENT (119, 209, 229)

Temperature

The temperature range used most often for tissue culture is between 20 and 27°C (68 and 81°F). There is some evidence to suggest that high temperatures should be avoided and can reduce cytokinin-induced shoot formation (60). Most tissue culture labs maintain a constant temperature; however, for some species (like lily), bulblet formation is enhanced by using alternating day/night temperature cycles (236).

Commercial growers also use temperature to suspend growth when demand for plantlets is down. Cultures are placed in refrigerated storage to reduce growth and eliminate the need and cost of subculturing (95, 188). These cultures are removed from refrigeration to resume multiplication of shoots as the demand for plants increases (Fig. 17–39, page 680).

Light

Light effects can be separated into the impact of **light irradiance** (expressed as photosynthetic active radiation—PAR or photon flux density—PFD), **duration** (photoperiod), and **quality** on growth and development (see Chapter 3). Typical irradiance levels for shoot multiplication during micropropagation are between 40 and 80 $\mu mol \cdot m^{-2} \cdot sec^{-1}$ at culture height, but irradiance inside the vessel may be much lower. Light distribution within the culture vessel was measured using a simulated microshoot constructed of a light sensor film (110). It was shown that PAR decreased toward the bottom of the vessel with reductions of 30 and 50 percent at the middle and lowest parts of the simulated microshoot, respectively. Also, the type of closure used can reduce light transmitted into the culture vessel (61). The number of lamps and the distance between lamps on light shelves impacts both micropropagation costs and uniformity and quantity of light irradiance distribution across the shelf. Regression models are available to describe the relationship between height and spacing of fluorescent lamps (29).

Although species differ in their light requirements, this range is a very low irradiance compared with outdoor and greenhouse irradiances, which are between 600 and 1200 $\mu mol \cdot m^{-2} \cdot sec^{-1}$. This low light level is used because sucrose is provided in the

Figure 17–39
During the slow part of the growing season, cultures can be placed in refrigerated storage to reduce or eliminate more costly subculturing. (a) "Restart" cultures emerging from cold storage. Shoots are etiolated from being in the dark storage. (b) Restart cultures being subcultured to re-initiate shoot multiplication.

medium of **heterotrophic** cultures rather than relying on photosynthesis for energy (96). Shoots in heterotrophic culture show low photosynthetic rates due to low CO_2 in the culture vessel and feedback inhibition of photosynthesis by the sugar in the medium. Higher light irradiance in heterotrophic cultures can result in loss of chlorophyll and leaf necrosis.

Photoperiod and Light Quality (Wavelength)

Very few studies have systematically studied the impact of photoperiod on culture development, but longer day lengths (12 to 16 hours) are most often incorporated. Light quality is a function of the lamps that illuminate the cultures and the type of vessels being used. In general, tissue culture labs use cool white fluorescent or growlight (added red light) lamps in the culture room. Light emitting diodes (LEDs) are an interesting alternative because they provide an efficient irradiance to heat production ratio compared to other lamps and can easily provide specific light qualities by mixing red and blue lamps. For banana micropropagation, a ratio of 90 percent red (peak emission 660 nm) to 10 percent blue (peak emission 450) at 60 $\mu mol \cdot m^{-2} \cdot sec^{-1}$ provided optimal plantlet growth (181).

Light quality is also affected by the transmission quality of the vessel and the type of closure for the vessel (46). Glass vessels do not transmit light wavelengths shorter than 290 nm, while polycarbonate vessels do not transmit light shorter than 390 nm. Although this is a measured difference, the important wavelengths for

---

BOX 17.8 GETTING MORE IN DEPTH ON THE SUBJECT
HETEROTROPHIC VERSUS AUTOTROPHIC GROWTH

An alternative to heterotrophic growth is autotrophic or photoautotrophic growth (119, 136). Various investigators have argued that many if not most of the problems in commercial micropropagation are related to the use of closed heterotrophic systems for producing plants that require a sugar-based medium. Toyoki Kozai and colleagues have investigated conditions that rely on photosynthesis as the major energy source rather than sugar in the medium, which would provide a photoautotrophic alternative to conventional micropropagation (137). Model systems using potato shoot-tip cultures have been very productive. These cultures are characterized by having little or no sucrose, increased CO_2 levels (3000 ppm), and increased light irradiance (100 to 200 $\mu mol \cdot m^{-2} \cdot sec^{-1}$) that provides enhanced photosynthesis for culture, while reducing costs and contamination. Numerous crops, including *Spathiphylum* (246), *Eucalyptus* (245), banana (181), and strawberry (182), have been successfully micropropagated using sugar-free autotrophic cultures. Part of this success is due to the development of novel containers using low-cost films with high gas permeability and use of agar substitutes like rockwool to support shoot growth (245). Photoautotrophic culture also offers a high potential for automation for some species (136).

photosynthesis and photomorphogenesis are between 400 and 800 nm.

Light quality can also alter the growth response of shoots *in vitro*. In general, red light tends to promote shoot proliferation as well as shoot elongation, possibly through its impact on shoot apical dominance and its interaction with endogenous cytokinin production or sensitivity (233). For example, geranium (*Pelargonium*) cultures exposed to incandescent (red) light showed increased shoot elongation compared to fluorescent (white) light, while blue light reduced shoot elongation (3). Red light also promoted shoot elongation in shortshoots of larch (138). In contrast, photosynthetic capacity was highest when birch (*Betula*) shoots were exposed to blue light compared to either white or red light (217). Blue light also stimulated chlorophyll production and increased leaf size. It was suggested that these qualities might be important during acclimatization of plantlets and could be promoted by a blue light treatment prior to removing plants from culture. In an interesting study using tomato mutants for phytochrome A and B, and narrow wave band light, it was determined that seedling explants require red light to be competent to form shoots in tissue culture (13). Blue light tended to inhibit shoot formation.

Light quality can also indirectly impact shoot development by causing changes to the media components, altering culture development (93).

In general, light inhibits root growth and there is some indication that excluding light from the rooting zone is beneficial to root formation. It has been suggested that one of the benefits of including activated charcoal to the *in vitro* rooting medium is a reduced light level. However, there can be an impact of light quality on rooting. Red light was shown to increase the number of roots formed per microcutting in Benjamin fig (62), and grape (28). In contrast, birch microcuttings exposed to blue light showed the fastest rooting, with the greatest number of roots, compared with red light (218).

Photoperiodic response of flowering can be produced *in vitro*. For example, carnation shoots can be induced to flower by treatment with 16 hours of light, but remain vegetative if kept at 12 hours (235). It is likely that other plants can respond similarly.

Gases

Oxygen, carbon dioxide, and ethylene impact shoot development in culture (209). Generally, commercial growers do not attempt to alter the levels of gases in culture. However, all closures and caps used for tissue culture are permeable to gases to some degree. There is usually some enhancement in growth by venting closures or providing filtered air exchange, due to increased CO_2 available for photosynthesis (even in heterotrophic cultures) and in some cases, reduced ethylene levels. Of course, elevated CO_2 concentrations are a key component of photoautotrophic systems. Even so, very high levels of CO_2, in excess of the compensation point for photosynthesis, can induce increased growth and development in a variety of species in tissue culture. Ambient CO_2 levels are about 380 ppm, but levels between 3,000 and 50,000 ppm have improved shoot and root growth of lettuce, tomato, citrus, kale, and radish (255). Obviously for health reasons, care should be taken in atmospheres containing very high CO_2.

The accumulation of ethylene within the culture vessel tends to reduce *in vitro* growth and development (141). Venting reduces vessel ethylene levels and can reduce ethylene's negative impact (111). The addition of ethylene action inhibitors such as silver nitrate ($AgNO_3$) can, in some cases, improve shoot formation (189) or alleviate culture problems such as hyperhydricity (289).

SPECIAL PROBLEMS ENCOUNTERED BY *IN VITRO* CULTURE

Hyperhydricity (96)

Hyperhydricity (originally called **vitrification**) is characterized by a translucent, water-soaked, succulent appearance that can result in cultures that deteriorate and fail to proliferate (Fig. 17–40, page 682) (40, 63). Physiologically, expression involves excess water uptake ("waterlogging") and inhibition of lignin and cellulose synthesis. Hyperhydricity appears to be a consequence of the difference between the water potential in the medium and developing shoots, as well as a low nitrate to ammonium ratio. Hyperhydricity can also be caused by hormonal imbalances, especially ethylene. In addition, polyamine levels are higher in hyperhydric compared to non-hyperhydric carnation leaves (194), which may relate the level of physiological stress.

Hyperhydricity is more prevalent if plants are grown in liquid media or with low agar concentrations, high humidity, and high ammonium concentrations (as in the MS medium). A few factors that have provided some relief from this problem include increasing the agar concentration, changing brands of agar, modification of inorganic ingredients in the medium, changing the cytokinin concentration, venting, and the addition of antivitrification agents (191, 285).

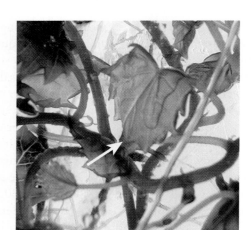

Figure 17–40
Tissues with hyperhydricity have a translucent, water-soaked appearance as seen in this grape culture.

Internal Pathogens

Even though micropropagation is considered to be a pathogen-free system, complete absence of pathogens should not be taken for granted. The original source material should be indexed for specific viruses even if the original selection is from shoot-tip cultures or micrografts.

For example, in the initial enthusiasm for orchid micropropagation, the importance of initial meristem treatment was not appreciated and shoot-tip culture was assumed to control viruses. As a result, many of the commercial sources of orchid cultivars were soon found to be infected, causing considerable economic loss.

Some internal bacteria can be very persistent and hard to eliminate (Fig. 17–4b). In some instances, internal pathogens are not apparent until after the culture has been grown for some time. In these cases, pathogens can inhibit growth and rooting but remain suppressed until the explant is subcultured. Examples of offending bacteria include *Bacillus subtilis, Erwinia,* and *Pseudomonas* (27, 132, 133).

Excessive Exudation

One of the most frequently encountered problems with establishing explants (especially woody perennials) in culture is the production of exudates by the explants (Fig. 17–41). The exudates are usually considered to be various phenolic compounds that oxidize to form a brown material in the medium that tends to be inhibitory to development (41, 74). Treatments to minimize this problem include treating the explant with an antioxidant (citric or ascorbic acid), including an adsorbent material in the medium (polyvinylpyrrolidone or activated charcoal), and frequent transfers to a new medium.

Shoot-Tip Necrosis

Actively growing shoot tips sometimes develop "tip dieback" (Fig. 17–42) usually caused by calcium deficiency that can be corrected by refining the basic nutrient medium to include more calcium (151, 226).

Tissue Proliferation

Tissue proliferation (TP) is the formation of gall-like growths on the stem of micropropagated plants (146). It may occur while plants are still in culture or may not be seen until after plants have been moved to the greenhouse or nursery. TP is most common in species in the heath family (Ericaceae) like *Rhododendron* and mountain laurel (*Kalmia*). The cause of TP is unknown, but apparently it is not pathogen-related. Similar gall-like structures have been seen naturally on *Rhododendron,* and it has been suggested that the tissue culture environment may promote the expression of

(a)

(b)

(c)

Figure 17–41
Exudation from explants in tissue culture is usually caused by phenolic exudation.
(a) Exudation from the explant causes dark coloration of the basal callus and medium.
(b) Including an anti-oxidant in the medium or (c) or activated charcoal can limit exudation and improve shoot growth.

PRINCIPLES OF TISSUE CULTURE AND MICROPROPAGATION CHAPTER SEVENTEEN 683

(a)

(b)

Figure 17–42
Shoot-tip necrosis (arrows) in cultures in (a) white oak (*Quercus alba*) and (b) eastern redbud (*Cercis canadensis*).

TP in species or cultivars having this tendency naturally. Some cultivars have a greater tendency to show TP, and care should be taken in micropropagating these plants.

Habituation

Habituation is the autotrophic growth in cultures that had previously required auxin or cytokinin for growth (156, 239). For example, shoot cultures (Fig. 17–43) may become habituated to cytokinin and continue to proliferate even after the culture has been transferred to a medium without any growth regulator (71). Cytokinin independence in tissue culture could be attributed to increased endogenous cytokinin production, reduction in cytokinin turnover, alteration in cytokinin signal transduction, or persistent epigenetic changes (194).

Figure 17–43
Habituation is the continued growth in tissue culture without the plant growth regulator inducement. Paw paw (*Asimina triloba*) cultures continually subcultured on a cytokinin medium over several years developed habituation for shoot formation (71).

BOX 17.9 **GETTING MORE IN DEPTH ON THE SUBJECT**
AGROBACTERIUM-INDUCED CELL GROWTH

Tissue that has been infected (transformed) with various strains of bacteria (*Agrobacterium tumefaciens* or *A. rhizogenes*) show growth on tissue culture medium without any growth regulators. *Agrobacterium* integrates a segment of bacterial DNA (T-DNA) into a chromosome of the plant's cell (131). Wild type *Agrobacterium* integrates segments of T-DNA that code for auxin and cytokinin, so infected tissue makes its own growth hormones for callus, shoot, or root development. The integration of T-DNA into the plant's genome is the basis for plant transformation techniques using *Agrobacterium* (see Chapter 2). *A. rhizogenes* incites the hairy root disease because it induces infected plants to make roots on above-ground parts. Infected plants can be used to initiate **hairy root cultures** that have the ability to produce roots in a liquid medium without growth regulators. The cultures are used for producing large amounts of roots that contain materials such as pharmaceutical compounds (90).

VARIATION IN MICROPROPAGATED PLANTS

Standards for micropropagated, vegetatively propagated plant cultivars are the same as standard propagated plants (i.e., true-to-cultivar, true-to-type, and free from serious pathogens) (see Chapter 16). The problem of deviation from these standards may be even more acute with micropropagation; however, because the very high multiplication rates magnify any problem, and potential economic losses can be greatly increased before detection and correction (Fig. 17–44).

Much field-testing has been carried out with micropropagated plants and considerable commercial experience has now been obtained with various crops (91, 235, 243, 283). Early experience with commercial products resulting from micropropagation showed specific instances of unexpected aberrations, and off-type performance in some cases resulted in some public resistance to these products. On the whole, most plants from micropropagation systems have been true-to-type and uniform in phenotypic appearance.

Variability can be categorized as in Chapter 16, although it is not always possible to assign a specific variant to a given category. Furthermore, variation may be species- or genotype-specific. The basic principle is that trueness-to-type must be tested through the same phenotypic and genotypic selection procedures as described in Chapter 16. Variation in tissue culture can occur through:

1. Genetic or chimeral variation
2. Transient phenotypic variation
3. Epigenetic (including rejuvenation) variation

Genetic or Chimeral Effects

Genetic changes prior to, or during, micropropagation are the most serious types of aberration because they are permanent and may be difficult to detect during the operation. As with conventional propagation, a problem may not appear until after the plants have been grown in a permanent location. It is important, therefore, to establish whether the problem is a preexisting one in the source material or due to conditions that occur during the propagation procedure.

Causes of Genetic Variation Genetic mutations that occur in tissue culture are referred to as **somaclonal variation** (17, 115, 221). In most cases such variations are undesirable, but they can be useful for developing mutants that may show desirable plant form, flower

(a)

(b)

(c)

Figure 17–44
Types of variation observed during micropropagation. (a) Micropropagation of chimeral plants requires judicious observation to maintain plants with the proper phenotype. A range of variants can be observed in a micropropagated chimeral hosta cultivar. (b) Fasciation such as this shoot clustering in a Rhododendron can occur in some plants. This may be a transient variation that is sensitive to the growing environmental conditions. (c) An increase in basal branching in these purple coneflower (*Echinacea*) plants is a common observation after micropropagation.

color, or resistance to disease. Systems that proliferate callus on the explant during any of the stages of development are susceptible to producing **mutant sectors,** or individual growing points, because callus proliferation can result in polyploid or aneuploid cells that produce aberrant mutant plants.

Shoots being produced through adventitious and axillary micropropagation systems can be susceptible to genetic variation, a problem recognized in Boston fern (*Nephrolepsis*) early in the development of micropropagation research (173). Axillary shoots are generally considered more genetically stable than adventitious shoots, but both can be negatively impacted by the stresses imposed by the culture environment, prolonged hormone exposure, by-products in the culture medium, and the time in culture (206). The probability of producing off-types increases with the age of the culture; therefore, new cultures should be initiated periodically from stock plants.

Many important horticultural crops are chimeras, as described in Chapter 16. Several cases in Chapter 16 are described in relation to conventional propagation concerning root and leaf cuttings that manifested reversions to underlying cell layers with different genotypes. Similarly, micropropagation with any method that includes adventitious shoot initiation will tend to result in chimeral breakdown (Fig. 17–44a). The result for chimeral plants producing variegated foliage is the production of plants that are all green, all white, and plants with varying degrees of variegation patterning. One interesting way to illustrate this phenomenon is with homogenized leaf tissue of chimeral African violet (*Saintpaulia ionantha* 'Crimson Frost')(190). Plants obtained from adventitious shoots from these homogenates formed plants that were like the original 'Crimson Frost' (33 percent), while the rest showed variation in leaf form, color, and flower shape. For plants with variegation patterns caused by internal virus infection, these patterns will be lost if the virus is eliminated during the tissue culture process.

Control of Genetic Variation Like conventional propagation, micropropagation systems that multiply by maintaining an intact shoot tip remain relatively stable and free from genetic mutation. Because of their genetic stability, axillary shoot proliferation culture is recommended for commercial micropropagation. Here it may be important to limit the number of subcultures from any single explant, and lowering the growth regulator levels during subculturing to reduce the amount of callusing may also be indicated. It may be that the best conditions for inducing rapid growth and high PRs are not the best conditions for production of true-to-type plants (55).

Systems that utilize adventitious shoot production or the multiplication of callus and cells may result in considerably more variation in the end product than the axillary shoot system. Where adventitious shoots are part of the system, the potential for variability of the specific genotype needs to be carefully evaluated through specific progeny tests of the end product. This precaution would apply to production from bulb scales, cotyledons, leaves, and flower stems. Chimeras would be difficult to duplicate by these methods.

In any micropropagation system, visual inspections should be practiced throughout the production cycle and any off-type or aberrant plant rogued prior to distribution. Maintaining progeny source records should be a regular aspect of production so that the origin of a specific problem can be traced to its source. Also, long-term evaluation is needed before introducing any plant into widespread distribution. Molecular DNA marker-based assessment of genetic fidelity is a sophisticated tool being utilized to track genetic changes during and after micropropagation (207). It has the advantages of providing rapid assessment of potential problems early in the micropropagation process and compliments field observations to ensure true-to-type progeny.

Transient Phenotypic Variation

One of the first observations with micropropagated plants was that in some cases, growth patterns tended to be different in the initial post-culture period compared to those plants propagated by conventional methods. Usually, these were temporary and disappeared as the plants continued to grow and/or were re-propagated. A particular variant may or may not be desirable, depending on the standards for trueness-to-type of the particular cultivar and its pattern of growth and development. Examples of changes in phenotype include:

1. Vigor
2. Changes in developmental stage
3. Branching

Vigor In many species, one common characteristic of micropropagated plants is enhanced vigor. For example, strawberry plants produced as a result of meristem-tip propagation invariably show enhanced vigor. The reason is not completely clear. One explanation has been that an unknown viruses or other pathogens have been eliminated; another explanation is that the plants have been rejuvenated to a more juvenile state, as described in a later section.

Changes in Developmental Stage Along with enhanced vigor, some species may show a specific developmental phase (25, 242). Strawberry (*Fragaria*) provides an example of variation in growth patterns. Micropropagated plants used directly for fruit production tended to produce somewhat smaller fruits and lower yields. The same plants tended to develop into an enhanced running stage and were thus useful for nursery plant production (see Chapter 19). Consequently, meristem-tip culture to control disease and shoot-tip multiplication is a useful combination to obtain "mother plants" for conventional nursery production.

Branching A common characteristic of many ornamentals propagated by micropropagation (235) is enhanced branching as compared to conventionally propagated plants (Fig. 17–44c). Enhanced branching is considered highly useful for some species, such as *Hosta, Begonia,* aster, chrysanthemum, rose (235), *Syngonium,* ferns, and other foliage plants (33), but undesirable for *Gypsophila, Kalanchoe, Gerbera,* and *Saintpaulia* (235). This branching characteristic has also occurred in strawberry, blackberry, apple (283), *Kalmia,* and *Rhododendron.*

This branching effect has been explained by a possible cytokinin as a carryover effect (see habituation, page 683) caused during the multiplication stage. Another explanation is an epigenetic effect related to phase change. Note that in some species (see Chapter 16), propagules taken from the more mature parts of a plant tend to be more branched with enhanced lateral orientations.

Epigenetic Variation

Variations due to epigenetic effects primarily involve phase changes involved with the juvenile-mature cycle (88). Refer to the clonal cycle described in Chapter 16 and the effect on reproductive capacity (flowering), trueness-to-type morphology, and regeneration (rooting potential). Micropropagated plants and the system of *in vitro* culture are not only sensitive to the initial status of the explant, but conditions during culture can bring about changes of rejuvenation.

Rejuvenation (Invigoration) Many, if not all, woody plants that are consecutively recultured in shoot-tip culture show some degree of rejuvenation as indicated by rooting potential, leaf morphology, and/or performance. This phenomenon has been observed in blackberry, gooseberry, grape, and apple (171, 283). Enhanced rooting potential makes it possible to produce clonal copies of plants that are difficult-to-root as conventional stem cuttings. Rooting potential may persist after micropropagation such that the plants produced may be useful later as stock plants in conventional cutting propagation (Fig. 17–45) (240). For example, birch (*Betula*) showed 31 percent rooting for cuttings from mature sources that were serially grafted, while 87 percent rooted from seedling cuttings. Microcuttings from tissue culture rooted at near the seedling percentage (95 percent), and conventional cuttings from stock plants derived from micropropagated birch continued to root at high percentages (75 percent).

Effect on Flowering Flowering problems have been reported as occurring in some herbaceous crops, such as chrysanthemum, *Dianthus, Kalanchoe,* and *Gypsophila* (235), manifested in unevenness and reluctance or delay in flowering. Scheduling flowering times becomes a problem, and excessive variation can limit the use of micropropagated material of some crops. Cultivar differences exist in these responses and selection of cultivars that resist variation in flowering may be the long-term solution. The explanation for this

(a)

(b)

Figure 17–45
Rejuventation (or invigoration) that occurs during micropropagation can be seen in the ability of cuttings taken from recently micropropagated plants to root easily under standard mist propagation. (a and b) Red maple (*Acer rubrum*) cuttings are commercially propagated from recently micropropagated plants.

phenomenon is variation in juvenility status among individual micropropagated plants. Future developments may include selection of cultivars specifically adapted to micropropagation procedures.

The effect of micropropagation on the age of flowering varies with genotype, method of handling, and source of explant. For example, fruit and nut cultivars propagated from shoot-tip culture have not demonstrated a significant increase in the age of bearing, compared with conventionally grafted trees (283), as might be expected if the plants remained in a juvenile condition after micropropagation. In fact, particularly precocious cultivars of micropropagated walnut have flowered and fruited at an unusually early age, even in the nursery row. This early flowering would be particularly favorable for the production of fruit and nut cultivars and for forest species grown for seed orchards, but it is a disadvantage for trees grown for timber.

True-to-Type Juvenile Phenotypes The production of a true-to-type juvenile phenotype is important in the propagation of forest species to be used for lumber, biomass production, and other cases where the vegetative structure is the product. In general, production of the mature phenotype persists from vegetative propagation, whether by conventional or micropropagation procedures. Selecting a superior ("elite") genotype of conifers for forest uses must be done at a mature age, but this creates problems not only in propagation but in the production of the true-to-type phenotype (15, 48). It is possible that the problem of propagation can be solved by explant selection from juvenile material, reversion of mature tissues, or development of suitable micropropagation techniques that will capture the genetic gains of clonal propagation, retain a juvenile phenotype, and provide an efficient economical mass-propagation system (15, 20).

DISCUSSION ITEMS

This chapter is a basic discussion of tissue culture procedures and the theories behind their practices. These concepts are essential to problem-solving for people involved in commercial micropropagation. Tissue culture propagation has become a standard propagation system for many common horticultural crops. People involved with these labs must have specialized training and a working knowledge of the basic concepts presented in this chapter. In addition, the emerging field of biotechnology relies heavily on regeneration of transformed (genetically engineered) plants from tissue culture. Increasingly, there are employment opportunities in laboratories for students trained in tissue culture propagation.

1. Compare and contrast zygotic embryo formation, somatic embryogenesis, and apomixis.
2. Why are orchid seeds germinated in tissue culture?
3. Compare procedures and advantages of rooting microcuttings in an *in vitro* vs. an *ex vitro* environment.
4. Compare axillary and adventitious (bud) shoot formation related to genetic stability, transformation, and species type.
5. What are advantages and disadvantages related to somaclonal variation?
6. Is all variation in tissue culture due to mutations?
7. Why is it often difficult to micropropagate variegated plants in tissue culture? (e.g., *Hosta*)

REFERENCES

1. Alfaro, F. P., and T. Murashige. 1987. Possible rejuvenation of adult avocado by graftage onto juvenile rootstocks *in vitro*. *HortScience* 22:1321–24.

2. Anderson, W. C. 1975. Propagation of rhododendrons by tissue culture. 1. Development of a culture medium for multiplication of shoots. *Comb. Proc. Intl. Plant Prop. Soc.* 25:129–35.

3. Appelgren, M. 1991. Effects of light on stem elongation of *Pelargonium* in vitro. *Scientia Hort.* 45:345–51.

4. Arnaud, Y., A. Franclet, H. Tranvan, and M. Jacques. 1993. Micropropagation and rejuvenation of *Sequoia sempervirens*: A review. *Ann. Sci. For.* 50:273–95.

5. Attree, S. M., D. Moore, V. K. Sawhney, and L. C. Fowke. 1991. Enhanced maturation and desiccation tolerance in white spruce [*Picea glauca* (Moench) Voss] somatic embryos: Effects of a non-plasmolysing water stress and abscisic acid. *Ann. Bot.* 68:519–25.

6. Azcón-Aguilar, C. A., and J. M. Barea. 1997. Applying mycorrhiza biotechnology to horticulture: Significance and potentials. *Scientia Hort.* 68:1–24.

7. Bajaj, Y. P. S. 1979. Establishment of germplasm banks through freeze storage of plant tissue culture and their implications in agriculture. In W. R. Sharp et al., eds. *Plant cell and tissue culture principles and applications.* Columbus: Ohio State Univ. Press. pp. 745–74.

8. Bajaj, Y. P. S., ed. 1986. *Biotechnology in agriculture and forestry, Vol. 1, Trees.* Berlin: Springer-Verlag.

9. Ball, E. 1946. Development in sterile culture of stem tips and subjacent regions of *Tropaeolum majus* L. and of *Lupinus albus* L. *Amer. J. Bot.* 3:301–18.

10. Barlass, M., and K. G. M. Skene. 1980. Studies on the fragmented shoot apex of grapevine. I. The regenerative capacity of leaf primordial fragments *in vitro*. *J. Exp. Bot.* 31:483–88.

11. Barlass, M., and K. G. M. Skene. 1980. Studies on the fragmented shoot apex of grapevine. II. Factors affecting growth and differentiation *in vitro*. *J. Exp. Bot.* 31:489–95.

12. Beasley, C. A. 1984. Culture of cotton ovules. In I. K. Vasil, ed. *Cell culture and somatic cell genetics of plants, Vol. 1. Laboratory procedures and their applications.* New York: Academic Press. pp. 232–40.

13. Bertram, L., and B. Lercori. 2000. Phytochrome A and Phytochrome B1 control the acquisition of competence for shoot regeneration in tomato hypocotyl. *Plant Cell Rpt.* 19:604–9.

14. Bilkey, P. C., B. H. McCown, and A. C. Hildebrandt. 1978. Micropropagation of African violet from petiole cross-sections. *HortScience* 13:37–8.

15. Bonga, J. M. 1986. Vegetative propagation in relation to juvenility, maturity, and rejuvenation. In J. M. Bonga and D. J. Durzan, eds. *Cell and tissue culture in forestry*, 2nd ed. Dordrecht: Martinus Nijhoff Publishers. pp. 387–411.

16. Boulay, M. 1987. *In vitro* propagation of tree species. In C. E. Green, D. A. Somers, W. P. Hackett, and D. D. Biesboer, eds. *Plant tissue and cell culture.* New York: Alan R. Liss. pp. 367–82.

17. Bouman, H., and G. J. de Klerk. 1997. Somaclonal variation. In R. L. Geneve, J. E. Preece, and S. A. Merkle, eds. *Biotechnology of ornamentals.* Biotechnol. Agric. Series No. 16. Wallingford, UK: CAB International.

18. Bourgin, J. P., and J. P. Nitsch. 1967. Obtention de *Nicotiana* haploides à partir d'etamines cultivées *in vitro*. *Ann. Phys. Veg.* 9:377–82.

19. Boxus, P. H., and P. Druart. 1986. Virus-free trees through tissue culture. In Y. P. S. Bajaj, ed. *Biotechnology in agriculture and forestry, Vol. 1, Trees.* Berlin: Springer-Verlag. pp. 24–30.

20. Boxus, P. H., M. Quoirin, and J. M. Laine. 1977. Large scale propagation of strawberry plants from tissue culture. In J. Reinert and Y. P. S. Bajaj, eds. *Applied and fundamental aspects of plant cell, tissue and organ culture.* Berlin: Springer-Verlag. pp. 131–43.

21. Brainerd, K. E., L. H. Fuchigami, S. Kwiatkowski, and C. S. Clark. 1981. Leaf anatomy and water stress of aseptically cultured 'Pixy' plum grown under different environments. *HortScience* 16:173–75.

22. Bridle, P., and C. F. Timberlake. 1997. Anthocyanins as natural food colours—selected aspects. *Food Chem.* 58:103–9.

23. Bueno, M. A., R. Astorga, and J. A. Manzanera. 1992. Plant regeneration through somatic embryogenesis in *Quercus suber*. *Physiol. Plant.* 85:30–4.

24. Burk, L. G., G. R. Gwynn, and J. R. Chaplin. 1972. Diploidized haploids from aseptically cultured anthers of *Nicotiana tabacum*. *J. Hered.* 63:355–60.

25. Cameron, J. S., and J. F. Hancock. 1986. Enhanced vigor in vegetative progeny of micropropagated strawberry plants. *HortScience* 21:1225–26.

26. Carlson, P. S., H. H. Smith, and R. D. Dearing. 1972. Parasexual interspecific plant hybridization. *Proc. Nat. Acad. Sci. USA* 69:2292–94.

27. Cassells, A. C., and E. A. Oherlihy. 2003. In S. M. Jain and K. Ishii, eds. *Micropropagation of woody trees and fruits.* Dordrecht: Kluwer Acad. Pub. pp. 103–28.

28. Chee, R., and R. M. Pool. 1989. Morphogenic responses to propagule trimming, spectral irradiance, and photoperiod of grapevine shoots recultured *in vitro*. *J. Amer. Soc. Hort. Sci.* 114:350–54.

29. Chen, C. 2005. Fluorescent lighting distribution for plant micropropagation. *Biosystems Engineering* 90:295–306.

30. Choi, Y. E., D. C. Yang, E. S. Yoon, and K. T. Choi. 1999. High-efficiency plant production via direct somatic single embryogenesis from preplasmolysed cotyledons of *Panax ginseng* and possible dormancy of somatic embryos. *Plant Cell Rpt.* 18:493–99.

31. Christianson, M. L. 1985. An embryogenic culture of soybean: Towards a general theory of somatic embryogenesis. In R. R. Henke, K. W. Hughes, M. J. Constantin, and A. Hollaender, eds. *Tissue culture in forestry and agriculture.* New York: Plenum Press. pp. 83–103.

32. Christianson, M. L., and D. A. Warnick. 1983. Competence and determination in the process of *in vitro* shoot organogenesis. *Develop. Bio.* 95:288–93.

33. Cocking, E. C. 1986. The tissue culture revolution. In L. A. Withers and P. G. Alderson, eds. *Plant tissue culture and its agricultural applications*. London: Butterworth. pp. 3–20.

34. Collins, G. B., and J. W. Grosser. 1984. Culture of embryos. In I. K. Vasil, ed. *Cell culture and somatic cell genetics of plants, Vol. 1*. New York: Academic Press. pp. 241–57.

35. Cooke, R. C. 1979. Homogenization as an aid in tissue culture propagation of *Platycerium* and *Davallia*. HortScience 14:21–2.

36. Cooper, D. C., and R. A. Brink. 1940. Somatoplastic sterility as a cause of seed failure after interspecific hybridization. *Genetics* 25:593–617.

37. Corredoira, E., A. Ballester, and A. M. Vieitez. 2003. Proliferation, maturation and germination of *Castanea sativa* Mill. Somatic embryos originated from leaf explants. *Ann Bot.* 92:129–36.

38. de Fossard, R. A. 1986. Principles of plant tissue culture. In R. H. Zimmerman, R. J. Griesbach, F. A. Hammerschlag, and R. H. Lawson, eds. *Tissue culture as a plant production system for horticultural crops*. Dordrecht: Martinus Nijhoff Publishers. pp. 1–13.

39. de Jeu, M. J., and A. Cadic. 2000. In vitro techniques for ornamental breeding. *Acta Hort.* 508:55–60.

40. Debergh, P., J. Aitken-Christie, D. Cohen, B. Grout, S. von Arnold, R. Zimmerman, and M. Ziv. 1992. Reconsideration of the term 'vitrification' as used in micropropagation. *Plant Cell Tiss. Organ Cult.* 30:135–40.

41. Debergh, P., and P. E. Read. 1991. Micropropagation. In P. C. Debergh and R. H. Zimmerman, eds. *Micropropagation: Technology and application*. Dordrecht: Kluwer Acad. Pub. pp. 1–14.

42. DeLanghe, E., and E. DeBruijne. 1976. Continuous propagation of tomato plants by means of callus cultures. *Scientia Hort.* 4:221–22.

43. Diaz-Barrita, A. J., M. Norton, R. A. Martinez-Peniche, M. Uchanski, R. Mulwa, and R. M. Skirvin. 2007. The use of thermotherapy and *in vitro* meristem culture to produce virus-free 'Chancellor' grapevines. *Intl. J. Fruit Sci.* 7:15–25.

44. Dodds, J. H., D. Silva-Rodriguez, and P. Tovar. 1992. Micropropagation of potato (*Solanum tuberosum* L.). In Y. P. S. Bajaj, ed. *Biotechnology in agriculture and forestry, Vol. 19. High tech and micropropagation III*. Berlin: Springer-Verlag.

45. Dodeman, V. L., G. Ducreux, and M. Kreis. 1997. Zygotic embryogenesis versus somatic embryogenesis. *J. Exp. Bot.* 48:1493–1509.

46. Dooly, J. H. 1991. Influence of lighting spectra on plant tissue culture. ASAE Paper No. 917530. *Amer. Soc. Agr. Eng.*, MI.

47. Driver, J. A., and G. R. L. Suttle. 1986. Nursery handling of propagules. In J. M. Bonga and D. J. Durzan, eds. *Cell and tissue culture in forestry, 2nd ed.* Dordrecht: Martinus Nijhoff Publishers. pp. 320–31.

48. Durzan, D. J. 1986. Special problems: Adult vs. juvenile explants. In D. A. Evans, W. R. Sharp, and P. V. Ammirato, eds. *Handbook of plant cell culture, Vol. 4*. New York: Macmillan. pp. 471–503.

49. Durzan, D. J., and P. K. Gupta. 1988. Somatic embryogenesis and polyembryogenesis in conifers. In A. Mizrahl, ed. *Biotechnology in agriculture*. New York: Alan R. Liss. pp. 53–81.

50. Economou, A. S., and P. E. Read. 1982. Effect of NAA on shoot production *in vitro* from BA-pretreated petunia leaf explants. *J. Amer. Soc. Hort. Sci.* 107:504–6.

51. El-morsy, A., and B. Millet. 1996. Rhythmic growth and optimization of micropropagation: The effect of excision time and position of axillary buds on *in vitro* culture of *Citrus aurantium* L. *Ann. Bot.* 78:197–202.

52. Estrada-Luna, A. A., F. T. Davies, Jr., and J. N. Egilla. 2000. Mycorrhizal fungi enhancement of growth and gas exchange of micropropagated guava plantlets (*Psidium guajava* L.) during *ex vitro* acclimatization and plant establishment. *Mycorrhiza* 10:1–8.

53. Etienne, H. 2005. Somatic Embryogenesis Protocol: Coffee (*Coffea arabica* L. and *C. canephora* P.) In S. M. Jain and P. K. Gupta. eds. *Protocol for somatic embryos in woody plants*. Dordrecht: Springer.

54. Etienne, H., and M. Berthouly. 2002. Temporary immersion systems in plant micropropagation (review). *Plant Cell Tiss. Organ Cult.* 69:215–31.

55. Evans, D. A., and J. E. Bravo. 1986. Phenotypic and genotypic stability of tissue cultured plants. In R. H. Zimmerman, R. J. Griesbach, F. A. Hammerschlag, and R. H. Lawson, eds. *Tissue culture as a plant production system for horticultural crops*. Dordrecht: Martinus Nijhoff Publishers. pp. 73–94.

56. Evans, D. A., W. R. Sharp, and C. E. Flick. 1981. Plant regeneration from cell cultures. *Hort. Rev.* 3:214–314.

57. Evans, D. A., W. R. Sharp, and H. P. Medina-Filho. 1984. Somaclonal and gametoclonal variation. *Amer. J. Bot.* 71:759–74.

58. Fiorino, P., and F. Loreti. 1987. Propagation of fruit trees by tissue culture in Italy. HortScience 22:353–58.

59. Fogh, J., ed. 1973. *Contamination in tissue culture*. New York: Academic Press.

60. Fonnesbech, M. 1974. Temperature effects on shoot and root development from *Begonia cheimantha* petiole segments grown *in vitro*. *Physiol. Plant.* 32:282–86.

61. Fujiwara, K., T. Kozai, Y. Nakajo, and I. Watanabe. 1989. Effects of closures and vessels on light intensities in plant tissue culture vessels. *J. Agric. Meteorol.* 45:143–49.

62. Gabryszewska, E., R. M. Rudnicki, and T. Blacquiere. 1997. The effects of light quality on the growth and development of shoots and roots of *Ficus benjamina in vitro*. *Acta Hort.* 418:163–67.

63. Gaspar, T., C. Kevers, P. Debergh, L. Maene, M. Paques, and P. Boxus. 1986. Vitrification: Morphological, physiological, and ecological aspects. In J. M. Bonga and D. J. Durzan, eds. *Cell and tissue culture in forestry, Vol. 1, General principles and biotechnology.* Dordrecht: Martinus Nijhoff Publishers. pp. 152–66.

64. Gautheret, R. J. 1934. Culture du tissu cambial. *C. R. Acad. Sci.* 198:2195–96.

65. Gautheret, R. J. 1939. Sur la possibilite de realiser la culture indefinie des tissus de tubercules de carotte. *C. R. Acad. Sci.* 208:118–30.

66. Gautheret, R. J. 1983. Plant tissue culture: A history. *Bot. Mag.* 96:393–410.

67. Gautheret, R. J. 1985. History of plant tissue and cell culture: A personal account. In I. K. Vasil, ed. *Cell culture and somatic cell genetics of plants, Vol. 2.* New York: Academic Press. pp. 1–59.

68. Geneve, R. L., and S. T. Kester. 1990. The initiation of somatic embryos and adventitious roots from developing zygotic embryo explants of *Cercis canadensis* L. cultured *in vitro*. *Plant Cell Tiss. Organ Cult.* 22:71–6.

69. Geneve, R. L., S. T. Kester, and S. El-Shall. 1990. *In vitro* shoot initiation in Kentucky coffeetree (*Gymnocladus dioicus* L.). *HortScience* 25:578.

70. Geneve, R. L. 2005. Comparative adventitious shoot induction in Kentucky coffeetree root and petiole explants treated with thidiazuron and benzylaminopurine. *In Vitro Cell. Dev. Bio.—Plant* 41:489–93.

71. Geneve, R. L., S. T. Kester, and K. W. Pomper. 2007. Autonomous shoot production in pawpaw [*Asimina triloba* (L.) Dunal] on plant growth regulator free media. *Propagation Ornamental Plants* 7:51–6.

72. Gengenbach, B. G. 1984. *In vitro* pollination, fertilization, and development of maize kernels. In I. K. Vasil, ed. *Cell culture and somatic cell genetics of plants, Vol. 1, Laboratory procedures and their applications.* New York: Academic Press. pp. 276–82.

73. George, E. F., and P. D. Sherrington. 1984. *Plant propagation by tissue culture: Handbook and directory of commercial laboratories.* Eversley, UK: Exegetics Ltd.

74. George, E. F., M. A. Hall, and G. J. De Klerk. 2007. *Plant propagation by tissue culture.* Dordrecht: Springer.

75. Georget, F., R. Domergue, N. Ferriere, and F. X. Cox. 2000. Morphohistological study of different constituents of a banana (*Musa* AAA, cv. Grande naine) embryogenic cell suspension. *Plant Cell Rpt.* 19:748–54.

76. Gingas, V. M. 1991. Asexual embryogenesis and plant regeneration from male catkins of *Quercus*. *HortScience* 26:1217–19.

77. Goldy, R. G., and U. Amborn. 1987. *In vitro* culturability of ovules from 10 seedless grape clones. *HortScience* 22(5):952.

78. Gray, D. J. 1987. Quiescence in monocotyledonous and dicotyledonous somatic embryos induced by dehydration. *HortScience* 22:810–14.

79. Gray, D. J., and A. Purohit. 1991. Somatic embryogenesis and development of synthetic seed technology. *Critical Rev. Plant Sci.* 10:33–61.

80. Gray, D. J. 1989. Effects of dehydration and exogenous growth regulators on dormancy, quiescence and germination of grape somatic embryos. *In Vitro Cell. Dev. Bio.—Plant* 25:1173–78.

81. Grosser, J. W., and R. G. Gmitter, Jr. 2005. Applications of somatic hybridization and cybridization in crop improvement, with citrus as a model. *In Vitro Cell. Dev. Bio.—Plant* 41:220–25.

82. Grout, B. W. W., and M. J. Aston. 1977. Transplanting of cauliflower plants regenerated from meristem culture. I. Water loss and water transfer related to changes in leaf wax and to xylem regeneration. *Hort. Res.* 17:1–7.

83. Guha, S., and S. C. Maheshwari. 1964. *In vitro* production of embryos from anthers of *Datura*. *Nature* (London) 204:497.

84. Gupta, P. K., and D. J. Durzan. 1986. Somatic polyembryogenesis from callus of mature sugar pine embryos. *Biotechnology* 4:643–45.

85. Gupta, P. K., D. J. Durzan, and B. J. Finkle. 1987. Somatic polyembryogenesis in embryogenic cell masses of *Picea abies* (Norway spruce) and *Pinus taeda* (loblolly pine) after freezing in liquid nitrogen. *Can. J. For. Res.* 17:1130–34.

86. Haberlandt, G. 1902. Kulturversuche mit isolierten Pflanzenzellen. Sitzungsber. Akad. der Wiss. Wien, Math.-*Naturwiss.* K1. 111:69–92.

87. Hackett, W. P. 1966. Applications of tissue culture to plant propagation. *Comb. Proc. Intl. Plant Prop. Soc.* 16:88–92.

88. Hackett, W. P. 1985. Juvenility, maturation, and rejuvenation in woody plants. *Hort. Rev.* 7:109–55.

89. Halperin, W. 1966. Alternative morphogenetic events in cell suspension. *Amer. J. Bot.* 53:443–53.

90. Hamill, J. D., and S. F. Chandler. 1994. Use of transformed roots for root development and metabolism studies and progress in characterizing root-specific gene expression. In T. D. Davis and B. E. Haissig, eds. *Biology of adventitious root formation.* New York: Plenum Press. pp. 163–80.

91. Hammerschlag, F. A. 1986. Temperate fruits and nuts. In R. H. Zimmerman, R. J. Griesbach, F. A. Hammerschlag, and R. H. Lawson, eds. *Tissue culture as a plant production system for horticultural crops.* Dordrecht: Martinus Nijhoff Publishers. pp. 221–36.

92. Han, B. H., B. W. Yae, H. J. Yu, and K. Y. Peak. 2003. Improvement of *in vitro* micropropagation of *Lilium* oriental hybrid 'Casablanca' by the formation of shoots with abnormally swollen basal plates. *Scientia Hort.* 123:351–59.

93. Hangarter, R. P., and T. C. Stasinopoulos. 1991. Repression of plant tissue culture growth by light is caused by photochemical change in the culture medium. *Plant Sci.* 79:253–57.

94. Harst, M. 1995. Development of a regeneration protocol for high frequency somatic embryogenesis from explants of grapevines (*Vitis* spp.). *Vitis* 34:27–9.

95. Hausman, J. F., O. Neys, C. Kevers, and T. H. Gaspar. 1994. Effects of *in vitro* storage at 4°C on survival and proliferation of poplar shoots. *Plant Cell Tiss. Organ Cult.* 38:65–7.

96. Hazarika, B. N. 2006. Morpho-physiological disorders in *in vitro* culture of plants. *Scientia Hort.* 108:105–20.

97. Henny, R. J. 1980. *In vitro* germination of *Maranta leuconeura* embryos. *HortScience* 15:198–200.

98. Henny, R. J., J. F. Knauss, and A. Donnan, Jr. 1981. Foliage plant tissue culture. In J. Joiner, ed. *Foliage plant production.* Englewood Cliffs, NJ: Prentice Hall.

99. Henry, Y., P. Vain, and J. de Buyser. 1994. Genetic analysis of *in vitro* plant tissue culture responses and regeneration capacities. *Euphytica* 79:45–58.

100. Hesse, C., and D. E. Kester. 1955. Germination of embryos of *Prunus* related to degree of embryo development and method of handling. *Proc. Amer. Soc. Hort. Sci.* 65:251–64.

101. Heuser, C. W., and J. Horker. 1976. Tissue culture propagation of day lilies. *Comb. Proc. Intl. Plant Prop. Soc.* 26:269–72.

102. Hicks, G. S. 1980. Patterns of organ development in plant tissue culture and the problem of organ determination. *Bot. Rev.* 46:1–23.

103. Holdgate, D. P., and J. S. Aynsley. 1977. The development and establishment of a commercial tissue culture laboratory. *Acta Hort.* 78:31–6.

104. Hu, C., and P. Wang. 1986. Embryo culture: Technique and applications. In D. A. Evans, W. R. Sharp, and P. V. Ammirato, eds. *Handbook of plant cell culture, Vol. 4.* New York: Macmillan. pp. 43–96.

105. Huang, S., and D. F. Millikan. 1980. *In vitro* micrografting of apple shoot-tips. *HortScience* 15:741–43.

106. Hussey, G. 1977. *In vitro* propagation of some members of the Liliaceae, Iridaceae and Amaryllidaceae. *Acta Hort.* 78:303–9.

107. Hussey, G., and A. Falavigna. 1980. Origin and production of *in vitro* adventitious shoots in the onion, *Allium cepa* L. *J. Exp. Bot.* 31:1675–86.

108. Hvoslef-Eide, A. K., and W. Preil. 2005. *Liquid culture systems for in vitro propagation.* Dordrecht: Springer.

109. Ibaraki, Y., and K., Kurata. 2001. Automation of somatic embryo production. *Plant Cell Tiss. Organ Cult.* 65:179–99.

110. Ibaraki, Y., and Y. Nozaki. 2005. Estimation of light intensity distribution in a culture vessel. *Plant Cell Tiss. Organ Cult.* 80:111–13.

111. Jackson, M. B., A. J. Abbott, A. R. Belcher, K. C. Hall, R. Butler, and J. Cameron. 1991. Ventilation in plant tissue cultures and effects of poor aeration on ethylene and carbon dioxide accumulation, oxygen depletion and explant development. *Ann. Bot.* 67: 229–37.

112. Jain, S. M. 2001. Tissue culture-derived variation in crop improvement. *Euphytica* 118:153–66.

113. Jain, S. M., and P. K. Gupta. 2005. *Protocol for somatic embryos in woody plants.* Dordrecht: Springer.

114. Jain, S. M., and K. Ishii. 2003. *Micropropagation of woody trees and fruits.* Dordrecht: Kluwer Acad. Pub.

115. Jain, S. M., D. S. Brar, and B. S. Ahloowalia. 1998. Somaclonal variation and induced mutations in crop improvement. *Curr. Plant Sci. Biotechnol. Agric.* Dordrecht: Kluwer Acad. Pub.

116. Jain, S. M., P. K. Gupta, and R. J. Newton. 1994. *Somatic embryogenesis in woody plants.* Dordrecht: Kluwer Acad. Pub.

117. James, D. J., and I. J. Thurbon. 1981. Shoot and root initiation *in vitro* in the apple rootstock M9 and the promotive effects of phloroglucinol. *J. Hort. Sci.* 56:15–20.

118. Jean, M., and M. Cappadocia. 1992. Effects of some growth regulators on *in vitro* tuberization in *Dioscorea alata* L. 'Brazo fuerte' and *D. abyssinica* Hoch. *Plant Cell Rpt.* 11:34–8.

119. Jeong, B. R., K. Fujiwara, and T. Kozai. 1995. Environmental control and photoautotrophic micropropagation. *Hort. Rev.* 17:125–72.

120. Jonard, R. 1986. Micrografting and its applications to tree improvement. In Y. P. S. Bajaj, ed. *Biotechnology in agriculture and forestry, Vol. 1, Trees.* Berlin: Springer-Verlag. pp. 31–48.

121. Jonard, R., J. Hugard, J. Machix, J. Martinez, L. P. Mosella-Chancel, J. L. Poessel, and P. Villemue. 1983. *In vitro* micrografting and its applications to fruit science. *Scientia Hort.* 20:147–59.

122. Jones, O. P. 1976. Effect of phloridzin and phloroglucinol on apple shoots. *Nature* 262:392–93.

123. Jones, R. O., R. L. Geneve, and S. T. Kester. 1994. Micropropagation of gas plant (*Dictamnus albus* L.). *J. Environ. Hort.* 12:216–18.

124. Kamo, K. 1990. Effect of phytohormones on plant regeneration from callus of *Gladiolus* cultivar Jenny-Lee. *In Vitro Cell. Dev. Bio.—Plant* 30:26–31.

125. Kartha, K. K. 1984. Elimination of viruses. In I. K. Vasil, ed. *Cell culture and somatic cell genetics of plants, Vol. 1.* New York: Academic Press. pp. 577–85.

126. Keller, W. A. 1984. Anther culture of *Brassica*. In I. K. Vasil, ed. *Cell culture and somatic cell genetics of plants, Vol. 1.* New York: Academic Press. pp. 302–10.

127. Kester, D. E., and C. O. Hesse. 1955. Embryo culture of peach varieties in relation to season of ripening. *Proc. Amer. Soc. Hort. Sci.* 65:265–73.

128. Kim, K. W., and A. Kako. 1984. Studies on clonal propagation in the *Cymbidium* floral organ culture *in vitro. J. Korean Soc. Hort. Soc.* 25:65–71.

129. Kim, K. W., and A. A. de Hertogh. 1997. Tissue culture of ornamental flowering bulbs (geophytes). *Hort. Rev.* 18:87–169.

130. Kitto, S. L., and J. Janick. 1985. Production of synthetic seeds by encapsulating asexual embryos of carrot. *J. Amer. Soc. Hort. Sci.* 110:277–80.

131. Klee, H., R. Horsch, and S. Rogers. 1987. *Agrobacterium*-mediated plant transformation and its further applications to plant biology. *Ann. Rev. Plant Physiol.* 38:467–81.

132. Knauss, J. F. 1976. A tissue culture method of producing *Dieffenbachia picta* cv. Perfection free of fungi and bacteria. *Proc. Fla. State Hort. Soc.* 89:293–96.

133. Knauss, J. F., and J. W. Miller. 1978. A contaminant, *Erwinia carotovora*, affecting commercial plant tissue culture. *In Vitro* 14:754–56.

134. Kohlenback, H. W. 1984. Culture of isolated mesophyll cells. In I. K. Vasil, ed. *Cell culture and somatic cell genetics of plants, Vol. 1, Laboratory procedures and their applications.* New York: Academic Press. pp. 204–12.

135. Kotte, W. 1922. Kulturversuche mit isolierten wurzelspitzen. *Beitr. Z. Allgem. Bot.* 2:413–34.

136. Kozai, T., F. Afreen, and S. M. A. Zobayed. 2005. *Photoautotrophic (sugar-free medium) micropropagation as a new micropropagation and transplant production system.* Dordrecht: Springer.

137. Kozai, T., K. Fujiwara, M. Hayashi, and J. Aitken-Christie. 1992. The *in vitro* environment and its control in micropropagation. In K. Kurata and T. Kozai, eds. *Transplant production systems.* Dordrecht: Kluwer Acad. Pub. pp. 247–82.

138. Kretzschmar, U. 1993. Improvement of larch micropropagation by induced short shoot elongation *in vitro. Silvae Genetica* 42:163–69.

139. Krikorian, A. D., R. P. Kann, S. A. O'Connor, and M. S. Fitter. 1986. Totipotent suspensions as a means of multiplication. In R. H. Zimmerman, R. J. Griesbach, F. A. Hammerschlag, and R. H. Lawson, eds. *Tissue culture as a plant production system for horticultural crops.* Dordrecht: Martinus Nijhoff Publishers. pp. 61–72.

140. Kumar, V., A. Ramakrishna, and G. A. Ravishankar. 2007. Influence of different ethylene inhibitors on somatic embryogenesis and secondary embryogenesis from *Coffea canephora* P ex Fr. *In Vitro Cell. Dev. Bio.—Plant* 43:602–7.

141. Kumar, V., D. Reid, and T. A. Thorpe. 1998. Regulation of morphogenesis in plant tissue culture by ethylene. In vitro *Cell. Dev. Biol—Plant* 69:244–52.

142. Kyte, L. 1987. *Plants from test tubes: An introduction to micropropagation* (rev. ed.). Portland, OR: Timber Press.

143. Lammerts, W. E. 1942. Embryo culture, an effective technique for shortening the breeding cycle of deciduous trees and increasing germination of hybrid seed. *Amer. J. Bot.* 29:166–71.

144. Langhans, R. W., R. K. Horst, and E. D. Earle. 1977. Disease-free plants via tissue culture propagation. *HortScience* 12:149–50.

145. Lee, C. W., R. M. Skirvin, A. I. Soltero, and J. Janick. 1977. Tissue culture of *Salpiglossis sinuata* L. from leaf discs. *HortScience* 12:547–49.

146. Linderman, R. G. 1993. Tissue proliferation. *Amer. Nurs.* 178:57–67.

147. Linsmaier, E. M., and F. Skoog. 1965. Organic growth factor requirements of tobacco tissue cultures. *Physiol. Plant.* 18:101–27.

148. Lutz, J. D., J. R. Wong, and J. Rowe. 1985. Somatic embryogenesis for mass cloning of crop plants. In R. R. Henke, K. W. Hughes, M. J. Constantin, and A. Hollaender, eds. *Tissue culture in forests and agriculture.* New York: Plenum Press. pp. 105–16.

149. Maene, L., and P. Debergh. 198. Liquid medium additions to established tissue cultures to improve elongation and rooting *in vivo*. *Plant Cell Tiss. Organ Cult.* 5:23–33.

150. McClelland, M. T., M. A. L. Smith, and Z. B. Carothers. 1990. The effects of *in vitro* and *ex vitro* root initiation on subsequent microcutting root quality in three woody plants. *Plant Cell Tiss. Organ Cult.* 23:115–23.

151. McCown, B. H. 1986. Woody ornamentals, shade trees, and conifers. In R. H. Zimmerman, R. J. Griesbach, F. A. Hammerschlag, and R. H. Lawson, eds. *Tissue culture as a plant production system for horticultural crops.* Dordrecht: Martinus Nijhoff Publishers. pp. 333–42.

152. McCown, B. H., and P. J. Joyce. 1991. Automated propagation of microtubers of potato. In I. K. Vasil, ed. *Scale-up and automation in plant propagation.* New York: Academic Press. pp. 95–109.

153. McCown, B. H., E. L. Zeldin, H. A. Pinkalla, and R. R. Dedolph. 1987. Nodule culture: a developmental pathway with high potential for regeneration, automated micropropagation, and plant metabolite production from woody plants. In J. W. Hanover and D. E. Keathley, eds. *Genetic manipulation of woody plants.* New York: Plenum Press. pp. 149–66.

154. McCown, D. D., and B. H. McCown. 1986. North American hardwoods. In J. M. Bonga and D. J. Durzan, eds. *Cell and tissue culture in forests, Vol. 3, Case histories: Gymnosperms, angiosperms and Palms* (2nd ed.). Dordrecht: Martinus Nijhoff. pp. 247–60.

155. McGrew, J. R. 1980. Meristem culture for production of virus-free strawberries. In R. H. Zimmerman, ed. *Proc. conf. on nursery production of fruit plants through tissue culture: Applications and feasibility.* U.S. Dept. of Agr. Sci. and Education Administration ARR-NE-11. pp. 80–5.

156. Meins, F., Jr. 1989. Habituation: heritable variation in the requirement of cultured plant cells for hormones. *Annu. Rev. Genet.* 23:395–408.

157. Meredith, C., and P. S. Carlson. 1978. Genetic variation in cultured plant cells. In K. W. Hughes, R. Henke, and M. Constantin, eds. *Propagation of higher plants through tissue culture.* Springfield, VA: U.S. Dept. of Energy, Tech. Inform. Center, pp. 166–76.

158. Merkle, S. A. 1997. Somatic embryogenesis in ornamentals. In R. L. Geneve, J. E. Preece, and S. A. Merkle, eds. *Biotechnology of ornamentals.* Biotechnol. Agric. Series No. 16. Wallingford, UK: CAB International. pp. 13–34.

159. Merkle, S. A., and P. J. Battle. 2000. Enhancement of embryogenic culture initiation from tissues of mature sweet gum trees. *Plant Cell Rpt.* 19:268–73.

160. Meyer, M. M., Jr. 1976. Culture of *Paeonia* embryos by *in vitro* techniques. *Amer. Peony Soc. Bul.* 217:32–5.

161. Meyer, M. M. 1980. *In vitro* propagation of *Hosta sieboldiana*. *HortScience* 15:737–38.

162. Meyer, M. M., L. H. Fuchigami, and A. N. Roberts. 1975. Propagation of tall bearded irises by tissue culture. *HortScience* 10:479–80.

163. Meyer, M. M., and G. M. Milbrath. 1977. *In vitro* propagation of horseradish with leaf pieces. *HortScience* 12:544–45.

164. Mneney, E. E., and C. H. Mantell. 2001. *In vitro* micrografting of cashew. *Plant Cell Tiss. Organ Cult.* 66:49–58.

165. Mølgaard, J. P., N. Roulund, V. Deichmann, L. Irgens-Moller, S. B. Andersen, and B. Farestveit. 1971. *In vitro* multiplication of *Saintpaulia ionantha* Wendl. *Scientia Hort.* 48:285–92.

166. Monteiro, A. C. B. A., E. N. Higashi, A. N. Gonçalves, and A. P. M. Rodriguez. 2000. A novel approach for the definition of the inorganic medium components for micropropagation of yellow passionfruit (*Passiflora edulis* Sims. f. *flavicarpa* Deg.). *In Vitro Cell. Dev. Bio.—Plant* 36:527–31.

167. Morel, G. M. 1960. Producing virus-free cymbidiums. *Amer. Orchid Soc. Bul.* 29:495–97.

168. Morel, G. M., and C. Martin. 1950. Guerison de dahlias atteints d'une maladie a virus. *C.R. Acad. Sci.* 235:1324–25.

169. Morel, G. M. 1964. Tissue culture: A new means of clonal propagation of orchids. *Amer. Orchid Soc. Bul.* 33:473–78.

170. Muir, W. H., A. C. Hildebrandt, and A. J. Riker. 1954. Plant tissue cultures produced from single isolated cells. *Science* 119:877–78.

171. Mullins, M. G., G. Y. Nair, and P. Sampet. 1979. Rejuvenation *in vitro*: Induction of juvenile characters in an adult clone of *Vitis vinifera* L. *Ann. Bot.* 44:623–27.

172. Murashige, T. 1966. Principles of *in vitro* culture. *Comb. Proc. Intl. Plant Prop. Soc.* 16:80–8.

173. Murashige, T. 1974. Plant propagation through tissue cultures. *Annu. Rev. Plant Physiol.* 25:135–66.

174. Murashige, T., and F. Skoog. 1962. A revised medium for rapid growth and bioassays with tobacco tissue cultures. *Physiol. Plant.* 15:473–97.

175. Nas, M. N., and P. E. Read. 2004. A hypothesis for the development of a defined tissue culture medium of higher plants and micropropagation hazelnuts. *Scientia Hort.* 101:189–200.

176. Navarro, L., and J. Juarez. 1977. Tissue culture techniques used in Spain to recover virus-free citrus plants. *Acta Hort.* 78:425–35.

177. Negueroles, J., and O. P. Jones. 1979. Production *in vitro* of rootstock/scion combinations of *Prunus* cultivars. *J. Hort. Sci.* 54:279–81.

178. Nestares, G., R. Zorzoli, L. Mroginski, and L. Picardi. 2008. Heritability of *in vitro* plant regeneration capacity in sunflower. *Plant Breeding* 121:366–68.

179. Ng, T. J. 1986. Use of tissue culture for micropropagation of vegetable crops. In R. H. Zimmerman, R. J. Griesbach, F. A. Hammerschlag, and R. H. Lawson, eds. *Tissue culture as a plant production system for horticultural crops.* Dordrecht: Martinus Nijhoff Publishers. pp. 259–70.

180. Nhut, D. T., J. A. T. Da Silva, and C. R. Aswath. 2003. The importance of the explant on regeneration in thin cell layer technology. *In Vitro Cell. Dev. Bio.—Plant* 39:266–76.

181. Nhut, D. T., N. T. Don, and M. Tanaka. 2007. Light-emitting diodes as an effective lighting source for *in vitro* banana culture. In S. M. Jain and H. Häggman, eds. *Protocols for micropropagation of woody trees and fruits.* Dordrecht: Springer. pp. 527–41.

182. Nhut, D. T., T. Takamura, H. Watanabe, A. Murakami, K. Murakami, and M. Tanaka. 2003. Responses of strawberry plantlets cultured *in vitro* under superbright red and blue light-emitting diodes (LEDs). *Plant Cell Tiss. Organ Cult.* 73:43–52.

183. Nitsch, J. P., and C. Nitsch. 1969. Haploid plants from pollen grains. *Science* 163:85–7.

184. Nobecourt, P. 1939. Sur la perennite et l'augmentation de volume des cultures de tissus vegetaux. *C. R. Soc. Biol.* 130:1270–71.

185. Oglesby, R. P., and J. L. Griffis, Jr. 1986. Commercial *in vitro* propagation and plantation crops. In R. H. Zimmerman, R. J. Griesbach, F. A. Hammerschlag, and R. H. Lawson, eds. *Tissue culture as a plant production system for horticultural crops.* Dordrecht: Martinus Nijhoff Publishers. pp. 253–57.

186. Onay, A., V. Pirinç, H. Yýldýrým, and D. Basaran. 2004. *In vitro* micrografting of mature pistachio (*Pistacia vera* var. *siirt*). *Plant Cell Tiss. Organ Cult.* 77:215–19.

187. Onishi, N., Y. Sakamoto, and T. Hirosawa. 1994. Synthetic seeds as an application of mass production of somatic embryos. *Plant Cell Tiss. Organ Cult.* 39:137–45.

188. Orlikowska, T. 1992. Effects of *in vitro* storage at 4°C on survival and proliferation of two apple rootstocks. *Plant Cell Tiss. Organ Cult.* 31:1–7.

189. Ozudogru, E. A., Y. Ozden-Tokatli, and A. Akgin. 2005. Effect of silver nitrate on multiple shoot formation of Virginia-type peanut through shoot tip culture. *In Vitro Cell. Dev. Bio.—Plant* 41:151–56.

190. Paek, K. Y., and E. J. Hahn. 1999. Variation in African violet 'Crimson Frost' micropropagated by homogenized leaf tissue culture. *HortTechnology* 9:625–28.

191. Pierik, R. L. M. 1999. *In vitro culture of higher plants.* 4th ed. Dordrecht: Martinus Nijhoff Publishers.

192. Pierik, R. L. M., and H. H. M. Steegmans. 1976. Vegetative propagation of *Freesia* through the isolation of shoots *in vitro*. *Neth. J. Agr. Sci.* 24:274–77.

193. Pierik, R. L. M., J. L. M. Jansen, A. Maasdam, and C. M. Binnendijk. 1975. Optimalization of *Gerbera* plantlet production from excised capitulum explants. *Scientia Hort.* 3:351–57.

194. Pischke, M. S., E. L. Huttlin, A. D. Hegeman, and M. R. Sussman. 2006. A transcriptome-based characterization of habituation in plant tissue culture. *Plant Physiol.* 140:1255–78.

195. Preece, J. E., and E. G. Sutter. 1991. Acclimatization of micropropagated plants to the greenhouse and field. In P. C. Debergh and R. H. Zimmerman, eds. *Micropropagation: Technology and application.* Dordrecht: Kluwer Acad. Pub. pp. 71–94.

196. Pullman, G. S., and A. Skryabina. 2007. Liquid medium and liquid overlays improve embryogenic tissue initiation in conifers. *Plant Cell Rpt.* 26:873–87.

197. Quak, F. 1977. Meristem culture and virus-free plants. In J. Reinert and Y. P. S. Bajaj, eds. *Applied and fundamental aspects of plant cell, tissue and organ culture.* Berlin: Springer-Verlag, pp. 598–615.

198. Raharjo, S. H. T., and R. E. Litz. 2005. Micrografting and *ex vitro* grafting for somatic embryo rescue and plant recovery in avocado (*Persea americana*). *Plant Cell Tiss. Organ Cult.* 82:1–9.

199. Rai, M. K., V. S. Jaiswal, and U. Jaiswal. 2008. Encapsulation of shoot tips of guava (*Psidium guajava* L.) for short-term storage and germplasm exchange. *Scientia Hort.* 118:33–8.

200. Raju, B. C., and J. C. Trolinger. 1986. Pathogen indexing in large-scale propagation of florist crops. In R. H. Zimmerman, R. J. Griesbach, F. A. Hammerschlag, and R. H. Lawson, eds. *Tissue culture as a plant production system for horticultural crops.* Dordrecht: Martinus Nijhoff Publishers. p. 138.

201. Ramirez-Parra, E., B. Desvoyes, and C. Gutierrez. 2005. Balance between cell division and differentiation during plant development. *Int. J. Dev. Bio.* 49:467–77.

202. Ramming, D. W. 1985. *In ovulo* embryo culture of early maturing *Prunus. HortScience* 20:419–20.

203. Rangan, T. S. 1984. Culture of ovaries. In I. K. Vasil, ed. *Cell culture and somatic cell genetics of plants, Vol. 1.* New York: Academic Press. pp. 221–26.

204. Rangan, T. S. 1984. Culture of ovules. In I. K. Vasil, ed. *Cell culture and somatic cell genetics of plants, Vol. 1.* New York: Academic Press. pp. 227–31.

205. Rangaswamy, N. S. 1977. Application of *in vitro* pollination and *in vitro* fertilization. In J. Reinert and Y. P. S. Bajaj, eds. *Applied and fundamental aspects of plant cell, tissue and organ culture.* Berlin: Springer-Verlag. pp. 412–25.

206. Rani, V., and S. N. Raina. 2000. Genetic fidelity of organized meristem-derived micropropagated plants: A critical reappraisal. *In Vitro Cell. Dev. Bio.—Plant* 36:319–30.

207. Rani, V., and S. N. Raina. 2003. Molecular DNA marker analysis to assess the genetic fidelity of micropropagated woody plants. In S. M. Jain and K. Ishii, eds. *Micropropagation of woody trees and fruits.* Dordrecht: Kluwer Acad. Pub. pp. 75–101.

208. Ranjit, M., and D. E. Kester. 1988. Micropropagation of cherry rootstocks: II. Invigoration and enhanced rooting of '46-1 Mazzard' by co-culture with 'Colt.' *J. Amer. Soc. Hort. Sci.* 113:150–54.

209. Read, P., and J. Preece. 2007. Micropropagation of ornamentals: the wave of the future? *Propagation Ornamental Plants* 7:150–59.

210. Redenbaugh, K. 1993. *Synseeds: Application of synthetic seeds to crop improvement.* Boca Raton, FL: CRC Press.

211. Redenbaugh, K., D. Slade, P. Viss, and J. A. Fujii. 1987. Encapsulation of somatic embryos in synthetic seed coats. *HortScience* 22:803–9.

212. Reed, S. 2005. Embryo Rescue. In R. N. Trigiano and D. J. Gray, eds. *Plant Development and Biotechnology.* Boca Raton, FL: CRC Press. pp: 235–39.

213. Reinert, J. 1959. Über die Kontrolle die Morphogenese und die Induktion von Adventiveembryonen an gewebekulturen aus karotten. *Planta* 53:318–33.

214. Robbins, W. J. 1922. Cultivation of excised root-tips and stem tips under sterile conditions. *Bot. Gaz.* 73:376–90.

215. Rodriguez, A. P. M., and W. A. Vendrame. 2003. Micropropagation of tropical woody species. In S. M. Jain and K. Ishii, eds. *Micropropagation of woody trees and fruits.* Dordrecht: Kluwer Acad. Pub. pp. 153–79.

216. Roest, S., and G. S. Bokelmann. 1973. Vegetative propagation of *Chrysanthemum cinerariaefolium* in vitro. *Scientia Hort.* 1:120–22.

217. Saebo, A., T. Krekling, and M. Appelgren. 1995. Light quality affects photosynthesis and leaf anatomy of birch plantlets *in vitro. Plant Cell Tiss. Organ Cult.* 41:177–85.

218. Saebo, A., G. Skjeseth, and M. Appelgren. 1995. Light quality of the *in vitro* stage affects the subsequent rooting and field performance of *Betula pendula* (Roth). *Scandinavian J. For. Res.* 10:155–60.

219. Sagawa, Y., T. Shoji, and T. Shoji. 1966. Clonal propagation of cymbidium through shoot meristem culture. *Amer. Orchid Soc. Bul.* 35:118–22.

220. Schwarz, O. J., A. R. Sharma, and R. M. Beaty. 2004. Propagation from nonmeristematic tissues: Organogenesis. In R. N. Trigiano and D. J. Gray, eds. *Plant development and biotechnology.* Boca Raton, FL: CRC Press.

221. Scowcroft, W. R., R. I. S. Brettell, S. A. Ryan, P. A. Davies, and M. S. Pallotta. 1987. Somaclonal variation and genomic flux. In C. E. Green, D. A. Somers, W. P. Hackett, and D. D. Biesboer, eds. *Plant tissue and cell culture.* New York: Alan R. Liss. pp. 275–88.

222. Senula, A., E. R. J. Keller, and D. E. Leseman. 2000. Elimination of viruses through meristem culture and thermotherapy for the establishment of an *in vitro* collection of garlic (*Allium sativum*). *Acta Hort.* 530:121–28.

223. Sharp, W. R., M. R. Sondahl, L. S. Caldas, and S. B. Maraffa. 1980. The physiology of *in vitro* asexual embryogenesis. *Hort. Rev.* 2:268–309.

224. Shepard, J. F., D. Bidney, and E. Shakin. 1980. Potato protoplasts in crop improvement. *Science* 208:17–24.

225. Simonton, W., C. Robacker, and S. Krueger. 1991. A programmable micropropagation apparatus using cycled medium. *Plant Cell Tiss. Organ Cult.* 27:211–18.

226. Singha, S., E. C. Townsend, and G. H. Singha. 1990. Relationship between calcium and agar

on Singha and shoot-tip necrosis of quince (*Cydonia oblonga* Mill.) shoots *in vitro*. *Plant Cell Tiss. Organ Cult.* 23:135–42.

227. Sinha, S., R. P. Roy, and K. K. Jha. 1979. Callus formation and shoot bud differentiation in anther culture of *Solanum surattense*. *Can. J. Bot.* 57:2524–27.

228. Skoog, F., and C. O. Miller. 1957. Chemical regulation of growth and organ formation in plant tissues cultured *in vitro*. *Symp. Soc. Exp. Bio.* 11:118–31.

229. Smith, M. A. L., and M. T. McClelland. 1991. Gauging the influence of *in vitro* conditions on *in vivo* quality and performance of woody plants. *In Vitro Cell. Dev. Bio.—Plant* 27:52–6.

230. Sorvari, S., M. László, M. G. Fári, and O. Toldi. 2006. Application of tissue culture and molecular farming in the production of pharmaceuticals. *Acta Hort.* 125:585–96.

231. Sriskandarajah, C., and M. G. Mullins. 1981. Micropropagation of Granny Smith apple: Factors affecting root formation *in vitro*. *J. Hort. Sci.* 56:71–6.

232. Stasolla, C., and E. C. Yeung. 2003. Recent advances in conifer somatic embryogenesis: improving somatic embryo quality. *Plant Cell Tiss. Organ Cult.* 74:15–35.

233. Stefano, M., and M. Rosario. 2003. Micropropagation of tropical woody species. In S. M. Jain and K. Ishii, eds. *Effects of light quality on micropropagation of woody species.* Dordrecht: Kluwer Acad. Pub. pp. 3–36.

234. Steward, F. C., M. O. Mapes, and K. Mears. 1958. Growth and organized development of cultured carrots. II. Organization in cultures from freely suspended cells. *Amer. J. Bot.* 454:705–8.

235. Stimart, D. P. 1986. Commercial micropropagation of florist flower crops. In R. H. Zimmerman, R. J. Griesbach, F. A. Hammerschlag, and R. H. Lawson, eds. *Tissue culture as a plant production system for horticultural crops.* Dordrecht: Martinus Nijhoff Publishers. pp. 301–15.

236. Stimart, D. P., and P. D. Ascher. 1981. Developmental response of *Lilium longiflorum* bulblets to constant or alternating temperatures *in vitro*. *J. Amer. Soc. Hort. Sci.* 106:450–54.

237. Stirn, S., A. P. Mordhorst, S. Fuchs, and H. Lorz. 1995. Molecular and biochemical markers for embryogenic potential and regenerative capacity of barley (*Hordeum vulgare* L.) cell cultures. *Plant Sci.* 106:195–206.

238. Stoltz, L. P. 1971. Agar restriction of the growth of excised mature iris embryos. *J. Amer. Soc. Hort. Sci.* 96:611–14.

239. Stoutemyer, V. T., and O. K. Britt. 1969. Growth and habituation in tissue cultures of English ivy, *Hedera helix*. *Amer. J. Bot.* 56:222–26.

240. Struve, D. K., and R. D. Lineberger. 1988. Restoration of high adventitious root regeneration potential in mature *Betula papyrifera* Marsh. Softwood stem cuttings. *Can. J. For. Res.* 18:265–69.

241. Sunderland, N. 1984. Anther culture of *Nicotiana tabacum*. In I. K. Vasil, ed. *Cell culture and somatic cell genetics of plants, Vol. 1.* New York: Academic Press. pp. 283–92.

242. Swartz, H. J. 1991. Post culture behavior: Genetic and epigenetic effects and related problems. In P. C. Debergh and R. H. Zimmerman, eds. *Micropropagation: Technology and application.* Dordrecht: Kluwer Acad. Pub. pp. 95–122.

243. Swartz, H. J., and J. T. Lindstrom. 1986. Small fruit and grape tissue culture from 1980 to 1985: Commercialization of the technique. In R. H. Zimmerman, R. J. Griesbach, F. A. Hammerschlag, and R. H. Lawson, eds. *Tissue culture as a plant production system for horticultural crops.* Dordrecht: Martinus Nijhoff Publishers. pp. 201–20.

244. Tanaka, K., Y. Kanno, S. Kudo, and M. Suzuki. 2000. Somatic embryogenesis and plant regeneration in chrysanthemum [*Dendranthema grandiflora* (Ramat.) Kitamura]. *Plant Cell Rpt.* 19:946–53.

245. Tanaka, M., D. T. T. Giang, and A. Murakami. 2005. Application of a novel disposable film culture system to photoautotrophic micropropagation of *Eucalyptus uro-grandis* (*urophylla x grandis*). *In Vitro Cell. Dev. Bio.—Plant* 41:173–80.

246. Tanaka, M., S. Nagae, S. Fukai, and M. Goi. 1992. Growth of tissue cultured *Spathiphyllum* on rockwool in a novel film culture vessel under high CO_2. *Acta Hort.* 314:19–46.

247. Tautorus, T. E., and D. I. Dunstan. 1995. Scale-up of embryogenic plant suspension cultures in bioreactors. In S. W. Jain, P. K. Gupta, and R. J. Newton, eds. *Somatic embryogenesis in woody plants: History, molecular and biochemical aspects and applications.* Dordrecht: Kluwer Acad. Pub. pp. 265–92.

248. Taylor, J., and L. A. Harrier. 2003. Beneficial influences of arbuscular mycorrhizal (AM) fungi on the micropropagation of woody and fruit trees. In S. M. Jain and K. Ishii, eds. *Micropropagation of woody trees and fruits.* Dordrecht: Kluwer Acad. Pub. pp. 129–49.

249. Teisson, C., and D. Alvard. 1995. A new concept of plant *in vitro* cultivation liquid medium: Temporary immersion. In M. Terzi, R. Cella, and

A. Falavigna, eds. *Current issues in plant molecular and cellular biology.* Dordrecht: Kluwer Acad. Pub. pp. 105–10.

250. Teixeira da Silva, J. A. 2003. Thin cell layer technology in ornamental plant micropropagation and biotechnology. *African J. Biotech.* 2:683–91.

251. Thomas, T. D., A. K. Bhatnagar, and S. S. Bhojwani. 2000. Production of triploid plants of mulberry (*Morus alba* L.) by endosperm culture. *Plant Cell Rpt.* 19:395–99.

252. Thomas, T. D., and R. Chaturvedi. 2008. Endosperm culture: a novel method for triploid plant production. *Plant Cell Tiss. Organ Cult.* 93:1–14.

253. Tisserat, B. 1991. Automated Systems. In Y. P. S. Bajaj, ed. *Biotechnology in agriculture and forestry, Vol. 17. High-tech and micropropagation I.* Berlin: Springer-Verlag. pp. 419–31.

254. Tisserat, B., E. B. Esan, and T. Murashige. 1979. Somatic embryogenesis in angiosperms. *Hort. Rev.* 1:1–78.

255. Tisserat, B., C. Herman, R. Silman, and R. J. Bothast. 1997. Using ultra-high carbon dioxide levels enhances plantlet growth *in vitro*. *HortTechnology* 7:282–89.

256. Toth, K. F., and M. C. Lacy. 1992. Micropropagation of celery (*Apium graveolens* var. Dulce). In Y. P. S. Bajaj, ed. *Biotechnology in agriculture and forestry, Vol. 19. High-tech and micropropagation III.* Berlin: Springer-Verlag.

257. Tran Thanh Van, M. 1973. *In vitro* control of *de novo* flower, bud, root and callus differentiation from excised epidermal tissues. *Nature* 245:44–5.

258. Tripepi, R. R. 1997. Adventitious shoot regeneration. In R. L. Geneve, J. E. Preece, and S. A. Merkle, eds. *Biotechnology of ornamentals*. Biotechnol. Agric. Series No. 16. Wallingford, UK: CAB International. pp. 45–72.

259. van Overbeek, J. 1942. Hormonal control of embryo and seedling. *Cold Spring Harbor Symp. Quantitative Biol.* 10:126–34.

260. Varma, A., and H. Schuuepp. 1995. Mycorrhization of the commercially important micropropagated plants. *Critical Rev. Biotechnol.* 15:313–28.

261. Vasil, I. K. 1985. Somatic embryogenesis and its consequences in the Gramineae. In R. R. Henke, K. W. Hughes, M. J. Constantin, and A. Hollaender, eds. *Tissue culture in forestry and agriculture.* New York: Plenum Press. pp. 31–47.

262. Vasil, I. K., and A. C. Hildebrandt. 1965. Differentiation of tobacco plants from single isolated cells in microcultures. *Science* 150:889–90.

263. Viéitez, F. J., and M. C. San-José. 1996. Adventitious shoot regeneration from *Fagus sylvatica* leaf explants *in vitro*. *In Vitro Cell. Dev. Bio.—Plant* 32:140–47.

264. Vieth, J., G. Laublin, C. Morisset, and M. Cappadocia. 1992. Multiplication *in vitro* of some iris plants from roots—histological aspects of somatic embryogenesis. *Can. J. Bot.* 70:1809–14.

265. Welander, M., L-H. Zhu, and X-Y. Li. 2007. Factors influencing conventional and semi-automated micropropagation. *Propagation Ornamental Plants* 7:103–11.

266. Wenzel, G., and B. Foroughi-Wehr. 1984. Anther culture of *Solanum tuberosum*. In I. K. Vasil, ed. *Cell culture and somatic cell genetics of plants, Vol. 1.* New York: Academic Press. pp. 293–301.

267. Wenzel, G., and B. Foroughi-Wehr. 1984. Anther culture of cereals and grasses. In I. K. Vasil, ed. *Cell culture and somatic cell genetics of plants, Vol. 1.* New York: Academic Press. pp. 31–127.

268. Wetzstein, H. Y., and H. E. Sommer. 1982. Leaf anatomy of tissue-cultured *Liquidambar styraciflua* (Hamamelidaceae) during acclimatization. *Amer. J. Bot.* 69:1579–86.

269. White, P. R. 1939. Potentially unlimited growth of excised plant callus in an artificial medium. *Amer. J. Bot.* 26:59–64.

270. White, P. R. 1963. *The cultivation of animal and plant cells.* 2nd ed. New York: Ronald Press.

271. White, P. R., and A. C. Braun. 1941. Crown-gall production by bacteria-free tumor tissues. *Science* 94:239–41.

272. Williams, E. G., and G. Maheswaran. 1986. Somatic embryogenesis: Factors influencing coordinated behaviour of cells as an embryonic group. *Ann. Bot.* 57:443–62.

273. Yantcheva, A., M. Vlahova, and A. Antanassov. 1998. Direct somatic embryogenesis and plant regeneration of carnation (*Dianthus caryophyllus* L.). *Plant Cell Rpt.* 18:148–53.

274. Yeung, E. C., T. A. Thorpe, and C. J. Jensen. 1981. *In vitro* fertilization and embryo culture. In T. A. Thorpe, ed. *Plant tissue culture: Methods and applications in agriculture.* New York: Academic Press. pp. 253–71.

275. Yusnita, S., R. L. Geneve, and S. T. Kester. 1990. Micropropagation of a white flowering form of eastern redbud. *J. Environ. Hort.* 8:177–79.

276. Zhang, C., C. M. Chi, and W. S. Hu. 1996. Application of image analysis to fed-cultures of somatic embryo development. *In Vitro Cell. Dev. Bio.—Plant* 32:190–98.

277. Zhang, S. B., and P. G. Lemaux. 2004. Molecular analysis of *in vitro* shoot organogenesis. *Critical Rev. Plant Sci.* 23:325–35.

278. Zhao, X. Y., Y. H. Su, Z. J. Cheng, and X. S. Zhang. 2008. Cell fate switch during *in vitro* plant organogenesis. *J. Integrative Plant Bio.* 50:816–24.

279. Zheng, Q., A. P. Dessai, C. S. Prakash, and Q. Zheng. 1996. Rapid and repetitive plant regeneration in sweet potato via somatic embryogenesis. *Plant Cell Rpt.* 15:381–85.

280. Zilis, M., D. Zwagerman, D. Lamberts, and L. Kurtz. 1979. Commercial propagation of herbaceous perennials by tissue culture. *Comb. Proc. Intl. Plant Prop. Soc.* 29:404–13.

281. Zilkah, A., E. Faingersh, A. Rotbaum, S. Spiegel, A. Stein, and M. Barba. 1995. Symptoms of prunus necrotic ringspot virus on micrografted *Prunus serrulata* cv. Shiorfugen shoot cultures. *Acta Hort.* 386:183–86.

282. Zimmerman, J. L. 1993. Somatic embryogenesis: A model for early development in higher plants. *Plant Cell* 5:1411–23.

283. Zimmerman, R. H. 1986. Propagation of fruit, nut and vegetable crops—overview. In R. H. Zimmerman, R. J. Griesbach, F. A. Hammerschlag, and R. H. Lawson, eds. *Tissue culture as a plant production system for horticultural crops.* Dordrecht: Martinus Nijhoff Publishers. pp. 183–200.

284. Ziv, M. 1986. *In vitro* hardening and acclimatization of tissue culture plants. In L. A. Withers and P. G. Alderson, eds. *Plant tissue culture and its agricultural applications.* London: Butterworth. pp. 187–96.

285. Ziv, M. 1991. Vitrification: Morphological and physiological disorders of *in vitro* plants. In P. C. Debergh and R. H. Zimmerman, eds. *Micropropagation: Technology and application.* Dordrecht: Kluwer Acad. Pub. pp. 45–70.

286. Ziv, M. 2000. Bioreactor technology for plant micropropagation. *Hort. Rev.* 24:1–30.

287. Ziv, M., A. H. Halevy, and R. Shilo. 1970. Organs and plantlets regeneration of *Gladiolus* through tissue culture. *Ann. Bot.* 34:671–76.

288. Ziv, M., and H. Lilien-Kipnis. 2000. Bud regeneration from inflorescence explants for rapid propagation of geophytes *in vitro*. *Plant Cell Rpt.* 19:845–50.

289. Zobayed, S. M. A., J. Armstrong, and W. Armstrong. 2001. Micropropagation of potato: Evaluation of closed, diffusive and forced ventilation on growth and tuberization. *Ann. Bot.* 87:53–9.

18
Techniques for Micropropagation

INTRODUCTION

Micropropagation is the production of new plants under the ultracontrolled environment within the culture vessel (i.e., test tube). Commercial micropropagation began in the United States in 1965 with orchid production. In the last quarter of the 20th Century, commercial micropropagation emerged as an important method for propagating horticultural plants. In 1997, it was estimated that the worldwide production of plants through micropropagation reached $15 billion dollars (49). In the United States, there are approximately 100 tissue-culture labs producing more than 100 million plants per year (16). However, it is still a small quantity compared with other forms of propagation, such as seeds or cuttings. Demand for micropropagated plants is high, but production is limited because of high labor costs. Interestingly, micropropagation in developed countries appears to have leveled off since the late 1980s primarily due to labor costs. In contrast, production in developing countries is increasing rapidly. For example, it is estimated that $5 million of micropropagated plants were produced in India in 1988. This total increased to $190 million by 1998 (49), a reflection of the lower cost of production and a demand for large quantities of disease-free propagules. The other interesting contrast is that in theUnited States, more than 80 percent of current production is for ornamental crops, while in developing countries the focus is on food, fiber, forestry, and medicinal crops. There is a trend for contract production of micropropagated plants from Asia and India for United States and European nurseries.

learning objectives

- Define the uses of micropropagation.
- Compare advantages and disadvantages of multiplying plants by micropropagation.
- Describe the procedure used for micropropagation.
- Describe the general laboratory facilities used for micropropagation.

USES FOR MICROPROPAGATION

Micropropagation has become an important part of commercial propagation for many plants (10, 14, 54). The advantages of micropropagation as a system have been reviewed by many authors (7, 15, 22, 24, 37, 46) and can be summarized as:

1. Mass propagation of specific clones.
2. Production of pathogen-free plants.
3. Clonal propagation of parental stock (inbred lines) for hybrid seed production.
4. Year-round nursery production.
5. Germplasm preservation.

Mass Propagation of Specific Clones

The controlled aspects of micropropagation permit the rapid propagation of individuals from a single plant. Multiplication rates can be very high, since plants in culture can theoretically be multiplied at an exponential rate by consecutive subculturing (e.g., one month apart). Although such theoretical rates are not usually maintained in practice, the actual rates that can be attained under proper management are impressive. A fourfold multiplication rate in cultures subcultured every 4 weeks will theoretically produce over one million plants in 10 months. Many commercial laboratories have the capacity to produce millions of micropropagated plants a year (35). Table 18–1 shows the approximate breakdown of the types of plants commercially micropropagated in the United States in 1996. Although the numbers may have changed since 1996, the relative proportion of micropropagated plants is similar today. However, because of the lower labor costs, there are many more micropropagated plants being imported into the United States at the present time compared to 1996.

Commercial micropropagation is particularly useful in the following situations:

1. Where the propagation rate is slow by conventional means.
2. New cultivars with high market demand.
3. Cultivars with high market value.
4. Plants that are difficult to propagate clonally.
5. Conservation of endangered species.

Plants Where Propagation Rate Is Relatively Slow For some species, including orchids, bulbs, ferns, many foliage plants, some herbaceous perennials, and palm, the only means for clonal propagation (other than micropropagation) is crown division or offshoots (see Chapter 15). This relatively slow process may not meet market demand. A good example is the herbaceous perennial hosta. Hosta plants are traditionally propagated by division, but division yields fewer than ten plants from the original mother plant per year. Micropropagated hosta can yield a similar multiplication rate every month!

New Cultivars with High Market Demand In some cases, such as bulbs, daylily, and hosta, there are commercial demands that require getting a new cultivar to market in as short a time as possible. Micropropagation is used until the number of stock plants of the new cultivar is sufficient for release, or can support other forms of propagation, like division or layering. For example, new herbaceous perennial cultivars (like *Heuchera* and *Pulmonaria*) can be very popular, and the nursery introducing the new cultivar can use micropropagation to take advantage of market demand while the novel plants still command a high introductory price.

Cultivars with High Market Value Micropropagation is more costly than other methods of clonal propagation. However, some crops have a high market value that makes micropropagation a viable alternative to conventional methods. In these cases, propagation costs are minimal compared to the value of the finished crop, and micropropagation may be the propagation method of choice for reasons other than cost. Examples include specialty woody plants such as valuable rootstocks (20), *Rhododendrons*, and specialty perennials such as daylily.

Plants That Are Difficult to Propagate Clonally Traditionally, grafting has been the only method to clonally propagate many woody plants that fail to root from cuttings (see Chapter 9). Micropropagation offers

Table 18–1
Yearly Production (1996) of Plants in the United States by Micropropagation

Crop type	Numbers in thousands	Percentage of total
Foliage plants	63,695	52.7
Trees and shrubs	15,294	12.7
Vegetables	12,862	10.6
Greenhouse crops	11,297	9.3
Herbaceous perennials	9,448	7.8
Fruits	3,721	3.1
Miscellaneous	4,545	3.8

Source: R. Zimmerman, USDA, Beltsville, MD.

a viable alternative for propagation of these plants on their own roots. An excellent example is the many cultivars of mountain laurel (*Kalmia latifolia*) that have only become available to the nursery trade because of micropropagation.

Conservation of Endangered Species Micropropagation can provide a rapid method for multiplication of endangered or threatened species, when only a few plants may be available as stock plants, or the collection of plants and seeds from wild plants needs to be minimized. Micropropagation can also be an alternative to collecting plants from natural habitats to supply commercial markets. For instance, over-harvesting bulbs of spring snowflake (*Leucojum*) from its natural populations for commercial sale has seriously reduced populations, and has been a driving force for the micropropagation of these plants. Several organizations, like the Conservation and Reproduction of Endangered Wildlife (CREW) at the Cincinnati Zoo and Botanical Garden and the Center for Plant Conservation at the Missouri Botanical Garden, are involved in developing and supporting efforts to conserve native plants, including the use of micropropagation.

Production of Pathogen-Free Plants

Production of propagation material free of fungal, bacterial, and systemic viruses and virus-like pathogens has become an essential aspect of propagation (see Chapter 16). Micropropagation provides a method to remove pathogens from a clone or breeding stock. Meristem culture is a routine method to re-initiate stock plants from vegetatively propagated plants as well as inbred parents used for breeding.

Propagators should not automatically assume systemic pathogens are absent unless culture indexes for bacteria, fungi, or viruses are made. Specific aspects of pathogen control are as follows:

1. Maintain germplasm and source material in a **pathogen-free** condition.
2. Include micropropagation as the initial step in production for **Foundation Clone Systems** as discussed in Chapter 16 (35, 39, 50).
3. Allow **movement of germplasm** materials across quarantine barriers (35).
4. **Facilitate the distribution** of commercial material in international trade. Trade has allowed the importation of specific plants between countries (37). For example, specialty producers of ferns in Israel and Australia supply many of the fern liners to United States growers.

Clonal Propagation of Parental Stocks for Hybrid Seed Production

This procedure has been used for hybrid seed production of many annuals such as petunia, impatiens, marigolds, asparagus, tomato, cucurbits, and broccoli (10, 31), where the inbred parental lines are multiplied by micropropagation.

Provide Year-Round Nursery Production

Most nursery operations are seasonal. Micropropagation has the potential for continuous year-round operation with production scheduled according to market demands. High-volume production requires high-volume distribution and the facilities to stockpile items. Combining production and cold storage facilities makes it possible to hold material for peak marketing periods (35).

Germplasm Preservation

The major method of germplasm preservation is seeds (see Chapter 6), but seed storage requires a relatively large facility with specialized storage conditions, and has not been useful for recalcitrant seeds that do not tolerate dry storage conditions. Also, seeds are not an option for preserving clonal germplasm. These reasons are what make preservation of vegetative tissue as explants an attractive alternative (19). **Cryopreservation,** often using antifreeze materials as an aid (35), has been used to ultrafreeze vegetative tissue in a fashion similar to that of seeds (see Chapter 6). Frozen material can be thawed and subsequently used to micropropagate the original plants. Proper freezing and storage conditions, as well as evaluation of stored material for genetic purity, are active areas of current research.

> **cryopreservation**
> The storage of seeds or vegetative organs at ultralow temperature, most commonly in liquid nitrogen at −196°C.

DISADVANTAGES OF MICROPROPAGATION

Micropropagation on a commercial scale has particular characteristics that may create problems that could limit use (30, 37), including:

1. A requirement for expensive and sophisticated facilities, trained personnel, and specialized techniques.
2. High labor costs.

3. A high-volume, more or less continuous distribution system, or adequate storage facilities to stockpile products, is required.
4. Pathogen contamination or insect infestation can cause high losses in a short time.
5. Variability and production of off-type individuals can be a risk in the products emerging from micropropagation. Careful roguing, field testing of new products, and continuing research and development are essential to decrease this risk (see Chapter 16).
6. Economics and marketing are key to the success of commercial operations. More companies fail for economic reasons than because of an inability to produce micropropagated plants.

GENERAL LABORATORY FACILITIES AND PROCEDURES

Facilities and Equipment

Facilities for micropropagation may be placed into three categories as determined by their scope, size, sophistication, and cost. These categories are: (a) research laboratories where precise work requires highly sophisticated equipment, (b) large commercial propagation facilities where several million plants can be mass-produced annually (3, 4, 5, 14, 18, 35), and (c) limited facilities for small research laboratories, individual nurseries, or hobbyists where a relatively small volume of material is handled (17, 47).

Regardless of size, the facility should include separate **preparation, transfer,** and **growing areas.** In addition, there may be a need for service areas, office, and cold storage. The facility should be separated from the regular nursery and greenhouse production area and entry should be restricted to avoid introducing contaminants into the culture rooms. Floors, benches, and tabletops should be kept scrupulously clean. In some commercial labs, technicians use clean lab coats, foot coverings, and hairnets.

There is much variation in the kinds of facilities and opportunities for cost cutting, but the basic principles of *in vitro* culture must be followed (8, 14, 17, 22, 47). In any case, careful consideration must be given to the costs and labor required (9, 35, 47).

Preparation Area The preparation area is where media are made and dispensed into containers. It has three basic functions: (a) cleaning containers, (b) preparation and sterilization of media, and (c) storage of containers and supplies (Figs. 18–1, 18–2, 18–3, and 18–4, page 704).

An efficient method of washing is required, either by hand or by machine. Normal washing is followed by rinsing in distilled or deionized water. A sink, running water, and electrical or gas outlets for heating are necessary; air or vacuum outlets are often useful. Table surfaces should be made of a material that can be cleaned easily.

The following equipment items are used in the preparation area (Figs. 18–2, 18–3, and 18–4):

1. Refrigerator to store chemicals, stock solutions, and small batches of media.
2. Scales or analytical balances; preferably top-loading.
3. Autoclave capable of reaching 120°C (250°F). A household pressure cooker can be used to sterilize small batches of media.
4. A pH meter. Indicator papers can be used as a substitute in less precise work.

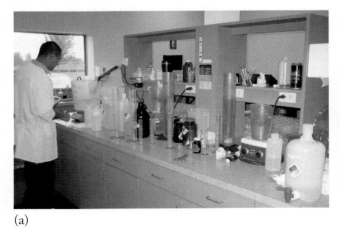

(a)

(b)

Figure 18–1
The preparation area is a laboratory wet lab including areas for weighing chemicals, preparing media, cleaning non-disposable glassware, and storage. (a) A large commercial lab. (b) A smaller lab that has converted a kitchen into the preparation area.

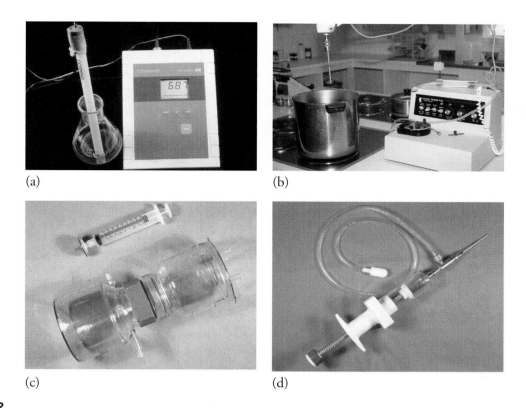

Figure 18–2
Common features found in the tissue culture preparation area. (a) Every lab must have a pH meter. (b) A peristaltic pump can significantly reduce the time to dispense medium. A mixer incorporates agar into the medium uniformly. (c) Filter apparatus should be available to sterilize chemicals that cannot be autoclaved. These may be disposable, like the presterilized 25-ml syringe and Acrodisc syringe tip filter, or permanent autoclavable units with replaceable filters to sterilize larger quantities of fluids. The permanent unit must use a vacuum to move fluid through the disposable paper filter. (d) For smaller operations that cannot afford a peristaltic pump, small plunger dispensers can be effective labor savers.

5. Gas or heating plate.
6. Stirrer and mixing device.
7. Filters for sterilizing non-autoclavable ingredients (particularly in research laboratories). Chemicals that are altered by autoclaving must be filter-sterilized. There are commercial devices available that come equipped with a cellulose acetate or cellulose nitrate membrane filter that is porous enough to allow liquid to pass but nonporous enough to prevent the movement of organisms such as bacteria. The devices require pressure from a syringe plunger or a gentle vacuum to pass solutions through the filter.

Figure 18–3
Autoclaves are used to sterilize media, containers, and tools. (a) A large, standard laboratory autoclave has the capacity for a large tissue culture lab. (b) Smaller units are cheaper and more suited to smaller labs.

(a)

(b)

Figure 18–4
The preparation area should be large enough to accommodate (a) media preparation and dispensing, and (b) storage of prepared media in containers.

8. Equipment to purify water, a glass still, an ion exchanger, reverse osmosis system, or a combination of equipment. In smaller operations, distilled water may be purchased and stored in plastic containers.
9. A vacuum pump or an ultrasonic cleaner. These items can help facilitate the decontamination process.
10. A media dispenser (often using peristaltic pumps) is valuable as a labor-saving device.
11. Storage for flasks, bottles, and other supplies.

Transfer Area The transfer area is the place where explants are inserted into culture, and where transfers or subcultures to fresh media take place. The key requirement is that the environment in the transfer area must be **sterile and free from any contaminating organisms.** Transfer is most conveniently done in an open-sided **laminar airflow hood** (Fig. 18–5) where filtered air is passed from the rear of the hood outward on a positive pressure gradient. Air passes through a prefilter to remove dust, and

> **laminar flow hood**
> A cabinet that filters air through a HEPA filter to eliminate fungal and bacterial spores from tissue culture work spaces.

(a)

(b)

Figure 18–5
The transfer area contains laminar flow transfer hoods used for all sterile operations. (a) A clean area with laminar flow hoods. Containers with media are wheeled into the transfer area for efficient subculturing. (b) During peak times of operation, hood space may need to be shared.

is drawn through a **high-efficiency particulate air (HEPA) filter** that removes microbial spores. Turn on the hood at least 30 minutes before use. With such equipment, transfers may be carried out in a room with other activities. Less expensive alternatives, such as an enclosed walk-in room or enclosed transfer boxes, can be used if they are carefully sterilized before use.

Ultraviolet (UV) germicidal lamps can be used to sterilize the interior of the transfer chamber. The lamps are turned on about 2 hours prior to using the chamber but must be turned off during operation. The light should be directed inward, not toward the worker's face, because UV may adversely affect the eyes.

Growing Area Cultures should be grown in a separate, lighted facility where both day length and light irradiance can be controlled, and where specific temperature regimes can be provided (Fig. 18–6). If different kinds of plants are to be propagated, it is useful to have several rooms, each programmed to meet the temperature and light needs of specific kinds of plants. Light requirements vary from about 30 to 80 $\mu mol \cdot m^{-2} \cdot sec^{-1}$ photosynthetically active radiation (PAR). This is accomplished with cool-white or "grow-light" fluorescent lamps above the cultures. A 16/8 (light/dark) period is most common. Temperatures of 21 to 27°C (70 to 81°F) are generally adequate, although some kinds of plants may need lower temperatures. The relative humidity in the growing area at this temperature range is about 30 to 50 percent, but depends on the type of closure. Dehydration of the medium and increase in salt concentration can occur during long culture periods if the relative humidity is too low. If humidity is too high, contamination may develop. Various kinds of rolling drum culture devices or shakers are sometimes used, which provide the added benefit of aeration in liquid culture systems.

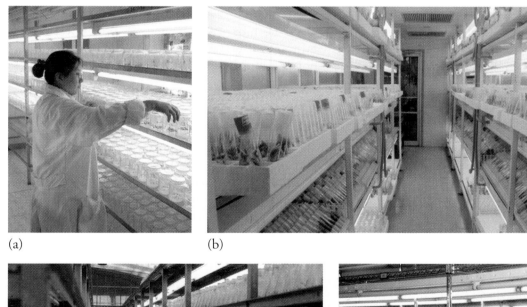

Figure 18–6
The culture growing area has lighted stainless steel racks that hold the culture vessels. (a) Technician checking on culture progress. (b) Racks containing test tubes. (c) Racks with orchids in Erlenmeyer flasks. (d) Cultures in Magenta jars.

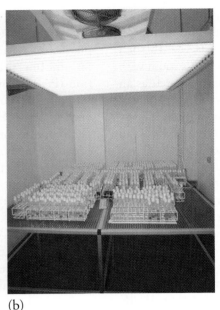

Figure 18–7
Growth rooms can be used for specialized production systems, such as photoautotrophic cultures where the plants are produced on sugar-free medium, and elevated carbon dioxide. (a) Growth rooms are large insolated rooms that minimize heat loss. Heat from the lamps may be the only heat source. (b) Light banks are adjustable to provide optimal light irradiance.

Growth rooms (Fig. 18–7) can be used for specialized production systems, such as photoautotrophic cultures where the plants are produced on sugar-free medium, high light (>200 $\mu mol \cdot m^{-2} \cdot sec^{-1}$ PAR), and elevated carbon dioxide (21). Similar systems have been proposed for rooting and acclimatizing microcuttings (34). Growth rooms are also constructed within standard greenhouse space for rooting microcuttings (Fig. 18–8), but must be modified for temperature and humidity control using enclosed mist or fog systems.

Containers for Growing Cultures Test tubes, Erlenmeyer flasks, Petri dishes, glass jars and plastic containers of various sizes are routinely used in research laboratories (Fig. 18–9). Glassware may be Pyrex glass but are often less-expensive soda glass and borosilicate, all of which can be sterilized and flamed during transfer operations. Various kinds of commercial plastic containers are also available that are less expensive, less likely to break, and often disposable.

Petri dishes can be made of glass or plastic and be reusable or disposable, respectively. Less-expensive larger glass containers commonly used in the past for home canning or commercial canning (Ball or Mason jars) are widely used for the later stages of development in large-scale propagation. Most of these items are available from commercial or food industry supply houses.

Various kinds of closures are necessary for the containers (Fig. 18–9). Nonabsorbent cotton plugs have been used in research laboratories, but are inconvenient in large commercial operations; metal or plastic covers or polyurethane or polypropylene plugs are more convenient. A second cover of aluminum foil or various plastic films is desirable for holding moisture and reducing infection, but these materials must still allow exchange of CO_2, O_2, and ethylene gases (37).

Parafilm (paraffined paper in a sterile roll) is used to seal Petri dishes and other containers. Parafilm or plastic films can also help reduce the spread of culture mites or thrips (Fig. 18–10, page 708), infestations of which can be devastating in a commercial or research lab. Mites move from one culture vessel to another, spreading contamination. Mite "trails" are evident as spots of fungal contamination on the surface of the agar. Thrips can enter the lab through unscreened windows or doorways. They can also be present on the original explant. A magnifying glass may be required to see the actual mites or thrips. To eliminate pests, destroy infested cultures in the autoclave and disinfect the culture room area.

Media Preparation

Ingredients Ingredients of the culture medium vary with the kind of plant and the propagation stage at which one is working. In general, certain standard mixtures are used for most plants, but exact formulations may need to be established by testing. Media can be made from the pure chemicals or purchased as commercial premixed culture media (Fig. 18–11, page 708). Empirical trials may be necessary to test available combinations of ingredients when dealing with a new kind of plant. A food colorant can be added to the medium

Figure 18–8
Rooting and acclimatization areas are usually modified greenhouse space where the humidity can be controlled. (a) A polyethylene tent with a mist system provides an enclosed space. (b) A growth room built inside a greenhouse with mist. (c) A growth room with insolated walls within a greenhouse with humidity controlled using (d) fog.

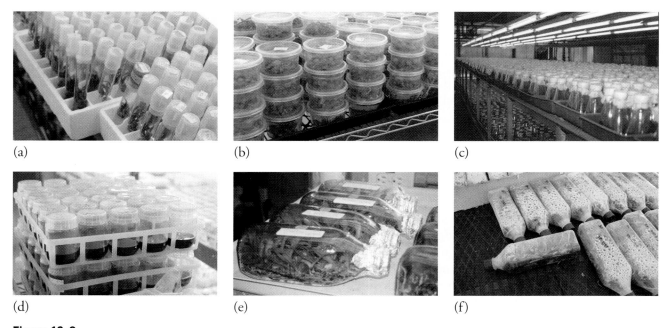

Figure 18–9
Cultures can be grown in: (a) Test tubes, (b) tubs, (c) flasks, (d) food jars, and (e and f) bottles. Container closures include plastic caps, cotton plugs, and bottle stoppers. Caps allow air to enter and leave the tissue culture vessel, but prevent contamination.

Figure 18–10
Culture vessels may be wrapped in plastic film to prevent cultures from drying and reduce the potential for insect and mite infestations.

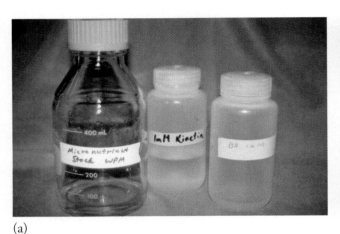
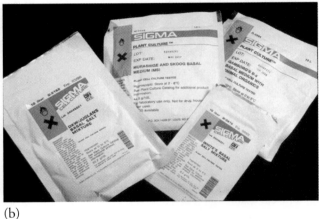

Figure 18–11
Media are usually prepared from: (a) Stock solutions or (b) premeasured commercial mixes.

to make it easy to tell each medium type apart (Fig. 18–12). Medium ingredients can be grouped into the following categories:

1. Inorganic salts
2. Organic compounds
3. Gelling agents

Inorganic Salts (Groups A to E, Table 18–2) Inorganic salts provide the macroelements (nitrogen, phosphorus, potassium, calcium, and magnesium) and microelements (boron, cobalt, copper, manganese, iodine, iron, and zinc). These salts can be made up in **stock solutions** by group and stored in a refrigerator. A stock solution is a concentrated solution (usually 100 times the concentration found in the final medium) from which portions are removed to make the final medium. Stock solutions save time because dry salts need only be weighed once. The reason for preparing different stock solutions is that certain kinds of chemicals, when mixed together, will precipitate and not remain in solution. Stock solutions should be stored in the refrigerator (not frozen) and are good for several months. Cloudy stock solution should be discarded.

Inorganic substances and many of the organic materials can be maintained under refrigeration in stock solutions of 100 times the required concentration (Fig. 18–11). Ten mL of this stock solution can be

stock solutions The concentrated solutions that are used to store inorganic and organic ingredients for tissue culture media preparation.

(a)

(b)

Figure 18–12
Media may be color coded with different colored caps or by placing food coloring in the medium. (a) A storage stock pile of media ready to be moved to the culture room for subculturing. (b) A recently subcultured group of Magenta jars with a color-coded medium.

Table 18–2
Stock Solutions (grams/liter) Used in Preparing Various Media for Micropropagation[a]

Group	Compound	Murashige and Skoog (MS)	Woody plant medium (WPM)	Anderson (AND)	Gamborg B5
A	NH_4NO_3	165.00	40.00	40.00	—
	KNO_3	190.00	—	48.00	250.00
	$Ca(NO_3 4H_2O)$	—	38.60	—	—
B	K_2SO_4	—	99.00	—	—
	$MgSO_4$	18.10	18.10	18.10	12.2
	$MnSO_4 \cdot H_2O$	1.69	2.23	1.69	1.00
	$ZnSO_4 \cdot 7 H_2O$	0.86	0.86	0.86	0.20
	$CuSO_4 \cdot 5 H_2O$	0.0025	0.0025	0.0025	0.0025
	NH_4SO_4	—	—	—	13.4
C	$CaCl_4 \cdot 2H_2O$	44.00	9.6	44.00	15.00
	KI	0.083	—	0.03	0.075
	$CoCl_2 \cdot 6 H_2O$	0.0025	—	0.0025	0.0025
D	KH_2PO_4	17.00	17.00	—	—
	H_3BO_3	0.62	0.62	0.62	0.30
	$Na_2MoO_4 \cdot 2 H_2O$	0.025	0.025	0.025	0.025
	$NaH_2PO_4 \cdot H_2O$	b	—	38.00	15.00
E	$FeSO_4 \cdot 7 H_2O$	2.78	2.78	5.57	2.78
	$Na_2 \cdot EDTA$	3.73	3.73	7.45	3.73
F	Thiamin · HCl	0.10	0.10	0.04	1.00
	Nicotinic acid	0.05	0.05	—	0.10
	Pyridoxine · HCl	0.05	0.05	—	0.10
	Glycine	0.20	0.20	—	—
G	Myo-inositol	10.00	10.00	10.00[c]	10.00

[a] These are 100 × the final solution. Use 10 ml of each stock solution for preparing 1 liter of the culture medium.
[b] Commonly added as a supplement directly to the culture medium at 85 to 255 mg/liter.
[c] Adenine sulfate added, 80 mg/liter.

added to each liter of medium to provide the required concentration. From the list in Table 18–2, the following materials can be combined without forming precipitates: (a) nitrates, (b) sulfates, (c) halides (i.e., chlorides and iodides), (d) phosphates, borates, and molybdates. Iron can be purchased as a chelated iron solution, or it can be prepared separately by combining the substances in group E, or alternatively supplied as inorganic iron—$FeCl_3 6H_2O$ (1 mg/l), $FeSO_4$ (2.5 mg/l), or Fe tartrate (10 mg/l). Iron solutions should be stored in a dark bottle.

There are many different formulations reported in the literature with detailed descriptions (14, 22, 37). In this chapter, we describe four representative media. The Murashige-Skoog (MS) and Linsmaier-Skoog media (23, 33) have been used extensively for a range of culture types and species, particularly herbaceous plants. For woody plants, a dilution of three to ten times inorganic salts or a shift to other inorganic media is desirable. The Woody Plant Medium (WPM) (25) was developed for woody plants and the Anderson (AND) medium was developed for rhododendrons (1, 22). The Gamborg B5 medium has been widely used for cell and tissue cultures (12).

Organic Compounds (Group F and G, Table 18–2) Various organic components are added to the medium including a **carbohydrate** as an energy source (usually sugar), **vitamins** to support plant enzyme reactions, **hormones** to direct organogenesis, and **miscellaneous materials** that facilitate plant growth and development.

Carbohydrates. **Sucrose** at 2 to 4 percent is used for most cultures, but concentrations as high as 12 percent might be used in some cases, as for young embryos. Glucose has sometimes been used for monocots and selected dicots (like strawberry), and fructose, maltose, and starch have been used occasionally; these materials are added at the time of making the medium.

Vitamins. Thiamin (0.1 to 0.5 mg/L) is almost always considered essential, and nicotinic acid (0.5 mg/L) and pyridoxine (0.5 mg/L) are usually added. Inositol at 100 mg/L is beneficial in many cultures and is usually added routinely. Pantothenic acid (0.1 mg/L) and biotin (0.1 mg/L) are sometimes beneficial. All of these substances are soluble in water and should be prepared as stock solutions, ready for dilution at 100 times the final concentration. Stock solutions should be stored in a refrigerator.

Hormones and Growth Regulators. The two most important hormones used to control organ and tissue development are auxins and cytokinins. They are usually stable and added prior to autoclaving. Gibberellins have sometimes been used to promote shoot elongation, but should be filter sterilized and not autoclaved.

Auxins. The natural auxin, indole-3-acetic acid (IAA), is usually used at concentrations of 1 to 50 mg/L. The other auxins are usually more stable, including α-naphthaleneacetic acid (NAA) (0.1 to 10 mg/L), indole-3-butyric acid (IBA) (0.1 to 10 mg/L), and 2, 4-dichlorophenoxyacetic acid (2,4-D) (0.05 to 0.5 mg/L).

Cytokinins. The cytokinins used for micropropagation include N^6-benzyladenine (BA), kinetin, N^6-isopentenyl-adenine (2iP), and zeatin. Of these, BA and 2iP are most commonly used because of cost and efficacy for shoot promotion, at a range of 0.01 to 10 mg/L. Thidiazuron and N-2-chloro-4-pyridyl N-phenylurea (CPPU) (29) have cytokinin activity and are often used in combination with traditional cytokinins (like BA) but at one-tenth the concentration. Adenine sulfate also has cytokinin activity and is added to 40 to 120 mg/L.

Gibberellins. Gibberellic acid (GA_3) is not considered essential for micropropagation but is sometimes used to promote shoot elongation in difficult plants.

All of these materials should be prepared in advance and maintained as stock solutions in a refrigerator at near-freezing temperature. Stock solutions of auxins and other organic materials deteriorate with time. Some organic compounds are relatively insoluble in water. A small quantity of organic solvent (not more than 0.5 percent of the final medium) is effective in dissolving most organic substances. Cytokinins are weak bases and can be dissolved in a dilute acid, then diluted to a final volume with water. Auxins are weak acids and can be dissolved in a dilute base or in alcohol. Use 0.3 mL of 1N HCl for cytokinin and NaOH (for auxin) for each 10 mg of the compound.

Miscellaneous. Additional organic materials may be added to the medium to prevent browning of the tissue or to effect shoot development. Citric acid (150 mg/L) or ascorbic acid (10 mg/L) can be used as an antibrowning agent (antioxidant). These materials should be filter-sterilized since autoclaving causes decomposition. Organic acids may be incorporated in the medium or used in washing steps prior to culture. Fine-grade activated charcoal, preferably prewashed, is used in some cases (0.1 to 1 percent) to adsorb and counteract inhibiting substances released by some tissues. Activated charcoal can reduce the effectiveness of growth regulators.

Various materials of unknown composition have been used to establish cultures when known substances fail to do so. Protein hydrolysates from casein or other proteins (30 to 3,000 mg/L) are sometimes helpful, primarily for providing organic nitrogen and amino acids. Coconut milk (endosperm of either green or ripe coconuts, 10 to 20 percent by volume) has been used but must be filter-sterilized for some uses. Malt (500 mg/L), yeast extract (50 to 5,000 mg/L), and such substances as tomato juice (30 percent v/v) and orange juice (30 percent v/v) also have been used; these substances are dissolved in water and added during preparation of the medium. Most of these materials are available from commercial sources.

Gelling Agents and Supports The most common form of tissue culture uses agar mixed with the nutrient medium to form a semi-solid base to support the explant. However, other systems are available that use agar substitutes, various fiber supports (like paper or cotton), or agitated liquid medium.

Agar. Agar is obtained from certain species of red algae and is commercially available in a powdered form. Its value in culture systems is (a) ability to melt when heated, (b) ability to change to a semisolid gel at room temperatures, and (c) relative biological inertness. It may contain certain impurities, notably salts, in its natural state, but most agars are prewashed or purified (USP grade). Quality may vary among commercial producers and can impact culture performance and water potential of the medium (2, 41). Agar performance may be affected by concentration and by pH. Usually a pH of 5.0 to 6.0 is obtained, but the pH of the medium may need to be adjusted through the addition of acids or bases. The agar may fail to solidify as the pH is lowered (4.5 or less).

The lowest concentration of agar that will support the explants is best, which is about 0.5 to 0.6 percent, a level that allows good contact between explant and medium and adequate movement of nutrients. At lower concentrations the explant may sink into the agar and aeration is impaired. As the concentration increases (up to 1.0 to 1.2 grams per liter), growth may be depressed because of higher osmotic pressure (more negative water potential).

Agar Substitutes. Gelrite (also called gellan gum and phytogel) is a polysaccharide derived from *Pseudomonas* bacteria (14, 22, 37) that makes a very clear gel upon heating in combination with cations (e.g., magnesium, calcium, potassium, and sodium for gelation). Gelrite requires a concentration less than agar but is often used in combination with agar, in a ratio such as 3:1 Gelrite to agar, with the most effective combination varying with the brand of agar (22). Gelrite should be avoided for species susceptible to hyperhydricity (vitrification).

Liquid. Nutrient solutions without agar have an advantage for some plants, but some type of support is required to keep explants and cultures from sinking, or a shaker or rotating drum must be used to provide aeration (Fig. 18–13). Filter paper bridges may be inserted into the liquid to support the culture with the paper acting as a wick. Plastic fabric supports and membrane rafts are available commercially that allow transfer of the liquid medium to the explant without submerging it in the liquid medium. Other solid support materials include cotton, vermiculite, and rockwool. Liquid media are useful for some plant species whose explants exude toxic substances from the cut surfaces because the liquid medium can be exchanged without reculturing.

(a)

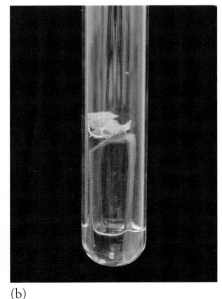
(b)

Figure 18–13
(a) Liquid cultures include those where explants or callus are submerged in the medium and agitated on a rotary shaker or (b) where the culture is supported on a filter paper bridge that wicks the liquid up to the culture.

Preparation of the Medium High water quality (i.e., low salts and/or low amounts of organic or chemical impurities) is important for growing cultures. Single distilled water may be adequate in commercial micropropagation, but it may be desirable to double distill the water or to pass it through deionizers and equipment to further purify it.

To prepare media, use a large Pyrex flask or beaker, and add a portion (one-half to two-thirds) of the final volume of water. Sugar may be added at this time or not until the end. Agitation helps dissolve the sugar. Add by pipette or a graduated cylinder the proper amounts of each ingredient from stock solutions. Agar is added next, and then the solution is made up to the prescribed volume by adding water. Adjust to the desired pH (usually about 5.7) with drop-by-drop additions of 1 N HCl or 1 N NaOH using a pH meter (or pH indicator strips) to measure.

Heating to melt the agar may be done in an autoclave at 121°C (218°F) for 3 to 7 minutes or on a hot plate, stirring continuously to prevent the agar from settling and burning. The hot solution is dispensed into containers and sterilized in an autoclave for 20 minutes at 121°C. The greater the volume of media to be sterilized, the longer the sterilization time (6).

Substances that are unstable during autoclaving must be filter-sterilized. The appropriate concentration of the material is added to the medium after autoclaving when the medium has cooled to 35 to 40°C (95 to 104°F) before dispensing into the final sterilized containers. Occasional agitation of the solution while dispensing is necessary to ensure even distribution of medium components into each container. The medium can be dispensed by pouring (Fig. 18–4a), using a metered pipette (Fig. 18–2d), or with a **peristaltic pump** (Fig. 18–2b). The medium should be used within 2 to 4 weeks, and refrigerated if kept longer.

> **peristaltic pumps** The equipment that is available from tissue culture supply companies, which make it easy to dispense media into tissue culture vessels.

MICROPROPAGATION PROCEDURES

As discussed in Chapter 17, there are basically four stages to the micropropagation process (Fig. 18–14):

- Stage I. Establishment
- Stage II. Shoot multiplication
- Stage III. Root formation
- Stage IV. Acclimatization

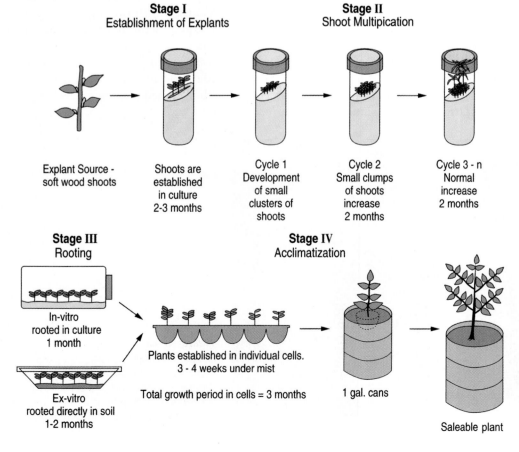

Figure 18–14
Schematic representation of micropropagation system applied to Rhododendron.
Courtesy of W. Anderson.

Common Terms Related to Micropropagation

Explant A plant part or tissue used to initiate the tissue culture process.

Subculture The act of subdividing the developing explants into smaller pieces and moving them to a new medium as occurs in the multiplication stage (Stage II) of micropropagation.

Transfer Moving the entire culture to a new medium. There is no subdivision, as there is in subculturing.

Microshoots Shoots developed during tissue culture.

Microcuttings Single microshoots moved to a medium to induce rooting. Microcuttings can be rooted either *in vitro* or *ex vitro*.

In vitro Growth and development in the tissue culture environment.

Ex vitro Growth and development outside the tissue culture environment.

Plantlets A plant derived from micropropagation that has both shoots and roots.

STAGE I—ESTABLISHMENT

The function of this stage is to disinfest the explant, establish it in culture, and then stabilize the explants for multiple shoot development.

Preparation for Establishing Cultures

Handling Stock Plants A principal consideration for handling stock plants is reducing the potential for contamination by fungi, viruses, and other pathogens. Pathogens are not automatically eliminated by *in vitro* technique unless this goal is built into the system. Selected "virus-free" or pathogen-indexed stock plants should be used. Otherwise special techniques, such as meristem-tip culture, heat treatment, and various culture-indexing tests, may be incorporated into the procedure.

Choice of Explant The kinds of explant and where and how they are collected varies with the purpose of the culture, the species, and often the cultivar (see Chapter 17). Shoot tips should be collected from plants in relatively active growth (Fig. 18–15). The size of the explant may vary from as small as a 1 to 5 mm long meristem tip for meristem culture to a piece of shoot several centimeters long. A single node bearing a lateral bud can also be used. For woody plants, a shoot tip from a dormant (but not resting) bud may be utilized but is often difficult to disinfest. Removal of the bud scales, plus cutting away the leaf scar, may reveal sterile tissue.

For woody plants, it may be helpful to collect branches in early spring before growth has begun and to bring the cut branches into a growth chamber or greenhouse to force growth (Fig. 18–15). A florist's cut-flower preservative in the forcing water can

(a)

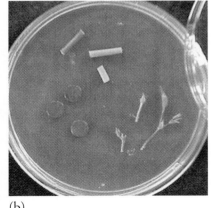
(b)

(c)

Figure 18–15
Some researchers suggest that there is a Stage 0 that precedes Stage I. It is the time when the donor plant is manipulated and the explant type selected. (a) Cut branches can be brought into the growth chamber or greenhouse to force shoot growth. These shoots are usually easier to disinfest, and growth regulators added to the solution may enhance later establishment in Stage I. (b) Explants can include leaf discs, internode stem sections, or single or multi-node explants. (c) Shoot tips being used as the initial explant in Stage I.

improve growth, and chemical additions of growth regulators can increase bud break (40). Explants collected from the expanding new growth are easier to disinfest than those collected directly from new growth outdoors.

Pieces of leaves with veins present, bulb scales, flower stems, and cotyledons also may be used for obtaining explants, as described in Chapter 17 for adventitious shoot initiation.

Disinfestation Disinfestation is the process of removing contaminants from the surface of the explant. A typical procedure would be to cut the plant material into small pieces and wash them in running tap water. For materials that are difficult to disinfest, a quick dip in alcohol may be helpful. Wrapping the shoots with small squares of sterile gauze will hold them intact during the treatment. In a laminar flow hood, the plant parts are placed into the disinfesting solution. The most common disinfesting solution contains 10 to 15 percent bleach (sodium or calcium hypochlorite) with a few drops of detergent (like Tween 20 or Alconox). Mercuric chloride ($HgCl_2$) and hydrogen peroxide (H_2O_2) are also used to disinfest tissue. $HgCl_2$ at 10 to 25 percent is usually more effective for heavily infested material taken from outdoor donor plants, but it is a toxic heavy metal that should be handled cautiously and disposed of properly. H_2O_2 (15 percent) is not commonly used, but can be effective in plants with hairy parts (51). The plant material is agitated in this solution for 10 to 15 minutes followed by three rinses in sterile water (Fig. 18–16). For very difficult to clean tissue (like roots or underground stems) antimicrobial or antibiotic agents may be necessary.

Microorganisms occur primarily as spores resting on surfaces of tables, hands, arms, clothing, and various objects which settle out of the air or are blown with the dust on air currents (11). Aseptic transfer procedures are done in a laminar flow cabinet equipped with a HEPA filter with a positive air pressure blowing outward from the rear of the hood to prevent movement of airborne spores into the chamber (Fig. 18–5). The sides and surfaces of the chamber should be washed with a disinfestant such as sodium or calcium hypochlorite or 95 percent alcohol. Hands and arms should be washed thoroughly, and gloves and a clean lab coat should be worn.

Supplies needed for a transfer operation are:

1. Device for sterilizing dissecting tools and containers, such as an automatic gas burner or an inexpensive glass alcohol lamp with a wick. A glass bead sterilizer (Fig. 18–17) and certain types of burner-incinerators are available commercially and can be used to avoid the fire dangers of alcohol or gas fumes.
2. Forceps to hold and manipulate tissue. Three sizes (47), 300 mm, 200 mm straight-tipped, and 115 mm curve-tipped, are particularly useful.
3. Dissecting needles with wood handles, or a metal needle holder with replaceable tips.
4. Dissecting scalpels. Metal scalpel holders with replaceable knife blades are most convenient.
5. Sterile Petri dishes can be used for holding explants and sectioning them. Small squares of sterile, moist

(a)

(b)

Figure 18–16
Explants being disinfested in a bleach solution. (a) The explants are placed in a beaker or tub containing the bleach solution. (b) The magnetic stir-bar in the solution spins to agitate the explants to ensure good contact between the solution and the explant. A timer is set to precisely stop the process before the bleach can seriously damage the explant tissue.

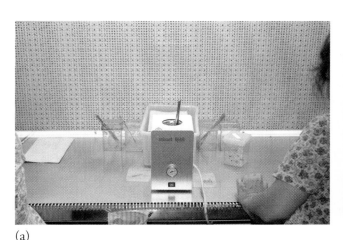  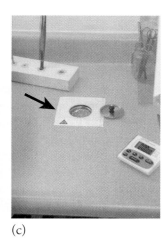

Figure 18–17
Following initial disinfestation procedures, all work needs to be done in a laminar flow hood, and tools sterilized between transfer or subculture of explants. (a) A glass bead sterilizer under the laminar flow hood with a tool being sterilized. (b) An innovative lab has mounted the glass bead sterilizer below the flow hood's table surface to save work space. (c) Top view of the table surface showing where the top of the glass bead sterilizer is located (arrow).

paper toweling inside each Petri dish are convenient and assist cleaning later (22). Sterilized paper towels placed directly on the surface of the laminar flow hood are commonly used as a working surface. The towels can be wetted with sterilized water to keep small explants from drying out.

6. A dissecting microscope; important for specific operations such as isolating shoot meristems or immature embryos. Equipment should be wiped with a disinfectant and placed in the laminar flow hood.

Implements must be sterilized in preparation of transferring explants. Although sterilization can be achieved by dipping tools into alcohol and flaming them before use, many propagators avoid this because of the fire hazard. Furthermore, simply dipping and flaming the scalpels may not kill the most resistant pathogens or spores; longer exposure to the flame, or use of a chemical sterilant, may be required (42). A Bacti-incinerator is a commercial device that heats forceps and tools to a very high temperature (about 1,600°F) when inserted into the core (22), but the manufacturer warns that sharp instruments should not be inserted into the core. An alternative to burners and the Bacti-incinerator is the **glass bead sterilizer** (Fig. 18–17). Instruments placed in the heated beads at 250°C are sterilized in 10 to 15 seconds. Tools should be dry and free of agar. Tools should not remain in glass beads for more than 1 minute, and glass beads must be replaced periodically. Regardless of how tools are sterilized, needles, scalpels, and forceps should be allowed to cool 30 to 60 seconds before using them in order to prevent damage to tender tissue. When working in an open area, the propagator usually flames the mouth of the container before opening it and after inserting the material to avoid entrance of contaminants.

Pretreatment of Explants Exudation of phenolics and other substances from the cut explant surface can inhibit initial shoot development. Using liquid media in the initial stage or frequent transfers on agar medium for several days may leach out the toxic materials. Likewise, using antioxidants, such as ascorbic acid and citric acid, in the preliminary washing may be useful.

Initial Shoot Development

Depending on the explant, shoots will initiate from: (a) the stimulation of axillary shoots; (b) the initiation of adventitious shoots on excised shoots, leaves, bulb scales, flower stems, cotyledons, and other organs, as described in Chapter 17; or (c) the initiation from callus on the cut surfaces. Which medium is selected varies with the species, cultivar, and kind of explant to be used. A basal medium includes the ingredients listed in Table 18–2, plus sucrose and sometimes other supplements. Control of development is most often achieved by manipulating the levels of auxin and cytokinin. If axillary shoots are desired, a moderate level of cytokinin (BA, kinetin, or 2iP) is used (0.5 to 1 mg/L). The auxin level is kept very low (0.01 to

0.1 mg/L) or omitted altogether. In some cases, auxin may be used alone without cytokinin.

A somewhat higher cytokinin level is needed to develop adventitious shoots on explants. For callus formation, increased auxin levels are needed, but must complement an appropriate level of cytokinin. Stage I requires 4 to 6 weeks before explants are ready to be moved to Stage II.

Some woody plant species can take up to a year to complete Stage I (26), a phenomenon called **stabilization.** A culture is stabilized when explants produce a consistent number of "normal" shoots after subculturing. "Abnormal" shoots usually are arrested in development and fail to elongate.

STAGE II—SHOOT MULTIPLICATION

In the multiplication stage each explant has expanded into a cluster of microshoots arising from the leaf axils or base of the explant (Fig. 18–18). The separate microshoots are transplanted into a new culture medium, which is a process called **subculturing.** During the multiplication stage, cultures are subcultured every 4 to 8 weeks (Fig. 18–19).

The kind of medium used depends on the species, cultivar, and type of culture. The basal medium is usually the same as in Stage I, but often the cytokinin and mineral supplement level is increased. Adjustments may need to be made after some experimentation (8, 32).

The optimum size of the microshoot and method of cutting apart varies, depending on the species. A minimum mass of tissue is generally necessary to produce uniform rapid multiplication in the next subculture. New cultures are initiated in two ways from these developing cultures. The elongated shoots are cut from the original culture and are subcultured as nodal explants. Shoots may have the leaves removed, are typically 2 to 4 nodes in length, and can be inserted into the medium in a vertical position or laid horizontally on the surface. Avoid pushing shoots too deeply into the agar and submerging the nodes. Elongated shoots laid horizontally on the agar surface perform as a type of layering and stimulate lateral shoots to develop. After the elongating shoots have been removed and subcultured, the original explant can be further subdivided and recultured. Tissue mass may be divided by vertically cutting it into sections, keeping some of the base with each piece. Sometimes one dominant shoot develops and inhibits elongation of others, in which case the shoot may be cut off and the base recultured. Clump-forming plants (like hosta or daylily) shoots form from a common base. Crowns are separated in a similar way to standard division (see Chapter 14), and each division is subcultured.

The number of microshoots produced ranges from 5 to 25 or more, depending on the species and conditions of culture. Multiplication may be repeated several times to increase the supply of material to a predetermined level for subsequent rooting and transplanting. Sometimes microshoots deteriorate with time, lose leaves, fail to grow, develop tip-burn, go dormant, or lose the potential to regenerate.

During multiplication, off-type propagules sometimes appear, depending on the kind of plant and method of regeneration. In some plants, such as Boston fern, restricting the multiplication phase to only three subcultures before reinitiating the cultures is recommended to avoid development of off-type microshoots. Other procedures that can reduce the potential for variability include reducing growth regulator concentration and avoiding

(a)

(b)

Figure 18–18
During repeated subculturing in the multiplication stage many shoots can arise from a single explant. (a) New shoot growth from a horizontally placed explant following a subculture. (b) Multiple shoots are formed during a 6-week culture cycle.

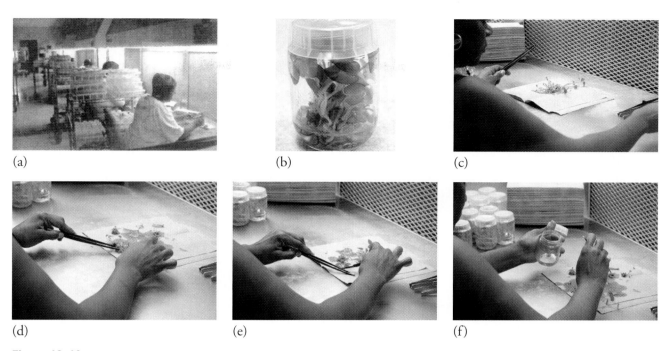

Figure 18–19
Subculturing in Stage II. (a) A row of laminar flow hoods with technicians at work. Media is on carts close to the workers to streamline work effort. (b) A five-week-old culture ready for subculturing. (c) The culture or a portion of the culture is removed from the jar and placed on a sterile paper towel. (d and e) A scalpel and forceps are used to cut and separate the larger culture into smaller pieces for (f) transfer to a new jar to complete the subculturing procedure.

basal callus that may give rise to adventitious shoots in subculturing (13).

STAGE III—ROOT FORMATION

Shoots developed during the multiplication stage do not usually have roots. There are some exceptions (like African violet and poplar) that spontaneously root on the multiplication medium, but for most other species, single shoots (microcuttings) must be moved to a medium or suitable environment to induce roots. Therefore, the purpose of Stage III is to prepare plantlets for transplanting from the tissue culture environment of the test tube to a free-living existence in the greenhouse and on to their ultimate location. Therefore, Stage III may not only involve rooting, but also conditioning the plantlet to increase its potential for acclimation and survival during transplanting, for example, by increasing agar and/or sucrose concentration. Light intensity is sometimes increased during Stage III. Some plants respond to an "elongation" phase between Stages II and III, achieved by placing the microshoot into an agar medium for 2 to 4 weeks without cytokinins (or at very low levels) and, in some cases, adding (or increasing) gibberellic acid (44), which reduces the influence of cytokinin. The microshoots are then rooted in a cytokinin- and GA-free medium.

Microcuttings can be rooted either *in vitro* or *ex vitro* (Fig. 18–20). For ***in vitro*** rooting, individual microcuttings are subcultured into new containers in a sterile medium with reduced or omitted cytokinin, an increased auxin concentration, and often a reduced inorganic salt concentration. For some plants, rooting

Figure 18–20
Stage III rooting can occur (a) *in vitro* on an agar solidified medium or (b) *ex vitro* in a soil-less greenhouse substrate.

Figure 18–21
A work-line for *ex vitro* microcutting rooting. (a) At each work station the technicians have their rooting flats, a syringe bottle to spray microcuttings periodically to keep them from drying out, and paper towels to (b) hold the explants that have been washed to remove any adhering media. (c) Microcuttings are inserted directly into the rooting substrate often using forceps to handle the small cuttings. (d and e) In some cases, the technician grades the microcuttings into large (red arrow) or small categories (white arrow) and places each grade in separate rooting flats.

is best if the microcutting is kept in the auxin medium for 1 to 5 days and then transferred to an auxin-free medium, or the microcutting may simply be dipped into a rooting (auxin) solution and inserted directly into an auxin-free medium. Some plants root best if placed in the dark during the auxin treatment period.

Microcuttings can also be rooted directly under **ex vitro conditions** (Figs. 18–21 and 18–22), which means microcuttings are "stuck" into plug cells (27, 53) containing a greenhouse rooting medium (e.g., two parts peat to one part vermiculite) and placed in a high humidity environment to form roots (Fig. 18–8). Microcuttings can be treated with auxin as a quick-dip, or talc, prior to sticking (Fig. 18–23, page 720) in the same procedure as that used during standard cutting propagation (see Chapter 10), except care must be taken to prevent the fragile leaves on the microcutting from drying out during the sticking operation. Frequent syringing or misting prevents damage from over-drying. *Ex vitro* rooting can be more efficient than *in vitro* rooting because it saves on labor and materials, and *ex vitro* rooted plantlets may resume normal growth quicker than *in vitro* rooted plantlets.

STAGE IV—ACCLIMATIZATION TO GREENHOUSE CONDITIONS

Once plantlets are well rooted, they must be acclimatized to the normal greenhouse environment (38, 52). *In vitro*-rooted plants are removed from the

Figure 18–22
A sequence showing the relationship between *ex vitro* rooting and Stage IV Acclimatization. (a) Cultures arrive from the tissue culture lab and must be removed from the container and separated from the agar medium. (b and c) Microshoots are cut from the culture clumps to provide uniform single-shoot microcuttings. (d) Microcuttings are inserted into the rooting medium. (e) Rooting and acclimatization take place in a polyethylene tent constructed within the greenhouse. Plastic lids cover the flats to provide a high-humidity and low-light environment for rooting. These lids are gradually vented and finally removed during acclimatization. Mist provides humidity control. (f) Fully acclimatized plants.

culture vessel and the agar is washed away completely to remove a potential source of contamination. Plantlets are transplanted into a standard pasteurized rooting or soil mix in small pots or cells in a more or less conventional manner (Fig. 18–24, page 720). Initially, plantlets should be protected from desiccation

Figure 18–23
Some microcuttings benefit from an auxin treatment during the rooting stage. (a and b) *Ex vitro* microcuttings being treated with an auxin talc prior to sticking.

Figure 18–24
Microcuttings rooted *in vitro* also arrive at the nursery in tubs containing an agar-based medium. (a) Cuttings are gently removed from the containers. Adhering medium can be washed off. (b) Rooted plantlets are placed in a greenhouse substrate. (c) A close-up of the rooted plantlet, ready to be transplanted.

Figure 18–25
Rooted plantlets need to be acclimatized before they can resume growth outdoors or in the greenhouse. (a) An enclosed mist system where the side walls roll up to gradually reduce humidity and raise light levels. (b) Once the plants have adjusted to a higher humidity they still may need additional shade before becoming completely acclimatized. Fully acclimatized and growing (c) hydrangea and (d) purple coneflower (*Echinacea*).

in a shaded, high-humidity tent or under mist or fog (Fig. 18–25). Several days may be required for new functional roots to form (43, 52).

Plantlets should be gradually exposed to a lower relative humidity and higher light irradiance. Any dormancy or resting condition that develops may need to be overcome as part of the establishment process.

In summary, micropropagation involves the vegetative propagation of plants through (1) establishment, (2) multiplication, (3) rooting, and (4) acclimatization (transplanting) of clones *in vitro* starting with small explants and ending with a rooted plant established in a container or in the ground. Table 18–3 (page 722) contains step-by-step micropropagation procedures for specific representative species. For more details about additional plant species, see individual genera in Chapters 19, 20, and 21.

Table 18-3
SPECIFIC MICROPROPAGATION PROTOCOLS FOR REPRESENTATIVE SPECIES

Species	Stage I. Establishment	Stage II. Multiplication	Stage III. Rooting	Stage IV. Acclimatization
Carnation (46) Carnation is typically started by meristem-tip culture to ensure that the material is pathogen-free. Once established, "clean" material can be readily propagated through axillary shoot proliferation.	Use shoots of about 12 nodes, removing leaves that are over 3 mm long. Wash with detergent. Place in 10 percent solution of bleach for 5 minutes. Rinse 2 times for 3 minutes each in sterile water. Remove stem tip with 1 to 2 leaves 1 mm or larger and place on surface of modified MS agar medium plus 0.2 mg/L kinetin and 0.2 mg/L NAA. Place in growth chamber at 22°C in continuous light.	After approximately 1 month, or when the culture has grown sufficiently, remove from the container, and cut into segments approximately 1 cm long. Transfer each to separate containers on agar with MS + 0.5 mg/L kinetin + 0.1 mg/L NAA. Continue to subculture to fresh medium.	Transfer microcuttings to a medium without kinetin or NAA. Roots form easily in vitro. Ex vitro rooting is an alternative. Microcuttings from Stage II are treated with IBA and placed in a rooting medium as a typical softwood cutting would be.	Remove rooted microshoots from container. Wash away solid media if present. Transplant into growing medium and hold under mist or under high humidity tent. New functional roots need to form to replace the older roots. Plug containers of the type used for seeds are useful. Once established, microplants are gradually exposed to lower humidity and higher light intensity.
Hosta (28) An example of an easily micropropagated herbaceous perennial.	Select top florets as an explant after the bottom florets have opened. Surface disinfest the entire flower stem in 10 percent bleach for 15 minutes followed by several rinses with sterile water. Remove florets and place on modified MS medium with BA at 0.5 to 2.5 mg/L and NAA at 0.5 to 2.5 mg/L. Callus and new shoots develop from the flower tissue in 8 to 12 weeks.	Transfer individual shoots to the same medium with 0.1 mg/L BA and 0.5 mg/L NAA to induce further shoot formation. Clumps of shoots can be subcultured every 4 to 6 weeks.	Shoots root readily in vitro on a medium without BA or NAA.	Wash agar from plantlets and transfer to greenhouse medium under high humidity. Gradually reduce humidity to acclimatize plants. Evaluate variegated cultivars for trueness-to-type.
Boston Fern (Nephrolepsis) (36, 43) Micropropagation is very effective because natural increase by rhizomes is slow.	Use 5- to 10-cm long actively growing runners. Surface disinfest in 20 percent bleach for 20 minutes; rinse 3 times in sterile water. Use one-half- or three-fourths-strength salts MS medium with 1.0 mg/L kinetin and 0.1 mg/L NAA. Cut runners into 2.5 cm sections. Growth should appear within 6 weeks.	Use the same medium. Establish at 6 to 8 weeks by subdividing clumps into 1 cm² sections. Subculture every 6 to 8 weeks but do not do more than 3 times to avoid excessive mutation from proliferating cultures.	Microcuttings form roots in vitro on one-half strength MS medium without growth regulators, or microcuttings can be rooted ex vitro directly into a rooting medium under high humidity or mist.	Gradually reduce humidity after plantlets are well rooted.
Cherry (Prunus) (45) Represents a woody plant that is sometimes difficult to propagate.	Collect 7- to 10-cm young shoots in spring. Wash vigorously in 10 percent bleach for 10 minutes. Place in running water for 2 minutes. Disinfest in 10 to 20 percent bleach for 20 minutes; rinse 3 times with sterile water. Cut above and below node and place upright in agar. Medium is one-half strength MS with 0.5 to 1.0 mg/L BA.	Place 1 or 2 node microshoots on one-half strength MS salts with BA (0.5 to 2.5 mg/L) and IBA (0.01 mg/L). Subculture microshoots every 6 to 8 weeks.	Root single microcuttings on a one-half strength MS medium with 0.2 mg/L IBA and no cytokinin.	When roots are well formed and new growth is evident from the plantlet, wash agar off roots and move to greenhouse potting mix under high humidity. Gradually reduce humidity to acclimatize to greenhouse conditions.
Rhododendron (1, 48) Woody ornamental that is extensively grown commercially by micropropagation.	Wash 5-cm shoot tips in 10 to 20 percent bleach; remove leaves and terminal bud; dip in 70 percent alcohol for 10 seconds, rinse; soak in 10 percent bleach for 20 minutes; rinse 3 times. Trim 1 cm from base leaving a 2- to 4-cm explant that is laid horizontally on an agar slant. A typical protocol includes: Anderson medium plus 2iP (5.0 mg/L), IAA (1.0 mg/L), and adenine (80 mg/L). Transfer to new medium every 2 weeks until growth starts, then every 5 weeks, but do not subculture until shoots are well developed.	Subculture individual shoots onto Anderson medium with 2iP.	Microcuttings can be rooted in vitro on one-half MS with 2 to 7 mg/L IBA plus activated charcoal or rooted ex vitro using 1:1 peat-perlite after treatment with IBA.	Reduce humidity after plantlets are well rooted.

DISCUSSION ITEMS

Micropropagation is a viable propagation option for many plant species, especially ornamentals. It is the standard system to propagate most foliage plants, many herbaceous perennials, and woody nursery crops, and it is an integral part of producing disease-free stock plants for vegetatively propagated greenhouse crops and many small and tree fruits. This chapter is a practical discussion of the techniques required to equip a lab and perform the procedures for micropropagation. Most horticultural students should become familiar with tissue culture even if their future employment does not involve performing micropropagation because they will be purchasing plants that were micropropagated.

1. Compare standard propagation systems with micropropagation for plants such as hosta, *Rhododendron*, spathiphyllum, ferns.
2. Compare the advantages and disadvantages of micropropagation.
3. Why are stock solutions used to prepare tissue culture media?
4. Which stage of micropropagation is the most difficult to complete successfully?

REFERENCES

1. Anderson, W. C. 1975. Propagation of rhododendrons by tissue culture. 1. Development of a culture medium for multiplication of shoots. *Comb. Proc. Intl. Plant Prop. Soc.* 25:129–35.

2. Berto, M., P. Curir, and P. Debergh. 1999. Influence of agar on *in vitro* cultures: II. Biological performance of *Ranunculus* on media solidified with three different agar brands. *In Vitro Cell. Dev. Biol.—Plant* 35:94–101.

3. Bridgen, M. P., and J. W. Bartok, Jr. 1987. Designing a plant micropropagation laboratory. *Comb. Proc. Intl. Plant Prop. Soc.* 37:462–67.

4. Broome, O. C. 1986. Laboratory design. In R. H. Zimmerman, R. J. Griesbach, F. A. Hammerschlag, and R. H. Lawson, eds. *Tissue culture as a plant production system for horticultural crops.* Dordrecht: Martinus Nijhoff Publishers. pp. 351–65.

5. Brown, D. C. W., and T. Thorpe. 1984. Organization of a plant tissue culture laboratory. In I. K. Vasil, ed. *Cell culture and somatic cell genetics of plants, Vol. 1.* New York: Academic Press. pp. 1–12.

6. Burger, D. 1988. Guidelines for autoclaving liquid media used in plant tissue culture. *HortScience* 23:1066.

7. Chu, I. Y. E. 1986. The application of tissue culture to plant improvement and propagation in the ornamental horticulture industry. In R. H. Zimmerman, R. J. Griesbach, F. A. Hammerschlag, and R. H. Lawson, eds. *Tissue culture as a plant production system for horticultural crops.* Dordrecht: Martinus Nijhoff Publishers. pp. 15–33.

8. de Fossard, R. A. 1976. *Tissue culture for plant propagators.* Armidale, Australia: Univ. of New England Printery.

9. Donnan, A., Jr. 1986. Determining and minimizing production costs. In R. H. Zimmerman, R. J. Griesbach, F. A. Hammerschlag, and R. W. Lawson, eds. *Tissue culture as a plant production system for horticultural crops.* Dordrecht: Martinus Nijhoff Publishers. pp. 167–73.

10. Dore, C. 1987. Application of tissue culture to vegetable crop improvement. In C. E. Green, D. A. Somers, W. P. Hackett, and D. D. Biesboer, eds. *Plant tissue and cell culture.* New York: Alan R. Liss. pp. 419–32.

11. Fogh, J., ed. 1973. *Contamination in tissue culture.* New York: Academic Press.

12. Gamborg, O. L., and L. R. Wetter, eds. 1975. *Plant tissue culture methods.* Saskatoon, Canada: Prairie Regional Laboratory.

13. Geneve, R. L. 1989. Variations within a clone in tissue culture production. *Comb. Proc. Intl. Plant Prop. Soc.* 39:458–62.

14. George, E. F., M. A. Hall, and G. J. De Klerk. 2007. *Plant propagation by tissue culture.* Dordrecht: Springer.

15. Hammerschlag, F. A. 1986. Temperate fruits and nuts. In R. H. Zimmerman, R. J. Griesbach, F. H. Hammerschlag, and R. H. Lawson, eds. *Tissue culture as a plant production system for horticultural crops.* Dordrecht: Martinus Nijhoff Publishers. pp. 221–36.

16. Hartman, R. D., and R. H. Zimmerman. 1999. Commercial micropropagation in the United States. *Curr. Plant Sci. Biotech. Agric.* 36:699–707.

17. Hartmann, H. T., and J. Whisler. 1977. Micropropagation exercises in teaching plant propagation. *Comb. Proc Intl. Plant. Prop. Soc.* 27:407–13.

18. Jones, J. B. 1986. Determining markets and market potential. In R. H. Zimmerman, R. J. Griesbach, F. A. Hammerschlag, and R. H. Lawson, eds. *Tissue culture as a plant production system for horticultural crops.* Dordrecht: Martinus Nijhoff Publishers. pp. 175–82.

19. Kartha, K. K. 1985. *Cryopreservation of plant cells and organs.* Boca Raton, FL: CRC Press.

20. Kester, D. E., and C. Grasselly. 1987. Almond rootstocks. In R. C. Rom and R. F. Carlson, eds. *Rootstocks for fruit crops.* New York: John Wiley & Sons. pp. 265–93.

21. Kozai, T., F. Afreen, and S. M. A. Zobayed. 2005. *Phytoautotrophic (sugar-free medium) micropropagation as a new micropropagation and transplant production system.* Dordrecht: Springer.

22. Kyte, L. 1987. *Plants from test tubes; an introduction to micropropagation.* (rev. ed.). Portland, OR: Timber Press.

23. Linsmaier, E. M., and F. Skoog. 1965. Organic growth factor requirements of tobacco tissue cultures. *Physiol. Plant.* 18:101–27.

24. Litz, R. E. 1987. Application of tissue culture to tropical fruits. In C. E. Green, D. A. Somers, W. P. Hackett, and D. D. Biesboer, eds. *Plant tissue and cell culture.* New York: Alan R. Liss. pp. 407–18.

25. Lloyd, G., and B. McCown. 1980. Commercially feasible micropropagation of mountain laurel, *Kalmia latifolia,* by use of shoot-tip culture. *Comb. Proc. Intl. Plant Prop. Soc.* 30:421–27.

26. McCown, B. H. 1986. Woody ornamentals, shade trees, and conifers. In R. H. Zimmerman, R. J. Griesbach, F. A. Hammerschlag, and R. H. Lawson, eds. *Tissue culture as a plant production system for horticultural crops.* Dordrecht: Martinus Nijhoff Publishers. pp. 333–42.

27. McCown, D. D. 1986. Plug systems for micropropagules. In R. H. Zimmerman, R. J. Griesbach, F. A. Hammerschlag, and R. H. Lawson, eds. *Tissue culture as a plant production system for horticultural crops.* Dordrecht: Martinus Nijhoff Publishers. pp. 53–60.

28. Meyer, M. M., Jr. 1980. *In vitro* propagation of *Hosta sieboldiana.* HortScience 15:737–38.

29. Mok, M. C., D. W. S. Mok, J. E. Turner, and C. V. Mufer. 1987. Biological and biochemical effects of cytokinin-active phenylurea derivatives in tissue culture systems. *HortScience* 22:1194–97.

30. Mudge, K. W., C. A. Borgman, J. C. Neal, and H. A. Weller. 1986. Present limitations and future prospects for commercial micropropagation of small fruits. *Comb. Proc. Intl. Plant Prop. Soc.* 36:538–43.

31. Murashige, T. 1974. Plant propagation through tissue cultures. *Ann. Rev. Plant Phys.* 25:135–66.

32. Murashige, T. 1977. Plant cell and organ cultures as horticultural practices. *Acta Hort.* 78:17–30.

33. Murashige, T., and F. Skoog. 1962. A revised medium for rapid growth and bioassays with tobacco tissue cultures. *Physiol. Plant.* 15:473–97.

34. Nas, M. N., and P. E. Read. 2003. *Ex vitro* survival of hybrid hazelnut shoots produced *in vitro*. *Acta Hort.* 616:215–19.

35. Paul, H., G. Kaigny, and B. S. Sangwan-Norreel. 2000. Cryopreservation of apple (*Malus xdomestica* Borkh.) shoot tips following encapsulation-dehydration or encapsulation-vitrification. *Plant Cell Rpt.* 19:768–74.

36. Pedhya, M. A., and A. R. Meha. 1982. Propagation of fern (*Nephrolepis*) through tissue culture. *Plant Cell Rpt.* 1:261–63.

37. Pierik, R. L. M. 1999. In vitro *culture of higher plants.* 4th ed. Dordrecht: Martinus Nijhoff Publishers.

38. Preece, J. E., and E. G. Sutter. 1991. Acclimatization of micropropagated plants to the greenhouse and field. In P. C. Debergh and R. H. Zimmerman, eds. *Micropropagation technology and application.* Dordrecht: Martinus Nijhoff Publishers. pp. 71–93.

39. Raju, B. C., and J. C. Trolinger. 1986. Pathogen indexing in large-scale propagation of florist crops. In R. H. Zimmerman, R. J. Griesbach, F. A. Hammerschlag, and R. H. Lawson, eds. *Tissue culture as a plant production system for horticultural crops.* Dordrecht: Martinus Nijhoff Publishers. p. 138.

40. Read, P., and J. Preece. 2007. Micropropagation of ornamentals: the wave of the future? *Propagation Ornamental Plants* 7:150–59.

41. Scholten, H. J., and R. L. M. Pierik. 1998. Agar as a gelling agent: Differential biological effects *in vitro*. *Scientia Hort.* 77:109–16.

42. Singha, S., G. K. Bissonnette, and M. L. Double. 1987. Methods for sterilizing instruments contaminated with *Bacillus* spp. from plant tissue cultures. *HortScience* 22:659.

43. Smith, R. H. 1992. *Plant tissue culture. Techniques and experiments.* New York: Academic Press.

44. Snir, I., and A. Erez. 1980. *In vitro* propagation of Malling Merton apple rootstocks. *HortScience* 15:597–98.

45. Snir, I. 1982. *In vitro* propagation of sweet cherry cultivars. *HortScience* 17:192–93.

46. Stimart, D. P. 1986. Commercial micropropagation of florist flower crops. In R. H. Zimmerman, R. J. Griesbach, F. A. Hammerschlag, and R. H. Lawson, eds. *Tissue culture as a plant production system for horticultural crops.* Dordrecht: Martinus Nijhoff Publishers. pp. 301–15.

47. Stoltz, L. P. 1979. Getting started in tissue culture: Equipment and costs. *Comb. Proc. Intl. Plant Prop. Soc.* 29:375–81.

48. Strode, R. A., P. A. Travers, and R. P. Oglesly. 1979. Commercial micropropagation of rhododendrons. *Comb. Proc. Intl. Plant Prop. Soc.* 29:439–43.

49. Suman, G., S. C. Gupta, and S. Govil. 1997. Commercialization of plant tissue culture in India. *Plant Cell Tiss. Organ Cult.* 51:65–73.

50. Swartz, H. J., and J. T. Lindstrom. 1986. Small fruit and grape tissue culture from 1980 to 1985: Commercialization of the technique. In R. H. Zimmerman, R. J. Griesbach, F. A. Hammerschlag, and R. H. Lawson, eds. *Tissue culture as a plant production system for horticultural crops.* Dordrecht: Martinus Nijhoff Publishers. pp. 201–20.

51. Welander, M., L-H. Zhu, and X-V Li. 2007. Factors influencing conventional and semi-automated micropropagation. *Propagation Ornamental Plants* 7:103–11.

52. Zimmerman, R. H. 1988. Micropropagation of woody plants: Post tissue culture aspects. *Acta Hort.* 227:489–99.

53. Zimmerman, R. H., and L. Fordham. 1985. Simplified method for rooting apple cultivars *in vitro*. *J. Amer. Soc. Hort. Sci.* 110:34–8.

54. Zimmerman, R. H., R. J. Griesbach, F. A. Hammerschlag, and R. H. Lawson, eds. 1986. *Tissue culture as a plant production system for horticultural crops.* Dordrecht: Martinus Nijhoff Publishers.

part five
Propagation of Selected Plant Species

CHAPTER 19 **Propagation Methods and Rootstocks for Fruit and Nut Species**
CHAPTER 20 **Propagation of Ornamental Trees, Shrubs, and Woody Vines**
CHAPTER 21 **Propagation of Selected Annuals and Herbaceous Perennials Used as Ornamentals**

The objective for the following three chapters is to provide information about propagation methods for plant species important to the major horticultural industries for which greenhouse and nurseries propagate plants, including fruit and nut crops as well as woody and herbaceous ornamental species used in the landscape. These chapters do not include all crop plants of importance. For instance, we do not include field crops, vegetables, lawn grasses, or all forest species. For all of the latter, special manuals are available and should be consulted. This part is organized into three separate chapters for fruits, woody ornamentals, and herbaceous annuals and perennials. Plants are listed alphabetically by genus names but are cross-referenced to common names in a separate plant index. Propagation of specific crops should also include a general knowledge of their taxonomy and something about their history and special needs; this information has been included where appropriate.

19

Propagation Methods and Rootstocks for Fruit and Nut Species

INTRODUCTION

Most fruit and nut crop cultivars are clones and do not reproduce true-to-type by seed, so propagation by vegetative methods is necessary. Many tree fruit and nut cultivars are propagated by budding or grafting onto rootstocks produced from seedlings, rooted cuttings, layers or micropropagated plants. Rootstocks are important. First, most fruit and nut tree cultivars are difficult to root by cuttings, except in a few species, although some can be propagated by layering. Second, rootstocks are used to resist critical soil pathogens and to adapt plants to specific areas and management conditions.

Clonal rootstocks are important and are propagated by cuttings, layering, and micropropagation. Micropropagation is used to mass-propagate some cultivars, but it is also often used to produce virus-tested sources.

Fruit and nut cultivars are examples of monocultures with the accompanying problems of vulnerability to introduce pathogens, which is why commercial propagation of fruit and nut crops has become very specialized, requiring continuing research and development. Nevertheless, large nurseries specializing in fruit and nut propagation produce millions of fruit and nut trees each year, mostly by vegetative methods.

Actinidia deliciosa. Kiwifruit (27). Chinese gooseberry. Large-fruited dioecious subtropical vine originating in China but now grown throughout the world. It requires male and female cultivars to ensure fruiting. *A. arguta* is a related species producing small fruits in a cluster. Kiwi vines are propagated commercially mostly by grafting cultivars to seedling rootstocks because grafted plants are believed to be more vigorous and to come into bearing sooner than those started as rooted cuttings. However, kiwis may also be propagated by leafy, semi-hardwood cuttings under mist, hardwood cuttings, and root cuttings.

Seed. Seedling plants have a long juvenile period and their sex cannot be determined until fruiting at 7 years or more. Seed should be taken from soft, well-ripened fruit, dried and stored at 5°C (41°F). After at least 2 weeks at this temperature, subject seed to fluctuating temperatures—10°C (50°F) night and 20°C (68°F) day—for 2 or 3 weeks before planting.

Cuttings. Leafy, semi-hardwood cuttings taken from apical and central parts of current season's growth in late spring and midsummer may be rooted under mist in coarse vermiculite with 6,000 ppm IBA. Hardwood cuttings, taken in midwinter and planted in a greenhouse, require higher IBA concentrations.

Grafting. Seedlings are grafted successfully by the whip graft in late winter using dormant scion-wood; T-budding in late summer is also successful. Seedlings are grown 1 year in the nursery. Collect dormant wood in previous winter and store. In California, graft in April but delay if bleeding is excessive. Use 1-bud scion with a whip graft. Wrap completely with budding rubber to make an air-tight seal. Complete operations by May 15.

Rootstocks. Most California growers use seedlings of 'Bruno' cultivar.

Micropropagation. Kiwifruit can be propagated by using short meristem tips or longer apical shoots as the initial explants.

Almond. *See Prunus dulcis.*

Anacardium occidentale. Cashew (8). These tender tropical evergreen trees are usually grown as seedlings. Clonal selections including precocious dwarf types (151) have been made, but vegetative propagation and transplanting are difficult.

Seed. Two to three are planted directly in place in the orchard and later thinned to one tree per location. Dormancy does not occur, but seeds may lack embryos. Test by placing seeds in water and discard those that float. Seeds germinate in 15 to 20 days. Growing in biodegradable containers will improve transplanting success.

Grafting. Rootstocks should be 1 to 2 months old but not tall and spindly. In softwood grafting, the scions are 8 to 10 cm (3 to 4 in) long, pencil thick, with bulging tips (20). Scions are green to brown in color with terminal leaves dark green. Leaf blades are clipped off 7 to 10 days before grafting, and grafting is done when the petioles abscise. One to two leaves are retained below the graft, which is done on the soft portion. A longitudinal cut about 3 to 4 cm (1 to 1.5 in) long is made. The scion wedge is inserted and the graft and tied with polyethylene grafting tape. Grafts are kept in shade for a week before transferring to the open.

Miscellaneous Asexual Methods. Cashew can also be propagated by rooting leafy cuttings, stooling (161), approach-grafting, and air layering (158, 161). In Nigeria, cashew is air-layered (2). T-budding and patch budding have been successfully used in limited trials.

Ananas comosus. Pineapple. Tropical terrestrial monocot of the Bromiliadaceae family which originated in South America. A large number of cultivars are grown in Hawaii, South America, Africa, the Philippines, Australia, and Southeast Asia, much of them in large industrial plantings (39). The plant consists of a thick stem with the flowering rosette on the top.

Offshoots. Three main types of vegetative propagules include offshoots (offsets) from the main stem either from below or above ground *(suckers)*, lateral offshoots from the fruiting stem (peduncle) just below the fruit *(slips)*, and the vegetative shoot emerging above the fruit *(crowns)*. All three types can be used for propagation by removing them from the plant and planting either by machine or by hand (see Fig. 14–21). Drying (curing) for several weeks to allow callusing on the cut surface will reduce decay after planting. Suckers are somewhat limited in number, producing a mature fruit in 15 to 17 months, and are rarely used as planting material. If used, they are removed about 1 month after peak fruit harvest. Slips, preferred and most abundant, are cut from the mother plant 2 to 3 months after peak harvest. A slip-produced plant will produce fruit in about 20 months. They can be stored for a relatively long time and retain vigor for replanting. Slips of the same size and age will produce uniform flowering. Crowns are removed at about the same time as fruit harvest and require approximately 24 months to produce fruit.

Micropropagation. Explants of axillary buds from crowns of mature fruits can be propagated in an MS medium with 24 percent coconut water followed by subculturing in reduced coconut water supplemented by BA (152).

Annona cherimola. Cherimoya, Chirimoya, Custard Apple. Sugar apple *(A. squamosa)*. Atemoyas *(A. squamosa × A. cherimola)*. Cherimoya *(A. cherimola)* is evergreen, sugar apple is deciduous, and atemoyas are semi-deciduous and somewhat tolerant of frost (20, 73). These tropical plants grow as small trees in Spain, Mexico, Ecuador, Peru, Chile, and southern Africa, with limited production in California, India, and Sri Lanka. In many areas, seedlings are grown, and some selections in Mexico and South America are reported to come nearly "true." Seedling plants are vigorous, have long juvenile periods and inferior fruits.

Seed. Viability is retained for many years if kept dry. Germination takes place a few weeks after planting. Seeds have been successfully propagated in containers transplanted from flats to small pots when 7.5 to 10 cm (3 to 4 in) high, then to small pots at 20 cm (8 in), and to larger pots or to open ground.

Grafting. Cultivars can be cleft grafted or T-budded. Topworking is successful with bark or cleft grafting carried out at the end of a dormant period. Seedling rootstocks are cherimoya, the related sugar apple (*A. squamosa*), which yields a dwarf plant, and custard apple (*Annona* hybrids). The latter two species are susceptible to cold, and root-rot is a problem on poorly drained soils.

Layering (20). Stooling (mound layering) has given promising results. Three-year-old seedlings were headed to ground level, shoots were girdled, and IBA applied.

Apple. See *Malus xdomestica*.

Apricot. See *Prunus armeniaca*.

Artocarpus heterophyllus. Jackfruit. A medium-sized tropical tree with fruit of unique flavor has many related genera and species including breadfruit (*Artocarpus altilis*). Although grown mostly by seed, cultivars are available. The seed has limited viability and must be germinated within 1 month. Presoaking the seed improves germination. Propagated by cuttings when stock plants are etiolated, and cuttings treated with IBA and rooted under mist. Grafting, including inarching, epicotyl grafting, and various budding methods, including chip budding, are successful (75). Can be propagated by air layering.

Asimina triloba. Common paw-paw and dwarf paw-paw (*A. parviflora*). See Chapter 20, page 780.

Atemoyas. See *Annona*.

Averrhoa carambola. Carambola (20, 159). This medium-sized evergreen tree originates from tropical and subtropical areas of Southeast Asia, but it is now grown in many parts of the world. The fruit is a distinctive five-cornered shaped fruit, used fresh for juice, or for decoration. Seeds will germinate within 7 days if sown immediately from the fruit. Storage can be maintained for about a week in the refrigerator if kept moist. A well-drained medium should be used for trays, and the seedlings transplanted when the first true leaves appear. Although propagated by seed, vegetatively propagated cultivars are available. Side-veneer grafting, chip budding, wedge grafting, whip-and-tongue, and approach grafts can be used. Standard air layering is highly successful. Side-veneer grafting can be used for topworking or wedge or side veneer on regrowth from stump trees.

Avocado. See *Persea*.

Banana. See *Musa*.

Berry (Youngberry, Boysenberry, Loganberry, Dewberry). See *Rubus*.

Blackberry. See *Rubus*.

Blueberry. See *Vaccinium*.

Butternut. See *Juglans cinerea*.

Cacao. See *Theobroma*.

Cactus (Prickly pear). See *Opuntia* spp.

Carambola. See *Averrhoa*.

Carica papaya. Papaya (159). These are large, tropical, fast-growing, short-lived herbaceous perennials. They come into fruiting within about 5 months and live 4 to 5 years. Plants are dioecious, require cross-pollination, and must be screened for sex form after coming into bearing. Some papaya seedling cultivars are highly inbred and come very true from seed (159). *Papaya ring spot* virus is a devastating aphid vectored disease for which transgenic resistant cultivars are now available (74).

Seed. Propagation is by seed from selected parents or controlled crosses. Seeds are equally useful if taken from stored or fresh fruit. Viability can be retained up to 6 years at 5°C (41°F) in sealed, moisture-proof bags (14). Seeds are sown in flats of soil or in seed beds in the open—and germinate in 2 to 3 weeks. Seedlings are first transplanted at about 10 cm (4 in) tall and then generally re-transplanted once or twice before they are put in their permanent location. Another technique is to plant 4 to 8 seeds per container and thin to 2 to 4 of the strongest when about 10 cm (4 in) tall. Soil pasteurization is recommended, since young papaya seedlings are very susceptible to damping-off organisms.

The usual practice in Florida is to plant seeds in midwinter and set the young plants in the field by early spring. Growth occurs during spring and summer, first fruits are mature by fall, and the plants bear all winter and the following season.

Miscellaneous Asexual Methods. Cuttings are used for specific cultivars with parthenocarpic fruit (20). Cuttings of branches with a "heel" are placed under mist with bottom heat. Budding is another method.

Micropropagation. Papaya can be micropropagated (135).

Carob. *See Ceratonia.*

Carya illinoinenesis. Pecan (86, 140). Pecans are native to the southwest United States and northern Mexico where many natural seedling groves exist. Commercial groves use selected cultivars grafted to pecan seedling rootstocks.

Seed. Pecan seeds of certain cultivars may show vivipary, which ruins the pecan crop. Seeds start to germinate in the hulls before harvesting and lose their viability in warm, dry storage. To maintain viability, seeds should be stored at 0°C (32°F) at 5 percent moisture immediately after harvest and until planted. Pecan seeds should be stratified for 12 to 16 weeks at about 1 to 5°C (34 to 41°F) to ensure good, rapid germination. Growth restrictions of the shell have been reported to be reduced by germination at high temperatures—30 to 35°C (86 to 95°F) (51, 139, 206).

Midwinter planting, with seedling emergence in the spring, is a successful procedure. Deep, well-drained, sandy soil should be used for growing pecan nursery trees. Young seedlings are tender and should be shaded against sunburn. In the summer, toward the end of the second growing season, the seedlings are large enough to bud.

Grafting. Cultivars are propagated by budding or grafting to two-year-old pecan seedling rootstocks. Patch budding in the nursery is the usual method. After the budded top grows for 1 or 2 seasons, the nursery tree is transplanted to a permanent location. Young pecan trees have a long taproot and must be handled carefully when digging and replanting.

Seedlings may be crown grafted with the whip graft in late winter or early spring. Pecan seedling trees, growing in place in the orchard, can be topworked by inlay bark grafting limbs 4 to 9 cm (1 1/2 to 3 1/2 in) in diameter. The four-flap or banana graft can be used with smaller-sized rootstock to graft more than one cultivar on the same rootstock (see Fig. 12–12).

Miscellaneous Asexual Methods. Pecans are difficult to root from cuttings but some rooting occurs with hardwood cuttings taken in the fall and treated with 10,000 ppm IBA (190). Mound (stool) layering for clonal rootstock production is done in Peru (see Fig. 14–12) (146). Plants have been propagated by root cuttings and by air layering when treated with IBA.

Micropropagation. Seedling shoots of pecans have been micropropagated and rooted *ex vitro* (88).

Rootstocks (86). Pecan seedlings from the 'Apache,' 'Riverside,' 'Elliott,' and 'Curtis' cultivars are reported to produce excellent grafted plants. Pecan roots are susceptible to *Verticillium wilt*. Experimental trials have indicated that *C. aquatics* might be a possible rootstock for wet soils. Although pecan cultivars will grow on hickory species rootstocks, the nuts generally do not attain normal size.

Carya ovata. Shagbark hickory (19, 138). Shagbark hickory is native to North America and grown as a nut tree and landscape ornamental.

Seed. Most trees are grown as seedlings because of difficulties grafting and transplanting, and slow growth of the trees. Some variation in seed requirements exists. Hickory seed will germinate when planted in the spring but should be kept in moist, cool storage until planting. Seed is orthodox, it can be stored dry at low temperatures (223); it has a stratification period of 3–4 months (52). Fall planting is successful if the soil in temperate climates is well mulched to prevent excessive freezing and thawing.

Grafting. Patch budding is done by nurseries in the commercial propagation of hickories, usually in late summer. Seedling rootstocks of *C. ovata* or pecan (*C. illinoensis*) are grown for 2 years or more before they are large enough to bud. The inlay bark graft is another option. Transplanting is a problem, so root pruning to force out lateral root growth is sometimes done.

Cashew. *See Anacardium.*

Castanea **spp.** Chestnut (19, 114, 150, 208). Chestnuts fall into three groups around the world: (a) European (*C. sativa*), (b) American (*C. dentata* and related species), and (c) Asiatic species (*C. mollissima* and *C. crenata*). The chestnut of commerce in edible nuts is the European (Spanish) sweet chestnut. It is produced in quantity from very old groves and new plantations throughout Europe, and large quantities of nuts are exported to the United States. The native American chestnut tree of eastern North America was very important for timber as well as nuts until the early 1900s when the fungal blight [*Cryphonectria (Endothia) parasitica*] was accidentally introduced from Asia. Essentially, all mature trees of the American chestnut have been killed to near the ground in natural woodlands of the eastern United States. The European chestnut is also susceptible but has not been as severely affected. Blight-resistant chestnut species in China and Japan provide an important source of resistance. Breeding programs both for cultivars and rootstocks are in progress and new, improved, blight-resistant materials are becoming available. In addition, a virus attacking the blight fungus has been discovered that has the promise of providing biocontrol of the disease.

Seed. Seeds are used for rootstocks, breeding, and occasionally for crops. Seeds are large, fleshy, recalcitrant, and should be prevented from drying following harvest and until planting. The nuts (seeds) are gathered as soon as they drop and are either planted in the fall or cold-moist stratified for 3 months, and later spring planted. Seeds are satisfactorily stored in tight tin cans, with one or two very small holes for ventilation, at 0°C (32°F) or slightly higher; this storage temperature also aids in overcoming embryo dormancy. Weevils in the nuts, which will destroy the embryo, can be killed by hot-water treatment at 49°C (120°F) for 30 minutes. After 1 year's growth, the seedlings should be large enough to transplant to their permanent location or be grafted to the desired cultivar.

Grafting and Budding (99). Clonal cultivars are propagated onto rootstocks using the splice, chip, whip, or inverted T-buds. Regular T-buds tend to "drown" from the excessive sap flow. A buried-inarch grafting technique for root chestnut cuttings has been reported (113). In this procedure, a nut that has started to germinate is cut off just above the root. A dormant scion with a wedge cut on the bottom is inserted into the inverted nut, which is then placed in a closed heated frame containing peat moss and vermiculite. After about 4 weeks of warm temperature the grafts are hardened-off and removed.

Cuttings. Generally difficult to root. Chinese chestnut can be propagated by rooting leafy cuttings under mist if treated with IBA (105), but the method has not been done on a large scale.

Stooling. This method of propagation is used in Europe for obtaining plants on their own roots, but it is expensive.

Rootstocks. Seedling rootstocks from seed of the cultivar being propagated are usually used. Because of hybridization, variability often results, which has been associated with graft incompatibility.

A chestnut hybrid (*C. sativa* × *C. castanea*) has been micropropagated using explants from mature trees (207).

Ceratonia siliqua. Carob, locust bean. Cultivated for its edible seed pods, also known as St. John's bread, it is a tough, low-maintenance plant, tolerant of dry, harsh conditions, that also has ornamental value. This subtropical evergreen tree is usually propagated by seeds, which germinate without difficulty when freshly harvested. However, if seeds dry out and the seed coat becomes hard, they should be softened by hot-water or sulfuric acid treatment. The taproot is easily injured, so it is best to sow seeds in air-pruning flats for more fibrous root development. Transplanting bare-root seedlings gives poor results, so seeds are best planted in their permanent location or started in containers for later transplanting. Selected cultivars are chip-budded in late spring, which is faster than grafting. Cuttings can be rooted if taken in mid-spring and treated with 7,500 ppm IBA. Air layering in late summer is successful.

Micropropagation. Micropropagation using explants from both seedling and mature trees has been done successfully (183).

Cherimoya. See *Annona*.

Cherry (Sour, Sweet). See *Prunus cerasus* (sour), *Prunus avium* (sweet).

Chinese Gooseberry. See *Actinidia*.

Citron. See *Citrus*.

Citrus **spp.** Citrus (48, 185). Includes cultivars of *C. aurantifolia* (lime), *C. limon* (lemon), *C. maxima* (pomelo), *C. medica* (citron), *C. reticulata* × *C. sinensis* (tangor), *C. xparadisi* (grapefruit), *C. reticulata* (mandarin orange), *C. sinensis* (sweet orange), *C. paradisi* × *C. reticulata* (tangelo), and related citrus species used for rootstock. In general, propagation is the same for all species of citrus. Members of this genus are readily intergrafted and grafted to other closely related genera such as *Fortunella* (kumquat) and *Poncirus* (trifoliate orange).

Seeds. Polyembryony occurs in seeds of most citrus species used as rootstocks due to nucellar embryony (see page 132). The sexual seedlings present within the embryo tend to be weak, variable, and are usually rogued out. The apomictic seedlings, which arise from the nucellus, are usually uniform and have the same genotype as the seed tree. As commercial plants, nucellar seedlings not only have a long juvenile period but are also vigorous, thorny, upright-growing, slow to come into bearing, and undesirable as an orchard tree. Vigor, thorniness, and delayed bearing become less pronounced in consecutive vegetative generations of propagation, particularly if propagules are taken from the upper part of older nucellar seedling trees. Nucellar seedlings are largely free of viruses and other systemic pathogens. Registered clonal selections of nucellar cultivars have been identified for all the commercial citrus cultivars and are the basis of citrus nursery source usage

[Citrus Clonal Protection Program (CCPP) http://ccpp.ucr.edu/] (see Chapter 16) (31). The vigor and uniformity of nucellar seedlings makes them ideal as rootstocks, and most rootstock species or cultivars are nucellar.

Citrus seeds generally have no dormancy but are injured if allowed to dry. Consequently, seeds should be planted immediately after being extracted from the ripe fruit. Seeds may be stored moist, in polyethylene bags at a low temperature [4°C (40°F)]; before storage they should be soaked for 10 minutes in water at 49°C (120°F) to aid in eliminating seed-borne diseases. Treatment of the seed with a fungicide is beneficial.

Grafting. The most important method used in commercial production of citrus is to bud or graft onto nucellar seedling rootstocks. T-buds, inverted T-buds, or modified cleft grafts can be used. In microbudding, a very small sliver of stem bearing the bud is inserted under the bark of the rootstock.

An alternate method used in commercial production of dwarf citrus is a cutting-graft in which the scion and the rootstock are grafted together, and placed in a mist bed or direct rooted into small liner pots so that the rootstock produces roots and the graft heals at the same time (50). Valencia sweet orange have been budded onto unrooted 'Swingle' citrumelo rootstock hardwood cuttings, which were then treated with 500 ppm IBA and rooted under mist (154). "Micro-budding" citrus to container-grown rootstock smaller than 2 mm diameter is being done to circumvent the typical nursery stage of T-budding to pencil-size rootstocks. After successful micro-buds take, plants go directly to the orchard. Tree growth is better than larger, conventional T-budded plants which have a typical 18-month nursery stage. Plants are more precocious, producing fruit 14–18 months after micro-budding. Microbudding has been used with grapefruit, oranges, lemons, and kumquat (see Fig. 13–22). See http://www.plantmanagementnetwork.org/pub/cm/news/2008/OrangeRevolution

Source Selection. Source selection of cultivars and rootstocks is essential in modern citrus nursery production. About 12 known viruses, a viroid causing "exocortis," and a mycoplasma-like organism causing "stubborn" disease can infect citrus. The most serious worldwide problem is *tristeza*, causing "quick decline" and/or "slow decline" in specific cultivar-rootstock combinations. Good indexing procedures have been worked out for most viruses (25, 122, 185). Citrus sometimes produce low-producing, inferior "bud-sport limbs," so it is desirable to select budsticks from a single tree, avoiding any off-type "sporting" branches (193).

Budwood should be taken only from known high-producing, disease-free trees, preferably registered clones (Citrus Clonal Protection Program http://ccpp.ucr.edu).

Rootstock Selection. Next to scion source selection, the choice of rootstock is critical in commercial citrus production. Rootstocks vary greatly in their relative susceptibility to diseases (*Phytophthora* sp.); viruses (particularly *tristeza*); interactions with scion; quality of fruit, size, and vigor; and tolerance to soil problems, such as chlorosis, salt, and boron excess.

Field Production of Nursery Plants. Avoid using soils infested with citrus nematodes (*Tylenchulus semipenetrans*) or burrowing nematodes (*Radopholus similis*), or soil-borne diseases, although citrus is resistant to *Verticillium* wilt. For nurseries, it is preferable to use virgin soil or at least a soil that has not been formerly planted with citrus. For small operations, raised seed beds enclosed by 12-in boards can be used. A soil mixture of sandy loam and peat moss is satisfactory. Treating the soil with a fumigant such as DD (dichloropropane-dichloropropene) will minimize the chances of nematode infestations. The seed bed and nursery site can be fumigated with methyl bromide to reduce the hazard of fungus infection. Planting should be delayed for 6 to 10 weeks following treatment to allow the fumigant to dissipate. Some soils in California, however, have remained toxic to citrus for a year following such treatment. The seed bed should be in a lathhouse, or some other provision should be made for screening the young seedlings from the full sun. The fruit of certain species, such as the trifoliate orange (*Poncirus trifoliata*) or its hybrids, matures in the fall. If the seeds are to be planted at that time, they should be held in moist storage at −1 to 4°C (30 to 40°F) 4 weeks before planting. Trifoliate orange seedlings, if grown during times of the year having short days, will respond strongly with increased growth when supplementary light is provided to lengthen the day. The same is true for seedlings of certain citrange and sweet orange cultivars (209).

The best time to plant the seeds is in the spring after the soil has warmed [above 15°C (60°F)]. Seeds should be planted in rows 5 to 7.5 cm (2 to 3 in) apart, and 2.5 cm (1 in) apart in the row. They are pressed lightly in the soil and covered with a 2 cm (3/4 in) layer of clean, sharp river sand. The sand prevents crusting and aids in the control of "damping-off" fungi. The soil should be kept moist at all times until the seedlings emerge. Either extreme, allowing the soil to become dry and baked or overly wet, should be avoided. PVC pipe with circulating hot water is placed below the seed bed to maintain a temperature of 27 to 29°C (80 to 85°F)—which hastens germination. By this method, seeds may be planted in the winter months, and the seedlings will be large enough to line out in the nursery in the spring. Many can be budded by fall or the following spring, a technique that often shortens the propagation time by 6 to 12 months.

After the seedlings are 20 to 30 cm (8 to 12 in) tall, they are ready to be transplanted from the seed bed to the nursery row, preferably in the spring after danger of frost has passed. The seedlings are dug with a spading fork after the soil has been wet thoroughly to a depth of 45 cm (18 in), because then they can be loosened and removed with little danger of root injury. All stunted or off-type seedlings, or those with crooked, misshapen roots, should be discarded.

The nursery site should be in a frost-free, weed-free location on a medium-textured, well-drained soil at least 61 cm (24 in) deep and with irrigation water available. Old citrus soils should be avoided unless fumigated before planting. The seedlings should be planted at the same depth as in the seed bed, and spaced 25 to 30 cm (10 to 12 in) apart in 90 to 120 cm (3 to 4 ft) rows.

Citrus seedlings are usually budded in the fall in Florida and California, starting in mid-September, early enough so that warm weather will ensure a good bud union, yet late enough so that bud growth does not start and the wound callus does not grow over the bud itself.

The best type of budwood is that next-to-the-last flush of growth, or the last flush after the growth hardens. A round bud-stick gives more good buds than an angular one. The best buds are those in the axils of large leaves. The bud-sticks are usually cut at the time of budding, the leaves removed, and the bud-sticks protected against drying. Bud-sticks may be stored for several weeks if kept moist and held under refrigeration at 4 to 13°C (40 to 55°F).

The T-bud method is extensively used for citrus. The bud piece is cut to include a sliver of wood beneath the bud. Fall buds are unwrapped in 6 to 8 weeks after budding, spring buds in about 3 weeks. In California and Texas the buds are usually inserted at a height of 30 to 45 cm (12 to 18 in), but in Florida the buds are inserted very low on the stock—2.5 to 5 cm (1 to 2 in) above the soil. Such low budding is often necessitated by profuse branching in rough lemon and sour orange rootstock seedlings, which is caused by partial defoliation due to scab and anthracnose spot.

Buds inserted in the fall are forced into growth by "lopping" or "crippling" the top of the seedlings 5 to 7.5 cm (2 to 3 in) above the bud just before spring growth starts, and consists of partly severing the top, allowing it to fall over on the ground. The top thus continues to nourish the seedling root, but forces the bud into growth. Lopping of spring buds is done when the bud wraps are removed—about 3 weeks after budding. If possible, the "lops" should be left until late summer, at which time they are cut off just above the bud union. Although lopping is necessary, it may make irrigation and cultivation difficult. An alternative practice is to cut the seedling completely off 30.5 to 35.5 cm (12 to 14 in) above the bud, then later cut it back immediately above the bud. However, this practice does not force the bud as well as lopping or cutting the seedling just above the bud.

Young citrus nursery trees may be dug "balled and burlapped" or bare-root. Bare-rooted trees should have the tops pruned back severely before digging. Transplanting of such trees is best done in early spring, but balled trees can be moved any time during spring before hot weather starts.

Container Production of Nursery Plants (56). Citrus seeds are mechanically extracted and washed, then germinated in special elongated square plastic pots in a greenhouse. After 3 or 4 months the seedling rootstock plants are large enough for inverted T-budding; finished budded trees can be developed in about 12 months when grown in plastic containers. Sometimes cleft grafting is used on the young seedlings after they are growing in their final containers (144). Citrus nursery trees grow well in containers and develop into good orchard trees upon field planting, providing they are not held in the container so long that root binding occurs (153).

Miscellaneous Asexual Methods. Citrus plants can be propagated by a variety of vegetative methods. Many citrus species can be propagated by rooting leafy cuttings, or by leaf-bud cuttings, although nursery trees are not commonly propagated in this manner (48). The Persian lime *(C. aurantifolia)* is propagated to some extent by air layering, as is the pummelo *(C. grande)* in southeast Asian countries. Citron *(C. medica)* is propagated from softwood or hardwood cuttings, because it tends to overgrow rootstock. Etrog *(C. medica)* must be grown on its own roots to be acceptable for Jewish holidays.

Micropropagation. Citrus can be micropropagated, but it is not a commercial method. Primary use is production of virus-free clones through micrografting procedures (177).

Rootstocks for Citrus (31, 48, 210). The principal characteristics of citrus are listed in the chart below. Information on recommended citrus rootstock for Florida can be found at http://news.ifas.ufl.edu/1999/04/14/new-florida-citrus-rootstock-selection-guide. *Sweet orange (C. sinensis)* is a good rootstock for all citrus cultivars, producing large, vigorous trees, but due to its high susceptibility to gummosis (*Phytophthora* spp.), it is not grown much today.

Sour Orange (C. aurantium). This is historically the most important rootstock species, used all over the world due to its vigor, hardiness, deep root system, resistance to gummosis diseases, and ability to produce high-quality, smooth, thin-skinned, and juicy fruit. Its high susceptibility to the *tristeza* virus, which produces quick decline, has eliminated its use in California and reduced its use in Florida, but this rootstock is still used in Texas, Mexico, Cuba, Venezuela, Honduras, Sicily, and Israel. The scion top may be tolerant to the virus, but a hyperreaction with the rootstock at the graft union kills the phloem tissues and the roots starve.

Satsuma oranges and kumquats, for which it is excellent. Trifoliate orange is commonly used as a stock for ornamental citrus and in home orchards for dwarfed trees. Trees on this stock yield heavily and produce high-quality fruits.

Trifoliate orange fruits produce large numbers of plump nucellar seeds that germinate easily. The upright-growing, thorny seedlings are easy to bud and handle, but their slow growth often necessitates an extra year in the nursery before salable trees are produced.

'Cleopatra' Mandarin (C. reshni). This is a good rootstock, particularly in Florida, for oranges, grapefruit, and other mandarin types, and has come into use in California and Texas as a replacement for sour orange rootstock. Its resistance to gummosis, comparative salt tolerance, and resistance to *tristeza* has justified its greater use. Its chief disadvantages are the slow growth of the seedlings, slowness in coming into bearing, and susceptibility to *Phytophthora parasitica* root rot.

Trait	Sweet Orange	Sour Orange	Rough Lemon	Citrange	Trifoliata	'Cleopatra'
Nucellar	70–90%		90–100%	nearly 100%	60%	80%
Incompatibility	OK	OK	some	OK	OK	'Eureka'
Vigor	large, vigorous	large, vigorous	very vigorous	semi-dwarf	dwarfing	
Fruit quality	very good	very good	tends to be low	good	very good	high resistant
Tristeza	high tolerant	very susceptible			resistant	susceptible
Phytophthora	very susceptible	resistant	susceptible		resistant	
Nematodes						
Soil	light, well drained		lighter soils, well drained		medium	salt-tolerant
Hardiness	medium	hardy	tender	hardy	hardy	

Rough Lemon (C. jambhiri). This has been the most important rootstock in Florida because its high vigor is well adapted to warm, humid areas with deep, sandy soils. Its primary drawback is the low quality of fruit that is produced, evidently a function of the high vigor, but it is also very cold-sensitive. Because of its susceptibility to blight and *Phytophthora,* its use is declining in Florida. However, the rootstock remains important in Arizona.

Trifoliate Orange (Poncirus trifoliata). This deciduous species is important because of its cold tolerance and tendency to dwarf the scion. It is the principal citrus rootstock in Japan and used in China as a stock for satsumas and kumquats. In northern Florida and along the Gulf Coast of Texas, it has been used as a stock for

Citranges (trifoliate orange × sweet orange Hybrids). These include several named cultivars—'Savage,' 'Morton,' 'Troyer,' and 'Carrizo,'—whose seedlings are used as rootstocks in California and Florida. 'Savage' is especially suitable as a dwarfing rootstock for grapefruit and the mandarins. 'Morton' is not as dwarfing but produces heavy crops of excellent-quality fruit. Seed production is so low, however, that getting quantities of nursery trees started is difficult. Sweet orange trees on 'Troyer' are vigorous, cold-hardy, resistant to gummosis, and produce high-quality fruit. This rootstock is widely used commercially in California particularly for replanting on old citrus soils. 'Troyer' is relatively fruitful and produces 15 to 20 plump seeds per fruit, facilitating the propagation of nursery trees.

Nucellar seedlings of citrange cultivars develop strong, single trunks, easily handled in the nursery. As with trifoliate orange, only exocortis-free buds should be used on citrange rootstocks; otherwise, dwarfing—and eventual low production—will result. 'Troyer' and 'Carrizo' are the two commercially most important citrus rootstocks in California and Arizona. 'Carrizo' is an important rootstock in Florida.

Rangpur Lime (C. aurantifolia × C. reticulate). This is the most widely used citrus rootstock in Brazil, producing vigorous, fruitful trees, which are resistant to *tristeza.* It is highly susceptible to *exocortis.* In Texas, it has been more salt-tolerant than other citrus rootstocks. Some strains of Rangpur lime are susceptible to *Phytophthora.*

Alemow (Citrus macrophylla). This rootstock is used in California for lemons because of its tolerance in high-boron areas. It is susceptible to *tristeza* when sweet orange scion cultivars are used, and to a rootstock necrosis first detected about 1960.

Coconut. See *Cocos.*

Cocos nucifera. Coconut (37). This tropical monocot is probably the most important nut crop in the world. There are two forms—tall and dwarf—with named seed-propagated cultivars of each, based on their origin in specific "seed gardens" collected from isolated groves. Some named cultivars maintain their characteristics quite dependably, but some of these are also susceptible to the highly lethal *golden disease.* Seeds should be collected from trees that produce large crops of high-quality nuts.

The nuts are usually germinated in seed beds. The nuts, still in the husk, are set at least 30 cm (12 in) apart in the bed and laid on their sides with the stem end containing the "eyes" slightly raised. The sprout emerges through the eye on the side that has the longest part of the triangular hull. As soon as this occurs (about a month after planting), the sprout sends roots downward through the hull and into the soil. In 6 to 18 months the seedlings are large enough to transplant to their permanent location.

Coconut is also tissue cultured for clonal propagation and safe germplasm exchange.

Coffea spp. Coffee (214). Coffee comes from three species, *Coffea arabica,* which is self-compatible, but also from *Coffea canephora* var. *robusta,* which is self incompatible, and *C. liberica.* Coffee comes relatively true from seed. *C. arabica* is the most important, such as seed propagated cultivars like 'Mondo Nova', 'Caturra' in Brazil, 'Villalbos' and 'Hibrido' in Central America, and 'Moko' in Saudi Arabia.

Seed. The most common method of propagation is by seed, preferably obtained from selected superior "mother" trees. Coffee seeds lose viability quickly and are subject to drying through the seed coverings. Seeds held at a moisture content of 40 to 50 percent and at 4 to 10°C (40 to 50°F) will keep for several months. There is no seed dormancy. Seeds are usually planted in seed beds under shade. Sometimes the seedlings are started in soil in containers formed from leaves, or in polyethylene bags, to facilitate transplanting. Germination takes place in 4 to 6 weeks. When the first pair of true leaves develop, the seedlings are transplanted to the nursery and set 30 cm (12 in) apart. After 12 to 18 months in the nursery the plant has formed 6 to 8 pairs of laterals and is ready to set out in the plantation.

Cuttings. Coffee can be propagated vegetatively by almost all methods. Leafy cuttings have potential for commercial use, but the resulting plants show strong topophysis. Cutting material should be taken only from upright shoot terminals in order to produce the desired upright-growing tree. Leafy cuttings of partially hardened wood treated with auxin can be rooted fairly easily with mist and partial shade (187).

Grafting. Coffee plants have been grafted using a hypocotyl-edonary grafting technique, primarily to utilize interspecific rootstock for nematode control (17).

Micropropagation. Coffee plants have been propagated by the formation of adventitious embryos in aseptic callus cultures (96).

Rootstocks. Seedlings of *C. canephora* var. *robusta* have been used for topworking to increase production in nematode-infested soils, but may not be desirable in uninfested areas (17).

Coffee. See *Coffea.*

Cola nitida. Kola. This tropical tree produces a caffeine-containing nut used to make refreshing beverages.

Seed. Plants are grown from seed sown directly in the field or in nurseries for later field planting. Seeds for planting should be harvested only when completely mature. Seedling plants have shown pronounced juvenility; 7 to 9 years are required for the seedlings to flower.

Miscellaneous Asexual Methods. Vegetative propagation using mature propagules can generate early production, but propagation is difficult. Rooting terminal leafy cuttings in polyethylene-covered frames without auxin has been successful. Air layering with IBA treatment is successful—commercial nut production is induced 12 to 18 months after planting the rooted layers. Patch budding is also a successful method of vegetative propagation.

***Corylus* spp.** Filbert, Hazelnut (125). Cultivars of hazelnut *(C. avellana)* are widely grown in Europe, but are called filberts in the Willamette valley of Oregon, where they are primarily grown in the United States. Native species in North America include *C. americana*.

Seed. Freshly harvested seeds of *C. avellana* are not dormant, but develop dormancy with drying in a complex interplay of inhibitors and growth promoters. Seed dormancy can be overcome with stratification for 3 months at 0 to 4°C (32 to 40°F). Seeds are mostly planted in the nursery in the fall.

Layering. The usual method of commercial filbert propagation is by simple layering in the spring using suckers arising from the base of vigorous young trees 4 to 8 years old. Special stool mother plants can be maintained in a bush form just for layering purposes. After one season's growth, a well-rooted plant 0.6 to 1.8 m (2 to 6 ft) tall may be ready to set out in the orchard. Old orchard trees are not suitable to use for layering. Sometimes suckers arising from the roots are dug and grown in the nursery row for a year or, if well rooted, planted directly in place in the orchard. Stool layering using girdling and spraying IBA at 750 mg/L^{-1} produces well-rooted liners ready for planting directly in the orchard (59).

Grafting. Using a whip graft in midwinter has been successful with the aid of a "hot grafting" procedure, (125). Commercial cultivars are grafted on seedling *C. avellana* rootstock.

Cuttings. Propagation is difficult, but some reports indicate that with careful timing in mid-June to mid-July while inactive, growth rooting can occur (52). Cuttings are treated with a 5,000 to 10,000 ppm IBA quick-dip or talc, and rooted under mist in a well-drained medium. Too much water results in decline, which is why fog is preferred to mist.

Micropropagation. Corylus can be micropropagated but is not a commercial procedure (7). Successful rooting is the chief limitation for large-scale micropropagation (10).

Crab Apple (Flowering). *See Malus in Chapter 20.*

Cranberry. *See Vaccinium.*

Currants. *See Ribes.*

Custard Apple. *See Annona.*

Cydonia oblonga. Quince. The quince is a small bushy tree that is grown for the jelly-making qualities of the fruit, due to its high pectin content, and as a dwarfing rootstock for pears; see the rootstock section for *Pyrus communis*.

Cuttings. Quince roots readily by hardwood cuttings using "heel" cuttings attached to one-year-old wood. Cuttings from two- to three-year-old wood also root easily. The burrs or knots found on this older wood are masses of adventitious root initials. Cuttings usually make sufficient growth in one season to transplant to their permanent location.

Grafting. Quince cultivars can be T-budded on rooted cuttings (such as 'Angers') or sometimes on quince seedlings.

Layering. Quince can be propagated readily by mound layering, which has been a traditional method of propagation.

Micropropagation. Methods for large-scale clonal production of 'Provence' quince has been described for use as a dwarfing pear rootstock (1).

Date Palm. *See Phoenix.*

Diospyros kaki. Japanese persimmon (180, 191). Subtropical tree species with many cultivars is grown widely in the Orient—and to a limited extent in the United States—for its large fruit. Cultivars are propagated by grafting onto seedling rootstocks. American persimmon *(D. virginiana)* is a hardy species that grows throughout the midwest and southern part of the United States. Some named cultivars are grown in home yards.

Seed. Seeds of *D. lotus, D. kaki,* and *D. virginiana* are stratified for 120 days at about 10°C (50°F). Dry seeds should be soaked in warm water for 2 days before stratification. Excessive drying of the seed is harmful, especially for *D. kaki*. Seeds are planted either in flats or in the nursery row. Young persimmon seedlings require shading.

Cuttings. Softwood cuttings with bottom heat, under mist, and treated with IBA will root, whereas hardwood cuttings do not root successfully.

Grafting. Crown grafting by the whip-and-tongue method or cleft can be done in early spring when both scionwood and rootstock are still dormant. Bench grafting may also be used; budding is less successful than grafting (180). *D. virginiana* is primarily produced by budding and grafting onto *D. virginiana* seedlings.

Rootstocks for Japanese Persimmon.

Diospyros lotus. This rootstock, widely used in California, is very vigorous and drought-resistant. It produces a fibrous type of root system that transplants easily. This rootstock is susceptible to crown gall (*Agrobacterium*) and *Verticillium*, will not tolerate poorly drained soils, and is highly resistant to oak root fungus (*Armillaria*). 'Hachiya' does not produce well on *D. lotus* stock because of excessive shedding of fruit in all stages (180). 'Fuyu' scions usually do not form a good union with *D. lotus*, although 'Fuyu' topworked on a compatible *D. kaki* interstock on *D. lotus* roots makes a satisfactory tree.

D. kaki. This rootstock is the most favored in Japan and is probably best for general use. It develops a good union with all cultivars, is resistant to crown gall (*Agrobacterium*) and oak root fungus (*Armillaria*), but is susceptible to *Verticillium*. Seedlings have a long taproot with few lateral roots, making transplanting somewhat difficult.

D. virginiana. Seedlings of this species are utilized in the southern United States due to their wide soil adaptation but have not proven satisfactory in some localities. 'Hachiya' growing on this rootstock in California is dwarfed, has sparse bloom, and yields poorly (180), but most other Oriental cultivars make a good union with this rootstock. However, diseases of an unknown nature carried in *D. kaki* scions will cause death of the *D. virginiana* roots. Otherwise, this rootstock seems to be tolerant of both drought and excess soil moisture. Its fibrous type of root system makes transplanting easy but increases suckering.

Micropropagation. Oriental persimmon cultivars can be micropropagated (42, 199) and produced from somatic embryogenesis (200).

Eriobotrya japonica. Loquat. This subtropical evergreen pome fruit is grown for its fruit which is used fresh, or in jams and jellies. The species is also used as a landscape plant. Many of the trees in use are unselected seedlings. Improved cultivars exist, but are difficult to propagate.

Seed. No treatment is apparently required, but seeds tend to germinate spontaneously. Should be sown soon after extraction, because heat and light tend to reduce quality of germination.

Grafting. Scions are T-budded or side-grafted to loquat seedlings. Established seedling trees may be topworked, using the cleft graft. Quince can also be used as a rootstock, producing a dwarfed tree. Inarching is also used, and topworking is easily done with bark grafting.

Layering. Air layering is successful and can be improved by auxin application (188). Three-month-old shoots are ringed and layered (20).

Euphoria longan. Longan (20). A tropical tree from southwest Asia and China. Seedlings have a long juvenile period and are variable. There are many vegetatively propagated cultivars in southern China.

Seed. Easily grown from seed.

Grafting. Cleft, whip, side, and inarch grafting are successful. The scions are semihard terminal shoots—about 0.5 cm (1/4 in) in meter, with leaves still attached. Grafts are made into a node subtended by a leaf and enclosed with a polyethylene bag, which is shaded. Seedling rootstocks of the cultivar to be grafted are used.

Cuttings. Leafy cuttings with thick hard terminal wood, about 25 cm (10 in) long, are taken in winter and rooted with bottom heat and mist.

Layering. Marcottage (air layering) is the preferred method. This practice is carried out in spring with well-matured, recently flushed growth. See the procedure described in Chapter 15. The layer is removed after 2 to 4 months when roots have turned from white to creamy brown. Establish in plastic bags or pots and transplant to the field after 6 to 12 months.

Feijoa sellowiana. Feijoa. Pineapple Guava (109). This small subtropical evergreen tree is grown for its edible fruit, useful for jelly and juice.

Seed. Seeds germinate without difficulty. They are started in flats of media and later transplanted to the nursery row.

Cuttings. Leafy softwood cuttings treated with auxin and started under closed frames can be rooted. Rooting under mist is usually quite difficult, otherwise layering can be successful.

Grafting. Scions are grafted to feijoa seedlings using a whip graft or a cleft graft (63).

Ficus carica. Fig. The cultivated fig is an ancient fruit crop from the Mediterranean area that grows into small trees (40, 123). Specific cultivars have been propagated by hardwood cuttings for centuries.

Seed. Seeds are used only for breeding new cultivars. The fruit is a fleshy receptacle bearing many small achenes inside. Some of these are sterile and others fertile. The heavier fertile seeds can be separated from the lighter ones by floating them in water. Seeds can be germinated easily in flats of soil mix.

Cuttings. Hardwood cuttings of 2- or 3-year-old wood or basal parts of vigorous 1-year shoots with a minimum of pith are suitable for cuttings. For best results cuttings should be prepared in early spring, well before bud-break, with the bases allowed to callus for around 10 days at about 24°C (75°F) in slightly moist bark or peat media before planting. IBA will enhance rooting. The cuttings are grown for one or two seasons in the nursery and then transplanted to their permanent location. Figs can also be grown in containers in the nursery. A method in European countries is to plant long cuttings, 0.9 to 1.2 m (3 to 4 ft), with their full length in the ground where the tree is to be located permanently. Sometimes two cuttings are set in one location to increase the possibility that at least one will survive. Pregirdling of fig shoots 30 days before removal for rooting is beneficial (20).

Grafting. The fig can be budded, using either T-buds inserted in vigorous 1-year-old shoots on heavily pruned trees, or patch-budded onto older shoots.

Layering. Figs are easily air layered. One-year-old branches, if layered in early spring, are usually well-rooted by midsummer.

Micropropagation. Figs can be micropropagated (173).

Fig. *See Ficus*

Fragaria xananassa. Strawberry (76, 85). The strawberry is an herbaceous perennial whose basic vegetative structure is a thickened vertical stem, known as a "crown." It produces a rosette of leaves and inflorescences in the spring, and runners (see Fig. 14–18) in the summer. In the normal annual cycle, the plant is induced to initiate flowering in the fall by warm days and shortened photoperiod, followed by chilling of the crown during winter to stimulate spring vegetative growth. Flowering and fruiting occur for about 6 weeks in spring. Runners are induced in the long days of summer and grow out from the main crown to start new plantlets at every other node. In some strawberry cultivars, runner formation is inhibited and the plant expands by increasing its crown ("everbearing types"). In recent years, "day-neutral" genotypes have been produced by breeding, which make flowering possible at any time of the year. Natural vegetative reproduction is by runner development or by crown division.

Modern commercial strawberry production has manipulated the strawberry plant annual growth cycle by combining breeding, nursery, and plant management and environment control into an integrated system, depending upon the region where grown (85). In areas of warm winters and hot or warm summers (e.g., California, Florida), the plants are planted as crowns, maintained in a *"hill" system* managed as an annual crop, and replanted each year after fumigating the previously planted ground. Annual production has increased more than tenfold and the fruiting season has been expanded from a few weeks in the spring to literally year-round production, depending on cultivar, management, and climate. In areas with shorter growing seasons and cold winters (northern or southern latitudes), the *matted row* is used, because it requires less inputs and produces over 3 to 5 seasons. Other systems utilize plastic high tunnels for production under controlled environments.

Strawberry has many disease problems that must be controlled both during propagation and production. These include fungal soil and foliage diseases, nematodes, viruses spread by aphids and nematodes, and mycoplasma diseases spread by aphids. Control involves cultivar resistance, pathogen-free propagation and planting material, chemical control, and sanitation.

Seeds. Seed propagation is used only in breeding programs for the development of new cultivars. The strawberry fruit is an expanded receptacle upon which are borne the true "fruit" or achenes. These are usually extracted with a blender. The achenes are quite hard, and scarifying with sulfuric acid for 15 to 20 minutes prior to planting improves germination (182).

Micropropagation. In the United States, meristem culture is used mostly in virus control programs. In Europe, on the other hand, millions of nursery plants are produced annually by micropropagation (47, 143). The first stage (establishment) is to excise a 0.1 to 0.5 mm meristem dome from surface-sterilized, newly formed runner tips. These are planted on a basal culture medium. Stage 2 (proliferation) is a transfer to a medium containing cytokinin to promote axillary buds development (may include several subcultures). Stage 3 is a transfer to an auxin medium for rooting, and stage 4 is acclimatization and transplantation (see Chapters 17 and 18). Long day and gibberellin treatment stimulate runnering.

Some genetic instability may occur as off-type individuals, apparently due to somaclonal variation. The effect can be minimized by regular inspection, rogueing, and limiting subcultures to 3 to 5 times.

Virus Control. Stock plants are heat pretreated at 36°C (97°F) for 6 weeks. From here, shoot tips are meristemed and individual stock plants are indexed by grafting to indicator plants (see Fig. 16–30), or tested by ELISA or DNA hybridization (see Chapter 16). Programs for certifying virus free, true-to-type plants are available in many countries [see UC Strawberry Links - http://www.innovationaccess.ucdavis.edu/strawberry/links.htm](41). Certification includes the following steps: I. Nuclear stock plants from virus-tested meristem cultured plants are grown in an insect proof greenhouse or screenhouse. II. The next generation provides registered foundation plants that are grown in fumigated fields to produce plants to be distributed to commercial nurseries. III. The next generation (daughter) plants that are produced are sold as certified plants to commercial strawberry producers.

Production of Planting Stock. Three different kinds of plants (crowns) are produced: (a) *actively growing green* plants, (b) *fully dormant crowns,* and (c) *semidormant* crowns. Active green plants are dug directly from production fields or by rooting runners as plug plants under mist in the greenhouse. Fully dormant plants are produced in strawberry nurseries located in areas (such as northern California) where they have a long, warm growing season and are exposed to adequate winter chilling. Nurseries are located in a favorable location, and isolated from other crops, including wild strawberries. Field nurseries are planted at relatively high elevations where there is little rainfall but lots of sunlight to maximize starch reserves. Land is fumigated the previous fall. Nursery plants are planted in the spring and subsequently maintained under irrigation for rapid growth and rooting of runners. Mature runner plants are dug with special machines in the fall when mature and desired dormancy is achieved. Immediately thereafter, the leaves and petioles are removed and the plants are packed with roots toward the center in unsealed cardboard or wood boxes lined with polyethylene (0.75 to 1.5 mil) to keep plants from drying out. These bare root crowns are stored until planting at slightly below freezing (−1°C). Semidormant plants with varying levels of chilling geared to different cultivars and planting dates are generated by digging plants from the field at different times in the autumn at different geographical locations (i.e., elevation and latitude).

Boxes of 1,000 or 2,000 plant units can be stored and kept in good condition up to one year at −2 and −1°C (28 and 30°F). During storage, the plants not only remain dormant but physiological changes occur that overcome endodormancy and enhance vigor when the plants are later removed and planted in the production area. The time schedule for digging, storage, and planting is geared to the individual cultivar, the production schedule at the production fields, and the environment at the growing site.

Crown Division. Everbearing cultivars that produce few runners are propagated by crown division. Certain cultivars may produce 10 to 15 strong crowns per plant by the end of the growing season. In the spring, such plants are dug and carefully cut apart; each crown may then be used as a new plant.

Garcinia mangostana. Mangosteen (159). This slow-growing evergreen tree of the Southeast Asian tropics produces a parthenocarpic, soft, sweet, and delicious fruit (20). There are no known wild progenitors; apparently Mangosteen originated as an interspecific hybrid of unknown species. Also, there are no known cultivars because the plants are reproduced apomictically. Plants are nucellar seedlings that may take 8 to 15 years to produce fruit. Seeds have low viability but can germinate in 4 to 6 days. Grafting with a top wedge is successful, but the rootstock must be compatible. Layering is improved by IBA treatment.

Gooseberries. *See Ribes.*

Grape. *See Vitis.*

Grapefruit. *See Citrus.*

Guava (Common or Lemon; Cattley or Strawberry). *See Psidium.*

Guava (Pineapple). *See Feijoa.*

Hickory (Shagbark). *See Carya.*

Jackfruit. *See Artocarpus.*

Juglans cinerea. Butternut. This tree is native to the eastern United States where it is grown mostly as a timber tree (19). Several butternut cultivars have been selected for nut production. *Butternut canker*, caused by a recently introduced fungus *Sirococcus clavigignenti-jugandacearun*, is now threatening its existence as a native species but propagation and maintenance of selected germplasm and the few cultivars may insure its preservation (170).

Seed. The very hard-shelled seeds are enclosed within a husk that must be removed after collection. Nuts may be stored at 3 to 4°C (37 to 39°F) at 80 to 90 percent humidity. Seeds have dormant embryos and should be either planted in the fall out-of-doors or stratified for several months and planted in the spring.

Grafting. Cultivars are propagated by chip budding or bench grafting (using the whip graft) on Persian *(J. regia)* or black *(J. nigra)* walnut seedling rootstocks. Butternut is also bark grafted onto *J. nigra* seedlings. Excessive sap "bleeding" is a problem in grafting, and necessary steps must be taken to overcome it.

Cuttings. There was enhanced rooting of softwood cuttings treated with 62 and 74 ppm K-IBA, and high survival of rooted cuttings overwintered in cold storage and acclimated to the field (171).

Juglans nigra. Black Walnut (71). This species is native to eastern North America where it is an important timber tree. It also produces excellent nuts, although they are very hard-shelled.

Seed. The hard-shelled nuts apparently are benefited by warm stratification to soften the shells, followed by cold stratification to overcome embryo dormancy. Early fall planting could provide both conditions. Seedlings are started in nurseries and only the strongest, most vigorous trees are set out in the plantation. Careful planting is necessary to obtain the essential rapid, early growth.

Cuttings. Cuttings are possible but difficult to root.

Grafting. Black walnut cultivars selected for their nut qualities are propagated by patch or ring budding or side grafting on *J. nigra* rootstocks (34, 189). This method is suitable for seedling stocks up to about 2.5 cm (1 in) in diameter.

Juglans regia. Persian or English Walnut (66). This important commercial nut is produced in California, Oregon, and in the Mediterranean region. Its origin is southwestern Asia, particularly from the area occupied by present-day Iran. "Carpathian walnuts" are a cold-hardy strain of *J. regia,* which originated in cold winter areas of the Carpathian Mountains of Poland and the Kiev and Poltava regions of the Ukraine. Carpathian cultivars have been developed that are winter-hardy in the eastern United States (79).

Seed. Nuts of *Juglans* species used as seedling rootstocks for *J. regia* should be either fall-planted, or stratified for about 2 to 4 months at 2 to 4°C (36 to 40°F) before they are planted in the spring. It is better to plant the seeds before they start sprouting in the stratification boxes, but with care, sprouted seeds can also be planted successfully. At the end of one season, the seedlings should be large enough to bud. Seedlings of walnut are very sensitive to waterlogged soil conditions under which they will not survive (33).

Grafting. *J. regia* is propagated by patch budding, T-budding, or whip grafting on one-year-old seedling rootstocks. Topworking-established seedling trees in the orchard that are 1- to 4-years old can be done by bark grafting in late spring. Patch budding in the spring or summer works well when topworking smaller sized material.

Micropropagation. *J. regia* cultivars (66) and clonal sources of Paradox walnut can be micropropagated but have difficulty in establishment.

Rootstocks
Northern California Black Walnut *(J. hindsii).* This rootstock is the most commonly used in California. The seedlings are vigorous and make a strong graft union. They are somewhat resistant to oak root fungus *(Armillaria mellea),* and root-knot nematode *(Meloidogyne* spp.), but are susceptible to crown and root rot *(Phytophthora* spp.), and the root-lesion nematode *(Pratylenchus vulnus).* Persian walnut trees on this rootstock are susceptible to a serious problem known as "blackline" caused by a pollen-transmissible virus, *cherry leaf roll,* which occurs in California, Oregon, England, Italy, and France. Symptoms include a breakdown of tissues in the cambial region at the graft union, leading to lethal girdling of the trees. The Persian walnut scion above the graft union is tolerant of the virus but the hypersensitive *J. hindsii* rootstock tissue at the union is killed. As a result the girdled top dies and the rootstock survives.

Persian Walnut *(J. regia).* Worldwide, this species is probably the most common rootstock for Persian walnut cultivars due to availability of seeds. Seedlings of this species as a rootstock produce good trees with an excellent graft union. The roots are susceptible to crown gall *(Agrobacterium),* oak root fungus *(Armillarea),* and salt accumulation in the soil. They are not as resistant to root-knot nematodes *(Meloidogyne)* as *J. hindsii.* Nurserymen object to the slow initial growth of the seedlings. This rootstock is used in Oregon, where blackline is a very serious problem. Seedlings of the 'Manregian' clone (of the Manchurian race of *J. regia*), imported by the USDA as P. I. No. 18256, are used as rootstocks. They are vigorous, cold-hardy, and—in Oregon—no more susceptible to oak root fungus than *J. hindsii* seedlings (166).

Paradox Walnut (J. hindsii × J. regia). First generation (F_1) hybrid seedlings are obtained from seed taken from *J. hindsii* trees, whose pistillate (female) flowers have been wind-pollinated with pollen from nearby *J. regia* trees. Seed source trees are found when seeds from a *J. hindsii* tree are planted and some of the seedling progeny are hybrids (an example of genotypic selection). The numbers vary widely, from none to almost 100 percent, depending upon the individual source tree. The hybrids are easily distinguished by their large leaves in comparison with the smaller-leaved, self-pollinated *J. hindsii* seedlings. Seed from Paradox trees should not be used for producing rootstocks, because of their great variability in all seedling characteristics. Although first-generation (F_1) seedlings are variable in some characteristics, most of them exhibit hybrid vigor and make excellent vigorous rootstocks for the Persian walnut. They are more resistant than either parent to crown rot *(Phytophthora),* and are tolerant of saline and heavy, wet soils. Paradox seedlings are susceptible to crown gall *(Agrobacterium).* Persian walnut on Paradox roots are just as susceptible to the blackline virus as *J. hindsii* seedlings. Trees on Paradox rootstocks grow and yield as well as, or better than, those on *J. hindsii* roots, and may produce large-size nuts with better kernel color. In heavy or low-fertility soils, trees on Paradox rootstock grow faster than those on *J. hindsii*.

Since Paradox seeds are difficult to secure in quantity, clonal selection efforts with accompanying vegetative propagation systems have been attempted but no mass propagation systems have yet evolved. Leafy cuttings under mist, hardwood cuttings, and trench layering can be rooted but with difficulty, and there are major problems when transplanting.

Eastern Black Walnut (J. nigra). This species has been recommended as a rootstock for Persian walnut cultivars in Europe and the former Soviet Union, but not in California. As compared with *J. regia* roots, it is reported to have greater tolerance to nematodes, oak root fungus, *Phytophthora,* and crown gall, but in California rootstock trials, trees on this stock showed poor yields.

Jujube. See *Zizyphus.*

Kiwifruit. See *Actinidia.*

Kola. See *Cola.*

Lemon. See *Citrus.*

Lime. See *Citrus.*

Litchi chinensis. Lychee. Evergreen subtropical tree fruit with delicious, juicy fruit that has been grown in China since ancient times. It is currently grown in most of the subtropical countries in the world. Many vegetatively propagated cultivars exist. It is typically propagated by air-layering mature tree branches. Growth can be enhanced by mycorrhizal fungi after air-layering (112).

Seed. The seeds are used in breeding, but have a long juvenile period of 10 to 15 years (20, 38); they are also used to produce seedling rootstocks. Litchi seeds germinate in 2 to 3 weeks if planted immediately upon removal from the fruit. They lose their viability within a month if not planted.

Cuttings. Cultivars are generally difficult to root from cuttings. Tip cuttings from a flush of growth in the spring have been rooted in fairly high percentages under mist in the full sun. Auxin is beneficial. Semi-hardwood cuttings from an active flush of new growth root more readily than those from dormant hardwood cuttings.

Grafting (20). Some success has been reported for splice grafting using subterminal portions of shoots for the scion. It is best to retain leaves on the rootstock. Approach grafting is also successful, as is chip and T-budding. Poor results may be attributed to graft incompatibility (159).

Layering (Marcottage). Air layering is the most important method of propagation and has been used in China for centuries. Layering can be done at any time of the year, but best results are obtained in spring or summer (137). Rooting takes place within 8 to 10 weeks with a success rate up to 100 percent. Large limbs with mature vegetative growth air layer more easily than small ones with recently flushed wood. Auxin applications may be beneficial. Stooling (mound layering) has also been reported to produce good results (20).

Loganberry. See *Rubus.*

Longan. See *Euphoria.*

Loquat. See *Eriobotrya.*

Lychee. See *Litchi.*

Macadamia spp. Macadamia nut. *M. integrifolia* and *M. tetraphylla* are the two principal species (84). These subtropical evergreen nut trees may be grown as seedlings, but clonal cultivars are preferred. Macadamia is highly resistant to *Phytophthora cinnamomi* and will tolerate heavy clay soil.

Seed. Fresh seeds should be planted in the fall as soon as they mature, either directly in the nursery or in sand boxes in a lathhouse. Seedlings are transplanted to the nursery or into poly containers after growing 10 to 15 cm (4 to 6 in) tall. Seeds should not be cracked because they are readily attacked by fungi. Only seeds that sink when placed in water should be used (84). Seeds retain viability for about 12 months at 4°C (40°F), but at room temperature viability starts decreasing after about 4 months. Scarifying or soaking the seed in hot water hastens germination.

Cuttings. Macadamia can be propagated by rooting leafy, semi-hardwood cuttings of mature, current season's growth. Tip cuttings 8 to 10 cm (3 to 4 in) work best. Treatment with IBA at 8,000 to 10,000 ppm is beneficial. Cuttings should be placed in a closed propagating frame or under mist for rooting. Bottom heat at 24°C (75°F) is beneficial. Cultivar differences exist in ease of rooting. *M. tetraphylla* cuttings root more readily than those of *M. integrifolia*. Rooted cuttings are not widely used in commercial plantings because of their shallow root system.

Grafting. One of the side graft methods is used. Leaves should be retained for a time on the rootstock. Rapid healing of the union is promoted if, prior to grafting, the rootstock is checked in growth by water or nitrogen deficiency to permit carbohydrate accumulation. Ringing the branches that are to be the source of the scions for 6 to 8 weeks before they are taken increases their carbohydrate content and promotes healing of the union (145). Wrap the entire scion with clear Parafilm or buddy grafting tape to prevent drying out; keep plants well hydrated after grafting. Budding has generally been unsuccessful (53). Seedling rootstock of *M. tetraphylla* is preferred for grafting *M. integrifolia* cultivars (84, 198). Macadamia can also be propagated by the cutting-graft method.

***Malus* spp.** Crab apple (Flowering). See Chapter 20.

***Malus xdomestica*.** Apple. This deciduous fruit crop is one of the oldest and most widely grown fruit crops in the world. Thousands of cultivars exist, many of which are chimeras and bud-sports. Viruses and virus-like organisms are serious problems for apples (see Chapter 16) such that essentially all commercial nurseries utilize virus-tested foundation clones and produce trees in a program of Registration and Certification (87).

Seed. Seeds are used for breeding and are the rootstock for many older orchards in the United States. At present, seedling rootstocks are used in limited parts of the world. The principal source of commercial seeds has been the pomace from processed apples. For spring sowing, seeds require stratification for 60 to 90 days at 2 to 7°C (35 to 45°F) to germinate. Some nurseries fall-plant seeds so that they receive the natural winter chilling. Soil should be raked over in the spring to avoid seedlings that become crooked in breaking through the crust. To obtain a branched root system, the seedlings may be undercut while small to prevent the development of a taproot. However, a straight root may be preferred for bench grafting. Seedlings that do not grow to a satisfactory size in 1 year should be eliminated.

Layering. The principal method of propagating the widely used clonal rootstocks is by mound layering (see Chapter 14). Large nursery areas in the Pacific Northwest, the eastern United States, and in northern Europe and England are devoted to stool beds.

Cuttings. Propagation of cultivars by hardwood cuttings is not successful, but certain clonal rootstocks can be propagated by special methods described in Chapter 10 (91). Softwood cuttings can be rooted under mist, but this method is not used commercially. Propagation of long cuttings [30 to 70 cm (12 to 28 in)] from semi-hardwood cuttings rooted under high pressure fog systems is being used as an alternative to replace more expensive mound layering systems for 'M9' and 'M27' dwarfing apple rootstock (195, 196).

Grafting. Root grafting either by the whip graft or by machine has been a traditional method of propagation (see Chapter 12). Seedling or layered liners are obtained from rootstock nurseries during the dormant season and stored at low temperatures until the time to graft. Plants are grafted during the dormant season and lined-out in the nursery in the spring as the outside temperatures become warm.

T-budding is a traditional method of field propagation, done either in the fall or in the spring. Dormant liners are obtained from rootstock nurseries and planted in late winter. In a mild winter area, such as California, rootstocks are budded (T- or chip) in April or early May with nursery trees produced by the end of the same season. In other areas, the rootstock may grow for the remainder of that season and are fall-budded to grow an additional year. In recent years, fall chip budding has become widely adopted in different parts of the world.

Micropropagation. Methods to micropropagate apple scions and rootstock have been developed (111, 126, 226). Nevertheless, commercial application is limited.

Micropropagated plants are more expensive than conventionally propagated ones and plants on their own roots are more vigorous and larger in size than those desired in commercial orchards. Rooting spur-type and dwarf genotypes may be possible (117). Clonal rootstock propagation is more feasible, and some commercial production occurs. Use of micropropagated source material has application to nursery production of apple rootstocks.

Rootstocks (64). All apple rootstocks, either seedling or clonal, are in the genus *Malus,* although the apple will grow for a time and even come into bearing while grafted on pear *(Pyrus communis)* rootstock. At one time, apple seedling rootstocks were widely used in the western and southeastern United States, but the trend has been almost entirely towards using clonal size-controlling rootstocks.

Seedling Rootstocks. Apples on seedling roots produce large-sized trees. Seedlings of 'Delicious,' 'Golden Delicious,' 'McIntosh,' 'Winesap,' 'Yellow Newtown,' and 'Rome Beauty' (but particularly 'Delicious') have been the most successful, being uniform with no incompatibility problems. In purchasing seed it is sometimes difficult to determine the seed source. In the colder portions of the United States—the Dakotas and Minnesota—the hardy Siberian crab apple *(Malus baccata)* and seedlings of such cultivars as 'Antonovka,' containing some *M. baccata* parentage are favored. In Poland, 'Antonovka' seedlings are the chief apple rootstock. Some nurseries in British Columbia, Canada, have used 'McIntosh' seedlings because of their winter-hardiness, upright growth in the nursery, and early fall shedding of leaves, but they tend to be somewhat susceptible to *hairy root,* a form of crown gall *(Agrobacterium),* and to powdery mildew *(Podosphaera teucotricha).* Trees tend to be variable in size and performance.

Apples with the triploid number of chromosomes, such as 'Gravenstein,' 'Baldwin,' 'Stayman,' 'Winesap,' 'Arkansas,' 'Rhode Island Greening,' 'Bramley's Seedling,' 'Jonagold,' 'Mutsu,' and 'Tompkins King,' produce seeds that are of low viability and are not recommended as a seed source. Seeds of 'Wealthy,' 'Jonathan,' or 'Hibernal' have given unsatisfactory results.

Apple roots are resistant to root-knot *(Meloidogyne)* and root-lesion nematodes *(Pratelynchus vulnus),* moderately susceptible to oak root fungus *(Armillaria),* and highly resistant to *Verticillium* wilt.

Clonal Rootstocks. Numerous clonal, asexually propagated apple rootstocks have been developed. Most are in the species *Malus xdomestica*. Rootstock breeding and testing programs are underway in various apple-producing countries to develop new dwarfing and semi-dwarfing rootstocks to replace them.

Malling and Malling-Merton Series. This rootstock selection program, which began in 1912 at the East Malling Research Station in England, included both a program of rootstock selection and standardization of propagation. Size of grafted cultivars (see Fig. 11–10) range from very dwarfed to very invigorated, although size control was also a function of the scion cultivar. Fruit size was large, especially on young dwarfed trees, often larger than on standard-sized trees. The Malling rootstocks are compatible with most apple cultivars, and trees grafted to these rootstocks have been planted in varying amounts all over the world. In 1928, a second phase of this program was instigated jointly by the John Innes Horticultural Institution and the East Malling Research Station (174). A new series of apple rootstocks, referred to as Malling-Merton (or MM), was produced that added resistance to woolly aphids, and to provide a further range in tree vigor.

Other improved cultivar characteristics associated with these rootstocks include high yield, precocity of flowering (with some rootstocks), well-anchored trees (with some rootstocks), freedom from suckering, and good propagation qualities. All of the Malling and Malling-Merton stocks are readily propagated, mostly by stool-bed layering. A number of rootstock clones, notably 'Malling 26,' 'MM 106,' and 'MM 111,' propagate well by hardwood cuttings (92). Virus-tested material of both of these groups of rootstocks has been developed by a further joint effort by East Malling and Long Ashton Research Stations in England. "Clean" material is distributed under the designations EMLA 7, EMLA 9, EMLA 26, EMLA 27, EMLA 106, and EMLA 111. In many cases, these "clean" rootstocks produce trees with 10 to 15 percent or more increased vigor than the older virus-infected stocks of the same cultivar.

Dwarfing Clonal Rootstocks

'Malling 27.' This most dwarfing of all the Malling stocks produces trees 1 1/2 to 2 m (4 to 6 ft) tall. They are one-half to two-thirds the size of those on 'Malling 9,' which makes it useful for high-density plantings. Virus-tested propagating material was released by the East Malling Research Station in 1970. 'Malling 27' can be used as an interstock to give a dwarfing effect.

'Malling 9' ('Jaune de Metz'). This rootstock originated as a chance seedling in France in 1879 and has been used widely in Europe for many years as an apple rootstock. It is a dwarfed tree itself and a valuable dwarfing rootstock much in demand for producing small trees for the home garden or for commercial high-density plantings. Recommended tree spacing is

1.8 × 3.6 m (6 × 12 ft). Such trees are seldom over 3 m (9 ft) tall when mature, usually starting to bear in the first year or two after planting. There are a number of clonal selections of 'Malling 9,' all dwarfing but somewhat different otherwise.

'Malling 9' has numerous thick, fleshy, brittle roots and requires a fertile soil for best performance. The trees require staking or trellising for support. It is resistant to collar rot *(Phytophthora cactorum)* but susceptible to mildew *(Podosphaera teucotricha),* crown gall *(Agrobacterium),* fire blight *(Erwinia amylovora),* and woolly apple aphid *(Eriosoma lanyerum).* Roots are sensitive to low winter temperatures. Propagation is by stooling.

'Malling 9' can be used as an *interstock* in double-worked trees, but the dwarfing of the scion cultivar is less than when it is used as the rootstock.

'Malling 26.' This rootstock was introduced in 1959 at the East Malling Research Station, originating from a cross between 'Malling 16' and 'Malling 9.' Better anchored than 'M 9,' it produces a tree somewhat larger and sturdier than 'Malling 9' but less so than 'Malling 7' or 'MM 106.' It still requires staking. Suggested tree spacing is 3 × 4.2 m (10 × 14 ft). It can be propagated by softwood cuttings under mist or by hardwood cuttings, but produces poorly in stool beds. Although quite winter-hardy, it does not tolerate heavy or poorly drained soils, and is especially susceptible to fire blight *(Erwinia amylovora)* but less to collar rot *(Phytophthora cactorum).*

Semi-dwarfing Clonal Rootstocks

'Malling 7a.' Malling 7 was originally selected at East Malling from a group of French rootstocks known as Doucin. This rootstock produces trees somewhat larger than those on 'Malling 26' roots. The 'Malling 7a' designation indicates a clonal selection free of certain viruses present in the original 'Malling 7,' the viruses having been removed by heat therapy. A virus-indexed EMLA clone was introduced in 1974. 'Malling 7a' has a stronger, deeper root system than 'Malling 9' and produces an early bearing, semi-dwarf tree. It is tolerant of excessive soil moisture, but susceptible to crown gall. Well-anchored staking is required for the first few years. Suggested tree spacing is 4.2 × 4.8 m (14 × 16 ft). This rootstock has the undesirable characteristic of suckering badly, and the trees are not very winter-hardy. 'Malling 7a' is easily propagated by stooling or by leafy cuttings under mist.

'Malling-Merton 106.' This clone originated as a cross between 'Northern Spy' and 'Malling 1,' producing trees two-thirds to three-fourths the size of trees on seedling rootstocks. Although once popular, its planting has decreased because of crown rot *(Phytophthora)* problems. On good soils, with some scion cultivars, it can produce a large tree. The roots are well anchored and do not sucker. Suggested planting distances are 4.2 × 5.4 m (14 × 18 ft). In some areas 'MM 106' is susceptible to mildew *(Podsphaera)* and highly susceptible to collar rot *(Phytopthora),* which may be its chief weakness. It has not been affected by fire blight. It grows well in the nursery but drops its leaves later in the fall than most other understocks and is not resistant to early fall freezes. Hardwood and softwood cuttings root easily and stool beds are quite productive.

Vigorous Clonal Rootstocks

'Malling-Merton 111.' This stock originated as a cross between 'Northern Spy' and 'Merton 793' ('Northern Spy' × 'Malling 2'). A virus-indexed EMLA clone was introduced in 1969. Grafted trees are about 75 to 80 percent the size of seedlings in most orchards where they are grown. It produces more precocious bearing than those trees on seedlings, but less than those on 'MM 106,' 'M 7,' and 'M 26.' It does well on a wide range of soil types. It is susceptible to mildew but not to collar rot or woolly aphid. Stool beds are highly productive, with heavy root systems developing. Hardwood cuttings root well with proper treatment, as do softwood cuttings under mist. 'MM 111' is more winter-hardy than 'Malling 7' or 'MM 106.' Suggested planting distances are 4.8 × 6 m (16 × 20 ft). It shows excessive vigor in some situations.

Clonal Rootstocks from Miscellaneous Sources

'Alnarp 2.' This rootstock, developed at the Alnarp Fruit Tree Station in Sweden, is widely used there for its winter-hardy properties. The roots are well anchored and give about 20 percent size reduction as compared to trees on seedling roots. It is susceptible to woolly apple aphid and to fire blight.

'Robusta No. 5' [M. robusta (M. baccata × M. prunifolia)]. This vigorous, very hardy clonal apple rootstock, propagated by stooling or stem cuttings, originated in 1928 at the Central Experimental Farm, Ottawa, Canada. It is apparently resistant to fire blight *(Erwinia amylovora)* and crown rot *(Phytophthora),* and seems to be compatible with most apple cultivars. It is extensively used as an apple rootstock in eastern Ontario and Quebec, as well as in the New England states. It is the best rootstock for use where extreme winter-hardiness is required. However, its low chilling requirement produces the unfavorable habit of starting growth too early in the spring following three or four warm days in late winter. It is not a dwarfing rootstock.

Polish Series. These rootstocks were the result of a breeding program started in Poland in 1954. Six clones (P-1, P-2, P-14, P-16, P-18, and P-22) are available (64) with a range of tree size potential and other characteristics. As a group, they were selected for hardiness.

Budagovski Series. This material originated from a breeding program in Russia to produce very hardy apple rootstocks (64). 'Budagovski 9' is very dwarfing and is recommended as an interstock in cold regions. 'Budagovski 118' (75 percent size), 'Budagovski 490' (65 percent), and 'Budagovski 491' (20 percent) are other promising rootstocks of the series.

Mandarin. *See Citrus.*

Mangifera indica. Mango (35, 159). Most plantings of this tropical evergreen fruit tree are seedlings that may be sexual (monoembryonic) or nucellar (polyembryonic) in origin—both conditions may occur simultaneously in the seed. Growth of several shoots from one seed does not necessarily indicate nucellar embryos, since certain cultivars develop shoots from below ground, arising in the axils of the cotyledons of one embryo, which may or may not be of zygotic origin (9). Polyembryonic cultivars commonly occur in mango. Monoembryonic cultivars should not be propagated by seed, as they do not come true.

Seed. Mango seeds are used for rootstocks, although seedlings of polyembryonic cultivars are apomictic. Seedlings have long juvenile periods and are overly vigorous. Seeds should be planted within a week of maturation. Storage is possible within the fruits or in polyethylene bags at about 21°C (70°F) for at least 2 months. Low-temperature (below 10°C, 50°F) storage and excessive drying should be avoided. By removing the tough endocarp that surrounds the seed and planting in a sterilized medium, good germination should occur within 2 to 3 weeks. Seedlings should be transplanted to pots or into nursery rows soon after they start to grow.

Cuttings. Cuttings are difficult, but some success is reported from India using etiolation, IBA treatment, mist, and bottom heat (142).

Grafting. Mangos are commonly propagated in Florida by veneer grafting or by chip budding. A week after budding, the top of the rootstock is removed two to three nodes above the bud—with final removal back to the bud when the bud shoot is 7.5 to 10 cm (3 to 4 in) long. The best budwood is prepared from hardened terminal growth 6 to 10 mm (1/4 to 3/8 in) in diameter. The leaves are removed, with the exception of two or three terminal ones. The buds swell in 2 to 3 weeks, and are then ready to use. If the buds are to be used on stocks older than 3 weeks, ringing the base of the shoots from which the buds are to be taken about 10 days before they are used increases their carbohydrate supply and seems to promote graft union formation.

Budding is best done when the rootstock seedlings are 2- to 3-weeks old—in the succulent red stage. Four to six weeks after budding, the inserted bud should start growth (137). T-budding has also been successful. Patch, shield, and Forkert budding is done in India (142).

Approach grafting, termed *inarching* in India, has been used since ancient times in propagating the mango. Veneer grafting is also successful (156), as is saddle grafting (203). Veneer grafting is recommended in India (142) either for nursery propagation or top-working. The scion should be a terminal nonflowering shoot of 3 to 4 months' maturity. The scions are defoliated 7 to 10 days before they are removed for grafting, keeping a part of the petioles attached, which helps the buds to swell. This defoliation is done from March to September in northern India. In India, assorted rootstock seedlings are mostly used. Monoembryonic rootstocks can be multiplied by air layering or by cuttings, and subsequently clonally propagated by stooling (155).

A method called epicotyl/stone grafting has been described in India (142). Seeds are germinated in a sand bed covered with leaf mold. Eight- to fifteen-day-old germinated seedlings are lifted, decapitated about 5 cm above the stone, and wedge-shaped scions are inserted into a vertical split. Buds are wrapped with poly tape, and grafted plantlets are planted in poly bags or outdoors in a trench, after which they are then shifted to the field.

Layering. Air layering is successful, especially when etiolated shoots are treated with IBA at 10,000 ppm; such treatments have given high rooting and survival (157). Pot-layering and stooling are also suggested in India (142).

Mango. *See Mangifera.*

Mangosteen. *See Garcinia.*

Manikara zapota. Sapodilla, Nispero. Tropical fruit tree originating in Mexico and Central America but grown throughout the tropics (20, 29, 149). *M. chicle* is a medium-sized evergreen tree that produces chicle, an ingredient in chewing gum. Many cultivars exist in different parts of the world; 'Prolific' and 'Brown Sugar' are grown in Florida. Although seedling may be readily grown, the plants are variable and have a very long juvenile period of 5 to 8 years. Seedlings are used as rootstocks. Fruits are depulped, washed, and dried. Soak seed overnight and sow in seed beds or pots. Germination takes about 4 weeks. Seedlings are transferred to small pots when 15 cm high. While side veneer grafting is the most important graft, budding can also be done. Air layering is done in India. Cuttings are difficult.

***Morus* spp.** Mulberry. Mulberry cultivars grown for their fruit are mostly *M. alba* or *M. nigra*, but some of *M. rubra* occur. Seeds are easily grown but produce weedy trees.

Cutting propagation is readily started by hardwood cuttings 20 to 30 cm (8 to 12 in) long and planted in early spring. Leafy softwood or semi-hardwood cuttings treated with 8,000 ppm IBA root easily under mist (52).

Mulberry can be commercially micropropagated (107).

Mulberry. *See Morus.*

***Musa* spp.** Banana. Edible bananas apparently originated in the tropical lowlands of Malaysia as seedless, triploid mutations with year-long production. They since have become not only an essential food resource around the world, but perhaps the most important fruit export to the rest of the world. The banana "tree" is a large herbaceous perennial monocot whose "stem" (pseudostem) grows horizontally as a rhizome (see Chapter 15 and Fig. 15–32), consisting of compressed, curved bases of leaf stalks arranged spirally at nodes. Lateral buds develop at the nodes of a new segment of the stem and grow into upright "suckers" that soon develop their own roots and a large leaf base. The older fruiting section (i.e. "parent") is cut after the fruit is harvested. A banana plant may live for many years, but as a succession of new plants, each arising as a sucker (daughter shoot) from the rhizome; any given sucker fruits only once, then dies.

Asexual Methods. Edible clones rarely produce seeds, and commercial propagation is asexual. Two primary methods are division of the rhizome (locally called corms) and suckers. A large rhizome is cut into pieces, which are termed heads, suckers, or pups, weighing 3 to 4.5 kg (7 to 10 lb), depending on the cultivar. Larger suckers are the preferred planting material. Each head should contain at least two buds capable of growing into suckers. Before planting, the pieces are pared to remove old roots and disease and are immersed in water at 52°C for 20 minutes, or treated with pesticides to control nematodes and borers. Each sucker produces two branches in the first crop.

Fairly large "sword" suckers 90 to 180 cm (3 to 6 ft) high with well-developed roots are also used, but the leaves must be shortened considerably to reduce water loss after the sucker is cut from the parent plant. Suckers are removed with a sharp cutting tool inserted vertically about halfway between the parent stalk and stem of the sucker. These sword suckers produce only one bunch of fruit in the first crop, but they are often preferred, because of the large size of the bunch.

Micropropagation (106). Millions of banana plants are now produced annually around the world by micropropagation from sources free of *Fusarium* wilt. These are started from explants obtained from a decapitated shoot apex from "clean" banana suckers. In 1 year, about 1 million pathogen-free plantlets can be produced for commercial plantings. Similarly, mosaic-free plants obtained after meristem tip culture and thermotherapy (20) can be propagated. These protocols have been standardized for commercial production (179). A majority of intensive, commercial banana production utilizes micropropagation. Field tests have shown that micropropagated plants yield somewhat better than sucker-derived plants (24). A low percentage of off-type plants of 'Lady Finger' could be controlled by early detection and roguing in the production process (192). Because of the clonal nature of banana and the difficulty in breeding, biotechnology has provided much promise in banana improvement and propagation to control its many problems (80).

Nectarine. *See Prunus persica.*

Nephelium lappaceum. Hairy Litchi. A tropical evergreen fruit tree similar to litchi and longan (20). Seedlings are variable and have a long juvenile period. Cultivars are grafted to seedling rootstocks. Seeds are sown immediately upon ripening because drying causes loss of viability. Seeds are removed from the fruit, washed, and sown horizontally. Two leaved seedlings are produced within 2 weeks. The greatest success is with patch budding or Forkert budding onto 1- to 2-year-old rootstocks of the same cultivar. Inarching is successful, as is air layering.

Olea europaea. Olive (90, 186). Many cultivars of this ancient crop originating in Southwest Asia exist. They are mostly propagated by budding or grafting onto seedling or clonal rootstocks, hardwood or semi-hardwood cuttings, or "suckers" from old trees.

Seed. Seeds are used primarily for rootstocks. Germination is somewhat difficult due to the hard stony endocarp ("pit") that surrounds the seed. Seeds of small-fruited cultivars germinate more easily than those of large-fruited ones. Since germination is sometimes prolonged over 1 or 2 years, a common practice is to plant more seeds than will be needed as seedlings to offset the low germination percentage. Clipping the end of the endocarp will hasten germination. Softening the endocarp with sulfuric acid or sodium hydroxide, then warm-stratifying at 15°C (59°F) will improve germination, as will stratifying the seeds (still in the endocarp) at 10°C (50°F), removing the endocarp, and planting at 20°C (68°F) (186). Removing the seed from the pit, removing the seed coats, and germinating at 20°C in Petri dishes on moist filter paper will give high germination if the fruits are collected shortly after pit hardening.

Cuttings. Hardwood cuttings may be made from two- to three-year-old wood about 2.5 cm (1 in) in diameter and 20 to 30 cm (8 to 12 in) long. All leaves are removed. IBA aids rooting. An older technique is to soak the basal ends of the cuttings in 15 ppm IBA for 24 hours, followed by storage in moist sawdust at 15 to 21°C (60 to 70°F) for a month preceding spring planting in the nursery (90).

Semi-hardwood cuttings are made from vigorous, 1-year-old wood about 6 mm (1/4 in) in diameter. Cuttings are taken in early summer or midsummer, and trimmed to 10 to 15 cm (4 to 6 in) long with two to six leaves retained on the upper portion of the cutting. Cuttings are treated with a 4,000 ppm IBA quick-dip and rooted under mist. Cultivars vary greatly in their ease of rooting (93).

Grafting. Seedlings tend to grow slowly and may take a year or two to become large enough to be grafted or budded. Successful methods include T-budding, patch budding, whip grafting, and side-tongue grafting. A widely used method in Italy is bark grafting small seedlings in the nursery row in the spring. The stocks are cut off several centimeters above ground, and one small scion is inserted in each seedling, followed by tying and waxing. After grafting or budding, it takes 1 or 2 more years to produce a tree large enough for transplanting to the orchard.

Rootstocks. *Olea europaea* seedlings are commonly used, although they vary considerably in tree vigor and size and are susceptible to *Verticillium* wilt (93). Using rooted cuttings of a strong-growing cultivar such as 'Mission' may be more desirable because of uniformity and vigor. Clonal rootstocks 'Oblonga' or 'Swan Hill' are resistant to *Verticillium* wilt (90). Other *Olea* species do not make satisfactory rootstocks for the edible olive. (72)

Olive. *See Olea.*

***Opuntia* spp.** Prickly pear cactus (61, 130). The most important species in the *Cactaceae* family, produced on five continents. It is a multi-purpose crop, used for production of fruits, vegetables, forage for animal feed, and as an ornamental plant. Also used as a raw-industrial material to produce wine, candies, flour, and as the host for the cochineal insect (*Dactylopius*) to obtain the natural-dye carminic acid, which is used for coloring fabrics, food, and cosmetics. Can be propagated from sexually produced or apomictic seed. Primarily, vegetatively propagated from single or multiple cladodes (see Fig. 10–14); also with small portions of cladodes comprising two or more areoles, or by using fruits as propagules. Wounding and applying auxin can enhance rooting and propagation rates of cladodes (130). Can also be grafted and micrografted (60). Micropropagation has higher propagation rates and reduced requirements for space than other clonal regeneration techniques, and can produce healthy, pathogen-free plants (61).

Orange (Mandarin, Sour, Sweet). *See Citrus.*

Papaya. *See Carica.*

Passiflora edulis. Passion fruit (54). The genus has about sixty species that produce edible fruit grown in tropical and subtropical areas. In Hawaii, Brazil, Sri Lanka, Kenya, and New Zealand, production is based on seedlings of the yellow form (*P. edulis* f. *flavicarpa*). The plant grows as a vine.

Seed. Must be extracted from freshly harvested fruit and removed from fruit and juice after fermentation for about 3 days. Seeds are washed, dried, and kept in sealed containers in a refrigerator where they will remain viable for at least three months. Germination takes place in 2 to 3 weeks after planting in pots, whereupon they are transplanted in the two- to three-leaf stage.

Grafting. Cleft-grafting is done when seedlings are 30 to 40 cm high. Rootstock is cut off about 15 to 20 cm above ground and split. The scion is a small piece of vine wood about 8 to 10 cm long with larger leaves removed. The wedge at the scion base is inserted into the rootstock cleft and the union is bound with plastic budding tape. Callusing takes 4 to 6 weeks, at which time plants are transplanted to the field. Scions of purple-fruited hybrid cultivars are grafted to seedling rootstocks of golden passion fruit (*P. edulis* forma *flavicarpa*), which are resistant to *Fusarium* and to nematodes (202).

Passion Fruit. *See Passiflora.*

Peach. *See Prunus persica.*

Pear. *See Pyrus.*

Pecan. *See Carya.*

Persea americana. Avocado (21, 159). This subtropical and tropical tree has been grown in South America for centuries, apparently originating in three areas which identify the three important horticultural races. These areas are the highlands of central Mexico (Mexican race), the highlands of Guatemala (Guatemalan race), and coastal regions of Guatemala (West Indian race). Many cultivars exist and are grafted onto seedling or clonal rootstocks. A major limitation in avocado production is susceptibility to *Phytophthora cinnamomi*.

Seed. Seeds are used for the production of seedling rootstocks. To avoid *sun-blotch viroid,* the seeds (and budwood) must be taken from source trees that have been registered by state certifying agencies as free of the disease (218). Seeds are generally planted shortly after removal from fruit taken from the tree (not picked from the ground), care being taken not to allow them to dry out. Seeds can be stored 6 to 8 months if packed in dry peat moss and held at 5°C (41°F) with 90 percent relative humidity

Seeds are immersed in hot water at 49 to 52°C (120 to 125°F) for 30 minutes before planting to eliminate infection from avocado root rot (*Phytophthora cinnamomi*) (225). Germination of the seeds is hastened by removing the brown seed coats and cutting a thin slice from the apical and basal end of each seed before planting. The seed coats can be removed by wetting the seeds and allowing them to dry in the sun. Seeds are then sown in nursery rows or into containers.

Cuttings. Semi-hardwood stem cuttings from mature trees of the Mexican race have been rooted under intermittent mist using bottom heat at 25°C (77°F), provided seven or more leaves are retained on a 20-cm (8-in) cutting. If the leaves drop, rooting ceases (178). Cuttings taken from mature trees of the Guatemalan and West Indian races are difficult to root. Avocado cultivars started as rooted cuttings eventually make satisfactory trees, but in general, grow poorly in their initial stages.

Rooting has been produced by etiolation (see Chapters 9 and 14). Frolich (69) first showed the advantage of etiolation by covering the base of developing shoots with Velcro or black electrical tape to keep the attached shoot base in darkness while allowing the terminal part in light to develop five to six leaves. Shoots with etiolated bases were then detached and rooted in a propagating case (70). See Chapters 9 and 10 for discussions on banding and etiolation systems for stock plants. Brokaw (22) developed this concept into the commercial practice known as the *nurse-seed etiolation* procedure, which with further modifications have become the standard procedure for propagating avocado cultivars on clonal rootstocks (16).

Grafting. Nursery trees are commercially propagated by tip grafting (splice or whip graft) (201), or wedge or cleft grafting on avocado seedlings or clonal rootstocks (22). T-budding is used occasionally in Florida and some of the Caribbean countries—but the usual nursery practice is to graft mature tip scions, either as side-veneer or cleft grafts, on young succulent rootstock seedlings.

Scion and budwood must be selected properly. The best buds are usually near the terminal ends of a completed growth cycle with fully matured, leathery leaves. The leaves should be removed when the budwood is taken to prevent drying.

Nursery Field Propagation Systems. Seeds of the Mexican race, which ripen their fruit in the fall, may be planted in beds in late fall or early winter. They should be placed with the large basal end down, just deep enough to cover the tips. If grown in a warm area, the sprouted seeds are ready to line-out in the nursery row the following spring. By summer or fall, the seedlings are usually large enough to permit T-budding; if not, they can be budded the following spring.

Four to six weeks after budding, the seedling rootstock is cut off 20 to 25 cm (8 to 10 in) above the bud, or bent over a few inches above the bud. The remaining portion of the seedling above the bud is not cut off until the bud shoots have completed a cycle of growth. The new shoots are usually staked and tied. In digging from the nursery, the trees are "balled and burlapped" for removal to their permanent location following the first or second growth cycle or just as the first flush starts.

Container Production of Nursery Plants. Container production has become standard in the industry because of the greater ease in controlling *Phytophthora* (11) and its use in Certification programs (see Chapter 16) (16).

(a) Seedling rootstock production. Seeds are planted in polyethylene bags and the resulting seedlings

are cleft or wedge grafted about 9 cm above the seed 2 to 4 weeks after germination. Scions are taken from freshly cut terminal shoots showing strong plump buds. The plastic containers, with the grafts, are placed on raised benches in plastic-covered houses, which facilitates sanitation procedures to avoid *P. cinnamomi*. After 4 to 6 weeks, the grafts are moved to a 50 percent shade house to acclimatize, then are transplanted into large containers for moving to an outdoor area (172).

In Florida, young, succulent West Indian seedlings (137), grown in gallon containers, are grafted when they are 15 to 25 cm (6 to 10 in) high and 6 to 10 mm (1/4 to 3/8 in) in diameter. The scions are shoot terminals, 5 to 7.5 cm (2 to 3 in) long, with a plump terminal bud, taken just as it resumes growth.

(b) Nurse seed/etiolation method (21). A seed is planted about one-third of the way from the bottom of a 300 × 70 (12 × 2.8 in) polyethylene bag, the top of which has been folded down on the bottom half of the bag. The seedling shoot is grafted close to the surface using a single strong shoot. When scion shoot growth starts, the plant is placed in darkness. When new etiolated shoots have grown 300 to 400 mm (12 to 16 in) they are removed from the dark, a metal or plastic "C" ring is clamped around the base of the shoot, and the bag is extended to its full length and filled with moist potting medium. The ring gradually constricts and eventually kills the "nurse" seedling, usually within one year. The rooted scion is the new clonal rootstock which can be grafted to the fruiting scion, usually at 200 to 250 mm (8 to 10 in).

This method has since been modified (16) for nursery use by encouraging more than one seedling shoot, which is individually grafted, each to be enclosed in a micro-container of plastic. Each shoot is grafted, the new shoot is allowed to grow, subsequently wounded and treated with 7,000 ppm IBA, surrounded by rooting media, and placed in the dark. After the grafted shoot has reached 300 mm (12 in) the plant is returned to the light and grafted to a fruiting scion. When this shoot reaches 55 mm (2 in) the nurse seedling is pruned off and the rooted clone/fruiting cultivar grown on a salable nursery tree. The entire process takes 16 to 20 months. Other changes include increasing temperature from 20 to 25°C (68 to 75°F) to 27°C (81°F) for rooting and surrounding stem with a clear plastic cup to monitor rooting.

Rootstocks

Mexican Race (P. americana var. drymifolia). Seedlings of this race are preferred in California for their cold-hardiness and their partial resistance to *Phytophthora cinnamomi,* lime-induced chlorosis, and *Dothiorella* and *Verticillium.* They are, however, susceptible to injury from high salinity. In Florida, where seedlings of large diameter are preferred for grafting, the Mexican types are little used, owing to their thin shoots.

Guatemalan Race (P. americana). These are occasionally used in California when there is a scarcity of Mexican seeds. Guatemalan seedlings are often more vigorous initially than Mexican seedlings, but are more susceptible to diseases and to injury from cold.

West Indian Race (P. americana). These seedlings are susceptible to frost injury in California but are widely used in Florida. The large seed produces a pencil-sized shoot suitable for side grafting in 2 to 4 weeks after germination.

Resistant Rootstocks. Research toward the selection of disease-resistant avocado rootstocks began in the 1940s (225). This research led to the introduction of clonal selections 'Duke 6' and 'Duke 7,' the latter becoming commercially important. These were individual seedling selections from the original 'Duke' cultivar. More recently 'Thomas,' from a survivor tree in an orchard, has become an important rootstock, although a number of others are undergoing orchard testing. 'Toro Canyon' is a nursery-selected clonal selection (148).

Persimmon (American, Japanese). See *Diospyros.*

Phoenix dactylifera. Date palm (100, 101, 102, 160). This monocotyledonous plant is propagated by seeds, offshoots, and more recently by micropropagation (somatic embryogenesis). Date palms are dioecious. In a commercial planting, most of the trees are female, but a few male trees are necessary for cross-pollination.

Seed. Date seeds germinate readily. However, seedling populations have a long juvenile period and about half of the trees are males. Seedling female trees produce fruits of variable and generally inferior types.

Offshoots (100, 101, 102). Superior clonal cultivars exist, some since ancient times. The natural method of vegetative propagation is from offshoots which arise from axillary buds near the juvenile base of the tree. Roots develop on the base after 3 to 5 years and are ready to be removed. To promote rooting, soil should be added to the base of the offshoot. Large, well-rooted offshoots, weighing 18 to 45 kg (40 to 100 lb), are more likely to grow than smaller ones. Offshoots higher on the trunk can be induced to root by placing moist rooting medium against the base of the offshoot in a box or polyethylene tube. Unrooted offshoots arising even higher on the stem can be cut off and rooted in the nursery, but percentages may be reduced. A single date palm may yield 10 to 25 offshoots during its first 10 to 15 years.

Considerable skill is required to cut off the date offshoot properly. Soil is dug away from the rooted offshoot, but a ball of moist earth as thick as possible should remain attached to the roots. The connection with the parent tree should be exposed on each side by removing loose fiber and old leaf bases. A special chisel, with a blade flat on one side and beveled on the other, is used to sever the offshoot. The first cut is made to the side of the base of the offshoot close to the main trunk. The beveled side of the chisel is toward the parent tree, which gives a smooth cut on the offshoot. A single cut may be sufficient, but usually one or more cuts from each side are necessary to remove the offshoot. The offshoot should never be pried loose; it should be cut off cleanly. After removal, it should be handled carefully, replanted as soon as possible, and roots should be prevented from drying out (see Fig. 14–20).

Micropropagation. Commercial micropropagation has become an important alternative because of its greater efficiency and speed (163). Two general methods can be used. One is production of somatic embryos from callus generated from shoot tips; this technique produces a larger number of plants but has greater danger of variability. The second method is rooting axillary shoots produced from the shoot apex of offshoots. The general procedure is described in Chapters 17 and 18.

Pineapple. *See Ananas.*

Pistacia vera. Pistachio (83, 116). This popular nut crop originated in the deserts of southwest Asia where it has been a food staple for centuries. In recent years, a major pistachio industry has developed in California.

Seed. Seeds are used in commercial production for rootstocks. Obtain fruits when the hulls turn blue-green. Pinkish seeds are blank. The hulls contain a germination inhibitor and should be removed. *P. terebinthus* seed should be held in moist sand at 5 to 10°C (41 to 50°F) for 6 weeks before sowing and then germinated at 20°C (68°F). With early planting of seed, the more vigorous seedlings may reach budding size by fall.

Seed may be affected by chalcid larvae.

Nursery and Field Production. Because of their long taproot, best results are obtained by planting seedling rootstocks in their permanent orchard location, then T-budding them 0.6 m (2 ft) or more above ground after the seedlings are well established. Soak seed for 24 hours at room temperature in loosely rolled damp burlap in the dark at 20 to 30°C (70 to 90°F). When germinated, plant seedlings in double Jiffy pots with one-third each of humus, fumigated sand, and loam. When danger of frost is past, and seedlings are 4 to 6 inches in height, transplant into 18 inch peat pots. Buds of *P. vera* cultivars are quite large, requiring a fairly large seedling to accommodate them. Budding is possible over a considerable period of time, but if done before mid-spring, when sap flow may be excessive, the percentage of takes is low. A marked improvement in bud union occurs as budding is extended through the summer and fall. Seedlings can be grown in long tubular, biodegradable pots ready for planting in the orchard.

Rootstocks.

P. atlantica. Seedlings of this species were used in earlier commercial orchards in California but have been found highly susceptible to *Verticillium* wilt *(Verticillium dahliae),* a serious soil-borne fungus of pistachio.

P. integerrima. Seedlings of this species show resistance to *Verticillium,* and is being used in areas having such problems.

P. terebinthus. Seedlings of this species may have some resistance to *Armillaria,* and are cold-tolerant, but grow slowly.

P. atlantica × *P. integerrima.* The F_1 hybrid seedlings between these species are vigorous and have been found to have both superior cold tolerance (as compared to *P. integerrima*) and resistance to *Vertillium.* In California, specific hybrid seed sources are available from the University of California under the name of "UCB-1" (68) and from some private nurseries under the name of Pioneer Gold. Some roguing for off-type seedlings should be practiced. Clonal rootstocks from selected seedlings may be propagated by softwood cuttings (5) or by micropropagation.

P. khinjuk. Used as rootstock in Turkey for its drought resistance, adaptability to poor soils, and resistance to *Meloidogyne* sp. It can be micropropagated (204).

Micropropagation. Cultivars of *P. vera* have been micropropagated (12, 164, 167), and somatic embryos developed into plantlets (165).

Pistachio. *See Pistacia.*

Plum (European). *See Prunus domestica.*

Plum (Japanese). See *Prunus salicina*.

Pomegranate. See *Punica*.

***Prunus* spp.** Stone-fruit and nut-producing trees include sweet cherry (*P. avium*), sour cherry (*P. cerasus*), apricot (*P. armeniaca*), almond (*P. dulcis*), peach and nectarine (*P. persica*), European plum (*P. domestica*), and Japanese plum (*P. salicina*). In addition, several hundred related species and species hybrids exist worldwide.

Prunus cultivars are highly sensitive to virus problems, and only virus-tested sources should be used for scions, clonal rootstocks, and seed orchards (see Chapter 16) (87). Registration and certification programs exist in many states of the United States, Canada, and in most European countries (28, 68). Likewise, stringent quarantine regulations exist for the movement of budwood across national boundaries (67).

Essentially, all *Prunus* cultivars are clones grafted to seedling or clonal rootstocks. Major rootstock problems include the following: (a) size control, (b) pest and disease susceptibility—particularly *Verticillium* wilt, crown rot (*Phytophthora* sp.), bacterial canker (*Pseudomonas*), and *Armillaria* root rot, (c) nematodes, as root knot (*Meloidogyne incognita, M. javanica*), lesion nematode (*Pratylenchus vulna*), and (d) soil characteristics, as calcareous and soil asphyxia.

Seed. Seed propagation is important for breeding and rootstock production. Prunus seeds have a hard protective endocarp (shell, pit, stone) that surrounds the seed. The embryo is nonendospermic and exhibits embryo dormancy that requires one to three months' cold stratification to germinate. Seeds of different species and sometimes individual cultivars have specific time requirements. In addition, the stone may be inhibiting either because of substances or because of hardness and/or lack of water uptake (cherry). Presoaking, stone removal, leaching in running water, or warm stratification preceding cold stratification may be helpful.

Cuttings. Propagation of *Prunus* cultivars or rootstocks varies greatly by species and cultivar. Producing own-rooted peach trees of specific cultivars by hardwood cuttings has been used to produce high density orchards (44). See cutting propagation methods for ornamental *Prunus* species in Chapter 20. For the most part, propagation by cuttings is more successful in producing clonal rootstocks that have been selected specifically for ease of rooting. Long cuttings [62 cm (24 in)] of *Prunus* 'Gisela 5' rootstock taken from semi-hardwood cuttings root well under high pressure fog systems (195, 196). Collecting hardwood cuttings (20 cm long and 9 to 11 mm wide) from the upper portion of parent shoots and quick-dipping in 1,000 ppm IBA yielded 50 percent rooting of 'Gisela 5' (62).

Direct rooting of hardwood cuttings in the nursery is a preferred method, and variations of this method are described in Chapter 10. Clones of many species can be rooted by softwood cuttings under mist (91, 147), although the method is not usually commercially practiced. Hard-to-root rootstock clones have been rooted commercially by trench layering ('F12/1' cherry) or mound layering ('Colt').

Budding. The most important method of propagation of stone fruits is to bud or graft cultivars to rootstocks in the nursery and transplant trees to the orchard. Seedling rootstocks are prepared either by stratifying the seeds during the winter and planting early in the spring, or planting in the fall and allowing natural chilling in the ground to bring about the moist-chilling process. Seedling liners that have been produced during the previous year may be purchased and lined-out in the nursery row for later budding.

T-budding is the usual practice and may be done in the fall, early spring or late spring, or summer depending on the length of the growing season. Chip budding may also be utilized in areas with a relatively short growing season. Trees are produced within 1, 2, or 3 years, depending upon the nursery site and conditions, the species, and the desires of the propagator.

In California, budding of plum, almond, or apricot to hardwood cuttings of 'Marianna' plum is done in the spring (April) directly onto hardwood cuttings in the nursery that had been inserted for rooting in the previous fall. Dormant budwood is stored over winter. Quick removal of the cutting top allows the new shoot to grow enough during the remainder of the season to produce a nursery tree in 1 year.

Micropropagation (55, 65, 108). Much research and development has been devoted to micropropagation of *Prunus* cultivars and rootstocks. Success has been achieved with plum, cherry, peach, and peach ×almond hybrid rootstocks. European micropropagation nurseries have produced millions of commercial plants of cultivars such as 'GF677' peach ×almond. Production has not been economical in the United States, and tissue–culture-based nursery systems have not replaced conventional propagation.

Embryo rescue and ovule culture are effective tools in breeding programs which make possible the propagation of early ripening cultivars or interspecific hybrids whose embryos abort when the fruits are ripe. Meristem culture is used for removing viruses from infected plants in initiating foundation clones.

Prunus cerasus. Sour cherry. *P. avium* (sweet cherry). Cultivars of both species are propagated by T-budding or chip budding on either seedling or clonal rootstocks. In addition to source selection in virus control programs, clonal selection is necessary to control genetic disorders such as *cherry crinkle leaf* and *deep suture* (194).

Nursery Production. Rootstock seedlings can be grown in a closely planted seed bed or nursery row for 1 year, dug, and then lined-out in the nursery row about 10 cm (4 in) apart and grown a second season before budding. Under favorable growing conditions, stratified seeds may be planted directly in the nursery in early spring, and the seedlings will be large enough for budding by late summer or early fall.

Rootstocks for Cherry (169). Historically, cherry rootstocks have been going through several generations of development. The first generation involved the selection and evaluation of the two most common rootstocks for sweet cherry and sour cherry cultivars throughout the world,—mazzard *(Prunus avium)* and mahaleb *(P. mahaleb)* seedlings.

Mazzard (P. avium). Considerable variation exists among sources of mazzard seeds. Mazzard seedlings used in the United States are virus-tested seed source clones selected in Oregon and Washington and need to be maintained in isolated seed orchards because infected plants can produce a certain proportion of infected seedlings. To germinate mazzard seeds it is beneficial to pre-soak them in water that is changed daily for about 8 days prior to stratification. A warm, moist stratification period at 21°C (70°F) for 4 to 6 weeks followed by a cold (4°C, 40°F) stratification period of 150 days should improve germination. When seeds show cracking of the endocarp with root tips emerging, they should be removed and planted (128).

Sweet cherry cultivars on mazzard roots produce an excellent graft union, as well as vigorous and long-lived trees, but are generally thought to grow too large for economical management. Mazzard roots do not grow particularly well on heavy, poorly aerated, wet soils, but will tolerate such conditions better than mahaleb. Mazzard seedlings have a shallow, horizontal root system. Mazzard rootstocks are susceptible to *Verticillium* wilt and moderately resistant to oak root fungus *(Armillaria).* Mazzard roots are immune to root-knot nematode *(Meloidogyne incognita),* resistant to *M. javanica,* but susceptible to the root-lesion nematode *(Pratylenchus vulnus).*

Mahaleb (P. mahaleb). This species is heterogeneous with considerable variability in many characteristics. Seeds are obtained from specific virus-tested clonal selections maintained in seed orchards. The seeds should be soaked in water for 24 hours and then stratified for about 100 days at 4°C (40°F). Mahaleb is the principal rootstock in the United States for 'Montmorency,' the leading sour cherry cultivar. Mahaleb rootstocks produce a somewhat dwarfed tree, particularly if budded high, [38 to 50 cm (15 to 20 in)] on the rootstock. Mahaleb roots are resistant to the *buckskin virus* and have been used to produce high-worked multitrunked trees with sweet cherry scions. Infected limbs can be cut off and regrafted below the union without losing the entire tree.

Mahaleb-rooted trees are not satisfactory in heavy, wet soils with high water tables, and are susceptible to *Phytophthora* root rot. Under dry, unirrigated conditions, Mahaleb is more likely to survive than Mazzard, presumably because of its deep, vertical rooting habit. Trees on Mahaleb roots are more cold-hardy than those on Mazzard. Sweet cherries often grow faster for the first few years on Mahaleb, produce precocious heavy bearing, and may result in dwarfing. Reports are that trees on Mahaleb roots are relatively short-lived in England but not in the United States. Mahaleb roots are more resistant to root-lesion nematode *(Pratylenchus vulnus)* than Mazzard. They are also resistant to the root-knot nematode *(Meloidogyne incognita)* but susceptible to *M. javanica.* Mahaleb roots are more resistant to bacterial canker *(Pseudomonas)* than Mazzard roots.

The ***second generation*** of rootstocks was the selection and release of the following two clonal rootstock cultivars by the East Malling Research Station, UK.

'F12/1.' This clone of *P. avium* was selected because it is vigorous and has high resistance to bacterial canker *(Pseudomonas).* Propagation is primarily by trench layering (see Chapter 14) but it has been produced by micropropagation. This rootstock is widely used in the UK, Australia, New Zealand, and South Africa. In Oregon, this stock has worked well as a root, trunk, and primary scaffold system onto which the scion cultivar is topworked.

'Colt' (Prunus avium × P. pseudocerasus). A semi-dwarfing clonal cherry 'Colt' rootstock was released in 1958 at the East Malling Research Station. A virus-tested strain, EMLA Colt, was released in 1977. This rootstock, used for both sweet and sour cherry cultivars, is easily propagated by hardwood cuttings as well as by mound layering. Tree size is about 60 to 70 percent of trees on F12/1 and 80 percent of trees on mazzard seedlings. It produces early heavy cropping, and is

resistant to bacterial canker (*Pseudomonas*), *stem pitting virus*, and replant problems. It is susceptible to crown gall (*Agrobacterium*), but size and yield have a high genotype × environmental interaction in that the size and vigor varies with location. It is somewhat cold-sensitive (30).

The third generation of cherry rootstocks began to emerge around the 1980s from research programs in Oregon to various stations in Europe (30). These have been extensively tested throughout the cherry growing regions of the world during the past few decades and some appear to be promising. Additional rootstock selections are being tested (224).

M × M Clones (P. avium × P. mahaleb). A series of rootstock clones identified by number were selected from large hybrid seedling populations in Oregon (197). 'M × M 14,' a semi-dwarfing clone, appears to be the most favored and is used in Europe.

G.M. Series. Two clonal rootstocks named 'Damil' (*Prunus dawyckansis*) and 'Camil' (*P. canescens*) were introduced among others by Fruit and Vegetable Research Station, Gembloux, Belgium (169). These produce about 60 percent of the size trees on F12/1. They are propagated by softwood cuttings or micropropagation.

Gisela Series. Five clonal rootstocks known by number (148-1, 148-2, 148-8, 148-9, and 195-1) were developed at the Justus Liebeg University, Giessen, Germany (169) which originated from progenies of crosses among various *Prunus* species, including *P. canescens*, *P. fruticosa*, and *P. avium*. Three rootstocks have been named 'Gisella 1' (17 to 45 percent of 'F12/1'), 'Gisella 5,' and 'Gisella 10' (40 to 60 percent of 'F12/1'). Trees on these rootstocks appear to be precocious, heavy bearing, well anchored, and hardy.

St. Lucie (P. mahaleb). Clonal selection was made at INRA, Grande Ferrade, France. Propagates from softwood and semi-hardwood cuttings. Grown in Europe. Adapts to a wide range of soils, including calcareous, but requires good drainage. Trees are compact and intermediate in vigor.

'Adara'. This is a Spanish selection of *P. cerasifera* which is vigorous and highly tolerant to calcareous and heavy soils (30).

Prunus armeniaca. Apricot. Cultivars of apricot are propagated commercially by T-budding or chip budding on various seedling *Prunus* rootstocks. Fall budding is the usual practice, but spring and June budding may be used. Bench grafting also has been successful (46).

Rootstocks for Apricot (46). Rootstocks include apricot seedlings, peach seedlings, and myrobalan plum (*P. cerasifera*) seedlings and recently clonal *Prunus* hybrids. Seeds of all these species require low-temperature 5°C (41°F) stratification before planting.

Apricot (P. armeniaca). Seedlings of local apricot cultivars are best on good, well-drained soils. Seeds of 'Royal' or 'Blenheim' produce excellent rootstock seedlings in California. In the eastern United States, seedlings of 'Manchurian,' 'Goldcot,' and 'Curtis' are recommended. In Europe local selections are used.

Apricot root is almost immune to the root-knot nematode (*Meloidogyne* spp.). In addition, it is somewhat resistant to the root-lesion nematode. It is susceptible to crown rot (*Phytophthora* spp.) and intolerant of poor soil-drainage conditions. Apricot roots are not as susceptible to crown gall (*Agrobacterium tumefaciens*), as are peach and plum roots. Apricot seedling roots are susceptible to oak root fungus (*Armillaria*) and highly susceptible to *Verticillium* wilt.

Peach (P. persica). In California, 'Nemaguard,' 'Nemared,' and 'Lovell' peach seedlings are satisfactory as rootstocks for apricot cultivars. See under **Peach** for characteristics. In the eastern United States and Canada, some apricot cultivars show incompatibility on peach seedlings (32, 127).

Myrobalan Plum (P. cerasifera). Although successful high-yielding apricot orchards grow on this rootstock, some instances have been observed where the trees have broken off at the graft union in heavy winds or where die-back conditions occur. Nurseries often have trouble starting apricots on myrobalan roots. Some of the trees fail to grow rapidly and upright or else have weak or rough unions. After those weaker trees are culled out, the remaining trees seem to grow satisfactorily. This rootstock is useful for apricot when the trees are to be planted in heavy soils or under excessive soil moisture conditions.

'Marianna 2624.' This clonal plum rootstock is being used to replace myrobalan plum seedlings on which apricot trees do well. See **plum** for characteristics.

'Citation.' This clonal rootstock is being used in new plantings in California.

Prunus dulcis. Almond (118). The almond is an ancient crop in the Mediterranean, Southwest Asia, and North Africa where at one time it was widely grown as seedling populations, topworking bitter individuals with better selections. Commercial almond orchards in the Mediterranean region have traditionally been T-budded in the fall onto almond seedling rootstocks, but more recently to 'GF 677' (*P. dulcis* × *P. persica*). In California, almond orchards followed the same practice initially, but as the industry came under irrigation, rootstocks shifted to peach seedlings. There is limited use for 'Marianna 2624' plum, and more recently peach × almond hybrids have been used. While success with rooting almond stem cuttings has been low, some progress has been made in clonal selection (118).

Virus-tested clonal selection should be used (28). In addition, genetic disorders such as *noninfectious budfailure* (119, 120) need to be controlled through clonal source selection and source tree management.

Rootstocks for Almond

Almond (P. dulcis). Seeds of bitter types or certain commercial cultivars are used. In California, 'Texas' ('Mission') seeds are commonly used. In the Mediterranean, seeds of 'Garrigues,' 'Atocha,' and 'Desmayo Royo' are typically used. Almond seeds require limited stratification of 3 to 4 weeks before planting. Almond roots are adapted to deep, well-drained soil. In poorly drained soils, almond roots are often unsatisfactory, owing to their susceptibility to waterlogging, infection by crown rot (*Phytophthora* spp.), and susceptibility to crown gall (*Agrobactium tumefasciens*). Their deep-rooting tendency is an advantage in orchards grown on unirrigated soils or where drought conditions occur. Almond seedlings are tolerant of high-lime soils and, of the rootstocks available for almonds, are the least affected by excess boron salts (91). Almond seedlings are susceptible to root-knot (*Meloidogyne*) and root-lesion nematodes as well as to oak root fungus (*Armillaria mellea*). Root-knot nematode-resistant almond selections 'Alnem 1,' 'Alnem 88,' and 'Alnem 201' have been selected in Israel.

Peach (P. persica). Peach seedlings are used as rootstocks for almond where irrigation is practiced, such as in California. 'Lovell' and 'Halford' are typically used in nonnematode soils but 'Nemaguard' and sometimes 'Nemared' are used in nematode-infested areas or where nematode concern exists. For characteristics of peach rootstocks see under **Peach.**

'Marianna 2624' Plum. This clonal rootstock is used in California where dense soils occur or where oak root fungus is present. Almond cultivars vary in their relative size and in their degree of incompatibility on this rootstock. In general, trees are about one-third smaller than those on the other available rootstocks. Not all almond cultivars are compatible with this rootstock. (See under **Plum** for characteristics.)

Peach-almond Hybrids (118). These first-generation hybrids of peach and almond can be produced as seedling populations by natural crossing in seed orchards between adjoining trees of the two species. Combinations of 'Titan' almond and 'Nemaguard' have been used. Seedling hybrid plants are identified in the nursery row by their high vigor, intermediate appearance between the parents, and red foliage color in certain combinations using a redleaf red parent.

Clonal hybrids have been developed. 'GF 677,' discovered as a chance seedling in France, is resistant to alkaline soils and is widely used in the calcareous soils of the Mediterranean region for both almonds and peaches. It is susceptible to all species of nematode. Commercial propagation by micropropagation is widely practiced in the Mediterranean. 'Hansen 536' and 'Hansen 2168' are hybrid rootstock clones released by the University of California in 1983 of which the first has achieved a commercial role. Both are immune to root knot nematodes (*Meloidogyne javanica* and *J. incognita acrita*). A new hybrid clone 'Nickels' has been released (121) which has the same nematode resistance potential as 'Nemaguard,' one of its parents, and a higher chilling requirement than 'Hansen 536' and 'Hansen 1268,' both of which are low. Rooting is possible by hardwood cuttings treated with IBA plus a fungicide, planted directly in the nursery row in the fall. Peach-almond hybrids can be rooted by semi-hardwood cuttings under mist if given IBA treatments (147). These rootstocks are noted for their vigor and excellent compatibility with scion cultivars. 'Hansen 536' has some susceptibility to crown rot (*Phytopthora*) and shows somewhat reduced survival in commercial orchards, whereas 'Nickels' has shown greater survival in early tests. All can be rooted well by micropropagation.

A number of additional clonal PA hybrid rootstocks have now been introduced in Europe, and additional programs are underway (175).

Prunus persica. Peach and Nectarine. Peach and nectarine cultivars are budded to peach seedling rootstocks. Most other species (almond, plum, apricot) tend to be incompatible to various degrees.

Cuttings (44). Some peach cultivars can be propagated by leafy, succulent softwood cuttings taken in spring or summer, treated with auxin, and rooted in a mist-propagating bed (44). Also, in areas with mild winters, some peach cultivars can be started from hardwood cuttings if they are treated with a 4,000 ppm IBA quick-dip, then set out in the nursery in the fall. Direct rooted peach cultivars have been used with high-density tree production systems, requiring large numbers of inexpensive nursery trees.

Budding. Nursery trees of peaches are propagated by T-budding or chip budding on seedling rootstocks. In California and other areas with a long growing season, trees are produced in 1 year by June budding. Seeds planted in early fall and allowed to stratify over winter germinate in early spring (March) to produce vigorous large plants by May. Fall budding may be used, but a 2-year production cycle is required.

Rootstocks for Peach (49, 129, 175)

Peach (P. persica). Peach seedlings are the most satisfactory rootstock for peach and should be used unless special conditions warrant other rootstocks. Seeds of 'Halford,' 'Lovell,' 'Elberta,' or 'Rutgers Red Leaf' are usually used, since they germinate well and produce vigorous seedlings. Seedlings of these cultivars are not resistant to root-knot nematodes, however, and where these are problematic, resistant peach stocks should be considered. Seeds from peach cultivars whose fruits mature early in the season should not be used because their germination percentage is usually low. It is best to obtain seeds from the current season's crops, since viability decreases with storage. In the eastern United States, seeds have historically been obtained from wild types such as Tennessee Naturals or Indian peaches.

Peach roots are susceptible to the root-knot nematode (*Meloidogyne* spp.), especially in sandy soils, as well as to root-lesion nematode (*Pratylenchus* vulnus). 'Nemaguard,' a *Prunus persica* × *P. davidiana* hybrid rootstock introduced by the USDA in 1959, produces seedlings that are uniform and resistant to both *M. incognita* and *M. javanica*, but not to the ring nematode (*Criconemella xenoplax*). Nemaguard roots give strong, well-anchored, high-yielding trees, although certain peach and plum cultivars have not done well on them and they have shown susceptibility to bacterial canker (*Pseudomonas syringae*) and to crown rot (*Phytophthora*). 'Nemaguard' is not hardy in colder peach-growing areas. 'Nemared' is a red-leaved selection from 'Nemaguard.' 'Guardian,' a peach rootstock clone whose seedlings are tolerant of "peach tree short life" syndrome (PTSL), has been released in the southeast United States (175).

Cold-tolerant rootstock cultivars have been sought. 'Siberian C,' a winter-hardy peach rootstock—withstanding 11°C (12°F) soil temperatures—introduced in 1967 by the Canadian Department of Agriculture Research Station at Harrow, Ontario, has not remained satisfactory. Similarly 'Harrow Blood' and Chinese cultivars ('Tzim Pee Tao,' 'Chi Lum Tao') have not been completely satisfactory. Future advancement may come from recent programs in Russia, Germany, and the Czech Republic (175).

Nursery trees on peach roots often make unsatisfactory growth when planted on soils previously planted to peach trees. Peach roots are susceptible to oak root fungus (*Armillaria*), crown rot (*Phytophthora*), crown gall (*Agrobactrium*), and *Verticillium* wilt.

Plum. Some plum rootstocks, including *Prunus insititia* ('St. Julien d'Orleans,' 'St. Julien Hybrid No. 1,' 'St. Julien GF 655.2'); *P. cerasifera* (myrobalan plums); *P. domestica* ('GF 43'); and *P. domestica* × *P. spinosa* ('Damas GF 1869'), are used for budding peach cultivars to produce somewhat smaller trees and to grow on heavy, dense soil. In the UK, rootstocks such as these have been compatible with all peach and nectarine cultivars worked on them, and produce medium-sized to large trees. Plum rootstocks are especially adapted to wet, waterlogged soils.

Apricot (P. armeniaca). Apricot seedlings are occasionally used as a rootstock for the peach. The graft union is not always successful, but numerous trees and commercial orchards of this combination have produced fairly well for many years. Seedlings of the 'Blenheim' apricot seem to make better rootstocks for peaches than those of 'Tilton.' The apricot root is highly resistant to root-knot but not to root-lesion nematodes.

Almond (P. amygdalus). Almond seedlings have been used with limited success as a rootstock for peaches. There are trees of this combination that are growing well, but in general it is not a satisfactory combination. The trees are often dwarfed and tend to be short-lived.

Peach × Almond Hybrids. These clonal rootstocks, such as 'GF556' and 'GF667,' are important in France and other Mediterranean countries. They are excellent rootstocks for peach cultivars and can be propagated by rooting semi-hardwood cuttings under mist (44). Micropropagation is used commercially for 'GF667'.

Western Sand Cherry (P. besseyi). When used in limited experiments as a dwarfing rootstock for several peach cultivars, the bud unions appeared excellent, but about 40 percent of the nursery trees failed to survive. The remainder grew well, however, and developed into typical dwarf trees with healthy, dark-green foliage. The trees bore normal-sized fruit in the second or third year after transplanting to the orchard.

Nanking Cherry (P. tomentosa). May be suitable for some peach cultivars as a dwarfing rootstock.

Newer rootstock selections (175). During the 1980s and 1990s another generation of clonal rootstocks has begun to appear, particularly in Europe, many of which are being propagated by micropropagation.

Prunus domestica. European plums and prunes. *P. salicina.* Japanese plums. Plum cultivars are propagated by T-budding or chip budding in the fall on seedling rootstocks or, with certain rootstocks, on rooted cuttings or layers. Budding can also be done in the spring. Plums tend to be easier to root than other species, being propagated by hardwood cuttings (52) and some by leafy, softwood cuttings under mist (91). Bench grafting in winter and planting the grafts in the nursery in spring gives good results using the whip graft (49). Japanese plum *(P. salicina)* nursery trees have been produced by micropropagation (55).

Rootstocks for Plum (162)

Myrobalan Plum (P. cerasifera). This widely used plum rootstock is particularly desirable for the European plums *(P. domestica)* (which includes the commercially important prune cultivars), but it is also satisfactory for Japanese plums. However, some cultivars of plum—'President,' 'Kelsey,' 'Stanley,' and 'Robe de Sergeant'—are not entirely compatible with this stock. Cultivars that are Japanese-American hybrids *(P. salicina × P. americana)* are best worked on American plum seedlings *(P. americana).*

Myrobalan roots are adapted to a wide range of soil and climatic conditions. They will endure fairly heavy soils and excess moisture, and are resistant to crown rot but susceptible to root-knot nematodes and oak root fungus. They grow well on light sandy soils.

Myrobalan seeds require stratification for about 3 months at 2 to 4°C (36 to 40°F). Then they may be planted thickly in a seed bed for one season and then transplanted to the nursery and grown for a second season before budding; or the seeds may be planted directly in the nursery row and grown there for one season, the seedlings being budded in late summer or fall.

Certain vigorous myrobalan selections are propagated by hardwood cuttings. One of these, 'Myro 29C,' is immune to root-knot nematodes. A selection, 'Myrobalan B,' was developed at the East Malling Research Station and is propagated by hardwood cuttings. It is particularly valuable in producing vigorous trees, although there are cultivars not completely compatible with it. Most plum cultivars in England belong to the *P. domestica* species.

Marianna Plum (P. cerasifera × P. munsoniana). This clonal rootstock originated in Texas as an open-pollinated seedling of the myrobalan plum, apparently a hybrid with *P. munsoniana*. It is propagated by hardwood cuttings. Some plums have grown well on it; others have not. An exceptionally vigorous seedling selection of the parent 'Marianna' plum, made by the California Agricultural Experiment Station in 1926, is widely used under the identifying name 'Marianna 2624.' It is adaptable to heavy, wet soils and is immune to root-knot nematodes *(Meloidegyne),* resistant to crown rot *(Phytopthora),* crown gall *(Agrobactium),* oak root fungus *(Armillaria),* and *Verticillium* wilt, but is susceptible to bacterial canker *(Pseudomonas).* It suckers badly from roots and is sensitive to "brownline" virus. Propagated as rooted cuttings, the rootstock has a shallow root system for the first few years, but older trees develop a deeper root system. Hardwood cuttings root well if fall-planted (in mild winter climates) after IBA treatments (52).

Peach (P. persica). Some plum and prune orchards in California are on peach seedling rootstocks. The stock has proved satisfactory for light, well-drained soils or where bacterial canker has been a problem. However, peach rootstocks should be avoided if the trees are to be planted on a site formerly occupied by a peach orchard. In some areas, plum on peach roots tend to overbear and develop a die-back condition under low potassium conditions. Peach is not satisfactory as a rootstock for some plum cultivars, including 'Sugar' prune and 'Robe de Sergeant.'

Apricot (P. armeniaca). Apricot seedlings can be used as a plum stock in nematode-infested sandy soils for those cultivars that are compatible with apricot. Japanese plums tend to do better than European plums on apricot roots.

Almond (P. dulcis). Some plum cultivars can be grown successfully on almond seedlings. The 'French' prune does very well on this rootstock. The trees grow faster and bear larger fruit than when myrobalan roots are used. Plum cultivars on almond roots tend to overbear, sometimes to the detriment of the tree. This stock probably should not be used for plums except where plantings are to be made on well-drained, sandy soils, high in lime or boron.

'Brompton' and 'Common' Plum (P. domestica). These two clonal plum rootstocks are chiefly used in England. 'Brompton' seems to be compatible with all plum cultivars and tends to produce medium-to-large

trees. 'Common' plum, which produces small-to-medium trees, has shown incompatibility with some cultivars. Propagation is by hardwood cuttings.

'St. Julien,' 'Common Mussel,' 'Damas,' and 'Damson' (P. insititia). The first three stocks are used mostly in England. 'Damas C' produces medium-to-large trees, whereas 'Common Mussel' generally produces small-to-medium trees. The latter stock seems to be compatible with all plums. 'St. Julien A' produces small-to-medium trees for all compatible scion cultivars. Results have been variable with these stocks, since they vary widely in type. At the East Malling Research Station, 'St. Julien,' 'Mussel,' and 'Damas' clones have been selected and listed as A, B, C, D, and so forth. 'St. Julien' has been used to some extent as a plum stock for the *P. domestica* cultivars in the United States. 'Pixy' has been released by the East Malling Research Station as a dwarfing plum rootstock.

In England, the following clonal rootstocks are recommended for plums: for *vigorous* trees, 'Myrobalan B'; for *semi-vigorous* trees, 'Brompton'; for *intermediate* trees, 'Marianna,' 'Pershore'; for *semi-dwarf* trees, 'Common' plum, 'St. Julien A'; for *dwarf* trees, 'St. Julien K.'

Florida Sand Plum (P. angustifolia). This species may be useful as a dwarfing rootstock for compatible plum cultivars. In California, after 16 years, 'Giant,' 'Burbank,' and 'Beauty' on this stock were healthy and very productive, with a dwarf type of growth.

Japanese Plum (P. salicina). Seedlings of this species are used as plum rootstocks in Japan but apparently not elsewhere. European plums *(P. domestica)*, when topworked on Japanese plum rootstocks, result in very short-lived trees; the reverse combination, though, produces compatible unions.

Western Sand Cherry (P. besseyi). This rootstock has produced satisfactory dwarfed plum trees of the Japanese and European types, but poor bud unions and shoot growth developed when it was used as a rootstock for cultivars of *P. insititia*.

Hybrid Prunus Rootstocks. A number of clonal rootstocks originating as interspecific hybrids of a number of species have been developed and introduced. Among these are the following:

'Citation' (Interspecific Hybrid of Peach and Plum). Compatible with apricot and plum. Induces early bearing in the orchard, and early defoliation and dormancy in the nursery. Tolerant of waterlogging and resistant to root-knot nematodes (*Meloidegyne*). Susceptible to crown gall (*Agrobacterium*) and oak root fungus (*Armillaria*). Not suitable for peach and almond.

'Viking' (Interspecific among Peach, Almond, Plum, and Apricot). Extremely vigorous and precocious. May have resistance to root-knot nematodes. Intolerant of wet soil conditions.

'Myran'(Prunus cerasifera × P. salicina × 'Yunnan' peach). Compatible with almond and peach cultivars. Tolerates wet soil conditions and has some tolerance to *Armillaria*. Propagated by hardwood cuttings.

'Ishtara'(P. myrobalan × P. myrobalan × P. persica). Introduced by INRA, France. Used for prunes in Europe but compatible for apricot, almond, plum, and some cultivars of peach.

Psidium spp. Guava. Cattley or Strawberry guava *(P. cattleianum)*, and Common or Lemon Guava *(P. guajava)* (20, 141, 159). The Cattley guava has no named cultivars, and nursery plants are propagated by seed. This species comes nearly true from seed—large-fruited, superior trees are used as the seed source—but it is difficult to propagate by vegetative methods. Indigenous to tropical America, the common guava *(P. guajava)* is widely grown in the world as a tropical and subtropical fruit tree. The common guava has several cultivars, but they are not grown extensively because of difficulties in asexual propagation.

Seed. Seeds are short-lived and must be planted immediately. They are grown in poly bags or in a nursery, and transplanted to the field when 6 to 8 weeks old. Most trees of these two species are propagated by seeds, which germinate easily and in high percentages. Seedlings are somewhat susceptible to damping-off organisms, and should be started in sterilized soil or otherwise treated with fungicides. When about 4 cm (1 1/2 in) high, the seedlings should be transplanted into individual containers. In 6 months the plants should be about 30 cm (12 in) high and can be transplanted to their permanent location.

Cuttings. Cuttings are somewhat difficult to root. Best results are obtained by treating softwood and semi-hardwood cuttings with IBA and rooting under mist (168).

Grafting. For large-scale propagation of common guava cultivars, grafting or budding is necessary. Chip budding has been successful when done any time during the summer using greenwood buds from selected cultivars inserted into seedling rootstocks about 5 mm in thickness. Plastic wrapping tape is used to cover the buds. The rootstock is cut off above the inserted bud after about 3 weeks (110). Inarching is an important method of propagation, but labor is expensive. Side-veneer grafting is also possible, using scions from terminal growth flushes with well-developed axillary buds. Several species of *Psidium* are used as seedling rootstock: *P. cujanillis, P. molle, P. cattleianum,* and *P. guineense,* but they tend to sucker from the base.

Layering. For the common guava, air layering is one of the most important commercial propagation methods, as described in Chapter 14. With both species, simple and mound layering are also effective methods of starting new plants; the layers may be tightly wrapped with wire just below the point where roots are wanted, or can be ringed, and treated with 500 ppm IBA in lanolin prior to earthing up. Two series of shoots can be produced in one year (141).

Micropropagation. Can be micropropagated with explants from mature trees (6).

Pummelo. *See Citrus.*

Pomelo. *See Citrus.*

Punica granatum. Pomegranate. Small tree or shrub of subtropical and tropical Mediterranean and southwest Asian areas. Both deciduous and semi-deciduous types exist and are adapted to semiarid climates. Seedlings are grown but cultivars can be readily propagated by hardwood cuttings. Cuttings are taken from fully mature 1-year-old wood or suckers from the base of the tree. Cuttings are trimmed to 20 to 25 cm (8 to 10 in) long, and treated with IBA. Softwood cuttings root successfully under mist. Suckers are produced; these may be dug during the dormant season, leaving a piece of root attached, and then planted.

***Pyrus* spp.** Pear. Over 20 species of pears exist around the world (136). The most important edible types are the European pear (*P. communis,* including a number of related species in Europe) and the Asiatic pears (mostly *P. pyrifolia,* but also complex hybrids). Pears are ancient fruit crops in Europe, North America, China, and Japan, respectively. Many named cultivars are hundreds of years old, and many of the related species are involved in rootstock selection.

The history of the pear illustrates the problems of monocultures and the changing patterns of disease and insect vectors. European pear cultivars are highly susceptible to the bacteria fire blight *(Erwinia amylovorus).* "Pear decline" is a graft union failure caused by a mycoplasma-like organism (98) which was spread by the pear psylla *(Psylla pyricola)* first in Europe (184) and then in western North America (15, 18). Certain pear species are highly susceptible and, when used as a rootstock, they cause decline or death of the tree. Other rootstock species produce a serious defect on the fruit known as "black-end" or "hard-end." Serious viruses and virus-like diseases affect the pear, and production must be carried through a virus control program (181).

Seed. Seeds are used for rootstocks and in breeding programs. Pear seeds must be stratified for 60 to 100 days at about 4°C (40°F). Then they are planted thickly about 13 mm (1 1/2 in) deep in a seed bed, where they are allowed to grow during one season. The following spring they are dug, the roots and top are cut back, and then they are transplanted to the nursery row, where they are grown a second season, ready for budding in the fall.

Budding. Essentially all pear cultivars are propagated by budding to rootstocks in the fall (T- or chip budding). Pear trees can also be started by whole-root grafting, using the whip-and-tongue method.

Cuttings. Some pear cultivars, such as 'Old Home' and 'Bartlett,' can be propagated by hardwood cuttings or by leafy cuttings under mist if treated with IBA (91, 94, 216). Own-rooted 'Bartlett' trees have shown excellent production with large, well-shaped fruit, are resistant to pear decline (78, 91), and with age become partially dwarfed—a desirable attribute for high-density plantings. However, rooting percentages are not high, and cutting propagation is not widely practiced. Long cuttings [23 to 62 cm (9 to 24 in)] of *Pyrus* 'Pyrodwarf' dwarfing rootstock from semi-hardwood cuttings root well under high pressure fog systems (195, 196).

Micropropagation. Pear cultivars and rootstocks can be micropropagated (36, 227). Commercial production has occurred primarily in Italy and France where own-rooted 'Bartlett' ('Williams') have been grown for commercial orchards. Micropropagation has been used to enhance the rooting potential of hard-to-root clones in the establishment of stock blocks.

Rootstocks for Pear (77, 136, 217)

Pear Seedlings. Seed selection is very important if rootstocks of a specific species are desired. Various *Pyrus* species hybridize freely, some bloom at the same time, and cross-pollination is necessary for seeds to develop. Use isolated groups of trees as the seed source of a known species and avoid collecting seeds from several different species, since hybrids are likely to be produced.

French Pear (P. communis). French pear seedlings are generally grown from seeds of 'Winter Nelis' or 'Bartlett.' This rootstock is moderately vigorous and winter-hardy, produces moderately productive, uniform trees with a strong, well-anchored root system, and is resistant to pear decline. French Pear forms an excellent graft union with all pear cultivars and will tolerate relatively wet (but not waterlogged) and heavy soils. French pear roots are resistant to *Verticillium* wilt, oak root fungus (*Armillaria mellea*), root-knot (*Meloidogyne*), and root-lesion nematodes (*Pratalynchus vulnus*) and crown gall (*Agrobacterium*).

French pear rootstocks are susceptible to pear root aphid (*Eriosoma pyricola*) and fire blight (*Erwinia amylovora*). Millions of pear trees on French pear roots have died from fire blight, due to the high susceptibility of this stock. Seedlings of the 'Kieffer' pear—a hybrid between *P. communis* and *P. pyrifolia*—have been satisfactorily used for many years in Australia as a pear rootstock.

Blight Resistant French Pear Clonal Rootstocks and Interstocks. Pear selections resistant to blight include 'Old Home' (94) and 'Farmingdale' (176), and have been used as an intermediate stock grafted on seedling rootstocks. The desired cultivar is topworked after the trunk and primary scaffold branches develop. A blight attack in the top of the tree will only go to the resistant interstock, which can then be regrafted after blight has been cut out. Thirteen clonal selections from an 'Old Home' × 'Farmingdale' progeny have been made in Oregon to be propagated by hardwood cuttings to provide blight resistance and give a range of dwarfing that will be of potential value in pear production.

P. calleryana. This stock is blight-resistant and produces vigorous trees with a strong graft union; fruit quality is good, with no black-end. 'Bartlett' orchards with this rootstock have produced well in California, and it is popular in the southern part of the United States for 'Keiffer' and other hybrid pears. In areas with severe winters it lacks winter-hardiness. Where good pear psylla control has been practiced, trees on *P. calleryana* roots are resistant to pear decline. However, trees with *P. calleryana* roots show less resistance to oak root fungus (*Armillaria*) than those with *P. communis* roots. Seedlings of a selection of this species, known as 'D-6,' are widely used as a pear rootstock in Australia.

P. ussuriensis. This east Asian stock has been used in the past to some extent; many pear cultivars on this root develop black-end, although not to the extent found with *P. pyrifolia* roots. It produces small trees that are susceptible to pear decline, and should not be used in areas where this problem may occur.

P. betulaefolia. This species has vigorous seedlings, resistance to leaf spot and pear root aphid, tolerance to alkali soils, an adaptability to a wide range of climatic conditions, good resistance to pear decline, and produces large high-yielding trees.

Cydonia oblonga. Quince (215). This species has been used for centuries as a dwarfing stock for pear and is widely used as a pear rootstock in Western Europe. Some cultivars fail to make a strong union directly on the quince, hence double-working with an intermediate stock such as 'Hardy' or 'Old Home' is necessary. Cultivars that require such a compatible interstock when worked on quince roots include 'Bartlett,' 'Bosc,' 'Winter Nelis,' 'Seckel,' 'Easter,' 'Clairgeau,' 'Conference,' 'Guyot,' 'Clapp's Favorite,' 'Farmingdale,' and 'El Dorado.' A selection of 'Bartlett' ('Swiss Bartlett'), compatible with 'Quince A' rootstock, originated in Switzerland, presumably as a bud mutation from an incompatible form. The following pears also appear to be compatible when worked directly on quince: 'Anjou,' 'Old Home,' 'Hardy,' 'Packham's Triumph,' 'Gorham,' 'Comice,' 'Flemish Beauty,' 'Duchess,' and 'Maxine.'

Quince roots are resistant to pear root aphids and nematodes, but are susceptible to oak root fungus, fire blight, and calcareous soil, and are not winter-hardy in areas where extremely low temperatures occur. In some areas, trees on quince roots have developed pear decline, but in others they have not, possibly because different quince rootstocks were used. The black-end trouble has not developed with pears on quince roots. There are a number of quince cultivars, most of which are easily propagated by hardwood cuttings or layering. 'Angers' quince is commonly used as a pear rootstock because its cuttings root readily, it grows vigorously in the nursery, and it does well in the orchard. 'Provence' quince roots produce larger pear trees than 'Angers' and are more winter-hardy.

The East Malling Research Station has selected several clones of quince suitable as pear rootstocks and designated them as 'Quince A,' 'B,' and 'C.' 'Quince A' ('Angers') has proved to be the most satisfactory stock. 'Quince B' (Common quince) is somewhat dwarfing, whereas 'Quince C' produces very dwarfing but highly productive trees. It is important to use only virus-tested quince stock.

'Provence' quince (LePage Series C and BA-29C) originated in France. These are winter-hardy and, when used as rootstocks for pears, give trees one-half to two-thirds the size of standard pear trees. The BA-29C series is a virus-free selection of LePage Series C.

Pyrus pyrifolia (P. serotina). Asiatic pear. Cultivars of Asiatic pear are propagated on seedling rootstocks of this species in Japan. This rootstock was once widely used in the United States as a pear rootstock from about 1900 to 1925, but is no longer recommended due to its high susceptibility to pear decline and to the physiological defect black-end (or "hard-end"), which may occur when 'Bartlett,' 'Anjou,' 'Winter Nelis,' and other cultivars are grafted to it (95).

Quince. *See Cydonia.*

Rambutan. *See Nephelium.*

Raspberry. *See Rubus.*

***Ribes* spp.** Currants and gooseberries. Specific species, hybrids, and cultivars are found in the northern hemisphere, where they grow as bushes, and produce small berries used in making jams, jellies, and pies. Cultivars of red and white currants are in *Ribes sativum*; black currants are in *R. nigrum*. Flowering current is *R. odoratum*. American gooseberries include *R. grossularia* and its hybrids, while European gooseberries are in *R. uvacrispa*. All of the species are alternate hosts of white pine blister rust, and their planting is restricted by law. Propagation is similar for cultivars of all of these species.

Cuttings. Currants are propagated by collecting hardwood cuttings 20 to 25 cm (8 to 10 in) long in late fall, precallusing them at low temperatures over winter, and planting in the spring. Cultivars of gooseberry can also be rooted but with more difficulty.

Layering. Mound layering is used for American gooseberry. Shoots usually root well after one season. They are then cut off and transferred to the nursery row for a second season's growth before they are set out in their permanent location. The slower-rooting layers of European cultivars may have to remain attached to the parent plant for two seasons before they develop enough roots to be detached.

Micropropagation. Gooseberries can be easily micropropagated and the rooted plantlets stored under refrigeration for as long as 130 days with 100 percent survival (212).

***Rubus* spp.** Raspberry, Blackberry, Youngberry, Boysenberry, Loganberry, Dewberry. This group includes a complex of species growing worldwide from which a large number of commercially important cultivars have been produced. Commercial propagation requires vegetative methods. Seeds are used in breeding programs. Raspberry and blackberry produce clusters of drupelets surrounding a hard-coated achene.

Asexual Methods

Upright Type. Blackberries (*Rubus*; subgenus *Eubatus*). Reproduce by suckers removed in the spring with roots attached and replanted to a new location or to a nursery row for an additional year (115). Also propagate by root cuttings, conventional stem cuttings, one-node cuttings, and leaf-bud cuttings under mist with auxin (26). They are commercially micropropagated (see below).

Trailing Types. Youngberry, Boysenberry, Loganberry, or Dewberry (*Rubus* spp.). These do not produce many suckers but reproduce naturally by tip layering (see Chapter 14). They can be propagated by root cuttings, but some thornless forms are periclinal chimeras and revert to the nonmutated (thorny) type. Thornless cultivars that do not revert to thorny types have been developed through biotechnology (see Box 16.7, page 612). Boysenberry types can also be propagated by conventional stem cuttings, one-node stem cuttings (228), or leaf-bud cuttings rooted under mist with auxin hormones. They are commercially micropropagated (see below).

Black Raspberry (R. occidentalis). Usual method is by tip layering but can be rooted from leaf-bud cuttings in about 3 weeks under mist.

Purple Raspberry (R. occidentalis × R. idaeus) (115). Tip layering is the usual method of propagation but roots less easily than the black raspberry.

Red Raspberry (R. strigosus × R. idaeus). Propagated usually by removing 1-yearold suckers in early spring with a piece of old root attached (97). Young, green suckers of new wood may also be dug in the spring shortly after they appear above ground. Leave piece of old root attached. Survival requires removal during cool weather and irrigation following transplanting. Sucker production is stimulated by inserting a spade deeply at intervals in the vicinity of old plants to cut off roots, or mulch with straw or sawdust.

Viruses and crown gall can be problems with red raspberries, so suckers should be taken only from clean plants obtained from nurseries specializing in certified plants. Root cuttings (97) should be thick; use 15 cm (6 in) root lengths; for thin roots use 5 cm (2 in) long pieces. Cuttings are planted very shallow—13 mm (1/2 in). Strong nursery plants may be produced in one year.

Cuttings. Leafy softwood cuttings can be made in early spring from young sucker shoots just emerging from the soil. The shoots should have a 2.5 to 5 cm (1 to 2 in) etiolated section from below the surface, and be placed in a propagating frame. Shoot cuttings taken directly from the canes are difficult to root, although success has been reported with pre-etiolated shoots, treated with IBA, and rooted under ventilated high-humidity fog (103).

Micropropagation. Blackberry and raspberry (23, 213) cultivars are commercially micropropagated by shoot-tip cultures on a large enough scale to allow mass propagation. In practice, micropropagation is used as part of a system to produce virus-free stock plants in virus control programs (see Chapter 16).

Sapodilla (Nispero). See *Manikara*.

Strawberry. See *Fragaria xananassa*.

Tangelo. See *Citrus*.

Tangerine. See *Citrus*.

Theobroma cacao. Cacao. The dried, partly fermented fatty seeds of this native South American tree are used in the production of cocoa, chocolate, and cocoa butter. Large commercial cacao plantings have been established in West Africa and South America. Production is mostly from seedling trees, which are highly variable.

Seed. Seeds are obtained from selected high-yielding clones as well as from hybrid seeds resulting from cross-pollination between parents that are propagated vegetatively. Freshly harvested mature seeds are planted immediately, since cacao seeds quickly deteriorate after harvesting, normally losing all capacity to germinate within a week after removal from the pod. Longer seed life can be produced by keeping seeds from drying, and storing at 24 to 29°C (75 to 85°F) (13).

Cacao is often directly field planted, 3 to 4 seeds to a planting site, retaining all seedlings surviving as branches of one tree. Seedlings may be started in a nursery bed and later transplanted to their permanent site. Seedlings may also be started in poly bags, baskets, bamboo or paper cylinders, or clay pots, or containers from which they are removed later and planted. The germination temperature should be about 26°C (80°F).

Miscellaneous Asexual Methods. Air-layering cacao is quite successful. Grafting is done using patch budding, T-budding, or top wedge-grafting on seedling rootstocks. Top wedge-grafting using orthotropic budwood that contains one or two buds of selected seedlings is being used for accelerated hybrid clone selection (58). There is 70 to 90 percent rooting success with semi-hardwood cuttings rooted close to 100 percent relative humidity and low light–15 to 20 percent normal light (World Cocoa Foundation, http://search.worldcocoafoundation.org). Cocoa is commercially micropropagated (205).

***Vaccinium* spp.** Blueberry and Cranberry. Lowbush blueberry *(V. angustifolium)*, Rabbiteye blueberry *(V. ashei)*, Highbush blueberry *(V. corymbosum* L. and *V. austral)*, Cranberry *(Vaccinium macrocarpon)* (57). The blueberry and cranberry are native to North America and are grown, respectively, as a small bush or trailing vine.

Seed. Seed propagation is used for breeding. There is no pregermination treatment for blueberry, but with cranberry, optimum seed germination occurs after a 3-month cold stratification. Seeds are removed from ripe berries and spread over a well-drained, acid-type soil mix containing one-third peat moss. Seeds are covered with a layer of finely ground sphagnum moss and kept moist until they germinate, usually in 3 to 4 weeks. Seedlings 2 cm (3/4 in) tall are transferred to peat pots.

Miscellaneous Asexual Methods
Lowbush Blueberry (V. angustifolium) (82). Cultivars of this species are probably best propagated by leafy softwood cuttings under intermittent mist with bottom heat, using sand and peat moss (1:1) as a rooting medium. Cuttings taken in late spring and early summer from actively growing shoots root well, some clones giving almost 100 percent rooting. Rooted cuttings are transferred to peat pots for further growth and overwintering. Cuttings made from rhizomes also can be rooted; they are best taken in early spring or late summer and fall, avoiding the midsummer rest period of the rhizome buds.

Rabbiteye Blueberry (V. ashei). Cultivars of this species can be propagated by hardwood cuttings as well as by leafy softwood cuttings taken in midsummer, treated with a 10,000 ppm IBA talc, and rooted under mist. Using supplementary light to give a 16-hour daylength may improve root production (43). Micropropagation is also possible (227). The inlay bark graft has been used with 'Tif Blue' blueberry *(V. ashei)* on a farkelberry *(V. arboreum)* rootstock, which tolerates a more basic soil pH; this allows the acid-loving blueberry to be produced in a site with higher soil pH (see Fig. 11–2).

Highbush Blueberry (V. corymbosum L. and *V. austral) (57).* Dormant hardwood stem cuttings are used. The blueberry can also be started by leaf-bud cuttings. Cuttings should be spaced about 5 cm (2 in) apart in the rooting bed and set with the top bud just showing. When leaves appear, the frame should be raised slightly to allow for ventilation. Either mist or frequent watering to maintain a high humidity is required. Roots start to form in about 2 months. Both softwood and hardwood cuttings treated with a 8,000 ppm IBA talc root well (52). To take advantage of different soil types, *V. corymbosum* can be grafted on *V. ashei* and *V. arboreum* rootstock. Grafting has been by cleft, whip, side graft, and T-budding (52).

Cranberry (V. macrocarpon) (45, 81). This vine type of evergreen plant produces trailing runners upon which are numerous short upright branches. Propagation is by cuttings made from either runners or upright branches. Cutting material is obtained by mowing the vines in early spring before new growth has started. The cuttings are then set directly in place in their permanent location without previous rooting at distances of 6 to 18 inches apart each way. Two to four cuttings are set in sand in each "hill." The cuttings are 13 to 25 cm (5 to 10 in) long and set deep enough so that only an inch is above ground. A more rapid method of starting a cranberry bog is to scatter the cuttings over the ground and work them into the soil with a special disk-type planter. This is justifiable when there is an abundance of cutting material and a scarcity of labor for setting the cuttings by hand. Water is applied to the bog immediately after planting. The cuttings root during the first year and make some top growth, but the plants do not start bearing until 3 or 4 years later.

Vitis spp. Grape (219, 222). The grape is an ancient crop used for fresh fruit, raisins, and making wine. Although there are many species, cultivated grape cultivars come into several groups. The first group, European grapes, includes cultivars originating from *V. vinifera*, which tend to lack hardiness but are easy to root by hardwood cuttings. Many ancient clonal cultivars exist today and are grown in Europe, California, and many countries of the world with a Mediterranean climate. The second group includes American type cultivars originating from *V. labruska* (and hybrids). These cultivars tend to be winter-hardy and differ in fruit characteristics from *V. vinifera*. The third type of grape cultivar originated from *Vitis rotundifolia*, the muscadine grape from southern United States. Cultivars of this origin have a distinct flavor and are often difficult to root.

Historically, grape cultivars as own-rooted clones are highly vulnerable to pests such as nematodes (*Meloidogyne* spp.) and phylloxera *(Dacylosphaera vitifoleae)*, as well as systemic virus and virus-like organisms. Since these are propagation-related problems, the worldwide grape industry is highly dependent upon pathogen control, "clonal selection," "clean stock" programs, and resistant rootstocks (68).

Grapevines are propagated by seeds, cuttings, layering, budding, or grafting (4). Grape propagation methods have been modernized by the use of virus-indexed "clean" planting stock, mist propagation techniques for leafy cuttings, and rapid machine-grafting procedures. Root-knot nematodes can be eradicated from grapevine rootings by dipping them in hot water [52 to 55°C (125 to 130°F)] for 5 to 3 minutes, respectively (131).

Seeds. Seeds are used in breeding programs to produce new cultivars. Grape seeds are not difficult to germinate. Best results with *vinifera* grape seeds are obtained after a moist stratification period at 1 to 4°C (33 to 40°F) for about 3 months before planting (89).

Hardwood Cuttings. Grape cultivars and clonal rootstocks have traditionally been propagated by dormant hardwood cuttings, which root readily. Cutting material should be collected during the winter from healthy, vigorous, mature vines. Well-developed current season's canes should be used; which are medium in size and have moderately short internodes. Cuttings 8 to 13 mm (1/3 to 1/2 in) in diameter and 36 to 46 cm (14 to 18 in) long are generally used and planted in the spring deep enough to cover all but one bud. One season's growth in the nursery should produce plants large enough to transplant to the vineyard. Auxins have not been needed to root hardwood grape cuttings. Muscadine grapes are normally hard to root, but some success has occurred.

Leafy Cuttings. Leafy greenwood grape cuttings of *V. vinifera* cultivars root profusely under mist in about 10 days if given relatively high [27 to 30°C (80 to 85°F)] bottom heat and treated with IBA. Scarce planting stock (such as virus-indexed material) can be increased very rapidly using one-node stem cuttings, then consecutively taking additional cuttings from the shoots arising from each node on the rooted cutting. Several consecutive cycles of cuttings can be taken to increase the supply of rooted cuttings to very high numbers in a short period.

Layers. Grape cultivars difficult to start by cuttings can be propagated by simple, trench, or mound layering (see Chapter 15). Layering has been used to root muscadine grapes.

Micropropagation. Grapes have been produced by several *in vitro* techniques, including somatic embryo formation and fragmented shoot-tip cultures (124, 227).

Grafting. Budding or grafting onto resistant rootstocks is necessary to increase vine life, plant vigor, and yield specifically where pests, such as phylloxera *(Dacylosphaera vitifoliae)* or root-knot nematodes (*Meloidogyne* spp.) are present. Bench grafting (211) is widely used where scions are grafted either onto rooted or unrooted disbudded rootstock cuttings. Methods include the whip graft, or better, machine grafting (see Chapter 12). Grafting is done in late winter or early spring from completely dormant scion and rootstock material. The stocks are cut to 31 to 36 cm (12 to 14 in) with the lower cut just below a node and the top cut 2.5 cm (1 in) or more above a node. All buds are removed from the rootstock to prevent subsequent suckering. Scionwood should have the same diameter as the stock.

After grafting with a one-bud scion, the union is stapled together or wrapped with budding rubber. The grafts are placed for 3 to 4 weeks in boxes or plastic bags with well-aerated, moist wood shavings or peat moss at about 26.5°C (80°F) for callusing. When callusing is complete, the grafts are removed from the callusing boxes or plastic bags and any roots and the scion shoot are carefully trimmed back to an 18 mm (1/2 in) stub. The scion is dipped into a temperature-controlled container of melted (low melting point) paraffin or rose wax to a depth of 2.5 cm (1 in) below the graft union and then quickly into cool water. The paraffined bench grafts can then be planted into 5.0 × 5.0 × 25 cm (2 × 2 × 10 in) milk cartons, containers, or planting tubes that contain a mixture of perlite and pumice, or perlite and peat moss. The containers or tubes are placed upright in flats, and the flats are set on pallets and moved into a heated greenhouse for 6 to 8 weeks. Following this operation, the growing bench grafts are transferred to a 50 percent shade screen house for about 2 weeks for hardening-off prior to planting in the nursery or vineyard.

The bench grafts are planted deeply so that the tops of the cartons or planting tubes are at least 5 to 7 cm (2 to 3 in) below the soil level to ensure that water will get inside the cartons or tubes. At no time, however, should bench grafts be planted in the vineyard with the graft union at or below ground level.

Greenwood Grafting. Greenwood grafting is a simple and rapid procedure for propagating *vinifera* grapes on resistant rootstocks (89). A one-budded greenwood scion is splice-grafted during the active growing season on new growth arising either from a 1-year-old rooted cutting or from a cutting during midseason of the second year's growth.

Field Budding (134). An older method of establishing grape cultivars on resistant rootstocks is to field bud onto rapidly growing, well-rooted cuttings that had been planted in their permanent vineyard location the previous winter or spring. T-budding can then be done in late spring using dormant budwood held under refrigeration. Shortly after budding, the "trunks" should be cut with diagonal slashes at the base to allow "bleeding" to take place (3). An alternate method is to chip bud in late summer or early fall as soon as fresh mature buds from wood with light brown bark can be obtained and before the stock goes dormant. In areas where mature buds cannot be obtained early in the fall, growers may store under refrigeration the bud-sticks collected in the winter and bud them in late spring or early summer.

The bud is inserted in the stock 5 to 10 cm (2 to 4 in) above the soil level, preferably on the side adjacent to the supporting stake, tied in place with budding rubber, or poly budding tape, but not waxed. The bud is then covered with 13 to 25 cm (5 to 10 in) of well-pulverized, moist soil to prevent drying. In areas of hot summers, or in soils of low moisture, variable results are likely to be obtained, and bench or nursery grafted vines should be used. If the buds are tied with white 13 mm (1/2 in) plastic tape, it is unnecessary to mound them over with soil.

Top-Grafting Grapevines. Cultivars of mature grapevines can be changed by cutting off the tops of the vines in early spring 30 to 53 cm (12 to 21 in) below the lower wire. Stocks are side whip-grafted, using a two-bud scion of the desired cultivar. The scions are wrapped with 1-inch white plastic tape and the cut surfaces are covered with grafting wax.

The vines may also be T-budded with the inverted T and wrapped with white plastic tape after the bark "slips" in late spring. No grafting compound is needed. Both methods give highly successful takes.

Rootstocks for American Hybrid Grapes (104). 'Ramsey' ('Salt Creek') and 'Dog Ridge.' Widely used as rootstocks for grapes in the southern United States, they have also been used successfully to resist nematodes on low fertility sites in California.

The northeastern grape-growing regions are largely own-rooted, but in areas where *V. vinifera* cultivars are grown, or where more vigor is desired, scions are primarily grafted to 'Couderc 3309,' 'Teleki 5C,' and '101–14 Mgt.' Other stocks that are occasionally used include 'Couderc 1616,' and 'Kober 5BB.' Some inland areas still derive scion vigor from American hybrid rootstocks, such as 'Cynthiana' or 'Lenoir.'

Rootstocks for Vinifera Grapes (132, 133, 220, 221).

The rootstock AXR#1 (*V. vinifera* × *V. rupestris*), although widely used in California through the early 1980s, does not possess sufficient resistance to phylloxera and its use is declining within the state. The replanting effort is proceeding with a wide range of new rootstocks, although current evaluation data indicating which rootstocks are best suited for which areas is not complete. The following list describes the most widely used rootstocks at this time. Most of these rootstocks were developed in Europe about 100 years ago and were designed to resist phylloxera and grow well under European conditions.

'Rupestris St. George.' This pure *V. rupestris* rootstock was one of the original rootstocks developed to address the phylloxera crisis in France. It is only moderately resistant to phylloxera but has not collapsed to this pest. It is very susceptible to nematodes, but roots and grafts well. 'St. George' is best suited to shallow soils, or soils with moderate fertility, and can produce overly vigorous scions on fertile soils or on sites with excessive soil moisture.

'Teleki 5C.' This widely used *V. berlandieri* × *V. riparia* rootstock (104) is low to moderate in vigor and well adapted to fertile soils. It resists phylloxera and many nematode species, but is sensitive to water stress and performs best with adequate irrigation. Its cuttings root relatively easily and its low-to-moderate vigor produces good fruit quality.

'3309 Couderc.' This *V. riparia* × *V. rupestris* hybrid is widely used in California, as is the similar '101–14 Mgt.' These two rootstocks are suited to fertile soils and need adequate irrigation. Both are phylloxera resistant, but are susceptible to nematodes. They root and graft well and produce vines of moderate vigor (scions on '101–14' are relatively more vigorous) with good fruit quality.

'110 Richter.' This *V. berlandieri* × *V. rupestris* rootstock is widely used for drought tolerance on sites with shallow soils and limited rainfall. The similar rootstock '1103 Paulsen' is also used on these sites, and is slightly more vigorous. Both rootstocks are phylloxera resistant, but susceptible to nematodes. They are tolerant of limestone soils, and root and graft well.

'Ramsey.' This pure *V. champinii* rootstock is used on soils with low fertility and nematode problems. It resists most root knot nematode strains but can produce an overly vigorous vine on more fertile soils leading to problems with fruit set and fruit quality. Its phylloxera resistance is only moderate, but phylloxera pressure is usually low on the sandy soils for which 'Ramsey' is best suited. 'Dog Ridge' is a *V. champinii* rootstock with similar characteristics, but slightly less vigor. Both rootstocks root with difficulty, but graft well once rooted.

'Freedom.' This widely used rootstock was developed in the 1960s at the USDA-Fresno station as a cross between a seedling of open-pollinated 'Couderc 1613' × (seedling of open pollinated 'Dog Ridge'). Both rootstock parents are female vines and the pollen source is unknown. 'Couderc 1613' is ¼ *V. vinifera* and is moderately susceptible to phylloxera; this fact and the open pollinated nature of this cross cast suspicion on the phylloxera resistance of 'Freedom' and the similar rootstock 'Harmony.' However, 'Freedom' has excellent nematode resistance and is used on many sites throughout California. 'Freedom' imparts relatively high vigor to the scion and is best suited for sandy or low fertility soils. It can be difficult to root and graft well.

'039–16.' This *V. vinifera* × *V. rotundifolia* rootstock was released by UC Davis in 1989 to resist fanleaf degeneration (220), a virus vectored by the dagger nematode, *Xiphinema* index. Although this rootstock allows the nematode to vector the virus, it moderates the virus' negative effect on fruit set. Because it is half *V. vinifera*, 039–16's phylloxera resistance is questionable. It is only recommended for sites where fanleaf degeneration is severe, and no alternative exists. '039–16' can impart excessive vigor to the scion and is very sensitive to water stress. Its *M. rotundifolia* percentage makes this rootstock difficult to root, but once rooted it will graft moderately well.

Walnut (Black). *See Juglans nigra.*

Walnut (Paradox). *See J. hindsii* × *J. regia.*

Walnut (Persian or English). *See Juglans regia.*

Zizyphus iujuba. Jujube, Chinese date. Deciduous temperate-zone tree adapted to hot, arid regions. Other evergreen tropical species with cultivars also exist in India. The leading jujube cultivars in the United States are 'Lang' and 'Li.' Seeds should be stratified at about 4°C (40°F) for several months before planting. Can be rooted by hardwood stem cuttings and root cuttings. T-budding is done on jujube seedling rootstock.

REFERENCES

1. Al Maarri, K., Y. Arnaud, and E. Miginiac. 1986. *In vitro* micropropagation of quince (*Cydonia oblonga* Mill.). *Scientia Hort.* 28:315–21.

2. Aliyu, O. M. 2007. Clonal propagation in cashew (*Anacardium occidentale*): Effect of rooting media on the rootability and sprouting of air-layers. *Tropic. Sci.* 47:65–72.

3. Alley, C. J., and A. T. Koyama. 1978. Vine bleeding delays growth of T-budded grapevines. *Calif. Agr.* 32:6.

4. Alley, C. J. 1980. Propagation of grapevines. *Calif. Agr.* 34:29–30.

5. Almehdi, A. A., D. E. Parfitt, and H. Chan. 2001. Clonal propagation of pistachio rootstock by rooted stem cuttings. (abst.). *HortScience* 36:460.

6. Amin, M. N., and V. S. Jaiswal. 1987. Rapid clonal propagation of guava through *in vitro* shoot proliferation on nodal explants of mature trees. *Plant Cell Tiss. Organ Cult.* 9:235–43.

7. Anderson, W. C. 1984. Micropropagation of filberts, *Corylus avellana*. *Comb. Proc. Intl. Plant Prop. Soc.* 33:132–37.

8. Argles, G. K. 1976. *Anacardium occidentale* (cashew). In R. J. Garner and S. A. Chaudri, eds. *The propagation of tropical fruit trees*. East Malling, England: FAO and Commonwealth Agricultural Bureaux.

9. Arndt, C. H. 1935. Mango polyembryony and other multiple shoots. *Amer. J. Bot.* 22:26.

10. Bacchetta, L., M. Aramini, C. Bernardini, and E. Rugini. 2008. *In vitro* propagation of traditional Italian hazelnut cultivars as a tool for the valorization and conservation of local genetic resources. *HortScience* 43:562–66.

11. Baker, K. F. 1957. *The U.C. system for producing healthy container grown plants*. Calif. Agr. Exp. Sta. Man. 23.

12. Barghchi, M., and P. G. Alderson. 1985. *In vitro* propagation of *Pistacia vera* L. and the commercial cultivars, Ohadi and Kalleghochi. *J. Hort. Sci.* 60:423–30.

13. Barton, L. V. 1965. Viability of seeds of *Theobroma cacao* L. *Contrib. Boyce Thomp. Inst.* 23:109–22.

14. Bass, L. N. 1975. Seed storage of *Carica papaya* L. *HortScience* 10:232–33.

15. Batjer, L. P., and H. Schneider. 1960. Relation of pear decline to rootstocks and sievetube necrosis. *Proc. Amer. Soc. Hort. Sci.* 76:85–97.

16. Bender, G. S. and A. W. Whiley. 2002. Propagation. In A. W. Whiley, B. Schaffer, and B. N. Woilstenholme, eds. *The Avocado: Botany, Production and Uses*. Wallingford, UK: CAB International.

17. Bertrand, B., H. Etienne, and A. Eskes. 2001. Growth, production and bean quality in *Coffea arabica* as affected by interspecific grafting: Consequences for rootstock breeding. *HortScience* 36:269–73.

18. Blodgett, E. C., H. Schneider, and M. D. Aichele. 1962. Behavior of pear decline disease on different stock-scion combinations. *Phytopathology* 52:679–84.

19. Bonner, F. T. 2001. *Woody Plant Seed Manual*. http://wpsm.net.

20. Bose, T. K., and S. K. Mitra. 1990. *Fruits: Tropical and subtropical*. Calcutta, India: Naya Prokash.

21. Brokaw, W. H. 1977. Subtropical fruit tree production; avocado as a case study. *Comb. Proc. Intl. Plant Prop. Soc.* 27:113–21.

22. Brokaw, W. H. 1988. Avocado clonal rootstock propagation. *Comb. Proc. Intl. Plant Prop. Soc.* 38:97–102.

23. Broome, O. C., and R. H. Zimmerman. 1978. *In vitro* propagation of blackberry. *HortScience* 13:151–53.

24. Bush, J. N., Y. Kawamitsu, S. Yonemori, and S. Murayama. 2000. Field performance of *in vitro*-propagated and sucker-derived plants of banana (*Musa* spp.). *Plant Production Sci.* 3:124–28.

25. Calavan, E. C., E. E. Frolich, J. B. Carpenter, C. N. Roistacher, and D. W. Christiansen. 1964. Rapid indexing of exocortis of citrus. *Phytopathology* 54:1359–62.

26. Caldwell, J. D. 1984. Blackberry propagation. *HortScience* 19:193–95.

27. Caldwell, J. D., D. C. Coston, and K. H. Brock. 1988. Rooting of semi-hardwood 'Hayward' kiwifruit cuttings. *HortScience* 23:714–17.

28. Calif. Dept. of Food and Agr. 2001. *California Fruit and Nut Tree: Registration and Certification Program*. http://fpms.ucdavis.edu/Tree/treeR&C.html.

29. Calif. Rare Fruit Growers. 2001. *Fruit facts*. http://www.crfg.org/.

30. Calleson, O. 1998. Recent developments in cherry rootstock research. In J. Ystaas, and O. Calleson, eds. *Proc. Third Intl. Cherry Symp. Acta. Hort.* No. 468.

31. Cameron, J. W., and R. K. Soost. 1953. Nucellar lines of citrus. *Calif. Agr.* 7:8, 15, 16.

32. Carlson, R. F. 1965. Growth and incompatibility factors associated with apricot scion/rootstock in Michigan. *Mich. Quart. Bul.* 48(1):23–9.

33. Catlin, P. B., and E. A. Olsson. 1986. Response of eastern black walnut and northern California black walnut to waterlogging. *HortScience* 21:1379–80.

34. Chase, S. B. 1947. Budding and grafting eastern black walnut. *Proc. Amer. Soc. Hort. Sci.* 49:175–80.

35. Chaudri, S. A. 1976. *Mangifera indica*—Mango. In R. J. Garner and S. A. Chaudri, eds. *The propagation of tropical fruit trees.* Hort. Rev. 4. East Malling, England: FAO and Commonwealth Agricultural Bureaux.

36. Chevreau, E., B. Thibault, and Y. Arnaud. 1992. Micropropagation of Pear (*Pyrus communis* L.). In Y. B. S. Bajaj, ed. *Biotechnology in agriculture and forestry*, 18:244–61. Berlin: Springer-Verlag.

37. Child, R. 1964. *Coconuts.* London: Longmans, Green.

38. Cobin, M. 1954. The lychee in Florida. *Fla. Agr. Exp. Sta. Bul.* 546.

39. Collins, J. L. 1960. *The pineapple—history, cultivation, utilization.* London: Leonard-Hill.

40. Condit, I. J. 1969. *Ficus, the exotic species.* Berkeley: Univ. Calif. Div. Agr. Sci.

41. Converse, R. A. 1972. The propagation of virus-tested strawberry stocks. *Comb. Proc. Intl. Plant Prop. Soc.* 22:73–6.

42. Cooper, P. A., and D. Cohen. 1984. Micropropagation of Japanese persimmon (*Diospyros kaki*). *Comb. Proc. Intl. Plant Prop. Soc.* 34:118–24.

43. Couvillon, G. A., and F. A. Pokorny. 1968. Photoperiod, indolebutyric acid, and type of cutting wood as factors in rooting of rabbiteye blueberry (*Vaccinium ashei* Reade), cv. Woodward. *HortScience* 3:74–5.

44. Couvillon, G. A., and A. Erez. 1980. Rooting, survival, and development of several peach cultivars propagated from semi-hardwood cuttings. *HortScience* 14:41–3.

45. Cross, C. E., J. E. Demoranville, K. H. Deubert, R. M. Devlin, J. S. Norton, W. E. Tomlinson, and B. M. Zuckerman. 1969. Modern cultural practice in cranberry growing. *Mass. Agr. Exp. Sta. Ext. Ser. Bul.* 39.

46. Crossa-Raynaud, P., and J. M. Anderson. 1987. Apricot rootstocks. In R. C. Rom and R. F. Carlson, eds. *Rootstocks for fruit crops.* New York: John Wiley.

47. Damiano, C. 1980. Strawberry micropropagation. In R. H. Zimmerman, ed. *Proc. Conf. Nurs. Prod. Fruit Plants through Tissue Cult.* USDA, SEA, ARR-NE-11.

48. Davies, F. S. 1986. The navel orange. *Hort. Rev.* 8:129–80.

49. Davis, B. H. 1976. Bench grafting plum and apricot as compared to T-budding. *Comb. Proc. Intl. Plant Prop. Soc.* 26:253–55.

50. Dillon, D. 2004. Mist propagation of citrus. *Comb. Proc. Intl. Plant Prop. Soc.* 54:370–71.

51. Dimalla, G. G., and J. V. Staden. 1978. Pecan nut germination: A review for the nursery industry. *Scientia Hort.* 8:1–9.

52. Dirr, M. A., and C. W. Heuser, Jr. 2006. *The reference manual of woody plant propagation.* 2nd ed. Portland, OR: Timber Press.

53. Doggrell, L. M. 1976. Commercial propagation of macadamias. *Comb. Proc. Intl. Plant Prop. Soc.* 26:391–93.

54. Doncaster, T. 1981. The whys and hows of passion fruit growing in Queensland. *Comb. Proc. Intl. Plant Prop. Soc.* 31:616–17.

55. Druart, P. 1992. *In vitro* culture and micropropagation of plum *(Prunus* sp.*).* In Y. P. S. Bajaj, ed. *Biotechnology in agriculture and forestry.* Berlin: Springer-Verlag.

56. Durling, D. L. 1990. International citrus nursery production as it relates to *Phytopthora*, virus and growing media. *Comb. Proc. Intl. Plant Prop. Soc.* 40:51–4.

57. Eck, P. 1988. *Blueberry science.* New Brunswick, NJ: Rutgers Univ. Press.

58. Efron, Y., P. Epaina, and J. Marfu. 2006. Guidelines for accelerated clone development. In A.B. Eskes and Y. Efron, eds. *Global Approaches to Cocoa Germplasm Utilization and Conservation. Final report of the CFC/ICCO/IPGRI project on "Cocoa Germplasm Utilization and Conservation: a Global Approach" (1998–2004).* Amsterdam: CFC/ICCO/IPGRI.

59. Erdogan, V., and D. C. Smith. 2005. Effects of tissue removal and hormone application on rooting of hazelnut layers. *HortScience* 40:1457–60.

60. Estrada-Luna, A. A., C. Lopez-Peralta, and E. Cardenas-Soriano. 2002. *In vitro* micrografting and histology of the graft union formation of selected species of prickly pear cactus (*Optuntia* spp.). *Scientia Hort.* 92:317–27.

61. Estrada-Luna, A. A., J. J. Martinez-Hernandez, M. E. Torres-Torres, and F. Chable-Moreno. 2008. *In vitro* micropropagation of the ornamental prickly pear cactus *Optunita lanigera* Salm-Dyck and effects of sprayed GA_3 after transplantation to *ex vitro* conditions. *Scientia Hort.* 117:378–85.

62. Exadaktylou, E., T. Thomidis, B. Grout, G. Zakynthinos, and C. Tsipouridis. 2009. Methods to improve the rooting of hardwood cuttings of 'Gisela 5' cherry rootstock. *HortTechnology* 19:254–59.

63. Fankhauser, I. 1985. Propagation of feijoa by bench grafting. *Comb. Proc. Intl. Plant Prop. Soc.* 35:401–3.

64. Ferree, D. C., and R. F. Carlson. 1987. Apple rootstocks. In R. C. Rom and R. F. Carlson, eds. *Rootstocks for fruit crops*. New York: John Wiley.

65. Fiorino, P., and F. Loreti. 1987. Propagation of fruit trees by tissue culture in Italy. *HortScience* 22:323–58.

66. Forde, H. I., and G. H. McGranahan. 1996. Walnuts. In J. Janick and J. N. Moore, eds. *Fruit Breeding. Volume III*. New York: John Wiley & Sons.

67. Foster, J. A. 1988. Regulatory actions to exclude pests during the international exchange of plant germplasm. *HortScience* 23:60–6.

68. Foundation Plant Materials Service. 2001. *FPMS Test Index Page*. http://fpms.ucdavis.edu.

69. Frolich, E. F. 1966. Rooting citrus and avocado cuttings. *Comb. Proc. Intl. Plant Prop. Soc.* 16:51–4.

70. Frolich, E. F., and R. G. Platt. 1972. *Use of the etiolation technique in rooting avocado cuttings*. Calif. Avoc. Soc. Yearbook, 1971–1972.

71. Funk, D. T. 1979. Black walnuts for nuts and timber. In R. A. Jaynes, ed. *Nut tree culture in North America*. Hamden, CT: North American Nut Growers Assn.

72. Gascó, A., A. Nardini, F. Raimondo, E. Gortan, A. Motisi, M. A. Lo Gullo, and S. Salleo. 2007. Hydraulic kinetics of the graft union in different *Olea europaea* L. scion/rootstock combinations *Environ. Exp. Bot.* 60:245–50.

73. George, A. P., and R. J. Nissen. 1987. Propagation of *Annona* species: A review. *Scientia Hort.* 33:75–85.

74. Golsalves, D., S. Ferreira, R. Manshardt, M. Fitch, and J. Sligthom. 2000. *Transgenic virus resistant papaya: New hope for controlling papaya ringspot virus in Hawaii*. http://apsnet.org/education/feature/papaya/Top.htm.

75. Gonzalez, J. M. 2008. Jack fruit propagation testing four graft types. *HortScience* 43:1180.

76. Goode, D. Z., Jr., D. T. Mason, and E. G. Ing. 1965. Cold storage of strawberry runners at different temperatures. *Exp. Hort.* 12:38–41.

77. Griggs, W. H., D. D. Jensen, and B. T. Iwakiri. 1968. Development of young pear trees with different rootstocks in relation to psylla infestation, pear decline, and leaf curl. *Hilgardia* 39:153–204.

78. Griggs, W. H., J. A. Beutel, W. O. Reil, and B. T. Iwakiri. 1980. Combination of pear rootstocks recommended for new Bartlett plantings. *Calif. Agr.* 34:20–4.

79. Grimo, E. 1979. Carpathian (Persian) walnuts. In R. A. Jaynes, ed. *Nut tree culture in North America*. Hamden, CT: North American Nut Growers Assn.

80. Haicour, R., V. B. Trang, D. Djailo, A. Assani, F. Bakry, and F. X. Cote. 1998. Banana improvement through biotechnology—ensuring food security in the 21st century. *Cahiers Agricultures*. 468–75 (in French).

81. Hall, I. V. 1969. *Growing cranberries*. Can. Dept. Agr. Publ. 1282 (rev.).

82. Hall, I. V., L. E. Aalders, L. P. Jackson, G. W. Wood, and L. H. Lockhart. 1975. *Lowbush blueberry production*. Can. Dept. Agr. Publ. 1477.

83. Hall, W. J. 1975. Propagation of walnuts, almonds, and pistachios in California. *Comb. Proc. Intl. Plant Prop. Soc.* 25:53–7.

84. Hamilton, R. Z., and W. Yee. 1974. *Macadamia: Hawaii's dessert nut*. Univ. Hawaii Cir. 485.

85. Hancock, J. F. 1999. *Strawberries*. Wallingford, UK: CAB International.

86. Hanna, J. D. 1987. Pecan rootstocks. In R. C. Rom and R. F. Carlson, eds. *Rootstocks for fruit crops*. New York: John Wiley.

87. Hansen, A. J. 1985. An end to the dilemma: Virus free all the way. *HortScience* 20:852–59.

88. Hansen, K. C., and J. E. Lazarte. 1984. In vitro propagation of pecan seedlings. *HortScience* 19:237–39

89. Harmon, F. N. 1954. A modified procedure for greenwood grafting of *Vinifera* grapes. *Proc. Amer. Soc. Hort. Sci.* 64:255–58.

90. Hartmann, H. T., K. Opitz, and J. A. Beutel. 1980. Olive production in California. Univ. Calif. Div. Agr. Sci. Publ.: Leaflet 2474.

91. Hartmann, H. T., and C. J. Hansen. 1958. Rooting pear, plum rootstocks. *Calif. Agr.* 12:4, 14–45.

92. Hartmann, H. T., C. J. Hansen, and F. Loreti. 1965. Propagation of apple rootstocks by hardwood cuttings. *Calif. Agr.* 19:4–5.

93. Hartmann, H. T., W. C. Schnathorst, and J. Whisler. 1971. Oblonga, a clonal olive rootstock resistant to verticillium wilt. *Calif. Agr.* 25:12–5.

94. Hartmann, H. T., W. H. Griggs, and C. J. Hansen. 1963. Propagation of own-rooted Old Home and Bartlett pears to produce trees resistant to pear decline. *Proc. Amer. Soc. Hort. Sci.* 82:92–102.

95. Heppner, M. J. 1927. Pear black-end and its relation to different rootstocks. *Proc. Amer. Soc. Hort. Sci.* 24:139–42.

96. Herman, E. B., and G. J. Hass. 1975. Clonal propagation of *Coffea arabica* L. from callus culture. *HortScience* 10:588–89.

97. Heydecker, W., and M. Marston. 1968. Quantitative studies on the regeneration of raspberries from root cuttings. *Hort. Res.* 8:142–46.

98. Hibino, H., G. H. Kaloostian, and H. Schneider. 1971. Mycoplasmalike bodies in the pear psylla vector of pear decline. *Virology* 43:34–40.

99. Hilton, H. 1986. Budding European (Spanish) chestnut (*Castanea sativa* Mill.). *Comb. Proc. Intl. Plant Prop. Soc.* 36:72–6.

100. Hodel, D. R., and D. R. Pittenger. 2003a. Studies on the establishment of date palm (*Phoenix dactylifera* 'Deglet Noor') offshoots. Part 1. Observations on root development and leaf growth. *Palms* 47:191–200.

101. Hodel, D. R., and D. R. Pittenger. 2003b. Studies on the establishment of date palm (*Phoenix dactylifera* 'Deglet Noor') offshoots. Part ll. Size of offshoot. *Palms* 47:201–5.

102. Hodel, D. R., D. V. Johnson, and R. W. Nixon. 2007. *Imported and American Varieties of Dates (Phoenix dactylifera) in the United States*. Oakland: Univ. Calif. ANR Publications.

103. Howard, B. H., E. Tal, and S. K. Miles. 1987. Red raspberry propagation from leafy summer cuttings. *J. Hort. Sci.* 62:485–92.

104. Howell, G. S. 1987. *Vitis* rootstocks. In R. C. Rom and R. F. Carlson, eds. *Rootstocks for fruit crops,*. New York: John Wiley.

105. Hunter, A. G., and J. D. Norton. 1985. Rooting stem cuttings of Chinese chestnut. *Scientia Hort.* 26:43–5.

106. Hwang, S. C., C. L. Chen, J. C. Lin, and H. L. Lin. 1984. Cultivation of banana using plantlets from meristem culture. *HortScience* 19:231–33.

107. Ivanicka, J. 1987. *In vitro* micropropagation of mulberry. *Morus nigra* L. *Scientia Hort.* 32:33–9.

108. Ivanicka, J. 1992. Micropropagation of cherry. In Y. P. A. Bajaj, ed. *Biotechnology in agriculture and forestry*. Berlin: Springer-Verlag.

109. Ivey, I. D. 1979. Feijoas: Selection and propagation. *Comb. Proc. Intl. Plant Prop. Soc.* 29:161–68.

110. Jaffee, A. 1970. Chip grafting guava cultivars. *The Plant Propagator* 16(2):6.

111. James, D. J., and I. J. Thurbon. 1981. Shoot and root initiation *in vitro* in the apple rootstock M.9 and the promotive effects of phloroglucinol. *J. Hort. Sci.* 56:15–20.

112. Janos, D. P., M. S. Schroeder, B. Schaffer, and J. H. Crane. 2001. Inoculation with arbuscular mycorrhizal fungi enhances growth of *Litchi chinensis* Sonn. trees after propagation by air-layering. *Plant Soil* 233:85–94.

113. Jaynes, R. A. 1961. Buried-inarch technique for rooting chestnut cuttings. North. Nut Grow. Assoc. *52nd Ann. Rpt.* pp. 37–9.

114. Jaynes, R. A. 1979. Chestnuts. In R. A. Jaynes, ed. *Nut tree culture in North America*. Hamden, CT: Northern Nut Growers Assn.

115. Jennings, D. L. 1988. *Raspberries, blackberries, their breeding, diseases, and growth*. London: Academic Press.

116. Joley, L. E. 1979. Pistachios. In R. A. Jaynes, ed. *Nut tree culture in North America*. Hamden, CT: Northern Nut Growers Assn.

117. Jones, O. P. 1993. Propagation of apple *in vitro*. In A. Ahuja, ed. *Micropropagation of woody plants*. Dordrecht: Kluwer Acad. Pub.

118. Kester, D. E., and C. Grasselly. 1987. Almond rootstocks. In R. C. Rom and R. F. Carlson, eds. *Rootstocks for fruit crops*. New York: John Wiley.

119. Kester, D. E., and T. M. Gradziel. 1996. Almonds. In J. Janick and J. N. Moore, eds. *Fruit breeding: Volume III. Nuts*. New York: John Wiley & Sons.

120. Kester, D. E., and T. M. Gradziel. 1996. Genetic Disorders. In W. C. Micke, ed. *Almond Production Manual*. Univ. of Cali. Div. of Agr. and Nat. Res. Publ. 3364.

121. Kester, D. E., R. Asay, and T. M. Gradziel. 2002. 'Nickels' almond x peach rootstock. *HortScience* 36:855–57.

122. Klotz, L. J., E. C. Calavan, and L. G. Weathers. 1972. Virus and virus-like diseases of citrus. *Calif. Agr. Exp. Sta. Circ. 559.*

123. Krezdorn, A. H., and G. W. Adriance. 1961. *Fig growing in the South*. Washington, DC:U.S. Govt. Printing Office.

124. Krul, W. R., and J. Meyerson. 1980. *In vitro* propagation of grape. In R. H. Zimmerman, ed. *Proc. Conf. Nurs. Prod. Fruit Plants through Tissue Culture*. USDA, SEA, ARR-NE-11.

125. Lagerstedt, H. B. 1979. Filberts. In R. A. Jaynes, ed. *Nut tree culture in North America*. Hamden, CT: Northern Nut Growers Assn.

126. Lane, W. D. 1992. Micropropagation of apple (*Malus domestica* Burkh.). In Y. P. S. Bajaj, ed. *Biotechnology in Agriculture and Forestry*. Berlin: Springer-Verlag.

127. Lapins, K. 1959. Some symptoms of stock-scion incompatibility of apricot varieties on peach seedling rootstock. *Can. J. Plant Sci.* 39:194–203.

128. Lawyer, E. M. 1978. Seed germination of stone fruits. *Comb. Proc. Intl. Plant Prop. Soc.* 28:106–9.

129. Layne, R. E. C. 1987. Peach rootstocks. In R. C. Rom and R. F. Carlson. *Rootstocks for fruit crops.* New York: John Wiley.

130. Lazcano, C. A., F. T. Davies, Jr., V. Olalde-Portugal, J. C. Mondragon, S. A. Duray, and A. A. Estrada-Luna. 1999. Effects of auxin and wounding on adventitious root formation of prickly-pear cactus cladodes (*Opuntia amyclaea* T.). *HortTechnology* 9:99–102.

131. Lear, B., and L. Lider. 1959. Eradication of root-knot nematodes from grapevine rootings by hot water. *Plant Dis. Rpt.* pp. 314–17.

132. Lider, L. 1958. Phylloxera-resistant grape rootstocks for the coastal valleys of California. *Hilgardia* 27:287–318.

133. Lider, L. 1960. Vineyard trials in California with nematode-resistant grape rootstocks. *Hilgardia* 30:123–52.

134. Lider, L. 1963. Field budding and the care of the budded grapevine. Calif. Agr. Ext. Ser. *Leaflet 153.*

135. Litz, R. E., and V. S. Jaiswal. 1991. Micropropagation of tropical and subtropical fruits. In P. C. deBergh and R. H. Zimmerman, eds. *Micropropagation: Technology and application.* Dordrecht: Kluwer Acad. Pub.

136. Lombard, P. B., and M. N. Westwood. 1987. Pear rootstocks. In R. C. Rom and R. F. Carlson, eds. *Rootstocks for fruit crops.* New York: John Wiley.

137. Lynch, S. J., and R. O. Nelson. 1956. Current methods of vegetative propagation of avocado, mango, lychee, and guava in Florida. *Cieba (Escuela Agricola Panamericana, Tegucigalpa, Honduras).* 4:315–77.

138. MacDaniels, L. H. 1979. Hickories. In R. A. Jaynes, ed. *Nut tree culture in North America.* Hamden, CT: Northern Nut Growers Assn.

139. Madden, G. E., and H. W. Tisdale. 1975. Effects of chilling and stratification on nut germination of northern and southern pecan cultivars. *HortScience* 10:259–60.

140. Madden, G. E. 1979. Pecans. In R. A. Jaynes, ed. *Nut tree culture in North America.* Hamden, CT: Northern Nut Growers Assn.

141. Majumdar, P. K., and S. K. Mukherjee. 1968. Guava: A new vegetative propagation method. *Indian Hort.* Jan.–Mar.

142. Majumdar, P. K., and D. K. Sharma. 1990. Mango. In T. K. Bose and S. K. Mitra, eds. *Fruits: Tropical and subtropical.* Calcutta, India: Naya Prokash.

143. Martinelli, A. 1992. Micropropagation of strawberry (*Fragaria* spp.). In Y. P. S. Bajaj, ed. *Biotechnology in agriculture and forestry.* Berlin: Springer-Verlag.

144. Maxwell, N. P., and C. G. Lyons. 1979. A technique for propagating container-grown citrus on sour orange rootstock. *HortScience* 14:56–7.

145. McPhee, D. 2007. Improving take rates in grafting macadamias. *Comb. Proc. Intl. Plant Prop. Soc.* 57:206–8.

146. Medina, J. P. 1981. Studies of clonal propagation of pecans at Ica, Peru. *Plant Propagator* 27:10–1.

147. Mehlenbacher, S. A. 1986. Rooting of interspecific peach hybrids by semi-hardwood cuttings. *HortScience* 21:1374–76.

148. Menge, J. A., B. S. McKee, F. B. Guillemet, and E. C. Pond. 1999. Results of recent tests for root rot tolerance in avocado rootstocks in California. *Calif. Avocado Society Yearbook* 83:75–81.

149. Mickelbart, M. V. 1996. Sapodilla: A potential crop for subtropical climates. In J. Janick, ed. *Progress in new crops.* Alexandria, VA: ASHS Press.

150. Miller, G., D. D. Miller, and R. A. Jaynes. 1996. Chestnuts. In J. Janick and J. N. Moore, eds. *Fruit breeding Volume III. Nuts.* New York: John Wiley & Sons.

151. Montealegre, J. C., N. F. Childers, S. A. Sargent, L. de M. Barros, and R. E. Alves. 1999. Cashew (*Anacardium occidentale*, L.) nut and apple: A review of current production and handling recommendations. *Fruit Var. J.* 53:2–9.

152. Moore., G. A., M. G. Dewald, and M. H. Evans. 1992. Micropropagation of pineapple (*Ananas comosus* L.). In Y. P. Bajaj, ed. *Biotechnology in agriculture and forestry.* Berlin: Springer-Verlag.

153. Moss, G. I., and R. Dallies. 1981. Rapid propagation of citrus in containers. *Comb. Proc. Intl. Plant Prop. Soc.* 31:262–74.

154. Mourao Filho, F. A. A., E. A. Girardi, and H. T. Zarate do Couto. 2009. 'Swingle' citrumelo propagation by cuttings for citrus nursery tree production or inarching. *Scientia Hort.* 120:207–11.

155. Mukherjee, S. K., and P. K. Majumdar. 1963. Standardization of rootstock of mango. I. Studies on the propagation of clonal rootstocks by stooling and layering. *Indian J. Hort.* 20:204–9.

156. Mukherjee, S. K., and P. K. Majumdar. 1964. Effect of different factors on the success of veneer grafting in mango. *Indian J. Hort.* 24:46–50.

157. Mukherjee, S. K., and N. N. Bid. 1965. Propagation of mango. II. Effect of etiolation and growth regulator treatment on the success of air layering. *Indian J. Agr. Sci.* 35:309–14.

158. Nagabhushanam, S., and M. A. Menon. 1980. Propagation of cashew (*Anacardium occidentale* L.) by etiolation, girdling, and stooling. *Plant Propagator* 26:11–3.

159. Nakasone, H. Y., and R. E. Paull. 1988. *Tropical fruits.* Wallingford, UK: CAB International.

160. Nixon, R. W. 1969. Growing dates in the United States., *USDA Agr. Inf. Bul.* 207.

161. Northwood, P. J. 1964. Vegetative propagation of cashew (*Anacardium occidentale*) by the air layering method. *E. Afr. Agr. For. J.* 30:35–7.

162. Okie, W. R. 1987. Plum rootstocks. In R. C. Rom and R. F. Carlson, eds. *Rootstocks for fruit crops.* New York: John Wiley.

163. Omar, M. S., M. K. Hameed, and M. S. Al-Rawi. 1992. Micropropagation of date palm (*Phoenix dactylifera* L.). In Y. P. S. Bajaj, ed. *Biotechnology in agriculture and forestry.* New York: John Wiley.

164. Onay, A. 2000. Micropropagation of pistachio from mature trees. *Plant Cell Tiss. Organ Cult.* 60:159–62.

165. Onay, A., E. Tilkat, H. Yildirum, and V. Suzerer. 2007. Indirect somatic embryogenesis from mature embryo cultures of pistachio (*Pistacia vera* L.). *Prop. Ornamental Plants* 7:68–74.

166. Painter, J. H. 1967. Producing walnuts in Oregon. *Oregon Agricultural Extension Bulletin 795.*

167. Parfitt, D. E., and A. A. Almehdi. 1994. Use of high CO_2 atmosphere and medium modifications for the successful micropropagation of pistachio. *Scientia Hort.* 56:321–29.

168. Pennock, W., and G. Maldonado. 1963. The propagation of guavas from stem cuttings. *J. Agr. Univ. Puerto Rico* 47:280–90.

169. Perry, R. L. 1987. Cherry rootstocks. In R. C. Rom and R. F. Carlson, eds. *Rootstocks for fruit crops.* New York: John Wiley.

170. Pijut, P. M., and M. J. Barker. 1999. Propagation of *Juglans cinerea* (butternut). *HortScience* 34:458–59.

171. Pijut, P. M., and M. J. Moore. 2002. Early season softwood cuttings effective for vegetative propagation of *Juglans cinerea*. *HortScience* 37:697–700.

172. Platt, R. G. 1976. Current techniques of avocado propagation. In J. W. Sauls, R. L. Phillips, and L. K. Jackson, eds. *Proc. 1st Inter. Trop. Short Course: The avocado.* Gainesville, FL: Fruit Crops Dept., Univ. of Fla.

173. Pontikis, C.A., and P. Melas. 1986. Micropropagation of *Ficus carica* L. *HortScience* 21:153.

174. Preston, A. P. 1966. Apple rootstock studies: Fifteen years' results with Malling-Merton clones. *J. Hort. Sci.* 41:349–60.

175. Reighard, G., R. Andersen, J. Anderson, W. Autio, and T. Beckman. 2004. Eight-year performance of 19 peach rootstocks at 20 locations in North America. *J. Amer. Pomol. Soc.* 58:174–202.

176. Reimer, R. C. 1925. Blight resistance in pears and characteristics of pear species and stocks. *Oreg. Agr. Exp. Sta. Bul.* 214.

177. Reuther, W. , ed. *The citrus industry*, Vol. 1, 1967; Vol. 2, 1968; Vol. 3, 1973; Vol. 4 1978. Berkeley: Univ. Calif. Div. Agr. Sci.

178. Reuveni, O., and M. Raviv. 1981. Importance of leaf retention to rooting of avocado cuttings. *J. Amer. Soc. Hort. Sci.* 106:127–30.

179. Rout, G. R., S. Samantaray, and P. Das. 2000. Biotechnology of the banana: A review of recent progress. *Plant Biol.* 2:512–24.

180. Ryugo, K., C. A. Schroeder, A. Sugiura, and K. Yonemori. 1988. Persimmons for California. *Calif. Agr.* 42:7–9.

181. Schneider, H. 1970. Graft transmission and host range of the pear decline causal agent. *Phytopathology* 60:204–7.

182. Scott, D. H., and D. P. Ink. 1948. Germination of strawberry seed as affected by scarification treatment with sulfuric acid. *Proc. Amer. Soc. Hort. Sci.* 51:299–300.

183. Sebastian, K. T., and J. A. McComb. 1986. A micropropagation system for carob. *Scientia Hort.* 28:127–31.

184. Shalla, T., and L. Chiarappa. 1961. Pear decline in Italy. *Bul. Calif. Dept. Agr.* 50:213–17.

185. Shamel, A. D., C. S. Pomeroy, and R. E. Caryl. 1929. *Bud selection in Washington Navel orange: Progeny tests for limb variation. USDA Tech Bul. 123.*

186. Sibbett, G. S., L. Ferguson, J. L Coviello, and M. Lindstrand. 2005. *Olive Production Manual.* 2nd ed. Berkeley, CA: ANR Publications, University of California (System).

187. Singh-Dhaliwal, T., and A. Torres-Sepulveda. 1961. Recent experiments on rooting coffee stem cuttings in Puerto Rico. *Univ. Puerto Rico Agr. Exp. Sta. Tech. Paper 33.*

188. Singh, S., and D. V. Chugh. 1961. Marcotting with some plant regulators in loquat (*Eriobotrya japonica* Lind.). *Indian J. Hort.* 18:123–29.

189. Sitton, B. G. 1931. Vegetative propagation of the black walnut. *Mich. Agr. Exp. Sta. Tech. Bul. 119.*

190. Smith, I. E., B. N. Wolstenholme, and P. Allan. 1974. Rooting and establishment of pecan [*Carya illinoensis* (Wang) K. Koch.] stem cuttings. *Agroplantae.* 6:21–8.

191. Smith, J. C., D. J. Jordon, and F. H. Wood. 1986. Persimmon propagation: Research highlights from Ruakura. *Comb. Proc. Intl. Plant Prop. Soc.* 35:323–28.

192. Smith, M. K., S. D. Hamill, V. J. Doogan, and J. W. Daviells. 1999. Characterization and early detection of an off-type from micropropagated 'Lady Finger' bananas. *Aust. Jour. Expt. Agr.* 39:1017–23.

193. Soost, R. K., J. W. Cameron, W. P. Bitters, and R. G. Platt. 1961. Citrus bud variation. *Calif. Citrograph.* 46:176, 188–93.

194. Southwick, S. M., and J. Uyemoto. 1999. *Cherry crinkle- leaf and deep suture disorders.* University of California Publ. 8007.

195. Spethmann, W. 2004. Use of long cuttings to reduce propagation time of rose and fruit rootstocks and street trees. *Comb. Proc. Intl. Plant Prop. Soc.* 54:223–31.

196. Spethmann, W. 2007. Increase of rooting success and further shoot growth by long cuttings of woody plants. *Prop. Ornamental Plants* 7:160–66.

197. Stebbins, R. L., and H. R. Cameron. 1984. Performance of 3 cherry varieties, *Prunus avium* L. cultivars on 5 clonal rootstocks. *Fruit Var. J.* 38:21–3.

198. Storey, W. B. 1976. *Macadamia tetraphylla*—the preferred rootstock. *Calif. Macadamia Soc. Yearbook.* 22:101–7.

199. Sugiura, A., R. Tao, H. Murayama, and T. Tomana. 1986. *In vitro* propagation of Japanese persimmon. *HortScience* 24:1205–7.

200. Sugiura, A., Y. Matsuda-Habu, M. Gao, T. Esumi, and R. Tao. 2008. Somatic embryogenesis and plant regeneration from immature persimmon (*Diospyros kaki* Thunb.) embryos. *HortScience* 43:211–14.

201. Teague, C. P. 1966. Avocado tip-grafting. *Comb. Proc. Intl. Plant Prop. Soc.* 16:50–1.

202. Teulon, J. 1971. Propagation of passion fruit (*Passiflora edulis*) on a fusarium-resistant rootstock. *Plant Propagator* 17:4–5.

203. Thomas, C. A. 1981. Mango propagation by saddle grafting. *J. Hort. Sci.* 56:173–75.

204. Tilkat, E., C. Isikalan, and A. Onay. 2005. *In vitro* propagation of Kinjuk pistachio (*Pistacia Khinjuk*) stocks through seedling apical shoot tip culture. *Prop. Ornamental Plants* 5:124–28.

205. Traore, A., S. N. Maximova, and M. J. Guiltinan. 2003. Micropropagation of *Theobroma cacao* L. using somatic embryo-derived plants. *In Vitro Cell. Dev. Biol. Plant.* 39:332–37.

206. van Staden, J., B. N. Wolstenholme, and G. G. Dimala. 1976. Effect of temperature on pecan seed germination. *HortScience* 11:261–62.

207. Vietez, A. M., A. Ballester, M. L. Vieitez, and E. Vieitez. 1983. *In vitro* plantlet regeneration of mature chestnut. *J. Hort. Sci.* 58:457–63.

208. Wallace, R. D., and L. G. Spinella. 1992. *World chestnut industry conference.* Aluchua, FL: Chestnut Marketing Association.

209. Warner, R. M., Z. Worku, and J. A. Silva. 1979. Effect of photoperiod on growth responses of citrus rootstocks. *J. Amer. Soc. Hort. Sci.* 104:232–35.

210. Webber, H. J. 1948. Rootstocks: Their character and reactions. In H. J. Webber and L. D. Batchelor, eds. *The citrus industry*, Vol. II. Berkeley: Univ. of Calif. Press.

211. Weinberger, J. H., and N. H. Loomis. 1972. *A rapid method for propagating grapevines on rootstocks.* USDA Agr. Res. Ser. ARS-W-2.

212. Welander, M. 1985. Micropropagation of gooseberry, *Ribes grossularia. Scientia Hort.* 26:267–72.

213. Welander, M. 1985. *In vitro* culture of raspberry (*Rubus ideaus*) for mass propagation. *J. Hort. Sci.* 60:493–99.

214. Wellman, F. L. 1961. *Coffee: Botany, cultivation, utilization.* London: Leonard-Hill.

215. Westwood, M. N., and A. N. Roberts. 1965. Quince rootstock used for dwarf pears. *Oreg. Orn. and Nurs. Dig.* 9(1):1–2.

216. Westwood, M. N., and L. A. Brooks. 1963. Propagation of hardwood pear cuttings. *Comb. Proc. Intl. Plant Prop. Soc.* 13:261–68.

217. Westwood, M. N., H. R. Cameron, P. B. Lombard, and C. B. Cordy. 1971. Effects of trunk and rootstock on decline, growth, and performance of pear. *J. Amer. Soc. Hort. Sci.* 96:147–50.

218. Whitsell, R. H., G. E. Martin, B. O. Bergh, A. V. Lypps, and W. W. Brokaw. 1989. *Propagating avocados: Principles and techniques of nursery and field grafting.* 1989, Univ. of Calif. Div. of Agr. and Nat. Res. Publ. 21461 p. 30.

219. Winkler, A. J., J. A. Cook, L. A. Lider, and W. M. Kliewer. 1974. Propagation. Chapter 9 in *General viticulture* (2nd ed.) Berkeley Univ. of Calif. Press.

220. Wolpert, J. A., M. A. Walker, and E. Weber, eds. 1992. *Proceedings of rootstock seminar. A worldwide perspective.* Davis, CA: Amer. Soc. Enol. and Vit. Press.

221. Wolpert, J. A., M. A. Walker, E. Weber, L. Bettiga, R. Smith, and P. Verdegal. 1994. Use of phylloxera-resistant rootstocks in California: Past, present and future. *Grape Grower,* Part 1 January: 16–19, Part 2 February: 4–7, Part 3 March: 10–17.

222. Wolpert, J. P., M. A. Walker, E. Weber, and D. Roberts. 1995. *Proc. Intern. Symp. on Clonal Selection.* Davis, CA: Amer. Soc. Enol. and Vit.

223. Young, J. A., and C.G. Young. 1986. *Seeds of wildland plants.* Portland, OR: Timber Press.

224. Ystaas, J. Y., ed. 1998. Proc. Third Intern. Cherry Symposium. *Acta Hort.* Leuven, Belgium: IHS.

225. Zentmyer, G. A., A.O. Paulus, and R. M. Burns. 1962. Avocado root rot. *Calif. Agr. Exp. Sta. Circ. 511.*

226. Zimmerman, R. H., and D. C. Broome. 1980. *Apple cultivar micropropagation.* USDA, Agr. Research Results ARR-NE-11 pp. 54–63.

227. Zimmerman, R. H. 1991. Micropropagation of temperate zone fruit and nut crops. In P. C. DeBergh and R. H. Zimmerman, eds. *Micropropagation: Technology and application.* Dordrecht: Kluwer Acad. Pub.

228. Zimmerman, R. H., G. J. Galleta, and O. C. Broome. 1980. Propagation of thornless blackberries by one-node cuttings. *J. Amer. Soc. Hort. Sci.* 105:405–7.

20

Propagation of Ornamental Trees, Shrubs, and Woody Vines

INTRODUCTION

Ornamental trees, shrubs, and woody vines are perennial plants. A few ornamental shrubs and vines are used as annuals and planted in landscapes for one season in hardiness zones where they will not survive winter temperatures. This chapter describes propagation systems that include seed, cuttings, grafting, and micropropagation where appropriate for the listed species. Extensive references are included for more in-depth details of propagation.

As a general rule, shrubs and vines are propagated by cuttings, whereas trees are produced by seed or selected cultivars grafted onto seedling rootstocks. There are exceptions, such as tree species that can be propagated by cuttings or micropropagated.

In any commercial propagation system it is ***important to conduct small trials*** before propagating on a large scale. The propagation techniques and references listed are to serve as a guide. Propagators must develop their own procedures and chemical treatments that work best for their particular propagation systems.

Abelia xgrandiflora. Abelia, Glossy Abelia. Commercially propagated with semi-hardwood cuttings. Can be rooted easily under mist in spring, summer, or fall. Rooting is enhanced by applying talc or quick-dips of 1,000 to 2,000 ppm IBA, or IBA-NAA combinations totaling 1,000 to 2,000 ppm have produced superior results (131). Hardwood cuttings also may be rooted in fall or late winter but less successfully than with semi-hardwood cuttings. Abelia is commercially micropropagated.

***Abies* spp.** Fir.

Seed. Seed propagation is not difficult, but fresh seed should be used, since most species lose their viability after 1 year in ordinary storage. Embryo dormancy is generally present; fall planting or stratification at about 4°C (40°F) for 1 to 3 months is required for good germination. Bulk presoaking *A. procera* (noble fir) seed in water should be avoided because of imbibition damage. It is best to allow seeds to slowly uptake water on moist filter paper and then stratify at 4°C (40°F) for a minimum of 3 weeks (237). Alternatively, seeds can be placed for 5 to 10 days in moist perlite for imbibition, and then cold stratified. *Abies* seedlings are very susceptible to damping-off. They should be given partial shade during the first season, since they are injured by excessive heat and sunlight.

Cuttings. Fir cuttings are considered difficult-to-root, but *A. fraseri* hardwood cuttings—selected from young trees, and basally wounded and treated with IBA—can be rooted in high percentages (210). First order laterals root in higher percentages than primary axes (49). Cuttings taken from lateral branches root readily, but tend to become plagiotropic. A quick-dip of 5,000 ppm IBA and maintaining a bottom heat of 18 to 24°C (65 to 75°F) is best for rooting hardwood cuttings (49). This species, along with white fir (*A. concolor*), red fir (*A. magnifica*), and the California 'Silver Tip,' are important Christmas tree species.

Grafting. The side-veneer graft is commonly used. Japanese Momi fir (*A. firma*) is one of the few firs that will tolerate the heavy clay, wet soil conditions (low soil oxygen), and heat of the southeastern United States. Consequently, researchers at North Carolina State University recommend grafting desirable fir cultivars on *A. firma* rootstock, rather than the less tolerant *A. fraseri* or *A. balsamea* (114). Fraser fir (*A. fraseri*) is cleft grafted onto rootstocks of Turkish fir (*A. bornmuelleriana*) to reduce losses by phytophthora root rot. Cleft grafting is successful in North Carolina during early sprint (April) when scions are dormant and rootstocks are becoming active (206).

Micropropagation. *Abies* spp. have been regenerated via somatic embryogenesis (228).

***Abutilon* spp.** Abutilon, Flowering Maple, Chinese Bell Flower. Seeds germinate without difficulty. Can also be rooted by leafy cuttings under mist from summer through fall.

***Acacia* spp.** Acacia.

Seed. Generally propagated by seed. The impervious seed coats must be softened before planting by soaking in concentrated sulfuric acid for 20 minutes to 2 hours or by pouring boiling water over the seeds and allowing them to soak for 12 hours in the gradually cooling water. *A. cyanophylla* (beach acacia, golden willow), *A. farnesiana* (sweet acacia, West Indian blackthorn), and *A. koa* (koa acacia) seeds are scarified by soaking in warm water overnight. All but the youngest plants are difficult to transplant because of a pronounced taproot.

Cuttings. Generally considered difficult to root. Leafy cuttings of partially matured wood can be rooted under mist if treated with 8,000 ppm IBA talc. *A. redolens* and *A. subprosa* root during the fall with a 3,000 ppm IBA quick-dip. Cuttings with heels form vigorous new wood (99). *A. baileyana* can be rooted from semi-hardwood cuttings 10 to 13 cm (4 to 5 in) in length, quick-dipped in 5,000 ppm IBA (372). *A. koa* is difficult to root from stem cuttings, and better results have occurred with micropropagation.

Micropropagation. Many acacia species have been micropropagated, including *A. melanoxylon* has been reported (235, 295).

***Acer* spp.** Maple. Various methods of propagation are used—seeds, grafting, budding, cutting, layering, and micropropagation (62, 245). More propagation is being done via own-rooted cuttings and micropropagation.

Seeds. Most maples produce ripen seeds in the fall, while a few ripen seeds in the spring. Red maple (*A. rubrum*) and silver maple (*A. saccharinum*) produce mature seeds in the spring. Such spring-ripening seeds should be gathered promptly when mature and sown immediately without drying, since the seed does not store well. Typically, seed will germinate within a 7- to 10-day period. For selected species, excising embryos from fresh seed and incubating in cytokinin helps circumvent testa-imposed dormancy and the need to stratify—and results in germination within 1 week (443).

For species that produce mature seeds in the fall, stratification at 4°C (40°F) for 60 to 120 days, followed by spring planting, gives good germination. Fall planting out-of-doors may be done if the seeds are first soaked for a week, changing the water daily; seeds can also be soaked under running water. Dried seeds of the Japanese maple, *A. palmatum*, germinate satisfactorily if they are placed in warm water (about 43°C, 110°F) and allowed to soak for two days, followed by stratification for 60 to 120 days. Soaking dry seeds of *A. rubrum* (leach inhibitors) and *A. negundo* in cold running water for several days before planting may increase germination. Seeds of *Acer saccharum* var. *floridum* (formerly *A. barbatum*) and *A. negundo* are stratified. To avoid a hard seed coat, *Acer* seeds should not be allowed to dry out. Poor germination in trifoliate maples such as paperbark maple (*A. griseum*) is due to poor seed fill.

Cuttings (131, 436). Leafy Japanese maple (*A. palmatum*) cuttings, as well as those of other Asiatic maples, will root if they are made from tips of vigorous pencil-sized shoots in late spring and placed under mist. IBA at 8,000 to 20,000 ppm in talc or quick-dips has given good results in rooting leafy cuttings of various *Acer* species under mist and over bottom heat in the greenhouse. In general, it is best to wait until the terminal bud has formed, but before the last set of leaves has fully expanded. Semi-hardwood cuttings of *A. saccharum* var. *floridum* are propagated successfully with 8,000 ppm IBA talc. Softwood cuttings of *A. griseum* from stock blocks are rooted successfully with 8,000 ppm IBA talc or 5,000 ppm IBA quick-dip (212). Rooting is greatly enhanced by heavy pruning and later shading stock plants with 60 to 80 percent light-excluding saran (280). Etiolation of stock plants can improve rooting of softwood basal stem cuttings from 5-year-old plants of paper bark maple (*A. griseum*) (282). *A. macrophyllum* roots well from softwood cuttings (June and July in Washington) treated with 8,000 ppm IBA talc (206).

A. negundo semi-hardwood cuttings are rooted with 8,000 ppm IBA talc. Cuttings of selected *A. rubrum* and *A. xfreemanii* cultivars can be rooted from single-node cuttings treated with 3,000 to 8,000 ppm IBA talc (445). Single-node cuttings of *A. glabrum* subspecies *Douglasii* (Douglas maple) root with 8,000 ppm IBA talc; rooted cuttings should be overwintered before transplanting (410), but it is often difficult to overwinter rooted maple cuttings. To overcome this problem, new growth should be induced on the cuttings, shortly after rooting, by using supplementary lighting and supplying fertilizers. Rooting earlier in the season also allows more time to encourage new growth.

A. rubrum are primarily rooted by cuttings to avoid problems of graft incompatibility. Semi-hardwood cuttings are treated with 1,000 ppm IBA + 500 ppm NAA and rooted in small liner posts in 3 weeks (131); rooted liners need to be hardened-off out of doors, stored in a cooler during winter and spring planted. In Oregon, *A. rubrum* are rooted from cuttings taken from first generation tissue culture-produced plants.

In Alabama, cuttings of dwarf Japanese maple (*A. palmatum*) are taken in early spring, as soon as the new growth has hardened, and quick-dipped in 6,250 ppm IBA/2,500 ppm NAA. Liners are overwintered in a cool house, where temperatures do not drop below −3°C (25°F), and planted in the following spring. Hardwood cuttings of *A. palmatum* taken in midwinter have been rooted successfully in the greenhouse after wounding and treating with IBA (83). *A. truncatum* softwood cuttings taken from trunk sprouts of a 10-year-old tree had 79 percent rooting with a 5,000-ppm IBA ten-second quick-dip (312). Softwood cuttings taken in August (Kansas) from 3-year-old seedling stock root well with a 5,000 ppm IBA quick-dip. Growth after rooting via long-day manipulation is important for winter survival of transplanted rooted cuttings and cuttings left in the propagation bed before lifting in spring.

Sugar maple (*A. saccharum*) cuttings are best taken in late spring after cessation of shoot elongation, then rooted under mist. They are more difficult to root—IBA treatments give variable results (133).

Long, semi-hardwood cuttings of *Acer platanoides*, *A. pseudoplatanus*, and *A. campestre* ranging from 70 to 90 cm (28 to 35 in), rooted well when treated with 5,000 ppm IBA and propagated under high pressure fog systems. Rooting was enhanced and production time reduced with the larger propagules (389).

Softwood cuttings (June to August in Kansas) of *A. truncatum* root best with an IBA quick-dip of 1,000 ppm. Cuttings taken in August root readily, but are difficult to overwinter unless kept under long days to encourage some growth before winter (315).

Forcing epicormic shoots from woody stem segments [0.5 m (1.5 ft) long with a 2.5 to 25 cm (1 to 10 in) caliper] under mist and rooting the softwood cuttings can be used to propagate various *Acer* species: *A. ginnala, A. palmatum, A. rubrum, A. saccharum,* and *A. saccharinum* (332). Softwood cuttings are treated the same as if collected from a field-grown stock plant. Drench weekly with Banrot 40% WP fungicide.

Grafting. *A. palmatum* seedlings are used as the rootstock for Japanese maple cultivars. In Oklahoma, seedling rootstock of *A. palmatum* is selected from the more vigorous green-leaved cultivars, rather from

---

BOX 20.1 GETTING MORE IN DEPTH ON THE SUBJECT
USING ONLY END-USE AUXIN FORMULATIONS

In the United States, it is not legal for producers/growers to use technical-grade auxins. Only end-use formulations are permitted, such as Dip'N Grow, Woods Rooting Compound, and other commercial liquid and talc auxin formulations. Information is presented in Chapters 19, 20, and 21 for rooting with technical-grade auxins, since one can legally make recommended auxin concentrations by diluting end-use formulations. See Box 10.10, page 377 for examples of making recommended auxin concentrations with end-use formulation products for quick-dip applications.

red-leaved cultivars (167). Seedling rootstocks of *A. saccharum* are used for sugar maples, *A. rubrum* for red maple cultivars, *A. pseudoplatanus* for striped maple (*A. pennsylvanicum*), and *A. platanoides* for such Norway maple clones as 'Crimson King,' 'Schwedler,' and the pyramidal forms. There is some evidence of delayed incompatibility in using *A. saccharum* as a rootstock for red maple cultivars.

In Oregon, Japanese maples are direct seeded in liner pots in February, and by August the seedling rootstock is of graftable size. In other systems, seedling rootstock plants are grown for 1 year in a seed bed. In the fall or early spring they are dug and transplanted into small pots and grown in propagating frames through the second summer. In late winter, the rootstock plants are brought into the greenhouse preparatory to grafting. As soon as roots show signs of growth, the rootstock is ready for grafting. Dormant scions are taken from outdoor plants. The side-veneer graft is ordinarily used with *A. palmatum* 'Dissectum,' *A. pennsylvanicum*, *A. pseudoplatanus*, and *A. saccharinum* (181).

In Southern California, summer-grafted *A. palmatum* cultivars are side-grafted and wrapped with budding rubbers or poly tape for 3 weeks with 90 percent grafting success. It is important to completely remove the leaves of the scion, prior to grafting in the summer. In Canada, an apical splice graft is preferred to the traditional side-veneer graft for bench grafting cultivars of *A. palmatum* with containerized seedling rootstock (225).

See the published list of *Acer* rootstocks that can be used for interspecific grafting of rare and common *Acer* species (417). In general, scions of "milky" sap groups will not graft on "non-milky" rootstock, and conversely, "non-milky" scions will not take on "milky" rootstock (i.e., use *A. platanoides* rootstock for scions of *A. cappadocicum, A. catalpifolium, A. lobelii*, and other cultivars). These scions will not take on *A. pseudoplatanus* rootstock, which is "non-milky."

Maples are also T-budded and chip budded on one-year seedlings in the nursery row from mid- to late summer. In T-budding on active rootstock, the wood (xylem) is removed from the bud shield, which then consists only of the actual bud and attached bark. The seedling is cut back to the bud the following spring just as growth is starting. *A. platanoides* 'Crimson King' is more successfully chip budded than T-budded.

Micropropagation. Maple species are commercially micropropagated. The combination of cytokinins TDZ and BA enhanced micropropagation of sycamore maple (*A. pseudoplantanus*) (424).

***Actinidia* spp.** Gooseberry. See *Actinidia deliciosa* (kiwifruit). Fast-growing vigorous vines with fragrant flower and edible fruit. *A. hemsleyana* is a vigorous, narrow-leaved climbing vine that grows to 10 m (35 ft) and has attractive red soft bristles on the new growth. Gooseberry (*A. arguta*) can be propagated by seed, which is cold stratified for 3 months. It is best to use cuttings selected from female plants for flower and fruit characteristics. Can be propagated from single-node cuttings with talc applications of 8,000 ppm IBA (275). *Actinidia* is commercially micropropagated (448).

***Aesculus* spp.** Buckeye, Horse chestnut (131).

Seed. This tree may be propagated by seeds, but prompt sowing or stratification after gathering in the fall is necessary. If the seeds lose their waxy appearance and become wrinkled, their viability will be reduced. For best germination, seeds of *A. arguta* (Texas buckeye), *A. xcarnea* (red horse chestnut), *A. hippocastanum* (common horse chestnut), *A. indica, A. flava, A. sylvatica,* and *A. turbinata* should be stratified for 4 months at about 4°C (40°F) immediately after collecting. *A. californica* and *A. parvifolia* seed need no pretreatment. Though no dormancy has been reported for *A. pavia*, a one-month cold stratification is recommended (131). In southern California, seeds of *A. carnea* and *A. hippocastanum* are leached in a bag with running water for 10 days; seeds germinate within this leaching period. Seeds of buckeye are poisonous if eaten.

Cuttings. Cuttings of *A. parviflora* are taken from new growth on stock plants from June to July (Chicago). Leafy single node cuttings with two leaves and 2.5+cm (1+in) of stem tissue are treated with 1,250

BOX 20.2 GETTING MORE IN DEPTH ON THE SUBJECT
CAUTION IN HANDLING CONCENTRATED ACIDS FOR SCARIFICATION

Extreme care should be used when handling concentrated acids for scarification. Protective clothing, including safety glasses or face shields, should be worn. Use of the acid ideally should be under a laboratory hood, or at least in a well-ventilated area. The acid should be properly disposed of after use. It is important that seeds be *thoroughly washed* after acid treatment before storage or handling.

to 2,000 ppm IBA + 625 to 1,000 ppm NAA—the higher IBA/NAA level is used if the cutting wood is more lignified. The cuttings are inserted with buds set below the top surface of the propagation medium. Cuttings will not root under very moist medium conditions (247). Terminal stem cuttings of 15 cm (6 in) taken in May rooted better than those taken in July (North Carolina); a 2,500 ppm IBA quick-dip in ethanol gave the best results (45). *A. parviflora* can also be propagated by root cuttings.

Grafting. T-budding, cleft grafting, or bench grafting, using the whip graft, can be used for selected cultivars grafted on *A. hippocastanum* as the rootstock. Although normally propagated by seed, grafting is done to: (a) perpetuate superior sterile forms (*A. hippocastanum* 'Baumannii'), (b) more quickly produce a large-caliper tree, (c) avoid suckering problems, and (d) produce a plant with better flowers and foliage.

Agarita. See *Berberis trifoliata (Mahonia trifoliata)*.

Ailanthus altissima. Ailanthus, Tree of Heaven. Easily propagated by seed. Embryo dormancy is present in freshly harvested seed. Stratification at about 4°C (40°F) for 2 months aids germination. Seed propagation produces both male and female trees, but planting male trees should be avoided, since the staminate flowers produce an obnoxious odor. The more desirable female trees can be propagated by root cuttings planted in the spring. *A. malabarica* (110) and *A. altissima* (452) have been micropropagated.

Alberta magna. An attractive, flowering evergreen species, native to South Africa. It is used as a small landscape tree and has potential as a flowering pot plant. Seed propagation is poor and the seedlings have a long juvenile period. Cuttings taken from mature trees root poorly. Mature-phase scionwood has been successfully wedge-grafted onto micropropagated juvenile plants, which also speeds up flowering (34).

Albizia julibrissin. Albizzia, Silk tree, Mimosa. This species is propagated by seed. The impermeable seed coat is scarified by a 30-minute soak in sulfuric acid. Seed selection is important for obtaining plants with good flower color. Stem cuttings do not root, but root cuttings about 7.5 cm (3 in) long and 12 mm (0.5 in) or more in diameter taken and planted in early spring are successful (163). Juvenile shoots arising from root pieces can be removed and rooted with 1,000 to 3,000 IBA quick-dip or talc (123). Seedling material has been successfully micropropagated (361).

Alder. See *Alnus*.

Allamanda cathartica. Yellow Allamanda, Golden Trumpet. Easily propagated using semi-hardwood cuttings in summer and fall (Texas) with 1,000 to 2,000 ppm IBA quick-dip.

Almond (Dwarf, Flowering). See *Prunus*.

***Alnus* spp.** Alder.

Seed. *A. incana* (gray alder), *A. serrulata* (tag alder), *A. rhombifolia* (sierra alder), *A. rugosa* (speckled alder), *A. cordata* (Italian alder), and *A. glutinosa* (European alder). Seeds of this small deciduous tree should be thoroughly cleaned and can be planted immediately after fall harvest, but 3 to 5 months of stratification improves germination. Seed will also germinate if given a 24-hour soak in 0.2 percent potassium nitrate. Six weeks of cold stratification are optimal for seaside alder (*A. maritima*) (371). Light appears to be beneficial in germinating *Alnus* seed.

Cuttings. Softwood cuttings of *A. glutinosa* root with 3,000 ppm IBA talc. *A. incana* stem cuttings are rooted with 8,000 ppm IBA talc, and can also be grafted on potted seedling understock. Cultivars of *A. glutinosa* can also be side-veneer grafted (222). *A. maritima* can be rooted from softwood cuttings treated with 8,000 ppm IBA (370).

Micropropagation. Alder can be micropropagated.

***Amelanchier* spp.** Serviceberry (124, 131). These trees and shrubs are generally seed-propagated, however cutting propagation and micropropagation are also used.

Seed. Seeds show embryo dormancy, which can be overcome by stratification at 2°C (36°F) for 3 to 6 months. Seeds should not be allowed to dry out. Seeds of *A. alnifolia* (Saskatoon) and *A. laevis* (Allegheny serviceberry) require 15 to 30 minutes of sulfuric acid scarification prior to 3 months of cold stratification.

Cuttings. Many species are readily propagated by leafy softwood cuttings taken before the terminal bud has formed and when the new growth is several inches long, then rooted under mist. Talc and quick-dips of IBA from 1,000 to 10,000 ppm enhance rooting. At 40°N latitude, softwood to semi-hardwood cuttings of *A. laevis* taken in mid-May to mid-June root best with a K-IBA of 2,500 ppm or greater.

Micropropagation. The cultivars 'Prince Charles' and 'Princess Diana,' which are selections of *A. laevis* and *A. xgrandiflora*, respectively, are commercially micropropagated.

***Amorpha* spp.** Indigo bush (123). A large deciduous shrub in the legume family. They have an impermeable seed coat which requires acid scarification. *A. canescens* (leadplant amphora) seed is acid-scarified for 15 minutes followed by 2 to 8 weeks of stratification. Seed of *A. fruticosa* (indigo bush amorpha) can be fall planted or given 10 minutes of acid scarification; easily roots from untreated softwood cuttings. *A. canescens* is rooted with cuttings treated with 8,000 ppm IBA talc.

Anacho Orchid Tree. *See Bauhinia.*

Angelica Tree. *See Aralia.*

Anise. *See Illicium.*

Apricot (Japanese flowering). *See Prunus.*

***Aralia* spp.** Walking stick, Angelica tree. Suckering shrubs and trees. Seeds should be sown in early fall. Scarify with sulfuric acid for 30 to 40 minutes, and cold stratify for 3 months to overcome the double dormancy requirement. *A. spinosa* has a 3-month cold stratification requirement. Asexual propagation is with 10 to 13 cm (4 to 5 in) root cuttings, which are taken in fall and stored in cool storage until spring planting (123).

***Araucaria* spp.** These large evergreen trees are seed propagated.

Seed. Seeds of *A. bidwillii* (bunya-bunya), *A. columnaris* (Cook or New Caledonia pine), *A. araucana* (monkey puzzle tree), and *A. cunninghamii* (colonial pine) must be cleaned and then sown within a month after collection. There are no dormancy problems and germination time is 2 to 4 weeks. Plant seed of *A. heterophylla* (Norfolk Island pine) directly in pots, placing the seed vertical and burying to half the seed height. Keeping seeds moist is critical for the first 10 to 14 days after planting and can be accomplished by misting once each hour (Texas).

Cuttings. Can also be propagated by semi-hardwood cuttings. Cuttings of side branches will root but produce horizontally growing plants. Plants produced from terminal cuttings grow upright.

Micropropagation. *A. cunninghamii* and other species can be micropropagated (70).

Arborvitae. *See Thuja.*

***Arbutus* spp.** Pacific Madrone (*A. menziesii*), Texas madrone (*A. xalapensis*), Eastern strawberry tree (*A. andrachne*). *A. menziesii* is a Pacific coast evergreen tree that is usually propagated by seeds, which are stratified for 2 to 3 months at 2 to 4°C (35 to 40°F) (277). Seedlings are started in flats and then transferred to pots. They are difficult to transplant and should be set in their permanent location when not over 18 in tall. Propagation can be done by cuttings, layering, and grafting. *A. andrachne*, which is native to the East Mediterranean, can be seed propagated. While there are differences in seed provenance, a cold-moist stratification of 10 weeks or 24-hour soak with gibberellic acid enhances seed germination (404). Arbutus 'Marina' is a California hybrid that is more adaptable to varied landscapes than *A. menziesii*. However, it is difficult to produce by seed or cuttings. Air-layering is successful using stem diameters of 7 to 12 mm. A 1 to 2 cm (0.5 to 1 in) complete girdle was made, then painted with DipGel (gel thickener of Dip'N Grow rooting hormone) from 5,000 to 7,500 ppm, covered with moss and wrapped with a poly bag and aluminum foil (441).

A. xalapensis is native to Texas and southern New Mexico and propagated from fresh seed collected in the fall. The seed must be harvested at optimum ripeness and sown immediately to ensure high germination percentages. Rooting can be done with nonflowering stock material (405).

A. menziesii is commercially micropropagated, and *A. xalapensis* can be micropropagated on WPM (276).

***Arctostaphylos* spp.** Manzanita (*A.* spp.), bearberry (*A. uva-ursi*).

Seed. Seeds are water permeable, but benefit from sulfuric acid treatment for 3 to 6 hours (151). Seed of both species can also be scarified by inserting in boiling water, then immediately turning off the heat and soaking the seed in the gradually cooling water for 24 hours. This is followed by chilling stratification between 60 to 90 days for Manzanita or a combination of warm [60 days at 25°C (77°F)] then chilling [60 to 90 days at 5°C (41°F)] stratification for bearberry.

Cuttings. Cutting propagation is much more practical. Terminal cuttings are taken from November to February (California), submerged in 5 to 10 percent Clorox, given a 1,000 ppm IBA plus 500 ppm NAA quick-dip or 8,000 ppm IBA talc, and rooted under intermittent mist with bottom heat [21°C (70°F)]. The U.S. eastern form of this species is more difficult to root; cuttings taken from very short side branches from the middle of the plant root better than cuttings from long, end runners. Cuttings root best when taken in early March and treated with 8,000 ppm IBA (107). Colorado Manzanita (*Arctostaphylos xcoloradensis*) is considered difficult to propagate, but can be rooted with dormant cuttings. Cuttings are taken from previous year's growth from February to early April (Colorado). Cuttings are trimmed to 8 cm (3 in), quick-dipped in 1:10 Woods Rooting Hormone, lightly sprayed with water, paced in a Ziploc bag overnight in refrigeration to allow the cutting end to slightly dry down and not rot. The next morning cuttings are immersed in a 1:50 Zerotol (hydrogen peroxide) for disease control, removed, and excess solution drained off. Cuttings are stuck in small bottomless containers (bands) with a 3 perlite: 1 pear (v/v), bottom heated to 21°C (70°F) and rooted under mist (383). During propagation, a diluted concentrate is watered in containing the beneficial fungus, *Trichoderma harzianum,* for damping-off control, and a cocktail of ectomycorrhiza for enhanced stress resistance of the rooted cutting (383). Cuttings treated with 3,000 ppm Woods Hormone solution (1.03% IBA + 0/66% NAA) plus inoculum of arbuscular and arbutoid mycorrhizal fungi had enhanced rooting.

Micropropagation. *Arctostaphylos* is micropropagated.

***Ardisia* spp.** Coralberry, spiceberry (*A. crenata*), and Japanese (*A. japonica*). These medium-sized evergreen shrubs and ground covers (*A. japonica*) can be seed-propagated without stratification. *A. crenata* can be propagated by softwood cuttings. *A. japonica* is easily rooted with a 1,000 ppm IBA quick-dip; this stoloniferous plant can also be divided (123).

Aronia arbutifolia. Red Chokeberry (*Aronia arbutifolia*) and black (*A. melanocarpa*). Propagated by seed, which should be collected in autumn and then stratified for about 3 months at 5°C (41°F) before planting. Can also be started by leafy cuttings treated with a 4,000 ppm IBA quick-dip under mist or by suckers.

Ash. *See Fraxinus.*

Asimina triloba. Common pawpaw and dwarf pawpaw (*A. parviflora*) (177, 264). This fruit crop, small tree, and deciduous shrub are propagated primarily by seed. Seeds are recalcitrant and germination is reduced when seed moisture is less than 25 percent (157). Seed must be stored moist at chilling temperatures, but remains viable for only 2 years. Some seed lots require cold stratification (5°C, 41°F) for about 3 months to overcome dormancy. Seedling emergence occurs between 30 and 45 days after sowing (177). A southern ecotype of *A. triloba* easily germinates without stratification (131). Successful cutting propagation has not been reported. Chip budding is the most common method for propagating pawpaw cultivars with superior fruit characteristics onto seedling rootstock. Other clonal propagation techniques include grafting with the whip-and-tongue, cleft, or inlay bark graft (177, 264). Seedlings of *A. triloba* can be micropropagated (156), but microcuttings from mature sources have not been successfully rooted, nor acclimatized (176, 177).

Aspen. *See Populus.*

Aspen (Quaking). *See Populus.*

Australian Pine. *See Casuarina.*

Avocado. *See* Chapter 19.

Azalea. *See* Rhododendron.

***Baccharis* spp.** False willow, narrowleaf baccharis (*B. angustifolia*), and eastern baccharis (*B. halimifolia*). These small evergreen shrubs have excellent salt tolerance. Baccharis is seed-propagated with no seed pretreatment required. Germination occurs within 1 to 2 weeks. In Arizona, cuttings of male plants of *B. sarothodes* (Desert broom) are trimmed to 18 cm (7 in), dipped in 8,000 ppm IBA talc, stuck in a peat:perlite medium, and rooted under mist (357). *B. halimifolia* roots easily from softwood cuttings treated with 2,500 ppm K-IBA.

Bald Cypress. *See Taxodium.*

Bamboo. *See Bambusa.*

***Bambusa* spp.** Bamboo (*Arundinaria, Dendrocalamus, Gigantochloa, Phyllostachys, Pseudoasa, Sasa, Schizostachym, Sinarundinaria, Thamnocalamus, Thyrsostachys,* spp.) (344, 380). Bamboo has about 1,000 species in some 50 genera. Two classes of bamboo are commercially propagated in containers (Louisiana) to eliminate much of the hand labor and extensive field stock space needed in the old traditional method of hand digging and separating established clumps.

Division. *Clump-forming bamboo* (pachymorphs—with constricted rhizomes) are propagated by dividing the young, peripheral rhizomes in late spring or early summer; the culms (above-ground stem with branches and leaves) are cut back and the rhizome is containerized in a 5 bark to 1 sand medium with slow-release fertilizer and watered as needed. The following spring, rhizomes from the containerized stock plant are further divided and transplanted with attached culms and roots to new containers for further propagation or finishing-off as a marketable plant.

In Oregon, bamboo from 5-gallon containers are sawed vertically into quarters and then replanted into

5-gallon containers with 100 percent survival and faster reestablishment.

The *running-type bamboos* (leptomorphs—with vigorous rhizomes extending beyond parent plants) are more cold-tolerant but less desirable in the garden. They are propagated by the division technique described above for rhizomes less than 1 cm (1/2 in) thick, or by using 30-cm (1-ft) rhizomes 2 cm (3/4 in) thick, which are containerized and grown for 2 years as stock plants. Rhizomes 15 cm (6 in) long are then containerized and grown until they are a finished crop. It is important to prevent drying during transplanting. Best results are obtained from rhizomes taken and planted in late winter or early spring before the buds begin to elongate.

Micropropagation. Some 15 genera and 54 species of bamboo have been micropropagated, which offers potential for those bamboo species which are currently difficult to propagate or transplant (15, 334). Somatic embryogenesis and successful transplanting of plantlets has been reported (352).

Banksia **spp. (35).** Banksia are native Australian shrubs and small trees in the Proteaceae family. Two species with promise as a plantation-grown flower crop with export potential are *B. coccinea* (scarlet banksia) and *B. menziesii* (raspberry frost banksia).

Seed. These species are propagated by seed, although they are difficult to remove from the seed-containing structures. One method is to soak seed several days in water and then dry quickly. There are no pregermination treatments; however, 15°C is the optimum germination (20 days) temperature for *B. coccinea*.

Cuttings. *B. coccinea* can also be propagated from semi-hardwood cuttings using a quick-dip of 8,000 to 12,000 ppm IBA in 50 percent ethanol (35). Cuttings should be wounded and inserted in a porous mix under intermittent mist with 25°C (77°F) bottom heat. It is best to root cuttings directly in liner pots and avoid root disturbance. Graft incompatibility limits successful grafting of *Banksia* spp. to rootstock that are resistant to *phytophthora* and tolerant of high soil phosphorus and poorly drained sites.

Barberry. *See Berberis.*

Basswood. *See Tilia.*

Bauhinia **spp.** Orchid Tree. *Bauhinia lunarioides* (formerly *B. congesta*) Anacho Orchid Tree (219). This multi-stemmed deciduous shrub has emerald leaves, which are cleft to the petiole, and produces abundant white flowers from March through summer (Texas). Seeds of this species germinate without pregermination treatment. However, other *Bauhinia* spp. seeds have physical dormancy and must often need scarification. *Bauhinia variegata* and *B. xblakeana* (evergreen tree with thick leaves and striking purplish red flowers) are also important Bauhinia plants. *B. xblakeana* is a sterile hybrid propagated by air layering. *B. purpurea* has been micropropagated (254).

Bayberry. *See Myrica.*

Bay Laurel. *See Laurus nobilis.*

Beaucarnea recurvata. Ponytail palm. *See palms.*

Beech. *See Fagus.*

Berberis **spp.** Barbarry.

Seed. Propagated without difficulty by fall-sowing or by spring-sowing seeds that have been stratified for 2 to 6 weeks at 4°C (40°F). Some species require up to 2 to 3 months of stratification, or seed can be sown in fall. It is important to remove all pulp from seeds. Seedlings are susceptible to damping-off and some to black stem rust (Oregon).

Cuttings. Berberis are generally propagated by cuttings. Semi-hardwood cuttings taken from spring to fall can be rooted under mist (431). Some propagators have success with hardwood cuttings (leaves still attached) taken from October to November. IBA from 2,000 to 8,000 ppm aids rooting in some species. *B. thunbergii* 'Atropurpurea nana' is propagated with 5 to 6 in (14 cm) semi-hardwood cuttings in May and June (Texas), which are collected when the new growth is firm at the cutting base (25). The stem of the cutting should have a greenish-yellow color; avoid using brown wood. Cuttings are quick-dipped into 1,870 ppm IBA with 50 percent alcohol and direct stuck into 5 cm (2 1/4 in) liner pots. Timing, application of mist, and the hardening-off process are critical. In Georgia, cuttings of *B. thunbergii* cultivars and other *Berberis* species are taken from May 15 through July; the 8 to 10 cm (3 to 4 in) cuttings are wounded 6 to 12 mm (1/4 to 1 1/2 in) from the base and quick-dipped with 1,500 to 3,500 ppm IBA. In Alabama, golden barberry (*B. koreana* × *B. thunbergii*) can be rooted from softwood medial stem cuttings 7 to 10 cm (3 to 4 in) long, treated with a 1,250 ppm IBA quick-dip and propagated under 70 to 80 percent shade (302).

Greenhouse grafting of some selected types is also practiced, and layering is done occasionally. Division of crowns is useful for small quantities of plants.

Micropropagation. *B. thunbergii* 'Atropurpurea' is micropropagated (411).

Berberis trifoliata. (*Mahonia trifoliata*). Agarita (219). A dense evergreen rounded shrub with holly-like leaves that is an early spring bloomer in Texas with yellow fragrant flowers in March. This drought-resistant plant also has edible red berries and grows best in full sun. Seeds require a 2- to 3-month cold stratification. *B. trifoliata* can be micropropagated (297).

***Betula* spp.** Birch.

Seed. Seeds germinate without pretreatment. Birch seed needs stratification to germinate in the dark, but will generally germinate without stratification if germinated in the light. Stratification and germination in the light accelerates germination rate compared to light alone. The key is to plant seeds shallow and not too deep (131).

Cuttings. Birch is considered difficult to propagate by cuttings, but leafy, semi-hardwood cuttings root under mist if taken in midsummer. Softwood cuttings of *B. alleghaniensis*, *B. lenta*, *B. nigra*, *B. papyrifera*, and *B. pendula* are rooted with 8,000 ppm IBA talc. Softwood cuttings (July in Ohio) of *B. nigra* root best with a 1,000 ppm IBA dip. Softwood one-node cuttings of 18-year-old *B. papyrifera* collected in mid-July (Vermont) treated with 4,000 to 6,000 ppm IBA in 50 percent ethanol had 40 to 60 percent rooting; problems occurred with survival of post-rooted cuttings (319).

Grafting. Some of the selected, weeping forms are grafted on *Betula pubescens* or *B. pendula* seedlings (117). T-budding and chip budding have been used in field-grown rootstocks of *B. pendula* (109). In the selection of birches that will tolerate the wet soils and high temperatures of the southeast United States, whitespire Japanese birch (*B. platyphylla*)—which has heat and drought resistance, but is intolerant of poorly drained soils—has proven ideal when grafted to rootstock of river birch (*B. nigra*) or European birch rootstock (*B. pendula*) (339, 340). Other white birches, such as *B. pendula* 'Youngii' have been successfully grafted to *B. nigra* rootstock, which is tolerant of poorly drained soils, high humidity, and high temperature (22).

Micropropagation. Birch is commercially micropropagated (62, 224, 284). *B. platyphylla* 'Fargo' (Asian white birch) can be micropropagated with WPM and BA (95).

Birch. *See Betula.*

Bittersweet. *See Celastrus.*

Black gum. *See Nyssa.*

Black Haw. *See Bumelia.*

Black Tupelo. *See Nyssa.*

Boston Ivy. *See Parthenocissus.*

Bottlebrush. *See Callistemon.*

***Bougainvillea* spp.** Bougainvillea. This showy, tropical, woody, evergreen vine grows outdoors only in mild climates and is propagated by leafy cuttings taken at any time of the year. Rooting is aided by supplemental bottom heat. Difficult-to-root cultivars should be treated with IBA. In Australia, semi-hardwood (wood changing from green to brown) 4- to 6-node cuttings are cut to 10 to 13 cm (4 to 5 in) length (207). IBA at 4,000 to 16,000 ppm is applied, depending on the cultivar and the time of year. Bottom heat is used to maintain a propagation bed surface temperature of 25 to 27°C (77 to 81°F). Intermittent mist is used, but bougainvilleas do not tolerate an overly wet environment. Can be micropropagated by shoot-tip culture (93).

Boxwood. *See Buxus.*

Boxwood (Oregon). *See Paxistima.*

Breath of Heaven. *See Diosma (Ericoides).*

Broom. *See Cytisus.*

Buckeye. *See Aesculus.*

Buckeye (Mexican). *See Ungnadia.*

Buckthorn. *See Rhamnus.*

***Buddleia* spp.** Butterfly Bush. Seeds require no pretreatment. Seeds started in the greenhouse in early spring will provide flowering plants by fall, although reproduction is not genetically true by seed. Generally propagated by softwood or semi-hardwood cuttings taken in the summer, treated with 8,000 ppm IBA, and rooted under mist. *Buddleia* is micropropagated (322).

Buffaloberry. *See Shepherdia.*

Bumelia lanuginosa. Bumelia, Black Haw, or False Buckthorn. This deciduous small tree is propagated by seed. Seed is fall-collected and stratified at 4°C (40°F) for 2 months. Seed is sown in the spring in outdoor beds.

Bunya-Bunya. *See Araucaria.*

Butia capitata. Pindo palm. *See* palms.

Butterfly Bush. *See Buddleia.*

***Buxus* spp.** Boxwood (*Buxus sempervirens* and *B. microphylla*). Seeds are rarely used because of the very slow growth of the seedlings. Cuttings are commonly used—either softwood taken in spring or summer, semi-hardwood taken in late summer and fall, or hardwood cuttings taken in the winter (Texas). Quick-dips of 2,500 to 5,000 ppm IBA enhance rooting (424). An Oregon nursery uses a light application of liquid fertilizer (about 50 ppm N) after callus formation to help speed root development. Young liner plants should always be grown in containers or transplanted with balls of soil around their roots.

Cactus (Prickly Pear). *See Opuntia* spp. in Chapter 19, page 748.

***Callistemon* spp.** Bottlebrush. Although seeds germinate without difficulty, seedlings should be avoided because many of them prove worthless as ornamentals. The preferred method of propagation is by semi-hardwood cuttings taken from selected cultivars, which root under mist quite easily.

Calluna vulgaris. Heather. *See Erica* (Heath).

Calocedrus decurrens. Incense Cedar. Propagated by seed. Germination is promoted by a stratification period of about 8 weeks at 0 to 4°C (32 to 40°F). Can be rooted from cuttings treated with quick-dips of 2,500 ppm NAA, or 2,500 ppm NAA plus 2,500 ppm IBA. *C. decurrens* can be grafted on *Thuja*.

***Calycanthus* spp. (123).** Sweetshrub, Allspice. Stoloniferous shrubs. *C. floridus* is common to the eastern and southern United States. *C. floridus* 'Athens' ('Katherine') is very fragrant with yellow flowers, while *C. occidentalis* has reddish flowers. Seed should be collected when receptacles change from green to brown, and immediately planted. Seed collected later should be cold stratified for 3 months. Good rooting success is reported with summer semi-hardwood cuttings using 2,000 to 3,000 ppm IBA, or a 10,000 ppm K-IBA quick-dip.

***Camellia* spp.** Camellias can be propagated by seed, cuttings, grafting, layering, and micropropagation (274, 359), but they do not come true from seed. Seedlings are used in breeding new cultivars, as rootstocks for grafting, or in growing hedges where foliage is the only consideration. To perpetuate cultivars, cuttings, grafts, and micropropagation are used.

Seed. In the fall when the capsules begin to turn reddish brown and split, seeds should be gathered before the seed coats harden and the seeds become scattered. The seeds should not be allowed to dry out and should be planted before the seed coats harden. If the seeds must be stored for long periods, they will keep satisfactorily mixed with ground charcoal and stored in an airtight container placed in a cool location. After the hard seed coats develop, scarification is accomplished by pouring boiling water over the seeds and allowing them to remain in the cooling water for 24 hours. It takes 4 to 7 years to bring camellias into flowering from seeds.

Cuttings. Most *C. japonica, C. oliefera, C. sasanqua, C. reticulata,* and *C. sinensis* cultivars and hybrids are produced commercially from cuttings. Cuttings are best taken from midsummer to mid-fall from the flush of growth after the wood has matured somewhat and changed from green to light brown in color. Tip cuttings are used, 8 to 15 cm (3 to 6 in) long, with two or three terminal leaves. Rooting is much improved if the cuttings are treated with IBA quick-dips of 3,000 to 5,000 ppm, 6,000 to 8,000 ppm IBA plus 2,500 ppm NAA, or 3,000 to 8,000 ppm IBA talc (27, 38, 131). *C. reticulata* hybrids root best with 8,000 to 20,000 ppm IBA talc (354). In Alabama, semi-hardwood cuttings 8 to 10 cm (3 to 4 in) in length are collected in September and October from stock blocks of *C. japonica, C. sasanqua, C. xhiemalis,* and quick-dipped in 10,000 ppm IBA (365). Wounding the base of the cuttings before they are treated is also likely to improve rooting. Cuttings root best either in a polyethylene closed frame or under mist.

Camellias also may be started as leaf-bud cuttings which are handled as stem cuttings. In this case, high auxin concentrations should be avoided because this may inhibit development of the single bud.

Grafting. Camellias are frequently grafted since some cultivars (e.g., 'Pink Pagoda') are poor rooters and grow poorly on their own roots. Grafting is also used for multiplying new cultivars faster and changing out cultivars of older established plants. Vigorous seedlings or rooted cuttings of either *C. japonica* or *C. sasanqua* can be used as rootstocks for grafting. Any of the side-graft methods are suitable. *C. japonica, C. sinensis,* and *C. reticulata* can be container-grafted by using a whip graft.

Micropropagation. Camellias are commercially micropropagated.

Camphor Tree. *See Cinnamomum.*

***Campsis* spp.** Trumpet Creeper. This vine is usually propagated by cuttings, but seeds can also be used. With the latter method, stratification for 2 months at 4 to 10°C (40 to 50°F) hastens but does not increase germination. Both softwood and hardwood cuttings root readily and can be quick-dipped with 1,000 ppm IBA. *C. radicans* can be started by root cuttings. One

Oregon nursery takes root pieces [5 cm length × 0.6 cm width (2 × 1/4 in)], which are direct-stuck vertically with 0.3 cm (1/8 in) of the root showing. Maintain the correct polarity of the root cutting.

Cape Jasmine. See Gardenia.

Caragana **spp.** Peashrub.

C. arborescens (Siberian Peashrub). The most common peashrub for northern gardens. Seed germination is erratic, so best to soak in water for 24-h; 4-weeks stratification or fall planting results in good germination (131). *C. arborescens* is propagated from softwood cuttings, similar to *C. pygmaea* (pygmy peashrub). Untreated cuttings taken in May–July will root. A 3,000 ppm IBA-talc can slightly enhance rooting (131). Weekping and other specialty cultivars are top grafted on seedling rootstock.

C. pygmaea [Pygmy (Dwarf) peashrub]. Shrub with bright yellow, pealike pendulous flowers. Propagated by softwood tip cuttings, which are treated with 1,000 ppm IBA + 500 ppm NAA quick-dip. Rooted cuttings are transplanted the following spring.

Carob. See *Ceratonia* in Chapter 19.

Carpinus **spp.** Hornbeam.

Seed. For seed propagation, collect seeds while the wings are still soft and pliable. Check to make sure that seeds have embryos, which may be absent after stressful growing conditions. Do not allow seeds to dry out. Sow outdoors in autumn or stratify (3 to 4 months), overwinter, and sow in spring. If seeds of *C. caroliniana* are allowed to dry, the seed coat hardens and inhibits germination. Consequently, double dormancy occurs and seeds must be scarified and then warm stratified for 2 months, followed by 2 months of cold stratification (61).

Cuttings. Summer stem cuttings of *C. betulus* 'Fastiagata' root with a quick-dip of 20,000 ppm IBA, *C. caroliniana* 'Pyramidalis' with 16,000 ppm IBA, and *C. japonica* with 3,000 ppm IBA talc (131). Wounding cuttings has been recommended. Softwood cuttings may root better than semi-hardwood cuttings. Stock plant etiolation, shading, and stem banding improved rooting of *C. betulus* softwood cuttings (281).

Grafting. Cultivars may be side-veneer grafted or budded on seedlings of the same species. *C. betulus* is used as the principle understock, since the majority of hornbeam cultivars are from European hornbeam (*C. betulus*), while grafting onto American hornbeam (*C. caroliniana*) results in significant overgrowth of the understock by the scion. However, *C. japonica* or *C. cordata* should not be grafted onto (*C. betulus*) (444).

Carya **spp.** Hickory, Water Hickory (*Carya aquatica*), bitternut hickory (*C. cordiformis*), pignut hickory (*C. glabra*), shellbark (*C. laciniosa*), nutmeg hickory (*C. myristicae-formis*), shagbark hickory (*C. ovata*), and mockernut hickory (*C. tomentosa*). Some of these tree species have edible fruits. Seeds (nuts) are fall-collected and will erratically germinate without pretreatment. Most species have a 3- to 4-month cold stratification requirement for uniform germination. Few reports exist on successful rooting or grafting of hickory, and cutting propagation is difficult. However, Medina (294) reported high rooting success with pecans using mounding and stooling techniques. Pecan (*C. illinoiensis*) is successfully grafted by patch budding, the inlay bark graft, and four-flap (banana graft). Some of these asexual propagation techniques could be applied to other *Carya* species (see Chapter 19).

Castanea **spp.** Chestnut. American chestnut (*Castanea dentata*), Chinese chestnut (*C. mollissima*), and sweet chestnut (*C. sativa*).

Seed. These deciduous medium-sized trees are predominantly seed propagated. Seeds are harvested in early fall and stratified for 2 to 3 months. It is important that seeds not dry out.

Cuttings. Mature cuttings are very difficult to root. Rooting success has occurred with juvenile cuttings of *C. mollissima* (8,000 ppm IBA talc or a 7,000 ppm IBA quick-dip in 95 percent alcohol) (131). Limited success has occurred with hardwood cuttings quick-dipped with up to 12,000 ppm IBA and with air layering. Leafy softwood summer cuttings have rooted with 1,250 ppm IBA five-second quick-dip in a high-humidity ventilated fog system; pretreatment with a blanching tape also improved rooting (327). Grafting success has occurred with cleft graftage, inlay bark graftage, and chip budding.

Micropropagation. Juvenile (349) and mature (418) explants of *Castanea* have been successfully micropropagated (375, 449, 450).

Casuarina **spp.** Ironwood (*C. cunninghamiana*), Australian pine (*C. equisetifolia*), and longleaf casuarina (*C. glauca*). These large evergreen trees from Australia are easily propagated from seeds. Seeds require no pretreatment, and after harvesting and cleaning, seeds are stored and spring-sown in outdoor beds. Seed germination of *C. cunninghamiana* varies from 30 to 70 percent. Poor germination results from shriveled, empty, and insect damaged seeds. There was a 91 percent germination when seed coats were removed (377). It is recommended to inoculate seedling roots with the nitrogen-fixing bacteria, *Frankia*. *C. cunninghamiana* can be micropropagated (377).

***Catalpa* spp.** Catalpa.

Seed. Catalpa seeds germinate readily without any pretreatment. They are stored dry, overwintered at room temperature, and planted in late spring. Catalpa species can also be propagated in summer by softwood cuttings rooted under mist. Hardwood cuttings of *C. bignonioides* and *C. speciosa* root with 8,000 ppm IBA talc (131). The terminal bud of the cuttings should be removed.

Grafting. *C. bignonioides* 'Nana' is often budded or grafted high on stems of *C. speciosa*, giving the "umbrella tree" effect. A strong shoot is forced from a 1-year-old seedling rootstock, which is then budded with several buds in the fall at a height of 1.8 m (6 ft).

***Ceanothus* spp.** Ceanothus can be propagated by seed, cuttings, layering, and sometimes grafting.

Seed. Seeds must be gathered shortly before the capsules open or they will be lost. Those of *C. arboreus*, *C. cuneatus*, *C. jepsoni*, *C. megacarpus*, *C. oliganthus*, *C. rigidis*, and *C. thyrsiflorus* have only seed coat dormancy. Germination is aided by placing seeds in hot water [82 to 87°C (180 to 190°F)] and allowing them to cool for 12 to 24 hours. To obtain germination in other *Ceanothus* species, which have both seed-coat and embryo dormancy, the seed should be immersed in hot water (as previously described) and then stratified at 2 to 4°C (35 to 40°F) for 2 to 3 months.

Cuttings. Ceanothus hybrid cultivars 'Concha,' 'Frosty Blue,' 'Joyce Coulter,' and 'Victoria' are all cutting-grown. Semi-hardwood cuttings can be rooted under mist at any time from spring to fall. Rooting is enhanced with 1,000 to 3,000 ppm (and up to 5,000 ppm) IBA/NAA quick-dips (387). Terminal softwood cuttings taken from vigorously growing plants in containers root when treated with 1,000 ppm IBA.

Cedar. See *Cedrus*.

Cedar (Incense). See *Calocedrus*.

***Cedrus* spp.** Cedar.

Seed. Seeds germinate if not permitted to dry out. No dormancy conditions occur, but soaking the seeds in water several hours before planting may be helpful. Also, a 1-month cold stratification period will improve the germination rate.

Cuttings. *C. ailantica* and *C. libani* are difficult to root. *C. deodara* can be rooted by wounding cuttings and using bottom heat; cuttings are collected in late fall to early winter and quick-dipped with 5,000 ppm IBA.

Grafting. Side-veneer grafting of selected forms on 1- or 2-year-old potted seedling stocks may be done in the spring (or winter in southern California). Scions should be taken from vigorous terminal growth of current season's wood rather than from lateral shoots. In winter grafting, both lateral and terminal scions from the previous season's growth can be used, producing identical results. *C. atlantica* selections and other species are grafted on *C. deodara* seedling rootstock. Rooted cuttings of *C. deodara*, which are a year or older, also make acceptable rootstock material.

***Celastrus* spp.** Bittersweet. These dioecious twining vines have male flowers on one plant and female on another, and the two types must be near each other to produce the plant's attractive berries.

Seed. To propagate, seeds must be removed from the berries and then fall-planted or stratified for about 3 months at 4°C (40°F) before planting.

Cuttings. Clones of known sex can be propagated by softwood cuttings taken in midsummer, or by hardwood cuttings taken in winter. Semi-hardwood summer cuttings root well when treated with 8,000 ppm IBA talc (131). IBA-treated softwood cuttings can also be used.

***Celtis* spp.** Hackberry. Seeds are ordinarily used, sown either in the fall or stratified for 2 or 3 months at about 4°C (40°F), and planted in the spring (288). Clones of two species, *C. occidentalis* and *C. laevigata* (sugarberry), can be started by cuttings, but the rooting percentage is low. Grafting and chip budding also have been used with *C. occidentalis* and *C. laevigata* as rootstock for other species.

Cephalotaxus harringtonia. Japanese plum yew. This ornamental species also has anticarcinogenic attributes. Can be propagated by seed, which requires a 3-month cold stratification period. Rooting by cuttings is difficult. An IBA talc or quick-dip of 5,000 to 10,000 ppm is reported to enhance rooting (131). Rooting of cuttings varies with cultivars. The species can be micropropagated. Micropropagated shoots can be rooted *ex vitro* under mist with or without auxin (229). See *Taxus* spp.

Ceratonia siliqua. Carob. See *Ceratonia* in Chapter 19, page 732.

***Cercis* spp.** Redbud.

Seed. Seed propagation is successful, but seed treatments are necessary because of dormancy resulting from an impervious seed coat plus a dormant embryo. Probably the most satisfactory treatment is a 30- to 60-minute soaking period in concentrated sulfuric acid, or an 82°C (180°F) hot-water soak; scarification treatments must be followed by stratification for 3 months at 2 to 4°C (35 to 40°F). A 1-hour sulfuric acid scarification followed by 35 days of cold stratification is optimum for Mexican redbud (*C. canadensis* var. *mexicana*) (407). Gibberellic acid and ethephon can overcome dormancy in nonchilled dormant seed (174). Outdoor fall-sowing of untreated seeds also may give good germination. Seed provenance is important, since ecotypic variation affects survival, dormancy, and plant growth potential (e.g., Florida versus Canadian seed sources).

Cuttings. Propagating *Cercis* by cuttings is difficult. Leafy softwood cuttings of some *Cercis* species root under mist if taken in spring or early summer. *C. canadensis* var. *mexicana* roots readily when 10 to 15 cm (4 to 6 in) long, leafy terminal cuttings are taken 4 weeks after bud-break and treated with a 10,000 ppm K-IBA 5-second dip (406). Softwood and semi-hardwood cuttings of *C. Canadensis* 'Flame' rooted best with, respectively, 5,000 and 10,000 ppm K-IBA (447).

Grafting. T-budding in midsummer on *C. canadensis* seedlings is used commercially for *Cercis* cultivars such as *C. canadensis* var. *texensis* 'Oklahoma' (429, 430). Chip budding redbuds is commercially done. Stick budding is used in California. Apical wedge grafts are used with *C. canadensis* 'Forest Pansy' using a hot-pipe callusing system (303).

Micropropagation. Cercis is commercially micropropagated. Both *C. canadensis* and *C. canadensis* var. *mexicana* have been micropropagated and somatic embryos have been produced from micropropagated zygotic embryo explants (172, 175, 451).

***Chaenomeles* spp.** Quince (Flowering). Flowering quince is easily started by seeds, which should be fall-planted or stratified for 2 or 3 months at 4°C (40°F) before sowing. Clean the seeds from the fruit. Softwood cuttings treated with 1,000 to 5,000 ppm IBA root readily. Quince is commercially micropropagated.

***Chamaecyparis* spp.** Chamaecyparis, False cypress. *C. thysoides* (Atlantic white cedar).

Seed. In the fall, cones are collected and dried, and the seeds knocked free. The seeds are stratified at about 4°C (40°F) for 2 to 3 months. Some species such as *C. nootkatensis, C. thyoides,* and *C. praecox* germinate better with warm followed by cold stratification. Optimal seedling growth of *C. thyoides* occurs under long-day with day/night temperatures of 30/22°C (86/72°F) (240).

Cuttings. Cuttings of most species are not difficult to root, particularly if juvenile forms are used. Cuttings may be taken in fall and rooted in a cold frame, or in winter and rooted under mist or a polytent using bottom heat. Quick-dips of 3,000 to 8,000 ppm IBA are used. *C. thyoides* roots well with hardwood or softwood cuttings 24 cm (9 in) long. At high auxin levels, more primary roots are produced, while at lower to moderate auxin levels, more secondary roots are produced, which is more desirable (211).

Micropropagation. *C. nootkatensis* (Alaska yellow cedar) is micropropagated (271).

***Chamelaucium* spp.** Waxflower. Small- to medium-sized shrubs with waxy flowers. In Australia, terminal stem cuttings 8 to 10 cm (3 to 4 in) long with growing points intact are dipped in a 2 percent sodium hypochlorite solution for disease prevention. Leaves are stripped from the basal 2 cm of the stem and cuttings quick-dipped in 2,000 ppm IBA in ethanol solution (183). Grafting is done on *Phytophthora*-resistant rootstock using a splice graft wrapped with Parafilm or Buddy tape (318).

Cherry (Flowering). *See Prunus.*

Chestnut. *See Castanea.*

Chilopsis linearis. Desert willow. This small tree is a member of the *Bignoniaceae* family, and has funnel-form flowers in open terminal panicles that range in color from white to lavender to burgundy. Can be propagated by seed or cuttings (408). Seeds will germinate immediately after harvest, or they can be stored under refrigeration. Germination takes 1 to 3 weeks. Hardwood cuttings root easily. Softwood and semi-hardwood cuttings treated with less than 5,000 ppm IBA can be rooted either under mist or in a high-humidity chamber. Air-layering is optimal during most active growth period (July and August in Arizona) (42).

***Chionanthus* spp.** Fringe Tree (*Chionanthus virginicus* and *C. retusus*).

Seed. Seed propagation can be used but is very slow. Manually nicking the seed coat helps. There is a double dormancy requirement, with a 3-month warm stratification period (watch for the radicle to emerge) followed by a 3-month cold stratification period, after which seeds can be sown (131).

Cuttings. Cutting propagation of the Chinese fringe tree (*C. retusus*) is very difficult, but by taking softwood cuttings in late spring, treating cuttings with 8,000 to 10,000 ppm IBA talc and propagating under mist, excellent rooting percentages can be obtained. Rooting improves when stock plants are kept juvenile or when serial propagation is used.

Cuttings of *C. virginicus* are very difficult to root, but some success has been reported in Alabama by taking cuttings in early spring from hardened-off new growth and quick-dipping with 1,250 ppm IBA. Wounding appears to be helpful, and soft-tip cuttings have been reported to root in limited percentages (20). Serially propagating those cuttings that root can increase rooting success.

Micropropagation. *C. virginicus* can be micropropagated via embryo culture (89).

Chokeberry (Red). *See Aronia.*

Cinquefoil. *See Potentilla.*

Cinnamomum camphora. Camphor Tree. Seed propagation is the most common propagation method, unless cultivars have been selected. Can also be propagated by semi-hardwood cuttings taken in spring and rooted under mist.

Citrus **spp.** See Chapter 19, page 732. There are a number of citrus cultivars for landscapes and home gardens, all of which are asexually grown by cuttings or grafted or budded onto seedling rootstock. *Citrus sinensis* 'Valencia' and 'Washington' are grown by cuttings (California).

Grafting. In Texas, citrus varieties are grafted on *Poncirus trifoliata* 'Rubidoux' rootstock. Some grafted cultivars include 'Improved Myer's Limon' (*C. limon*), which is grown by cuttings in California, 'Satsuma' (*C. reticulata*), 'Tomango' and 'Louisiana Sweet' (*C. sinensis*), 'Bloomsweet,' 'Rio Red' and 'Star Ruby' (cutting-grown) (*C. xparadisi*), and 'Kumquat' (*Fortunella miewa*).

T-budding in Texas is done in the spring and fall. Dormant budwood is collected in February and refrigerated. Budding begins as soon as the bark begins slipping and ends in mid-April when the weather gets hot. In the fall, budwood is taken from stems that are round—compared to the normal triangular shape of citrus stems. The fall budding is from October until mid-November when the bark stops slipping. Budding rubbers and clear poly tape are used for tying.

Clematis **spp.** Clematis can be propagated by seed, cuttings, grafting (358), division of roots, or layering (153).

Seed. Seeds of some clematis species have embryo dormancy, so stratification for 1 to 3 months at 4°C (40°F) is needed. Some species require a warm-cold stratification.

Cuttings. Clematis is probably best propagated by cuttings taken from young plants; these root under mist in about 5 weeks. Young wood with short internodes taken in the spring gives satisfactory results, but partially matured wood taken in late spring to late summer is more commonly used. Leaf-bud cuttings taken in midsummer will also root readily under mist. Semi-hardwood and hardwood cuttings can be treated with 3,000 to 8,000 ppm IBA talc, and softwood cuttings with a 3,000 to 5,000 ppm IBA quick-dip. Internodal cuttings (leaf-bud cuttings with one leaf intact) and the stem node inserted above the rooting medium root best with 1,000 ppm IBA quick-dip or 4,000 ppm talc (336). Basal cuttings can also be taken from unwanted flower stems, treated with 3,000 ppm IBA powder, set in a mix of 1 river sand: 1 perlite (v/v) and rooted under mist at 25°C (77°F) (369).

Micropropagation. Clematis is commercially micropropagated.

Clethra **spp.** Sweet pepperbush. Clethra are low-growing shrubs or small trees with attractive foliage, bark, and fragrant flowers; they can be propagated by seed sown in flats under mist. *C. alnifolia* is rooted from softwood cuttings (June to September in Delaware, USA) and treated with 1,000 ppm K-IBA (69, 128). Other species are rooted with 1,000 to 5,000 ppm IBA.

Coccoloba **spp. (317).** Sea grape. Some of these species are useful landscape plants with high salt tolerance. *C. uvifera* (sea grape) and *C. diversifolia* (pigeon plum) are primarily seed-propagated. Seeds are collected from August to November (Florida), and the seed coats peeled off. No other pregermination is needed, but it is important that seeds not dry out before planting. Some nursery producers propagate *C. uvifera* by cuttings and air layering.

Conospermum mitchellii **(Proteacea).** Victorian Smokebush. Can be rooted from softwood cuttings treated with 3,000 ppm IBA (321).

Coralberry. *See Ardisia.*

Cornus **spp.** Dogwood.

Seed. Seeds have various dormancy conditions; those of the popular flowering dogwood (*C. florida*) require either fall planting or a stratification period of about 4 months at 4°C (40°F). Best germination is obtained if the seeds are gathered as soon as the fruit starts to color and sown or stratified immediately. If allowed to dry out it is best to remove seeds from the fruit and soak in water. Seed germination of *C. canadensis* is enhanced with a 4-month cold stratification. Low precipitation during seed formation reduces seed germination of red-osier dogwood (*C. stolonifera*), which must also be stratified (1). Other species require a warm-moist stratification to soften the seed coat, such as 2 months in moist sand at diurnally fluctuating temperatures (21 to 30°C, 70 to 85°F), followed by cold-moist stratification of 4 to 6 months at 0 to 4°C (32 to 40°F). With some species, the warm stratification period may be replaced by mechanical scarification or soaking in sulfuric acid.

Cuttings. Some dogwoods can be started easily by cuttings. *C. kousa* cuttings collected in mid-June through July (Massachusetts) root well when wounded on one side and treated with either 8,000 ppm IBA talc or quick-dips of 2,500 ppm IBA and NAA (165). Summer softwood cuttings of *C. alba* root readily when treated with a 1,000 to 3,000 IBA quick-dip or talc. Cuttings of *C. florida* are best taken in late spring or early summer from new growth after flowering, then rooted under mist. In Alabama, two-node cuttings of *C. florida* 'Stokes Pink' and 'Weaver's White' are taken from slightly stiff new growth early in the spring and quick-dipped with 10,000 ppm IBA. *C. florida* 'Spring Grove' roots well from softwood cuttings (June, July in Ohio) treated with 8,000 ppm IBA in talc. In Florida, semi-hardwood cuttings taken in May and late August are quick-dipped in 10,000 ppm K-IBA for 3 or 10 seconds, respectively (106). *C. florida* 'Rubra' can be rooted successfully if cuttings are taken in early summer after the second growth flush, treated with 3,000 ppm IBA, and propagated under mist (364).

To ensure survival through the following winter in cold climates, the potted cuttings should be kept in heated cold frames or polyhouses to hold the temperature between 0 and 7°C (32 and 45°F). Rooted cuttings that had shoot growth in the fall, but were not fertilized, had the best overwinter survival (182).

Grafting. Chip- and T-budding are major ways that *C. florida*, *C. Kousa* and their hybrids are propagated in the southeast United States. Selected types, such as the red flowering dogwood, *C. florida* 'Rubra' and the weeping forms (difficult to start by cuttings), are often propagated by T-budding in late summer or by whip grafting in the greenhouse in winter on *C. florida* seedling rootstock. *C. kousa* and *C. florida* can be reciprocally grafted (165).

Micropropagation. Dogwood is micropropagated for breeding programs aiming to incorporate resistance to dogwood anthracnose and powdery mildew in flowering dogwood. Other dogwood species have been micropropagated (144, 147).

Corylopsis **spp.** Winterhazel. Difficult to propagate by seeds, which have a warm and cold stratification requirement. Softwood to semi-hardwood cuttings can be treated with a 1,000 ppm IBA quick-dip and rooted under mist. IBA talc at 3,000 ppm has also been used. Buttercup winterhazel (*C. pauciflora*) is commercially micropropagated.

Corylus **spp.** Filbert, Hazelnut. European filbert (*C. avellana*) is grown for fruit production and ornamental characteristics (*C. avellana* 'Contorta' and *C. fargesii*).

Seed. Seeds should be stored under refrigerated conditions immediately after cleaning (131). Seeds have a 2- to 6-month cold stratification requirement. *C. fargesii* has a 3-month cold stratification requirement (4).

Cuttings. In Oregon, cuttings in an active stage of growth are taken in mid-June to mid-July and treated with 5,000 to 10,000 ppm IBA quick-dip or talc. Cuttings are very sensitive to overwatering and should be kept in an active growth stage after rooting. Etiolation of stock plants and use of softwood cuttings enhanced rooting of *C. maxima* (281). Stooling and collection of hardwood cuttings in mid-February (England), and treating cuttings with 10,000 ppm IBA at 21°C (70°F), can enhance rooting. *C. avellana* is commercially propagated by mound layering in Oregon. Stooled shoots are ringed at the base, a 1,000 to 2,500 ppm IBA paste applied, and stools covered with soil *or* sawdust. IBA at 750 ppm sprayed on the lower 15 cm of each sucker (stool) and in combination with girdling enhance root quality (152). *C. fargesii* has not been successfully rooted.

Grafting. Commercial fruit and ornamental cultivars are grafted using a whip graft on *C. avellana* seedling rootstock. Chip building is successful and budded plants are more vigorous than grafted plants (131). *C. fargesii* can be grafted with a modified veneer graft (side wedge graft) (4, 412).

Micropropagation. Corylus is micropropagated (7).

Cotinus coggygria. Smoke Tree or Smoke Bush. Smoke trees should not be propagated by seeds, since many of the seedlings are male plants, lacking the showy flowering panicles. Only vegetative methods should be used, with cuttings taken from plants known to produce large quantities of the desirable fruiting clusters. Also, the purple foliage color will only come true from vegetative reproduction.

Cuttings. Tip cuttings taken from spring growth (May in Oregon) should be very soft and flexible, 5 to 8 cm (2 to 3 in) long. Treated with a 1,000 to 3,000 ppm IBA quick-dip, cuttings should root in about 5 weeks. After the mist is discontinued, the rooted cuttings are left in place, undisturbed, and transplanted the following spring. *C. coggygria* 'Royal Purple' can be rooted from softwood cuttings [13 to 19 cm (5 to 7.5 in) long] taken in late May (Oklahoma) and treated with 10,000 ppm IBA + 5,000 ppm NAA (140).

***Cotoneaster* spp.** Cotoneaster.

Seed. Seeds of most Cotoneaster species should be soaked for about 90 minutes in concentrated sulfuric acid, rinsed, and then stratified for 3 to 4 months at about 4°C (40°F).

Cuttings. Leafy cuttings of many species taken in spring or summer will root under mist without much difficulty. Evergreen/semi-evergreen types root better than deciduous, and are treated with a 1,000 to 3,000 ppm IBA quick-dip. Semi-hardwood cuttings of *C. buxifolius* root best with a 4,000 to 6,000 ppm IBA quick-dip in 100 percent perlite medium.

Grafting. Cotoneaster can be budded high onto pear trees to produce a "tree" cotoneaster. A blight-resistant pear rootstock, such as 'Old Home,' should be used. *C. bullatus* and *C. actifollus* rootstocks are more commonly used.

Cottonwood. *See Populus.*

Crab Apple (Flowering). *See Malus.*

Crape Myrtle. *See Lagerstroemia.*

***Crataegus* spp.** Hawthorn.

Seed. Hawthorns tend to reproduce true by seed. Seeds have a pronounced dormancy and generally more from acid scarification and stratification. Scarify seed with sulfuric acid, and then cold stratify for 5 months at about 4°C (40°F). Some species germinate better with a warm treatment prior to cold stratification. Since hawthorn develops a long taproot, transplanting is successful only with very young plants. Air pruning of seedling roots may be beneficial.

Grafting. Selected clones may be T-budded or root-grafted on seedlings of *C. crus-galli*, or *C. coccinea* for the American (entire-leaf) types and on seedlings of *C. laevigata* or *C. monogyna* for the European (cut-leaf) types.

Cryptomeria japonica. Japanese Cryptomeria, Sugi. Can be propagated either by seeds or by cuttings.

Seed. Seeds should not dry out. Seeds should be cold stratified for 3 months or given a 3-month warm stratification followed by a 3-month cold stratification.

Cuttings. Chinese and Japanese foresters have been commercially propagating Sugi by cuttings for more than 500 years, and detailed methods for taking and rooting cuttings were published in the 1600s (348). Cuttings, 5 to 15 cm (2 to 6 in) long, should be taken from greenwood at a stage of maturity at which they break with a snap when bent. Root with bottom heat; keep the cuttings shaded and cool. After roots start to form, in about 2 weeks, give more light; transplant to pots when roots are about 13 mm (1/3 in) long. Stem cuttings of *C. japonica* 'Yoshino' can be rooted year-round. Hardwood cuttings taken in January (North Carolina) can be rooted when tips of first-order laterals or proximal halves of first-order laterals are treated with 3,000 to 9,000 IBA quick-dip applications (238, 239).

xCupressocyparis leylandii. Leyland Cypress. This bigeneric hybrid of *Cupressus macrocarpa* and *Chamaecyparis nootkatensis* is propagated by cuttings and rooted under mist with bottom heat (440). Cuttings can be taken any time from late winter to autumn and should be treated with IBA at 3,000 ppm, or 8,000 ppm in December–January (Georgia) (331). In California, 6,000 ppm IBA was optimal (54). Hardwood cuttings of 'Castlewellan Gold' are propagated in winter (Greece) and treated with 9,000 ppm IBA (413). Leyland cypress is micropropagated (400).

***Cupressus* spp.** Cypress. Seeds have embryo dormancy, so stratification for about 4 weeks at 2 to 4°C (35 to 40°F) is used. Cuttings can be rooted if taken during winter months. Treatments with 2,000 to 8,000 ppm IBA enhance rooting. In California, quick-dips of 6,000 to 8,000 ppm IBA or 6,000 IBA + 6,000 NAA aid rooting (54). *C. arizonica* (Arizona cypress) can be rooted by hardwood cuttings treated with 4,000 ppm IBA (399). Side-veneer grafting of selected forms on seedling *Cupressus* rootstocks in the spring is often practiced.

Currant (Red Flowering). *See Ribes.*

Cycads. *See Cycas.*

***Cycas* spp.** Cycads. Species of *Cycas*, *Zamia*, and *Encephalartos*.

Seed. Many are seed-propagated. The seed of most cycads germinate without difficulty but sometimes slowly over a span of several months. In general, soaking the seed in water for a few days and subsequently in 1,000 ppm gibberellic acid for 24 to 48 hours enhances germination (118). Species of *Cycas, Zamia,* and *Encephalartos* have seed germination problems that include an impermeable seed coat (the sclerotesta), an immature embryo, and physiological dormancy. Scarification with sulfuric acid, followed by a gibberellic acid treatment, resolves most of these dormancy problems. Scarifying the seeds by cracking them is not recommended. Seeds of *Zamia floridana* germinate well when scarified with sulfuric acid for 1 hour, soaked in gibberellic acid for 48 hours, and then placed under intermittent mist for 6 weeks; *Z. furfuracea* seeds germinate best with a 15-minute acid treatment, 24-hour gibberellic acid soak, and placement under mist for 6 weeks (118).

Seeds of *Cycas* float in water whether viable or nonviable. The embryo is immature at the time of seed collection. For improved germination, Sago palm (*C. revoluta*) seed should be stored at room temperature and scarified for one hour with concentrated sulfuric acid.

Cycad seedlings should be root-pruned by severing the roots at the root-shoot juncture, dipping the cut end of the leafy portion in IBA, after which they are potted in liner pots and placed under mist for 2 weeks to encourage branched root systems and more rapid growth and development of the seedlings.

Vegetative propagation. Slow growth rates limit the potential for vegetative propagation. The American *Zamia* can be regenerated from its underground tuberous root and stem tissue. Cycads do not have lateral buds to produce side shoots. However, when taxa such as *Cycas* are injured, side shoots known as "pups" regenerate from callus. The pups are removed and their severed base is quick-dipped in 2,000 to 10,000 ppm IBA; rooting can take up to 1 year (118).

Micropropagation. The South African cycad, *Stangeria eriopus,* has been micropropagated (311). However, there is no commercial micropropagation of cycads.

Cypress. See *Cupressus.*

***Cytisus* spp.** Broom.

Seed. Seeds of many of the species germinate satisfactorily if gathered as soon as mature and treated with sulfuric acid for 30 minutes to soften the hard seed coats before planting. Since the various *Cytisus* species crossbreed readily, stock plants for seed sources should be isolated.

Cuttings. Hardwood cuttings taken in late February or early March root well with 2,500 ppm IBA. Semi-hardwood cuttings can be rooted easily under mist in midsummer if treated with 3,000 to 8,000 ppm IBA and given bottom heat.

***Daphne* spp.** Daphne can be propagated by seed, stem and leaf-bud cuttings, layering, and grafting (81, 90).

Seed. Seeds should be sown at once after harvest or, if dried, scarified and given a moist-chilling period before sowing. Berries are very poisonous if eaten.

Cuttings. Daphne is probably best propagated by leafy cuttings in perlite and peat moss (4:1) mix under mist; a well-drained 100 percent medium perlite is also recommended for cuttings (134). In England, cuttings are taken in mid- to late May, toward the end of flowering. Cuttings are also taken in summer from partially matured current season's growth. Rooting can be enhanced with a 1,000 ppm IBA + 500 ppm NAA, or a 2,500 ppm IBA quick-dip (134). The daphnes do not transplant easily and should be moved only when young.

Micropropagation. *D. odora* is micropropagated (346).

Davidia involucrata. Dove tree. Seeds require a 5 to 6 months' warm stratification followed by 3 months of cold stratification (166). Polyethylene bags containing a ratio of 1 part sand to 1 part peat is a suggested system for the pretreatment of *Davidia* seeds. Rooting is variable. Leaf-bud cuttings treated with 3,000 ppm IBA talc rooted 85 percent. Wounded cuttings with four leaves that were treated with 8,000 ppm IBA talc and propagated under mist had 50 percent rooting (131). Disturbance and overwintering of rooted cuttings can be a problem.

Dawn Redwood. See *Metasequoia.*

Delonix regia. Royal Poinciana. This spectacular tropical flowering tree is propagated by seed. Germination is rapid when seeds are treated to soften seed coats, as by pouring boiling water over seeds or by soaking in concentrated sulfuric acid for 1 hour.

***Deutzia* spp.** Deutzia is easily propagated either by hardwood cuttings lined-out in the nursery row in spring, or by softwood cuttings taken in the spring or summer and treated with IBA talc or quick-dip at 1,000 to 3,000 ppm.

Diosma ericoides. Diosma, Breath of Heaven. Propagated by leafy cuttings taken in summer and rooted under mist.

***Diospyros* spp.** Japanese persimmon (*D. kaki*), common persimmon (*D. virginiana*), and Texas persimmon (*D. texana*). Japanese persimmon is an important fruit crop and propagation is covered in Chapter 19. Texas and common persimmon are propagated from seed. No pretreatment is required for Texas persimmon, but germination percentages can be low (219). Common persimmon requires 3 months of cold stratification at 4°C (40°F) for 5 months for good germination. *D. kaki* is also propagated by shield (T-budding) or chip budding and side-veneer or side-tongue grafted on *D. virginiana* rootstock. Other cultivars of *D. kaki* are chip-budded on *D. kaki* rootstock (379). *D. kaki* is micropropagated (104).

***Disanthus cercidifolius* (123).** A multistemmed shrub of potential in the Hammamelidaceae family with red fall color and purple flowers. Seeds require a warm stratification for 5 months followed by 3 months of cold stratification. Cuttings collected in July (Georgia) root best when treated with a 10,000 ppm IBA alcohol quick-dip and propagated under mist in a peat-perlite media. Successfully overwintered in a polyhouse if potted up immediately after rooting, fertilized, and allowed to harden-off naturally in the fall. *Disanthus* is commercially micropropagated.

Dogwood. *See Cornus.*

Douglas-fir. *See Pseudotsuga.*

Dove Tree. *See Davidia.*

***Elaeagnus* spp.** Elaeagnus, Russian olive, Silverberry, Silverthorn. Seeds planted in the spring germinate readily following a stratification period of 3 months at 4°C (40°F). Removal of the pit (endocarp) for silverberry seeds (*E. commutata*) resulted in about 90 percent germination of unstratified seeds, since a germination inhibitor is apparently present in the pit; if not fall-planted, seeds should be cold stratified for 3 months. Seeds of the Russian olive, *E. angustifolia,* should be treated with sulfuric acid for 30 to 60 minutes before fall planting or stratification; this deciduous species is commercially seed-propagated, and only limited rooting success (28 percent) has been obtained with 3,000 ppm IBA talc (131). Leafy cuttings of the evergreen species root readily. Hardwood and semi-hardwood cuttings of *E. pungens* root well when treated with 8,000 to 20,000 ppm IBA talc (38).

Elderberry. *See Sambucus.*

Elm. *See Ulmus.*

English Ivy. *See Hedera.*

***Enkianthus* spp.** Enkianthus are attractive shrubs with pronounced fall color. Seed propagation requires no pretreatment, and germination occurs within 2 to 3 weeks after sowing (131, 307). Cuttings root quite easily. Leafy cuttings taken in mid-June (Massachusetts) are treated with talc or quick-dips of IBA at 5,000 to 8,000 ppm, and rooted under mist. Rooted cuttings should be allowed to harden-off and overwinter before they are disturbed. An extended photoperiod is helpful in ensuring overwintering success. Enkianthus is micropropagated.

Epigaea repens. Mayflower, trailing arbutus. This creeping, evergreen shrub has fragrant white to pink flowers and is propagated by seed collected in late spring (Massachusetts). Immature seed is white, and dark when mature. The cleaned seed is stored at 4°C (40°F) until sowing; there is no stratification requirement (350).

***Erica* spp.** Heath. The propagation of the closely related genera, Heather (*Calluna vulgaris*), is quite similar (132).

    Seed. Seeds may be germinated in flats in the greenhouse in winter or in a shaded outdoor cold frame in spring.

    Cuttings. Leafy, partially matured cuttings taken at almost any time of year, but especially in early summer, root readily under mist in a glasshouse or polyethylene-covered cold frame. While most species root readily, IBA (1,000 ppm quick-dip or 4,000 ppm talc) speeds up rooting.

    Micropropagation. *Erica* and *Calluna* are micropropagated.

Eriobotrya japonica. Loquat. These large evergreen shrubs or small trees can be propagated by seed, without pretreatment. Considered difficult-to-root. Side grafting and budding is another form of propagation. *E. japonica* has been successfully micropropagated. See Chapter 19.

***Eriostemon* spp.** Eriostemon. An Australian evergreen shrub with long-lasting white or pink flowers and fragrant foliage. Can be propagated by seed and stem cuttings. Stem cuttings can be slow to root or root poorly, and ease of rooting is clonally variable. *E. myoporoides* and *E.* 'Stardust' can be micropropagated (14).

***Escallonia* spp.** Escallonia is easily started by leafy cuttings taken after a flush of growth. Cuttings root well under mist and respond markedly to treatment with 1,000 ppm IBA. It is best to direct-root in liner pots, since transplanting is difficult.

***Eucalyptus* spp.** Eucalyptus.

Seeds. Eucalyptus is largely propagated by seeds planted in the spring. Mature capsules are obtained just before they are ready to open. No dormancy conditions occur in most species, so seeds are able to germinate immediately following ripening. Seeds of some species, however—for example, *E. dives, E. niphophila,* and *E. pauciflora*—require stratification for about 2 months at 4°C (40°F) for best germination. Eucalyptus seedlings are very susceptible to damping-off. Seeds are usually planted in flats of pasteurized soil placed in a shady location or flats covered with a white plastic (46). From flats, they are transplanted into small pots, from which they are later lined-out in a nursery row. The roots of young trees will not tolerate drying, so the young plants should be handled as container-grown stock. Seeds may be sown directly into containers in which the seedlings are grown until planting in their permanent location (188).

Cuttings. Eucalyptus is difficult to start from cuttings, but good rooting can be obtained from some species. For example, leafy cuttings of *E. camaldulensis* root when taken in early spring from shoots arising from the base of young trees, wounded, treated with a 4,000 ppm IBA + 4,000 ppm NAA, and propagated under mist with bottom heat at 21°C (70°F) (155). *E. grandis* stock plants are hedged and new developing shoots are used as cuttings. Four-node cuttings are prepared and treated with IBA. Leaf retention is very important, so avoid using older shoots with leaves that will abscise before rooting occurs (82). Clones of *E. cladocalyx, E. tereticornis, E. grandis, E. camaldulensis,* and *E. trabutii* will root when three- to four-node cuttings are treated with a quick-dip of 4,000 to 8,000 K-IBA (356). Rooting is best when cuttings are taken from rapidly growing stock plants (April to October in California). Apical minicuttings are taken from intensively managed mother plants to optimize shoot production and rooting of cuttings (385). Mother plants are maintained in cell trays or small pots and flood irrigated. Apical minicuttings are taken every 7 to 10 days, treated with auxin and rooted under mist. Plants produced from mini-cuttings, compared to semi-hardwood cuttings, have better form, a more robust and developed root system, reduced production time as juvenile apical shoot grows more rapidly, and the area required for hedging mother plants for traditional semi-hardwood production is reduced (385).

Grafting. *E. ficifolia* has been grafted successfully by a side-wedge method, using young vigorous Eucalyptus seedling rootstocks growing in containers and placed under very high humidity following grafting. Use of scions taken from shoots that had been girdled at least a month previously increased success. Top cleft grafting is used in Australia. Semihardwood scions are grafted onto seedling rootstocks and the grafted ramets are initially placed under intermittent mist (385).

Micropropagation. Eucalyptus are routinely micropropagated, even with mature explant tissue (36, 116, 236, 265, 296).

***Euonymus* spp.** Euonymus.

Seed. Generally cold stratification at 0 to 10°C (32 to 50°F) for 3 to 4 months is required for satisfactory seed germination (171). Some species (*E. atropurpureus* and *E. europaeus*) require several months of warm stratification prior to cold stratification for germination. Remove seeds from fruit (colorful aril) and prevent drying.

Cuttings. Euonymus is easily started by cuttings. Deciduous species are started from hardwood cuttings in late winter/early spring, while evergreen types are propagated with leafy semi-hardwood cuttings. *E. alatus* roots best from stem cuttings collected before growth is fully hardened in the spring. IBA can be applied as a 1,000 to 3,000 ppm quick-dip or from 3,000 to 8,000 ppm talc (131). The production cycle of *E. fortunei* 'Sarcoxie' is significantly shortened when summer-rooted cuttings are exposed to a minimum of 500 chilling hours and then winter-forced in a heated greenhouse (96).

Micropropagation. *E. alatus* has been commercially micropropagated, and somatic embryogenesis has been successful with other Euonymus species (228).

***Euphorbia* spp.** Poinsettia (*E. pulcherrima*). This species is generally propagated by leafy cuttings under mist. A quick-dip of 500 to 1,000 ppm IBA + NAA enhances rooting. Stock plants of self-branching cultivars should be used as a source of cuttings. It is best to root cuttings in small container liners or root cubes, so that the roots of rooted cuttings are not disturbed during transplanting to larger containers. Cuttings can be rooted in the greenhouse from spring to fall. Specialists or their licensed propagators will sell rooted liners to greenhouse producers for either stock plants or for shifting up to a larger pot for finishing. Scarlet plum (*E. fulgens*), which is a medium-sized shrub from Mexico used for cut-flower production in greenhouses, is generally propagated by cuttings. Micropropagation of this species is faster and requires less greenhouse space than conventional cuttage (453).

***Fagus* spp.** Beech.

Seed. Seeds germinate readily in the spring from fall planting or after being stratified for 3 months at about 4°C (40°F). Seeds should not be allowed to dry out. *F. sylvatica* seeds are sometimes hollow and nonviable.

Cuttings. Cuttings taken from seedling plants will root at different percentages due to clonal influences. Overwintering survival of rooted cuttings (Denmark) is a problem (263). Etiolation of stock plants and covering shoot bases with black adhesive tape enhances rooting.

Grafting. Selected clones are grafted by the cleft, whip, or side-veneer method on seedling rootstock of European beech (*F. sylvatica*). In Oregon, *F. sylvatica* cultivars are field-grafted on seedling rootstock using a "stick bud" method; the scionwood generally contains two buds and is inserted into a T-cut in the rootstock (149). Cleft grafting is used with American beech (*F. americana*), but the species is difficult to graft with only a 25 to 30 percent success rate (338). Grafting success increased when the scion diameter was slightly greater than rootstock diameter.

False Cypress. See *Chamaecyparis*.

xFatshedera (Fatsia × Hedera). See *Hedera helix*.

Feijoa sellowiana. See *Feijoa* in Chapter 19.

***Ficus* spp.** Fig, rubber plant. A wide range of ornamental trees, vines, and ornamental potted plants. For the edible, common fig, *see Ficus carica* in Chapter 19. Many species form aerial roots, and are easily rooted by cuttings and layering.

Cuttings. Single-node propagation of *Ficus* spp. produces many more plants per stock plant than the common method of air layering (329). *F. benjamina* (weeping fig) is easily started by leafy semi-hardwood cuttings taken in spring or early summer and rooted under mist. *F. elastica* (rubber plant) is propagated by cuttings taken from 5 to 27 cm (6 to 12 in) shoots; single buds or "eyes" can be removed and rooted. These cuttings are made in spring, inserted in sand or a similar medium, and held in a warm greenhouse. *F. pumila* (creeping fig) is propagated with 10 to 13 cm (4 to 5 in) long cuttings. The juvenile form, which is a climbing vine with aerial rootlets, roots easily all year round; 1,000 to 1,500 ppm IBA quick-dips enhance rooting (113). The mature form lacks aerial roots, but can be rooted successfully with 2,000 to 3,000 ppm IBA. *F. lyrata* (fiddleleaf fig) can be propagated by cuttings, or by air layering (234).

Air layering. Shoots of trees growing outdoors in the tropics are air layered and the rooted air layers are shipped to wholesale nurseries for finishing off the crop. In subtropical regions of the United States, *Ficus* stock blocks are maintained for air layering (see Fig. 14–5). Indoor plants that become too "leggy" also can be air layered.

Micropropagation. *Ficus* are commercially micropropagated (328).

Fig. See *Ficus*.

Fig (Common). See *Ficus carica* in Chapter 19.

Filbert. See *Corylus*.

Fir. See *Abies*.

Firethorn. See *Pyracantha*.

***Forsythia* spp.** Forsythia is easily propagated by hardwood cuttings set in the nursery row in early spring or by leafy softwood cuttings taken during late spring and rooted under intermittent mist.

Fothergillia gardenii. Dwarf Fothergillia. Generally difficult to propagate by seed—based on lack of availability, and a 6-month warm stratification, followed by a 3-month cold stratification requirement. Easy to root by cuttings, which are treated with 4,000 ppm IBA (131). Cultivars of Fothergillia are commercially micropropagated.

Franklinia alatamaha. Franklinia can be propagated by seed that is sown immediately (if fresh). Otherwise, seed needs a 1-month stratification period if stored. Roots easily by cuttings treated with 1,000 ppm IBA. Root early in the season to allow growth of rooted liners prior to overwintering. Franklinia is commercially micropropagated.

***Fraxinus* spp.** Ash.

Seed. Seeds of most species germinate if stratified for 1 to 3 months followed by cold stratification for up to 6 months. In some cases, seeds of some species like *F. anomala* and southern accessions of *F. pennsylvanica* and *F. Americana* require only a 3-month cold stratification treatment.

Cuttings. Ashes are difficult to propagate from stem cuttings. *F. greggii* (littleleaf ash) roots readily when 10 to 15 cm (4 to 6 in) long—leafy terminal cuttings are taken 16 weeks after bud-break and treated with a 17,000 ppm K-IBA quick-dip (406). Rooting is enhanced with early, softwood cutting material. *F. pennsylvanica* 'Summit' has been rooted in low percentages when taken from softwood cuttings in April and May in Pennsylvania (20). *F. anomala* cuttings taken from containerized seedling stock plants had high rooting when treated with auxins from 5,000 to 20,000 ppm.

Grafting. *F. excelsior, F. ornus,* and *F. pennsylvanica* seedlings are used as rootstocks for grafting or budding ash cultivars.

Micropropagation. Ashes are commercially micropropagated (12, 30).

Fringe Tree. *See Chionanthus.*

Fuchsia **spp.** Fuchsia is easily rooted by leafy cuttings maintained under humid conditions (422). Roots develop in 2 to 3 weeks.

Gardenia jasminoides. Gardenia, Cape jasmine. Leafy terminal cuttings are treated with 3,000 ppm IBA and rooted under mist from spring to fall. Gardenias are difficult to transplant and should be moved only when small. Gardenia is commercially micropropagated (146).

Garrya elliptica **(345).** Garrya 'James Roof' is an ornamental tree producing catkins 300 mm long. Seeds generally require cold stratification at 4°C (40°F) for 1 to 3 months. In some cases, a 1-day gibberellic acid (100 ppm) treatment following stratification has improved germination. It is generally considered difficult to propagate asexually. Tip nodal cuttings 10 cm long on side shoots taken with a heel with well-developed terminal buds are collected from late summer to December (England). Wounding is optional and cuttings are treated with 8,000 ppm IBA talc and propagated under mist or plastic film. Direct rooting in small liner pots is desirable to avoid root disturbance problems. Rooted cuttings should complete their first spring flush of growth before transplanting. *Garrya* is micropropagated.

Genista **spp.** Genista. Landscape plants that range from creeping ground covers to massing shrubs. These species are not difficult to propagate.

Seed. Seed propagation is similar to that of other legumes with hard seed coats. Scarify with a 30-minute treatment of concentrated sulfuric acid or a boiling-water treatment. Seeds are placed in boiling water which is immediately allowed to cool to room temperature (162). Some species have no scarification requirement. Seeds of *G. tinctoria* and some other species also require a 3-month cold stratification of 5°C (41°F). When in doubt, immerse seeds in water overnight, and then separate those that have imbibed water and swelled from those that did not. Cold stratify swollen seeds for 3 months or until radicle growth first becomes visible. Soak those seeds that did not swell in concentrated sulfuric acid for 30 minutes, then wash them thoroughly, and cold stratify in the same way (162).

Cuttings. Softwood cuttings taken in early summer (Indiana) root easily without auxin, while semi-hardwood cuttings taken in late summer and hardwood cuttings taken in fall root best with 8,000 ppm IBA in talc (162). Bottom heat should not be allowed to drop below 18°C (65°F). Seedlings or rooted cuttings should be grown in containers, since field-dug plants do not transplant well.

Micropropagation. *G. monosperma* has been micropropagated (353).

Ginkgo biloba. Ginkgo. Seed propagation should not be used except for rootstock production. The "fruits" are collected in mid-fall and the pulp removed. A warm and cold stratification of 1 to 2 months, each, is required. Seedlings produce either male or female trees, but the sex cannot be determined until the trees flower—after about 20 years. The plumlike "fruits" on the female trees have a very disagreeable odor, so only male trees are used for ornamental planting. Cutting-produced male liners do not grow as quickly as grafted plants. Softwood cuttings taken in early summer are treated with 8,000 ppm IBA talc or quick-dip and rooted under mist. Commercial propagation is also by T-budding or chip budding, using buds from male trees inserted into ginkgo seedlings. Male cultivars can also be cleft or whip grafted on container seedling rootstock during winter.

Gleditsia triacanthos. Common Honeylocust. Readily propagated either by seeds planted in the spring or by budding. In seed propagation, soaking the seed in sulfuric acid for 1 to 2 hours, gives good germination. Very difficult to propagate by cuttings—can be rooted by root cuttings. The thornless honey locust, *G. triacanthos,* var. *inermis,* and the thornless and fruitless 'Moraine' locust are usually propagated by T-budding on seedlings of the thorny type.

Gold Tree. *See Tabebuia argentea.*

Golden Chain. *See Laburnum.*

Golden Trumpet. *See Allamanda.*

Goldenrain Tree. *See Koelreuteria.*

Gordonia **spp. (131).** Loblolly bay, Black laurel. These medium-sized trees can be propagated by seed, which germinates readily without pregermination requirements. Semi-hardwood cuttings are easily rooted with a 2,500 to 3,000 ppm IBA quick-dip or 3,000 ppm IBA talc. These species transplant easily after rooting.

***Grevillea* spp.** These Australian native shrubs or trees are propagated by seed, cuttings, or graftage. Cuttings of the low-growing species root readily, but larger-growing species, such as *G. robusta,* silk oak, are best propagated by seed. *G. johnsonii* is difficult to root, but fall cuttings (Australia) treated with 4,000 ppm IBA-ethanol 5-second quick-dip have 70 percent rooting (150). With *G. asplenifolia,* single-node cuttings with reduced leaf areas work best, whereas retaining all leaves on larger two- to three-node cuttings is optimal for *G. juniperina.* A number of Grevillea species can also be grafted using a whip graft (56). *G. robusta* is the primary rootstock, since it has resistance to *Phytophthora cinnamomi.* Specialized weeping forms of Grevillea are produced by using approach grafting of two independent, containerized Grevillea plants (108). *G. scapigera* (68) and *G. robusta* (337) can be micropropagated.

Gymnocladus dioicus. Kentucky Coffeetree. An attractive landscape tree that is dioecious. Male trees are more desirable than female trees, which have long seed pods that abscise and detract from the ornamental value of the plant. Can be propagated by seed, but undesirable female plants are produced. Seeds require 2 to 4 hour acid scarification, but no stratification needed to germinate. Cutting propagation is very difficult. Can be propagated by root cuttings. Seedlings (173) and mature male trees have been successfully micropropagated (386).

Hackberry. *See Celtis.*

***Halesia* spp.** Silverbell. A small, valuable landscape tree, native to the southeastern United States. It has a striped bark, bell-shaped flowers, and interesting fruit (two- to four-winged drupe). Seed propagation of *Halesia* requires complex stratification regimes and success is often limited. *H. carolina* is warm stratified for 2 to 4 months, followed by 2 to 3 months of cold stratification. Spring and summer cuttings root well when treated with 1,000 to 10,000 ppm IBA quick-dips (123). Silverbell is commercially micropropagated (60).

***Hamamelis* spp.** Witch Hazel. Propagated by seed, budding, or grafting.

Seed. *Hamamelis* requires a 3-month warm stratification followed by a 3-month cold stratification. The treated seed can be germinated in the spring.

Cuttings. Cutting propagation is difficult but possible. Leafy cuttings of *H. mollis, H. virginiana, H. japonica,* and *H. vernalis* are treated with 8,000 ppm IBA talc or a 10,000 IBA quick-dip and rooted under mist. Cuttings of *H. xintermedia* 'Arnold Promise' should be taken as early as possible in spring—cuttings taken in late spring and summer will root, but not survive the winter. Collect three- to four-node cuttings with basal portion firm, quick-dip with 10,000 ppm K-IBA, or use an 8,000 ppm talc. For softer tissue use 2,500 to 5,000 ppm IBA (130). To improve overwintering survival, it is best to induce a growth flush after rooting and prior to winter. The rooted cuttings need to be left undisturbed in an unheated house (avoid freezing) and not transplanted until leaves have started to expand in the spring (148, 257).

Grafting. Cultivars of *H. mollis, H. xintermedia,* and *H. japonica* are propagated by budding or grafting (side-veneer graft) on *H. virginiana* seedlings; however, rootstocks tend to sucker and compete with scions (293, 335). In Oregon, chip budding works well. Bud as low as possible on the rootstock.

Micropropagation. Witch hazel is commercially micropropagated (62).

Hawthorn. *See Crataegus.*

Heath. *See Erica.*

Heather. *Calluna. See Erica* (Heath).

***Hebe* spp.** Hebe, Veronica. Propagated by seed, by leafy cuttings in summer under mist, or by layering. Hebe is commercially micropropagated.

Hedera helix. English Ivy. English ivy is readily propagated by rooting cuttings of the juvenile (nonfruiting, lobed-leaf) form. For *H. canariensis* (Algerian ivy), cuttings are taken after first growth flush and treated with a 1,000 ppm IBA quick-dip (131). *H. helix* is also sometimes grafted onto Fatshedera (*Fatsia japonica x Hedera helix*) as a rootstock.

Hemlock. *See Tsuga.*

Heptacodium miconioides. Seven Son Flower. A small deciduous flowering tree in the *Caprifoliaceae* family, which is native to China. It has potential as a new nursery crop because of its exfoliating bark, vigorous growth, and fragrant, white, late-summer flowers. In the fall it has beautiful rose to crimson fruit display that is as ornamental as the flowers. Best rooting occurred with basal and middle softwood cuttings and basal semi-hardwood cuttings (266). A quick-dip of 5,000 ppm K-IBA enhanced rooting. Shoots from lateral cuttings appear to exhibit plagiotrophic growth, but vertical shoots will rapidly grow from the base of the plant. Plagiotrophic growth can be reversed by growing rooted cuttings for a season or two and then cutting back to one bud.

Heteromeles arbutifolia. Toyon, Christmas Berry. Usually propagated by seed, which is stratified for 3 months or sown in the fall to obtain outdoor winter chilling of the seed. It can be rooted from softwood tip cuttings that are taken in mid-spring, treated with 8,000 ppm IBA talc, and placed under mist (185). It can also be propagated by layering.

Hibiscus spp. Hibiscus.

Seed. There are no pregermination requirements. Seed propagation is not commonly used for propagation.

Cuttings. Softwood and hardwood cuttings of *H. rosa-sinensis* (Chinese hibiscus) are not difficult to root; however, there are cultivar differences (242). Softwood and semi-hardwood cuttings are rooted under mist in liner pots with peat:perlite, peat blocks, or rock wool to prevent root damage during transplanting. In Australia, soft-tip cuttings 11 cm (4 in) long are taken from December to April and treated with 5,000 ppm IBA, and rooted under mist. Cultivars that root rapidly receive maximum benefit from propagation medium temperatures of 26 to 30°C (79 to 86°F) (77).

H. syriacus (Hibiscus, Shrub-althea, Rose of Sharon) are propagated, either by hardwood cuttings in the nursery row in spring or by softwood cuttings in midsummer under mist. Lateral shoots make good cutting material. Softwood cuttings respond well to treatment with a 1,000 ppm IBA quick-dip. Deciduous, hardwood cuttings have rooted well when treated with 8,000 ppm IBA talc. Softwood cuttings root faster than hardwood.

Grafting. Vigorous cultivars of *H. rosa-sinensis* which are resistant to soil pests and can be started easily by cuttings—such as 'Single Scarlet,' 'Dainty,' 'Euterpe,' or 'Apple Blossom'—are also used as rootstocks. Some clones develop into much better plants when grafted on these rootstocks than on their own root system from cutting propagation. Whip grafting in the spring or cleft grafting or side grafting in late spring or early summer is successful. Scions of current season's growth, about pencil size, are grafted on rooted cuttings of about the same size (373).

In New Zealand, simultaneous bench grafting and rooting is done with a grafting tool that makes matching "V" cuts in the scion and unrooted rootstock. After wrapping the graft union with floral tape, the grafted rootstock is quick-dipped with 2,000 ppm IBA and rooted under mist (32).

Micropropagation. Hibiscus is commercially micropropagated.

Hickory. *See Carya.*

Holly. *See Ilex.*

Honey Locust (Common). *See Gleditsia.*

Honeysuckle. *See Lonicera.*

Hornbeam. *See Carpinus.*

Horse Chestnut. *See Aesculus.*

Hovenia dulcis. Japanese Raisin Tree. Used for ornamental, medicinal, and fruit production. Seeds should be collected in the fall (Delaware). Fruits are dried, cracked, and added to water to separate the seeds. Discard seeds that float. The remaining seeds are scarified with sulfuric acid for 45 minutes, stratified by storage at 5°C (41°F) for 90 days, and planted in seedling flats (169).

Huckleberry (Evergreen). *See Vaccinium.*

Hydrangea spp. Most species are propagated by cuttings and not seeds. Seed propagation is used for plant breeding purposes. While seeds germinate without pretreatment, cold stratifycation enhances germination (131). Hydrangea are generally easy to root. Softwood and semi-hardwood cuttings are treated with 1,000 ppm IBA, and hardwood cuttings and more difficult-to-root species with 3,000 to 5,000 ppm IBA as a talc or quick-dip (131). In Oregon, soft-tip cuttings of *H. pedularis* are taken late May to early June, treated with 1,000 ppm IBA, and rooted under mist. In Michigan, softwood cuttings [10 to 15 cm (4 to 6 in)] at 1 cutting per tray cell are propagated from April to September, and hardwood cuttings [15 to 20 cm (6 to 8 in)] at 2 to 3 per tray cell from November to March are used (105). Prophylactic fungicides applied at 2 days after sticking, and then rotated on 21-day cycle, include Terrachlor-75WP, Cleary 3336, BannerMax, Terrazole, and Chipco (105). Cuttings are quick-dipped in 500 ppm IBA:330 ppm NAA. In Alabama, *H. quercifolia* roots well from spring to early summer. Half of the leaf surface on one-node cuttings are trimmed off and the cuttings are dipped in 2,500 ppm IBA. Rooted liners do not overwinter well in a warm greenhouse, so they must be kept cool during the winter. Florists' hydrangea have been rooted with 5,000 to 10,000 ppm IBA talc under mist (16).

Forcing epicormic shoots from woody stem segments [0.5 m (1.5 ft) long with a 2.5 to 25 cm (1 to

10 in) caliper] under mist and rooting the softwood cuttings can be used to propagate *H. quercifolia* oak leaf hydrangea (332). This species is unique in that individual stem segment continue to produce softwood shoots for several successive months. Softwood cuttings are treated the same as if collected from a field-grown stock plant. Drench weekly with Banrot 40% WP fungicide.

Hydrangea is commercially micropropagated (17).

Hypericum **spp.** St. John's wort. Easily started by softwood cuttings taken in late summer from the tips of current growth and rooted under high humidity or mist. *Hypericum* is commercially micropropagated.

Ilex **spp.** Holly. Can be propagated by seeds, cuttings, grafting, budding, layering, division, and micropropagation (161, 381). *Cutting propagation is the preferred method for producing superior clones.* Most hollies are dioecious. The female plants produce the very desirable decorative berries if male plants are nearby for pollination. For most hollies, any male will pollinate the female flower as long as the two are in bloom at the same time and planted fairly close together. In seed propagation, sex cannot be determined until the seedlings start blooming at 4 to 12 years.

Seeds. The only reason to grow seedlings is for rootstock graftage, or for breeding and selection work. Germination of holly seed is very erratic; species such as *I. crenata, I. cassine, I. glabra, I. vomitoria, I. amelanchier,* and *I. myrtifolia* germinate promptly and should be planted as soon as they are gathered. Seeds of other species, *I. aquifolium* (English holly), *I. cornuta* (Chinese holly), *I. verticilliata, I. decidua,* and most *I. opaca* (American holly), should be collected and cleaned as soon as the fruit is ripe in the fall and then stored at about 4°C (40°F) until spring in a mixture of moist sand and peat moss; they do not germinate until a year or more after planting even though stratified, due to rudimentary embryos at the time of harvest. For deciduous hollies (*I. verticilliata, I. decidua*), stratify seeds for 3 to 4 months, and if no germination occurs, place seeds under warm stratification at 20 to 30°C (68 to 86°F) for after-ripening and embryo development for 2 to 3 months—followed by a 2- to 3-month cold stratification period (125).

Cuttings. This is the method most used commercially, permitting large-scale production of superior clones. Semi-hardwood tip cuttings from well-matured current season's growth produce the best plants. Cuttings taken from flat, horizontal branches of *I. crenata* tend to produce prostrate plants (plagiotropic) and those from upright growth produce upright (orthotropic) plants.

Timing is important—best rooting is usually obtained from mid to late summer, but cuttings may be successfully taken into the following spring. Wounding the base of the cuttings helps induce root formation. The wounding induced by stripping off the lower leaves may be sufficient.

The use of auxins, particularly IBA at relatively high concentrations (8,000 to 10,000 ppm), is essential in obtaining rooting of some cultivars, such as *I. opaca* 'Savannah' (10,000 ppm IBA liquid), whereas 2,500 ppm IBA is sufficient for medium-difficult species such as *I. cornuta* (38). Semi-hardwood cuttings of *I. crenata* root best with 8,000 ppm talc, whereas hardwood cuttings should be treated with a 10,000 ppm IBA dip. Deciduous hollies such as *I. serrata, I. decidua,* and *I. verticillata* are best rooted in early July (Indiana) with 10 to 13 cm (4 to 5 in) cuttings having four to five nodes; the bottom leaf is stripped off and the cutting quick-dipped in 5,000 ppm IBA. Hollies can be sensitive to solvents used as IBA carriers. They usually exhibit chlorosis, followed by defoliation, e.g., *I. vomitoria* is sensitive to alcohol quick-dips. It is best to switch to water-based K-IBA products.

I. vomitoria 'Nana' are direct-stuck into 6-cm (2 1/4-in) liner pots using 3,000 ppm K-IBA spring-summer (Alabama) and 5,000 ppm K-IBA fall-winter (139). Semi-hardwood cuttings taken in early fall (Alabama) are strip-wounded at the base by tearing off the lowest branch and given a three-second dip with 2,000 ppm IBA (189, 190). Using NAA on *I. vomitoria* will burn the stem and cause defoliation. Bottom heat at 21 to 24°C (70 to 75°F) is beneficial. A table of liquid auxin solutions for Ilex taxa has been published (40).

Micropropagation. Ilex are commercially micropropagated.

Illicium **spp.** Anise. Medium-sized, fragrant, evergreen shrubs that are fragrant. Seed propagation requires no pretreatment; however, most species are propagated by semi-hardwood cuttings, with good to poor rooting, depending on the species. Summer cuttings treated with a 3,000 ppm IBA quick-dip are rooted in peat-perlite media with mist (123).

Incense Cedar. *See Calocedrus.*

Indigo Bush. *See Amorpha.*

Ironwood. *See Casuarina.*

Jacaranda mimosifolia. Jacaranda. Propagation is by softwood cuttings, grafting, or by seed. Easily propagated by seed taken from capsules after blooming (432). Seeds should be sown fresh for best germination results; fresh seed soaked in water for 24 hours prior to sowing. A white flowering form is grafted in California.

Jamesia americana. Cliffbush. A native shrub of United States Rocky Mountain states with peeling bark with white and pink flowers. Seeds require a 4-month cold stratification period (23). Seedlings should not be overwatered and prophylactic application of Daconil or similar fungicide is recommended. Can be propagated by softwood cuttings. Semi-hardwood cuttings should be wounded and direct stuck into liner pots. No auxins are applied, but bottom heat is recommended. Go light on the mist. Cuttings root in about 3 weeks with a 50 to 60 percent success rate (23).

Japanese Raisin Tree. *See Hovenia.*

Japanese Ternstroemia (Cleyera). *See Ternstroemia.*

Jasmine (Asiatic). *See Trachelospermum.*

Jasmine ('Maid of Orleans,' 'Sambac Jasmine,' and 'Grand Duke'). *See Jasminum.*

Jasmine (Star, Confederate). *See Trachelospermum.*

***Jasminum* spp.** Jasmine. Propagated without difficulty by leafy semi-hardwood cuttings taken in late summer and rooted under mist. *J. sambac* 'Maid of Orleans,' 'Sambac Jasmine,' and 'Grand Duke' are propagated by single- or double-node cuttings that root within 4 to 5 weeks. Cuttings of *J. nudiflorum* root best when treated with a 3,000 ppm IBA quick-dip (115). Layers and suckers also can be used.

***Juglans* spp.** Walnut. See Chapter 19.

***Juniperus* spp.** Juniper. Junipers are divided into ground cover types, bushes, and upright pyramidal types. They are generally propagated by cuttings, but in some cases difficult-to-root species such as the upright types are grafted onto seedlings or select cutting-grown clonal rootstock.

Seeds. Seedlings of the red cedar, *Juniperus virginiana* or *J. chinensis*, are ordinarily used as rootstocks for grafting ornamental clones. Seeds should be gathered in the fall as soon as the berry-like cones become ripe. For best germination, seeds should be removed from the fruits, then treated with sulfuric acid for 30 minutes before being stratified for about 4 months at 4°C (40°F). Rather than the acid treatment, 2 to 3 months of warm [21 to 30°C (70 to 85°F)] stratification, or summer planting, could be used. As an alternative for cold stratification, the seed may be sown in the fall. Germination is delayed at temperatures above 15°C (60°F). Viability of the seed varies considerably from year to year and among different lots, but it never germinates much more than 50 percent. Treated seed is usually planted in the spring, either in outdoor beds or in flats in the greenhouse. Two or three years are required to produce plants large enough to graft. For various species germination has been improved by treating with different stimulants, such as citric acid, hydrogen peroxide, and gibberellic acid.

Cuttings. The spreading, prostrate types of junipers are more easily rooted than upright kinds. Auxins enhance rooting, with easy-to-root types (*J. horizontalis*) treated at lower concentrations and more difficult-to-root species (*J. virginiana*) treated at higher auxin levels. Cuttings are made 5 to 15 cm (2 to 6 in) long from new lateral-growth tips stripped off older branches. Sometimes a small piece of old wood—a heel—is left attached to the base of the cutting. In other cases, good results are obtained when the cuttings are just clipped without the heel from the older wood. Terminal growth of current season's wood also roots well.

Cuttings to be rooted in the greenhouse can be taken at any time during the winter (209) or rooted outdoors on heated beds (southern California). In more temperate areas, exposing the stock plants to several hard freezes seems to stimulate better rooting. Optimum time for taking cuttings is when stock plants have ceased growth (late fall-winter propagation period is more successful than during summer). For propagating in an outdoor cold frame, cuttings are taken in late summer or early fall. There may be advantages to using bottom heat.

Lightly wounding the base of the cuttings is sometimes helpful, as is the use of IBA. Recommendations have included 2,500 IBA quick-dip (Alabama) for medium-difficult juniper species to 3,000 to 8,000 ppm IBA liquid or talc. For upright junipers, one large California nursery uses combinations of 3,000 to 6,000 ppm K-IBA (102).

Maintaining a minimum bottom heat of 16 to 18°C (60 to 65°F) is critical during the first 6 weeks of propagation in order for the basal wound of cuttings to callus. In southern California, heat is withheld for 6 weeks to allow callusing, then bottom heat is raised to 21 to 24°C (70 to 75°F) to encourage rooting (102). In the southern United States (Gulf Coast), many growers use no bottom heat during the rooting process and take advantage of the ambient temperatures. Hardwood cuttings can be rooted in outdoor field beds. Optimal rooting of dwarf Japanese juniper (*J. procumbens* 'Nana') occurred with an IBA quick-dip or talc at 8,000 ppm; rooted cuttings were potted up after

15 weeks and unrooted, but callused cuttings were re-treated with IBA and restuck in a peat:perlite medium to allow for further rooting (55).

In North Carolina, optimum rooting of eastern red cedar (*J. virginiana*) occurred with hardwood cuttings collected in January, then wounded, treated with 5,000 ppm IBA, and rooted under mist in a propagation house (204). Unlike many coniferous species, eastern red cedar has no problem with plagiotropism (undesirable prostrate growth); cuttings from lateral branches retain an upright growth habit after rooting.

Grafting (18, 326). Vigorous seedling rootstocks with straight trunks, about pencil size, are dug in the fall from the seedling bed and potted in small pots set in peat moss in a cool, dry greenhouse. Seedlings potted earlier—in the spring—may also be used for grafting in the wintertime. About 2 to 3 weeks before grafting, the rootstock is brought into the greenhouse to induce root activity before grafting takes place. Seedlings can give variable graft results, and later, variable growth of the grafted plant. In Oregon, *J. virginiana* 'Skyrocket,' which is clonally propagated from cuttings and has a very columnar form resembling a rocket, is widely used as a rootstock or standard for patio tree grafts.

The scions should be selected from current season's growth taken from vigorous, healthy plants and preferably of the same diameter as the rootstock to be grafted. Scion material can be stored under high humidity at −1 to 4°C (30 to 40°F) for several weeks until used.

Side-veneer or other side-graft methods are ordinarily used. The unions are best tied with budding rubber strips. The grafted plants are set in a greenhouse bench filled with peat moss deep enough to cover the union. The temperature around the graft union should be kept as constant as possible at 24°C (75°F) with a relative humidity of 85 percent or more around the tops of the plants. A lightly shaded greenhouse should be used to avoid injuring the grafts. Adequate healing will take place in 2 to 8 weeks, after which the temperature and humidity can be lowered. The rootstock plant is then cut off above the graft union to allow the scion to develop.

Kalmia **spp.** Mountain Laurel. Can be propagated readily by seed germinated at about 20°C (68°F). Germination of *K. latifolia* (mountain laurel) seed is enhanced by cold stratification for 8 weeks, or by a 12-hour soak in 200 ppm gibberellic acid (230, 231). Seeds of *K. latifolia* should not be covered during sowing, since they require light for germination (continuous light is optimal) and the seeds are extremely small [1.4 million seeds/28 g (1 oz)] (278). Rooting of *Kalmia* is highly variable among species and cultivars. Some respond to higher IBA concentration (8,000 ppm), but others do not (164). *Kalmia* can also be propagated by cleft or side grafting, or by layering.

K. latifolia cultivars are commercially micropropagated, which alleviates rooting problems and is probably the most efficient propagation method (62, 272).

Kentucky Coffeetree. See *Gymnocladus*.

Koelreuteria **spp.** Goldenrain Tree. This tree is usually propagated by seed. Seeds germinate best if the seed coats are scarified for about 60 minutes in sulfuric acid or by mechanical scarification, followed by stratification for about 90 days at 2 to 4°C (35 to 45°F) to overcome embryo dormancy. In southern California, stratification is avoided by pouring 83°C (180°F) water on seeds and soaking overnight. Can be propagated from root cuttings, and shoots that develop from the root cuttings can also be taken as softwood cuttings in the spring and treated with 1,000 to 3,000 ppm IBA and rooted under mist.

Laburnum **spp.** Laburnum, Golden Chain. Propagated by seed which are scarified for 2 hours in sulfuric acid. Seeds are poisonous if eaten. Some cultivars are propagated by grafting or budding on laburnum seedling rootstock.

Lagerstroemia indica. Crape Myrtle. Seeds will germinate readily without pregermination treatment. However, *this species is commercially propagated by cuttings* because it is easily rooted from softwood or hardwood cuttings. Most softwood cuttings are much easier to root prior to flower initiation. An IBA quick-dip of 1,000 to 1,250 ppm will aid root formation. Most cultivars are easy to transplant; however, some dwarf cultivars must be transplanted with a ball of soil. Hardwood cuttings of field-grown plants are gathered after the first hard frost, sawed into 8-inch (20-cm) cuttings, graded, bunched, stored over winter, and then planted in open fields in spring (March in Alabama) without auxin treatment. Eighty percent rooting is achieved (72, 73) using this method. Crape myrtle can be micropropagated (454).

Larch. See *Larix*.

Larix **spp.** Larch. Most of these deciduous conifers are propagated easily by fall-planted seeds (202). Cones should be collected before they dry and open on the tree. Several species have empty or improperly developed seeds. Seeds of some species have some embryo dormancy, so for spring planting, stratification for 1 month at about 4°C (40°F) is advisable. Cutting propagation is best done by rooting leafy, semi-hardwood cuttings in late summer under mist. IBA at a 8,000 ppm quick-dip or 2,000 ppm talc promotes rooting, but only at a low percentage. Cutting material should be taken from young trees only.

Selected cultivars are side-veneer grafted and weeping forms are top-grafted (whip and tongue) on established seedling rootstock during the winter.

Adult larch trees have been successfully micropropagated by shoot-tip culture and plantlets that are acclimatized (251).

Laurel (Cherry, English). *See Prunus.*

Laurus nobilis. Bay Laurel. It is normally propagated by seed. Seed germination is best with mechanical scarification, followed by a 30-day cold stratification. Rooting of stem cuttings should be carried out in the summer (Israel), which is the period of most active growth. Semi-hardwood cuttings from hedged trees root better than softwood cuttings (342).

***Leptospermum* spp.** Tea Tree. Some species of this Australian native must be propagated by seed. Cultivars of other species, such as *L. scaparium,* can be readily propagated by cuttings.

***Leucophyllum* spp.** *L. frutescens, L. candidum* (Texas Sage, Purple Sage). Easily propagated from seed, but cuttings are preferred. Softwood, succulent cuttings from 8 to 10 cm (3 to 4 in) are treated with 8,000 ppm IBA talc, and propagated under mist in a well-drained propagation medium (e.g., 100 percent perlite). Cuttings will root in 3 to 4 weeks under high light and a propagation medium temperature of 20 to 22°C (70 to 72°F).

Rooting is enhanced when leaf tissue of stockplants is greater than 2 percent nitrogen. For *L. candidum* 'Silver Cloud,' stock plants are cut back to force new growth (244). In June, softwood cuttings are treated with 7,500 ppm IBA, stuck in perlite media, maintained at 25 to 30°C (77 to 86°F) and rooted under mist. Go light on the mist. Forced softwood cuttings collected later in summer (Arizona) are treated with 2,500 ppm IBA (244). Diseases after rooting cause major problems with survivability. It is difficult to transplant, so the best practice is to direct-stick (root) in small liner pots.

***Leucospermum* spp.** Leucospermum, Protea. *L. conocarpodendron* x*L. cuneiforme* 'Hawaii Gold,' which is a cut-flower protea. Leucospermum is micropropagated (256). See *Grevillea* for other proteaceous plants that can be micropropagated.

***Leucothoe* spp.** Switch Ivy, Drooping Leucothoe (*L. fontaneisana*), Coast Leucothoe (*L. axillaris*). Evergreen shrubs with white, pitcher-shaped flowers. Because of the small seed size and light requirement to germinate, seeds should not be covered with media during propagation. Seeds can be germinated at 25°C with continuous light (50). Shrubs can be asexually propagated by stem cuttings from June to December (southern United States). Treating with 1,000 to 3,000 ppm IBA will hasten rooting but is optional (131). Terminal cuttings 10 to 13 cm (4 to 5 in) long are taken, and leaves stripped from the basal end 2.5 cm (1 in). Cuttings root in 10 to 12 weeks and can be transplanted after rooting. Leucothoe is commercially micropropagated.

Leyland Cypress. *See xCupressocyparis.*

***Ligustrum* spp.** Privet. Seed propagation is easily done. The cleaned seed should be stratified for 2 to 3 months at 0 to 10°C (32 to 50°F) before planting. Hardwood cuttings of most species planted in the spring root easily, as do softwood cuttings in summer under mist. Japanese privet (*L. japonicum*) is somewhat difficult to start from cuttings, but good results were obtained with actively growing terminal shoots rather than more mature wood. Treat with a 2,500 ppm IBA quick-dip (38).

Lilac. *See Syringa.*

Linden. *See Tilia.*

***Lindera* spp.** Spicebush. *L. obtusiloba* (Japanese spicebush) is propagated by cuttings quick-dipped with 4,000 ppm NAA or 8,000 ppm IBA (131). *L. benzoin* [(American) spicebush] is native to the U.S. eastern seaboard, west to Kansas. Propagation by cuttings is difficult, but propagation by seed is successful with a cold stratification treatment of 3 months (330). Lindera is micropropagated.

Liquidambar styraciflua. Liquidambar, American Sweet Gum. Propagation is usually by seeds, which are collected in the fall. Seeds are stratified for 1 to 3 months at about 4°C (40°F) to overcome seed dormancy. Sweet gum can also be propagated by stem cuttings treated with 4,000 to 8,000 ppm IBA—overwintering may be a problem. Girdling of 10-year-old *L. formosana* prior to taking cuttings, and IBA treatment enhanced rooting (195). Selected clones are grafted or T-budded from spring through fall onto *L. styraciflua* seedlings. Sweet gum is commercially micropropagated (58, 59, 270, 401).

The Mouth and Foot Painting Artists.

***Malus* spp.** Crab Apple (Flowering). Crab apples can be produced from seed, rooted as cuttings, grafted or budded, and some cultivars are commercially produced from micropropagation.

Seed. Four species of crab apples—*M. toringoides, M. hupehensis, M. sikkimensis,* and *M. florentina*—will reproduce true from seed. Seed is also used for rootstock production for grafting. Seed should be cold stratified for 60 days, but some species require longer durations up to 120 days.

Cuttings. Selected forms of all other crab apple species, such as *M. sargentii, M. floribunda,* and *M.* 'Dolgo,' should be *clonally propagated by asexual methods,* which results in a stronger, more uniform crop. *M.* 'Hopa,' 'Almey,' and *M. xeleyi* are commercially propagated with July semi-hardwood cuttings (Florida), which are wounded at the base, quick-dipped in 5,000 ppm K-IBA, and rooted under mist within 4 to 6 weeks (106). Hardwood cuttings of crab apples are difficult to root. Establishing selected crab apple cultivars and species on their own root system with softwood cuttings is cheaper and avoids rootstock suckering, crooks in the trunk, and graft incompatibility associated with traditional grafting and budding systems (66). Softwood cuttings taken in late spring root when treated with 3,000 to 8,000 ppm IBA in talc (428) or 2,500 to 10,000 ppm IBA liquid under mist (123).

Grafting. Nursery trees are commonly propagated either by root grafting, using the whip graft, or by T-budding and chip budding (see Fig. 13–10) rootstock in the nursery row. Budding is done either in spring or summer (428). Summer budding (late summer in Oregon) is considered by most nursery people to be the faster and most desirable method of propagating crab apples. Various seedling rootstocks have been used, such as *M. xdomestica* (common apple), *M. baccata, M. ioensis,* and *M. coronaria.* Hardiness and suckering can be a problem with some seedling rootstock.

To eliminate suckering, clonal apple rootstock—'EMLA 111' and semi-dwarfing 'M7a'—are used. They also give better anchorage than own-rooted crabs and give a moderate improvement in root-hardiness (Oregon). While T-budding generally works well on domestic apple rootstock, chip budding is preferable with 'EMLA 111' (428). Crab apples are commercially micropropagated (382).

***Mandevilla* spp.** Mandevilla 'Alice DuPont.' Take one-node cuttings and trim off half of the leaf surface and dip cuttings in 2,500 ppm IBA. In Alabama, cuttings are rooted in July, rooted liners planted in October, and plants finished off and sold in 3.7-liter (l-gal) containers in April.

Manzanita. *See Arctostaphylos.*

Maple. *See Acer.*

***Melaleuca* spp.** The seeds of these native Australian species are almost dustlike but germinate easily and can be handled like eucalyptus seed.

Mesquite. *See Prosopis.*

Metasequoia glyptostroboides. Metasequoia, Dawn Redwood. Seeds germinate without difficulty, and both softwood and hardwood cuttings will root (98). The leafless hardwood cuttings may be lined-out in the nursery row in early spring. Leafy cuttings root easily under mist if taken in summer and treated with 3,000 to 8,000 IBA talc or quick-dip.

Cultivars can be grafted and weeping forms are top grafted on seedling rootstocks.

Mimosa. *See Albizzia.*

Mock Orange. *See Philadelphus.*

Monkey Puzzle Tree. *See Araucaria.*

***Morus* spp.** Mulberry. *M. alba* (fruitless or white mulberry) and *M. nigra* (black mulberry). Some mulberry trees produce only male flowers and hence do not bear fruits. Seeds need to be separated from fruits; some seeds have dormant embryos. Seeds collected in early summer, cleaned, and sown have high percent germination. Stored seeds, cold stratified for 1 to 3 months, can be sown in the spring. Roots easily from semi-hardwood (summer cuttings) and hardwood cuttings treated with 8,000 ppm talc or 10,000 to 15,000 ppm IBA quick-dip. *M. nigra* is more difficult to propagate, is clonally regenerated using hardwood cuttings treated with 4,000 to 5,000 ppm IBA and bottom heat or softwood cuttings treated with 4,000 to 8,000 ppm IBA (421). *M. nigra* can be T-budded and chip budded (421). *M. alba* 'Pendula' is grafted on a standard of *M. alba* 'Tatarica' to produce a small weeping tree (131). The fruitless *Morus alba* 'Chaparral' is also grafted. Mulberry is commercially micropropagated, and *M. nigra* microcuttings have high rooting and a 90 percent survival (227).

Mountain Ash. *See Sorbus.*

Mountain Laurel. *See Kalmia.*

Mulberry (Fruitless). *See Morus.*

Mussaenda. *See Mussaenda* spp.

***Mussaenda* spp.** Mussaenda. Ornamental shrubs indigenous to the Pacific Islands, Asia, and Africa; they are important ornamental species in the Philippines. Low pollen fertility and poor fruit production limit seed propagation. They are difficult to propagate by cuttings, but rooting of *M. erythrophylla* 'Ashanti Blood' stem cuttings was enhanced with 3,000 ppm IBA (390).

***Myrica* spp.** Bayberry, Wax myrtle. Propagated by seed or by cuttings. With northern wax myrtle (*M. pennsylvanica*) and southern wax myrtle (*M. cerifera*), the wax must be removed from the seed for germination to occur. The wax coating may be removed by rubbing the seed over a screen. One propagator uses a hot water soak with several drops of mild detergent to help remove the wax. A 3-month cold stratification is recommended. Kinetin and gibberellic acid treatments are reported to enhance germination of northern wax myrtle (*M. pennsylvanica*).

M. cerifera is fairly easy to root under mist using a 2,000 ppm IBA quick-dip (Texas). *M. pennsylvanica* can be easily rooted from softwood cuttings (Pennsylvania) taken early before the terminal bud has formed—a 5,000 ppm IBA quick-dip is beneficial.

Myrtle. *See Myrtus.*

***Myrtus* spp.** Myrtle. Crop production of *M. communis* takes longer by seed than by cuttings. Seedling plants produce a special swollen root (lignotuber) that permits myrtle to survive drought conditions, disease, and insect damage. Plants produced by cuttings do not produce the lignotuber, and are not as drought-resistant as seed-propagated plants. The species is most commonly propagated from 8 to 15 cm (3 to 6 in) shoot-tip cuttings taken in August (Pennsylvania). A quick-dip of 5,000 ppm IBA enhances rooting. Seasonal variation occurs in the rooting of cuttings. Rooting is much higher in cuttings taken during December–February (Israel) than May–August (250). Micropropagation has been obtained with explants from mature field-grown plants (308).

Nandina domestica. Nandina, Heavenly Bamboo.

Seed. Can be propagated by seed. The embryos in the mature fruits are rudimentary but will develop in cold storage. Seeds can be collected in late fall, held in slightly moist storage at 4°C (40°F), and then planted in late summer. Germination occurs in about 60 days. Germination tends to take place in autumn regardless of planting date. No cool moist stratification period is necessary.

Cuttings. Dwarf nandina ('Compacta nana,' 'Purpurea Nana,' 'Gulf Stream,' 'Harbour Dwarf,' and 'Moon Bay') are easy to root (26). Shoot tips (without brown wood) are trimmed to 4 cm (1.5 in) lengths, stripped of bottom leaves, quick-dipped in 1,250 ppm IBA and 500 ppm NAA, and stuck directly into 6 cm (2.2 in) liner pots. Cuttings can be rooted any time of the year (Texas), except during the spring growth flush. Winter rooting requires bottom heat. 'Harbour Dwarf' and 'Compacta' can be propagated by separation of suckers at the base; this technique often results in suckering in the liner pot with new shoots, more rapidly producing a salable plant. Suckering and more rapid plant development in the liner pot can also be promoted by rooting a cutting with one node under soil; when the liner is transplanted, it is buried deep enough to cover the next node up the stem, which results in suckering from one or both of the nodes.

Micropropagation. Nandinas are commercially micropropagated (62).

Nerium oleander. Oleander. Seedlings reproduce fairly true-to-type, although a small percentage of plants with different flower colors will appear. The seeds should be collected in late fall after a frost has caused the seed pods to open. Rubbing the seeds through a coarse mesh wire screen removes most of the fuzzy coating. The seeds are then planted immediately in the greenhouse in flats without further treatment. Germination occurs in about 2 weeks. Leafy cuttings root easily under mist if taken from rather mature wood during the summer and treated with a 3,000 ppm IBA quick-dip. Plant parts are very poisonous. Can be micropropagated (310).

Norfolk Island Pine. *See Araucaria.*

Nyssa sylvatica. Black Gum, Black Tupelo, Pepperidge Tree. One of the best trees for dependable fall color, even in mild climates. Propagated by seed. A high seed germination occurs with a 3-month cold stratification. Softwood cuttings root when treated with 8,000 ppm IBA talc and 21°C (70°F) bottom heat. *N. aquatica* and *N. ogeche* can be propagated from subterminal, softwood cuttings treated with a liquid application of 2,400 ppm NAA and 130 ppm IBA. *Nyssa* can be grafted. There have been a number of new cultivars developed that are budded onto seedling rootstock.

Oak. *See Quercus.*

Olea europaea. Olive. A fruitless cultivar, 'Swan Hill,' is available for use as a patio or street tree; it is usually grafted on *O. oblonga*. Cuttings are difficult to root. They should be placed under mist and treated with IBA at 2,000 to 3,000 ppm, or they can be grafted on easily rooted cultivars used as a rootstock (197).

O. europaea 'Wilsonii' and 'Majestic Beauty' are also fruitless and root more easily.

Oleander. *See Nerium.*

Olive. *See Olea.*

***Opuntia* spp.** Cactus (Prickly pear). *See Opuntia* spp. in Chapter 19.

Orchid (anacho) Tree. *See Bauhinia.*

Osage-Orange. *See Maclura.*

***Osmanthus* spp.** Sweet olive, Fragrant tea olive (*O. fragrans*). Seeds are slow and difficult to propagate. Semi-hardwood cuttings or those with firm wood are used. IBA at 3,000 to 8,000 talc or quick-dip enhances rooting. Softwood cuttings of *Osmanthus xfortunei* treated with 2,500 IBA root well (47). In Alabama, *O. fragrans* cuttings root best when taken from semi-hardwood, new-growth stock in early August and quick-dipped in 15,000 ppm IBA—the bigger the caliper of the cutting, the better it roots. Semi-hardwood and hardwood cuttings of *O. heterophyllus* 'Ilicifolius' and hardwood cuttings of 'Rotundifolius' root in high percentages without auxin (48).

Oxydendrum arboreum. Sourwood. Sourwood is propagated commercially by seeds. Fall-harvested seeds need no pretreatment, and flats containing seeds are generally placed under continuous light. Stratification of 2 to 3 months speeds germination and decreases the light required for germination (28). Propagation by cuttings is extremely difficult. However, soft-tip cuttings treated with a 2,000 to 3,000 ppm IBA quick-dip can be rooted (Pennsylvania). Sourwood is commercially micropropagated (19).

Pachysandra terminalis. Japanese spurge. This evergreen ground cover for shady areas spreads naturally by rhizomes. Can be propagated easily by division or by leafy cuttings under mist without auxin. *Pachysandra* is commercially micropropagated.

Paeonia suffruticosa. Tree Peony. Seed propagation is complicated by epicotyl dormancy. Seeds are sown in pots and exposed to a warm-moist stratification of 18 to 21°C (65 to 75°F) for root/hypocotyl growth, then cold stratified to break epicotyl dormancy. Further epicotyl growth occurs under warm conditions. Selected cultivars are propagated by grafting in late summer on herbaceous peony (*P. lactiflora*) roots as the rootstock. The grafts are covered and callused in a sand-peat medium in a greenhouse until fall, when they are potted. In Oregon, peonies are cleft grafted in early August with a fresh, single-edge razor blade for each graft. The newly grafted units are packed in plastic-lined crates containing slightly moist peat, and maintained at 27°C (80°F) for 2 to 4 weeks until the graft union has formed. The species has been micropropagated (57).

Palms. There are numerous species and genera of ornamental palms. See *Phoenix dactylifera* (Date Palm) in Chapter 19.

Seed (67, 86). They are propagated with fully mature seed, as indicated by color changes from green to red, yellow, black, etc., depending on species. It is best to use only fresh seed, which should be planted as soon as possible after harvesting and not allowed to dry out. Generally, palm seeds remain viable for only a short period. Seed should be collected from trees and not taken from the ground. It is recommended that the fleshy seed coat be removed, in part because the coat often contains an inhibitor. Soak the cleaned seed for 3 days prior to planting, changing the water daily (341). Any seeds that float should be discarded. Presoaking seed for 1 to 2 days in water containing a fungicide and insecticide can also speed germination time and the percentage of germination. Palm seeds are susceptible to surface molds and should be protected by dusting with a fungicide.

For storing palm seed, fresh, cleaned seed should be conditioned for 2 days to 85 to 90 percent RH, then dusted with a seed protectant fungicide (e.g., thiram), tightly sealed in heavy polyethylene bags, and stored at 23°C (73°F) (135).

Seed germination of some species can be accelerated by scarification, followed by soaking in gibberellic acid at 1,000 ppm for 72 hours—and seed flats placed over bottom heat at 27°C (81°F) (305). *Rhapidophyllum hystrix* seed can be scarified under sterile conditions by removing the seed coat and cap that cover the cavity containing the embryo (79). High germination was obtained from scarified seeds of *R. hystrix*, which were stored for 12 months at 5°C (41°F) and 100 percent relative humidity; high viability of the stored seeds was obtained when the seeds' moisture content was 14 to 36 percent (78). Successful germination also occurs with unscarified seeds of *R. hystrix*—but the trick is to fully hydrate the seeds and expose them to alternating temperatures. Seeds are soaked for 7 days with aerated and/or running tap water at 30°C (86°F), then seeds are germinated in moist peat moss with alternating 40°C (104°F) for 6 hours and 25°C (77°F) for 18 hours daily; the fully hydrated embryos can readily penetrate the thick-walled seed coats, but need the high temperature stimulus to promote embryo growth (78, 79).

Seeds of most species germinate in 1 to 3 months, especially if bottom heat, 28°C (80°F), is maintained, but some may take as long as 1 to 2 years. Some palm species have limited optimum temperatures for seed germination, and temperatures above and below the optimum level contribute to irregular and low percentage germination. Four native Florida palms had optimum germination temperatures of 35°C (95°F) (75). As a general rule, germination temperatures should be maintained between 29 and 35°C (85 and 95°F). With *Butia capitata,* seeds require 90 to 150 days after-ripening at 5 to 25°C (41 to 77°F) before sowing (76). After ripening, it is followed by a 40°C (104°F) germination temperature, increased seed germination and reduced germination time. Removing the endocarps of *B. capitata* was reported to be more effective in enhancing germination rate than after-ripening storage of endocarps (63). Since each endocarp contains 1 to 3 seeds, removing the endocarp also increased the overall number of seedlings [i.e., from 100 endocarps more than 100 seedlings would emerge (63)]. Endocarp removal was not successful when seeds were germinated in poly bags filled with damp sphagnum peat but worked fine in seed flats containing 1 sphagnum peat:1 perlite (v:v).

Division. Multiclumping palms, such as some date palm species, can be asexually propagated by removing and rooting offshoots or basal suckers.

Micropropagation. The date palm (*Phoenix dactylifera*) (409), ponytail palm (*Beaucarnea recurvata*) (360), and Canary Island date palm (*Phoenix canariensis*) (221) can be micropropagated.

Parrotia persica. Persian Parrotia. Lack of available seed is a problem, as is a long warm/cold stratification requirement. Easily propagated from semi-hardwood cuttings treated with 1,000 to 4,000 ppm IBA. It is commercially micropropagated.

***Parthenocissus* spp.** Virginia Creeper (*P. quinquefolia*), Boston Ivy (*P. tricuspidata*). These two ornamental vines can be propagated by seeds planted in the fall or stratified for 2 months at about 4°C (40°F) before planting in the spring. Leafy softwood cuttings taken in late summer root easily under mist when treated with a 3,000 ppm IBA talc, as do hardwood cuttings planted outdoors in early spring. Grafting of named cultivars on *P. quinquefolia* seedlings or rooted cuttings is done by some nurseries using the whip or cleft graft.

Passiflora xalatocaerulea. Passion Vine. This tender, subtropical vine is propagated by leafy cuttings under glass or mist. *Passiflora mollissima, P. tricuspsis,* and *P. nov* sp. can be seed propagated. Presoaking seed for 24 hours with 400 ppm gibberellic acid and removing the apical point enhanced germination (120). A combination of mechanical and physiological dormancy occur in these three species.

Paxistima myrsinites. Oregon Boxwood. An evergreen shrub native to the Pacific Northwest. Readily propagated from semi-hardwood cuttings from mid-summer until bud-break in spring (Vancouver, B.C.). Cuttings are treated with 8,000 ppm IBA talc, placed under mist, and maintained with a minimum basal temperature of 21°C (70°F). Can also be rooted in fall and winter under contact polyethylene film (410).

Pawpaw. *See Asimina.*

Pear (Callery cvs. Bradford, Capital, Whitehouse, etc.). *See Pyrus.*

Pear (Evergreen). *See Pyrus.*

Penstemon fruticosus. Purple Haze Penstemon. A low, compact, evergreen to semi-evergreen ground cover (subshrub) with tubular, mauve-purple flowers. Propagated with softwood and semi-hardwood cuttings (June through September in Vancouver); 3,000 ppm IBA talc is optional. Requires a well-drained container soil. Penstemon is commercially micropropagated.

Peony (Tree). *See Paeonia.*

Persian Parrotia. *See Parrotia.*

Persimmon (Texas). *See Diospyros.*

***Philadelphus* spp.** Mock Orange. The many cultivars of mock orange are best propagated by cuttings, which root easily. Hardwood cuttings (late winter) treated with 2,500 to 8,000 ppm IBA or softwood cuttings treated with a 1,000 ppm IBA quick-dip can be rooted under mist. Removing rooted suckers arising from the base of old plants is an easy means of obtaining a few new plants.

Phoenix dactylifera. Date palm. See Chapter 19.

Photina arbutifolia. *See Heteromeles.*

***Photinia* spp.** Photinia. Large evergreen shrubs with small pome fruits. *P. xfraseri* (Fraser or red-tip photinia) can be propagated by seed exposed to a 2-month cold stratification. Semi-hardwood cuttings are best rooted with a 10,000 ppm IBA quick-dip (38) or 8,000 ppm talc. Wounding cuttings, trimming leaves of cuttings, and 3,000 to 8,000 ppm IBA talc or quick-dips are optional for *P. xfraseri* 'Red Robin,' which is a hybrid of *P. glabra* and *P. serrulata* (97, 122). Photinia is commercially micropropagated.

Picea **spp.** Spruce.

Seed. Ordinarily propagated without difficulty by seed, most species of spruce have embryo dormancy, requiring a 1- to 3-month cold stratification at 4°C (40°F) for good germination. Spruce can also be fall-planted and naturally stratified. Seeds of *P. abies, P. engelmannii, P. glauca* var. *albertiana,* and *P. omorika* germinate well without stratification. Colorado blue spruce (*P. pungens* 'Glauca') grown from seed produces trees with a slight bluish cast. Only a small percentage of the seedlings have the very desirable bright blue color. Several exceptionally fine blue seedling specimens have been selected as clones and are perpetuated by grafting. Three of the best-known grafted blue spruces are Koster blue spruce (*P. pungens* 'Koster'), the compact 'Moerheim' blue spruce (*P. pungens* 'Moerheimii'), and *P. pungens* 'Hoopsii'—which is considered the bluest form.

Cuttings. Selected clones of spruce are difficult to propagate by cuttings, but there are instances in which good percentages of cuttings have rooted (180, 226), especially from selected young source trees. In general, pyramidal forms are difficult-to-root, and are usually grafted or seed propagated. Dwarf and more prostrate, spreading forms are easier to root from cuttings.

Taking cuttings from vigorous containerized trees gives good results. Cuttings taken in spring, midsummer, and mid-autumn have been rooted. For cuttings, it is best to use only shoot terminals, which should be gathered in early morning when the wood is turgid. Wounding, mist, and high light irradiance during rooting are helpful. *P. glauca* and *P. pungens* have been successfully rooted with 3,000 to 10,000 ppm IBA treatments (131). In making cuttings of upright-growing types, terminal shoots should be selected rather than lateral branches, since the latter, if rooted, tend to produce prostrate, sprawling plants rather than the desired upright form.

Grafting. The Koster blue spruce is propagated commercially by grafting scions on Norway spruce (*P. abies*) seedlings. In Canada, bench grafting of *P. abies* on seedling rootstock is done with a side-veneer or apical-wedge graft. Most grafting is done in winter, but there are advantages to late summer graftage. In Oregon, blue spruce (*P. Pungens* 'Hoopsi') is grafted on containerized, dormant *P. abies* rootstock and overwintered in unheated structures; soaking the scion bases in 200 ppm IBA for 3 minutes increased grafting success (33).

Micropropagation. Clones of *P. glauca* and *P. pungens* can be vegetatively micropropagated by somatic embryogenesis (87, 228). Meristem micrografting has been done with *P. abies in vitro* (298).

Pieris **spp.** Pieris. *P. floribunda* and *P. japonica* reproduce readily by seed with no treatment necessary. During sowing, seeds should not be covered with media, since they require light to maximize germination and are relatively small [210,000 seeds per 28 g (1 oz)] (393). Some Pieris species can be started easily by cuttings with a 1,000 ppm IBA quick-dip, but *P. floribunda* cuttings are difficult to root. Rooting under mist or polyethylene-covered frames is enhanced by wounding, IBA talc, or quick-dips at 5,000 to 8,500 ppm, and bottom heat. Rockwool propagating sheets are reported to reduce transplant shock (13). Pieris is commercially micropropagated (393).

Pileostegia viburnoides. A self-clinging evergreen climber in the Hydrangeaceae family. Produces panicles of creamy-yellow flowers. Can be rooted from cuttings using talc applications of 8,000 ppm IBA.

Pine. *See Pinus.*

Pinus **spp.** Pine.

Seed. Pines are ordinarily propagated by seed. Considerable variability exists among the species in regard to seed dormancy conditions. Seeds of many species have no dormancy and will germinate immediately upon collection, whereas others have embryo dormancy. With the latter, stratification at 0 to 4°C (32 to 40°F) for 1 to 3 months will increase germination. Moist perlite as a stratification medium will enhance overall germination. *P. cembra* may have immature embryos, so a warm stratification of 21 to 27°C (70 to 80°F) for 2 to 3 months, followed by a cool stratification for 3 months at 2°C (36°F), will aid germination.

Species whose seeds have no dormancy conditions and can be planted without treatment include *Pinus aristata, P. banksiana, P. canariensis, P. caribaea, P. clausa, P. contorta, P. coulteri, P. edulis, P. halepensis, P. jeffreyi, P. latifolia, P. monticola, P. mugo, P. nigra, P. palustris, P. pinaster, P. ponderosa, P. pungens, P. radiata, P. resinosa, P. roxburghii, P. sylvestris, P. thunbergii, P. virginiana,* and *P. wallichiana.* However, if seeds of the above species have been stored for any length of time, it is advisable to give them a cold stratification period before planting. Pine seeds can be stored for considerable periods of time without losing viability if held in sealed containers between −15 and 0°C (5 and 32°F). Seeds should not be allowed to dry out.

Cuttings. Pine cuttings are difficult to root, although those of mugo pine (*Pinus mugo*) root easily if taken in early summer (246), and selected clones of *P. radiata* are commercially rooted. Success is more likely if cutting material is taken in winter from low-growing lateral shoots on young trees. Treatment with IBA is beneficial (269).

Considerable study has been given to the rooting of cuttings of Monterey pine (*P. radiata*) because of its importance as a timber crop in New Zealand and Australia, and as a Christmas tree species. There are clonal variations in the commercial rooting of this species. Best rooting is from cuttings taken in early winter. Wounding, plus a 4,000 ppm IBA quick-dip, enhances rooting. A more symmetrical root system could be induced by clipping the ends of the original roots and allowing a root system to develop from the secondary roots.

Rooting of *P. strobus* was improved when stock plants were etiolated and shoot bases covered with black adhesive tape. Accelerated growth techniques of supplementary lighting, elevated CO_2, temperature manipulation, and optimum fertility enhance rooting in *P. echinata, P. thunbergiana,* and *P. elliottii* (194).

Rooting needle fascicles (179). Pines can be propagated asexually by rooting needle fascicles (needle leaves held together by the scale leaves, containing a base and a diminutive shoot apex). Rooting is best when the fascicles are taken from trees less than 4 years old. IBA treatments are helpful. By selecting certain seedlings whose cuttings root easily and by using critical timing in taking cuttings, it is possible to select clones in which cutting propagation is commercially feasible, as has been shown to be true for the mugo pine (179). Cuttings of the Scotch pine have been rooted by a unique method of forcing out interfascicular shoots from young stock plants. These shoots, when made into cuttings, root with a high percentage of success.

Grafting. Clonal regeneration of pines is usually done via grafting. Side-veneer grafting is used for propagating selected clones. Well-established 2-year-old seedlings of the same or closely related species should be used as rootstocks. Scions should be of new growth, taken from firm, partly matured wood. Winter grafting in greenhouses works well in Oregon. *P. cembra* is side-veneer grafted on *P. strobus,* and selected forms of *P. densiflora* are grafted on seedling *P. densiflora* or *P. sylvestris.*

Micropropagation. Selected pine species can be micropropagated (62, 374, 442, 446).

Pistache (Chinese). *See Pistacia.*

Pistacia chinensis. Chinese Pistache. Commercially propagated from seed, which results in highly variable plant form and fall leaf color. Seeds should be collected from relatively large fruits, blue-green in color, in mid-fall. Pulp must be removed. Soak fruits in water and then rub over a screen. Seeds that float in water have aborted embryos and should be discarded. Stratification at 4 to 10°C (40 to 50°F) for 10 weeks gives good germination. Seedlings exhibit a wide range of variability.

Propagation by cuttings is very difficult. Some rooting success was obtained with softwood cuttings treated with a quick-dip of 5,000 ppm IBA + 5,000 ppm NAA (142). T-budding selected clones on seedling *P. chinensis* rootstocks in late summer is used to produce uniform, superior trees (233), but they can also be propagated by mound layering (141). Male Chinese pistache cuttings should be collected from green softwood or red semi-hardwood stems when about 380 to 573 degree days have accumulated after orange budbreak (143). Cuttings taken in May (Oklahoma) and treated with 8,750 ppm IBA had the greatest rooting. Limited success has occurred with the micropropagation of *P. atlantica* (324).

***Pittosporum* spp.** Pittosporum. Can be started by seeds or cuttings. The seeds are not difficult to germinate; dipping a cloth bag containing the seeds in boiling water for several seconds may hasten germination. Cultivars are propagated with leafy semi-hardwood cuttings. An IBA quick-dip or talc application of 1,000 to 3,000 ppm is beneficial to rooting of *P. tobira*—and 6,000 to 8,000 ppm IBA for *P. tenuifolium.*

Plane Tree. *See Platanus.*

***Platanus* spp.** Plane Tree, Sycamore.

Seed. Seeds are ordinarily used in propagation, but they should not be allowed to dry out. The best procedure is to allow the seeds to overwinter in the seed balls right on the tree. They may be collected in late winter or early spring and planted immediately. Germination usually occurs promptly. If the seeds are collected in the fall, then stratification at about 4°C (40°F) should be used.

Cuttings. The hybrid London plane tree, *Platanus xacerifolium,* can be propagated by hardwood cuttings taken and planted in the nursery in autumn or by leafy softwood cuttings taken in midsummer and rooted under mist. Auxin treatments do not always enhance rooting (304). *P. xacerifolia* (London Plane tree) cultivars, such as 'Bloodgood,' 'Columbia,' and 'Liberty,' can be budded on seedling rootstock; the same cultivars can also be rooted from cuttings treated with 8,000 ppm IBA (131). Softwood cuttings propagated in June (Ontario, Canada) root well when wounded 1 cm (1/2 in) above the bottom node and treated with 5,000 ppm IBA (368).

Platycladus orientalis. *See Thuja.*

***Plumbago* spp.** Plumbago seeds sown in late winter usually germinate easily. Leafy cuttings taken from partially matured wood can be rooted without difficulty under mist. Root cuttings also can be used, and old plants can be divided. *Plumbargo* has been micropropagated (255).

***Plumeria* spp.** Plumeria. Leafy cuttings 15 to 20 cm (6 to 8 in) long of this tender tropical Hawaiian shrub will root readily under mist if treated with 2,500 ppm IBA.

Plums (Flowering, Mexican). *See Prunus.*

***Podocarpus* spp.** Podocarpus. These evergreen trees and shrubs have foliage resembling the related yews (*Taxus*) and make good container plants. Generally they are seed-propagated. They can be propagated by stem cuttings taken in late summer and early fall. Rooting of *P. marophyllus* is slow, even with a 3,000 to 8,000 ppm IBA talc application.

Poinsettia. *See Euphorbia.*

Poplar. *See Populus.*

***Populus* spp.** Poplar, Cottonwood, Aspen.

Seed. These trees can be propagated by seeds; they should be collected as soon as the capsules begin to open in spring, and planted at once, because they lose viability rapidly, and should not be allowed to dry out. However, if held in sealed containers near 0°C (32°F), seeds of some species can be stored for as long as 3 years. There are no dormancy conditions and seeds germinate within a few days after planting. The seedlings are highly susceptible to damping-off fungi and will not tolerate excessive heat or drying. Poplars are difficult to propagate in quantity by seed.

Cuttings (323). Hardwood cuttings of Populus root easily (except the aspens) when propagated in the spring. Treatment with IBA is likely to improve rooting. Leafy softwood cuttings (of some species at least) taken in midsummer also root well. *P. tremuloides*, quaking aspen, can be propagated by removing root pieces, root suckers, and layering—inducing adventitious shoots to form from these propagules in vermiculite and then rooting these adventitious shoots as stem cuttings under mist with IBA treatments.

Micropropagation. Many *Populus* species are micropropagated (435).

***Potentilla* spp.** Potentilla, Cinquefoil. Propagation is usually by cuttings, but seed and division can be used. Cuttings are taken from early summer through fall (168). Rooting is best under light mist with bottom heat; mist should be reduced as soon as rooting occurs. Softwood cuttings of *P. fruticosa* root readily with 1,000 ppm IBA. Potentilla is commercially micropropagated.

Privet. *See Ligustrum.*

***Prosopis* spp.** Mesquite. A landscape shrub or small tree for arid and semi-arid regions. Seed germination is hindered by external seed dormancy due to a water-impermeable seed coat. Mechanical scarification and hot-water treatment enhance seed germination of *P. alba* and *P. flexuosa* (85). Auxin has been used to enhance the rooting of *P. alba* cuttings (249). A thornless variety of Chilean mesquite (*P. chilensis*) is propagated by cuttings, and also by air-layering (42). Prosopis has been micropropagated (9).

***Protea* spp.** Protea. This South African native, popular for its cut flowers, is propagated by seed, cuttings, and grafting. Cuttings are difficult to root, but some species respond to auxin treatments (320).

***Prunus* spp.** Flowering Apricot (*P. mume*), Flowering Cherries (cultivars of *P. campanulata, P. sargentii, P. serrulata, P. sieboldi, P. subhirtella,* and *P. yedoensis*), Plums (*P. cerasifera, P. mexicana*), Almonds (*P. dulcis, P. glandulosa, P. tenella*), Cherry Laurel, English Laurel (*P. caroliniana, P. laurocerasus*). Prunus are classified as evergreen and deciduous shrubs, flowering and fruiting trees.

Seeds. Seed from most *Prunus* species require cold stratification for 1 to 3 months, but some species (like *P. cerasifera, P. cerasus,* and *P. virginiana*) may require longer stratification durations up to 5 months. If cross-pollination with other species can be avoided, *P. sargentii, P. campanulata,* and *P. yedoensis* will reproduce true from seed.

Cuttings. As a general rule, ornamental Prunus shrubs are propagated by cuttings (e.g., *P. xcistena*, cultivars of *P. laurocerasus, P. lusitanica*), while ornamental Prunus trees are grafted (e.g., cultivars of *P. cerasifera, P. serrulata*). Leafy cuttings of some of the flowering cherry species can be rooted under mist in high percentages if treated with IBA, but subsequent survival and overwintering are sometimes difficult. Rooting is enhanced in *P. tenella* semi-hardwood cuttings collected in July (Ireland), quick-dipped in 1,250 ppm K-IBA + 1,250 K-NAA; IBA alone was ineffective (243).

P. serrulata 'Kwanzan' is successfully own-rooted from softwood cuttings by wounding and using a 10,000 ppm IBA and 5,000 ppm NAA quick-dip (273). Dwarf flowering almond (*P. glandulosa*) is easily rooted from softwood summer cuttings with 1,000 ppm IBA. In Georgia, cuttings of *P. mume* are collected early in the season, when growth is just starting to firm, and treated with 3,000 ppm K-IBA (126); with later, firmer wood cuttings, no auxin, or a low dosage of 1,000 ppm, is used. Mexican plums (*P. mexicana*) are propagated by softwood cuttings (May and June in Texas), treated with 1,000 to 1,500 ppm IBA, or they are produced as seedlings. *P. caroliniana* are rooted from softwood or semi-hardwood cuttings treated with 3,000 to 8,000 ppm IBA. *P. laurocerasus* are rooted from semi-hardwood cuttings treated with 1,000 to 3,000 ppm IBA.

In Florida, *P. campanulata* (Taiwan flowering cherry) is propagated from green, softwood cuttings, preferably terminal tips, taken in April. Cuttings are trimmed to 10 to 20 cm (4 to 8 in) in length, quick-dipped in 1,000 to 3,000 ppm IBA, and stuck into thirty-eight-cell trays. Cuttings are misted only to keep the tips from wilting, and the mist should be reduced as soon as roots are 2.5 to 5 cm (1 to 2 in) in length (285).

Grafting. Cultivars of oriental cherry (*P. serrulata*) are grafted onto seedling or clonal rootstocks of the Mazzard cherry (*P. avium*); T-budding, either in the fall or in the spring, is also done. *Prunus dropmoreana* is a suitable rootstock for *P. serrulata* 'Kwanzan.' In England, *P. mume* is bench grafted in February and chip budded in August onto *P. cerasifera* rootstock (126). Seedling 'Nemaguard' peach rootstock is used in California. Mature scions of micropropagated *P. avium* can be top cleft grafted onto seedling rootstock (367).

Micropropagation. Prunus species are commercially micropropagated (58, 367, 388).

Pseudotsuga menziesii. Douglas-fir. This important lumber and Christmas tree species is normally seed-propagated, but clonal regeneration with cuttings is becoming important with Christmas tree selection and the production of elite timber trees (347). Limited seed from elite, control-pollinated plants (full-sib families) are collected and sown. Each seedling is used as a stock plant (ortet) and is sheared, and cuttings (ramets) are taken and rooted. The rooted cuttings are combined so that many clones are mixed together, thus avoiding monoculture production. **Bulking** refers to the mixing of many elite clones together in the nursery and plantation—for example, a few clonal copies are made of a large number of genotypes; when a large number of copies are produced from only one or a few genotypes, it's called **cloning** (347, 348).

Seeds. Seeds exhibit varying degrees of embryo dormancy (43). For prompt germination, it is best to sow the seeds in the fall or stratify them in moist perlite for 2 months at about 4°C (40°F).

Cuttings. Douglas-fir cuttings are rather difficult to root, but by taking them in late winter, treating them with IBA, and rooting them in a sand and peat-moss mixture, it is possible to obtain fairly good rooting. Cuttings from young trees root much more easily than those from old trees, and cuttings from certain source trees are easier to root than those from others. Cuttings are taken in December and January from sheared or hedged stock blocks, and treated with 5,000 ppm NAA + 5,000 ppm IBA. One timber company uses a 3-year propagation regime with greenhouse production of stock plants (ortets) from elite seed; the cuttings of the ortets are rooted under fog, then transplanted and finished off in an outdoor bare-root nursery (347). Rooting of cuttings for Douglas-fir Christmas trees is strongly related to clonal variation and seasonal fluctuation. NAA is more effective than IBA in stimulating rooting (333). Cuttings taken from stockplants in a juvenile condition, maintained by hedging, serial propagation, and grafting have a higher rooting success. Rooting decreases markedly with increasing stockplant age. Some success treating softer cuttings with 5,000 ppm IBA and more lignified cuttings with 10,000 ppm IBA (84). Plagiotropism (horizontal growth) can be a problem with rooted cuttings and grafts. Douglas-fir is a difficult species to propagate by cuttings.

Micropropagation. Micropropagation was successful with cotyledon explants (91).

Pygmy Pea Shrub. See *Caragana*.

Pyracantha **spp.** Pyracantha, Firethorn. Seeds require a 3-month cold stratification, however, propagation is almost always by cuttings. Partially matured, leafy, current-season's growth is taken from late spring to late fall and rooted either in the greenhouse or under mist for good results. Treatments with a 2,500 ppm IBA quick-dip are beneficial (38). Semi-hardwood cuttings collected in fall rooted well when wounded and treated with 8,000 ppm IBA talc or quick-dip (131).

Pyrus calleryana. Bradford Pear (*P. calleryana* 'Bradford'). Can be rooted with 8,000 ppm IBA talc, quick-dips of 10,000 ppm IBA + 5,000 ppm NAA, or a 10,000 ppm K-IBA. Semi-hardwood cuttings are taken June to August (after the terminal bud has formed), and rooted under mist (2, 131). There are some reports of poor performance of 'Bradford,' which is being replaced by better-performing *P. calleryana* cultivars, such as 'Capital,' 'Cleveland Select,' 'Redspire,' and 'Whitehouse.' These cultivars are grafted on seedling *P. calleryana* rootstock by T-budding. Bench grafting can also be done with a whip-and-tongue graft. 'Bradford' is commercially micropropagated (62).

Pyrus kawakamii. Evergreen Pear. Propagated by cuttings or, more commonly, by grafting on *P. calleryana* seedling rootstock. Bench grafting is done in midwinter using the cleft graft. The grafts can be planted in containers or in the nursery row.

***Quercus* spp.** Oak.

Seed. Seed propagation is generally practiced. Wide variations exist in the germination requirements of oak seed, particularly between the black or red oak (acorns maturing the second year) and white oak (acorns maturing the first year) groups. Seeds of the white oak group have little or no dormancy and, with few exceptions, are ready to germinate as soon as they mature in the fall. Seeds of the following species will germinate without a low-temperature stratification period: *Quercus agrifolia, Q. alba, Q. arizonica, Q. bicolor, Q. chrysolepis, Q. douglassii, Q. garryana, Q. lobata, Q. macrocarpa, Q. montana, Q. petraea, Q. prinus, Q. robur, Q. stellata, Q. suber, Q. turbinella,* and *Q. virginiana*. Germination of *Q. nigra* is maximized when the fruit (nut) covering is removed (3). Seeds of most species of the black oak group have embryo dormancy, requiring either stratification [0 to 2°C (32 to 35°F)] for 1 to 3 months, or fall planting to allow natural stratification.

Acorns are often attacked by weevils, but soaking in water held at 49°C (120°F) for 30 minutes will rid the acorns of this damaging pest. However, research with *Q. virginiana* indicates no commercial advantage with heat treatment for weevil control. In fact, there can be a loss of seed viability.

Seeds are usually floated in water, and those seeds that float are discarded. With seeds of stratified *Q. rubra*, the presence of a split pericarp following a ten-day aerated water-soak treatment is a nondestructive method for identifying high-quality seed before sowing (103). Acorns of many species tend to lose their viability rapidly when stored dry at room temperature. Seeds of some northern species can be stored for several years without losing viability by holding them at 1 to 3°C (34 to 37°F) in polyethylene bags. The seeds should have a 60 to 70 percent moisture level at the start of storage (154). Consider treating seeds with an insecticide or heat treatment to prevent weevil damage during long-term storage.

Oaks are strongly taprooted and seedling roots will quickly encircle containers. To obtain lateral root branching—which makes the seedlings more adaptable to transplanting—the acorns can be planted in bottomless flats or small liner containers. The bottoms of the containers are covered by a screen mesh or placed on an open wire bench, which allows for air-pruning of the taproot. The tip of the taproot, upon contacting this mesh, is air-pruned and killed, forcing development of many lateral roots. Copper-based paints are being used to coat the inside of liner pots and larger production containers to prevent root girdling (397). A system for speeding up red oak whip production in containers has been described (396).

Cuttings. Attempts to propagate oaks by cuttings or layering is frequently difficult, and species specific. Rooting of oaks depends on the species, with no consistent patterns between red and white oak groups (138). Evergreen types of *Quercus* species tend to root better than deciduous forms (201). Cuttings from younger trees root more readily than those from older trees.

Some success has been obtained in rooting leafy softwood cuttings of *Quercus robur* 'Fastigiata' under outdoor mist in midsummer after treatment with IBA at 20,000 ppm (159). Long [90 cm (35 in)], semi-hardwoood cuttings of *Q. robur* 'Fastigiata Koster' rooted well when treated with 5000 ppm IBA and propagated under high pressure fog systems. Rooting was enhanced and production time reduced with the larger propagules (389). Etiolating *Q. bicolor* and *Q. macrocarpa* stockplants can enhance rooting of softwood cuttings (5).

Commercial propagation of *Q. virginiana* and *Q. laurifolia* has been done by taking July semi-hardwood cuttings (Florida) at 6 cm (2 to 3 in), wounding, quick-dipping in 12,000 ppm K-IBA salt, and rooting under mist; rooting takes 7 to 9 weeks (106). *Q. shumardii* cuttings taken in July (Florida) are quick-dipped in 10,000 ppm K-IBA and also root in 7 to 9 weeks. Some success has occurred with serial propagation of rooted *Q. virginiana* maintained under accelerated growth techniques in a greenhouse and used as stock plants for future propagules (300). *Q. virginiana* can be rooted from rhizomic shoots produced at the crown of the tree, which are treated with a five second quick-dip of IBA in 50 percent ethanol (425).

A Florida nursery that has success with propagating *Q. virginiana* 'Cathedral Oak' and other clonal selections collects cutting wood early in the morning from 2- to 4-year old trees, since juvenility optimizes adventitious root formation (343). Rooting percentages decrease as the tree ages, even with rootable selections (clones). Cuttings are taken from branches in the first 2 m (6 to 7 ft of the stock plant), collecting only long and straight terminal tips that are 15 to 30 cm (6 to 12 in). Cuttings are taken from the first and second flush, just when the new wood and foliage hardens. The wood should have turned from green to gray (May through August in northeast Florida). During the processing, the lower 3 to 4 cm (1 to 1.5 in) of the cutting base is removed, the remaining cutting base is wounded vertically 3 to 4 cm (1 to 1.5 in) by dragging the cutting clippers along the side of the base. Only the basal 2.5 to 5.0 cm (1 to 2 in) of the cutting is quick-dipped in a 10,000 ppm K-IBA + 6,000 ppm K-NAA; these potassium salt formulations of auxin are dissolved in 3 parts distilled water and 1 part isopropyl alcohol; other clones may require 8,000 to 12,000 ppm K-IBA and 4,000 to 8,000 ppm K-NAA. Propagation media is Canadian peat and perlite (1:4, v/v). Sticking depth is 2.5 to 4 cm (1 to 1.5 in) with one cutting per container. Bottomless and/or liner containers 6.4 cm (2.5 in) in diameter and 7.6 cm (3 in) or more in depth are used for maximizing fibrous and branched root systems. The propagation polyhouses are 50 percent shaded. Mist is controlled by solar misting controllers, to automatically compensate for changes in light intensity. Once 50 to 60 percent of the cuttings have initiated roots, the mist is reduced 20 percent. *Some drought stress stimulates the callused cuttings to form primorida and start to root* (343). When a majority of cuttings have rooted, the mist is reduced another 50 percent for 4 weeks to harden-off cuttings. Then, rooted liners are watered as needed until the root system is finished as desired for potting-up. Sanitation is important with clippers periodically disinfected. During the first week of propagation, cuttings are drenched with Subdue and Captan, then biweekly with Phyton 27 (fungicide and bactericide) (343). Liquid fertilizer is applied when the cuttings begin to root.

Q. phillyreoides (ubame oak) can be rooted from softwood cuttings treated with a 3,000 ppm IBA talc (290). There are strong clonal differences in rooting success. *Q. palustris* × *Q. phellos* can be rooted from semi-hardwood stem cuttings (July in South Carolina) treated with 10,000 ppm K-IBA (201).

Girdling of stock-plant shoots prior to collecting cuttings and treating cuttings with a rooting powder of auxins, sucrose, and fungicide increased rooting (76 percent) and survival of 19- to 57-year-old water oak (196). Rooting of *Q. bicolor* and *Q. macrocarpa* was enhanced when cuttings were taken from pot-bound (root-restricted) and etiolated stock plants (29).

Forcing epicormic shoots from woody stem segments [0.5 m (1.5 ft) long with a 2.5 to 25 cm (1 to 10 in) caliper] under mist and rooting the softwood cuttings can be used to propagate *Q. bicolor* and *Q. rubra* (332). Softwood cuttings are treated the same as if collected from a field-grown stock plant. Drench weekly with Banrot 40% WP fungicide.

Grafting. Bench grafting of potted seedling rootstocks in the greenhouse in late winter or early spring is moderately successful. Side or whip-grafting is ordinarily used, with dormant 1-year-old wood for scions. Seedlings in place in the nursery row are occasionally crown grafted in the spring, after the rootstock plants start to leaf out. Scions are taken from wood gathered when dormant and stored under cool, moist conditions until used. Various grafting methods are satisfactory—whip, cleft, or bark. A modified bark graft has been successfully used for topworking established rootstock in the field (403). Budding generally has been unsatisfactory. In grafting, only seedlings of the black oak group should be used for scion cultivars of the same group and, in the same manner, only seedlings of the white oak group should be used as rootstocks for other members of the white oaks. The use of seedlings of the same species is preferable. Although some distantly related species of oak will unite satisfactorily, incompatibility symptoms usually appear later.

Quercus virginiana is the recommended rootstock for the southeastern United States. Researchers at North Carolina State University report that it tolerates the heavier wet clay soils, and temperature of the South.

Root grafting has been done with a white oak *Q. macrocarpa* scion on a *Q. robur* root piece (rootstock) (268). Root grafting with a side-veneer graft on *Q. robur* has been reported to produce plants with more uniform growth and no suckering; this technique was not successful with *Q. rubra* or *Q. palustris* (267).

Micropropagation. Single-node-stem sections of *Q. shumardii* were successfully micropropagated (37). Juvenile and mature explants of *Q. robur* have been micropropagated and rooted successfully *in vitro* (419, 420). Embryoid germination and plant regeneration have been done with *Q. bicolor* micropropagated from male catkins (178).

Quince, Flowering. See *Chaenomeles*.

Redbud. See *Cercis*.

Redwood (Coast redwood, Giant Sequoia, Sierra Redwood). See *Sequoia*.

Redwood (Dawn). See *Metasequoia*.

Raphiolepis indica. Raphiolepis, India Hawthorn. Seed germination is easy after the pulp has been removed. The cultivars of *R. indica* are commercially propagated by cuttings. *R. indica* 'Jack Evans' is best rooted when quick-dipped in 50 percent ethanol solutions of 5,000:2,000 ppm or 9,000:1,000 ppm IBA:NAA (258). Auxin response will vary with cultivar. Semi-hardwood cuttings taken from lower portions of stems rooted slightly better than those from upper portions (California). The patio tree form, *R. indica* 'Monrey,' is top grafted on clonal rootstock propagated from cuttings.

***Rhamnus* spp.** Buckthorn. Can be propagated by planting seed out-of-doors in the fall. Macerate fruits and clean seeds. Some species may germinate better if scarified for 20 minutes with sulfuric acid before sowing. *R. frangula* cultivars are propagated by softwood cuttings treated with 8,000 ppm IBA under mist.

Rhapidophyllum hystrix. Needle palm. *See Palms.*

***Rhododendron* spp.** Rhododendron, Azalea (415, 437). Azaleas are no longer a separate genus and are incorporated within the genus *Rhododendron*. *Rhododendron* are normally propagated by cuttings or tissue culture, but seeds and grafting are an option. All parts of *Rhododendron* and azalea plants are poisonous if ingested.

Seeds. Seedlings may be used as rootstocks for grafting or for propagation of ornamental species. *Rhododendron ponticum* is one of the principal rootstocks for grafting. The seed should be collected just when the capsules are beginning to dehisce, and may be stored dry and planted in late winter or early spring in the greenhouse. Seed to be kept for long periods should be put in sealed bottles and held at about 4°C (40°F). A good germination medium is a layer of shredded sphagnum moss or vermiculite over a mixture of sand and peat. The very small seeds are sifted on the surface of the medium and watered with a fine spray. The flats should be covered with glass or propagated under mist (438). Careful attention must be given to provide adequate moisture and ventilation as well as even heat: 15 to 21°C (60 to 70°F). The plants grow slowly, taking 3 months to reach transplanting size. After two or three true leaves form, they are moved to another flat and spaced 2.5 to 5 cm (1 to 2 in) apart, where they remain through the winter in a cool greenhouse or in cold frames. In the spring, the plants are set out in the field in an acid soil, and by fall they are ready to be dug and potted preparatory to grafting in the winter. Seeds of *R. carolinianum* should not be covered during propagation because of their extremely small size [825,000 seeds per 28 g (1 oz)] and light requirement for germination (52). Likewise, seeds of *R. chapmanii* (Chapman's *Rhododendron*) are easy to germinate and should be dusted on the surface of the propagation media because of their small size and light requirement to germinate (10). Seeds of *R. catawbiense* and *R. maximum* should be subjected to light to maximize germination (51).

Cuttings (184, 223, 415). Rooting cuttings is the chief method of *Rhododendron* propagation. Cuttings are best taken, midsummer to fall, from stock plants grown in full sun. However, stem cuttings—or leaf-bud cuttings—of some hybrids taken in midwinter will root well. Any flower buds should be removed from the cuttings. Treatments with IBA at relatively high concentrations are required—an IBA talc or quick-dip of 8,000 to 20,000 ppm works well, depending on cultivars. Most standard large-flowered evergreen *Rhododendron* types root readily with 8,000 ppm IBA or less; slightly more difficult cultivars respond to 10,000 ppm, and some of the more difficult red cultivars respond to 20,000 ppm IBA. Wounding the base of the cutting on both sides is a strong stimulus to rooting in *Rhododendrons* (see Fig. 10–28). A rooting medium of two-thirds sphagnum peat moss and one perlite or 1 peat:1 perlite (v/v) is suitable. Bottom heat at 24°C (75°F) should be used. *Rhododendron* cuttings are best rooted under mist in the greenhouse or in a closed polytent, and should be lifted soon after roots are well formed (about 3 months) or the roots will deteriorate.

In California, *Rhododendrons* are direct-rooted into liners (small rose pots) (362). After rooting and transplanting (into peat moss, with added fertilizers), the cuttings should be held at 4°C (40°F) for about 20 days, after which the night temperature can be raised to a minimum of 18°C (65°F). Supplementary light at this stage to extend the daylength will give good growth response. Plants started from cuttings usually develop rapidly and are free of the disadvantage of suckering from rootstock, which occurs with grafted plants.

In Alabama, spring cuttings of Azalea cultivars are cut to 7 to 10 cm (3 to 4 in) and tender tops removed. Basal ends are quick-dipped in 3,000 ppm K-IBA or 5,000 ppm K-IBA for more difficult-to-root cultivars (189).

For native azaleas, butter-soft cuttings are taken in early spring (Louisiana). Cuttings are harvested, dropped into water, placed in Ziploc bags, and labeled. Cuttings are kept refrigerated until stuck, and no rooting compounds are used. Cuttings are placed under mist. After rooting it is important to keep rooted cuttings well watered in thirty-six-cell flats under shade, fertilize them once per week, and leave roots undisturbed until shifting up into larger container during the following spring after new growth has occurred (232).

Grafting. A side-veneer graft is most successful. The best scionwood is taken from straight, vigorous current season's growth. After grafting, the plants are kept in closed frames under high humidity at a temperature near 21°C (70°F) until the union has healed. Then the plants should be moved to cooler conditions—10 to 15°C (50 to 60°F)—and the top of the rootstock removed above the graft union. After the plant has hardened, it is transplanted to the nursery row and grown for 2 years, after which it is ready to dig as a salable nursery plant. A modified chip-bud method using leaf buds in early summer (Rhode Island) was successful for grafting difficult-to-root *Rhododendron* cultivars (292).

Micropropagation. Many *Rhododendron*, evergreen, and deciduous azaleas are commercially micropropagated (6, 62, 145, 241, 262). There have been some micropropagation problems with reproducing *Rhododendrons* that are true-to-type. See Chapters 17 and 18 for discussion.

***Rhus* spp.** Sumac.

Seed. Commonly propagated by seeds, which are collected in the fall. For prompt germination the seeds should be scarified in concentrated sulfuric acid for 1 to 6 hours, depending upon the species—then either fall-planted out-of-doors or stratified for 2 months at about 4°C (40°F) before planting. Seeds of *R. virens* (evergreen sumac) need to be acid scarified with concentrated sulfuric acid for 50 minutes and then cold-stratified for 73 days (219, 407). To ensure fruiting, only plants bearing both male and female flowers should be propagated asexually.

Cuttings. Leafy softwood cuttings of some species, such as *R. aromatica*, taken in midsummer, root well if treated with 10,000 ppm IBA. For those species that sucker freely, such as *R. typhina* and *R. copallina*, root cuttings are planted in the nursery row in early spring. In Minnesota, root cuttings of *R. typhina* are trimmed to 10 cm (4 in), with a 6- to 19-mm (1/4- to 3/4-in) diameter and the ends dipped in talc containing 3,000 ppm IBA; the root cuttings are allowed to callus at 10 to 16°C (50 to 60°F)—then are stored and field-planted. It is important to maintain proper polarity of the root cuttings.

Ribes sanguineum. Red Flowering Currant. An attractive flowering deciduous shrub native to the Pacific Northwest. Easily propagated by softwood cuttings taken during summer to early September (Vancouver, B.C.), or by hardwood cuttings made with a "heel" of 2-year-old wood (410). Both cutting types do best with basal heat [21°C (70°F)] and 8,000 ppm IBA talc. Currant is commercially micropropagated.

Robinia pseudoacacia. Black Locust. Readily propagated by seeds, which are soaked in concentrated sulfuric acid for 1 hour, followed by thorough rinsing in water, before planting. A hot-water scarification followed by a 24-hour soak prior to sowing can also be used. Black locust can be propagated by root cuttings and by grafting using a whip, side-veneer, cleft, or wedge ("V") graft, (316, 427) or chip budding (215). Black locust is micropropagated (88).

***Rosa* spp.** Rose. All rose cultivars selected are propagated by asexual methods (199, 253). T-budding on vigorous rootstocks is most common, although the use of softwood or hardwood cuttings, chip budding, simultaneous budding and rooting (stenting), layering, the use of suckers, and micropropagation are also practiced.

Seed. Seed propagation is used in breeding new cultivars, in producing plants in large numbers for conservation projects or mass landscaping, and in growing seedling rootstock of certain species, such as *R. canina* and *R. multiflora*. Some commercial rose-bush producers prefer propagating rootstock by seed to avoid virus transmission through conventional asexual techniques (80).

As soon as the rose fruits ("hips") are ripe but before the flesh starts to soften, they should be collected and the seeds extracted. It is best to stratify them immediately at 2 to 4°C (35 to 40°F). Six weeks is sufficient for *Rosa multiflora*, but others—*R. rugosa* and *R. hugonis*—require 4 to 6 months, and *R. blanda* 10 months. *Rosa canina* germinates best if the seeds are held at room temperature for 2 months in moist vermiculite and then transferred to 0°C (32°F) for an additional 2 months.

Hybrid rose seeds usually respond best to a stratification temperature of 1 to 4°C (34 to 40°F) for 60 to 90 days, although some seeds may germinate with no cold stratification treatment. Germination is probably prevented in rose seeds by inhibitors occurring in the seed coverings, as well as by the mechanical restriction imposed by the pericarp (fruit) covering the seed. Acid scarification as well as macerating enzymes have promoted germination in some cases substituted for the need for warm stratification. The seeds may be planted either in the spring or in the fall in seed beds or in the nursery row. In areas of severe winters, seedlings are likely to be winter-killed if they are smaller than 10 cm (4 in).

Hardwood Cuttings. Hardwood cuttings are widely used commercially in the propagation of rose rootstocks, and to some extent for propagating the strong-growing polyanthas, pillars, climbers, and hybrid perpetuals on their own root systems. The Meidiland series of hybrid roses are own-rooted and require no budding (203). The hybrid teas and other similar ever-blooming roses also can be started by cuttings, but more winter-hardy and nematode-resistant plants are produced if they are budded on selected vigorous rootstocks.

In mild climates (Texas, California, Spain), the cuttings are taken and field propagated in the nursery in the fall (191). In areas with severe winters, cuttings may be made in late fall or early winter, tied in bundles, and stored in damp peat moss or sand at about 4°C (40°F) until spring, when they are planted in the nursery row; rootstocks are ready to bud by the following spring, summer, or fall. The cuttings are made into 15 to 20 cm (6 to 8 in) lengths from previous season's canes of 6 to 9 mm (1/4 to 1/8 in) diameter. Commercially, large bundles of canes are run through band saws to cut them to the correct length. Disbudding ("de-eying") is done in rootstock propagation; all axillary buds except the top two or three are removed to prevent subsequent sucker growth in the nursery row. The use of auxins and other rooting pretreatments are of little benefit in the commercial propagation of *Rosa multiflora* hardwood cuttings (111).

Softwood Cuttings. Softwood cuttings are made from current season's growth, from early spring to late summer, depending upon the time the wood becomes partially mature. Rooting is fairly rapid, occurring in 10 to 14 days. At the end of the season the cuttings may be transplanted to their permanent location, potted, and overwintered in a cold frame, or transferred to the nursery row for another season's growth. Some desired cultivars are budded to rooted rootstock. Cultivars of most miniature roses are easily propagated by softwood or semi-hardwood cuttings under mist.

Budding. T-budding is the method ordinarily used. The buds are inserted into 5- to 10-mm (3/16- to 3/8-in) diameter rootstock plants. In mild climates, budding can be done during a long period, from late winter until fall, but mostly in the spring. Early buds will make some growth during the summer and produce a salable plant by fall. Some propagators partially break over the top or "cripple" the rootstock about 2 weeks after budding to force the bud out. After the bud has reached a length of 10 to 20 cm (4 to 8 in), the top of the rootstock is entirely removed. In areas with shorter growing seasons, budding is done during the summer. Buds inserted late in the summer either make little growth or remain dormant until the following spring. In this case, the rootstock is cut off just above the bud in late winter or early spring, forcing the inserted bud into growth. Shoots from buds started in the fall are cut back to 13 mm (1/2 in) in the spring. The shoot then grows through the following entire summer, producing a well-developed plant by fall. After the shoot has grown about 15 cm (6 in), it is generally cut back to 5 to 7.5 cm (2 or 3 in) to force out side branches.

In California, Texas, and Spain, commercial landscape rose bush producers collect budwood in late fall (prior to digging 2-year-old rose bushes), store the budwood at 29°F (−2°C), and then T-bud in the following spring (112). The budwood is collected in late fall after the flowers are shed and the thorns become dark. The leaves are removed by hand, or "sweated-off" under high humidity, and the thorns are left intact. Sticks 15 to 25 cm (6 to 10 in) long are put up in bundles of 30 or 40 each. The bundles are wrapped as tightly as possible in moist paper and then placed in polyethylene bags.

From spring to early summer, the budwood is taken out of storage, the thorns are removed, and shield buds from the budwood are T-budded by inserting them into the medial, de-eyed portion of the original cuttings (rather than into new growth arising from the rootstock).

There has been an interest in simultaneously grafting (budding) and rooting landscape roses through bench chip budding and direct rooting in the field (112)—and using the stenting system in grafting and rooting greenhouse cut-flower roses (see Figs. 12–52 and 12–53) (414).

Rootstocks and Interstocks for Rose Cultivars (65). Most rose rootstock clones have been in use for many years, propagated by cuttings. Many of the clones are virus-infected, and will infect the cultivar top after budding. However, these clonal rootstocks are available with the viruses eliminated by heat treatments. Holding potted rose plants at a dry heat of 37 to 38°C (98 to 100°F) for 4 to 5 weeks will rid infected rootstocks of viruses. Again, seed propagation of rootstocks helps avoid most virus problems.

Rosa multiflora. This is a useful rootstock, especially in its thornless forms for landscape roses. Several cultivars have been developed, some giving better bud unions and bud development than others. Cuttings of this species root readily, develop a vigorous root system, and do not sucker excessively. It is adaptable to a wide range of soil and climatic conditions. Seedlings are used in the eastern part of the United States, Great Britain, and Australia, and cuttings are used on the Pacific Coast and Texas. The bark often becomes so thick late in the season that budding is impossible.

Rosa canina. Dog rose. Although this species has not done well under American conditions, the rootstock is commonly used in Europe. It is usually propagated by seed, which are difficult to germinate. While it is generally difficult to root, long [50 to 150 cm (20 to 59 in)] semi-hardwood cuttings of *R. canina* rooted well when treated with 5000 ppm IBA and propagated under high pressure fog systems. Rooting was enhanced and production time reduced with the larger propagules (389). The prominent thorns make it difficult to handle, and it also tends to sucker. Young plants on this rootstock grow slowly, but they are long-lived. *Rosa canina* is adaptable to drought and alkaline soil conditions, and is used as greenhouse rootstock for cutflower rose production in the Netherlands.

Rosa chinensis. 'Gloire des Rosomanes,' 'Ragged Robin.' This old French rootstock is popular in California for outdoor roses, resisting heat and dry conditions well. It is also resistant to nematodes and does not sucker if the lower buds on the cuttings are removed. This rootstock grows steadily through the summer, permitting budding at any time, under irrigated conditions. The fibrous root system is easy to transplant but requires good soil drainage. In some areas, however, it is difficult to propagate and is injured by leaf spot.

Rosa 'Dr. Huey.' This is the principal rootstock in Arizona and the southern San Joaquin Valley, California, rose districts, replacing 'Ragged Robin' to a large extent. It has also performed well in Australia (351). It is useful for late-season budding because of its thin bark. It is very vigorous and well adapted to irrigated conditions, and its cuttings root readily. It is very good as a rootstock for weak-growing cultivars. Defects include injury from subzero temperatures and susceptibility to black spot, mildew, and verticillium.

Rosa xnoisettiana 'Manetti.' This is an old rootstock, very popular for greenhouse forcing roses. It is also of value for dwarf roses and for planting in sandy soils. It is easily propagated by cuttings, produces a plant of moderate vigor, and is resistant to some strains of verticillium.

Rosa odorata (Odorata 22449). Tea rose. This rootstock is excellent for greenhouse forcing roses. Cuttings root easily under suitable conditions, and produce a large symmetrical root system. It is adapted to both excessively dry and wet soil conditions. Since it is not cold-hardy, it should be used only in areas with mild winters. Some propagation stock of this clone is badly diseased and does not root well. The plants are not adaptable to cold-storage handling, and this variety is more susceptible than *R. manetti* to verticillium.

Rosa IXL (Tausendschon/Veilchenblau). This interstock is used primarily as a trunk for tree roses. It is very vigorous and has no thorns. The canes tend to sunburn and are somewhat susceptible to low-temperature injury.

Rosa Multiflore de la Grifferaie. This interstock is useful as a trunk for tree roses, producing desirable straight canes. It is vigorous, extremely hardy, and resistant to borers, but very susceptible to mite injury.

Rosa rugosa. This form, which is used as a rootstock, bears single, purplish-red flowers. It is propagated by cuttings for bush roses, and by seed for tree roses. The root system is shallow and fibrous and tends to sucker badly, but the plants are very long-lived. It is also used as the upright stem in producing standard (tree) roses.

Propagating Tree (Standard) Roses. A satisfactory method of producing this popular form of rose is to use *Rosa multiflora* as the rootstock, which is budded in the first summer to *IXL* or, preferably, the *Grifferaie* interstock (see Fig. 11–12). These plants are trained to an upright form and kept free of suckers. In the second summer, at a height of about 0.9 m (3 ft), three or four buds of the desired flowering cultivar are inserted into the interstock trunk. During the winter, the cane above the inserted buds is removed. The buds develop the following summer, as do buds from the rootstock, which must be removed. In the fall, the plants may be dug and moved to their permanent location. Tree roses are sometimes dug as balled and burlapped plants because an extensive root system is formed during the 2 years in which the top is being developed.

Propagation of Miniature Roses (299). Soft or semi-hardwood cuttings are taken year round and rooted under mist, after dipping in an IBA quick-dip or talc. Miniature rose cultivars are especially bred for their ease of rooting.

Micropropagation. Selected rose cultivars are commercially micropropagated (198, 423).

Rose of Sharon. *See Hibiscus.*

Rosmarinus officinalis. Rosemary. This ground cover with aromatic leaves is easily propagated by seed and leafy cuttings under mist.

Rosemary. *See Rosmarinus.*

Rosewood, Arizona. *See Vauquelinia.*

Royal Poinciana. *See Delonix.*

Rubber Plant. *See Ficus.*

Russian Olive. *See Elaeagnus.*

Sage (Texas, Purple). *See Leucophyllum.*

***Salix* spp.** Willow.

Seed. Seeds must be collected as soon as the capsules mature (when they have turned from green to yellow) and planted immediately, since they retain their viability for only a few days at room temperature. Even under the most favorable conditions, maximum storage is 4 to 6 weeks. Willows are difficult to propagate in quantity by seed.

Cuttings. *Willows root so readily by either stem or root cuttings that there is little need to use other methods.* Hardwood cuttings planted in early spring root promptly. Summer cuttings from hardened new growth also perform well.

Grafting. Top grafting of *Salix* standards (*S. caprea* 'Kilmarnock,' *S. purpurea* 'Pendula,' and *S. integra* 'Hakuro-nishiki') is done with a whip-and-tongue or side-veneer graft (261). Other miniature standard *Salix* are also top-grafted on unrooted rootstock (279). The rootstock is *Salix xsmithiana,* which are clonally propagated from hardwood cuttings. After bench grafting, the unrooted, grafted cuttings are dipped in IBA talc, stuck directly in pots, and rooted under mist in a greenhouse (February in Canada) (261).

Micropropagation. *Salix* can be micropropagated (303).

***Sambucus* spp.** Elderberry. Seed propagation is difficult because of complex dormancy conditions involving both the seed coat and embryo. Probably the best treatment is a warm [21 to 30°C (70 to 85°F)], moist stratification period for 2 months, followed by a cold [4°C (40°F)] stratification period for 3 to 5 months. These conditions could be obtained naturally by planting the seed in late summer, after which germination should occur the following spring. Since softwood cuttings root easily if taken in spring or summer, this method is generally practiced. Cuttings are treated with 7,500 ppm IBA. Disease control can be a problem, so preventative fungicides are used.

Sapindus drummondii. Soapberry (Western). Western soapberry is a deciduous landscape tree with a bright yellow fall color and a low water requirement that performs well on highly calcareous soils. For maximum germination, fall or winter collection of seeds followed by a 60-minute acid scarification and 3-month cold stratification is recommended (301). May and June cuttings can be rooted when treated with 16,000 ppm IBA (131).

***Sarcococca* spp.** Sarcococca, Fragrant Sweetbox (*S. ruscifolia*). Attractive, broad-leaf evergreen ground cover that tolerates shade conditions. Sarcococca species can be propagated from firm cuttings. A 3,000 ppm IBA quick-dip enhanced rooting of *S. hookerana* (129).

Sassafras albidum. Sassafras. An eastern North American native tree with spectacular fall color. Seeds require a long period (4 months) of cold stratification and should be sown in the nursery early in the fall or use stratified seeds for spring planting. Sassafras can be propagated from root cuttings. Shoots from the root cuttings can be rooted. Can also be propagated by layering and by chip-budding.

Schizophragma hydrangeoides. A vine in the Hydrangeaceae family that climbs to 12 m (40 ft) on the bark of trees and bears a mass of creamy-white flowers with reddish new shoots. Can be propagated from seed stratified at 1 to 3°C (34 to 38°F) for 10 to 12 weeks. The seeds are very small and can be difficult to handle. Can be propagated asexually by nodal tip cuttings or single nodal cuttings with two opposite buds using talc applications of 5,000 ppm IBA (275). The most easily rooted cuttings are those obtained from nonflowering shoots near the base of the plant, which often have preformed root initials. Schizophragma is commercially micropropagated.

Sciadopitys verticillata. Umbrella pine. A columnar, slow-growing, conical, evergreen conifer with attractive, peeling red-brown bark. Seed germination is poor and seedling growth is very slow. While there are many cultivars, rooting is difficult, and there are advantages of using mycorrhiza in combination with auxin application to enhance rooting (136). A 24-hour soak in water (to remove latex sap at the cut ends of the stems), before applying auxin enhanced rooting success and root mass, particularly with hardwood cuttings (50 percent rooting), compared to softwood cuttings (40 percent rooting) (136); a quick-dip of 2000 ppm IBA + 1000 ppm NAA was used. Cuttings from shade-grown sources had greater rooting than when taken from full-sun trees. Age or height of stock plants did not affect rooting.

Sea Grape. *See Coccoloba.*

Senicio cineraria. *See Senicio (see Chapter 21).*

Sequoia spp. Coast Redwood (*Sequoia sempervirens*), Giant Sequoia or Sierra Redwood (*Sequoiadendron giganteum*). Both genera are ordinarily propagated by seed; however, some cultivars are propagated by cuttings.

Seed. Seeds of *Sequoia sempervirens* are mature at the end of the first season, but those of *Sequoiadendron giganteum* require two seasons for maturity of the embryos. Cones are collected in the fall and allowed to dry for 2 to 4 weeks, after which the seeds can be separated. Seeds may be kept for several years in sealed containers under 4°C (40°F) storage without losing viability. Stratification for 10 weeks at about 4°C (40°F) promotes germination of *Sequoiadendron giganteum* seed. Seeds of *Sequoia sempervirens* will germinate without a stratification treatment, however, a 24-hour water soak followed by 4 weeks of stratification can improve germination.

Fall planting also may be done, sowing the seed about 3 mm (1/8 in) deep in a well-prepared seed bed. The young seedlings should be given partial shade for the first 60 days.

Cuttings. Introduction of vegetatively propagated ornamental cultivars of *S. sempervirens* has resulted in vast improvement over highly variable seed-propagated specimens. Hardwood cuttings (February in southern California) were trimmed to 12 cm (5 in) in length so that the outer tissue on the main stem of the cutting was brown at the base and green above. Cuttings were propagated in 9 perlite:1 peat media, placed in cutting flats on outdoor heated beds [21°C (70°F)] and rooted with mist under full sun for 5 months. The cultivar 'Majestic Beauty' rooted best with 3,000 ppm IBA and 3,000 ppm NAA liquid; 'Santa Cruz' rooted best with 16,000 ppm IBA talc, and 'Soquel' rooted best with 6,000 ppm IBA and 6,000 ppm NAA quick-dip (53). *Sequoiadendron giganteum,* grown for Christmas trees, roots easily if cuttings are taken from young trees and treated with IBA under mist (158).

Micropropagation. Both genera are micropropagated (41, 220).

Sequoiadendron giganteum. See *Sequoia.*

Serviceberry. See *Amelanchier.*

Seven Son Flower (Tree). See *Heptacodium.*

Shepherdia spp. Buffaloberry. Hardy deciduous and evergreen shrubs. Seeds of *S. canadensis* and *S. rotundifolia* have been reported to exhibit seed-coat and internal dormancy. Seeds should be scarified in sulfuric acid for 30 minutes, followed by a minimum 30-day cold stratification. *S. canadensis* can be rooted with hardwood cuttings treated with 3,000 ppm IBA. *S. rotundifolia* can be micropropagated with WPM (252).

Shrub-Althea. See *Hibiscus.*

Silk Tree. See *Albizzia.*

Silverbell. See *Halesia.*

Skimmia spp. Skimmia. Hardy, woody ornamental shrubs in the citrus family (Rutaceae).

Seed. *S. japonica* ssp. *Reevesiana,* though bisexual, can be propagated by seed since over 99 percent of the progeny will be male (259). As soon as the fruit becomes ripe, the fruit is pulped and the seeds collected via flotation. Seeds are then sown in seedling flats and allowed to naturally stratify during the winter in a non-heated greenhouse (England).

Cuttings. *S. japonica* (Japanese skimmia) and *S. reevesiana* (Reeves skimmia) are easy-to-room, and are propagated from semi-hardwood cuttings 6 to 8 cm (2.5 to 3 in) long, wounded 1.5 cm (0.5 in) from the base and treated with 8,000 ppm IBA talc or quick-dip solution (259). Cuttings are rooted in a low polyethylene tent, not mist, and given bottom heat [20°C (68°F)].

Smoke Tree. See *Cotinus.*

Snowberry. See *Symphoricarpos.*

Soapberry (Western). See *Sapindus.*

Sophora spp. Sophora. For uniform seed germination, seed of Texas Mountain Laurel (*S. secundiflora*) must be scarified, regardless of the seed age (young, soft seeds with yellow seed coats or mature, firm seeds with red seed coats) (426). Scarifying with sulfuric acid from 35 to 120 minutes, or nicking the seed coat with a saw, work well (395). *S. japonica* (syn. *Styphnolobium japonicum*) (Japanese pagoda tree) is generally seed-propagated. Seed must be cleaned by soaking fruit for 48 hours; generally, no scarification or stratification treatments are needed (131). Cuttings are not a successful method for propagating *S. secundiflora* and *S. japonica.* However, cultivars of *S. microphylla* are rooted successfully by taking semi-hardwood cuttings in winter (New Zealand), which are stripped of lower leaves, wounded, dipped in 8,000 ppm IBA talc, and inserted into 1 bark:1 pumice medium under mist with 20°C (68°F) bottom heat (71). *S. microphylla* can be grafted by T-budding or with a side graft. *Sophora japonica* (syn. *Styphnolobium*) can be T-budded in summer, side-grafted on seedling rootstock during late spring. Weeping forms of *S. japonicum* are top grafted on seedling rootstock. *S. secundiflora* can be micropropagated, which offers potential for clonal selection (170).

Sorbus spp. Sorbus, Mountain Ash.

Seed. Seeds should be collected as soon as the fruits mature; fleshy parts are removed to eliminate inhibitors. Seed is usually cold stratified at 1 to 4°C (34 to 40°F) with the best germination occurring at the cold stratification temperatures. However, germination at these temperatures can take up to 4 months. For *S. glabrescens,* the best germination was a 4-month cold stratification [1°C (34°F)] followed by germination at 10°C (50°F). For some species, a period of 3- to 5-month warm dry storage prior to the cold stratification/germination period can shorten the germination time. Exposing stratified seeds to warm germination temperatures above 25°C (77°F) can induce secondary dormancy (402).

Cuttings. Cuttings are generally difficult to root; however, 6 to 10 cm (2 to 4 in) softwood cuttings of *S. aucuparia* and *S. hybrida*—which had been severed from forced stock plants at the junction between the growth of the current year and that of the previous year—root well (193). A talc formulation of 3,000 ppm IBA improved rooting of the softwood cuttings.

Grafting. Either fall chip budding or bench grafting (whip grafting) is successful. Selected cultivars are best worked on seedlings of their own species, although *S. aria, S. aucuparia,* and *S. cuspidata* (European mountain ash) seedlings seem satisfactory as a rootstock for other species.

Micropropagation. *S. aucuparia* (88) and *S. domestica* are micropropagated (11).

Spiceberry. *See Ardisia.*

Spicebush (Japanese, American). *See Lindera.*

Spiraea spp. Spirea. Usually propagated by cuttings, although some species, such as *S. thunbergi,* are more easily started by seeds, which should not be allowed to dry out. Leafy softwood cuttings taken in midsummer are generally successful. Treatments with auxin at 1,000 ppm aid rooting. Some species, such as *S. xvanhouttei,* can be started readily by hardwood cuttings, planted in early spring. Spirea is commercially micropropagated.

Spruce. *See Picea.*

Spurge (Japanese). *See Pachysandra.*

St. John's Wort. *See Hypericum.*

Stangeria eriopus. *See Cycas.*

Stewartia spp. Stewartia. Desiccation causes a very rapid decline in seed viability, so seed should be harvested when the capsules are still green and indehiscent (309). For Japanese stewartia (*S. pseudocamellia*)—harvest seed capsules when they turn from green to brown, maintain seed under nondesiccating conditions. Seed dormancy is overcome by a 3- to 5-month warm stratification followed by a 3- to 7-month cold stratification (123, 309).

Due to the difficulty of seed propagation, softwood and semi-hardwood cuttings are the most common form of propagation (306). Semi-hardwood cuttings collected in early summer root well when treated with 5,000 to 8,000 ppm IBA, or 2,500 ppm IBA × 2,500 ppm NAA. Problems occur with winter survival. Rooted cuttings should be hardened-off and not transplanted during the fall. It is best not to fertilize until the cuttings leaf out the following spring. In Louisiana, *S. malacondendron* cuttings are taken from new growth in mid-May. The 15 cm (6 in) cuttings are treated with 1,000 ppm IBA talc and rooted under mist. Plants in rooted thirty-six-cell trays are placed under shade and fertilized weekly for cuttings to break growth. Roots are left undisturbed until the following spring when the rooted liners are shifted up into larger containers after new growth has occurred (232). In Oregon, 15 to 20 cm (6 to 8 in) cuttings taken in mid-July are treated with 2,000 ppm IBA + 1,000 ppm NAA and direct-stuck in small liner pots under mist. Rooted cuttings are not transplanted until after one full year of growth (149). *S. pseudocamellia* can be propagated by softwood and semi-hardwood stem cuttings dipped in low IBA (100 ppm) with 0.1 M ascorbic or caffeic acid; rooted cuttings can be overwintered successfully (398).

Micropropagation. *S. pseudocamellia* can be micropropagated (289).

Sumac. *See Rhus.*

Sweetshrub. *See Calycanthus.*

Swietenia mahagoni. Mahogany. A salt-tolerant, semi-deciduous landscape tree in southern Florida. Seeds are collected when mature in late winter and germinate easily without soaking or pregermination treatments (317). While seed propagation is the preferred method, propagation by cuttings is important for clonal propagation and as a research tool. The inclusion of older growth in stem cuttings is necessary for successful rooting. Good rooting was obtained from a 20-year-old tree (South Florida). Cuttings are trimmed to 10 to 20 cm (4 to 8 in) with four to five fully expanded leaves, succulent terminal growth removed, treated with 1,000 ppm IBA + 300 ppm NAA, and rooted under mist (216).

Sycamore. *See Platanus.*

***Symphoricarpos* spp.** Snowberry. Seed propagation is difficult because of a hard, impermeable endocarp and a partially developed embryo at harvest. Give seeds a 3- to 4-month warm, moist stratification followed by cold stratification at 5°C (41°F) for 6 months before planting. Softwood and semi-hardwood cuttings of *S. albus* root easily when treated with 1,000 to 3,000 ppm IBA. *Symphoricarpos xchenaulti* semi-hardwood cuttings root when treated with 8,000 ppm IBA talc (131).

***Syringa* spp.** Lilac. *Syringa vulgaris* cvs. and other *Syringa* species are either micropropagated or propagated from cuttings.

Seeds. Seedlings are used mostly as rootstock for grafting or in hybridization. Lilac cultivars will not reproduce true from seed. Seeds require fall planting out-of-doors or a stratification period of 40 to 60 days at about 4°C (40°F) for good germination.

Cuttings. Cuttings can be rooted, if attention is given to proper timing. Ordinarily, good rooting of lilacs can be obtained only if terminal leafy cuttings are taken within a narrow period shortly after growth commences in the spring. When the new, green shoots have reached a length of 10 to 15 cm (4 to 6 in) they should be cut off and trimmed into cuttings. Since they are very succulent at this state it is difficult to prevent wilting. Rooting can be obtained in a polyethylene-covered bed in the greenhouse with bottom heat, or in indoor and outdoor mist beds.

Chinese lilac (*S. chinensis*) roots well in late spring (before the terminal bud has set), when treated with a 4,000 ppm IBA quick-dip. Korean lilac (*S. patula*) cuttings rooted well when taken from plants forced in a greenhouse and treated with 8,000 ppm talc (131). Softwood cuttings of *S. henryi, S. josikaea,* and *S. villosa* root easily with a 3,000 ppm IBA talc. With *S. vulgaris,* optimal rooting occurs with softwood cuttings treated with 1,500 to 3,000 ppm IBA, depending on the cultivar. One nursery buys in unrooted microcuttings (stage II) of *S. vulgaris,* roots them in a growing chamber. The rooted cuttings are planted in small liner pots and then treated with 1000 ppm Florel (an ethylene-generating chemical) at 10 to 15 cm (4 to 6 in) to encourage more basal buds to develop and increase propagule numbers (137). Treated plants are overwintered in unheated greenhouses covered with black and white poly to exclude light. Temperature is gradually raised in late winter (Minnesota) and the first harvest of softwood cuttings is taken in April. Cuttings are quick-dipped in 750 ppm IBA, stuck in 128-cell trays and rooted under mist. Up to 4 crops are harvested and rooted during the year, a new juvenile stand of *S. vulgaris* is treated with Florel Brand Pistal, and the process is repeated for the next year (137).

Cuttings of hybrid *S. vulgaris* cultivars showed improved rooting with prior etiolation of stock plants, and the period over which lilac cuttings could be propagated successfully was lengthened considerably. IBA talc of 3,000 to 8,000 ppm enhanced rooting.

Grafting. Today, grafting is less popular due to the expense, graft incompatibility problems, suckering of the rootstock, reduced plant life, and poor growth renewal. Also, tissue culture production of *Syringa* is much more broadly utilized. Grafting or budding on rootstock of California privet (*Ligustrum ovalifolium*) or Amur privet (*L. amurense*) cuttings, or on lilac or green ash (*Fraxinus pennsylvanica*) seedlings, has been used in lilac propagation (100, 192, 433).

Micropropagation. Lilacs are commercially micropropagated (208, 287).

Tabebuia argentea. Gold Tree, Silver Trumpet Tree. Has a cork-like light-colored bark with spectacular golden yellow flowers. It is generally seed-propagated without any pregermination treatment (317). It is important that the propagation medium be drenched with fungicides to control damping-off. This species can also be produced by cuttings and air layering.

Tamarisk. *See Tamarix.*

***Tamarix* spp.** Tamarisk, salt cedar. Seed has no dormancy and germinates within a few days at warm temperature [20°C (68°F)]. Tamarix (*T. aphylla*) is a salt tolerant plant and has become an invasive plant in the southwest United States. These plants are easily rooted by hardwood cuttings, usually made about 30 cm (12 in) long and planted deeply. Softwood cuttings taken in early summer also will root readily under mist. Cuttings of *T. ramosissima* (*T. pentandra*) will root during any season.

Taxodium distichum. Bald Cypress. Bald cypress is propagated by seeds, which should be fall-planted or stratified at 5°C (41°F) for 90 days. Soaking the seeds for 24 to 48 hours enhances germination. Difficult to root by cuttings. Softwood cuttings treated with 1000 ppm IBA and hardwood cuttings with 8000 ppm (131). Bench grafting is done using a whip and tongue graft. *T. mucronatum* (Mexican bald cypress or Montezuma cypress), which is the national tree of

Mexico, can be propagated by seed, which has a 2-month cold stratification requirement. Seed should be cleaned of its resinous coating prior to germination. Some limited success with cuttings collected from lower branches, wounded and treated with 3,000 to 8,000 ppm IBA (391).

***Taxus* spp.** Yew. These ornamental trees and shrubs are also valued for the anticancer drug, taxol, which is found in various parts of the plant.

Seeds. Seedling propagation is little used, because of variation in the progeny, complicated seed dormancy conditions, and the slow growth of seedlings. Seedling propagation is almost entirely confined in commercial practice to the Japanese yew (*T. cuspidata*), which comes fairly true from seed if isolated plants can be located as sources of seed. Seed imported from Japan is believed to produce uniform offspring. Plant growth habit and basal branching is better with seedling-grown plants than cutting-grown.

For good germination, seeds should be given a warm 20°C (68°F) stratification period in moist peat moss or other medium for 3 months, followed by 4 months of cold stratification [5°C (41°F)]. Seedling growth is very slow. Two years in the seed bed, followed by 2 years in a lining-out bed, then 3 or 4 years in the nursery row are required to produce a salable-size plant of Japanese yew.

Cuttings. Most clonal selections of yews are propagated by cuttings, which root without much difficulty (355, 378). Taxus cuttings can be rooted outdoors in cold frames or in the greenhouse under mist, the latter giving much faster results. For the cold frame, fairly large cuttings, 20 to 25 cm (8 to 10 in) long, are made in early fall from new growth with a section of old wood at the base. Many species and cultivars respond best when they are stripped of needles at the base of the cutting (wounding) and treated with 8,000 ppm IBA talc or an IBA quick-dip of 5,000 to 10,000 ppm. Cuttings may be kept in closed frames through the winter. Rooting takes place slowly during the following spring and summer.

In Ohio and Rhode Island (291), spring-rooted Taxus were equal to or superior to cuttings taken in late fall, rooted on bottom heat, and spring-planted. With bottom heat, there is no need to strip needles from the basal end of the cuttings, and cuttings propagated on bottom heat are ready to plant sooner than those without heat. Cuttings are initially quick-dipped in 2,000 to 2,500 ppm IBA and 1,000 to 1,250 ppm NAA combinations. Mycorrhizal fungi in the rooting substrate enhances rooting of stem cuttings of Hick's yew (*Taxus xmedia* 'Hicksii') (366).

For greenhouse propagation, cuttings should be taken in early winter, after several frosts have occurred, and rooted under mist in a 90 percent sand and 10 percent peat mixture with bottom heat at about 21°C (70°F) and an air temperature of 10 to 13°C (50 to 55°F). Rooting in the greenhouse takes only about 2 months, but cuttings should not be dug too soon. Allow time for secondary roots to develop from the first-formed primary roots. There may be advantages of a 2-month cold period after cuttings have rooted and before planting out liners. Cuttings from male plants (at least in *T. cuspidata expansa*) root more readily than cuttings from female plants (those that produce fruits).

Side or side-veneer grafting is practiced for those few cultivars that are especially difficult to start by cuttings, with easily rooted cuttings used as rootstock. Taxus can be micropropagated (94).

Telopea speciosissima. Waratah. Seeds of this Australian native shrub with beautiful chrysanthemum-like flowers germinate easily, but the seedlings are difficult to transplant and grow outside their native environment. Waratah requires soil of extremely low phosphorus content.

Ternstroemia gymnanthera. Japanese ternstroemia (cleyera). An evergreen shrub with alternate, glossy, dark green leaves. New growth is red, and changes to dark green as it matures. The shrub grows upright but full and rounded, and a mature specimen can attain a height of 2.4 to 3 m (8 to 10 ft) with a 2 m (6 ft) spread. Fruit is collected in mid-September (Mississippi) (217). The pulp of the fruit is removed and seed is washed. Seed is sown directly and emerge after 30 to 60 days without pregermination treatment (217). Semi-hardwood cuttings are started in late summer.

***Thuja* spp.** Arborvitae. American (*Thuja occidental*), Oriental [Platycladus orientalis (*Thuja orientalis*)], Korean (*T. koraiensis*), and Western Arborvitae (*T. plicata*).

Seeds. Germination is relatively easy, but stratification of seeds for 60 days at about 4°C (40°F) may be helpful. *T. plicata* seed generally does not require stratification. *T. occidentalis* are fall-planted and *Platycladus orientalis* are spring-planted.

Cuttings. Hardwood cuttings of *T. occidentalis* can be rooted in midwinter under mist in the greenhouse. Best rooting is often found with cuttings taken from older plants that are no longer making rapid growth. The cuttings should be about 20 cm (6 in) long and may be taken either from succulent, vigorously growing terminals or from more mature side growth several years old. Wounding and treating with 3,000 to 8,000 ppm IBA quick-dip or talc is beneficial. No shading should be used. Cuttings may also be made in midsummer and rooted out-of-doors in a shaded, closed frame.

Cuttings of *P. orientalis* are often more difficult to root than those of *T. occidentalis*. Small, soft cuttings several centimeters long, taken in late spring, can be rooted in mist beds if treated with 8,000 ppm IBA talc. Cuttings of 10 to 15 cm (4 to 6 in) can be taken in summer or winter, treated with a 5,000 ppm IBA quick-dip, and rooted under mist (Texas). *T. plicata* roots easily from fall cuttings treated with 3,000 ppm IBA talc (123). In southern California, conifers are rooted outdoors in full sun; cuttings are taken in November and December before spring rains, since these species have cultural problems with too much water.

Thuja × 'Green Giant' ('Green Giant' arborvitae) can be rooted from softwood, semi-hardwood, or hardwood terminal cuttings (tips of first order laterals) or laterals cuttings (187). Lateral cuttings (side shoots removed from those portions of terminal cuttings inserted into the rooting media) root better than terminal cuttings. Rooting of hardwood cuttings was enhanced with 3,000 ppm IBA (187).

Grafting. The side graft is used in propagating selected clones of *P. orientalis*, with 2-year-old potted *P. orientalis* seedlings as the rootstock. Grafting is done in late winter in the greenhouse. After making the grafts, the potted plants are set in open benches filled with moist peat moss just covering the union. The grafted plants should be ready to set out in the field for further growth by mid-spring. Selected cultivars of *T. occidentalis* can also be side-veneer grafted. *P. orientalis* is generally own-rooted, rather than grafted.

***Tilia* spp.** Linden, Basswood.

Seed. Seeds are used primarily for seedling rootstock for budding and grafting; it is difficult to get good germination and seedling production. The seeds have a dormant embryo plus an impermeable seed coat which, in some species, is surrounded by a hard, tough pericarp. Such seeds are slow and difficult to germinate (160, 416). Removing the pericarp, either mechanically or by soaking the seeds in concentrated nitric acid for 1/2 to 2 hours, rinsing thoroughly and drying, then soaking the seeds for about 15 minutes in concentrated sulfuric acid to etch the seed coat, followed by stratification for 4 months at 2°C (35°F), may give fairly good germination; otherwise, warm [15 to 27°C (60 to 80°F)], stratification for 4 to 5 months, followed by an equal period of cold stratification at 2 to 4°C (35 to 40°F), can be used. Collecting the seed from the tree just as the seed coats turn completely brown (but before the seeds drop and the seed coats become hard and dry), followed by immediate planting, has given good germination.

Cuttings. *T. americana* and *T. cordata* softwood and semi-hardwood cuttings, wounded, and treated with 20,000 to 30,000 IBA quick-dips will root successfully (Kansas). Cuttings should be very soft and taken before the terminal bud hardens (Pennsylvania). Leaf abscission under mist is a problem. Suckers arising around the base of trees cut back to the ground have been successfully mound layered, and softwood cuttings taken from stump sprouts have been rooted. Long [62 cm (24 in)], semi-hardwoood cuttings of *T. cordata* rooted well when treated with 5,000 ppm IBA and propagated under high pressure fog systems. Rooting was enhanced and production time reduced with the larger propagules (389).

Budding. T-budding or chip budding (England) in late summer on seedling rootstocks of the same species gives good results (160).

Micropropagation. *T. cordata* is commercially micropropagated (88).

Toyon. See Heteromeles.

***Trachelospermum asiaticum*.** Asiatic jasmine. An important ground cover that can be rooted any time of the year (Texas). Cuttings are quick-dipped in 3,500 ppm IBA and rooted under mist or under heavy shade in outdoor beds. They root easily from the nodes and can be direct-stuck in any suitable rooting media. In Alabama, four to five cuttings are stuck in the center of a small pot, which results in less time to produce a full liner plant. *T. asiaticum* can be micropropagated (8).

***Trachelospermum jasminoides*.** Star or Confederate Jasmine. Leafy cuttings of partially matured wood root easily, especially when placed under mist and treated with 1,000 to 3,000 ppm IBA quick-dip or talc.

Tree of Heaven. See Ailanthus.

Tree Tea. See Leptospermum.

Trumpet Creeper. See Campsis.

***Tsuga* spp.** Hemlock.

Seed. Hemlocks are propagated by seed without difficulty. Seed dormancy is variable; some lots exhibit embryo dormancy, but others do not. To ensure good germination it is advisable to stratify the seeds for 2 to 4 months at about 4°C (40°F). Fall planting outdoors generally gives satisfactory germination in the spring. The seedlings should be given partial shade during the first season.

Cuttings. The preferred method of propagating hemlock cultivars is by cuttings. *T. canadensis* cultivars can be rooted with either hardwood cuttings (bottom heat plus 5,000 ppm each of IBA and NAA, or a 10,000 ppm IBA quick-dip) or softwood cuttings treated with 8,000 ppm IBA talc plus wounding the cuttings. Softwood cuttings have a lower rooting percentage, but once rooted have a greater growth rate than hardwood cuttings (119). Moisture stress will cause needles to drop, so do not store the cuttings prior to rooting. Rooting varies widely among cultivars.

Grafting. Weeping cultivars of *T. canadensis* have been grafted. Bench grafting is done with seedling rootstock in small liner pots which are brought into the greenhouse and maintained at 18°C (65°F) for 6 weeks. Dormant scions are collected as needed during the winter and a side-tongue graft is used (131).

Tulip Tree. See *Liriodendron*.

***Ulmus* spp.** Elm.

Seeds. Seed propagation is commonly used. Elm seed loses viability rapidly if stored at room temperature, but can be kept for several years in sealed containers at 0 to 4°C (32 to 40°F). Seed that ripens in the spring should be sown immediately, and germination usually takes place promptly. For those species that ripen their seed in the fall, either fall planting or stratification for 2 months at about 4°C (40°F) should be used. To obtain tree uniformity, selected clones are propagated by budding on seedling rootstocks of the same species.

Cuttings. Softwood cuttings of several elm species can be rooted under mist when taken in early summer. Semi-hardwood cuttings have been rooted without mist or auxin treatments (363). Softwood cuttings taken from new growth arising from cut-off stumps root readily if treated with IBA and placed under mist. *Ulmus* 'Frontier' (*U. capinifolia* × *U. parvifolia*), *U. parvifolia* 'Pathfinder,' and *U. wilsoniana* 'Prospector'—all of which are highly adaptable to tough urban sites and have resistance to Dutch elm disease—are propagated with softwood cuttings, treated with an 8,000 ppm IBA quick-dip.

U. parvifolia can be rooted by softwood cuttings, treated with 5,000 ppm IBA or 1,250 ppm, each, of IBA + NAA. However, hardwood cuttings treated with 10,000 ppm IBA are preferred, since there is greater growth of the rooted cutting in the same season, and winter protection is not required, as it is with softwood cuttings (Kansas) (314). *U. hollandica* can be propagated by hardwood cuttings taken in late winter, treated with IBA at 1,500 ppm, then placed in a bin over bottom heat for 6 weeks before planting (439). Long [62 cm (24 in)], semi-hardwoood cuttings of *U.* 'Regal' rooted well when treated with 5000 ppm IBA and propagated under high pressure fog systems. Rooting was enhanced and production time reduced with the larger propagules (389).

Micropropagation. *Ulmus* is commercially micropropagated.

Umbrella Pine. See *Sciadopitys*.

Ungnadia speciosa. Mexican buckeye. An outstanding native of central Texas and northern Mexico which grows as a tree or a large multistemmed shrub. Propagation is by seed, which is collected in late summer and needs no pregermination treatment (219). Seeds turn from green to dark brown and become mature in July (Texas). Seedlings require full sun (392).

***Vaccinium* spp.** Blueberry, Cranberry, Huckleberry. Evergreen huckleberry (*V. ovatum*) is similar to Japanese holly, but with reddish-colored new growth and pale pink flowers in profusion.

Seed. Most *Vaccinium* species have no seed pregermination requirements. Exceptions are lingonberry or mountain cranberry (*V. vitis-idaea*)—its seed is separated from the fruit and cleaned, then cold stratified for 90 days at 4°C (40°F). A higher percentage of seed will germinate with stratification than without (350). *Vaccinium* seed reguire light for germination.

Cuttings. *V. ovatum* is propagated by cuttings from fully matured shoots taken in fall and winter; cuttings made from previous year's growth taken the third week in April root readily (Vancouver, B.C.). Basal heat [21°C (70°F)] and 3,000 to 4,000 ppm IBA talc enhance rooting (410). Rabbiteye Blueberry (*V. ashei*) and Highbush Blueberry (*V. corymbosum*) are propagated by terminal softwood cuttings treated with 8,000 ppm IBA talc or 4,000 ppm NAA. American cranberry (*V. macrocarpon*) is easily rooted with 1,000 ppm IBA talc (131). Many blueberry species root best when very soft cuttings are taken and not treated with auxins.

Micropropagation. Blueberry cultivars are commercially micropropagated. Cascade huckleberry (*V. deliciosum*), mountain huckleberry (*V. membranaceum*), and oval-leaf bilberry (*V. ovalifolium*) are also micropropagated (24).

***Vauquelinia* spp.** Arizona rosewood (*V. californica*). This is a drought-tolerant, rosaceous evergreen shrub, which is one of the most popular landscape plants used in semi-arid regions of the southwestern United States. Seed propagation is highly variable (384). Cuttings taken from 6- to 10-year-old stock plants root best when propagated from May to June (Arizona) under mist and bottom heat [32°C (90°F)]. An IBA quick-dip at 8,000 ppm is of some benefit, but clonal variability (0 to 95 percent rooting) and season timing are the most important factors.

Veronica. *See Hebe.*

***Viburnum* spp.** Viburnum. This large group of desirable shrubs can be propagated by a number of methods, including seeds, cuttings, grafting, and layering (44). At least one species (*V. dentatum*) is readily started by root cuttings.

Seeds. The viburnums have rather complicated seed dormancy conditions. Seeds of some species, such as *V. sieboldi*, will germinate after a single ordinary low-temperature [4°C (40°F)] stratification period, but for most species, a period of 2 to 9 months at high temperatures [20 to 30°C (68 to 86°F)], followed by a 2- to 4-month period at low temperatures [4°C (40°F)] is required. The initial warm temperatures cause root formation, and the subsequent low temperature causes shoot development. Cold stratification alone will not result in germination. Such rather exacting treatments may best be given by planting the seeds in summer or early fall (at least 60 days before the onset of winter), thus providing the initial high-temperature requirement; the subsequent winter period fulfills the low-temperature requirement. The seeds should then germinate readily in the spring. Often, collecting the seeds early, before a hard seed coat has developed, will hasten germination. Viburnum seed can be kept up to 10 years if stored dry in sealed containers and held just above freezing. *V. lantana, V. opulus,* and *V. rhytidophyllum* are commonly propagated by seed.

Cuttings. Although some viburnum species (*V. opulus, V. dentatum,* and *V. trilobum*) can be propagated by hardwood cuttings, softwood cuttings rooted under mist are successful for most species. Soft, succulent cuttings taken in late spring root faster than those made from more mature tissue in midsummer, but the latter are more likely to develop into sturdy plants that will survive through the following winter. Treatments with 8,000 ppm IBA talc and a 2,500 ppm IBA quick-dip have been recommended (38). In Illinois, cuttings of *V. carlesii* taken in June root best when treated with 10,000 ppm K-IBA. Cuttings are direct-rooted in small liner pots in a heated quonset house, left in place, and allowed to go dormant and maintained throughout the winter at a minimum temperature of 2°C (28°F) (325).

One of the chief problems with viburnum cuttings is to keep them growing after rooting. Cuttings made from succulent, rapidly growing material often die within a few weeks of being potted. This problem may be overcome by not transplanting the cuttings too soon, and allowing a secondary root system to form, which will better withstand the transplanting shock. It may help to fertilize the rooted cuttings with a nutrient solution about 10 days before the cuttings are to be removed. It is best not to fertilize rooted cuttings of *V. carlesii* until new growth has started. Placing the rooted cuttings under supplementary lighting to increase daylength is also helpful. Cuttings of some species root more easily than others. *Viburnum carlesii,* for example, is difficult to root, while *V. burkwoodii, V. rhytidophylloides, V. lantana, V. sargentii,* and *V. plicatum* root readily.

Grafting (21). Selected types of viburnum are often propagated by grafting on rooted cuttings, layers, or seedlings of *V. dentatum* or *V. lantana*. Often, grafted viburnums will develop into vigorous plants more quickly than those started as cuttings. *V. opulus* 'Roseum' (Snowball) is dwarfed when grafted onto *V. opulus* 'Nanum' cuttings. It is important that all buds be removed from the rootstock so that subsequent suckering from the rootstock does not occur. The rootstocks are potted in the fall and brought into the greenhouse, where they are grafted in midwinter by the side-graft method, using dormant scionwood. After grafting, the potted plants are placed in a closed, glass-covered or poly frame with the unions buried in damp peat moss.

Grafting can be done in late summer also, using potted rootstock plants and scion material that has stopped growing and become hardened. *V. carlesii* 'Compactum' is grafted on *V. dentatum* rootstock during mid-September (Rhode Island) using a conventional side-veneer graft (213). The grafted plants are buried in slightly damp peat moss in closed frames in the greenhouse until the unions heal, after which they are moved to outdoor, poly-covered cold frames for hardening-off over the winter.

Micropropagation. Viburnums are commercially micropropagated.

Victorian Smokebush. *See Conospermum.*

Virginia Creeper. *See Parthenocissus.*

Walking Stick. *See Aralia.*

Waratah. *See Telopea.*

Waxflower. *See Chamelaucium.*

Wax Myrtle. *See Myrica.*

***Weigela* spp.** Weigela. This shrub is easily propagated either by hardwood cuttings planted in early spring or by softwood tip cuttings under mist taken any time from late spring into fall. Rooting is promoted by treating softwood cuttings with a 1,000 to 3,000 ppm IBA quick-dip, and hardwood cuttings with a 8,000 ppm IBA talc (England). Weigela is commercially micropropagated.

Willow. *See Salix.*

Willow, False Willow. *See Baccharis.*

Winterhazel. *See Corylopsis.*

***Wisteria* spp.** Wisteria. It may take 8 to 12 years for seed-propagated wisteria to flower, which is why it is vegetatively propagated. Seeds and pods are poisonous. Wisterias may be started by softwood cuttings under mist taken in midsummer. IBA often aids rooting. Some species can be started by hardwood cuttings set in the greenhouse in the spring. *W. sinensis* can be propagated with 6 to 8 cm (2.5 to 3 in) root cuttings without auxins (283). It is important to maintain polarity of root cuttings and to plant vertically with tops of cuttings level with the surface of a porous rooting medium. Simple layering of the long canes is quite successful. Choice cultivars are often grafted on rooted cuttings of less desirable types. Suckers arising from roots of such grafted plants should be removed promptly. Wisteria is commercially micropropagated.

Witch Hazel. *See Hamamelis.*

Xylosma congestum. This is propagated by rooting leafy cuttings taken in late summer or early fall, using the first and second subterminal cuttings on the shoot. Cuttings are treated with a 5,000 ppm IBA quick-dip and propagated under mist with bottom heat [21°C (70°F)]. Rooted cuttings should be hardened in a cool, humid greenhouse.

Yellow Poplar. *See Liriodendron.*

Yew. *See Taxus.*

Zamia. *See Cycas.*

REFERENCES

1. Acharya, S. N., C. B. Chu, R. Hermesh, and G. B. Schaalje. 1992. Factors affecting red-osier dogwood seed germination. *Can. J. Bot.* 70:1012–6.

2. Ackerman, W. L., and G. A. Seaton. 1969. Propagating the Bradford pear from cuttings. *Amer. Nurs.* 130:8.

3. Adams, J. C., and K. W. Farrish. 1992. Seed coat removal can enhance the germination rate of water oak seed. *Amer. Nurs.* 176:141.

4. Aiello, A. S., and S. Dillard. 2007. *Corylus fargesii*: A new and promising introduction from China. *Comb. Proc. Intl. Plant Prop. Soc.* 57:391–95.

5. Amissah, N. J., and N. Bassuk. 2007. Effect of light and cutting age on rooting in *Quercus bicolor*, *Quercus robur*, and *Quercus macrocarpa* cuttings. *Comb. Proc. Intl. Plant Prop. Soc.* 57:286–92.

6. Anderson, W. C. 1978. Rooting of tissue-cultured rhododendrons. *Comb. Proc. Intl. Plant Prop. Soc.* 28:135–39.

7. Anderson, W. C. 1983. Micropropagation of filberts, *Corylus avellana*. *Comb. Proc. Intl. Plant Prop. Soc.* 33:132–37.

8. Apter, R. C., F. T. Davies, and E. L. McWilliams. 1993. *In vitro* and *ex vitro* adventitious root formation of Asian jasmine (*Trachelospermum asiaticum*). II. Physiological comparisons. *J. Amer. Soc. Hort. Sci.* 118:906–9.

9. Arce, P., and O. Balboa. 1991. Seasonality in rooting of *Prosopis chilensis* cuttings and *in vitro* micropropagation. *Forest Ecol. Manag.* 40:163–73.

10. Arocha, L. O., F. A. Blazich, S. L. Warren, M. Thetford, and J. B. Berry. 1999. Seed germination of *Rhododendron chapmanii*: Influence of light and temperature. *J. Environ. Hort.* 17:193–96.

11. Arrillaga, I., T. Marzo, and J. Segura. 1991. Micropropagation of juvenile and adult *Sorbus domestica* L. *Plant Cell Tiss. Organ Cult.* 27:341–48.

12. Arrillaga, I., V. Lerma, and J. Segura. 1992. Micropropagation of juvenile and mature ash. *J. Amer. Soc. Hort. Sci.* 117:346–50.

13. Artlett, E. G. 1985. Propagation of *Pieris japonica*. *Comb. Proc. Intl. Plant Prop. Soc.* 35:128–29.

14. Ault, J. R. 1994. *In vitro* propagation of *Eriostemon myoporoides* and *Eriostemon* 'stardust.' *HortScience* 29:686–88.

15. Bag, N., S. Chandra, L. M. S. Palni, and S. K. Nandi. 2000. Micropropagation of devringal [*Thamnocalamus spaihiflorus* (Trin.) Munro]—a

temperate bamboo, and comparison between *in vitro* propagated plants and seedlings. *Plant Sci.* 156: 125–35.

16. Bailey, D. A., and T. C. Weiler. 1984. Rapid propagation and establishment of florists' hydrangea. *HortScience* 19:850–52.

17. Bailey, D. A., G. R. Seckinger, and P. A. Hammer. 1986. *In vitro* propagation of florists' hydrangea. *HortScience* 21:525–26.

18. Bakker, D. 1992. Grafting of junipers. *Comb. Proc. Intl. Plant Prop. Soc.* 42:439–41.

19. Banko, T. J., and M. A. Stefani. 1989. *In vitro* propagation of *Oxydendrum arboreum* from mature trees. *HortScience* 24:683–85.

20. Barnes, H. W. 1988. Rooting responses and possibilities of *Fraxinus, Osmanthus, Chionanthus*. *Plant Propagator* 34:5–6.

21. Barnes, H. W. 1992. Grafting viburnums: New ideas and techniques. *Comb. Proc. Intl. Plant Prop. Soc.* 42:436–38.

22. Barnes, H. W. 2006. Grafting white barked birches onto *Betula nigra*: Practice and possibilities. *Comb. Proc. Intl. Plant Prop. Soc.* 56:407–10.

23. Barnes, H. W. 2006. Propagation of *Jamesia americana*. *Comb. Proc. Intl. Plant Prop. Soc.* 56:411–12.

24. Barney, D. L., O. A. Lopez, and E. King. 2007. Micropropagation of cascade huckleberry, mountain huckleberry, and oval-leaf bilberry using woody plant medium and Muashige and Skoog medium formulations. *HortTechnology* 17: 279–84.

25. Barr, B. 1985. Propagation of *Berberis thunbergii* 'Atropurpurea Nana.' *Comb. Proc. Intl. Plant Prop. Soc.* 35:711–12.

26. Barr, B. 1987. Propagation of dwarf nandina cultivars. *Comb. Proc. Intl. Plant Prop. Soc.* 37:507–8.

27. Barr, B. 1994. Propagation of Camellias by cuttings. *Comb. Proc. Intl. Plant Prop. Soc.* 44:454–56.

28. Barton, S. S., and V. P. Bonaminio. 1986. Influence of stratification and light on germination of sourwood [*Oxydendrum arboreum* (L.) DC]. *J. Environ. Hort.* 4:8–11.

29. Bassuk, N. L. 2001. A taxing taxon. *Amer. Nurs. Mag.* 193:30–1.

30. Bates, S., J. E. Preece, N. E. Navarrete, J. W. Van Sambeek, and G. R. Gaffney. 1992. Thidiazuron stimulates shoot organogenesis and somatic embryogenesis in white ash (*Fraxinus americana* L.). *Plant Cell Tiss. Organ Cult.* 31:21–9.

31. Batson, D. 1998. Southern magnolia propagation and production. *Comb. Proc. Intl. Plant Prop. Soc.* 48:624–26.

32. Bayly, R. B. 1989. Propagating *Hibiscus* by cuttings and grafting. *Comb. Proc. Intl. Plant Prop. Soc.* 39:278–80.

33. Beeson, R. C., Jr., and W. M. Proebsting. 1991. Propagation tips for blue spruce. *Amer. Nurs.* 172:86–90.

34. Ben-Jaacov, J., A. Ackerman, and E. Tal. 1991. Vegetative propagation of *Alberta magna* by tissue culture and grafting. *HortScience* 26:74.

35. Bennell, M., and G. Barth. 1986. Propagation of *Banksia coccinca* by cuttings and seed. *Comb. Proc. Intl. Plant Prop. Soc.* 36:148–52.

36. Bennett, I. J., J. A. McComb, C. M. Tonkin, and D. A. J. McDavid. 1994. Alternating cytokinins in multiplication in media stimulated *in vitro* shoot growth and rooting of *Eucalyptus globulues* Labill. *Ann. Bot.* 74:53–8.

37. Bennett, L. K., and F. T. Davies. 1986. *In vitro* propagation of *Quercus shumardii* seedlings. *HortScience* 21:1045–47.

38. Berry, J. B. 1984. Rooting hormone formulations: A chance for advancement. *Comb. Proc. Intl. Plant Prop. Soc.* 34:486–91.

39. Berry, J. B. 1991. Cleft grafting of *Magnolia grandiflora*. *Comb. Proc. Intl. Plant Prop. Soc.* 41:345–46.

40. Berry, J. B. 1994. Propagation and production of *Ilex* species in the southeastern United States. *Comb. Proc. Intl. Plant Prop. Soc.* 44:425–29.

41. Berthon, J. Y., S. B. Tahar, G. Gasper, and N. Boyer. 1990. Rooting phases of shoots of *Sequoiadendron giganteum in vitro* and their requirements. *Plant Physiol. Biochem.* 28:631–38.

42. Bhattacharya, S. 2003. Asexual reproduction of trees by air-layering. *Comb. Proc. Intl. Plant Prop. Soc.* 53:387–88.

43. Bhella, H. S. 1975. Some factors affecting propagation of Douglas fir. *Comb. Proc. Intl. Plant Prop. Soc.* 25:420–24.

44. Bhella, H. S. 1980. Vegetative propagation of *Viburnum*, cvs. Alleghany, Mohican, and Onondaga. *Plant Propagator* 26:5–9.

45. Bir, R. E., and H. W. Barnes. 1994. Stem cutting propagation of bottlebrush buckeye. *Comb. Proc. Intl. Plant Prop. Soc.* 44:499–502.

46. Blakemore, S. D. 1993. Propagation of acacias and eucalypts. *Comb. Proc. Intl. Plant Prop. Soc.* 43:317–19.

47. Blazich, F. A., and J. R. Acedo. 1987. Propagation of *Osmanthus x fortunei* by softwood cuttings. *J. Environ. Hort.* 5:70–1.

48. Blazich, F. A. 1989. Propagation of *Osmanthus heterophyllus* 'llicifolius' and 'Rotundifolius' by stem cuttings. *J. Environ. Hort.* 7:133–35.

49. Blazich, F. A., and L. E. Hinesley. 1995. Fraser fir. *Amer. Nurs.* 181:54–67.

50. Blazich, F. A., S. L. Warren, J. R. Acedo, and R. O. Whitehead. 1991. Seed germination of *Leucothoe fontanesiana* as influenced by light and temperature. *J. Environ. Hort.* 9:72–5.

51. Blazich, F. A., S. L. Warren, J. R. Acedo, and W. M. Reece. 1993. Seed germination of *Rhododendron catawbiense* and *Rhododendron maximum*: Influence of light and temperature. *J. Environ. Hort.* 9:5–8.

52. Blazich, F. A., S. L. Warren, M. C. Starrett, and J. R. Acedo. 1993. Seed germination of *Rhododendron carolinianum*: Influence of light and temperature. *J. Environ. Hort.* 11:55–8.

53. Blythe, G. 1984. Cutting propagation of *Sequoia sempervirens*. *Comb. Proc. Intl. Plant Prop. Soc.* 34:204–11.

54. Blythe, G. 1989. Cutting propagation of *Cupressus* and *Cupressocyparis*. *Comb. Proc. Intl. Plant Prop. Soc.* 39:154–60.

55. Blythe, G. 1994. Cutting propagation of *Juniperus procumbens* 'Nana.' *Comb. Proc. Intl. Plant Prop. Soc.* 44:409–13.

56. Boorman, D. 1991. How to commercially graft grevilleas for profit or preservation. *Comb. Proc. Intl. Plant Prop. Soc.* 41:57–8.

57. Bouza, L., M. Jacques, and E. Miginiac. 1994. *In vitro* propagation of *Paeonia suffruticosa* Andr. cv. 'Mme de Vatry': Developmental effects of exogenous hormones during the multiplication phase. *Scientia Hort.* 57:241–45.

58. Brand, M. H. 1988. *In vitro* adventitious shoot formation on mature-phase leaves and petioles of *Liquidambar styraciflua* L. *Plant Sci.* 57:173–79.

59. Brand, M. H. 1991. The effect of leaf source and development stage on shoot organogenic potential of sweetgum (*Liquidambar styraciflua* L.) leaf explants. *Plant Cell Tiss. Organ Cult.* 24:1–7.

60. Brand, M. H., and R. D. Lineberger. 1986. *In vitro* propagation of *Halesia Carolina* L. and the influence of explanation timing on initial shoot proliferation. *Plant Cell Tiss. Organ Cult.* 7:103–13.

61. Bretzloff, L.V., and N. E. Pellet. 1979. Effect of stratification and gibberellic acid on the germination of *Carpinus caroliniana* Walt. *HortScience* 14:621–22.

62. Briggs, B. A., and S. M. McCulloch. 1983. Progress in micropropagation of woody plants in the United States and Canada. *Comb. Proc. Intl. Plant Prop. Soc.* 33:239–48.

63. Broschat, T. K. 1998. Endocarp removal enhances *Butia capitata* (Mart.) Becc. (pindo palm) seed germination. *HortTechnology* 8:586–87.

64. Browse, P. M. 1986. Dormancy control in *Magnolia* seed germination. *Comb. Proc. Intl. Plant Prop. Soc.* 36:116–20.

65. Buck, G. J. 1951. Varieties of rose understocks. *Amer. Rose Ann.* 36:101–16.

66. Bunge, B. 1988. Propagation and production of crab apples on their own roots. *Comb. Proc. Intl. Plant Prop. Soc.* 38:542–44.

67. Bunker, E. J. 1975. Germinating palm seed. *Comb. Proc. Intl. Plant Prop. Soc.* 25:377–78.

68. Bunn, E., and K. W. Dikon. 1992. *In vitro* propagation of the rare and endangered *Grevillea scapigera* (Proteaceae). *HortScience* 27:261–62.

69. Bunting, B. A. 1985. Propagation of *Clethra alnifolia*. *Comb. Proc. Intl. Plant Prop. Soc.* 35:712–13.

70. Burrows, G. E., D. D. Doley, R. J. Haines, and D. K. Nikles. 1988. *In vitro* propagation of *Araucaria cunninghamii* and other species of *Araucariacae* via axillary meristems. *Austral. J. Bot.* 36:665–76.

71. Butcher, S. M., and S. M. N. Wood. 1984. Vegetative propagation and development of *Sophora microphylla* Ait. *Comb. Proc. Intl. Plant Prop. Soc.* 34:407–16.

72. Byers, D. 1983. Selection and propagation of crape myrtle. *Comb. Proc. Intl. Plant Prop. Soc.* 33:542–45.

73. Byers, D. 1997. *Crape myrtle: A grower's thoughts*. Auburn, AL: Owl Publications, Inc.

74. Byrnes, B. 2002. Air layering: A rooting alternative. *Comb. Proc. Intl. Plant Prop. Soc.* 52:450–52.

75. Carpenter, W. J. 1988. Temperature affects seed germination of four Florida palm species. *HortScience* 23:336–37.

76. Carpenter, W. J. 1988. Seed after-ripening and temperature influence *Butia capitata* germination. *HortScience* 23:702–3.

77. Carpenter, W. J. 1989. Medium temperature influences the rooting response of *Hibiscus rosa-sinensis* L. *J. Environ. Hort.* 7:143–46.

78. Carpenter, W. J., E. R. Ostmark, and J. A. Cornell. 1993. Embryo cap removal and high temperature exposure stimulate rapid germination of needle palm seeds. *HortScience* 28:904–7.

79. Carpenter, W. J., E. R. Ostmark, and K. C. Ruppert. 1995. Promoting the rapid germination of needle palm seeds *N. Amer. Plant Propagator* 7:20–3.

80. Carter, A. R. 1969. Rose rootstocks—performance and propagation from seed. *Comb. Proc. Intl. Plant Prop. Soc.* 19:172–80.

81. Carter, A. R. 1979. Daphne propagation. *Comb. Proc. Intl. Plant Prop. Soc.* 29:248–51.

82. Carter, A. R., and M. U. Slee. 1992. The effects of shoot age on root formation of cuttings of *Eucalyptus grandis* W. Hill ex Maiden. *Comb. Proc. Intl. Plant Prop. Soc.* 42:43–7.

83. Carville, L. 1975. Propagation of *Acer palmatum* cultivars from hardwood cuttings. *Comb. Proc. Intl. Plant Prop. Soc.* 25:39–42.

84. Cash, R. C. 2007. Vegetative propagation of cuttings of Douglas-fir (*Pseudotsuga menziesii*). *Comb. Proc. Intl. Plant Prop. Soc.* 57:270–73.

85. Catalan, L. A., and R. E. Macchiavelli. 1991. Improving germination in *Prosopis flexuosa* D. C. and *P. alba* Griseb with hot water treatments and scarification. *Seed Sci. Technol.* 19:253–62.

86. Caulfield, H. W. 1976. Pointers for successful germination of palm seed. *Comb. Proc. Intl. Plant Prop. Soc.* 26:402–5.

87. Cervelli, R. 1994. Propagation of ornamental varieties of spruce (*Picea* spp.) through somatic embryogenesis. *Comb. Proc. Intl. Plant Prop. Soc.* 44:300–3.

88. Chalupa, V. 1987. Effect of benzylaminopurine and thidiazuron on *in vitro* shoot proliferation of *Tilia cordata* Mill., *Sorbus aucuparia* L., and *Robinia pseudoacacia* L. *Biolog. Plantar (Praha)* 29:425–29.

89. Chan, C. R., and R. D. Marquard. 1999. Accelerated propagation of *Chionanthus virginicus* via embryo culture. *HortScience* 34:140–41.

90. Chandler, G. P. 1969. Rooting daphnes from cuttings. *Comb. Proc. Intl. Plant Prop. Soc.* 19:205–6.

91. Chang, T. Y., and T. H. Vogue. 1977. Regeneration of Douglas-fir plantlets through tissue culture. *Science* 198:306–7.

92. Chase, H. H. 1964. Propagation of oriental magnolias by layering. *Comb. Proc. Intl. Plant Prop. Soc.* 14:67–9.

93. Chaturvedi, A., K. Sharma, and P. N. Prasad. 1978. Shoot apex culture of *Bougainvillea glabra* 'Magnifica.' *HortScience* 13:36.

94. Chee, P. B. 1994. *In vitro* culture of zygotic embryos of *Taxus* species. *HortScience* 29:695–97.

95. Cheng, Z. M., J. P. Schmurr, and W. Dai. 2000. Micropropagation of *Betula platyphylla* 'Fargo' via shoot tip culture and regeneration from leaf tissues. *J. Environ. Hort.* 18:119–22.

96. Chong, C. B. 1988. Dormancy requirement and greenhouse forcing of three *Euonymus* cultivars. *Comb. Proc. Intl. Plant Prop. Soc.* 38:580–83.

97. Christie, C. B. 1986. Factors affecting root formation on *Photinia* 'Red Robin' cuttings. *Comb. Proc. Intl. Plant Prop. Soc.* 36:490–94.

98. Chu, K., and W. S. Cooper. 1950. An ecological reconnaissance in the native home of *Metasequoia glyptostroboides*. *Ecology* 31:260–78.

99. Coate, B. 1983. Vegetative propagation of *Acacia iteaphylla*. *Comb. Proc. Intl. Plant Prop. Soc.* 33:118–20.

100. Coggeshall, R. C. 1977. Propagating French hybrid lilacs by softwood cuttings. *Comb. Proc. Intl. Plant Prop. Soc.* 27:442–44.

101. Connor, D. A. 1985. Propagation of *Mahonia* species and cultivars. *Comb. Proc. Intl. Plant Prop. Soc.* 35:279–81.

102. Connor, D. A. 1985. Propagation of upright junipers. *Comb. Proc. Intl. Plant Prop. Soc.* 35:719–21.

103. Cooper, C., D. K. Struve, and M. A. Bennett. 1991. Pericarp splitting after aerated water soak can be used as an indicator of red oak seed quality. *Can. J. For. Res.* 21:1694–97.

104. Cooper, P. A., and D. Cohen. 1984. Micropropagation of Japanese persimmon (*Diospyros kaki*). *Comb. Proc. Intl. Plant Prop. Soc.* 34:118–24.

105. Corbett, M. P. 2002. Propagating hydrangeas year-round. *Comb. Proc. Intl. Plant Prop. Soc.* 52:415–16.

106. Covan, D. A. 1986. Softwood cutting propagation of oaks, magnolias, crab apples and dogwoods. *Comb. Proc. Intl. Plant Prop. Soc.* 36:419–21.

107. Cross, J. 1984. Propagation tips on some less widely grown plants. *Comb. Proc. Intl. Plant Prop. Soc.* 34:565–68.

108. Crossen, T. 1990. Approach grafting *Grevillea*. *Comb. Proc. Intl. Plant Prop. Soc.* 40:69–72.

109. Currie, R. 1990. *Betula* propagation. *Comb. Proc. Intl. Plant Prop. Soc.* 40:315–17.

110. D'Silva, I., and L. D'Souza. 1992. Micropropagation of *Ailanthus malabarica* D. C. using juvenile and mature tree tissues. *Silvae Genet.* 41:333–39.

111. Davies, F. T., Jr. 1985. Adventitious root formation in *Rosa multiflora* 'Brooks 56' hardwood cuttings. *J. Environ. Hort.* 3:55–6.

112. Davies, F. T., Y. Fann, J. E. Lazarte, and D. R. Paterson. 1980. Bench chip budding of field roses. *HortScience* 15:817–18.

113. Davies, F. T., Jr., and J. N. Joiner. 1980. Growth regulator effects on adventitious root formation in leaf bud cuttings of juvenile and mature *Ficus pumila*. *J. Amer. Soc. Hort. Sci.* 105:91–5.

114. Davis, R. G. 1995. A consortium of conifers. *Amer. Nurs.* 182:20–7.

115. Davis, T. 1995. *Jasminum nudiflorum*. *Nurs. Manag. Prod.* 11:26.

116. de Fossard, R. A., and H. de Fossard. 1988. Micropropagation of some members of the Myrtaceae. *Acta Hort.* 227:346–51.

117. Deering, T. 1979. Bench grafting of *Betula* species. *Plant Propagator* 25:8–9.

118. Dehgan, B. 1996. Permian permanence: Cycads. *Amer. Nurs.* 183:66–81.

119. Del Tredici, P. 1985. Propagation of *Tsuga canadensis* cultivars: Hardwood versus softwood cuttings. *Comb. Proc. Intl. Plant Prop. Soc.* 35:565–69.

120. Delanoy, M., M. P. Van Damme, X. Scheldeman, and J. Beltran. 2006. Germination of *Passiflora mollissima* (Kunth) L. H. Bailey, *Passiflora tricuspis* Mast. and *Passiflora nov* sp. seeds. *Scientia Hort.* 110:198–203.

121. Dick, L. 1990. Magnolia production. *Comb. Proc. Intl. Plant Prop. Soc.* 40:318–20.

122. Dirr, M. A. 1983. Comparative effects of selected rooting compounds on the rooting of *Photinia x fraseri*. *Comb. Proc. Intl. Plant Prop. Soc.* 33:536–40.

123. Dirr, M. A. 1985. Ten woody plants that deserve a longer look. *Comb. Proc. Intl. Plant Prop. Soc.* 35:728–34.

124. Dirr, M. A. 1987. Native amelanchiers: A sampler of northeastern species. *Amer. Nurs.* 166:64–83.

125. Dirr, M. A. 1988. To know them is to love them. *Amer. Nurs.* 163:23–41.

126. Dirr, M. A. 1991. The Japanese apricot: Flowers in the winter and style in the landscape. *Nurs. Manag.* 7:40–2.

127. Dirr, M. A. 1993. *Loropatalum chinense* var. *rubrum*: As a specimen, in a grouping in masses or in a container, the shrub has exciting potential for the landscape. *Nurs. Manag.* 9:24–5.

128. Dirr, M. A. 1994. Confessions of a Clethraphile. *Nurs. Manag.* 10:14–21.

129. Dirr, M. A. 1994. Lustrous leaves, fragrant flowers and shiny red fruit make *Sarcococca* a sweet addition to the landscape. *Nurs. Manag.* 10:24–5.

130. Dirr, M. A., and A. E. Richards. 1989. Cutting propagation of *Hamamelis x intermedia* 'Arnold Promise.' *N. Amer. Plant Prop.* 1:9–10.

131. Dirr, M. A., and C. W. Heuser, Jr. 2006. *The reference manual of woody plant propagation.* Portland, OR: Timber Press.

132. Dodge, M. H. 1986. Heath and heather propagation. *Comb. Proc. Intl. Plant Prop. Soc.* 36:563–67.

133. Donnelly, J. R., and H. W. Yawney. 1972. Some factors associated with vegetatively propagating sugar maple by stem cuttings. *Comb. Proc. Intl. Plant Prop. Soc.* 22:413–31.

134. Donovan, D. M. 1990. *Daphne cneorum*: Its propagation by softwood cuttings. *N. Amer. Plant Prop.* 2:13.

135. Donselman, H. 1990. Ornamental palm propagation and container production. *Comb. Proc. Intl. Plant Prop. Soc.* 40:236–41.

136. Douds, D. D., G. Becard, P. E. Pfeffer, L. W. Doner, T. J. Dymant, and W. M. Kayser. 1995. Effect of vesicular-arbuscular mycorrhizal fungi on rooting of *Sciadopitys verticillata* Sieb. & Zucc. cuttings. *HortScience* 30:133–34.

137. Drahn, S. R. 2003. Multiplying *Syrina* propagules …Easy as one, two, ethylene. *Comb. Proc. Intl. Plant Prop. Soc.* 53:518–21.

138. Drew, J. J., and M. A. Dirr. 1989. Propagation of *Quercus* L. species by cuttings. *J. Environ. Hort.* 7:115–17.

139. Duck, P. H. 1985. Propagation of *Ilex vomitoria* 'Nana.' *Comb. Proc. Intl. Plant Prop. Soc.* 35:710–11.

140. Dunn, D. E. 1999. Timing and auxin concentration affect *Cotinus coggygria* 'Royal Purple' rooting. *Comb. Proc. Intl. Plant Prop. Soc.* 49:510–13.

141. Dunn, D. E., and J. C. Cole. 1995. Propagation of *Pistacia chinensis* by mound layering. *J. Environ. Hort.* 13:109–12.

142. Dunn, D. E., J. C. Cole, and M. W. Smith. 1996. Position of cut, bud retention and auxins influence rooting of *Pistacia chinensis*. *Scientia Hort.* 67:105–10.

143. Dunn, D. E., J. C. Cole, and M. W. Smith. 1996. Timing of *Pistacia chinensis* Bunge. rooting using morphological markers associated with calendar date and degree days. *J. Amer. Soc. Hort. Sci.* 121:269–73.

144. Durkovic, J. 2008. Micropropagation of mature *Cornus mas* 'Macrocarpa.' *Trees* 22:597–602.

145. Economou, A. S., and P. E. Read. 1984. *In vitro* shoot proliferation of Minnesota deciduous azaleas. *HortScience* 19:60–1.

146. Economou, A. S., and K. M. J. Spanoudaki. 1985. *In vitro* propagation of *Gardenia*. *HortScience* 20:213.

147. Edson, J. L., D. L. Wenny, and A. Leege-Brusven. 1994. Micropropagation of Pacific dogwood. *HortScience* 29:1355–56.

148. Effner, W. S. 1992. Successful cutting propagation of *Hamamelis x intermedia* 'Arnold Promise' and *Hamamelis mollis* 'Brevipetala.' *Comb. Proc. Intl. Plant Prop. Soc.* 42:497–98.

149. Ekstrom, D. E., and J. A. Ekstrom. 1988. Propagation of cultivars of *Stewartia*, *Acer palmatum*, and *Fagus sylvatica* for open ground production. *Comb. Proc. Intl. Plant Prop. Soc.* 38:180–84.

150. Ellyard, R. K., and P. J. Ollerenshaw. 1984. Effect of indolebutyric acid, medium composition and cutting type on rooting of *Grevillea johnsonii* cuttings at two basal temperatures. *Comb. Proc. Intl. Plant Prop. Soc.* 34:101–8.

151. Emery, D. E. 1985. *Arctostaphylos* propagation. *Comb. Proc. Intl. Plant Prop. Soc.* 35:281–84.

152. Erdogan, V., and D. C. Smith. 2005. Effect of tissue removal and hormone application on rooting of hazelnut layers. *HortScience* 40:1457–60.

153. Evison, R. J. 1977. Propagation of clematis. *Comb. Proc. Intl. Plant Prop. Soc.* 27:436–40.

154. Farmer, R. E., Jr. 1975. Long term storage of northern red and scarlet oak seed. *Plant Propagator* 20, 21:11–4.

155. Fazio, S. 1964. Propagating *Eucalyptus* from cuttings. *Comb. Proc. Intl. Plant Prop. Soc.* 14:288–90.

156. Finneseth, C. H., and R. L. Geneve. 1997. Adventitious bud and shoot formation in pawpaw [*Asimina triloba* (L.) Dunal] using juvenile seedling tissue. *Comb. Proc. Intl. Plant Prop. Soc.* 47:638–40.

157. Finneseth, C. H., D. R. Layne, and R. L. Geneve. 1998. Morphological development of the North American pawpaw during germination and seedling emergence. *HortScience* 33:802–5.

158. Fins, L. 1980. Propagation of giant sequoia by rooting cuttings. *Comb. Proc. Intl. Plant Prop. Soc.* 30:127–32.

159. Flemer, W., III. 1962. The vegetative propagation of oaks. *Comb. Proc. Intl. Plant Prop. Soc.* 12:168–71.

160. Flemer, W., III. 1980. Linden propagation review. *Comb. Proc. Intl. Plant Prop. Soc.* 30:333–36.

161. Fleming, R. A. 1978. Propagation of holly in southern Ontario. *Comb. Proc. Intl. Plant Prop. Soc.* 28:553–57.

162. Flint, H. 1992. *Genista*—golden treasures for the landscape. *Amer. Nurs.* 76:57–61.

163. Fordham, A. J. 1972. Vegetative propagation of *Albizia*. *Amer. Nurs.* 128:7, 63.

164. Fordham, A. J. 1977. Propagation of *Kalmia latifolia* by cuttings. *Comb. Proc. Intl. Plant Prop. Soc.* 27:479–83.

165. Fordham, A. J. 1984. *Cornus kousa* and its propagation. *Comb. Proc. Intl. Plant Prop. Soc.* 34:598–602.

166. Fordham, A. J. 1987. *Davidia involucrata* var. vilmoriniana—dove tree and its propagation by seeds. *Comb. Proc. Intl. Plant Prop. Soc.* 37:343–45.

167. Foster, S. 1992. Propagation of Japanese maples by softwood cutting and grafting. *Comb. Proc. Intl. Plant Prop. Soc.* 42:373–74.

168. Freeland, K. S. 1977. Propagation of potentilla. *Comb. Proc. Intl. Plant Prop. Soc.* 27:441–42.

169. Frett, J. J. 1989. Germination requirements of *Hovenia dulcis* seeds. *HortScience* 24:152.

170. Froberg, C. A. 1985. Tissue culture propagation of *Sophora secundiflora*. *Comb. Proc. Intl. Plant Prop. Soc.* 35:750–54.

171. Fuller, C. W. 1979. Container production of *Euonymus alata* 'Compacta.' *Comb. Proc. Intl. Plant Prop. Soc.* 29:360–62.

172. Geneve, R. L., and S. T. Kester. 1990. The initiation of somatic embryos and adventitious roots from developing zygotic embryo explants of *Cercis canadensis* L. cultured *in vitro*. *Plant Cell Tiss. Organ Cult.* 22:71–6.

173. Geneve, R. L., S. T. Kester, and S. El-Shall. 1990. *In vitro* shoot initiation in Kentucky coffeetree. *HortScience* 25:578.

174. Geneve, R. L. 1991. Seed dormancy in eastern redbud (*Cercis canadensis*). *J. Amer. Soc. Hort. Sci.* 116:85–8.

175. Geneve, R. L., S. T. Kester, and S. Yusnita. 1992. Micropropagation of Eastern Redbud (*Cercis canadensis* L.). *Comb. Proc. Intl. Plant Prop. Soc.* 42:417–20.

176. Geneve, R. L., S. T. Kester, and K. W. Pomper. 2007. Autonomous shoot production in pawpaw [*Asimina triloba* (L.) Dunal] on plant growth regulator free media. *Prop. Ornamental Plants* 7:51–6.

177. Geneve, R. L., K. W. Pomper, S. T. Kester, J. N. Egilla, C. L. H. Finneseth, S. B. Crabtree, and D. R. Layne. 2003. Propagation of pawpaw—A review. *HortTechnology* 13:428–33.

178. Gingas, V. M. 1991. Asexual embryogenesis and plant regeneration from male catkins of *Quercus*. *HortScience* 26:1217–18.

179. Girouard, R. M. 1971. Vegetative propagation of pines by means of needle fascicles—a literature review. *Inform. Rpt. Dept. Environ.* 1971, Can. For. Ser.: Quebec.

180. Girouard, R. M. 1973. Rooting, survival, shoot formation, and elongation of Norway spruce stem cuttings as affected by cutting types and auxin treatments. *Plant Propagator* 19:16–7.

181. Goddard, A. N. 1970. Grafting Japanese maples. *Plant Propagator* 16:6.

182. Goodman, M. A., and D. P. Stimart. 1987. Factors regulating overwinter survival of newly propagated tip cuttings of *Acer palmatum* Thumb 'Bloodgood' and *Cornus florida* L. var. rubra. *HortScience* 22:1296–98.

183. Gordon, I. 1992. Cutting propagation of *Chamelaucium* cultivars. *Comb. Proc. Intl. Plant Prop. Soc.* 42:196–99.

184. Goreau, T. 1980. Rhododendron propagation. *Comb. Proc. Intl. Plant Prop. Soc.* 30:532–37.

185. Greever, P. T. 1979. Propagation of *Heteromeles arbutifolia* by softwood cuttings or by seed. *Plant Propagator* 25:10–1.

186. Griffin, J. J., F. A. Balzich, and T. G. Ranney. 1999. Propagation of *Magnolia virginiana* 'Santa Rosa' by semi-hardwood cuttings. *J. Environ. Hort.* 17:47–8.

187. Griffin, J. J., F. A. Balzich, and T. G. Ranney. 2000. Propagation of *Thuja* x 'Green Giant' by stem cuttings: Effects of growth stage, type of cuttings and IBA treatment. *J. Environ. Hort.* 17:47–8.

188. Grossbechler, F. 1981. Mass production of eucalyptus seedlings by direct sowing method. *Comb. Proc. Intl. Plant Prop. Soc.* 31:276–79.

189. Gwaltney, T. 1992. Dwarf yaupon, weeping yaupon and azalea propagation at Flowerwood Nursery, Inc. *Comb. Proc. Intl. Plant Prop. Soc.* 42:369–72.

190. Gwaltney, T. 1997. Propagation of weeping yaupon holly. *Comb. Proc. Intl. Plant Prop. Soc.* 47:390–92.

191. Hambrick, C. E., F. T. Davies, and H. B. Pemberton. 1991. Seasonal changes in carbohydrate/nitrogen levels during field rooting of *Rosa multiflora* 'Brooks 56' hardwood cuttings. *Scientia Hort.* 46:137–46.

192. Hand, N. P. 1978. Propagation of lilacs. *Comb. Proc. Intl. Plant Prop. Soc.* 28:348–50.

193. Hansen, O. B. 1990. Propagating *Sorbus aucuparia* L. and *Sorbus* hybrida L. by softwood cuttings. *Scientia Hort.* 42:169–75.

194. Hare, R. C. 1974. Chemical and environmental treatments promote rooting of pine cuttings. *Can. J. For. Res.* 4:101–6.

195. Hare, R. C. 1976. Rooting of American and Formosan sweetgum cuttings taken from girdled and nongirdled cuttings. *Tree Planters' Notes* 27:6–7.

196. Hare, R. C. 1977. Rooting of cuttings from mature water oak. *S. Jour. Appl. For.* 1:24–5.

197. Hartmann, H. T. 1967. 'Swan Hill': A new fruitless ornamental olive. *Calif. Agr.* 21:4–5.

198. Hasegawa, P. M. 1979. Factors affecting shoot and root initiation from cultured rose shoot tips. *J. Amer. Soc. Hort. Sci.* 105:216–20.

199. Hasek, R. F. 1980. Roses. In R. A. Larson, ed. *Introduction to floriculture*. New York: Academic Press.

200. Head, R. H. 1995. Propagation of *Magnolia grandiflora* cultivars. *Comb. Proc. Intl. Plant Prop. Soc.* 45:591–93.

201. Head, R. H. 1999. Cutting-grown *Quercus*: The future. *Comb. Proc. Intl. Plant Prop. Soc.* 49:514–16.

202. Heit, C. E. 1972. Propagation from seed: Growing larches. *Amer. Nurs.* 135:14–5, 99–110.

203. Henderson, J. C., and F. T. Davies. 1990. Drought acclimation and the morphology of mycorrhizal *Rosa hybrida* L. cv. 'Ferdy' is independent of leaf elemental content. *New Phytologist* 115:503–10.

204. Henry, P. H., F. A. Blazich, and L. E. Hinesley. 1992. Vegetative propagation of eastern red cedar by stem cuttings. *HortScience* 27:1272–74.

205. Hesselein, R. 2005. Chip budding hard-to-root magnolias. *Comb. Proc. Intl. Plant Prop. Soc.* 55:384–86.

206. Hibbert-Frey, H., J. Frampton, F. A. Blazich, and L. E. Hinesley. 2010. Grafting Fraser fir (*Abies fraseri*): Effect of grafting date, shade, and irrigation. *HortScience* 45:617–20.

207. Higginbotham, R. 1992. Bougainvillea propagation. *Comb. Proc. Intl. Plant Prop. Soc.* 42:37–8.

208. Hildebrandt, V., and P. M. Harney. 1983. *In vitro* propagation of *Syringa vulgaris* 'Vesper.' *HortScience* 18:432–34.

209. Hill, J. B. 1962. The propagation of *Juniperus chinensis* in greenhouse and mist bed. *Comb. Proc. Intl. Plant Prop. Soc.* 12:173–78.

210. Hinesley, L. E., and F. A. Blazich. 1980. Vegetative propagation of *Abies fraseri* by stem cuttings. *HortScience* 15:96–7.

211. Hinesley, L. E., F. A. Blazich, and L. K. Snelling. 1994. Propagation of Atlantic white cedar by stem cuttings. *HortScience* 29:217–19.

212. Hoogendoorn, D. P. 1984. Propagation of *Acer griseum* from cuttings. *Comb. Proc. Intl. Plant Prop. Soc.* 34:570–73.

213. Hoogendoorn, D. P. 1988. Grafting *Viburnum carlesii* 'Compactum.' *Comb. Proc. Intl. Plant Prop. Soc.* 38:563–66.

214. Hooper, V. 1990. Selecting and using *Magnolia* clonal understock. *Comb. Proc. Intl. Plant Prop. Soc.* 40:343–46.

215. Howard, B. H., and W. Oakley. 1997. Budgrafting difficult field-grown trees. *Comb. Proc. Intl. Plant Prop. Soc.* 47:328–33.

216. Howard, F. W., S. D. Verkade, and J. D. DeFilippis. 1988. Propagation of West Indies, Honduran and hybrid mahoganies by cuttings, compared with seed propagation. *Proc. Fla. State Hort. Soc.* 101:296–98.

217. Howell, J. 2007. Liner production of *Ternstroemia gymnanthera*. *Comb. Proc. Intl. Plant Prop. Soc.* 57:648–49.

218. Howkins, A. 1997. Propagation of magnolias: Rooting techniques, post-rooting care, and overwintering of rooted cuttings. *Comb. Proc. Intl. Plant Prop. Soc.* 47:282–84.

219. Hubbard, A. C. 1986. Native ornamentals for the U.S. southwest. *Comb. Proc. Intl. Plant Prop. Soc.* 36:347–50.

220. Hunter, S. A., and N. O'Donnell. 1988. Outline of a system for *in vitro* propagation of *Sequoia sempervirens*. *Comb. Proc. Intl. Plant Prop. Soc.* 38:268–72.

221. Huong, L. T. L., M. Baiocco, B. P. Huy, B. Mezzetti, R. Santilocchi, and P. Rosati. 1999. Somatic embryogenesis in Canary Island date palm. *Plant Cell Tiss. Organ Cult.* 56:1–7.

222. Huss-Danell, K., L. Eliasson, and I. Ohberg. 1980. Conditions for rooting of leafy cuttings of *Alnus incana*. *Physiol. Plant.* 49:113–16.

223. Hyatt, D. W. 2006. Propagation of deciduous azaleas. *Comb. Proc. Intl. Plant Prop. Soc.* 56:542–47.

224. Iliev, I., P. Kitin, and R. Funada. 2001. Morphological and anatomical study on *in vitro* root formation on silver birch (*Betula pendula* Roth). *Prop. Ornamental Plants* 1:10–9.

225. Intven, W. J., and T. J. Intven. 1989. Apical graftage of *Acer palmatum* and other deciduous plants. *Comb. Proc. Intl. Plant Prop. Soc.* 39:409–12.

226. Iseli, J., and D. Howse. 1981. New cultivars of *Picea pungens glauca*—their attributes and propagation. *Plant Propagator* 27:5–8.

227. Ivanicka, J. 1987. *In vitro* micropropagation of mulberry, *Morus nigra* L. *Scientia Hort.* 32:853–63.

228. Jain, S. M., P. M. Gupta, and R. J. Newton. 2000. Somatic embryogenesis in woody plants. Dordrecht: Kluwer Acad. Pub.

229. Janick, J. A., Whipkey, S. L. Kitto, and J. Frett. 1994. Micropropagation of *Cephalotaxus harringtonia*. *HortScience* 39:120–22.

230. Jaynes, R. A. 1976. Mountain laurel selections and how to propagate them. *Comb. Proc. Intl. Plant Prop. Soc.* 26:233–36.

231. Jaynes, R. A. 1988. *Kalmia, the laurel book II*. Portland, OR: Timber Press.

232. Jenkins, M. Y. 2008. Rooting native azaleas and stewartia. *Amer. Nurs.* 207:22–5.

233. Joley, L. 1960. Experiences with propagation of the genus *Pistacia*. *Comb. Proc. Intl. Plant Prop. Soc.* 10:287–92.

234. Jona, R., and I. Gribaudo. 1987. Adventitious bud formation from leaf explants of *Ficus lyrata*. *HortScience* 22:651–53.

235. Jones, C., and D. Smith. 1988. Effect of 6-benzylaminopurine and 1-naphthylacetic acid on *in vitro* axillary bud development of mature *Acacia melanoxylon*. *Comb. Proc. Intl. Plant Prop. Soc.* 38:389–93.

236. Jones, N. B., and J. V. Staden. 1994. Micropropagation and establishment of *Eucalyptus grandis* hybrids. *S. Afr. J. Bot.* 60:122–26.

237. Jones, S. K., Y. K. Samuel, and P. G. Gosling. 1991. The effect of soaking and prechilling on the germination of noble fir seeds. *Seed Sci. Technol.* 19:287–93.

238. Jull, L. G., S. L. Warren, and F. A. Blazich. 1993. Effects of growth stage, branch order and IBA treatment on rooting stem cuttings of 'Yoshino' *Cryptomeria*. *Comb. Proc. Intl. Plant Prop. Soc.* 43:404–7.

239. Jull, L. G., S. L. Warren, and F. A. Blazich. 1994. Rooting of 'Yoshino' *Cryptomeria* stem cuttings as influenced by growth stage, branch order and IBA treatment. *HortScience* 29:1532–35.

240. Jull, L. G., F. A. Blazich, and L. E. Hinesley. 1999. Seedling growth of Atlantic white-cedar as influenced by photoperiod and day/night temperature. *J. Environ. Hort.* 17:107–13.

241. Kavanagh, J. M., S. A. Hunter, and P. J. Crossan. 1986. Micropropagation of catawba hybrid *Rhododendron* 'Nova Zembla,' 'Cynthia' and 'Pink Pearl.' *Comb. Proc. Intl. Plant Prop. Soc.* 36:264–72.

242. Kelety, M. M. 1984. Container-grown hibiscus: Propagation and production. *Comb. Proc. Intl. Plant Prop. Soc.* 34:480–86.

243. Kelly, J. C. 1986. Propagation of *Prunus tenella* 'Firehill' from cuttings. *Comb. Proc. Intl. Plant Prop. Soc.* 36:279–81.

244. Kelly, J. J. 2004. Auxin, bottom heat, and time of propagation affect the adventitious rooting of *Leucophyllum candidum* 'Silver Cloud.' *Comb. Proc. Intl. Plant Prop. Soc.* 54:335–36.

245. Kerns, H. R., and M. M. Meyer. 1986. Tissue culture propagation of *Acer xfreemanii* using thidiazuron to stimulate shoot tip proliferation. *HortScience* 21:1209–10.

246. Kiang, Y. T., O. M. Rogers, and R. B. Pike. 1974. Rooting mugho pine cuttings. *HortScience* 9:350.

247. King, P. C. 1993. Fog and air circulation techniques to propagate *Aesculus parviflora* and trifoliate maples. *Comb. Proc. Intl. Plant Prop. Soc.* 43:470–73.

248. King, S. M. 1988. Tissue culture of osage-orange. *HortScience* 23:613–15.

249. Klass, S., J. Wright, and P. Felker. 1987. Influence of auxins, thiamine and fungal drenches on the rooting of *Prosopis alba* clone B2V50 cuttings. *J. Hort. Sci.* 62:97–100.

250. Klein, J. D., S. Cohen, and Y. Hebbe. 2000. Seasonal variation in rooting ability of myrtle (*Myrtus communis* L.) cuttings. *Scientia Hort.* 83:71–6.

251. Kretzschmar, U., and D. Ewald. 1994. Vegetative propagation of 140-year-old *Larix decidua* trees by different *in vitro* techniques. *J. Plant Physiol.* 144:627–30.

252. Krishnan, S., and H. Hughes. 1991. Asexual propagation of *Shepherdia canadensis* and *S. rotundifolia*. *J. Environ. Hort.* 9:218–20.

253. Krussmann, G. 1981. *The complete book of roses*. Portland, OR: Timber Press.

254. Kumar, A. 1992. Micropropagation of a mature leguminous tree—*Bauhinia purpurea*. *Plant Cell Tiss. Organ Cult.* 31:257–59.

255. Kumar, K. S., and K. V. Bhavanandan. 1988. Micropropagation of *Plumbago rosea* Linn. *Plant Cell Tiss. Organ Cult.* 15:275–78.

256. Kunisaki, J. T. 1989. *In vitro* propagation of *Leucospermum* hybrid, 'Hawaii Gold.' *HortScience* 24:686–87.

257. Lamb, J. G. D. 1976. The propagation of understocks for *Hamamelis*. *Comb. Proc. Intl. Plant Prop. Soc.* 26:27–30.

258. Lane, B. C. 1987. The effect of IBA and/or NAA and cutting wood selection on the rooting of *Rhaphiolepis indica* 'Jack Evans.' *Comb. Proc. Intl. Plant Prop. Soc.* 37:77–82.

259. Lane, C. 1997. The genus *Skimmia* and its production at Hadlow College. *Comb. Proc. Intl. Plant Prop. Soc.* 47:169–70.

260. Lane, C. G. 1993. Magnolia propagation. *Comb. Proc. Intl. Plant Prop. Soc.* 43:163–66.

261. Langendoen, J. 1998. Simultaneous top grafting of *Salix* standards and hardwood rooting of the understock. *Comb. Proc. Intl. Plant Prop. Soc.* 48:394–96.

262. Lapichino, G., T. H. H. Chen, and L. H. Fuchigami. 1991. Adventitious shoot production from a *Vireya* hybrid of *Rhododendron*. *HortScience* 26:594–96.

263. Larsen, O. N. 1985. Clonal propagation of *Fagus sylvatica* L. by cuttings. *Comb. Proc. Intl. Plant Prop. Soc.* 35:438–42.

264. Layne, D. R. 1996. The pawpaw [*Asimina triloba* (L.) Dunal]: A new fruit crop for Kentucky and the United States. *HortScience* 31:777–84.

265. Le Roux, J. J., and J. V. Staden. 1991. Micropropagation of *Eucalyptus* species. *HortScience* 26:199–200.

266. Lee, C. C., and T. E. Bilderback. 1990. Propagation of *Heptacodium miconioides*, seven son flower, by softwood and semi-hardwood cuttings. *Comb. Proc. Intl. Plant Prop. Soc.* 40:397–401.

267. Leiss, J. 1984. Root grafting of oaks. *Comb. Proc. Intl. Plant Prop. Soc.* 34:526–27.

268. Leiss, J. 1988. Piece root grafting of oaks: An update. *Comb. Proc. Intl. Plant Prop. Soc.* 38:531–32.

269. Libby, W. J., and M. T. Conkle. 1966. Effects of auxin treatment, tree age, tree vigor, and cold storage on rooting young Monterey pine. *For. Sci.* 12:484–502.

270. Lin, X., B. A. Bergmann, and A. M. Stomp. 1995. Effect of medium physical support, shoot length and genotype on *in vitro* rooting and plantlet morphology of sweetgum. *J. Environ. Hort.* 13:117–21.

271. Ling, C. H., and L. K. C. Clay. 1988. Commercial conifer micropropagation. *Comb. Proc. Intl. Plant Prop. Soc.* 38:209–15.

272. Lloyd, G., and B. McCown. 1980. Commercially feasible micropropagation of mountain laurel, *Kalmia latifolia*, by use of shoot-tip culture. *Comb. Proc. Intl. Plant Prop. Soc.* 30:421–27.

273. Lohnes, J. P., B. C. Van Duyne, and C. H. Case. 1988. Propagating 'Kwanzan' cherries. *Amer. Nurs.* 167:69–85.

274. Macdonald, A. B. 1974. Camellia propagation. *Comb. Proc. Intl. Plant Prop. Soc.* 24:152–54.

275. Macdonald, A. B. 1990. Ornamental Asian climbing vines for the Pacific Northwest. *Comb. Proc. Intl. Plant Prop. Soc.* 40:464–66.

276. Mackay, W. A. 1996. Micropropagation of Texas madrone, *Arbutus xalapensis* H. B. K. *HortScience* 31:1028–29.

277. Maleike, R., and R. L. Hummel. 1991. Germination of Madrona seed. *Comb. Proc. Intl. Plant Prop. Soc.* 41:283–85.

278. Malek, A. A., F. A. Blazich, S. L. Warren, and J. E. Shelton. 1989. Influence of light and temperature on seed germination of mountain laurel. *J. Environ. Hort.* 7:161–62.

279. Marczynski, S. 1998. Top grafting of *Salix*. *Comb. Proc. Intl. Plant Prop. Soc.* 48:337–41.

280. Maynard, B., W. Johnson, and T. Holt 1996. Stock plant shading to increase rooting of paperbark maple cuttings. *Comb. Proc. Intl. Plant Prop. Soc.* 46:611–13.

281. Maynard, B. K., and N. L. Bassuk. 1987. Stock plant etiolation and blanching of woody plants

prior to cutting propagation. *J. Amer. Soc. Hort. Sci.* 112:273–76.

282. Maynard, B. K. 1990. Rooting softwood cuttings of *Acer griseum*: Promotion by stockplant etiolation, inhibition by catechol. *HortScience* 25:200–2.

283. McConnell, J. F. 1993. Wisteria propagation by root cuttings. *Comb. Proc. Intl. Plant Prop. Soc.* 43:304.

284. McCown, B., and R. Amos. 1979. Initial trials with micropropagation of birch selections. *Comb. Proc. Intl. Plant Prop. Soc.* 29:387–93.

285. McCoy, C. J. 2003. Taiwan flowering cherry (*Prunus campanulata*) propagation: Seedling or Cuttings? *Comb. Proc. Intl. Plant Prop. Soc.* 53:270–71.

286. McCracken, P. 1995. The controversy continues: Comparisons of propagation techniques of *Magnolia grandiflora*. *Comb. Proc. Intl. Plant Prop. Soc.* 45:587–90.

287. McCulloch, S. M. 1989. Tissue culture propagation of French hybrid lilacs. *Comb. Proc. Intl. Plant Prop. Soc.* 39:105–8.

288. McDaniel, J. D. 1964. A look at some hackberries. *Comb. Proc. Intl. Plant Prop. Soc.* 14:143–46.

289. McGuigan, P. J., F. A. Blazich, and T. G. Ranney. 1997. Microprapation of *Stewartia pseudocamellia*. *J. Environ. Hort.* 15:65–8.

290. McGuigan, P. J., F. A. Blazich, and T. G. Ranney. 1997. Propagation of *Quercus phillyreoides* by stem cuttings. *Amer. Nurs. Mag.* 185:72–3.

291. McGuire, J. J. 1991. Cold storage of rooted Taxus cuttings and subsequent summer regrowth. *J. Environ. Hort.* 9:36–7.

292. McGuire, J. J., W. Johnson, and C. Dawson. 1987. Leaf–bud or side-graft nurse root grafts for difficult-to-root *Rhododendron* cultivars. *Comb. Proc. Intl. Plant Prop. Soc.* 37:447–49.

293. Meacham, G. E. 1997. Homing in on Hamamelidaceae. *Amer. Nurs. Mag.* 185:34–9.

294. Medina, J. P. 1981. Studies of clonal propagation of pecans at Ica, Peru. *Plant Propagator* 27:10–1.

295. Meyer, H. J., and J. V. Staden. 1987. Regeneration of *Acacia melanoxylon* plantlets *in vitro*. *S. Afr. Tydskv. Planik.* 53:206–9.

296. Mokotedi, M. E. O., M. P. Watt, N. W. Pammenter, and F. C. Blakeway. 2000. *In vitro* rooting and subsequent survival of two clones of a cold-tolerant *Eucalyptus grandis* x *E. nitens* hybrid. *HortScience* 35:1163–65.

297. Molinar, F., W. A. Mackay, M. M. Wall, and M. Cardenas. 1996. Micropropagation of Agarita (*Berberis trifoliata* Moric.). *HortScience* 31:1030–32.

298. Monteuuis, O. 1994. Effect of technique and darkness on the success of meristem micrografting of *Picea abies*. *Silvae Genet.* 43:2–3.

299. Moore, R. S. 1981. Miniature rose production. *Comb. Proc. Intl. Plant Prop. Soc.* 30:54–60.

300. Morgan, D. L. 1985. Propagation of *Quercus virginiana* cuttings. *Comb. Proc. Intl. Plant Prop. Soc.* 35:716–19.

301. Munson, R. H. 1984. Germination of western soapberry as affected by scarification and stratification. *HortScience* 19:712–13.

302. Murphree, B. H., J. L. Sibley, D. J. Eakes, and J. D. Williams. 2000. Shade influences propagation of golden barberry 'Bailsel.' *HortTechnology* 10:752–53.

303. Murphy, N. 2005. Propagation of *Cercis canadensis* 'Forest Pansy.' *Comb. Proc. Intl. Plant Prop. Soc.* 55:273–76.

304. Myers, J. R., and S. M. Still. 1979. Propagating London plane tree from cuttings. *Plant Propagator* 25:8–9.

305. Nagao, M. A., K. Kanegawa, and W. S. Sakai. 1980. Accelerating palm seed germination with gibberellic acid, scarification, and bottom heat. *HortScience* 15:200–1.

306. Nair, A., and D. Zhang. 2010. Propagation of stewartia: Past research endeavors and current status. *HortTechnology* 20:277–82.

307. Nicholson, R. 1987. *Enkianthus*—a worthy genus waits to emerge from companion-plant status. *Amer. Nurs.* 166:91–7.

308. Nobre, J. 1994. *In vitro* shoot proliferation of *Myrtus communis* L. from field-grown plants. *Scientia Hort.* 58:253–58.

309. Oleksak, B. A., and D. K. Sturve. 1999. Germination of *Stewartia pseudocamellia* seeds is promoted by desiccation avoidance, gibberellic acid treatment and warm and cold stratification. *J. Environ. Hort.* 17:44–6.

310. Oliphant, J. L. 1992. Micropropagation of *Nerium oleander*. *Comb. Proc. Intl. Plant Prop. Soc.* 42:288–89.

311. Osborne, R., and J. V. Staden. 1987. *In vitro* regeneration of *Stangeria eriopus*. *HortScience* 22:13–26.

312. Pair, J. C. 1986. Propagation of *Acer truncatum*, a new introduction to the southern great plains. *Comb. Proc. Intl. Plant Prop. Soc.* 36:403–13.

313. Pair, J. C. 1986. New plant introductions with great stress tolerance to conditions in the southern great plains. *Comb. Proc. Intl. Plant Prop. Soc.* 36:351–55.

314. Pair, J. C. 1992. Evaluation and propagation of lacebark elm selections by hardwood and softwood cuttings. *Comb. Proc. Intl. Plant Prop. Soc.* 42:431–35.

315. Pair, J. C. 1995. *Acer tuncatum. Nurs. Manag. Prod.* 11:27.

316. Parr, G. 1983. Summer grafting of golden robinia. *Comb. Proc. Intl. Plant Prop. Soc.* 33:164–67.

317. Patel, S. I. 1983. Propagation of some rare tropical plants. *Comb. Proc. Intl. Plant Prop. Soc.* 33:573–80.

318. Pearce, B. J. 2006. Propagation of *Chamelaucium uncinatum* cultivars by grafting. *Comb. Proc. Intl. Plant Prop. Soc.* 56:88–90.

319. Pelleh, N. E., and K. Alpert. 1985. Rooting softwood cuttings of mature *Betula papyrifera*. *Comb. Proc. Intl. Plant Prop. Soc.* 35:519–25.

320. Perry, D. K. 1987. Successfully growing Proteaceae. *Comb. Proc. Intl. Plant Prop. Soc.* 37:112–15.

321. Perry, F., and S. J. Trueman. 1999. Cutting propagation of Victorian smokebush, *Conospermum mitchellii* (Proteaceae). *S. Afr. J. Bot.* 65:243–44.

322. Phelan, S., A. Hunter, and G. Douglas. 2005. Bacteria detection and micropropagation of ten *Buddleia* cultivars. *Prop. Ornamental Plants* 5:164–69.

323. Phipps, H. M., D. A. Belton, and D. A. Netzer. 1977. Propagating cuttings of some *Populus* clones for tree plantations. *Plant Propagator* 23:8–11.

324. Picchioni, G. A., and F. T. Davies. 1990. Micropropagation of *Pistacia atlantica* shoots from axillary buds. *N. Amer. Plant Propagator* 2:14–5.

325. Pickerill, J. D. 1992. Overwintering rooted cuttings of *Viburnum carlesii*. *Comb. Proc. Intl. Plant Prop. Soc.* 42:468–69.

326. Pinney, J. J. 1970. A simplified process for grafting junipers. *Amer. Nurs.* 131:7, 82–4.

327. Ponchia, G., and B. H. Howard. 1988. Chestnut and hazel propagation by leafy summer cuttings. *Acta Hort.* 227:236–41.

328. Pontikis, C. A., and P. Melas. 1986. Micropropagation of *Ficus carica* L. *HortScience* 21:153.

329. Poole, R. T., and C. A. Conover. 1984. Propagation of ornamental *Ficus* by cuttings. *HortScience* 19:120–21.

330. Poston, A. L., and R. L. Geneve. 2006. Propagation of spicebush (*Lindera benzoin*). *Comb. Proc. Intl. Plant Prop. Soc.* 56:458–60.

331. Powell, J. C. 1985. Production of xCupressocyparis leylandii. *Comb. Proc. Intl. Plant Prop. Soc.* 35:722–23.

332. Preece, J. E., and P. Read. 2007. Forcing leafy explants and cuttings from woody species. *Prop. Ornamental Plants* 7:138–44.

333. Proebsting, W. E. 1984. Rooting of Douglas-Fir stem cuttings: Relative activity of IBA and NAA. *HortScience* 19:845–56.

334. Prutpongse, P., and P. Gavinlertvatana. 1992. *In vitro* micropropagation of 54 species from 15 genera of bamboo. *HortScience* 27:453–54.

335. Purcell, G. V. 1973. The budding of *Hamamelis*. *Comb. Proc. Intl. Plant Prop. Soc.* 23:129–32.

336. Putland, L. 1994. The best way to propagate *Clematis*. *Nurs. Manag.* 10:55–8.

337. Rajesekaran, P. 1994. Production of clonal plantlets of *Grevillea robusta* in *in vitro* culture via axillary bud activation. *Plant Cell Tiss. Organ Cult.* 39:277–79.

338. Ramirez, M., M. J. Krasowski, and J. A. Loo. 2007. Vegetative propagation of American beech resistant to beech bark disease. *HortScience* 42:320–24.

339. Ranney, T. G., and R. E. Bir. 1994. Comparative flood tolerance of birch rootstocks. *J. Amer. Soc. Hort. Sci.* 119:43–8.

340. Ranney, T. G., and E. P. Whitman. 1995. Growth and survival of 'Whitespire' Japanese birch grafted on rootstocks on five species of birch. *HortScience* 30:521–22.

341. Rauch, F. D. 1994. Palm seed germination. *Comb. Proc. Intl. Plant Prop. Soc.* 44:304–7.

342. Raviv, M., and E. Putievsky. 1989. The effect of in-tree position and type of cutting on rooting of bay laurel (*Laurus nobilis* L.) stem cuttings. *N. Amer. Plant Propagator* 1:14–5.

343. Reeves, B. 2002. Propagation of *Quercus virginiana* by cuttings. *Comb. Proc. Intl. Plant Prop. Soc.* 52:448–49.

344. Richard, M. A. 1984. Propagation of bamboo by vegetative means. *Comb. Proc. Intl. Plant Prop. Soc.* 34:440–43.

345. Ridgway, D. 1984. Propagation and production of *Garrya elliptica*. *Comb. Proc. Intl. Plant Prop. Soc.* 34:261–65.

346. Ripphausen, F. 1989. Propagation and nutrition of daphne cuttings and tissue culture plantlets. *Comb. Proc. Intl. Plant Prop. Soc.* 39:305–8.

347. Ritchie, G. A. 1993. Production of Douglas-Fir, *Pseudotsuga menziesii* (Mirb.) Franco, rooted cuttings for reforestation by Weyerhaeuser company. *Comb. Proc. Intl. Plant Prop. Soc.* 43:284–88.

348. Ritchie, G. A. 1994. Commercial application of adventitious rooting to forestry. In T. D. Davis and B. E. Haissig, eds. *Biology of adventitious root formation.* New York: Plenum Press.

349. Rodríquez, R. 1982. *In vitro* propagation of *Castanea sativa* Mill. through meristem-tip culture. *HortScience* 17:888–89.

350. Rogers, C. S. 1994. Perennial preservation. *Amer. Nurs.* 179:64–7.

351. Ross, D. M. 1977. Rose rootstock, 'Dr. Huey,' in South Australia. *Comb. Proc. Intl. Plant Prop. Soc.* 27:562–63.

352. Rout, G. R., and P. Das. 1994. Somatic embryogenesis and *in vitro* flowering of 3 species of bamboo. *Plant Cell Reports* 13:683–86.

353. Ruffoni, B., M. Rabaglio, L. Semeria, and A. Allavena. 1999. Improvement of micropropagation of *Genista monosperma* Lam. by abscisic acid treatment. *Plant Cell Tiss. Organ Cult.* 57:223–25.

354. Rumbal, J. 1992. The development of cutting propagation of *Camellia reticulata* hybrids. *Comb. Proc. Intl. Plant Prop. Soc.* 42:295–96.

355. Sabo, J. E. 1976. Propagation of *Taxus* in Northern Ohio. *Comb. Proc. Intl. Plant Prop. Soc.* 26:174–76.

356. Sachs, R. M., C. Lee, J. Ripperda, and R. Woodward. 1988. Selection and clonal propagation of *Eucalyptus. Calif. Agr.* 42:27–31.

357. Sallee, K. 1989. Desert broom—drought-tolerant shrub of Southwest deserves respect in the landscape. *Nurs. Manag.* 5:106–9.

358. Salter, C. E. 1970. *Clematis armandii* grafting. *Comb. Proc. Intl. Plant Prop. Soc.* 20:330–32.

359. Samartin, A. 1984. *In vitro* propagation of *Camellia japonica* seedlings. *HortScience* 19:225–26.

360. Samyn, G. L. J. 1993. *In vitro* propagation of ponytail palm: Producing multiple-shoot plants. *HortScience* 28:225.

361. Sankhla, D., T. D. Davis, and N. Sankhla. 1994. Thidiazuron-induced *in vitro* shoot formation from roots of intact seedlings of *Albizzia julibrissin. J. Plant Growth Reg.* 14:267–72.

362. Santana, C. 1993. Direct stick *Rhododendron* production. *Comb. Proc. Intl. Plant Prop. Soc.* 43:293.

363. Saul, G. H., and L. Zsuffa. 1978. Vegetative propagation of elms by green cuttings. *Comb. Proc. Intl. Plant Prop. Soc.* 28:490–94.

364. Savella, L. 1984. Propagating pink dogwoods from rooted cuttings. *Comb. Proc. Intl. Plant Prop. Soc.* 30:405–7.

365. Sawada, S. T. 1996. Camellia propagation and production. *Comb. Proc. Intl. Plant Prop. Soc.* 46:635–38.

366. Scagel, C. F., K. Reddy, and J. M. Armstrong. 2003. Mycorrhizal fungi in rooting substrate influences the quantity and quality of roots on stem cuttings of Hick's yew. *HortTechnology* 13:62–6.

367. Scaltsoyiannes, A., P. Tsoulpha, I. Iliev, and K. Theriou. 2005. Vegetative propagation of *Prunus avium* L. genotypes selected for wood production. *Forest Genet.* 12:145–54.

368. Schmidt, D. 1998. Softwood propagation of *Platanus x hispanica* 'Columbia.' *Comb. Proc. Intl. Plant Prop. Soc.* 48:479–80.

369. Scholes, C. A. 2004. *Clematis* Propagation. *Comb. Proc. Intl. Plant Prop. Soc.* 2004:131–34.

370. Schrader, J. A. 2000. Propagation of *Alnus maritima* from softwood cuttings. *HortScience* 35: 293–95.

371. Schrader, J. A., and W. R. Graves. 2000. Seed germination and seedling growth of *Alunus maritima* from its three disjunct populations. *J. Amer. Soc. Hort Sci.* 125:128–34.

372. Schwarz, J. L., P. L. Glocke, and M. Sedgley. 1999. Adventitious root formation in *Acacia baileyana* F. Muell. *J. Hort. Sci. Biotech.* 74:561–65.

373. Scott, A. 1976. A successful technique for grafting hibiscus. *Comb. Proc. Intl. Plant Prop. Soc.* 26:389–91.

374. Sen, S., M. E. Magallanes-Cedeno, R. H. Kamps, C. R. McKinley, and R. J. Newton. 1994. *In vitro* micropropagation of Afghan pine. *Can. J. For. Res.* 24:1248–52.

375. Serres, R., P. Read, W. Hackett, and P. Nissen. 1990. Rooting of American chestnut microcuttings. *J. Environ. Hort.* 8:86–8.

376. Sharma, J., G. W. Knox, and M. L. Ishida. 2006. Adventitious rooting of stem cuttings of yellow-flowered magnolia cultivars is influenced by time after budbreak and indole-3-butyric acid. *HortScience* 41:202–6.

377. Shen, X., W. S. Castle, and F. G. Gmitter. 2009. Micropropagation of a *Casuarina* hybrid (*Casuarina equisetifolia* L. × *Casuarina glauca* Sieber ex Spreng) following facilitated seed germination. *Plant Cell Tiss.Organ Cult.* 97:103–8.

378. Shugert, R. 1985. *Taxus* production in the U.S.A. *Comb. Proc. Intl. Plant Prop. Soc.* 35:149–53.

379. Simkhada, E. P., and H. Gema. 2003. The effect of time, cultivar and grafting method on graft compatibility in Persimmon (*Diospyros*). *Comb. Proc. Intl. Plant Prop. Soc.* 53:430–35.

380. Simon, R. A. 1986. A survey of hardy bamboos: Their care, culture and propagation. *Comb. Proc. Intl. Plant Prop. Soc.* 36:528–31.

381. Simpson, R. C. 1981. Propagating deciduous holly. *Comb. Proc. Intl. Plant Prop. Soc.* 30:338–42.

382. Singha, S. 1982. *In vitro* propagation of crab apple cultivars. *HortScience* 17:191–92.

383. Skogerboe, S. 2003. Asexual propagation of *Arctostaphylos xcoloradensis*. *Comb. Proc. Intl. Plant Prop. Soc.* 53:370–71.

384. Smith, E. Y., and C. W. Lee. 1983. Propagation of Arizona rosewood by stem cuttings. *HortScience* 18:764–65.

385. Smith, H. 2003. Improved plantation eucalypts: The role of vegetative propagation. *Comb. Proc. Intl. Plant Prop. Soc.* 53:124–27.

386. Smith, M. A. L., and A. A. Obdeidy. 1991. Micropropagation of a mature, male Kentucky coffeetree. *HortScience* 26:1426.

387. Smith, M. N. 1985. Propagating *Ceanothus*. *Comb. Proc. Intl. Plant Prop. Soc.* 35:301–5.

388. Snir, I. 1982. *In vitro* propagation of sweet cherry cultivars. *HortScience* 17:192–93.

389. Spethmann, W. 2007. Increase of rooting success and further shoot growth by long cuttings of woody plants. *Prop. Ornamental Plants* 7:160–68.

390. St. Hilaire, R., and C. A. Fierro. 2000. Three *Mussaenda* cultivars propagated by stem cuttings exhibit variation in rooting in response to hormone and rooting conditions. *HortTechnology* 10:780–84.

391. St. Hilaire, R. 2003. Propagation of *Taxodium mucronatum* from softwood cuttings. *Desert Plants* 19:29–30.

392. Stanford, G. 1982. *Ungnadia speciosa* (Mexican buckeye). *Plant Propagator* 28:5–6.

393. Starrett, M. C., F. A. Blazich, and S. L. Warren. 1993. Seed germination of *Pieris floribunda*: Influence of light and temperature. *J. Environ. Hort.* 10:121–24.

394. Starrett, M. C., F. A. Blazich, J. R. Acedo, and S. L. Warren. 1993. Micropropagation of *Pieris floribunda*. *J. Environ. Hort.* 11:191–95.

395. Strong, M. E., and F. T. Davies. 1982. Influence of selected vesicular-arbuscular mycorrhizal fungi on seedling growth and phosphorus uptake of *Sophora secundiflora*. *HortScience* 17:620–21.

396. Struve, D. K., M. A. Arnold, and D. H. Chinery. 1987. Red oak whip production in containers. *Comb. Proc. Intl. Plant Prop. Soc.* 37:415–20.

397. Struve, D. K., M. A. Arnold, R. Beeson, J. M. Ruter, S. Svenson, and W. T. Witte. 1994. The copper connection: The benefits of growing woody ornamentals in copper-treated containers. *Amer. Nurs.* 179:52–61.

398. Struve, D. K., and L. M. Lagrimini. 1999. Survival and growth of *Stewartia pseudocamellia* rooted cuttings and seedlings. *J. Environ. Hort.* 17:53–6.

399. Stubbs, H. L., F. A. Blazich, T. G. Ranney, and S. L. Warren. 1997. Propagation of 'Carolina Sapphire' smooth Arizona cypress by stem cuttings: Effects of growth stage, type of cutting and IBA treatment. *J. Environ. Hort.* 15:61–4.

400. Sturrock, J. W., and J. D. Ferguson. 1989. Macro and micro propagation of Leyland cypress. *Comb. Proc. Intl. Plant Prop. Soc.* 39:285–90.

401. Sutter, E., and P. Barker. 1985. *In vitro* propagation of mature *Liquidambar styraciflua*. *Plant Cell Tiss. Organ Cult.* 5:13–21.

402. Taylor, C. W., and W. A. Gerrie. 1987. Effects of temperature on seed germination and seed dormacy in *Sorbus glabrescens* Cardot. *Acta Hort.* 215:185–92.

403. Tietje, W. D., J. H. Foott, and E. L. Labor. 1990. Grafting California oaks. *Calif. Agr.* 44:30–1.

404. Tilki, F., and S. Guner. 2007. Seed germination of three provenances of *Arbutus andrachne* L. in response to different pretreatments, temperature and light. *Prop. Ornamental Plants* 7:175–79.

405. Tipton, J. L. 1981. Asexual propagation of juvenile *Arubuts xalapensis* in a high humidity chamber. *Plant Propagator* 27:11–2.

406. Tipton, J. L. 1990. Vegetative propagation of Mexican redbud, larchleaf goldenweed, littleleaf ash, and evergreen sumac. *HortScience* 25:196–98.

407. Tipton, J. L. 1992. Requirements for seed germination of Mexican redbud, evergreen sumac, and mealy sage. *HortScience* 27:313–16.

408. Tipton, J. L. 1995. Easy-to-grow *Chilopsis* lends an ethereal look to desert vistas. *Nurs. Manag. Prod.* 11:14–5.

409. Tisserat, B. 1979. Propagation of the date palm *in vitro*. *J. Exp. Bot.* 30:1275–83.

410. Tubesing, C. E. 1985. Cutting propagation of *Paxistima myrsinites*, *Vaccinium ovatum*, *Ribes sanguineum* and *Acer glabrum* subsp. *Douglasii*. *Comb. Proc. Intl. Plant Prop. Soc.* 35:293–96.

411. Uno, S., and J. E. Preece. 1987. Macro-and cutting, propagation of 'Crimson Pygmy' barberry. *HortScience* 22:488–91.

412. Upchurch, B. L. 2006. Grafting with care. *Amer. Nurs.* 203:18–22.

413. Vakouftsis, G., S. Kostas, T. Syros, A. Economou, M. Tsaktsira, A. Scaltsoyiannes, and D. Metaxas. 2008. Rooting of *xCupressocyparis Leylandii* 'Castlewellan Gold' by cuttings as influenced by IBA,

harvesting season, type of cuttings and rooting medium. *Prop. Ornamental Plants* 8:125–32.

414. Van de Pol, P. A., and A. Breukelaar. 1982. Stenting of roses: A method for quick propagation by simultaneously cutting and grafting. *Scientia Hort.* 17:187–96.

415. Van Veen, T. 1971. The propagation and production of rhododendrons. *Amer. Nurs.* 133:15–6, 52–8.

416. Vanstone, D. E. 1978. Basswood (*Tilia americana*) seed germination. *Comb. Proc. Intl. Plant Prop. Soc.* 28:566–69.

417. Vertrees, J. D. 1991. Understock for rare *Acer* species. *Comb. Proc. Intl. Plant Prop. Soc.* 41:272–75.

418. Vieitez, A. M., A. Ballester, M. L. Vieitez, and E. Vieitez. 1983. *In vitro* plantlet regeneration of mature chestnut. *J. Hort. Sci.* 58:457–63.

419. Vieitez, A. M., M. C. San-Jose, and E. Vieitez. 1985. *In vitro* plantlet regeneration from juvenile and mature *Quercus robur* L. *J. Hort. Sci.* 60:99–106.

420. Vieitez, A. M., M. Concepción Sanchez, J. B. Amo-Marco, and A. Ballester. 1994. Forced flushing of branch segments as a method for obtaining reactive explants of mature *Quercus robur* trees for micropropagation. *Plant Cell Tiss. Organ Cult.* 37:287–95.

421. Vural, U., H. Dumanoglu, and V. Erdogan. 2008. Effect of grafting/budding techniques and time on propagation of black mulberry (*Morus nigra* L.) in cold temperate zones. *Prop. Orn. Plants* 8:55–8.

422. Wallis, J. S. 1976. The propagation and training of standard fuchsias. *Comb. Proc. Intl. Plant Prop. Soc.* 26:346–48.

423. Wang, H. C., and N. A. Reichert. 1992. *In vitro* propagation of modern roses. *Comb. Proc. Intl. Plant Prop. Soc.* 42:398–403.

424. Wang, Y. T. 1989. Effect of water salinity, IBA concentration and season of rooting on Japanese boxwood cuttings. *Acta Hort.* 246:191–98.

425. Wang, Y. T., and R. E. Rouse. 1989. Rooting live oak rhizomic shoots. *HortScience* 24:1043.

426. Wang, Y. T. 1991. Enhanced germination of *Sophora secundiflora* seeds. *Subtrop. Plant Sci.* 44:37–40.

427. Ward, P. W. 1993. The production of *Robinia pseudoacacia* 'Frisia.' *Comb. Proc. Intl. Plant Prop. Soc.* 43:357–58.

428. Warren, K. 1989. A crab apple system. *Amer. Nurs.* 170:31–5.

429. Warren, P. 1973. Propagation of *Cercis* cultivars by summer budding. *Plant Propagator* 19:16–7.

430. Warren, P. 1995. *Cercis canadensis* ssp. *texensis* 'Oklahoma' 1995. *Nurs. Manag. Product.* 11:13.

431. Wasley, R. 1979. The propagation of *Berberis* by cuttings. *Comb. Proc. Intl. Plant Prop. Soc.* 29:215–16.

432. Watkins, J. V. 1972. Jacaranda. *Horticulture* 50:22–3.

433. Wedge, D. 1977. Propagation of hybrid lilacs. *Comb. Proc. Intl. Plant Prop. Soc.* 27:432–36.

434. Weilhelm, E. 1999. Micropropagation of juvenile sycamore maple via adventitious shoot formation by use of thidiazuron. *Plant Cell Tiss. Organ Cult.* 57:57–60.

435. Welander, M., E. Jansson, and H. Lindquist. 1989. *In vitro* propagation of *Populus xwilsocarpa*—a hybrid of ornamental value. *Plant Cell Tiss. Organ Cult.* 18:209–19.

436. Wells, J. S. 1980. How to propagate Japanese maples. *Amer. Nurs.* 151:14, 117–20.

437. Wells, J. S. 1981. A history of rhododendron cutting propagation. *Amer. Nurs.* 154:14–5, 114–26.

438. Wells, J. S. 1985. *Plant propagation practices.* Chicago: American Nurseryman Publishing Co.

439. Whalley, D. N. 1975. Propagation of Commelin elm by hardwood cuttings in heated bins. *Plant Propagator* 21:4–6.

440. Whalley, D. N. 1979. Leyland cypress—rooting and early growth of selected clones. *Comb. Proc. Intl. Plant Prop. Soc.* 29:190–202.

441. Whitlow, C., and D. Hannings. 2005. Propagation of *Arbutus* 'Marina' by air-layering. *Comb. Proc. Intl. Plant Prop. Soc.* 55:469–72.

442. Whitten, M. 1994. A better Virginia pine—send in the clones. *Nurs. Manag.* 10:39–42.

443. Wiegrefe, S. J. 1994. Overcoming dormancy in maple seeds without stratification. *N. Amer. Plant Propagator* 6:8–11.

444. Wiegrefe, S. J. 2003. Graft compatibilities of hornbeam (*Carpinus*) species and hybrids. *Comb. Proc. Intl. Plant Prop. Soc.* 53:549–54.

445. Wilkins, L. C., W. R. Graves, and A. M. Townsend 1995. Development of plants from single-node cuttings differs among cultivars of red maple and Freeman maple. *HortScience* 30:360–62.

446. Wisniewski, L. A., L. J. Frampton, and S. E. McKeand. 1986. Early shoot and root quality effects on nursery and field development of tissue-cultured loblolly pine. *HortScience* 21:1185–86.

447. Woodridge, J. M., F. A. Blazich, and S. L. Warren. 2007. Propagation of selected clones of eastern redbud (*Cercis canadensis*) by stem cuttings. *Comb. Proc. Intl. Plant Prop. Soc.* 57:616–21.

448. Woodward, S. 1993. Micropropagation, rooting and survival of *Actinidia kolomikta*. *N. Amer. Plant Propagator* 5:9–12.

449. Xing, Z., M. F. Satchwell, W. A. Powell, and C. A. Maynard. 1997. Micropropagation of American chestnut: Increasing rooting rate and preventing shoottip necrosis. *In Vitro Cell, Dev. Biol. Plant* 33:43–89.

450. Xing, Z., W. A. Powell, and C. A. Maynard. 1999. Development and germination of American chestnut somatic embryos. *Plant Cell Tiss. Organ Cult.* 47:47–55.

451. Yusnita, S., R. L. Geneve, and S. T. Kester. 1990. Micropropagation of white flowering eastern redbud (*Cercis canaensis* var. alba L.). *J. Environ. Hort.* 8:177–79.

452. Zenkteler, M., and B. Stefaniak. 1991. The *de novo* formation of buds and plantlets from various explants of *Ailanthus altissima* Mill. culture *in vitro*. *Biolog. Plant.* 33:332–36.

453. Zhang, B., and L. P. Stoltz. 1989. Shoot proliferation of *Euphorbia fulgens in vitro* affected by medium components. *HortScience* 24:503–4.

454. Zhang, Z. M., and F. T. Davies. 1986. *In vitro* culture of crape myrtle. *HortScience* 21:1044–45.

21
Propagation of Selected Annuals and Herbaceous Perennials Used as Ornamentals

INTRODUCTION

Herbaceous plants are classified as annuals, biennials, or perennials, although the differences among these types may not be obvious. They may also be classified as hardy, semi-hardy, or tender. In general, the propagation procedures for such plants depend on their categories and the locality where they are to be grown. In the following list of plants, seed germination data are given for some species, including suggested approximate temperatures that should give the most rapid and complete germination (223). Temperatures provided are for media temperature and it should be recognized that the medium can be several degrees cooler than the air temperature. In addition, media temperature can be severely lowered if irrigated with cold water, which will slow or retard germination. In addition, many seeds from herbaceous plants can show some dormancy (125). The propagation methods indicated serve as a guide, but some variation from these methods may be necessary with individual cultivars (24, 27).

Acanthus mollis. Bear's breeches. Hardy perennial used in containers or perennial bed for its bold foliage. Seed germination is at 18 to 21°C (65 to 70°F). Seeds may benefit from scarification. Plants are commonly multiplied by dormant divisions of the crown. It can also be propagated by root cuttings.

***Achillea* spp.** Yarrow. Hardy perennial used in garden beds or as cut flowers. Seed germination is at 18 to 21°C (65 to 70°F) with light. Fast propagation is common from summer softwood cuttings that respond to auxin (266). Propagation by division is easy and necessary for good garden performance (286). *A. filipendulina* can be micropropagated (109).

Achimenes* spp. *(92). Cupid's Bower. Tender perennial commonly used in hanging baskets. Seeds germinated in a warm greenhouse [24 to 27°C (75 to 81°F)] are used for propagating this species. Plants grow from small, scaly rhizomes, which can be divided for propagation. Commercial propagation is from softwood cuttings in early spring (322) under mist using bottom heat. Leaf cuttings have also been successful (213).

Aconitum* spp. *(286). Monkshood. Hardy perennial grown in garden beds or as a cut flower. Seeds often show dormancy and must be moist-chilled below 5°C (41°F) for 6 weeks before planting. Considered difficult to propagate by seed, and fresh seed may be less dormant. Propagation is most often accomplished by division of the tuberous roots. Once established they should not be transplanted. All parts of the plants are poisonous. Can be micropropagated (318).

Adiantum. See Fern.

***Aegopodium* spp.** Goutweed. A hardy perennial used as a groundcover. Easily propagated by division.

African violet. See *Saintpaulia ionantha.*

***Agapanthus* spp.** Lily-of-the-Nile. Tender perennial. Grown for blue lily-like flowers. Some selections come true from seeds. The thick rhizomes can also be divided to produce new plants. Plants have been micropropagated (156).

Agastache foeniculum. Anise hyssop. A long-blooming hardy perennial. Seeds germinate at 18 to 21°C (65 to 70°F). Propagated by dormant division of the crown or stem cuttings that root easily under mist.

***Agave* spp.** Many species of succulents, including the century plant. Plants are perennial. Seeds should be sown in sandy soil when mature at 15°C (59°F). Reproduces vegetatively by offsets from base of plant, or the production of plantlets on the flower stalk of some species; these are removed along with roots and repotted in spring. Some species produce bulbils that can be used for propagation. Agave are commercially mass-produced by tissue culture for the distillery market (244).

Ageratum houstonianum. Ageratum (223). Half-hardy annual. Blue and white flowering bedding plants. Taller forms are grown as cut flowers. Seed germination is at 24 to 29°C (75 to 85°F) and may benefit from light. Ageratum may also be propagated by mist propagated stem cuttings with bottom heat [21 to 24°C (70 to 75°F)] (99).

***Aglaonema* spp.** Chinese Evergreen. An important foliage plant that is easily propagated by canes (long stems), shoot cuttings, division, or seeds. Canes should be treated as delicate leaf cuttings. Rooting is enhanced with IBA and bottom heat [24 to 29°C (75 to 80°F)] (239). Can also be micropropagated (69).

Agrostemma githago. Corncockle. Hardy annual used in the garden bed or container. Seeds germinate within 1 week at 20°C (68°F).

***Ajuga* spp.** Bugle flower. Hardy perennial. Blue or pink flowers on a spreading groundcover. Naturally layers itself by stolons. Can be propagated by seeds, cuttings, divisions, or tissue culture. Seeds germinate within 2 weeks at 20°C (68°F). Division is the most common form of multiplication. Variegated foliage types are from division or cuttings (286).

Alcea rosea. Hollyhock. Half-hardy biennial. Seed germination is at 20°C (68°F). Where winters are not too severe, sow seeds in summer, transplant in fall for bloom the following year. For annual production, sow seeds in a warm greenhouse in winter and transplant outdoors.

Alchemilla mollis. Lady's mantle. Short-lived hardy perennial grown for its interesting gray-green hairy foliage. Seed germination is at 18 to 21°C (65 to 70°F) with light. Plants can also be divided. Plants have been micropropagated (108).

***Allium* spp. (94).** Ornamental onions; also onion, chives, and garlic. Often propagated by seed that is germinated at 18 to 21°C (65 to 70°F). Some ornamental species may benefit from chilling stratification for 2 weeks and germination at a cooler temperature 15°C (59°F). Plants grow from bulbs, which produce offsets. Clumps can also be divided. Many species produce aerial bulbils. Can be micropropagated (331).

***Aloe* spp.** Succulents of the lily family. Propagated by seed in well-drained sandy soil. Germination is at 20 to 24°C (68 to 75°F). Plants produce offshoots that can be detached and rooted. Plants with long stems can be made into cuttings, which should be exposed to air for a few hours to allow cut surfaces to suberize. Some species are commercially micropropagated. *Aloe barbadensis* can be micropropagated from shoot explants (206, 221).

***Alstroemeria* spp.** Parrot lily (92). Half-hardy perennial. Grown commercially as a cut flower and pot plant. Propagated by division of the fleshy rhizome. Rhizomes are also multiplied in tissue culture to produce disease-free plants (47, 216). *Alstroemeria* as pot plants are propagated from seed. Fresh seeds are dormant, but moistened 1-year-old seeds germinate after 4 weeks of 18 to 25°C (65 to 77°F) followed by 4 weeks of 7°C (45°F) conditions (173).

Alyssum. See *Aurinia* or *Lobularia.*

Alyxia olivaeformis. Maile. An important foliage plant, native to Hawaii. It is propagated almost exclusively by seeds which need to be depulped (295). Rooting of single-node maile stem cuttings is improved by removing one-half leaf surface area, placing greenhouse-grown cuttings in water prior to treatment, and propagating in a shady cloth-covered greenhouse.

A five-second quick-dip in 3,000 to 8,000 ppm IBA improves rooting (296).

Amaranthus caudatus. Love-lies-bleeding (223). Half-hardy annual. Seed germination is at 21 to 24°C (70 to 75°F). Light may increase germination (18). Sow in warm greenhouse for later transplanting or sow out-of-doors when frost danger is past.

A. tricolor. Joseph's coat. Same propagation methods as *A. caudatus*. Sensitive to excess water.

Amaryllis belladonna. Belladonna lily (92). Perennial. Grows from bulbs outdoors in mild areas or in pots in cold climates. Propagate by bulb cuttings, separation of bulbs, or tissue culture (73, 95).

Amsonia tabernaemontana. Willow amsonia. Hardy perennial. Pale blue flowers above willowlike foliage. Propagation is from seed or summer softwood cuttings. Seed should be stratified at 2 to 5°C (34 to 40°F) for 4 to 6 weeks.

Anaphalis margaritacea. Pearly everlasting. Hardy perennial used in garden beds or rock gardens. Propagation is by seed or division. Seeds germinate within 1 week at 18°C (65°F).

Anchusa capensis. Bugloss. Hardy annual or biennial. Seeds germinate within 1 week at 20 to 22°C (68 to 72°F) with light. Sow seeds in summer for bloom next year or plant in greenhouse in winter for later transplanting to garden.

A. azurea. Perennial. Selected clones best propagated by root cuttings or clump division (97).

Anemone spp. Windflowers. Tender perennials.

A. coronaria. Poppy enemone (209). Seed germination is at 15°C (59°F) and may be sensitive to higher temperatures. Plants develop from tubers. There is an export market for *Anemone* tubers (128).

A. blanda. Greek windflower (92). Hardy perennial produced from a tuber. Propagated from seed or by division of the tuber into sections.

A. japonica. Japanese anemone and *A. xhybrida*. Hardy perennial. Since seeds do not come true, cultivars are propagated by division or by root cuttings (102). Roots are dug in fall and cut into 5-cm (2-in) pieces, which are laid in flats or in a cold frame, then covered with 2.5 cm (1 in) of soil. Plants can be potted after shoots appear.

Anigozanthus spp. Kangaroo paw (189). This native perennial Australian genus is used for cut flower production and for containerized plants. Seed supplies are often scarce, and germination rates of available seed are usually low and variable for many species. Hot-water and chemical pretreatment can be used to improve germination. Some hybrids are sterile and do not set seed at all. Clumps of rhizomes can be divided, but the rate of multiplication is low and unreliable. The most effective means of commercial propagation is through micropropagation (201).

Angelonia angustifolia. Summer snapdragon. A tender perennial commonly grown as an annual bedding plant. Can be propagated from seed or more commonly shoot tip cuttings. Seed is germinated at 22 to 24°C (72 to 76°F). Auxin-treated (2,500 ppm IBA) cuttings should be rooted under mist with bottom heat 24°C (75°F) (99).

Anthemis spp. Golden Marguerite, camomile. Hardy perennial. Seed is germinated at 20°C (68°F). Plants can be divided or propagated by stem cuttings.

Anthurium andraeanum. Anthurium. Remove offshoots with attached roots from the parent plant or root two- or three-leaved terminal cuttings under mist. *Anthurium* can be micropropagated using a vegetative bud explant (185). Seed propagation is at 25°C (77°F) and is a lengthy process requiring 1.5 to 3 years for flowering, and cultivars do not come true from seed (143).

Antirrhinum majus. Snapdragon. Tender perennial, treated as an annual. Seed is germinated with 27°C (80°F) days and 24°C (75°F) nights (65). Chilling seeds at 5°C (40°F) can improve germination. Light seeds for the first 3 days, and then provide dark to allow radicle growth, and move back to light when seedlings emerge. Softwood cuttings root readily. *A. majus* is tissue cultured (224, 235).

Aquilegia spp. Columbine. Hardy perennials. Seed is germinated at 21 to 24°C (70 to 75°F) in light. A short period of stratification for 3 to 4 weeks, moist-chilling at 5°C (41°F), can improve germination (116), but may not be necessary for all species. *A. chysantha* germinates better with alternating day/night temperature cycles of 25°C (76°F) day and 20°C (68°F) night (89).

Arabis spp. Rockcress. Hardy perennials. Seed is germinated at 18 to 21°C (65 to 70°F) and may respond to light (18). Softwood cuttings taken from new growth immediately after bloom root readily. Plants can be divided in spring or fall.

Arctotis stoechadifolia. African daisy. Half-hardy annual. Seed is germinated at 20°C (68°F). Sow indoors for later transplanting.

***Arisaema* spp.** Jack-in-the-pulpit. Perennials developing from rhizomes or tubers that are used in woodland or rock gardens. Propagated by division of the tuber or rhizome. Seeds require 60 to 90 days of stratification prior to sowing.

***Armeria* spp.** Thrift. Hardy evergreen perennials. Seeds emerge within 2 weeks at 15 to 21°C (59 to 70°F). Can also be propagated by clump division in spring or fall.

***Artemisia* spp.** Hardy perennial. *A. ludoviciana* can be used as a foliage plant and is propagated by division or stem cuttings. *A. schmidtiana* (wormwood) is a hardy perennial used as a specimen plant and is propagated by stem cuttings, rather than by division.

Arum italicum 'Pictum.' Painted arum. Hardy perennial produced from a tuber. Evergreen foliage and naked red seed heads are attractive for the perennial garden. Propagated by division of the tuber. Seed requires stratification.

Aruncus dioicus *(A. sylvester)*. Goat's beard. A hardy perennial used as a specimen or border plant. Seeds benefit from a cold stratification treatment of 5°C (40°F) for 4 weeks. Seeds germinate at 16°C (60°F). Usually propagated by dormant crown divisions.

***Asarum* spp.** Ginger. Perennial plants grown in the woodland or rock garden. Can be propagated by seed sown in the fall (125) or by division of the creeping rhizome.

Asclepias tuberosa. Butterfly weed. Hardy perennial. Seed is germinated at 21 to 24°C (70 to 75°F). Fresh seed may need chilling. Vegetative propagation is from 3-cm-long root cuttings (105) or stem cuttings taken before the plants flower. Plants should not be disturbed once established.

A. curassavica. Bloodflower. Tropical perennial. Propagated by seed or by rooting softwood cuttings. Long taproot makes division difficult.

Asparagus asparagoides. *A. plumosus* (fern asparagus), *A. sprengeri* (Sprenger asparagus). Tender perennials. Propagated by seeds at 24 to 30°C (75 to 86°F). Sow seeds soon after they ripen, since seeds are short-lived (27). Cuttings can be made of young side shoots taken from old plants in spring; clumps can be divided.

Asplenium nidus. Bird's nest fern. *See* Fern.

***Aster* spp.** Hardy perennials. Seed is germinated at 18 to 21°C (65 to 70°F). Cultivars are propagated by lifting clumps in fall and dividing into rooted sections, discarding the older parts. If stem cuttings are used, the stems must be from juvenile material and rooted under long days or the resultant plant will be short and quickly flower (100). *A. frikartii* 'Monch' has been micropropagated commercially.

***Astilbe* spp.** Astilbe. Hardy perennials. Propagated by division in early spring when 2.5 cm (1 in) tall and then again following flowering (29). Seed germination is slow and produces a mixed progeny. Seed is germinated at 16 to 21°C (60 to 70°F) in the light (286).

***Astrantia* spp.** Masterwort. Perennial with unusual and attractive flowers. Generally propagated by division. Seeds require cold stratification.

***Athrium* spp.** Painted fern. *See* Fern.

Aubrieta deltoidea. Aubrieta. Hardy perennials, sometimes treated as annuals. Seed is germinated at 18 to 21°C (65 to 70°F). Clumps are difficult to divide; cuttings may be taken immediately after blooming.

Aurinia saxatilis (formerly *Alyssum saxatile*). Goldentuft (223). Short-lived hardy perennial grown for its early yellow flower display. Seed is germinated at 15 to 21°C (60 to 70°F). Sow in summer for bloom the following year. Germination may be stimulated by light or exposure of moist seeds at 15°C (50°F) for 5 days (18). Propagate by division or by softwood cuttings in spring. Double forms must be propagated by cuttings or division.

***Baptisia* spp.** False Indigo. Hardy perennial. *Baptisia* is a hard-seeded legume and requires scarification. Seeds germinate in 3 weeks at 21°C (70°F). Stem cuttings can be rooted and they respond to auxin treatment (266). Plants can be divided, and micropropagation is also possible (21).

***Begonia* spp.** Begonia (262, 263). Tropical perennials. Seeds, which are very fine and need light, emerge in 2 to 4 weeks at 22°C (72°F). Best seed germination is at 28°C (82°F) for 5 to 7 days followed by 25°C (78°F) until seedlings emerge (223). Sow on moist, light medium with little or no covering. Begonia species, tuberous begonias, and wax begonias are propagated by seed, but other types are propagated vegetatively.

Fibrous-Rooted Begonias. Wax begonias, Christmas begonias, and others are propagated by leaf cuttings or softwood cuttings taken from young shoots in spring or summer. The cytokinin PBA was more effective in bud and shoot development from leaf cuttings than BA or kinetin (88).

Rhizomatous Types. Various species and cultivars, including Rex begonia plants, are divided or their rhizomes are cut into sections. Propagation is usually by leaf cuttings, but stem cuttings also will root. Treatment of leaf cuttings with a cytokinin increases the number of plantlets produced per leaf (320). *B. evansiana* produces small tubercles, which are detached and planted.

Tuberous Begonias (130). In addition to seed propagation, these can be grown from tuberous stems, which are divided into sections, each bearing at least one growing point. Leaf, leaf-bud, and short-stem cuttings (preferably with a piece of tuberous stem attached) will root readily. Can be micropropagated (220, 294).

Begonia can also be micropropagated using leaf petioles (212), petiole explants (243, 264, 265, 320), and somatic embryos (229).

Belamcanda chinensis. Blackberry lily. Summer-blooming hardy perennial with orange blossoms are held above iris-like foliage. Commercially propagated by seed that emerge in 3 to 4 weeks at 18°C (65°F). Division is also possible while plants are dormant.

Bellis perennis. English daisy. Hardy perennial often treated as an annual or biennial. Seed is germinated at 21 to 24°C (70 to 75°F) and may respond to light (18). Plants may also be multiplied by division.

Bergenia cordifolia. Hardy perennial with evergreen cabbage-like foliage and attractive spring flowers. Can be propagated by seed or division. Seeds require chilling stratification of 5°C (41°F) for 6 to 8 weeks. Germination is within 2 weeks at 21 to 24°C (70 to 75°F) in the light. *Bergenia* is commercially micropropagated (220, 294).

Bleeding heart. See *Dicentra*.

***Boltonia* spp.** Boltonia. Autumn-blooming, hardy perennial resembling asters. Seed is germinated at 20°C (68°F). Commercially propagated by division while plants are dormant.

Bouvardia ternifolia (153). Scarlet bouvardia. An outstanding perennial with scarlet tubular flowers that bloom from midsummer to frost in Texas and New Mexico. It is propagated by semi-hardwood cuttings throughout the growing season.

Brachycome iberidifolia. Swan River daisy. Annual plant grown in hanging baskets or as bedding plants. Seeds germinate in 1 week at 21°C (70°F) in light.

Brassica oleracea. Flowering cabbage or kale. Cool season plants grown for fall display of their colorful foliage. Seed is germinated at 21°C (70°F).

***Brodiaea* spp.** (syn. *Triteleia*) (134). Perennial plants grown as cut flowers and produced from a corm. *Brodiaea* is propagated from seed that requires stratification for 8 weeks at 3°C (37°F) (133), or from cormels. Plants can also be propagated through liquid tissue cultures to develop corms (157).

Bromeliads. About 2,000 species of tropical herbs or subshrubs in 45 genera, of which the pineapple (*Ananas*) is the best known. Propagation is mainly by seeds or by asexual division of lateral shoots, but micropropagation has been used successfully with some species (148). Conditions vary for the successful micropropagation of *Cryptanthus* (178), *Guzmania, Puya* (308), *Tillandsia,* and *Vriesea* species (205).

***Browallia* spp.** Amethyst flower (223). Tender, blue-flowered perennial often treated as an annual. Can be used as flowering pot plant indoors in winter. Seed is germinated at 24°C (75°F). Softwood cuttings can be taken in fall or spring.

Brunnera macrophylla. Siberian bugloss. Hardy perennial with blue forget-me-not flowers that appear in the spring followed by large green leaves. Can be propagated by seed, division, or root cuttings. Commercially micropropagated (240).

Cactus (70, 234). Large group of many genera, species, and some cultivars. Tender to semi-hardy perennials. Seed propagation can be used for most species, but seeds often germinate slowly. Sow fungicide-treated seed in well-drained sterile mixture and water sparingly, but do not allow medium to dry out. Pieces of stem can be broken off and rooted as cuttings (306), or small offsets, which root readily, can be removed. Allow offsets to dry for a few days to heal (suberize) cut surfaces before rooting. High humidity during rooting is unnecessary, but bottom heat is beneficial. Grafting is used to provide a decay-resistant stock for certain kinds and to produce unusual growth forms. For example, the pendulous *Zygocactus truncatus* is sometimes grafted on tall erect stems of *Pereskia aculeata*. Intergeneric grafts are usually successful. A type of cleft graft or splice graft is used. The stem of the stock is cut off, and a wedge-shaped piece is removed. The scion is prepared by removing a thin slice from each side of the base, which is fitted into the opening in the stock. The scion is held in place with a pin or a thorn. A grafting adhesive can be used to adhere scions to stocks of transversally cut (tip-grafted) cactus (328). The completed graft is held in a warm greenhouse until healed (66, 106, 328). The development of cacti shoots by micropropagation can be extremely rapid in comparison with greenhouse-germinated seedlings (20), and where poor branching limits propagation by traditional vegetative propagation methods (76, 154, 190).

Caladium hybrids (323). This tropical perennial, grown for its strikingly colorful foliage, produces tubers. Propagation is by removing the tubers from the parent plant at the end of the 4- to 5-month dormancy period just before planting. Commercially, tubers are cut into 2-cm pieces (chips), each containing at least 2 buds ("eyes"). Caladiums do best out-of-doors when planted after the minimum night temperature is above 18°C (65°F) or as pot plants maintained with night temperatures of 18 to 21°C (65 to 70°F) and day temperatures of 24 to 29.5°C (75 to 85°F). Dried caladium seed has a short storage life. Seeds require light and temperatures between 25 to 30°C (68 to 86°F) (57). Can be micropropagated (132).

***Calibrachoa* spp.** A tender perennial grown as an annual bedding plant. Flowers resemble small petunias. Commonly propagated from stem cuttings under mist and bottom heat 24°C (75°F) (99).

Calamagrostis acutiflora. Feather reed grass. Hardy perennial. Upright grass with attractive flowering plumes. Propagation is by division in late spring.

***Calathea* spp.** Tropical perennials grown as indoor foliage plants. Propagation is from division or commercially by micropropagation (237).

***Calceolaria* spp.** Pocketbook plant. Tender perennials often grown as annuals and seasonal pot plants. Seed is germinated at 21°C (70°F) with light. Germination percentages in some species can be low. In some cases, propagation is also possible by softwood cuttings.

Calendula officinalis. Pot marigold. Hardy annual grown as a bedding plant or cut flower. Plants can provide winter bloom in mild climates from seed sown in late summer. Seed is germinated at 21°C (70°F) in the dark.

Calla. See *Zantedeschia* spp.

Callistephus chinensis. China aster (297). Annual grown as a bedding plant or cut flower. Seeds germinate at 20°C (68°F).

Caltha palustris. Marsh marigold. A hardy perennial used around water gardens or ponds. Seeds require chilling stratification and may benefit from temperatures below freezing. Following stratification, sow seeds at 18°C (64°F). Propagation is by division.

***Camassia* spp.** Quamash (92). Hardy perennials produced from bulbs. Blue or white spikes emerge in the spring above grass-like foliage. Propagation is by offsets of the bulb.

Campanula carpatica. Tussock, Bellflower. Hardy perennial. Seed is germinated at 18 to 21°C (65 to 70°F) and responds to light. Plants can also be divided and stem cuttings root easily.

C. isophylla. Falling stars. Perennial often grown as an annual pot plant for indoor use or hanging baskets. Seeds are used for hybrid cultivars and germinate in 3 weeks at 18 to 21°C (65 to 70°F) with light. Cuttings can also be rooted after treatment with 1,000 ppm IBA from vegetative stock plants kept under short days (100).

C. lactiflora. Bellflower. Hardy perennial. Seeds germinate best after 2 to 4 weeks of stratification at 4°C (40°F).

C. medium. Canterbury bells. Hardy biennial. Seeds, which germinate in 2 to 3 weeks at 21°C (70°F), are sown in late spring or early summer for bloom the following year.

C. persicifolia. Peach bells. Hardy perennial. Seeds germinate in 2 to 3 weeks at 18 to 21°C (65 to 70°F) in light. Small offsets can be detached and rooted.

Many of the named cultivars of *Campanula* species cannot be produced by seed, so division or cuttings are used. Cuttings are produced from the rhizomatous growth of stock plants, and rooting occurs from the etiolated base. Cuttings are placed in peat-perlite media and given basal heat under glass or in a tunnel (England) (269).

***Canna* spp.** Canna (92). Tender perennial. Cultivars do not come true from seed. Seeds, which have hard coats, must be scarified before planting. Seeds germinate in 2 weeks at 21 to 24°C (70 to 75°F). Cultivars are propagated by dividing the rhizome, keeping as much stem tissue as possible for each growing point. In mild climates, rhizomes are divided after the shoots die down in the fall or before growth starts in the spring. In cold climates, the plants are dug in fall, stored over winter, and divided in spring for transplanting outdoors when frost danger is over. Plants can be micropropagated (182).

Capsicum annuum. Christmas or ornamental pepper. Tender annual used as a bedding plant or seasonal pot plant. Plants are most commonly propagated by seeds germinated at 21 to 24°C (70 to 75°F) that emerge within 2 weeks. Stem cuttings also root readily.

Carnation. See *Dianthus caryophyllus.*

Catananche caerulea. Cupid's dart. Hardy perennial. Seed is germinated at 18 to 21°C (65 to 70°F). Plants may be divided in the fall.

Catharanthus roseus. Vinca (65). Tender annual. Vinca is a major bedding plant grown from seed. Optimum temperature for germination is 24 to 27°C (75 to 78°F) in the dark. Do not keep seeds too moist. Vinca is the commercial source of the cancer drugs vincristine and vinblastine.

Celosia argentea* and *C. spicata. Cockscomb. Tender annual. Both plumed and cockscomb (fasciated) cultivars are available as bedding plants and cut flowers. Seed is germinated at 24°C (75°F).

***Centaurea* spp.** Tender and hardy perennials. Seeds of *C. cyanus* (cornflower or bachelor button) and *C. moschata* (sweet sultan) emerge in 1 to 2 weeks at 18 to 21°C (65 to 70°F).

C. hypoleuca (knapweed), *C. macrocephala* (Globe centaurea), and *C. montana* are hardy perennials propagated by division or seed. Seeds of *C. montana* are germinated at 21 to 24°C (70 to 75°F). *C. macrocephala* has been micropropagated (149).

Centranthus ruber. Red valerian. Hardy perennial. Rose or white-colored flowers are produced throughout the summer. Seed is germinated at 15 to 18°C (60 to 65°F). Stem cuttings are also possible.

Cephalotus follicularis. Australian pitcher plant. Perennial carnivorous plant. Usually propagated from IBA-treated leaf or stem cuttings. Easily propagated by tissue culture (1).

Cerastium tomentosum. Snow-in-summer. Hardy perennial. Seed is germinated at 20°C (68°F). Easily propagated by division in the fall or by softwood cuttings in summer.

***Ceratostigma* spp.** Plumbago. Hardy perennials with bright blue flowers on trailing stems. Propagation is from softwood cuttings under mist, or division while dormant.

Cheiranthus cheiri (synonym is *Erysimum asperum*). Wallflower. Semi-hardy perennial often treated as a biennial. Seeds germinate in 1 week at 16°C (60°F) and may respond to light (18). Choice plants may be increased by cuttings taken in early summer.

***Chelone* spp. (286).** Turtlehead. A hardy perennial used in wet areas. Propagation is by division or by cuttings. Seeds may benefit from 2 to 4 weeks chilling stratification.

***Chionodoxa* spp.** Glory-of-the-snow. Hardy perennial bulb. Primarily grown from seed. Ripe seeds are stored through the summer at 17°C (63°F) and sown outdoors in September to stratify over winter (92). Bulb cuttage (see *Hyacinthus*) and offsets are successful for vegetative propagation.

Chlorophytum comosum. Spider plant. Propagated mainly by planting miniature plants developing at ends of stolons. Stolon formation is under daylength control; short days (12 hours or less daily) promote stolon production (131). It can also be propagated by division.

Chrysanthemum. See *Dendranthema*.

Chrysogonum virginianum. Goldenstar. Trailing hardy perennial grown for its yellow flowers. Propagation is by division in spring or fall.

***Cimicifuga* spp.** Cohosh. Hardy perennials for the border or woodland garden. White flower spikes appear in late summer. Propagated by seed or division. Seed is germinated at 18°C (65°F) in light. Plants are divided in spring.

Clarkia amoena (synonym *Godetia*). Hardy annuals grown as cut flowers or pot plants (8). Sow seeds in early spring; they germinate at 21°C (70°F).

Cleome spinosa. Spiderflower. Tender annual. Germinate seeds at 21 to 22°C (70 to 72°F) with light. Seeds may benefit from brief chilling stratification for 5 days. Can also be propagated by division.

***Clivia* spp. (92).** Tropical perennial grown indoors for its colorful floral display. Plants can be propagated by seeds sown as soon as they are ripe at 21°C (70°F). It is more common to divide plants after they finish blooming. Can be micropropagated (115).

Cobaea scandens. Cup and saucer vine. Tropical vine grown for its large purple flowers. Propagation is from seeds that germinate in 1 week at 21°C (70°F).

Codiaeum variegatum. Croton (247). Tropical perennial. Propagated by leafy terminal cuttings in spring or summer. Tall "leggy" plants can be propagated by air layering. Stem cutting root number (root initiation) was unaffected by bottom heat (28°C, 83°F) or increasing light intensity; however, root length was increased (314). During shipping of cuttings, exposure to light results in shorter roots. Unrooted cuttings can be shipped for 5 to 10 days at 15 to 30°C (59 to 86°F) (313).

Colchicum autumnale. Autumn crocus, saffron (92). Hardy perennial that grows from corms. Seeds are sown as soon as they are ripe in the fall but may require chilling over winter to germinate. Several years are required for plants to reach flowering size. Propagation is from offsets of the corm (see *Crocus*). Can be micropropagated (141).

Coleus. *See Solenostemon.*

Consilida ambigua. Larkspur. Upright annual plants similar to delphinium in bloom. Seeds germinate at 18°C (65°F) and can self-sow in the garden.

Convallaria majalis. Lily-of-the-valley. Hardy perennial that grows as a rhizome, whose end develops a large underground bud, commonly called a "pip." In fall, the plants are dug, and the pip, with attached roots, is removed and used as the planting stock. Digging should take place in early autumn, with replanting completed by late autumn. Single pips may be stored in plastic bags in the refrigerator, then planted in late winter for spring bloom. Microproagation has also been successful (309).

Convolvulus spp. Morning glory. Trailing annuals used in hanging baskets or as bedding plants. Seeds benefit from hot water scarification and germinate at 21°C (70°F).

Cordyline spp. Ti plant (*C. terminalis*) is easily propagated by cuttings and by micropropagation (90, 184). Other *Cordyline* species can be seed-propagated with seeds germinating in approximately 4 months with 30°C (86°F) days and 20°C (68°F) nights. A vegetative propagation technique for *C. australis, C. kaspar,* and *C. pumillo* by division of the underground stems of stock plants has been described (238).

Coreopsis spp. Hardy annuals and perennials. *C. grandiflorum* (tickseed) is a short-lived perennial usually grown from seed. Seedlings emerge within 2 weeks when sown at 18 to 24°C (65 to 75°F) in light. Growth regulator treatment or seed priming (61, 256) can improve seed germination.

C. verticillata can be propagated by cuttings and is hardy to zone 3 (81). Perennial clumps can be divided in spring or fall.

Cortaderia selloana. Pampass grass. A half-hardy, ornamental grass with large feathery plumes. Propagated by clump division. Micropropagation is possible from immature flower parts (250).

Corydalis spp. Garden perennials. Propagated by seeds that that should be sown soon after they are ripe because they have a short storage life. Sow outdoors or provide 6 weeks warm [24°C (75°F)], moist conditions followed by 6 to 8 weeks' chilling [4°C (45°F)] and germination at 10°C (50°F). Commonly multiplied from dormant crown divisions.

Cosmos bipinnatus **and** ***C. sulphureus.*** Half-hardy annuals grown as cut flowers or garden plants. Seeds usually germinate within 1 week at 21°C (70°F).

Crassula argentea. Jade plant. Succulent perennial grown as an indoor pot plant. Can be propagated at any time by leaf or stem cuttings.

Crinum spp. Tender bulbs grown for seasonal flower display. Propagation is from offsets of the bulb or bulb cuttage techniques such as bulb chipping (see Chapter 15). Plants can be micropropagated (267, 301).

Crocosmia spp. Crocosmia (92). Hardy perennial used for cut flowers or border plants. Plants are propagated from offsets of the corm.

Crocus spp. *(33).* Hardy perennials that grow from corms. Seeds germinate as soon as ripe in summer; several years are required for plants to flower. When leaves die in fall, plants are dug and corms and cormels are separated and planted. Tissue culture propagation is from corm fragments, isolated buds, or flower parts (111, 145).

Crossandra infundibuliformis. Flame flower. Tender perennial grown as a specialty pot plant for indoor use. Can be propagated by seeds or cuttings. Seeds emerge sporadically over 2 to 3 weeks at 27 to 29°C (80 to 85°F), but show better germination at alternating 29 to 21°C (85 to 70°F) day/night regimes (100). Most commercial plants are grown from stem cuttings treated with 3,000 ppm IBA and rooted under mist or tents using bottom heat at 24°C (75°F). Micropropagation is possible from axillary shoot explants (129).

Cucurbita pepo var. ***ovifera.*** Ornamental gourds. Tender annuals. Seed is germinated at 20° to 30°C (68 to 86°F).

Cuphea spp. Mexican heather; bat-faced cuphea. Tender perennials grown as bedding plants or in baskets. Cuttings root readily without hormone under mist with bottom heat at 24°C (75°F).

Cyclamen spp. *(92, 193, 199, 203).* Tender and half-hardy perennials. Plants grow from a large tuberous stem. Cyclamen is propagated best by seeds that germinate in 3 to 4 weeks in the dark at temperatures about 20°C (68°F), no higher than 22°C. Seeds are planted from midsummer to midwinter. Germination is best in a medium of peat moss to which pulverized limestone and mineral nutrients have been added for a pH of 6 to 6.5 (321). Seedlings under commercial production conditions flower after 30 weeks (100). The tuberous stem can be divided for the production of a few plants identical to the parent. Short shoots of cyclamen with two to three leaves are easily rooted in 3 weeks when given a 10-second dip of 3,000 to 5,000 ppm K-IBA under intermittent mist, 21°C (70°F) basal heat, and 1 perlite:1

vermiculite rooting medium (193). Tissue culture is used to multiply F_1 hybrids (259).

Cymbalaria muralis. Kenilworth ivy. Semi-hardy perennial. Seed is germinated at 12°C (54°F). Self-seeds readily. Softwood cuttings or clump division may be used.

Cynara cardunculus. Cardoon. Large bold perennial grown for its thistle-like flowers that appear in late summer. Seeds germinate in 3 weeks at 24°C (75°F). Plants can also be multiplied by division while dormant and by root cuttings.

Cynoglossom amabile. Chinese forget-me-not. Hardy biennial grown as an annual. Seed is germinated at 20°C (68°F) and may respond to light (18).

Cypripedium. Lady slipper. *See* Orchids.

Dahlia (93). Tender perennials consisting of hundreds of cultivars. Commercially propagated by seed, division, or stem cuttings. Seed is germinated at 26°C (80°F) when planted indoors for later transplanting outdoors. Large plants can be dug in the fall before frost and stored over winter at 2 to 10°C (30 to 50°F) and covered with a material such as soil or vermiculite to prevent shriveling. In spring, when new sprouts begin to appear, divide the clumps so that each tuberous root section has at least one or two buds. Plant outdoors when danger of frost is over. Dahlias are often commercially propagated by softwood cuttings (93) from stems forced in the greenhouse in early spring. Tissue culture is used to recover virus-free plants (319).

Daylily. *See Hemerocallis* spp.

***Delphinium* spp.** Hardy perennials, usually propagated by seeds, which germinate at 18 to 24°C (65 to 75°F). 'Giant Pacific' germinates better with alternating 26°C (75°F) day and 21°C (70°F) night temperatures. There may be some benefit to chilling dry seed at 3°C (35°F) for 1 week prior to sowing seeds (59, 223). Seeds are short-lived and should be used fresh or stored in containers at low temperature and reduced moisture. Seeds can be sown outdoors in spring or summer to produce plants that flower the following year. Delphiniums can be propagated by softwood cuttings or root cuttings (97). Clumps can be divided in spring or fall, but such plants tend to be short-lived. Plants can be micropropagated (227).

Dendranthema xgrandiflorum (formerly *Chrysanthemum xmorifolium*). Garden and greenhouse chrysanthemum and *Argyranthemum frutescens* (formerly *C. frutescens*, Marguerite). Hardy and semi-hardy perennials. Shoots 6 to 8 cm (2.5 to 3 in) long are rooted as softwood cuttings under mist and with 1,500 ppm IBA usually as a talc (99). The best source of new cuttings is a mother block (or increase block) grown in an isolated area away from the producing area kept under long days (43, 101). Such plants are grown in programs designed to keep them pathogen- and virus-free and true-to-type (27). Disease-indexed stock plants are regenerated through micropropagation (53). Unrooted cuttings can be held for as long as 30 days at 0.5°C (33°F). Chrysanthemum is readily micropropagated by shoot-tip and petal-segment explants (26, 171, 253).

Dianthus caryophyllus (36). Carnation. Tender to semi-hardy perennial grown as an annual that has many cultivars used in the florist's trade. Seeds germinate easily but are used primarily for breeding. Carnations are readily propagated by softwood cuttings (144) with mist and auxin treatment. The best source of cuttings is a mother (or stock) block isolated from the producing area. This block originates from cuttings taken from stock plants maintained under a program designed to keep them pathogen- and virus-free and true-to-type. As with chrysanthemum, carnation stock blocks are periodically replenished with meristem-tip culture for disease-free plants. Conventional cutting propagation then proceeds with these clean stock blocks (see Chapter 16). Commercially, rooted cuttings are produced by propagation specialists for sale to growers. Lateral shoots ("breaks") that arise after flowering are removed and used as cuttings. Cuttings root in 2 to 4 weeks. Carnations can also be micropropagated on a large scale using shoot-tip explants (104). Tissue culture is also possible from petal explants (166, 219) and somatic embryos have been recovered from leaf explants (327).

D. barbatus Sweet William. Short-lived perennial that is grown as a biennial. Seeds are germinated in the summer and seedlings emerge in 1 to 2 weeks at 21°C (70°F). Seedlings are overwintered in frames and flower the following spring. Plants can also be propagated by cuttings or division.

D. chinensis, D. plumarius,* and *Related Species. Garden pinks. Hardy perennials, although some kinds are grown as annuals or biennials. Seeds germinate readily at 15 to 21°C (60 to 70°F). Softwood cuttings are taken in early summer and rooted to produce next year's plants. Layering and division also can be used.

D. gratinanopolitanus. Cheddar pinks. A popular herbaceous perennial. Seeds germinate in 1 to 2 weeks at 21°C (70°F). The cultivars may be propagated by stem cuttings or division.

Diascia **spp.** Twinspur. Annual plants grown as bedding or container plants. Seed germination is in 1 week at 21°C (70°F) in light. Shoot tip cuttings root easily without auxin or with 2,500 ppm IBA (99).

Dicentra **spp.** Bleeding heart. Hardy perennials. Seeds are sown in late summer or fall for overwintering at low temperatures; alternatively, seeds should be stratified for 6 weeks below 5°C (41°F) before planting. Seeds will germinate in 3 to 4 weeks at 10 to 13°C (50 to 55°F). Divide clumps in spring or fall. Stem cuttings can be rooted if taken in spring after flowering. Root cuttings about 7.5 cm (3 in) long can be taken from large roots after flowering. *D. exima* is commercially propagated from seeds or division. *D. spectabilis* is usually propagated by division of the woody rhizome or from stem cuttings. It can also be micropropagated.

Dictamnus albus. Gas plant. Hardy perennial. Seed germination is difficult and inconsistent. Stratify seeds at 1 to 5°C (34 to 41°F) for 3 to 4 months. Cultivars can be propagated from root cuttings. Gas plant is easily micropropagated from shoot explants (164). Plants should not be disturbed after establishment; thus, division is not done. Contact with the leaves of gas plant can produce severe dermatitis.

Dieffenbachia **spp.** Dumbcane. Tropical perennial. Cut stem into 5-cm (2-in) segments, with one or two nodes per section, and place horizontally, half exposed in sand or other well-drained media. New shoots and roots will develop from nodes. If plant gets tall and "leggy," the top may be cut off and rooted as a cutting, or the plant may be air layered. Leaves and stem are poisonous and may cause rashes on skin.

Digitalis **spp.** Foxglove. Hardy biennial or perennial plants. Seed is germinated at 15 to 18°C (60 to 65°F) in light. Sow seeds outdoors in spring, transplant to a nursery row at 9-inch spacing, and then transplant to a permanent location in fall. Perennial species is increased by clump division.

Dimorphotheca **spp.** Cape or African marigold. Half-hardy annual with daisy-like flowers. Seeds germinate within 1 week at 21°C (70°F).

Dionaea muscipula. Venus flytrap. Carnivorous plants, which have unique appearance, unusual mode of life, and are in demand by plant collectors. Although plants can be grown from seed or division, they are usually propagated by tissue culture, from leaves, adventitious buds, and peduncle explants (32, 214).

Dodecatheon maedia. Shooting star. Hardy perennial. Unique flowers in white or purple resemble tiny darts. Easily divided when plants are dormant (41). Seed is sown in autumn to stratify over winter for spring emergence.

Doronicum **spp.** Leopard's bane. Hardy perennial grown for its early yellow, daisy-like flowers in spring. Seed is germinated at 20°C (68°F). Divide plants in spring or fall.

Dorotheathus bellidiformis (formerly *Mesembryanthemum criniflorum*). Livingstone daisy. Drought-tolerant annual grown as a bedding plant. Seedlings emerge within 2 weeks at 18°C (65°F) in light.

Dracaena **spp.** Variable group of tropical perennial foliage plants which are available in bush, cane, tree, and stump forms. Seed is germinated at 30°C (86°F). Some species are propagated from leaf-bud cuttings that are treated with IBA and rooted under intermittent mist. *D. fragrans,* which is an important cane form used for interiorscapes, is propagated from cane stem cuttings which are cut into 30- to 183-cm (1- to 6-ft) sections waxed on the distal (top) end; basal ends are treated with IBA and placed in a porous medium without intermittent mist under shade (indoors or field-propagated). Branching of canes during field propagation is done by cutting one-third to one-half way through the cane, which results in the development of lateral buds anywhere from directly below to 15 cm (6 in) below the cut (75). Micropropagation is from stem explants (74, 90, 310).

Drosera **spp.** Sundew. Easily grown carnivorous plants that produce leaves with conspicuous glandular hairs that trap insects. *Drosera* can be propagated by seed, root cuttings (68), or easily from tissue culture (9). Non–tuber-forming plants can also be propagated from IBA-treated leaf cuttings.

Dryoptris **spp.** *See* Fern.

Dusty miller. *See Senecio* spp. or *Tanacetum ptarmiciflorum.*

Dyssodia tenuiloba. Dahlberg daisy. Tender annual. Yellow daisylike flowers on compact edging plants. Propagated from seed germinated at 18 to 21°C (65 to 70°F) in light. Germination can be erratic due to a percentage of dormant seeds.

D. pentacheta. Perennial. A low-growing Texas wildflower suitable for xeriscapes. Propagated from softwood cuttings treated with IBA under mist or by tissue culture (330).

Echeveria. *See* Succulents.

Echinacea **spp.** Purple coneflower. Hardy perennial. Purple, orange, yellow, or white flowers typical of the sunflower family are attractive as garden plants or as cut flowers. Seed is germinated at 21 to 25°C (70 to 78°F). Germination can be erratic and improved by seed priming or chilling stratification at 15°C (59°F) (256, 316, 317). Garden plants can be divided. *Echinacea* hybrids are commercially micropropagated (71, 139).

Echinops **spp.** Globe thistle (223). Hardy perennials. Unique metallic blue flowers on thistlelike plants. Seed is germinated at 15 to 20°C (60 to 68°F). Plants may be divided in spring. Root cuttings, 5- to 7.5-cm (2- to 3-in) long, may be made in the fall and planted in sandy soil in a cold frame.

Echium **spp.** Viper's bugloss. Annual and perennial plants grown as bedding or container plants. Plants produce large blue spikes of flowers. *E candicans* (Pride of Madeira) is a drought tolerant plant often seen in seaside gardens. Propagation is from seed sown at 15 to 20°C (60 to 68°F). Larger plants can be propagated by stem cuttings.

Epimedium **spp.** Barrenworts. Hardy perennials. Popular ground cover for shady areas. Propagated by division in the spring (174). Can be micropropagated (6).

Epiphyllum **spp.** Leaf-flowering cactus. Tender perennial. Seeds do not germinate well when fresh but will after 6 to 12 months' storage if planted in a warm greenhouse. Propagated readily by leaf cuttings (botanically, modified stems called phyllocades) or by grafting to *Optunia*. Can be micropropagated (190). *See* Cactus.

Epipremnum aureum. Golden pothos. Among the most important commercially produced foliage plants. Producers cut the long vines into single-node, leaf-bud cuttings for propagation. Leaf-bud cutting propagation is enhanced with light intensity of about 2,000 ft-c for stock plants and basal heat of 28°C (83°F) (312). Stock plants should be maintained at four to five nodes (14 to 15 leaves), with a 3-cm or longer internode section below the node, and a fraction of the old aerial root retained on the cuttings for most rapid axillary shoot development (315). Maintained in the juvenile phase by cutting propagation. Mature phase has a much larger leaf and flowers.

Episcia **spp.** Flame violet. Tropical trailing plants grown indoors in hanging baskets. Seed propagation is at 24°C (75°F), but plants root easily from stem cuttings using mist and bottom heat or plantlets taken from ends of the runners.

Eranthis hyemalis. Winter aconite (92). Hardy perennial produced from tubers. Early yellow flowering plants popular in the rock garden. Commercially propagated from seed (see *Chionodoxa*).

Eremurus bungei. Foxtail lily (92). Tender perennial produced from a tuberous root. Tall flowering spikes are commercially grown as a cut flower. Propagated from seed or more commonly by division. Seeds apparently have an after-ripening requirement that is satisfied by dry storage at 30°C (86°F) for 2 months (329). Plants are commercially propagated by division of the tuberous roots.

Erigeron **spp.** Fleabanes. Hardy perennial. Blue, pink, or white daisy-like flowers with yellow centers. Can be propagated by seed, division, and stem cuttings. Usually commercially propagated by seed that germinates at 21 to 24°C (70 to 75°F).

Eryngium **spp.** Sea-holly. A diverse species of perennials, used as specimen and border plants or as cut flowers. Propagation by division is possible, but a long taproot makes transplanting difficult. Root cutting propagation is the commercial method for species not coming true from seed (286). Seeds have a warm-cold stratification requirement of 4 weeks at 21°C (70°F), followed by 6 weeks of 3°C (38°F) and then a warm temperature of 18 to 23°C (65 to 75°F).

Erythronium **spp.** Dog's tooth violet or trout lily (92). Hardy woodland perennials with recurved lily-like flowers. Plants are used in rock or alpine gardens as well as being naturalized. Plants naturally multiply by seeds that are spread by ants and by producing new plants at the end of slender stolons. Propagation is by division.

Eschscholzia californica. California poppy. Hardy annual grown en masse for its flower display. Seed is germinated at 21°C (70°F). Seeds can be sown outdoors in fall in mild climates or in early spring in colder areas. Tends to self-sow. Seedlings may be difficult to transplant because of a long taproot.

Eucomis **spp.** Pineapple lily (92). Summer-blooming perennial bulb. Propagation is by offsets or from seeds. Micropropagation by twin scaling offers commercial potential (19).

Euphorbia **spp.** Euphorbia, spurge. Perennials used as border and sometimes specimen plants. Plants are also grown for latex production. Propagation is easy by division. The thick seeds of *E. epithymoides* (*polychroma*) take 15 to 20 days to germinate at 18 to 21°C (65 to 70°F) (286). *In vitro* techniques have also been developed from stem explants (241).

Eustoma grandiflorum. Lisianthus (223). Annual. Grown as a pot plant or cut flower. Seed is germinated at 22°C (72°F). Plants have been micropropagated (112).

Evolvulus glomeratus. Prostrate plant with blue flowers grown as a bedding plant or in hanging baskets. Propagation is by stem cuttings using mist.

Exacum affine. Exacum or Persian violet. A popular greenhouse-grown pot plant with fragrant flowers, multiple blooms, and good postharvest quality. Can be propagated from seed or cuttings. Seeds germinate within 2 weeks at 21 to 24°C (70 to 75°F). Herbaceous stem cuttings root easily with bottom heat [22°C (72°F)]. Exacum can also be micropropagated (299).

xFatshedera lizei. Tree ivy. An intergeneric cross between *Hedera helix* and *Fatsia japonica*. Propagated by auxin-treated stem cuttings using mist and bottom heat [24°C (75°F)] or by air layering.

***Felicia* spp.** Blue daisy. Tender, trailing plants most often grown for hanging baskets. Propagation is from stem cuttings using mist with bottom heat [21°C (70°F)].

***Festuca* spp.** Blue fescue. Hardy perennial grass. A clump-forming grass with blue foliage. A warm season grass, fescue should be divided in the spring or fall. *F. ovina glauca* comes relatively true from seeds that germinate within 1 week at 21°C (70°F).

Fern (233, 251, 291). Many genera and species. Spores are collected from the spore cases on lower sides of fronds. Examine these sporangia with a magnifying glass to be sure they are ripe but not empty. Place fronds with the spores in a manila envelope and dry for a week at 21°C (70°F). Screen them to separate spores from the chaff. Transfer to a vacuum-tight bottle and store in a dry, cool place. Sow spores evenly on top of sterilized moist substrate (e.g., two-thirds peat moss, one-third perlite in flats), paying particular attention to sanitation. Leave 1 inch space on top and cover with a pane of glass. Use 18 to 24°C (65 to 75°F) air temperature; bottom heat may be helpful. Keep moist, preferably using distilled water to avoid salt injury.

Spores germinate and produce mosslike growth composed of many small gametophytic prothalli. Fertilization requires free water on the prothallus to allow motile male gametes to reach the female archegonium. After fertilization occurs, a small leafy sporophyte appears in 2 to 3 months on the surface of the prothallus. Transplant the developed sporophyte to a greenhouse substrate for further growth into the fern plant. Procedures have also been developed for *in vitro* fern propagation from spores (177, 188).

Several vegetative propagation methods are possible. Many ferns grow from rhizomes, which can be divided. In few species, like sword fern (*Nephrolepsis tuberosa*), underground tubers are produced on thin stolons and can be used for propagation. Certain other ferns (e.g., *Cystopteris bulbifera, Polystichum setiferum*) produce small plantlets (also called bulblets) along the leaf surface or leaf tip (*Camptosorus rhizophyllum*). These can be removed and used to produce a new plant (291).

Cultivars of the sterile (non-propagatable by spores) Boston fern group (*Nephrolepsis*) are now largely micropropagated starting with rhizome tips (7, 51). This micropropagation technique is also applicable to other fern genera, such as *Adiantum* (maidenhair fern), *Alsophila* (Australian tree fern), *Pteris* (brake fern), *Microlepia, Playcerium* (staghorn fern), and *Woodwardia* (chain fern).

Ficus. See Chapter 20 (82, 91).

***Filipendula* spp.** Queen of the prairie. Hardy perennials used in garden beds or naturalized areas. Propagation is from dormant divisions or root cuttings.

Fragaria. Pink Panda. Perennial strawberry plant grown for its ornamental pink flowers rather than fruit, which it rarely produces. Propagation is from rooted plantlets produced on runners.

***Freesia* spp. (158).** Tender perennials produced from a corm. Commercially produced as a cut flower. Seeds planted in fall germinate in 4 to 6 weeks and will bloom the next spring. A germination temperature of about 18.5°C (65°F) is best (127). Plants are commercially propagated from cormels that are planted in spring and dug in fall. Micropropagation is also used to produce corms, as well as disease-free plants (22, 236).

***Fritillaria* spp.** Checker lily (92). Perennials produced from non-tunicate bulbs. Interesting group of spring-flowering bulbs of which *F. imperialis* (crown imperial) and *F. melaegris* are the best known. Propagated from offsets, bulb scaling, and bulb cuttage (chipping). Micropropagation is also successful (183).

Fuchsia xhybrida. *Fuchsia magellanica* hybrids. Fuchsias are tender perennials treated as annuals that are utilized as hanging baskets, containers, or trained to tree form on standards. The most effective way to propagate cultivars is by cuttings. Softwood cuttings root easily under mist or high humidity with bottom heat at 21°C (70°F). Auxin (1,000 ppm IBA) can improve rooting. For optimum cutting production it is best to maintain stock plants on short day, 10-hour photoperiods (311).

***Gaillardia* spp.** Blanketflower (223). Annuals and hardy perennials. Seed is germinated at 21 to 24°C (70 to 75°F) in light (18). Perennial kinds can be started from root cuttings or may be divided in spring or fall but are not long-lived.

Galanthus **spp.** Snowdrop (92). Hardy perennial produced from an annual tunicate bulb. Bulbs are planted in the fall for bloom the following spring. Offsets are removed when bulbs are dug. *G. nivalis* and *G. elwesii* are mostly propagated by seed sown as soon as they are ripe in the spring. Bulb cuttage (chipping) and twin scale micropropagation can also be used (73, 156).

Galtonia candicans. Summer hyacinth (92). Tender bulb with small pendant flowers that are fragrant. Propagated by offsets of the bulb. Plants can also be micropropagated (326).

Gasteria. *See* Succulents.

Gaura lindheimeri. Hardy perennial that produces wispy flowers on long stems. Seeds germinate within 2 weeks at 21°C (70°F). Can also be multiplied by dormant division of the crown.

Gazania* spp. *(223). Tender perennial often grown as an annual. Seed is germinated at 21°C (70°F). Divide clumps after 3 or 4 years.

Gentiana **spp.** Gentian. Many species, mostly hardy perennials, although some are annuals and biennials. Plant fresh seed in the fall to overwinter outdoors. Seeds germinate in 1 to 4 weeks at 20°C (68°F). However, seeds need to be stratified for 3 weeks at 2°C (36°F) or treated with 300 ppm GA_3 (39). Seedlings are very delicate and should not be transplanted until roots are established during the first month. Cutting propagation is used for some white cultivars, which have poor seed germination. Micropropagated liners are now becoming available (147).

Geranium. *See Perargonium.*

Geranium **spp.** Cranesbill. Hardy perennials. True geraniums are popular herbs and perennial garden plants. Geraniums can be propagated by seed, division, stem, or root cuttings. Seeds germinate in 2 to 4 weeks at 21°C (70°F), but fresh seeds may have a hard seed coat that requires scarification. Commercial propagation is most often by division when plants are dormant or from root cuttings taken in the winter. Root cuttings are sensitive to rot if overwatered. One approach is to allow buds to form on root pieces by holding them in "sweat" boxes (polyethylene tents) at near 100 percent humidity prior to planting in a potting medium.

Gerbera jamosonii. Transvaal daisy. Tender perennial. Seed is germinated at 20°C (68°F); it is important to use fresh seed (64). Alternately, basal shoots from the rhizome can be used as cuttings. Commercially, micropropagation from shoot tips is used for rapid, large-scale multiplication (187, 218).

Geum **spp.** Avens. Hardy perennials used in garden beds and cut flowers. Common perennial easily propagated by seeds. Seed is germinated at 18 to 21°C (65 to 70°F). Can also be propagated by crown division in spring or fall.

Gladiolus (78). Tender perennial grown from a corm. Popular cut flower. Seed propagation is used for developing new cultivars. Seeds are planted in spring either indoors for later transplanting or outdoors when danger of frost is over (see Chapter 15). Commercial propagation is from cormels or division of the corm, leaving at least one bud (eye) per piece (202). *In vitro* techniques using buds or liquid-shake culture have improved multiplication rates (25, 196, 332).

Gloriosa **spp.** Gloriosa lily (92). Tender perennial. Vines are produced from tuberous stems. Unique flowers with recurved petals are grown as cut flowers or container plants. Propagation is from daughter tubers that form at the shoot base of the original tuber. Micropropagation is possible from tuber explants (114).

Gloxinia. *See Sinningia speciosa.*

Godetia. *See Clarkia.*

Gomphrena globosa. Globe amaranth (223). Button-like white or purple flowers make this a popular cut flower and bedding plant. Propagate from seeds that emerge in 1 to 2 weeks at 24 to 25°C (75 to 78°F).

Gunnera **spp.** Tender perennials producing extremely large leaves on creeping rhizomes. Propagation is from seed or division. Seeds will germinate in 3 weeks at 18°C (65°F) with light. Division of the creeping rhizome is also possible while plants are dormant.

Guzmania **spp.** *See* Bromeliads.

Gypsophila **spp.** (*G. elegans*). Baby's breath. Annual. Grown as a cut flower. Seed germinates in 2 to 3 weeks at 21 to 26°C (70 to 79°F).

G. paniculata. Hardy perennial. Started by seed as above. Plants can be divided in spring and fall. Terminal stem cuttings treated with 3,000 ppm IBA can be rooted under mist (99). Double-flowered cultivars can be difficult to root and can be propagated by seeds that yield approximately 60 percent double plants. Double-flowering types have also been wedge-grafted on seedling *G. paniculata* (single-flowering) roots or crowns. Grafting can be done in summer and fall, using outdoor-grown plants for rootstocks and placing them in a cold frame for healing of the graft; grafting is also done in winter and early spring, using greenhouse-grown stock plants. *G. paniculata* can be micropropagated using shoot-tip explants (186).

Haworthia. *See* Succulents.

***Hedychium* spp.** Ginger lily. Tropical plants producing attractive spikes of flowers from creeping rhizomes. Propagation is from division of the rhizome.

Helenium autumnale. Sneezeweed. Hardy perennial. Sunflower-like blooms in unique colors produced in late summer in the perennial garden. Seed is germinated at 22°C (72°F) with light. Cultivars are increased by division. Separate rooted shoots in spring, line-out in nursery, then transplant in fall and winter.

Helianthemum nummularium. Sunrose. Hardy perennial. Drought-tolerant spring blooming groundcover. Seed is germinated at 21 to 24°C (70 to 75°F). Cultivars are propagated by softwood cuttings taken from young shoots in spring. Transplant to pots and place in permanent location the following winter or spring. Division of clumps is also possible, but plants tend to be short-lived.

Helianthus annuus. Sunflower. Hardy annual. Popular as a cut flower. Seeds germinate within a few days at 20 to 30°C (68 to 86°F). Some sunflower cultivars show dormancy and benefit from 6 weeks of chilling stratification (46). *H. decapetalus* and other hardy perennial species can be propagated by seeds or increased by division.

Helichrysum bracteatum (current name *Bracteantha bracteata*). Strawflower (223). Annual. Popular cut flower for drying. Propagated by seeds that emerge in 1 to 2 weeks at 21 to 24°C (70 to 75°F).

H. petiolare. (Licorice plant). A trailing plant with interesting foliage grown as a companion plant in hanging baskets. Propagated from stem cuttings treated with 3,000 ppm IBA under mist (99).

Heliconia* spp. *(85). These tropical ornamental herbaceous perennials are prized for their showy inflorescences. Commercially produced as a cut flower, they are easily propagated by division of the rhizomes. Micropropagated from rhizome buds (222).

***Heliopsis* spp.** Heliopsis. Hardy perennial with sunflower-like flowers. Seed is germinated at 18 to 21°C (65 to 70°F). Divide clumps in fall.

***Heliotropium* spp.** Heliotrope. Tender perennial usually grown as an annual. Seed germinates within 1 week at 21°C (70°F) and may respond to light (18). Plants will also root from softwood tip cuttings from stock plants grown under short days to remain vegetative (99).

***Helleborus* spp.** Hellebore, Christmas and lenten rose. Perennials used as border and woodland gardens. One of the earliest plants to bloom in the perennial garden. Propagation by seed is very slow. *H. lividus* has a combinational dormancy requirement of warm stratification [21°C (70°F)] for 8 to 10 weeks followed by cold stratification of 3°C (37°F) for 8 to 10 weeks; seeds still may take up to 2 years to germinate. Division is the most common method, carried out by carefully separating the crown (286). Roots and leaves are poisonous. Hellebore can be micropropagated (52).

Hemerocallis* spp. *(87). Daylily. Hardy perennial. Seed propagation is used only to develop new cultivars and requires 6 weeks of stratification; germination takes 3 to 7 weeks at 16 to 21°C (60 to 70°F) (286). Divide clumps in fall or spring, separating into rooted sections, each with about three offshoots. Clones can also be micropropagated using flower petals and sepals as explants (10, 207).

***Hepatica* spp.** Hardy perennial used in the rock garden or wildflower garden. Propagated by seed or division. Seeds have epicotyl dormancy (see Chapter 7) and may take 2 years to emerge. Division is in early spring before growth begins. Plants can be micropropagated (225).

Hesperis matronalis. Sweet rocket. A biennial or short-lived perennial grown in garden borders or in mass display. Seeds germinate in 1 week at 21 to 26°C (70 to 75°F).

***Heuchera* spp.** American alumroot, coralbells. Perennials used as border plants for their foliage and flowers. Propagated by dormant crown division. Seed is germinated at 18°C (65°F) in light. Commercially micropropagated from stem explants (279).

***Hibiscus* spp.** Mallow. Hardy perennials. Large saucer-shaped flowers in vibrant colors. Cultivars are propagated by division while dormant. Seed is germinated at 21 to 26°C (70 to 75°F).

***Hippeastrum* spp.** Amaryllis (228). Tender bulbous perennial. Garden plant and popular indoor flowering bulb. Propagated by offsets or micropropagation. Bulb offsets will flower the second year. Bulb cuttings (chipping) can be made in late summer. Dry membranous seeds are borne in dehiscing capsules. Seeds germinate under warm conditions at 20 to 30°C (68 to 86°F). Seedlings take 2 to 4 years to produce flowers. Micropropagated by the twin scaling method (73, 151).

***Hosta* spp.** Plantain-lily. Herbaceous perennials which are used for massed plantings or as specimen plants for their foliage and flowers. Propagated by clump division in spring. One producer removes the apical dominance of the crown (terminal) bud, slices (divides) the remaining clumps into quarters, places the quarters outdoors in trays (England) which are winter-protected, and then plants when they begin to shoot. Offset formation has also been increased by spraying crowns with cytokinin (123). It takes 3 years to produce a flowering plant from seed. Micropropagation is being used with new cultivars to speed up propagation (208).

Houttuynia cordata. An aggressive hardy perennial plant growing by rhizomes. Tolerates wet conditions and is often confined to a shallow pond. Propagation is from division or softwood cuttings.

***Hoya* spp.** Wax flower. Tropical vines grown in indoor hanging baskets for its attractive foliage and flowers. Propagation is from single node stem cuttings that benefit from bottom heat.

Hummulus lupulus. Hops. Tender perennial vine grown as an annual for quick coverage of a support. The yellow foliage 'Aureus' is most common in gardens. Propagated from stem cuttings.

Hunnemannia fumariifolia. Goldencup. Tender perennial often grown as an annual. Seeds germinate in 2 to 3 weeks at 20°C (68°F). For bloom first year, sow seeds early indoors then transplant outdoors when danger of freezing is over.

***Hyacinthus* spp.** Hyacinth (226). Hardy, spring-flowering perennial; bulbs are planted in the fall. Removal of offset bulbs gives small increase. For commercial propagation, new bulbs are obtained by scoring or scooping mature bulbs (see Chapter 15). Micropropagation, using segments of the bulb, leaf, inflorescence, or stem as an explant, is successful (23, 155). Seeds may be planted outdoors in fall, but up to 6 years are required to produce blooms.

***Hymenocallis* spp.** Spider lily (92). Tender bulb producing large white flowers with extended tepals, suggesting the common name. Propagated by offsets from the bulb. Plants can also be micropropagated (325).

***Hypericum* spp.** St. John's wort. Hardy herbaceous and woody perennials grown for their saucer-shaped yellow flowers. Also grown as a cut flower. Seed is germinated at 21°C (70°F). Plants are also propagated by division of the crown or creeping rhizome. Stem cuttings will also root easily.

Hypoestes phyllostachya. Polka dot plant (223). Tender annual. Grown as a bedding plant or indoor plant because of unique spotted foliage. Propagated from seeds that emerge in 1 to 2 weeks at 21 to 24°C (70 to 75°F). Stem cuttings root easily without auxin.

***Iberis* spp.** Candytuft. Hardy annual and perennial species. Seed is germinated at 15 to 18°C (60 to 65°F) in light. Cultivars are propagated by softwood cuttings in summer or plants are divided in fall.

***Impatiens* spp.** Touch-me-not, balsam. Impatiens are important bedding plants and include impatiens (*I. walleriana*), New Guinea impatiens (*I. hawkeri*), balsam (*I. balsamina*), and several new hybrids. They can be propagated by seeds or cuttings. Seed germination is at 24 to 25°C (75 to 78°F) and benefits from light (63). Light seeds for the first 3 days and then move to darkness until seedling emergence. Plants can be started by terminal cuttings treated with 2,500 ppm IBA under mist with bottom heat [22°C (72°F)]. Impatiens have also been micropropagated (283, 284).

***Incarvillea* spp.** Incarvillea. Hardy perennial producing large gloxinia-like flowers in the spring. Seed is germinated at 20°C (68°F). Divide in fall or, preferably, in spring. Basal cuttings taken in the spring can also be rooted.

***Ipomoea* spp.** Morning glory, ornamental sweet potato. Tender perennials grown as an annual. Seeds germinate within 1 week at 18 to 21°C (65 to 70°F). Notch seed coats or soak seeds overnight in warm water before planting.

   I. batatas. Sweet potato vine. Popular annual vine with bright yellow or dark purple foliage used in containers or for bedding. Cuttings are easily rooted without auxin under mist.

***Iresine* spp.** Bloodleaf. Tender perennial used as an indoor plant or outdoors as a container or bedding plant. Seeds germinate at 24°C (75°F) with light. Softwood cuttings root easily without auxin.

***Iris* spp. (96).** Perennials. There are several different groups of hardy or semi-hardy iris, which grow either from rhizomes or from bulbs. Rhizomes are divided after bloom. Discard the older portion and use only the vigorous side shoots. Leaves are trimmed to about 15 cm (6 in).

   Bulbous species follow a typical spring-flowering, fall-planting sequence. The old bulb completely disintegrates, leaving a cluster of various-sized new bulbs. These are separated and graded, the largest size being used to produce flowers, the smaller for further growth.

   Seeds, which are used to propagate species and to develop new cultivars, should be planted as soon as ripe after being given a moist-chilling period; germination is often irregular and slow. Removal of embryo from the seed and growing it in artificial culture has yielded prompt germination in some cases. Iris can be micropropagated, which greatly hastens production of new cultivars over the customary division of rhizomes (155, 161, 305).

***Ixia* spp.** Corn lily (92). Tender, summer- or fall-flowering perennials grown from corms. In cold climates, these are dug in fall and stored over winter. Small cormels are removed and planted in the ground or in flats to reach flowering size, as is done with gladiolus. Can be micropropagated (293).

***Ixora* spp. (245).** Ixora. Tender perennial used for indoor gardens. Several species are used in Hawaii as landscape and flowering pot plants. Many species are easy to propagate by cuttings. Difficult-to-root *I. acuminata* three-node cuttings had optimal rooting when given a 5-second dip of IBA-NAA, both at 2,500 ppm.

***Justicia* spp.** Shrimp plant (*J. brandegeeana*) and Brazilian plume (*J. carnea*) are tropical plants grown as landscape plants where hardy and indoor flowering pot plants. Propagation is from softwood cuttings with bottom heat [24°C (75°F)].

Kalanchoe blossfeldiana. A tropical succulent perennial grown as an indoor flowering pot plant. Plants can be propagated by cuttings or tissue culture, but they are more commonly propagated from terminal cuttings. Cuttings are rooted easily under mist or poly tent with bottom heat [22°C (72°F)]. Disease-free cuttings are available from commercial specialists that have been cleaned in tissue culture (31). Plants can also be micropropagated (270).

K. diagremontiana. (formerly *Bryophyllum*). A succulent plant grown for its interesting habit of forming plantlets (often called foliar embryos) along the margin of the leaf. These plantlets can be removed for propagation.

Kangaroo Paw. See *Anigozanthus* spp.

Kirengeshoma palmata. Yellow wax bells. Herbaceous perennial producing pendulous yellow blossoms for shady garden beds. Propagated by crown division when plants are dormant.

***Knautia* spp.** Herbaceous perennial plants, many of which were formerly in *Scabiosa*. Used in perennial beds. Propagation is from dormant crown division or stem cuttings from new growth in spring.

***Kniphofia* hybrids (*K. tritoma*) (286).** Torch lily or poker plant. Perennials used as specimen plants, borders, and cut flowers. Seeds germinate at 18 to 24°C (65 to 75°F) in light. Many of the cultivars are propagated by crown division.

***Lachinelia* spp.** Cape cowslip (92). Tender bulbous perennials most often seen as container plants in display areas. Propagated by bulb offsets or micropropagation (176).

Lamiastrum galeobdolan. Yellow archangel. Hardy perennial used as a ground cover for shade gardens. Grown for the attractive variegated foliage. Only cultivars are grown and these are usually propagated by softwood cuttings. Division is also possible.

Lamium maculatum. Spotted deadnettle. Hardy perennial with trailing stems that are used as a ground cover for shade. Propagation is by stem cuttings treated with 2,500 ppm IBA under mist. Plants can also be divided.

Lantana sellowiana, L. camara. Lantana. Tender perennials treated as landscape plants where hardy and as annuals where tender. Lantana has become invasive in some tropical countries. Seeds germinate at 20°C (68°F). Softwood cuttings root easily under mist.

Lathyrus latifolius. Perennial pea vine. Hardy perennial. Seeds germinate at 13 to 18°C (55 to 65°F). Clumps may be divided.

L. odoratus. Sweet pea. Hardy annual grown for garden display or cut flowers. Seed germinates in 2 weeks at 20°C (68°F). Notching seed or soaking in warm water may hasten germination. Plant outdoors in fall where winters are mild, in spring where winters are severe.

***Lavandula* spp.** Lavender. Half-hardy perennial native to Mediterranean. Essential oils from these plants are important for the perfume industry. Seeds germinate at 18 to 24°C (65 to 75°F) in light. Stem cuttings are treated with 2,500 ppm IBA and rooted under mist with bottom heat. Divide clumps in the fall. Micropropagation is from hypocotyl explants (55) or leaf-derived callus (300).

Lavatera trimestris. Tender herbaceous perennials used as annuals. Plants produce showy hibiscus-like flowers. Seeds germination is at 21°C (70°F).

Leonotis leonurus. Lion's ear. Tender perennial usually grown as a container plant, greenhouse displays, or in garden beds. Propagation is from stem cuttings taken prior to flowering.

Leontopodium alpinum. Edelweiss (223). Hardy perennial. Mounded plants with silvery leaves and unique flowers. Seeds germinate at 20 to 22°C (68 to

72°F). Division is also possible. Plants can also be multiplied in tissue culture (146).

***Lespedeza* spp.** Bushclover. Considered an herbaceous perennial in more northern latitudes or a semi-woody shrub in the southern United States. Excellent landscape shrub for massing and screening with its bluegrass foliage and purple flowers. Seeds can be direct-sown after harvest or scarified with a 15-minute acid treatment if stored. Roots easily from softwood cuttings with 1,000 ppm IBA.

Leucanthemum xsuperbum (formerly *Chrysanthemum superbum*). Shasta daisy. Hardy perennial but often treated as a biennial, since it is short-lived. Propagated by seeds at 18 to 21°C (65 to 70°F) that emerge in about 1 week. Division is from side shoots that have roots. Can be commercially micropropagated.

Leucocorne ixioides. Glory of the sun. Tender bulb grown as a pot plant or for cut flowers. Propagation is from bulb offsets.

***Leucojum* spp. (92).** Spring snowflake. Hardy perennial bulb. Bulbs have been collected for sale from native populations to the point of endangerment. Plants produced from seed take 4 to 5 years to flower. Propagation is normally done by separating bulbs, which is done by digging bulbs after foliage has turned brown. Can be micropropagated (73, 277).

Lewisia cotyledon. Semi-hardy perennial commonly seen in rock, alpine, or dish gardens. Seeds germinate at 21°C (70°F) (2). Cuttings from the basal rosette are possible. Leaf cuttings can also be used to produce new plants.

***Liatris* spp.** Gayfeather (203). Hardy perennials used in garden beds, naturalized prairies, as well as cutflowers (16). Seeds may benefit from stratification at 4°C (34°F) for 6 weeks and will germinate at 21°C (70°F), but flower development can take 2 years. Asexual propagation is by woody corms or rhizomes, which are divided in the spring (286). *L. spicata* has been micropropagated (289).

***Ligularia* spp.** Large herbaceous perennials producing yellow composite flowers and ornamental foliage. Seeds germinate at 18°C (65°F) in dark. Plants can also be divided.

***Lilium* spp. (30, 204).** Lily. Hardy perennials. These are spring- and summer-flowering plants grown from scaly bulbs; most have a vertical axis, but in some species growth is horizontal with a rhizomatous structure. Lilies include many species, hybrids, and named cultivars. Seed propagation is used for species and for new cultivars. Seeds of different lily species have different germination requirements (252).

Immediate seed germinators include most commercially important species and hybrids (*L. amabile, L. concolor, L. longiflorum, L. regale, L. tigrinum,* Aurelian hybrids, Mid-Century hybrids, and others). Germination is epigeous; shoots generally emerge 3 to 6 weeks after planting at moderately high temperatures. Treat seeds with a fungicide to control *Botrytis*. Sow 3/4 inches deep in flats during winter or outdoors in a seed bed in early spring. Dig the small bulblets in fall, sort for size, and replant with similar sizes together. Plants normally grow 2 years in a seed bed and 2 years in a nursery row before producing good-sized flowering bulbs.

Another group consists of the slow seed germinators of the epigeal type (*L. candidum, L. henri,* Aurelian hybrids, and others), in which seed germination is slow and erratic; the procedures used are essentially the same as described above. The most difficult group to propagate are the slow seed germinators of the hypogeous type (*L. auratum, L. bolanderi, L. canadense, L. martagon, L. parvum, L. speciosum,* and others). Seeds of this group require 3 months under warm conditions for the root to grow and produce a small bulblet, then a cold period of about 6 weeks, followed by another warm period in which the leaves and stem begin to grow. This sequence can be provided by planting the seeds outdoors in summer as soon as they are ripe, or by planting seeds in flats and then storing under appropriate conditions to provide the required temperature sequence. These procedures are described in Chapter 15. *L. longiflorum* can also be propagated by leaf cuttings.

Vegetative methods of propagation include natural increase of the bulbs, such as bulblet production on underground stems (either naturally or artificially), aerial stem bulblets (bulbils), or scaling. Outer and middle scales are used for scale propagation to increase the number of forcible commercial bulbs (200).

Lilies can be micropropagated from bulb scales (288) and pedicels (198). *L. longiflorum* can also be propagated by leaf cuttings.

***Limonium* spp.** Statice. *L. sinuatum* is a perennial herb, native to the eastern Mediterranean, which is grown commercially around the world as a cut flower for both fresh- and dry-flower arrangements. *L. latifolia* is a hardy perennial. Plants are propagated by seeds that germinate at 21°C (70°F). Statice has been micropropagated (54, 138).

***Linaria* spp.** Toadflax, butter and eggs. Hardy annual and perennials. Seeds germinate in 1 to 2 weeks at 18°C (65°F) in light. Perennial species may take 2 years to produce bloom from seed. Perennial types can be divided in spring or fall.

***Linum* spp.** Flax. Hardy annual and perennial species with blue or yellow flowers. Used in perennial beds and rock gardens. Seeds germinate in 2 to 3 weeks at 18 to 24°C (65 to 75°F). Divide clumps of perennial species in fall or spring.

***Liriope* spp.** Lily turf. Hardy perennial. Vigorous ground cover. Can be propagated by seed or division. Seeds have morphological dormancy and require warm stratification [21 to 30°C (70 to 85°F)] for germination (110). However, commercial propagation is by division in spring or autumn. Plants can be micropropagated (119).

Lisianthus. See *Eustoma*.

Lithodora diffusa. Semi-hardy prostrate perennial grown for intense blue flowers. Propagated by stem cuttings in summer or from division.

***Lithops* spp.** Living stones. Interesting succulents with plants adapted to mimic stones. Propagation is from seeds sown at 18 to 24°C (65 to 75°F). Some species produce offsets that can be divided.

Lobelia erinus. Lobelia. Tender perennial grown as an annual. Seeds germinate at 24 to 26°C (75 to 80°F), but seedling growth is slow. May respond to light (18). Mature plants, if potted in the fall and kept in greenhouse over winter, can be used to provide new growth for cuttings to be taken in late winter. Commercial cuttings are treated with 2,500 ppm IBA under mist.

***Lobelia cardinalis* (cardinal flower), *L. siphilitica* (blue cardinal flower) and *hybrids*.** Hardy and half-hardy perennials. Seeds germinate at 20 to 24°C (68 to 75°F). Species self-seeds. Divide clumps in fall or spring.

Lobularia maritima. Sweet alyssum (223). Perennial grown as a hardy annual. Seed germinates in 1 to 2 weeks at 26 to 28°C (75 to 82°F) in light and blooms appear in 6 weeks. Often seeded with more than one seed per plug for better pack development.

Lunaria annua. Honesty or money plant. Biennial, sometimes grown as an annual. Seeds germinate at 20°C (68°F).

L. rediviva. Hardy perennial. Propagated at 20°C (68°F). Also increased by dormant division.

***Lupinus* spp. and *hybrids*.** Hardy annuals and perennials. All lupines have physical seed dormancy and require scarification. Seeds germinate at 20°C (68°F).

L. texensis. (Texas bluebonnet) *L. texensis* has been micropropagated from cotyledonary node explants (302). Sow seeds in spring or summer, or propagate by cuttings taken in early spring with a small piece of root or crown left attached.

***Lychnis* spp.** Campion. Mostly hardy perennials, but some are grown as annuals or biennials. Seeds germinate at 20°C (68°F) or propagated from dormant crown divisions.

***Lycoris* spp.** Spider lily, surprise lily (92). Tender and semi-hardy perennials from a tunicate bulb. Propagation is by bulb offsets, which are removed when the dormant bulbs are dug. These are replanted to grow larger. Bulb cuttings can also be used for increase. Micropropagation by twin scaling has also been developed (152).

***Lysimachia* spp.** Loosestrife. Hardy perennial. Vigorous plants with striking white or yellow flowers. Trailing plants are grown for their colorful foliage. Can be propagated from seed, division, or softwood cuttings. Seeds germinate at 18 to 21°C (65 to 70°F). Terminal stem cuttings root easily under mist without auxin.

Lythrum salicaria. Purple loosestrife. Hardy perennial. Long-blooming plants produce purple flowering spikes. *Lythrum* can be an invasive weed in wet habitats and some states prohibit its use. Propagation is easy from softwood cuttings. Division is possible but difficult, because the roots are woody.

Macleaya cordata. Plume poppy. Aggressive, hardy perennial from a creeping rhizome. Produces bold foliage and large plumes of small white flowers. Propagation is from division of the rhizome.

Malva alcea. Hollyhock mallow; and *M. moschata.* Musk mallow. Drought-tolerant perennials utilized in border plantings. Seeds germinate within 2 weeks at 21°C (70°F). Plants can also be propagated by division.

Mammillaria. See *Cactus*.

Maranta leuconeura. Prayer plant. Tender perennials grown as indoor potted plants. They are normally propagated asexually by stem cuttings because seeds are difficult to germinate. Cuttings are rooted in humidity tents for 4 to 6 weeks. *M. leuconeura* 'Kerchoviana' can be micropropagated (103).

Marigold. See *Tagetes* spp.

Matteuccia. Ostrich fern. See Fern.

Matthiola incana. Common stock (223); *M. longipetala bicornis.* Evening scented stock. Perennials grown as biennial or annual plants most often used as cut flowers or for bedding. Seed germinates at 18 to 21°C (65 to 70°F) and may respond to light (18). Seeds are sown in summer or fall for winter bloom, in late winter indoors for spring bloom, or outdoors in spring for summer bloom.

Mazus reptans. Hardy perennial that forms a creeping mat of foliage and purple-blue flowers. Propagated from division or cuttings of the creeping stem.

Meconopsis **spp.** A group of alpine annual and short-lived perennials. Poppy-like flowers can be white, red, yellow, or blue depending on the species. Must have cool summers to survive. Seeds germinate at 13°C (55°F). Most often the seeds are sown in a cold frame in fall for spring emergence. Perennial plants can also be divided before growth begins in spring.

Melampodium paludosum. Annual. Rounded plants produce yellow daisy-like flowers throughout the growing season. Seeds germinate at 18°C (65°F). Self-sowing can be a problem in the garden. Plants also root from softwood cuttings.

Mertensia **spp.** Hardy perennial plants usually grown in woodland gardens but also at home in the perennial bed. *M. virginiana* (currently *M. pulmonarioides*) Virginia blue bells produces nodding blue flowers early in the spring. Propagation is most often from division in dormant plants.

Mesembryanthemum **spp.** See Succulents.

Mimulus **spp.** Monkey flower. Includes many species of tender to hardy plants. Mostly perennials but sometimes grown as annuals. Seeds germinate at 15 to 21°C (60 to 70°F). Softwood cuttings taken from young shoots can be rooted under mist.

Miscanthus sinensis. Maiden grass. There are many cultivars of this popular perennial grass. They make excellent specimens in the landscape with their showy feathery inflorescences. Cultivars are propagated by division. *Miscanthus* is a warm-season grass and should be divided in late spring. *Miscanthus* can be micropropagated (124).

Molluccella laevis. Bells of Ireland. Half-hardy annual grown as a garden plant or cut flower. Seed germinates at 10°C (50°F) but may do better with 30°C (86°F) days alternating with 10°C (50°F) nights (223). Plants can be difficult to transplant.

Monarda didyma. Bee balm (286). A perennial garden plant native to eastern North America. Can be propagated by seed, which germinates at 16 to 21°C (60 to 70°F). May benefit from a brief (1 week) chilling stratification. Can also be propagated by softwood cuttings or by crown division.

Monstera deliciosa. Swiss cheese plant. Cut-leaf philodendron. Easily propagated by rooting sections of the main stem, by stem cuttings, or by air layering. Seeds germinate at 21 to 29°C (70 to 85°F).

Muscari **spp.** Grape hyacinth (255). Hardy perennials from tunicate bulbs. Plants bloom in spring; bulbs become dormant in fall when they are lifted and divided by removing bulb offsets. Seed propagation can also be used. Bulb scooping and scoring produce bulb offsets. Micropropagation is easy from leaves, scales, or flower parts (231).

Mussaenda erythrophylla. Red Flag Bush. Tropical evergreen perennials used as a landscape plant where hardy and as a container plant in other regions. Flowers are subtended by colorful bracts. Propagation is from stem cuttings with bottom heat (275). Plants can be micropropagated (84).

Myosotis sylvatica. Forget-me-not. Hardy biennial or short-lived perennial. Grown for its blue or pink flowers in the spring. Plants may naturalize from seeds. Seeds germinate at 20°C (68°F). Sow in summer and transplant to permanent location the following spring. *M. scorpioides* is a perennial started from seed; division in spring is also used.

Narcissus. Daffodil (135). Hardy perennials; spring-flowering tunicate bulbs. Vegetative propagation procedures are described in Chapter 15. Commercial propagation is mainly from natural bulb offset production. Vegetative techniques include twin scaling, chipping (almost the same method as twin scaling, except that the bulb is cut across the root plate into 8 to 16 pieces with up to 2 bulbils developing per section), and micropropagation (73, 136, 150).

Nasturtium. See *Tropaeolum majus*.

Nelumbo lutea. American lotus; *N. nucifera.* Sacred Lotus. Aquatic plants for water gardens. Lotus produces one of the longest-lived seeds, which have physical dormancy and must be scarified. Seeds germinate readily at 24°C (75°F). Lotus is usually propagated vegetatively through rhizome division. Rhizome cultures have been established *in vitro* from excised embryos (167).

Nemesia strumosa. Capejewels. Half-hardy annual used as bedding plants. Seeds germinate at 13 to 18°C (55 to 65°F) in darkness. Temperatures above 18°C (65°F) can inhibit germination. Terminal stem cuttings root easily under mist without auxin.

***Nemophila* spp.** Baby blue eyes. Low-growing native North American annuals used as bedding plants. Seeds germinate at 18°C (65°F) and require darkness to germinate.

Neoregilia. *See* Bromeliads.

***Nepenthes* spp.** A large group of carnivorous plants producing pitchers to trap insects. Traps are formed at the tips of tendrils that extend from the midrib of the leaves. Propagation is from seeds at 27°C (81°F), leaf or stem cuttings (68), and tissue culture (248).

***Nepeta* spp.** Catmint. Hardy perennial. Seeds germinate at 20°C (68°F). Softwood cuttings of non-flowering side shoots taken in early summer root readily. Plants may be divided in spring, using newest parts and discarding older portion of clumps.

***Nerine* spp. (303).** A tender, perennial tunicate bulb. Lanscape bulb also used as a cut flower crop. Propagation is from offsets or bulb cuttage, but mostly by twin scaling. Tissue culture is possible from scales, lateral buds, and the young flower stalk (86).

***Nicotiana* spp.** Flowering tobacco. Half-hardy annuals with bright fragrant flowers. Seeds germinate at 24°C (75°F) and may respond to light (18). Stem cuttings root readily and plants are easily micropropagated.

***Nierembergia* spp.** Cupflower (223). Tender perennial, sometimes grown as an annual. Seeds germinate at 21°C (70°F). Softwood cuttings removed from new growth in spring root readily. Clumps can be divided.

Nigella damascena. Love-in-a-mist. Herbaceous annuals grown as bedding plants or as cut flowers for the interesting seed pods. Seeds germinate at 18 to 24°C (65 to 75°F).

Nolana paradoxa. Prostrate, annual plant with flowers that resemble small petunia blossoms. Seeds germinate at 20 to 22°C (68 to 72°F). Can also be propagated from stem cuttings.

***Nymphaea* spp.** Water lily. Consists of numerous species and many named cultivars. Plants grow as rhizomes. Clumps are divided in spring. Seeds are used to grow species and to develop new cultivars. Tropical water lily hybrids and species grow from tubers. Seeds do not reproduce hybrids. Both kinds of seeds are planted 2.5 cm (1 in) deep in sandy soil, then immersed in water 3- to 4-inch deep. Hardy species should be started at 16°C (60°F), tropical species at 21 to 27°C (70 to 80°F). Vegetative propagation is either from small tubers that can be removed from old tubers in fall or from small epiphyllous plantlets growing from the leaf (268). Micropropagation is also possible by using epiphyllous plantlets as explants (162). *N.* 'Gladstone' production can be extended by photoperiod control (169).

Ocimum basilicum. Basil. Annual plants usually reserved for the herb garden, but numerous colored leaf forms are available for the flower garden. Seeds germinate in 1 week at 21°C (70°F). Stem cuttings also root easily.

***Oenothera* spp.** Evening-primrose. Hardy perennials, but some kinds are biennial. Seeds germinate at 21 to 26°C (70 to 80°F). Plants can also be increased by dormant crown divisions.

Ophiopogon japonicus. Mondo grass. Hardy perennial. Evergreen ground cover. Propagation is by division in the spring or fall. Plants can be micropropagated (119).

***Opuntia* spp.** Prickly pear cactus. Hardy perennial. The only cactus hardy to the northern United States. Propagation is from seed, division, or cuttings (191). *See* Cactus.

Orchids (217, 261, 285). Many genera, hybrids, and cultivars are cultivated, and many more are found in nature. Some, such as *Aerides, Arachnis, Phalaenopsis, Renanthera,* and *Vanda,* exhibit a monopodial habit of growth. This means they are erect and grow continuously from the shoot apex and can be propagated by tip cuttings. Adventitious roots are produced along the stem and inflorescences are produced laterally from leaf axils. Most others including *Brassovola, Calanthe, Cattleya, Cymbidium, Laelia, Miltonia, Odontoglossom, Oncidium,* and *Phalus* have a sympodial habit of growth, are procumbent, and do not grow continuously from the apex. Their main axis is a rhizome in which new growth arises from offshoots or "breaks." Pseudobulbs are usually present on plants of this type. Many orchids are epiphytes (i.e., air plants), typically growing on branches of trees. Others (*Cypripedium* and *Paphiopedilum*) are terrestrial and grow in the soil (12).

Epiphytic Orchids. Seed propagation is mainly used for hybridization. Many important cultivars are seedling hybrids, either from species or between genera, resulting from controlled crosses of carefully selected parents. Many such important crosses are between tetraploid and diploid parents to produce triploids. These offspring are sterile and are not usable as parents. Seedling variation occurs, since orchids are heterozygous. Five to seven years are required for a seedling plant to bloom. Orchid flowers are hand-pollinated. A seed capsule requires 6 to 12 months to mature.

A single capsule will contain many thousands of tiny seeds with relatively underdeveloped embryos. *In vitro* culture is universally used for seed propagation. (The procedure is described in Chapter 18.) Knudson's C medium is usually used. Arditti (12, 13) has summarized the many experiences of testing various nutritional and other factors for orchid seed germination. Orchid seed can be stored for many years if held in sealed containers over calcium chloride at about 2°C (36°F). Vegetative methods for orchids are generally slow, difficult for many genera, and usually too low-yielding for extensive commercial use. Sympodial species are increased by division of the rhizome while it is dormant or just as new growth begins. Four or five pseudobulbs are included in each section. "Back-bulbs" and "greenbulbs" can be used for some genera.

Orchids with long canelike stems, such as *Dendrobium* and *Epidendrum,* sometimes produce offshoots ("keiki") that produce roots. Offshoots can also be produced if the stem is cut off and laid horizontally in moist sphagnum or some other medium. Flower stems of *Phaius* and *Phalaenopsis* can be cut off after blooming and handled in the same way. A drastic method of inducing offshoots is to cut out or mutilate the growing point of *Phalaenopsis,* remove the small leaves, and treat the injured portion with a fungicide. Offshoots may then be produced. Monopodial species can be propagated by long (30 to 37 cm) tip cuttings with a few roots already present. Air layering is also possible. Vegetative propagation by proliferation of shoot-tip (meristem) cultures *in vitro* has revolutionized orchid propagation, particularly for *Cymbidium, Cattleya,* and some other genera (14, 217). The shoot growing point is dissected from the plant and grown on a special, sterile medium; a proliferated mass of tissue and small protocorms develops, which can be divided periodically. Many thousands of separate protocorms can be developed in this way within a matter of months, each of which will eventually differentiate shoots and roots to produce an orchid plant (14). The procedure is described in Chapter 18. *Vanilla planifolia,* which is an orchid, essential for its oil, that grows as a vine, is normally propagated commercially by cuttings; nodal stem explants of this species were successfully micropropagated with BA and microshoots rooted *ex vitro* (179).

Terrestrial Orchids. These orchids can be difficult to propagate because they require a symbiotic relationship with an appropriate mycorrhizal fungus. *In vitro* seed germination using techniques similar to those used with epiphytic orchids has been successful both with and without fungal assistance (11). However, these procedures have not been extensively used commercially. Sowing seeds in pasteurized potting mixes containing mycorrhizal fungi has been successful and offers commercial potential (242).

***Origanum* spp.** Oregano, Dittany of Crete. Tender perennials usually produced as annuals. This genus contains the culinary herbs oregano and marjoram, but also contains showy ornamentals grown as bedding or container plants. Seeds germinate at 21°C (70°F).

***Ornithogalum* spp.** Star of Bethlehem (92). Herbaceous perennials that grow from tunicate bulbs. They are grown as garden plants, but are most important as cut flowers and seasonal pot plants. Plants have been propagated by seeds, leaf cuttings, bulb divisions, and tissue culture. Emerging leaves can be taken as leaf cuttings with plant regeneration taking place under mist (see Chapter 15). Micropropagation is also practiced (307, 325).

***Osmunda* spp.** *See* Fern.

***Osteospermum* spp.** Tender annual and perennial plants grown as bedding or container plants. Flowers are daisy-like with some interesting cultivars that have spoon-type ray florets. Seeds germinate at 21 to 24°C (70 to 75°F). Tip cuttings (99) are treated with 3,000 ppm IBA under mist with bottom heat 22°C (72°F).

***Oxalis* spp.** Shamrock or wood sorrel (92). Herbaceous perennials used in the garden or grown as a seasonal pot plant. Oxalis is an interesting and diverse group of plants. Some members of this genus are perhaps the only dicot species that produces a bulb. Others produce tubers or rhizomes. Propagation is usually from offsets of the bulbs or division of the tubers or rhizomes. Can be micropropagated (170).

***Pachysandra* spp.** *See* Chapter 20.

***Paeonia* spp.** Paeonia hybrids (*P. hybrida*), fernleaf peony (*P. tenuifolia*), tree peony [*P. suffructicosa* (*P. arborea*)]. Hardy perennials native to China used for specimen plants in borders and as cut flowers. Seed propagation is difficult, taking 5 to 7 years to produce a flowering plant from seed (286). Germination may take 1 to 2 years to meet epicotyl dormancy requirement. Seeds are sown in fall for cold stratification requirement during the winter. Roots develop during the first summer, and shoots develop the second spring. Plants developed from seed are generally not true-to-type. Another method is to collect seeds before they become black and completely ripe. Do not allow them to dry out; sow in pots, which should be buried in the ground for 6 to 7 weeks. Roots will develop; dig up and plant in a protected location or under mulch over winter. Best propagation method for herbaceous peonies is to divide clumps in fall; each tuberous root should have at least one bud or "eye," preferably three to five.

P. suffruticosa. Wedge grafted in late summer onto a herbaceous (*P. lactiflora*) understock (254). Peonies can be micropropagated (3, 45, 50).

Pansy. *See Viola.*

Papaver nudicaule. Iceland poppy. Hardy perennial often grown as a biennial. Seeds germinate in 1 to 2 weeks at 18 to 24°C (65 to 75°F). Sow in permanent location in summer for bloom next year.

P. orientale. Oriental poppy. Hardy perennial. Very fine seeds, which may respond to light, should be covered very lightly (18). Seeds germinate in same time as *P. nudicaule.* Cultivars are propagated by root cuttings. Dig when leaves die down in fall, cut into 7.5- to 10-cm (3- to 4-in) sections, and lay horizontally in a flat covered by 2.5 cm (1 in) of a sandy greenhouse substrate. Root cuttings are transplanted in spring. Or dig plants in spring, prepare root cuttings, and plant directly in a permanent location.

P. rhoeas. Corn poppy, Shirley poppy. Hardy annual. Seed germinates at 13°C (55°F). Sow in late summer for early spring bloom or in early spring for summer bloom.

***Pelargonium* spp.** Geranium. A group of tender perennials grown as bedding plants, hanging baskets, and seasonal pot plants. Plants are propagated by seeds or cuttings.

Pelargonium xdomesticum. Regal Geranium. Grown primarily as a cool-season conservatory plant or as a seasonal pot plant. If used in the garden, summer temperatures must be cool to keep plants in bloom. Propagation is from stem cuttings under mist with bottom heat [21 to 24°C (70 to 75°F)] from disease-free stock plants.

P. xhortorum. Geranium. Started by cuttings and by seed. Traditionally propagated by cuttings, which root easily with bottom heat, but *Pythium* and *Botrytis* infection can be serious problems. Pathogen-free stock, identified by culture indexing, should be used and can be supplied by specialists (298). There may be practical value in applying ABA in the shipment and storage of geranium cuttings (17).

In the mid-1970s, large-scale seed propagation of geraniums began with the introduction of certain cultivars that would grow from seed to flower in 14 to 16 weeks. Seeds germinate best at about 21 to 24°C (70 to 75°F) in medium greenhouse substrate (15). Plant growth regulators (like cycocel) are utilized in greenhouse to produce compact, early-flowering, well-branched plants (258). *In vitro* propagation has been developed and can be used for virus elimination in stock plants (67, 98).

P. peltatum. Ivy geranium. Popular as a window box or hanging basket plant. Propagation is from terminal stem or leaf-bud cuttings handled as described for other geraniums.

***Pennisetum* spp.** Fountain grass. Perennial grasses with feathery inflorescences. They are propagated by division in late spring to early summer (81). Seeds germinate easily, and volunteers in the garden can be a nuisance.

***Penstemon* spp.** Beardtongue. Semi-hardy to hardy perennials, sometimes handled as annuals. Seeds germinate at 18 to 21°C (65 to 70°F), but growth can be slow and uneven; seeds may respond to light (18). Some species benefit from 8 weeks of stratification at 15°C (59°F) (5). Plants started indoors in early spring and transplanted outdoors later may bloom the first year. Plants are usually short-lived.

Softwood cuttings taken from non-flowering side shoots of old plants root readily. Make cuttings in fall to obtain plants for next season. Clumps may be divided. *Penstemon* can also be micropropagated from lateral buds used as explants (197).

Pentas lanceolata. Star-cluster pentas. Tender perennial with clusters of white, pink, lavender or red star-shaped flowers grown as a pot or bedding plant. Can be propagated from seeds or cuttings. Seeds germinate at 21 to 22°C (70 to 72°F). Tip cuttings from non-flowering plants are treated with 1,000 ppm IBA + 500 ppm NAA and rooted under mist with bottom heat at [20 to 22°C (68 to 72°F)] (99).

***Peperomia* spp.** Tender perennials grown as indoor house plants and occasionally as bedding plants. Softwood stem, leaf-bud, or leaf cuttings root readily. Plants can also be divided. Peperomia can also be micropropagated from excised leaf explants (142) and petioles (243).

Pericallis xhybrida. Florist's cineraria (formerly *Senecio cruentus*) is a cool-season crop with true blue or lavender flowers grown as a seasonal pot plant. Seeds germinated at 20 to 22°C (68 to 72°F) will emerge in 2 weeks in light (168). Plants are most often produced from seeds because vegetatively propagated plants can show reduced vigor and flower size. Plants have been micropropagated (77, 126).

Periwinkle. *See Vinca minor.*

***Persicaria* spp.** Fleeceflower or knot weeds. This genus now contains plants that were formerly *Polygonum* and *Tovara*. Versatile perennials and annuals used as ground covers, in rock gardens, planters, and hanging baskets. Generally propagated by division or seed. *P. capitatum* 'Magic Carpet,' which is an excellent annual ground cover, is propagated by seed at 21 to 27°C (70 to 80°F). Other species (*P. affine, P. amplexicaulis,* and *P. virginiana*) are more commonly propagated by division.

Pervoskia atriplicifolia. Russian sage. Hardy perennial. Large shrub-like perennial with silvery foliage and blue flowering spikes. Propagated from softwood cuttings.

Petunia xhybrida. Petunia (65). Tender perennial often grown as an annual. Seeds germinate at 24 to 25°C (75 to 78°F). Give light for the first 3 days then move to darkness until seedlings emerge. Several cultivars are propagated from stem cuttings that root easily under mist. Petunia is easily micropropagated from leaf segments (287).

***Philodendron* spp.** Tropical vines. Seeds germinate readily at about 25°C (77°F) if sown as soon as they are ripe and before they become dry. Vining types are propagated by single node leaf-bud or stem cuttings. Larger plants may be multiplied by air layering. Non-vining types are propagated from seeds or, more commonly, tissue culture (274).

Phlomis russeliana. Jerusalem sage. Hardy perennial grown for pale yellow flowers on stiff, upright stems. Most commonly propagated by crown division or stem cuttings.

***Phlox* spp.** Phlox. Colorful herbaceous annual or perennial garden plants. Propagated by seeds or cuttings.

Phlox divaricata. Sweet William. Hardy perennial. Expose seeds to cold during winter or chilling stratification before planting. Softwood cuttings taken in spring root easily. Divide clumps in spring or fall.

Phlox drummondii. Annual phlox (65). Hardy annual. Seed germinate at 15 to 18°C (60 to 65°F). Initial seed germination occurs in either light or darkness, but light inhibits radicle growth, so it is common to germinate phlox in the dark (62). Start indoors for later outdoor planting, or outdoors after frost.

Phlox paniculata. Garden phlox. Hardy perennial. Plants do not come true from seed. Sow seeds as soon as ripe in fall to germinate the next spring. Seed will germinate at 20°C (68°F). Grow plants one season and transplant in fall. Softwood cuttings taken from young shoots in spring or summer root easily, but are subject to damping-off if kept too wet. Garden phlox are often commercially propagated from root cuttings (195). Dig clumps in fall; remove all large roots to within 5 cm (2 in) of crown (which is replanted). Cut roots into 2-inch lengths and place in flats of sand; cover 13 mm (0.5 in) deep. Transplant next spring. Divide clumps in fall or spring. Phlox can be micropropagated using shoot explants (257, 292).

Phlox subulata. Moss pink. Hardy perennial. Evergreen ground cover with profusion of blooms in early spring. Propagation is by division of dormant plants or softwood stem cuttings taken after plants finish blooming in the spring. Cuttings are treated with 1,000 ppm IBA under mist. Plants can be micropropagated (257).

Physalis alkekengi. Chinese lantern. Hardy perennial grown for showy fruits shaped like orange paper lanterns. Seeds emerge in 2 weeks at 16 to 21°C (60 to 70°F). Can be divided in the autumn.

Physostegia virginiana. False dragonhead or lionsheart. A herbaceous perennial that is used as a field-grown cut flower crop (16). Seeds will germinate at 18 to 21°C (65 to 70°F). This species can also be propagated by division.

***Pinguicula* spp.** Butterwort. Perennial carnivorous plants. Glandular hairs are produced on the leaves of these rosette plants to trap insects. An attractive flowering carnivorous plant. Propagation is from seed or tissue culture (1).

Platycodon grandiflorus. Balloon flower. Hardy perennial. Long-lasting blue flowers that resemble balloons when in bud. Most commonly propagated by seeds that emerge in 1 to 2 weeks at 20 to 21°C (68 to 70°F) with light. Plants can also be divided in spring.

***Plectranthus* spp.** Spurflowers. Tender herbaceous perennials related to coleus. Plants are primarily grown for their interesting foliage, but some have showy flowers. Propagation is primarily from stem cuttings with bottom heat (21 to 24°C; 70 to 75°F) that root easily without auxin.

Poinsettia *(Euphorbia pulcherrima).* *See* Chapter 20.

***Polemonium* spp.** Jacob's ladder. Hardy perennial. In spring, blue or white flowers appear on plants with a leaflet pattern that resembles a ladder. Seeds germinate at 21°C (70°F). Plants can also be multiplied by crown division and stem cuttings.

Polianthes tuberosa. Tuberose (34). Tender bulbous perennial. Propagated by removing offsets at planting time. The small bulbs take more than 1 year to flower. Divide clumps every 4 years. Plants can be micropropagated (44).

Poliomintha longiflora (153). Mexican oregano. Has small evergreen leaves that smell like the spice oregano and is a striking landscape plant that produces light lavender flowers throughout the summer. It is easily rooted from stem cuttings.

Polygonatum **spp.** Solomon's seal. Hardy perennials. Used in wildflower and perennial gardens. Propagation is by seed or division. Cultivars are propagated by dividing rhizomes in the spring.

Portulaca **spp.** Moss rose, flowering purslane. Half-hardy annual used as a bedding or container plant. Seeds germinate at 24 to 27°C (75 to 80°F) and may respond to light (18). Can also be propagated by tip cuttings.

Potentilla **spp.** Cinquefoil. Hardy perennial. Creeping perennials with attractive blooms. Can be propagated by seed, division, or stem cuttings. *P. nepalensis* is a common garden perennial and is grown from seed germinated at 21°C (70°F).

Pothos. See *Epipremnum*.

Primula **spp.** Primrose. A group of cool-season perennials used in the rock or perennial garden. Some species and hybrids used as seasonal pot plants. Most primroses are propagated by seed. Seed germination can be erratic. Seeds generally require light, and germination temperature should be below 21°C (70°F). Erratic germination may be related to seed dormancy, and germination has been enhanced by a pregermination treatment with gibberellic acid (100). Micropropagation is possible (35).

P. xpolyantha **and** ***P. vulgaris (syn P. aucale).*** Hardy perennials grown outdoors or used as seasonal pot plants. Seeds are slow to germinate and can take up to 30 days. Germination is at 16 to 18°C (60 to 65°F), but some species may require lower temperatures (80, 107). It is best to collect and sow seeds as soon as they are ripe in the fall. Clumps can be divided just after flowering.

Primula malacoides. Fairy primrose. Seeds will germinate in 2 to 3 weeks at 15 to 18°C (59 to 65°F).

Primula obconica. Primrose. Tender perennial grown as an annual. Seeds germinate well in a cool greenhouse or after 3 to 4 weeks at 18 to 20°C (65 to 68°F). The very tiny seeds respond to light and should not be covered.

Primula sinensis. Chinese primrose. Seeds will germinate within 3 weeks at 20°C (68°F). In these species, double-flowering cultivars do not produce seeds but are propagated by cuttings taken in spring, or by division.

Pulmonaria **spp.** Lungwort. Hardy perennial grown for their attractive spotted or silvery foliage on plants adapted to the shade garden. Common companion plant to hostas. Can be propagated by division, root cuttings (48), or tissue culture. Tissue culture propagation and the introduction of new cultivars has increased the availability of lungworts (290).

Pulsatilla **spp.** Pasque flower (formerly *Anemone pulsatilla*). Hardy perennial with showy anemone-like flowers and feathery seed heads. Seed germination is at 20°C (68°F) but may be sensitive to high temperature. Plants can be divided or propagated from root.

Puschkinia scilloides. Striped squil (92). Hardy perennial bulb. Propagation is the same as *Scilla*.

Ranunculus **spp.** Buttercup (210). A group of herbaceous perennials produced on tuberous roots or rhizomes. Plants are used as annual or perennial garden plants, seasonal pot plants, and cut flowers. Commonly propagated by seeds germinated at 20°C (68°F). Divide perennial types while dormant in spring or fall.

Ranunculus asiaticus. Turban or Persian ranunculus. Tender perennial. Plants are most commonly propagated from seeds that germinate at 15 to 17°C (58 to 62°F). Plants can also be multiplied by division of the tuberous root, or micropropagated (172, 211).

Ratibida **spp.** Prairie coneflower. Hardy perennial producing yellow coneflowers. Most often used in native prairie plantings and wildflower mixes. Seed germination is at 21 to 24°C (70 to 75°F) and may benefit from several weeks of chilling stratification. Plants can be divided.

Reseda odorata. Mignonette. Hardy annual. Seeds should not be covered; Seed germination is at 12°C (54°F), and seeds may respond to light (18).

Rhipsalidopsis **spp.** Easter cactus. Propagation same as *Epiphyllum*.

Ricinus communis. Castor bean. Soak seeds in water for 24 hours or nick with file before planting. Seed germination is at 21°C (70°F). Seeds are poisonous.

Rodgersia **spp.** Featherleaf, Rodgersflower. Perennial for moist border with attractive foliage and ornamental flowers. Propagation is by seeds or division.

***Rudbeckia* spp.** Black-eyed Susan, coneflower. Hardy annual, biennial, and perennial species. Seeds germinate in 2 to 3 weeks at 21 to 24°C (70 to 75°F). Perennial kinds are propagated by crown division. Plants often re-seed naturally. Seed priming can improve germination (113).

Rumex sangineus. Red-veined dock. Semi-hardy perennial grown for the interesting red-veined pattern on leaves. Propagation is from seed germinated at 20°C (68°F).

Saintpaulia ionantha. African violet (159, 181, 324). Tropical perennial. Propagation is by seed, division, or cuttings. The very fine seeds, which germinate at 30°C (86°F), should not be covered. Seedlings are subject to damping-off. Vegetative methods are necessary to maintain cultivars. Leaf cuttings (blade and petiole) are easily propagated using a rooting medium at 25°C (75°F) and air temperature at 18°C (64°F) (260). Variegated leaf cultivars that do not come true from leaf cuttings should be propagated by crown divisions or tissue culture (271). Rapid, large-scale propagation can be accomplished by *in vitro* culture techniques using leaf petiole sections (40, 83, 165, 281).

Salpiglossis sinuata. Painted tongue. Semi-hardy annual. Seed germination is at 21 to 22°C (70 to 72°F) but can be slow and uneven. Can be micropropagated (194).

***Salvia* spp.** Sage (65). Annual, biennial, and perennial species. Salvia are used in borders, containers, or as cut flowers (16). Seed germination is at 24 to 25°C (75 to 78°F) and may respond to light (18). Softwood cuttings from non-flowering plants root readily under mist and may benefit from 1,000 ppm IBA.

S. greggii. Easily propagated by semi-hardwood cuttings (153). Plants can be divided, but such divisions are slow to recover.

S. officinalis. Basal cuttings root better than apical cuttings. Flowering reduces rooting and removal of flowers enhances propagation. Rooting is enhanced with basal dips of 1,000 ppm K-IBA salt (246). Rooting ability was highest in spring (Israel).

S. splendens. Scarlet sage. Tender perennial grown as an annual. Germinate seeds at 24 to 25°C (75 to 78°F), then grow at 13°C (55°F) night temperature. Seeds soaked for 6 days at 6°C (43°F) can promote germination (56). Softwood cuttings taken in fall root readily.

Sandersonia aurantiaca. Climbing, tender, tuberous plant with orange flowers like inverted lanterns. Grown as a container plant or cut flower. Propagation is by division of the tuberous root or by micropropagation (114).

Sanguinaria canadensis. Bloodroot. Ephemeral, North American native producing white flowers in early spring. Seeds have morphophysiological dormancy and require cycles of warm followed by chilling stratification. It can take several years to produce a seedling.

Sansevieria trifasciata* and *S. 'hahnii' (Dwarf Form). Bowstrip hemp. Snakeplant. Tropical perennial. Plants grow from a rhizome, which can be readily divided. Leaves may be cut into sections, several inches long, and inserted into a rooting medium; a new shoot and roots will develop from the base of leaf cutting. The variegated form, *S. trifasciata* 'Laurentii,' is a chimera, which can be maintained only by division. *S. trifasciata* has been micropropagated (42).

Santolina chamaecyparissus (incana). Lavender cotton. A perennial native to the Mediterranean that is used as carpet bedding or a low hedge. Must be trimmed to maintain compact growth. Seeds germinate at 18 to 21°C (65 to 70°F). It is easily propagated by stem cuttings.

Sanvitalia procumbens. Creeping zinnia. Hardy annual. Seed germination is at 20°C (68°F). Plants are easily propagated using stem cuttings treated with up to 2,500 ppm IBA under mist (99).

Saponaria ocymoides. Rock soapwort. Hardy short-lived perennial used in rock gardens or as a bedding plant. Seed germination is at 21°C (70°F) in the dark.

S. officinalis. Bouncing Bet. Hardy perennial. Seeds germinate at 20°C (68°F). The plant spreads rapidly by an underground creeping stem, which can be divided.

***Sarracenia* spp.** Pitcher plant. Perennial carnivorous plants native to bog ecosystems. Leaves are modified to form spectacular pitchers that entrap insects. Also grown for cut foliage for the florist industry. Propagation is by division of the rhizome (68).

***Saxifraga* spp.** Many interesting unusual species and hybrids. Tender or hardy perennials. Seeds germinate easily at 21°C; they are preferably sown when ripe. Seed germination is at 21 to 24°C (70 to 75°F). Some hybrids and cultivars are maintained only by vegetative methods and are propagated once flowering is finished in June or July (England) (269). Cuttings are small and slow growing—it takes one year to grow a liner.

Cuttings are rooted in a cold frame or seed tray. Most plants grow as small rosettes and are easily propagated by making small cuttings involving single rosettes. Small plants from stolons root readily. Plants can be divided in spring or fall.

S. stolonifera. Strawberry geranium. A tender perennial grown indoors or as a bedding plant. It can be reproduced by removing plantlets from the runners.

***Scabiosa* spp.** Pincushion flower. Annual and hardy perennial garden plants. Seed germination is at 18 to 21°C (65 to 70°F). Perennial kinds can be divided.

Scaevola aemula. Fan flower. Tender perennial grown for its trailing habit and blue flowers. Most commonly used as a hanging basket. Propagation is from stem cuttings that are rooted under moderate mist with bottom heat at 20 to 22°C (68 to 72°F). Auxin is not required but up to 2,500 ppm IBA can be used to improve uniformity (99). Micropropagation is possible (38).

Scarlet sage. *See Salvia splendens.*

***Schefflera* (synonym *Brassaia*) *arboricola* and *S. actinophylla* (137).** Umbrella tree. Tropical perennial plants used in warm climate landscapes and as an important indoor foliage plant. Schefflera is an important foliage plant that can be propagated easily by seeds, cuttings, or air layering. Basal cuttings develop more roots and longer shoots and require less time to break lateral buds than do apical cuttings (137). As single-node cutting length increased to 20 cm (8 in) so did rooting, bud-break, and shoot growth. Seeds germinate at 22 to 24°C (72 to 75°F).

***Schizanthus* spp.** Butterfly flower. Tender annual grown as a seasonal pot plant. Seed germination is at 18 to 22°C (64 to 72°F) in the dark. Seeds are sensitive to high temperatures (18). Sow seeds in fall for early spring blooms indoors, or sow in early spring to be transplanted outdoors for summer blooms.

Schlumbergera truncata. Christmas cactus. Propagation same as *Epiphyllum*.

***Scilla* spp.** Squill (92). Includes several kinds of bulbous hardy and half-hardy spring-flowering perennials. Dig plants when leaves die down in summer and remove the bulblets. *S. autumnalis* is planted in spring and blooms in fall. Can be micropropagated (155).

***Scutellaria* spp.** Skullcap. Annual or herbaceous perennials used as bedding plants, in containers or in the perennial bed. Seed germination is at 21 to 24°C (70 to 75°F) in light. Perennial types can also be divided or multiplied by stem cuttings.

***Sedum* spp. (269).** Sedum is composed of a wide range of species including herbaceous perennials, evergreens, and monocarps. Many of the *Sedum* species can be raised by seed, but generally this method is limited to herbaceous perennials only; seeds germinate at 15 to 18°C (65°F) in light. *S. acre* should have alternating day [29°C (86°F)] and night [21°C (70°F)]. The mat-forming species are propagated by division, since the creeping shoots root into the ground as they travel and mats are easily pulled apart. Direct sticking cuttings into containers is done, since many species root so readily.

Sempervirens. *See* Succulents.

Senecio cineraria. Dusty miller (223). Semi-hardy perennials usually grown as annuals for their silvery foliage. Seeds germinate at 24 to 26°C (75 to 80°F). Stem-tip cuttings root rapidly if treated with IBA and placed under mist with bottom heat.

Sidalcea malviflora. Sidalcea (286). A native western U.S. perennial used as a border plant. Seeds germinate at 18 to 21°C (65 to 70°F). Cultivars are commonly propagated by division.

***Silene* spp.** Catchfly, campion. Annual and short-lived perennials grown for bright pink, red, or white flowers. Seeds germinate at 18 to 21°C (65 to 70°F). Perennial types are divided, or basal stem cuttings can be rooted.

Sinningia speciosa. Gloxinia. Tropical perennial. Commonly grown from seeds, which are very fine and require light. Sow uncovered, in well-drained peat moss medium; they emerge in two weeks at 20°C (68°F). Vegetative methods are required to reproduce cultivars. Plant grows from a tuber on which a rosette of leaves is produced. The root can be divided as described for tuberous begonia. Softwood cuttings or leaf cuttings taken in spring from young shoots starting from the tubers root easily. Gloxinia can also be micropropagated using leaf explants (163).

Snapdragon. *See Antirrhinum majus.*

***Solanum* spp.** A diverse group of tender, annual, herbaceous, and woody perennials related to potato. *S. pseudocapsicum* (Christmas cherry) is grown as a seasonal container plant. It is propagated by seeds, which germinate at 18 to 21°C (65 to 70°F) in light.

S. crispeum and *S. jasminoides* are vines and *S. rantonnettii* is a woody shrub usually grown as a container standard. All are commonly propagated by softwood stem cuttings rooted under mist.

Solenostemon scutellarioides. (formerly *Coleus blumei*) (27). Tender perennials. Used for containers and as bedding plants. Plants are grown for their wide selection colorful foliage. Seeds germinate at 21 to 24°C (70 to 75°F) in light. Coleus is commonly propagated from cuttings that root easily under mist without auxin.

***Solidago* spp.** Goldenrod. Hardy perennial. Grown as a cut flower and for the perennial garden. Plants can be grown from seeds or division, but are usually propagated from stem cuttings. Seeds germinate at 20 to 22°C (68 to 72°F) and some species benefit from 10 weeks of chilling stratification (46). Stem cuttings are taken from basal vegetative growth in the spring and rooted under mist without auxin. Division is from dormant plants.

xSolidaster. An intergeneric hybrid between *Solidago* and *Aster* that is also grown as a cut flower from stem cuttings. Plants can also be divided.

***Spathophyllum* spp.** Spatheflower. Tropical evergreen plants grown in indoor containers or as bedding plants in tropical regions. Plants can be propagated by crown division or commercially by micropropagation (117).

Spider plant. *See Chlorophytum comosum.*

Spigelia marilandica. Pinkroot. Herbaceous perennial with tubular red flowers. Underused perennial due to difficulty in propagation. Plants can be multiplied by division or stem cuttings. Stem cuttings root under mist when taken from greenhouse-grown stock plants (118). Plants can also be micropropagated.

***Stachys* spp.** Lamb's ear (*S. byzantina*), big betony (*S. grandiflora*). Hardy perennials used as border plants or ground covers. Propagated by clump division or by seed. Seed germinates at 21°C (70°F).

***Stapelia* spp.** Carrion flower. Tender perennial succulent that produces large star-shaped flowers that have an unpleasant odor, which attracts flies as pollinators. Propagated by seeds sown at 21°C (70°F), by division or stem cuttings. See Succulents for detailed information.

Sternbergia lutea. Autumn daffodil. Small fall flowering bulb with bright yellow flowers. Propagated by offsets of the bulb.

Stock. *See Matthiola.*

Stokesia laevis. Stokes aster. Hardy perennial. Seeds germinate at 21 to 24°C (70 to 75°F) and the plants bloom the first year. Commonly propagated from root cuttings or division.

Strelitzia reginae. Bird-of-paradise. Tropical perennial producing the recognizable bird of paradise flower. Grown as a greenhouse ornamental and a cut flower. Seed propagation of this tropical perennial is undesirable due to juvenility and genetic variation. Seeds germinate in 6 to 8 weeks using bottom heat at 37°C (98°F) (72). This species grows from a rhizome, which can be divided in the spring; however, division is limited by a low rate of multiplication with 0.5 to 1.5 divisions per branch per year. A technique has been developed to overcome the strong apical dominance, which inhibits branching of axillary buds into propagules. A triangular incision with a knife is made at the base of a separated branch 8 to 12 mm above the basal plate to reach and remove the apex from adult plants (304). After 2 to 6 months, 2 to 30 lateral shoots develop from each fan (separated branch). During the next 6 months, newly formed laterals root and can be divided into individual plants (see Chapter 15).

Streptocarpus xhybridus. Cape primrose is a herbaceous perennial in the Gesneriaceae. It is generally propagated by seed or leaf cuttings. Seeds germinate at 21°C (70°F). Seeds are tiny and should not be allowed to dry out. Leaf cuttings are made by splitting the leaf down the midrib to form two leaf segments. The base of the leaf segment is removed and the remaining leaf cutting is placed into a peat-lite medium held at 21°C (70°F). New plants appear in several months. Cape primrose has been successfully micropropagated using explants of leaf discs, shoot apices, stem, petiole, pedicel segments, and corolla flower parts (230).

Succulents (28, 106, 234, 282). This loosely defined horticultural group includes many genera, such as *Agave, Aloe, Crassula, Echevaria, Euphorbia, Gasteria, Haworthia, Hoya, Kalanchoe, Mesembryanthemum, Portulaca, Sedum, Sempervirens,* and *Yucca*. These are plants with fleshy stems and leaves that store water, or plants that are highly drought-resistant. Most are half-hardy or tender perennials. Seed propagation is possible, although young plants are often slow to develop and to produce flowers. It is best to germinate seeds indoors at high day temperatures (29 to 35°C, 85 to 95°F). Seedlings are susceptible to damping-off.

Cuttings of most species root readily—either stem, leaf-bud, or leaf—in a 1:1 peat-perlite medium. They should be exposed to the open air or inserted into dry sand for a few days to allow callus to develop over the cut end. Some protection from drying is needed during rooting. Some species can be reproduced by removing offsets. Grafting is possible as described for cacti. Many of the succulents can be micropropagated (249, 270, 276, 287).

***Sutera* spp.** Bacopa. Trailing annual mainly grown in hanging baskets or mixed containers. Plants are propagated from cuttings treated with 2,500 ppm IBA under mist with bottom heat at 18 to 20°C (65 to 68°F). Overuse of mist can be a problem if rooting substrate becomes saturated (99).

Sweet alyssum. *See Lobularia maritima.*

Sweet pea. *See Lathyrus.*

***Symphytum* spp.** Comfrey. Hardy perennial. Grown as an herb or ornamental garden plant. Cultivars are propagated mainly from root cuttings or division. Micropropagation is from stem explants (140).

Syngonium podophyllum. Arrowhead vine. Tropical perennial vine grown as an indoor foliage plant. Seeds germinate at 24 to 27°C (75 to 80°F). Plants also root easily from single eye leaf-bud or stem cuttings as well as tissue culture.

***Tagetes* spp.** Marigold (223). Tender annuals. Plants are used as bedding plants and cut flowers. Seeds germinate readily within 1 week at 21 to 24°C (70 to 75°F). Can be sown in place in spring after frost in mild climates. *T. erecta* (African marigold) has been micropropagated from leaf segments (180). Somatic embryos have also been developed (37).

Tanacetum coccineum (formerly Chrysanthemum coccineum). Painted daisy. Hardy perennial for the flower border. Can be propagated by seeds that emerge in 2 weeks at 16 to 21°C (60 to 70°F). Most often propagated by division or basal stem cuttings.

T. parthenium (formerly Chrysanthemum parthenium). Feverfew. Hardy perennial usually grown as an annual. Seeds germinate in about 1 week at 20°C (68°F) in light. Plants easily self-seed and can be divided.

T. ptarmiciflorum. Dusty miller. Tender perennial grown as an annual. Propagated from seeds that emerge within 2 weeks at 22 to 24°C (72 to 75°F) in light.

***Teucrium* spp.** Germander. Herbaceous and woody perennials grown mainly for their green or silvery foliage. Often pruned into garden borders or used in knot gardens. Seeds germinate at 18 to 21°C (65 to 70°F). Commonly propagated by stem cuttings using mist or high humidity with bottom heat.

***Thalictrum* spp.** Meadow rue. Hardy perennial. Seeds germinate at 20°C (68°F). May benefit from 1 week of chilling stratification. Plants can be multiplied by dormant divisions in spring or fall.

Thermopsis caroliniana. Hardy perennial. Seeds have hard seed coats and require scarification. Seeds germinate at 21°C (70°F). Plants can be divided, but it is best to leave them undisturbed for garden performance.

***Thunbergia* spp.** Clockvine. Tender annual and perennial vines. *T. alata* (black-eyed Susan) and *T. gregorii* are grown as summer annuals on a garden support or as a ground cover. *T. grandiflora* (Blue trumpet vine) and *T. mysorensis* are usually used as perennial vines in greenhouse displays for spring or fall blooms. Seeds germinate at 21 to 24°C (70 to 75°F), but seedlings grow slowly. Softwood cuttings taken from new shoots root readily with bottom heat.

***Thymus* spp.** Thyme. Hardy creeping perennials grown as ground covers for small spaces or in containers. Leaves are fragrant. Seeds germinate at 21°C (70°F). Germination may be promoted by light. Can also be increased by division or by softwood cuttings taken in summer. Plants have been micropropagated (121).

Ti. *See Cordyline terminalis.*

Tiarella cordifolia. Foamflower. Hardy perennial grown as a ground cover in the shade. Can be propagated from seed or division. Seeds emerge within 2 weeks at 18°C (65°F). *Tiarella* is commercially micropropagated from shoot explants (175).

Tigridia pavonia. Tiger flower (92). Tender bulbous perennials. Plant bulbs in spring and dig in fall when leaves die. Increase by removal of small bulblets. Easily started by seed.

***Tillandsia* spp.** *See* Bromeliads.

Tithonia rotundifolia. Mexican sunflower. Tender perennial with large composite flowers grown as an annual. Seeds germinate at 21°C (70°F).

Tolmiea menziesii. Piggyback plant. Tender perennial used as an indoor foliage plant. New plantlets form on upper surface of leaves. Plantlets can be removed and rooted. Also propagated from stem (rhizome) cuttings.

Torenia fournieri. Wishbone flower (272). *Torenia* is a tender perennial grown as an annual for semi-shady garden areas. Seed germination is at 21 to 23°C (70 to 75°F) with light. Plants are also propagated from terminal cuttings treated with 2,500 ppm IBA under mist (99).

Trachymene coerulea. Laceflower. Tender perennial grown as an annual for display or cut flowers. Seeds germinate at 20°C (68°F).

***Tradescantia* spp.** Spiderwort. Hardy perennial grown for their foliage color or their long-blooming blue, pink, or white flowers held above grass-like foliage. Seeds germinate at 21°C (70°F). Cultivars are propagated from dormant divisions in spring or autumn. Stem cuttings root easily under mist or high humidity without auxin.

***Tricyrtis* spp.** Toad lily. Hardy perennial. Unique flowers produced in the early autumn. Plants prefer shade. Propagation is from seed, division, or stem cuttings. Seeds require stratification. *Tricyrtis* has become widely available partly due to its ease of micropropagation.

***Trillum* spp.** Trillum, wake robin. Hardy herbaceous perennials grown in the woodland garden. Plants produce white, yellow, or red flowers with three pigmented petals. Seeds are covered with an aril called an elaisome that attracts ants, which disseminate the seed. Seeds should be sown as soon as the seed ripens. Maintain even moisture levels until seedlings emerge. Seeds have morphophysiological dormancy where the radicle emerges after the first chilling stratification and the epicotyl emerges after a second chilling stratification, which means two seasons are required for seedling emergence (49, 273). Many trillium produce shoots from rhizomes. The rhizome can be notched behind the apical shoot bud to induce numerous new buds near the wound site. After 1 year, these new shoots can be separated, including a piece of the rhizome (41). Some trillium species have been micropropagated (232).

***Trollius* spp.** Globeflower. Hardy perennial grown for their striking orange-yellow flowers. Seeds show endogenous dormancy and should be stratified or planted in fall to produce plants that flower by next spring (49). Plants are also increased by crown division.

Tropaeolum majus. Nasturtium. Tender perennial grown as an annual. Seeds germinate at 18 to 21°C (65 to 70°F). Plants can be difficult to transplant and are often seeded in place. Double-flowering kinds must be propagated vegetatively, usually by softwood cuttings.

Tulipa* spp. and *Hybrid Cultivars. Tulip (192). Hardy perennials from tunicate bulbs. Tulips are used for seasonal bedding plants, pot plants, and cut flowers. Plant bulbs in fall for spring blooms. Seeds are used to reproduce species and for breeding new cultivars. They germinate readily after chilling stratification. Vegetative methods include removal of offset bulbs in the fall. Different bulb sizes are planted separately, since the time required to produce flowers varies with size. For details of procedure, see Chapter 15. Tissue culture of tulip is possible but results are variable and success rates are low. Additional work is needed to develop a commercial micropropagation system (4).

Tweedia caerulem. (synonym *Oxypetalum*). Hardy perennial grown for its pale blue flowers that make long-lasting cut flowers. Seeds germinate within 2 weeks at 21°C (70°F) in light.

Valeriana. See *Centranthus.*

***Verbascum* spp.** Mullein. Hardy perennials and biennials grown for their showy spikes of flowers. Seeds germinate at 18 to 21°C (65 to 70°F). Propagate named cultivars by root cuttings taken in early spring before vegetative growth begins.

***Verbena* spp.** Verbena. Tender perennials grown as an annual. Seeds germinate at 24 to 27°C (75 to 80°F), but germination can be erratic (65). Seeds are sensitive to overwatering (60). Hardy kinds can be propagated by division or from softwood cuttings taken in summer. Stem cuttings are treated with 1,000 ppm IBA and rooted under mist (99).

***Veronica* spp.** Speedwell. Diverse group of hardy perennials producing white or blue flowers on upright or spreading plants depending on the species. Seeds germinate at 18 to 24°C (65 to 75°F) with light. Plants are increased by division in spring or fall or by softwood cuttings taken in the spring or summer. Plants can be micropropagated (280).

Vinca. See *Catharanthus.*

Vinca major. Tender perennial grown as a ground cover. Propagated by division or by softwood cuttings treated with 1,000 ppm IBA under mist (99).

V. minor. Periwinkle. Hardy perennial. Seeds germinate at 20°C (68°F). Easily propagated by softwood cuttings or by division. *V. minor* can also be micropropagated (278).

***Viola* spp.** Violets. Many hardy perennial kinds. Species violets can be propagated by seeds, but germination may be slow and seeds are best exposed to chilling stratification before planting. Many species produce seeds in inconspicuous, enclosed (cleistogamous) flowers near the ground, whereas the conspicuous, showy flowers produce few or no seeds. These plants can also be reproduced by cuttings or by division (80).

Viola cornuta. Horned violet, tufted pansy. Hardy perennial. Seeds germinate at 18 to 21°C (65 to 70°F). Seeds of some cultivars need light. Vegetative propagation is by cuttings taken from new shoots obtained by heavy cutting back in the fall. Clumps may also be divided.

V. odorata. Sweet violet. Tender to semi-hardy perennials. Grows by rhizome-like stems, which can be separated from others on the crown and treated as a cutting with some roots present.

V. tricolor. Johnny-jump-up. Hardy or semi-hardy, short-lived perennial. Usually propagated by seeds as described for *V. cornuta* but may also be increased by division.

Viola *xwittrockiana*. Pansy. Short-lived perennial grown as an annual. Popular as an autumn and early spring bedding plant. Propagation is by seed. Seeds germinate at 18 to 21°C (65 to 70°F) with light. Pansy seed experiences thermoinhibition at temperatures above 30°C (86°F) and fails to germinate. These temperatures are common in summer greenhouses when pansy seed is normally sown, but seed priming alleviates this problem (58). Pansy seed is commonly sold as primed seed from the seed company.

Vriesea. See Bromeliads.

***Yucca* spp.** Yucca. Tender to semi-hardy perennials. Seeds germinate at 20°C (68°F) but require several years to flower. Plants are monocots; some are essentially stemless and grow as a rosette, while others have either long or short stems. Offsets growing from the base of the plant can be removed and handled as cuttings; sometimes entire branches or the top of the plant can be detached a few inches below the place where leaves are borne and replanted in sandy soil. Sections of old stems can be laid on sand or other medium in a warm greenhouse, and new side shoots that develop can be detached and rooted. *Y. elephantipes* is rooted by long canes (239).

***Zantedeschia* spp.** Calla (120). Tropical perennials used as cut flowers and seasonal pot plants. Plants are propagated by seed, division, and tissue culture. Plants grow by thickened rhizomes (or tuber) that produce offsets or rooted side shoots; these are removed and planted. Calla are also commercially micropropagated (79, 160).

Zebrina pendula. Wandering Jew. Tender perennial. Easily propagated at any season by stem cuttings.

***Zephyranthes* spp.** Zephyr lily (92). Tender perennial produced from a bulb. Used as a seasonal pot plant or annual display. Multiplied by offsets of the bulb.

***Zinnia elegans* and other species.** Zinnia (223). Half-hardy hot-weather annuals used as bedding plants or cut flowers. Greenhouse germination is rapid (within 1 week) for seeds sown at 21°C (70°F).

REFERENCES

1. Adams, R. M., S. S. Koenigsberg, and R. W. Langhans. 1979. *In vitro* propagation of *Cephalotus follicularis* (Australian pitcher plant). *HortScience* 14:512–13.

2. Aelbrecht, J. 1989. The effect of different treatments on the germination of *Lewisia* hybrid seeds. *Acta Hort.* 252:239–45.

3. Albers, M. R. J., and B. P. A. M. Kunneman. 1992. Micropropagation of *Paeonia. Acta Hort.* 314:85–92.

4. Alderson, P. G., and A. G. Taeb. 1990. Influence of culture environment on shoot growth and bulbing of tulip *in vitro. Acta Hort.* 266:91–4.

5. Allen, P. S., and S. E. Meyer. 1990. Temperature requirements of three *Penstemon* species. *HortScience* 25:191–93.

6. Al-Matar, M., J. M. Al-Khayri, M. S. Brar, and G. L. Klingaman. 1999. *In vitro* regeneration of long spur barrenwort (*Epimedium grandiflorum* Morr.) from rachis explants. *In Vitro Cell. Dev.—Plant* 35:245–48.

7. Amaki, A., and H. Higuchi. 1992. Micropagation of Boston ferns (*Nephrolepis* spp.). In Y. P. S. Bajaj, ed. *Biotechnology in Agriculture and Forestry. Vol. 20. High-tech and micropropagation IV.* Berlin: Springer-Verlag. pp. 484–94.

8. Anderson, R. A., R. L. Geneve, and L. Utami. 1991. Gotta cut godetia. *Greenhouse Grower* 9(1):46–8.

9. Anthony, J. L. 1992. *In vitro* propagation of *Drosera* spp. *HortScience* 27:850.

10. Apps, D. A., and C. W. Heuser. 1975. Vegetative propagation of *Hemerocallis*—including tissue culture. *Comb. Proc. Intl. Plant Prop. Soc.* 25:362–67.

11. Arditti, J. 1967. Factors affecting the germination of orchid seeds. *Bot. Rev.* 33:1–97.

12. Arditti, J. 1977. *Orchid biology.* Ithaca, NY: Cornell Univ. Press.

13. Arditti, J. 1982. Orchid seed germination and seedling culture—a manual. In J. Arditti, ed. *Orchid biology: Reviews and perspectives II.* London: Cornell Univ. Press. pp. 242–370.

14. Arditti, J, and R. Ernst. 1993. *Micropropagation of orchids.* New York: John Wiley.

15. Armitage, A. M. 1986. *Seed propagated geraniums.* Portland OR: Timber Press.

16. Armitage, A. M. 1987. The influence of spacing on field grown perennial crops. *HortScience* 22:904–7.

17. Arteca, R. N., D. S. Tsai, and C. Schiangnhauler. 1985. Abscisic acid effects on photosynthesis and transpiration in geranium cuttings. *HortScience* 20:370–72.

18. Assn. Off. Seed Anal. 1993. Rules for seed testing. *J. Seed Tech.* 16:1–113.

19. Ault, J. R. 1995. *In vitro* propagation of *Eucomis autumnalis, E. comosa* and *E. zambesiaca* by twin scaling. *HortScience* 30:1441–42.

20. Ault, J. R., and W. J. Blackmon. 1987. *In vitro* propagation of *Ferocactus acanthodes* (Cactaceac). *HortScience* 22:126–27.

21. Ault, J. R., and K. Havens. 1999. Micropropagation of *Baptisia* 'Purple Smoke.' *HortScience* 34:353–54.

22. Bach, A. 1987. The capability of *in vitro* regeneration of various cultivars of *Freesia hybrida. Acta Hort.* 212:715–18.

23. Bach, A. 1992. Micropagation of Hyacinths (*Hyacinthus orientalis* L.). In Y. P. S. Bajaj, ed. *Biotechnology in agriculture and forestry. Vol. 20. High-tech and micropropagation IV.* Berlin: Springer-Verlag. pp. 144–59.

24. Bailey, L. H., E. Z. Bailey, and Staff of Bailey Hortorium. 1976. *Hortus third.* New York: Macmillan.

25. Bajaj, Y. P. S., M. M. S. Sidhu, and A. P. S. Gill. 1992. Micropagation of *Gladiolus.* In Y. P. S. Bajaj, ed. *Biotechnology in agriculture and forestry. Vol. 20. High-tech and micropropagation IV.* Berlin: Springer-Verlag. pp. 135–43.

26. Bajaj, Y. P. S. 1992. Micropagation of *Chrysanthemum.* In Y. P. S. Bajaj, ed. *Biotechnology in agriculture and Forestry. Vol. 20. High-tech and micropropagation IV.* Berlin: Springer-Verlag. pp. 69–80.

27. Ball, V., ed. 1991. *The Ball red book: Greenhouse growing.* 15th ed. Reston, VA: Reston Publ.

28. Bayer, M. B. 1982. *The new Haworthia handbook.* Pretoria: National Botanic Gardens of South Africa.

29. Beattie, D. J. 1993. Propagation of *Astilbe. Comb. Proc. Intl. Plant Prop. Soc.* 43:509–10.

30. Beattie, D. J., and J. W. White. 1993. *Lilium*—hybrids and species. In A. De Hertogh and M. Le Nard, eds. *The physiology of flowering bulbs.* Amsterdam: Elsevier. pp. 423–54.

31. Bech, K., and K. Husted. 1996. Verification of ELISA as a reliable test method for kalanchoe mosaic potyvirus (KMV) and establishment of virus-free *Kalanchoe blossfeldiana* by meristem-tip culture. *Acta Hort.* 432:298–305.

32. Beebee, J. D. 1980. Morphogenetic response of seedlings and adventitious buds of the carnivorous plant *Dionaea muscipula* to aseptic culture. *Bot. Gaz.* 141:396–400.

33. Benschop, M. 1993. Crocus. In A. De Hertogh and M. Le Nard, eds. *The physiology of flowering bulbs.* Amsterdam: Elsevier. pp. 257–72.

34. Benschop, M. 1993. Polianthes. In A. De Hertogh and M. Le Nard, eds. *The physiology of flowering bulbs.* Amsterdam: Elsevier. pp. 589–602.

35. Benson, E. E., J. E. Danaher, I. M. Pimbley, C. T. Anderson, J. E. Wake, S. Daley, and L. K. Adams. 2000. *In vitro* micropropagation of *Primula scotica*: A rare Scottish plant. *Biodivers. Conserv.* 9:711–26.

36. Besemer, S. 1980. Carnations. In R. A. Larson, ed. *Introduction to floriculture.* New York: Academic Press.

37. Bespalhok, E., and K. Hattori. 1998. Friable embryogenic callus and somatic embryo formation from cotyledon explants of African marigold (*Tagetes erecta* L.). *Plant Cell Rept.* 17:870–75.

38. Bhalla, P. L., and K. Sweeney. 1999. Direct *in vitro* regeneration of the Australian fan flower, *Scaevola aemula* R. Br. *Scientia Hort.* 79:65–74.

39. Bicknell, R. A. 1984. Seed propagation of *Gentiana scabra. Comb. Proc. Intl. Plant Prop. Soc.* 34:396–401.

40. Bilkey, P. C., B. H. McCown, and A. C. Hildebrandt. 1978. Micropropagation of African violet from petiole cross-sections. *HortScience* 13:37–8.

41. Blanchette, L. 1998. Asexual propagation of *Anemonella, Dodecatheon,* and *Trillium. Comb. Proc. Intl. Plant Prop. Soc.* 48:327–29.

42. Blazich, F. A., and R. T. Novitzky. 1984. *In vitro* propagation of *Sansevieria trifasciata. HortScience* 19:122–23.

43. Borowski, E., P. Hagen, and R. Moe. 1981. Stock plant irradiation and rooting of chrysanthemum cuttings in light or dark. *Scientia Hort.* 15:245–53.

44. Bose, T. K., B. K. Jana, and S. Moulilk. 1987. A note on the micropropagation of tuberose from scale stem section. *Indian J. Hort.* 44:100–1.

45. Bouza, L., M. Jacques, and E. Miginiac. 1994. *In vitro* propagation of *Paeonia suffruticosa* Andr. cv. 'Mme. De Vatry': Developmental effects of exogenous hormones during the multiplication phase. *Scientia Hort.* 57:241–51.

46. Bratcher, C. B., J. M. Dole, and J. C. Cole. 1993. Stratification improves seed germination of five native wildflower species. *HortScience* 28:899–901.

47. Bridgen, M. P., J. J. King, C. Pedersen, M. A. Smith, and P. J. Winski. 1992. Micropropagation of *Alstroemeria* hybrids. *Comb. Proc. Intl. Plant Prop. Soc.* 42:427–30.

48. Bridgen, M. P., and J. Todd. 1998. Genotypic and environmental effects on root cutting propagation of *Pulmonaria* species and cultivars. *Comb. Proc. Intl. Plant Prop. Soc.* 48:319–23.

49. Brumback, W. E. 1985. Propagation of wildflowers. *Comb. Proc. Intl. Plant Prop. Soc.* 35:542–48.

50. Buchheim, J. A. T., and M. M. Meyer, Jr. 1992. Micropropagation of peony (*Paeonia* spp.). In Y. P. S. Bajaj, ed. *Biotechnology in agriculture and forestry. Vol. 20. High-tech and micropropagation IV.* Berlin: Springer-Verlag. pp. 269–85.

51. Burr, R. W. 1975. Mass production of Boston fern through tissue culture. *Comb. Proc. Intl. Plant Prop. Soc.* 25:122–24.

52. Burrell, C. C. 2006. *Hellebores: a comprehensive guide.* Portland, OR: Timber Press.

53. Bush, S. R., E. D. Earle, and R. W. Langhans. 1976. Plantlets from petal segments, petal epidermis, and shoot tips of the periclinal chimera, *Chrysanthemum morifolium* 'Indianapolis,' *Amer. J. Bot.* 63:729–37.

54. Butcher, S. M., R. A. Bicknell, J. F. Seelye, and N. K. Borst. 1986. Propagation of *Limonium peregrinum*. *Comb. Proc. Intl. Plant Prop. Soc.* 36:448–50.

55. Calvo, M. C., and J. Segura. 1989. *In vitro* propagation of lavender. *HortScience* 24:375–76.

56. Carpenter, W. J. 1989. *Salvia splendens* seed pregermination and priming for rapid and uniform plant emergence. *J. Amer. Soc. Hort. Sci.* 114:247–50.

57. Carpenter, W. J. 1990. Light and temperature govern germination and storage of *Caladium* seed. *HortScience* 25:71–4.

58. Carpenter, W. J., and J. F. Boucher. 1991. Priming improves high-temperature germination of pansy seed. *HortScience* 26:541–44.

59. Carpenter, W. J., and J. F. Boucher. 1992. Temperature requirements and germination of *Delphinium xcultorum* seed. *HortScience* 27:989–92.

60. Carpenter, W. J., and S. Maekawm. 1991. Substrate moisture level governs the germination of verbena seed. *HortScience* 26:1468–72.

61. Carpenter, W. J., and E. R. Ostmark. 1992. Growth regulator and storage temperature govern germination of *Coreopsis* seed. *HortScience* 27:1190–93.

62. Carpenter, W. J., E. R. Ostmark, and J. A. Cornell. 1993. The role of light during *Phlox drummondii* Hook. seed germination. *HortScience* 28:786–88.

63. Carpenter, W. J., E. R. Ostmark, and J. A. Cornell. 1994. Light governs the germination of *Impatiens wallerana* Hook. f. seed. *HortScience* 29:854–57.

64. Carpenter, W. J., E. R. Ostmark, and J. A. Cornell. 1995. Temperature and seed moisture govern germination and storage of *Gerbera* seed. *HortScience* 30:98–101.

65. Carpenter, W. J., and S. W. Williams. 1993. Keys to successful seeding. *Grower Talks* 57:34–44.

66. Carter, F. M. 1973. Grafting cacti. *Horticulture* 51:34–5.

67. Cassells, A. C. 1992. Micropropagation of commercial *Pelargonium* species and hybrids. In Y. P. S. Bajaj, ed. *Biotechnology in Agriculture and Forestry: Vol. 20. High-tech and micropropagation IV.* Berlin: Springer-Verlag. pp. 286–306.

68. Cheers, G. 1992. *Lett's guide to carnivorous plants of the world.* London: Charles Lett and Co.

69. Chen, W. L. 2007. Elimination of *in vitro* contamination, shoot multiplication, and *ex vitro* rooting of *Aglaonema*. *HortScience* 42:629–32.

70. Chidamian, C. 1984. *Book of cacti and other succulents.* Portland, OR: Timber Press.

71. Choffe, K. L., J. M. R. Victor, S. J. Murch, and P. K. Saxena. 2000. *In vitro* regeneration of *Echinacea purpurea* L.: Direct somatic embryogenesis and indirect shoot organogenesis in petiole culture. *In Vitro Cell. Dev. Biol.—Plant* 36:30–6.

72. Chopping, N. 1986. Replacing the bird: Pollination in the genus *Strelitzia*. *Comb. Proc. Intl. Plant Prop. Soc.* 36:204–7.

73. Christie, C. B. 1985. Propagation of amaryllids: A brief review. *Comb. Proc. Intl. Plant Prop. Soc.* 35:351–57.

74. Chua, B. U., J. T. Kunisaki, and Y. Sagawa. 1981. *In vitro* propagation of *Dracaena marginata* 'Tricolor.' *HortScience* 16:494.

75. Cialone, J. 1984. Developments in *Dracaena* production. *Comb. Proc. Intl. Plant Prop. Soc.* 34:491–94.

76. Clayton, P. Q., J. F. Hubstenberger, and G. C. Phillips. 1990. Micropropagation of members of the *Cactaceae* subtribe *Cactinae*. *J. Amer. Soc. Hort. Sci.* 115:337–43.

77. Cockrel, A. D., G. L. McDaniel, and E. T. Graham. 1986. *In vitro* propagation of florist's cineraria. *HortScience* 21:139–40.

78. Cohat, J. 1993. *Gladiolus.* In A. De Hertogh and M. Le Nard, eds. *The physiology of flowering bulbs.* Elsevier, Amsterdam. pp. 297–320.

79. Cohen, D. 1981. Micropropagation of *Zantedeschia* hybrids. *Comb. Proc. Intl. Plant Prop. Soc.* 31:312–16.

80. Colborn, L. N. 1986. Primroses and violets: What's new. *Comb. Proc. Intl. Plant Prop. Soc.* 36:245–49.

81. Connor, D. M. 1985. Plants for the discriminating propagator. *Comb. Proc. Intl. Plant Prop. Soc.* 35:274–77.

82. Conover, C. A., and R. T. Poole. 1978. Production of *Ficus elastica* 'Decora' standards. *HortScience* 13:707–8.

83. Cooke, R. C. 1977. Tissue culture propagation of African violets. *HortScience* 12:549.

84. Cramer, C. S., and M. P. Bridgen. 1997. Somatic embryogenesis and shoot proliferation of *Mussaenda* cultivars. *Plant Cell Tissue Org. Cult.* 50:135–38.

85. Criley, R. A. 1988. Propagation of tropical cut flowers: *Strelitzia, Alpinia,* and *Heliconia. Acta Hort.* 226:509–17.

86. Custers, J. B. M., and J. H. W. Bergervoet. 1992. Differences between *Nerine* hybrids in micropropagation potential. *Scientia Hort.* 52:247–56.

87. Darrow, G. M., and F. G. Meyer, eds. 1968. *Day lily handbook.* American Horticultural Society, Vol. 47, No. 2.

88. Davies, F. T., Jr., and B. C. Moser. 1980. Stimulation of bud and shoot development of Rieger begonia leaf cuttings with cytokinins. *J. Amer. Soc. Hort. Sci.* 105:27–30.

89. Davis, T. D., D. Sankhla, N. Sankhla, A. Upadhyaya, J. M. Parsons, and S. W. George. 1998. Improving seed germination of *Aquilegia chrysantha* by temperature manipulation. *HortScience* 28:798–99.

90. Debergh, P. 1976. An *in vitro* technique for the vegetative multiplication of chimaeral plants of *Dracaena* and *Cordyline. Acta Hort.* 64:17–9.

91. Debergh, P., and J. DeWall. 1977. Mass propagation of *Ficus lyrata. Acta Hort.* 78:361–64.

92. De Hertogh, A., and M. Le Nard. 1993. *The physiology of flowering bulbs.* Amsterdam: Elsevier.

93. De Hertogh, A., and M. Le Nard. 1993. Dahlia. In A. De Hertogh and M. Le Nard, eds. *The physiology of flowering bulbs.* Amsterdam: Elsevier. pp. 273–84.

94. De Hertogh, A., and K. Zimmer. 1993. Allium. In A. De Hertogh and M. Le Nard, eds. *The physiology of flowering bulbs.* Amsterdam: Elsevier. pp. 187–200.

95. De Bruyn, M. H., D. I. Ferreira, M. M. Slabbert, and J. Pretorius. 1992. In *vitro* propagation of *Amaryllis belladonna. Plant Cell Tiss. Organ Cult.* 31:179–84.

96. De Munk, W. J., and J. Schipper. 1993. *Iris*—bulbous and rhizomatous. In A. De Hertogh and M. Le Nard, eds. *The physiology of flowering bulbs.* Amsterdam: Elsevier. pp. 349–80.

97. Desilets, H. 1985. Propagation of herbaceous perennials by root cuttings. *Comb. Proc. Intl. Plant Prop. Soc.* 35:548–55.

98. Desilets, H., Y. Desjardins, and R. B. Belanger. 1993. Clonal propagation of *Pelargonium xhortorum* through tissue culture: Effects of salt dilution and growth regulator concentration. *Can. J. Plant Sci.* 73:871–78.

99. Dole, J. M., and J. L. Gibson. 2006. *Cutting propagation. A guide to propagating and producing floriculture crops.* Batavia, IL: Ball Pub.

100. Dole, J. M., and H. F. Wilkins. 1999. *Floriculture: Principles and species.* Upper Saddle River, NJ: Prentice Hall.

101. Druge, U., S. Zerche, R. Kadner, and M. Ernst. 2000. Relation between nitrogen status, carbohydrate distribution and subsequent rooting of chrysanthemum cuttings as affected by pre-harvest nitrogen supply and cold-storage. *Ann. Bot.* 85:687–701.

102. Dubois, J. B., F. A. Blazich, S. L. Warren, and B. Goldfarb. 2000. Propagation of *Anemone xhybrida* by root cuttings. *J. Environ. Hort.* 18:79–83.

103. Dunston, S., and E. Sutter. 1984. *In vitro* propagation of prayer plants. *HortScience* 19:511–12.

104. Earle, E. D., and R. W. Langhans. 1975. Carnation propagation from shoot tips cultured in liquid medium. *HortScience* 10:608–10.

105. Ecker, R., and A. Barzilay. 1993. Propagation of *Asclepias tuberosa* from short root segments. *Scientia Hort.* 56:171–74.

106. Edinger, P., ed. 1970. *Succulents and cactus.* Menlo Park, CA: Lane.

107. Erickson, D. M. 1985. Propagating and growing primroses in the Pacific Northwest. *Comb. Proc. Intl. Plant Prop. Soc.* 35:219–21.

108. Evenor, D. 2001. Micropropagation of *Achemilla mollis. Plant Cell Tiss. Organ Cul.* 65:169–72.

109. Evenor, D. and M. Reuveni. 2004. Micropropagation of *Achillea filipendulina* cv. Parker. *Plant Cell Tiss. Organ Cult.* 79:91–3.

110. Fagan, A. E., M. A. Dirr, and F. A. Pokorny. 1981. Effects of depulping, stratification, and growth regulators on seed germination of *Liriope muscari. HortScience* 16:208–9.

111. Fakhrai, F., and P. K. Evans. 1989. Morphogenetic potential of cultured explants of *Crocus chrysanthus* Herbert cv. E. P. Bowles. *J. Exp. Bot.* 40:809–12.

112. Farina, E., and B. Ruffoni. 1993. The effect of temperature regimes on micropropagation efficiency and field performance of *Eustoma grandiflorum*. *Acta Hort.* 387:73–80.

113. Fay, A. M., M. A. Bennett, and S. M. Still. 1994. Osmotic seed priming of *Rudbeckia fulgida* improves germination and expands germination range. *HortScience* 29:868–70.

114. Finne, J. F., and J. Van Staden. 1989. *In vitro* propagation of *Sandersonia* and *Gloriosa*. *Plant Cell Tiss. Organ Cult.* 19:151–58.

115. Finne, J. F., and J. van Staden. 1995. *In vitro* culture of *Clivia miniata*. In Kim, K. W. and A. A. de Hertogh. 1997. Tissue culture of ornamental flowering bulbs (geophytes). *Hort. Rev.* 18:87–169.

116. Finnerty, T. L., J. M. Zajicek, and M. A. Hussey. 1992. Use of seed priming to bypass stratification requirements of three *Aquilegia* species. *HortScience* 27:310–13.

117. Fonnesbech, M., and A. Fonnesbech. 1979. *In vitro* propagation of *Spathiphyllum*. *Scientia Hort.* 10:21–5.

118. Foster, S., and S. L. Kitto. 2000. Vegetative propagation of *Spigelia marilandica* (Indian pinks) from shoot-tip cuttings. *HortScience* 35:447–48.

119. Frett, J. J., and M. A. Dirr. 1983. Tissue culture propagation of *Liriope muscari* and *Ophiopogon jaburan*. *HortScience* 18:431–32.

120. Funnell, K. A. 1993. *Zantedeschia*. In A. De Hertogh and M. Le Nard, eds. *The physiology of flowering bulbs*. Amsterdam: Elsevier. pp. 683–704.

121. Furmanowa, M., and O. Olszowska. 1991. Micropropagation of thyme (*Thymus vulgaris* L.). In Y. P. S. Bajaj, ed. *Biotechnology in agriculture and forestry. Vol. 19. High-tech and micropropagation IV.* Berlin: Springer-Verlag. pp. 230–43.

122. Furmanowa, M., and L. Rapczewska. 1993. *Bergenia crassifolia* (L.) Fritsch (*Bergenia*): Micropropagation and arbutin contents. In Y. P. S. Bajaj, ed. *Biotechnology in agriculture and forestry. Vol. 21. Medicinal and aromatic plants IV.* Berlin: Springer-Verlag. pp. 18–33.

123. Garner, J. M., G. J. Keever, D. J. Eakes, and J. R. Kessler. Benzyladenine-induced offset formation in *Hosta* dependent of cultivar. *Comb. Proc. Intl. Plant Prop.* 42:605–10.

124. Gawel, N. J., C. D. Robacker, and W. L. Corley. 1990. *In vitro* propagation of *Miscanthus sinensis*. *HortScience* 25:1291–93.

125. Geneve, R. L. 1999. Seed dormancy in vegetable and flower seeds. *J. Seed Technol.* 20:236–50.

126. Gertsson, U. E. 1992. Micropropagation of cineraria (*Senecio xhybridus* Hyl.). In Y. P. S. Bajaj, ed. *Biotechnology in agriculture and forestry. Vol. 20. High-tech and micropropagation IV.* Berlin: Springer-Verlag. pp. 396–406.

127. Gilbertson-Ferriss, T. L., and H. F. Wilkins. 1977. Factors influencing seed germination of *Freesia refracta* Klatt cv. Royal Mix. *HortScience* 12:572–73.

128. Gill, L. M. 1984. Anemone tuber production in southwest England. *Comb. Proc. Intl. Plant Prop. Soc.* 34:290–94.

129. Girija, S., A. Ganapathi, and G. Vengadesan. 1999. Micropropagation of *Crossandra infundibuliformis* (L.) Nees. *Scientia Hort.* 82:331–37.

130. Haegeman, J. 1993. *Begonia*—Tuberous hybrids. In A. De Hertogh and M. Le Nard, eds. *The physiology of flowering bulbs*. Amsterdam: Elsevier. pp. 227–38.

131. Hammer, P. A. 1976. Stolon formation in *Chlorophytum*. *HortScience* 11:570–72.

132. Han, B. H., J. S. Kim, S. L. Choi, and K. Y. Paek. 1991. *In vitro* mass propagation of *Begonia rex* Putx. and *Caladium bicolor* Vent. *Korean J. Plant Tissue Cult.* 18:95–101.

133. Han, S. S. 1993. Chilling, ethephon and photoperiod affect cormel production of *Brodiaea*. *HortScience* 28:1095–97.

134. Han, S. S., and A. H. Halevy. 1993. *Triteleia*. In A. De Hertogh and M. Le Nard, eds. *The physiology of flowering bulbs*. Amsterdam: Elsevier. pp. 611–16.

135. Hanks, G. R. 1993. *Narcissus*. In A. De Hertogh and M. Le Nard, eds. *The physiology of flowering bulbs*. Amsterdam: Elsevier. pp. 463–558.

136. Hanks, G. R., and A. R. Rees. 1979. Twin-scale propagation of *Narcissus*. A review. *Scientia Hort.* 10:1–14.

137. Hansen, J. 1986. Influence of cutting position and stem length on rooting of leaf-bud cuttings of *Schefflera arboricola*. *Scientia Hort.* 28:177–86.

138. Harazy, A., B. Leshem, A. Cohen, and H. D. Rabinowitch. 1985. *In vitro* propagation of statice as an aid to breeding. *HortScience* 20:361–62.

139. Harbage, J. F. 2001. Micropropagation of *Echinacea angustifolia*, *E. pallida*, and *E. purpurea* from stem and seed explants. *HortScience* 36:360–64.

140. Harris, P. J. C., C. G. Grove, and A. J. Harvard. 1989. *In vitro* propagation of *Symphytum* species. *Scientia Hort.* 40:275–81.

141. Hayashi, T., K. Oshida, and K. Sano. 1988. Formation of alkaloids in suspension cultured *Colchicum autumnale*. *Phytochemistry* 27:1371–74.

142. Henny, R. J. 1978. *In vitro* propagation of *Peperomia* 'Red Ripple' from leaf discs. *HortScience* 13:150–51.

143. Higaki, T., and D. P. Watson. 1973. *Anthurium* culture in Hawaii. *Univ. Hawaii Coop. Ext. Ser. Circ.* 420.

144. Holley, W. D., and R. Baker. 1963. *Carnation production.* Dubuque, Iowa: William C. Brown Co., Publishers.

145. Homes, J., M. Legeos, and M. Jaziri. 1987. *In vitro* multiplication of *C. sativus* L. *Acta Hort.* 212:675–76.

146. Hook, I. L. I. 1993. *Leontopodium alpinuin* Cass. (edelweiss): *In vitro* culture, micropropagation, and the production of secondary metabolites. In Y. P. S. Bajaj, ed. *Biotechnology in agriculture and forestry. Vol. 21. Medicinal and aromatic plants IV.* Berlin: Springer-Verlag. pp. 217–32.

147. Hosokawa, K., Y. Oikawa, and S. Yamamura. 1998. Mass propagation of ornamental gentian in liquid medium. *Plant Cell Rept.* 17:747–51.

148. Hosoki, T., and T. Asahira. 1980. *In vitro* propagation of bromeliads in liquid culture. *HortScience* 15:603–4.

149. Hosoki, T., and D. Kimura. 1997. Micropropagation of *Centauria macrocephala* Pushk. ex Willd. by shoot-axis splitting. *HortScience* 32:1124–25.

150. Houghton, W. J. 1984. New narcissus and their propagation. *Comb. Proc. Intl. Plant Prop. Soc.* 34:294–96.

151. Huang, C. W., H. Okubo, and S. Uemoto. 1990. Comparison of bulblet formation from twin scales and single scales on *Hippeastrum hybridum* cultured *in vitro*. *Scientia Hort.* 46:151–60.

152. Huang, F. H., G. L. Klingaman, and H. H. Chen. 1992. Micropropagation of surprise lily (*Lycoris squamigera*). In Y. P. S. Bajaj, ed. *Biotechnology in agriculture and forestry Vol. 20. High-tech and micropropagation IV.* Berlin: Springer-Verlag. pp. 198–212.

153. Hubbard, A. C. 1986. Native ornamentals for the U. S. southwest. *Comb. Proc. Intl. Plant Prop. Soc.* 36:347–50.

154. Hubstenberger, J. F., P. W. Clayton, and G. C. Phillips. 1992. Micropropagation of cacti (Cactaceae). In Y. P. S. Bajaj, ed. *Biotechnology in agriculture and forestry. Vol. 20. High-tech and micropropagation IV.* Berlin: Springer-Verlag. pp. 49–68.

155. Hussey, G. 1975. Totipotency in tissue explants and callus of some members of the Lilliaceae, Iridaceae, and Amaryllidaceae. *J. Exp. Bot.* 26:253–62.

156. Hussey, G. 1980. Propagation of some members of the Liliaceae, Iridaceae and Amaryllidaceae by tissue culture: Petaloid monocotyledons. In C. D. Brickell, D. F. Cutler, and M. Gregory, eds. *Linnean Soc. Symp. Ser. 8.* London: Academic.

157. Ilan, A., M. Ziv, and A. H. Halevy. 1995. Propagation of corm development of *Brodiaea* in liquid cultures. *Scientia Hort.* 63:101–12.

158. Imanishi, H. 1993. *Freesia.* In A. De Hertogh and M. Le Nard, eds. *The physiology of flowering bulbs.* Amsterdam: Elsevier. pp. 285–96.

159. Jackson, H. C. 1975. Propagation and culture of African violets. *Comb. Proc. Intl. Plant Prop. Soc.* 25:269–71.

160. Jamieson, A. C. 1988. New Zealand callas. *Grower Talks* 51:56–60.

161. Jeham, H., D. Courtois, C. Ehret, K. Lerch, and V. Petiard. 1994. Plant regeneration in *Iris pallida* Lam. and *Iris germanica* L. via somatic embryogenesis from leaves, apices and young flowers. *Plant Cell Rpts.* 13:671–75.

162. Jenks, M., M. Kane, F. Morousky, D. McConnell, and T. Sheehan. 1990. *In vitro* establishment and epiphyllous plantlet regeneration of *Nymphaea* 'Daubeniana.' *HortScience* 25:1664.

163. Johnson, B. B. 1978. *In vitro* propagation of gloxinia from leaf explants. *HortScience* 13:149–50.

164. Jones, R. O., R. L. Geneve, and S. T. Kester. 1994. Micropropagation of gas plant (*Dictamnus albus* L.). *J. Environ. Hort.* 12:216–18.

165. Jungnickel, F., and S. Zaid. 1992. Micropropagation of African violets (*Saintpaulia* spp. and cvs.). In Y. P. S. Bajaj, ed. *Biotechnology in agriculture and forestry. Vol. 20. High-tech and micropropagation IV.* Berlin: Springer-Verlag. pp. 357–95.

166. Kallak, H., M. Reidla, I. Hilpus, and K. Virumae. 1997. Effects of genotype, explant source and growth regulators on organogenesis in carnation callus. *Plant Cell Tiss. Organ Cult.* 51:127–35.

167. Kane, M. E., T. J. Sheehan, and F. H. Ferwerda. 1988. *In vitro* growth of American lotus embryos. *HortScience* 23:611–13.

168. Kaukovirta, E. 1973. Effects of temperature and light intensity on the growth of *Senecio xhybridus* Hyl. seedlings. *Acta Hort.* 31:63–9.

169. Kelly, J. W., and J. J. Frett. 1986. Photoperiodic control of growth in water lilies. *HortScience* 21:151.

170. Khan, M. R. I., J. K. Heyes, and D. Cohen. 1988. Plant regeneration from oca (*Oxalis tuberosa* M.): The effect of explant type and culture media. *Plant Cell Tiss. Organ Cult.* 14:41–50.

171. Khehra, K., C. Lowe, M. R. Davey, and J. B. Power. 1995. An improved micropropagation system for chrysanthemum based on Pluronic F-68-supplemented media. *Plant Cell Tiss. Organ Cult.* 41:87–90.

172. Kim, Y. S., E. H. Park, S. O. Yoo, J. S. Eun, and W. S. Song. 1991. *In vitro* propagation of ranunculus (*Ranunculus asiaticus* L.). I: Embryogenic callus induction and somatic embryogenesis. *J. Korean Soc. Hort. Sci.* 32:401–10.

173. King, J. J., and M. P. Bridgen. 1990. Environmental and genotypic regulation of *Alstroemeric* seed germination. *HortScience* 25:1607–9.

174. King, P. 1997. Production of *Epimediums* by division. *Comb. Proc. Intl. Plant Prop. Soc.* 47:521–22.

175. Kitto, S. L., and A. Hoopes. 1992. Micropropagation and field establishment of *Tiarella cordifolia*. *J. Environ. Hort.* 10:171–74.

176. Klesser, P. J., and D. D. Kel. 1976. Virus diseases and tissue culture of some South African flower bulbs. *Acta Hort.* 59:71–6.

177. Knauss, J. F. 1976. A partial tissue culture method for pathogen-free propagation of selected fern from spores. *Proc. Fla. State Hort. Soc.* 89:363–65.

178. Koh, Y. C., and F. T. Davies, Jr. 1997. Micropropagation of *Cryptanthus* with leaf explants with attached intercalary meristems, excised from greenhouse stock plants. *Scientia Hort.* 70:301–7.

179. Kononowicz, H., and J. Janick. 1984. *In vitro* propagation of *Vanilla planifolia*. *HortScience* 19:58–9.

180. Kothari, S. L., and N. Chandra. 1984. *In vitro* propagation of African marigold. *HortScience* 19:703–5.

181. Kramer, J. 1971. *How to grow African violets*. 4th ed. Menlo Park, CA: Lane.

182. Kromer, K. D., and K. Kukulczanka. 1985. *In vitro* cultures of meristem tips of *Canna indica* (L.). *Acta Hort.* 251:147–53.

183. Kukulczanka, K., K. Kromer, and B. Cvzanstka. 1989. Micropropagation of *Fritillaria melengris* L. through tissue culture. *Acta Hort.* 251:147–53.

184. Kunisaki, J. T. 1975. *In vitro* propagation of *Cordyline terminalis* (L.) Kurth. *HortScience* 10:601–2.

185. Kunisaki, J. T. 1980. *In vitro* propagation of *Anthurium andreanum* Lind. *HortScience* 15:508–9.

186. Kusey, W. E., Jr., P. A. Hammer, and T. C. Weiler. 1980. *In vitro* propagation of *Gypsophila paniculata* L. 'Bristol Fairy.' *HortScience* 15:600–1.

187. Lalibert, S., L. Chretien, and J. Vieth. 1985. *In vitro* plantlet production from young capitulum explants of *Gerbera jamesonii*. *HortScience* 20:137–39.

188. Lane, B. C. 1981. A procedure for propagating ferns from spore using a nutrient agar solution. *Comb. Proc. Intl. Plant Prop.* 30:94–7.

189. Lawson, G. M., and P. B. Goodwin. 1985. Commercial production of kangaroo paws. *Comb. Proc. Intl. Plant Prop. Soc.* 35:57–65.

190. Lazarte, J. E., M. S. Gaiser, and O. R. Brown. 1982. *In vitro* propagation of *Epiphyllum chrysocardium*. *HortScience* 17:84.

191. Lazcano, C. A., F. T. Davies, Jr., A. A. Estrada-Luna, S. A. Duray, and V. Olalde-Portugal. 1999. Effect of auxin and wounding on adventitious root formation of prickly-pear cactus cladodes. *HortTechnology* 9:99–102.

192. Le Nard, M., and A. A. De Hertogh. 1993. *Tulipa*. In A. De Hertogh and M. Le Nard, eds. *The physiology of flowering bulbs*. Amsterdam: Elsevier. pp. 617–82.

193. Lee, C. I., and H. C. Kohl. 1985. Note on vegetative reproduction of *Cyclamen indicum*. *The Plant Propagator* 31:4.

194. Lee, C. W., R. M. Skirvin, A. I. Soltero, and J. Janick. 1977. Tissue culture of *Salpiglossis sinuata* L. from leaf discs. *HortScience* 12:547–49.

195. Leiss, J. 1998. Propagation of *Phlox paniculata* from root cuttings. *Comb. Proc. Intl. Plant Prop. Soc.* 48:326–27.

196. Lilien-kipnis, H., and M. Kochba. 1987. Mass propagation of new *Gladiolus* hybrids. *Acta Hort.* 212:631–38.

197. Lindgren, D. T., and B. McCown. 1992. Multiplication of four penstemon species *in vitro*. *HortScience* 27:182.

198. Liu, L., and D. W. Burger. 1986. *In vitro* propagation of Easter lily from pedicels. *HortScience* 21:1437–38.

199. Lyons, R. E., and R. E. Widmer. 1980. Origin and historical aspects of *Cyclamen persicum* Mill. *HortScience* 15:132–35.

200. Matsuo, E., and J. M. van Tuyl. 1986. Early scale propagation results in forcible bulbs of Easter lily. *HortScience* 21:1006–7.

201. McComb, J. A., and S. Newton. 1981. Propagation of kangaroo paws using tissue culture. *HortScience* 16:181–83.

202. McKay, M. E., D. E. Bythe, and J. A. Tommerup. 1981. The effects of corm size and division of the mother corm in gladioli. *Aust. J. Expt. Agric. Animal Husbandry* 21:343–48.

203. McMillan, R. N. 1980. Cyclamen production problems. *Comb. Proc. Inter. Plant Prop. Soc.* 29:173–76.

204. McRae, E. A. 1978. Commercial propagation of lilies. *Comb. Proc. Intl. Plant Prop. Soc.* 28:166–69.

205. Mekers, O. 1977. *In vitro* propagation of some Tillandsioideae (Bromeliaceae). *Acta Hort.* 78:311–20.

206. Meyer, H. J., and J. Van Stadden. 1991. Rapid *in vitro* propagation of *Aloe barbandensis* Mill. *Plant Cell Tiss. Organ Cult.* 26:167–71.

207. Meyer, M. M., Jr. 1976. Propagation of daylilies by tissue culture. *HortScience* 11:485–87.

208. Meyer, M. M., Jr. 1980. *In vitro* propagation of *Hosta sieboldianum*. *HortScience* 15:737–38.

209. Meynet, J. 1993. *Anemone*. In A. De Hertogh and N. Le Nard, eds. *The Physiology of flowering bulbs*. Amsterdam: Elsevier. pp. 211–18.

210. Meynet, J. 1993. *Ranunculus*. In A. De Hertogh and M. Le Nard, eds. *The physiology of flowering bulbs*. Amsterdam: Elsevier. pp. 603–10.

211. Meynet, J., and A. Cuclos. 1990. Culture in vitro de la renoncule des fleuristes (*Ranunculus asiaticus* L.). *Agronomie* 10:157–62.

212. Mikkelsen, B. P., and K. C. Sink, Jr. 1978. *In vitro* propagation of Rieger Elatior begonias. *HortScience* 13:242–44.

213. Miller, C. T. 2005. Photoperiod and stock plant age effects on shoot, stolon, and rhizome formation response from leaf cuttings of *Achimenes*. *Acta Hort.* 673:349–54.

214. Minocha, S. C. 1985. *In vitro* propagation of *Dionaea muscipula*. *HortScience* 20:216–17.

215. Moe, R. 1993, *Liatris*. In A. De Hertogh and M. Le Nard, eds. *The physiology of flowering bulbs*. Amsterdam: Elsevier. pp. 381–90.

216. Monette, P. L. 1992. Micropropagation of Inca Lily (*Alstroemeria* spp.). In Y. P. S. Bajaj, ed. *Biotechnology in agriculture and forestry. Vol. 20. High-tech and micropropagation IV*. Berlin: Springer-Verlag. pp. 1–18.

217. Morel, G. M. 1966. Meristem culture: Clonal propagation of orchids. *Orchid Digest* 30:45–9.

218. Murashige, T., M. Serpa, and J. B. Jones. 1974. Clonal multiplication of *Gerbera* through tissue culture. *HortScience* 9:175–80.

219. Nakano, M., Y. Hoshino, and M. Mii. 1994. Adventitious shoot regeneration from cultured petal explants of carnation. *Plant Cell Tiss. Organ Cult.* 36:15–20.

220. Nakano, M., D. Niimi, D. Kobayashi, and A. Watanabe. 1999. Adventitious shoot regeneration and micropropagation of hybrid tuberous begonia (*Begonia xtuberhybrida* Voss). *Scientia Hort.* 79:245–51.

221. Natali, L., I. C. Sanchez, and A. Cavallini. 1990. *In vitro* culture of *Aloe barbadensis* Mill.: Micropropagation from vegetative meristems. *Plant Cell Tiss. Organ Cult.* 20:71–4.

222. Nathan, M. J., C. J. Goh, and P. P. Kumar. 1992. *In vitro* propagation of *Heliconia psittacorum* by bud culture. *HortScience* 27:450–52.

223. Nau, J. 1993. *Ball culture guide: The encyclopedia of seed germination*. Batavia, IL: Ball Pub.

224. Newbury, H. J., E. A. B. Aitken, N. J. Atkinson, and B. V. Ford-Lloyd. 1992. Micropropagation of snapdragon (*Antirrhinum majus* L.). In Y. P. S. Bajaj, ed. *Biotechnology in agriculture and forestry. Vol. 20. High-tech and micropropagation IV*. Berlin: Springer-Verlag. pp. 19–33.

225. Nomizu, T., Y. Niimi, S. Kasahara. 2003. *In vitro* micropropagation of 'Yukiwariso' (*Hepatica nobilis* Schreber var. japonica f. magna) by leaf segment culture. *J. Jap. Soc. Hort. Sci.* 72:205–11.

226. Nowak, J., and R. M. Rudnicki. 1993. *Hyacinthus*. In A. De Hertogh and M. Le Nard, eds. *The physiology of flowering bulbs*. Amsterdam: Elsevier. pp. 235–48.

227. Ohki, S., and S. Sawaki. 1999. The effects of inorganic salts and growth regulators on *in vitro* shoot proliferation and leaf chlorophyll content of *Delphinium cardinale*. *Scientia Hort.* 81:149–58.

228. Okubo, H. 1993. *Hippeastrum (Amaryllis)*. In A. De Hertogh, and M. Le Nard, eds. *The physiology of flowering bulbs*. Amsterdam: Elsevier. pp. 321–34.

229. Peck, D. E. 1984. *In vitro* propagation of *Begonia* x *tuberhydra* from leaf sections. *HortScience* 19:395–97.

230. Peck, D. E. 1984. *In vitro* vegetative propagation of cape primrose using the corolla of the flower. *HortScience* 19:399–400.

231. Peck, D. E., and B. G. Cumming. 1986. Beneficial effects of activated charcoal on bulblet production in cultures of *Muscari armeniacum*. *Plant Cell Tiss. Organ Cult.* 6:9–14.

232. Pence, V. C., and V. G. Soukup. 1993. Factors affecting the initiation of mini-rhizomes from *T. erectum* and *T. grandiflorum* tissues *in vitro*. *Plant Cell Tiss. Organ Cult.* 35:229–35.

233. Perl, P. 1977. *Ferns*. Alexandria, VA: Time-Life Books.

234. Perl, P. 1978. *Cacti and succulents*. Alexandria, VA: Time-Life Books.

235. Pfister, J. M., and J. M. Widholm. 1984. Plant regeneration from snapdragon tissue culture. *HortScience* 19:852–54.

236. Pierik, R. L. M., and H. H. M. Steegmans. 1976. Vegetative propagation of *Freesia* through

isolation of shoots *in vitro*. *Netherlands J. Agric. Science* 24:274–77.

237. Piqueras, A., J. M. van Huylenbroeck, B. H. Han, and P. C. Debergh. 1998. Carbohydrate partitioning and metabolism during acclimatization of micropropagated *Calathea*. *Plant Growth Regul.* 26:25–31.

238. Platt, G. C. 1985. Propagation of the cordylines by vegetative means. *Comb. Proc. Intl. Plant Prop. Soc.* 35:364–66.

239. Poole, R. T., and C. A. Conover. 1987. Vegetative propagation of foliage plants. *Comb. Proc. Intl. Plant Prop. Soc.* 37:503–7.

240. Prakash, J. 2009. Micropropagation of ornamental perennials: progress and problems. *Acta Hort.* 812:289–94.

241. Preece, J. E., and K. P. Ripley. 1992. *In vitro* culture and micropropagation of *Euphorbia sp.* In Y. P. S. Bajaj, ed. *Biotechnology in agriculture and forestry. Vol. 20. High-tech and micropropagation IV.* Berlin: Springer-Verlag. pp. 91–112.

242. Quay, L., J. A. McComb, and K. W. Dixon. 1995. Methods for *ex vitro* germination of Australian terrestrial orchids. *HortScience* 30:182.

243. Ramachandra, S., and H. Khatamian. 1989. Micropropagation of *Peperomia* and *Begonia* using petiole segments. *HortScience* 24:153.

244. Ramcurez-Malagcon, R. 2008. *In vitro* propagation of three *Agave* species used for liquor distillation and three for landscape. *Plant Cell Tiss. Organ Cult.* 94:201–7.

245. Rauch, F. D., and R. M. Yamakawa. 1980. Effects of auxin on rooting of *Ixora acuminata*. *HortScience* 15:97.

246. Raviv, M., E. Putieusky, and D. Sanderovich. 1984. Rooting stem cuttings of sage (*Salvia officinalis* L.). *The Plant Propagator* 30:8–9.

247. Raward, I. D. 1975. Propagation of *Codiaeum* (croton) by tip cuttings. *Comb. Proc. Intl. Plant Prop. Soc.* 25:386.

248. Redwood, G. N., and J. C. Bowling. 1990. Micropropagation of *Nepenthes* species. *Bot. Gardens Microprop. News* 1:19–20.

249. Richwine, A. M., J. L. Tipton, and G. Thompson. 1995. Establishment of *Aloe, Gasteria,* and *Haworthia* shoot cultures from inflorescence explants. *HortScience* 30:1443–44.

250. Robacker, C. D., and W. L. Corley. 1992. Plant regeneration of pampas grass from immature inflorescences cultured *in vitro*. *HortScience* 27:841–43.

251. Roberts, D. J. 1965. Modern propagation of ferns. *Comb. Proc. Intl. Plant Prop. Soc.* 15:317–21.

252. Rockwell, F. F., G. C. Grayson, and J. de Graff. 1961. *The complete book of lilies.* Garden City, NY: Doubleday.

253. Roest, S., and G. S. Bokelmann. 1975. Vegetative propagation of *Chrysanthemum morifolium* Ram *in vitro*. *Scientia Hort.* 3:317–30.

254. Rogers, R. W. 1995. Divide and conquer: Propagation of herbaceous and tree peonies. *Comb. Proc. Intl. Plant Prop. Soc.* 45:325–26.

255. Rudnicki, R. M., and J. Nowak. 1993. *Muscari*. In A. De Hertogh and M. Le Nard, eds. *The physiology of flowering bulbs.* Amsterdam: Elsevier. pp. 455–62.

256. Samfield, D. M., J. M. Zajicek, and B. G. Cobb. 1988. The effects of osmoconditioning on herbaceous perennial seed germination at different temperatures. *HortScience* 23:750.

257. Schnabelrauch, L. S., and K. C. Sink. 1979. *In vitro* propagation of *Phlox subulata* and *Phlox paniculata* tissue culture. *HortScience* 14:607–8.

258. Schwartz, M. A., R. N. Payne, and G. Sites. 1985. Residual effects of chlormequat on garden performance in sun and shade of seed and cutting propagated cultivars of geraniums. *HortScience* 20:368–70.

259. Schwenkel, H. G., and J. Grumwaldt. 1988. *In vitro* propagation of *Cyclamen persicum* Mill. *Acta Hort.* 226:659–62.

260. Scott, M., and M. E. Marson. 1967. Effects of mist and basal temperature on the regeneration of *Saintpaulia ionantha* Wendl. from leaf cuttings. *Horticultural Res.* 7:50–60.

261. Sheehan, T. J. 1983. Recent advances in botany, propagation and physiology of orchids. *Hort. Rev.* 5:279–315.

262. Sheerin, P. 1974. Propagation of various types of begonia. *Comb. Proc. Intl. Plant Prop. Soc.* 24:292–93.

263. Shoemaker, C. A., and W. H. Carlson. 1992. Temperature and light affect seed germination of *Begonia semperflorens-cultorum*. *HortScience* 27:181.

264. Simmonds, J. 1992. Micropropagation of *Begonia* spp. In Y. P. S. Bajaj, ed. *Biotechnology in agriculture and forestry. Vol. 20. High-tech and micropropagation IV.* Berlin: Springer-Verlag. pp. 34–48.

265. Simmonds, J., and T. Werry. 1987. Liquidshake culture for improved micropropagation of *Begonia xhiemalis*. *HortScience* 22:122–24.

266. Singletary, S. R., and S. A. Martin. 1998. IBA-K induced rooting in perennials. *Comb. Proc. Intl. Plant Prop. Soc.* 48:480–85.

267. Slabbert, M. M., M. H. De Bruyn, D. I. Ferreira, and J. Pretorius. 1995. Adventitious *in vitro*

plantlet formation from immature floral stems of *Crinum macowanii*. *Plant Cell Tiss. Organ Cult.* 43:51–7.

268. Slocum, P. D. 1985. Propagating water lilies and aquatics. *Brooklyn Botanic Garden Record* 41:25–8.

269. Small, D. J. 1986. Propagation of choice alpines. *Comb. Proc. Intl. Plant Prop. Soc.* 36:241–44.

270. Smith, R. H., and A. E. Nightingale. 1979. *In vitro* propagation of *Kalanchoe*. *HortScience* 14:20.

271. Smith, R. H., and R. E. Norris 1983. *In vitro* propagation of African violet chimeras. *HortScience* 11:436–37.

272. Solem, M. 1988. *Torenia*. *Grower Talks* 52:18.

273. Solt, S. 1998. Commercial propagation of *Trillium*. *Comb. Proc. Intl. Plant Prop. Soc.* 48:329–32.

274. Sriskandarajahi, S., and R. M. Skirvin. 1991. Effect of NH_4NO_3 level on philodendron root development *in vitro*. *HortTechnology* 1:37–8.

275. St. Hilaire, R., C. A. Fierro-Berwart, and C. A. Perez-Mumoz. 1996. Adventitious root formation and development in cuttings of *Mussaenda erythrophylla* L. Schum. & Thonn. *HortScience* 31:1023–25.

276. Standifer, L. C., E. N. O'Rourke, and R. Porche-Sorbet. 1984. Propagation of *Haworthia* from floral scapes. *The Plant Propagator* 30:4–6.

277. Stanilova, M. I., V. P. Ilcheva, and N. A. Zagorska. 1994. Morphogenetic potential and *in vitro* micropropagation of endangered plant species *Leucojum aestivum* L. and *Lilum rhodopaeum* Delip. *Plant Cell Rpts.* 13:451–53.

278. Stapfer, R. E., and C. W. Heuser. 1985. *In vitro* propagation of periwinkle. *HortScience* 20:141–42.

279. Stapfer, R. E., and C. W. Heuser. 1986. Rapid multiplication of *Heuchera sanguinea* Engelm. 'Rosamundi' propagated *in vitro*. *HortScience* 21:1043–44.

280. Stapfer, R. E., C. W. Heuser, and C. F. Deneke. 1985. Rapid multiplication of *Veronica* 'Red Fox' propagated *in vitro*. *HortScience* 20:866–67.

281. Start, N. D., and B. G. Cumming. 1976. *In vitro* propagation of *Saintpaulia ionantha* Wendl. *HortScience* 11:204–6.

282. Stefanis, J. P., and R. W. Langhans. 1980. Factors affecting the production and propagation of xerophytic succulent species. *HortScience* 15:504–5.

283. Stephens, L. C., S. L. Krell, and J. L. Weigle. 1985. *In vitro* propagation of Java, New Guinea and Java New Guinea Impatiens. *HortScience* 20:362–63.

284. Stephens, L. C., J. L. Weigle, S. L. Krell, and K. Han. 1991. Micropropagation of *Impatiens*. In Y. P. S. Bajaj, ed. *Biotechnology in agriculture and forestry. Vol. 20. High-tech and micropropagation IV.* Berlin: Springer-Verlag. pp. 160–72.

285. Stewart, J., and E. Hennessy. 1981. *Orchids of Africa*. Boston: Houghton Mifflin.

286. Still, S. M. 1994. *Manual of herbaceous ornamental plants*. 4th ed. Champaign, IL: Stipes Publ.

287. Stimart, D. P. 1986. Commercial micropropagation of florist flower crops. In R. H. Zimmerman et al., eds. *Tissue culture as a plant production system for horticultural crops.* Dordrecht: Martinus Nijhoff Publishers.

288. Stimart, D. P., and P. D. Ascher. 1978. Tissue culture of bulb scale sections for asexual propagation of *Lillium longiflorum* Thumb. *J. Amer. Soc. Hort. Sci.* 103:182–84.

289. Stimart, D. P., and J. F. Harbage. 1989. Shoot proliferation and rooting *in vitro* propagation of *Liatris spicata*. *HortScience* 24:835–36.

290. Stimart, D. P., J. C. Mather, and K. R. Schroeder. 1998. Shoot proliferation and rooting *in vitro* of *Pulmonaria*. *HortScience* 33:339–41.

291. Stokes, P. 1984. Hardy ferns. *Comb. Proc. Intl. Plant Prop. Soc.* 34:332–33.

292. Sul, I. W., and S. S. Korban. 1998. Influence of bombardment with BA-coated microprojectiles on shoot organogenesis from *Phlox paniculata* L. and *Pinus pinea* L. tissues. *In Vitro Cell Dev. Biol.—Plant* 34:300–2.

293. Sutter, E. G. 1986. Micropropagation of *Ixia viridifolia* and a *Gladiolus xhomoglossum* hybrid. *Scientia Hort.* 29:181–89.

294. Takayama, S. 1990. Begonia. In P.V. Ammirato, D.A. Evans, W.R. Sharp, and Y.P.S. Bajaj, eds. *Handbook of plant cell culture.* McGraw-Hill. pp. 253–83.

295. Tanabe, M. J. 1980. Effect of depulping and growth regulators on seed germination of *Alyxia olivaeformis*. *HortScience* 15:199–200.

296. Tanabe, M. J. 1982. Single node stem propagation of *Alyxia olivaeformis*. *HortScience* 17:50.

297. Thompson, P. A., and S. A. Cox. 1979. Germination responses of half-hardy annuals. 2. China aster (*Callistephus chinensis*). *Seed Science and Technol.* 7:201–7.

298. Thorn-Horst, A., R. K. Horst, S. H. Smith, and W. A. Oglevee. 1977. A virus-indexing tissue culture system for geraniums. *Flor. Rev.* 160(4148):28–9, 72–4.

299. Torres, K. C., and N. J. Natarella. 1984. *In vitro* propagation of *Exacum*. *HortScience* 19:224–25.

300. Tsuro, M., M. Koda, and M. Inowe. 1999. Comparative effect of different types of cytokinin for shoot formation and plant regeneration from leaf-derived callus of lavender (*Lavandula vera* DC). *Scientia Hort.* 81:331–36.

301. Ulrich, M. R., F. T. Davies, Jr., Y. C. Koh, S. A. Duray, and J. N. Egilla. 1999. Micropropagation of *Crinum* 'Ellen Bosanquet' by tri-scales. *Scientia Hort.* 82:95–102.

302. Upadhyaya, A., T. D. Davis, D. Sankhla, and N. Sankhla. 1992. Micropropagation of *Lupinus texensis* from cotyledonary node explants. *HortScience* 27:1222–23.

303. Van Brenk, G., and M. Benschop. 1993. *Nerine*. In A. De Hertogh and M. Le Nard, eds. *The physiology of flowering bulbs*. Amsterdam: Elsevier. pp. 559–88.

304. Van de Pol, P. A., and T. F. Van Hell. 1988. Vegetative propagation of *Streilitzia reginae*. *Acta Hort.* 226:581–86.

305. Van der Linde, P. C. G., and J. A. Schipper. Micropropagation of iris with special reference to *Iris xhollandica* Tub. In Y. P. S. Bajaj, ed. *Biotechnology in agriculture and forestry. Vol. 20. High-tech and micropropagation IV.* Berlin: Springer-Verlag. pp. 173–97.

306. Van Dyk, M., and R. Currah. 1983. Vegetative propagation of prairie forbs native to southern Alberta, Canada. *The Plant Propagator* 28:12–4.

307. Van Rensburg, J. G. J., B. M. Vcelar, and P. A. Landby. 1989. Micropropagation of *Ornithogalum maculatum*. *South African J. Bot.* 55:137–39.

308. Varadarajan, G. S., U. Varadarajan, and R. D. Locy. 1993. Application of tissue culture techniques maintain a rare species, *Puya tuberosa*. *J. Bromeliad Soc.* 43:12–8.

309. Verron, P., M. Le Nard, and J. Cohat. 1995. *In vitro* organogenic competence of different organs and tissues of lily of the valley 'Grandiflora of Nantes'. *Plant Cell Tiss. Organ Cult.* 40:237–42.

310. Vinterhalter, D. V. 1989. *In vitro* propagation of green-foliaged *Dracaena fragrans* Ker. *Plant Cell Tiss. Organ Cult.* 17:13–9.

311. Von Hentig, W. U., M. Fisher, and K. Köhler. 1984. Influence of daylength on the production and quality of cuttings from *Fuchsia* mother plants. *Comb. Proc. Intl. Plant Prop. Soc.* 34:141–49.

312. Wang, Y. T. 1987. Effect of warm-medium, light intensity, BA and parent leaf on propagation of golden pothos. *HortScience* 22:597–99.

313. Wang, Y. T. 1987. Effect of temperature, duration and light during simulated shipping on quality and rooting of croton cuttings. *HortScience* 22:1301–2.

314. Wang, Y. T. 1987. Influence of light and heated medium on rooting and shoot growth of two foliage plant species. *HortScience* 23:346–47.

315. Wang, Y. T., and C. A. Boogher. 1988. Effect of nodal position, cutting length, and root retention on the propagation of golden pothos. *HortScience* 23:347–49.

316. Wartidininghsih, N., and R. L. Geneve. 1994. Source and seed quality influence germination in purple coneflower (*Echinacea purpurea*). *HortScience* 29:1443–44.

317. Wartidininghsih, N., and R. L. Geneve. 1994. Osmotic priming or chilling stratification improve seed germination of purple coneflower (*Echinacea purpurea*). *HortScience* 29:1445–48.

318. Watad, A. A., M. Kochba, A. Nissim, and V. Gaba. 1995. Improvement of *Aconitum nepellus* micropropagation by liquid culture on floating membrane rafts. *Plant Cell Rpt.* 14:345–48.

319. Watelet-Gonad, M. C., and J. M. Favre. 1981. Miniaturization et rajeunissement chez *Dahlia variabilis* (variete Television) cultive *in vitro*. *Ann. Sciences Naturelles Bot.*, Paris. pp. 51–67.

320. Welander, T. 1978. *In vitro* organogenesis in explants from different cultivars of *Begonia x hiemalis*. *Physiol. Plant.* 41:142–45.

321. Widmer, R. E. 1980. Cyclamens. In R.A. Larson, ed. *Introduction to floriculture*. New York: Academic Press.

322. Wikesjö, J. 1982. Growing hybrid *Achimenes* as a pot plant. *Florists' Review* 168:34, 120–22.

323. Wilfret, G. J. 1993. *Caladium*. In A. De Hertogh and M. Le Nard, eds. *The physiology of flowering bulbs*. Amsterdam: Elsevier. pp. 239–48.

324. Wilson, H. V. 1980. *Saintpaulia* species. *Amer. Hort.* 58:35–9.

325. Yanagawa, T., and I. Ito. 1988. Differences in the capacity for bulblet regeneration between bulb-scale explants excised from different parts of *Hymenocallis* and *Ornithogalum* bulbs. *J. Japan. Soc. Hort. Sci.* 57:454–61.

326. Yanagawa, T., and Y. Sakanishi. 1980. Studies on the regeneration of excised bulb tissues of various tunicated-bulbous ornamentals. 1. Regenerative capacity of the segments from different parts of bulb scales. *J. Japan. Soc. Hort. Sci.* 48:495–502.

327. Yantcheva, A., M. Vlahova, and A. Antanassov. 1998. Direct somatic embryogenesis and

plant regeneration of carnation (*Dianthus caryophyllus* L.). *Plant Cell Rpt.* 18:148–53.

328. Zieslin, N., and A. Keren. 1980. Effects of rootstock on cactus grafted with an adhesive. *HortScience* 21:153–54.

329. Zimmer, K. 1985. *Eremurus*. In A. H. Halevy, ed. *Handbook of flowering*, Vol. II. Boca Raton, FL: CRC Press. pp. 466–69.

330. Zimmerman, T. W., F. T. Davies, Jr., and J. M. Jajicek. 1991. *In vitro* and macropropagation of the wildflower *Dyssodia pentacheta* (D.C.) Robins. *HortScience* 26:1555–57.

331. Ziv, M., N. Hertz, and Y. Biran. 1983. Vegetative reproduction of *Allium ampeloprasum* L. *in vivo* and *in vitro*. *Israel J. Bot.* 32:1–9.

332. Ziv, M., and J. Lilien-Kipnis. 1990. *Gladiolus*. In P. V. Armmirato, D. A. Evans, W. R. Sharp, and Y. P. S. Bajaj, eds. *Handbook of plant cell culture*. New York: McGraw-Hill. pp. 461–78.

subject index

A

Abiotic stress, 421
Abscisic acid (ABA), 38, 41–42
   adventitious root, bud, and shoot formation, 295, 298–299
   dormancy control, 236–237, 238
   germination, 228, 229, 236
   scion-rootstock relationships, 456
   seed development, 129, 130, 135
Accelerated growth techniques (AGT), 89, 90, 330–331
Acclimation. *See* Hardening-off
Acclimatization stage of micropropagation, 658–659, 718–723
Acids, safe use of, 258
Acid scarification, 222
Acrylic greenhouse coverings, 66
Action plans, 398–399
Additive heritability, 153
Adhesion, 425, 426
Adult phase of seedling life cycle, 15–16, 18
   clones, 594, 616
   retention of, 19
Advection, 326
Adventitious bud formation, 281–287
   callus formation, 287–289
   hormonal control, 293–299
   leaf cuttings, 290–292, 360–361
   root cuttings, 292–295
Adventitious buds, 280
Adventitious embryony, 132
Adventitious organs, 280
Adventitious root formation, 281–287, 302–304. *See also* Cutting propagation
   biochemical basis for, 299–304
   biotechnological advances in, 304–307, 545
   bulbs, 564
   callus formation, 287–289
   growth regulators, 295–299
   hormonal control, 293–299
   leaf cuttings, 290–292, 360–361
   managing and manipulating, 305–307
   photosynthesis, 331
   predictive indices, 317, 318
   root cuttings, 292–295
Adventitious roots, 280
Adventitious shoot formation, 281–287, 285, 291. *See also* Axillary shoot formation; Cutting propagation
   biotechnological advances, 305–306
   callus formation, 287–289
   hormonal control, 293–299
   leaf cuttings, 290–292, 360–361
   managing and manipulating, 305–307
   root cuttings, 292–295
   shoot organogenesis, 285, 289–290, 298
   tissue culture propagation, 646, 651, 659, 663–666, 685
   tuber propagation, 583–584
Adventitious shoots, 280
Adventive embryony, 132, 133
Aeration
   chilling stratification, 225–226
   germination, 216–217
After-ripening technique, 225, 228–230, 231, 237
Agar, 653, 711
Agar substitutes, 711
Aggregates, as media components, 80
Aging, 308, 602, 613–616
Aging tests, 183
Agricultural development, 3
Agricultural seeds, commercial production, 162–166
Agriculture, 2, 3–5
Agrobacterium-induced cell growth, 683
Agrobacterium-mediated transformation, 36
Agrobacterium rhizogenes, 383, 545, 683
Agrobacterium tumefaciens, 36, 383, 683
AGT (accelerated growth techniques), 89, 90, 330–331
Air atomizers, 392–393
Air layering, 541, 542–545, 793
Air-root pruning, 72, 101–102
Albino plants, 604
Aleurone layer, 208
Algal growth in misting systems, 395
Almond leaf scorch, 622
American Nursery and Landscape Association (ANLA), 12
American Seed Trade Association (ASTA), 11
American Society for Horticultural Science (ASHS), 11
Amplification, 35
α-amylase, 209
Analog environmental controls, 63, 65
Anaphase, 24, 28
Androecious plants, 142
Aneuploidy, 598, 600
Angiosperms, 112
   fertilization, 120–122
   life cycle, 114
   ploidy levels, 122
   polyembryony, 132–133
   reproductive structure, 117–118
ANLA (American Nursery and Landscape Association), 11
Annuals, 14. *See also* Herbaceous plants
   categories of, 154–159
   genetic variability, 150–152
   seed production systems, 153–154
Anther culture, 666, 667
Anthesis, 566
Anticlinal division, 284, 286, 605, 606
Antisense technology, 36–37
Anvil mist nozzles, 385–387
AOSA (Association of Official Seed Analysts, Inc.), 11, 176
AOSCA (Association of Official Seed Certifying Agencies), 11
Aphid control, 92
Apical grafting, 467–477
Apomictic cultivars, 16–18
Apomicts, 131
Apomixis, 130–134
   breeding systems, 141, 146–147, 148
   life cycles, 16–18
Apoplastic connections, 429
Apospory apomixis, 133
Approach graftage, 435, 465–466, 485–487, 488–489
Arils, 118
Aryl amids, 376
Aryl esters, 376
Aseptic conditions, 644, 704–705. *See also In vitro* conditions
Aseptic seed culture, 667, 668
Asexual propagation. *See also* Vegetative propagation
   Anacardium occidentale, 729
   Carica papaya, 730
   Carya illinoinensis, 731
   Citrus spp., 734
   Cola nitida, 736–737
   Musa spp., 747
   Pinus spp., 808
   Rubus spp., 761
   Theobroma cacao, 762
ASHS (American Society for Horticultural Science), 11
Association of Official Seed Analysts, Inc. (AOSA), 11, 176
Association of Official Seed Certifying Agencies (AOSCA), 11
ASTA (American Seed Trade Association), 11
Aster yellows, 622
Autografting, 448
Autotrophic growth, 680
Autotrophic micropropagation, 90
Auxins, 38, 39–40. *See also* Indole-3-acetic acid (IAA); Indole-3-butyric acid (IBA); α-naphthalene acetic acid (NAA); Plant growth regulators
   adventitious root, bud, and shoot formation, 293–297
   application methods, 377–381
   cutting propagation, 319–320, 346–347, 373–381, 382, 406
   cytokinin interaction, 41, 455–456
   dormancy control, 239
   end-use formulations, 776
   gene regulation, 33, 34
   grafting, 431, 439, 448–449, 455
   polarity and, 293

Auxins (cont'd)
  seed development, 134
  shelf life, 320
  stem cutting treatments, 377
  tissue culture propagation, 649, 654, 710, 721
Auxin synergists, 300
Avocado sunblotch, 622
Axillary branching, 661
Axillary shoot formation, 646, 659–663, 685

B

BA (benzyladenine), 40, 296
Back breaks, 588
Back bulbs, 588–590
Bags, polyethylene, 76–77
Bailey, Liberty Hyde, 6–8, 189
Balled and burlapped stock, 407–408
Banana graft, 466, 474–476, 477
Banding, 312, 367, 370, 371
Bare-root grafts, 503
Bare-root liners
  handling, 406–407, 408
  harvesting, 265, 406–407
  transplanting, 275–276
Bark, as media component, 80–81
Bark, budding and grafting, 418, 432, 513
Bark grafting, 467, 480–483, 484
Basal cuttage, 574–576
Basal plates, 563
Base-pairs, 624. *See also* Nucleotides
Base temperature, 216
Base water potential, 216
Basic seed, 154
Batch cultures, 671
Bedding plants, transplanting, 275–276
Bell jars, 70
Belt seeder, 254
Bench budding, 514. *See also* Chip budding
Bench grafting, 435, 469, 483
  aftercare, 502–503
  machines, 495–496
  rootstock handling, 500–501
  systems, 505, 506–507
Bench systems, 56, 57, 58
Bending, 538
Beneficial microorganisms, 95, 185–186, 382–383
Benzyladenine (BA, BAP), 40, 296
Best practices management (BMP), 98, 100
Biennials, 15. *See also* Herbaceous plants
  categories of, 154–159
  genetic variability, 150–152
  seed production systems, 153–154
Biocontrol agents, 95
  beneficial microorganisms, 95, 382–383
  biotic factors, 49, 50
  in cutting propagation, 382–383
  in Integrated Pest Management, 94–95
  seed protectants, 185–186
Biofungicides, 94
Biological control. *See* Biocontrol agents
Bioreactors, 672

Biotechnology, 34–45. *See also* Genetically modified organisms
  aesexual propagation, 304–307
  air layering, 545
  apomixis, 134
  cell and tissue culture technology, 34
  clonal selection, 597, 599–601
  DNA-based marker technology, 34–35, 623
  potato production, 635
  recombinant DNA technology, 35–38, 599–601
  seed reserves, 128
  woody plant propagation, 280
Biotic factors, 49, 50. *See also* Biocontrol agents
Biotic stress, 421
Black clothing, 61
Blanching
  cutting propagation, 367, 371
  layering, 538
Bleeding, 436, 526
Blender, 172
BMP (best practices management), 98, 100
Boron salts, 85
Botanical gardens, 634
Botanical varieties, 147, 148–149
Bottom heat, 273, 350, 385, 401
Bracing, 467, 489–490, 492
Branching, enhanced, 686
Brassinosteroids, 38, 42–43, 239–240
Breeder's seed, 153, 154
Breeding systems, 140–147
Bridge grafting, 438, 467, 488–489, 491, 492
Broadcasting fertilizers, 54, 401
Bromeliads, propagation of, 844
Bud-break suppression, 319–320
Budders, 512, 527, 528
Budding, 415, 418, 512. *See also* Chip budding; T-budding
  advantages, 419–420, 512–513
  Castanea spp., 732
  Citrus spp., 734
  conditions, 513
  cost, 415
  crippling, 439, 440, 503, 517
  double-working, 505–506, 533–534
  fruit and nut species, 728
  grafting combined with, 513
  Prunus persica, 756
  Prunus spp., 752
  Pyrus spp., 759
  rootstocks, 512, 513, 522
  Rosa spp., 815, 816
  Tilia spp., 822
  timing of, 513–519
  tools, 530
  top-budding, 532–533
  types, 519–532
Budding and grafting strips, 493
Budding-off, 572
Budding rubbers, 493
Buddy Tape, 494

Bud-forcing methods, 439, 440
Bud-mutation, 598–599
Buds, adventitious, 280. *See also* Adventitious bud formation
Bud-sports, 598–599
Budsticks, 499, 514–516, 518
Bud takes, 512
Budwood, 499, 514–516, 518
Bulb chipping, 576–577
Bulb cuttings, 576–577
Bulbils, 132, 564
Bulblets, 563, 572–573
Bulbs, 563–576
  basal cuttage, 574–576
  bulb cuttings, 576–577
  bulblet formation, 572–574
  growth patterns, 565–566
  micropropagation, 577, 645
  specialized stems and roots, 561–563, 569–572
  spring-flowering, 566–567
  structure, 563–564
  summer-flowering, 567–569
  tissue culture propagation, 645
  winter-flowering, 569
Bulb scales, 563, 666
Bulb splitting, 572
Bulking (micropropagation), 810
Bulking up (forestry propagation), 617
Burr knots, 282
Business-to-business niche, 366

C

Calcined clay, as media component, 80
Calcium, 84–85
Callus, 419
Callus bridge formation, 425–428, 431–432, 436, 448
Callus cultures, 670–671
Callus formation, 287–289
  in hardwood cuttings, 350
  tissue culture propagation, 645, 646–647, 670–671
  in unrooted cuttings, 330
  wounding response, 283
Cambial peroxidase, 449
Capillary mat watering systems, 272
Carbohydrates in media, 710
Carbohydrates in storage reserves
  cutting propagation, 313, 327, 329, 330
  germination, 208–209
  grafting, 454
Carbon dioxide, 89–90
  cutting propagation, 313–314, 402
  seedling germination, 217
  tissue culture propagation, 681
Carpal tunnel syndrome, 501
Causal agents, 621. *See also* Pathogens
Cell and tissue culture technology, 34, 599. *See also* Tissue culture propagation
Cell biology, propagation and, 21–33
Cell expansion, 122, 126–128

Cell suspensions, 670, 671–672
Cellular recognition, 448
Cell wall, 22, 204
Centripetal foggers, 391–392
Certification, disease testing and, 632
Certified seed, 153–154, 157
Chasmogamous flowers, 141
Chemical dormancy, 223–224
Chemical root pruning systems, 74
Chemical treatments. *See also* Auxins;
        Cytokinins; Gibberellins;
        Hormones; Plant growth
        regulators
    cutting propagation, 319–320,
        346–347, 373–381, 382
    in genetic purity tests, 181, 182
    in Integrated Pest Management, 91–94
    minor-use, 93
    pathogens, 621
    root pruning, 74
    scarification, 258
    seed protectants, 185
    seed separation, 174
Chemostats, 671
Chilling stratification, 225–226,
    230, 237, 238
Chimeral breakdown, 612
Chimeras, 602
    patterns within clones, 603–613
    propagating, 292, 612, 684–685
Chip budding, 512, 519–521
    Acer spp., 517, 777
    combined with grafting, 513
    criteria, 522
    graft union formation, 432–433
    photos, 517, 523, 524
    Rosa spp., 523
    timing of, 514
Chlorine, 98–100
Chloropicrin, 92–93
Chloroplasts, 22
Chromosomes, 22, 31
Chronological aging, 615–616
Chrysanthemum mottle, 622
Chrysanthemum stunt, 622
Circling, 101–102
Citrus exochortis, 622
Citrus greening, 534, 622
Cleaning
    plant material, 98
    pots, 73
    seeds, 168–171, 172
Clean stock programs, 632–634
Cleft graft, 466, 472–474
Cleistogamy, 141
Clines, 149
Clonal cultivars, 16–18, 594, 595–601
Clonal fixing, 596, 612
Clonal forestry, 595, 617
Clonal regeneration. *See* Cloning
Clonal repositories, 634
Clonal rootstock, 419, 728
Clonal seed, 148, 149–150

Clonal selection, 594, 595
    anticlinal division, 605, 606
    cell and tissue culture technology, 599
    history, 594–595
    monocultures, 596–597
    mother blocks, 597, 617
    mutations, 597, 598–613
    ontogenetic aging, 602, 613–615
    pathogen detection, 625–630
    pedigreed production systems, 632–634
    rootstock, 419, 728
    seed, 148, 149–150
    source characteristics, 623–625
    source management, 630–635
Clones, 280, 344
    as cultivars, 595–601
    genetic variation, 601–613
    grafting and budding, 420, 439–440
    pathogens, 619–622
    phase variation management, 613–619
Cloning, 279, 594
Closed-case propagations systems. *See*
    Enclosure systems
Club leaf, 622
Coated seeds, 186, 253
Coconut fiber, as media component, 81, 369
Codons, 31
Coir, as media component, 81
Cold frames, 68–70, 269, 390, 391
Cold-hardiness, 452
Cold storage of cuttings, 405
Cold test, 184
Cold waxes, 495
Coleoptilar stage of seed development,
    123–125
Collenchyma cells, 25
Combinatorial dormancy, 234–235
Commercial plantings, 153, 631
Commercial seed production, 150–152,
    162–175
Commercial seed sources, 154, 162–166
Competence-to-root, 306, 311
Competency, 44, 660, 673
Competing sinks, 315, 454
Compost, 81
Compound layering, 539–542
Computer modeling of propagation
    environments, 90, 390
Condensation, 88
Conditional dormancy, 235–236
Conditioned seed storage, 193, 194
Cone of juvenility, 618
Conifers, propagation of, 146, 551
Conjugation, 39, 296
Contact polyethylene systems, 69, 391
Contact systems for cutting propagation, 326
Container-grown plants
    alternatives to, 102
    budding, 527
    Citrus spp., 734
    fertilization, 101
    irrigation, 100
    media mixes, 81–83

    Persea americana, 749–750
    production, 409
    root development in, 101–102
    woody plant seedlings, 268, 269
Containers, propagation, 70–77
    liner plants, 70–75
    micropropagation cultures, 706–708
    post-liner production, 75–77
    seedling liners, 268, 269, 274, 275
Contractile roots, 565, 578
Contracts, 22
Controlled-release fertilizers (CRFs),
    83–84, 401, 402
Conversion stage of somatic
    embryogenesis, 679
Cooling. *See* Heating and cooling systems
Cool test, 184
Copyrights, 22
Cormels, 578, 579
Corms, 562, 577–579
Cornell Peat-Lite mixes, 82
Corpus, 605
Correlative effects
    adventitious root, bud, and shoot
        formation, 293–299
    grafting, 431, 455
Cost-benefit analysis, 398–399
Costing variables, 398–399
Costs of propagation
    cutting propagation, 398–399
    grafting, 415
    labor, 280–281
    vegetative propagation, 597
Cotyledonary node, 207
Cotyledons
    adventitious shoot formation, 664
    gymnosperms, 126
    seedling emergence, 207
    storage reserves, 114–116
Cotyledon stage of seed development,
    123, 126
Crepe rubber sheets, 494
CRFs (controlled-release fertilizers),
    83–84, 401, 402
Crinkle, 603
Crippling
    aftercare, 503
    bud forcing, 439, 440, 517
    Citrus spp., 734
Critical photoperiod, 51
Cross cut, 501–502
Crossing-over, 27
Cross-pollination, 141–146, 150–151
Crown division, 555–558
Crown grafting, 472, 504–505
Cryopreservation, 192, 675, 701
Culms, 584–585, 587
Cultivars
    apomictic, 16–18
    categories of, 147–150
    changing established, 422–424, 504,
        505, 512, 532–533
    clonal, 18, 594, 595–601

Cultivars (cont'd)
  combining, 420–422
  herbaceous, 147–148
  hybrid, 148, 154
  legal protection, 21, 22
  open-pollinated, 147–148
  resistance, 91
  seedling, 15–16
  seed-propagated, 147–150
  synthetic, 148
  taxonomy, 19–21
  true-to-name, 623
  true-to-type, 616, 623–624
Cultural control, 95. See also Integrated Pest Management
Culture indexing, 626, 653
Culture media, 653, 655–656
Cutting-grafts, 505, 507
Cutting propagation, 280–281. See also Adventitious bud formation; Adventitious root formation; Adventitious shoot formation; Cutting propagation of specific plants; Stock plants for cutting propagation
  accelerated growth techniques (AGT), 330–331
  auxin treatments, 319–320, 373–381
  banding, 312, 367, 370, 371
  bulbs, 576–577
  chemical treatments, 382
  corms, 562, 577–579
  cutting length, 353–354
  cutting types, 344–345
  deciduous plants, 315, 316, 345–350, 353–356
  direct sticking, 394–395
  disease control, 381–383
  environmental control, 70, 323–331, 350, 383–393, 401–402
  etiolation, 312, 313, 367, 369, 370, 371
  evergreens, 316, 346–347, 350–356
  facility preparation, 393–394
  fertilization, 401, 402
  hardening-off, 403–404
  Integrated Pest Management, 370, 402
  leaf-bud cuttings, 281, 346–347, 357–359, 573
  leaf cuttings, 346–347, 360–361, 576
  leafy cuttings, 383–393
  management, 305–306, 396–401
  nutrition, 320–321, 401, 402
  pathogen avoidance, 98
  polarity, 293, 361–362
  post-propagation care, 404–406
  pruning, 364, 365–367, 374
  rhizomes, 552, 562, 584–587
  root cuttings, 346–347, 361–363
  rooted cuttings, 403–406
  rooting, 401–403
  rooting media, 367–373
  sanitation, 395–396, 402
  seasonal timing, 315–318, 346–347
  shrubs, 774
  standard operating procedures, 398
  stem cuttings, 281, 345–357, 572–573, 584
  storage of cuttings, 318, 404–406
  tubers, 582
  unrooted cuttings, 318–319, 330, 357, 403–406
  weed control, 402–403
  woody vines, 774
  wounding, 321–323, 373, 374
Cutting propagation of specific plants
  Abies spp., 774
  Acacia spp., 775
  Acer spp., 775–776
  Actinidia deliciosa, 728
  Aesculus spp., 777–778
  Alnus spp., 778
  Amelanchier spp., 778
  Araucaria spp., 779
  Arctostaphylos spp., 779–780
  Banksia spp., 781
  Berberis spp., 781
  Betula spp., 782
  Camellia spp., 783
  Carpinus spp., 784
  Castanea spp., 732, 784
  Ceanothus spp., 785
  Cedrus spp., 785
  Celastrus spp., 785
  Cercis spp., 786
  Chamaecyparis spp., 786
  Chionanthus spp., 787
  Clematis spp., 787
  Coffea spp., 736
  Cornus spp., 787–788
  Corylus spp., 737, 788
  Cotinus coggygria, 789
  Cotoneaster spp., 789
  Cryptomeria japonica, 789
  Cydonia oblonga, 737
  Cytisus spp., 790
  Daphne spp., 790
  Diospyros kaki, 737
  Erica spp., 791
  Eucalyptus spp., 792
  Euphoria longan, 738
  Fagus spp., 793
  Feijoa sellowiana, 738
  Ficus carica, 739
  Ficus spp., 793
  Fraxinus spp., 793–794
  Genista spp., 794
  Hamamelis spp., 795
  Hibiscus spp., 796
  Ilex spp., 797
  Juglans cinerea, 741
  Juglans nigra, 741
  Juniperus spp., 798–799
  Litchi chinensis, 742
  Macadamia spp., 743
  Magnolia spp., 801–802
  Malus xdomestica, 743
  Malus spp., 803
  Mangifera indica, 746
  Myrtus spp., 804
  Olea europaea, 748
  Persea americana, 749
  Picea spp., 807
  Pinus spp., 807–808
  Platanus spp., 808
  Populus spp., 809
  Prunus persica, 755–756
  Prunus spp., 752, 809–810
  Pseudotsuga menziesii, 810
  Psidium spp., 758
  Pyrus spp., 759
  Quercus spp., 811–812
  Rhododendron spp., 813–814
  Rhus spp., 814
  Ribes spp., 761
  Rosa spp., 815, 816
  Rubus spp., 761
  Salix spp., 817
  Sequoia spp., 818
  Skimmia spp., 818
  Sorbus spp., 819
  Syringa spp., 820
  Taxus spp., 821
  Thuja spp., 821–822
  Tilia spp., 822
  Tsuga spp., 823
  Ulmus spp., 823
  Viburnum spp., 824
  Vitis spp., 763
Cyanogenic glucoside, 447
Cybrids, 670, 672, 675
Cyclic irrigation, 100
Cylinder seeders, 271
Cytokinesis, 24–25
Cytokinins, 38, 40–41. See also Plant growth regulators
  adventitious root, bud, and shoot formation, 295, 296–298
  auxin interaction, 41, 455–456
  dormancy control, 239
  graft union formation, 439, 455
  light-activated germination, 228
  in micropropagation media, 710
  seed development, 135
  tissue culture propagation, 649, 654, 655
Cytoplasm, 22

D

Damping-off, 91, 217–218
Darwin, Charles, 6, 214
Davis, Ben, 401
Daylength. See Photoperiod
Day-neutral plants, 51
Deciduous plants, propagation of, 315, 316, 345–350, 353–356
Dedifferentiation, 281, 299–300
Deep physiological dormancy, 230
Deep suture, 603
Deflection mist nozzles, 385–387
Dehiscing, 166, 167–172

Deionization, 85, 86
Delayed fruiting, 146
Deletion, 598, 600
De novo adventitious roots, 282, 283
De novo meristems, 425
Dessication
  cutting propagation, 323–325
  grafting, 465
Detached scale propagation, 573
Detached scion graftage, 465, 466
  apical grafting, 467–477
  bark grafting, 480–483
  root grafting, 483–485
  side grafting, 477–480
Deterioration of seeds, 191
Determinate stems, 585
Determination (organogenesis), 44, 660
Development stage of somatic
    embryogenesis, 678
Dewinging, 172
Dichogamy, 143
Dicotyledonous plants, 112, 113
  germination, 205
  seed development, 122–123
Dictyosomes, 427–428
Differentiation
  bulbs, 566
  cells, 25–26
  embryonic, 122–126
  vascular, 425, 448
Dihaploids, 666
Dihydrozeatin, 40
Dioecious plants, 142
Dioecy, 142
Diploid plant regeneration, 663–666
Diplospory apomixis, 133
Direct-cooled refrigeration, 405
Direct fall planting, 346, 349–350
Direct root formation, 287, 288
Direct rooting, 50, 70–72, 394–395,
    403–404
Direct-seeded nursery rows, 251–255,
    264–266
Direct solution soaking method, 378
Direct spring planting, 350
Direct sticking, 50, 70–72, 394–395,
    403–404
Disease control. *See also* Pathogen
    management
  cutting propagation, 381–383
  grafting, 424, 438–439
  seed germination, 217–218
Disease indexing, 424, 626, 627, 628
Disease resistance, 452–453
Disinfestation of explants, 652–653,
    714–715
Displacement, cellular, 606–607
Distal end of shoots and roots, 437
Division
  Bambusa spp., 780–781
  cells, 23–27, 28
  palms, 806
  pseudobulbs, 587
  specialized stems and roots, 561
  tubers, 581, 583
DNA (deoxyribonucleic acid), 22, 31–32
DNA-based marker technology, 34–35, 623
DNA fingerprinting, 181
DNA sequencing, 34–35, 623
Dormancy, 201, 218–219. *See also*
    Germination
  budding, 515
  cycle, 235–236
  ecological advantages of, 219
  hormone control of, 236–240
  primary, 218–219, 220–235
  quiescent seeds, 130
  release mechanisms, 231–235, 239
  secondary, 219, 221, 235–236
  types of, 220–231
Double budding test, 627
Double-cross hybrids, 148
Double dormancy, 234
Double fertilization, 120–121
Double-nose bulbs, 571
Double phase culture, 656
Double pruning, 367
Double-shield budding. *See* T-budding
Double-working
  budding, 505–506, 533–534
  grafting, 422, 423
Drill, 253
Drop layering, 541, 551
Droppers, 565
Drum priming, 214, 215–216
Drum seeders, 271
Drying process
  conifer cones, 172
  fruits, 168–169, 172
  maturation drying, 122, 129–130, 201
  seeds, 175
Duct tape, 494
Duplication, 598, 600
Dutch elm disease, 545
Dwarfing clonal rootstocks, *Malus
    xdomestica,* 744–745
Dwarfing scion cultivars, 451, 453, 454, 455
Dynamic control systems, 388–390

E

Ebb and flood system, 57–59, 272
Ecotypes, 149
Edaphic environmental factors, 49, 50
Elaiosomes, 117, 118
Electrical impedance, 317
Electrical tape, 494
Electric heat pasteurizers, 96
Electrolyte leakage test, 184
Electrophoresis, 33, 449, 628
ELISA tests, 628
Elite stock plants, 634
Elite trees, 149
Elongation of flowering shoots, 566
Embryo, 112–113
  differentiation, 122–126
  foliar, 290, 361
  sac development, 118, 119, 120, 133
  types, 115, 233–234
Embryo culture, 667–669
Embryogenesis, 122–126
Embryoids, 673
Embryonic phase of seedling life cycle,
    15, 594
Embryo rescue, 667–669
Enclosed mist systems, 384, 385
Enclosed poly tents, 326, 391, 470, 503
Enclosure systems, 88, 390–391
  cold frames, 68–70, 269, 390, 391
  contact polyethylene systems, 69, 391
  cutting propagation, 326, 391
  enclosed mist systems, 384, 385
  grafting, 470, 503
  hot frames, 67–70, 390, 391
  indoor polytents, 326, 391, 470, 503
  lathhouses, 69
  low tunnels, 67–70, 390
  nonmisted enclosures, 326
Encrusted seeds, 186
Endangered species, micropropagation
    of, 701
Endogenous dormancy, 224–231,
    233–234
Endogenous rooting inhibitors, 300
Endoplasmic reticulum (ER), 429
Endosperm, 112
  differentiation, 122
  double fertilization, 120, 121
  germination, 205–206, 228
  ploidy levels, 114, 122, 125
  storage reserves, 114, 116
Endosperm culture, 666
Endospermic seeds, 114, 116
Environment. *See* Propagation
    environment
Environmental controls, 63, 65, 66, 390
Environmental variation, 601, 602
Enzyme production, 204
Epicormic shoots, 618, 619
Epicotyl, 207, 234
Epigenetic control, 602
Epigenetic variation, 309, 616, 686–687
Epigeous germination, 208
Epigeous-type bulblets, 573
Ergonomics of cutting propagation, 398
Establishment stage of micropropagation,
    650–654, 713–716
Ethylene, 38, 42
  adventitious root, bud, and shoot
    formation, 295, 298
  dormancy control, 238–239
  seed development, 135
  tissue culture propagation, 681
Etiolation
  cutting propagation, 312, 313, 367,
    369, 370, 371
  layering, 538
Etiology, 621
Evergreens, propagation of, 316–317,
    346–347, 350–356

Excess vigor, 146
Excised-embryo test, 177–178, 179
Exogenous dormancy, 220, 221–224
Exons, 31–32
Expanded foam blocks, 75
Explants, 713
   disinfestation, 652–653, 714–715
   establishment of, 650
   material selection, 675–676, 713–714
   pretreatment of, 715
   thin layer, 664, 665
Extraction of commercial seeds, 168, 172–175
Exudation, excessive, 682
Ex vitro conditions, 657–658, 713, 717–718
Eyes, 579–580

F

F_1 hybrids, 148
F_2 cultivars, 148
Facultative apomicts, 131, 146, 148
Fall seeding, 262–263
Families, grafting between, 441
Family selection, 150
Fastidious bacteria, 622
Federal Insecticide, Fungicide, and Rodenticide Act (FIFRA, 1988), 93
Federal Seed Act (1939), 175
Fermentation, 173–174
Ferns, propagation of, 851
Fertigation, 53, 101
Fertilization
   angiosperms, 120–121, 122
   biology of, 26, 110–112
   broadcasting, 54
   gymnosperms, 119–120, 121–122, 125, 126
   postplant, 83–85
   seed formation, 118–122
   systems, 83–85, 101
Fertilizers, 82–84, 401, 402
Fiber blocks, 75
Fiberglass greenhouse coverings, 67
Fiber pots, 73–74
Fibers (cell differentiation), 25
Fibrous roots, 578
Field budding, *Vitis* spp., 764
Field grafting, 469, 503–506
Field nurseries, 250, 255–264
Field planting of bulbs, 573
Field-propagated plants
   Citrus spp., 733–734
   handling, 406–409
   Persea americana, 749
   Pistachia vera, 751
Field seeding, 250–255, 264–266
FIFRA (Federal Insecticide, Fungicide, and Rodenticide Act, 1988), 93
First-generation seed, certified, 154
Fixing, 29, 140
   clonal, 596, 612
   self-pollination, 141
Flat production, 266, 267

Flats, 71
Fleshy roots, 581, 583–584
Flexible greenhouse covering materials, 64–66
Flipped bud, 519
Float bed watering systems, 272
Floor ebb and flood watering systems, 57–59, 272
Flowering, epigenetic variation and, 686–687
Flowering, time of, 16, 17
Flowering plants. *See* Angiosperms
Flowers
   cleistogamy, 141
   structure, 117–122, 144
   tissue culture propagation, 645
   woody plants, 17
Flower seeds, commercial production of, 162–166
Fluid drilling, 188–189
Fluoride, 85
Flute budding, 522, 533
Foam blocks, 75
Fog systems, 88
   controllers, 393
   cutting propagation, 326–327, 395–396
   droplet sizes, 391
   long cuttings, 353
Foliar embryos, 290, 361
Foliar leaching, 385
Food reserves. *See* Storage reserves
Form, 149
Foundation blocks, 633–634
Foundation plants, 634
Foundation seed, 153, 154
Four-flap graft, 466, 474–476, 477
Fractional scale-stem cuttage, 576
Fragmentation restriction enzymes, 35
Frameworking, 533
Free nuclear stage, 125, 126
Fruit, drying, 168–169, 172
Fruit chimeras, 609
Fruit development, 134, 135, 136, 451–452
Fruiting, delayed, 146
Fruiting precocity, 451
Fruit-processing industries, 166
Fruit set, 451
Fruit species propagation, 728
Fumigation, chemical, 92
Fungi
   beneficial, 94, 95, 659
   pathogenic, 90–91
Fungicides
   cutting propagation, 382
   damping-off, 218
   soil drenches, 93–94
Funiculus, 127, 129

G

GA (gibberellic acid), 38, 41, 456
Gable-roof greenhouses, 55
Gametes, 27, 110–111

Gametophyte production, 112–114, 119. *See also* Pollen
Gametophytic apomixis, 131, 132, 133
Gametophytic self-incompatibility, 143–145
Gas exchange, 53. *See also* Carbon dioxide; Ethylene; Oxygen
Gas-fired infrared heaters, 60, 88
Gel electrophoresis, 33, 449, 628
Gelling agents in micropropagation media, 711
Gel seeders, 254–255
Gene expression, 32–33. *See also* Biotechnology
   adventitious root formation, 306
   fixing, 29, 140, 141, 596, 612
   hormone influence, 33, 34, 135
   seed filling, 128
   transcription and translation, 31, 32–33
Gene guns, 675
Genera, grafting between, 440–441
Generative nuclei, 119
Genes, 22, 27–34
Gene silencing, 36–37, 456
Genetically modified organisms (GMOs), 35–38, 151, 181–182
Genetic code, 31
Genetic drift, 152
Genetic improvement, 674
Genetic inheritance, 27–29
Genetic purity, testing for, 181
Genetic resources, conserving, 192, 193
Genetics. *See also* Biotechnology; Gene expression
   cell biology, 22–27, 28
   grafting limitations, 439–441
   history, 30–31
Genetic transformation, 674–675
Genetic variability, controlling. *See also* Chimeras; Phenotypic variation
   micropropagation process, 684–687
   seed production, 150–153
   tissue culture propagation, 685
   uniformity of populations, 596
Genome, 34
Genome sequencing, 34–35, 623
Genome-wide gene expression, 37
Genotype, 140, 601, 602
Genotypic selection, 152–153, 624, 633
Genus, 18
Geophytes, 561
Germination, 200–209. *See also* Dormancy
   cell expansion, 130–131
   disease control, 217–218
   enhancement, 186–189
   environmental factors, 211–217
   facilities, 269–270
   inhibitors, 223, 224
   measuring, 209–211
   models, 216
   plug production requirements, 268
   primary seed dormancy, 218–219, 220–235

rate of, 210–211
regulating, 218–219
secondary seed dormancy, 219, 221, 235–236
seed ripening, 136
seed treatments and, 184–189
storage and, 191–192
storage reserves, 208–209
uniformity, 211
Germination stage of somatic embryogenesis, 679
Germplasm preservation, 675, 701
Germplasm storage, 192, 194
Gibberellic acid (GA), 38, 41, 456
Gibberellins, 38, 41
adventitious root, bud, and shoot formation, 295, 298, 299, 308
characteristics, 38
dormancy control, 236, 237–238
germination, 209, 228–229
gibberellic acid, 38, 41, 456
in micropropagation media, 710
seed development, 134–135
Girdling
cutting propagation material, 312–313, 365, 367, 368
layering, 538
Glass bead sterilizers, 715
Glass-covered greenhouses, 64
Globular stage, 123, 125
Glyoxysomes, 209, 210
Graft chimeras, 609–612
Graft compatibility, 441, 448–449, 464
Grafter, 472
Graft failure, 441
Graft incompatibility, 434, 441–450, 464
anatomical flaws, 443, 445
causes and mechanisms, 446–449
correcting, 450
external symptoms, 442–443
localized, 443–445, 447
pathogen-induced, 445, 446
predicting, 449–450
translocatable, 445–446
Grafting of specific plants, 415, 417–418. *See also* Budding; Grafting process; Grafting techniques
Abies spp., 775
Acer spp., 776–777
Actinidia deliciosa, 729
Aesculus spp., 777–778
Anacardium occidentale, 729
Annona cherimola, 730
Betula spp., 782
Camellia spp., 783
Carpinus spp., 784
Carya illinoinensis, 731
Carya ovata, 731
Castanea spp., 732
Catalpa spp., 785
Cedrus spp., 785
Cercis spp., 786
Citrus spp., 733, 787

Coffea spp., 736
Cornus spp., 787–788
Corylus spp., 737, 788
Cotoneaster spp., 789
Crataegus spp., 789
Cydonia oblonga, 737
Diospyros kaki, 737–738
Eriobotrya japonica, 738
Eucalyptus spp., 792
Euphoria longan, 738
Fagus spp., 793
Feijoa sellowiana, 738
Ficus carica, 739
Fraxinus spp., 794
Hamamelis spp., 795
Hibiscus spp., 796
Juglans cinerea, 741
Juglans nigra, 741
Juglans regia, 741
Juniperus spp., 799
Litchi chinensis, 742
Macadamia spp., 743
Magnolia spp., 801–802
Malus xdomestica, 743
Malus spp., 803
Mangifera indica, 746
Olea europaea, 748
Passiflora edulis, 748–749
Persea americana, 749
Picea spp., 807
Pinus spp., 808
Prunus spp., 810
Psidium spp., 758–759
Quercus spp., 812
Rhododendron spp., 813–814
Salix spp., 817
Sorbus spp., 819
Syringa spp., 820
Thuja spp., 822
Tsuga spp., 823
Viburnum spp., 824
Vitis spp., 763–764
Grafting process, 491
aftercare, 502–504
automation, 496, 498, 499
correlative effects, 431, 455
cost, 415
dessication, 465
disease control, 424, 438–439
disease resistance, 452–453
environmental conditions, 435–436, 470, 503
fruit and nut species, 728
genetic limitations, 439–441
history, 415–417
machines, 495–496, 497, 499
natural, 424–425
ornamental trees, 774
polarity, 437–438, 487
reasons for, 419–420
requirements, 464–465
scion-rootstock relationships, 450–457
study of plant processes and, 424

systems, 504–509
timing, 504
tools, 492–495
wound repair, 423–424, 428–431
Grafting techniques, 464
approach graftage, 435, 465, 466, 485–489
craftmanship, 501–502
crippling, 439, 440, 503, 517
cross cuts, 501–502
crown grafting, 472, 504–505
detached scion graftage, 465, 466–485
double-working, 422, 423
repair graftage, 465, 466, 487–490, 491
rootstock handling, 500–501
scion insertion, 471
scionwood selection and handling, 496–500
species-specific recommendations, 434–435, 440
Grafting waxes, 494–495
Graft union formation, 425–432. *See also* Graft compatibility; Graft incompatibility
endoplasmic reticulum, 429
factors influencing, 433–439, 455
T- and chip budding, 432–433, 512
vascular cambium, 426, 464–465
Green bulbs, 589, 590
Greenhouses, 54–70. *See also* Intermittent mist; Irrigation systems; Light manipulation; Misting systems; Shading
bench systems, 56, 57, 58
construction types, 55–59
covering materials, 61, 62, 63–67
disinfection of, 97–98
environmental controls, 63, 65, 66, 390
heating and cooling systems, 59–63, 88
sources, 58
ventilation systems, 59
Greenwood grafting, *Vitis* spp., 764
Grow bags, 76, 102
Growing area, micropropagation, 705–706
Growth habit, 451
Growth potential, 203, 224, 231
Growth regulators. *See* Plant growth regulators (PGRs)
Growth retardants and inhibitors, 299
Growth rooms, 269–270
Gutter-connected greenhouses, 55
Gymnosperms, 112
fertilization, 119–120, 121–122, 125, 126
life cycle, 113
seed development, 125–126
storage tissue, 114, 115
Gynoecious plants, 142

H

Habituation, 683
Hairy root cultures, 683
Hale, Stephen, 424

Hand pollination, 154, 156, 157
Haploids, 675
Hardening-off, 100
    cutting propagation, 50, 403–404
    micropropagation, 658
    seed propagation, 275
Hard pruning, 366–367
Hardwood stem cuttings, 345–351
    direct planting, 346, 349–350
    material selection, 348–349
    propagation systems, 346–347
    Rosa spp., 815
    storage, 405
    Vitis spp., 763
Harvesting
    bare-root liners, 265, 406–407
    bulbs, 570–571
    field-grown seedlings, 263–264
Harvest maturity, 166–172
Heart stage, 123
Heating and cooling systems, 59–63, 88
Heat treatment
    hardwood cuttings, 350
    pathogen elimination, 629
    seed protectants, 186
Hedge rows, 618–619
Hedging, 367
Heeling-in, 407, 572
Height control in flat and plug production, 273–274
Heirloom varieties, 148
Herbaceous plants, 840. *See also* Annuals; Perennials
    commercial seed production, 150–152, 162–166
    cultivars, 147–148
    cutting propagation, 346–347, 356–357
    genetic variability, 150–152
    perennials, 15
    seed propagation, 153–159
Herbaceous stem cuttings, 356–357, 377
Herbicides, 402–403
Heteroblastic phase changes, 615
Heterogeneous seedlings, 140
Heterografts, 449
Heterosis, 142
Heterostyly, 143, 144
Heterotrophic growth, 680
Heterozygous populations, 140–141
HIG (hole insertion graft), 467, 476, 478
High additive heritability, 153
High-pressure fogging equipment, 392, 393
High-pressure sodium vapor lamps, 89
High-temperature scarification, 258
Hilling up, 547
Hilum, 129, 222, 232
Histodifferentiation, 122–126
Histogens, 605
Hole insertion graft (HIG), 467, 476, 478
Homoblastic phase changes, 614
Homogenous seedlings, 140
Homoptera, 621
Homozygous populations, 141

Hormones, 38–44. *See also* Abscisic acid; Auxins; Cytokinins; Ethylene; Gibberellins; Plant growth regulators
    adventitious root, bud, and shoot formation, 295–299, 302–304, 660
    cutting propagation, 319–320, 373–382
    dormancy control, 236–240
    embryo culture, 669
    grafting, 431, 448–449, 455
    layering, 538
    physiological dormancy, 228–229
    plant development, 38–45
    seed development, 133–135
    tissue culture propagation, 660, 671, 710
Horticultural literature, 6–8
Hot air convection heating systems, 88
Hot frames, 67–70, 390, 391
Hot-pipe callusing system, 502–503
Hot water distribution, 88
Hot waxes, 495
Humidity control
    cutting propagation, 323–327
    enclosure systems, 52, 88, 326
    grafting, 435–436
    seed storage, 193–194
Hybrid cultivars
    seed production, 148, 154, 155
    seed sources, 150
    taxonomy, 20
Hybrid lines, 148
Hybrid vigor, 142
Hydroponic systems, 70
Hydrothermal time, 216
Hydrotime, 216
Hyperhydricity, 681–682
Hypocotyl, 207, 208, 664
Hypocotyl hook, 208
Hypogenous germination, 208

I

IAA. *See* Indole-3-acetic acid
IBA. *See* Indole-3-butyric acid
I-budding, 522, 532
Illumination. *See* Light intensity
Imbibition, 201, 203–204
Impaction pin atomizers, 392
Improved seed sources, 149–150
Inarching, 467, 487–490
Inbred lines, 142, 148
Inbreeding depression, 142
Incision (layering), 538
Incompatibility. *See* Graft incompatibility
Increase blocks, 634
Incubation of bulbs, 573
Independent assortment of chromosomes, 27
Indeterminate stems, 585–586
Indicator plants, 626
Indirect organogenesis, 280
Indirect root formation, 287, 288
Indole-3-acetic acid (IAA), 39–40
    adventitious root formation, 295–296, 297
    cutting treatment, 319, 320
    seed development, 134

Indole-3-butyric acid (IBA), 40
    adventitous root formation, 295–296, 297
    cutting treatment, 319, 320, 374–377
    layering, 543
Indoor polytent systems, 326, 391, 470, 503
Induced mutations, 599
Induction of flowering, 566, 569
Induction stage of somatic embryogenesis, 676–677
Inflorescences, immature, 664
Infrared radiation heating systems, 60, 88
In-ground fabric containers, 76, 102
Inheritance, genetic, 27–29
Initials, 607
Initial shoot development, 646, 651, 659–666, 685, 715–716
Initiation codons, 32
Initiation of flowering, 566–567
Inlay approach graft, 467, 487, 489
Inlay bark graft, 467, 482–483, 486
Inorganic salts, 708–710
Insecticidal sprays and drenches, 94
Integrated Pest Management (IPM), 91–98, 370, 402
Integuments, 119
Intermediate physiological dormancy, 230
Intermediate stem section. *See* Interstock
Intermediate stock. *See* Interstock
Intermittent mist, 52, 384
    cutting propagation, 326, 344, 384–385
    fungal spores, 91
    nutrient leaching, 54, 321
    system components, 386
International Code of Botanical Nomenclature, 19
International Fruit Tree Association, 11
International Plant Names Index, 20
International Plant Propagators Society (IPPS), 11
International Seed Testing Association (ISTA), 11, 176
International Society for Horticultural Science (ISHS), 11
Interphase, 23–24, 28
Interstem. *See* Interstock
Interstock, 418
    bridge graft system, 488
    double-working, 422–423, 505, 533
    effect on scion and rootstock, 453
    mutual compatibility, 445
    Rosa spp., 815–816
Interval irrigation, 100
Introns, 32
Inversion, 598, 600
Inverted T-incision, 522, 525–526
Invigoration. *See* Rejuvenation
In vitro conditions, 644, 713
    explant establishment, 650
    problems, 681–683
    root formation, 657–658, 717–718, 721
In vitro micrografting, 663

IPM (Integrated Pest Management), 91–98, 370, 402
IPPS (International Plant Propagators Society), 11
Irradiance. *See* Light irradiance
Irrigation systems
  computer-controlled, 100
  fertigation, 53, 101
  field nurseries, 256
  floor ebb and flood, 57–59, 272
  recycled water, 86–88, 98
ISHS (International Society for Horticultural Science), 11
Isolation, 150–151
Isolation layer, 426, 428
Isopentenyladenine (2iP), 296
Isozymes, 623
ISTA (International Seed Testing Association), 11

J

Jacketed coolers, 405
Japanese tube graft, 466, 469–472
Jarvis, B.C., 300, 301
Jasmonates, 38, 43
Jumping genes, 604
June budding, 514–516, 519, 529
June yellows, 603
Juvenile phase of seedling life cycle, 15, 18
  clones, 594, 616–619
  retaining, 19
Juvenile phenotypes, 616

K

Kains, M. G., 8
Kinetin, 40, 296
Knives, 492, 493, 530

L

Lag phase of germination, 201, 203, 204
Laminar flow hoods, 704–705, 715
Laminate bulbs, 564, 565, 570, 571
Landraces, 109, 147
Late embryogenesis abundant (LEA) proteins, 129, 130
Latent primary meristems, 290
Latent roots, 282–283
Lathhouses, 69
Layered meristems, 605
Layering of specific plants, 537
  Annona cherimola, 730
  Corylus spp., 737
  Cydonia oblonga, 737
  Eriobotrya japonica, 738
  Euphoria longan, 738
  Ficus carica, 739
  Ficus spp., 793
  Litchi chinensis, 742
  Magnolia spp., 801–802
  Malus xdomestica, 743
  Mangifera indica, 746
  mother plants, 538, 546, 550
  Psidium spp., 759
  Ribes spp., 761
  Vitis spp., 763
Layering process. *See also* Air layering
  etiolation, 538
  natural, 551–558
  plant management, 539
  procedures, 539–551
  success of, 537–539
Leaf-bud cuttings, 281, 357–359
  bulb propagation, 573
  propagation systems, 346–347
Leaf cuttings, 360–361
  adventitious bud, shoot, and root formation, 290–292
  bulb propagation, 576
  propagation systems, 346–347
Leaf scorch, 622
Leaf-to-stem ratio, 354–355
Leaf variegation, 609–611
Leafy cuttings
  environmental conditions, 383–393
  Vitis spp., 763
Leaky seeds, 203
LEA (late embryogenesis abundant) proteins, 129, 130
Leaves
  adventitious shoot formation, 663–664
  effect on rooting of cuttings, 294
LeBlanc, Charlotte, 356
Legal protection of cultivars, 21, 22
Lens, 222
Leptomorphs, 585–586
LFR (low fluence response), 228
Life cycles, 14–18
  angiosperms and gymnosperms, 113, 114
  apomictic cultivars, 16–18
  clonal cultivars, 18, 616
  reproductive, 110–112
  seedling cultivars, 15–16, 594
Light duration. *See* Photoperiod
Light-emitting diodes (LEDs), 89
Light exclusion, 538
Light intensity, 49, 50
  cutting propagation, 330
  light sources, 52
Light irradiance, 49, 50
  cutting propagation, 311, 328
  light sources, 52
  measurement of, 51
  tissue culture propagation, 679–680
Light manipulation, 51–52. *See also* Photoperiod
  cutting propagation, 311–312, 328–331
  lighting systems, 65
Light pruning, 367
Light quality, 52, 88
  cutting propagation, 311, 329
  germination, 217, 227
  measurement of, 51
  photodormancy, 227
  tissue culture propagation, 679, 680–681
Light-sensitive seeds, 217
Lignification process, 447
Lignin, 285, 288
Linear embryos, 233
Liner plants, 49, 50. *See also* Cutting propagation of specific plants; Layering
  bare-root, 265, 275–276, 406–407, 408
  carbohydrate reserves, 329
  cleaning, 98
  containers, 70–77
  harvesting field-grown seedlings, 263–264
  media, 78–83
  micropropagated, variation in, 684–687
  overwintering rooted, 404
  post-propagation care, 100–102
  potted rootstock, 503
  production, 50, 84–85
  seedlings, 207–208, 630
  tissue-culture produced, 364
  transplanting, 264, 274–276
  transporting, 634–635
Lines, 148
Lipids in storage reserves, 209
Liquid cultures, 655–656, 711
Liquid fertilizers, 83
Liquid slurries, 185
Local seed, 152
Long-day plants, 51
Long-lived seeds, 190–191
Lopping, 439, 440, 503, 517
Low additive heritability, 153
Low fluence response (LFR), 228
Low tunnels, 67–70, 390

M

Maceration, 172–173
Macrosclerid cells, 221, 222
Magnesium, 84–85
Magnetic Resonance Imaging (MRI), 449–450
Maintenance stage of somatic embryogenesis, 677–678
Malling apple rootstock, 548, 744–745
Manual grafting techniques, 501–502
Marcottage. *See* Air layering
Markers, 456
Market-led propagation systems, 400–401
Matriconditioning, 214, 215
Matric potential, 202
Matrix seed priming, 214, 215
Maturation drying, 122, 129–130, 201
Maturation stage of somatic embryogenesis, 678–679
Mature phase. *See* Adult phase of seedling life cycle
Mature phenotypes, 616
Maturity, harvest, 166–172
MB (methyl bromide), 92–93, 529
Mechanical scarification, 257–258
Mechanical seeders
  for direct seeding, 253–255
  for flat and plug production, 270–274

Media, 77–85. *See also specific propagation methods*
   characteristics, 77–78
   components, 78–81
   herbaceous plant germination, 269, 270
   mixes, 81–83
   pH, 86
   sterilization, 95–96
   substrates, 367–373
   temperature, 327
   water management, 370
Media substrates, 367–373
Medium-lived seeds, 190
Megaspore mother cells, 112, 117, 119
Megasporocytes, 112, 117, 119
Megasporogenesis, 117–118
Meiosis, 23, 26–27, 28, 122
Mendelian genetics, 30
Mericlinal chimeras, 605, 607
Meristem aging, 308
Meristematic cells, 291
Meristem culture, 659–660
Meristemoids, 283
Meristems
   adventitious root, bud, and shoot formation, 28, 290–292
   layered, 605
   unlayered, 607
Mesomorphs, 586
Messenger RNA. *See* mRNA
Metaphase, 24, 28
Metering system, 253
Microbudding, 522, 534–535
Microclimatic conditions, 49, 50, 85–90. *See also* Gas exchange; Humidity; Light quality; Photoperiod; Temperature control; Water
Microcuttings, 650, 713
Micrografting, 505, 507–509, 629, 662–663
Micromist systems, 391
Micropropagated plants, variation in, 684–687
Micropropagation of specific plants. *See also* Explants
   Abies spp., 775
   Acacia spp., 775
   Acer spp., 777
   Actinidia deliciosa, 729
   Alnus spp., 778
   Amelanchier spp., 778
   Ananas comosus, 729
   Araucaria spp., 779
   Arctostaphylos spp., 779–780
   Bambusa spp., 780–781
   Berberis spp., 781
   Betula spp., 782
   bulbs, 577
   Camellia spp., 783
   Carica papaya, 730
   carnation, 722
   Carya illinoinensis, 731
   Castanea spp., 784
   Ceratonia siliqua, 732
   Cercis spp., 786
   Chamaecyparis spp., 786
   chimeras, 612, 684–685
   Chionanthus spp., 787
   Citrus spp., 734
   Clematis spp., 787
   Coffea spp., 736
   Cornus spp., 787–788
   Corylus spp., 737, 788
   Cycas spp., 790
   Cydonia oblonga, 737
   Daphne spp., 790
   Diospyros kaki, 738
   Erica spp., 791
   Eucalyptus spp., 792
   Ficus carica, 739
   Ficus spp., 793
   Fragaria xananassa, 739
   Fraxinus spp., 794
   Genista spp., 794
   Hamamelis spp., 795
   Hibiscus spp., 796
   hosta, 722
   Ilex spp., 797
   Juglans regia, 741
   Magnolia spp., 801–802
   Malus xdomestica, 743–744
   Musa spp., 747
   Myrtus spp., 804
   Nephrolepsis (Boston fern), 722
   palms, 806
   Phoenix dactylifera, 751
   Picea spp., 807
   Pinus spp., 808
   Pistachia vera, 751
   Populus spp., 809
   Prunus (cherry), 723
   Prunus spp., 752, 810
   pseudobulbs, 590
   Pseudotsuga menziesii, 810
   Psidium spp., 759
   Pyrus spp., 759
   Quercus spp., 812
   Rhododendron spp., 813–814, 723
   Ribes spp., 761
   Rubus spp., 761
   Salix spp., 817
   Sequoia spp., 818
   Sorbus spp., 819
   Syringa spp., 820
   Tilia spp., 822
   tubers, 582
   Ulmus spp., 823
   Viburnum spp., 824
   Vitis spp., 763
Micropropagation process, 699
   acclimatization stage, 50, 658–659, 718–723
   advantages, 699–701
   automated systems, 655–656
   clonal rootstock, 419
   containers for culture growth, 706–708
   culture media, 653, 655–656, 706–712
   disadvantages, 701–702
   facilities and equipment, 702–706
   genetic variation, 684–687
   history, 649
   phase variation management, 619
   plantlets, 650, 659–666
   procedures, 712–722
   production rates, U.S., 700
   stages of, 650–659, 712–713, 713–723
   uses, 699–701
Micropyle, 119
MicroRNAs, 33
Microsatellites, 623, 624–625
Microscreening, 92
Microshoots, 650, 713
Microspore cultures, 666
Microsporogenesis, 117
Microtubers, 582, 662
Mineral media components, 367–368, 370
Mineral nutrition, 53–54. *See also* Fertilization; Fertilizers
   cutting propagation, 320–321, 401
   fertigation, 53, 101
   grafting, 454
Mineral wool, 80, 369, 372
Minitubers, 582, 662
Minor-use chemicals, 93
Misting systems. *See also* Fog systems
   controls, 388–390
   intermittent mist, 384–385
   nozzles, 385–387
   potential problems, 395–396
   safety cautions, 389
   vs. fog and micromist systems, 391
Mitochondria, 22, 204
Mitosis, 23–26
Modeling. *See* Plant modeling
Moderate pruning, 367
Modified stem cuttings, 344
Modified stooling, 365
Moist-chilling. *See* Stratification
Moisture content of seeds. *See* Seeds, moisture content of
Moisture stress. *See* Water stress
Molecular genetics, 35–36
Monocotyledonous plants, 112, 113
   seed development, 123–125
   storage tissue, 114
Monocultures, 596–597
Monoecious plants, 142
Monoecy, 142–143
Morphological dormancy, 224, 233–234
Morphological expression, 16
Morrill Act (1862), 6
Mosaic of clonal blocks, 597
Mosaics, 603
Mother blocks, 597, 617
Mother bulbs, 567–568
Mother plants, 538, 546, 550, 634
Mound layering, 537, 541, 545–548
Movable thermal curtains, 61

MRI (Magnetic Resonance Imaging), 449–450
mRNA (messenger RNA), 32–33
  germination process, 204
  storage protein accumulation, 128
Mutagenesis, 35–36
Mutagenic agent, 599
Mutants, 598, 602, 612–613
Mutant sectors, 685
Mutation breeding, 599
Mutations in clonal selection, 597, 598–613
Mutually incompatible interstock, 445
Mycorrhizal fungi, 94, 95
Mycostop, 94

N

α-naphthalene acetic acid (NAA), 40
  adventitious root formation, 295
  application methods, 374–377, 379
  cutting propagation, 319, 320
1-naphthalene acetic acid (NAA), 40
Natural grafting, 424–425
Naturalization, 264–266
Natural layering, 551–558
Natural selection, 6
Necrosis, shoot-tip, 682, 683
Necrotic layer, 426, 428
Needle fascicles, 664
Needle seeders, 271
Network (GRIN) Taxonomy for Plants, 20
Nitrogen efficiency, 452
Nitrogen levels, 84, 313–314
Nitrogenous compounds, 240
Nodal cultures, 661
Nondeep physiological dormancy, 225–230
Non-dormant seeds, 218
Non-endospermic seeds, 114–116
Noninfectious bud-failure, 602–603, 631
Nonmisted enclosures, 326
Nonproductive syndrome, 631
Nonrecurrent apomixis, 132–133
Nonrepeats, 624
Nontunicate bulbs, 564–565
Normal seedlings, 176
Noxious weeds, 175, 180
NP-IBA (phenyl amide), 319
Nucellar budding, 132
Nucellar embryony, 132
Nucellus, 116
Nuclear plants, 633
Nucleotides, 31, 33
Nucleus, 22
Nurse-root grafting, 437–438, 484
Nursery adhesive tape, 494
Nursery grafting, 503–504
Nursery industry, history of, 8–11
Nursery production
  Citrus spp., 733–734
  Fragaria xananassa, 740
  Persea americana, 749
  Prunus cerasus, 753
Nursery row selection, 158–159
Nutrient leaching, 321

Nutrition. *See* Fertigation; Fertilization; Mineral nutrition
Nut species, propagation of, 728

O

Oasis, 75
Obligate apomicts, 131, 146, 148
OCG (one-cotyledon graft), 466, 469–472
Offsets (offshoots), 537
  Ananas comosus, 729
  bulbs, 563, 564, 569–572
  natural layering, 552, 553–555
  Phoenix dactylifera, 750–751
  pseudobulbs, 587
Ohio Production System (OPS), 101
Oil bodies, 209
Omega graft, 497
One-cotyledon graft (OCG), 466, 469–472
Ontogenetic aging, 308, 602, 613–615. *See also* Phase changes
Open bench. *See* Open case systems
Open case systems, 503
Open mist systems, 385
Open-pollinated cultivars, 147–148
Open seed storage, 193–194
OPS (Ohio Production System), 101
Organ formation, 44–45
Organic compounds, 710
Organic media components, 367–369
Organogenesis, 670, 671
  indirect, 280
  root and shoot, 287–290, 298
  shoot, 289–290, 298
Organs, 25, 26
Ornamental trees, 774
Ortets, 598, 616
Orthodox seeds, 129, 189–191, 201
Orthotropic plants, 314, 315, 620
Osmoconditioning, 214, 215
Osmotic potential, 202, 203, 213
Osmotic priming, 214, 215
Outplanting, 503
Ovary culture, 669
Overwintering, 404
Ovule, 117–118, 121, 127
Ovule culture, 669
Own-rooted plants, 452
Oxygen, 216–217, 681

P

Pachymorph, 585
Pad and fan systems, 62
Paper pots, 71, 74–75
PAR (Photosynthetic active radiation), 51, 679
Parafilm tape, 494
Parasexual hybridization, 670, 672
Parenchyma cells, 25, 283, 284. *See also* Callus formation
Parthenocarpy, 134, 135
Particle bombardment, 36
Pasteurization, media, 95–96
Patch budding, 522, 526–532, 533

Patents, 22, 623
Pathogen-free plants, micropropagation of, 701
Pathogen-induced incompatibility, 445, 446
Pathogen management, 90–100. *See also* Disease control
  cutting propagation, 382–383
  fungicides, 93–94, 218, 382
  micropropagation, 651
  vegetative propagation, 619–622, 624–630
  virus control, *Fragaria xananassa*, 740
  virus indexing, 424, 626, 627, 628
Pathogens, 90–91
  clonal vulnerability, 597
  fungi, 90–91
  graft incompatibility and, 445, 446
  internal, 682
  systemic, 604
  viruses, 446, 621
Patterned genes, 604
PCR (Polymerase chain reaction), 35
Peach yellows, 622
Pear decline, 622
Peat, as media component, 79, 372
Peat blocks, 75
Peat-lite media, 82
Pedigreed production systems, 632–634
Pedigreed stock system, 153
Pegging down, 549
Pelleted seeds, 186, 187
Peptide hormones, 38, 44
Peptides, 33
Perennials, 15. *See also* Herbaceous plants
  genetic variability, 150–152
  seed production, 153–159
Perennial seed sources, 154–159
Pericarp, color of, 233
Periclinal chimeras, 605, 607
Periclinal division, 605, 606
Perisperm, 114, 129, 206
Peristaltic pumps, 712
Perlite, as media component, 80
Pesticides, 92, 383
Pest management. *See* Integrated pest management
Pest resistance, 452–453
Pests, 90–91, 438–439, 597. *See also* Pathogens
PFD (photon flux density), 679
Phase changes, 308
  clonal development, 594, 596
  seedling life cycle, 15–16, 18
  vegetative propagation, 613–619
Phenotype, 140
Phenotypic selection, 152, 624, 633
Phenotypic variation, 601–603, 685–687
Phenyl amide (NP-IBA), 319
Phenyl-IAA (P-IAA), 319
Phenyl-IBA (P-IBA), 319
Phenyl thioester (P-ITB), 319
Phloem, 25
Phony peach, 622

Phosphorus, 84
Photoautotrophic growth, 680
Photodormancy, 225–228
Photoelectric cell, 389–390
Photometric sensors, 51
Photon flux density (PFD), 679
Photoperiod, 51
   cutting propagation, 311–312, 328–330
   germination, 217, 228, 235
   manipulation of, 61, 65, 88–89
   tissue culture propagation, 679, 680–681
Photosynthates, 538
Photosynthesis, 329–330, 395
Photosynthetic active radiation (PAR), 51, 679
Photosynthetic lighting, 89
Photosynthetic photon flux (PPF), 51, 52
Physical abrasion, 222
Physical dormancy, 220, 221–223
Physiological age, 308
Physiological dormancy, 224–231
Physiological dwarfing, 230–231
Physiological maturity, 129
Phytochrome, 226, 227, 228–229
Phytohormones, 38. See also Hormones
Phytoplasma, 446, 621–622
P-IAA (Phenyl-IAA), 319
P-IBA (Phenyl-IBA), 319
Piece-root graftage, 483, 487
Piecework, 502, 398
Pierce's disease, 622
Pips, 587
P-ITB (phenyl thioester), 319
Plagiotropic plants, 314, 315, 620
Plant collections, 634
Plant exchanges, 5–6
Plant growth regulators (PGRs), 38. See also Auxins; Cytokinins
   adventitious root, bud, and shoot formation, 295–299
   graft union formation, 439
   micropropagation, 654, 710
Plant hormones. See Hormones
Planting density, 262
Planting stock, *Fragaria xananassa,* 740
Plantlets, propagation of, 650, 659–666, 713
Plant material, cleaning, 98
Plant modeling, 90, 390
Plant patents, 22, 623
Plant residency period, 400
Plant rolls, polyethylene, 76–77
Plant Variety Protection Act (1970), 22, 159
Plant wastage, controlling, 399–400
Plasmalemma, 428
Plasmids, 304
Plasmodesmata, 426–427
Plastic graft clips, 493, 494
Plastic-lined boxes, 573
Plastic pots, 71–73, 75–77
Plate seeder, 254
Ploidy, altered, 666
Ploidy levels, 114, 122

Plug production, 266–268
Plugs, 266, 268
Plum leaf scald, 622
Pneumatic humidifier nozzles, 392–393
Point mutation, 598, 600
Polarity
   cutting propagation, 293, 361–362
   grafting, 437–438, 487
Pollen, 117, 121
Pollen culture, 666
Pollen sterility, 150
Pollen tube, 119, 125
Pollination, 118–119, 125
   cross-pollination, 141–146, 150–151
   hand-pollination, 154, 156, 157
   self-pollination, 141
Polyacrylamide gel electrophoresis, 33, 449, 628
Polyamines, 38, 43–44, 299
Polycarbonate greenhouse coverings, 66–67
Polyembryony, 130–133
Polyethylene, 390
   bags and plant rolls, 76–77, 494, 573
   greenhouse coverings, 64–66
Polymerase chain reaction (PCR), 35
Polymer film-coated seeds, 185, 186, 187
Polymorphic seeds, 219
Polymorphism, 143
Polypeptide hormones, 44
Polypeptides, 33
Polyploidy, 598, 600
Polytents. See Indoor polytent systems
Post-liner production
   containers for, 75–77
   in-ground alternatives, 102
Postplant fertilizers, 82
Postplanting care, 255, 263
Post-propagation care of cuttings, 404–406
Potassium, 84
Potato production, 581, 582, 635
Potato spindle-tuber, 622
Pot-in-pot systems, 76, 102
Pots
   liner plants, 71–75
   post-liner production, 75–77
   root development and, 101–102
   sterilization, 73
Potted rootstock liners, 503
PPF (photosynthetic photon flux), 51, 52
PR (propagation ratio), 655
Precision seeders, 253–254
Precocious germination, 130–131
Predictive indices of rooting, 317, 318
Preformed meristems, 290
Preformed roots, 281–283
Pregermination, 188–189
Preincorporated fertilizer, 54, 401
Preparation area, micropropagation, 702–704
Preplant fertilizers, 82
Pressure jet mist nozzles, 385–386, 387
Pressure potential, 202–203
Primary seed dormancy, 218–219, 220–235

Priming, 187–188, 214–216
Production-led propagation systems, 400
Production material, as nursery sources, 631
Production systems, 50, 72, 84–85
Production windows, chip budding, 521
Proembryo stage, 123, 125, 126
Progeny, 624
Progymnosperms, 112
Propagating frames, 69–70
Propagation, history of, 1–12
Propagation, vegetative. See Vegetative propagation
Propagation biology
   genetics, 22–27, 28
   life cycles, 14–18
   reproduction, 111–122
   taxonomy, 18–21
Propagation by seed. See also Dormancy; Germination; Seed development; Seedling production; Seed production
   breeding systems, 140–147
   genetic variability, 150–153
   plant categories, 147–150
   seed storage, 189–195, 215
   trees, 774
Propagation containers. See Containers, propagation
Propagation environment, 49–50. See also Enclosure systems; Environmental controls; Greenhouses; *specific environmental factors*; *specific propagation methods*
   aseptic conditions, 644, 704–705
   biotic factors, 49, 50
   computer modeling, 90, 390
   containers, 70–77
   controls, 63, 65, 66, 390
   edaphic factors, 49, 50
   gas exchange, 681
   media and nutrition, 77–85
   micropropagation, 704–706, 707
   post-propagation care, 100–102
   seedling germination, 211–217
Propagation generations, 598
Propagation organizations, 11
Propagation ratio (PR), 655
Propagation unit systems, 372–373
Propagules, 49, 594. See also Explants
Prophase, 24, 28
Protected culture, 344, 350
Protein bodies in storage reserves, 208
Protein electrophoresis, 181
Proteins, 33
Protein synthesis, 204, 215
Proteomic analysis, 306
Provenance, 149
Proximal end of shoots and roots, 437
Prunasin, 447
Pruning for cutting propagation, 364, 365–367, 374
Prunus necrotic ring spot virus, 625
Pseudobulbs, 562, 587–590

Pseudocorms, proliferation of, 662
Pseudogamy, 133
Psylla, 622
Pulse irrigation, 100
Pumice, as media component, 80
Pups. *See* Suckers
Pure stand, 152
Purity determination, 180, 181–182
Pyranometric sensors, 51

Q

Quality in propagation, 400
Quantum sensors, 51
Quarantines, 634–635
Quick-dips, 377, 378–380
Quiescence, 513, 514
Quiescent seeds, 130
Quonset-type greenhouses, 57

R

Radical protrusion, 201, 204–207
Radicle, 207
Raffia, 493
Ramets, 598, 616
Randomly Applied Polymorphic DNA (RAPD), 623
Random seeders, 253
Reasonable Expectancy (RE) programs, 398
Recalcitrant plants, 299, 356
Recalcitrant seeds, 129
  germination, 201
  ripening and dissemination, 136
  storage, 189–190, 191–192, 194
Recombinant DNA technology, 35–38, 599–601
Recombination, 27
Record keeping
  cutting propagation, 396–398
  grafting, 502
Recycled water systems, 86–88, 98
Reforestation, 264–266
Refrigerated stratification, 259–261
Refrigeration systems, 405
Regeneration, potential for, 16
Regermination, 189
Registered seed, 154
Registered source plants, 634
Regulator sequences, 32
Rejuvenation
  cutting propagation, 308–310
    layering, 538
    phase variation management, 619, 620
    somatic embryogenesis, 675
    tissue culture propagation, 686
Relative storability index, 190
Repair graftage, 465, 466, 487–490, 491
Repeats, SSR testing, 624–625
Replacement, cellular, 607
Repositories, clonal, 634
Reproduction
  cell biology, 26–27, 110
  seedless plants, 111–112
  seed plants, 112–122

Reproductive bulb development, 565–566
RE (Reasonable Expectancy) programs, 398
Resistance of cultivars, 91
Respiration. *See* Gas exchange
Rest period, 294
Restriction Fragment Length Polymorphism (RFLP), 623
Restriction fragments, 35
Retractable roof greenhouses, 55–57
Reverse osmosis, 85, 86
Reversion
  chimeral, 612, 613
  mature to juvenile phase, 619
Rhizocaline, 293, 301
Rhizomes
  growth pattern, 586–587
  natural layering, 552
  propagation of, 562, 587
  structure, 584–586
Rhizosphere, 94, 382
Rigid greenhouse covering materials, 66–67
Rind graft, 467, 480–483, 484
Ring budding, 522
RNA (ribonucleic acid), 22, 32
RNA interference, 37
RNA polymerase, 32
Rockwool, as media component, 80, 369, 372
Rockwool blocks, 73, 75
Roguing, 151
Rolled-towel test, 177, 178
Root axis, 207
Root cuttings, 361–363
  adventitious bud, shoot, and root formation, 281, 292–293, 664
  propagation systems, 346–347
Root development in containerized plants, 101–102. *See also* Adventitious root formation
Rooted cuttings, 403–404, 404–406
Root emergence, 287, 288–289
Root formation stage of micropropagation, 657–658, 717–718
Root grafting, 483–484, 483–485, 502
Rooting. *See also* Adventitious root formation; Direct sticking
  cutting propagation, 367–373, 401–403
  hardwood cuttings, 350
  semi-hardwood cuttings, 354
Rooting bioassays, 300
Rooting co-factors, 300
Rooting inconsistency, 382
Rooting inhibitors, 300, 301
Rooting morphogen, 301
Rooting needle fasicles, *Pinus* spp., 808
Rooting treatments, 374–376
Root initials
  formation of, 283, 288, 299–300, 327
  preformed and latent roots, 281–283
Root morphogens, 299
Root organogenesis, 287–289
Root primordia, 283, 284, 288, 289, 299

Root pruning systems
  air, 72, 101–102
  chemical, 74
Root regeneration potential (RRP), 407
Roots, adventitious, 280. *See also* Adventitious root formation
Root spiraling, 101–102
Rootstocks, 418
  Actinidia deliciosa, 729
  budding, 512, 513, 522
  Carya illinoinensis, 731
  Castanea spp., 732
  Citrus spp., 733, 735–736
  Coffea spp., 736
  composite plants, 421–422
  Diospyros kaki, 737
  double-working, 505
  fruit and nut species propagation, 728
  growth activity, 436–437
  handling, 500–501
  Juglans regia, 741
  Malling apple, 548
  Malus xdomestica, 548, 744–746
  nut species, 728
  Olea europaea, 748
  ornamental trees, 774
  Persea americana, 750
  Pistachia vera, 751
  Prunus spp., 753–758
  Pyrus spp., 759–761
  Rosa spp., 815–816
  scion relationships, 451–457
  seedling, 419, 744
  Vitis spp., 764–765
Root zone heating systems, 60
Round bulbs, 571
RRP (root regeneration potential), 407
Rudimentary embryos, 233
Runners, 552, 553

S

Saddle graft, 466, 474, 476, 487
Salicylate, 299
Salicylic acid, 38, 43
Salinity, measurement of, 85
Sampling, for seed testing, 176
Sand, as media component, 78–79
Sanitation methods, 96–98, 395, 402
Saturation, 88
Saw-kerf graft, 466, 474, 475, 497
Scaffold management, 368
Scaling, 573–574
Scaly bulbs, 564, 565, 570, 571
Scarification, 222, 257–258, 777
Scheduling, cutting propagation, 399
Scientific literature, 6–8
Scion blocks, 498–499
Scion rooting, 419, 484
Scion-rootstock relationships, 450–457
Scions, 418
  composite plants, 420
  double-working, 505
  insertion technique, 471

Scionwood
    budding, 514–516, 518
    grafting, 499
    selection and handling, 496–500
Sclereids, 25, 285
Sclerenchyma ring, 284–285
Scooping bulbs, 575
Scoring bulbs, 575–576
Scouting systems, 91
Scrapage, controlling, 399–400
Screen balance, 389
Scutellar stage, 123, 125
Scutellum, 123, 208
Sealed-container seed storage, 193, 194
Seasonal timing
    budding, 513–519
    cutting propagation, 315, 316–318, 346–347
    grafting, 504
    layering, 538, 542
    top-budding, 514, 533
Secondary meristems, 290–292
Secondary phloem, 431
Secondary seed dormancy, 219, 221, 235–236
Secondary xylem, 431
Second-generation seed, certified, 154
Sectorial chimeras, 605–606, 607
Seed, 112, 200
    Abies spp., 774
    Acacia spp., 775
    Acer spp., 775
    Actinidia deliciosa, 728
    Aesculus spp., 777
    Ailanthus altissima, 778
    Albizia julibrissin, 778
    Alnus spp., 778
    Amelanchier spp., 778
    Anacardium occidentale, 729
    Annona cherimola, 729–730
    Araucaria spp., 779
    Arctostaphylos spp., 779
    Banksia spp., 781
    Berberis spp., 781
    Betula spp., 782
    Camellia spp., 783
    Carica papaya, 730
    Carpinus spp., 784
    Carya illinoinensis, 731
    Carya ovata, 731
    Castanea spp., 784
    Catalpa spp., 785
    Ceanothus spp., 785
    Cedrus spp., 785
    Celastrus spp., 785
    Cercis spp., 786
    Chamaecyparis spp., 786
    Chionanthus spp., 786–787
    Citrus spp., 732–733
    Clematis spp., 787
    Coffea spp., 736
    Cola nitida, 736
    Cornus spp., 787–788
    Corylus spp., 737, 788
    Cotoneaster spp., 789
    Crataegus spp., 789
    Cryptomeria japonica, 789
    Cycas spp., 789–790
    Cytissus spp., 790
    Daphne spp., 790
    Diospyros kaki, 737
    Erica spp., 791
    Eriobotrya japonica, 738
    Eucalyptus spp., 792
    Euphoria longan, 738
    Fagus spp., 793
    Feijoa sellowiana, 738
    Ficus carica, 738–739
    Fragaria xananassa, 739
    Fraxinus spp., 793
    Genista spp., 794
    Hamamelis spp., 795
    Hibiscus spp., 796
    Ilex spp., 797
    Juglans cinerea, 740–741
    Juglans nigra, 741
    Juglans regia, 741
    Juniperus spp., 798
    Litchi chinensis, 742
    Macadamia spp., 742–743
    Magnolia spp., 801
    Malus xdomestica, 743
    Malus spp., 803
    Mangifera indica, 746
    Myrtus spp., 804
    Olea europaea, 747–748
    palms, 805–806
    Passiflora edulis, 748
    Persea americana, 749
    Phoenix dactylifera, 750
    Picea spp., 807
    Pinus spp., 807
    Pistachia vera, 751
    Platanus spp., 808
    Populus spp., 809
    Prunus spp., 752, 809
    Pseudotsuga menziesii, 810
    Psidium spp., 758
    Pyrus spp., 759
    Quercus spp., 811
    Rhododendron spp., 813
    Rhus spp., 814
    Rosa spp., 814
    Salix spp., 817
    Sequoia spp., 818
    Skimmia spp., 818
    Sorbus spp., 819
    Syringa spp., 820
    Taxus spp., 821
    Theobroma cacao, 762
    Thuja spp., 821
    Tilia spp., 822
    Tsuga spp., 822–823
    Ulmus spp., 823
    Viburnum spp., 824
    Vitis spp., 763
Seed, foundation, 153, 154
Seed bank, 219
Seed bed preparation, 251–252, 256–257
Seed certification, 153–154, 157
Seed classification, 116
Seed coat. *See* Seed coverings
Seed coating, 186, 253
Seed collecting, 166
Seed collection zones, 154–155, 158
Seed coverings, 116–117
    germination, 203, 205, 224, 231–233
    water gap, 222
Seed development
    flower parts and, 118–122
    gymnosperms, 125–126
    plant hormones and, 133–135
    stages, 122–130
    unusual, 130–133
Seed dissemination, 136–137
Seeders
    gel, 254–255
    mechanical, 253–255, 270–274
Seed exchanges, 166
Seed filling, 122, 126–128
Seed hoppers, 253
Seedling chimeras, 607–608
Seedling cultivars, 15–16, 594
Seedling emergence, 207–208, 219
Seedling liners
    harvesting field-grown, 263–264
    transplanting, 264, 274–276
Seedling production
    containers, 268, 269
    control methods, 268–274
    field nurseries, 251, 255–264
    field seeding, 250–255, 264–266
    flats, 266, 267
    grow-out tests, 184
    pathogen elimination, 630
    plugs, 266–268
    protected-conditions seeding, 266–274
    tissue culture propagation, 646–647, 666–669
    transplanting, 264, 274–276
Seedling progeny tests, 149, 151–153
Seedling rootstock, 419, 744
Seedling selection, 597
Seed orchards, 157–158, 166
Seed plants, 112
Seed potatoes, 581
Seed priming, 187–188, 214–216
Seed production. *See also* Drying process; Seed sources
    breeding programs, 140–147
    commercial, 150–152, 162–175
    conditioning, 168–171, 172
    extraction, 168, 172–175
    genetic variability, 150–153
    handling, 163, 166–175
    hybrid cultivars, 148, 154, 155
    maturity and ripening, 166–175
    micropropagation, 701
    open-pollinated cultivars, 147–148

ripening, 136–137
seed collection areas, 155–157
seed sizing, 186–188
storage, 189–195, 215
synthetic seeds, 673–679
systems, 153–159
treatments, 184–189, 252–253
Seed propagation. *See* Propagation by seed
Seed protectants, 185–186
Seeds
health tests, 182–184
local, 152
long-lived, 190–191
orthodox, 129, 189–191, 201
quality of, 250, 252
quiescent, 130
recalcitrant, 129, 136, 189–192, 194, 201
structure, 115–118, 200, 201
synthetic, 673–679
testing requirements, 175–184
viability, 176–180, 189, 191
Seeds, moisture content of
germination, 225, 230
storage, 191–192
testing, 167
Seed selection, 140–147
Seed sources
commercial, 154, 162–166
woody perennials, 149–150, 152, 154–159
Selected families, 150
Selected tree seed, 157
Select seed, 154
Self-incompatibility, 143
Self-pollination, 141
Semi-dwarfing clonal rootstocks, *Malus xdomestica,* 745
Semi-dwarfing scion cultivars, 451
Semi-hardwood stem cuttings, 346–347, 351–354
Separation of DNA fragments, 35
Separation of specialized stems and roots, 561
Serial propagation, 308, 619
Serology, 626–628
Serpentine layering, 541, 543
Sexual incompatibility, 143–145
Sexual reproduction, 27
Shade houses, 69
Shading
cutting propagation, 50, 367, 370, 371
field nurseries, 263
Shield budding. *See* T-budding
Shoot apex culture, 628–629
Shoot apices, fragmented, 664
Shoot axis, 207
Shoot guide clips, 494, 517
Shoot multiplication stage of micropropagation, 654–656, 716–717
Shoot organogenesis, 285, 289–290, 298
Shoot-root relationships, 450–457

Shoots. *See also* Adventitious shoot formation
adventitious, 280
structure of, 314–315
tissue culture propagation, 645
Shoot tip grafting, 629, 663
Shoot-tip necrosis, 682, 683
Short-day plants, 51
Short-lived seeds. *See* Recalcitrant seeds
Shrubs, 276, 774
Side-insertion graft (SIG), 467, 477–480, 483
Side-stub graft, 467, 477–478, 479
Side-tongue graft, 467, 478, 480
Side-veneer graft, 467, 478–479, 481, 483
SIG (side-insertion graft), 467, 477–480, 483
Silicon tubing, 493, 494, 496
Simple layering, 539, 541, 542
Simple morphophysiological dormancy, 234
Simple Sequence Repeats (SSRs), 623, 624–625
Single-cross hybrids, 148
Single-eye cuttings, 281. *See also* Leaf-bud cuttings
Single-node cuttings. *See* Leaf-bud cuttings
Single-nose bulbs, 571
Singulator device, 254
Site selection for field nurseries, 251–252, 256–257
SIVB (Society for *In Vitro* Biology), 12
Slabs (bulbs), 571
Slant graft, 466, 469–472
Slice cut, 501
Slipping, 513
Slips, 554, 583–584
Societies, horticultural, 4–5. *See also* Propagation organizations
Society for *In Vitro* Biology (SIVB), 11
Society of Commercial Seed Technologists, 180
Softwood stem cuttings, 354–355
auxin treatments, 377
etiolation, 367, 369
forcing, 618
length of, 346
propagation systems, 346–347
Rosa spp., 815, 816
storage, 405
Soil, as media component, 78
Soil structure, 78
Soil temperatures for pasteurization, 96
Solar heating systems, 60
Solid matrix priming, 214, 215
Somaclonal decline, 602–603
Somaclonal variation, 599, 604, 670, 674, 684
Somatic embryogenesis, 646–647, 670, 673–679
Somatic embryos, 645
Somatic hybridization, 670, 672
Somatoplastic sterility, 122

Source-identified tree seed, 157
Source plants for vegetative propagation
characteristics, 623–625
cuttings, 363–367
management, 630–635
micropropagation, 650–654
Southern Nursery Association (SNA), 12
Sowing procedures, 252, 255, 261–263
Spatulate embryos, 233
Specialized stems and roots, 561–563. *See also* Bulbs; Corms; Pseudobulbs; Rhizomes; Tubers
Species, 18–19
Species, grafting between, 434, 440. *See also* Graft compatibility; Graft incompatibility
Sphaeroblasts, 308–309, 618
Sphagnum moss peat, as media component, 79–80
Spliced approach graft, 467, 485–486, 489
Splice graft, 466, 469–472
Splints, 494
Split-base treatments, 322
Split graft, 466, 472–474
Splits (bulbs), 571
Spoons, 571
Sporophyte, 110–111
Sporophytic apomixis, 131–132, 133
Sporophytic self-incompatibility, 143, 145–146
Spring budding, 514, 515, 517, 518–519
Spring patch budding, 531–532
Spring seeding, 263
Sprinkler systems, 391
Spurs, 454
SSRs (simple sequence repeats), 623, 624–625
Stabilization, 653–654, 716
Stage 0 in micropropagation, 651–652, 713
Standard germination test, 176–177
Standard operating procedures (SOPs), 398
Starch. *See* Carbohydrates in storage reserves
Static control systems, 388
Stecklings, 164
Stem cuttings, 281, 345
bulb propagation, 572–573
hardwood, 345–351
herbaceous, 356–357
modified, 344
semi-hardwood, 351–354
softwood, 354–355
tuber propagation, 584
Stem structure and rooting, 288
Stenting, 505, 507
Sterilization methods, 73, 95–96
Stimulative parthenocarpy, 135
Stock. *See* Rootstocks
Stock blocks, 345, 631–632
Stock plants for cutting propagation
cleaning, 98
cutting material selection, 307–308, 314–318, 345, 363–364

Stock plants for cutting propagation (cont'd)
  environmental control, 50, 310–314
  maintenance, 307–308
  manipulation techniques, 364–367
  rejuvenation, 308–310
Stock plants for micropropagation, 675–676, 713–714
Stock seed, 153
Stock solutions, 708–710
Stolons, 552, 553
Stool beds, 540, 546–547
Stooling
  Castanea, 732
  cutting propagation, 50, 365
  layering, 538, 545–548
Stool shoots, 538, 546, 619, 661–662
Storage
  cuttings, 318, 404–406
  scionwood, 499–500
  seeds, 189–195, 215
Storage reserves, 114–116
  cell expansion, 126–128
  genetic engineering, 128
  germination, 208–209
  metabolism, 204
Stratification, 258–259
  chilling, 225–226
  embryo growth potential, 231
  hormone-controlled dormancy, 239
Strip tests, 181–182
Strophiole, 222
Stump sprouts, 618
Stunt disease, 622
Subculturing, 655, 713, 716, 717
Suberization, 357, 581
Subirrigation watering systems, 272, 393, 394
Substrates, media, 367–373
Succulents, propagation of, 866
Suckers, 292, 537
  budding, 518
  clonal selection, 618
  grafting, 444
  natural layering, 552, 554–555, 556
  rhizome propagation, 587
Summer budding, 514–518
Summer patch budding, 530
Summer seeding, 262
Sun tunnels, 67–70, 390
Supplemental carbon dioxide. *See* Carbon dioxide
Surfactants in media, 370
Suspension cultures, 671
Suspensor, 123, 125, 135
Suspensor tier stage, 125, 126
Symplastic connections, 429
Synchronized germination, 219
Synergid cells, 120–121, 132–133
Synseeds, 673–679
Synthetic cultivars, 148
Synthetic lines, 148
Synthetic seed production, 673–679
Systemic pathogens, 604, 621–622

T

TAG (tongued approach graft), 467, 486, 489
Target cells, 44
Taxonomy, 18–21
T-budding, 512–513
  Acer spp., 777
  Citrus spp., 787
  Cornus spp., 789
  Crataegus spp., 789
  criteria for, 522
  double-working, 506
  graft union formation, 432–433
  June budding, 519
  methods, 521–527
  spring budding, 518
  wood-in and wood-out, 519, 528–530
T-DNA, 304–305
Telophase, 24, 28
Temperature control, 53
  chilling stratification, 226
  cutting propagation, 310, 327–328
  flat and plug production, 272–273
  germination, 211–212, 216, 227, 230
  grafting, 435
  heating and cooling systems, 59–63, 88
  propagation facilities, 88
  seed storage, 192, 193–194
  tissue culture propagation, 679, 680
Template seeders, 270–271
Terminal/top insertion graft, 467, 476, 478
Termination codons, 32
Tested tree seed, 157
Tetrazolium test, 178–179, 180
Thermal curtains, 61
Thermal time, 216
Thermodormancy, 235
Thermogradient tables, 212
Thermoinhibition, 212–213
Thermotherapy
  grafting, 424
  pathogen elimination, 629–630, 660
  seed protectants, 186
Thidiazuron (TDZ), 40, 296
Thin layer explants, 664, 665
Thorniness, 146
Thornless mutations, 612
Three-way cross hybrids, 148
Tiers, 472, 512, 527, 528
Timers, mist-system, 388
Tip layering, 537, 551–553
Tissue, homogenized, 664
Tissue-culture produced liners, 364
Tissue culture propagation, 644. *See also* Micropropagation
  callus cultures, 670–671
  cultures and suspensions, 670–672
  environmental control, 679–681
  genetic variation, 684–685
  history, 644–649
  problems, 681–683
  seedling plant reproduction, 666–669
  somatic embryogenesis, 646–647, 670, 673–679
  systems, 649–650, 655–656, 670–672
  techniques, 646–647
  technology, 34, 599
Tissue culture technology, 34, 599
Tissue overlay culture, 656
Tissue proliferation (TP), 682–683
Tissues, 25
Tobacco mosaic, 622
Tongued approach graft (TAG), 467, 486, 489
Tools, disinfecting, 97–98
Top-budding. *See* Topworking
Top-cross hybrids, 148
Top-dressing, 401
Top-grafting, 436–437
  aftercare, 504
  Vitis spp., 764
Topographical test, 178
Topophysis, 315, 620
Topworking, 422–423, 424
  aftercare, 504
  budding, 512, 532–533
  frameworking, 533
  grafting, 486, 505
  timing of, 514, 533
Torpedo stage, 123
Totipotency, 44, 645, 648
TP (tissue proliferation), 682–683
Trademark names, 623
Trademarks, 22
Trade secrets, 22
Transcription, 31, 32
Transcriptome analysis, 37
Transfer (micropropagation), 704–705, 713
Transfer cells, 127
Transfer RNA, 33
Transformation technology, 36
Transgenic clones, 600
Transient phenotypic variation, 685–687
Transition phase of seedling life cycle, 15, 18, 594
Translation, 31, 32–33
Translocatable incompatibility, 445–446
Translocation, 598, 600
Transmitting cells, 119
Transplant shock, 275
Transposons, 604
Trees, ornamental, 774
Trees, transplanting, 276
Tree seed certification, 157
Trench layering, 537, 541, 550
Triacylglycerides, 209
Trichoderma fungi, 94, 95, 383
Triploids, 666
True potato seed (TPS), 635
True-to-name cultivars, 623
True-to-type cultivars, 616, 623–624
True-to-type juvenile phenotypes, 687
Tube nucleus, 119
Tubercles, 581
Tuberization, 580
Tuberous roots and stems

growth pattern, 582–583
propagation, 583–584
structure, 581–582
Tubers
growth pattern, 580–581
propagation, 562, 581, 635
structure, 579–580
Tunica, 605
Tunicate bulbs, 564, 565, 570, 571
Tunics, 578
Turgor, 52, 202
Twin-scaling, 576
Tying materials, grafting, 492–493

U

Ultrasonic humidifier nozzles, 392–393
Unclassified seeds, 116
Understock. See Rootstocks
Undifferentiated embryos, 233–234
Uniformity of populations, 596
Unlayered meristems, 607
Unrooted cuttings (URCs), 318–319
  environmental controls, 330
  handling, 357, 403–406
  wholesale production, 366
USDA online resources, 20
Utility patents, 22

V

Vacuum seeder, 254
Vapor pressure deficit (VPD), 324–326
Var, 149
Variegation, 604, 609–611
Varietas, 149
Varieties. See Cultivars
Vascular cambium, 418–419
  graft union formation, 426, 464–465
  wound repair, 428–431
Vascular differentiation, 425, 448
Vascular plant reproduction, 110–112
Vegetable crops
  commercial seed production, 162–166
  field seeding, 251–255
  grafting, 499
  transplanting seedlings, 275–276
Vegetative apomixis, 133
Vegetative bulb development, 565–566, 569
Vegetative parthenocarpy, 135
Vegetative progeny test, 624
Vegetative propagation. See also Asexual propagation; specific propagation methods
  biotechnological advances in, 304–307
  cost, 597
  Cycas spp., 790
  of fruit and nut species, 728
  history, 595
  phase variation management, 613–619
  Pseudotsuga menziesii, 810
Ventilation systems, 59
Vermiculite, as media component, 80
Vernalization, 164, 569

Very low fluence response (VLFR), 228
Viability of seeds
  determining, 176–180
  storage and, 189, 191
Vigor
  of micropropagated plants, 685–686
  of seedlings, 146
  of seeds, 180–182, 183–184, 191
Vigorous clonal rootstocks, *Malus xdomestica,* 745
Vigorous scion cultivars, 451
Vines, woody, 774
Viral diseases. See Virus indexing
Virginia Tech System, 83
Viroids, 622
Virus contamination, 438–439
Virus control, *Fragaria xananassa,* 740
Viruses, 446, 621
Virus-free designation, 630
Virus indexing, 424, 626, 627, 628
Virus-tested designation, 630
Visualization of DNA fragments, 35
Vitamins in media, 710
Vitrification, 681–682
Viviparous seeds, 130–131, 201
VLFR (very low fluence response), 228
VPD (vapor pressure deficit), 324–326

W

Wardian cases, 5, 69
Water, recycled, 86–88, 98
Water gap, 222
Watering systems. See Irrigation systems; Misting systems
Water management. See also Enclosure systems; Irrigation systems; Misting systems
  cutting propagation, 310, 323–327, 370
  germination, 213–214, 216
  grafting, 435–436
  propagation facilities, 52, 86–88
  systems for flat and plug production, 271–272, 273
Water potential, 52, 201–203
Water quality
  in misting systems, 395
  pathogen suppression, 95
  salinity, 85–86
Watersprouts, 554, 618, 619
Water stress
  bare-root liners, 407
  germination, 213, 214
Wavelength. See Light quality
Wedge graft, 466, 474, 475, 497
Weeds
  cutting propagation, 402–403
  history, 3
  noxious, 175, 180
  in restoration plantings, 266
  roguing, 151
  seed dormancy, 219
Wetting agents in media, 370

Wheel seeder, 254
Whip-and-tongue graft, 466, 467–469, 483
Whip graft, 466, 469–472
Whirl-type mist nozzles, 385–386, 387
Whitewash, 63
Whole-root graftage, 483, 487
Wood, grafting and budding, 418, 432, 513
Wood containers, 77
Wood-in and wood-out, 519, 528–530
Woody plants. See also Hardwood stem cuttings; Semi-hardwood stem cuttings; Softwood stem cuttings
  adventitious root formation, 280, 283, 284, 286, 298
  clonal selection, 596
  commercial seed production, 166
  conifer propagation, 146, 551
  container-grown, 268
  cutting stock management, 308–309, 314–315, 328
  deciduous plant propagation, 315, 316, 345–350, 353–356
  evergreen propagation, 316–317, 346–347, 350–356
  field seeding, 256
  flower development, 17
  genetic variability, 152–153
  micropropagation, 651
  perennials, 15
  perennial seed propagation, 148–150, 152–153
Woody vines, 774
Worker Protection Standard (WPS), 92
World checklist of plant families, 20
Wound cambium, 430
Wound-induced meristems, 290–292
Wound-induced roots, 281–283, 284
Wounding, 321–323, 373, 374
Wounding-related compounds (WRCs), 321
Wounding response, 283, 426
Wound periderm, 283
Wound repair, 423–424, 428–431
Wound repair xylem and phloem, 430–431
WPS (Worker Protection Standard), 92
Wrapping materials, grafting, 492–493

X

X-disease, 622
X-ray analysis, 179
Xylem, 430–432, 513

Y

Yield efficiencies, fruit, 451

Z

Zeatin, 40, 296
Zeatin riboside, 296
Zygote, 21–22
  double fertilization, 120
  formation, 110, 112
  ploidy levels, 122

plant index: scientific names

Note: The Scientific and common name indexes contain names as they are used in the text. To find all references to a plant, use both indexes.

A

Abelia, 315
Abelia xgrandiflora, 400, 774
Abies, 172, 347, 350, 415, 420, 452, 457, 504, 774–775
Abies amabilis, 17
Abies balsamea, 775
Abies bornmuelleriana, 775
Abies concolor, 775
Abies firma, 452, 775
Abies fraseri, 314, 620, 775
Abies magnifica, 775
Abies procera, 774
Abutilon, 775
Abutilon striatum, 622
Acacia, 221, 616, 663, 775
Acacia baileyana, 775
Acacia cyanophylla, 775
Acacia farnesiana, 775
Acacia koa, 308, 775
Acacia melanoxylon, 775
Acacia redolens, 775
Acacia subprosa, 775
Acalypha, 293
Acanthus, 132
Acanthus mollis, 840
Acer, 189, 300, 315, 320, 347, 354, 357, 370, 404, 436, 490, 500, 516, 517, 528, 604, 662, 775–777
Acer campestre, 776
Acer cappadocicum, 777
Acer catalpifolium, 777
Acer ginnala, 776
Acer glabrum, 776
Acer griseum, 775, 776
Acer lobelii, 777
Acer macrophyllum, 776
Acer negundo, 775, 776
Acer palmatum, 356, 465, 482, 501, 504, 775–777
Acer pennsylvanicum, 777
Acer pentaphyllum, 444
Acer platanoides, 353, 370, 408, 421, 432, 517, 776, 777
Acer pseudoplatanus, 17, 353, 354, 444, 776, 777
Acer rubrum, 329, 440, 686, 775, 776, 777
Acer saccharinum, 191, 194, 775, 776, 777
Acer saccharum, 776, 777
Acer truncatum, 776
Achillea, 840
Achillea filipendulina, 840
Achimenes, 840
Aconitum, 840–841
Actinidia, 233, 777
Actinidia arguta, 728, 777
Actinidia deliciosa, 305, 365, 454, 473, 728–729, 777
Actinidia hemsleyana, 777
Adiantum, 851
Aegopodium, 841
Aerides, 859
Asclepias, 118
Aesculus, 190, 194, 504, 777–778
Aesculus arguta, 777
Aesculus californica, 777
Aesculus xcarnea, 777
Aesculus flava, 777
Aesculus hippocastanum, 777, 778
Aesculus indica, 777
Aesculus parviflora, 365, 777–778
Aesculus pavia, 777
Aesculus sylvatica, 777
Aesculus turbinata, 777
Agapanthus, 841
Agastache foeniculum, 841
Agathis australis, 284
Agave, 133, 841, 866
Aglaonema, 841
Agrostemma githago, 841
Ailanthus altissima, 365, 778
Ailanthus malabarica, 778
Ajuga, 552, 553, 841
Alberta magna, 778
Albizia julibrissin, 365, 778
Albizia lophantha, 222
Albuca, 576
Alcea rosea, 841
Alchemilla mollis, 841
Allamanda cathartica, 778
Allium, 133, 145, 150, 168, 217, 564, 567, 569, 629, 841
Alnus, 146
Alnus cordata, 778
Alnus glutinosa, 778
Alnus incana, 778
Alnus maritima, 778
Alnus rhombifolia, 778
Alnus rugosa, 778
Alnus serrulata, 778
Aloe, 841, 866
Aloe barbadensis, 841
Alsophila, 851
Alstroemeria, 577, 841
Alyssum saxatile, 843
Alyxia olivaeformis, 841–842
Amaranthus, 217
Amaranthus caudatus, 842
Amaranthus gangeticus tricolor, 842
Amaryllis, 570
Amaryllis belladonna, 842
Amelanchier, 172, 778
Amelanchier alnifolia, 778
Amelanchier xgrandiflora, 778
Amelanchier laevis, 778
Amorpha, 779
Amorpha canescens, 779
Amorpha fruticosa, 779
Amsonia tabernaemontana, 842
Anacardium occidentale, 729
Ananas comosus, 729
Anaphalis margaritacea, 842
Anchusa azurea, 842
Anchusa capensis, 842
Andropogon, 134, 146
Anemone, 234, 569, 842
Anemone blanda, 842
Anemone coronaria, 234, 842
Anemone xhybrida, 842
Anemone japonica, 365, 842
Anemone pulsatilla, 863
Angelonia angustifolia, 842
Anigozanthus, 842
Annona cherimola, 729–730
Annona squamosa, 233, 729, 730
Annona squamosa xA. cherimola, 729
Anthemis, 842
Anthurium andraeanum, 842
Antirrhinum majus, 842
Aquilegia, 842
Aquilegia chysantha, 842
Arabidopsis, 129, 134, 205, 206, 228, 232, 236, 237, 239, 295–296, 297, 305
Arabidopsis thaliana, 456
Arabis, 842
Arachis, 141
Arachnis, 859
Aralia, 779
Aralia spinosa, 365, 779
Araucaria, 620, 779
Araucaria araucana, 779
Araucaria bidwillii, 779
Araucaria columnaris, 779
Araucaria cunninghamii, 420, 779
Araucaria heterophylla, 779
Arbutus, 779
Arbutus andrachne, 779
Arbutus menziesii, 779
Arbutus xalapensis, 779
Arctostaphylos, 779
Arctostaphylos xcoloradensis, 779–780
Arctostaphylos uva-ursi, 779
Arctotis stoechadifolia, 842
Ardisia, 780
Ardisia crenata, 780
Ardisia japonica, 780
Argyranthemum frutescens, 848
Arisaema, 843
Armeria, 843

Aronia, 363
Aronia arbutifolia, 780
Aronia melanocarpa, 780
Artemisia, 843
Artemisia ludoviciana, 843
Artemisia schmidtiana, 843
Artocarpus altilis, 365, 730
Artocarpus heterophyllus, 370, 730
Arum italicum, 843
Aruncus dioicus, 843
Aruncus sylvester, 843
Arundo donax, 556
Asarum, 843
Asclepias curassavica, 843
Asclepias tuberosa, 843
Asimina, 163, 552
Asimina parviflora, 780
Asimina triloba, 194, 683, 730, 780
Asparagus, 132, 142, 143, 144, 556, 557, 701
Asparagus asparagoides, 843
Asparagus plumosus, 843
Asparagus sprengeri, 843
Asplenium nidus, 851
Aster, 843
Aster frikartii, 843
Astilbe, 843
Astrantia, 843
Athrium, 851
Atriplex, 224
Aubrieta deltoidea, 843
Aurinia saxatilis, 843
Avena, 141
Averrhoa carambola, 730

B
Baccharis, 780
Baccharis angustifolia, 780
Baccharis halimifolia, 780
Baccharis sarothodes, 780
Bambusa, 780–781
Banksia, 781
Banksia coccinea, 781
Banksia menziesii, 781
Baptisia, 843
Bauhinia, 781
Bauhinia xblakeana, 781
Bauhinia lunarioides, 781
Bauhinia purpurea, 781
Bauhinia variegata, 781
Beaucarnea recurvata, 806
Begonia, 148, 297, 298, 328, 380, 584, 686, 843–844
Begonia evansiana, 581
Begonia rex, 290, 361
Begonia xtuberhybrida, 582
Belamcanda chinensis, 844
Bellis perennis, 844
Benincasa hispida, 499
Berberis, 136, 172, 367, 381, 541, 551, 781–782
Berberis koreana x B. thunbergii, 781
Berberis thunbergii, 781, 782
Berberis trifoliata, 782

Bergenia cordifolia, 844
Betula, 190, 227, 408, 415, 436, 452, 490, 662, 681, 782
Betula alleghaniensis, 782
Betula lenta, 782
Betula nigra, 356, 782
Betula papyrifera, 370, 782
Betula pendula, 524, 782
Betula platyphylla, 452, 782
Betula populifolia, 452
Betula pubescens, 17, 782
Boltonia, 844
Bougainvillea, 401, 782
Bouteloua curtipendula, 134, 146
Bouvardia ternifolia, 844
Boxus, 541
Brachycome iberidifolia, 844
Bracteantha bracteata, 853
Brassaia arboricola, 865
Brassica, 134, 143, 145, 168, 190, 204, 666, 667, 672
Brassica napus, 128
Brassica oleracea, 844
Brassovola, 859
Brodiaea, 844
Bromus, 148
Broussonetia papyrifera, 365
Browallia, 844
Brunnera macrophylla, 844
Bryophyllum, 290, 297, 361, 855
Bryophyllum daigremontianum, 362
Bryophyllum tubiflorum, 370
Buddleia, 782
Bumelia lanuginosa, 782
Butia capitata, 806
Buxus, 782–783
Buxus microphylla, 782–783
Buxus sempervirens, 782–783

C
Cactus, 844
Caladium, 562, 579, 581, 845
Calamagrostis acutiflora, 556, 845
Calanthe, 859
Calathea, 845
Calceolaria, 217, 845
Calendula officinalis, 845
Calibrachoa, 845
Callistemon, 783
Callistephus chinensis, 845
Calluna vulgaris, 791
Calocedrus, 172
Calocedrus decurrens, 783
Caltha palustris, 845
Calycanthus, 783
Calycanthus floridus, 783
Calycanthus occidentalis, 783
Camassia, 845
Camellia, 351, 357, 604, 783
Camellia xhiemalis, 783
Camellia japonica, 783
Camellia oliefera, 783
Camellia reticulata, 435, 783

Camellia sasanqua, 783
Camellia sinensis, 783
Campanula carpatica, 845
Campanula isophylla, 845
Campanula lactiflora, 845
Campanula medium, 845
Campanula persicifolia, 845
Campsis, 783–784
Campsis radicans, 365, 783–784
Camptosorus, 290
Camptosorus rhizophyllum, 851
Canna, 845
Capsella bursa-pastoris, 122–123, 124
Capsicum, 141
Capsicum annuum, 451, 845
Caragana, 784
Caragana arborescens, 784
Caragana pygmaea, 784
Carica papaya, 293, 730
Carpinus, 190, 504, 784
Carpinus betulus, 353, 370, 784
Carpinus caroliniana, 194, 784
Carpinus cordata, 784
Carpinus japonica, 784
Carya, 190, 192, 194, 308, 418, 483, 486, 504, 505, 517, 531, 541, 547, 548, 784
Carya aquatica, 784
Carya cordiformis, 784
Carya glabra, 784
Carya illinoinensis, 731, 784
Carya laciniosa, 784
Carya myristicae-formis, 784
Carya ovata, 731, 784
Carya tomentosa, 784
Castanea, 167, 190, 194, 300, 312, 545, 731–732, 784
Castanea crenata, 731
Castanea dentata, 731, 784
Castanea mollissima, 370, 440, 784
Castanea sativa, 731, 784
Casuarina, 784
Casuarina cunninghamiana, 784
Casuarina equisetifolia, 784
Casuarina glauca, 784
Catalpa, 785
Catalpa bignonioides, 785
Catalpa speciosa, 785
Catananche caerulea, 845
Catharanthus roseus, 846
Cattleya, 587, 859
Ceanothus, 235, 785
Ceanothus arboreus, 785
Ceanothus cuneatus, 785
Ceanothus jepsoni, 785
Ceanothus megacarpus, 785
Ceanothus oliganthus, 785
Ceanothus rigidis, 785
Ceanothus thyrsiflorus, 785
Cedrus, 504, 785
Cedrus atlantica, 785
Cedrus deodara, 785
Cedrus libani, 785

Celastrus, 785
Celastrus scandens, 365
Celosia argentea, 846
Celosia spicata, 846
Celtis, 785
Celtis laevigata, 785
Celtis occidentalis, 785
Centaurea, 846
Centaurea cyanus, 846
Centaurea hypoleuca, 846
Centaurea macrocephala, 846
Centaurea montana, 846
Centaurea moschata, 846
Centranthus ruber, 846
Cephalotaxus harringtonia, 785
Cephalotus follicularis, 846
Cerastium tomentosum, 846
Ceratonia siliqua, 732
Ceratostigma, 846
Cercis, 205, 224, 235, 315, 504, 786
Cercis canadensis, 127, 179, 222, 231, 232, 655, 657, 679
Cercis canadensis var. mexicana, 786
Cercis canadensis var. texensis, 786
Chaenomeles, 786
Chaenomeles japonica, 365
Chaenomeles speciosa, 365
Chamaecyparis, 350, 786
Chamaecyparis lawsonia, 616
Chamaecyparis nootkatensis, 232, 441, 786, 789
Chamaecyparis pisifera, 616
Chamaecyparis praecox, 786
Chamaecyparis thyoides, 786
Chamelaucium, 786
Chasmanthe, 576
Cheiranthus cheiri, 846
Chelone, 846
Chenopodium album, 228
Chilopsis linearis, 786
Chionanthus, 259, 347, 354, 786–787
Chionanthus retusus, 316, 786, 787
Chionanthus virginicus, 786, 787
Chionodoxa, 846
Chlerodendrum trichotomum, 365
Chlorophytum, 552, 553
Chlorophytum comosum, 846
Chrysanthemum, 284, 300, 312, 327, 328, 347, 380, 405, 595, 596, 597, 626, 629, 633, 660, 676, 686
Chrysanthemum coccineum, 867
Chrysanthemum frutescens, 848
Chrysanthemum morifolium, 406
Chrysanthemum xmorifolium, 848
Chrysanthemum parthenium, 867
Chrysanthemum superbum, 556, 856
Chrysogonum virginianum, 846
Cichorium, 431
Cimicifuga, 846
Cimicifuga racemosa, 234
Cinnamomum camphora, 370, 787
Citrullis, 145, 174
Citrullus lanatus, 452, 499

Citrullus vulgaris, 499
Citrus, 17, 132–134, 136, 146, 194, 347, 419, 440, 451–452, 486, 507, 513, 534, 541, 548, 595, 617, 629, 650, 663, 666, 672, 681, 732–736, 787
Citrus aurantifolia, 542, 732, 734
Citrus aurantifolia x C. reticulate, 736
Citrus aurantium, 446, 452, 609, 735
Citrus grande, 734
Citrus jambhiri, 735
Citrus limon, 599, 732, 787
Citrus macrophylla, 736
Citrus maxima, 732
Citrus medica, 282, 609, 732, 734
Citrus paradisii, 599, 787
Citrus xparadisii, 732
Citrus paradisii x C. reticulata, 732
Citrus reshni, 735
Citrus reticulata, 732, 787
Citrus reticulata x C. sinensis, 732
Citrus sinensis, 441, 446, 452, 535, 732, 787
Clarkia amoena, 846
Clematis, 347, 380, 787
Cleome spinosa, 846
Clerodendrum, 118
Clethra, 363, 787
Clethra alnifolia, 787
Clivia, 846
Cobaea scandens, 846
Coccoloba, 787
Coccoloba diversifolia, 787
Coccoloba uvifera, 787
Cocus nucifera, 129, 736
Codiaeum variegatum, 318, 846
Coffea, 736
Coffea arabica, 129, 315, 736
Coffea liberica, 736
Coffea robusta, 736
Cola nitida, 736–737
Colchicum autumnale, 846
Coleus, 190, 217, 282, 604
Coleus blumei, 866
Comptonia, 363
Comptonia peregrina, 365
Conospermum mitchellii, 787
Consilida ambigua, 847
Convallaria, 234, 562, 584, 585, 587
Convallaria majalis, 587, 847
Convolvulus, 847
Cooperia, 576
Cordyline, 847
Cordyline australis, 847
Cordyline kaspar, 847
Cordyline pumillo, 847
Cordyline terminalis, 847
Coreopsis, 190, 847
Coreopsis grandiflorum, 847
Coreopsis verticillata, 847
Cornus, 320, 367, 504, 529–530, 787–788
Cornus canadensis, 788
Cornus florida, 596, 787–788
Cornus kousa, 788

Cornus stolonifera, 552, 553, 788
Cortaderia selloana, 847
Corydalis, 847
Corylopsis, 788
Corylopsis pauciflora, 788
Corylus, 167, 190, 194, 238, 239, 537, 539, 541, 653, 737, 788
Corylus americana, 370
Corylus avellana, 305, 788
Corylus colurna, 369
Corylus fargesii, 788
Corylus maxima, 370, 788
Cosmos bipinnatus, 847
Cosmos sulphureus, 847
Cotinus coggygria, 370, 788–789
Cotoneaster, 174, 232, 258, 381, 402, 789
Cotoneaster actifollus, 789
Cotoneaster bullatus, 789
Cotoneaster buxifolius, 789
Cotoneaster divaricatus, 236
Cotton, 190
Crambe maritima, 362
Crassula, 290, 866
Crassula argentea, 370, 847
+Crataegomespilus durdarii, 609
Crataegus, 172, 232, 408, 789
Crataegus coccinea, 789
Crataegus crus-galli, 789
Crataegus laevigata, 789
Crataegus monogyna, 609, 789
Crataegus phaenopyrum, 408
Crepis, 133
Crinum, 575, 847
Crocosmia, 847
Crocus, 562, 577, 847
Crossandra infundibuliformis, 847
Cryptanthus, 844
Cryptomeria japonica, 595, 616, 789
Cucumis, 148, 448, 499
Cucumis melo, 452, 499
Cucumis sativus, 228, 229, 452, 499
Cucurbita, 142, 148, 416, 417, 430, 443, 448, 451, 469, 471, 476, 477, 499, 701
Cucurbita ficifolia, 421, 452, 499
Cucurbita maxima, 452
Cucurbita pepo, 847
Cucurma, 580
Cunninghamia lanceolata, 595
Cuphea, 847
xCupressocyparis leylandii, 789
Cupressus, 789
Cupressus arizonica, 789
Cycas, 789–790
Cycas revoluta, 790
Cyclamen, 190, 233, 562, 584, 847–848
Cyclamen persicum, 582
Cydonia, 347, 419, 445, 447, 541, 549
Cydonia oblonga, 441, 618, 737, 760
Cymbalaria muralis, 848
Cymbidium, 589, 859
Cynara cardunculus, 848
Cynodon dactylon, 553
Cynoglossom amabile, 848

Cypressus macrocarpa, 789
Cypripedium, 859–860
Cystopteris bulbifera, 851
Cytissus, 609, 790

D

Dactylis, 148
Dahlia, 300, 315, 562, 581–584, 629, 848
Daphne, 790
Daphne genkwa, 365
Datura, 232, 666
Datura stramonium, 441
Davidia involucrata, 790
Delonix regia, 790
Delphinium, 92, 380, 848
Dendranthema xgrandiflorum, 848
Dendrobium, 553, 587, 860
Deutzia, 354, 790
Dianthus, 143, 595, 596, 626, 629, 633, 686
Dianthus barbatus, 848
Dianthus caryophyllus, 284, 358, 848
Dianthus chinensis, 848
Dianthus gratinanopolitanus, 848
Dianthus plumarius, 848
Diascia, 849
Dicentra, 365, 849
Dicentra exima, 849
Dicentra spectabilis, 849
Dictamnus albus, 849
Dieffenbachia, 347, 380, 651, 849
Digitalis, 92, 849
Dimorphotheca, 849
Dionaea muscipula, 849
Dioscorea, 595
Dioscorea alta, 580
Dioscorea batatas, 581
Diosma ericoides, 790
Diospyros, 596, 791
Diospyros kaki, 737–738, 791
Diospyros lotus, 738
Diospyros texana, 791
Diospyros virginiana, 738, 791
Disanthus cercidifolius, 791
Dodecatheon maedia, 849
Doronicum, 849
Dorotheathus bellidiformis, 849
Dracaena, 357, 358, 367, 383, 541, 849
Dracaena fragrans, 849
Dracaena marginata, 545
Drosera, 849
Duchesnea indica, 553
Dwarf rhododendron, 328–329
Dyssodia pentacheta, 849
Dyssodia tenuiloba, 849

E

Echevaria, 866
Echinacea, 169, 224, 238, 684, 721, 849–850
Echinops, 850
Echium, 850
Echium candicans, 850
Eleagnus, 367, 791
Eleagnus angustifolia, 791
Eleagnus commutata, 791
Eleagnus pungens, 791
Eleutherococcus sieboldianus, 365
Encephalartos, 789–790
Enkianthus, 791
Ephedra, 122
Epidendrum, 860
Epigaea repens, 791
Epimedium, 850
Epiphyllum, 850
Epipremnum aureum, 347, 356–357, 850, 863
Episcia, 850
Equisetum, 111
Eranthis hyemalis, 850
Eremurus, 569
Eremurus bungei, 850
Erica, 791
Erigeron, 850
Eriobotrya japonica, 194, 441, 738, 791
Eriostemon, 791
Eriostemon myoporoides, 791
Eryngium, 850
Erysimum asperum, 846
Erythronium, 566, 850
Escallonia, 791
Eschscholzia californica, 92, 365, 850
Eucalyptus, 146, 300, 308, 420, 595, 616, 619, 620, 680, 792
Eucalyptus camaldulensis, 792
Eucalyptus cladocalyx, 792
Eucalyptus dives, 792
Eucalyptus ficifolia, 792
Eucalyptus globus, 310, 318
Eucalyptus grandis, 792
Eucalyptus niphophila, 792
Eucalyptus pauciflora, 792
Eucalyptus robusta, 614
Eucalyptus tereticornis, 792
Eucalyptus trabutii, 792
Eucomis, 575, 850
Eugenia, 483
Euonymus, 347, 351, 363, 402, 792
Euonymus alatus, 354, 792
Euonymus atropurpureus, 792
Euonymus europaeus, 792
Euonymus fortunei, 792
Euonymus japonicus, 370
Euphorbia, 850, 866
Euphorbia epithymoides, 850
Euphorbia fulgens, 792
Euphorbia pulcherrima, 331, 792, 866
Euphoria, 542, 595
Euphoria longan, 738
Eustoma grandiflorum, 850
Evolvulus glomeratus, 851
Exacum affine, 851

F

F12/1 cherry, 752, 753, 754
Fagus, 142, 146, 190, 194, 436, 490, 504, 792–793
Fagus americana, 793
Fagus sylvatica, 17, 239, 650, 793
Fallopia baldschuanicum, 370
x*Fatshedera lizei*, 851
Fatsia, 233, 793
Fatsia japonica, 851
Fatsia japonica x Hedera helix, 795
Feijoa sellowiana, 738
Felicia, 851
Festuca, 851
Festuca ovina, 556
Ficus, 298, 300, 345, 347, 380, 489, 541, 542, 595, 596, 793
Ficus aurea, 217
Ficus benjamina, 543, 545
Ficus carica, 365, 543, 738–839
Ficus elastica, 543
Ficus lyrata, 793
Ficus macrophylla, 543
Ficus pumila, 281, 282, 284–287, 289, 316, 614, 615, 616
Filipendula, 851
Forsythia, 327, 345, 347, 354, 365, 367, 541, 662, 793
Fortunella, 732
Fortunella miewa, 787
Fothergillia gardenii, 793
Fragaria, 172, 537, 552, 553, 600, 603, 686, 851
Fragaria xananassa, 739–740
Franklinia alatamaha, 793
Franxinus excelsior, 17, 234
Fraxinus, 232, 490, 793
Fraxinus americana, 793
Fraxinus anomala, 793, 794
Fraxinus excelsior, 436, 794
Fraxinus greggii, 793
Fraxinus ornus, 794
Fraxinus pennsylvanica, 793–794, 820
Freesia, 562, 577, 629, 851
Fritillaria, 577, 851
Fritillaria imperialis, 851
Fritillaria melaegris, 851
Fuchsia, 794
Fuchsia xhybrida, 851
Fuchsia magellanica, 851

G

Gaillardia, 851
Galanthus, 852
Galanthus elwesii, 852
Galanthus nivalis, 852
Galtonia candicans, 852
Garcinia, 132, 134
Garcinia mangostana, 740
Gardenia jasminoides, 794
Garrya elliptica, 794
Gasteria, 866
Gaura lindheimeri, 852
Gaylussacia, 172
Gazania, 852
Genista, 794
Gentiana, 852

Geranium, 347, 362, 365, 607, 626, 629, 660, 852
Gerbera, 686
Gerbera jamosonii, 852
Geum, 852
Ginkgo, 122, 504
Ginkgo biloba, 794
Ginseng, 233
Gladiolus, 562, 577, 578, 579, 629, 676, 852
Gleditsia, 223
Gleditsia triacanthos, 129, 794
Gloriosa, 852
Gloxinia, 562
Glycine, 147
Gomphrena globosa, 852
Gongora quinquenervis, 589
Gordonia, 794–795
Gossypium, 132, 141
Grevillea, 297, 795
Grevillea asplenifolia, 795
Grevillea johnsonii, 795
Grevillea juniperina, 795
Grevillea robusta, 795
Grevillea scapigera, 795
Gunnera, 852
Guzmania, 844
Gymnocladus dioicus, 653, 664, 795
Gypsophila, 686, 852
Gypsophila elegans, 852
Gypsophila paniculata, 852

H

Haemanthus, 576
Halesia, 795
Halesia carolina, 795
Hamamelis, 132, 320, 404, 504, 795
Hamamelis xintermedia, 795
Hamamelis japonica, 795
Hamamelis mollis, 795
Hamamelis vernalis, 444, 795
Hamamelis virginiana, 795
Haworthia, 866
Hebe, 795
Hedera, 347, 490, 616, 617, 619
Hedera canariensis, 795
Hedera helix, 17, 18, 19, 285, 289, 299, 300, 308, 356, 614, 795, 851
Hedychium, 853
Helenium autumnale, 853
Helianthemum nummularium, 853
Helianthus, 145
Helianthus annuus, 305, 441, 853
Helianthus decapetalus, 853
Helianthus tuberosus, 579
Helichrysum bracteatum, 853
Helichrysum petiolare, 853
Heliconia, 853
Heliopsis, 853
Heliotropium, 853
Helleborus, 853
Helleborus lividus, 853
Hemerocallis, 556, 853
Hepatica, 853

Hepatica acutiloba, 234
Heptacodium miconioides, 795–796
Hesperis matronalis, 853
Heteromeles arbutifolia, 796
Heuchera, 599, 700, 853
Hevea, 422, 489–490, 663
Hevea brasilensis, 309
Hibiscus, 352, 380, 428, 796, 853
Hibiscus rosa-sinensis, 300, 370, 796
Hibiscus syriacus, 796
Hippeastrum, 564, 569, 576, 577, 629, 853
Hollyhock, 190
Hordeum, 141
Hosta, 552, 554, 556, 557, 610, 665, 686, 700, 722, 853–854
Houttuynia cordata, 854
Hovenia dulcis, 796
Hoya, 854, 866
Hummulus lupulus, 854
Hunnemannia fumariifolia, 854
Hyacinthus, 562, 566, 567, 570, 575, 576, 577, 854
Hydrangea, 282, 300–304, 365, 366, 796–797
Hydrangea petiolaris, 796
Hydrangea quercifolia, 355, 796–797
Hymenocallis, 576, 854
Hypericum, 797, 854
Hypericum calycinum, 365
Hypoestes phyllostachya, 854

I

Iberis, 854
Ilex, 142, 234, 239, 315, 347, 351, 402, 797
Ilex amelanchier, 797
Ilex aquifolium, 797
Ilex cassine, 797
Ilex cornuta, 797
Ilex crenata, 797
Ilex decidua, 797
Ilex glabra, 797
Ilex myrtifolia, 797
Ilex opaca, 797
Ilex verticilliata, 797
Ilex vomitoria, 797
Illicium, 797
Impatiens, 163, 183, 189, 268, 330, 701, 854
Impatiens balsamina, 854
Impatiens hawker, 854
Impatiens walleriana, 854
Incarvillea, 854
Ipomoea, 854
Ipomoea batatas, 581, 584, 595, 854
Iresine, 854
Iris, 205, 224, 552, 562, 566, 570, 574, 577, 584, 585, 629, 668, 676, 854–855
Ixia, 855
Ixora, 855
Ixora acuminata, 855

J

Jacaranda mimosifolia, 797–798
Jamesia americana, 798
Japanese holly, 320
Jasminum, 282, 798
Jasminum nudiflorum, 798
Jasminum sambac, 798
Jeffersonia, 118, 234
Jerusalem sage, 580
Juglans, 142, 146, 190, 192, 194, 236, 435, 483, 504, 528, 541, 549, 616, 619
Juglans cinerea, 740–741
Juglans hindsii, 158, 435, 438, 489, 741
Juglans hindsii xJ. regia, 150, 158, 742
Juglans nigra, 741, 742
Juglans regia, 159, 435, 438, 446, 489, 741–742
Juniperus, 172, 174, 317, 322, 328, 347, 350, 352, 405, 616, 798
Juniperus chinensis, 351, 798
Juniperus excelsa, 351
Juniperus horizontalis, 798
Juniperus procumbens, 798
Juniperus virginiana, 798, 799
Justicia, 855
Justicia brandegeeana, 855
Justicia carnea, 855

K

Kalanchoe, 217, 686, 866
Kalanchoe blossfeldiana, 855
Kalanchoe diagremontiana, 855
Kalmia, 682, 701, 799
Kalmia latifolia, 369, 799
Kirengeshoma palmata, 855
Knautia, 855
Kniphofia, 855
Kniphofia tritoma, 855
Koelreuteria, 235, 799
Koelreuteria paniculata, 365

L

+*Laburnocytissus*, 609
Laburnum, 609, 799
Lachenalia, 576, 577
Lachinelia, 855
Laelia, 587, 859
Laelia anceps, 589
Lagenaria siceraria, 499
Lagerstroemia, 354
Lagerstroemia indica, 799
Lamiastrum galeobdolan, 855
Lamium maculatum, 855
Lantana camara, 855
Lantana sellowiana, 855
Larix, 149, 305, 504, 617, 799–800
Larix decidua, 17
Larix decidua x L. leptolepsis, 150
Lathyrus latifolius, 855
Lathyrus odoratus, 855
Laurus nobilis, 800
Lavandula, 381, 855

Lavatera trimestris, 855
Lavendula, 224, 381
Lens, 204
Leonotis leonurus, 855
Leontopodium alpinum, 855–856
Leptospermum, 800
Leptospermum scaparium, 800
Lespedeza, 856
Lettuce, 123, 190, 224, 227, 228, 229, 235, 239, 681
Leucanthemum xsuperbum, 856
Leucocorne ixioides, 856
Leucojum, 701, 856
Leucophyllum, 800
Leucophyllum candidum, 800
Leucophyllum frutescens, 800
Leucospermum, 800
Leucospermum conocarpodendron x L. cuneiforme, 800
Leucothoe, 800
Leucothoe axillaris, 800
Leucothoe fontaneisana, 800
Lewisia cotyledon, 856
Liatris, 562, 856
Liatris spicata, 856
Ligularia, 856
Ligustrum, 315, 354, 800
Ligustrum amurense, 820
Ligustrum japonicum, 401, 800
Ligustrum ovalifolium, 820
Lilium, 143, 145, 234, 290, 562, 564, 567, 568, 570, 572, 573, 577, 856
Lilium amabile, 856
Lilium auratum, 856
Lilium bolanderi, 856
Lilium bulbiferum, 572
Lilium canadense, 572, 856
Lilium candidum, 292, 572, 856
Lilium chalcedonicum, 572
Lilium concolor, 572, 856
Lilium hansonii, 572
Lilium henryi, 572, 856
Lilium hollandicum, 565
Lilium longiflorum, 292, 567, 568, 569, 572, 856
Lilium maculatum, 572
Lilium martagon, 856
Lilium pardalinum, 572
Lilium parryi, 572
Lilium parvum, 856
Lilium regale, 572, 856
Lilium sargentiae, 572
Lilium speciosum, 856
Lilium sulphureum, 572
Lilium superbum, 572
Lilium testaceum, 572
Lilium tigrinum, 572, 856
Limonium, 856
Limonium latifolia, 856
Limonium sinuatum, 856
Linaria, 856–857
Lindera, 800

Lindera benzoin, 800
Lindera obtusiloba, 800
Linum, 141, 224, 668, 857
Liquidambar, 159, 504, 658, 675
Liquidambar formosana, 800
Liquidambar styraciflua, 365, 800
Liriodendron, 504
Liriodendron tulipifera, 801
Liriope, 365, 366, 556, 557, 857
Lisiathus, 850
Litchi chinensis, 742
Lithodora diffusa, 857
Lithops, 857
Lobelia cardinalis, 857
Lobelia erinus, 857
Lobelia siphilitica, 857
Lobularia maritima, 857
Lonicera, 801
Lonicera hirsute, 801
Lonicera oblongifolia, 801
Lonicera tatarica, 801
Loropetalum chinense, 801
Lotus, 145
Lunaria annua, 857
Lunaria rediviva, 857
Lupinus, 857
Lupinus texensis, 857
Lychnis, 857
Lycopersicon, 441, 499, 599
Lycopersicon esculentum, 441, 499, 611
Lycopodium, 111
Lycoris, 576, 857
Lysimachia, 857
Lythrum salicaria, 857

M

Macadamia, 507, 742–743
Macleaya cordata, 857
Maclura pomifera, 801
Magnolia, 320, 347, 352, 354, 373, 374, 404, 616, 801–802
Magnolia acuminata, 802
Magnolia grandiflora, 544, 801, 802
Magnolia kobus, 802
Magnolia xsoulangiana, 801, 802
Magnolia sprengeri, 802
Magnolia stellata, 801
Magnolia virginiana, 802
Mahonia, 136, 802
Mahonia aquifolia, 802
Mahonia bealei, 802
Mahonia japonica, 802
Mahonia lomarifolia, 802
Mahonia nervosa, 802
Mahonia pinnata, 802
Mahonia repens, 802
Mahonia wagneri, 802
Malus, 17, 133, 135, 143, 146, 236, 237, 292, 296, 308, 315, 347, 354, 365, 422, 423, 432, 438, 440, 441, 442, 446, 452, 453, 454, 456, 457, 483, 505, 507, 517, 525, 537, 540, 541, 547–550, 599, 616, 618, 803
Malus baccata, 744, 803
Malus coronaria, 803
Malus xdomestica, 370, 743–746, 803
Malus xeleyi, 803
Malus florentina, 803
Malus floribunda, 803
Malus hupehensis, 618, 803
Malus ioensis, 803
Malus sargentii, 803
Malus sikkimensis, 803
Malus sylvestris, 370
Malus toringoides, 803
Malva alcea, 857
Malva moschata, 857
Malva rotundifolia, 191
Mammillaria (cactus), 844
Mandevilla, 803
Mangifera, 132, 133, 134, 146, 595
Mangifera indica, 370, 435, 537, 746
Manihot esculenta, 581
Manikara zapota, 746–747
Maranta, 669
Maranta leuconeura, 857
Matteuccia (fern), 851
Matthiola incana, 857–858, 866
Matthiola longipetala bicornis, 857–858
Mazus reptans, 858
Meconopsis, 858
Melaleuca, 803
Melampodium paludosum, 858
Mentha, 552, 553
Mertensia, 858
Mertensia pulmonarioides, 858
Mertensia virginiana, 858
Mesembryanthemum, 866
Mesembryanthemum criniflorum, 849
Mespilus germanica, 609
Metasequoia glyptostroboides, 803
Metilotus alba, 441
Microlepia, 851
Miltonia, 587, 859
Miscanthus sinensis, 858
Molluccella laevis, 858
Monarda didyma, 858
Monstera, 542
Monstera deliciosa, 858
Morus, 541, 747, 803
Morus alba, 803
Morus nigra, 803
Musa, 587, 589, 595, 596, 747
Muscari, 570, 576, 577, 858
Mussaenda, 804
Mussaenda erythrophylla, 804, 858
Myosotis scorpioides, 858
Myosotis sylvatica, 858
Myrica, 804
Myrica cerifera, 804
Myrica pennsylvanica, 365, 804
Myrtus, 804
Myrtus communis, 804

N

Nandina domestica, 584
Narcissus, 562, 564–567, 569–570, 571–572, 576, 577, 858
Nelumbo lutea, 858
Nelumbo nucifera, 191, 858
Nemesia strumosa, 858
Nemophila, 235, 858
Neoregilia (bromeliad), 844
Nepenthes, 859
Nepeta, 859
Nephelium, 542, 595
Nephelium lappaceum, 747
Nephrolepis, 553, 685, 722, 851
Nephrolepis tuberosa, 851
Nerine, 576, 577, 629, 859
Nerium oleander, 804
Nicotiana, 150, 151, 206, 297, 305, 859
Nicotiana tabacum, 441, 667
Nierembergia, 859
Nigelia, 217
Nigella damascena, 859
Nolana paradoxa, 859
Nymphaea, 859
Nyssa aquatica, 804
Nyssa ogeche, 804
Nyssa sylvatica, 194, 804

O

Ocimum basilicum, 859
Odontoglossum, 587, 859
Oenothera, 859
Olea europaea, 317, 455, 747–748, 804–805
Olea oblonga, 804
Oncidium, 859
Oncidium longifolium, 589
Ophiopogon japonicus, 859
Opuntia (cactus), 748, 844, 859, 850
Opuntia ficus-indica, 359
Origanum, 860
Ornithogalum, 569, 577, 860
Oryza, 141
Osmanthus, 805
Osmanthus fragrans, 805
Osmanthus heterophyllus, 805
Osmunda (fern), 851
Osteospermum, 860
Oxalis, 860
Oxydendrum arboreum, 805
Oxypetalum caerulem, 868

P

Pachysandra terminalis, 805
Paeonia, 860–861
Paeonia arborea, 860
Paeonia hybrida, 860
Paeonia lactiflora, 805, 861
Paeonia suffruticosa, 805, 860, 861
Paeonia tenuifolia, 860
Pancratium, 576
Pandanus utilis, 281
Papaver nudicaule, 861
Papaver orientale, 361, 365, 861
Papaver rhoeas, 861
Paphiopedilum, 859
Parrotia persica, 806
Parthenium, 133
Parthenocissus, 806
Parthenocissus quinquefolia, 806
Parthenocissus tricuspidata, 806
Paspalum notatum, 134, 146, 148
Passiflora, 663
Passiflora xalatocaerulea, 806
Passiflora edulis, 748–749
Passiflora tricuspsis, 806
Paulownia, 226
Paxistima myrsinites, 806
Pelargonium, 633, 681, 861
Pelargonium xdomesticum, 861
Pelargonium xhortorum, 861
Pelargonium peltatum, 861
Pennisetum, 133, 861
Pennisetum ciliare, 146, 148
Pennisetum glaucus, 148
Penstemon, 861
Penstemon fruticosus, 806
Pentas lanceolata, 861
Peperomia, 292, 861
Pereskia aculeata, 844
Pericallis xhybrida, 861–862
Persea, 194, 294, 321
Persea americana, 370, 662, 749–750
Persicaria, 862
Persicaria affine, 862
Persicaria amplexicaulis, 862
Persicaria capitatum, 862
Persicaria virginiana, 862
Pervoskia atriplicifolia, 862
Petunia, 143, 148, 156, 190, 206, 268, 329, 330, 604, 701
Petunia xhybrida, 862
Phacelia, 217
Phaius, 859
Phalaenopsis, 859
Phaseolus, 141
Philadelphus, 806
Philodendron, 862
Philodendron oxycardium, 543
Phlomis russeliana, 862
Phlox, 217, 365, 862
Phlox divaricata, 862
Phlox drummondii, 862
Phlox paniculata, 862
Phlox subulata, 862
Phoenix canariensis, 806
Phoenix dactylifera, 750–751, 806
Photina arbutifolia. See Heteromeles
Photinia, 806
Photinia xfraseri, 806
Photinia glabra, 806
Photinia serrulata, 806
Physalis alkekengi, 862
Physostegia virginiana, 862
Picea, 350, 420, 504, 663, 807
Picea abies, 17, 149, 314, 807
Picea alba, 149
Picea engelmannii, 807
Picea glauca, 230, 232, 807
Picea omorika, 807
Picea pungens, 159, 421, 438, 807
Picea sitchensis, 17, 314, 370, 427, 436, 448
Pieris, 807
Pieris floribunda, 807
Pieris japonica, 807
Pileostegia viburnoides, 807
Pinguicula, 862
Pinus, 112, 113, 126, 149, 164, 172, 305, 308, 350, 370, 383, 425, 490, 504, 617, 807–808
Pinus aristata, 17, 807
Pinus banksiana, 172, 807
Pinus canariensis, 807
Pinus caribaea, 149, 368, 420, 807
Pinus cembra, 807, 808
Pinus clausa, 807
Pinus contorta, 302–304, 306, 807
Pinus coulteri, 807
Pinus densiflora, 808
Pinus echinata, 149, 808
Pinus edulis, 807
Pinus elliottii, 370, 420, 808
Pinus halepensis, 807
Pinus jeffreyi, 807
Pinus latifolia, 807
Pinus monticola, 17, 807
Pinus mugho, 149
Pinus mugo, 807
Pinus murrayana, 149
Pinus nigra, 807
Pinus palustris, 807
Pinus pinaster, 807
Pinus ponderosa, 149, 807
Pinus pungens, 807
Pinus radiata, 283, 289, 308, 420, 442, 549, 551, 595, 596, 617, 807–808
Pinus rigida x P. taeda, 150
Pinus roxburghii, 807
Pinus strobus, 149, 370, 808
Pinus sylvestris, 17, 149, 228, 289, 308, 807, 808
Pinus thunbergiana, 808
Pinus thunbergii, 807
Pinus virginiana, 807
Pinus wallichiana, 807
Pistacia atlantica x P. integerrina, 751
Pistacia integerrima, 751
Pistacia khinjuk, 751
Pistacia terebinthus, 751
Pistacia vera, 370, 663, 751
Pistacia, 142, 318, 517
Pistacia atlantica, 808
Pisum, 141
Pittosporum, 347, 351, 808
Pittosporum tenuifolium, 808
Pittosporum tobira, 366, 808
Platanus, 808

Platanus xacerifolia, 808
Platanus xhispanica, 808
Platanus occidentalis, 370
Platycladus orientalis, 441, 821, 822
Platycodon grandiflorus, 862
Playcerium, 851
Plectranthus, 862
Plumbago, 365, 380, 809
Plumeria, 809
Poa, 133, 146
Poa bulbosa, 133
Poa pratensis, 134, 148
Podocarpus, 315, 620, 809
Podocarpus marophyllus, 809
Polemonium, 862
Polianthes tuberosa, 863
Poliomintha longiflora, 863
Polygonatum, 234, 552, 863
Polygonum, 862
Polystichum setiferum, 851
Poncirus trifoliata, 132, 441, 732, 735
Populus, 146, 189, 282, 300, 345, 595, 809
Populus alba, 365
Populus tremula, 365
Populus tremuloides, 365, 809
Populus trichocarpa x P. deltoides, 150
Portulaca, 224, 863, 866
Potentilla, 345, 809, 863
Potentilla fruticosa, 809
Potentilla nepalensis, 863
Primula, 144, 217, 863
Primula malacoides, 863
Primula obconica, 863
Primula xpolyantha, 863
Primula sinensis, 863
Primula vulgaris, 863
Prosopis, 809
Prosopis alba, 809
Prosopis chilensis, 809
Prosopis flexuosa, 809
Protea, 809
Prunus, 17, 117, 141, 143, 236, 240, 320, 345, 347, 354, 381, 404, 419, 422, 446, 447, 449, 452, 455, 456–457, 486, 504, 507, 513, 517–519, 595, 599, 605, 626, 631, 663, 669, 722, 752
Prunus amygdalus, 440, 756
Prunus angustifolia, 758
Prunus armeniaca, 440, 752, 754, 756, 757
Prunus avium, 297, 451, 452, 603, 752, 753, 810
Prunus avium x P. mahaleb, 754
Prunus avium x P. pseudocerasus, 753–754
Prunus besseyi, 756, 758
Prunus calleryana, 760
Prunus campanulata, 809, 810
Prunus canescens, 754
Prunus caroliniana, 809, 810
Prunus cerasifera, 446, 754, 756, 757, 809, 810

Prunus cerasifera x P. munsoniana, 440, 757
Prunus cerasifera x P. salicina x 'Yunnan' peach, 758
Prunus cerasus, 752, 753–754, 809
Prunus dawyckansis, 754
Prunus domestica, 370, 440, 752, 756–758
Prunus domestica x P. spinosa, 756
Prunus dropmoreana, 810
Prunus dulcis, 143, 442, 446, 517, 518, 519, 545, 602, 631, 663, 752, 754–755, 757, 809
Prunus dulcis x P. persica, 755
Prunus glandulosa, 365, 809
Prunus hybrida, 452
Prunus insititia, 756, 758
Prunus laurocerasus, 809, 810
Prunus lusitanica, 809
Prunus mahaleb, 451, 753, 754
Prunus mexicana, 809, 810
Prunus mume, 809, 810
Prunus myrobalan x P. myrobalan x P. persica, 758
Prunus persica, 440, 608, 752, 754–757
Prunus persica x P. davidiana, 754, 755, 756, 810
Prunus salicina, 440, 752, 758
Prunus salicina x P. americana, 757
Prunus sargentii, 809
Prunus serrulata, 424, 809, 810
Prunus sieboldi, 809
Prunus subhirtella, 809
Prunus tenella, 809
Prunus tomentosa, 757
Prunus ussuriensis, 760
Prunus virginiana, 809
Prunus yedoensis, 809
Pseudotsuga, 663
Pseudotsuga menziesii, 17, 149, 294, 308, 405, 420, 425, 440, 449, 810
Psidium cattleianum, 758
Psidium guajava, 758
Psidium spp., 758
Psilotum, 111
Pteris, 851
Pulmonaria, 700, 863
Pulsatilla, 863
Punica granatum, 759
Puschkinia scilloides, 863
Puya, 844
Pyracantha, 367, 381, 810
Pyrus, 17, 146, 353, 354, 599, 600, 616, 759–761
Pyrus betulaefolia, 441, 759–760
Pyrus calleryana, 365, 760, 810–811
Pyrus communis, 441, 759–760
Pyrus kwakamii, 811
Pyrus pyrifolia, 441, 759–761
Pyrus serotina, 760–761
Pyus, 300, 347, 353, 419, 422, 432, 434, 445, 447, 451, 483, 505, 517, 595, 597, 619, 631

Q

Quercus, 129, 167, 186, 190, 192, 194, 308, 370, 404, 408, 415, 420, 425, 452, 541, 675, 676, 811–812
Quercus agrifolia, 811
Quercus alba, 811
Quercus arizonica, 811
Quercus bicolor, 811, 812
Quercus chrysolepis, 811
Quercus douglassii, 811
Quercus garryana, 811
Quercus laurifolia, 811
Quercus lobata, 811
Quercus macrocarpa, 811, 812
Quercus montana, 811
Quercus nigra, 313, 811
Quercus palustris, 812
Quercus petraea, 811
Quercus phillyreoides, 812
Quercus prinus, 811
Quercus robur, 17, 353
Quercus rubra, 440, 811, 812
Quercus shumardii, 811, 812
Quercus stellata, 811
Quercus suber, 811
Quercus turbinella, 811
Quercus virginiana, 811–812

R

Ranunculus, 233, 569, 863
Ranunculus asiaticus, 863
Raphiolepis indica, 399, 812–813
Ratibida, 863
Renanthera, 859
Reseda odorata, 863
Rhamnus, 813
Rhamnus frangula, 813
Rhaphiolepis indica, 812–813
Rhapidophyllum hystrix, 805
Rhipsalidopsis, 863
Rhizophora mangle, 131
Rhododendron, 233, 314–315, 347, 351, 357, 370, 373, 402, 404, 474, 483, 507, 541, 551, 612, 682, 686, 700, 712, 722, 813–814
Rhododendron catawbiense, 318, 320, 322, 813
Rhododendron chapmanii, 813
Rhododendron xlimbatum, 611
Rhododendron maximum, 813
Rhododendron ponticum, 483, 813
Rhus, 235, 347, 814
Rhus aromatica, 814
Rhus copallina, 365, 814
Rhus glabra, 365
Rhus typhina, 365, 814
Ribes, 282, 761
Ribes grossularia, 761
Ribes nigrum, 761
Ribes odoratum, 761
Ribes sanguineum, 814
Ribes sativum, 761

Ribes uvacrispa, 761
Ricinus, 118
Ricinus communis, 863
Rieger begonia, 291, 360
Robina hispida, 365
Robinia, 221
Robinia pseudoacacia, 365, 524, 814
Rodgersia, 863
Romneya, 233
Rosa, 17, 133, 172, 258, 284, 347, 370, 434, 440, 507, 508, 516, 522, 527, 528, 529, 541, 595, 597, 686, 814–816
Rosa blanda, 365, 814
Rosa canina, 814, 816
Rosa chinensis, 816
Rosa 'Dr. Huey', 816
Rosa hugonis, 814
Rosa IXL, 816
Rosa multiflora, 316, 345, 348, 349, 523, 814, 815–816
Rosa Multiflore de la Grifferaie, 816
Rosa nitida, 365
Rosa xnoisettiana, 816
Rosa odorata, 816
Rosa rugosa, 814
Rosa virginiana, 365
Rosmarinus officinalis, 816
Rubus, 133, 347, 357, 365, 537, 551, 552, 612, 686, 761–762
Rubus idaeus, 370
Rubus occidentalis, 357, 761
Rubus occidentalis x R. idaeus, 761
Rubus strigosus x R. idaeus, 761
Rudbeckia, 133, 864
Rumex sangineus, 864

S

Saccharum, 595
Saintpaulia, 217, 290, 291, 347, 360, 610–611, 664, 674, 686, 717
Saintpaulia ionantha, 685, 864
Salix, 189, 282, 298, 595, 817
Salix caprea, 817
Salix integra, 817
Salix purpurea, 817
Salix tetrasperma, 300
Salpiglossis, 190, 664
Salpiglossis sinuata, 864
Salvia, 864
Salvia greggii, 864
Salvia officinalis, 864
Salvia splendens, 864
Sambucus, 345, 817
Sandersonia, 170
Sandersonia aurantiaca, 864
Sanguinaria, 234
Sanguinaria canadensis, 864
Sansevieria, 290, 347, 358, 361, 585, 586
Sansevieria 'Hahnii', 864
Sansevieria trifasciata, 864

Santolina chamaecyparissus, 864
Santolina incana, 864
Sanvitalia procumbens, 864
Sapindus crummondii, 817
Saponaria ocymoides, 864
Saponaria officinalis, 864
Sarcococca, 817
Sarcococca hookerana, 817
Sarcococca ruscifolia, 817
Sarracenia, 864
Sassafras albidum, 365
Saxifraga, 864–865
Saxifraga sarmentosa, 552
Saxifraga stolonifera, 865
Scabiosa, 865
Scaevola aemula, 865
Schefflera, 544, 545
Schefflera actinophylla, 865
Schefflera arboricola, 356, 370, 865
Schizanthus, 865
Schizophragma hydrangeoides, 817
Schlumbergera truncata, 865
Sciadopitys verticillata, 817
Scilla, 576, 865
Scutellaria, 865
Sedum, 289, 290, 292, 347, 430, 448, 865, 866
Sedum telephoides, 450
Selaginella, 111
Sempervirens, 865, 866
Senecio, 365
Senecio cineraria, 865
Senecio cruentus, 861–862
Sequoiadendron, 172, 617, 618
Sequoiadendron giganteum, 17, 818
Sequoia sempervirens, 17, 297, 619, 663, 818
Shepherdia, 818
Shepherdia canadensis, 818
Shepherdia rotundifolia, 818
Sidalcea malviflora, 865
Silene, 865
Sinningia speciosa, 865
Skimmia, 818
Skimmia japonica, 818
Skimmia reevesiana, 818
Solanum, 133, 141, 415, 865
Solanum crispeum, 865
Solanum integrifolium, 499
Solanum jasminoides, 865
Solanum melongena, 499
Solanum nigrum, 441, 611
Solanum pennellii, 450
Solanum pseudocapsicum, 865
Solanum rantonnettii, 865
Solanum torvum, 499
Solanum tuberosum, 441, 579, 580, 595, 600, 632, 633, 635
Solenostemon scutellarioides, 866
Solidago, 866
Solidago solidaster, 866

Sophora, 221, 818
Sophora japonica, 818–819
Sophora microphylla, 818
Sophora secundiflora, 818, 819
Sorbus, 818
Sorbus aria, 819
Sorbus aucuparia, 819
Sorbus cuspidata, 819
Sorbus domestica, 819
Sorbus glabrescens, 818
Sorbus hybrida, 819
Sorghum halepense, 586, 587
Spathiphyllum, 680, 866
Spigelia marilandica, 866
Spiraea, 345, 354, 363, 819
Spiraea thunbergi, 819
Spiraea xvanhouttei, 819
Sprekelia, 576
Stachys, 553, 866
Stachys byzantina, 866
Stachys grandiflora, 866
Stangeria eriopus, 790
Stapelia, 866
Sternbergia lutea, 866
Stewartia, 356, 819
Stewartia malacondendron, 819
Stewartia pseudocamellia, 819
Stewartia pseudocamillia, 401
Stokesia, 347
Stokesia laevis, 365, 866
Strelitzia, 92
Strelitzia reginae, 587, 588, 866
Streptocarpus, 360
Streptocarpus xhybridus, 866
Striga, 239
Styphnolobium japonicum, 365, 818–819
Succulents, 866
Sutera, 867
Swietenia mahagoni, 819–820
Symphoricarpos, 820
Symphoricarpos xchenaulti, 820
Symphoricarpos xchenaultii, 365
Symphytum, 867
Syngonium, 686
Syngonium podophyllum, 867
Syringa, 205, 224, 316, 347, 354, 404, 820
Syringa chinensis, 820
Syringa henryi, 820
Syringa josikaea, 820
Syringa patula, 820
Syringa villosa, 820
Syringa vulgaris, 312, 313, 365, 367, 370, 820

T

Tabebuia argentea, 820
Tagetes, 148, 867
Tamarix, 133, 820
Tamarix aphylla, 820
Tamarix pentandra, 820

Tamarix ramosissima, 820
Tanacetum coccineum, 867
Tanacetum parthenium, 867
Tanacetum ptarmiciflorum, 867
Taraxacum, 133
Taxodium distichum, 820–821
Taxodium mucronatum, 820–821
Taxus, 234, 317, 347, 350, 352, 405, 821
Taxus cuspidata, 315, 821
Taxus xmedia, 821
Telopea speciosissima, 821
Ternstroemia gymnanthera, 821
Teucrium, 867
Thalictrum, 867
Theobroma cacao, 762
Thermopsis caroliniana, 867
Thuja, 347, 350, 352, 373, 405, 821
Thuja x'Green Giant,' 822
Thuja koraiensis, 821
Thuja occidentalis, 401, 821–822
Thuja orientalis, 821, 822
Thuja plicata, 821, 822
Thunbergia, 867
Thunbergia alata, 867
Thunbergia grandiflora, 867
Thunbergia gregorii, 867
Thunbergia mysorensis, 867
Thymus, 867
Tiarella cordifolia, 867
Tigridia pavonia, 867
Tilia, 235, 370, 432, 822
Tilia americana, 822
Tilia cordata, 353, 370, 421, 822
Tilia platyphyllos, 432
Tilia tomentosa, 370
Tillandsia, 844
Tithonia rotundifolia, 867
Tobacum, 599
Tolmiea, 290, 361, 362
Tolmiea menziesii, 867
Torenia fournieri, 867
Tovara, 862
Trachelospermum asiaticum, 822
Trachelospermum jasminoides, 822
Trachymene coerulea, 867
Tradescantia, 867–868
Tricyrtis, 868
Trifolium, 145, 669
Trifolium pratense, 145
Trifolium repens, 145
Trillum, 868
Tripterospermum, 312, 329, 339
Triteleia, 844
Triticum, 141
Trollius, 868
Tropaeolum majus, 868
Tsuga, 347, 350, 425, 822–823

Tsuga canadensis, 823
Tsuga heterophylla, 17
Tulipa, 562, 563, 565, 566, 567, 569–571, 577, 868
Tulipa gesneriana, 132
Tweedia caerulem, 868

U

Ulmus, 189, 353, 383, 545, 823
Ulmus alata, 444
Ulmus capinifolia xU. parvifolia, 823
Ulmus carpinifolia, 365
Ulmus hollandica, 823
Ulmus parvifolia, 354, 823
Ulmus wilsoniana, 823
Ungnadia speciosa, 823
Urceolina, 576

V

Vaccinium, 587, 762–763, 823
Vaccinium angustifolium, 584, 762
Vaccinium arboreum, 416
Vaccinium ashei, 415, 416, 762, 823
Vaccinium atrococcum, 315
Vaccinium austral, 762–763
Vaccinium corymbosum, 315, 762–763, 823
Vaccinium deliciosum, 823–824
Vaccinium itis-idaea, 823
Vaccinium macrocarpon, 762, 763
Vaccinium membranaceum, 823–824
Vaccinium ovalifolium, 824
Vaccinium ovatum, 823
Vanda, 859
Vauquelinia, 824
Vauquelinia california, 824
Verbascum, 868
Verbascum blattaria, 191
Verbascum thapsus, 191
Verbena, 868
Veronica, 868
Viburnum, 174, 234, 347, 363, 367, 381, 404, 541, 824
Viburnum burkwoodii, 824
Viburnum carlesii, 824
Viburnum dentatum, 824
Viburnum lantana, 824
Viburnum opulus, 824
Viburnum plicatum, 824
Viburnum rhytidophylloides, 824
Viburnum xrhytidophyllum, 824
Viburnum sargentii, 824
Viburnum sieboldi, 824
Viburnum trilobum, 824
Vicia faba, 128
Vigna, 282, 284, 296, 300
Vinca, 190

Vinca major, 868
Vinca minor, 868
Viola, 141, 148, 190, 224, 268
Viola cornuta, 868
Viola odorata, 869
Viola tricolor, 869
Viola xwittrockiana, 869
Viscum album, 217
Vitis, 17, 135, 419, 434, 437, 474, 483, 496, 541, 543, 595, 598, 763–765
Vitis berlandieri x V. riparia, 765
Vitis berlandieri x V. rupestris, 765
Vitis champinii, 765
Vitis labruska, 763
Vitis riparia x V. rupestris, 765
Vitis rotundifolia, 537, 763
Vitis rupestris, 765
Vitis vinifera, 763, 764
Vitis vinifera xrotundifolia, 765
Vitis vinifera xrupestris, 765
Vriesea (bromeliad), 844

W

Watsonia, 564
Weigela, 347, 354, 825
Wisteria, 616, 825
Wisteria sinensis, 825
Woodwardia, 851

X

Xanthium pennsylvanicum, 219
Xylosma congestum, 825

Y

Yucca, 866, 869
Yucca elephantipes, 869

Z

Zamia, 789–790
Zamia floridana, 790
Zamia furfuracea, 790
Zantedeschia, 869
Zea, 142, 167
Zea mays, 123, 135, 148, 155
Zebrina pendula, 869
Zelkova serrata, 421
Zephyranthes, 869
Zingiber, 585
Zingiber officinale, 586
Zinnia elegans, 869
Zizania, 190
Zizyphus jujuba, 765
Zygocactus truncatus, 844

plant index: common names

A

Abelia, 315, 400, 774
Abutilon, 775
Acacia, 221, 616, 663, 775
African daisy, 842
African marigold, 849
African violet, 217, 290, 291, 347, 360, 610–611, 664, 674, 685, 686, 717, 864
Agarita, 782
Agave, 133, 841, 866
Ailanthus, 778
Albizzia, 778
Alder, 146
Alemow, 736
Alfalfa, 145, 148, 152, 190
Algerian ivy, 795
Allamanda, 778
Allegheny serviceberry, 778
Allspice, 783
Almond, 143, 440, 442, 446, 517, 518, 519, 545, 602, 631, 663, 752, 754–757, 809
Aloe, 841, 866
Alyssum, 190, 217
Amaryllis, 564, 569, 576, 577, 629, 853
American alumroot, 853
American arborvitae, 401, 821–822
American beech, 793
American bittersweet, 365
American chestnut, 731, 784
American cranberry, 762, 763
American gooseberry, 761
American holly, 797
American hornbeam, 194, 784
American hybrid grape, 764
American lotus, 858
American spicebush, 800
American sweet gum, 365, 800
Amethyst flower, 844
Amur privet, 820
Anacho orchid tree, 781
Anemone, 233, 234, 562, 569, 842
Angelica tree, 779
Anise, 797
Anise hyssop, 841
Annual phlox, 862
Anthurium, 842
Apple, 17, 133, 135, 143, 146, 236, 237, 292, 296, 308, 347, 354, 365, 370, 422, 423, 432, 438, 440, 441, 442, 452, 453, 454, 456, 457, 483, 505, 507, 537, 540, 541, 547–550, 595, 597, 599, 616, 618, 629, 631, 662, 663, 666, 686, 743–746
Apricot, 354, 440, 442, 518, 519, 595, 752, 754, 756, 757

Aralia, 347, 362
Arborvitae, 347, 350, 352, 373, 405, 821
Arizona cypress, 789
Arizona rosewood, 824
Arrowhead vine, 867
Artemisia, 843
Artichoke, 579
Arum, 843
Ash, 17, 232, 234, 436, 490, 793, 794
Asian white birch, 782
Asiatic chestnut, 370, 440, 731, 784
Asiatic jasmine, 822
Asiatic pear, 759, 760–761
Asparagus, 132, 142, 143, 144, 556, 557, 701
Aspen, 809
Aster, 686, 843
Astilbe, 843
Atemoyas, 729
Atlantic white cedar, 786
Aubergine, 499
Aubrieta, 843
Aurelian lilies, 856
Australian pine, 784
Australian pitcher plant, 846
Australian tree fern, 851
Autumn crocus, 846
Autumn daffodil, 866
Avens, 852
Avocado, 194, 294, 321, 370, 662, 749–750
AXR#1 grape, 765
Azalea, 316, 328–329, 351, 367, 405, 813–814

B

Baby blue eyes, 235, 858
Baby's breath, 852
Baccharis, 780
Bachelor button, 846
Bacopa, 867
Bahia grass, 134, 146, 148
Bald cypress, 820–821
Balloon flower, 862
Balsam, 854
Bamboo, 562, 584, 586, 595, 780–781
Banana, 554, 584, 587, 589, 595, 596, 597, 629, 660, 666, 680, 747
Banksia, 781
Barberry, 172, 541, 551, 781–782
Barley, 141, 190
Barrenwort, 850
Bartlett pear, 422, 447, 506, 759
Basil, 217, 859
Basswood, 822
Bat-faced cuphea, 847
Bayberry, 365, 804
Bay laurel, 800
Beach acacia, 775

Bean, 141
Bearberry, 779
Beardtoungue, 861
Bear's breeches, 840
Bee balm, 858
Beech, 17, 190, 194, 239, 404, 420, 436, 490, 792–793
Beet, 190
Begonia, 148, 190, 217, 291, 297, 298, 328, 347, 360, 380, 584, 652, 686, 843–844
Belladonna lily, 842
Bellflower, 845
Bells of Ireland, 858
Bergenia, 844
Bermuda grass, 190, 537, 553
Big betony, 866
Birch, 17, 190, 227, 408, 415, 436, 452, 490, 662, 681, 782
Bird-of-paradise, 92, 587, 588, 866
Birdsfoot trefoil, 145
Bird's nest fern, 851
Bitternut hickory, 784
Bittersweet, 785
Blackberry, 347, 357, 365, 537, 551, 612, 686, 761–762
Blackberry lily, 844
Black chokeberry, 780
Black cohosh, 234
Black-eyed Susan, 864, 867
Black gum, 804
Black haw, 782
Black laurel, 794–795
Black locust, 365, 814
Black mulberry, 803
Black nightshade, 441, 611
Black raspberry, 357, 761
Black tupelo, 804
Black walnut, 741, 742
Blanketflower, 851
Bleeding heart, 365, 849
Bloodflower, 843
Bloodleaf, 854
Blood lily, 576
Bloodroot, 234, 864
Blueberry, 315, 415, 416, 587, 762–763, 823
Blue cardinal flower, 857
Blue daisy, 851
Blue fescue, 556, 851
Bluegrass, 133, 190
Blue trumpet vine, 867
Boltonia, 844
Boston fern, 553, 685, 722, 851
Boston ivy, 806
Bottlebrush, 783
Bottle-brush buckeye, 365, 777–778
Bottle gourd, 499

Bougainvillea, 401, 782
Bowstrip hemp, 864
Boxwood, 541, 782–783
Boysenberry, 347, 357, 761–762
Bradford pear, 810–811
Brake fern, 851
Brazilian plume, 855
Breadfruit, 365, 556, 730
Breath of heaven, 790
Bristlecone pine, 17, 807
Broad bean, 128
Broccoli, 190, 701
Brodiaea, 844
Bromegrass, 148
Bromeliads, 844
Brompton plum, 757–758
Broom, 790
Brussels sprouts, 629
Buckeye, 190, 194, 777–778
Buckthorn, 813
Budagovski apple series, 746
Buffaloberry, 818
Buffelgrass, 146, 148
Bugle flower, 841
Bugleweed, 552, 553
Bugloss, 842
Bumelia, 782
Bunya-bunya, 779
Bushclover, 856
Buttercup, 863
Buttercup winterhazel, 788
Butterfly bush, 782
Butterfly flower, 865
Butterfly weed, 118, 843
Butternut, 740–741
Butterwort, 862
Buttonbush, 235

C

Cabbage, 143, 145, 168, 190
Cacao, 762
Cactus, 357, 421, 431, 844, 850
Cactus pear, 359
Caladium, 845
California poppy, 92, 365, 850
California privet, 820
Calla, 869
Camellia, 351, 357, 604, 783
Camomile, 842
Camphor tree, 787
Campion, 857, 865
Canary Island date palm, 806
Candytuft, 854
Canna, 845
Canterbury bells, 845
Cape cowslip, 576, 577, 855
Cape jasmine, 794
Capejewels, 858
Cape marigold, 849
Cape primrose, 360, 866
Carambola, 730
Cardinal flower, 857
Cardoon, 848
Caribbean pine, 420, 807

Carnation, 143, 190, 284, 285, 358, 405, 406, 595, 596, 626, 629, 633, 660, 686, 722, 848
Carob, 732
Carrion flower, 866
Carrot, 233, 648, 649, 672
Cascade huckleberry, 823–824
Cashew, 663, 729
Cassava, 581, 629, 660
Castor bean, 118, 863
Casuarina, 784
Catalpa, 785
Catchfly, 865
Catmint, 859
Cattley guava, 758
Cauliflower, 190, 629
Ceanothus, 785
Cedar, 504, 785
Celery, 235
Century plant, 841
Chain fern, 851
Chamaecyparis, 350, 786
Chapman's rhododendron, 813
Checker lily, 851
Cheddar pinks, 848
Chenille, 293
Cherimoya, 729–730
Cherry, 143, 353, 419, 491, 541, 547, 549, 595, 603, 722
Cherry laurel, 809, 810
Chestnut, 167, 190, 194, 300, 312, 545, 731–732, 784
Chickpea, 239
Chilean mesquite, 809
Chili pepper, 451
China aster, 845
Chinese arborvitae, 441
Chinese bell flower, 775
Chinese chestnut, 370, 440, 784
Chinese date, 765
Chinese evergreen, 841
Chinese fir, 595
Chinese forget-me-not, 848
Chinese fringe tree, 316, 786, 787
Chinese hibiscus, 796
Chinese holly, 797
Chinese lantern, 862
Chinese lilac, 820
Chinese pistache, 318, 808
Chinese primrose, 863
Chinese witch hazel, 801
Chirimoya, 729–730
Chives, 841
Christmas begonia, 843
Christmas berry, 796
Christmas cactus, 865
Christmas cherry, 865
Christmas pepper, 845
Christmas rose, 853
Chrysanthemum, 284, 300, 312, 327, 328, 347, 380, 405, 595, 596, 597, 626, 629, 633, 660, 676, 686, 848
Cineraria, 861–862
Cinnamon vine, 581

Cinquefoil, 863
Citranges, 735–736
Citron, 282, 609, 732, 734
Citrus, 17, 132–134, 136, 146, 194, 347, 419, 440, 451–452, 486, 507, 513, 534, 541, 548, 595, 617, 629, 650, 663, 666, 672, 681, 732–736, 787
Clematis, 347, 370, 380, 541, 543, 787
Cleyera, 821
Cliffbush, 798
Clockvine, 867
Clover, 145, 190, 669
Coast leucothoe, 800
Coast redwood, 619, 663, 818
Cocklebur, 219
Cockscomb, 846
Coconut, 129, 736
Coffee, 736
Cohosh, 846
Cole crop, 134, 666, 667, 672
Coleus, 190, 217, 282
Colonial pine, 779
Colorado manzanita, 779–780
Colorado spruce, 159, 807
Columbine, 842
Comfrey, 867
Common fig, 365, 738–839
Common guava, 758
Common honeylocust, 794
Common horse chestnut, 777, 778
Common Mussel plum, 756, 758
Common pawpaw, 194, 683, 730, 780
Common persimmon, 791
Common stock, 857–858
Coneflower, 133, 864
Confederate jasmine, 822
Conference pear, 432
Conifers, 146, 551
Cook pine, 779
Coralbell, 853
Coralberry, 780
Coreopsis, 190, 847
Corn, 123, 125, 135, 142, 146, 148, 155, 167, 181, 190, 666
Corncockle, 841
Cornflower, 846
Corn lily, 855
Corn poppy, 861
Corydalis, 847
Cotoneaster, 174, 232, 258, 381, 402, 789
Cotton, 132, 141, 190
Cottonwood, 146, 809
Couderc 3309 grape, 765
Crab apple, 599, 618, 803
Crab apple, flowering, 315, 354, 365, 446, 517, 525, 803
Cranberry, 762–763, 823
Cranesbill, 852
Crape myrtle, 345, 347, 354, 799
Creeping fig, 281, 282, 284–287, 289, 316, 614, 615, 616
Creeping zinnia, 864
Crimson King maple, 432

Crocosmia, 847
Crocus, 562, 577, 847
Croton, 318, 541, 542, 846
Crown imperial, 851
Cryptomeria, 789
Cucumber, 135, 148, 190, 206, 228, 229, 231, 421, 452, 499
Cucurbits, 142, 416, 417, 430, 443, 448, 451, 469, 471, 476, 477, 499, 701
Cup and saucer vine, 846
Cupflower, 859
Cupid's bower, 840
Cupid's dart, 845
Currant, 282, 761
Custard apple, 729–730
Cut-leaf philodendron, 858
Cycads, 789–790
Cyclamen, 148, 190, 233, 562, 582, 584, 847–848
Cypress, 789

D

Daffodil, 562, 564–567, 569–570, 571–572, 576, 577, 858
Dahlberg daisy, 849
Dahlia, 300, 315, 562, 581–584, 629, 848
Daisy, 133
Damson plum, 756, 758
Dandelion, 133
Daphne, 365, 790
Dasheen, 562, 629
Date palm, 554, 750–751, 806
Dawn redwood, 803
Daylily, 552, 556, 562, 599, 700, 853
Delicious apple, 744
Delphinium, 848
Desert broom, 780
Desert willow, 786
Deutzia, 354, 790
Devil's walking stick, 365
Dewberry, 551, 761–762
Dieffenbachia, 347
Diosma, 790
Dittany of Crete, 860
Dog Ridge grape, 764
Dog rose, 814, 816
Dog's tooth violet, 850
Dogwood, 345, 347, 354, 367, 404, 529–530, 552, 553, 613, 787–788
Douglas-fir, 17, 149, 294, 308, 405, 420, 425, 440, 449, 663, 810
Douglas maple, 776
Dove tree, 790
Drooping leucothoe, 800
Dumbcane, 849
Dusty miller, 865, 867
Dwarf flowering almond, 809
Dwarf fothergillia, 793
Dwarf Japanese juniper, 798
Dwarf Japanese maple, 776
Dwarf pawpaw, 780
Dwarf rhododendron, 328–329

E

Easter cactus, 863
Easter lily, 292, 567, 568, 569, 572
Eastern baccharis, 780
Eastern black walnut, 741, 742
Eastern larch, 617
Eastern redbud, 127, 179
Eastern red cedar, 798, 799
Eastern strawberry tree, 779
Eastern white pine, 149, 808
Edelweiss, 855–856
Elaeagnus, 791
Elderberry, 345, 817
Elm, 189, 353, 354, 383, 545, 823
Empress tree, 226
English daisy, 844
English holly, 416, 797
English ivy, 19, 347, 490, 614, 795
English laurel, 809, 810
English oak, 353
English walnut, 446, 741–742
Enkianthus, 791
Epiphytic orchids, 562, 859–860
Eriostemon, 791
Escallonia, 791
Etrog, 734
Eucalyptus, 146, 300, 308, 310, 318, 420, 595, 616, 619, 620, 680, 792
Euonymous, 347, 351, 354, 363, 402, 792
Euphorbia, 850, 866
European alder, 778
European aspen, 365
European beech, 650, 793
European birch, 782
European chestnut, 731
European filbert, 788
European gooseberry, 761
European larch, 149, 305
European mountain ash, 819
European pear, 759–760
European plum, 440, 752, 756, 757–758
European prune, 756, 757–758
Evening-primrose, 859
Evening scented stock, 857–858
Evergreen huckleberry, 823
Evergreen pear, 811
Exacum, 851

F

Fairy primrose, 863
Falling stars, 845
False buckthorn, 782
False cypress, 786
False dragonhead, 862
False indigo, 843
False willow, 780
Fan flower, 865
Farkelberry, 416
Farmingdale pear, 760
Fatshedera, 795
Fatsia, 233, 793, 851
Featherleaf, 863

Feather reed grass, 556, 845
Feijoa, 738
Fern, 111, 851
Fern asparagus, 843
Fernleaf peony, 860
Fescue, 190, 851
Feverfew, 867
Fibrous-rooted begonia, 843
Fiddleleaf fig, 793
Field corn, 190
Field pea, 141
Fig, 298, 300, 345, 347, 365, 380, 489, 541, 542, 595, 738–839, 793
Filbert, 190, 194, 238, 239, 537, 539, 541, 737, 788
Fir, 17, 172, 347, 350, 415, 420, 452, 457, 504, 774–775
Firethorn, 810
Fiveleaf aralia, 365
Flame flower, 847
Flame violet, 850
Flax, 141, 224, 666, 668, 857
Fleabane, 850
Fleeceflower, 862
Florida sand plum, 758
Florist's cineraria, 861–862
Flowering almond, 365
Flowering apricot, 809, 810
Flowering cabbage, 844
Flowering cherry, 424, 809, 810
Flowering crab apple, 315, 354, 365, 446, 517, 525, 803
Flowering currant, 814
Flowering dogwood, 596, 787–788
Flowering maple, 622, 775
Flowering purslane, 863, 866
Flowering quince, 365, 786
Flowering tobacco, 859
Foamflower, 867
Forget-me-not, 858
Forsythia, 327, 345, 347, 354, 365, 367, 541, 662, 793
Fothergillia, 793
Fountain grass, 861
Foxglove, 92, 849
Foxtail lily, 850
Fragrant sweetbox, 817
Fragrant tea olive, 805
Franklinia, 793
Fraser fir, 314, 620, 775
Fraser photinia, 806
Freedom grape, 765
Freesia, 562, 577, 629, 851
French pear, 441, 759–760
Fringe tree, 259, 354, 786–787
Fruitless mulberry, 803
Fuchsia, 613, 794, 851

G

Garden bean, 141
Garden chrysanthemum, 848
Gardenia, 794
Garden phlox, 862

Garden pinks, 848
Garlic, 567, 629, 841
Garrya James Roof, 794
Gas plant, 849
Gayfeather, 856
Genista, 794
Gentian, 233, 852
Geranium, 347, 362, 365, 607, 626, 629, 633, 660, 681, 861
Germander, 867
GF 677 almond, 755
Giant reed, 556
Giant sequoia, 818
Ginger, 585, 586, 843
Ginger lily, 853
Ginger root, 562
Ginkgo, 794
Ginseng, 233
Gisela cherry, 752, 754
Gladiolus, 629
Globe amaranth, 852
Globe centaurea, 846
Globeflower, 868
Globe thistle, 850
Gloire des Rosomanes, 816
Gloriosa lily, 852
Glory bower, 118, 365
Glory-of-the-snow, 846
Glory of the sun, 856
Glossy abelia, 774
Gloxinia, 562, 865
Goat's beard, 843
Golden barberry, 781
Golden chain, 799
Goldencup, 854
Golden Marguerite, 842
Golden passion fruit, 748–749
Golden pothos, 850
Golden rain tree, 235, 365, 799
Goldenrod, 866
Goldenstar, 846
Golden trumpet, 778
Goldentuft, 843
Golden willow, 775
Gold tree, 820
Gooseberry, 233, 629, 686, 761, 777
Goutweed, 841
Granadilla, 663
Granny Smith apple, 454
Grape, 17, 135, 419, 434, 437, 474, 483, 496, 541, 543, 595, 597, 598, 629, 631, 663, 686, 763–765
Grapefruit, 452, 535, 599, 732, 787
Grape hyacinth, 570, 576, 577, 858
Grass, 133
Gray alder, 778
Greek windflower, 842
Green ash, 793–794, 820
Green bean, 190
Greenhouse chrysanthemum, 848
Grevillea, 297, 795
Guava, 758
Guayule, 133

H

Hackberry, 785
Hairy litchi, 747
Hale's Early peach, 446
Halesia, 795
Hawmedlar, 609
Hawthorn, 172, 232, 408, 609, 789
Hazelnut, 167, 190, 236, 305, 537, 541, 653, 737, 788
Heath, 380, 791
Heather, 233, 791
Heavenly bamboo, 584
Hebe, 795
Heliopsis, 853
Heliotrope, 853
Hellebore, 853
Hemlock, 17, 347, 350, 425, 822–823
Herbaceous peony, 805
Hibiscus, 352, 380, 428, 796
Hickory, 192, 194, 784
Hick's yew, 821
Highbush blueberry, 315, 762–763, 823
Holly, 142, 234, 239, 315, 347, 351, 402, 797
Hollyhock, 190, 841
Hollyhock mallow, 857
Honesty plant, 857
Honey locust, 129, 223, 516
Honeysuckle, 801
Hoop pine, 420
Hoopsii blue spruce, 807
Hops, 629, 854
Hornbeam, 190, 784
Horned violet, 868
Horse chestnut, 777–778
Horse radish, 347, 664
Horsetails, 111
Hosta, 552, 554, 556, 557, 610, 665, 686, 700, 722, 853–854
Huckleberry, 172, 823
Hyacinth, 562, 566, 567, 570, 575, 576, 577, 854
Hydrangea, 282, 300–304, 365, 366, 721, 796–797

I

Iceland poppy, 861
Impatiens, 163, 183, 189, 268, 330, 701, 854
Incarvillea, 854
Incense cedar, 172, 783
India hawthorn, 399, 812–813
Indian lotus, 191
Indigo bush, 779
Iris, 205, 224, 552, 562, 566, 570, 574, 629, 854–855
Irish potato, 635
Ironwood, 784
Ishtara plum, 758
Italian alder, 778
Ivy, 17, 18, 285, 289, 299, 300, 308, 356, 619
Ivy geranium, 861
Ixora, 855

J

Jacaranda, 797–798
Jackfruit, 370, 730
Jack-in-the-pulpit, 843
Jack pine, 172, 807
Jacob's ladder, 862
Jade plant, 847
Japanese anemone, 365, 842
Japanese ardisia, 780
Japanese birch, 452, 782
Japanese cryptomeria, 789
Japanese flowering quince, 365
Japanese holly, 320
Japanese maple, 356, 465, 482, 504, 775–777
Japanese Momi fir, 452, 775
Japanese pagoda tree, 365, 818–819
Japanese persimmon, 737–738, 791
Japanese plum, 440, 752, 758
Japanese plum yew, 785
Japanese privet, 800
Japanese raisin tree, 796
Japanese skimmia, 818
Japanese spicebush, 800
Japanese spurge, 805
Japanese stewartia, 819
Japanese ternstroemia, 821
Japanese yew, 315, 821
Jasmine, 282, 798
Jaune de Metz apple, 744
Jerusalem artichoke, 579
Jerusalem sage, 862
Jimson weed, 441
Johnny-jump-up, 869
Johnson grass, 586, 587
Jonagold apple, 441, 744
Joseph's coat, 842
Jujube, 765
Juniper, 172, 174, 317, 322, 328, 347, 350, 352, 405, 616, 798

K

Kale, 681, 844
Kangaroo paw, 842
Kenilworth ivy, 848
Kentucky bluegrass, 134, 146, 148, 190
Kentucky coffee tree, 653, 664, 795
Kieffer pear, 760
King Ranch bluestem, 134, 146
Kiwifruit, 305, 365, 454, 473, 663, 728–729, 777
Knapweed, 846
Knot weed, 862
Koa acacia, 308, 775
Kola, 736–737
Korean arborvitae, 821
Korean lilac, 820
Koster spruce, 421, 438
Kumquat, 732, 733, 735, 787

L

Laburnum, 799
Laceflower, 867
Lady slipper, 859–860

Lady's mantle, 841
Lamb's ear, 866
Lantana, 855
Larch, 17, 305, 504, 799–800
Larkspur, 92, 380, 847
Laurel, 799
Lavender, 855
Lavender cotton, 864
Lawson cypress, 616
Leadplant amorpha, 779
Leadwort, 365, 380
Leaf-flowering cactus, 850
Lemon, 357, 599, 732, 735, 787
Lemon guava, 758
Lenten rose, 853
Lentil, 204
Leopard's bane, 849
Lettuce, 123, 190, 224, 227, 228, 229, 235, 239, 681
Leucospermum, 800
Leyland cypress, 789
Liatris, 562
Licorice plant, 853
Ligustrum, 401, 800, 820
Lilac, 205, 224, 316, 347, 354, 404, 820
Lily, 121, 143, 145, 234, 290, 292, 562, 564, 565, 567, 568, 570, 572, 577, 629, 666, 856
Lily-of-the-Nile, 841
Lily-of-the-valley, 234, 562, 584, 585, 587, 847
Lily turf, 857
Lime, 542, 732, 734, 736
Linden, 235, 370, 432, 822
Lingonberry, 823
Lion's ear, 855
Lionsheart, 862
Liquidambar, 800
Liriope, 365, 366, 556, 557
Lisianthus, 850
Litchi, 194, 537, 542, 595
Littleleaf ash, 793
Liverwort, 234
Livingstone daisy, 849
Living stones, 857
Lobelia, 857
Loblolly bay, 794–795
Loblolly pine, 149, 308, 617
Locust bean, 732
Lodgepole pine, 149
Loganberry, 761–762
London plane tree, 808
Longan, 542, 595, 738
Longleaf casuarina, 784
Loosestrife, 857
Loquat, 194, 441, 738, 791
Love-in-a-mist, 859
Love-lies-bleeding, 842
Lovell peach, 754
Lowbush blueberry, 584, 762
Lungwort, 863
Lycopods, 111

M

M 26 apple, 745
Macadamia nut, 507, 742–743
Madrone, 779
Magnolia, 320, 347, 352, 354, 373, 374, 404, 616, 801–802
Mahaleb, 451, 753, 754
Mahogany, 819–820
Mahonia, 136, 802
Maiden grass, 858
Maiden-hair fern, 851
Maile, 841–842
Malling apple series, 453, 454, 456, 744–745
Mallow, 165, 853
Mandarin orange, 732, 735, 787
Mandevilla Alice DuPont, 803
Mango, 132, 133, 134, 146, 370, 435, 537, 595, 746
Mangosteen, 132, 134, 740
Mangrove, 131
Manzanita, 779
Maple, 17, 189, 300, 315, 320, 347, 354, 357, 370, 404, 436, 490, 500, 516, 517, 528, 604, 662, 775–777
Marguerite, 848
Marianna plum, 440, 757
Marigold, 148, 190, 701, 867
Marsh marigold, 845
Masterwort, 843
Mayflower, 791
Mazzard cherry, 452, 753, 810
McIntosh apple, 438, 454, 744
Meadow rue, 867
Medlar, 609
Melon, 148, 190, 206, 416, 417, 430, 443, 452, 499
Mesquite, 809
Metasequoia, 803
Mexican bald cypress, 820–821
Mexican buckeye, 823
Mexican heather, 847
Mexican oregano, 863
Mexican plum, 809, 810
Mexican redbud, 786
Mexican sunflower, 867
Mid-Century lilies, 856
Mignonette, 863
Millet, 190
Mimosa, 778
Miniature roses, 816
Mini-watermelon, 452, 499
Mint, 552, 553
Mistletoe, 217
Mockernut hickory, 784
Mock orange, 806
Moerheim blue spruce, 807
Mondo grass, 859
Money plant, 857
Monkey puzzle tree, 779
Monkshood, 840–841
Monterey pine, 283, 308, 420, 442, 549, 551, 595, 596, 617, 807–808

Montezuma cypress, 820–821
Moraine locust, 794
Morning glory, 190, 847, 854
Moss pink, 862
Moss rose, 863, 866
Mother-in-law's tongue, 586
Mountain ash, 818
Mountain cranberry, 823
Mountain huckleberry, 823–824
Mountain laurel, 369, 682, 701, 799
Mugho pine, 149
Mulberry, 541, 747, 803
Mullein, 868
Mung bean, 282, 284, 296, 300
Muscadine grape, 537, 763
Musk mallow, 857
Mussaenda, 804
Mustard, 224
Myran plum, 758
Myrobalan plum, 446, 754, 756, 757
Myrtle, 804

N

Nandina, 584
Nanking cherry, 757
Narcissus, 629
Narrowleaf baccharis, 780
Nasturtium, 868
Nectarine, 422, 519, 752, 755–757
Needle palm, 805
Nemaguard peach, 754, 755, 756, 810
Nemared peach, 754, 755, 756
Nerine, 576, 577, 629, 859
New Caledonia pine, 779
New Guinea impatiens, 854
Newport plum, 452
Nispero, 746–747
Noble fir, 774
Nootka cypress, 441
Norfolk Island pine, 315, 620, 779
Northern California black walnut, 158, 435, 438, 489, 741
Northern wax myrtle, 804
Norway maple, 353, 370, 408, 421, 432, 517, 776, 777
Norway spruce, 17, 149, 314
Nutmeg hickory, 784

O

Oak, 17, 129, 142, 146, 167, 186, 190, 192, 194, 308, 370, 404, 408, 415, 420, 425, 452, 541, 675, 676, 811–812
Oats, 141, 190
Old Home pear, 422, 447, 506, 759, 760, 789
Oleander, 804
Olive, 146, 308, 316, 317, 345, 347, 455, 747–748, 804–805
Onion, 133, 145, 150, 163, 168, 190, 562, 567, 672, 841
Orange, 132, 441, 446, 452, 535, 609, 732, 735, 787
Orchard grass, 148

Orchid, 233, 234, 553, 562, 587, 589, 616, 629, 660, 662, 668, 682, 859–860
Orchid tree, 781
Oregano, 860
Oregon boxwood, 806
Oriental arborvitae, 821, 822
Oriental cherry, 810
Oriental magnolia, 802
Oriental pear, 365, 760
Oriental poppy, 361, 365, 861
Ornamental gourds, 847
Ornamental onions, 841
Ornamental pepper, 845
Ornamental sweet potato, 854
Osage-orange, 801
Ostrich fern, 851
Oval-leaf bilberry, 824
Oxalis, 860

P

Pacific madrone, 779
Painted arum, 843
Painted daisy, 867
Painted fern, 851
Painted tongue, 864
Palms, 261, 552, 805–806
Pampass grass, 847
Pansy, 148, 190, 268, 869
Papaya, 293, 730
Paper bark maple, 775, 776
Paper mulberry, 365
Paradox walnut, 150, 158, 742
Parrot lily, 841
Pasque flower, 863
Passion fruit, 748–749
Passion vine, 806
Pathos, 541
Pawpaw, 163, 552
Pea, 134, 135, 190, 296, 298
Peach, 141, 236, 347, 354, 422, 440, 446, 455, 486, 504, 513, 517, 519, 599, 605, 608, 631, 669, 752, 754, 755–757
Peach bells, 845
Peanut, 141, 190
Pear, 17, 146, 300, 347, 353, 354, 419, 422, 432, 434, 445, 447, 451, 483, 505, 517, 595, 597, 599, 600, 616, 619, 631, 759–761
Pearl millet, 148
Pearly everlasting, 842
Peashrub, 784
Pea vine, 855
Pecan, 190, 308, 418, 483, 486, 504, 505, 517, 531, 541, 547, 548, 731, 784
Penstemon, 806, 861
Pentas, 861
Peony, 860–861
Pepper, 141, 190, 205, 224, 232
Pepperidge tree, 804
Perennial pea vine, 855
Periwinkle, 868
Persian lime, 542, 732, 734

Persian parrotia, 806
Persian ranunculus, 863
Persian violet, 851
Persian walnut, 159, 435, 438, 489, 741–742
Persimmon, 596, 791
Petunia, 143, 148, 156, 190, 206, 268, 329, 330, 604, 701, 862
Philodendron, 347, 541–543, 862
Phlomis, 580
Phlox, 217, 365, 862
Photinia, 806
Pieris, 807
Pigeon plum, 787
Piggyback plant, 290, 361, 362, 867
Pignut hickory, 784
Pincushion flower, 865
Pindo palm, 806
Pine, 17, 112, 113, 126, 164, 172, 289, 305, 350, 370, 383, 425, 490, 504, 807–808
Pineapple, 347, 357, 554, 555, 629, 729
Pineapple guava, 738
Pineapple lily, 850
Pink panda, 851
Pinkroot, 866
Pistachio, 142, 370, 423, 517, 663, 751
Pitcher plant, 864
Pittosporum, 347, 351, 808
Plaintain-lily, 853–854
Plane tree, 808
Plum, 117, 143, 236, 345, 347, 354, 419, 422, 446, 518, 519, 595, 663, 756, 758, 809, 810
Plumbago, 846
Plume poppy, 857
Plumeria, 809
Pocketbook plant, 845
Podocarpus, 315, 620, 809
Poinsettia, 328, 330, 331, 347, 595, 792, 866
Poker plant, 855
Polish apple series, 746
Polka dot plant, 854
Pomegranate, 759
Pomelo, 732
Ponderosa pine, 149, 807
Ponytail palm, 806
Poplar, 189, 282, 300, 345, 595, 717, 809
Poppy, 233, 347
Poppy anemone, 234, 842
Potato, 441, 456, 553, 562, 579, 580, 582, 595, 600, 609, 629, 632, 633, 635, 660, 666, 667, 672
Potentilla, 345, 809
Pothos, 347, 356–357, 863
Pot marigold, 845
Prairie coneflower, 863
Prayer plant, 857
Prickly pear cactus, 748, 844, 859
Pride of Madeira, 850
Primrose, 144, 190, 217, 863
Primula, 233
Prince Charles serviceberry, 778

Princess Diana serviceberry, 778
Privet, 316, 345, 347, 800
Protea, 800, 809
Prune, 757–758
Pummelo, 734
Pumpkin, 284, 499
Purple coneflower, 169, 224, 238, 684, 721, 849–850
Purple haze penstemon, 806
Purple loosestrife, 857
Purple raspberry, 761
Purple sage, 800
Pygmy (dwarf) peashrub, 784
Pyracantha, 367, 381, 810

Q

Quaking aspen, 365, 809
Quamash, 845
Queen of the prairie, 851
Quince, 347, 419, 441, 445, 447, 541, 549, 618, 737, 760, 786

R

Rabbiteye blueberry, 762, 823
Radish, 204, 228, 681
Ragged Robin rose, 816
Ramsey grape, 764, 765
Rangpur lime, 736
Ranunculus, 233, 569, 863
Rape seed, 128, 190
Raphiolepis, 812–813
Raspberry, 133, 347, 357, 365, 537, 551, 552, 629, 761–762
Raspberry frost banksia, 781
Redbud, 205, 222, 224, 231, 232, 235, 315, 655, 657, 679, 786
Red cedar, 351, 798, 799
Red chokeberry, 780
Red clover, 145
Red currant, 761
Red fir, 775
Red flag bush, 858
Red flowering currant, 814
Red horse chestnut, 777
Red maple, 329, 440, 686, 775, 776, 777
Red oak, 440, 811, 812
Red-osier dogwood, 788
Red raspberry, 761
Red-tip photinia, 806
Red valerian, 846
Red-veined dock, 864
Redwood, 17, 172
Reeves skimmia, 818
Regal geranium, 861
Rhizomatous begonia, 843
Rhododendron, 233, 314–315, 318, 320, 322, 347, 351, 357, 370, 373, 402, 404, 474, 483, 507, 541, 551, 612, 682, 686, 700, 712, 722, 813–814
Rhubarb, 629
Rice, 141, 190, 666
Richter 110 grape, 765

River birch, 356, 782
Robusta No. 5 apple, 745
Rockcress, 842
Rock soapwort, 864
Rodgersflower, 863
Rome Beauty apple, 744
Rose, 17, 133, 172, 258, 284, 316, 345, 347, 348, 349, 365, 370, 434, 440, 507, 508, 516, 522, 523, 527, 528, 529, 541, 595, 597, 686, 814–816, 815–816
Rose acacia, 365
Rosemary, 816
Rose of Sharon, 796
Roses, climbing, 543
Rosewood, 824
Rough lemon, 735
Royal poinciana, 790
Rubber plant, 793
Rubber tree, 309, 422, 489–490, 663
Rubra linden, 432
Rupestris St. George grape, 765
Russian olive, 791
Russian sage, 862
Rye, 666

S
Sacred lotus, 858
Saffron, 846
Sage, 864
Sago palm, 790
Salpiglossis, 190, 664, 864
Salt cedar, 820
Salt Creek grape, 764
Santa Rosa sweetbay, 802
Sapodilla, 746–747
Sarcococca, 817
Saskatoon serviceberry, 778
Sassafras, 365, 556
Saucer magnolia, 802
Scarlet banksia, 781
Scarlet bouvardia, 844
Scarlet plum, 792
Scarlet sage, 864
Schefflera, 544, 545, 865
Scotch pine, 17, 149, 289, 807, 808
Scots pine, 228, 289, 308
Screwpine, 281
Sea grape, 787
Sea-holly, 850
Sea kale, 362
Seaside alder, 778
Sedum, 289, 290, 292, 347, 430, 448, 865, 866
Sequoia, 17, 617
Serviceberry, 172, 778
Seven son flower, 795–796
Shagbark hickory, 731, 784
Shamrock, 860
Shasta daisy, 190, 556, 856
Shellbark hickory, 784
Shepherd's purse, 122–123, 124

Shining sumac, 365, 814
Shirley poppy, 861
Shirofugen cherry, 627
Shooting star, 849
Shortleaf pine, 149, 808
Shrimp plant, 855
Shrub-althea, 796
Siberian bugloss, 844
Siberian crab apple, 744, 803
Siberian peashrub, 784
Sidalcea, 865
Side oats grama, 134, 146
Sierra alder, 778
Sierra redwood, 818
Silk oak, 795
Silk tree, 365, 778
Silverbell, 795
Silverberry, 791
Silver maple, 191, 194, 775, 776, 777
Silverthorn, 791
Silver trumpet tree, 820
Sitka spruce, 17, 314, 427, 436, 448
Skimmia, 818
Skullcap, 865
Slash pine, 149, 368, 420
Smoke bush, 788–789
Smoke tree, 788–789
Smooth-leaved elm, 365
Smooth sumac, 365
Snake plant, 290, 358, 864
Snapdragon, 842
Sneezeweed, 853
Snowball viburnum, 824
Snowberry, 365, 820
Snowdrop, 852
Snow-in-summer, 846
Soapberry, 817
Solomon's seal, 234, 552, 863
Sophora, 221, 818
Sorbus, 818
Sour cherry, 752, 753–754, 809
Sour orange, 446, 452, 609, 735
Sourwood, 805
South African cycad, 790
Southern wax myrtle, 804
Soybean, 147, 190
Spatheflower, 866
Speckled alder, 778
Speedwell, 868
Spiceberry, 780
Spicebush, 800
Spiderflower, 846
Spider lily, 854, 857
Spider plant, 552, 553, 846
Spiderwort, 867–868
Spinach, 190
Spirea, 345, 347, 354, 363, 819
Spotted deadnettle, 855
Sprenger asparagus, 843
Spring snowflake, 701, 856
Spruce, 230, 347, 350, 420, 551, 663, 807
Spurflowers, 862

Spurge, 850
Squash, 148
Squill, 865
Staghorn fern, 851
Staghorn sumac, 365, 814
Star-cluster pentas, 861
Star jasmine, 822
Star of Bethlehem, 860
Statice, 190, 856
Stewartia, 356, 819
St. John's wort, 365, 797, 854
St. Julien plum, 756, 758
St. Lucie cherry, 754
Stock, common, 866
Stocks, 190
Stokes aster, 365, 866
Stone fruits, 17, 629
Strangling fig, 217
Strawberry, 134, 172, 537, 552, 553, 600, 603, 622, 627, 629, 631, 680, 686, 739–740
Strawberry geranium, 552, 865
Strawberry guava, 758
Strawflower, 853
Striped maple, 777
Striped squil, 863
Succulents, 866
Sugar apple, 233, 729–730
Sugar beet, 190
Sugarberry, 785
Sugar cane, 584, 595, 629, 672
Sugar maple, 776, 777
Sugi, 595, 789
Sumac, 235, 347, 814
Summer hyacinth, 852
Summer snapdragon, 842
Sundew, 849
Sunflower, 117, 145, 190, 236, 239, 305, 441, 650–651, 853
Sunrose, 853
Surprise lily, 857
Swan River daisy, 844
Sweet acacia, 775
Sweet alyssum, 857
Sweet cherry, 297, 451, 454, 752
Sweet chestnut, 784
Sweet corn, 163, 183, 190
Sweet fern, 365
Sweet gum, 658, 675, 676
Sweet olive, 805
Sweet orange, 441, 446, 452, 535, 732, 787
Sweet pea, 855
Sweet pepperbush, 787
Sweet potato, 292, 347, 562, 581, 583–584, 595, 629
Sweet potato vine, 854
Sweet rocket, 853
Sweetshrub, 783
Sweet sultan, 846
Sweet violet, 869
Sweet William, 848, 862
Swiss cheese plant, 858

Switch ivy, 800
Sword fern, 851
Sycamore, 347, 353, 354, 808

T

Tag alder, 778
Tamarisk, 133, 820
Tangelo, 732
Tangor, 732
Taro, 629
Teak, 149
Tea tree, 800
Teleki 5C grape, 765
Terrestrial orchids, 860
Texas almond, 442
Texas bluebonnet, 857
Texas buckeye, 777
Texas madrone, 779
Texas mountain laurel, 818, 819
Texas persimmon, 791
Texas sage, 800
Thrift, 843
Thyme, 867
Tickseed, 847
Tiger flower, 867
Ti plant, 847
Toadflax, 856–857
Toad lily, 868
Tobacco, 150, 151, 206, 297, 305, 441, 599, 654, 666, 667, 672
Tomato, 134, 135, 141, 148, 163, 190, 224, 232, 236, 284, 415, 441, 451, 456, 495, 499, 500, 599, 611, 701
Torch lily, 855
Touch-me-not, 854
Toyon, 796
Trailing arbutus, 791
Transvaal daisy, 852
Tree ivy, 851
Tree-of-heaven, 365, 778
Tree peony, 805, 860, 861
Tree roses, 816
Trifoliate orange, 132, 441, 732, 735
Trillium, 234, 868
Trout lily, 566, 850
Trumpet creeper, 783–784
Trumpet vine, 365
Tuberose, 863
Tuberous begonia, 562, 582, 583, 584, 844
Tucson side oats grama, 134, 146

Tufted pansy, 868
Tulip, 562, 563, 565, 566, 567, 569–571, 577, 672, 868
Tulip tree, 801
Tupelo, 194
Turban ranunculus, 863
Turkish fir, 775
Turnip, 168
Turtlehead, 846
Tussock, 845
Twinleaf, 118, 234
Twinspur, 849

U

Ubame oak, 812
Umbrella pine, 817
Umbrella tree, 865

V

Vanilla orchid, 439
Venus flytrap, 849
Verbena, 868
Veronica, 795
Vetch, 190
Viburnum, 174, 234, 347, 363, 367, 381, 404, 541, 824
Victorian smokebush, 787
Viking plum, 758
Vinca, 846
Vinifera grapes, 764–765
Violet, 141, 224, 868–869
Viper's bugloss, 850
Virginia blue bells, 858
Virginia crab, 446
Virginia creeper, 806

W

Wake robin, 868
Walking fern, 290
Walking stick, 779
Wallflower, 846
Walnut, 142, 146, 190, 192, 194, 236, 423, 435, 483, 504, 528, 541, 549, 616, 619, 663
Wandering Jew, 869
Waratah, 821
Washington hawthorn, 408
Water hickory, 784
Water lily, 859
Watermelon, 145, 174, 476, 478, 499
Water oak, 313, 811
Wax begonia, 843

Wax flower, 786, 854, 866
Wax myrtle, 804
Weeping fig, 543, 545
Weigela, 347, 354, 825
Western arborvitae, 821, 822
Western sand cherry, 756, 758
Western soapberry, 817
West Indian blackthorn, 775
Wheat, 141, 190, 666
White clover, 145
White currant, 761
White fir, 775
White gourd, 499
White mulberry, 803
White oak, 811, 812
White pine, 314
White poplar, 365
Whitespire Japanese birch, 452, 782
White spruce, 149, 232
White sweet clover, 441
Wild rice, 190
Willow, 189, 282–283, 298, 300, 347, 595, 817
Willow amsonia, 842
Windflower, 234, 569
Winesap apple, 438, 744
Winter aconite, 850
Winterhazel, 788
Wishbone flower, 867
Wiskferns, 111
Wisteria, 345, 504, 541, 543, 616, 825
Witch hazel, 795
Witchweed, 239
Wood sorrel, 860
Wormwood, 843

Y

Yam, 580, 595
Yarrow, 840
Yellow allamanda, 778
Yellow archangel, 855
Yellow cedar, 232
Yellow Newtown apple, 744
Yellow poplar, 801
Yellow wax bells, 855
Yew, 234, 317, 347, 350, 352, 405, 821
Youngberry, 761–762
Yucca, 866, 869

Z

Zephyr lily, 869
Zinnia, 190, 869